Development

McGraw-Hill Book Company

New York St. Louis San Francisco Auckland Bogotá Hamburg
Johannesburg London Madrid Mexico Montreal New Delhi
Panama Paris São Paulo Singapore Sydney Tokyo Toronto

Development

SECOND EDITION

Gerald Karp Department of Microbiology and Cell Science, University of Florida

N. J. Berrill Swarthmore, Pennsylvania

DEVELOPMENT

2 3 4 5 6 7 8 9 0 D O D O 8 9 8 7 6 5 4 3 2 1

This book was set in Primer by Progressive Typographers.
The editors were Marian D. Provenzano and James S. Amar; the designer was Hermann Strohbach; the production supervisor was Charles Hess.
The drawings were done by J & R Services, Inc.
R. R. Donnelley & Sons Company was printer and binder.

Cover photos courtesy of Karl Illmensee,
University of Geneva, Switzerland.

Front: Photomicrograph showing a single malignant teratocarcinoma cell (seen within the pipette) about to be injected microsurgically into a 4½ day mouse embryo.
Back: The injected cell is now seen within the embryo attached to its inner cell mass. These experiments are discussed in Chapter 18, page 608.

Library of Congress Cataloging in Publication Data

Karp, Gerald.
Development.

Authors' names in reverse order in previous ed.
Includes bibliographies and index.
1. Developmental biology. 2. Embryology. I. Berrill, Norman John, date joint author. II. Title.
[DNLM: 1. Embryology. QS604 K18d]
QL955.B37 1981 591.3 80-19599
ISBN 0-07-033340-8

Contents

Preface

The second edition of *Development,* while retaining similar emphases to its predecessor, contains an extensively revised content. Much of the change reflects the findings of research carried out in the past five years which we have attempted to integrate into the framework of each chapter. Also, in the case of many topics we have gone back to the older literature with the aim of reexamining the subject in order to present a more informative treatment. We have, in addition, chosen to expand the coverage of certain topics which were slighted somewhat in the first edition. Some of the major changes incorporated into the second edition are as follows.

The topic of positional information has been added to the introductory chapter, in connection with budding in hydra, so that questions concerning the spatial as well as the temporal aspects of development can be considered at the outset. In Chapter 2, the techniques of DNA cloning and sequencing are introduced in anticipation of their use in studies discussed in later chapters. In Chapter 5, coverage of mammalian spermatogenesis has been expanded. Our understanding of fertilization has increased considerably during the past few years and new insights into sperm-egg recognition, the role of calcium in sperm-egg responses, and in vitro mammalian fertilization are fully discussed in Chapter 5. Chapter 7 has undergone major reorganization and now includes sections on the experimental analysis of the early development of sea urchins, amphibians, spiralians, ascidians, cephalopods, insects, and mammals. The subject of insect development has been greatly expanded: included in this part of Chapter 7 are discussions of compartmentalization, homeotic genes, maternal-effect mutations, and imaginal disc determination. The section on mammals includes an expanded discussion of the presumptive fates of early embryonic regions, in vitro development of mammalian embryos, and experiments designed to determine the prospective potency of embryonic parts. Chapter 8 on the molecular biology of early development has been enlarged to include the most important recent findings on the molecular aspects of sea urchin, amphibian, and mammalian development. An extensive consideration is given the recent series of studies on the analysis of RNA sequences present in the nuclei and cytoplasm of sea urchin embryos and adult tissues. The general findings of these studies are presented in the body of the chapter and the experimental approaches described in an appendix. The control of histone synthesis in sea urchins and amphibians is discussed and contrasted, and attempts to distinguish between transcriptional and

posttranscriptional level control are made. In Chapter 11 the topic of the histogenesis of the nervous system has been added as well as an expanded discussion of the regionalization of the brain and development of the cranial and spinal nerves. In the chapter on organogenesis, discussion of the formation of gut structures has been increased as has the treatment of the development of the circulatory, excretory, and reproductive systems. Chapter 13 on limb development has been largely rewritten to include a more current analysis of cell-type determination, positional information, and the proposed role of the apical ectodermal ridge. In Chapter 14 on the development of the sense organs and the nervous system, the topics of nerve outgrowth and the development of behavior have been expanded. In Chapter 16 on tissue assembly and morphogenesis, the subjects of cell-cell recognition, histogenesis, and the role of the cell surface have been enlarged as has the topic of branching morphogenesis. Genetic aspects of the appearance of pigmentation have been added. Finally, the discussion of cancer has been totally rewritten and combined with that of aging, both of which are presented in a developmental context.

In addition to overhauling the text, the illustration program has also been reexamined. Those figures from the first edition felt to be inappropriate have been dropped and a large number of new photographs and line drawings have been added. Altogether, this second edition contains a much more thorough illustration program to accompany the text description.

Many people have participated in the preparation of this book and without their help its completion would not have been possible. We are indebted to the following developmental biologists who read and commented on large parts of the manuscript: Fred Wilt, University of California, Berkeley; Douglas P. Easton, State University College at Buffalo; Bonnie Joy Sedlak, State University of New York, Purchase; Marilyn S. Kerr, Syracuse University; Wayne Dougherty, San Diego State University; Kenneth R. Barker, Canisius College; and Charles Lambert, California State University, Fullerton. We are particularly grateful to Johannes Holtfreter for his interest and suggestions on discussions of the experimental analysis of amphibian development.

It has been a pleasure once again to work with the tireless staff at McGraw-Hill. Thanks there to Marian Provenzano for coordinating the many aspects of manuscript preparation and to James Amar who somehow kept track of all the paragraphs, figures, and legends despite the fact that we were continually moving them around. Finally, a special thanks is due those persons who have contributed micrographs for our use; it is these photographs which bring the study of development to life.

Gerald Karp
N. J. Berrill

Development

Chapter 1 Introduction to development

1. Nature of development

What is development? No one has completely defined it, any more than the organism itself has been fully defined. In all cases, a cell or a group of cells becomes separate from the organism as a whole, either physically or physiologically, and progressively becomes a new complete organism or a new part thereof. A fern spore settles and develops into a fern gametophyte. An insect egg may become a caterpillar which transforms into a pupa and emerges as a butterfly. The stump of a salamander leg regenerates a new limb. A microscopic cell, the human egg, proliferates and develops into a giant creature able to contemplate its own nature and origin. Development is a clearly ordered process whereby the structural and functional organization of the system, whether of a single cell or of a multicellular organism, becomes progressively expressed. Apart from energy supply and utilization, the fundamental phenomena appear to be the production of differentiated structural substances and the organization of such materials into patterns distinctive of cell, tissue, or organism. The end point is highly structured matter in action. The aim of developmental biology is to discover what processes are involved in producing this culmination, how they are related, and how they are controlled.

Life on earth has passed through three recognizable stages: the evolution of the cell and its capacity for self-replication, which occupied the first several billion years of earthly history; the evolution of multicellular organisms and their capacity for self-creation by development from single cells, which has been going on some hundreds of millions of years; and the emergence of mind with a creative imagination capable of contemplating the cell and its development, and everything else, which has happened only very recently. The three constitute a continuum: development is a phenomenon of cells, and mind is a product of development.

Development and evolution may both be regarded as a process of attaining successively higher levels of organization of matter by means of self-assembly and directed assembly. During evolution, matter progressively evolved from prebiotic assemblies to prokaryotic cells to eukaryotic cells to multicellular organisms, collectively constituting ecosystems of increasing magnitude and complexity, successive generations being linked in time by replication. In develop-

ment, as in evolution, the organism passes from one level of organization to a higher level with the emergence of new properties at each such translation. But whereas evolution has occupied several billion years of planetary time, development expresses a comparable phenomenon in hours, days, or weeks.

All cells arise from preexisting cells. There is no other way for a cell to be produced, unless scientists eventually succeed in synthesizing cells from simple elements or molecules in the laboratory. At this late stage in the history of the planet, however, a single cell may be a bacterial or a protistan organism which through division gives rise to successive new generations. Or it may be a constituent cell of a multicellular organism—animal or plant—and by multiplication contribute to the growth of the whole. In such a system it may become a relatively very large cell indeed, an egg, and by a series of divisions give rise to an integrated cell assembly, a multicellular organism of the next generation.

An egg, any egg, at the moment of activation is awesome in its potential. It is essentially always a single cell, however large. Small eggs may develop into prodigious organisms, particularly in mammals, although only with major nutritive assistance from the maternal organism. Large eggs, enormously rich in yolk, may develop independently and directly into near replicas of the adult, as in birds, reptiles, sharks, and some others. Very small eggs, entirely self-contained and unassisted, develop into adult-type creatures only when the adult itself is of minute dimensions, as in rotifers, where the egg divides into about a thousand cells and that number suffices to make a rotifer; otherwise such development is indirect, the small egg becoming only what is possible with the material it is endowed with, attaining adult size and character by devious and diverse means at later times. Yet whatever the course of development or the final destiny, all eggs undergo various progressive changes simultaneously in a strictly coordinated manner.

Following activation, an animal egg undergoes successive cleavages to become a multicellular organism consisting of a multitude of small more or less differentiated cells cohering and functioning together as a whole. Such an organism must be viable, i.e., capable of self-maintenance, by the time the various processes released during activation have run their course. The fertilized egg has an obvious unity to begin with, as a single cell with but one zygote nucleus and a single unbroken surface layer of plasma membrane and cortical material. The division of the egg into cells does not alter the fact that the developing organism remains an integral protoplasmic system. This needs to be kept in mind, in spite of the elusive nature of the integration, particularly since analytical studies tend to break the whole developmental process into separate components that appear to be independent.

The central problem of development is the developmental origin of complex structures. At the level of the single cell, the phenomenon is known as *differentiation*, or more particularly, *cytodifferentiation*, and it underlies the specialization both of single-cell organisms and of constituent cells of multicellular organisms. In multicellular development there is, in addition, cohesive cell multiplication and an orderly transformation of the cell population into successive forms of increasing complexity, i.e., *progressive organization*, over and above intracellular differentiation. The problem has long been recognized.

Whatever new qualities or features appear during the course of development, much of what happens clearly derives from the nature of the cell as we know it. Cells grow and divide and have a complex structural organization. They exhibit both process (which is temporal) and pattern (which is spatial). Contemporary

cell biology is mostly the study of temporal processes, resulting in the spectacular advances of present and recent time. Pattern and organization are spatial phenomena, of equal importance in our understanding of the organism and its development, but so far mostly eluding analytical study.

2. The organism as a whole

As the developing organism proceeds from level to level in the organizational scale, something new emerges at each step. This is the principle underlying the phenomenon of epigenesis. A simple example is that when an egg cell or a spore undergoes successive divisions, the property of cohesiveness becomes apparent, together with all the complexities of the multicellular situation. The nature and origination of intercellular cohesion and the consequences of such cohesion and its variability thus become vitally important to an understanding of the developmental process. At a later stage in the development of most animals, some tissue shifts to a location internal to the remainder, so that environmental conditions become different for the two components and in addition there is opportunity for interactions between the two. Each new circumstance leads to new possibilities, and development is primarily the orderly and ordered sequence of transformations.

The organism as a whole, although clearly recognizable, is practically indefinable. An organism is more than an object; it is a happening. A frog, for instance, is all that takes place during its life span. It begins with a cell that becomes an egg, is fertilized, and then progressively transforms to become a tadpole larva equipped with adhesive suckers and external gills, only to change into one with internal gills and no suckers. Growth continues, but at some point, depending on the species, metamorphosis intervenes and what was a tadpole becomes a froglet different in almost every character. The froglet grows to adult size, attains sexual maturity, undergoes tissue aging, and ends its life span. All this is "frog." The potential lies in the cell that becomes an egg, either in the genes alone or together with cytoplasmic properties not yet identified.

So it is with human development. The cell that becomes the human organism is one of many thousands dormant in the ovary from before birth. Of these, only about 200 in each ovary ripen during the years between puberty and menopause, and but a few become fertilized and develop. A human egg that does develop is therefore from 15 to 50 years old when it starts. During development and growth it forms transient structures that are integral parts of the organism. The fetal membranes, or amnion, which enclose the fetus, and the placenta and cord, which sustain the fetus in the womb, are cast off or cut at birth. Yet they are as much a part of the developing organism as the embryonic heart or brain is, just as milk teeth are at a later time. In all cases, an organism changes with time, sometimes dramatically, as in insect metamorphosis, sometimes with little outward sign, but always with incessant loss and replacement of vital cells and substance.

In animals, the eggs and sperm are the exclusive means of reproduction in most groups, although in many cases the egg is able to develop without being fertilized by a sperm. The egg is always a relatively large cell specialized to a greater or lesser degree to undergo development. In a few groups, new individuals can form without the intervention of an egg, i.e., by asexual reproduction. In animals, asexual reproduction consists of the development of tissue fragments usually containing many small somatic cells, in certain cases remarkably few. In plants, asexual reproduction is widespread, both by small groups of tissue cells and by asexual spores.

3. Historical perspectives

Books have been written on the history of developmental biology, although under the older name of *experimental embryology*. A chronological anthology of the more classical articles and biographical material on the most outstanding of the pioneering investigators is the "Foundations of Experimental Embryology," compiled by Benjamin Willier and Jane Oppenheimer.

The central problem has always been to understand the genesis of organismic form. Developmental biology began with the descriptive studies of as many creatures as possible, inspired by Darwin's theory of evolution and Haeckel's concept that individual development recapitulates in a condensed form the evolutionary history of the species. These studies have left us with descriptive records of great value, since the early observers, with relatively simple microscopes, made meticulously accurate and detailed accounts that are still unmatched.

Reaction to the evolutionary interpretation of developmental phenomena appeared in the 1880s. Wilhelm Roux proposed a detailed program of research which he called *developmental mechanics*, a program designed to discover the laws governing the generation of form (*morphogenesis*), in direct analogy with those previously expressed as the laws of motion. Complementing Roux' approach were the observations of Hans Driesch, his great contemporary, that embryonic parts have the capacity to generate wholes, and that the potentialities for forming a particular structure, such as a limb, are more widely distributed in developing tissue than the area actually destined to form that structure. Driesch concluded that spatial position in an embryonic system is as important as developmental age (the temporal factor) in determining the fate of embryonic tissues. This is a "field" concept, from which the contemporary notions of morphogenetic fields (Section 13.1) and positional information (Section 1.5.C) have arisen. Mechanical causality and spatial fields, two of the most powerful analytical concepts developed in science, became available to developmental biologists. The profitable outcome of this combined analytical approach has been surveyed by the leading exponents in a massive work entitled "Analysis of Development," covering the period from 1885 to 1955, edited by Benjamin Willier, Paul Weiss, and Viktor Hamburger, with major emphasis on the spectacular work on the vertebrate embryo by Hans Spemann, Hilde Mangold, and Johannes Holtfreter in Germany and Ross Harrison and associates in the United States.

In all of this, however, the role of the nucleus was hardly considered. Meanwhile, August Weismann had advanced his idea of the separation of the organism into germ plasm, whose continuity establishes the material basis of inheritance, and its carrier, the soma or body, whose form is dictated by the germ plasm. By 1900, Edmund B. Wilson had produced his monumental classic "The Cell in Development and Heredity," and Thomas Hunt Morgan soon introduced the fruit fly as the main research organism to be used in the new science of genetics. Both men, however, were very conscious that genetics and experimental embryology were but two aspects of the same thing, the cell and its development. Nevertheless, although this became widely recognized, progress for the next half century took two paths—genetics became virtually confined to the study of the connection between nucleus and characters of the phenotype, while experimental embryology remained restricted to cytoplasmic events in development without much consideration for the nuclei, which were considered to be identical throughout the organism and therefore not related to the differences between one cell and another or one part and another.

Since the midcentury, biochemical embryology has proceeded on many fronts, but the light remains focused on the molecular events associated with the storage and utilization of information within the cells and the embryo. When the struc-

ture of DNA and its manner of duplication became known, particularly its action in controlling enzyme production in bacteria, both geneticists and embryologists recognized the challenge to discover the mechanism by which genes regulate cell differentiation. In other words, how does a developing multicellular organism organize its numerous cells so that they express their genes selectively and thereby become different from one another? This dominant and insistent question pervades present developmental biology.

Advances in biochemical or molecular biology have gone hand in hand with exploitation by the electron microscope, which has portrayed the cell's anatomy as an amazingly complex, fine structure of membranes and granules within the cell and its organelles, differing from cell to cell, yet fundamentally the same in all. We are supplied, however, with static pictures of frozen moments in the ongoing life of the individual cell. Studies of living cells, living developing systems, and living organisms are essential accompaniments to the inevitably disruptive procedures of chemical analysis and electron microscopy. Otherwise the essence of the vital phenomenon is lost. At the cell level, the insight afforded by the new knowledge of molecular and cellular phenomena is great indeed. At the level of the developing multicellular system, the problems recognized and formulated by the nineteenth-century pioneers remain the outstanding problems of today. Inasmuch as developmental biology, to date, is a harmless, fascinating, and profoundly enjoyable activity, long may the challenge remain.

4. Application in developmental biology

Knowledge and understanding of developmental processes are important to human welfare:

1. Many kinds of abnormalities can arise during human development, and comprehension is the first step toward alleviation. These may be caused by genetic defects, drugs, viruses, hormones, or any disturbance of normal circumstances. The kind of abnormality produced depends less on the nature of the effective agent than on the stage of development on which it acts. In human development, whatever processes are most active at a particular time will be the ones most affected.
2. Malignant tumors are in a real sense abnormal forms of tissue development, and the nature and control of malignancy relate to the normal processes of growth and differentiation.
3. The nature of aging is elusive, and any understanding, and therefore control, of the process depends on analytical insight into cell, tissue, and organismal growth throughout the life span. The aging process is itself one aspect of development, and in humans and other mammals it can be said to begin within the womb.
4. The control of reproduction and development—at least in livestock and potentially in humans—involves the use of hormones, low-temperature sperm storage, and storage and transfers of early embryos; in effect, seed stocks are maintained and genetic improvement can be made without dependence on the quality of the foster mothers. These techniques are readily applicable to humans, and further human evolution may well depend on their exploitation, even though the social implications are alarming.

Developmental biology consequently holds out both a promise and a threat to the human species. The more that is understood concerning the manifold and integrated processes of development, the more likely we are to ensure the quality of development and growth, not only in the womb, but also throughout life. An un-

derstanding of development leads to an understanding of aging, and much alleviation of the afflictions of age might follow. Sooner or later it will be understood what it is that enables the human egg and the eggs of certain other creatures to endure in a developing, growing, and maintaining state for nearly a century, in contrast with the short-lived existence of most others. This understanding would be essential if a true extension of the human life span were to be attained, as distinct from merely warding off some of the more deleterious changes that now characterize our aging period.

The threat lies in the prospect of success itself. The more we can do technologically, the more certain it is we will undertake to do it. The technological control of development has already begun in the procedures of in vitro fertilization of mammalian eggs (including human) and their reintroduction into a suitable host mother; freezing, thawing, and transferring mammalian embryos, with the opportunity to vary the conditions experimentally before implantation into a foster mother; and nuclear transplantation, which permits numbers of genetically identical nuclei from one embryo to be introduced into a variety of different enucleated egg recipients, with the goal of producing genetically identical individuals. Each of these techniques has great benefits for animal husbandry, and the temptation exists to meddle where meddling should not be possible.

Why, then, study development at all? The benefits to humanity are there but are not overwhelming. Progress in medicine clearly depends on progress in cell biology and, to a lesser extent, in developmental biology. Public funding of any activity, scientific or other, is generally forthcoming only insofar as public officials can see beneficial application, actual or potential. Yet developmental biology and most other branches of biology exist because a few individuals during the late nineteenth and early twentieth centuries had time and opportunity to be caught up in the esthetic fascination and intellectual challenge of developmental phenomena as such, usually without research funds or assistants.

Few people who have watched—through a microscope—a fertilized sea urchin egg become an elaborately sculptured, free-swimming larva within 2 or 3 days, or who have seen a small ascidian egg become an active tadpole between dawn and dusk, are likely to forget it. The process of self-creation of organisms is an unceasing attraction and a perpetual mental challenge, and those who could devote time and attention to a materially rather unrewarding activity were fortunate. Molecular and cell biology have now progressed to such a high technological level that expensive equipment and technical assistance are necessary. At the developmental level, for the most part, this phase of sophistication has not arrived, and the scope for individual imagination and investigation remains as broad as ever. Human beings are by nature wondering, problem-solving creatures with esthetic sense, fascinated by stars, by the presence of life on earth, by the nature of matter, and by the mysterious mind-brain that they feel themselves to be. The phenomenon of development—the process of becoming, at all levels and in all forms of life—is an equally essential part of the whole. A developmental biologist, like all biologists, like all human beings, grows from a single, microscopic cell to become a giant, multicellular complex now self-consciously exploring the process of becoming, together with that of all else. The procedure is both objective and subjective and brings its own reward.

5. Primitive systems

Most of this book is concerned with the differentiation of cells and the development of animal eggs into embryos, tissues, and organs, as well as with various associated topics. Chapters 19 and 20 relate to regenerative and asexual develop-

ment as seen in animals. It is therefore appropriate here to emphasize that single somatic or nonsexual cells of plants are capable of developing into complex, multicellular forms without benefit of the internal reserves that eggs possess. This is seen most clearly in the development of spores, which are specialized for reproduction only in having a protective coat that ruptures under conditions compatible with development. For example, single spores liberated from fern fronds develop in water into *gametophytes*, small multicellular plants that produce eggs and sperm. Such chlorophyll-containing cells are able to grow and multiply as a multicellular unit by virtue of photosynthesis and do not depend on previously built-up reserves.

A. Cell division and morphogenesis in ferns

The fern spore cell, moreover, has but one set of chromosomes; it is *haploid*. Nevertheless it follows a progressive course, demonstrating that the developmental potential may be present in all cells not overly specialized for other functions. Fern spores develop readily in water that contains a mixture of inorganic salts and is exposed to light. The first cell division gives rise to one *rhizoidal* and one *protonema* cell (Fig. 1.1). The protonema cell then begins a phase of successive cell divisions in one direction, resulting in the so-called filamentous protonema.

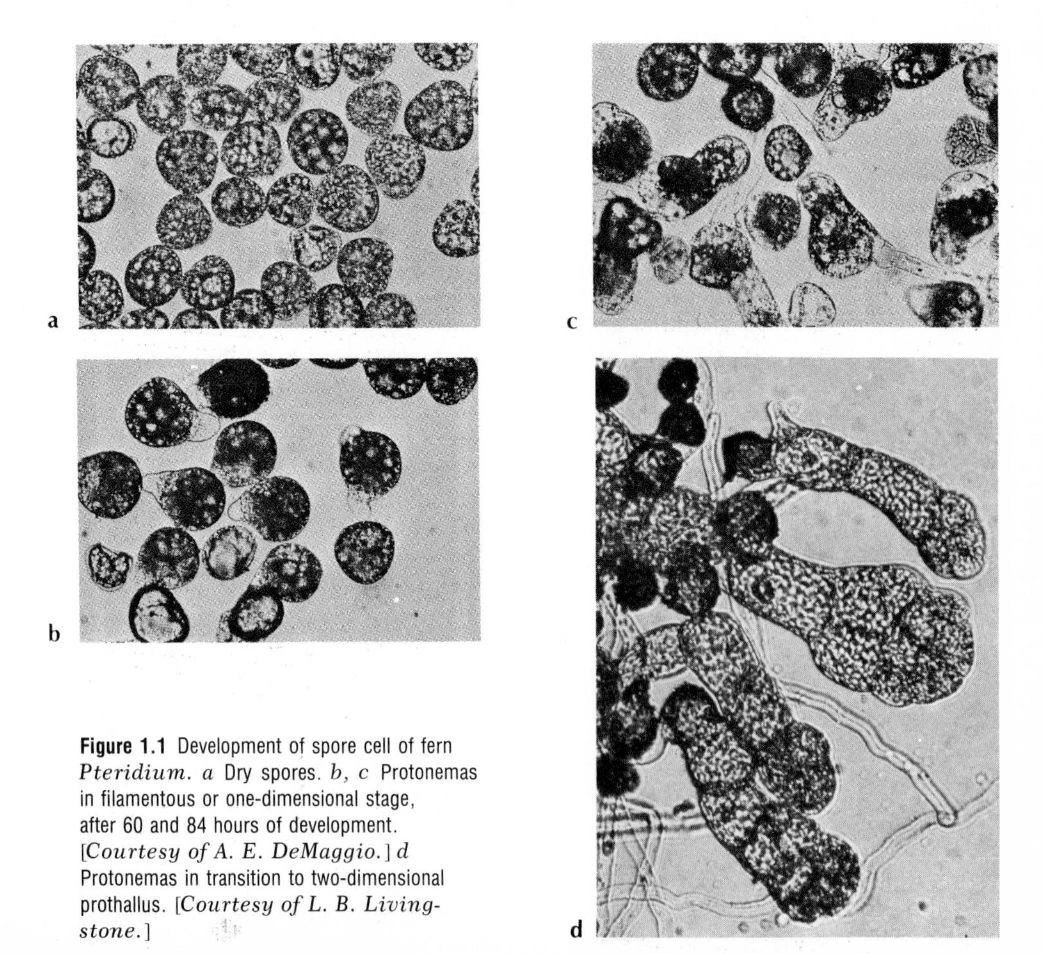

Figure 1.1 Development of spore cell of fern *Pteridium*. *a* Dry spores. *b, c* Protonemas in filamentous or one-dimensional stage, after 60 and 84 hours of development. [*Courtesy of A. E. DeMaggio.*] *d* Protonemas in transition to two-dimensional prothallus. [*Courtesy of L. B. Livingstone.*]

This growth pattern is *primary one-dimensional growth.* After a definite time lapse, the direction of cell division of the protonema apical cell is converted from longitudinal to transverse. The subsequent division of these cells takes place in all directions but in the same plane, forming a platelike, monocellular layer. This growth pattern is *two-dimensional.* The platelike organism resulting from two-dimensional growth develops first into a heart-shaped *prothallium* and then into a mature prothallium that contains antheridia and archegonia, or male and female sex-cell tissue. As a rule, the primary one-dimensional growth proceeds to a seven-cell stage. This takes about 25 days, while about 100 days are required for the completion of the mature prothallium, the gametophyte. Fertilized egg cells develop in a different manner, under different circumstances, to become the large, spore-producing fern that is more familiar and with which we are not concerned here.

The comparatively simple process just described is a good example of a developmental system suitable for analysis; one point of interest is the nature and cause of the switch from one-dimensional to two-dimensional growth. It can be induced prematurely, for instance, by transferring spores growing in red light to a white-light environment, the switch being critically dependent on exposure to the blue-light component. It is also known that a sharp increase in protein concentration occurs whenever two-dimensional growth takes place, and it is associated with a change in the nucleotide composition of the dividing cells. Accordingly, a developmental control system exists that combines an environmental agent with gene activation in the cells.

B. Self-assembly and differentiation in slime molds

Another primitive eukaryotic developmental system is now under intensive investigation, that of the cellular slime molds, of which the species *Dictyostelium discoideum* has received the most attention. Slime molds lie near the taxonomic base of the eukaryotic ladder, classified among the Protozoa by many biologists. Their life cycle has provided developmental biologists with events that resemble many of the more complex processes seen in the development of higher animals. The life cycle, shown in Fig. 1.2, is as follows: As spores germinate, each liberates a small amoeba which divides repeatedly. The population of amoebas continues to divide as long as an adequate supply of bacteria (the colon bacillus *Escherichia coli* is routinely employed in laboratory cultures) permits them to feed and grow. When the bacterial food supply is exhausted and the population of amoebas is sufficiently dense, the amoebas begin to stream together to form cell masses, or aggregates. These assume a sluglike form and move about for a while, eventually becoming upright. The anterior cells of the slug give rise to stalk cells, and the posterior cells give rise to spore cells which form an ovoid mass at the tip of the stalk, the whole being known as a *sorocarp.*

An organism with such a life cycle obviously lends itself to the study of many aspects of cell and developmental biology. Primarily three developmental phases predominate:

1. The phase of *aggregation,* whereby a local population of free amoebas congregate at a center and become an integrated mass (Fig. 1.3*a*)
2. The phase of the slug, or pseudoplasmodium, with the degree of *differentiation* that its cells exhibit (Fig. 1.3*b*)
3. The phase of morphogenesis, whereby a slug transforms into a mature fruiting body, the sorocarp, a process known as *culmination* (Fig. 1.3*c*)

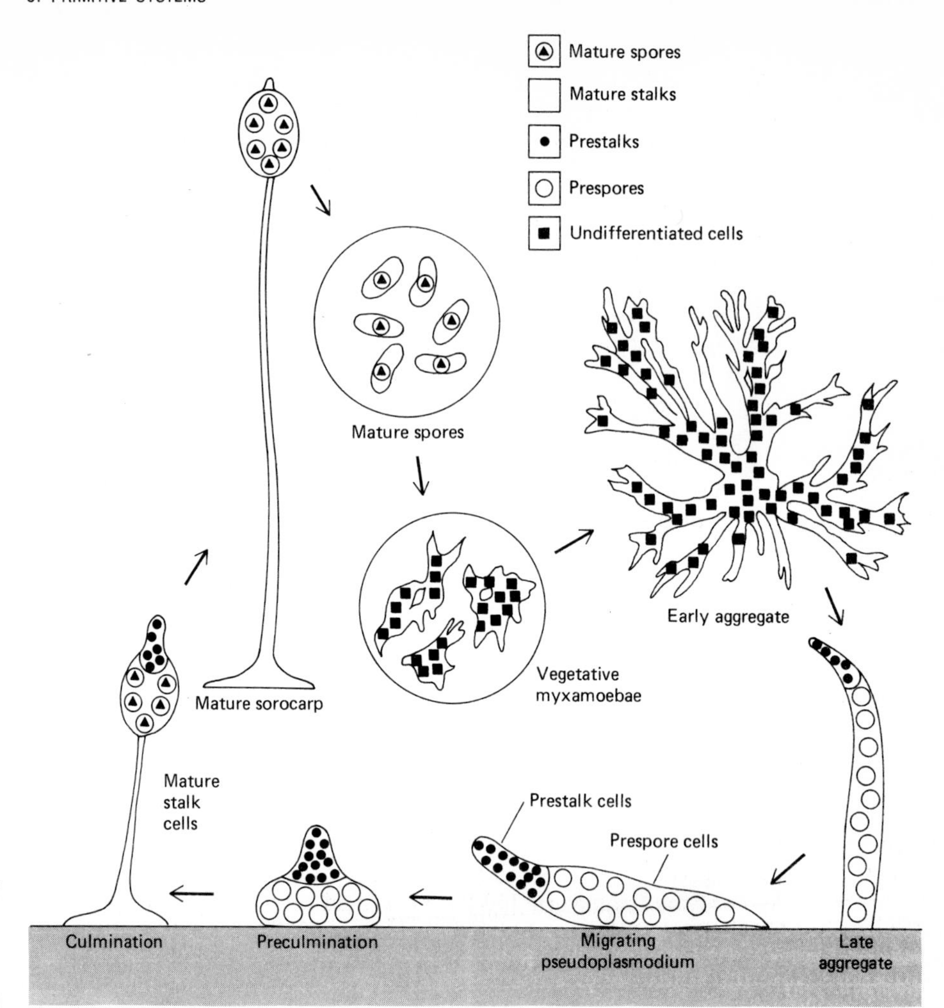

Figure 1.2 Life cycle of the slime mold *Dictyostelium discoideum*. [*Courtesy of J. H. Gregg.*]

Each phase presents problems of its own, although all relate to the attainment of the multicellular state and its subsequent self-assembly into an organized and differentiated reproductive structure.

The aggregation phase represents a profound change in the life-style of the organism, reflecting the transformation of a population of single cells into a single, multicellular structure. Why does it happen? What is the mechanism? What is the consequence? Evolution has selected adaptations that ensure survival and reproduction. In many animals, environmental hardship triggers a stage in the life cycle that is resistant to a deleterious environment and contains the potential for the establishment of the next generation, should times improve. In the cellular slime molds, the spore provides that capability, and factors in the environment trigger a series of rapid developmental activities to ensure the production of the necessary spore cells.

Why should cells living as isolates suddenly aggregate? The aggregation phenomenon has been shown to result from concentration gradients produced by the secretion of certain cells of the nucleotide cyclic adenosine monophosphate (cy-

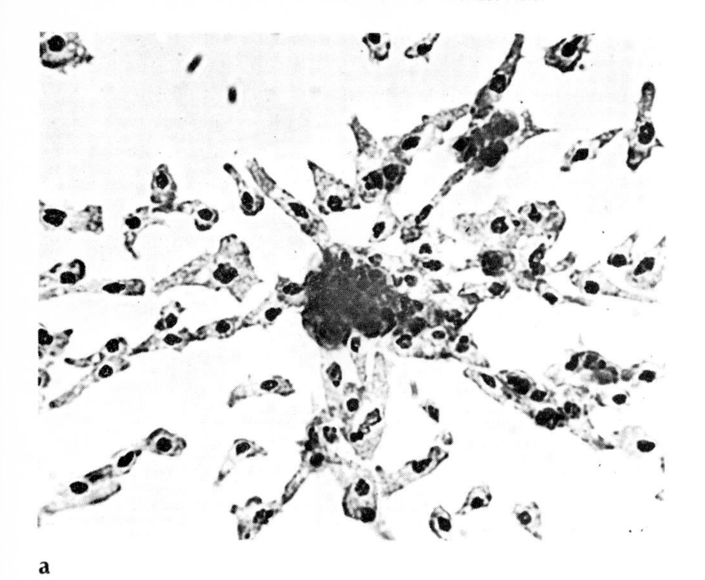

a

b

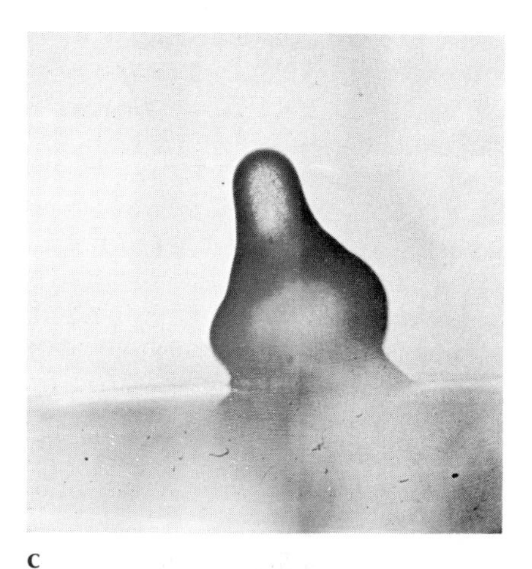

c

Figure 1.3 Stages in the life cycle of the slime mold. *a* Aggregation of amoebas at a center, showing alignment and head-tail contacts of amoebas in migration. [*Courtesy of J. T. Bonner.*] *b* Side view of slug (pseudoplasmodium) near close of migrating period, showing tip lifting from substratum and a trail of mucus behind. [*Courtesy of David Francis.*] *c* Early culmination phase and beginning of stalk formation in *Dictyostelium discoideum.* [*Courtesy of K. B. Raper.*] *d* Mature sorocarp. [*Courtesy of W. F. Loomis.*]

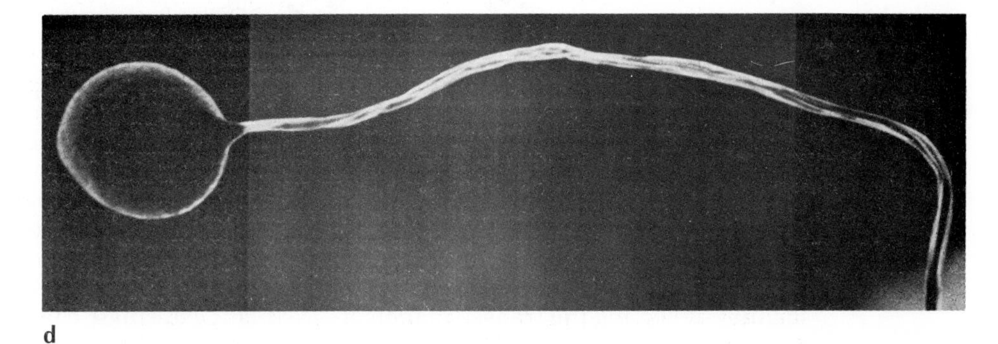

d

clic AMP). Dispersed cells within a certain distance of a source of cyclic AMP move in the direction of increasing concentration, a condition termed *chemotaxis.* Centers of aggregation may contain more than 100,000 cells or as few as a dozen. The concepts of chemotaxis and chemical gradients as controlling agents in development have prevailed in various forms since the early days of developmental biology.

The attainment of the multicellular state provides the opportunity for specialization so that different cells can accomplish different tasks. In a higher organism, one cell, the fertilized egg, gives rise to a multitude of cell types; in the slime-mold pseudoplasmodium, only two types are clearly recognizable: the pre-

spore cells, which will eventually form the spores at the top of the fruiting body, and the prestalk cells, which will eventually form a slender stalk bearing the mature spores at the apex (Fig. 1.3*d*). What determines which cells will become each type? What characteristics do each of these cells acquire, and how does this differentiation come about? How stable are the changes? That is, can the process of differentiation be reversed? This set of questions is just as valid for the differentiation of the cells of a human embryo as it is for those of a slime mold. We cannot provide all the answers in either instance, but it is hoped that the analysis of basic processes in simpler organisms can provide knowledge applicable to all forms of development.

The fate of a given cell in a pseudoplasmodium can be traced to its position within that cellular mass. Those at the leading end of the migratory slug become prestalk cells, while the remainder differentiate along the prespore line (Fig. 1.2). This observation tells us about the fate of the cells in different regions of the slug, but not about the underlying basis for a cell becoming one or the other type. Is the fate of a cell fixed by its relative position within the pseudoplasmodium, or alternatively, is the relative position of a cell fixed by its predetermined fate? If the latter possibility were the case, then we might conclude that two types of amoebas were present within a given population, one destined to differentiate eventually into stalk cells and the other into spores. During aggregation and subsequent slug formation, these two types of cells would "sort out" from one another such that the prespore group headed to the rear of the slug and the prestalk group toward the front. The alternate hypothesis, although more complex in some ways, is likely to prove closer to the truth. In this case, aggregating cells would consist of a homogeneous population of amoebas which aggregates in a random manner to form the slug. Then, during the formation of the slug, those cells which happened by circumstance to reside in the anterior portion of the aggregate would become prestalk cells, those in the rear would become prespores.

How can the relative position of a cell or group of cells within a mass determine the direction in which that cell differentiates? It must be assumed that cells located at different positions within the whole come to be influenced in different ways; i.e., they receive different types of stimulation. It can be said that cells within the mass receive some type of *positional information* which "tells" them of their relative location. Once so informed, it is up to the individual cell to interpret this information and respond in an appropriate manner. In the case of the pseudoplasmodium, the response consists of a shift in phenotype in one of the two available directions. This decision has readily observed consequences, since there are many recorded differences between prestalk and prespore cell types. The two types of cells can be distinguished biochemically on the basis of their stainability, specific protein content, antigenic properties, biosynthetic capacities, etc. Presumably the differentiation into prespore and prestalk cells reflects the activation of different sets of genes which then provide each cell with the specific templates needed to carry out its own differentiation. The underlying biochemical differentiation manifests itself morphologically, as seen in the electron microscope. The prespore cells, for example, develop a specific type of membrane-bound vacuole that is not found in prestalk cells.

Even though the cells of the slug have become differentiated in anticipation of their respective positions within the mature fruiting body, their fate is not yet sealed irreversibly. The prespore region of the slug can be isolated from the prestalk region, and each can form a complete fruiting body; under these conditions each cell can shift its path of differentiation to produce the missing type. We can conclude from this experiment that differentiation, in this instance at least, is re-

versible to a relatively late stage and that each part of the slug has an awareness of the whole. Not only can a prespore cell become a prestalk cell and vice versa, but each can halt its differentiation and revert to the single-celled, ameboid stage. For this to happen, a part of the slug is removed, reduced to single cells by mechanical disruption, and provided with a food source, whose absence may have triggered the initial aggregation. If there is no food source, the isolated cells reaggregate and return to the formation of a fruiting body. The cells of the slime mold are remarkably plastic in their differentiation. This plasticity, however, is not long-lasting. The transition from the ameboid to the prestalk or prespore state is an early step in differentiation. At a later stage in the developmental cycle, a point is reached in the life of a given prestalk or prespore cell at which it becomes committed to undergo the final steps in the differentiation of a mature stalk or spore cell, respectively. At this point a cell is said to be fully "determined" and can no longer be directed along alternate pathways.

Of primary interest to developmental biologists is the nature of the agents that cause a cell or group of cells to proceed along one path of differentiation as opposed to another. In some cases, specific chemical substances produce complex responses in competent cells, i.e., cells capable of responding to the stimulus. This is clearly the case for the aggregation response by starved amoebas to cyclic AMP. More recently, another such "on-off switch" has been found to operate at a later stage in slime-mold development. Under normal circumstances, the pseudoplasmodium which forms at the site of aggregation undergoes a period of migration over the substrate before settling down to the business of constructing a fruiting body. Presumably this migratory phase has evolved as a dispersal mechanism geared to moving the cells to a more favorable environment. Regardless of the selective advantages of slug migration, it is not a required step in the sequence of events leading to culmination. The transition from aggregation to migration appears to be triggered by a critical concentration of a ubiquitous substance, ammonia, which is excreted by the amoebas as they aggregate. If the ammonia is removed from the environment, a feat that is readily accomplished by addition of an appropriate enzyme, fruiting-body formation is initiated directly following aggregation at the site over which the amoebas had collected. It is evident that low-molecular-weight molecules such as cyclic AMP or ammonia are not substances in which developmental information can be stored. Rather, each serves as a specific trigger to elicit a preprogrammed reaction within the responding cells.

One of the consequences of the multicellular state in a pseudoplasmodium, or in any other cell mass, is the potential for intercellular communication that results from cell contact. The outer edge of all cells forms a complex structure capable of transmitting and receiving a wide variety of stimulatory signals. Cells kept in isolation, and therefore free of cell contact, cannot undergo differentiation into prespore and prestalk cells. Certain changes in the plasma membrane occur as an early event in the differentiation of slime-mold amoebas. There are changes, for example, in the size of intramembrane particles and appearances and disappearances of specific membrane proteins. Whereas the cell surfaces of vegetative amoebas fail to be adherant to one another, those of starved amoebas become very sticky. The increased adhesiveness of aggregation-competent amoebas is believed due to the synthesis of new cell-surface materials. Whether or not a change at the cell surface is the first step in differentiation in this or any other system remains uncertain, but many examples show that information received at the cell periphery has a profound effect on the internal activities of the cell.

The latter stages in the life cycle of *Dictyostelium* involve the construction of the fruiting body and provide an example of *morphogenesis*, the development of form and structure. The analysis of morphogenesis includes the study of changes in cell shape, the interaction between cells and extracellular materials such as cellulose and glycosaminoglycans, and the assembly of the parts of this complex structure.

The genetic control of morphogenesis is a complex subject, studied in various ways. Biochemical analysis has provided evidence for the role of specific messenger RNAs and proteins in these events, and genetic mutants have been isolated in which morphogenesis proceeds toward a greatly deranged fruiting body.

Events in the life cycle of the cellular slime mold have been included in this introductory chapter to provide an overview of some basic questions of concern to developmental biologists, regardless of the particular organism under study. The processes of differentiation and morphogenesis provide the focus for research in developmental biology and are discussed throughout this book.

C. Positional information: studies on *Hydra*

One of the most simple and best-studied of multicellular organisms is *Hydra*, a small freshwater coelenterate. In gross structure it is essentially a tube, or column, with a mouth, oral cone (*hypostome*), and ring of tentacles at one end, comprising the "head," and an adhesive basal disc, or "foot," at the other. The wall of the column consists of two epithelial layers (Fig. 1.4), back to back, the *epidermis* and *gastrodermis*, separated by a thin layer of collagenous material to which they both adhere. Small *interstitial cells* are distributed among the bases of the epithelial layers, especially the epidermis. They represent an unspecialized reserve capable of giving rise to nerve cells, sting cells (*nematocysts*), reproductive cells, or epithelial cells as need arises. Buds form as outgrowths from the middle region of the column, each of which successively develops a head at its distal end and finally separates from the parent (Fig. 1.5).

Hydra has been a classic subject for experimental analysis ever since its discovery more than two centuries ago, when the Abbé Trembley in Holland observed and experimented with hydras, employing very simple means. He suc-

Figure 1.4 Histology of *Hydra*. *a* Section through the gastric region of the body wall. *b* An epitheliomuscular cell. [*After L. H. Hyman, 1940.*]

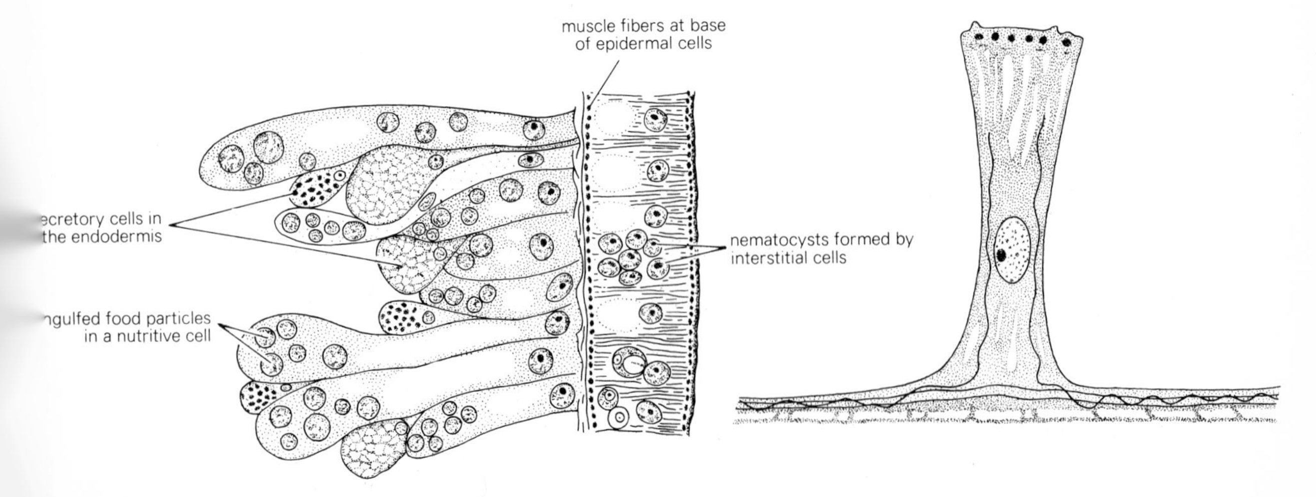

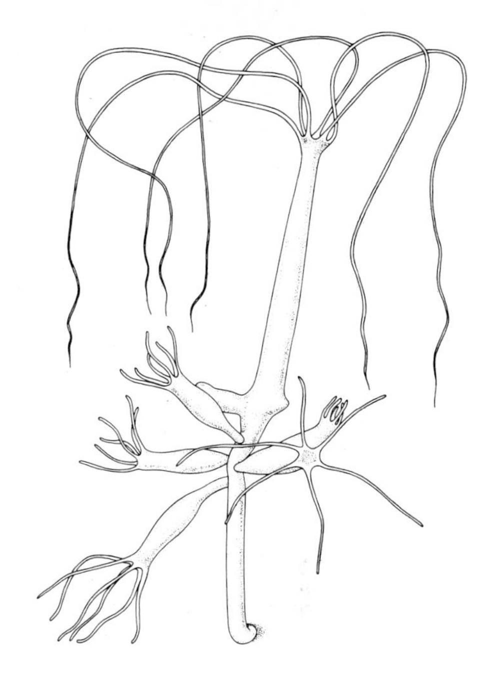

Figure 1.5 Hydra with tentacles expanded and with a number of buds forming on the body column. Successively forming buds arise in a helical arrangement, seemingly according to available space on the narrow column in the budding region. [*After P. Brien and A. Decoen, 1949.*]

ceeded in cutting the small animal in two and observed the regeneration of the amputated parts. He managed to insert a hair through the mouth of a hydra and pull the creature inside out, subsequently observing its recovery to the normal state. Recovery usually occurs by action of the musculoepithelial tissue comprising the epidermis, which reinverts the animal. In many cases, the normal state is reattained through the interpenetration of the two layers; i.e., the large gastrodermal cells independently move outward, and the large epidermal cells move inward, thereby coming to occupy their original relative positions. In this case, specialized cells reassume positions in accordance with their particular differentiation. There is a reestablishment of an inside-outside spatial relationship, a phenomenon also observed in tissue recombinations of amphibian embryos (Section 16.1).

The reestablishment of normal body form in *Hydra* can occur after disturbances much more severe in nature than that involved in turning the animal inside out. Suspensions of hydra cells produced by mechanical disruption of hydra tissues in a suitable culture medium reaggregate spontaneously in dense cell suspensions, but especially when the dissociated cells are centrifuged together. Clearly, the capacity to form or re-form the organized shape and the cell and tissue activities we recognize as a hydra is a property of any assemblage of hydra cells that consist of a minimum number and diversity. Within such an aggregate, local centers self-assemble as hypostomes of specific, innate dimensions, each of which organizes adjacent tissue in competition with other centers, a phenomenon reminiscent of aggregation in slime molds.

A hydra exhibits both polarity and individuality. If segments of the column are cut from the middle, each piece regenerates a new head from the end that was nearest the host head and a new foot from the end that was nearest the host foot. It is evident that the head-to-foot (distal-proximal) polarity of the whole individual is present in pieces throughout its length, and furthermore, that the original polarity of the tissue is retained in isolation. When heads and bases of hydras are

amputated and grafted together in series, the giant tube thus formed, although regenerating a new head with tentacles, does not remain a single giant individual (Fig. 1.6). Each portion asserts its individuality, regenerates its own head and basal part, and becomes separate.

Further insight into the control of form in *Hydra* has come from experiments in which a piece from one animal is grafted to a lateral position on the column of a second intact individual. A number of these types of experiments are shown in Fig. 1.7. Transplantation of the head from one hydra to the stalk of another (Fig. 1.7*a*) will lead to the formation of a second individual, regardless of where on the column the graft is placed. In contrast, if a piece is taken from the region below the mouth (the suboral region) and grafted to a second individual near its mouth (Fig. 1.7*b*), donor tissue will be resorbed; i.e., no second individual will form in attachment to the host. The presence of the nearby head on the intact

Figure 1.6 Polarity and individuality in *Hydra*. *a* Head of hydra (shaded) grafted to foot of decapitated hydra: decapitated hydra regenerates new head from its anterior cut surface, grafted head grows new column of its own, and the two individual hydras eventually separate. *b* A series of five decapitated hydras are grafted tandem to form a single long column. The original anterior region of each decapitated hydra reconstitutes a new head (hypostome and tentacles), commencing with the most posterior original individual, the process proceeding anteriorly according to their original polarity. [*After P. Tardent*, Arch. Entwickl.-mech., **146:**640 (1954).]

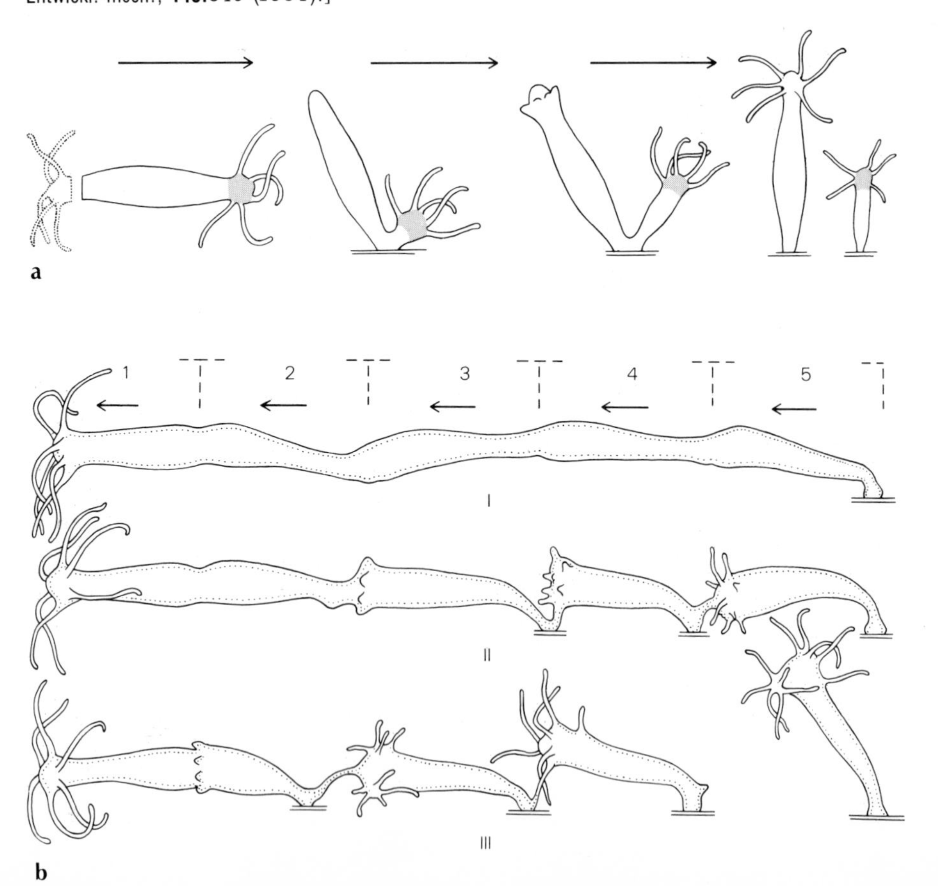

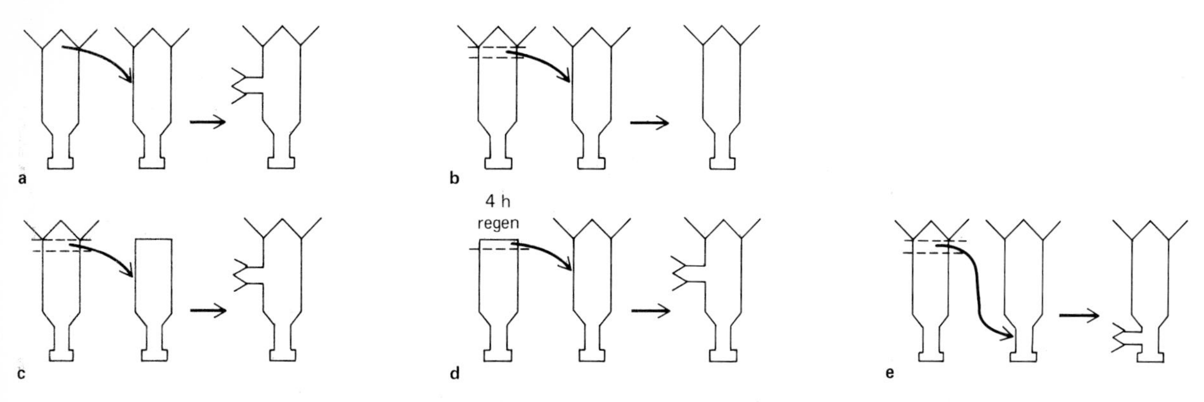

Figure 1.7 Diagram to illustrate some of the main results obtained from lateral grafts of distal regions. A piece of head will induce a secondary axis when grafted into an intact hydra *a*. A suboral region similarly placed will not do so *b* unless the host's head is removed *c* or it is allowed to regenerate for about 4 hours *d*. The suboral region will induce an axis if grafted into the peduncle *e*. [*From L. Wolpert et al.*, 25th Symp. Soc. Exp. Biol., **25**:*398* (*1970*).]

host is capable of suppressing the development of a head from this suboral graft. However, there are several ways to overcome the host's resistance to the formation of a second individual from the graft of a suboral region on its column. For example, if the head of the host animal is removed prior to the grafting procedure (Fig. 1.7*c*), the graft is now able to develop into a head in its new location (along with the regeneration of a new head on the previously decapitated host). The removal of the head prior to grafting appears to remove the suppressive capability of the host column (Fig. 1.7*d*). Alternatively, the graft can be made farther from the head of the host animal (Fig. 1.7*e*). In this position, near the base of the recipient, the suppressive power of the distant host head appears to be too weak to cause graft resorption. Still another way to promote graft development is to decapitate the *donor* individual several hours prior to removal and grafting of the suboral region onto the host. In the absence of the donor head, the suboral region of the column gains headlike properties and, following grafting, develops into a second individual.

What is the basis for these experimental results? The concept of gradients is prominent in all attempts made so far to explain the nature and control of morphological and other qualities distributed along an axis. Metabolic gradients of an oxidative-reductive nature, underlying and responsible for the differentials in visible structure, were postulated by C. M. Child early in this century, and the notion has been supported by a vast array of circumstantial evidence, yet has never been widely embraced. In *Hydra* an obvious gradient exists in the capacity for head regeneration along the length of the column, occurring most readily at the head end and diminishing toward the base. There is also a gradient in the density of nerve cells and of associated neurosecretory granules, the density being greatest at the head end and decreasing in a proximal direction. Nerves and their secretory products have in fact been proposed as the agents involved in control of growth and regeneration in this organism. Interstitial cells also have been strongly implicated in this proposal, since they are the source of new nerve cells.

The concept of positional information, as presently formulated by Lewis Wolpert and associates, has been developed with particular regard to these problems of form presented in *Hydra*. It is an attempt to introduce a more rigorous and quantitative framework for considering pattern formation and regulation and is discussed in this book in a number of places in relation to various subjects. The basis of the proposed mechanism is that cells have their position specified with respect to certain boundary regions (such as the tip of the head and/or bottom of the foot in a hydra), as in a coordinate system. As a result of its position in the system, each cell would possess a particular positional value, which would be interpreted according to that cell's genetic potential and developmental history, thereby leading to its subsequent differentiation. In this way, the phenotype of each cell will be based on its relative position within the group.

Once the idea of position specification is accepted, our attention is drawn to the specification of the boundary regions and the properties of the system that enable position to be specified. According to Wolpert, a hydra may be regarded, along its main axis, as a bipolar field. He defines a *field* as that set of cells which have their position specified with respect to the same coordinate system or boundary regions. The hydra field is bipolar, since the two boundary reference regions appear to be the head and foot ends. The variation in positional value along the hydra may be represented by a gradient in some cellular property, which decreases steadily from the head to the foot end. For this gradient to provide positional information, a fixed value must be assigned to the head and foot ends, and the values must be reestablished when a head or foot end is removed. Moreover, in order to explain the suppressive effects of the head region revealed in the type of experiments illustrated in Fig. 1.7, it is suggested that the head end produces a second substance, one that acts as an inhibitor of head formation and also falls off with increasing distance from the head.

Two simple mechanisms have been proposed by which stable gradients of a substance might be generated: (1) a simple diffusion mechanism involving continuous production of a substance by a source and continuous removal by a sink (either the sink or the source must be located at one end of the axis); and (2) a mechanism involving polarized, or unidirectional, transport of a substance coupled with back diffusion. At the present time, little evidence is available for deciding which, if either, of these mechanisms might prevail, or what the nature of the graded substance might be. The discovery that neurosecretory granules, which contain a specific hormone of peptide character, are associated with regenerative properties fits well into the proposed conceptual scheme. It is, however, the only tangible candidate for a specific substance so far that might be involved. The potential role of nerve cells in regenerative events in hydras is discussed below.

Whatever the nature of a gradient along a polar axis, the polarity itself comes first. Even though the polarity of a hydra is one of its most basic properties, it can be reversed under certain circumstances. In one type of experiment (Fig. 1.8), the head and basal discs from a vitally stained hydra were grafted to the opposite ends of the column of an unstained individual whose own head and foot had been removed. Then at 12-hour intervals, the unstained column was cut out from various of these composite individuals and allowed to regenerate from its ends. Results showed that up until 24 hours, regeneration occurred according to the original polarity of the column. After 36 hours, regeneration occurred according to the new *reversed* polarity; i.e., a head formed at that site from which a foot had originally been removed, and vice versa.

The role of nerve cells in regenerative activities was then studied by carrying out similar experiments on nerve-free animals. It has been found that treatment

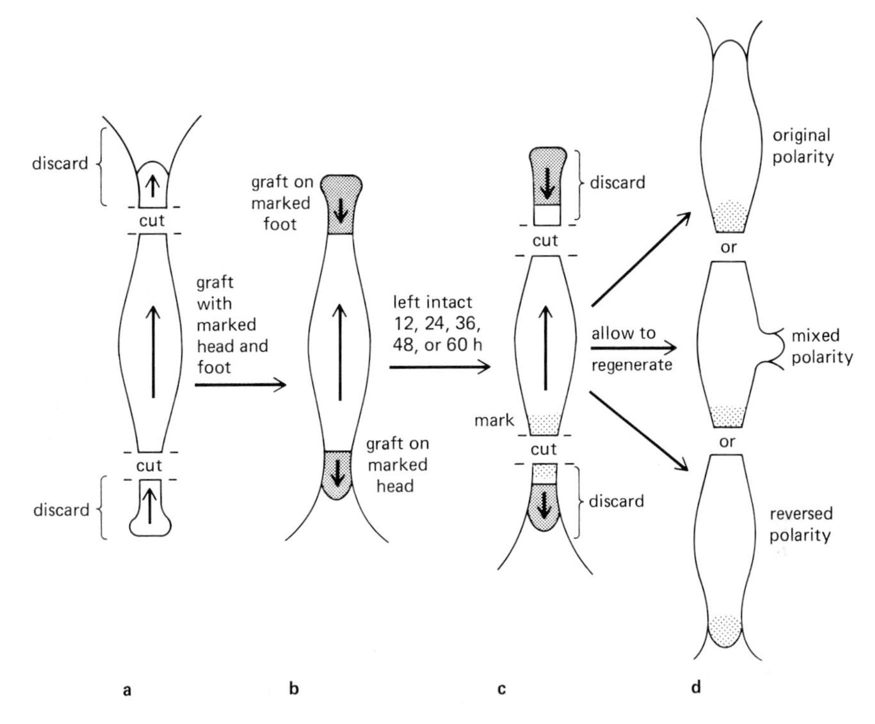

Figure 1.8 Diagram of polarity-reversal experiment. [*After B. A. Marcum, R. D. Campbell, and J. Romero,* Science **197**:773 *(1977); copyright © 1977 by the American Association for the Advancement of Science.*]

of hydras with colchicine leads to the formation of animals lacking all nonepithelial cells (interstitial, nematocyst, gland, and nerve cells) (Fig. 1.9). These individuals consist entirely of epitheliomuscular cells of the epidermis and epithelial digestive cells of the gastrodermis. Although lacking spontaneous behavior in the absence of nerve cells, they continue to bud normally and give rise to asexually propagated clones. Members of these clones were shown to consist only of the two types of epithelial cells, and they were used experimentally only after the twenty-fifth generation of asexual propagation. The experiments showed that the nerve-free hydras regenerated normally, in addition to their being able to produce buds. Moreover, reversal experiments demonstrated that full-polarity reversal occurs with the same time course as in normal hydras, beginning after 24 hours and complete by 48 hours. These experiments strongly suggest that hydra polarity is determined by epithelial cells, although by as yet unknown means. This is further substantiated by experiments of the type shown in Fig. 1.9*c*. It is found that a small graft of hypostomal tissue of a nerve-free hydra has the same capacity as a piece from a normal hydra to induce formation of a complete individual when grafted to the side of an intact host.

The positional information theory remains just that. It has undoubted descriptive value. It stimulates thought and experimentation, and it enters into discussions concerning spatial or morphogenetic aspects of development. Above all, perhaps, it consistently calls attention to the probable existence of a great and still unknown property of all cells and all developing systems, denoted by the

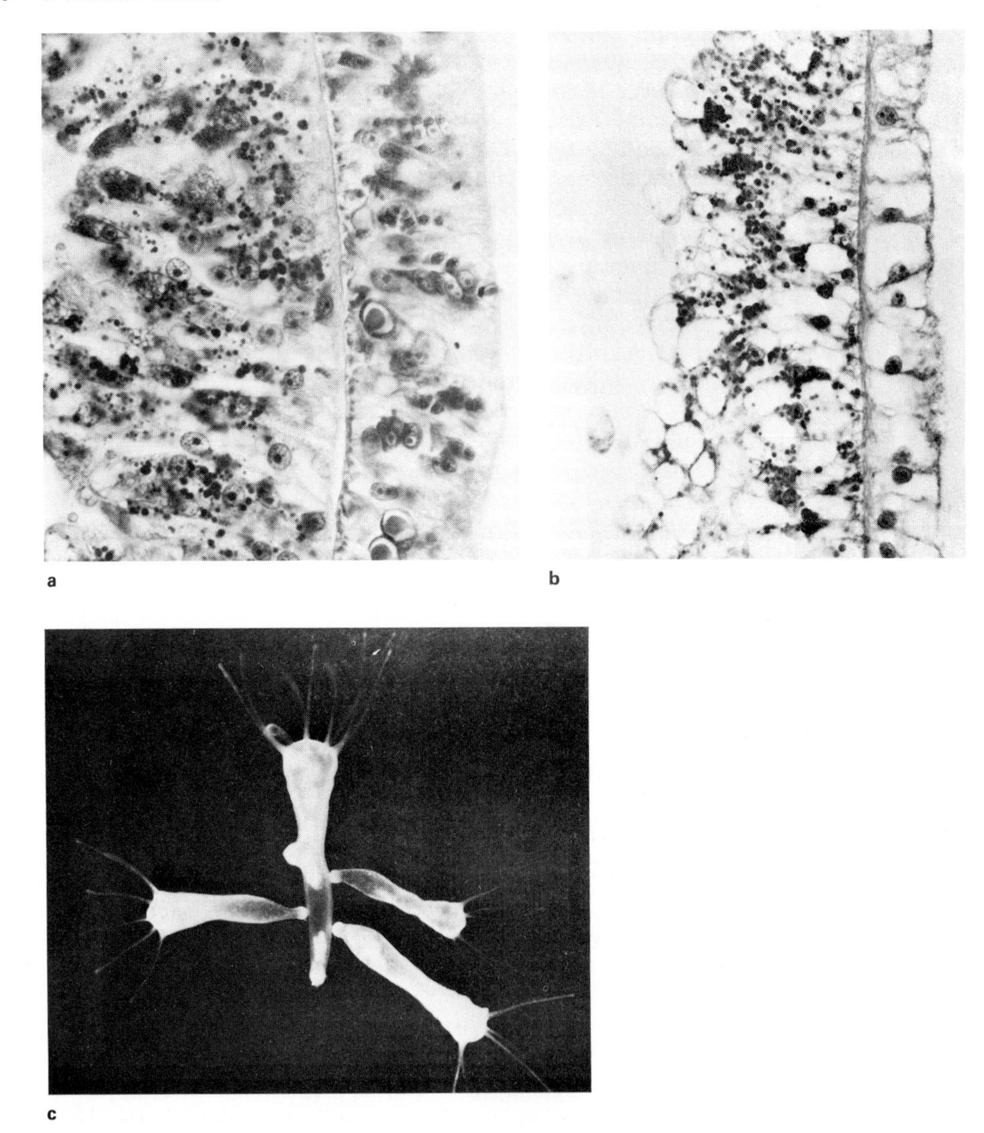

Figure 1.9 Cross section through the body wall of a normal a and nerve-free b hydra. c A nerve-free hydra with several buds on its column. [*Courtesy of R. D. Campbell and B. A. Marcum.*]

term *biological field,* more akin to physics than chemistry. This is highlighted by a recent symposium sponsored by the Society for Developmental Biology (1978), where much of the discussion concerned three questions: (1) What is polarity, how does it originate, and where does it reside? (2) What is the relationship between polarity and the origin of morphogenetic fields? and (3) How does pattern formation occur within a morphogenetic field, and what does *position effect* mean? Developmental biology may accordingly be viewed in two contrasting ways: the molecular and the field approach. Both are necessary to a full under-

standing of the cell and its development; neither can stand alone. Because of the breakthrough symbolized by the double helix, the molecular approach now dominates. We have learned little, however, of how the information stored in this one-dimensional linear sequence is translated into the formation of three-dimensional form and pattern. The field approach, although still pursued, has been relatively neglected, partly because of the promise of rich rewards from the molecular, but more significantly because of the intrinsic difficulty of the subject. Lack of insight, however, does not eliminate the problem.

We have tried in this first chapter to introduce the subject of developmental biology and discuss a few problems of developmental interest as they apply to simple eukaryotic organisms. Many of the general questions posed for these model systems in the past few pages will resurface in following chapters, primarily as they relate to embryonic development in higher organisms. First, however, we will take a side trip and consider the basic findings of molecular and cellular biology, the subjects of Chapters 2 and 3, respectively.

Readings

Black, S., 1978. On the Thermodynamics of Evolution, *Persp. Biol. Med.*, **21**:348–356.

Bonner, J. T., 1967. "The Cellular Slime Molds," 2d ed., Princeton University Press.

Campbell, R. D., 1979. "The Development of Nerve-Free Hydra," 37th Symp. Soc. Dev. Biol., Academic.

Child, C. M., 1941. "Problems and Patterns of Development," Chicago University Press.

Gierer, A., 1977. Biological Features and Physical Concepts of Pattern Formation Exemplified by Hydra, *Curr. Topics Dev. Biol.*, **11**:17–59.

Goodwin, B. C., 1977. Mechanics, Fields and Statistical Mechanisms in Developmental Biology, *Proc. Roy. Soc. Lon.*, **B199**:407–414.

Gurdon, J. B., 1977. Genes and the Structure of Organisms, *Proc. Roy. Soc. Lon.*, **B199**:399–406.

Lash, J., and J. R. Whittaker, 1974. "Concepts of Development," Sinauer.

Locke, M. (ed.), 1970. "Control Mechanisms of Growth and Differentiation," 25th Symp. Soc. Exp. Biol., Academic.

Loomis, W. F., 1975. "*Dictyostelium discoideum,* A Developmental System," Academic.

———, 1979. Biochemistry of Aggregation in Dictyostelium, *Dev. Biol.*, **70**:1–12.

MacWilliam, H. K., and J. T. Bonner, 1979. The Prestalk-Prespore Pattern in Cellular Slime Molds, *Differentiation,* **14**:1–22.

McMahon, D., and C. West, 1976. Transduction of Positional Information During Development, in G. Poste and G. L. Nicolson, "The Cell Surface in Animal Embryogenesis," Elsevier.

Muller, K., and G. Gerisch, 1978. A Specific Glycoprotein as the Target Site of Adhesion Blocking Fab in Aggregating Dictyostelium Cells, *Nature,* **274**:445–449.

Needham, J., 1942. "Biochemistry and Morphogenesis," Macmillan.

Newell, P. C., 1978. Progress on Slime Mould Morphogens, *Nature,* **271**:302–303.

Raghavan, V., 1974. Control of Differentiation in the Fern Gametophyte, *Amer. Sci.*, **62**:465–475.

Schaller, H. C., 1978. "Action of a Morphogenetic Substance in Hydra," 5th Symp. Soc. Dev. Biol., Academic.

Shostak, S., 1974. The Complexity of Hydra: Homeostasis, Morphogenesis, Controls, and Integration, *Quart. Rev. Biol.*, **49**:287–310.

Soll, D. R., 1979. Timers in Developing Systems, *Science,* **203**:841–849.

Spemann, H., 1938, "Embryonic Development and Induction," Yale University Press (reissued 1968).

Subtelny, S., and I. R. Konigsberg (eds.), 1979. "Determinants of Spatial Organization," 37th Symp. Soc. Dev. Biol., Academic.

Willier, B. H., and J. M. Oppenheimer, 1974. "Foundations of Experimental Embryol-

ogy," Prentice-Hall. Contains classical articles by Roux, Driesch, Wilson, Boveri, Harrison, Child, Spemann and Mangold, Holtfreter, Loewenstein et al.[1]

WILLIER, B. H., P. A. WEISS, and V. HAMBURGER (eds.), 1955. "Analysis of Development," Saunders.

WILT, F. H., and N. K. WESSELLS, 1967. "Methods in Developmental Biology," Crowell.

WOLPERT, L., 1969. Positional Information and the Spatial Pattern of Cellular Differentiation, *J. Theor. Biol.*, **25**:1–47.

[1] For reference throughout this text, the articles in this chronological anthology are listed here. **Contents:** 1888, Contributions to the Developmental Mechanics of the Embryo. On the Artificial Production of Half-Embryos by Destruction of One of the First Two Blastomeres, and the Later Development (Postgeneration) of the Missing Half of the Body (W. Roux). 1892, The Potency of the First Two Cleavage Cells in Echinoderm Development. Experimental Production of Partial and Double Formations (Hans Driesch). 1898, Cell-lineage and Ancestral Reminiscence (Edmund B. Wilson). 1902, On Multipolar Mitosis as a Means of Analysis of the Cell Nucleus (Theodor Boveri). 1907, The Living Developing Nerve Fiber (Ross G. Harrison). 1908, Observations on Oxidative Processes in the Sea Urchin Egg (Otto Warburg). 1913, The Mechanism of Fertilization (Frank R. Lillie). 1914, Susceptibility Gradients in Animals (C. M. Child). 1916, The Theory of the Free-Martin (Frank R. Lillie). 1924, Induction of Embryonic Primordia by Implantation of Organizers from a Different Species (Hans Spemann and Hilde Mangold). 1939, Tissue Affinity, A Means of Embryonic Morphogenesis (Johannes Holtfreter). 1954, *In Vitro* Experiments on the Effects of Mouse Sarcomas 108 and 37 on the Spinal and Sympathetic Ganglia of the Chick Embryo (Rita Levi-Montalcini, Hertha Meyer and Viktor Hamburger). 1969, Ionic Communication between Early Embryonic Cells (Shizuo Ito and Werner R. Loewenstein). 1973, Positional Information in Chick Limb Morphogenesis (D. Summerbell, J. H. Lewis and L. Wolpert).

Chapter 2 Molecular biology and assembly

Current research in developmental biology leans heavily on the findings and technology of those who have examined the mysteries of metabolic function in the simpler viral and bacterial systems. Many of these studies represent some of the finest endeavors in the history of science. Their elegant yet simple design has allowed questions to be answered concerning organisms too small to be seen with the light microscope. Their methods and tools have been adopted by biologists working at higher levels of organization, and we are rapidly finding ourselves in possession of an extensive knowledge of the workings of the complex eukaryotic cell. The developmental biologist must utilize this core of basic information on cellular structure and function in an attempt to understand the processes involved in embryonic differentiation and morphogenesis. Cellular and molecular biology are starting points for the study of development, and one must become familiar with the basic techniques and core of knowledge in these areas in order to appreciate developmental research. That is the goal of this chapter and the one that follows.

Although many molecular processes found in microorganisms are also found in animal cells, many are not. The bacterial cell is a simple unit by mammalian standards, one that has evolved a direct responsiveness to the foreign environment around it. The cells of higher plants and animals are, instead, responsive to the needs of a multicellular organism, and the levels of regulation are vastly more complex. This is reflected in the complexity of the genome, in the architecture of the cytoplasm and cell surface, and in the many intercellular activities. During the course of these two chapters we will consider selected properties of both prokaryotic and eukaryotic cells.

The most important questions at the molecular level concern the storage and utilization of genetic information. It is clear to all of us that the basic characteristics of any organism are transmitted from generation to generation in the form of some type of genetic information. In some unexplained way, the geometry of a spider's web, the score of a bird's song, the human capability to recall an event from childhood, and the complex regulatory processes that maintain the homeo-

static activities of the cell must all be *coded for* within the genetic material passed on within the body of a sperm and an egg. In many ways, we have only scratched the surface in our attempts to understand how this vast genetic inheritance is utilized in the activities of the organism.

1. The concept of the gene

The credit for the concept of a "unit of inheritance" belongs to the Austrian monk Gregor Mendel, who in 1866 published the results of his studies on the breeding of different races of pea plants. Starting with pure stocks that had recognizably different characteristics, he artificially cross-pollinated the plants and carefully recorded the characteristics of the offspring over several generations. From these data, Mendal reached several conclusions that now form the foundation of the science of genetics. He concluded that a pair of discrete "factors" governed each trait and that they segregated upon formation of the gametes. We know this pair of factors as the maternally and paternally derived alleles on homologous chromosomes that first come together at fertilization and later segregate during meiosis.

Mendel's findings, published in an obscure periodical, were rediscovered in 1900, a time when several investigators were beginning to open doors in the science of genetics. During the first half of the twentieth century, genetic studies by many biologists, most notably T. H. Morgan, on a number of organisms (particularly the fruit fly) were being carried out. These studies provide much of our present knowledge of genetic inheritance. The geneticists of this period discovered that new genes could appear as mutations of existing genes and that crossing-over and recombination could redistribute maternal and paternal characteristics. They realized that genes occur in a linear sequence, and they described many complex *linkage groups*, i.e., groups of genes that segregate together. Another branch of genetics, *cytogenetics*, was making great strides in explaining the physical basis of the gene theory. The chromosome was singled out as the carrier of the linear array of genes and the physical basis of the linkage groups. The duplication and separation of the chromosomes and their crossing-over activities during meiosis were observed and described.

One type of cell of particular interest was discovered and greatly exploited for its usefulness to cytogenetics. This special cell, found in several tissues of the larval fruit fly, contains giant chromosomes which, except for their unusual size, reflect the nature of the genome in every cell of the fly. These chromosomes result from duplication of genetic material and failure of the duplicates to separate during mitosis. This process continues until there are over 1,000 identical units together in perfect genetic register. The giant chromosomes have a banded appearance under the light microscope (Fig. 2.1), and individual bands have been correlated with specific genes. Thus the cytogeneticist could examine the chromosomes of an individual fly and be presented with a visual array of the genes of that fly. Similarly, visible alterations in these chromosomes led to explanations of gross genetic change. Missing genes (*deletions*) were observed, as were a variety of ways in which the chromosomal pieces could be broken and reorganized.

In the past quarter century, much of genetic theory has been explained at the molecular level. The single, most basic question in molecular genetics has been the nature of the genetic material. This is the point of departure of this still-young science. During the first half of this century, considerable information became available on the composition of macromolecules. Proteins were the logical candidates for the role of genetic carrier. In contrast with nucleic acids, proteins were known to mediate complex reactions and were composed of a variety of different

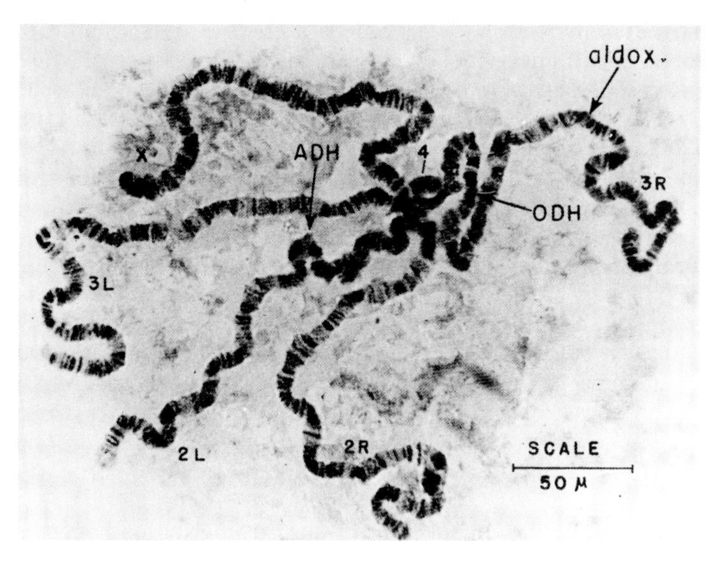

Figure 2.1 The giant polytene chromosomes of *Drosophila melanogaster*, prepared from a squash of a larval salivary gland. Several genetic loci are indicated. [*Courtesy of H. Ursprung.*]

building blocks. There are approximately 20 different amino acids in a protein, but only 4 different nucleotides in a nucleic acid molecule. Presumably this greater complexity could better satisfy the requirements of the job.

We know now, of course, that protein is not the genetic carrier. The first report to point directly to deoxyribonucleic acid (DNA) was by Oswald Avery, Maclyn Macleod, and Colin McCarty in 1944 on transformation in pneumococcus. Two strains of the bacteria had been isolated: one produced colonies having a smooth (S) appearance and was able to cause pneumonia in a suitable host; the other grew into rough (R) colonies as a result of a defect in its capsule and was nonvirulent. When a cell-free extract of the S bacteria was added to the medium in which the R strain was growing, a few of the latter grew into smooth colonies and were virulent. They had been *transformed.* From the time of transformation, the progeny of that cell continued to have the properties of the S strain; i.e., the transformation was a stable genetic change. Avery and his coworkers purified the contents of cells after their disruption in detergent and found that among cellular constituents, only the purified DNA was capable of causing the transformation.

To further corroborate this finding, they treated the cell extract with various enzymes that digest specific macromolecules. Treatment of the homogenate with ribonuclease or protease had no effect, indicating that transformation could occur in the absence of RNA or protein; these could not be the genetic material. Enzymes that destroyed DNA, however, eliminated the ability of the homogenate to transform. As a result of these and later experiments, it was shown that in order for transformation to occur, DNA fragments entered the recipient cell intact and substituted in the bacterial chromosome for the original DNA, which was eliminated; a new genetic microorganism was produced.

Even though the chemical identity of the gene was clearly established by 1952, the elucidation of the actual structure of DNA was yet to come. To appreciate the analysis of DNA structure by James Watson and Francis Crick, we will consider some of the facts available at that time.

2. The structure of DNA

DNA was known to be a very long, fibrous molecule with a backbone composed of alternate sugar and phosphate groups joined by 3′,5′-phosphodiester linkages. To each sugar was attached one of four possible nitrogenous bases. The bases were of two types: the pyrimidines, cytosine (C) and thymine (T), and the purines, adenine (A) and guanine (G). It was known that the amount of purine equaled the amount of pyrimidine, and specifically that adenine content was equal to thymine content and guanine to cytosine. However, the percentage of adenine plus thymine or guanine plus cytosine could vary from source to source. With this type of information in mind, Watson and Crick in 1953 proposed, from the analysis of x-ray diffraction studies and the construction of feasible models, that DNA had the following structure (Fig. 2.2): DNA was made of two chains of nucleotides coiled around a common axis, with the sugar-phosphate backbone on the outside

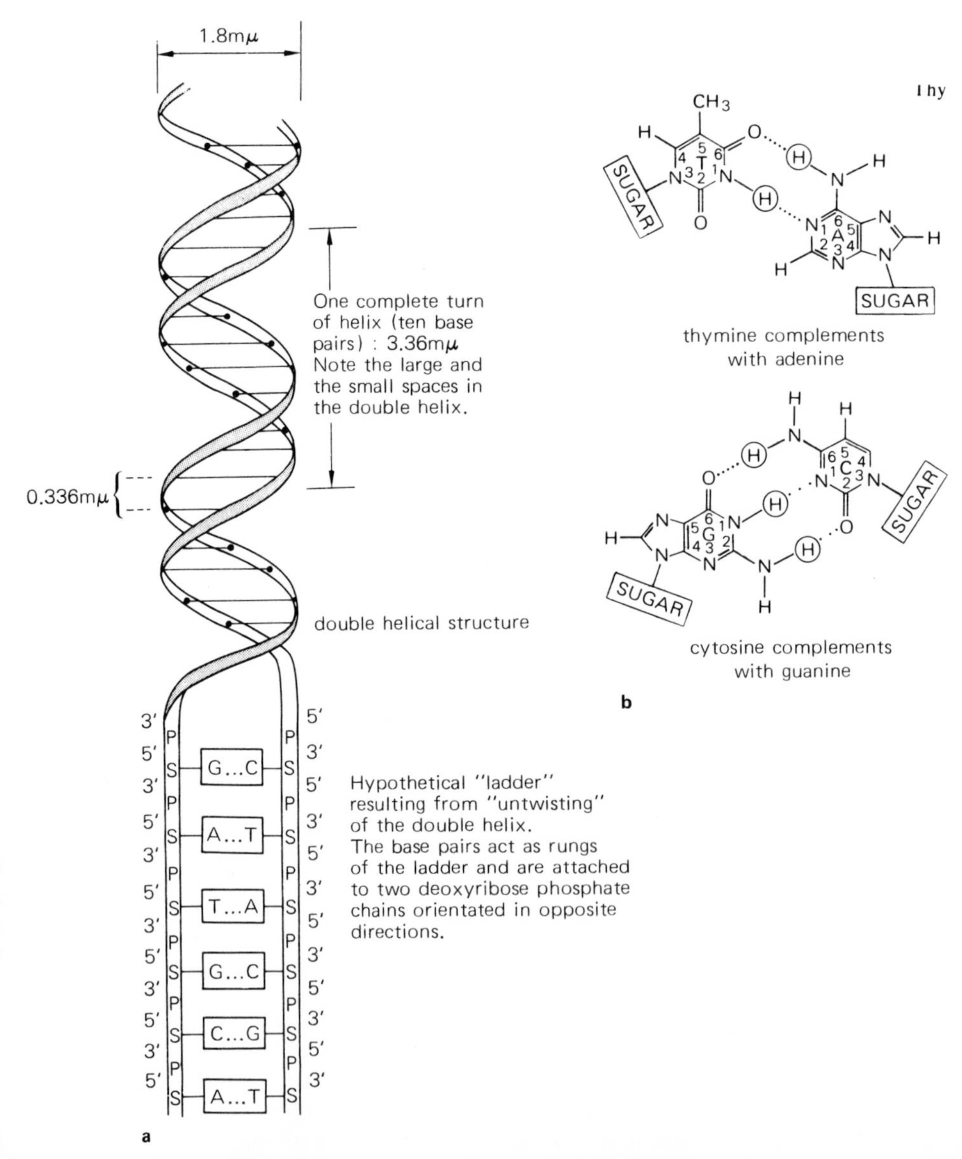

Figure 2.2 Structure of DNA. *a* Nature of the double helix. *b* Hydrogen bonding between the base pairs in DNA. [*After N. A. Edwards and K. A. Hassal, "Cellular Biochemistry and Physiology," McGraw-Hill, 1971.*]

and the bases pointing in toward the axis. The two chains were held together by hydrogen bonds, which occur between each base of one chain and an associated base on the other chain. The 20-angstrom (A) width of the fiber requires that a pyrimidine from one chain is always paired with a purine of the other chain. Consideration of the structures of the bases suggested that there were further restrictions on the association of nucleotides. It was suggested that adenine was the only purine capable of bonding to thymine and guanine was the only purine capable of bonding to cytosine. Therefore, the only possible pairs were AT and GC, which fit perfectly with the previous base-composition analysis.

With this model Watson and Crick speculated on the relationship between DNA structure and several basic genetic functions. They proposed that the information in DNA was coded for by the linear sequence of the base pairs. They theorized that a mutation could be accounted for by a chance mistake in the formation of the sequence during duplication. Finally, they suggested that during duplication, each chain would act as a template for the formation of the companion chain. By this mechanism, the requirement for self-duplication of the genetic material was met.

One of the most important aspects of the Watson-Crick model was the proposed *complementarity* between hydrogen-bonded nucleotides. For example, adenine is complementary to thymine, AGC is complementary to TCG, and one entire chain is complementary to the other. If the base sequence of one chain were to become known, then the base sequence of the complementary chain would automatically be available. The concept of complementarity of nucleic acids—in DNA chains or in RNA chains—is of overriding importance in nearly all the activities and mechanisms in which this class of macromolecules is involved. Not only was the elucidation of the structure of DNA significant in its own right, but it also provided the stimulus for investigation of all the activities in which the genetic material must take part.

DNA is a highly asymmetric molecule. Although the double helix is only 20 A in diameter, the single circular DNA molecule comprising the *Escherichia coli* chromosome is approximately 1,100 microns (μm) in length, while molecules nearly an inch long have been isolated from animal cells. It is generally believed that each chromosome of an eukaryotic cell contains a single unbroken DNA molecule. Since the center-to-center distance between adjacent nucleotides on a chain is 3.4 A, there are approximately 4 million base pairs in an *E. coli* chromosome. It is estimated that the *E. coli* chromosome contains in the neighborhood of 5,000 genes, each being composed of a discrete sequence of nucleotides representing a small section of the total continuous nucleotide sequence of the molecule.

Since the information content of a nucleic acid molecule is determined by the specific sequence of its component nucleotides, it has been imperative that molecular biologists develop techniques for the analysis of nucleic acids based on their nucleotide sequences. Consider two strands of DNA whose length and overall base composition are identical, but whose base sequence is quite different. How can one go about distinguishing between these two molecules? They cannot be separated by centrifugation, electrophoresis, gel filtration, or any other bulk biochemical technique that one might use to separate two proteins. Differences in the base sequence of a molecule hundreds of nucleotides long can involve very subtle distinctions. The way to distinguish between molecules of different sequence is to use complementary molecules as probes. Techniques of this type involve the association of specific complementary single-stranded nucleic acids to form stable double-stranded duplex, or *hybrid*, molecules. The hybrids formed

can be of a DNA–DNA, DNA–RNA, or even RNA–RNA nature. The technique, termed *molecular hybridization*, stemmed from an early finding that single-stranded bacterial DNA molecules (formed by heating a solution of DNA to 100°C to separate the two strands) were capable of reassociation under appropriate conditions to form stable double-stranded molecules that were essentially indistinguishable from native DNA. It appears that the *reannealing* process is highly specific; only those strands which are truly complementary to one another are capable of complete association.

Utilizing a technique based on nucleic acid hybridization, Roy Britten and David Kohne made an important discovery in 1966. They found that within the genome of higher organisms, some DNA sequences were present as many copies, while other sequences were present in only one copy per haploid set of chromosomes. The techniques used by Britten and Kohne in their studies are relatively complex, but they must be understood before the significance of the findings can be appreciated. (The reader also will have to understand these techniques to follow the series of investigations of gene activity that will be described in later chapters.) To begin the analysis, a solution of DNA is prepared and the large DNA molecules are broken into fragments, in this case of approximately 400 nucleotides in length. DNA is generally fragmented into smaller pieces of uniform length by forcing a solution through a tiny orifice at high hydrostatic pressure. The double stranded DNA fragments are then denatured into their component single strands by heating the solution to approximately 100°C in the appropriate salt solution.

Each strand of a DNA duplex releases a strand of complementary base sequence. Two of these types of fragments are considered here: first, a fragment with the same sequence of several hundred nucleotides that is present repetitiously in many places on several different chromosomes; and second, a fragment with a unique nucleotide sequence found in only one place on one of each set of chromosomes. Clearly, the solution of single-stranded DNA will have many more fragments with the repetitious sequences than with the unique sequences. In solution these DNA fragments are continually colliding with one another, and inevitably those with a complementary base sequence will come into contact. At 100°C, H bonds cannot form between the colliding complementary strands, but if the temperature of the solution is dropped to about 60°C, stable double-stranded fragments (reannealed DNA) can result from successful collisions. The sequences most common within the DNA and present at the greatest concentration will have the greatest number of successful collisions and will reanneal most rapidly. Therefore, the longer a solution of single-stranded fragments is allowed to remain at 60°C in the appropriate solvent (e.g., 0.12 M phosphate), the greater will be the percentage of the DNA present in a double-stranded reannealed form. Sequences present at very low concentration will reanneal only after very long periods.

This relationship is shown by the reannealing curves of Fig. 2.3 (see the appendix to Chapter 8 for further discussion). A reannealing curve for mammalian DNA has several components. First there is a fraction representing about 10 percent of the total DNA. It reanneals so rapidly its progress can be followed only in dilute solutions where the concentrations of the reactants are very low. This is called *satellite DNA;* it consists of great numbers of very short nucleotide sequences and has been localized primarily in the centromeres of mammalian chromosomes. Another fraction (representing about 20 percent of the total DNA sequences in the mouse) reanneals over a part of the curve that indicates that the sequences are present in numerous copies in the genome. This is the *repeated*

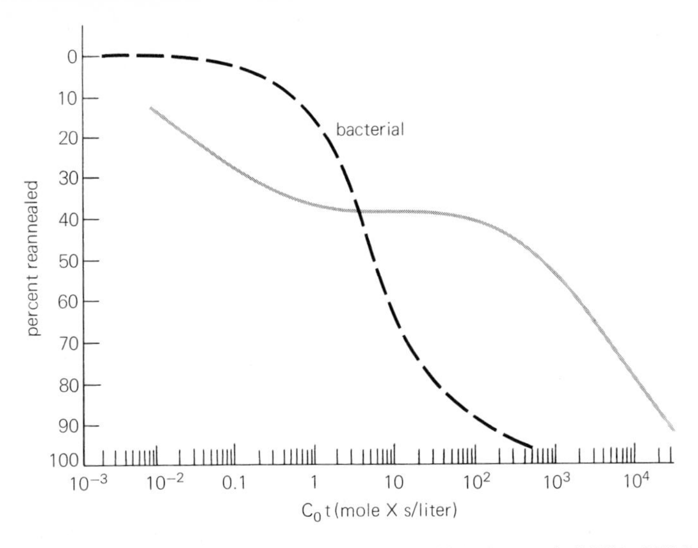

Figure 2.3 Reannealing curves of heat-denatured (single-stranded) DNA. DNA is fragmented, converted to single strands, allowed to incubate at a suitable temperature (e.g., 60°C) in a suitable medium (e.g., 0.12 *M* phosphate buffer) for various times, and the percentage of DNA that has reannealed is determined. The term C_0t is a measure of the concentration-time variable. Either the greater the concentration or the longer the time, the greater will be the number of successful collisions between complementary strands, and the greater will be the percentage of the DNA that has reannealed. A solution containing a high concentration of DNA incubated for a short time will have the same C_0t as one of low concentration incubated for a correspondingly longer time; both will have the same percentage reannealed. In this figure the dashed line represents the reannealing of bacterial DNA, and the solid line represents mammalian DNA. The mammalian-reannealing curve starts with a significant percentage already double-stranded, reflecting the reannealing of satellite DNA at very low C_0t. The first part of the mammalian curve represents the reannealing of the repeated DNA sequences. The last part of the curve (from 50 to 100 percent reannealed) represents the nonrepeated sequences. The reannealing curve for the bacterial DNA has a simpler shape. Bacterial DNA has essentially no repeated DNA sequences and therefore reanneals with the same kinetics as single-copy (unique) mammalian DNA. The bacterial curve is to the left of the single-copy mammalian DNA because the sequences are at a much higher concentration. The bacterial genome is small and, therefore, a comparable amount of bacterial DNA will have a greater number of genomes and thus a greater number of each sequence. [*After R. J. Britten and D. Kohne, 1968.*]

fraction; its function remains unknown, with the exception of those DNA sequences known to code for histone proteins, ribosomal RNA, and transfer RNA, which together constitute a small percentage of the repeated DNA. The bulk of this DNA is believed to have important regulatory functions.

The remaining fraction of sequences (representing 70 percent of the DNA of the mouse) appears to be present in only one copy, or at best a very few copies, per haploid genome. This is the *unique (single-copy) fraction,* which has been shown to include the DNA that codes for a number of different messenger RNAs (including those for hemoglobin, silk fibroin, ovalbumin, etc.). Of known mRNAs, only those for the histones have been shown to be synthesized from repeated DNA templates, which are generally found to be highly clustered within the genome. Calculations indicate that one gene copy per haploid set of DNA can

turn out sufficient numbers of mRNA molecules in the time allowed to account for the synthesis of protein, even when one protein greatly dominates the cell's synthetic activity.

The evidence from the analysis of different-sized DNA fragments from both the sea urchin and *Xenopus*, an amphibian, suggests that approximately 50 percent of the genome contains closely interspersed repeated and unique sequences. The repeated sequences average 300 nucleotides in length, while the unique sequences average about 800 nucleotides. A significant fraction (about 20 percent in the sea urchin) is made of essentially uninterrupted unique DNA sequences. In the sea urchin, about 6 percent of the genome apparently contains relatively long regions of repeated DNA. The remaining DNA is made of repeated sequences interspersed with unique sequences of considerable length (4,000 nucleotides). Indications are, therefore, that the organization of the genome is very complex and its significance is yet to be understood.

In the past several years, a number of techniques of great importance in nucleic acid methodology have been developed. Enzymes, termed *restriction endonucleases*, have been isolated which are capable of cleaving a DNA molecule at sites of specific nucleotide sequence. These enzymes allow one to fragment DNA from complex eukaryotic genomes into a predictable collection of specific shorter pieces. Second, advances in DNA sequencing techniques have made it possible to rapidly determine the exact base sequence of large DNA fragments. As a result of this latter methodology, the sequences of many genetic regions are becoming known. Third, techniques of DNA cloning have been developed which allow investigators to (1) isolate DNA fragments of particular sequence present among large numbers of contaminating fragments and (2) greatly amplify the number of copies of a given DNA fragment. Cloning procedures involve considerable methodology, and a bit of background information is necessary for an understanding of the general technique.

In many bacterial cells, small DNA molecules occur which exist independently of the main bacterial chromosome. Included in this group are viral chromosomes (such as that of the bacteriophage lambda) and *plasmids*, which are small circular bacterial DNAs that maintain a semiautonomous existence in the cell along with the main bacterial chromosome. Plasmids contain a variety of genes coding for nonessential functions, among them ones that confer resistance to various antibiotics. Molecular cloning techniques take advantage of the fact that bacterial cells will pick up small circular DNA molecules (such as plasmids) from the medium and replicate them along with their own chromosomal DNA. In order to clone eukaryotic DNA fragments, viral or plasmid DNA is made to contain a piece of eukaryotic DNA (making it a *recombinant DNA molecule*), which also will be picked up by the bacterial cells and replicated. Since the uptake of a plasmid is a rare event, most cells in the culture will remain devoid of extrachromosomal genetic elements, while a small percentage of the cells now will contain the recombinant molecule. Since the uptake of the plasmid is accompanied by a gain in resistance to particular antibiotics, those members of the population which lack the plasmid (and the attached eukaryotic fragment) can be selectively eliminated by addition of the antibiotic. The resistant bacteria can then be grown as colonies from which the cloned DNA can be recovered.

There are basically two different types of eukaryotic DNA sequences that have been cloned by these procedures. In one case, the DNA of an organism is fragmented by use of restriction endonucleases and the various fragments are joined to individual plasmids, thereby producing a population of recombinant DNA molecules. When this population of recombinant DNAs is incubated with host bacte-

ria, different cells can be expected to pick up plasmids containing different eukaryotic fragments. After a period of exposure, the plasmid-deficient cells are destroyed, and the remainder are spread sparsely over a culture dish so that individual cells produce distinct colonies of cells of identical genetic composition. In most of these experiments, a particular genetic region, such as that containing a globin gene or the ovalbumin gene, is sought. Bacterial colonies containing plasmids with the desired sequence can be identified by in situ hybridization, and the particular eukaryotic DNA fragment can be isolated in a purified state for use in subsequent experiments or sequence analysis. It is possible in these types of experiments to isolate, in addition to the coding regions of important genetic loci, the noncoding portions which flank the structural genes and are believed to be important in the control of gene expression.

In the other type of approach, the DNAs being cloned are ones that have been synthesized in vitro from RNA templates; they are termed *cDNAs*. In one type of experiment, a purified mRNA is isolated, such as the globin mRNA from reticulocytes, and used as a template for the production of a homogeneous population of complementary single-stranded DNA molecules. These single-stranded molecules can be made double-stranded, combined with the plasmid DNA, and cloned so as to greatly increase the amount of this specific sequence available for further work. In other experiments, a mixture of mRNAs can be used to produce a heterogeneous population of cDNAs, from which individual sequences can be isolated after amplification within bacterial colonies. These techniques have been of great value in the study of nucleic acids and will undoubtedly find large application by developmental biologists studying various aspects of gene function. A few of the prospects of DNA cloning in the study of development are considered in the paper by Igor Dawid and W. Wahli (1979).

3. DNA replication

The term *replication* describes the process of the doubling of DNA or DNA synthesis. The details will not concern us here, but a few of the most important points will be mentioned. Replication is semiconservative; i.e., each of the strands of the parent molecule remains intact and becomes one of the strands of each of the two daughter molecules. Each strand serves as a template during the copy process for the polymerization of the complementary strand. The separation of strands, the unwinding, and the polymerization are accomplished by an enzyme, DNA polymerase, in association with several other proteins. The copy process is precise, although a rare mistake is made when an incorrect nucleotide is incorporated into the strand, producing a mutation, i.e., a change in nucleotide sequence. This is a major cause of mutation, another cause being a modification of a nucleotide base that was already incorporated into the DNA molecule.

DNA synthesis is not a continual process in the life of a eukaryotic cell; it is restricted to a period between mitoses and is sandwiched between two periods when DNA is not synthesized, G_1 and G_2. In other words, in a dividing population of eukaryotic cells, there is a period following mitosis when DNA synthesis has not yet begun. This period is G_1 (gap_1). In the following period, the S phase, DNA synthesis begins and continues until the chromosomal DNA is doubled. Studies indicate that DNA synthesis is a carefully controlled event; certain chromosomes are predictably replicated in early S phase and others replicated in late S phase. Within each chromosome, many points of replication are found along each DNA molecule. After DNA synthesis is completed, the cell enters G_2 (gap_2) preparatory to its next mitosis.

4. Transcription

The chromosomes are the seats of information storage and as such have virtually no direct effect on the cell's activities. To the present time, only one way has been uncovered in which DNA can manifest itself in cell function. This is indirectly via the synthesis of complementary RNA molecules, which is the process of *transcription*. In general, RNA differs structurally from DNA in three ways: (1) the sugar is ribose rather than deoxyribose; (2) thymine is not incorporated into RNA but is replaced by uracil, which differs by lacking a methyl group; and (3) RNA is generally a single-stranded molecule, although base pairing is still possible and many RNAs have double-stranded regions.

As research continues, the number of types of RNA discovered continues to grow. The three RNAs first identified, messenger RNA, transfer RNA, and ribosomal RNA, are the best studied and most important. As will be discussed in detail later, messenger RNA (mRNA) is the one that contains within its linear sequence of nucleotides the information for the specific sequence of amino acids of one polypeptide chain. Transfer RNAs (tRNAs) are the intermediates in the manufacture of proteins and are responsible for bringing the proper amino acid, as specified by the bases of the mRNA, into the growing polypeptide chain. Ribosomal RNA (rRNA) is found as part of the ribosome. Ribosomal RNA is initially transcribed as one long RNA molecule, which is then broken into several fragments by a complex process within the nucleolus. In eukaryotic cells, two large (the 28 S and 18 S) and one small (the 5.8 S) fragment are retained as ribosomal RNA. The 28 S and 5.8 S rRNAs join with a 5 S species, which is synthesized from an entirely distinct genetic site, in the formation of the 60 S ribosomal subunit. The 18 S RNA becomes the only nucleic acid in the smaller 40 S subunit. Ribosomal RNAs are believed to be involved in numerous ribosome functions, including assembly, structural maintenance, and protein synthesis itself. The tRNAs and mRNAs also are synthesized from larger RNA precursors. The interaction of these RNAs in protein synthesis will be reviewed later.

Transcription is accomplished by the activity of an enzyme, RNA polymerase, of which the eukaryotic cell has more than one type. RNA polymerases catalyze the formation of a molecule of RNA whose base sequence is complementary to one strand, the sense strand, of the DNA. In a bacterial cell, the *single* type of polymerase consists of a core enzyme, made of several polypeptide chains, which functions in association with another polypeptide, termed the *sigma factor*. An enzyme made of more than one polypeptide chain is said to have a subunit structure. Depending on the particular enzyme, the subunits may be identical or diverse in nature.

The function of the polymerase is complex. In prokaryotes, it is clear that the chromosome is composed of a unbroken molecule of DNA without protein or other material within the DNA chain. From this chromosome-sized DNA, small RNA molecules of defined length and base sequence must be produced. In order to accomplish this task, the polymerase must recognize where it should bind on the DNA molecule, which strand it should use as a template, and where transcription should begin. In bacterial polymerase activity, these recognition functions are accomplished by the sigma factor. There are special regions of the DNA, the promoter regions, with which the sigma factor of the polymerase first makes contact. The promoter sites of the chromosome contain specific nucleotide sequences to which the polymerase can bind, thereby preventing the enzyme from attaching to the incorrect strand or the internal region of a gene. The transcription process requires that the polymerase move along the DNA template strand in the 3′ to 5′ direction, separating the strands as it travels and fabricating a com-

plementary antiparallel RNA strand which grows from its 5′ terminus in a 3′ direction. As the polymerase moves along, it must insert the proper nucleotide into the growing (*nascent*) chain at each site. The reaction being catalyzed can be written as

$$\text{RNA}_n + \text{NPPP} \rightarrow \text{RNA}_{n+1} + \text{PP}_i$$

in which ribonucleoside triphosphates (NPPP) are polymerized into a chain of nucleotides with the simultaneous release of pyrophosphate (PP_i). Presumably the enzyme selects the proper ribonucleoside triphosphate (UTP, GTP, CTP, or ATP) for incorporation on the basis of its ability to form complementary hydrogen bonds with the nucleotide in the DNA strand being transcribed. Once the polymerase has moved past a particular site, the double helix is reformed; the RNA chain does not remain associated with its template as a DNA-RNA hybrid. It has been estimated that bacterial RNA polymerase molecules are capable of incorporating approximately 50 nucleotides into a growing RNA molecule per second. Just as transcription is initiated at specific points in the chromosome, it also is terminated when a specific nucleotide sequence is reached. In some cases, a protein termed the *rho factor* is required for termination, an event followed by chain release.

The details of transcription are presented because they illustrate the dynamic processes that occur at the molecular level. Enzymes are catalysts; they are constructed in such a way as to facilitate a specific chemical reaction without being themselves altered in the process. These proteins are inanimate objects, capable of performing these complex acts in unison with one another as a result of their geometric shapes and the reactivities of the component amino acids. It must be presumed that each act produces the necessary change in the substrate and the enzyme itself, whereby the next act must inevitably follow. Polymerization follows nucleotide selection, and movement must then occur before the process can be repeated. These events occur in conjunction with conformational changes within the proteins, changes which involve subtle alterations in the relative positions of the polypeptide chains. The development of techniques to visualize transcription in the electron microscope has provided a fitting visual demonstration of this important process. The photographs of "genes in action," such as that shown in Fig. 2.4, represent precisely the type of portrait expected on the basis of a large body of biochemical information.

The process of DNA-dependent RNA synthesis leads to the production of messenger RNA molecules that contain the same information as that stored within the genetic material. The ultimate role of these RNA messages is to serve as a template for the production of a polypeptide chain, i.e., a continuous sequence of amino acids joined by peptide bonds. Whereas the formation of RNA is a relatively straightforward enzymatic process, the production of the corresponding polypeptide chain is a vastly more complex activity. This added complexity is not surprising when one considers that protein synthesis involves the utilization of information in one language, that composed of a nucleotide alphabet, to form the words of a very different language, one which uses an amino acid alphabet. Before considering this process of *translation*, it is necessary to briefly consider events that might occur between synthesis of an RNA molecule and its use as a template in protein synthesis.

5. Translation

In a bacterial cell, the synthesis of an mRNA strand is followed by the immediate translation of that strand into protein. In other words, the processes of transcrip-

tion and translation are closely linked both in time and in space. In contrast, in the cells of higher organisms there are numerous intermediate steps between the formation of an RNA molecule and its translation. The eukaryotic messenger RNA is a molecule containing a sequence of nucleotides which code for a single polypeptide chain; i.e., it is said to be *monocistronic*. In addition to the coding sequence, there are considerable stretches of nucleotides on both the 3′ and 5′ sides of the template section. The function of these noncoding sequences is not well understood, although certain portions are believed to be involved in specific steps of protein synthesis. The terminal sections of the molecule contain unusual features. The 5′ terminus of virtually all eukaryotic messages contains a methylated guanine present in an inverted orientation and separated from the next-to-last nucleotide by a triphosphate. This 7-methylguanine "cap" functions in the initial steps of protein synthesis. The 3′ end of the majority of eukaryotic mRNAs contains a tail of adenosine residues referred to as *poly-A*. These adenosines are not coded for by poly-T stretches in the DNA, but rather are added enzymatically one at a time after the RNA has been synthesized. These poly-A fragments are in the order of 200 nucleotides long and have been implicated primarily as a determinant in the stability of the message. Since neither the 5′ cap nor the 3′ tail are part of the initial RNA transcript, they represent posttranscriptional modifica-

Figure 2.4 Micrograph of a nucleolus and of ribosomal cistrons transcribing rRNA (in the newt *Triturus viridescens*). *a* Thin section of extrachromosomal nucleolus, showing compact fibrous core surrounded by a granular cortex. *b* Portion of a nucleolar DNA molecule in the process of transcription. RNA molecules (in the form of ribonucleoprotein complexes) in various stages of completion are evident in the lateral extensions from the central thread of DNA. RNA polymerase molecules can be seen at the base of each matrix fibril. [*Courtesy of O. L. Miller, Jr. and Barbara R. Beatty, Biology Division, Oak Ridge National Laboratory.*]

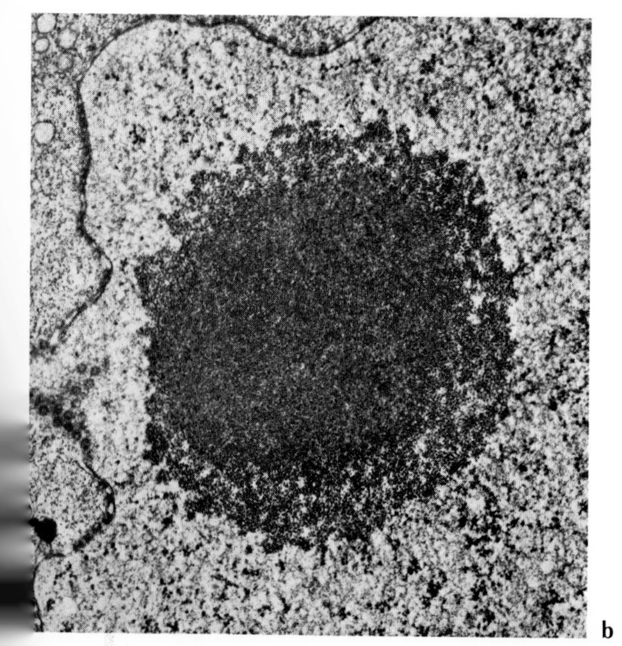

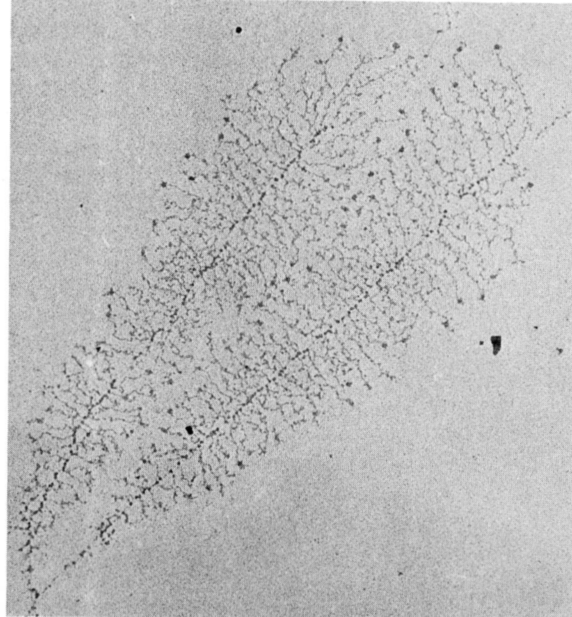

tions. Whereas the 5′-methylguanine cap seems always to be added in the nucleus to the end of newly synthesized RNA, the 3′ poly-A can be added either within the nucleus or in the cytoplasm. Cytoplasmic polyadenylation of preexisting mRNA is particularly evident in eggs soon after fertilization. Despite considerable investigation, the role of poly-A in RNA metabolism remains uncertain. The best indications suggest that the poly-A extension is involved in protecting the message from nuclease degradation, thereby extending its lifetime in the cytoplasm.

As described in the previous paragraph, the structure of the eukaryotic mRNA is well understood; in fact a variety of these molecules have been sequenced; i.e., the precise order of nucleotides is known. However, mRNAs are not synthesized in the same forms as that existing in the cytoplasm, but rather arise by a series of ill-defined processing steps from larger precursor molecules. If one incubates mammalian cells in a radioactive RNA precursor for a very brief period of time (for example, 2 minutes), extracts the RNA, and analyzes the radioactive products, one finds that radioactivity appears in a diverse array of RNA molecules whose size is considerably larger than the size of the mRNA population. If the labeling period is extended to an hour or more, radioactivity in these large RNA transcripts drops, indicating their relatively short life span in the cell. This rapidly labeled population of RNA molecules is referred to as *heterogeneous nuclear RNA* (*hnRNA*)—"heterogeneous" because it contains molecules of widely different molecular weight and "nuclear" because they are restricted to the nucleus. The events occurring between the formation of the *primary transcript* and the smaller mRNA product are referred to as *processing steps*. The nature of these steps, which are only recently becoming better understood, involve the removal of specific sections of RNA from the precursor molecule. Not only are pieces removed from the 3′ and 5′ ends of the precursor, but pieces are actually taken out from within the coding stretch of the molecule, a step that is followed by the splicing of the two newly formed ends together. The ovalbumin mRNA of the chick, for example, is formed by the removal of seven distinct pieces, termed *intervening sequences*, that are present as interruptions of the template sequence within the primary transcript. Intermediates in the processing of specific mRNAs have been isolated, and therefore, the mechanism responsible for these highly specific activities may soon be understood.

The information content of an mRNA molecule is coded in its linear sequence of nucleotides, specifically in the form of a continual sequence of triplets of nucleotide bases (*codons*). Because there are 4 possible nucleotides, there are 64 possible codons, but only 20 amino acids to be coded for. Three of the possible codons specify termination, and the remaining 61 triplets, including the initiation codon AUG, specify amino acids. Obviously there are amino acids with several different codons; i.e., the code is *degenerate*.

There is no obvious steric relationship between a triplet sequence of nucleotides in RNA and any given amino acid; thus the RNA cannot serve as a direct template for polypeptide formation. Rather, an intermediate must serve to translate the words in nucleic acid language into an amino acid alphabet. This intermediate function is performed by the transfer RNAs. One end of each tRNA molecule has a triplet sequence (*anticodon*) complementary to a specific triplet in the mRNA. The other end of the tRNA molecule becomes charged with a specific amino acid for which that mRNA triplet must code. Since the sequence of amino acids polymerized in a forming polypeptide chain is dependent on the sequence of tRNAs that interacts (via the codon-anticodon complementarity) with the mRNA template, it is essential that each tRNA is charged with the appropriate amino

acid. The specificity in the linkage between the amino acid and the tRNA is accomplished by a collection of enzymes, termed *amino acid activating enzymes,* that catalyze the reaction whereby the two molecules are linked. Even though amino acids generally have more than one species of tRNA to which they are attached, a single amino acid activating enzyme is capable of "charging" all the tRNAs specific for a particular amino acid. It is apparent, therefore, that the ability of each enzyme to recognize specific tRNA and amino acid molecules is of paramount importance in the entire process of information utilization within the cell.

As mentioned earlier, protein synthesis is an extremely complex process. The components required include the mRNA template, the ribosome, charged tRNAs, GTP, and a considerable variety of soluble, i.e., nonribosomal, protein factors. The events occurring during polypeptide formation are very similar in eukaryotic and prokaryotic cells, although the structure of the components can be quite different. The major differences are found in comparison of the ribosome with the mRNA. Ribosomes of prokaryotic cells are smaller and contain fewer proteins than those of eukaryotic cells. Whereas the mRNAs of eukaryotic cells are monocistronic and contain poly-A and 7-methylguanine termini, the prokaryotic message is typically polycistronic, i.e., contains the information for more than one polypeptide chain and lacks these terminal modifications.

The synthesis of a polypeptide can be divided into three rather distinct activities: the *initiation* of the chain, its *elongation,* and finally, its *termination.* The first step in protein synthesis occurs by the association of the small subunit of the ribosome with a specific initiation codon of the message, an AUG located near the 5′ end of the coding section. This triplet, which codes for methionine, serves as the first codon of the message to be read and ensures that the ribosome will translate the message in the proper reading frame. All polypeptide chains to be synthesized begin with methionine (or formylmethionine in prokaryotic cells) at their amino terminus. This methionine (or formylmethionine) may be removed in a subsequent step, so that the second amino acid to be incorporated comes to reside at the amino end of the chain. The attachment of the small subunit of the ribosome to the mRNA is followed by the association of the large subunit and the entry into the ribosome of the methionyl tRNA (formylmethionyl tRNA in prokaryotes), whose anticodon forms a complementary interaction with the AUG codon. Three soluble protein factors are required in prokaryotes for the events of initiation to occur, and one molecule of GTP is hydrolyzed.

Each ribosome is believed to contain two binding sites for tRNAs, the A site and the P site. At first the P site is occupied by the charged methionyl tRNA and the A site is open (Fig. 2.5). The first step in elongation occurs with the association of the appropriate tRNA with the second codon of the message in the A site of the ribosome. Peptide bond formation is accomplished by the transfer of the methionine (or formylmethionine in prokaryotes) on the tRNA of the P site to the amino acid attached to the tRNA that has just entered the ribosome. The deacylated tRNA of the P site is displaced from the ribosome, which then moves one codon along the mRNA in the 3′ direction. This last step, termed *translocation,* is accompanied by the movement of the tRNA dipeptide to the P site of the ribosome, still hydrogen-bonded to the second codon of the mRNA. Translocation opens the A site of the ribosome for the entry and association of the third aminoacyl tRNA, whose anticodon is complementary to the third codon of the message, and the cycle is repeated. Each elongation cycle is accompanied by the hydrolysis of two molecules of GTP and the involvement of at least three distinct soluble protein elongation factors in prokaryotes.

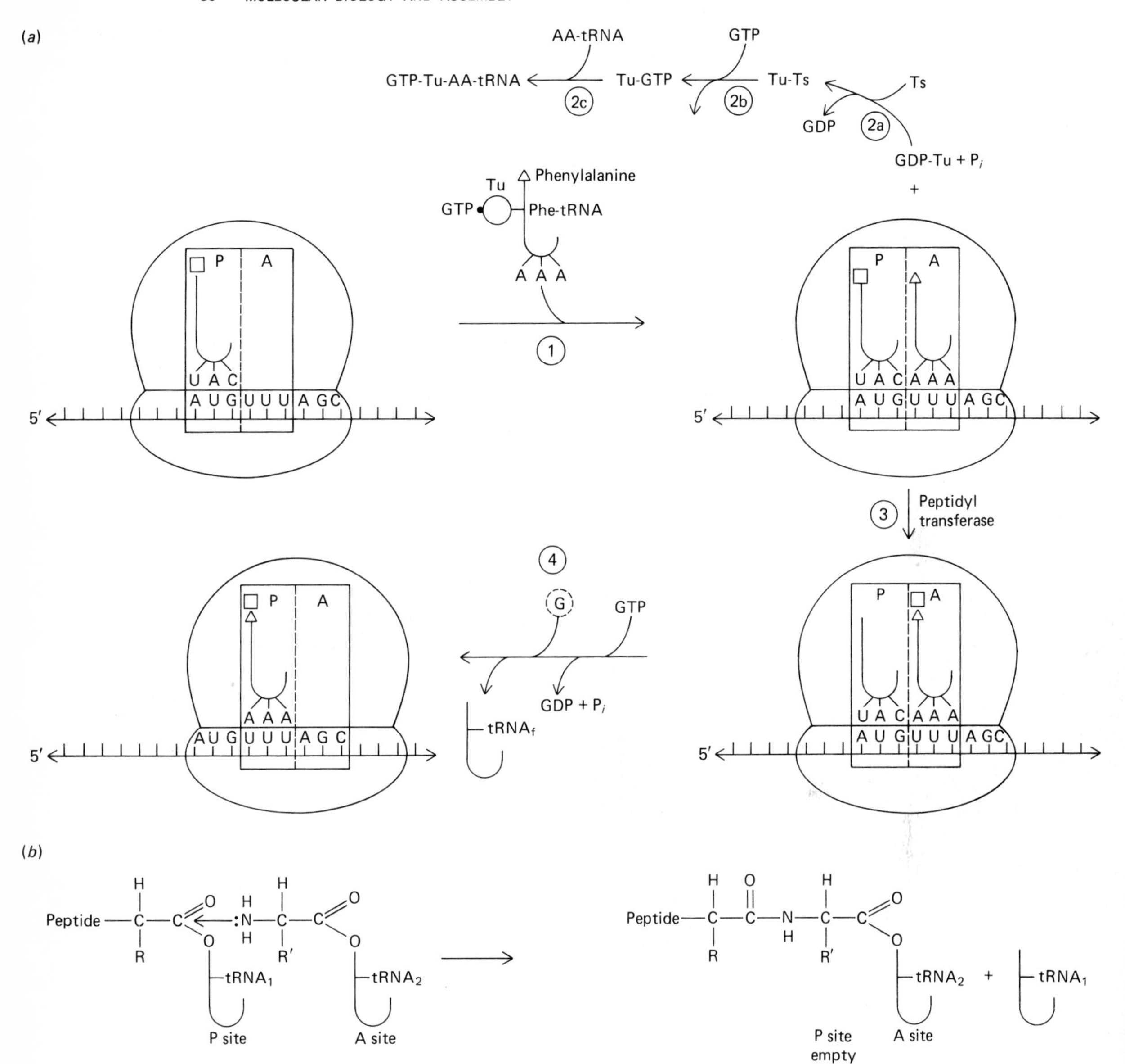

Figure 2.5 *a* Steps in the elongation of a polypeptide chain: (1) the entrance of the aminoacyl-tRNA into the empty A site of the ribosome; (2) the binding of the tRNA is accompanied by the release of GDP-Tu, which is recycled as shown; (3) peptide-bond formation is accomplished by the transfer of the nascent polypeptide chain from the tRNA at the P site to the aminoacyl-tRNA of the A site; and (4) the binding of factor G and the hydrolysis of GTP result in the release of the tRNA from the P site and the translocation of the ribosome relative to the mRNA. *b* Peptide bond formation and the subsequent displacement of the deacylated tRNA of the P site.

The last step in protein synthesis occurs when the ribosome reaches one of the three termination codons, UAA, UGA, or UAG. When this happens, no new tRNA associates with the ribosome-mRNA complex. Instead, the newly synthesized polypeptide chain is released from the ribosome and its connection with the last tRNA is severed. The events of termination also require the hydrolysis of GTP and the involvement of soluble protein factors.

Protein synthesis is carried out by numerous ribosomes attached to the same mRNA at the same time, just as RNA synthesis is accomplished by numerous polymerase molecules operating simultaneously at different sites along the DNA template. The complex formed by the mRNA and its associated ribosomes is termed a *polyribosome* or *polysome;* it represents the functional unit of protein synthesis (Fig. 2.6). Typically, the number of attached ribosomes reflects the length of the mRNA chain, although various factors can modify this relationship. Polysomes involved in the synthesis of small polypeptides, such as the histones or globins, generally contain about 5 or fewer ribosomes (Fig. 2.6*a*), while those involved in the synthesis of a very large polypeptide, such as that of myosin (Fig. 2.6*b*), can contain as many as 40 or more ribosomes.

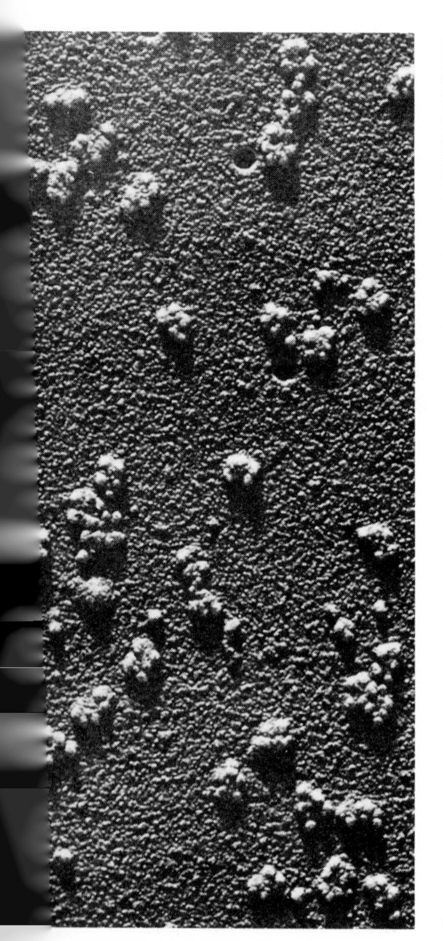

(*a*)

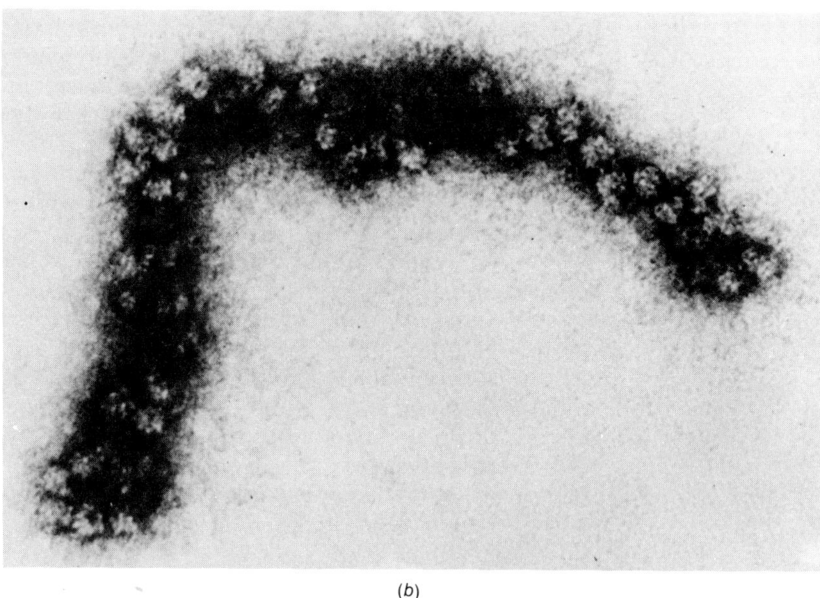

(*b*)

Figure 2.6 *a* Electron micrograph of metal-shadowed polyribosomes isolated from reticulocytes engaged in hemoglobin synthesis. [*Courtesy of A. Rich.*] *b* Electron micrograph of a negatively stained polyribosome engaged in the synthesis of the heavy chain of the myosin molecule. [*Courtesy of S. M. Heywood, R. M. Dowben, and A. Rich.*]

6. The basis for selective gene expression

One of the key words in developmental biology is *differentiation*, the process whereby the cells of the body acquire different properties that enable them to carry out their specific functions. To a large extent, differentiation is reflected in the production of specialized proteins. It is generally believed that the production of a specific set of proteins within a particular differentiated cell occurs as a consequence of the synthesis of a specific set of RNAs. According to this view, the specific activities of a specialized cell reflect the selective utilization of only a fraction of the genetic material of that cell. Liver cells would transcribe a particular battery of genes involved in liver function, and so forth. If selective gene transcription is the key to differentiation, we have pushed the question back one level and must then search for the regulatory mechanisms responsible for causing only certain genes to be transcribed in a given cell. This search has been particularly fruitful in microorganisms and viruses, and recent studies on eukaryotic cells are beginning to reveal important insights into the mechanisms operating in the control of gene expression in higher organisms. We will begin with a review of the best understood mechanism for transcriptional control, that of the operon which functions in prokaryotic cells.

A. The operon concept

In this section, the best studied case of genetic activation is reviewed: the activation by lactose of the synthesis of mRNAs which code for a series of enzymes needed for lactose utilization in *Escherichia coli*. It was the analysis of this system which formed the basis for the operon theory proposed in 1961 by François Jacob and Jacque Monod. Every aspect of this theory has been substantiated, and it has been extended to several other genetic units in bacterial and viral systems.

Every bacterial cell lives in direct contact with its environment, a constantly changing chemical milieu. At certain times a given molecule may become available for use, while at other times that compound is absent. It is of obvious selective advantage for these cells to utilize their available space and resources in the most efficient way, and mechanisms have evolved that allow them to respond to specific environmental changes by selective gene expression. There are two broad classes of response by a bacterial cell to the introduction of metabolic substrates to its medium. If a substance such as lactose, which the cell can use to provide energy or metabolic intermediates, becomes available, the cell responds by the synthesis of the hydrolytic enzymes necessary to utilize the substance. The enzymes are said to be *induced*, and lactose is termed the *inducer*. However, if a substance such as an amino acid becomes available, for which the cell is normally expending energy and materiel to produce, the enzymes involved in the synthesis of this amino acid are no longer needed; their synthesis is *repressed*.

The genetic and biochemical basis for the phenomena of induction and repression was explained by the formulation of the operon. An *operon* (Fig. 2.7) is a functional genetic unit whose genes are coordinately controlled. The elements of the operon are the structural genes, i.e., DNA that codes for the specific enzymes required, and the operator gene (o gene), i.e., DNA at which the synthesis of the mRNAs of the structural genes is controlled. Outside the physical region of the operon is the regulator gene (i gene), from which an mRNA is transcribed. The mRNA of the i gene is subsequently translated into a *repressor* (or *corepressor*) protein that will physically interact with the DNA of the o gene to control the transcription of the adjacent structural genes.

The lac operon contains a collection of genes which code for three adjacent inducible enzymes involved in the uptake and breakdown of the disaccharide lactose. In the absence of lactose, these enzymes are not needed and there is less

than one mRNA for them per cell. This shutdown results from the specific attachment of the repressor protein of the i gene to the operator site, thereby inhibiting the movement of RNA polymerase to the structural genes. If lactose, or another inducer, is added to the culture, after a lag of a few minutes the enzymes are found within the cell. The repressor-DNA interaction serves to illustrate the potential for proteins to recognize and bind to specific nucleotide sequences in DNA (or RNA). The activation of enzyme synthesis results from the specific interaction of the inducer with the repressor protein, since the complex (inducer-repressor) is no longer able to bind to the operator DNA. The inactivation of the repressor leads to synthesis of the mRNAs for the three structural genes which code for β-galactosidase, galactoside permease, and galactoside acetylase. The mRNAs for the polypeptide chains of these three enzymes are synthesized as one long, polycistronic messenger that is translated as three distinct polypeptides, each portion of the mRNA having its own initiation and termination codon. In this way, all the enzymes that will be needed together can be synthesized together, although the translation of each portion of the message need not occur at the same rate. For example, it has been found that more β-galactosidase chains are synthesized than either of the other two proteins. In the induced state there are 35 to 50 β-galactosidase mRNAs (of the approximate 1,000 total) per cell and over a thousand times more enzyme than in the absence of inducer. As is the case for most interactions within the cell, the complexes that form between the repressor and operator and between the inducer and repressor are reversible. As a result, when the lactose supply becomes exhausted and its intracellular concentration drops, the repressors are freed to interact once again with the operator and block transcription of the structural genes.

Figure 2.7 The mechanism of repression and induction in the lac operon. In the upper diagram it is seen that, in the absence of inducer, the product of the i gene acts to repress transcription of the structural genes as a consequence of its binding to the operator site. In the lower diagram, the presence of the inducer leads to the formation of an inactive repressor complex, thereby allowing transcription of the structural genes to proceed.

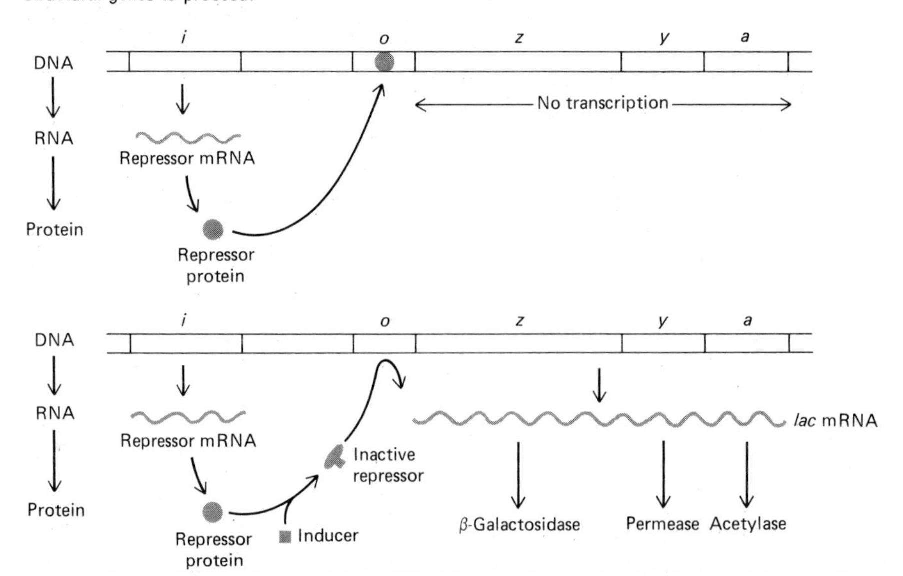

The operons of repressible enzymes work in a somewhat reversed fashion. One of the best studied is the tryptophan operon, which consists of the tandem array of five structural genes that code for the enzymes making up the pathway for tryptophan biosynthesis. In the absence of added tryptophan, the enzymes are needed, the product of the i gene is unable to attach to the operator gene, and the long polycistronic mRNA that codes for all the enzymes is produced and translated. If tryptophan becomes available, the enzymes are no longer needed. The tryptophan molecule complexes with the i gene protein, called a *corepressor* in repressible systems, and this complex is now capable of attachment to the operator gene, which then shuts down mRNA synthesis in this operon.

B. Control in eukaryotes

It is important for the life of any cell that it utilize its store of genetic information in an appropriate manner. In the case of the bacterial cell, the proteins being synthesized are largely determined by the conditions in which the cell is growing. The problems confronting a eukaryotic cell are many times more complex. In addition to having a genome containing much more information, the cells of a higher organism can exist in a wide variety of different phenotypic states. There are hundreds of different types of cells in a vertebrate body, each far more complex than the bacterial cell and each having a distinct battery of proteins which allow it to carry out its specialized set of activities. The mechanisms underlying the control of gene expression in eukaryotic cells are still poorly understood. Since the chain of events leading to the synthesis of a particular protein consists of a number of discrete steps, there are several levels at which control might be exercised. It is the goal of this section to briefly explore the possibilities available for regulation in eukaryotic cells and provide some estimate of their relative overall importance.

VARIATION IN THE NUMBER OF AVAILABLE DNA TEMPLATES: The union of a sperm and an egg at the time of conception brings together two sets of chromosomes, each containing a haploid amount of genetic information. It is possible that between the time of zygote formation and the differentiation of specific cell types, there occur changes in the nature of the genetic information passed to successive generations of cells during mitosis. There are two general directions that such genetic alterations might take. On the one hand, there may occur the selective loss of DNA sequences which are not required for the course of differentiation of the recipient cell and its progeny. If this were true, cells taken from the liver, for example, would be deficient in genetic information utilized only by kidney cells. On the other hand, it is possible that differentiation leads to a selective increase in those DNA sequences of particular importance for the activities of the recipient cell and its descendants. Results from an extensive variety of studies indicate that neither selective gene loss nor selective gene amplification occurs as a general method for regulating gene expression. Liver cells, for example, contain all the genetic information that was originally present within the egg and sperm at fertilization. Similarly, these same cells do not contain additional copies of liver-specific genes, such as those which code for serum albumin or tyrosine aminotransferase. Even though a given differentiated cell may be actively producing large quantities of a particular species of protein, there is no need for an increased presence of that DNA template, despite the fact that the template, representing a nonrepeated DNA sequence, is present in only one copy per haploid set of chromosomes.

It should be noted, however, that there are certain important exceptions to the statement that all cells contain the same genetic information. Surprisingly, gene loss does occur during the development of certain insects and nematodes (Section 8.1.A). The evolutionary basis for these exceptions is obscure. Just as there are cases of selective gene loss, there are also cases of selective gene amplification, the most important involving the DNA that codes for ribosomal RNA. During the development of most oocytes, cell volume enlarges greatly, with a corresponding increase in nuclear volume and in the number or size of the nucleoli. If one extracts the DNA from a population of amphibian oocytes, approximately 50 percent of it consists of the genes for ribosomal RNA. The selective amplification of rDNA is needed to fill a cell of great volume with sufficient ribosomes in the allotted time (Section 4.3.A). In contrast with the synthesis of rRNA for ribosomes, the synthesis of proteins, such as those present in the same amphibian oocyte ribosomes, can occur with the normal number of DNA templates. The difference between the situation for the rRNA and the ribosomal proteins (or other proteins, such as hemoglobin in a reticulocyte) can be appreciated if one considers that each ribosomal RNA molecule is present in only one ribosome while each messenger RNA molecule is capable of serving as the template for the production of hundreds or thousands of polypeptide chains. The additional step involved in the translation of a message, such as that for the ribosomal proteins of the oocyte, makes the presence of a large number of DNA templates unnecessary.[1]

TRANSCRIPTIONAL LEVEL CONTROL: As previously noted, selective gene transcription is believed to be widespread in differentiated cells of higher organisms, just as it has been demonstrated to be important in prokaryotic cells. Despite this strong belief in the predominance of transcriptional level control, direct evidence of it is difficult to obtain and is not, therefore, overwhelming. There have been several different approaches to this question. One of these approaches involves the examination of populations of RNA molecules present in different types of cells. Regulation at the transcriptional level requires that different cells synthesize different, but likely overlapping populations of RNA molecules. The difficulty arises when one considers that mRNAs of eukaryotic cells are derived from much larger primary transcripts. Therefore, in order to determine whether different DNA sequences are actually being transcribed in different cells, one must examine their respective populations of primary transcripts with this question in mind. Since it is impossible to isolate a population consisting solely of primary transcripts, one begins this experiment already at a disadvantage, since total nuclear populations must be used. If, as is likely the case, the initial processing steps occur very rapidly during and after synthesis, the actual level of primary transcript may be so low within the hnRNA population as to go undetected. Even if we assume that studies of hnRNA populations do tell us of the extent of transcription in a given cell type, a quick search of the literature reveals the relative scarcity of these types of experiments.

Most studies of differential gene expression have examined differences in populations of mRNA molecules, i.e., molecules taken from cytoplasmic polyribosomes, and thus do not bear directly on the question of transcriptional level control. It is possible, for example, that two different types of cells have very different

[1] There now appears to be a clear-cut example of selective gene amplification, specifically those genes in the follicle cells of *Drosophila* egg chambers which code for certain abundant proteins of the egg shell (reported by A. C. Spradling and A. P. Mahowald in Proc. Natl. Acad. Sci. 77, 1096, 1980). In this case, large amounts of these proteins must be produced in a very short period of time.

populations of mRNAs (as is generally found) but very similar populations of mRNA precursors, i.e., hnRNAs. If this were the case (discussed further in Sections 8.3.C and 15.7), then the differences in protein synthesis occurring in these two cells would be the result of differences in the manner in which the two populations of hnRNA molecules were processed to form their respective populations of mature messengers. At the time of this writing, there are reports of large differences in hnRNA populations among cells, as well as reports of basic similarities. It has been reported, for example, that hnRNAs of mammalian brain cells are much more complex than hnRNAs of other tissues. If one disregards the possibility of the existence of hnRNAs below the level of detectability in nonbrain tissues, this finding supports the concept of differential gene expression. In contrast, studies on the sea urchin (discussed in Chapter 8) indicate that hnRNA populations from various embryonic stages and adult tissues may be very similar. The matter remains to be settled.

The mechanisms responsible for selective gene activation and repression in higher cells remain uncertain. The search for operons in eukaryotic cells has not been productive, and alternate control mechanisms are believed to function. Several models have been presented, the most extensive one by Roy Britten and Eric Davidson in 1969. The Britten-Davidson model is theoretical and complex (to be expected in a model that attempts to explain genetic regulation in higher organisms), and a detailed presentation is beyond the scope of this book. In the model, the redundant sequences are assigned regulatory roles in a hierarchy of genetic function. Certain sequences act as sensors to external stimuli, such as hormones. Others produce secondary signals that could act at receptor sites in the DNA to turn batteries of genes on or off for the production of messengers as required by the activities of each cell at a given time. The potential role of the repetitive sequences as transcriptional level regulators has been recently reexamined (see Davidson and Britten, 1979).

Whatever the role of the various types of DNA sequences, the selective transcription of chromosomal DNA must be dependent on the nature of the chromosomal-associated proteins. The proteins of chromatin are generally divided into two major groups, the histones and the nonhistones. The *histones* represent a collection of well-defined basic proteins, while the *nonhistone* chromosomal proteins include a large number of uncharacterized species. There are five distinct classes of histones: the H1 (lysine-rich), the H2a and H2b (slightly lysine-rich), and the H3 and H4 (arginine-rich) classes. The interaction between these histone molecules and DNA leads to a highly defined organization within the chromatin. Most of the DNA of an interphase chromosome does not exist in an extended state, but rather is found wrapped around the outside of balls of histone. Each of these balls, or *nucleosome core particles,* consists of eight molecules of histone, two each of the following: H2a, H2b, H3, and H4. Approximately 1½ turns, or 140 base pairs, of DNA is curled around each of these 100-A particles; the DNA is associated with the histones by ionic bonds between the positively charged amino acids and the negatively charged phosphate residues. The DNA present between the nucleosomes can vary in length, but typically exists as a stretch of 55 to 60 base pairs. This internucleosomal or *linker* DNA is found to be associated with the remaining class of histone, the H1.

Are the histones simply structural molecules or do they play a role in the control of gene expression? Despite a long history of experimentation, the answer to this fundamental question in the study of the molecular biology of eukaryotic cells remains very unclear. A variety of early experiments indicated that purified DNA was a much poorer template for RNA synthesis (with added RNA polym-

erase) when histones were complexed to it than when present in the naked state. Based primarily on these types of experiments, it became generally accepted that histones acted as genetic repressors. However, support for this concept has been indirect and definitive evidence has not been forthcoming.

If histones do serve as genetic repressors within eukaryotic cells, they must act in conjunction with other types of molecules, since by themselves they comprise only a handful of different protein species. It is currently believed that the specificity required for transcriptional level control is provided by the nonhistone proteins, which as a group consist of at least several hundred different species in a given type of vertebrate cell. It is envisioned that the nonhistone proteins, presumably as a result of their affinity to particular nucleotide sequences in the DNA, somehow attach themselves to specific sites within the chromatin, thereby interfering with the repressive action of the histones in that region and facilitating the transcription of that genetic locus. It might be supposed, then, that the activation of a particular genetic locus for transcription would be accompanied by the removal of the histones from that section of DNA. However, biochemical and electron microscopic studies indicate that transcriptionally active genes remain in the nucleosomal configuration, thereby continuing their association with histones. In order to account for the presence of potentially repressive histone molecules in close proximity to active DNA, it has been suggested that the histones are modified when present on active genes. Histones are known to undergo reversible acetylation, methylation, ribosylation, and phosphorylation, which might well inhibit their ability to block the access of RNA polymerase molecules to the DNA template. Although it is important to keep in mind that this is a very simplified scheme for a complex and obscure process, this type of hypothesis allows investigators to design various types of experiments which might provide support or refutation for the proposal.

Transcriptional level control need not be solely an on-or-off phenomenon; it also might involve the rate at which various transcripts are synthesized. Molecular-hybridization studies have indicated that RNA populations contain species that are present at greatly different concentrations. Examination of chromatin under the electron microscope has provided direct visual support for the concept that some of the differences in the abundance of RNA sequences reflects differences in the rates of their transcription. For example, when chromatin of cells of the silk gland are examined, it is found that there is one length of DNA, clearly nonribosomal in nature, that is very densely covered by nascent RNA chains. All the other stretches of nonribosomal DNA are either inactive or contain a much lower nascent RNA density. It is believed that this particularly active segment of the chromosome represents the single copy of the fibroin gene that is responsible for the large amount of this protein produced by these cells. In this and other cases, there is good evidence that the rate of initiation of transcription need not be the same for all DNA templates.

POSTRANSCRIPTIONAL LEVEL CONTROL: After a primary RNA transcript has been synthesized, numerous steps precede translation. Most important, the smaller mRNA must be carved out of the larger hnRNA, a process which often involves the removal of sections of RNA from within the coding region of the message and the splicing of the newly exposed ends. During the processing steps, poly-A tracts are added, the RNA is capped, and exit to the cytoplasm must occur. All these processes occur through the association of the RNA precursor with specific proteins, and each represents a site at which regulation might be exerted. Once the mRNA is in the cytoplasm, decisions concerning translation must be made. Fac-

tors operating to regulate the use of the mRNA template as a template for protein synthesis are said to operate in *translational level control.* As discussed earlier, the synthesis of a protein is complex, involving a wide variety of different components, any of which could serve in a regulatory capacity. In the best-studied case of translational level control, namely, the regulation of globin synthesis by hemin (the iron-containing porphyrin group of the hemoglobin molecule), control is executed via the phosphorylation and dephosphorylation of one of the elongation factors required for protein synthesis.

Translational level control is particularly important during early development since the embryo begins its life with a large store of mRNAs made during oogenesis but intended for translation after fertilization. In the best-studied case, that of the sea urchin (discussed at length in Chapter 8), the mRNA of the unfertilized egg appears to be associated with certain protein components which interfere with the ability of the message to interact with ribosomes. In some unknown manner, fertilization leads to the removal of the inhibitory condition and the subsequent translation of the preexisting messages.

Once a given mRNA has been translated, we must consider the question of how many times it will perform this function. Some mechanism has to be available for a cell ro rid itself of RNAs that are no longer needed. Differentiation encompasses the loss of previous function as well as the gain of new properties. It appears that cells can distinguish, to some extent at least, between different mRNA molecules. Some have a long half-life—once produced, they remain intact for days or weeks—while others have a much shorter half-life, possibly as little as an hour or so. The hemoglobin mRNA is an example for a very stable message. Once the mRNAs for a protein are destroyed, no further synthesis of that protein can occur in the absence of new transcription of that mRNA.

In the foregoing pages, a brief summary of some of the molecular activities of the cell has been given. These processes do not occur simply in a homogeneous solution, but are integrally connected with the structures of the cell itself. The basic nature of the eukaryotic cell will be described in Chapter 3, but first we will consider some of the basic aspects of macromolecular organization.

7. Protein structure and formation

We have considered in the previous section the manner in which the cell manufactures its polypeptide chains. However, a very brief examination of protein structure and function indicates that these macromolecules are much more than simply a large polymer of amino acid building blocks. Analysis of the three-dimensional nature of proteins indicates that the polypeptide chain is twisted and folded in a variety of complex ways. However, the structure of a given species of protein is precisely constant from molecule to molecule.

Whereas nucleic acids are assigned the roles of information storage and transfer, proteins mediate the cell's many activities. Proteins, as a group, can assume the innumerable geometric and electrostatic configurations needed for these many functions. Proteins can be categorized on the basis of shape, charge, amino acid composition, function, etc., and they serve in several capacities. Many are enzymes constructed to catalyze metabolic reactions. Others have structural roles, such as certain of the ribosomal proteins, microtubule proteins, collagen, keratin, and the proteins of silk. Many have a regulatory function, including genetic repressors and hormones. There are proteins with specialized functions not easily categorized. Some bind molecules, e.g., hemoglobin and hormone receptors; others transport molecules through the blood or through a membrane. Antibodies are proteins. There are proteins with elastic properties, contractile proper-

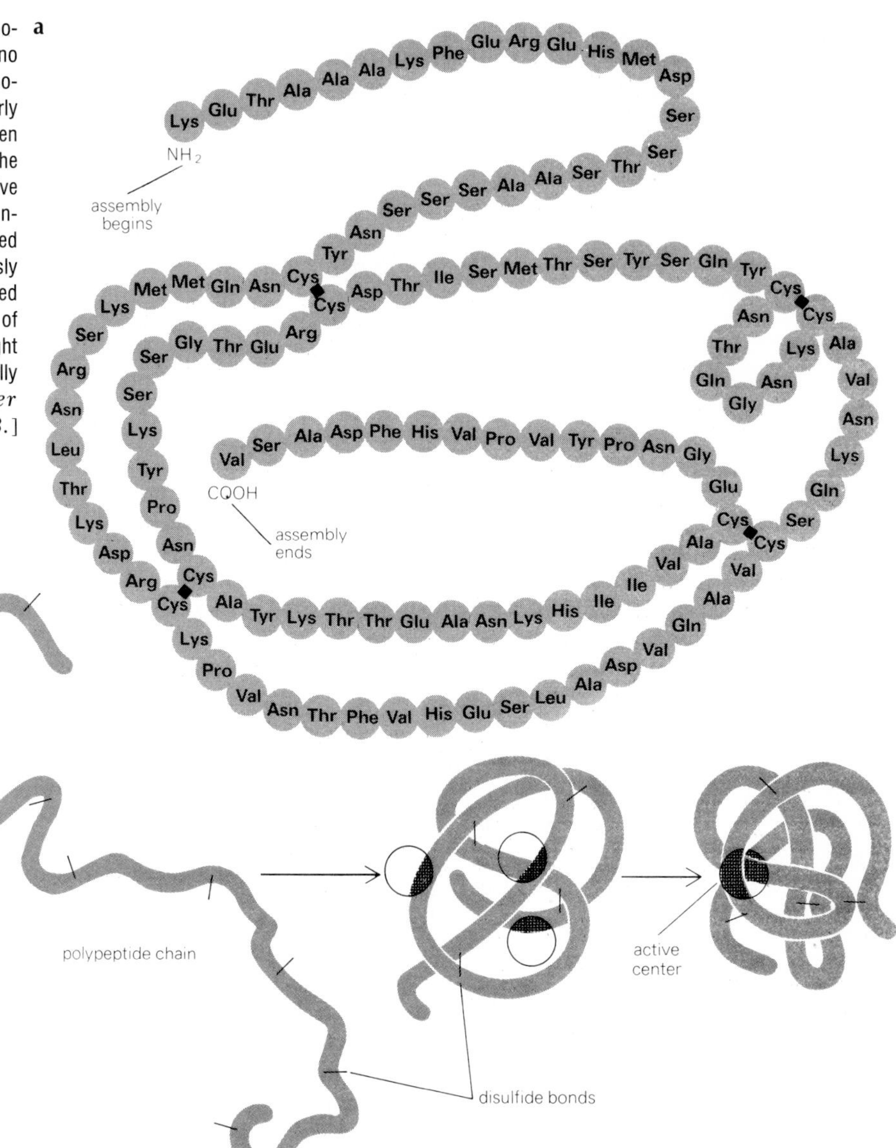

Figure 2.8 *a* The enzyme ribonuclease, consisting of 124 amino acids of 19 different types. Ribonuclease was utilized in early studies of the relationship between the amino acid sequence and the shape of the protein. *b* Native protein molecule in natural configuration, converted to an extended polypeptide chain, spontaneously reconverts to the original folded condition. Separated regions of the extended chain are thus brought together to form a specifically structured active center. [*After C. B. Anfinsen, 1968.*]

ties, and antiviral properties; and undoubtedly there are many unknown proteins with unknown properties.

The major question to be considered in this book is the manner in which form and function initially arise within the embryo. With this in mind, it is worthwhile to discuss how the structure of such complex molecules as proteins is generated. The overall shape of a polypeptide chain of a protein molecule is dependent on the interaction of amino acids that may be located at some distance from one another in the linear arrangement but are brought very close to one another when the molecule becomes folded. This is clearly illustrated in the diagram of the enzyme ribonuclease (Fig. 2.8). The interactions between amino acids within a protein

are mediated by the weak secondary bonds, namely, ionic bonds, hydrogen bonds, and hydrophobic interactions.

How is the tertiary structure, i.e., the three-dimensional conformation, generated? Since it is impossible to isolate polypeptide chains before they undergo their folding process inside the cell, some indirect procedure must be employed to study this question. The approach that has been taken is to isolate and purify intact proteins from cells, then treat them with agents that cause them to lose tertiary structure, and then observe the capabilities that such disorganized polypeptides possess in *re-forming* meaningful spatial organizations. The unfolding or disorganization of a protein is called *denaturation,* and it can be brought about by a wide variety of agents that disrupt noncovalent bonds. Denaturation can be either reversible or irreversible depending on the protein and the conditions under which it is tested. The first and best understood case of reversible denaturation is that of ribonuclease, studied by Christian Anfinsen and coworkers. Ribonuclease is a small enzyme which consists of one polypeptide chain of 124 amino acids with 4 disulfide bonds linking various parts of the chain (Fig. 2.8). Denaturation is accomplished by subjecting the protein to a solution of mercaptoethanol, a reducing agent that breaks the covalent disulfide (—SS—) bonds and converts them to sulfhydryl (—SH) groups of cysteine, and 8 *M* urea, which breaks noncovalent bonds. If the urea and mercaptoethanol are removed in an appropriate manner, active enzyme molecules are re-formed which are indistinguishable from those present at the beginning of the experiment.

The results of these types of experiments suggest that the formation of the tertiary structure of a protein is a self-directed process, one that derives automatically from its primary structure, i.e., its linear sequence of amino acids. In other words, once a given amino acid polymer is constructed, given the appropriate environment, the secondary interactions occur spontaneously, causing the protein to fold into the proper conformation; no additional information is required. The term *self-assembly* is used to refer to this type of spontaneous formation of higher structure. It would appear that of the vast number of possible conformations that could be formed, that of the native protein is thermodynamically favored above all the others.

The most important generating forces in the folding of most proteins arise from hydrophobic interactions. Certain amino acids have structures which pay little regard to the solvent water. These amino acids tend to associate in the center of the protein causing the exclusion of water from that region. The hydrophobic core of the protein is surrounded by those parts of the molecule which contain the hydrophilic amino acids with which water readily associates. The primary folding of the polypeptide is believed to result from the withdrawal of the hydrophobic side chains from the aqueous solvent. This association of hydrophobic amino acid residues in the core of the molecule is driven by the increase in entropy resulting from the consequent disorganization of the surrounding water molecules. Once the primary folding processes occur, various side chains are brought into close enough proximity to form other weak types of interactions. Disulfide bridge formation is more responsible for stabilizing a three-dimensional structure already present than for causing it to be generated in the first place.

The importance of the shape of a polypeptide chain goes beyond its own activity to its interaction with its neighbors. As mentioned earlier, many proteins are composed of more than one polypeptide chain. As in the determination of a folded structure, the association of polypeptide chains is also a spontaneously occurring process. The same types of procedures that denature the coiling of a single chain dissociate proteins made of more than one chain; these include high salt concen-

trations, changes in pH, detergents, and the presence of urea. Such agents interfere with noncovalent associations, which are responsible for holding subunits together. In certain cases, several different types of proteins come together, and again there is evidence that these interactions can occur simultaneously as a direct result of the structure of each individual protein in the complex. These interactions clearly illustrate the potential for complex organization with a minimum of outside input. In a direct way, the primary sequence of amino acids is responsible for the far-reaching consequences of protein organization and, therefore, of protein function. This type of spontaneous development of structure will be considered later in the formation of even more complex products by self-assembly.

It is apparent from the preceding that the genetic code can exert a major impact on organizational features. The genome determines the character and diversity of a multitude of component structural elements and associated catalysts. Any changes in production processes that result in changes in the primary products will be reflected in the final construction, for better or worse. Faulty material means faulty construction. However, changes in quantities of construction materials or changes in delivery schedules may greatly affect the kind and direction of the assembly processes, even though they are largely self-controlled.

8. Self-assembly

In the preceding discussion of protein structure it was noted that specific parts of polypeptide chains interact spontaneously with one another to form more thermodynamically stable associations. *Given the appropriate amino acid sequence*, a region of a newly synthesized polypeptide chain can spontaneously curl into an alpha helix; an entire polypeptide can fold onto itself to form a more globular structure; separate chains can associate to form a multisubunit protein; and separate enzymes can associate to form a multienzyme complex. The term *self-assembly* is used to describe the spontaneous ordering of components into a more complex structure—all the information and energy required for the formation of the whole is contained in the parts.

If distinct proteins can become associated with each other in a highly predictable and ordered fashion, it should be possible for different types of biological molecules to do the same. Cellular organelles are composites of various macromolecules: ribosomes consist of RNA and protein, chromosomes of DNA and protein, membranes of protein and lipid, and so forth. Can these structures be derived by the spontaneous association of less complex components, or does their association require the guidance of outside agents (e.g., the enzymes needed to catalyze the formation of covalent bonds)? How far can a process of self-assembly be carried to explain biological organization? In the following sections we will examine this question by presenting several examples of molecular morphogenesis.

A. Tobacco mosaic virus

Although simple in construction and unable to perform the most basic metabolic activity, virus particles contain all the information required for attachment to a cell, penetration of the cell, and control of that cell's activities for its own purposes. It is strikingly clear how a handful of viral genes is capable of rerouting the activities of an infected cell causing it to direct its energy and resources toward the production of new virus particles.

Tobacco mosaic virus (TMV) is of particularly simple construction, being made of approximately 2,200 identical protein subunits organized into a helix around one long RNA molecule of approximately 6,400 nucleotides which contains the genetic information. If the components are gently dissociated and then

brought together under the proper conditions, they reassemble to form complete and infectious viral particles. The entire particle, therefore, can be formed by self-assembly. The manner in which these viruses are assembled in vitro has been closely scrutinized and demonstrates the complex interactions and conformational changes that macromolecules can undergo.

Although the structural unit of the TMV particle is a 4 S subunit, the unit of assembly is a specific complex of 34 subunits. This complex (see Fig. 2.10) has the shape of a disk made of two layers of subunits, 17 per layer, arranged in a ring. If a preparation of these disks is combined with the RNA, complete virus particles are formed in a matter of minutes. If, however, one starts with a preparation of individual 4 S subunits, the process takes several hours; the conversion from a single subunit to the disk, which is a required intermediate in the assembly process, is the slower event. Initiation begins with the association of a disk with a specific region of the RNA molecule. Although it was previously believed that assembly began at one end of the RNA molecule, it is now evident that the first disk attaches at approximately 15 percent of the distance from the 3′ end, with assembly then proceeding in both directions (although at different rates). The use of a disk, instead of the smaller subunit, as the unit of assembly provides certain advantages. All 17 subunits of a ring are able to interact in a coordinated manner with a considerable stretch of RNA nucleotides. Consequently, the initiation of assembly can occur in a specific manner following recognition of the required viral sequence; no other type of RNA can be incorporated into the virus. Moreover, the simultaneous interaction of the RNA with 17 subunits around the ring helps to ensure that the considerable energy barrier against initiation is surmounted.

In order to learn more of the nature of the specific interaction between the initiation site of the RNA molecule and the protein disk, the following experiment was performed. Purified viral RNA was incubated for brief periods with relatively small numbers of disks. Under these conditions, viral assembly is initiated but has little opportunity to progress to more advanced stages. When these early assembly intermediates are treated with ribonuclease, most of each RNA molecule is degraded leaving only a small fragment which consists of that portion of the genome which escapes destruction as a result of its association with the protec-

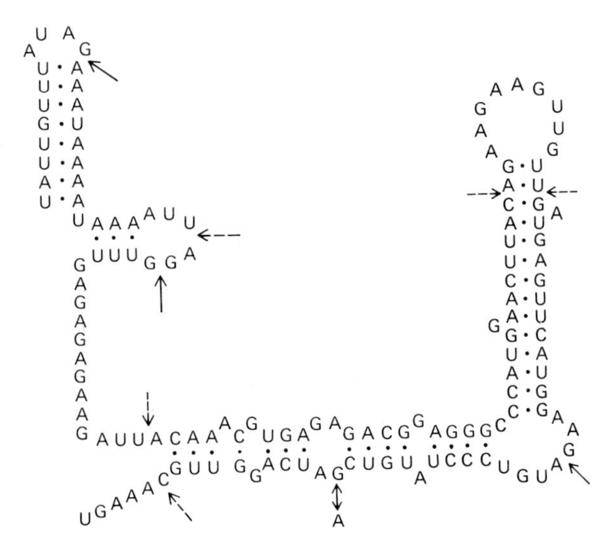

Figure 2.9 The nucleotide sequence and postulated secondary structure of TMV RNA in the region in which assembly is initiated. Solid arrows mark points very labile to partial RNAase T1 digestion; dashed arrows are points most susceptible to pancreatic RNAase digestion. [*From D. Zimmern*, Cell, **11**:477 *(1977); copyright by M.I.T.*]

tive protein overcoat. Sequence analysis of this RNA fragment, which represents the initiation site for TMV assembly, has revealed an interesting feature; this portion of the RNA has the potential to form a double-stranded hairpin loop (Fig. 2.9). It is believed that initiation begins when the end of this RNA loop reaches into the central hole of a disk (Fig. 2.10*a*) making contact with the inner surface of the ring of subunits at a site between the two layers.

We will return to the story of RNA and protein interaction shortly, but first we need to consider how the assembly of a helical TMV particle can result from the

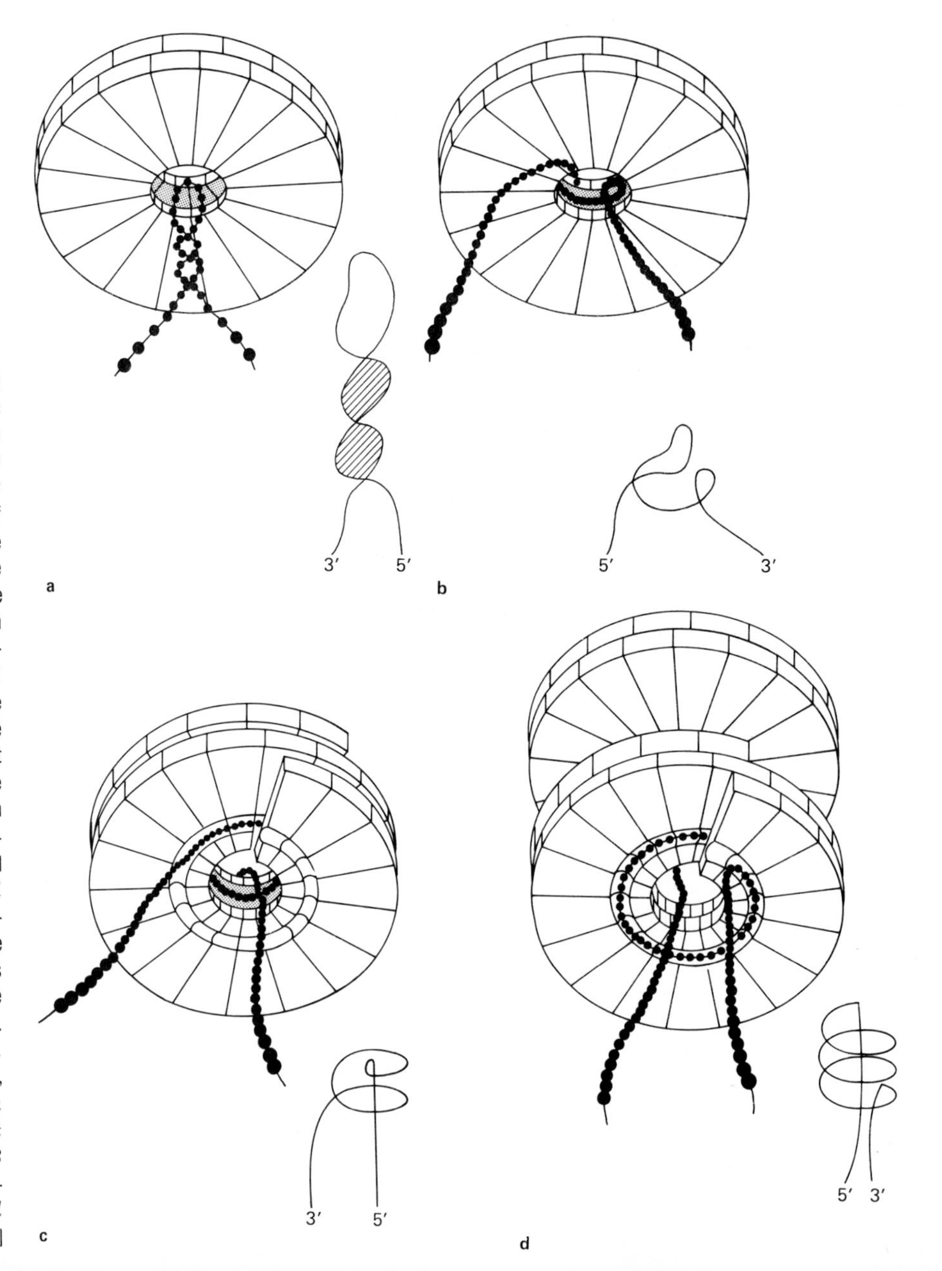

Figure 2.10 *a* Nucleation of the tobacco mosaic virus (TMV) begins with the insertion of the hairpin loop formed by the initiation region of the viral RNA into the central hole of the protein disk. *b* The loop intercalates between the two layers of subunits and binds around the first turn of the disk, opening up the base-paired stem as it does so. *c* Some feature of the interaction causes the disk to translocate into the helical lock-washer form. *d* This structural transformation closes the jaws made by the rings of subunits, trapping the viral RNA inside. [*From P. J. G. Butler and A. Klug, "Assembly of a Virus"; copyright © November 1978 by Scientific American, Inc. All rights reserved.*]

ordered accumulation of disks. A pile of disks would merely form a cylinder, while a helix is a continuing spiral that requires that none of the disks have a closed shape. The answer to this puzzle is seen in a comparison of parts *b* and *c* of Fig. 2.10 and reemphasizes the complex conformational changes that protein molecules can make. The transition to a helical unit occurs because of a shift in the way the subunits are organized within the disk; a dislocation converts the two-layered disk into approximately two turns of the spiral. This structure is reminiscent of a lock washer found in a hardware store.

Recent analysis of TMV disks by x-ray diffraction has provided considerable information on the molecular changes that occur during the disk to lock-washer transition, a transition in which the spatial relationship between the two layers of subunits undergoes marked alteration. In the disk the two layers are in close contact with one another at the outer surface of the rings, but are quite far apart at their inner surface. Consequently, there is a wedge-shaped gap between the subunits of the two layers much like an open set of jaws. The distance between the layers at their inner surface appears to be maintained primarily as a result of ionic repulsion between amino acid residues lining the central hole. The inner surface of the subunits is largely hydrophobic in nature but punctuated by patches of charged residues. There are, for example, three aspartate and three glutamate residues in each subunit, forming what is referred to as a *carboxyl cage*. It appears that disk stability is maintained by repulsion between carboxyl groups. In fact, lowering the pH of the medium from 7 to 5 can result in a triggering of the disk to lock-washer transition causing assembly of helical TMV rods that lack genetic material. Normally the dislocation occurs in the cell only after the disk has combined with the RNA molecule, an event which somehow reduces the electrostatic repulsion between the layers and facilitates the change. The requirement for RNA in this transition ensures that particles will never form without genetic material. We can now return to the events of the initiation of assembly.

The proposed scheme of initiation events in TMV assembly is shown in Fig. 2.10. Contact between the double-stranded RNA loop and the inner surface of the disk is believed to cause the nucleotides in the stem region to unpair and wind around the ring within the gap between the layers. Three nucleotides of RNA fit between each vertical pair of subunits, 51 per ring, with the negative phosphate groups apparently binding to positively charged arginine residues in the region. Once the RNA is bound within the open circular crevice between the layers at their inner surface, the disk transforms to the lock washer, an event that closes the "jaws" with the RNA inside.

As shown in Fig. 2.11, the addition of the second and subsequent disks also is believed to occur by insertion of a loop of RNA into the central hole of an incoming disk. Electron micrographs (Fig. 2.12) of particles caught at intermediate stages in the assembly process indicate that both tails of RNA extend from the same end of the forming rod. To account for this observation, it is proposed that following the addition of each disk and its conversion to a helical unit, one of the tails of RNA is drawn farther up the central hole so that it extends above the last disk as a loop capable of interacting with the next disk, thereby repeating the process (Fig. 2.11). Whereas the interaction of the first disk with the RNA molecule requires the presence of a particular recognition sequence, thereby guaranteeing the formation of RNA-containing particles, later interactions do not require such sequence specificity. Once the disks have consumed the entire RNA molecule, the assembly process automatically comes to an end and the particle is complete.

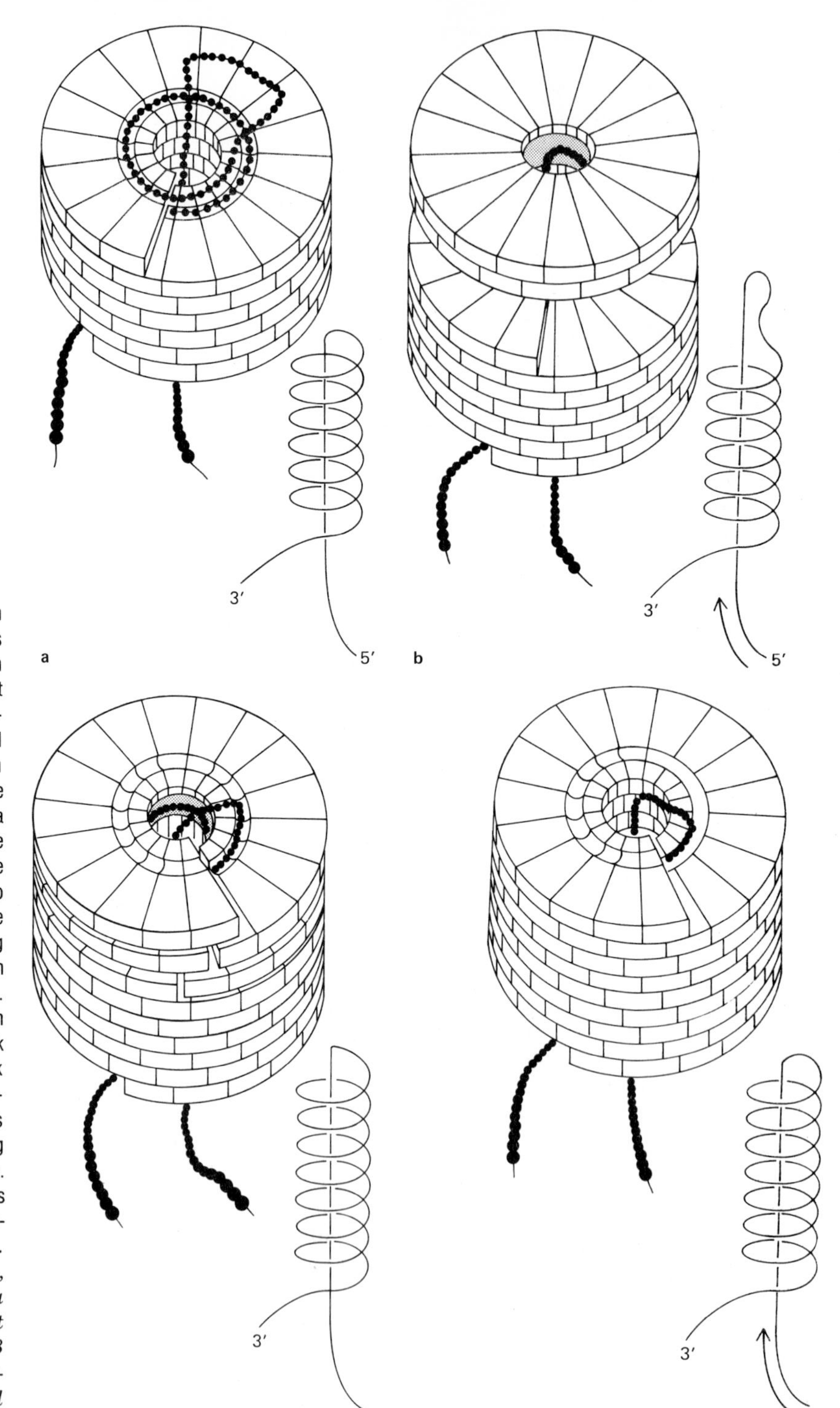

Figure 2.11 Elongation of TMV also proceeds by the addition of protein disks. *a* As a result of the mode of initiation, the longer RNA tail is doubled back through the central hole of the growing rod forming a traveling loop at the growing end of the particle. *b* The loop inserts itself into the center of an incoming disk and binds within open jaws of the ring. *c* This interaction converts the new disk into a helical lock washer. *d* The transformed disk then stacks onto the rod, providing two turns of the helix. The process repeats until assembly is complete. [*From P. J. G. Butler and A. Klug, "Assembly of a Virus"; copyright © November 1978 by Scientific American, Inc. All rights reserved.*]

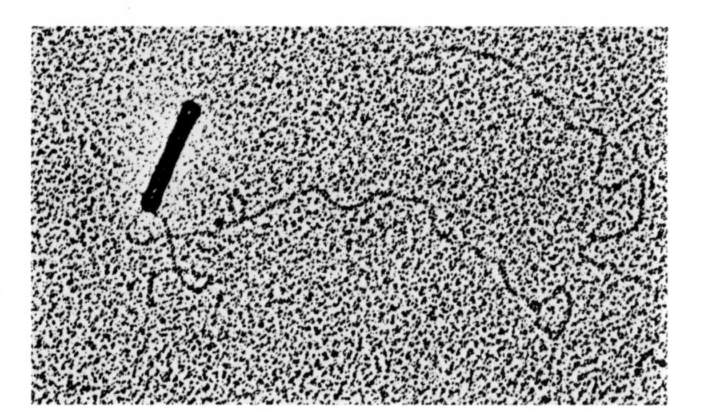

Figure 2.12 Electron micrograph of an incompletely reconstituted TMV particle showing both RNA tails protruding from the same end of the rod. [*Courtesy of G. Lebeurier, A. Nicolaieff, and K. E. Richards.*]

B. The ribosome

Bacterial ribosomes are organelles of much greater structural complexity than TMV particles, yet they too can be reconstituted in vitro from their purified components. We will begin this discussion with a consideration of the 30 S subunit, a particle containing one RNA molecule (the 16 S rRNA of about 1,600 nucleotides) and 21 different proteins. Unlike the TMV particle, the 30 S subunit has an irregular embryolike exterior shape (Fig. 2.13) and a highly asymmetric interior organization. A number of laboratories in recent years have been concentrating their efforts on determining the position of various proteins within the ribosome. Many techniques have been employed in this regard, the most successful one involving the use of antibody molecules capable of attaching to only one of the ribosomal proteins. In order to carry out this determination, 30 S ribosomal subunits are mixed with a preparation of antibodies specific for one of the proteins. If this protein is present at the surface (and all the 30 S ribosomal proteins appear to have some part of their structure at the surface), then the antibody should be able to attach to it and its location should be apparent under examination of the complex in the electron microscope. In some cases, antibodies prepared against purified ribosomal proteins bind at more than one site on the subunit's surface. This finding indicates that certain of the proteins are highly extended, possibly even branched in nature, and capable of spanning the width or even the length of the particle. These types of studies, as well as others, have revealed the close association that ribosomal proteins have with one another and with the ribosomal RNA.

As mentioned earlier, mixture of the 21 proteins of the small subunit with the 16 S rRNA under the appropriate conditions leads to the formation of fully active 30 S subunits. Earlier work had suggested that only a few of the ribosomal proteins actually bound to the RNA, the remainder binding in a secondary manner to the RNA-protein complex that was formed in the initial assembly step. More re-

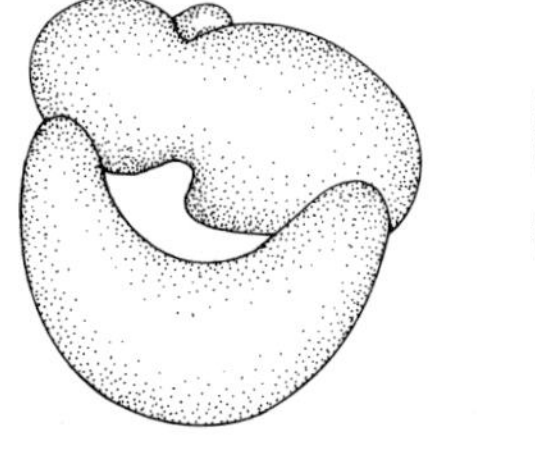

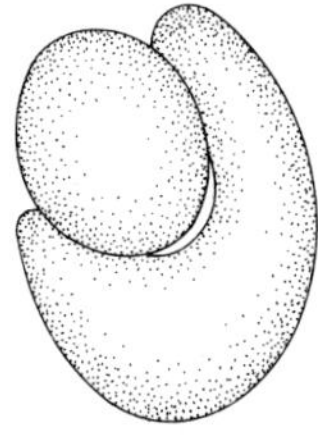

Figure 2.13 Two views of the *E. coli* ribosome showing the configuration of the two subunits and their steric relationship to each other.

cent studies suggest that most, if not all, of the proteins of the 30 S subunit do have one or more specific sites on the RNA molecule to which they bind. It appears that the binding properties of the ribosomal RNA molecule are greatly affected by its secondary structure, which is, in turn, affected by the extraction procedure used in its purification. It would appear from these observations that protein binding sites on the rRNA can appear and disappear according to the conformation of the RNA molecule. Sequence analysis of the 16 S rRNA indicates that much of the molecule has the potential to exist in a double-stranded state. It is very likely, therefore, that the attachment of one group of proteins to the RNA changes the conformation of the RNA molecule such that additional sites become available for the binding of other proteins. Regardless of the precise binding characteristics of the molecule, the 16 S rRNA is required for both the assembly and function of the subunit. Chemical modification of only a few nucleotides is sufficient to destroy all activity. In some cases these RNA-modified inactive subunits are indistinguishable from normal ones by all criteria based strictly on structure.

The development of in vitro reconstitution systems for ribosomal subunits has led to further analysis of the role of the various proteins. In one approach, mutant bacteria that contain an altered protein are selected, and the effect on ribosomal function is studied. Another approach involves chemical modifications of one specific, purified ribosomal protein, its incorporation into a reconstituted ribosome, and the analysis of any malfunction. A similar technique is to reconstitute subunits in the absence of one of the ribosomal proteins and to determine the effect. The results of these studies suggest that some ribosomal proteins are required exclusively for the assembly process, some for the maintenance of the proper conformation of the subunit, and others for specific functional activities of the ribosome. Some proteins may have more than one role. For example, the omission of the ribosomal protein S11 increases the translational errors when the reconstituted subunits are tested. The exact mechanism for this is unknown, but it suggests that this protein has a functional role. Many omissions produce ribosomes devoid of activity, but the interdependence of structure and function makes it difficult to assign specific roles to these proteins.

Like the small subunit, the 50 S particle also can be reconstituted from its purified components, thereby demonstrating its ability to self-assemble. One of the proteins of the large subunit, L24, appears to function only during assembly and has no role in the activities of the completed particle. This particular protein binds to the 23 S rRNA in a very tight manner and appears to be required for a major conformational change (from a 33 S to a 41 S intermediate), which occurs during assembly within the cell. Even though the 70 S ribosome has been reconstituted in vitro, one should not assume that the pathway of assembly followed in the test tube is exactly the same as that which occurs within the bacterium. Although there appear to be definite parallels between the in vitro and in vivo pathways, the fact that assembly in the cell seems to begin before the RNA molecule is completely transcribed and processed into the mature RNA species (the 16 S and 23 S rRNAs are initially synthesized as part of a single precursor molecule) requires that some differences must exist.

With the successful assembly of the bacterial ribosome in vitro, the question arises as to whether the ribosomes of eukaryotic cells are also capable of reconstitution outside the cell. The eukaryotic ribosome is basically similar to those of prokaryotes, despite the fact that they contain about 25 additional proteins and an extra RNA molecule in their large subunit. Some of the extra proteins are required for the attachment of eukaryotic ribosomes to cytoplasmic membranes,

while others are believed important in translational level control. Although proteins from eukaryotic ribosomes have been purified and some of their amino acid sequences explored, functional particles have not been assembled in vitro. Since the formation of ribosomes within the eukaryotic cell involves the transient association of several nucleolar proteins (which do not end up in the final particle) and the removal of approximately half the nucleotides of the large rRNA precursor, it is very likely that the components of the mature ribosome no longer possess the information to reconstitute themselves as an active organelle.

C. T_4

The tobacco mosaic virus is a relatively simple virus whose coat is made of one protein and whose genetic material contains only a few genes. Many viruses, including many of those which attack bacterial cells, the bacteriophages, are much more complex. An electron micrograph of one of these is shown in Fig. 2.14, and a diagram is presented in Fig. 2.15. The double-stranded DNA of the phage is contained within the polyhedral head. A short neck connects the head with a tail complex, which consists of a springlike contractile sheath surrounding a hollow central core and attached to a hexagonal base plate. Six slender fibers and six short spikes protrude from the base plate. Altogether the virus is a highly organized structure, not unlike a landing module for the moon.

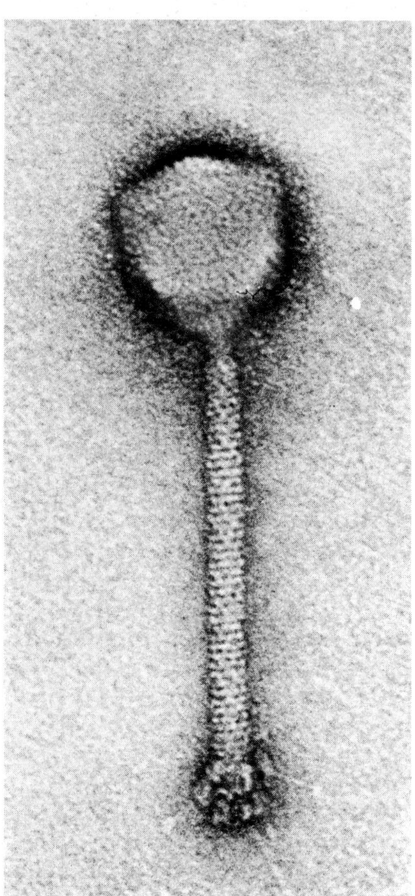

Figure 2.14 Bacteriophage particle showing shell, sheath-enclosing core, and plate with anchoring fibrils. [*Courtesy of A. K. Kleinschmidt.*]

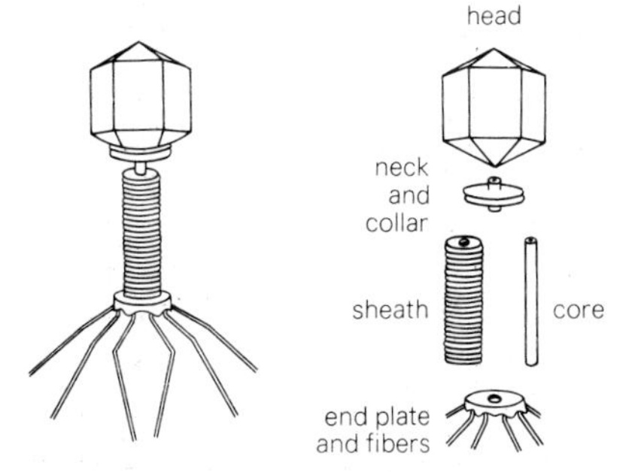

Figure 2.15 The parts and structure of the T bacteriophage. [*From J. R. Whittaker, "Cellular Differentiation"; copyright © 1968 Dickinson Publishing Company, Inc., Belmont, California.*]

Before discussing the assembly process, a brief review of the events that occur during bacteriophage infections will be presented. First, the virus becomes anchored to the bacterial wall by specific receptors on its tail fibers so that the tail and head stand perpendicular to the surface as if guyed by the tail fibers. The response which begins with contact of the tail fibers with the bacterial cell soon spreads to other parts of the virus. The base plate, which settles down on the cell surface, undergoes a striking change in conformation from its original hexagonal shape to that of a star. During this process, the bonds between the base plate and the hollow core are loosened and a subsequent contraction of the overlying sheath forces the core through the wall of the bacterium; this allows the DNA of the head to pass into the cell. Once inside, a series of intricate biochemical steps result in the appearance of the first complete virus particle about 13 minutes after infection. The process continues for another 12 minutes until about 200 such particles have been completed and the raw materials of the bacterial cell have been exhausted. A viral enzyme, a lysozyme, then appears and attacks the cell wall, liberating the new viral particles. The known details of this process could easily take up the remainder of this book and would provide a fascinating account of new sigma factors, carefully controlled gene activation, altered host enzymes and ribosomes, and complex assembly steps. Instead we will try to keep to the original question concerning the assembly process.

Studies of T_4 construction indicate there are three independent assembly lines. Each produces one finished part of the virus—the head, the tail, and the tail fibers—and these subsequently combine to form the finished virus particle. This is illustrated in Fig. 2.16. Each of the three assembly lines consists of steps in which specific genes are known to be involved.

Some of the best insights into the nature of self-assembling systems have been gleaned from the study of T_4 tail formation. The tail section (base plate, core, and sheath) consists of at least 16 proteins, all of which are synthesized in the host cell during the same stage of infection. Despite the presence of all these proteins in solution, their assembly proceeds in a stepwise, orderly manner. Each protein is added sequentially to the growing assemblage; those added late in the process simply remain as independent molecules until needed. In order to explain the lack of interaction among soluble proteins prior to their incorporation into the tail assembly, it is suggested that each protein is activated as it is incorporated into a substrate structure. This activation event is believed to occur as a result of conformational changes in the molecules as they attach to the growing structure, thereby making the structure reactive toward the binding of the next protein. If any of the proteins of the assembly pathway are missing, the intermediate accumulates and subsequent proteins remain unassembled. This concept can be illustrated using the polymerization of the tail core as an example. The initiation of tail core and sheath polymerization can occur only on the surface of a completed base plate; in its absence, tail core and sheath subunits remain soluble owing to the absence of a suitable site to which they can bind. Once formation of the core begins, soluble subunits are able to associate with core protein molecules only at the growing tip of the core, ones that have been activated by previous incorporation into the tail complex. The basis for termination of polymerization at the required point is unclear. It would appear from this type of analysis that the order of assembly is regulated by protein-protein interactions of the assembly path itself.

If we return to the original question of whether or not the morphogenesis of this virus can be accounted for simply by self-assembly, the answer appears to be that it cannot. Of the 150 to 175 estimated genes in T_4, 55 have been identified as necessary for viral morphogenesis. However, fewer than 40 different species of

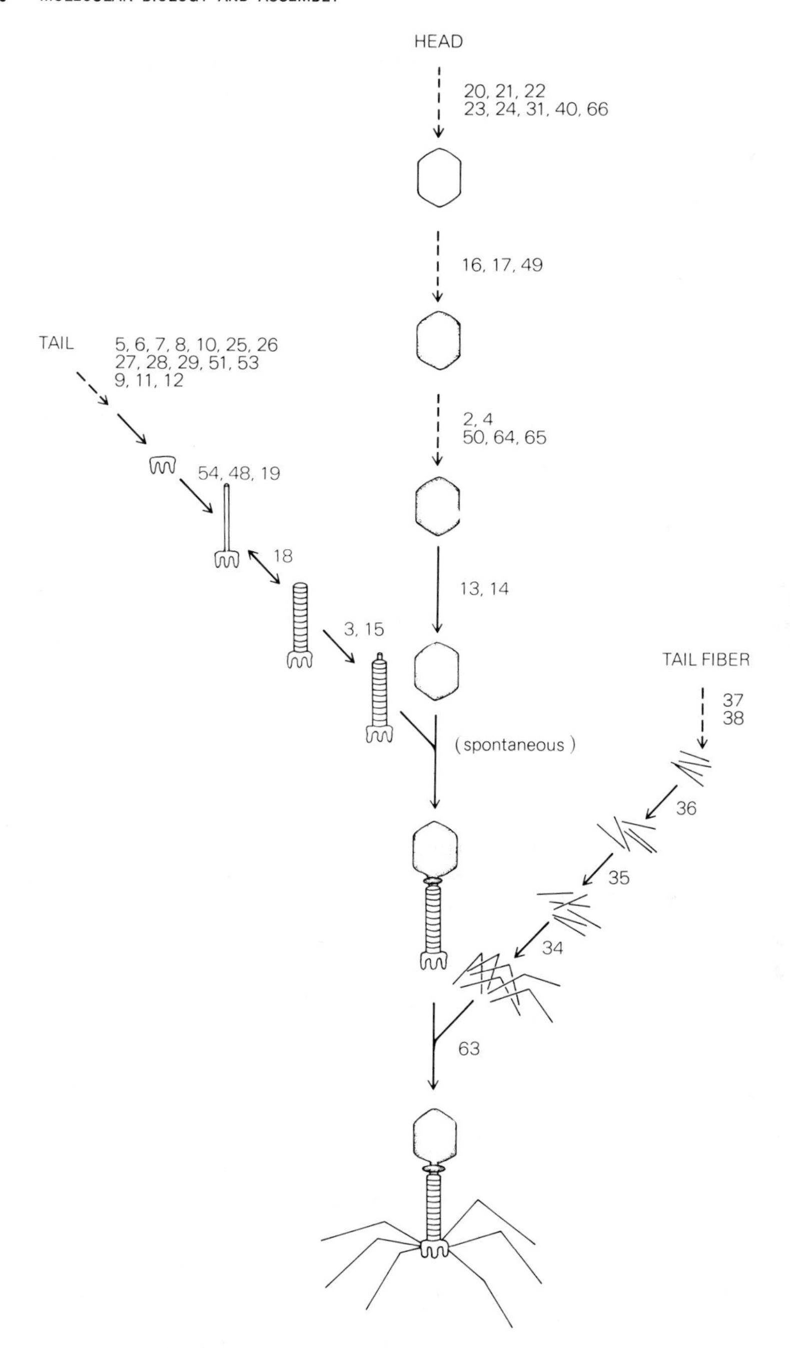

Figure 2.16 Pathway of morphogenesis of T_4 virus has three main branches leading independently to the formation of heads, tails, and tail fibers, which subsequently combine to form complete virus particles. The numbers refer to products of identified genes. [*After W. B. Wood, R. S. Edgar, et al.,* Fed. Proc. Amer. Soc. Exp. Biol., **27:***1160* (*1968*).]

protein have been identified as part of the virus particle. This does not mean that all the other genes required for morphogenesis do not code for structural proteins —they may be present in nondetectable quantities per virus—but it does seem clear that there are genes (at least 7) whose products are not contained within the virus but instead have some morphogenetic role. In other words, the assembly is not entirely coded for by the structure of the components, but requires the outside intervention of morphogenetic gene products. Mutations in some of these genes may be responsible for many bizarre results, including giant virus particles, polysheaths composed of many sheaths stuck together, etc. There appear to be several types of morphogenetic gene products; these include proteolytic enzymes that destroy previously incorporated proteins, transient scaffolding proteins, and proteins that seem to accelerate the formation of noncovalent bonds.

To illustrate the roles of nonstructural accessory proteins, we can briefly examine the assembly of tail fibers. Six of the seven genes now known to be required for this branch of assembly are shown in Fig. 2.16. Of these seven gene products, only four (34, 35, 36, and 37) have been accounted for as structural elements. Two other gene products (38 and 57) appear to be required for tail fiber formation, and the third (63) appears to catalyze the noncovalent attachment of the tail fiber to the fiberless base plate. Other examples of morphogenetic genes have come from studies of head assembly. For example, mutations in gene 31 cause the head proteins to aggregate randomly and adhere to the bacterial envelope. There is evidence that this gene product interacts with a component coded for by the bacterial genome that is also required to promote polymerization of head proteins to form a head intermediate. In this case, several bacterial mutants have been isolated in which genetically normal virus cannot proliferate as a result of the deficiency in this host protein needed for virus assembly.

Although many of the steps in T_4 construction occur by self-assembly, some do not, suggesting that structures in the order of complexity of this virus begin to rely on outside agents to put themselves together. Throughout this book we will discuss biological architecture, for this is the essence of differentiation. In the study of suborganisms such as T_4, it is believed that principles will be uncovered that we can apply to the construction of more complex cellular elements. Considering the current effort and past success in the analysis of many viral systems, we should soon possess the blueprints for many of their constructions. Whether or not this information will be useful as models in the analysis of embryonic development remains to be seen.

We have taken the approach in this chapter that an understanding of the organization of even the most complex multicellular organisms can be best gained by dividing events into two parts: (1) the biosynthesis of the appropriate molecules and their control, and (2) the assembly of these molecules into organized structures. The first events are known to be under the direction of the genome, and the second seem largely under thermodynamic control; i.e., the organized structures result simply from the search of each molecule for its position of lowest chemical potential.

Readings

ANFINSEN, C. B., 1973. Principles that Govern the Folding of Protein Chains, *Science*, **181**:223–230.

AVERY, O. T., C. M. MACLEOD, and M. MCCARTY, 1944. Studies on Chemical Nature of the Substance Inducing Transformation of Pneumococcal Types, *J. Exp. Med.*, **79**:137–157.

BALDWIN, R. L., 1975. Intermediates in Protein Folding Reactions and the Mechanism of Protein Folding, *Ann. Rev. Biochem.*, **44**:453–476.

BLOOMER, A. C., J. N. CHAMPNESS, G. BRICOGNE, R. STADEN, and A. KLUG., 1978. Protein Disk of Tobacco Mosaic Virus at 2.8 A Resolution Showing the Interaction within and between the Subunits, *Nature*, **276**:362–368.

BRIMACOMBE, R., 1978. Ribosome Structure, *Ann. Rev. Biochem.*, **47**:217–249.

———, 1978. "The Structure of the Bacterial Ribosome," 28th Symp. Soc. Gen. Microbiol. Cambridge University.

BRITTEN, R. J., and E. H. DAVIDSON, 1969. Gene Regulation for Higher Cells: a Theory, *Science*, **165**:349–357.

———, and D. E. KOHNE, 1968. Repeated Sequences in DNA, *Science*, **161**:529–540.

BUTLER, P. J. G., and C. H. DURHAM, 1977. Tobacco Mosaic Virus Protein Aggregation During Virus Assembly, *Adv. Prot. Chemistry*, **31**:187–251.

BUTLER, P. J. G., and A. KLUG, 1978. The Assembly of a Virus, *Sci. Amer.*, **239**:52–58 (Nov.).

CASJENS, S., and A. KING, 1975. Virus Assembly, *Ann. Rev. Biochem.*, **44**:555–611.

CHAMBON, P., 1977. The Molecular Biology of the Eukaryotic Chromosome is Coming of Age, *Cold Spring Harbor Symp. Quant. Biol.*, **42**:1209–1234.

COHEN, S. N., 1977. Recombinant DNA: Fact and Fiction, *Science*, **195**:654–657.

CRICK, F. C., 1979. Split Genes and RNA Splicing, *Science*, **204**:264–271.

DAVIDSON, E. H., and R. J. BRITTEN, 1973. Organization, Transcription, and Regulation in the Animal Genome, *Quart. Rev. Biol.*, **48**:565–613.

———, 1979. Regulation of Gene Expression: Possible Role of Repetitive Sequences, *Science*, **204**:1052–1059.

DAWID, I. B., and W. WAHLI, 1979. Application of Recombinant DNA Technology to Questions of Developmental Biology, *Develop. Biol.*, **69**:305–328.

DICKERSON, R. E., and I. GEIS, 1969. "The Structure and Action of Proteins," Harper & Row.

DICKSON, R. C., J. ABELSON, W. M. BARNES, W. S. REGNIKOFF, 1975. Genetic Regulation: The Lac Control Region, *Science*, **187**:27–35.

DUPRAW, E. J., 1968. "Cell and Molecular Biology," Academic.

HENDRIX, R. W., 1979. Bacteriophage Assembly, *Nature*, **277**:172–173.

JACOB, F., and J. MONOD, 1961. On the Regulation of Gene Activity, *Cold Spring Harbor Symp. Quant. Biol.*, **26**:193–211.

KENDREW, J. C., 1961. Three-Dimensional Structure of a Protein, *Sci. Amer.*, **205**:96–110 (Dec.).

KORNBERG, A., 1974. "DNA Synthesis," Freeman.

———, and J. O. THOMAS, 1974. Chromatin Structure, *Science*, **184**:865–871.

LEHNINGER, A. L., 1976. "Biochemistry; The Molecular Basis of Cell Structure and Function," 2d ed., Worth.

LEWIN, B., 1974. "Gene Expression," 3 vols., Wiley.

LODISH, H. F., 1976. Translational Control of Protein Synthesis, *Ann. Rev. Biochem.*, **45**: 39–72.

MENDEL, G., The Birth of Genetics, suppl. to *Genetics*, **35**(5): part 2, reprinted in 1950.

MILLER, O. L., and B. A. HAMKALO, 1972. Visualization of RNA Synthesis on Chromosomes, *Int. Rev. Cytol.*, **33**:1–25.

MOORE, J. A., 1977. "Heredity and Development," 2d ed., Oxford University Press.

MURIALDO, H., and A. BECKER, 1978. Head Morphogenesis of Complex Double Stranded Deoxyribonucleic Acid Bacteriophages, *Microbiol. Revs.*, **42**:529–576.

NEURATH, H., and R. L. HILL (eds.), 1975. "The Proteins," 3d ed., Academic.

O'FARRELL, P. H., 1975. High Resolution Two-Dimensional Electrophoresis of Proteins. *J. Biol. Chem.*, **250**:4007–4021.

PARDUE, M. L., and J. G. GALL, 1970. Chromosomal Localization of Mouse Satellite DNA, *Science*, **168**:1356–1358.

SANGER, F., and E. O. P. THOMPSON, 1953. The Amino Acid Sequence in the Glycyl Chain of Insulin, *Biochem. J.*, **53**:353–374.

SINSHEIMER, R. L., 1977. Recombinant DNA, *Ann. Rev. Biochem.*, **46**:415–438.

STENT, G., and R. CALENDAR, 1978. "Molecular Genetics," 2d ed., Freeman.

STRYER, L., 1975. "Biochemistry," Freeman.

Symposium on the Genetic Code, 1966. *Cold Spring Harbor Symp. Quant. Biol.* **31**:1–742.

TANFORD, C., 1978. The Hydrophobic Effect and the Organization of Living Matter, *Science*, **200**:1012–1018.

TAYLOR, J. H., 1965. "Selected Papers on Molecular Genetics," Academic.

WATSON, J. D., 1976. "Molecular Biology of the Gene," 3rd ed., Benjamin.

———, and F. H. C. CRICK, 1953*a*. The Structure of DNA, *Cold Spring Harbor Symp. Quant. Biol.*, **18**:123–131.

———, ———, 1953*b*. Genetical Implications of the Structure of Deoxyribonucleic Acid, *Nature*, **171**:964–967.

WOOD, W. B., 1978. Bacteriophage T_4 Assembly and the Morphogenesis of Subcellular Structure, *Harvey Lects.*, **73**:203–223.

———, and R. S. EDGAR, 1967. Building a Bacterial Virus, *Sci. Amer.* **217**:60–74 (July).

WOOL, I. G., 1979. The Structure and Function of Eukaryotic Ribosomes, *Ann. Rev. Biochem.*, **48**:719–754.

Chapter 3 The eukaryotic cell

The cell is the structural and functional unit of living tissue. It is a complete entity with a well-defined perimeter and a predictable internal organization. It is the smallest unit that can maintain itself in a living state. In isolation, cells can divide, metabolize, grow, and take part in every basic activity required for the maintenance of life. Whereas the previous chapter was concerned primarily with information storage and utilization, this chapter will serve as a brief survey of the fundamental components of eukaryotic cell structure and function, with a particular emphasis on those aspects which relate to developmental phenomena.

1. The plasma membrane

The cell is bounded by the *plasma membrane,* a structure of extreme thinness (approximately 75 A) and delicacy. It is of the utmost importance, for it maintains a highly specialized environment within. In this capacity, the plasma membrane is a barrier, but a highly selective one. Many substances are kept out, numerous substances are allowed in. Among the substances that enter the cell, some do so simply by diffusion in response to a concentration difference across the membrane. Others are actively carried through the membrane in order to be maintained at a higher concentration on the inner side. Still other materials, including the fluid they are suspended in, are brought inside as a result of the formation of pinocytotic vesicles at the surface, which are then brought inward. In a similar manner, fusions of plasma and vesicle membranes are capable of extruding materials from the intracellular space to the outside world. One important consequence of the selective permeability of the plasma membrane and its high electric resistance is its capacity to separate charged ions and therefore establish a potential difference. This voltage is critical for the so-called excitable cells, the neurons and muscle cells, but it is also of great importance in the ability of all cells to respond to their environment.

In its role of screening the substances that are allowed entrance to the cell's interior, the plasma membrane can determine what types of regulatory influences can be exerted on the cell from the outside. In a more direct sense, several regulatory molecules that have dramatic effects on the activities of the entire cell

appear to have their primary site of action at the cell surface. One of these regulatory molecules is the hormone insulin, whose presence has profound effects on cell metabolism, but whose site of interaction with the cell appears to be at the membrane itself. Similarly, there is evidence that the nerve-growth factor, a protein that produces dramatic changes in the differentiation of developing neurons, does not penetrate the membrane but instead binds to it to initiate its effect.

Membranes serve as supportive structures as well as walls upon which a variety of proteins can be hung. In actuality these proteins are integral parts of the membrane itself, serving both structural and functional roles. Each type of cellular membrane has its own characteristic composition. For example, many of the proteins of the plasma membrane are required in the dealings of this membrane with its environment: the membranes of the mitochondria include a sequence of enzymes involved in oxidative metabolism; the membranes of the Golgi complex possess enzymes that function in the addition of sugar groups to newly synthesized proteins; and so on.

Membranes also serve to compartmentalize the cell. They are very thin, continuous sheets, never have free edges, and accordingly separate the cell interior into cytoplasmic compartments. In this capacity, membranes ramify throughout the cytoplasm (Fig. 3.1). Taken together, membranes are active participants in virtually all cellular processes and may include a large part of the cell's total enzymatic machinery. Observations that membranes have a similar appearance under the electron microscope, are interconnected, and appear capable of giving rise to one another led to the concept of a "unit membrane," i.e., that all membranes are essentially alike. This concept remains valid if provisions are made for specialization of the various cellular membranes within a broader framework of similar overall structure.

A. Molecular structure of membrane

All membranes are lipid-protein assemblies in which the components are held together in a thin sheet by noncovalent bonds. The proportion of lipid to protein varies considerably depending on the type of cellular membrane, the type of organism, and the type of cell. In addition to lipid and protein, most membranes contain small amounts of carbohydrate, present as oligosaccharide chains covalently linked to protein and lipid molecules. There are several basic types of membrane lipids, all having one point in common: they are *amphipathic;* i.e., they have hydrophilic and hydrophobic portions within one lipid molecule. Most

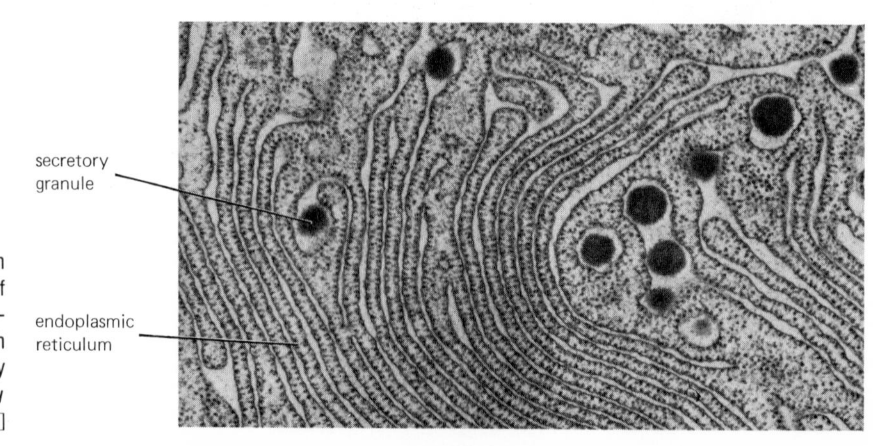

Figure 3.1 Electron micrograph of part of a cell, showing endoplasmic reticulum with ribosomes and secretory granules. [*Courtesy of K. Porter.*]

membrane lipids contain phosphate, the main exception being cholesterol. Most of the phospholipids of membranes are phosphoglycerides; their structure is based on a glycerol backbone. Glycerol is a three-carbon polyalcohol upon which substitutions for the hydroxyl groups can be made enzymatically. In the case of the glycerides present within the membrane, two of the hydroxyls are esterified to fatty acids, the third to a phosphate. One of a variety of low-molecular-weight groups, including choline, ethanolamine, serine, and inositol, also is esterified to the phosphate moiety. All these groups are hydrophilic, and together with the phosphate to which they are attached, this end of the phospholipid (called the *head group*) forms a highly water-soluble territory within the molecule. The fatty acid chains are long hydrocarbons, i.e., chains containing only hydrogen and carbon, and are very hydrophobic, thus the asymmetric, amphipathic nature of these molecules (Fig. 3.2). Since the nature of the fatty acid chains can be highly variable, both with regard to chain length and number of double bonds, phospholipid molecules as a group can have a wide range of properties.

Our current concept of the molecular organization of cellular membranes is shown in Fig. 3.3. The core of the membrane consists of a lipid bilayer, i.e., a double row of phospholipid molecules, organized so that the hydrophobic ends of each lipid molecule are pointed inward to the center of the bilayer and their hydrophilic ends are pointed toward the bilayer's outside edges. The proteins of the membrane are present as discrete particles, many of which penetrate deeply into, or entirely through, the lipid bilayer. In this model, the lipid bilayer is continuous in the sense that no region of lipids would be walled off from the remainder of the bilayer by a protein barrier; however, the integrity of the bilayer as an unbroken sheet would be interrupted by membrane proteins.

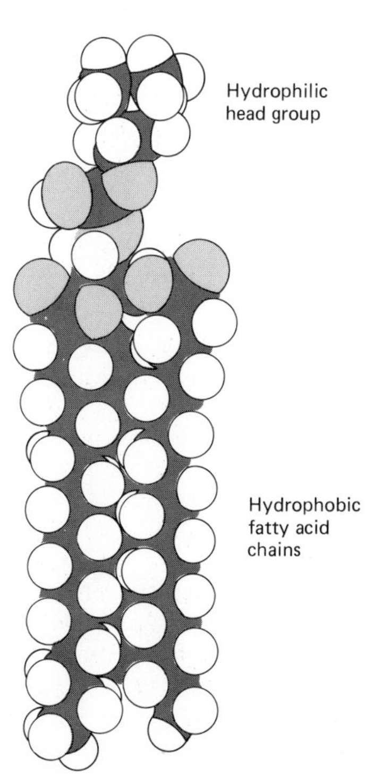

Figure 3.2 A space-filling model of a phospholipid, phosphatidylcholine.

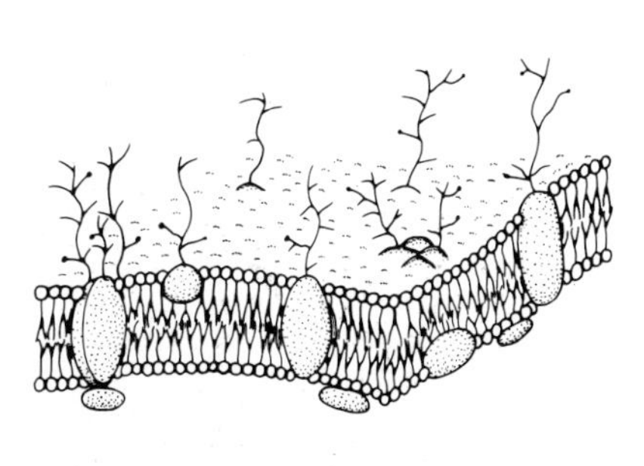

Figure 3.3 The fluid-mosaic model of the plasma membrane. Integral proteins are seen to penetrate into and completely through the lipid bilayer. Oligosaccharide chains are seen to protrude from the external leaflet into the extracellular space. Peripheral proteins are seen on the cytoplasmic face of the membrane associated with the hydrophilic head groups of the phospholipids and the surfaces of integral proteins.

In the fluid-mosaic model shown in Fig. 3.3, two types of membrane proteins can be distinguished on the basis of the intimacy of their relationship to the lipid bilayer. Those proteins which penetrate into the lipid bilayer are referred to as *integral membrane proteins*. Integral proteins are ones that contain major hydrophobic portions, i.e., regions composed primarily of hydrophobic amino acids, capable of penetration into the hydrophobic lipid bilayer. In addition to having a capacity to form hydrophobic associations with lipids, integral membrane proteins also possess hydrophilic portions which allow them to protrude beyond the edges of the bilayer. As a consequence, interactions with water-soluble substances (ions, small-molecular-weight substrates, hormones, etc.) can occur at the membrane surface, while the barrier properties afforded by the continuous lipid phase also can be maintained. In the case of the best-studied cells, integral proteins that project from the plasma membrane to the extracellular space are actually glycoproteins. The oligosaccharide chains that are covalently attached to the protruding proteinaceous base (Fig. 3.3) have been shown to play an important role in the interactions between the cell and its environment. In addition to the integral proteins, there exists a class of membrane proteins that are not so tightly involved in the membrane's architecture. These are the *peripheral proteins* associated with the membrane by weak bonds to either the hydrophilic head groups of the lipids or the hydrophilic portions of the integral proteins protruding from the bilayer.

The diagram of membrane structure shown in Fig. 3.3 portrays the membrane as it might appear frozen in time. However, in the living state, the membrane is anything but a static structure. The dynamic nature of the membrane is nicely illustrated by employing the technique of cell fusion, whereby cells of two different species are joined together to produce one cell with a common cytoplasm and a continuous plasma membrane—in the present case, human and mouse cells. Certain proteins of the cell membrane are readily identifiable as being derived from a human cell rather than of mouse origin. The locations of human and mouse surface proteins were mapped at various times after fusion, and it was found that they did not remain in their original positions. Shortly after fusion the two types of proteins were still present in either one or the other half of the united membrane, but in 40 minutes they were essentially intermixed. These results suggest that the proteins of the membrane are free to move—to actually diffuse—within the membrane itself. In other words, the components of the membrane are not in a static state, but rather are capable of movement in lateral directions within the plane of the membrane. The results also suggest that the lipid phase of the membrane is in a fluid state, through which the proteins are suspended and capable of migration. The length of time it takes the human and mouse proteins to approach an equilibrium state is a measure of the rate of diffusion of the molecules, which is in turn a measure of the mass of the protein and the fluidity (or viscosity) of the membrane. Calculations indicate that the viscosity of the lipid solvent phase is approximately 1,000 times that of water.

Earlier studies of the turnover of membrane components, i.e., the continual synthesis and destruction of the membrane constituents, indicated the dynamic nature of the membrane in a metabolic sense. The fluid-mosaic model illustrates its dynamic nature in an organizational sense. Membrane fluidity seems to provide the perfect compromise between a very rigid, ordered structure, in which mobility would be lacking, and a completely fluid nonviscous liquid, in which the molecules would not be oriented and mechanical support would be lacking. Most important, fluidity allows for interactions within the plane of the membrane. The interactions can be lipid-lipid, lipid-protein, or protein-protein in nature. In any

case, the interacting molecules can come together, carry out the necessary reaction, and either remain together or move apart depending on the conditions. Similarly, it is believed that the association of integral membrane proteins could lead to the formation of hydrophilic "pores" in the membrane, through which low-molecular-weight materials could diffuse or be transported.

The cell-fusion experiment just discussed (and many others) suggests that the potential for migration is present among integral membrane proteins, but does not indicate that all membrane proteins are free to drift around randomly on the lipid "sea." In fact, there is extensive evidence indicating that protein mobility can be greatly restricted. Restraints imposed on protein mobility can be of an intrinsic (i.e., within the membrane) or extrinsic (i.e., external to the membrane itself) nature. For example, the physical state of the lipid and the size of the protein aggregate represent important factors in this regard. The relationship between integral and peripheral proteins also can be a determining factor. In the case of the erythrocyte, for example, integral proteins are held in place as a result of their connections to a network of peripheral proteins on the membrane's inner surface. Similarly, there is good evidence that certain of the cytoskeletal components of the cytoplasm can become hooked to the inner surface of certain integral membrane proteins and either facilitate or inhibit lateral mobility. The consequence of these various types of restraints is significant, since it allows various types of associations within the membrane to be stabilized and maintained. As a result, different parts of the membrane can become specialized for particular functions. Examples of membrane specializations can be seen in the organization of the gap junctions in Fig. 3.6 and that of the sperm plasma membrane in Fig. 4.7.

In order to reveal the existence of regional specializations within the membrane, special types of techniques have had to be developed. One of the most important of these techniques is freeze-fracture replication, by which pieces of tissue are frozen and fractured along planes that run through the middle of cellular membranes. Once the membrane is split, the exposed surfaces can be coated with a metallic layer that fits over the hills and valleys present within the membrane, features that reflect the presence and absence of integral membrane proteins. Once the metallic replica of the surface is formed, the tissue that provided the template can be discarded, and the replica can be viewed in the electron microscope to reveal the topographical organization of the membrane's integral proteins. Figures 3.6 and 4.7 contain electron micrographs prepared in this manner.

2. The cell coat and associated materials

Although the plasma membrane exists at the outer edge of the cytoplasm, there is usually extracellular material on the outer surface of the membrane itself. It is generally difficult to determine how much of this material represents a layer referred to as the *cell coat*, which consists of the hydrophilic carbohydrate-containing portions of the integral membrane proteins projecting to the outside of the cell, and how much represents a truly independent extracellular layer. In many cases, the glycoproteins of the outer leaflet of the plasma membrane have similar staining properties to the extracellular materials beyond the membrane (both stain with ruthenium red, for example), and the two layers tend to merge with one another.

The most commonly encountered extracellular materials are the *glycosaminoglycans* (or *mucopolysaccharides*) and the protein *collagen*. Glycosaminoglycans are built up of long chains of repeating disaccharides in which one of the

two sugars of the repeating unit is an amino sugar, either *N*-acetylglucosamine or *N*-acetylgalactosamine. Glycosaminoglycans (such as hyaluronic acid, heparan sulfate, chondroitin sulfate, and keratin sulfate) are extremely acidic molecules owing to the presence of large numbers of carboxyl and/or sulfate groups. In most cases, glycosaminoglycans exist in combination with proteins, the complex being termed a *proteoglycan* (or *mucoprotein*). Glycosaminoglycans are secreted into the extracellular space, where they generally exist as amorphous materials of high viscosity, ones that become greatly hydrated and occupy very large volumes for a given weight of material.

Collagen molecules, however, are composed of long polypeptide chains organized into ropelike fibers. These collagen fibers are readily identified in the extracellular space because of their banded appearance following various types of staining procedures. Collagen fibers are characterized by their extremely high tensile strength. In addition to its general presence in the extracellular space of many types of tissues, collagen is a major component of skin, cartilage, bones, tendon, and the cornea. In many cases, collagen and glycosaminoglycans interact with one another in the extracellular space to form a variety of loosely organized structures, including basement membranes and connective tissue capsules. In addition to glycosaminoglycans and collagen, a variety of glycoproteins of ill-defined character are also associated with the surfaces of cells. One of these, *fibronectin*, has been extensively studied. When present, this large-molecular-weight glycoprotein exists as a loose fibrous meshwork that clings to the cell and to the neighboring substrate should it be present.

Materials present on the outside of cells, both in the cell coat and beyond, are of vital importance in determining the manner in which cells interact with one another, as well as other properties, such as adhesion to substrates, migratory activity, and possibly, malignancy. Certain of the macromolecules at the cell surface act as recognition sites for cell-to-cell identification. The nature and distribution of these molecules are believed to vary from one type of cell to the next and facilitate proper interactions. There are clearly differences in the surfaces of cells taken from different individuals, and these form the basis for the graft-rejection mechanism discussed in Chapter 17.

3. Cell junctions

One of the primary consequences of the multicellular state is that cells have evolved mechanisms for intercellular interaction. The types of contacts that cells can form with one another are of great importance to the cooperative activity they might show, and it is the cell surface that must mediate these contacts. Among the various types of contacts cells can make with one another, three types of *junctional complexes* are commonly observed with the electron microscope. These three types of junctions have different morphologies and apparent functions.

The *desmosome* (Fig. 3.4) is a specialized junction with considerable space between adjoining plasma membranes (about 150 A) and extensive modifications in the adjacent cytoplasm, including the presence of associated filaments (Fig. 3.4*a*). Desmosomes are believed to serve primarily as sites of strong mechanical adhesion between adjacent cells. The *tight junction* (Fig. 3.5) brings the two membranes together in their closest contact and is believed to serve as a barrier to prevent materials from passing between cells. Tight junctions are found to completely encircle cells (Fig. 3.5*b*) and to seal off the extracellular space on the two sides of the junction. In the third type of junction, the *gap junction* (Fig.

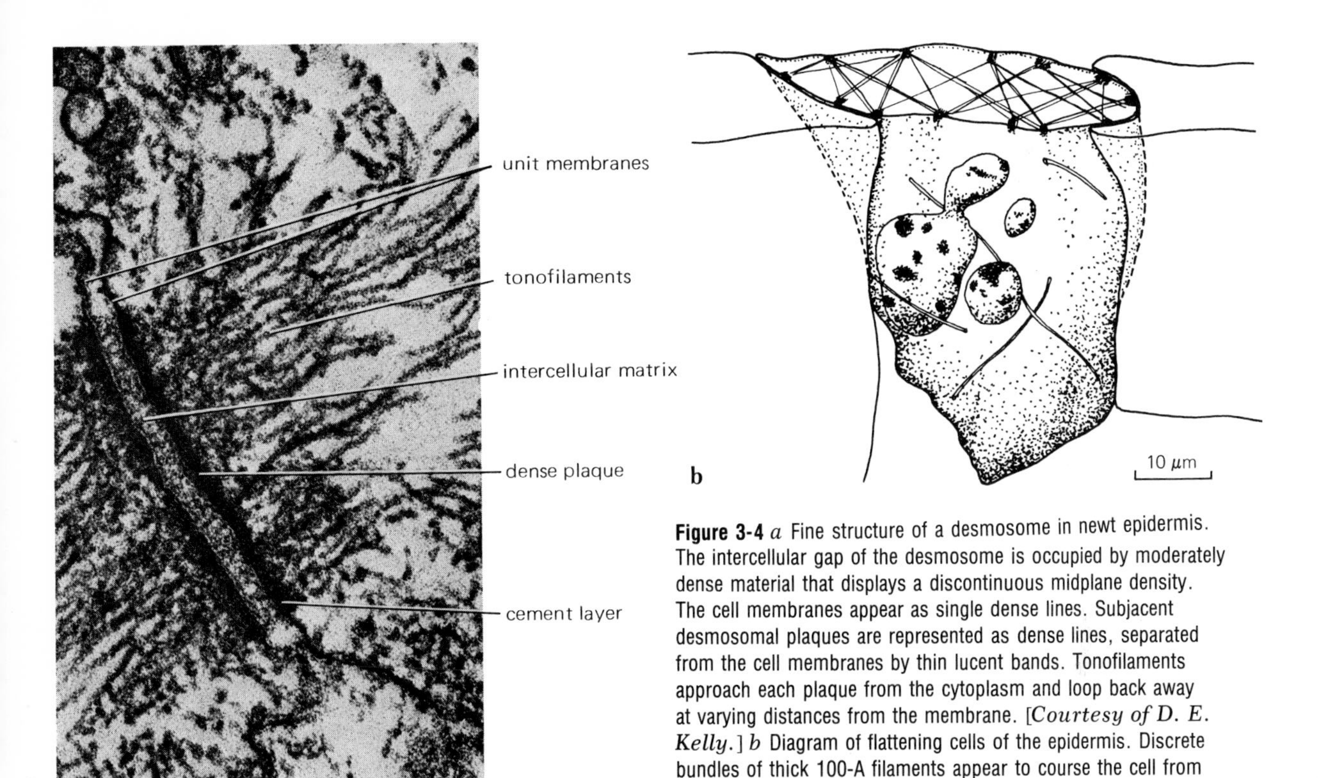

Figure 3-4 *a* Fine structure of a desmosome in newt epidermis. The intercellular gap of the desmosome is occupied by moderately dense material that displays a discontinuous midplane density. The cell membranes appear as single dense lines. Subjacent desmosomal plaques are represented as dense lines, separated from the cell membranes by thin lucent bands. Tonofilaments approach each plaque from the cytoplasm and loop back away at varying distances from the membrane. [*Courtesy of D. E. Kelly.*] *b* Diagram of flattening cells of the epidermis. Discrete bundles of thick 100-A filaments appear to course the cell from desmosome to desmosome. [*After B. Burnside,* Amer. Zool., **13**:*997* (*1973*).]

Figure 3.5 The tight junction. *a* Model of a tight junction indicating that the membranes are held together along lines of attachment composed of proteinaceous particles. [*From L. A. Staehlin,* Int. Rev. Cyt., **39**:*207* (*1974*).] *b* Electron micrograph of this type of junction. [*Courtesy of D. Friend.*] *c* Scanning electron micrograph of the apical surface of an epithelium showing the encircling nature of the junction. [*Courtesy of D. Tarin.*]

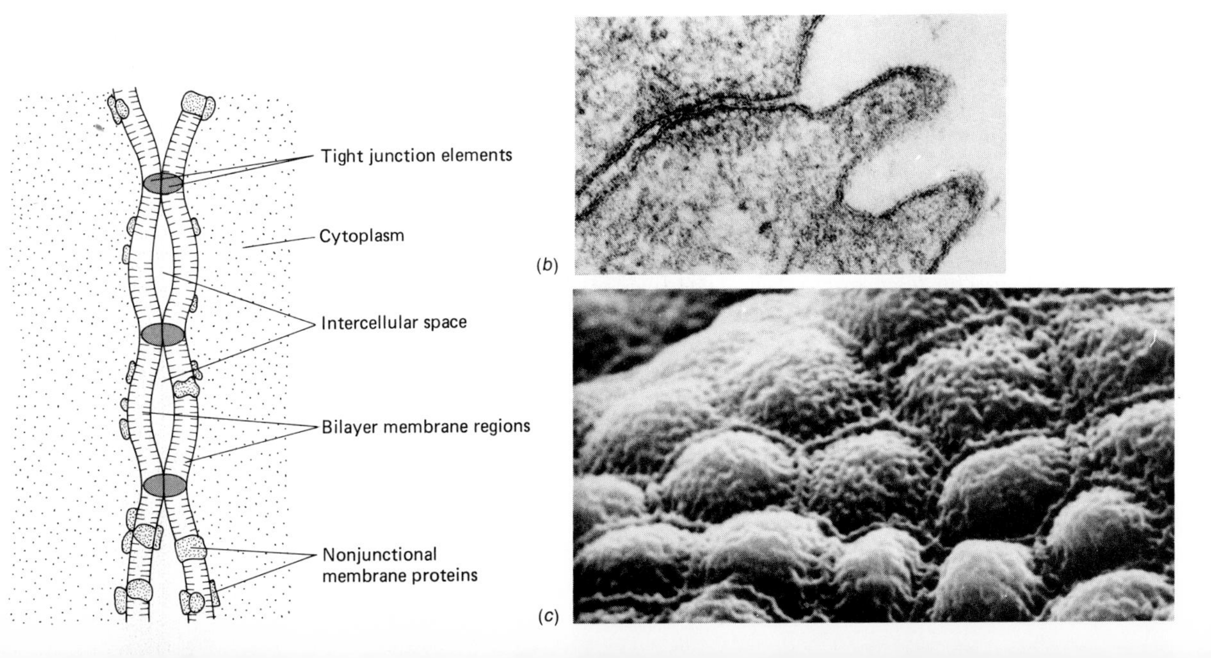

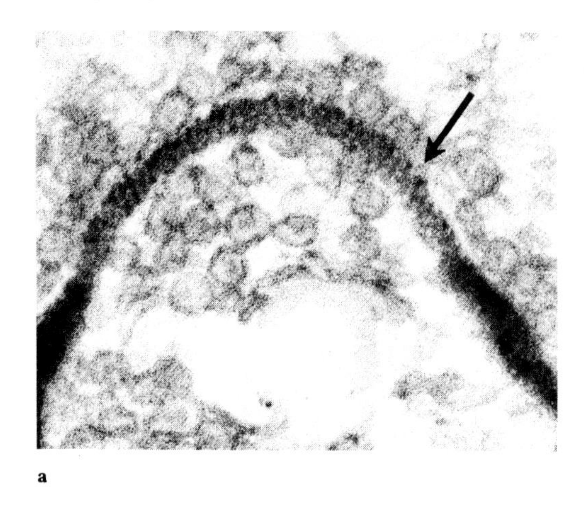

a

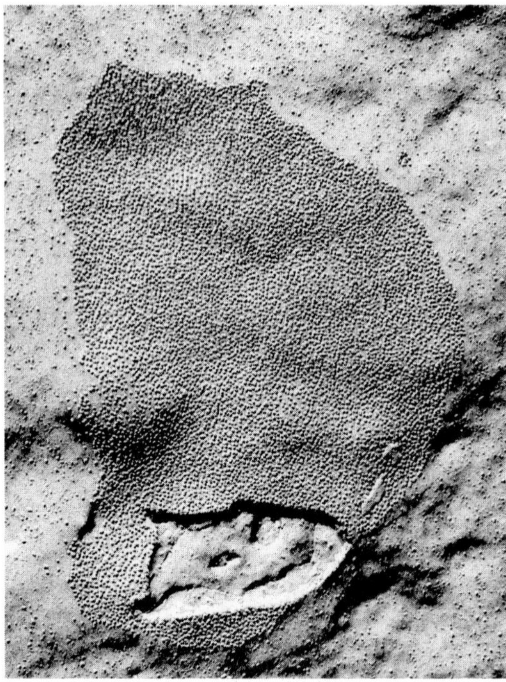

b

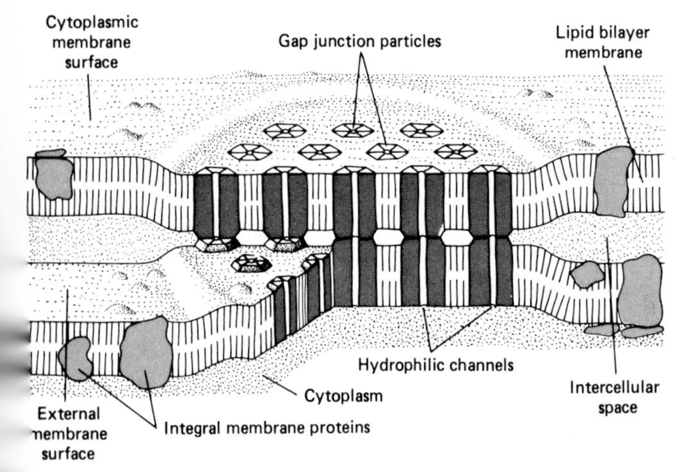

c

Figure 3.6 The gap junction. *a* Electron micrograph of a section through a gap junction. Very dense staining is seen in the extracellular space at the edge of the junction, and a repeating pattern is discernible within the junction (arrow). [*Courtesy of B. W. Payton.*] *b* Freeze-fracture replica through a gap junction showing the concentrated presence of membrane particles. [*Courtesy of D. F. Albertini.*] *c* Three-dimensional reconstruction of a gap junction. [*From L. A. Staehlin,* Int. Rev. Cyt., **39:**249 (1974).]

3.6), the membranes of adjoining cells are brought into close proximity (20 to 40 A), and there is evidence of fine connections passing between the cells through the extracellular space. Gap junctions, which appear as discrete patches on the cell surfaces, are believed to serve as sites through which small-molecular-weight materials can pass between adjoining cells.

There are two principal means to demonstrate the direct exchange of materials from one cell to another without their passing through the extracellular space

between (Fig. 3.7). In one technique, materials of different molecular weight are injected into one cell and are seen to diffuse rapidly and directly into the adjoining cell (Fig. 3.7*a*). A variety of tracer studies of this sort indicate that many substances, particularly of small molecular weight, pass readily in some channel across cell membranes. In some cases, particularly embryonic ones, molecules as large as several-hundred molecular weight (such as fluorescein) cannot make the journey; in other tissues, considerably larger molecules (possibly up to 10,000 molecular weight) can pass between. In the other method of analysis, electrodes are placed within two cells in a line of cells and the passage of ionic current (a measure of the ability of ions to move between cells) is determined (Fig. 3.7*b*). Cells between which a flow of current can be measured are said to be *electrically coupled.* Both tracer flow and electric coupling go hand in hand, and both are believed to be mediated by gap junctions. If substances are to be able to diffuse freely from one cell to another without appearing in the extracellular space, some form of pipeline must exist that directly connects the two cytoplasms. A model for the construction of a gap junction is shown in Fig. 3.6*c*. Tiny channels are believed to exist in a latticework within the patches where the membranes of two cells come together. The numerous interconnecting tubes are evident. The dimensions of the interconnecting channels seen in the electron microscope, together with a consideration of the molecular weights of materials that can pass between the cells, suggest that the diameter of these channels is on the order of 10 A.

One of the most remarkable features of these membrane complexes is the rapidity with which they can be disassembled and re-formed. Treatment of coupled cells with trypsin, removal of divalent ions, or micromanipulation readily causes the coupling to be lost. Cells can become coupled in minutes simply by putting their surfaces in contact with one another. Even cells of different tissues or of different species can be coupled one to another and gap junctions will form between them. These results reaffirm the dynamic nature of the cell membrane and

Figure 3.7 *a* Darkfield photomicrograph showing the passage of fluorescein from one cell into which it was injected (X) to the surrounding cells. [*Courtesy of R. Azarnia and W. R. Loewenstein.*] *b* Measurement of intercellular electrical communication through gap junctions. The photograph shows microelectrodes inserted into living cells of the Malphigian tubule of an insect. [*Courtesy of W. R. Loewenstein.*]

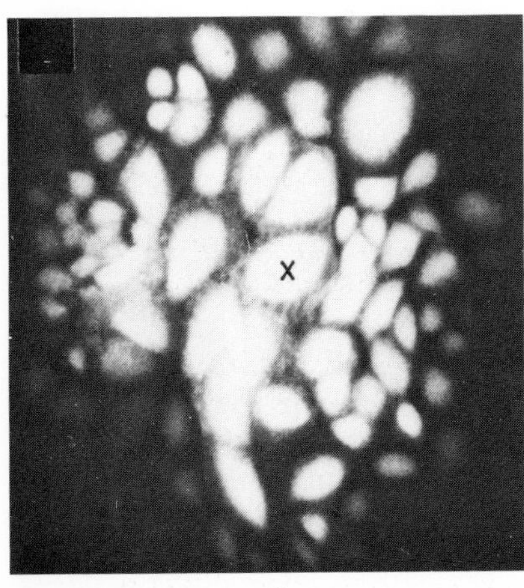

(*a*)

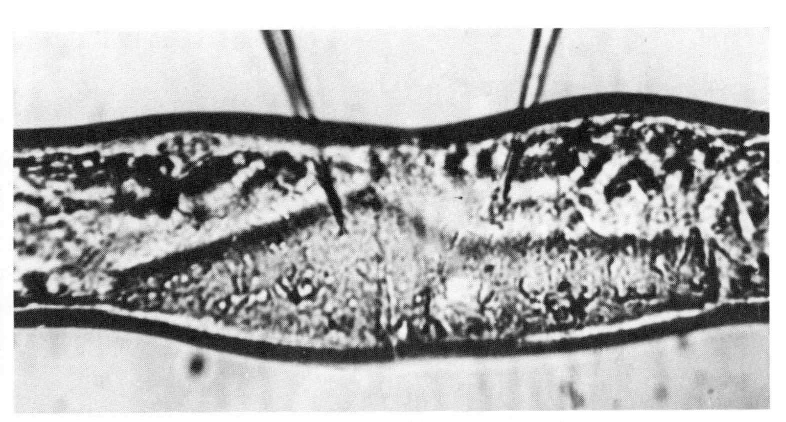

(*b*)

the spontaneity with which it can form itself or such derived structures. Presumably the same types of rules discussed in the section on self-assembly apply here as well; the major difference is the presence of a variety of lipid molecules in membranes, in addition to the ever-present protein. The reason for selecting the gap junction for special consideration among the numerous types of intercellular contacts is the potential it provides for intercellular communication. That a large number of cells can be put into intimate cytoplasmic contact with one another provides what is, in essence, one giant compartment. The possibility for embryonic cells to influence one another by this means is apparent. However, the process of diverse differentiation among a group of cells might be expected to be accompanied by a lack of such communication between cells whose development was continuing along different lines.

4. Intercellular controls

The characteristics of a tissue depend on the individual phenotypes of constituent cells and collective action among these cells. Collective action arises as a result of intercellular communication. There are several ways in which signals can be transmitted: by junctional transmission, as just described, by diffusion through intercellular fluid or matrix, and by way of systemic circuits. Apart from the direct junctional communication between electrically coupled cells, intercellular messages are generally sent in higher animals in one of two ways: by direct communication from cell to cell by way of nerves, and by indirect communication by way of chemical "messengers" within the circulatory system or diffusion within the intercellular matrix, i.e., by hormones or their equivalent. Important similarities between these two modes have long been obvious; for example, the transmission of signals across synapses in the nervous system is generally accomplished by chemical rather than electrical means. Synaptic transmission, in fact, is a highly specialized function carried out by a few unique low-molecular-weight neurotransmitters; hormonal control is involved in a complex and diverse set of functions carried out by many different chemical messengers of various sizes, structures, and chemical complexities. It has become increasingly apparent that a great variety of chemical messengers act on their target cells as a result of an interaction with the target cell membrane. Included within this group of messengers one finds neurotransmitters and other chemicals of the central nervous system, hormones, and growth-promoting substances. The best-studied hormones that act at the cell surface are ones whose main function is to control blood-sugar levels by regulating carbohydrate metabolism and the transport of sugars across cell membranes. These include the protein hormones glucagon and insulin and the catecholamine hormone epinephrine. All three of these hormones act on liver cells, each with a specific receptor molecule present on the cell surface, a component of the outer leaflet of the plasma membrane. The attachment of glucagon or epinephrine is transmitted through the plasma membrane as some type of signal capable of activating an enzyme, adenylate cyclase, believed present on the inner surface of the membrane. Upon activation, the enzyme functions to synthesize the cyclic nucleotide cAMP, which then diffuses into the cytoplasm.

This scenario in which an extracellular first messenger, such as glucagon, interacts with the membrane thereby leading to the synthesis of the intracellular second messenger, such as cyclic AMP, is repeated over and over again in many cell types. In each case it appears that the elevated levels of cyclic AMP serve to activate one or more specific protein kinases, enzymes which then add phosphate groups to specific proteins. Even though cyclic AMP serves as a second messenger in different types of cells, the response obtained can vary widely depending on

the nature of the cell involved. In the liver cell, increased cyclic AMP levels lead eventually to the activation of phosphorylase, the enzyme responsible for the breakdown of glycogen on the way to increasing blood glucose levels. In adipose tissue, elevated cAMP leads to the mobilization of energy reserves via eventual activation of an enzyme, triacylglycerol lipase, which hydrolyzes fat. There is evidence that increased cAMP levels are generally inhibitory to cell division, possibly via the activation of specific protein kinases that add phosphate groups to particular histone or nonhistone chromosomal proteins. The phosphorylation or dephosphorylation of certain proteins involved in cytoskeletal function also may occur in response to changes in cyclic AMP levels. Furthermore, cyclic AMP is not the only cyclic nucleotide that acts as a second messenger. Cyclic GMP, synthesized by the enzyme guanylate cyclase, seems also to mediate certain membrane-activated responses. Although the evidence remains fragmentary, cGMP may lead generally to reponses that are opposite to those elicited by cAMP. It has been proposed, for example, that elevated cGMP levels lead to glycogen synthesis in liver cells or increased potential for cell division in various types of cultured cells.

Another highly active agent involved in membrane-mediated responses is the calcium ion. We will have numerous occasions throughout this book to consider processes that are stimulated by increased levels of free calcium ions. In some cases, these responses are triggered by the release of calcium from some intracellular store, as occurs during muscle contraction. In other cases, the response occurs as a result of a sudden increase in the permeability of the plasma membrane toward this divalent ion. As in the case of the cyclic nucleotides, a single low-molecular-weight substance, Ca^{2+}, can have a variety of effects depending on the nature of the responding cell.

5. Cytoplasmic membranes

Coursing through the cytoplasm of most cells is an extensive series of interconnected cytoplasmic membranes. This series of membranes is found around spherical vesicles, lining long channels, organized into stacks of flattened disks, and in various other configurations. To a varying degree these membranous components can be categorized into several distinct types of organelles. The endoplasmic reticulum, the Golgi complex, and the lysosomes will be considered. Figure 3.1 shows a cell with an extensive *endoplasmic reticulum* (ER). In this cell, the endoplasmic reticulum is rough as a result of rows of ribosomes attached to the membranes. Presumably these ribosomes are organized into membrane-bound polyribosomes and are actively engaged in protein synthesis. Newly synthesized proteins come off the polyribosomes and can be carried through the enclosed channels of the ER to a variety of different destinations.

In some cases, the rough ER leads into a system of tubular membranes devoid of ribosomes, termed the *smooth endoplasmic reticulum.* Smooth ER is present to varying degrees in different cells and is particularly predominant in epithelial cells, steroid-producing cells, and striated-muscle cells. Cells responsible for the production of large quantities of protein—such as cells producing digestive enzymes in the pancreas—have a greatly expanded rough ER network; others, such as muscle and nephron cells, have very little rough ER. Rough ER is primarily involved in protein synthesis and transport, while smooth ER probably serves a variety of specialized functions in the cells in which it is best developed. Smooth ER has been implicated in the synthesis of nonprotein products, in glycogen metabolism, and as a channel for transport through the cell. As in the plasma

membrane, specific enzymes have been found to be associated with, and are part of, the membranes of the endoplasmic reticulum.

A. Golgi complex and lysosomes

Another distinct membranous structure is the *Golgi complex,* formed by a collection of flattened membranous cisternae and vesicles (Figs. 3.8 and 3.9). As with the ER, different types of cells have a varying degree of development of the Golgi, a variety of functions have been ascribed to this organelle. We will mention only a few. Golgi membranes are particularly well developed in secretory cells that pour their contents to the outside. In these cells, such as those of the pancreas, the newly synthesized products are collected within the membranes of the Golgi complex; there they appear to undergo a concentration process, producing vesicles with sufficient protein content to appear dense under the electron microscope after appropriate staining. In the pancreas the last stage in this process is the zymogen granule, which fuses to the plasma membrane, thereby releasing its products to the outside as a preliminary step in its passage to the digestive tract. Since the membrane of secretory vesicles ultimately contributes to the structure of the plasma membrane, the cisternae of the Golgi complex can be regarded as the source of much of the membrane of the cell.

The entire process from protein synthesis to expulsion from the cell can be followed by administering radioactively labeled amino acids and waiting various times before fixation of the tissue. As the time after injection is increased, the label moves farther from its site of synthesis in the rough ER, into the Golgi complex, then into the zymogen granules; and in about 2 hours, the label is seen to be expelled to the outside of the cells. Other functions of the Golgi complex include the attachment of carbohydrate molecules to newly synthesized proteins to produce glycoproteins and mucopolysaccharides. Certain of these carbohydrate-bearing proteins become sulfated, and the attachment of inorganic sulfate is

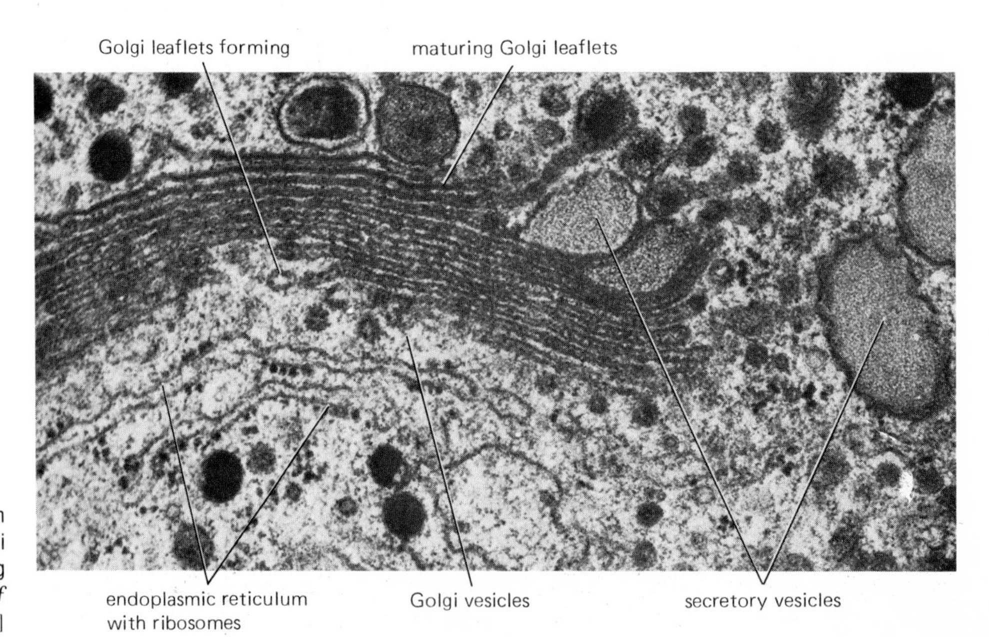

Figure 3.8 Electron micrograph of Golgi complex of a growing cell. [*Courtesy of O. Kiermayer.*]

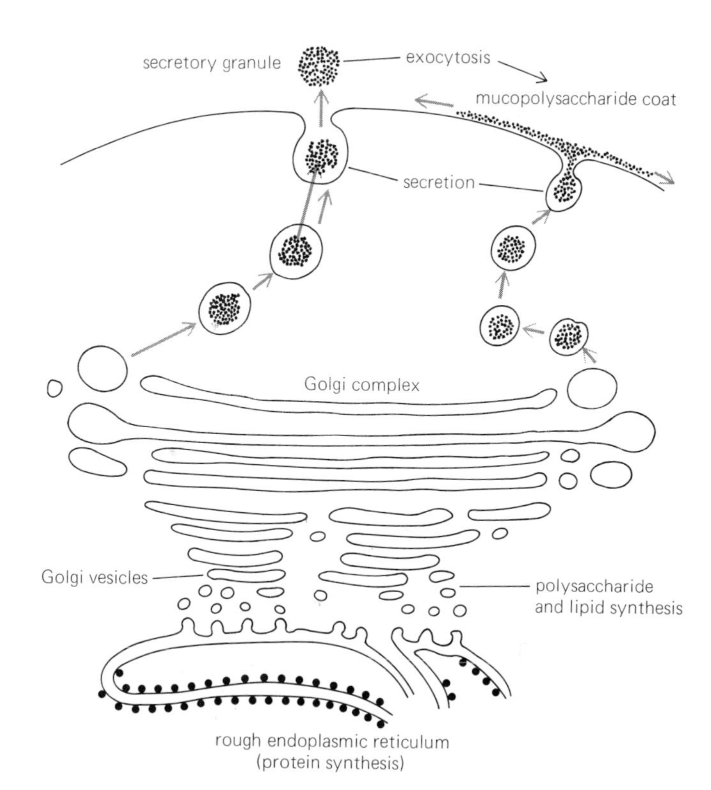

Figure 3.9 The cytoplasmic-membrane system. Rough endoplasmic reticulum at the bottom of the illustration is connected with smooth endoplasmic reticulum, which gives rise to smooth membranous Golgi vesicles that eventually fuse with the plasma membrane with which it becomes continuous. Proteins produced in the rough ER become part of the cellular membranes, and secretory products are released to the extracellular space.

accomplished within the confines of the Golgi complex. In many cells, such as mucous-secreting epithelia, these products are also transported to the outside via these membrane-bound vesicles. In at least two cases—the formation of the coelenterate nematocyst and of the acrosome at the tip of most sperm—the Golgi complex undergoes a complex transformation to form the two structures.

The other membranous organelle to be briefly considered is the lysosome. *Lysosomes* are membrane-bound vesicles formed in a manner similar to that of the zymogen granules described earlier. The Golgi complex, in its capacity of membrane-former, buds off these vesicles, which become filled with proteins produced by the endoplasmic reticulum (Fig. 3.9). In this case, the proteins are a collection of hydrolytic enzymes whose optimal activity occurs at acid pH. Included in the collection of enzymes are proteases, lipases, nucleases, and enzymes to break down polysaccharides and phosphate esters. It is this last capability of lysosomes—i.e., their acid phosphatase activity—that is typically used as the criterion for a vesicle's being considered a lysosome. As would be expected from its contents, lysosomes are involved in intracellular digestion.

In most cases lysosomes function by fusion with vesicles containing material to be digested. One of the properties of the cell membrane is its ability to surround particles or fluid on the outside of the cell and to enclose them within a vacuole that is pinched off inside the cell (Fig. 3.10). These vacuoles (termed *pinocytotic vacuoles* if the contents are more fluidlike and *phagocytotic vacuoles* if the contents are particulate) undergo fusion with the lysosomal vesicle, and their contents are digested by the hydrolytic enzymes. Lysosomes are particularly prominent in cells such as certain leukocytes and macrophages that have roles in intracellular digestion of material. In other cases, the hydrolytic enzymes gen-

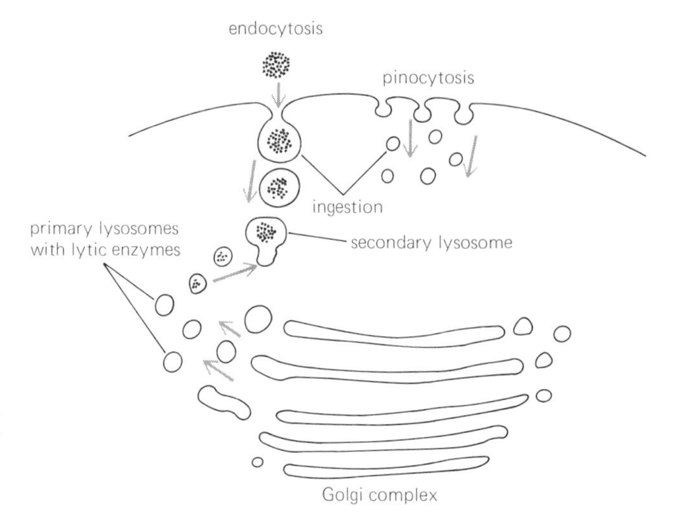

Figure 3.10 Formation of phagocytotic and pinocytotic vesicles and their subsequent fusion with primary lysosomes to form secondary lysosomes, in which digestion occurs.

erally ascribed to lysosomes are responsible for the death of the cell in which they are contained. In this capacity these organelles provide the cell with a capability for self-destruction, termed *autophagy*. This can result either from certain disease states characterized by tissue self-destruction or in the normal course of events of certain tissues, such as resorption of tail tissue in frog tadpole metamorphosis and progressive molding of structural bones during growth.

6. Microtubules and microfilaments

As the name implies, *microtubules* are small, hollow, cylindrical structures, rigid in nature, which have been found to occur in nearly every eukaryotic cell that has been scrutinized with the electron microscope. The tubule typically has an outer diameter of approximately 250 A and a wall diameter of approximately 50 A, leaving an internal (or lumen) diameter of 150 A. The structure of microtubules (Fig. 3.11) is strikingly similar in a very wide variety of organisms. Careful ultrastructural analysis of negatively stained microtubules has revealed that the wall is a polymer composed of globular subunits. Examination of a cross-section of the wall of a microtubule (Fig. 3.11*a* and *c*) nearly always reveals 13 subunits making up the complete circumference of the wall. When the surface of isolated microtubules is examined, the subunits are seen to be arranged in longitudinal rows, termed *protofilaments*, that are aligned parallel to the long axis of the tubule (Fig. 3.11*b*). When microtubules are caused to split open and flatten out, the 13 protofilaments making up the wall of the tubule can be seen, indicating the relatively tight association of the subunits with one another in these rows. If one traces the subunits around the wall, they are seen to spiral in a helical pattern (Fig. 3.11*d*).

Microtubules are composed totally of protein. They exist as polymers composed of dimeric subunits. Each subunit contains one molecule of α-tubulin and one of β-tubulin (Fig. 3.11*d*). These two proteins have very similar molecular weights and closely similar amino acid sequences. In addition to the tubulins, which account for 80 to 95 percent of the protein of the microtubule, a number of other proteins, termed *microtubule-associated proteins* (*MAPs*), are also present in the organelle (Fig. 3.11*b*) and are believed to be involved in both its assembly and function. Microtubules are found as part of a few highly organized struc-

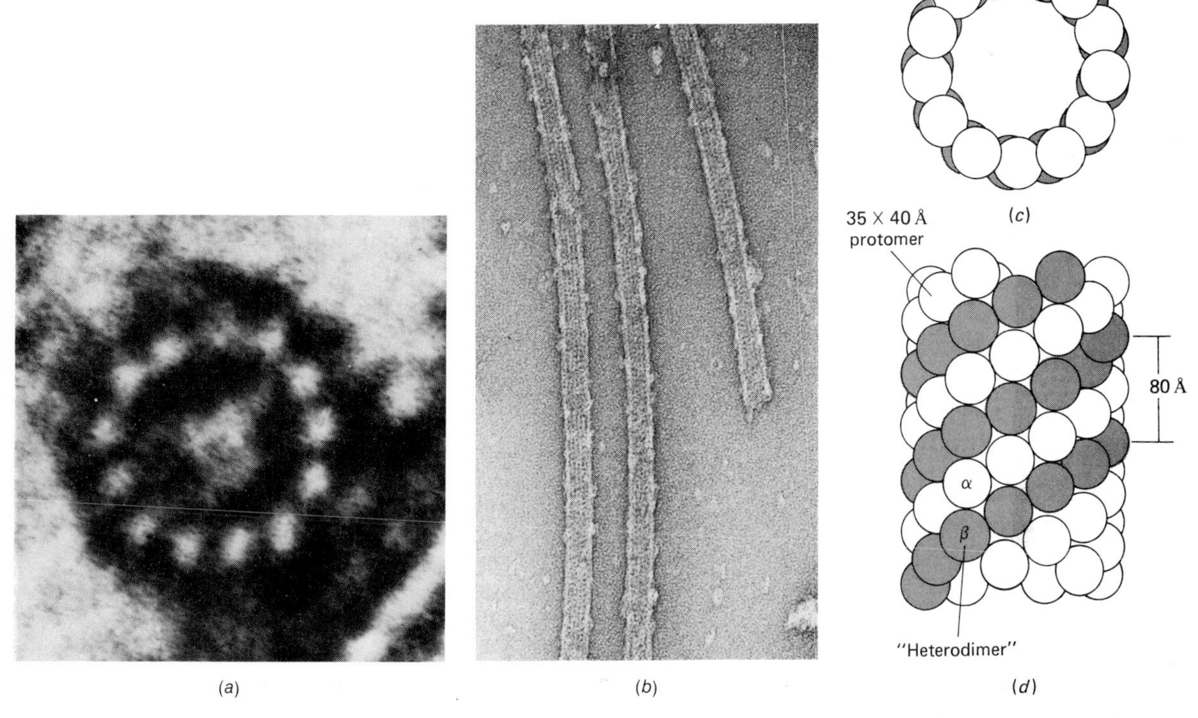

Figure 3.11 The structure of microtubules. *a* Electron micrograph of a cross section through a microtubule of a *Juniperus* root-tip cell revealing the 13 subunits arranged within the wall of the organelle. [*Courtesy of M. C. Ledbetter.*] *b* Electron micrograph of negatively stained microtubules from brain. Microtubule-associated proteins can be seen as regularly spaced projections on the surface of the microtubule. [*Courtesy of L. A. Amos.*] *c, d* Diagrams of a cross section and longitudinal section of a microtubule model. The model has 13 protofilaments composed of heterodimers. [*Reprinted from J. Bryan,* Federation Proceedings, **33:***156 (1974).*]

tures, such as cilia, flagella, centrioles, and the mitotic spindle, as well as in loosely organized networks or scattered through the cytoplasm. Their distribution in a given cell is best revealed by the use of fluorescent antitubulin antibodies, although in those cases where the microtubules are highly organized, they often can be seen with the polarization microscope.

In many cells, microtubules are capable of appearing and disappearing, as occurs normally for the mitotic apparatus before and after cell division. It is believed that under most conditions a dynamic equilibrium exists between the polymer and the subunits which make up the precursor pool. By control of this equilibrium the cell would regulate the assembly and disassembly of its microtubules. Factors having the potential to shift the equilibrium toward the polymeric or monomeric state include the concentrations of GTP, cyclic AMP, Ca^{2+}, Mg^{2+}, and certain of the MAPs. Each microtubule has a polarized nature and it is believed that assembly of the microtubule occurs at one of its ends and disassembly at the other. The relatively fragile condition of the microtubule is readily verified by the disruption of the structure by a variety of treatments—including hydrostatic pressure and cold temperature (treatments indicating that

polymerization is an endothermic, entropy-driven process similar to that of the assembly of certain viruses), as well as certain chemicals, such as colchicine, which binds to the dimers and the assembly end of the microtubule, thereby blocking the addition of more subunits. The microtubules of cilia and flagella are more stable structures, apparently not in equilibrium with the subunits and not sensitive to disruption by these treatments.

The nature of the microtubule assembly process within the cell is not well understood. In some cases, microtubules grow out from existing microtubular structures, as occurs in the formation of cilia and flagella as extensions of the preexisting basal body structure. In other cases, as in the mitotic spindle of animal cells, microtubules polymerize in the vicinity of specific types of organizing centers, including the pairs of centrioles at the mitotic poles and the kinetochores within the chromosomes. In many types of cells, microtubules appear to grow out of an electron-dense amorphous region of the cell whose properties have not been determined. The formation of microtubules from these formless regions is best seen in cultured cells after the existing microtubules are caused to depolymerize by treatment with cold or colchicine. When microtubules repolymerize in these cells, they do so from electron-dense regions just outside the nucleus.

Microtubules are believed to function in two interrelated activities: acting as a sort of cellular skeleton by providing structural support, and providing part of the machinery required in certain types of movements. For example, microtubules have been implicated in the movement of various granules within cells, and in an entirely different capacity, microtubules are believed to be of primary importance in the maintenance of cell shape. The natural shape of a free body with liquid properties is spherical, as seen in a suspended drop of water in oil or in a soap bubble. When free bodies are clustered together—again as seen most readily in a mass of soap bubbles—the exposed surfaces remain curved, but adjoining surfaces become more or less flat. Surface tension and mutually adhesive forces are responsible for these configurations. When other shapes appear, as in most cells, then other agencies must exist that either produce or maintain the distortions. Microtubules seem to be the most likely agent. They are commonly seen to be lined up parallel with and close to cell walls and in general to conform to whatever particular shape a cell may exhibit. A well-suited example of this is in the heliozoan protozoan *Echinosphaerium* (Fig. 3.12). This organism has numerous long axopodia (5 to 10 μm in diameter and up to 400 μm long) whose shape is determined by microtubules arranged parallel with the axopod (Fig. 3.12*c*). A cross section through an axopod (Fig. 3.13*a*), particularly near its base, reveals a remarkable structure, the *axoneme;* it consists of numerous microtubules arranged in two interlocking coils or spirals, the number of constituent tubules decreasing as sections are cut progressively toward the axopodial tip. Single tubules appear to traverse the entire length of the axopod.

Although microtubules are clearly involved in cell movements, it is not always evident whether or not they are playing an active, force-generating role. In cilia and flagella, the microtubules slide across one another in a manner analogous to sliding muscle filaments, thereby generating the force involved in the movement of the organelle. In this instance, the relative sliding is caused by a pair of dynein arms attached to one of the microtubules of each doublet of the cilium or flagellum (Fig. 3.13*b*). In the mitotic spindle, chromosomes are pulled toward the poles by the bundles of microtubules attached at their kinetochores. Whether the force for chromosome movement is generated by sliding of microtubules or by some other means, such as the action of an actomyosin system or the disassembly of the microtubules, remains to be established.

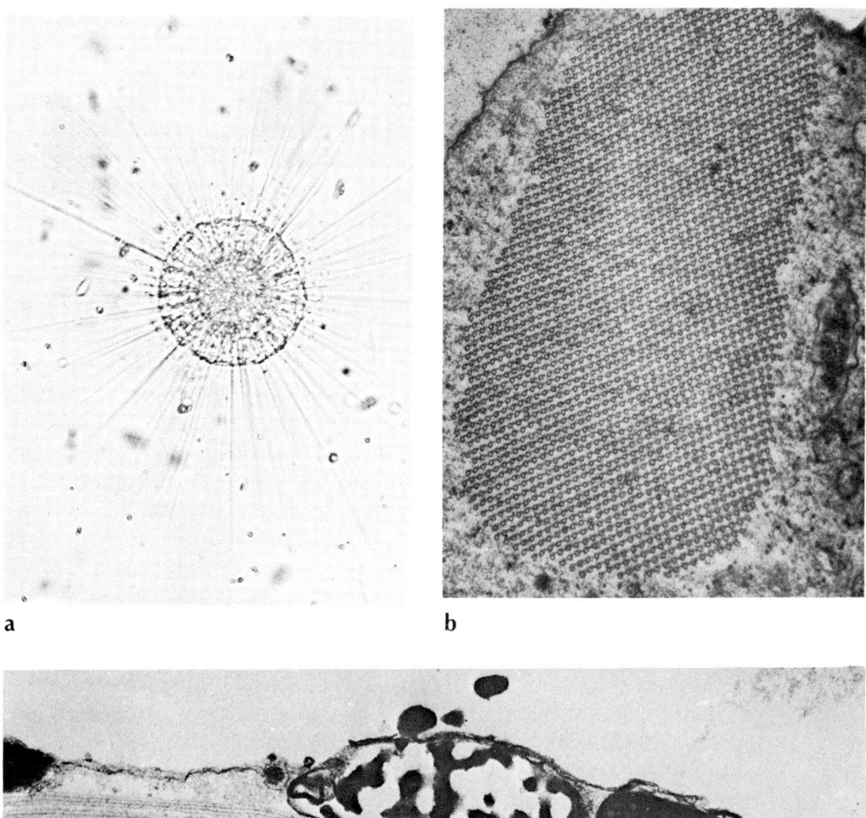

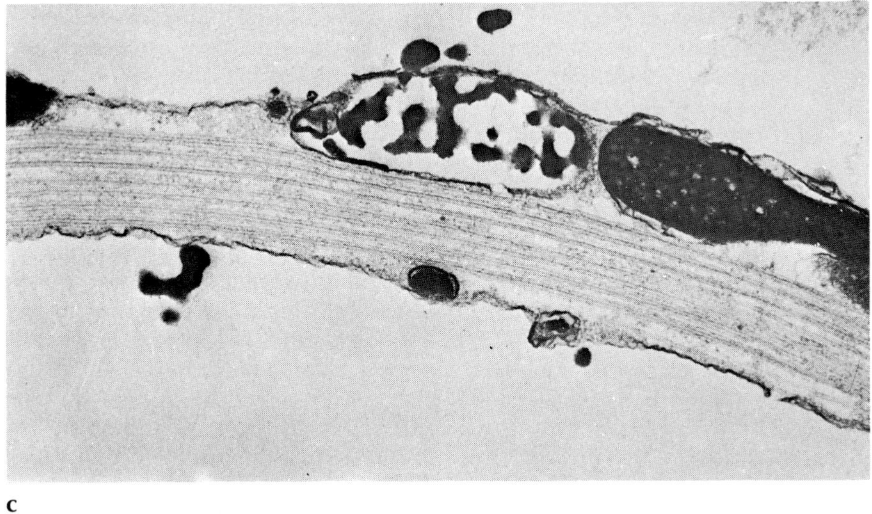

Figure 3.12 *a* The helizoan *Echinosphaerium*. Slender axopodial processes radiate from the cell body. Within each axopodium is a birefringent core, or axoneme. [*Courtesy of L. G. Tilney.*] *b* Microtubule array of an axopodium. Reassembly of axoneme structure after dispersion. [*Courtesy T. P. Fitzharris.*] *c* Longitudinal section shows a reforming axopodium after 10 minutes of recovery following dispersion of subunits by ultrasonic treatment. [*Courtesy of L. G. Tilney.*]

Embryonic development is characterized by a great deal of movement and mechanical activity, properties best revealed by time-lapse cinematography. Such diverse events as cleavage, gastrulation, neurulation, tubular gland formation, nerve outgrowth, and many other motile processes are believed to have a similar underlying mechanism, one involving the activity of certain contractile proteins. We are used to thinking of contractility of muscle cells, but contractility is a basic property of nonmuscle cytoplasm as well. Furthermore, the components of muscle and nonmuscle contractile systems are remarkably similar; they include actin, myosin, tropomyosin, alpha actinin, and probably troponin. Where

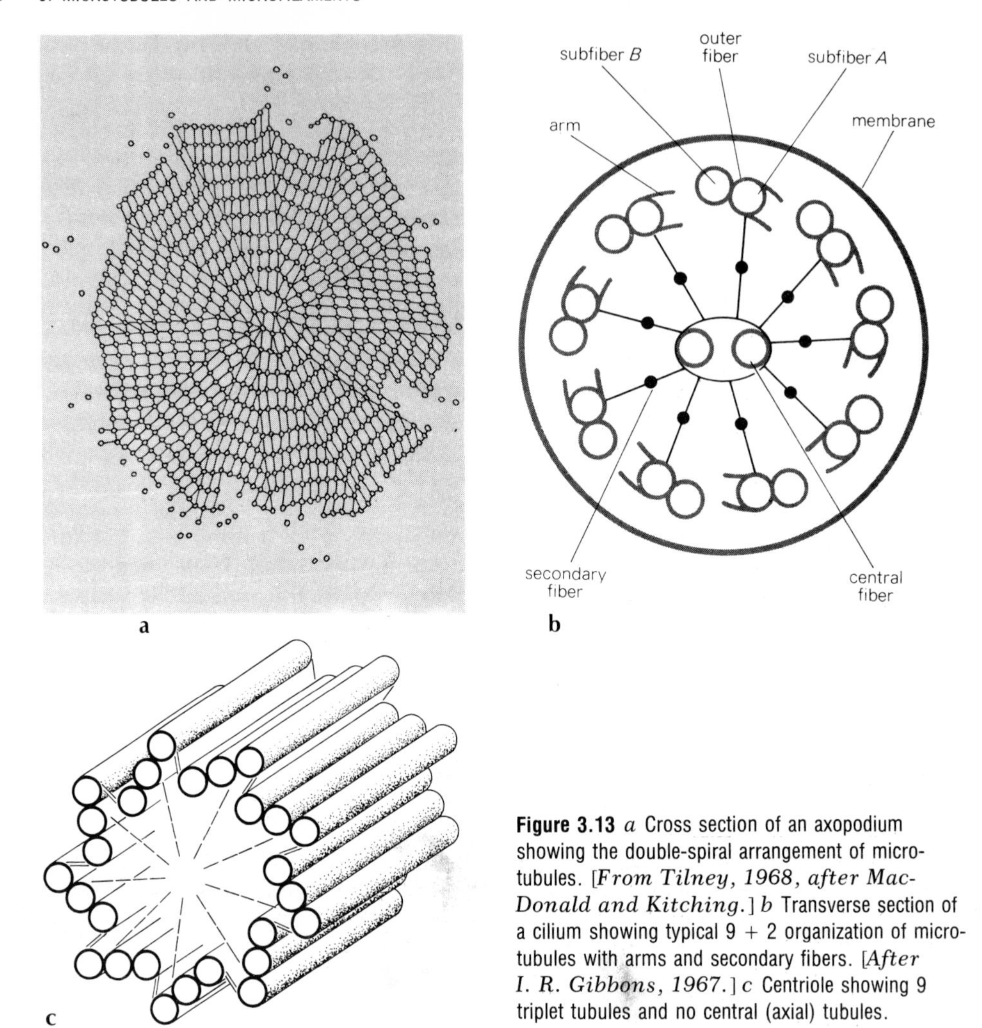

Figure 3.13 *a* Cross section of an axopodium showing the double-spiral arrangement of microtubules. [*From Tilney, 1968, after MacDonald and Kitching.*] *b* Transverse section of a cilium showing typical 9 + 2 organization of microtubules with arms and secondary fibers. [*After I. R. Gibbons, 1967.*] *c* Centriole showing 9 triplet tubules and no central (axial) tubules.

they have been best studied (in amoebas and cultured vertebrate cells), these proteins are similar in amino acid sequence, polymerization properties, and function to those of muscle tissue. Nonmuscle actin, for example, is capable of polymerization in vitro to form a 60-A double-helical filament capable of interaction with muscle or nonmuscle myosin. In fact, one of the best criteria for identifying an actin filament, regardless of its location, is its ability to bind to myosin (specifically the heavy meromyosin fragment). The complex which forms can then be identified directly in the electron microscope or, alternatively, by use of the fluorescent light microscope after association with fluorescent heavy meromyosin. Similarly, nonmuscle myosin molecules will polymerize in vitro to form the typical bipolar thick filament characteristic of myosin from muscle cells.

Although the proteins of muscle and nonmuscle cells are similar, their organization is very different. Myosin levels of nonmuscle cells are usually very low, and thick myosin filaments are not generally observed in electron micrographs. In contrast, actin levels are often very high, constituting as much as 10 percent of

the cell's protein, but the actin filaments are usually present in a relatively nonordered array. In some cells, for example, they exist as a loose network beneath the membrane, whereas in other cells, many of the actin filaments are organized into bundles, termed *stress fibers*. The thin filaments of nonmuscle cells exist in a dynamic state capable of rapid polymerization and depolymerization. The organization of actin filaments can change depending on the condition of the cell, whether it is in suspension or attached to the substrate, in interphase or mitosis or cytokinesis. The involvement of actin-containing filaments in a particular event is usually evidenced by the inhibition of that event by cytochalasin B, a substance capable of disrupting microfilaments in some undetermined manner. In many instances actin filaments are clearly involved in a contractile function, while in other instances they seem to serve in a support capacity, i.e., as part of a cytoskeleton. In the former case, it is likely that the actin filaments are functioning in conjunction with myosin molecules, even though thick filaments may not be apparent in the vicinity of the supposed action. Not all the actin of the nonmuscle cell is present in its polymerized form. The remainder exists as G-actin capable of undergoing polymerization if needed. As in the case of microtubules, the polymeric form of the actin molecule exists in a dynamic equilibrium with the monomers, an equilibrium that can be shifted in one direction or another depending on the conditions within the cell. In this latter regard, the concentrations of ATP, Mg^{2+}, and Ca^{2+} are particularly important.

Some of the best insights into the nature and function of nonmuscle actomyosin systems have come from studies carried out on in vitro preparations. Most cytoplasmic extracts appear to have inherent contractility. If sea urchin egg cytoplasm, for example, is slowly warmed to room temperature in the presence of Mg^{2+} and ATP (without Ca^{2+}), it becomes markedly gelated. Analysis of the gelated extract in the electron microscope indicates that it is made up of a huge network of interconnected (cross-linked) thin filaments. Protein analysis indicates that the actin is not involved by itself in the formation of the gel, but is present in association with one or more high-molecular-weight actin-binding proteins believed to be an important component in cross-linking the filaments. Addition of Ca^{2+} to these actin-containing gels leads to their contraction, a response which illustrates the basic contractility of nonmuscle cytoplasm. Examination of flattened unsectioned cells with the high-voltage electron microscope has revealed the presence within the cytoplasm of a fine filamentous latticework termed the *microtrabecular network*. It has been proposed that this network serves to interconnect all the cytoplasmic organelles, with the exception of the mitochondria, which do not appear to be contained within it. Furthermore, it appears that the lattice exists in a dynamic state and is capable of undergoing movements generated by actomyosin components of the structure, as well as being subject to disassembly and reassembly so as to give the cell control over its very fabric.

7. Cell locomotion and contact inhibition

Embryonic development is characterized by a diverse array of cellular movements. In some cases cells move as part of a sheet of cells; in others they move as individuals, often from one part of the embryo to a very distant part. The movement of single embryonic cells is accomplished with the aid of various types of cytoplasmic projections that reach out over the substrate, make the necessary attachments, and cause the remainder of the cell to move forward. In some cases, the extended processes are thin and elongated; these are referred to as *pseudopodia* or *filopodia*. More often, examination of the leading edge of a migratory cell reveals the presence of a broad, flattened projection termed a *lamellipo-*

dium. Lamellipodia are particularly apparent at the leading edge of flattened cells moving over the surface of a culture dish, as opposed to cells moving through the three-dimensional space of an embryonic structure. Cells having lamellipodia move over the substrate without pseudopodial formation or any evidence of cytoplasmic flow. Phase and interference microscopy show that the leading edge of a moving fibroblast consists of an exceedingly thin, fanlike membrane, 5 to 10 μm wide and closely applied to the substrate (Fig. 3.14). This membrane undergoes continual folding movements that beat inward, i.e., appear as ruffles. Electron microscopic examination shows the lamellipodia to be filled with microfilaments, particularly near those points involved in adhesion. If the microfilaments are disrupted by the addition of cytochalasin B, movement is brought to a stop. As a lamellipodium stretches forward and adheres to the substrate, the elongated cell can be seen to be stretched and under tension. After a time, the trailing edge of the cell is ripped loose from its attachment and it springs forward. Once the rear of the cell has moved forward, a new lamellipodium can be formed and the cell can move off in the same or some different direction. Unlike the movements of cells across the surface of a culture dish, migrations within the embryo are highly directed in nature. Cells migrate from one region to another along predictable routes over difficult terrain.

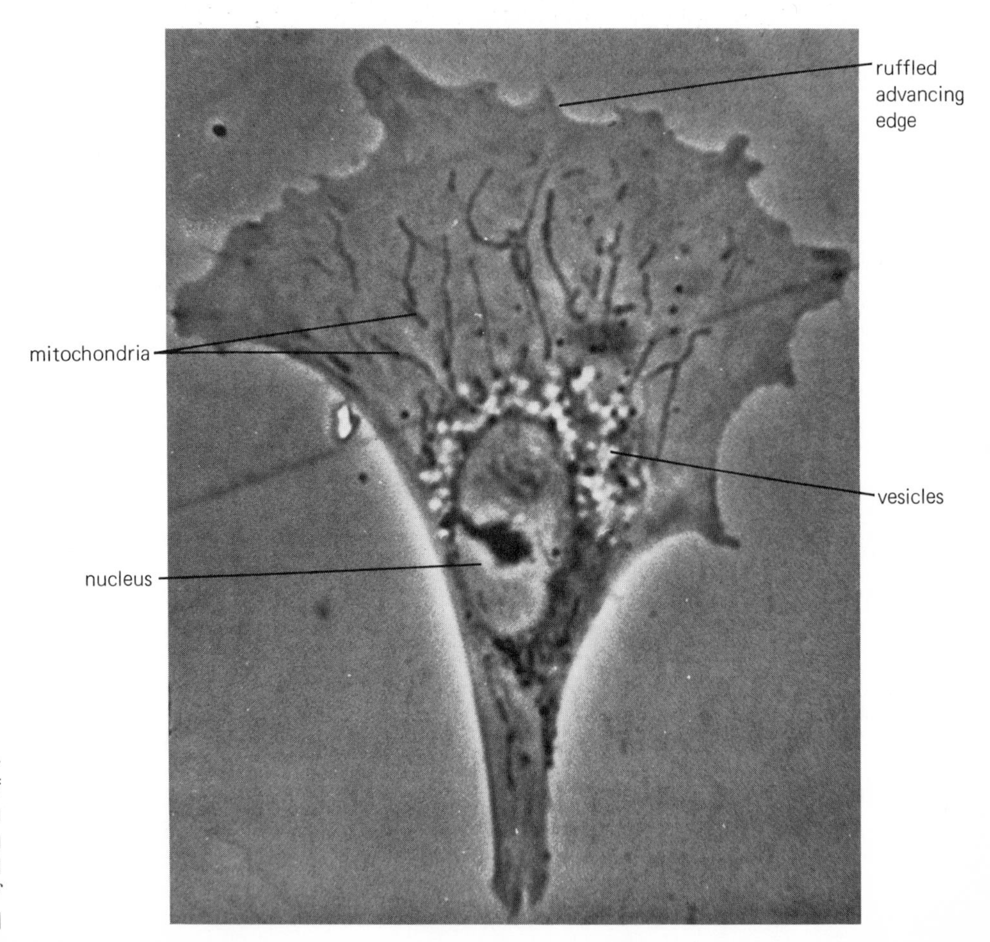

Figure 3.14 Phase-contrast micrograph of a fibroblastic cell in locomotion, showing broad advancing ruffled edge. [*Courtesy of N. K. Wessels.*]

An important feature of the locomotory properties of cells can be seen when one cultured fibroblast makes contact with another; contact between these cells is followed by a dramatic cessation of their locomotor activities. If one observes the ruffled lamellipodia of the two cells, one sees an immediate inhibition of its undulatory behavior at those points at which contact is made. This behavior of cells is termed *contact inhibition of movement*. The initial contact between cells can be followed by a more stable cohesion, or alternatively, the cells can break the contact and move off in different directions. Regardless of the outcome, cells generally do not move directly over one another's membranes. As we will see in various places in this text, the composition of the extracellular material can be of considerable importance in the locomotory behavior of the neighboring cells. Consideration of intercellular contacts, recognition, and adhesion will be discussed in Chapter 16.

8. Centrioles and basal bodies

Apart from any capacity the microtubule components of the various structures may have for self-assembly, disassembly, and reassembly, the basal bodies and centrioles serve as nucleation centers for microtubule production. The fine structures of basal bodies and centrioles are so much alike that these bodies can be regarded as a single type of organelle serving two possible functions in the cell. All may be called *centrioles*.

A typical centriole is a cylinder about 200 nm (nanometers) wide and about twice as long. Nine evenly spaced fibrils run the length of the centriole; each fibril appears in cross section as a band of three microtubules (Fig. 3.13*c*), designated the A, B, and C subfibrils. Each band of three microtubules is inclined at an angle to the surface of the centriole. At one end, identified as the proximal or old end, the central part is usually occupied by a characteristic pinwheel structure that consists of a central cylindrical hub and nine delicate spokes extending toward the peripheral fibers.

Characteristically, therefore, centrioles have nine sets of triplet tubules but no central tubules, whereas ciliary and flagellar axonemes have nine sets of double tubules plus central tubules. It is significant that when outer doublets of the ciliary or flagellar axoneme are isolated, depolymerized, and allowed to redevelop in vitro without the basal body, only single microtubules are formed.

The primary questions about centrioles—whether mitotic and/or axonemic—concern their role in the assembly process of the associated physically active structure and the process by which they are replicated. During mitotic cell division, the centriole at each pole of the mitotic spindle duplicates, so that each "daughter" cell has a pair of centrioles. The daughter centriole always arises close to the old centriole, with its long axis at right angles to that of the old (see pair of centrioles in midpiece of sperm in Fig. 4.8). Through nearly a century since they were discovered in 1887—independently by Boveri and Van Beneden—centrioles have been considered to be self-replicating bodies, although there are cases where all evidence points to de novo formation. This paradoxical situation still remains unclear.

9. Nucleus

The nuclear compartment of the cell retains a unique composition, yet is in continual communication with the cytoplasm in which it is immersed. The organelles of the cytoplasm—mitochondria, endoplasmic reticulum, Golgi complex, lysosomes, and numerous vesicles and inclusion bodies—are not found within the confines of the nuclear membrane. Compared with the diverse functions of

the cytoplasm in the many different types of cells that have been studied, the *known* functions of the nucleus are essentially restricted to the storage and replication of genetic information, its selective transcription, and the assembly of ribosomes. Four prominent components make up the typical nucleus: the chromosomes, the nucleolus, the fluid matrix (nucleoplasm or nuclear sap), and the surrounding membrane. We will briefly consider the nature and function of each of these components.

DNA is contained within the chromosomes, with each probably having a single unbroken molecule, in close association with a great variety of proteins. Chromosomal proteins are generally divided into the histone and nonhistone categories, the possible role of each having been discussed in the previous chapter. The nucleosomal or "beads on a string" structure of chromatin discussed in Section 2.6.B is generally considered to represent its lowest level of organization; however, chromatin does not appear to exist in this relatively extended state within the cell. For example, if one examines electron micrographs of sections cut through nuclei, the chromatin fibers typically appear to be about 250 A in diameter. It seems that the 100-A nucleosomes are organized into some type of thicker filament in the interphase cell. Furthermore, this higher-order structure, which is believed to involve a type of helical coiling of the thinner 100-A filament, is dependent on the presence of the H1 internucleosomal histone molecules and divalent ions. Even though the overall character of the interphase and mitotic chromosomes are very different, the 250-A chromatin fiber is found in the chromosomes of both phases of the cell cycle. Figure 3.15*a* shows a whole mount of a highly condensed mitotic chromosome. A close look at this electron micrograph clearly reveals the fibrous nature of the mitotic chromosome. Further insight into the structure of mitotic chromosomes has come to light from studies in which all the histones and most of the nonhistone chromosomal proteins are removed. Under these conditions, a small residual group of nonhistone proteins form a type of backbone or scaffolding which is responsible for maintaining the basic structure of the mitotic chromosome (Fig. 3.15*b*). A closer look at these protein-depleted mitotic chromosomes indicates that the DNA filaments are organized into loops of at least 10 to 30 μm in length. The origin and terminus of each loop attach close to one another at sites on the nonhistone protein scaffold. It would appear that the nonhistone scaffolding proteins organize the long DNA molecule into shorter loops and the histones then organize the DNA of the loops into the thicker 250-A nucleoprotein fibers.

Along with the chromosomes, most nuclei contain either one or two dense structures, the *nucleoli* (Fig. 3.16), composed of DNA, RNA, and protein. The DNA of the nucleolus contains all the sequences in the cell which code for ribosomal RNA, which is synthesized in this organelle. The assembly of ribosomes, which occurs in the nucleolus, brings together molecules, synthesized in several sites within the cell. Of the four ribosomal RNAs found in the mature ribosome, three of them—the 28 S, 18 S, and 5.8 S in human ribosomes—are synthesized in the nucleolus, initially as part of a single large 45 S precursor. Of this large precursor, approximately half of its nucleotides will be discarded during the processing steps leading to the formation of the three rRNAs. The fourth ribosomal RNA, the 5 S rRNA, is coded by DNA sequences scattered over many of the chromosomes. In addition to the RNAs, eukaryotic ribosomes contain approximately 80 ribosomal proteins, all of which must be synthesized in the cytoplasm and shipped into the nucleolus through the nuclear envelope. The assembly of the ribosomal subunits from the components occurs in a stepwise manner characterized by the gradual association of proteins with the RNA as it undergoes its suc-

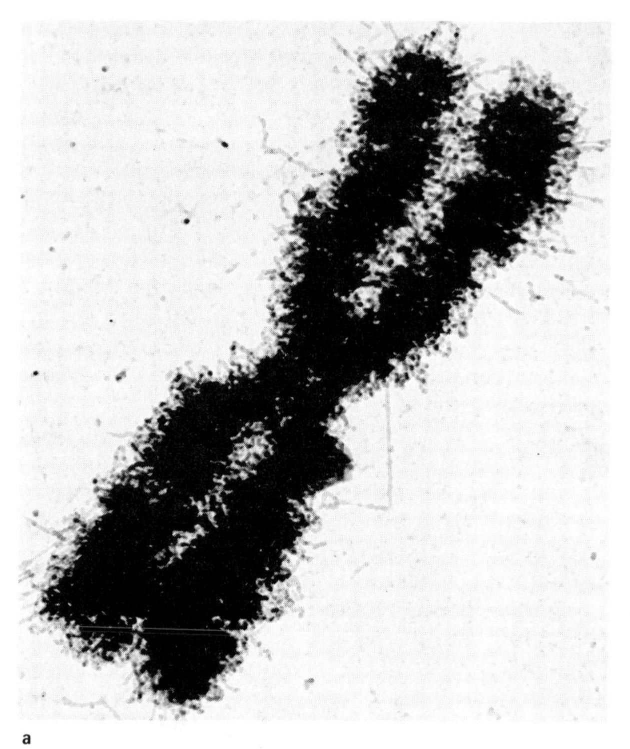

a

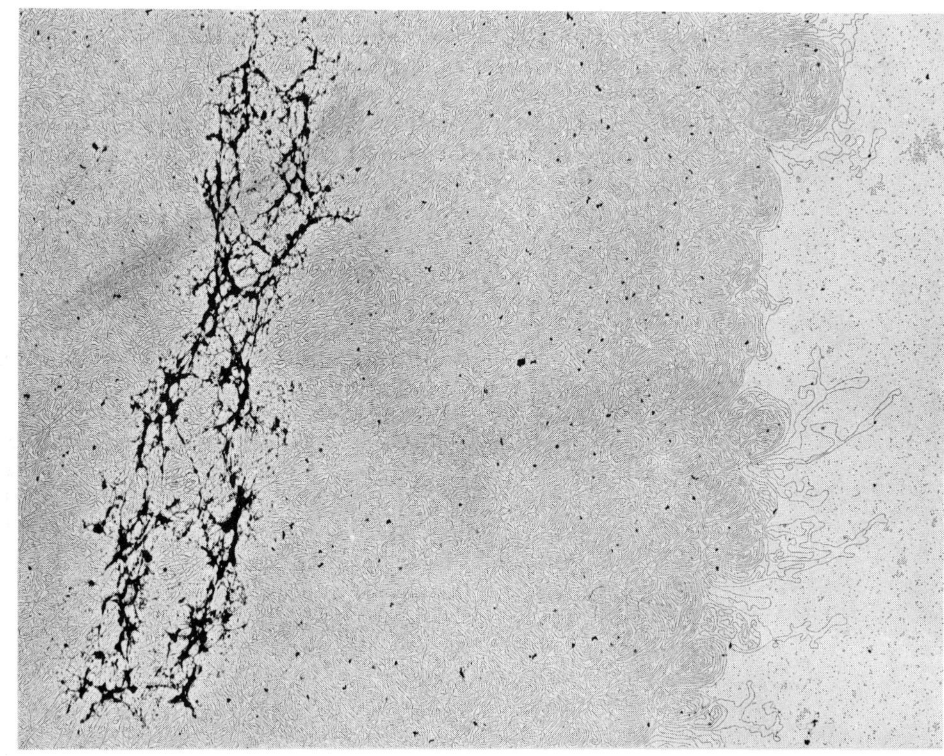

b

Figure 3.15 The mitotic chromosome. *a* Electron micrograph of a moderately condensed chromosome no. 1 from a human lymphocyte culture. [*Courtesy of G. F. Bahr.*] *b* Appearance of a mitotic chromosome after the histones and most of the nonhistone proteins have been removed. The residual proteins form a scaffold, which appears to consist of a dense network of fibers. Loops of DNA emerge from the scaffold. [*Courtesy of J. R. Paulson and U. K. Laemmli.*]

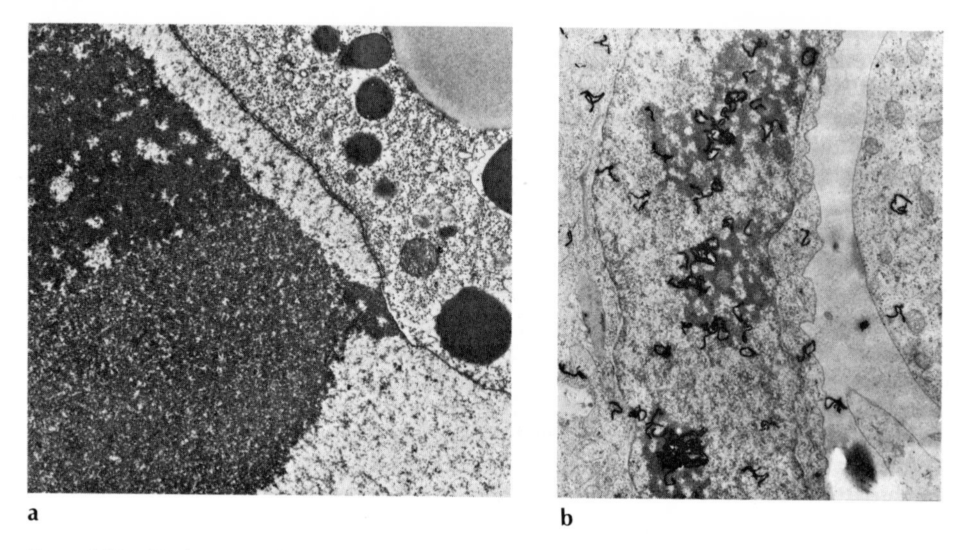

Figure 3.16 *a* Portion of a nucleolus of a molluscan oocyte showing both fibrous and granular regions. *b* Electron microscopic autoradiograph of a cell of a rabbit embryo after incubation of the embryo in [^{3}H]uridine. Nucleoli are seen within the nucleus and are found to contain numerous silver grains reflecting their role in ribosomal RNA synthesis.

cessive processing reactions. Other than its role in ribosome formation, no other functions of the nucleolus have been clearly established.

The nuclear sap, or *nucleoplasm,* is the fluid medium in which the chromosomes and nucleoli are found and through which regulatory molecules, synthetic products, substrates, enzymes, etc. must pass on their way to and from the cytoplasm. Analysis of the nucleoplasm reveals a wide variety of proteins, including enzymes capable of anaerobic energy metabolism (glycolysis), and an ionic composition that can be quite different from the cytoplasm that surrounds it.

Surrounding the nucleus is the *nuclear envelope,* a structure consisting of two distinct concentric membranes separated by a space of about 100 to 300 A. Scattered at frequent intervals along the nuclear envelope are "pores," which give the appearance that the nucleus is open at these sites to the cytoplasm (Fig. 3.17). More careful analysis using ultrastructural and physiological means indicates that these pores are complex structures and that they do not provide an open channel of communication between the cytoplasm and the nucleus. The pore is defined at its edges by the fusion of the two elements of the nuclear membrane; however, within the pore there is additional visible material (annular material) which appears to fill in much of the pore space. The number of pores present per unit surface of nuclear membrane is highly variable and can undergo change during a physiological or developmental process. For example, there is a reduction in nuclear pore density that accompanies the development of the spermatozoa.

The most important question for our purpose concerning the nuclear envelope is the constraint, if any, it places on communication between the two major cellular compartments. Evidence relating to this question has led to some controversy. A variety of techniques has been applied to determine just how porous the nuclear envelope might be. Observations of isolated nuclei that are osmotically active, i.e., capable of swelling when placed in hypotonic media, would

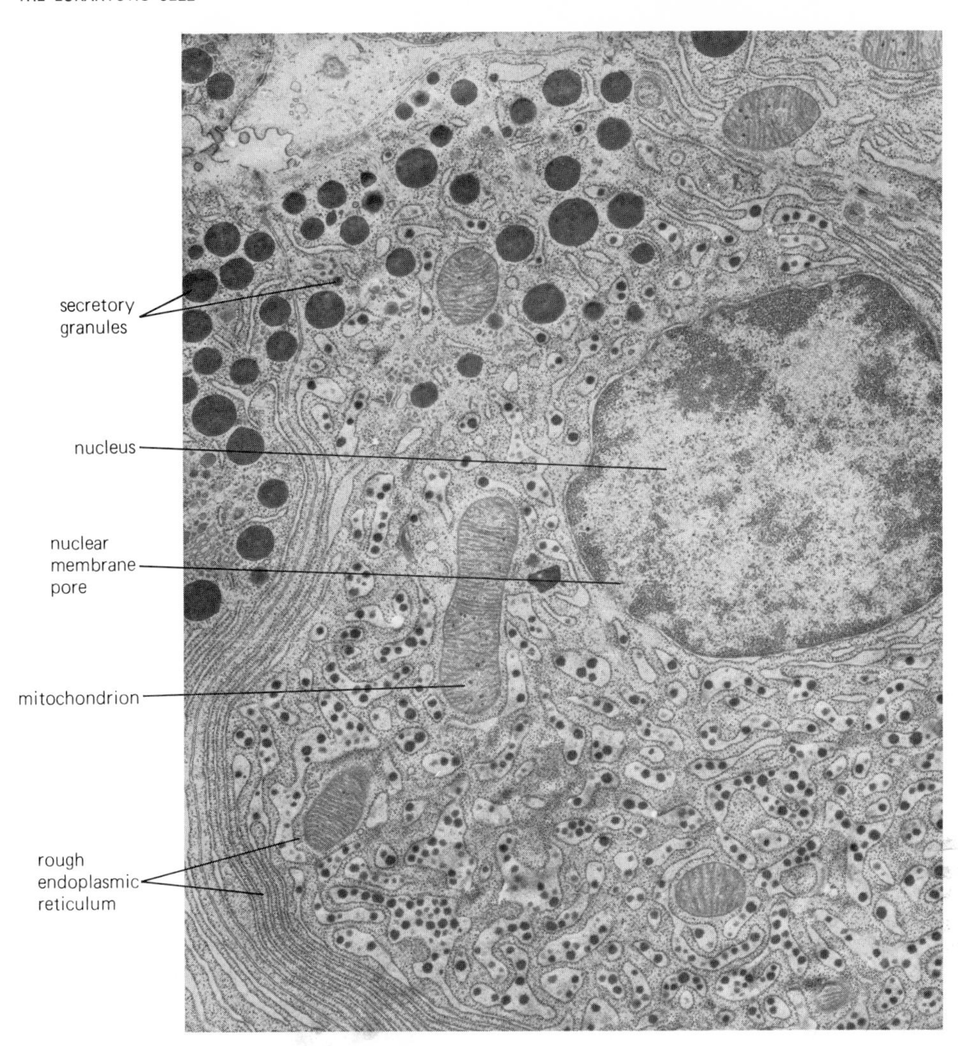

Figure 3.17 Part of a bat pancreas secretory cell showing nucleus, nuclear pores, mitochondria, endoplasmic reticulum, and secretory (zymogen) granules. [*Courtesy of K. Porter.*]

suggest that the nuclear envelope can act as a semipermeable membrane to restrict the free diffusion of small ions. Electrophysiological studies of the large nuclei of *Drosophila* larval salivary gland cells indicate that a potential difference is maintained across the nuclear envelope in these cells, which therefore must restrict the movement of small ions across itself. However, in other cases, such as the nuclear envelope of frog oocytes, no such potential is found; this suggests that there may be species differences to consider with respect to nuclear membrane permeability.

In another series of experiments, particles of varying diameter were injected into the cytoplasm of cells (amoebas and oocytes); the ability of the various-sized particles to penetrate into the nuclear space was determined in the electron microscope. In these studies it was found that smaller particles could penetrate the envelope and that penetration was restricted to the center of the pores (Fig. 3.18). Particles of about 125 A or greater remained in the cytoplasm, however, suggesting that this is the upper limit for penetration, even though the diameter of the pore itself is considerably greater. That particles of 100 A can enter the nu-

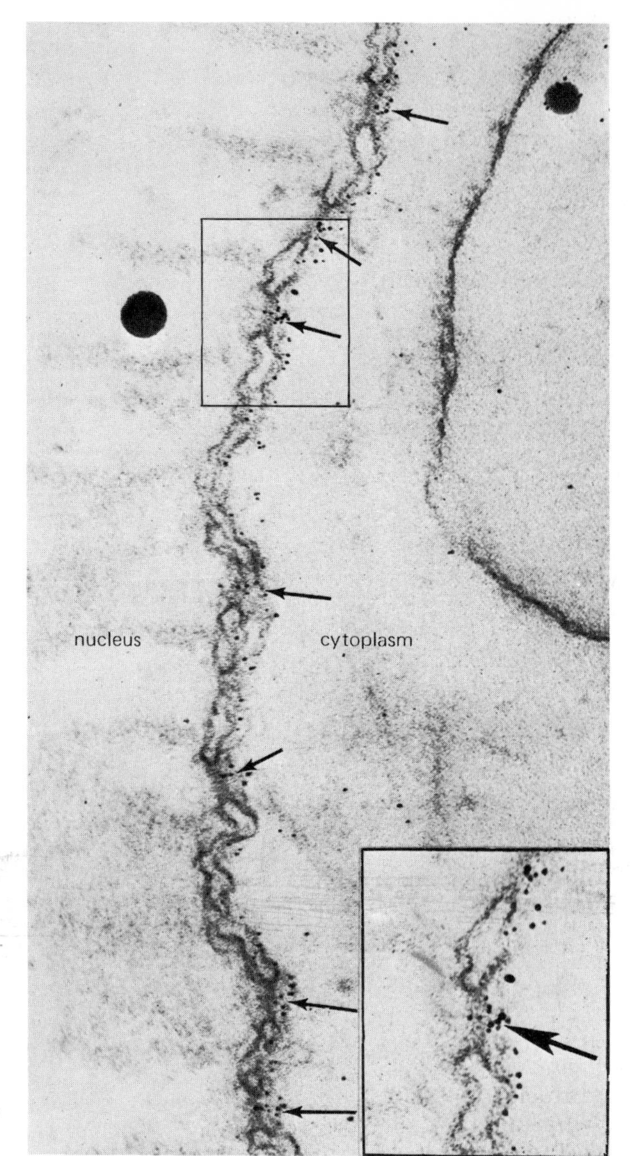

Figure 3.18 Electron micrograph of the nuclear-cytoplasmic border of an amoeba after injection with coated gold particles. These particles are seen to penetrate through the nuclear envelope via the center of the pores. Insert shows a portion of the figure at higher magnification. Arrows correspond to nuclear pores. [*Courtesy of C. Feldherr.*]

cleus through the pores does not prove that the pores in the envelope are wide-open gates in these cells. Mechanisms might exist for capture and penetration of certain materials, maintaining the exclusion of others. Whatever the mechanism, some facility must exist for passage of materials—including particles at least the size of the ribosomal subunits—through the envelope.

10. Mitochondria

One of the most characteristic structures of the cell is the *mitochondrion,* present in all cells except the bacteria. Each mitochondrion is typically about 1 μm long and one-third as wide. The number of mitochondria per cell can vary greatly; a renal tubule cell contains about 300, a rat liver cell about 1,000, and an

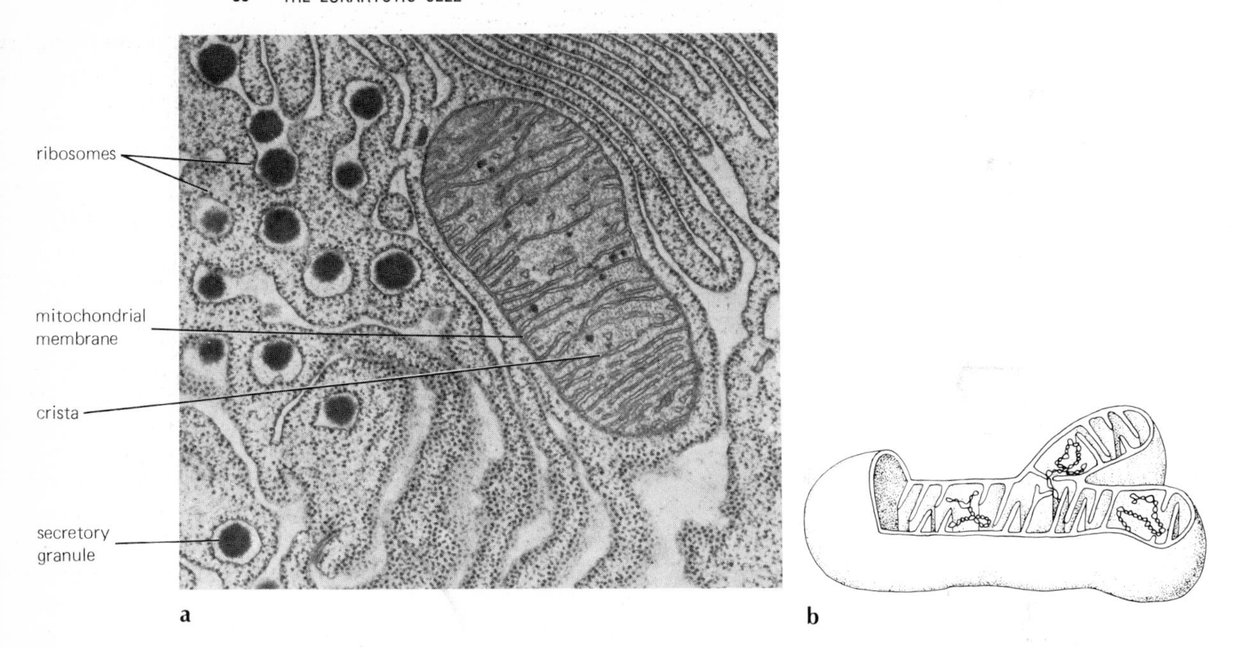

Figure 3.19 *a* Electron micrograph of a mitochondrion, together with secretory granules and rough endoplasmic reticulum. [*Courtesy of K. Porter.*] *b* Model of a mitochondrion showing outer membrane, inner membrane extending inward as folds, or cristae, and possible arrangement of circular DNA within the mitochondrion. The DNA molecules may be attached to portions of the membrane. [*After M. M. K. Nass; copyright © 1969 by the American Association for the Advancement of Science,* Science, **165:**28 (*1969*).]

amphibian oocyte 100 times that number. The usual picture of the mitochondrion in section in the electron microscope (Fig. 3.19*a*) illustrates the main features of mitochondrial structure. The organelle is bounded by an outer, encircling membrane of typical trilaminar appearance. Within the outer membrane and separated from it by a space is an inner membrane, which is thrown into numerous folds that penetrate deeply into the mitochondrion and often connect with the inner membrane of the other side. Within the mitochondrion, bounded by the inner membrane and its foldings, is the fluid phase, or *matrix,* of the organelle. The electron microscope has provided a detailed examination of the structure of the mitochondrion, leading to the accepted model shown in Fig. 3.19*b*.

Observations of the mitochondria in living cells with cinematographic techniques have shown a dynamic picture of mitochondrial activity not obtainable in electron micrographs. In these movies the mitochondria are seen in constant motion undergoing branching, fusion, and division with great rapidity, revealing the type of spontaneous behavior characteristic of the plasma membrane previously described. It has been known for a long time that the seat of oxidative energy metabolism is housed in the mitochondrion, and recent biochemical analysis has localized many of the activities. Anaerobic oxidative metabolism (glycolysis) occurs outside the mitochondria; the products penetrate into the mitochondrial space, where further oxidative metabolism occurs. The intermembranous matrix contains the soluble enzymes of the Krebs cycle, during which electrons are removed from several of the intermediates and passed to a series of proteins and cofactors that constitute the electron-transport chain. The location of the compo-

nents of the electron-transport system is confined to the inner membrane, presumably organized in an assembly whereby electrons can be transferred from one component to the next on the path. A single mitochondrion from a liver cell has approximately 17,000 such respiratory assemblies, or about 650 per μm^2.

Another basic aspect of mitochondrial function has emerged in recent years: an entirely distinct information storage and utilization system operating in parallel with that of the nucleus. As in the nucleus, the code is in DNA, the utilization in RNA. Mitochondrial DNA occurs in small circles approximately 5 μm in circumference in most animal cells, sufficient to code for only about 20 proteins of 30,000 daltons, if that were its entire function. In actual fact, this small genome codes for a collection of transfer RNAs, a pair of ribosomal RNAs, and less than a dozen messenger RNAs, which are translated in the mitochondrion using special mitochondrial ribosomes smaller than those of the cytoplasm. All the proteins so far found to be synthesized in the mitochondrion (eight in human cells) represent hydrophobic subunits of inner-membrane proteins involved in electron transport and oxidative phosphorylation. All the proteins of the organelle's translational system, i.e., the ribosomal proteins, soluble protein factors, and aminoacyl tRNA synthetases, are synthesisized in the cytoplasm and transported into the mitochondrion. An interesting sidelight to mitochondrial study has been the proposal and speculation that they arose evolutionarily as a distinct prokaryotic organism that evolved a symbiotic relationship with eukaryotic cells. This proposal is discussed at length by Raff and Mahler (1972).

11. Mitosis and the cell cycle

The ability of a cell to divide into two cells is at the very foundation of the development of multicellular animals that arise from one cell, the fertilized egg. The division of a cell requires the contents to be parceled out, and the cellular events of mitosis ensure that each cell receives its share. The division of the genetic material is a precise event; each cell receives an entire set of intact, homologous chromosomes. The remaining contents appear to be divided in a less precise, but in most cases, approximately equivalent manner. The divisional process itself is a relatively short period, preceded and followed by much longer periods of time. Division occurs in two steps: nuclear division (*mitosis*) followed by cytoplasmic division (*cytokinesis*).

All cells have their individual period of existence, the *cell cycle*. They begin in the division of a parental cell; they cease at the end of a period of maturation and special functioning, or else lose their individuality in becoming a pair of daughter cells. Single-celled organisms, the *protists*, maintain various degrees of differentiated structure throughout the process of division, and only partial reconstruction of new individuals is evident. In multicellular organisms, cells fall roughly into three categories:

1. Cells that retain full capacity for division and exhibit little or no special differentiation, serving principally as limiting membranes and/or as reserves for tissue replacement.
2. Cells that exhibit considerable general differentiation—such as liver cells, the gastrodermal cells of hydras, and the photosynthetic cells of plants—but are still able to undergo division.
3. Cells that when mature have lost their capacity to divide, notably cells with extreme structural or chemical specialization.

If we consider a population of rapidly dividing cells, there are periods of mitosis alternating with periods of interphase in a continual cyclical process. This alter-

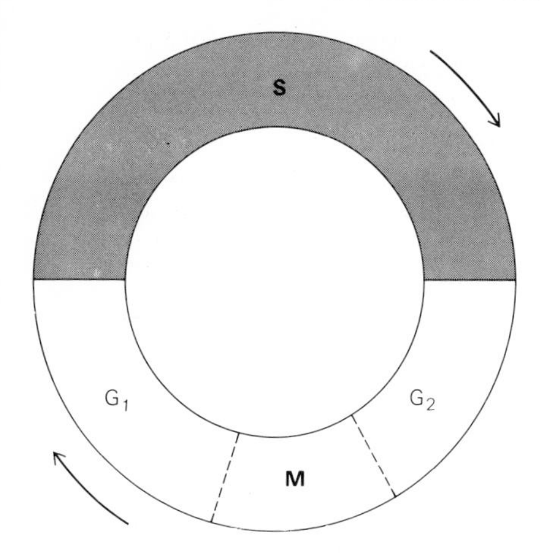

Figure 3.20 The cell cycle, or mitotic cycle, showing relative duration of phases in a growing cell. S, synthesis of DNA; G_1, phase prior to DNA synthesis; G_2, phase following DNA synthesis; M, the period occupied by mitosis (prophase through telophase).

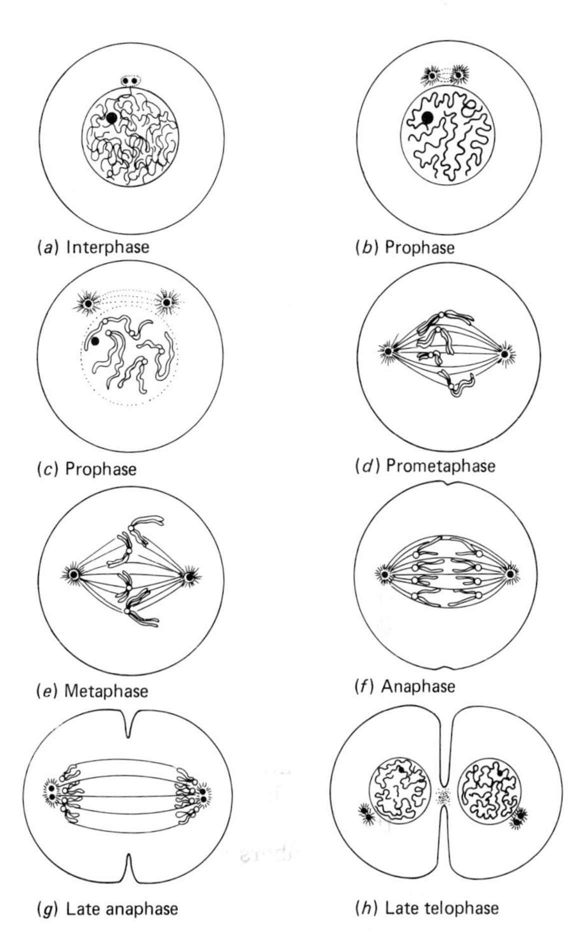

Figure 3.21 Diagram illustrating the successive cellular events in mitosis. [*From W. M. Copenhaver, D. E. Kelly, and R. L. Wood, "Bailey's Textbook of Histology," 17th ed., William and Wilkins, 1978.*]

nation of periods within the life of a population of cells is the cell cycle. The interphase period is generally divided into three parts, G_1, S, and G_2 (Fig. 3.20). During G_1, which follows immediately upon the completion of mitosis, preparatory synthetic steps begin that will ensure the following mitosis, if such is to occur. In cells that have reached a state in their differentiation where no further divisions will follow, the cell will remain in the G_1 state. During G_1, proteins are synthesized that are required for the synthesis of DNA during the ensuing S phase. DNA synthesis is restricted to the S phase, which is generally of a predictable length (several hours in mammalian cells in culture) and is followed by G_2. Throughout the interphase period, RNA synthesis and protein synthesis continue; some of these RNAs and proteins are required for the cell to undergo the following mitosis.

As mitosis begins, many dramatic changes in the cell's activities occur (Fig. 3.21). The precision of cell division relates primarily to the division of the chromosomal material that has previously undergone duplication and must be prepared for separation into two cells. In the interphase state, the chromosomes are diffuse, interwoven, and each is spread throughout the nucleus.

During *prophase,* the first stage in the continual sequence of mitotic events, the diffuse threads of the chromosome must become packed into the compact

structure of the metaphase chromosome. For example, the X chromosome in humans is estimated to contain nearly 5 cm of DNA double helix, which must become packed into a mitotic chromosome of less than 10 μm in length. The nature of the mechanism by which the chromatin threads become coiled and folded is unknown; presumably certain of the proteins associated with the DNA are of great importance for this process. Other events characteristic of prophase include the breakdown and disappearance of the nucleolus and the separation of two pairs of centrioles located just outside the nuclear membrane; this is followed by the migration of one of the pairs around the nucleus to the opposite end of the cell. Stretching between the centrioles, microtubules of the early spindle fibers can frequently be seen.

The junction between the prophase and the *metaphase* stages is characterized by the breakdown of the nuclear membrane, which is followed by the penetration of the growing spindle fibers into the center of the cell, the attachment of some of these fibers to the centromeres (*kinetochores*) of the chromosomes, and the alignment of the chromosomes at the cell's equator to form the metaphase plate. Spindle fibers are of two types, those which extend from pole to pole and those which extend from pole to chromosome. One theory suggests that chromosome movement at anaphase results from the sliding of the chromosome-attached fibers along the continuous pole-to-pole variety. Alternatively it has been proposed that the movement of the chromosomes results from the disassembly of the chromosomal spindle fibers at the poles. A third theory suggests that a contractile mechanism consisting of interacting actin and myosin molecules located in the region of the mitotic spindle is responsible for anaphase chromosome movement.

The initiation of *anaphase* is marked by the synchronous separation of the duplicated chromosomes (*chromatids*) from each other. Some mechanism within the chromosomes appears to cause the fused centromeres to split apart, because spindle fibers are not needed for the event. Once split, the two chromatids are pulled apart—the centromeres leading the way—by the action of the attached spindle fibers. During *telophase*, nuclear envelope and nucleolus re-form, and the chromosomes disperse to assume their typical interphase condition. The mechanisms underlying the events of cytokinesis are discussed in Chapter 6.

12. Cloning and synchronization

Cell cultures, particularly of mammalian cells, are often studied as synchronized populations, i.e., populations with all cells present in the same phase of the cell cycle. This situation can be brought about in several ways. Treatment of cells with millimolar concentrations of thymidine or with hydroxyurea, an inhibitor of DNA synthesis, will arrest cells at the G_1/S border, while treatment with colcemid, a microtubule depolymerizer, will arrest cells at metaphase. In any case, cells continue to collect at the particular stage until fresh medium is added, at which time all cells continue their growth beginning at the same point in the cycle.

A further degree of standardization of cell populations is attained by cloning, which eliminates the possible condition that a cell culture originating from a number of tissue cells consists of cells of subtly diverse kinds. Cloning consists of isolating a single cell from a culture and establishing a new culture which consists only of the descendants of that cell. The genetic and other constitutional characteristics are therefore, at least initially, the same for all the cells of that particular clone. Cloning and synchronization procedures are accordingly generally employed in the study of cell populations, whether of tissue cells of multi-

cellular organisms or of protistan organisms. Various pathways, however, are open to cells following division. They consist of continuing cycles of cell division maintaining a particular cell line, or of cycles leading to cell differentation and eventual death, or to differentiated cells (gametes) that relate to sexual reproduction and development.

13. Cell differentiation and the cell cortex

If a single fundamental process does exist which leads to cell differentiation, it is the genic control of synthesis of innumerable diverse structural and enzymatic proteins. All else follows from their subsequent interactions, except for possible genic control of the timing of synthesis initiations. The questions, however, remain: To what extent, if any, is essential biological information encoded and transmitted by materials and mechanisms other than the nucleic acid templates? To what extent does organized cell structure itself influence the course of differentiation?

Ciliate protozoans have long been studied with such thought as this in mind. They are highly appropriate organisms for several reasons: they are unicellular and readily multiply in laboratory cultures; they are usually comparatively large (some are even very large indeed); and they exhibit cortical structural patterns that are highly differentiated. The ciliate cortex, which may be as deep as 2 μm, contains rows of cilia, each cilium with its own basal body, or *kinetosome;* the rows of cilia typically orient with regard to body shape and locomotory axis and locally form specific assemblies, particularly feeding organelles. Individual rows of kinetosomes are known as *kineties* (Fig. 3.22). Ciliates also are characteristically equipped with two kinds of nuclei per individual: the micronucleus, which appears to be inactive most of the time but is necessary for mitotic division, and the macronucleus, which is essential to the general functioning of the cell-organism and for processes such as growth and regeneration.

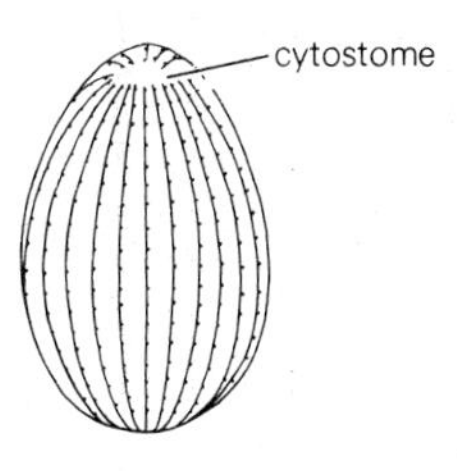

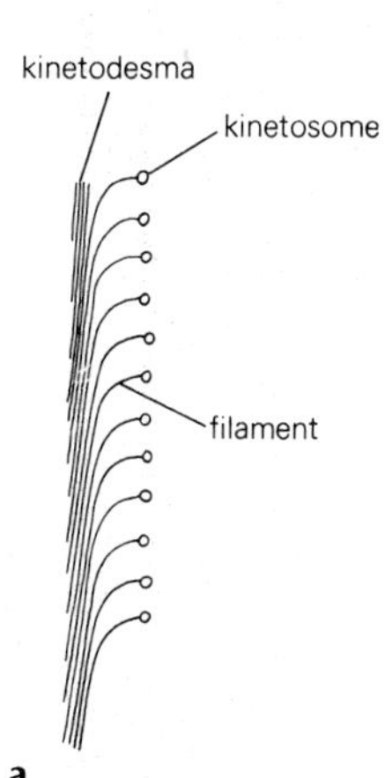

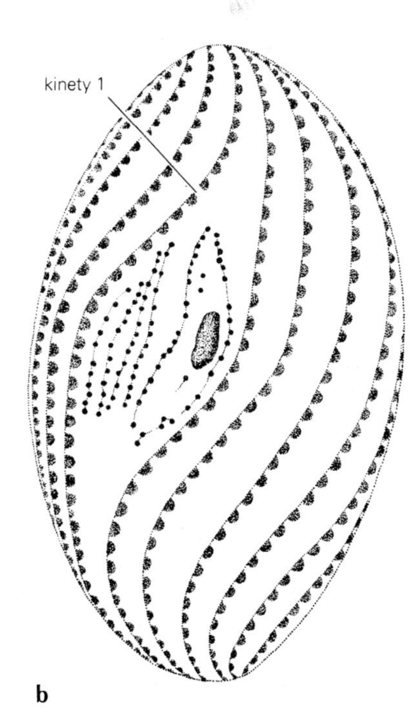

Figure 3.22 Kineties. *a* Kinety consisting of a row of kinetosomes, or basal bodies, with filaments constituting a kinetodesma. *b* Formation of successive rows of new kineties from kinety 1 to give rise to a "field," or oral primordium, associated with an expanding area of the cortex as a whole (typical of the ventral side of many ciliates).

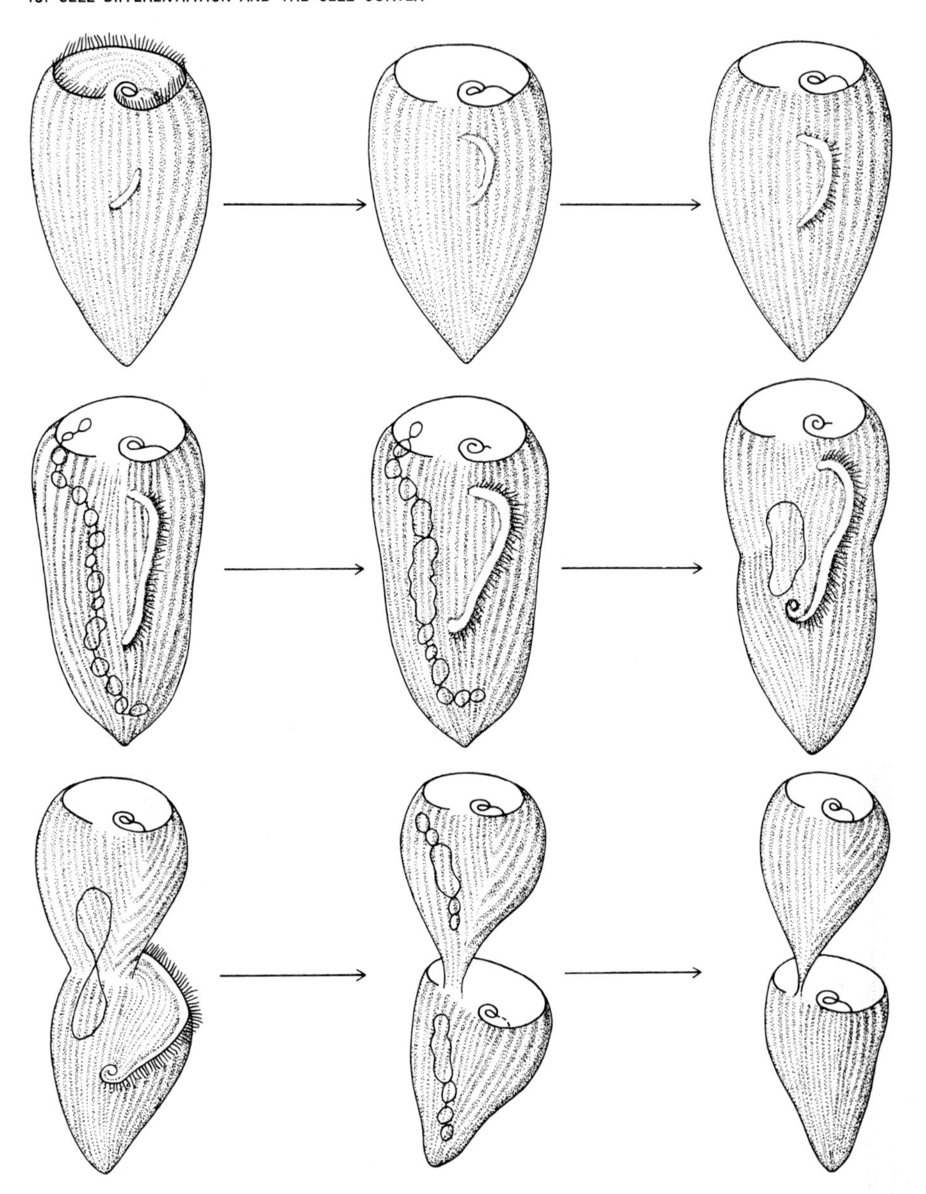

Figure 3.23 Stages in division of *Stentor*, showing formation and growth of a new primordium leading to development of a complete system of kineties and cytostome in posterior half of the individual and accompanied by constrictive separation of anterior and posterior daughter individuals. [*After V. Tartar, 1962.*]

A. Determination of pattern in *Stentor*

One of the favorite organisms of microsurgeons interested in the analysis of pattern determination is the large, cone-shaped ciliate *Stentor* (Fig. 3.23). In these protozoa about 100 pigmented stripes of graded width alternate in the cortex with rows of cilia extending from the narrow base of the cone to the wide saucerlike oral end. The broad end exhibits a whorl of concentric stripes and ciliary rows. Remarkably, very small fragments of a *Stentor* are capable of reconstitution. The

importance of the cortex in this connection is shown by certain experiments. When all or virtually all the endoplasm of a *Stentor* is withdrawn through a small incision in the surface, leaving only the cortex and the nucleus, endoplasm is promptly restored and normal growth and reproduction follow. However, if all the cortex is stripped off, leaving only endoplasm and nucleus, the cell becomes spherical, survives for a while, but finally dies. However, if a piece of cortex is left on the endoplasm, it gradually spreads around it, reconstituting the visible markers of its gradients in the form of gradation in stripe width, and eventually regenerates and reproduces normally.

Whether in normal division, experimental fragmentation, or cortical mincing, a new complex of feeding organelles arises in stentors from an oral primordium whenever such a complex needs to be formed. A new oral primordium develops in stentors in a certain region of the cortex determined or indicated by the mutual relationship of longitudinal stripes. Stripes vary in width, and at the broad end of an individual especially fine stripes are always seen to the right of the widest stripes. A new oral primordium always begins in the fine-stripe area next to the wide stripes. The pigmented stripes themselves, whatever their function, are in this connection probably no more than indicators of cortical geography and regional cortical differentiation. They indicate that a particular locality has properties of its own, and that an oral primordium develops in a precise situation in the system as a whole, specifically located in a normal individual but found elsewhere in modified individuals.

Many experiments have been performed to explore this situation, and always wherever wide-stripe areas are placed next to fine-stripe areas, an oral primordium, foreshadowing the development of oral membranellar structure, appears in the adjacent fine stripes or at the junction of the two contrasting areas. For example, if a piece of fine striping is implanted among the wide stripes at the back of the cell and regeneration of the *Stentor* is induced by cutting off the original broad end with its oral structure, a new primordium forms not only at the front end at the junction of broad and fine stripes, as expected, but also on the back side, where fine and broad stripes now lie close together, with the result that a doublet stentor is produced in place of a singlet (Fig. 3.24).

Moreover, if wide striping is experimentally placed to the right of a normal fine-stripe zone, the new oral primordium is correspondingly reversed and gives rise to membranes which coil to the left instead of to the right. The adjacent stripe pattern, therefore, seems to be involved in the manner in which the fine structures in the primordium are put together; reversed primordium sites produce reversed polarity in the band of oral membranes.

Figure 3.24 Grafting an extra fine-stripe zone into the wide-stripe region of a decapitated stentor leads to double oral regeneration, one from the graft and one from the host primordium. A second decapitation is followed by double regeneration from the posterior piece to form a doublet. [*After V. Tartar, 1962.*]

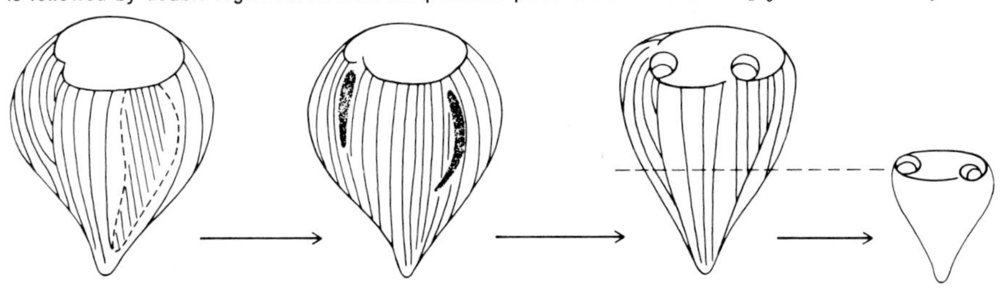

Apart from this question of the self-propagation of cortical structures and patterns, the very extensive work on *Stentor* indicates (1) that changes in the structural organization of the cell surface initiate the process of cell division; (2) that during cell division, the cell surface undergoes a series of changes that may control some events of organelle replication; and (3) that these regulatory changes do not seem to be caused by release of diffusible substances into the endoplasm. There is, accordingly, evidence that the cell surface controls the time of cell division and also plays a part in determining the replication of the macronucleus and the basal bodies during division.

B. Cortical inheritance

Inasmuch as the kinetosomes are apparently self-replicating bodies and form precise, complex patterns in the cortex, namely, characteristic infraciliatures, the question arises whether such cortical patterns perpetuate themselves more or less independently of nuclear genes. Although *Paramecium* is unsuitable for performing the sort of operations that have been so successful on *Stentor*, it has been possible by more indirect means to alter the precisely regular normal pattern of structure, to produce individuals with additional mouths, gullets, anuses, etc., without loss of viability. This is also true for certain other ciliates commonly employed in the laboratory, particularly *Tetrahymena* (Figs. 3.25 and 3.26). The altered structural pattern persists indefinitely, even through the sexual process of conjugation.

One of the simplest departures from the normal which has been induced is the inversion of one or more rows of the numerous longitudinal kineties, so that the inverted kineties show fibers emerging on the left and extending backward, instead of emerging on the right and extending forward. This partly deranged pattern has been perpetuated through several hundred generations as a result of replication of basal units with the same orientation as those of the original inverted rows.

Doublet individuals are frequently seen in cultures of various ciliates, i.e., two apparent individuals virtually complete in total structure and organelles except that they are conjoined like Siamese twins and usually possess a single macronucleus between them. In cultures of *Paramecium* and *Tetrahymena*, when doublets turn up as a result of abnormal fission or conjugation, the remarkable happening is that the doublets propagate as doublets by regular fission just as in singlets, and even undergo conjugative reproduction. Moreover, it is possible to mate doublets with singlets, so that a vast array of experimental possibilities is opened up.

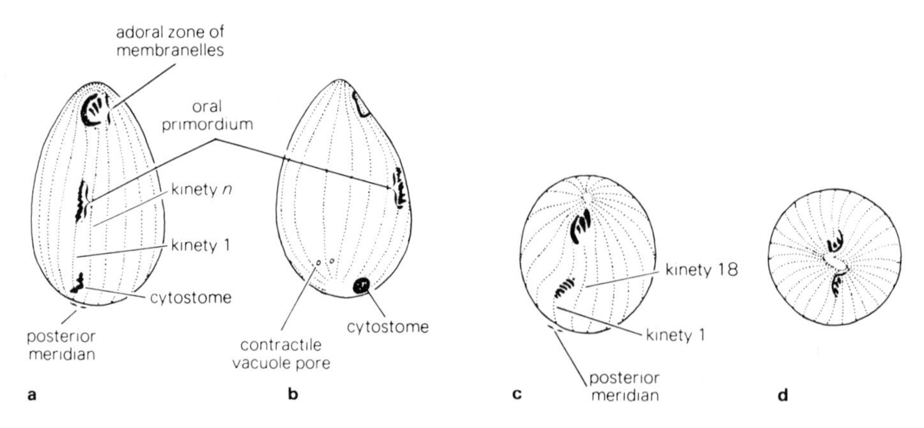

Figure 3.25 Semidiagrammatic view of silver-staining structures. *a* Ventral view. *b* Right lateral view. *c* Apical view. *d* Apical view of doublet cell. [*After* D. L. *Nanney* Science, **160**:*498 (1968); copyright © 1968 by the American Association for the Advancement of Science.*]

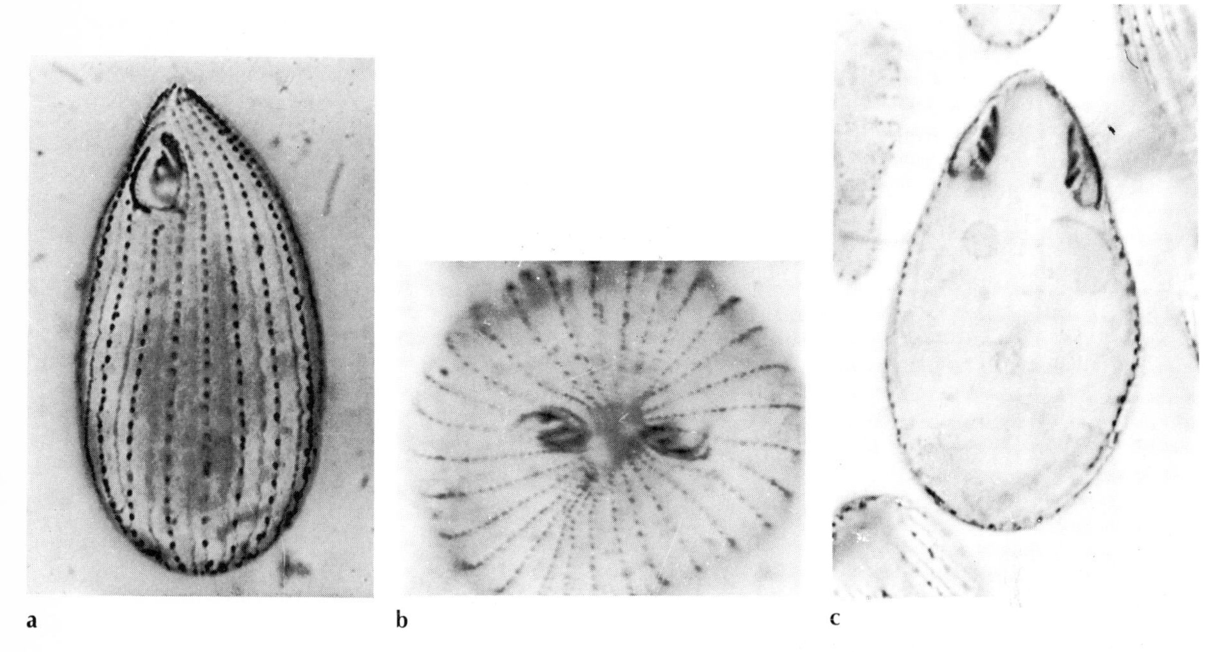

Figure 3.26 *Tetrahymena,* singlet and doublet. *a* Ciliate showing terminal cytostome and rows of kineties, each kinety consisting of a row of kinetosomes, or basal bodies, with filaments constituting a kinetodesma. *b* Polar view of a "duplex" form with two sets of oral structures and 29 instead of the normal 18 ciliary rows. *c* Longitudinal optical section of a duplex form. [*Courtesy of D. L. Nanney.*]

In such an instance, following conjugation between a doublet and a singlet, the singlet carried away a piece of a doublet's cortex at the time of separation and incorporated it as part of its own cortex. This particular freak was isolated and gave rise to a clone of like individuals, all of which were intermediates between singlets and doublets, with two sets of vestibules, two gullets, and the two ventral kinety fields, but with a single dorsal surface. That is, a piece of cortex pulled off from oral cortex of one cell and incorporated on the surface of another resulted in the inheritance and development of a complete additional oral region along the whole length of the animal on its ventral side. Accordingly, a small piece of cortex, as a natural graft, contains the genetic basis of a large but delimited part of the cell cortex.

Primarily, the question is the nature of the inheritable basis that enables a doublet to reproduce itself as a doublet and to behave generally as a singlet individual, although with duplicated cortical systems, in contrast with the self-replicating singlet individual. The difference between the two forms is the same kind of difference seen between normal individuals and those with one or more inverted kineties, but it is of a vastly more complex order. The question is essentially the same in the two cases: Is the inheritance of the new organization in any way under the control of, or determined by, the nucleus, or is the cortex as a whole, whether doublet or singlet, a truly self-perpetuating pattern, even though dependent on nuclear genes for the synthesis of some of its raw materials? We are concerned, in other words, with the nature of the inheritance of pattern as distinct from substance, and whether the innate tertiary or quaternary configura-

tive potentials of the gene-determined proteins are a sufficient or insufficient basis to work from. This question is crucial.

C. Positional information

Recent analyses of cortical patterning of kinetosomes and kineties in ciliate protozoa show that there are two types of developmental processes responsible for the positioning of these organelles. One is the propagation of ciliary rows through localized addition of new ciliary units along the axis of an existing row, which is the process responsible for the maintenance of the preexisting number of rows through many clonal generations. The position and orientation of the new units in the row are determined by the position and orientation of the preexisting units; the ciliary meridian can thus be considered to be a system in which the old structures serve as scaffolds for the new. This would be an example of a short-range positional effect. In *Tetrahymena* the normal number of ciliary rows is 19 to 20, which appears to be genetically specified, but the scaffolding mechanism of longitudinal extension of each row appears to be a nongenic perpetuation of a stable phenotypic feature. This is an example of an important emergent property, i.e., one that emerges from the specified structure itself.

Positional effects are more evident with regard to long-range positionings, particularly of major surface organelles, such as the primordium of a new oral apparatus during division and a new cytoproct (gullet) and contractile vacuole, which develop at a distance from the corresponding preexisting structures. In terms of a map, how does the cell control the *latitudes* at which structures develop, and how does the cell determine the relative *longitudes*? These may be called *global positional controls*, and they regulate the number of ciliary units produced in particular portions of the cell surface as well as the positioning of the organelles just mentioned. They represent a spatial framework which persists in spite of changes in cell length or size related to cell division and growth and left- and right-handedness as seen in doublets.

Global coordination is evident in several ways, notably: (1) as the number of ciliary rows increases, the number of ciliary units per row declines so that the *total* number of ciliary units in all rows is virtually constant; (2) when the number of ciliary rows is initially higher or lower than the stable number (19 to 20), the number declines or increases at a rate proportional to the degree above or below; and (3) a circumferential gradient system appears to exist that apportions basal-body proliferation among the different rows (meridians), each row being allotted a *unique* proportion of the total complement of ciliary units irrespective of the size of the total complement.

Positional information models have been proposed to account for situations such as that just described. A simple model postulates two orthogonal (perpendicular) gradients, one parallel to the long and short axis of the cell, longitudinal and latitudinal, respectively. The coordinates for development of specific structures would then be "read off" by the cell at appropriate levels of these two gradients, as postulated generally in the positional information hypothesis (Section 1.5.C). The most popular model for positional determination invokes the chemical measurement of at least one diffusible component. There is, however, a general difficulty in applying such models to cilates having a highly structured cortex in continuity with a cell interior characterized by rapid streaming movements (cyclosis). Many other incompatibilities also arise which cannot be dealt with here.

One possible answer to such complications is to discard the conceptual framework of Cartesian coordinates, adopt a polar coordinate system in which a cell

would establish a "preferred radius," normally coinciding with row number one, and then measure cortical angles clockwise from the preferred radius. This concept does away with the mysteries of a gradient horizontally wrapped around the cell, but it has various difficulties of its own, however. Despite the problems of constructing a detailed model of cell-surface patterning within an orthogonal framework, it does have the capacity to bring about the commonly observed asymmetry reversals resulting in mirror-image patterns.

Altogether we can conclude that long-range positional systems operate to determine the locations of new structural systems in the cell surface, as well as to suppress established systems. Moreover, we can conclude that there are gradients of positional information which provide each location in the cell-surface field with a unique positional value, whether or not their nature is known. Nuclear genes do play a significant role in determining cell-surface patterns, both by setting limits for the structural divergence that can be achieved through the operation of nongenic scaffolding mechanisms and also by influencing and probably specifying the system of positional value that underlies large-scale structural patterns. In many places throughout this book, the concept of diffusion gradients involving chemical "morphogens" with positional information will be invoked. It should be kept in mind at the outset that not one undisputed, convincing example of this phenomenon has been uncovered. Therefore, other mechanisms may well exist whereby cells can assess and regulate positional values.

14. An overview

Whether a cell is an egg, an organism, or a somatic cell, all components seem to be responsive to one another and constitute an open system with the immediate surroundings, with no part or property having a truly separate identity in the living state.

The cell is not a closed system whose properties are determined exclusively by its components. It is more than a "well-integrated chemical factory," although such a view of the cell may be necessary if we wish to investigate, for example, an enzyme reaction within it. No cell, nor any organism, exists in isolation from its immediate environment, and it cannot be conceived to do so in reality. In the case of cells, this becomes very clear when we consider the cell surface. Many cellular functions are directly influenced or controlled by macromolecules on the outside of the cell, either as components of the plasma membrane, as surface-associated materials, or as components of intercellular (extracellular) matrices. Many such macromolecules have been shown to have carbohydrate constituents (e.g., glycoproteins, mucopolysaccharides, proteoglycans, glycolipids). The Golgi complex functions both in synthesis or assembly of some of these carbohydrate-containing materials and in their transport to the cell surface as part of the secretory process. Such materials, with genetically controlled, specific carbohydrate groups, are involved in fundamental aspects of cellular function—recognition, motility, and association. For example, they are essential for recognition and adhesion between mating types in many unicellular organisms (probably for recognition by gametes of higher organisms) and in the structuring of multicellular forms and the association of embryonic cells during development.

Part of the Golgi complex function appears to be the assembly of membranes that have specific characteristics. It also plays an important role in the degradation of surface materials brought back into the cell by lysosomal enzymes, while secretion of lysosomal enzymes to the exterior results in the degradation of intercellular matrix materials.

The cycling and recycling of surface materials provide for the changing specificities in informational content essential for the control of differentiation and development. At both the cellular and multicellular (supracellular) levels, various developmental phenomena may be guided by cell-to-cell and cell-to-environment interactions in which carbohydrate-containing materials act as determinants. The informational potential of such materials located at the cell surface may explain characteristics of cell movement, morphogenesis, and adaptability to environmental stimuli during embryogenesis. Thus, altogether there is a close relationship between the genome and its regulators, the cytoplasmic membranes, the cell surface, the immediate external molecular and ionic environment, and the ultimate pattern of development and function of cells, both individually and collectively.

In this chapter we have dissected the cell into its most commonly found components and have described the most salient features of each. In a real sense, the cell is greater than the sum total of its parts, and an analytical approach can be misleading because it obliterates the interrelated nature of the structure and function of the organelles. Each organelle depends on the continuing function of the other cellular components; they do not function in isolation. For example, a mitochondrion depends on substrates provided by the cytoplasm and on proteins coded for on nuclear-produced RNA, translated on polyribosomes, and transported via cytoplasmic membrane channels. The cell's membranous components form a temporal and spatial continuum. Rough endoplasmic reticulum is believed to bud from the nuclear membrane; lysosomes and other membrane-bound vesicles form from Golgi complex; plasma membrane in many cells is a mosaic formed, in part, by the fusion of cytoplasmic membrane and conversely is an important contributor (via pinocytosis) to the formation of internal cytoplasmic structure. The developmental phenomena with which this book is concerned depend on the raw materials composing the cells. Differentiation, to a large extent, reflects specialized development of different organelles in different tissues, together with the accumulation of tissue-specific gene products.

Readings

BECK, J. S., 1980. "Biomembranes," McGraw-Hill.

BLAKE, C. C. F., 1978. Hormone Receptors, *Endeavour,* **2**:137–141.

BRINKLEY, B. R., and K. R. PORTER (eds.), "International Cell Biology 1976–1977," Rockefeller University Press.

BRYAN, J., 1976. Quantitative Analysis of Microtubule Elongation, *J. Cell Biol.,* **7**:749–767.

CAPALDI, R., 1974. A Dynamic Model of Cell Membranes, *Sci. Amer.,* **230**:26–33 (March).

CASE, R. M., 1978. Synthesis, Intracellular Transport, and Discharge of Exportable Proteins in the Pancreatic Acinar Cell and Other Cells, *Biol. Revs.,* **53**:211–354.

CLARKE, M., and J. A. SPUDICH, 1977. Nonmuscle Contractile Proteins, *Ann. Rev. Biochem.,* **46**:797–822.

DEROBERTIS, E., and E. M. DEROBERTIS, Jr., 1980. "Cell and Molecular Biology," 7th ed., Saunders.

DUPRAW, E. J., 1968. "Cell and Molecular Biology," Academic.

DUSTIN, P., 1978. "Microtubules," Springer-Verlag.

DYSON, R. D., 1978. "Cell Biology: A Molecular Approach," 2d ed., Allyn & Bacon.

EBASHI, S., 1976. Excitation-Contraction Coupling, *Ann. Rev. Physiol.,* **38**:293–313.

FAWCETT, D. W., 1966. "The Cell: Its Organelle and Inclusions," Saunders.

FRANKEL, J., 1979. "An Analysis of Cell-Surface Patterning in Tetrahymena," 37th Symp. Soc. Dev. Biol., Academic.

FRAZIER, W., and L. GLASER, 1979. Surface Components and Cell Recognition, *Ann. Rev. Biochem.,* **48**:491–523.

GIESE, A. C., 1979. "Cell Physiology," 4th ed., Saunders.

GOLDMAN, R., T. POLLARD, and J. ROSENBAUM (eds.), 1976. "Cell Motility," 3 parts, Cold Spring Harbor.

GOLDSTEIN, J. L., R. G. W. ANDERSON, and M. S. BROWN, 1979. Coated Pits, Coated Vesicles, and Receptor-Mediated Endocytosis, *Nature,* **279**:679–685.

HEAYSMAN, J. E. M., 1978. Contact Inhibition of Locomotion: A Reappraisal, *Int. Rev. Cytol.,* **55**:49–66.

HINKLE, P. C., and R. E. MCCARTY, 1978. How Cells Make ATP, *Sci. Amer.,* **238**:104–123 (March).

HITCHCOCK, S. E., 1977. Regulation of Cell Motility in Monmuscle Cells, *J. Cell Biol.,* **74**:1–15.

HOLTZMANN, E., 1976. "Lysosomes: A Survey," Springer-Verlag.

HUXLEY, H. E., 1969. The Mechanism of Muscle Contraction, *Science,* **164**:1356–1366.

INOUÉ, S., and R. E. STEPHENS, 1975. "Molecules and Cell Movement," Raven.

JENSON, W. A., and R. B. PARK, 1967. "Cell Ultrastructure," Wadsworth.

KAHN, C. R., 1976. Membrane Receptors for Hormones and Neurotransmitters, *J. Cell Biol.* **70**:261–286.

KARP, G., 1979. "Cell Biology," McGraw-Hill.

LAZARIDES, E., and J. P. REVEL, 1979. The Molecular Basis of Cell Movement, *Sci. Amer.,* **240**:100–113 (May).

LIMA-DE-FARIA, A., 1969. "Handbook of Molecular Cytology," Elsevier.

LINDAHL, U., and M. HOOK, 1978. Glycosaminoglycans and Their Binding to Biological Macromolecules, *Ann. Rev. Biochem.,* **47**:385–417.

LODISH, H. F., and J. E. ROTHMAN, 1979. The Assembly of Cell Membranes, *Sci. Amer.,* **240**:48–63 (Jan.).

LUFT, J. H., 1976. The Structure and Properties of the Cell Surface Coat, *Int. Rev. Cytol.,* **45**:291–382.

MANNHERZ, H. G., and R. S. GOODY, 1976. Proteins of Contractile Systems, *Ann. Rev. Biochem.,* **45**:427–465.

MARSHALL, J. M., et al. (eds.), 1980. "Motility in Cell Function," Academic.

MITCHELL, P., 1979. Keilin's Respiratory Chain Concept and Its Chemiosmotic Consequences, *Science,* **206**:1148–1159.

MORRE, D. J., and L. OLTRACHT, 1977. Dynamics of the Golgi Apparatus: Membrane Differentiation and Membrane Flow, *Int. Rev. Cytol. (suppl.),* **5**:61–188.

MUNN, E. A., 1975. "The Structure of Mitochondria," Academic.

NICKLAS, R. B., 1971. Mitosis, *Adv. Cell Biol.,* **2**:225–297.

NOVIKOFF, A. B., and E. HOLTZMANN, 1976. "Cells and Organelles," 2d ed., Holt, Rinehart and Winston.

OP DEN KAMP, J. A. F., 1979. Lipid Asymmetry in Membranes, *Ann. Rev. Biochem.,* **48**:47–71.

PETZELT, C., 1979. Biochemistry of the Mitotic Spindle, *Int. Rev. Cytol.,* **60**:53–92.

POLLARD, T. D., 1976. Cytoskeletal Functions of Cytoplasmic Contractile Proteins, *J. Supra. Struct.* **5**:317–334.

PORTER, K. R., and M. A. BONNEVILLE, 1968. "Fine Structure of Cells and Tissues," Lea and Febiger.

POSTE, G., and G. L. NICOLSON (eds.), "Cell Surface Reviews," Elsevier.

RAFF, R. A., and H. R. MAHLER, 1972. The Non-Symbiotic Origin of Mitochondria, *Science,* **177**:575–582.

RASH, J. E., and C. S. HUDSON (eds.), 1979. "Freeze-Fracture: Methods, Artifacts, and Interpretation," Raven.

REIJNGOUD, D.-J., and J. M. TAGER, 1978. The Permeability Properties of the Lysosomal Membrane, *Bioc. Biop. Acta,* **472**:419–449.

ROLAND, J. C., A. SZOLLOSI, and D. SZOLLOSI, 1977. "Atlas of Cell Biology," 2d ed., Little, Brown.

ROTHMAN, J. E., and J. LENARD, 1977. Membrane Asymmetry, *Science,* **195**:743–753.

SEGAL, H. L., and D. J. DOYLE (eds.), 1978. "Protein Turnover and Lysosome Function," Academic.

SHINITZKY, I., and P. HENKART, 1979. Fluidity of Cell Membranes—Current Concepts and Trends, *Int. Rev. Cytol.*, **60**:121–147.

SINGER, S. J., and G. L. NICOLSON, 1972. The Fluid Mosaic Model of the Structure of Cell Membranes, *Science*, **175**:720–731.

SLAVKIN, H. C., and R. C. GREVLICH (eds.), 1975. "Extracellular Matrix Influences on Gene Expression," Academic.

SNYDER, J. A., and J. R. MCINTOSH, 1976. Biochemistry and Physiology of Microtubules, *Ann. Rev. Biochem.*, **45**:699–720.

SONNEBORN, T. M., 1978. Local Differentiation of the Cell Surface of Ciliates: Their Determination, Effects, and Genetics, in G. Poste and G. L. Nicolson (eds.), "The Synthesis, Assembly, and Turnover of Cell Surface Components," Elsevier.

STAEHELIN, L. A., and B. E. HULL, 1978. Junctions between Living Cells, *Sci. Amer.*, **238**:140–152 (May).

Symposium on Chromatin, 1977. *Cold Spring Harbor Symp. Quant. Biol.*, **42.**

TARTAR, V., 1962. Morphogenesis in *Stentor*, *Adv. Morphogen.*, **2**:1–26.

TAYLOR, D. L., and J. S. CONDEELIS, 1979. Cytoplasmic Structure and Contractility in Ameboid Cells, *Int. Rev. Cytol.*, **56**:57–144.

TAYLOR, E. W., 1979. Mechanism of Actomyosin ATPase and the Problem of Muscle Contraction, *CRC Critical Revs. Biochem.*, **6**:103–164.

TEDESCHI, H., 1976. "Mitochondria: Structure, Biogenesis, and Transducing Functions," Springer-Verlag.

WHALEY, W. G., and M. DAUWALDER, 1979. The Golgi Apparatus, the Plasma Membrane, and Functional Integration, *Int. Rev. Cytol.*, **58**:199–245.

WEISMANN, G., and R. CLAIBORNE (eds.), 1975. "Cell Membranes," Hospital Practice.

Chapter 4 Gametogenesis

Gametogenesis is the prelude to sexual reproduction and encompasses two rather independent activities. In animals, the end product of gametogenesis is a specialized reproductive cell, the egg or the sperm, formed by a complex series of events to be examined in this chapter. The other aspect of gametogenesis centers on the requirement that the chromosome number be reduced from the diploid value to the haploid condition. Sexual reproduction involves the union of reproductive cells from two distinct organisms of the same species. If this is the case, then each reproductive cell must carry one-half the number of chromosomes of the parent cells or the chromosome number will double with each generation. The reduction in chromosome number is accomplished by *meiosis*, a process in which one round of replication is followed by two complete divisions rather than the single divisional event found in mitosis. Meiosis must therefore be incorporated into the steps leading to the formation of the reproductive cell, or *gamete*.

Sexual reproduction, by virtue of its requirement for a genetic contribution from two individuals, results in offspring with a new genetic identity distinct from that of either parent. To accomplish this end, the gametes have evolved as highly specialized cells with a surprising degree of similarity across the entire phylogenetic map. First we will examine *spermatogenesis*, the production of the male gamete, including meiosis, which can be applied to the female process as well. In the latter sections of this chapter we will examine *oogenesis*, the formation of the female gamete.

1. Spermatogenesis

In animals, with a few exceptions including nematodes and crustaceans, spermatogenesis produces a characteristic cell with a small, compact head and neck followed by a long, whiplike tail. The *spermatozoon* is a tiny cell which is packed with a variety of unusual organelles arranged in a highly ordered manner. The two principal functions of the spermatozoa in sexual reproduction are to activate the egg to begin development and to contribute a haploid set of chromosomes to the diploid offspring. In addition, the sperm typically (but not always) provides a centriole to the fertilized egg, which is used in the mitotic division that follows fertilization.

The structural unit of most vertebrate testes, and of many invertebrate gonads as well, is the seminiferous tubule (Fig. 4.1*a*), within which the process of spermatogenesis occurs. The mammalian seminiferous tubule is bounded on its outer surface by the *tunica propria*, which is composed of both cellular and noncellu-

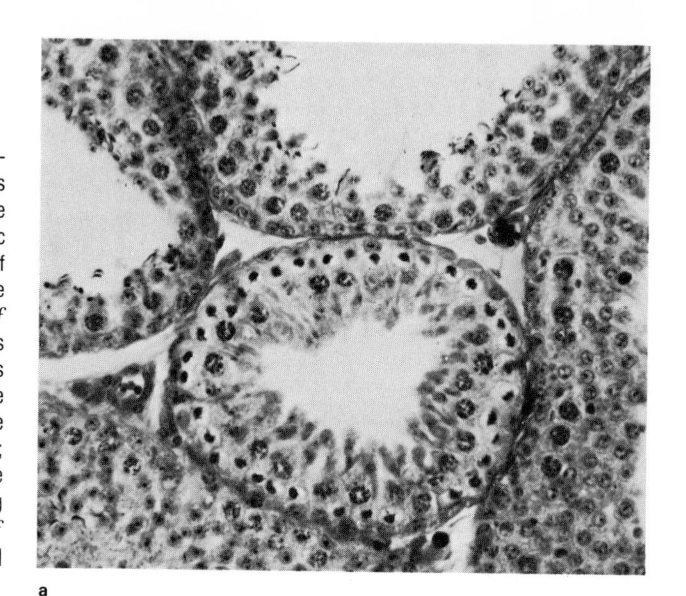

a

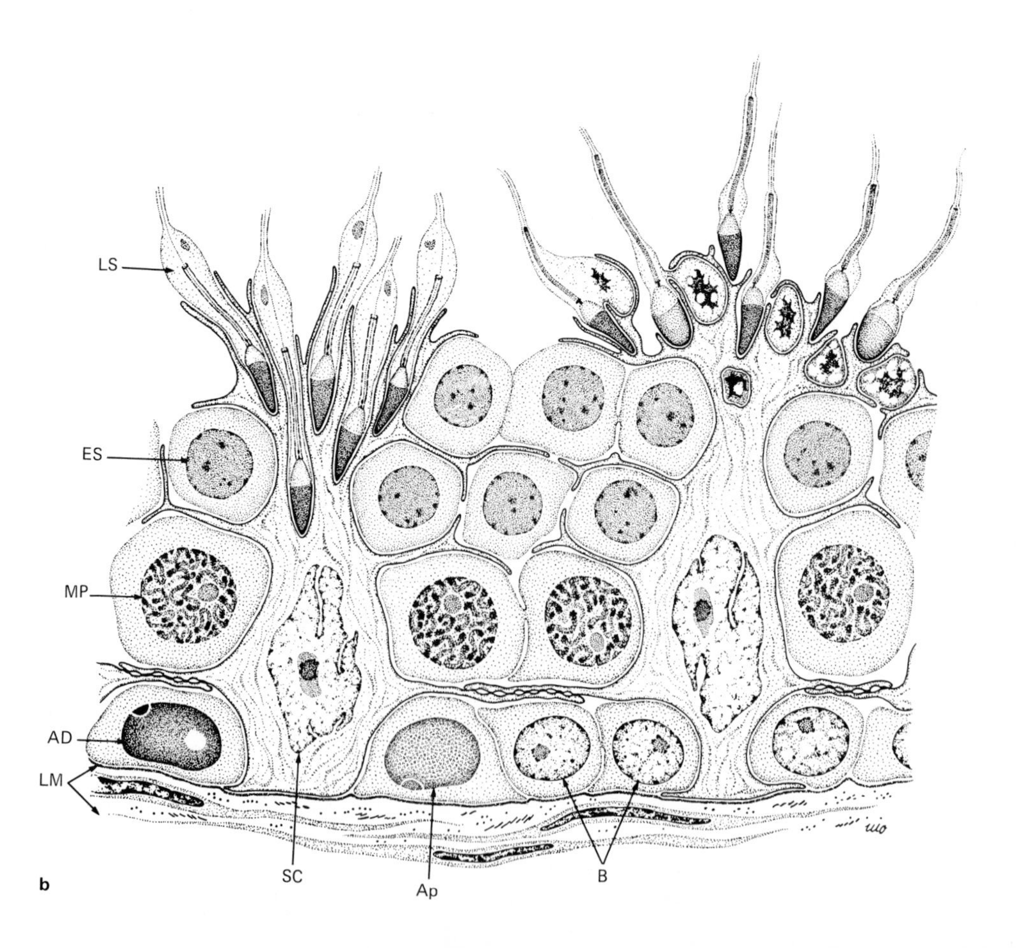

Figure 4.1 *a* Cross section through a group of seminiferous tubules of a mammalian testis. As germ cells differentiate, they move from the outer layer of the tubule toward its lumen. Note that within a given concentric layer of a tubule, all cells are at the same stage of differentiation and that stages present in one tubule are different from those of another tubule. [*Courtesy of M. R. Park.*] *b* Diagram of a portion of a seminiferous tubule showing a particular group of germ-cell stages and their relationship to the Sertoli cells that extend the entire width of the tubule wall. MP, middle pachytene spermatocytes; ES, early spermatids; LS, late spermatids; AD, type A dark spermatogonium; Ap, type A pale spermatogonium; B, type B spermatogonium; LM, limiting membrane; SC, Sertoli cell. [*Courtesy of Y. Clermont.*]

lar layers. Within the tubule, two very different types of cells can be found: the nondividing Sertoli cells (Fig. 4.1*b*) and the diverse population of germ cells in various stages on the path to becoming spermatozoa. Whereas individual Sertoli cells are seen to stretch from the tunica propria at their base to the lumen at their apical surface, the germ cells are arranged in five or six concentric layers within the tubule (Fig. 4.1*b*). Closer examination of the morphology of the germ cells of numerous tubules reveals certain important principles of tubule organization, which, in turn, provide important information on the process of spermatogenesis:

1. The position of individual germ cells within the tubule reflects the stage of spermatogenesis they have reached. Those cells closer to becoming mature spermatozoa are located nearer to the lumen of the tubule. Cells are continually migrating toward the inner free surface of the tubule into which they are released.
2. If all the cells of a given layer in a cross section of a tubule are examined, they are found to be at precisely the same stage of spermatogenesis. The cells of one concentric layer are said to be of the same generation; i.e., they began spermatogenesis at the same time and have progressed synchronously to their present stage.
3. If one notes the appearance of cells in one cross section, then one invariably finds this same combination of cells in cross sections of other tubules. In other words, the particular cellular associations undergoing spermatogenesis are not random, but are highly predictable; cells at specific steps of spermatogenesis always occur together. The reason for this constancy of associated stages is the constancy of time required for a cell to progress from one stage to that of the stage seen in the next layer. This time interval is termed the *cycle* (approximately 16 hours in the human male). If one were able to examine the contents of a particular site in a tubule from hour to hour, one would see a gradual progression in the stages of cells in each layer. After a period of time equivalent to one cycle, the initial grouping of stages would have reappeared, although each of the original cells would have moved upward one layer. The entire process of spermatogenesis in the human male occurs over a period of four cycles.
4. The same cellular associations are found over a short length of tubule. If one examines successive cross sections along a short segment of tubule, the cells of a given layer continue to appear in the same state. Therefore, the cells of that entire segment of tubule are undergoing spermatogenesis in a synchronous manner. The particular cellular associations in the rat testis are shown in Fig. 4.2. If one examines the cells of *each* layer from segment to segment along the length of a tubule, all the stages of each cycle can be seen (Fig. 4.2). The length of tubule one would have to examine before the complete series of stages repeated itself represents a *wave* of spermatogenesis. The wave is the equivalent in space to the cycle in time. The number of segments (corresponding to the number of cellular associations) that make up a wave is species-specific, as is the distance of tubule traversed for each wave.

With the organization of the tubule in mind, we can return to the events of spermatogenesis beginning with the steps leading to the formation of a meiotic cell, i.e., a primary spermatocyte. The premeiotic cells, or *spermatogonia,* are found in the outer layers of the tubule and represent the stock of cells from which the future gametes will originate. The spermatogonia arise from cells called the *primordial germ cells,* which are set aside very early during embryonic develop-

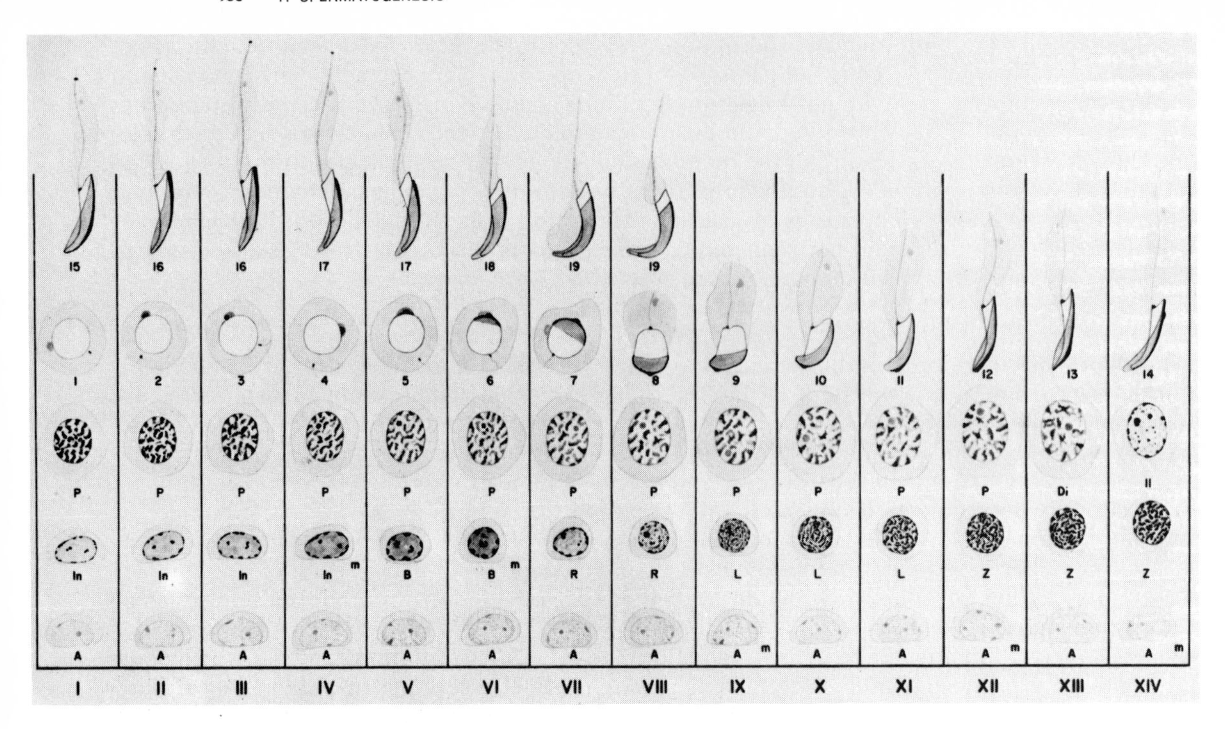

Figure 4.2 Composition of the 14 cellular associations observed in the seminiferous epithelium in the rat. Each column consists of the various cell types making a cellular association (identified by Roman numeral at the base). The cellular associations succeed each other in time in any given area of tubule according to the sequence indicated from left to right in the figure. The succession of 14 cellular associations makes up "the cycle of the seminiferous epithelium." The entire sequence of spermatogenesis can be seen by reading left to right through successive rows. A, type A spermatogonia; In, intermediate type spermatogonia; B, type B spermatogonia; R, resting primary spermatocytes; L, leptotene primary spermatocytes; Z, zygotene primary spermatocytes; P, pachytene primary spermatocytes; Di, diakinesis primary spermatocytes; II, secondary spermatocytes; 1–19, steps of spermiogenesis. The subscript m next to a spermatogonium indicates mitosis. [*From B. Perey, Y. Clermont, and C. P. Leblond,* Am. J. Anat., **108**:*49 (1961)*.]

ment and are destined to form the male gametes at a much later time. In many cases, cytoplasm of the primordial germ cells can be traced back to the fertilized egg itself. The primordial germ cells divide mitotically, find their way into the gonad, and become the spermatogonia. The spermatogonia of an adult testis form a heterogeneous population of cells distinguished primarily by the staining properties of their nuclei. Each of these various types is identifiable, and they are believed to be related to one another by mitotic division. Among the population of spermatogonia, two types of cells must be present, the *stem cell* and the *differentiating cell.* The latter is a cell that has taken one or more irreversible steps toward becoming a primary spermatocyte and ultimately a spermatozoon. In contrast, the stem cell is a less differentiated reserve cell, one that will continue to divide by mitosis to form future generations of spermatogonia, some of which will become stem cells and others of which will immediately begin the differentiation process. Gametogenesis in the male produces tremendous numbers of spermatozoa by a continual process. Reserve cells must be set aside in order to provide the cells needed for the future differentiation of gametes. If this did not occur, the population of germ cells would be rapidly depleted.

Through much of spermatogenesis, distinct bridges exist between clusters of reproductive cells, i.e., actual physical openings between cells where no plasma membrane is found. These intercellular openings arise during the mitotic divisions of the spermatogonia by an incomplete separation of the cells at each telophase. As a result, all the members of a clone—i.e., those cells originally derived from a single spermatogonium—remain in direct cytoplasmic contact with one another. No openings are seen to the Sertoli cells. It is believed that this direct communication between reproductive cells is responsible for the synchrony in the differentiation of the germ cells of a given generation.

2. Meiosis

Cells remain in the primary spermatocyte stage for a period of time during which there occur the steps of the first meiotic division. The first obvious step after the last mitotic division is the replication of the DNA, followed by a period recognizable as G_2. Then the prophase of the first meiotic division occurs, which can be divided into several stages. The events of meiosis described here for the primary spermatocyte occur in a similar manner in the primary oocyte of the female. In meiosis, two divisions follow one round of replication, and the chromosome number is reduced. Before meiosis, each cell contains a pair of each of the homologous chromosomes, while after meiosis, only one of each chromosome type is found per haploid cell.

The first stage in the meiotic prophase is called *leptotene* (Fig. 4.3). Since the chromosomes have been replicated, each is composed of two identical partners attached together at their centromere. Each partner is referred to as a *chromatid,* and the complex is called a *chromosome.* During leptotene, the very diffuse threads of the interphase chromatin become thicker and thicker as a result of some process believed to involve the coiling and supercoiling of the unit fiber. Important features of the leptotene chromosome include its attachment to the nuclear envelope and the presence of a lateral proteinaceous component which runs the length of each chromosome. This lateral fiber lies in the groove between sister chromatids.

When leptotene ends, the chromosomes are partially condensed and the homologs are ready to associate. The process by which homologs become joined to one another is termed *synapsis* and occurs during *zygotene,* the second stage of prophase I. The pairing of homologous chromosomes occurs via the formation of a new structure, the *synaptonemal complex.* The synaptonemal complex (SC) is a ladderlike structure (Fig. 4.4) composed of three parallel bars with many cross fibers connecting the central bar with the two lateral ones. The chromatin of the pairs of chromosomes is intimately associated with the lateral bars of the SC, which are separated by approximately 1,000 A. Observations on the assembly of the SC indicate that the lateral elements are derived directly from the lateral elements of the leptotene chromosomes. The role of the SC in meiosis remains uncertain. In the view of some researchers, the bars of the SC are involved simply in stabilizing the paired condition, whereas in other proposals, they are assigned a role in determining the precise chromosomal alignment. Once synapsis is complete, the complex formed by an SC and its four associated chromatids is termed a *tetrad* or *bivalent.*

It is during the next stage of prophase I, *pachytene,* that the important events of *crossing-over* are believed to occur. Unfortunately, the process cannot be directly visualized, and there are many questions remaining to be answered. It has long been felt that crossing-over involves the actual breakage and exchange of chromosomal material between one of the two chromatids of each homologous

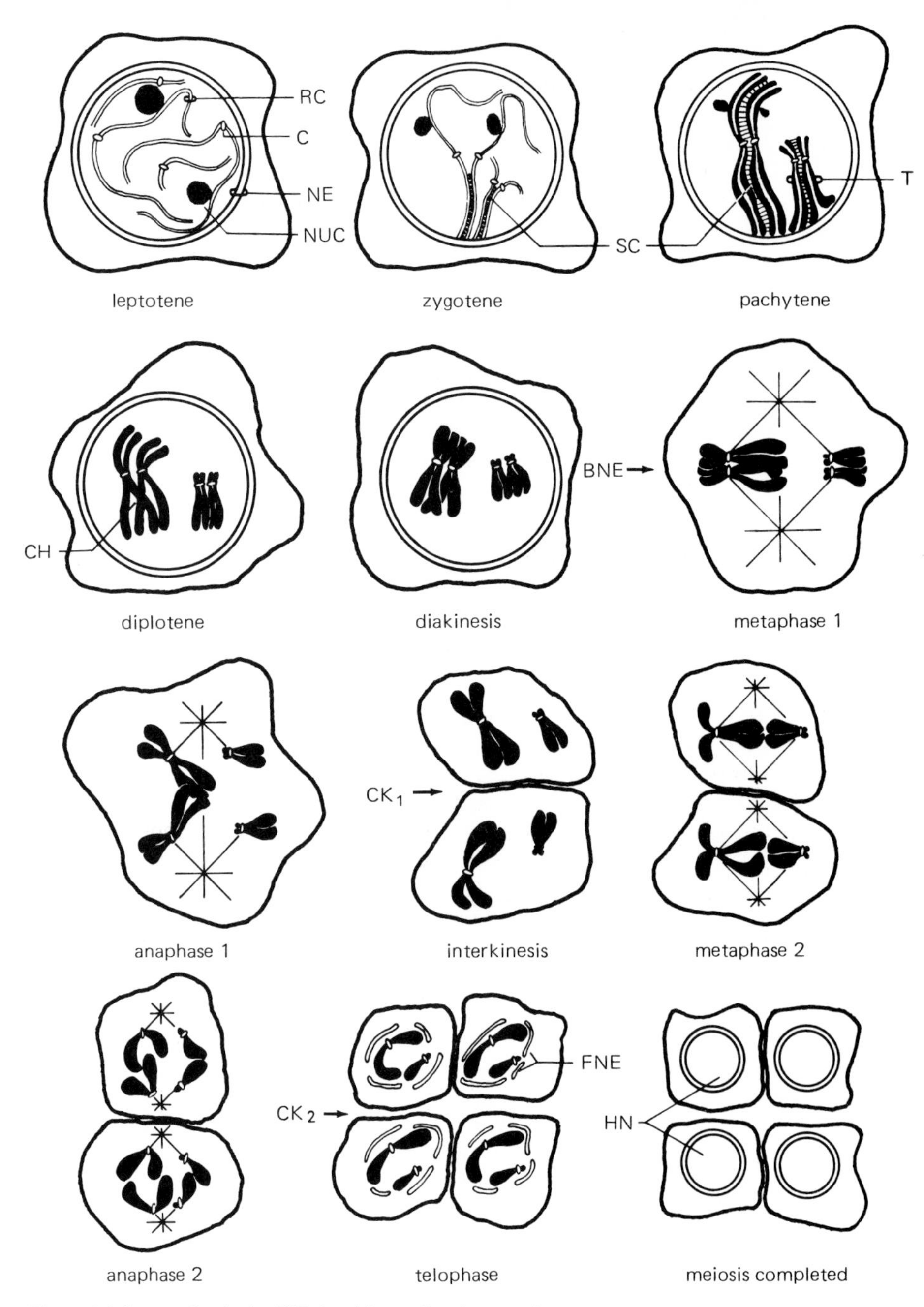

Figure 4.3 Stages of meiosis. BNE, breakdown of nuclear envelope; C, centromere; CH, chiasmata; CK_1 and CK_2, cytokinesis; FNE, formation of nuclear envelope; HN, haploid nuclei; NE, nuclear envelope; NUC, nucleolus; RC, replicated leptotene chromosomes; SC synaptonemal complex; T, tetrad. [*After F. Longo and E. Anderson, in "Concepts in Development," J. Lash and J. R. Whittaker (eds.), Sinauer, 1974.*]

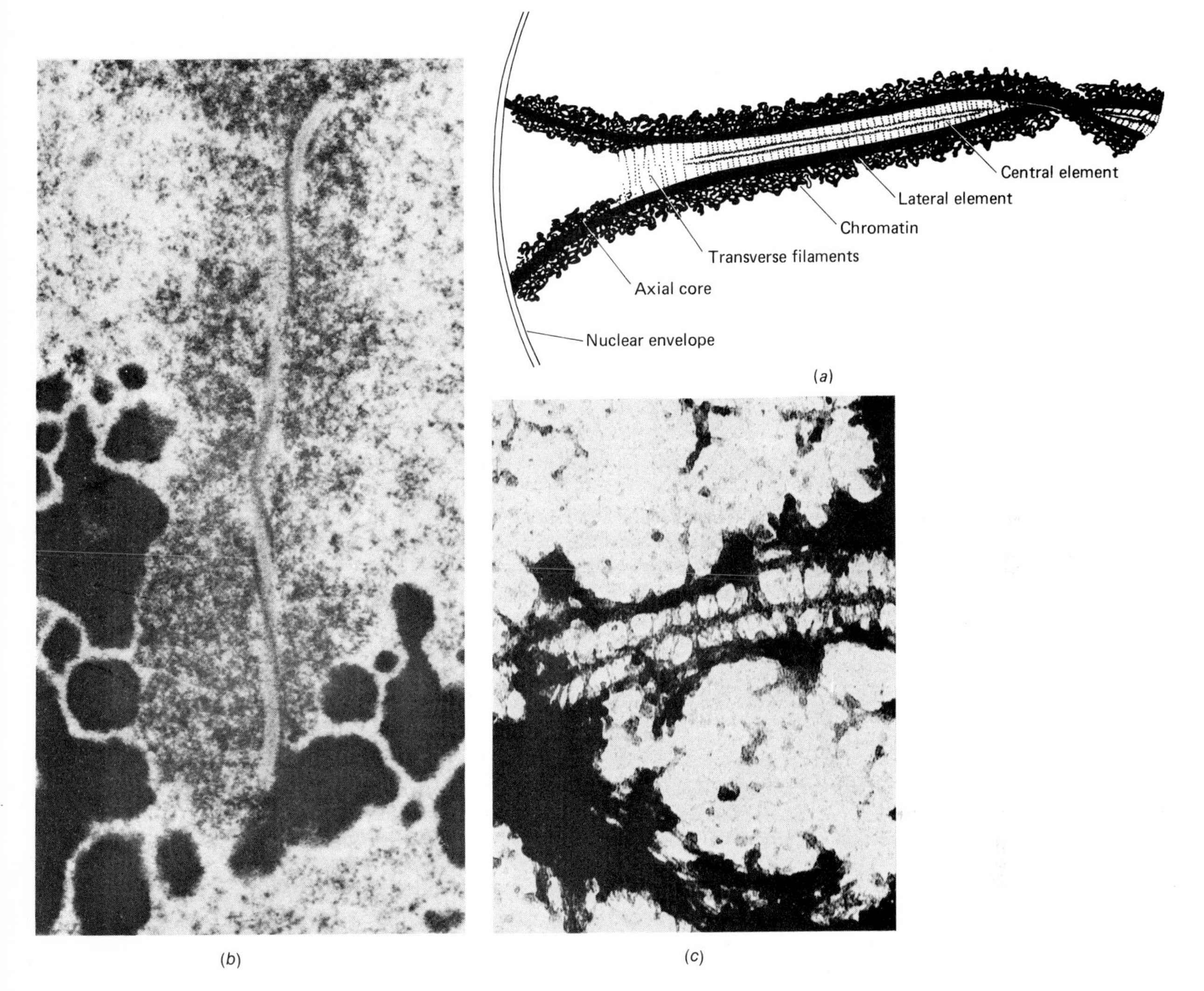

Figure 4.4 *a* Diagram of a synaptonemal complex (SC). The lateral elements are 1000 A apart. [*From P. B. Moens,* Int. Rev. Cytol., **35:***130 (1973).*] *b* Electron micrograph of a section through a bivalent in a grasshopper spermatocyte. The homologous chromosomes are held together by the SC, the dark-staining lateral elements and more opaque central element of which are clearly visible. The SC has a number of twists in it, which is commonly found at this stage (midpachytene). The very-dark-staining material at the periphery of the chromosome is part of the nucleolus, which is attached to this particular chromosome in this species. [*Courtesy of P. B. Moens.*] *c* Electron micrograph of the synaptonemal complex after treatment with deoxyribonuclease to remove the chromosomal fibers. The structure remaining represents the proteinaceous ladderlike SC itself. [*Courtesy of D. Comings and T. Okada.*]

chromosome of the tetrad. As a result of this exchange, there is a mixing of the genetic material (genetic recombination) that had originally come into the fertilized egg on different chromosomes. Direct evidence for the breakage and reunion of single strands of DNA has been obtained in studies on genetic recombination in viral and bacterial cells, and similar events are believed to occur during meiosis. One of the most remarkable aspects of genetic recombination is the pre-

cision by which it occurs. If each of the products of meiosis is to have a complete set of all the genes, it is essential that pieces of DNA are not added or removed from the chromosomes. The manner in which this precision is achieved is not understood, but the finding that DNA synthesis accompanies genetic recombination (discussed later) has provided a means whereby missing sections of single DNA strands that arise during recombination can be repaired.

After a period of tight apposition, the homologous chromosomes appear to separate and remain attached only at specific points termed *chiasmata*. It is only at this *diplotene* state, when the chromosomes have come apart from each other along most of their length, that the two chromatids of each chromosome are distinctly visible. The chiasmata of diplotene are thought to represent the points on the chromosome where crossing-over has occurred, although there does not appear to be a one-to-one correspondence between the number of chiasmata and the number of crossing-over events. During diplotene, in both spermatocytes and oocytes of many species, the chromosomes take on a distinct appearance and are called *lampbrush chromosomes*. This unusual chromosomal structure has been very important in genetic analysis and is discussed at length in Section 4.3.A. In both sexes, but particularly in the oocyte, the diplotene stage is an extended and active one.

Following diplotene, the chromosomes become thickened and the chiasmata disappear by sliding down the length of the chromosome. This is the last stage of prophase, termed *diakinesis*, being completed with the disappearance of the nucleolus and the nuclear membrane and the movement of the chromosomes to the metaphase plate.

The completion of meiosis is as follows: During anaphase I, the homologous chromosomes separate from each other. Since there is no interaction among the various tetrads, each one segregates into the two daughter cells in a manner independent of the other ones. It follows, therefore, that even in the absence of crossing-over there are a great number of genetically different gametes that can be produced. In humans, with a haploid number of 23, there can be 2^{23}, or nearly 10 million, different gametes in one individual. Crossing-over with recombination raises this value to an essentially infinite number of possibilities.

The first meiotic division produces two cells; each cell has only one of each of the original pair of homologous chromosomes, although at this point they have been mixed together by crossing-over. Telophase I occurs, and a short interphase, termed *interkinesis*, follows. These cells are now called *secondary spermatocytes*, a fleeting stage, and therefore, they are rarely found in tubules.

Interkinesis is followed by prophase II, a much simpler prophase than its predecessor. The chromosomes merely thicken and line up at the equatorial plane; the chromatids become separated in anaphase II as each centromere holding the pair is split in half. Four haploid cells, termed *spermatids*, are produced from each primary spermatocyte, and each will subsequently differentiate into a mature spermatozoon.

Although very little is known about the molecular activities that characterize the meiotic prophase, one series of experiments has uncovered some very interesting features and raised some very interesting questions. In the preceding discussion it was stated that replication of the DNA occurred prior to prophase; this does not appear to be the complete truth. If one provides a population of meiotic cells (the best-studied are those of the lily) with [^{3}H]thymidine, the cells take up the isotope and incorporate it into [^{3}H]DNA during first meiotic prophase. This occurs after the period of premeiotic replication when all the DNA, according to dogma, should have been already synthesized. In the case of the lily, only about

0.3 percent of the total DNA is made during first meiotic prophase, the remaining 99.7 percent being synthesized during the S period preceding the first meiotic prophase.

This DNA synthesis, which is seen in autoradiographs to involve all the chromosomes, occurs during zygotene and pachytene stages. It has been shown that the DNA synthesized during zygotene (Z-DNA) represents parts of the genome that had not been replicated during interphase 3 days earlier. In other words, for some reason a small amount of the DNA is left unreplicated until prophase. Reannealing studies indicate that the stretches of DNA synthesized during zygotene are approximately 5,000 to 10,000 base pairs in length and consist of single-copy sequences. Further analysis has revealed that the Z-DNA fragments are not

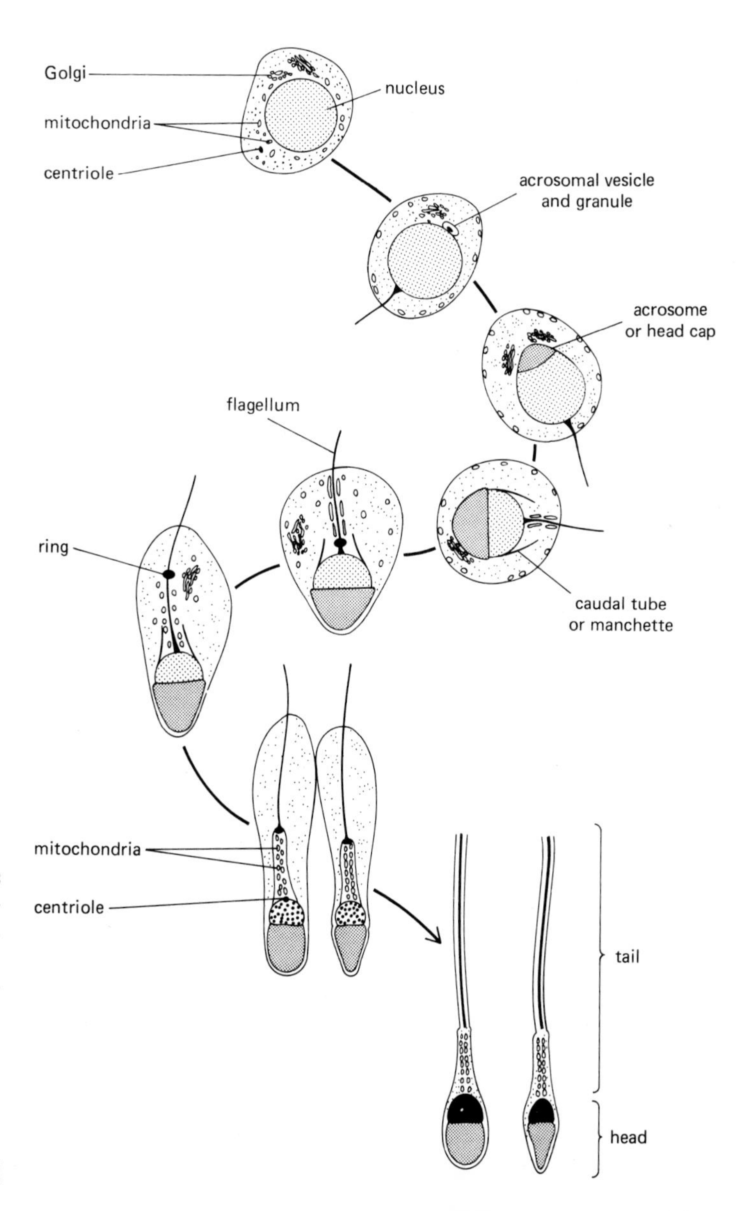

Figure 4.5 A schematic drawing of spermiogenesis in the human. [*From M. Dym in "Histology," 4th ed., L. Weiss and R. O. Greep (eds.), McGraw-Hill, 1977, after Y. Clermont and C. P. Leblond,* Am. J. Anat., **96:**229 *(1955).*]

completely replicated, but have gaps or nicks at each end. These stretches remain unconnected to the adjacent DNA until the gaps are finally closed 5 days later at the upcoming metaphase. Pachytene DNA (P-DNA) synthesis also has unusual properties. It is not replicated semiconservatively; instead, it is similar to a repair type of replication, where only one of the two strands is replicated in relatively short stretches. This latter type of replication is typically used by the cell to repair short damaged areas, such as those that occur after radiation. Unlike Z-DNA, P-DNA stretches are short (100 to 200 base pairs long) and consist of a special group of repeated sequences.

The question of primary importance concerns the role of meiotic prophase synthesis of DNA in the meiotic process. It has been suggested that Z-DNA synthesis is involved in the steps by which homologous chromosomes are precisely aligned with one another to form a tetrad. The two events occur at the same time during meiosis, and it has been shown that the inhibition of Z-DNA synthesis by treatment with deoxyadenosine stops the process of chromosome pairing and prevents the formation of the synaptonemal complex. It is proposed that delayed replication of Z-DNA leads to the availability of single-stranded stretches of DNA that pair with complementary stretches of the homologous chromosome, thereby facilitating pairing. It is speculated that the repair-type synthesis that occurs during pachytene is involved in the process of crossing-over believed to occur during this stage. In support of this hypothesis, it is found that inhibition of chiasmata formation during meiosis blocks the synthesis of DNA during zygotene. Although definitive conclusions concerning causality are difficult to make from studies utilizing inhibitors, the mechanism of genetic recombination during meiosis is likely to be one of the key elements in this synthetic puzzle.

We have followed spermatogenesis to a point where four relatively simple, spherical cells have been produced. The process (Fig. 4.5) that converts each spermatid into the characteristic structure of the spermatozoon is called *spermiogenesis* (or *spermioteliosis*), which represents the last phase of spermatogenesis. The final product is a beautifully constructed cell, one that shows remarkable uniformity of structure, with conspicuous exceptions, from lower invertebrates to mammals (Fig. 4.6). The main structural subdivisions are the head, midpiece, and tail. All are contained—as in living cells generally—by a continuous plasma membrane. Electron micrographs of freeze-fractured replicas and sections of spermatozoa are shown in Figs. 4.7 and 4.8. The whole cell is streamlined and pared down for action of a special sort and of limited duration. Its mission is to swim to the egg, to fuse with its surface, and to introduce its nucleus and centriole into the egg interior.

A. Structure and formation of the sperm

The head of a spermatozoon consists of two main parts, the nucleus and the acrosome. The nucleus of the sperm head is the ultimate in chromosome compactness. In progressing from the spermatid to the spermatozoon, the nucleus loses its entire fluid content, essentially all its RNA, and most of its protein. In most cases the typical histones of the spermatid are replaced by another class of small basic proteins, referred to as *sperm histones,* which are particularly rich in the amino acid arginine. In some species (as in the trout), these proteins contain tracts of two to six arginine residues which account for approximately two-thirds of the total amino acid residues of the polypeptide. Proteins of this latter type are termed *protamines*. These unusual arginine-rich proteins give the chromatin a smooth appearance in the electron microscope (one lacking the nucleosomal beads) and presumably facilitate the side-by-side association of chromatin fibers.

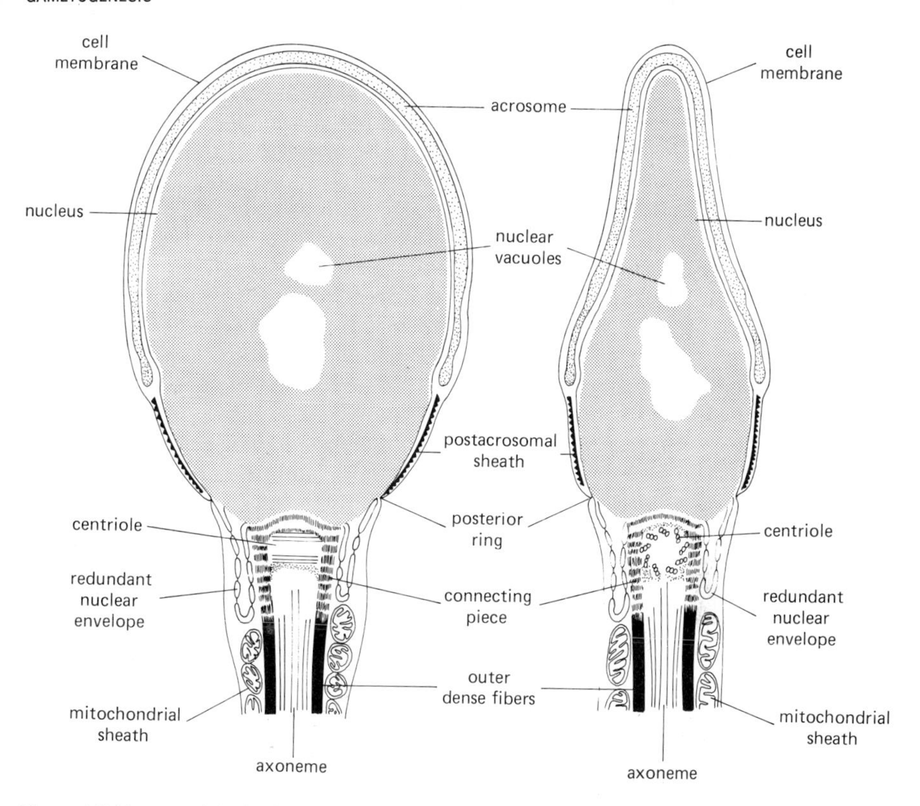

Figure 4.6 Diagram of the head and neck regions of a human spermatozoon. Longitudinal section parallel (left) and perpendicular (right) to the axis of the proximal centriole. [*From H. Pederson and D. Fawcett in "Human Semen and Fertility Regulation in the Male," E. S. E. Hafez (ed.), Mosby, 1976.*]

Figure 4.7 Surface replica after freeze-fracture treatment of a buffalo sperm. *a* Low-power micrograph showing acrosomal bulge (Ac) and the post-nuclear (or postacrosomal) sheath (S). *b* Higher magnification showing the postacrosomal sheath, which fuses with the egg. [*Courtesy of J. Koehler.*]

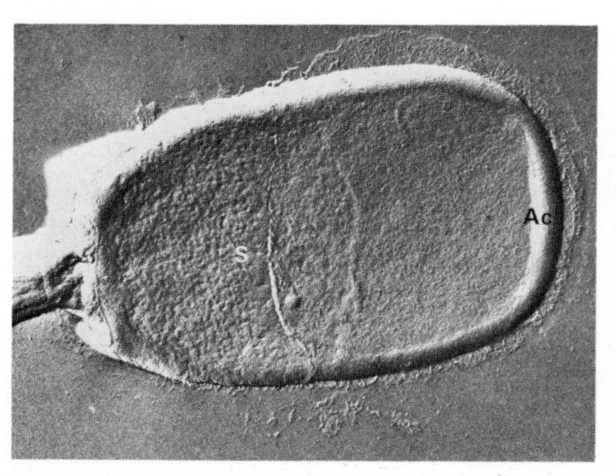

a

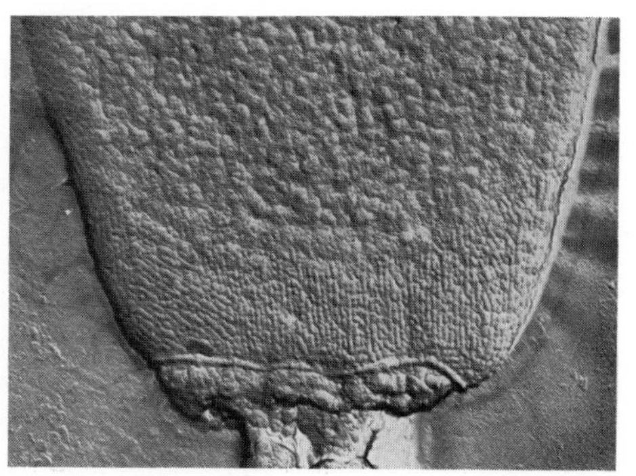

b

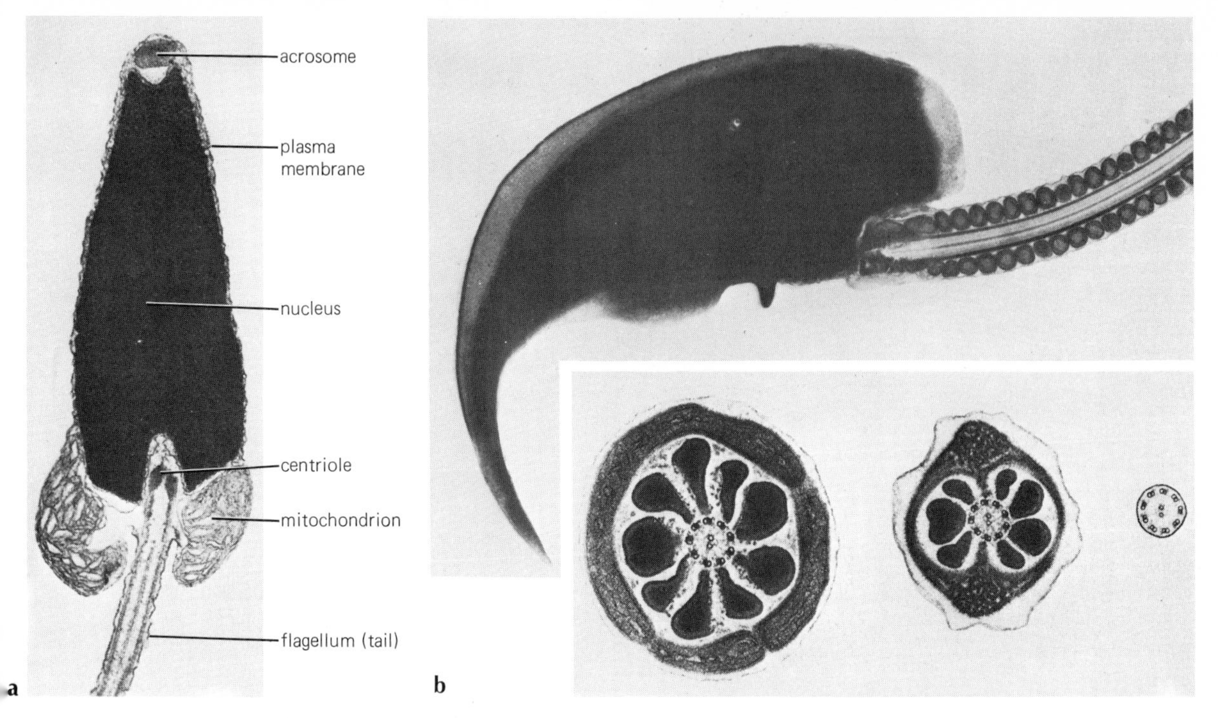

Figure 4.8 Electron micrographs comparing sea urchin sperm (*a*) with mammalian sperm (*b*). Mammalian sperm shown as head with curved acrosome, nucleus, and middle piece surrounded by mitochondria, together with successive cross sections of tail, showing typical internal 9 + 2 tubule organization and outer ring of coarse fibers. [*Courtesy of D. Fawcett.*]

In the final stages, the proteins of the chromatin are cross-linked to one another via disulfide bonds to provide for the final packaging of the genome in the least possible space. In some sperm heads the DNA is actually in a near crystalline state. Each chromosome of the spermatozoon is believed to remain in a small region of the nucleus rather than being spread throughout. For example, in the grasshopper, if the X chromosome, which replicates very late in the premeiotic S phase, is labeled with [^{3}H]thymidine, all the silver grains are localized in one region of the elongate sperm head.

The acrosome of the spermatozoon comes in various shapes, but generally forms an anterior cap over the sperm nucleus, as shown in Fig. 4.9. The acrosome is separated from the nucleus by a thin subacrosomal space, and the nuclear envelope in the region of the acrosome is reported to be devoid of nuclear pores. The sperm head is joined to the midpiece by a constricted region termed the *neck,* in which various types of connecting structures can be found (see Fig. 4.6). The neck and midpiece of the sperm contain a central portion consisting primarily of a pair of centrioles, the base of the tail apparatus, and an outer portion consisting of a mitochondrial sheath. The mitochondria of the outer sheath are organized into a tight helical array which spirals around the anterior section of the flagellum. The number of mitochondria can range from one or a few larger organelles to many smaller ones depending on the species. The mitochondria of the sperm midpiece supply the ATP required for the motility of the tail. Spermatozoa are launched, so to speak, like torpedoes with limited range. If they do not reach the egg within the alloted time, they exhaust their supplies and die. In

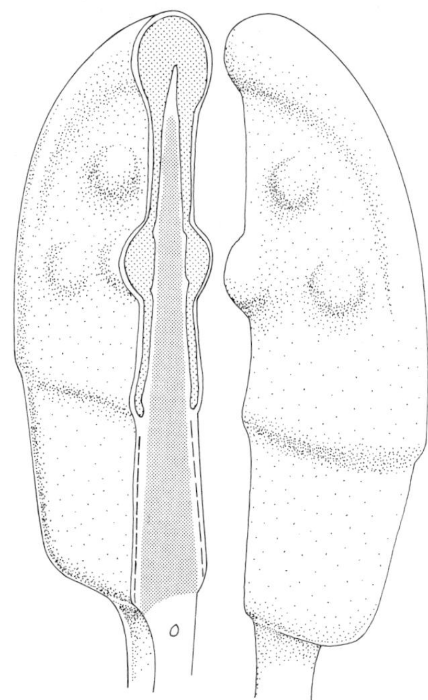

Fig. 4.9 Three-dimensional representation of the head of a spermatozoon from the rodent *O. degus*. [*From M. Berrios, J. E. Flechon, and C. Barros*, Am. J. Anat., **151**:*40 (1978)*.]

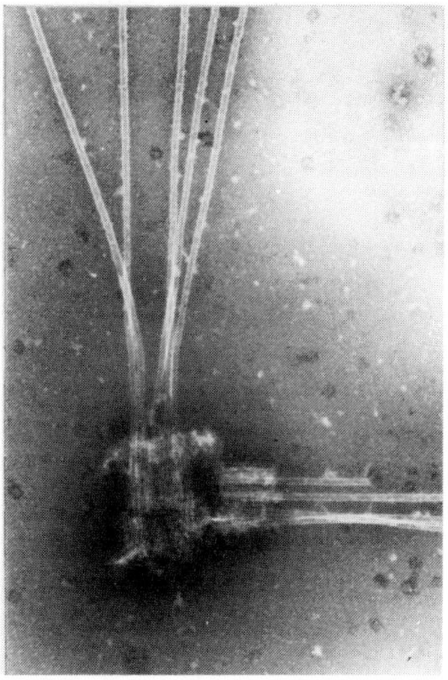

Figure 4.10 Microtubules formed by polymerization of pig-brain tubulin onto a centriole-procentriole pair from a Chinese hamster cell. [*Courtesy of R. R. Gould and G. G. Borisy.*]

mammals they are aided in traversing the long distance through the female reproductive tract by the contractions of the wall of the tube.

Within the neck of the sperm (Fig. 4.6), two centrioles are typically pressed against the posterior edge of the nucleus. Centrioles usually are seen in pairs, with one at right angles to the other. The anterior (or proximal) sperm centriole is typically donated to the egg and helps in forming the first mitotic spindle after fertilization. In the rat, however, this centriole disintegrates prior to fertilization, indicating that this donation to the egg cytoplasm is not an essential one. The posterior (or distal) centriole is responsible for the formation of the *axoneme,* i.e., the microtubule-containing core of the sperm tail. Although the manner in which subunits are polymerized into microtubules is not understood, centrioles are found to be an excellent seed or nucleating structure upon which polymerization can occur. The ability of centrioles to act as microtubule organizing centers can be demonstrated in vitro (Fig. 4.10) as well as in the cell. A cross section of the sperm tail appears very similar to that of any other flagellum (or cilium), although in the mammalian sperm there is an additional outer ring of nine more fibers of a much thicker nature (see Fig. 4.8). The function of the outer fibers is not known. The inner 9 + 2 group of microtubule-containing fibers is termed the *axial filament,* or *axoneme.* It is the sperm tail that whips against the medium to provide the mechanical force for locomotion. One of the last events in spermiogenesis is the elimination of nearly all the cytoplasm from the spermatozoon.

One of the most differentiated structures of the spermatozoon is its plasma membrane. The differentiation has been revealed in several ways. For example, the use of substances such as concanavalin A (which binds to exposed glucosyl and mannosyl residues) or colloidal iron (which binds to exposed anionic sites) indicates that the receptor sites for these substances are not uniformly dis-

tributed over the outer surface of the sperm plasma membrane. The membrane is a complex patterned mosaic. A similar conclusion concerning membrane differentiation has come from studies of freeze-fractured replicas, which allow one to examine the distribution of integral proteins within the membrane. The organization of the proteins within the plasma membrane of the acrosomal region can be quite different from their organization within the postacrosomal membrane (see Fig. 4.7). One of the characteristic organizations of the sperm membrane is a double row of particles overlying one of the dense outer fibers of the tail. The heterogeneity in structure of the continuous plasma membrane is believed to reflect the different roles that the membrane regions play in sperm function (described in the next chapter). The basis for the restriction of membrane components in one or another part of the sperm remains unknown.

If we reexamine the contents of the seminiferous tubule shown in Fig. 4.1*b*, one type of cell is seen to stretch all the way from the outer to the inner edge. Cells of this type are found in many different groups of animals; in mammals they are called *Sertoli cells* after their discoverer. The Sertoli cells are the only ones within the tubule that are not derived from the primordial germ cells, and the reproductive cells are pressed tightly up against the numerous processes that extend from these cells. The exact function of the Sertoli cells has proved difficult to elucidate, but they are generally implicated in a number of processes including (1) nutrition of the developing germ cells and regulation of the spermatogenic cycle, (2) support for the germ cells, and (3) release of mature spermatozoa into the lumen.

Each germ cell passes up the length of a Sertoli cell as it differentiates. A diagram of a Sertoli cell and its associated germ cells is shown in Fig. 4.11. A most important morphological feature of the seminiferous epithelium is the presence of tight junctions between the Sertoli cells toward their base. These junctions, which are remarkably wide, are seen to be composed of up to 50 parallel rows of membrane particles in freeze-fractured replicas. Earlier studies involving the injection of dyes and other substances had indicated that the lumen of the seminiferous tubule is isolated from the general circulation of the gonad. There was said to be a blood-testis barrier between the two fluid compartments. The tight junctions between the Sertoli cells near their base are responsible for this barrier. If one notes the position of the tight junctions between the Sertoli cells, it is apparent that the base of the Sertoli cells as well as the germ cells of the outer layers are separated from the apical portions of the Sertoli cells and the inner layers of germ cells. It is believed that as a consequence of their position, these junctions allow the differentiating spermatocytes and spermatids of the inner layer to be maintained in an environment quite different from that surrounding the spermatogonia and very early spermatocytes of the outer portion of the tubule. Micropuncture studies have supported this concept of distinct chemical environments within the tubule. The concentration of K^+, for example, is much higher within the lumen than within that portion directly bathed by the blood supply. Although no direct intercellular bridges are found between the Sertoli and germ cells, such as those existing among the germ cells themselves, it is generally believed that the Sertoli cells are responsible for providing the materials to the germ cells that are necessary for their differentiation.

The spermatozoa released to the lumen of the mammalian seminiferous tubule have still not fully completed their differentiation; they are not motile and they are not capable of fertilizing the egg. The finishing touches to the sperm occur during their transport through the tubules of the adjacent epididymis. During this period, subtle changes in the shape of the cell are made, chromatin con-

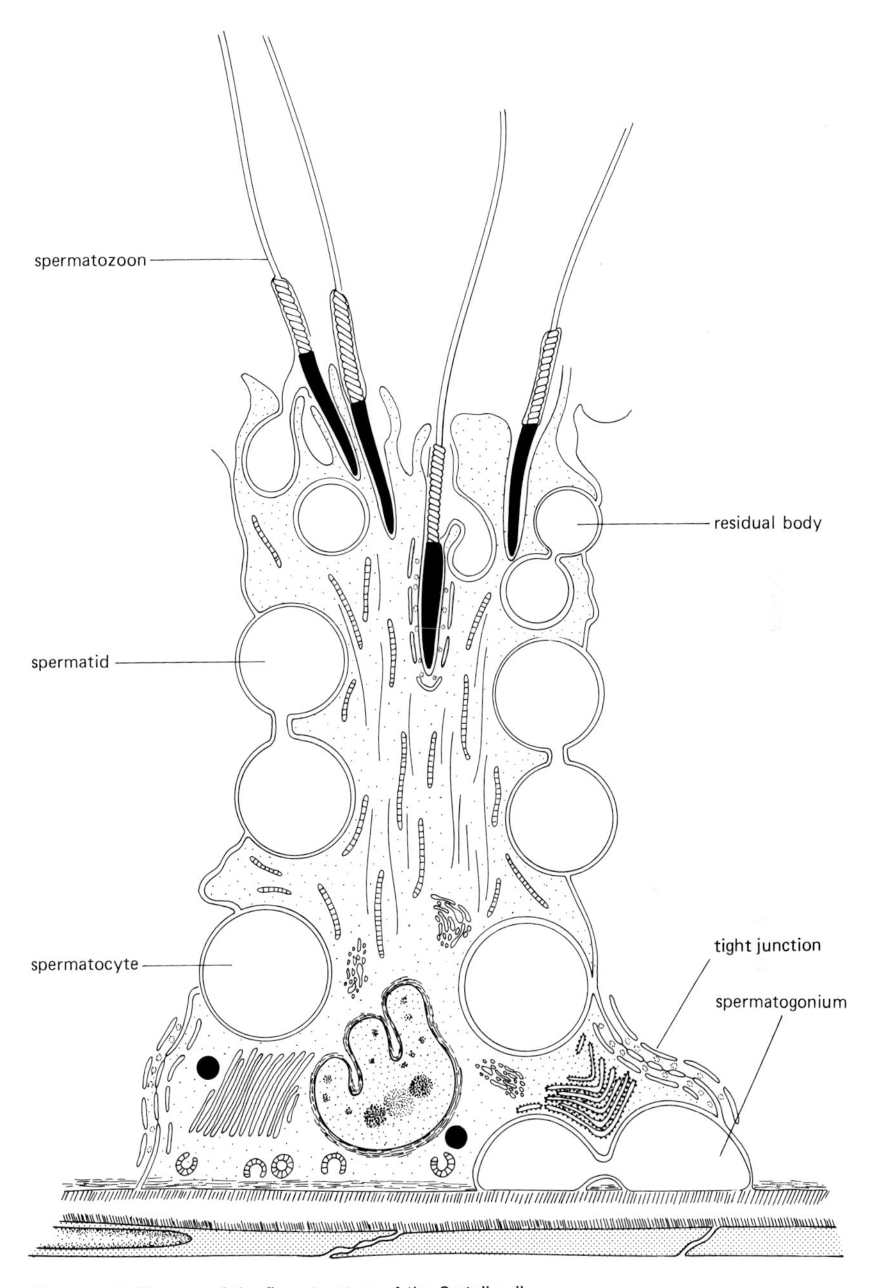

Figure 4.11 Diagram of the fine structure of the Sertoli cell showing the position of the tight junctions that divide the seminiferous epithelium into two compartments: a basal compartment containing the spermatogonia, preleptotene and leptotene spermatocytes, and an adluminal compartment containing the more advanced spermatocytes and spermatids. [*From M. Dym in "Histology," 4th ed., L. Weiss and R. O. Greep (eds.), McGraw-Hill, 1978; after M. Dym and D. Fawcett,* Biol. Reprod., **3**:*308 (1970).*]

densation is completed, the plasma membrane is modified, and the last bit of residual cytoplasm (present as a small drop in the midpiece region) is removed.

In plants, the final differentiation of the motile male gamete is quite different. There is no acrosome, for the sperm cell does not have to make its way through specialized egg membranes. More striking, in ferns and cycads, is the development of a ciliated band around the cell. The centrioles in the spermatid multiply until a large number are formed; these become arranged in rows, and each then produces a cilium. As in animal sperm differentiation, however, the residual cytoplasm is shed after the general structure of the mature cell has been formed.

Although in animals meiosis always precedes differentiation of spermatozoa, in plants it may be far removed in time and phase. Accordingly, we may conclude, provisionally at least, that although spermiogenesis follows meiosis during the production of spermatozoa, the two phenomena are essentially independent, and meiosis, as such, has no causal relationship to sperm differentiation.

B. Nuclear control of spermiogenesis

Analysis of reproductive processes provides a unique opportunity for the study of chromosomal, and therefore genetic, activity. In mammals, for example, the female of the species is XX and the male XY. Presumably all genetic differences between the sexes can ultimately be traced to this single chromosomal difference. Because in humans a person with one X chromosome and no Y chromosome (Turner's syndrome) develops as an immature female, it can be presumed that the presence of the Y chromosome is somehow responsible for male differentiations and therefore, directly or indirectly, for spermatogenesis. This does not mean that genetic information concerning sexual differentiation is restricted to the sex chromosomes; in fact, most of the necessary genetic functions are actually located on the other chromosomes, i.e., the *autosomes*. However, it does appear that the Y chromosome is necessary for their expression in males.

What evidence of a direct nature exists linking the genes on the Y chromosome to spermatogenesis? The best information comes from studies on *Drosophila*. Presumably, similar principles hold in other cases. As mentioned earlier, in the diplotene stages of gametogenesis, the chromosomes may assume a very unusual configuration termed the *lampbrush state*. Lampbrush chromosomes are fully described in the discussion of oogenesis, where they have been best studied. In brief, however, loops of chromatin extend out laterally from the backbone of the chromosome, with each loop containing an actively transcribed segment of DNA. In *Drosophila hydei*, the Y chromosome forms six giant loops in the primary spermatocyte stage. If any of these loops are absent, the end result is an abnormal, infertile sperm. This provides clear evidence that at least six genes on the Y chromosome are needed for the normal events of gametogenesis to occur.

How can the genes of the Y chromosome be needed for spermiogenesis when this process occurs after both meiotic divisions, a time when half the spermatids no longer possess the Y chromosome? To restate this paradox, X and Y chromosomes are separated in the first meiotic division; yet the spermatozoa that differentiates from the secondary spermatocyte receiving the X, not the Y, produce normal sperm. How, then, can the Y be needed? The answer to this puzzle seems to involve a phenomenon that recurs again and again in the analysis of developing systems, that of stable, stored mRNAs. The proteins synthesized during spermiogenesis are directed by mRNA templates that were made in the primary spermatocytes prior to the time of X and Y chromosome separation. The best-studied case of this type of *translational level control* (Section 2.6.B) operating during

spermatogenesis concerns the synthesis of the protamines of the trout sperm nucleus. The mRNAs for these proteins are synthesized prior to meiosis and then stored in an inactive state for approximately 1 month in small ribonucleoprotein particles for translation by the spermatid.

3. Oogenesis

The female gamete, the ovum or egg, arises by *oogenesis,* a process bearing little relation to its counterpart in the opposite sex apart from the meiotic events that occur in both. The growing egg, known as an *oocyte,* develops within the tissues of the ovary. To understand oogenesis, one must consider that its goal is to produce a cell capable of development after oogenesis has been completed. The events of oogenesis must be interpreted with the hindsight gained from the analysis of the postfertilization process.

The egg in all animals is large by comparison with the other cells of the body and is characterized by two important features, the presence of a blueprint for development and the means to construct an embryo from that blueprint. In other words, an egg has to be programmed and packaged during oogenesis. The program exists in the form of molecular information somehow coded within the protoplasmic structure of the egg. Together with its genetic material, the cytoplasmic information provides the egg with the potential to transform—through a predictable sequence of events—from a relatively simple, homogeneous-appearing cell into the complex preadult form. Since it is from the egg that this transformation springs, it must be within the organization of the egg that the directions are laid. The *packaging* refers to the presence within the egg of all the materials that will be needed to build the structures and provide the energy for the embryo up to the point where it can obtain its nutrition from an outside source. The basic features of oocyte organization are illustrated in the micrographs of Fig. 4.12, one taken from a molluscan source, the other from a mammal.

As in spermatogenesis, the oocytes arise from primordial germ cells that are set aside very early in development and multiply by mitosis to form a population of *oogonia,* from which the oocytes will arise. In most chordates (the exceptions are found among the amphibians and teleosts), all the oogonial divisions are completed and the meiotic prophase has begun before any of the oocytes become mature. In the human female, for example, all potential future eggs of an individual have entered diplotene of the first meiotic prophase by approximately the time of birth, and no reserve oogonia remain. Many of these primary oocytes will remain in this stage of meiosis for several decades. Oogenesis, therefore, is not necessarily a continuous process, as is spermatogenesis in the male. Rather, a primary oocyte population is maintained and a number of oocytes become mature as required. In the following section we will examine the events of oogenesis as they occur in amphibia, considering first the developmental program and then the nature of the packaging process.

A. Oogenesis in amphibia

The mature amphibian egg is a very large cell (up to about 2.5 mm in diameter) and therefore must undergo a tremendous growth phase to carry it from its initial small size of less than 50 μm in diameter. The oocyte begins its development in a manner analogous to that of a primary spermatocyte by entering into the first meiotic prophase. In *Xenopus,* the widely used African clawed frog, for example, the premeiotic S phase takes from 1 to 2 weeks, leptotene from 3 to 7 days, zygotene from 5 to 9 days, and pachytene about 20 days. The following stage, diplotene, is a very extended one, during which most of the growth of both the nucleus

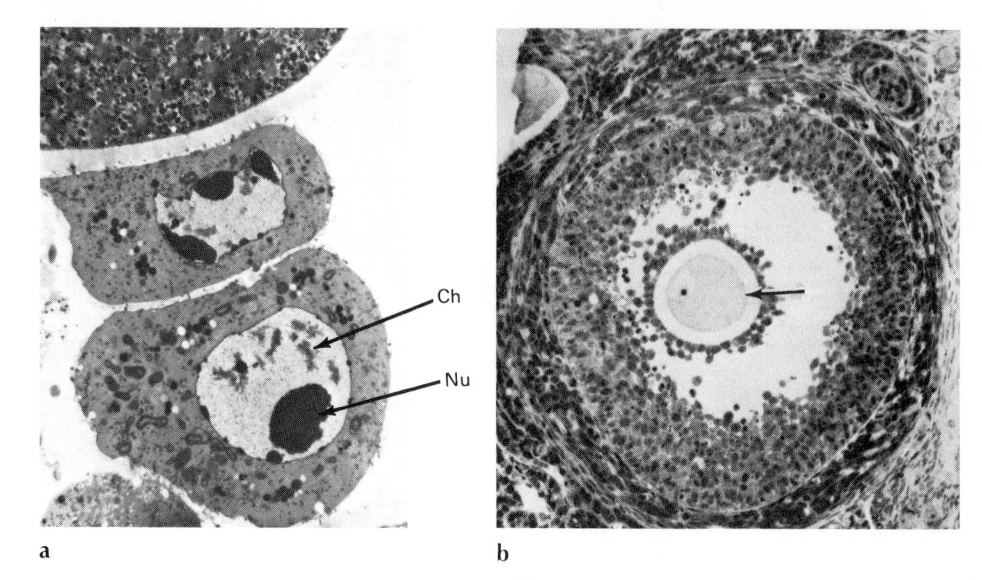

Figure 4.12 Structural organization of the oocyte. *a* Light micrograph of a section through molluscan oocytes. Note the large germinal vesicle (nucleus), the several nucleoli (Nu) (each active in rRNA synthesis), the chromosomal material (Ch), and the cytoplasm, which has not yet begun to accumulate yolk. The small portion of the much larger oocyte at the edge of the photograph illustrates the cytoplasm after it becomes filled with food-reserve materials. [*G. Karp.*] *b* Light micrograph of a portion of a mammalian ovary showing the disposition of the single large oocyte within the follicle. The arrow points to the boundary between the oocyte and the extracellular zona pellucida. The nature of the mammalian follicle is discussed later in the chapter. [*Courtesy of K. Selman.*]

and cytoplasm of the oocyte occurs. The very large oocyte nucleus is termed a *germinal vesicle.* One of the most characteristic cytological features of the mid-diplotene oocyte is the lampbrush state of the chromosomes. Lampbrush chromosomes are found in a wide variety of both invertebrates and vertebrates, including humans; those which have been best studied are found in the newt, *Triturus,* whose genome and corresponding chromosomes are extremely large. Figure 4.13 shows a micrograph of a homologous pair of lampbrush chromosomes and representative diagrams.

One investigative advantage of oocytes, amphibian in particular, is their large size. It is a simple procedure to take an oocyte, place it in a depression on a slide, and—with a pair of fine forceps—break open both the oocyte and its nuclear envelope, thereby liberating the chromosomes into a drop of culture medium. Lampbrush chromosomes consist of an axial backbone from which pairs of loops extend out in opposite directions. Loops arise in pairs because each member of the pair is part of one of the two chromatids that make up each replicated chromosome. The two homologous chromosomes remain attached to each other at the chiasmata. The backbone of the chromosomes contains DNA and tightly associated proteins and is transcriptionally inactive. The loops are made of one double-stranded stretch of DNA attached at both ends to the backbone at specific sites. At the base of each loop can be found a swelling (also containing DNA and protein) called a *chromomere,* from which the DNA of the loop is believed to be extended. A considerable amount of RNA and protein is associated with the DNA of the loops, giving them, as a group, a variety of appearances under the microscope. In *Triturus* there are approximately 5,000 loops present per haploid set of

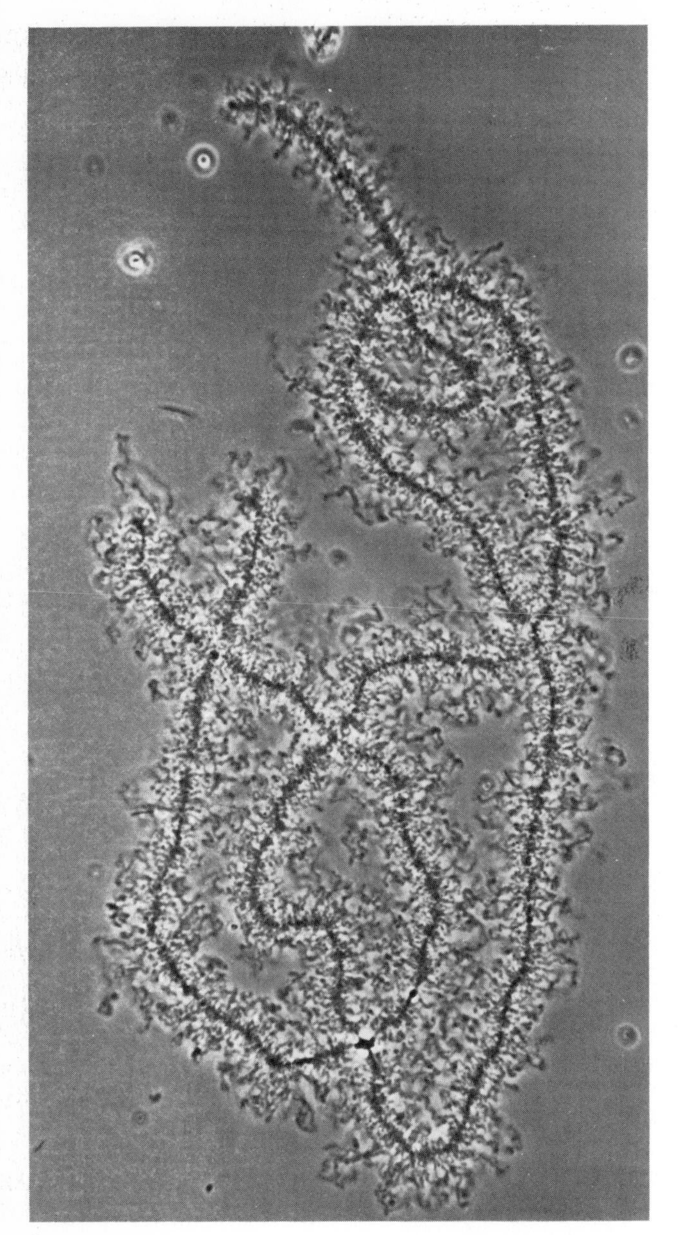

a

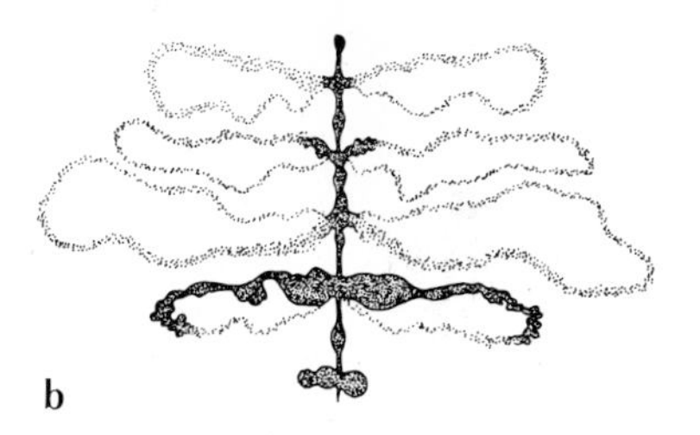

b

Figure 4.13 *a* Photomicrograph showing lampbrush chromosomes. [*Courtesy of J. G. Gall.*] *b* Diagrammatic representation of a section of a lampbrush chromosome showing paired loops emerging from the backbone. [*From J. G. Gall,* Nature, **198**:37 (*1963*).]

chromosomes and, therefore, 20,000 per oocyte nucleus. If a radioactively labeled precursor of RNA (such as [^{3}H]uridine) is added to isolated oocyte nuclei or ovarian tissue, the sites of RNA synthesis can be determined by autoradiography (see Chapter 21). When this is done, the autoradiograph (Fig. 4.14) shows that the loop structures are the sites of intense incorporation of [^{3}H]uridine into RNA.

The discovery that the lateral loops of lampbrush chromosomes are active sites of RNA synthesis has provided investigators with one of the best systems for the direct visual analysis of transcription in eukaryotic cells. Early studies with

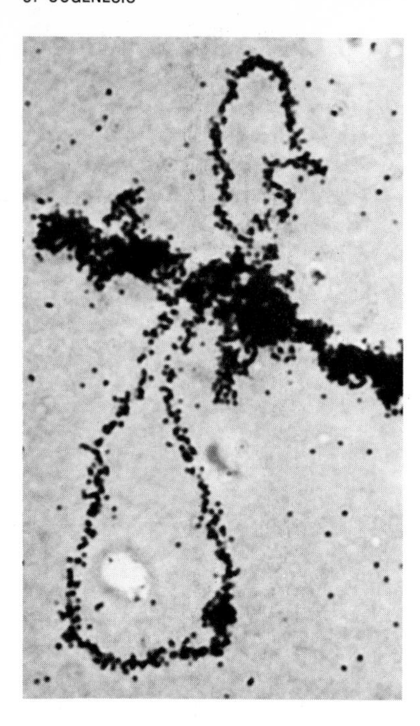

Figure 4.14 Autoradiograph of a pair of loops of a lampbrush chromosome of *Triturus viridescens* following incubation of a piece of ovary in [^{3}H]uridine for 5 hours prior to chromosome isolation. Most loops show this type of uniform labeling. Apparent labeling of the main chromosome axis is due to the presence of shorter loops, which contract and tangle about this region during fixation. [*Courtesy of J. G. Gall.*]

the light microscope had shown most loops to contain a ribonucleoprotein matrix that increased in thickness from one end of the loop to the other. Closer examination of loop matrix with the electron microscope reveals it to consist of aggregates of 200-A ribonucleoprotein (RNP) particles. These particles contain nascent RNA molecules of increasing length attached to the DNA template, upon which they are synthesized. A number of different transcription patterns appear among the loops (Fig. 4.15). In most cases, a loop is seen to have a single set of lateral fibrils that gradually increase in length as one progresses around the loop from one point of insertion to the other (Fig. 4.15*a*). It would appear that transcription of such a loop begins at one end of the exposed DNA and continues uninterrputed along its entire length. Under these conditions, the entire loop is said to contain a single *transcriptional unit.* Considering that the length of DNA in the loops of the large chromosomes of *Triturus* is typically 50 to 100 μm, the *primary transcripts* of these oocytes are tremendously long (well over 100,000 nucleotides). It may well be that a molecule of this size is ultimately responsible for the production of a single messenger RNA, typically less than 1,000 nucleotides in length. Other types of loop profiles containing more than one transcriptional unit are shown in Fig. 4.15*b* through *d*. Taken as a whole, lateral loops of the lampbrush chromosomes contain at least 5 percent of the DNA of the genome; the remainder is present in a condensed state primarily within the chromomeres.

Discussion of the activities of lampbrush chromosomes leads us to consider the broader question of the extent to which the genome is transcribed during oogenesis. We will begin this discussion by examining the products of transcription present in the egg at the end of oogenesis and then look backward with an eye toward their synthesis. It is estimated that the mature *Xenopus* oocyte contains approximately 4 μg of RNA. The most important question would seem to center on the percentage of this RNA that might be utilized in the programming of the egg as opposed to being part of the stockpile of ribosomal and transfer

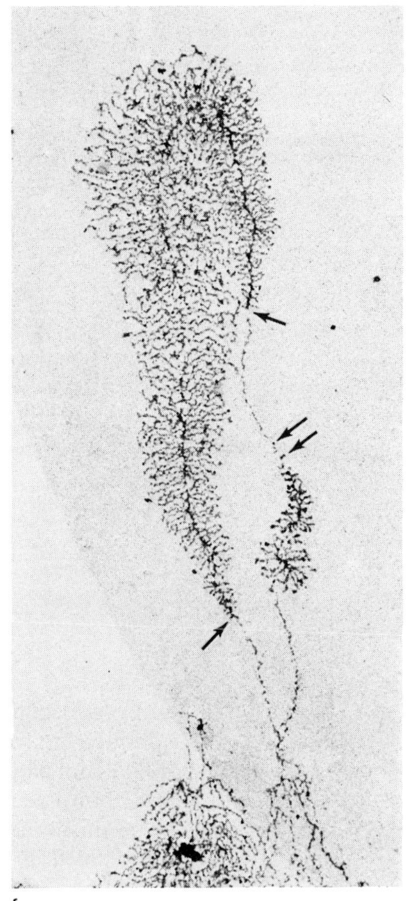

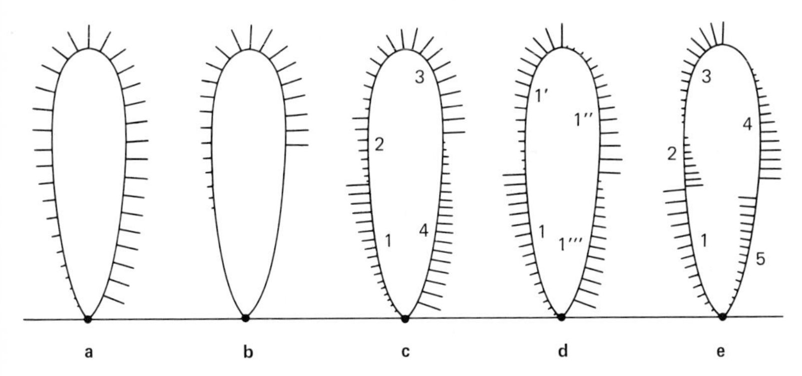

f

Figure 4.15 *a–e* Various alternatives for arrangements of transcriptional units within individual loops of lampbrush chromosomes. The numbers 1 through 1‴ denote units of equal length; 1 through 5, of different lengths. *f* Electron micrograph of an individual loop of a lampbrush chromosome isolated from an oocyte nucleus. The loop shown in this figure contains two matrix units in an end-to-end arrangement (their start regions denoted by arrows) and a small matrix unit (double arrow) separated by an apparently nontranscribed axial intercept. [*From U. Scheer, W. W. Franke, M. F. Trendelenburg, and H. Spring,* J. Cell, **22:***514 (1976).*]

RNAs. In other words, how much of this RNA is utilized as messenger RNA by the new embryo to support early development? This is a difficult question to answer, since the bulk of the mRNA stockpile of the mature occyte (termed *maternal mRNA*) is not being used as messenger at this stage. In most studies, the mRNA content of a cell is determined by measuring the amount of heterogeneous (or poly(A)-containing) RNA that is associated with the polysomes. In the case of the oocyte, however, only approximately 5 percent of the mRNA of the cell is present in polysomes, the remaining 95 percent being stored within cytoplasmic nonribosomal RNP particles for use after fertilization. How can we measure the population of oocyte messages? There are several approaches to the question:

1. Determination of the diversity of a population of RNA molecules by measurement of the percentage of the genome to which these molecules are complementary. Presumably those DNA sequences complementary to a given population of RNAs are the ones responsible for their production in the cell. The experimental procedure that is utilized in this type of determination is given in the appendix of Chapter 8. The most meaningful measurements are made using nonrepeated (single-copy) DNA; the interpretation of the hybridization data is more straightforward with single-copy

DNA, and it is from these DNA sequences that the great majority of the mRNAs are produced. It is estimated that the RNA of the mature oocyte is complementary to approximately 1.2 percent of the nonrepeated fraction of the genome, which in *Xenopus* corresponds to a *complexity* of approximately 27 million nucleotides. In other words, if one were able to line up one of each of the single-copy transcripts end to end, it would stretch to 2.7×10^7 nucleotides in length. Furthermore, a roughly similar value (between 25 and 50 million nucleotides) is found for the complexity of single-copy transcripts in the eggs of two species of sea urchins, an echuroid worm, and *Triturus*. It is significant that all these unrelated organisms appear to contain an RNA population of similar complexity. If we assume that an average sized mRNA is composed of 1,000 nucleotides, this population of RNA molecules would contain sufficient information to code for 25,000 to 50,000 different polypeptide chains. It would appear that this represents roughly the information necessary with which to begin development, whether the egg is a relatively small sea urchin egg, or a much larger *Xenopus* egg, or a *Triturus* egg whose genome is very large (approximately 7 times that of *Xenopus*). Even though the complexity of the single-copy transcripts is similar among these various organisms, the number of copies of the average stored RNA sequence is very different. *Xenopus,* for example, stores an average of nearly 2 million copies of each transcript, while sea urchin oocytes contain approximately one-thousandth of this number. Since the volume of the amphibian egg and its number of ribosomes are roughly 1,000 times that of the sea urchin, the transcription potential, i.e., the number of copies of a transcript per ribosome, is essentially equivalent. It is estimated that in both the sea urchin and the amphibian, approximately 0.5 to 2.0 percent of the total RNA of the oocyte is represented by RNA molecules complementary to nonrepeated DNA.

2. Determination of the poly(A)-containing RNA by measurement of the amount of RNA that is complementary to synthetic poly(U). Since virtually all these poly(A)-containing RNA molecules represent messengers (a small percentage representing a highly diverse group of polyadenylated hnRNAs), this simple measurement provides considerable information on the mRNA content of the egg. It is estimated that there is 40 to 70 ng of poly(A)-containing RNA in the mature *Xenopus* oocyte, a value that corresponds to approximately 1 to 2 percent of the total RNA present. A similar percentage is found on analysis of polyadenylated RNA in the sea urchin egg. The major problem with this type of measurement is that it tells us nothing about the mRNAs that lack poly-A tails and are, therefore, difficult to isolate. There is, in addition to the histone mRNAs, a considerable amount of poly(A)-mRNA in the oocyte that has largely escaped careful scrutiny.
3. Determination of the template activity of oocyte RNA as measured by its ability to direct the incorporation of amino acids in an in vitro protein-synthesizing system. As discussed in Chapter 21, this type of experiment allows one to directly measure the coding properties of a given population of RNA. Estimates from these experiments suggest that approximately 1 to 3 percent of the RNA stockpile has messenger activity, as measured under in vitro assay conditions.

Now that we have some idea of the developmental information that is stored in the mature amphibian oocyte, we can return to the question of its synthesis. Various types of labeling studies indicate that oocytes at virtually all stages of development, from the very small previttelogenic (prior to yolk deposition) oocytes

through to the fully grown post-lampbrush-stage oocytes, are very active in the synthesis of heterogeneous nuclear RNA (hnRNA). The RNA transcripts are similar to those synthesized by somatic cells; they are large and heterogeneous in molecular weight, highly complex in base sequence, DNA-like in base composition, have a short half-life, and are restricted to the nucleus. Presumably these are the types of RNA molecules visualized in electron micrographs of lampbrush chromosomes and presumably the types of molecules that give rise to mRNA species after extensive processing steps.

The most troubling aspects of the study of transcription during oogenesis concern the special function, if any, of the lampbrush chromosomes and the reason for so much RNA synthesis. Analysis of the transcripts present in oocytes of various stages indicate that by stage II (see Fig. 4.18), the *Xenopus* oocyte already contains a population of mRNA molecules essentially equivalent in both quantity and diversity to that present in later stages. Consequently, it is felt that the major transcriptional activity of oocytes during most of the period of growth and development serves primarily to maintain a steady-state level of mRNA. It is presumed that the mRNA produced through the later stages of oocyte development, including that of the lampbrush chromosome stages, which are particularly active, simply replaces that which is degraded. The basis for this turnover is not understood. The situation is even more puzzling in the sea urchin, where it is estimated that the thousand or so copies of the average single-copy transcript could be synthesized in a matter of days, yet oogenesis lasts for months with extensive periods of transcription. Similarly, it is not apparent why the lampbrush-chromosome configuration has evolved, although its widespread appearance among diverse groups points to its importance. The picture is particularly confusing when one considers the size of the *Triturus* chromosomes, with their giant loops and huge primary transcripts. These chromosomes remain transcriptionally active for 7 months, yet the store of mRNA in these oocytes is no greater than that of the *Xenopus* oocyte, which produces its mRNA on much less exposed DNA in a much shorter period of time. Certain of the earlier theories (see Callan, 1967) had suggested that the lampbrush chromosomes had a function unrelated to the transcriptional needs of oogenesis, although little evidence from molecular studies has supported these theories. The answers to these questions remain to be found.

The messenger RNA stockpile of the oocyte is small by comparison with the store of noninformational RNA, over 95 percent of which is ribosomal RNA. Since rRNA synthesis occurs only within nucleoli, we must examine the oocyte nucleoli to explain this accumulation of rRNA. Somatic cells of amphibians typically contain two nucleoli within a nucleus, one per haploid set of chromosomes. Since the growing amphibian oocyte remains in the first meiotic prophase and, accordingly, is tetraploid, four nucleoli could be expected to be present. Yet in *Triturus, Siredon,* and *Xenopus,* there are up to 600, 1,000 and 1,200 nucleoli, respectively.

If autoradiographs are made after incubation with [^{3}H]uridine, all these hundreds of nucleoli are labeled, which suggests that all of them are synthesizing rRNA. If each nucleolus is making rRNA, then presumably each nucleolus contains DNA that codes for this rRNA. Such DNA is termed *rDNA.* If one measures the quantity of rDNA in an oocyte nucleus as compared with the amount present in a somatic cell of the same animal, one finds approximately a thousand times as much in the oocyte (experiment described in Chapter 21). The tremendous increase in rDNA content has occurred as a result of a *selective gene amplification* of these particular sequences during the pachytene stage of meiosis I. Analysis of the DNA content of oocytes from a wide variety of organisms indicates that the

selective amplification of nucleolar DNA is widespread. There is no corresponding increase in the numbers of copies of other genes, including those for the 5 S and transfer RNAs.

Considering the rarity of selective gene amplification, can we justify the need for it in this present case? During the period of growth, the amphibian oocyte has increased in diameter from about 50 μm to about 1.5 mm. This corresponds to an increased volume of about 27,000-fold. At the end of oogenesis, the cytoplasm of this giant cell is seen to contain a considerable complement of ribosomes, approximately 1.1×10^{12}. If no amplification were to occur, it has been estimated that the nearly 2,000 rRNA genes in a replicated set of chromosomes, working at maximum synthetic capacity, would take approximately 500 years to accomplish this feat. Since the life span of a frog is well below this value, an alternate mechanism has evolved. Calculations suggest that no other species of RNA in these eggs is needed in sufficient numbers to require amplification of their DNA template. For example, even though as many 5 S RNAs are needed as 18 S and 28 S RNAs in the assembly of ribosomes, the 200,000 copies of these genes that are normally present in a tetraploid amount of DNA are sufficient to code for the quantity of RNA required. In order to accumulate the required amounts, the synthesis of the 5 S RNA (as well as the various tRNAs) begins much earlier in oogenesis than the synthesis of the 18 S and 28 S species. The early 5 S RNA transcripts are stored in a 42 S ribonucleoprotein particle until their utilization in ribosome formation much later in oogenesis. Interestingly, the 5 S RNA of *Xenopus* that is produced and stored during oogenesis has a somewhat different base sequence than the 5 S RNA produced in somatic cells. The reason for use of a different 5 S RNA gene during oogenesis remains unclear. An overall summary of nucleic acid syntheses during amphibian oogenesis is shown in Fig. 4.16.

Figure 4.16 Diagram of the relative rates of nucleic acid synthesis during amphibian oogenesis. Rates of synthesis of DNA, tRNA (4 S RNA), and rRNA are indicated very approximately on a low-to-high scale. [*From J. B. Gurdon, "The Control of Gene Expression in Animal Development," Harvard, 1976.*]

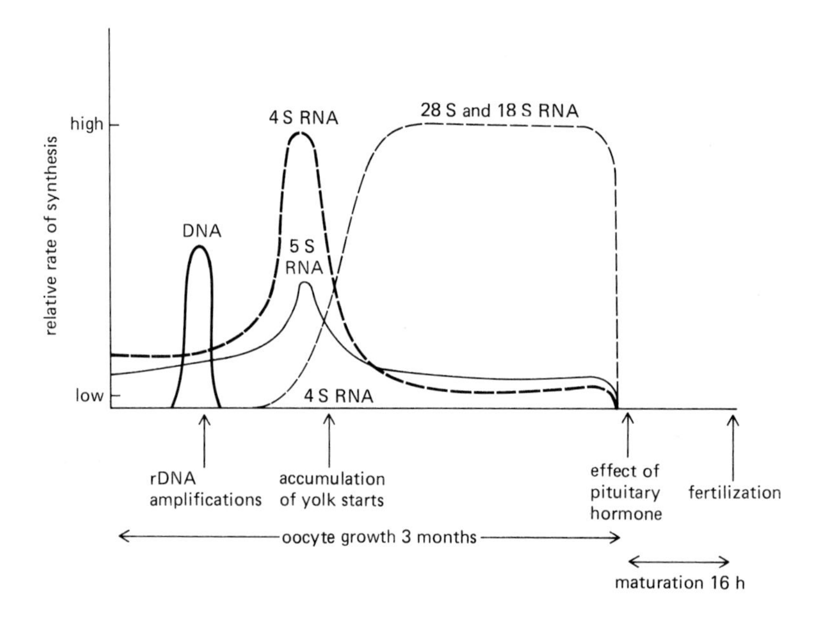

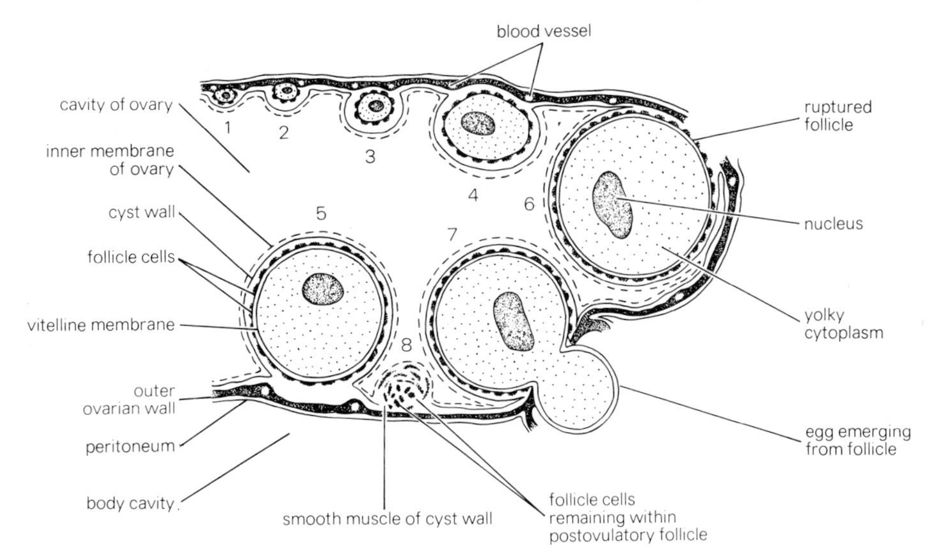

Figure 4.17 Lobe of frog ovary showing stages of growth of ovarian follicle, rupture of follicle, and ovulation. [*After C. D. Turner and J. T. Bagnara, "General Endocrinology," 6th ed., Saunders, 1976.*]

i. Vitellogenesis. The major changes that occur during oogenesis in amphibians are indicated diagramatically in Figs. 4.17 and 4.18. The most obvious change in the oocyte is the increase in its volume, approximately 100,000 times in *Rana* and 25,000 times in *Xenopus*. Although the germinal vesicle also increases greatly during this period, its relative volume cannot keep pace with that of the oocyte itself, and the nuclear/cytoplasmic ratio drops considerably. The major activities occurring within the oocyte are the synthesis of RNAs (previously described) and the deposition of yolk, termed *vitellogenesis*. Whereas RNA synthesis occurs throughout oogenesis, vitellogenesis begins in *Xenopus* in stage III (Fig. 4.18), becomes maximal during stage IV, and has decreased to very low levels in stage VI.[1] The major period of lampbrush chromosome activity occurs during the last part of stage II and stage III.

Vitellogenesis in vertebrates is usually a process involving the accumulation of material synthesized elsewhere in the body—particularly in the liver—and transported by the blood to the oocyte. *Yolk* is a general term that covers the major storage material of the egg. Yolk can be mostly protein in content (proteid yolk), mostly lipid (lipoid yolk), or both. Other materials also are often included under the term *yolk*.

Large molecules cannot directly penetrate the plasma membrane of the oocyte. Therefore, some other mechanism must prevail. In the amphibians, the yolk enters the oocyte by *micropinocytosis*. In this process, tiny specialized regions of the plasma membrane move inward as vesicles that take the extracellular yolk macromolecules into the egg from the outside. These small vesicles fuse with one another to form larger vesicles. In amphibians the yolk macromolecule synthesized and secreted by the liver is a lipophosphoprotein, vitellogenin, which undergoes conversion in the oocyte into two molecules. One of these, *phosvitin*, is a protein with a high phosphate content (8.4 percent in the form of phosphoserine residues); the other, *lipovitellin*, is a lipoprotein with a lipid content of 17.5 percent. Together these molecules are organized into large crystalline struc-

[1] Stage assignments used in this discussion for *Xenopus* are those of Dumont (Fig. 4.18); other staging schemes have been proposed and are commonly used.

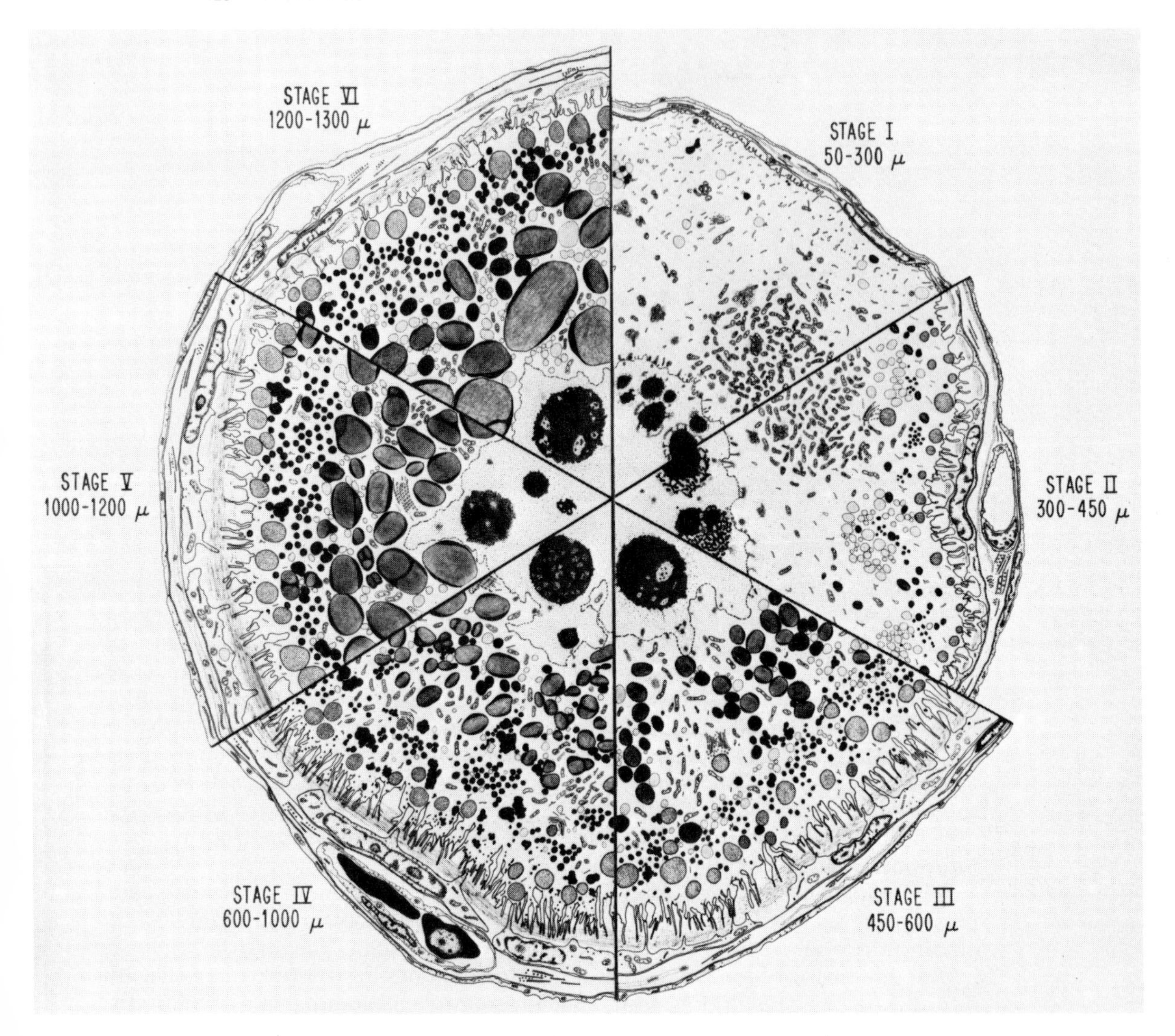

Figure 4.18 Diagrammatic representation of the six stages in the development of a *Xenopus* oocyte. Stage I oocytes are characterized by a very thin follicle cell covering and a large mitochondrial mass, few lipid droplets, and small Golgi complexes. During stage II, the follicle cells increase in thickness, the vitelline envelope forms, and cortical granules, premelanosomes, and some yolk appear. Vitellogenesis begins in stage III and continues until stage V, when it ceases. During stage VI many of the microvilli are lost, and the follicle cells decrease in thickness. [*From J. N. Dumont,* **136:***179 (1972).*]

tures called *yolk platelets* (Fig. 4.19). Surrounding the crystalline portion of the platelet is a less dense material that may be polysaccharide. The deposition of yolk into the crystalline platelets begins in the periphery; the oocyte becomes filled from the outside to near the edge of the nucleus, where the process stops. At the end of oogenesis, the yolk platelets are larger and more tightly packed at one end of the egg, the *vegetal pole.* During this period, the oocyte is active in its own protein synthesis, although this contribution is small compared with that reaching the oocyte from the outside.

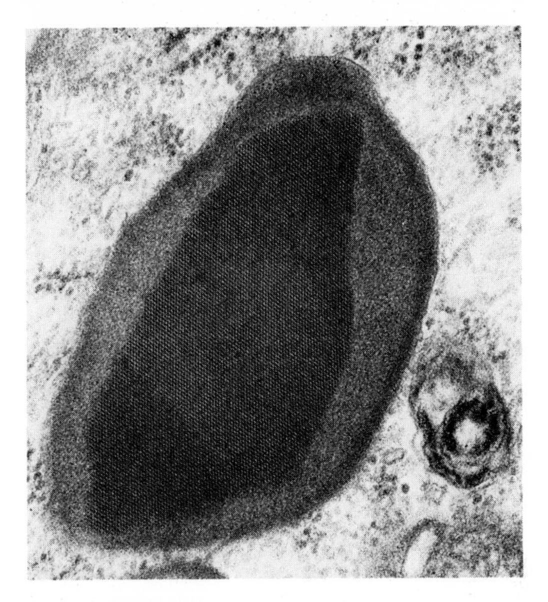

Figure 4.19 Electron micrograph of a yolk platelet of an amphibian oocyte. [*Courtesy of L. Opresko.*]

There has been an interesting sidelight to the story of yolk-platelet formation in amphibians. Considerable controversy has raged over the existence of DNA in yolk platelets. Recent evidence supports the claims of DNA being present in these cytoplasmic granules and indicates that, like the yolk protein, it finds its way into the yolk platelets by pinocytotic uptake from the serum. It is believed that the uptake of this DNA is fortuitous and that once in the yolk platelet the DNA plays no informational role.

The nature of yolk formation and the general structuring of the oocyte cytoplasm vary greatly from animal to animal. Various cytoplasmic organelles, including the endoplasmic reticulum, Golgi complex, and mitochondria, increase in amount during oogenesis and undergo modifications in structure in myriad different ways. As a result, it is dangerous to attempt to generalize in a brief discussion of these events for widely separate types of animals, particularly when the function of these structures is not always clear. One of the problems in analyzing results gained from the electron microscope is that one receives a series of completely static pictures of what we assume to be dynamic processes. As a result, we must attempt to put these various structures into action in our own mind's eye and formulate a theory as to what they are doing. Objects from widely separate species with similar appearances in the electron microscope become lumped together. If they are given the same name, the implication rapidly emerges that they have a similar function, which may not be justified.

In addition to yolk, other materials are included in the packaging of the oocyte. Carbohydrate is stored as glycogen and is present as dense cytoplasmic granules. Lipid accumulates as droplets with a surrounding protein coat; these structures are termed *lipochondria.* Ribosomes are stored, using the RNA made by the many nucleoli. Small pigment granules are incorporated into the surface layer of many amphibian eggs, although their distribution is not uniform. In addition, the cytoplasm contains a huge quantity of DNA—many times greater

than that of the nucleus—which is associated, primarily at least, with the mitochondria. In those vertebrates possessing very large, yolky eggs, such as reptiles and birds, yolk deposition continues far beyond that of amphibians, until the true cytoplasm is only a very small fraction of the finished product.

We can follow all these many examples of cytoplasmic packaging, but it is usually difficult to determine precisely the importance for development of each of these ingredients. In other words, at what point during embryonic development (if at all) does the stockpile become exhausted and the embryo become responsible for its own production of these materials? In the case of the store of ribosomes, an estimate can be provided. Mutants of *Xenopus* have been isolated whose chromosomes carry a deletion for the rDNA, have no nucleolus, and synthesize no rRNA. Heterozygous females produce oocytes with the typical number of ribosomes (even though the somatic cells of these individuals contain only one nucleolus). After meiosis, half these oocytes will be left with the chromosome lacking the rDNA. If such an egg is fertilized by a sperm that is also a mutant, the embryo that develops will have no rDNA; i.e., it is *anucleolate*.

If these embryos which lack the ability to manufacture ribosomes are followed, an estimate can be obtained of the extent to which development can proceed solely on the ribosomes inherited from oogenesis. In the normal embryo, nucleoli appear and rRNA synthesis begins at gastrulation, 8 hours after fertilization. In these anucleolate mutants, development proceeds normally to the swimming tadpole stage (fourth day), at which time they die. During this period, considerable protein synthesis and morphological changes occur; in the absence of any new ribosomes, development proceeds well beyond the stage of gastrulation, when new ribosomes normally appear.

ii. Accessory cells. In most animals the development of the oocyte takes place with the aid of another type of cell, an accessory cell. There are two main types: *nurse cells* and *follicle cells*.

The nurse cell is found only in invertebrates, including some coelenterates, annelids, and insects. Nurse cells are derived from the same oogonium that gives rise to the oocyte. For example, in *Drosophila*, an oogonium (termed a *cystoblast*) undergoes 4 mitotic divisions, producing 16 cells. One of these cells, as a result of its position among the group, becomes an oocyte, and the remaining 15 become nurse cells. Nurse cells are connected to the oocyte by direct cytoplasmic bridges, and it is through these openings that materials pass from the nurse cells into the oocyte, in which they are stockpiled. Nurse cells, therefore, bear much of the responsibility for the synthesis of materials for the unfertilized egg, although they do not synthesize yolk. In *Drosophila*, as in amphibians, yolk is produced outside the ovary, in this case by the fat body, and is taken up from the blood. The involvement of nurse cells in oogenesis provides for the extremely rapid growth of the oocyte. It is estimated that in *Drosophila*, an oocyte's volume can double every 2 hours. If [^{3}H]cytidine is provided to a female fly, it becomes incorporated into RNA. If autoradiographs (see Chapter 21) are made of the ovaries of these flies after a short exposure to this labeled RNA precursor, the highly polyploid nuclei of the nurse cells (Fig. 4.20) are strikingly labeled, indicating their intense RNA synthesis. In contrast, the oocyte nucleus is unlabeled. If several hours elapse after the injection of the isotope before the fly is killed, this labeled RNA has left the nurse-cell nucleus and is found in the nurse-cell cytoplasm, streaming into the oocyte through the intercellular bridges. It is estimated, for example, that approximately 2×10^{10} ribosomes are transferred from the nurse cells to the oocyte during oogenesis.

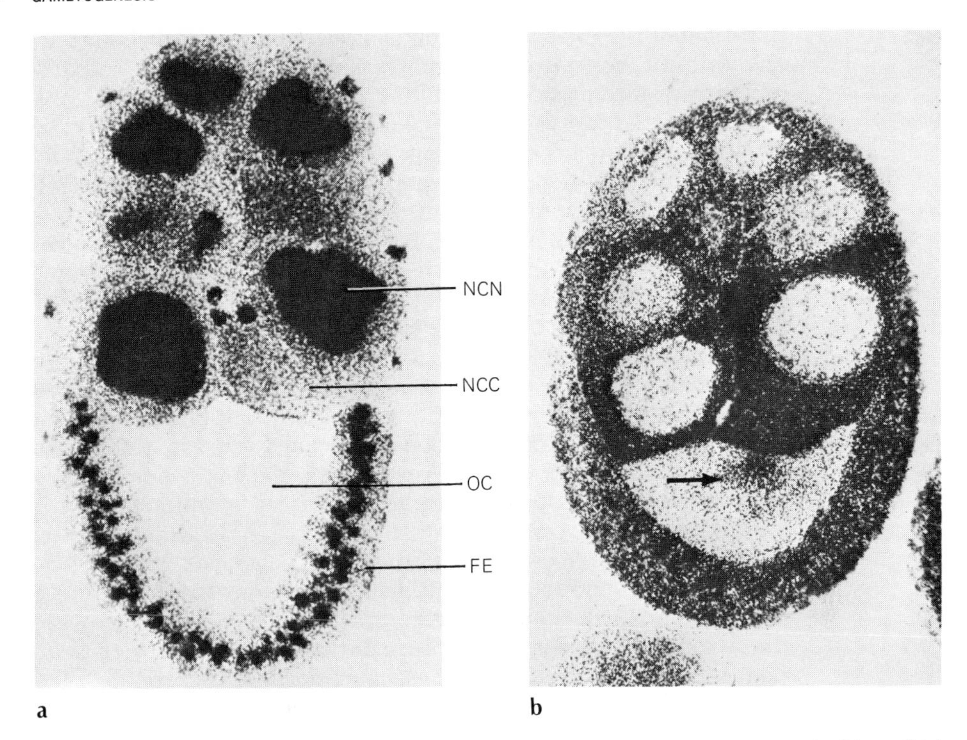

Figure 4.20 Autoradiographs of a follicle of the fly after incubation in [^{3}H]cytidine. *a* Tissue fixed immediately after incubation with the label. Nurse-cell nuclei are densely labeled, reflecting their role in providing newly synthesized RNA to the oocyte. The oocyte is unlabeled at this point. *b* A similar follicle fixed 5 hours after incubation with [^{3}H]cytidine. Nurse-cell nuclei no longer contain labeled RNA, which has moved out of the nucleus and has been replaced by newly synthesized RNA that is unlabeled. Labeled RNA is seen streaming into the oocyte cytoplasm through open channels between the two cells. NCN, nurse-cell nuclei. NCC, nurse-cell cytoplasm. OC, oocyte cytoplasm. FE, follicular epithelium. [*Courtesy of K. Bier.*]

Such autoradiographs elegantly illustrate the advantage that can be gained through the use of an isotope. The injection of the [^{3}H]cytidine provides a zero time for a process that is normally a continuum and has no starting point and is therefore difficult to follow with only the microscope as a tool. The pair of micrographs in Fig. 4.20 can tell what is being made, where it is being made, and the fate of the newly synthesized material. The longer one waits after the injection, the farther the process has continued from the zero point. In addition, if no new isotope becomes available, RNA synthesis will begin to produce molecules that are unlabeled, and one can estimate the time required to clear the previously synthesized molecules from the nucleus.

The term *follicle cell* refers to a heterogeneous collection of cells found in association with oocytes throughout the animal kingdom. Unlike nurse cells, follicle cells do not arise from oogonia and, with the exception of lizards, are not in direct cytoplasmic communication with the oocyte. In vertebrates, follicle cells generally form one or more layers which closely surround the developing oocyte. The follicle cells of vertebrates are believed to be involved in the formation of certain of the noncellular structures that surround the oocyte (such as the vitelline envelope of the amphibian) and in the production of steroid hormones. Where it has been studied, such as in the amphibian, projections of the follicle cells are seen to interdigitate to some extent with the microvilli of the oocyte, although it is not

certain what types of exchange occur between the cells (Fig. 4.21). In some species, such as the moth and squid, follicle cells have been shown to synthesize materials to be stored in the oocyte, but RNA does not appear to be among the products. In the absence of nurse cells, it appears that the oocyte must take care of its own transcriptional needs. As mentioned earlier, the bulk of the yolk materials for the oocytes of most species are synthesized in some nonovarian location in the body. Scanning electron micrographs of the ovarian follicle in *Xenopus* indicate that this blood-borne protein would have no trouble reaching the oocyte surface by passing between the overlying, but loosely packed follicle cells (Fig. 4.21). Studies employing various types of tracer substances further indicate that materials do not pass through the follicle cells themselves in traveling from the blood to the oocyte. In fact, a normal rate of growth of *Xenopus* oocytes can be obtained in vitro following the removal of the oocyte from its follicular investments. In this case, growth of the oocyte is accomplished by maintaining the naked oocytes in a defined medium containing the precursor to yolk protein vitellogenin. Under these conditions, oocytes can grow from 0.78 to 1.43 mm^3 in 28 days of culture and then undergo maturation by addition of progesterone (discussed later).

Whatever the extent of molecular organization present in the cytoplasm of an unfertilized egg, it must result during oogenesis either from the activities of surrounding tissue or from synthetic activities of the oocyte itself. The difficulty in trying to separate information from these two sources is that the same genes are responsible for it whether they happen to lie in the oocyte or, for example, in the liver. How can we distinguish, once it is inside the egg, whether a given RNA has entered from outside or was synthesized from within? Some way is needed to dis-

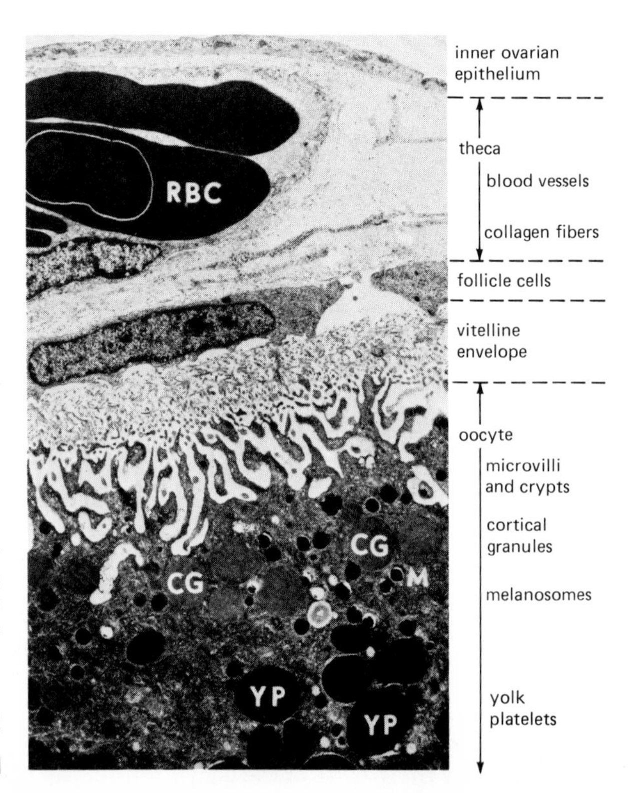

Figure 4.21 Electron micrograph illustrating the architectural relationships between a developing oocyte (stage IV) and its follicular investments. [*From J. N. Dumont and A. R. Brummett,* J. Morph., **155**:97 *(1978).*]

tinguish between what is being made by the oocyte and what is made by the ovary, the liver, etc. Since different species have different observable properties, we need an animal in which the oocyte is of one species and the remainder of the tissue of another. An animal composed of cells of more than one species is known as a *chimera*.

In amphibians, the necessary chimera has been produced by taking advantage of the following facts: (1) in amphibians, as in vertebrates in general, the gonad has two entirely different origins, the primordial germ cells and the remainder of the gonad; and (2) the primordial germ cells can be identified very early within the endoderm of the amphibian embryo and can be removed. To make the chimera, the region of the primordial germ cells of one species, *Xenopus laevis* (*X. laevis*), was removed and replaced with the primordial germ cells of another species, *Xenopus mulleri* (*X. mulleri*). When the ovary of this animal develops, all its oocytes will be from the *X. mulleri* donor, and all the remaining ovarian cells will be from the *X. laevis* host. Animals carrying the germ cells of another species are called *transmission hybrids*. When these transmission hybrids reach sexual maturity and produce eggs that can be fertilized, we can begin to analyze the resulting embryos with our original question in mind. If the eggs of this transmission hybrid are fertilized with the sperm from an *X. mulleri* male, then any deviation from the typical development of an *X. mulleri* embryo must be a result of the *X. mulleri* oocyte developing in a foreign environment. In this case, the resulting eggs, embryos, and adults were always completely *X. mulleri* in character and showed no effects of a developmental program arising from outside the oocyte during oogenesis. Although the number of distinguishing properties between these two closely related species is limited, the results suggest that the amphibian oocyte is responsible for its own program, to whatever extent such a program exists. If this conclusion is correct, the external tissue is left with the assignment of simply packaging the egg with necessary storage material.

B. Oogenesis in mammals

In the mammal, the retention of the developing embryo within the body of the mother makes the accumulation of large cytoplasmic stores unnecessary. Consequently, the mammalian oocyte remains much smaller than the eggs of lower vertebrates, in the order of 100 μm in diameter. One of the major aspects of reproductive capacity in mammals is its cyclical nature, a feature strikingly reflected in the growth and development of mammalian oocytes. Unlike other major organs of the body, the tissues of the ovary undergo dramatic cyclical changes, the results of which are very evident in the diagram in Fig. 4.22. In the following discussion we will concentrate on oogenesis as it occurs within humans. The process is broadly similar in other mammalian groups, even though nonprimate species have estrous cycles rather than menstrual cycles. As mentioned earlier in the chapter, all oocytes capable of participating in reproduction during a woman's life have already reached the diplotene (or *dictyate*) stage of meiosis by approximately the time of birth. Between the time of birth and the onset of reproductive maturity, the number of oocytes present in the ovary decreases from several million to several hundred thousand. Of the oocytes present within the ovaries of an adult woman, most are found within *primordial follicles*. A primordial follicle consists of a small oocyte (approximately 25 μm in diameter) surrounded by a single layer of flattened epithelial *granulosa cells* (follicle cells), which make up the *stratum granulosum* of the follicle.

The growth and development of a follicle can be divided into two phases. The first phase is characterized by the growth of the oocyte within the follicle, and the

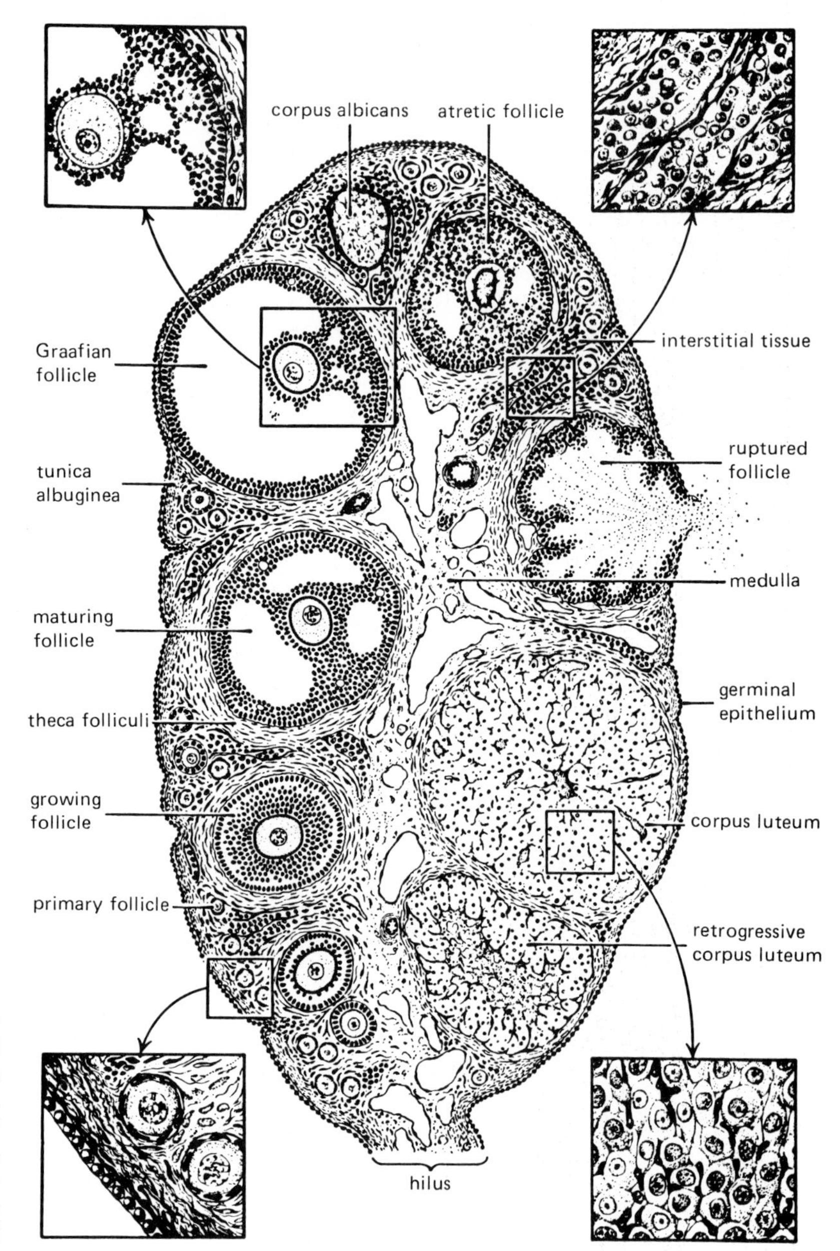

Figure 4.22 Diagrammatic representation of a section through a mammalian ovary. [*From C. D. Turner and J. T. Bagnara, "General Endocrinology," 6th ed., Saunders, 1976.*]

second by the tremendous increase in the size of the follicle itself (up to approximately 10 mm in diameter). The entry of primordial follicles into the pool of growing follicles occurs at a relatively constant rate and is independent of the presence or absence of circulating hormones. By the end of the first phase of follicular development, each follicle (see Fig. 4.12) is seen to be composed of three major components: an internal oocyte, the stratum granulosum, which now consists of several layers of granulosa cells, and an outer thecal capsule. The stratum granulosum is separated from the oocyte on its inner surface by a noncellu-

lar layer termed the *zona pellucida* and from the thecal layer on its outer surface by a noncellular basement membrane. It remains uncertain whether the zona pellucida, which plays an important role in fertilization, is formed by the oocyte, the granulosa cells, or as a result of the joint activity of both.

During the second phase of growth, which is strictly dependent on the secretion of pituitary gonadotropins, as described later, 15 to 20 of the thousands of growing follicles in the ovary are selected in some unknown manner during each menstrual cycle to undergo a remarkable growth process (Fig. 4.22). During the hormone-dependent phase of follicular growth, (1) the granulosa cells proliferate extensively; (2) the follicle enlarges greatly, becoming filled with a polysaccharide-containing fluid termed the *liquor folliculi;* and (3) the theca becomes differentiated into an inner glandular *theca interna* and an outer *theca externa* composed of connective tissue. At the end of this second growth phase, the oocyte is found suspended within the fluid-filled *antrum* of the follicle by a stalk of granulosa cells (Fig. 4.22). These mature follicles, termed *graafian follicles,* are so large that they can be seen as bulges on the outer surface of the ovary. It is from one of these enlarged follicles that, during each cycle, a single oocyte is released into the oviduct to be fertilized. This process of ovulation involves the rupture of the surface of the follicle and ovarian wall. Those graafian follicles which do not participate in ovulation undergo degeneration, a process called *atresia.*

Figure 4.23 indicates the relationship between the oocyte of a graafian follicle and the layer of granulosa cells (termed the *corona radiata*) immediately surrounding it. Electron microscopic studies indicate that the granulosa cells of the corona radiata extend processes through the zona pellucida to make contact with the microvilli of the oocyte. Morphological and physiological studies indicate that gap junctions exist between the oocyte and its innermost surrounding follicle cells, although it remains uncertain as to what types of substances are passed between the two types of cells. During the growth of the follicle, the oocyte remains in the diplotene stage while its germinal vesicle continues to synthesize RNA,

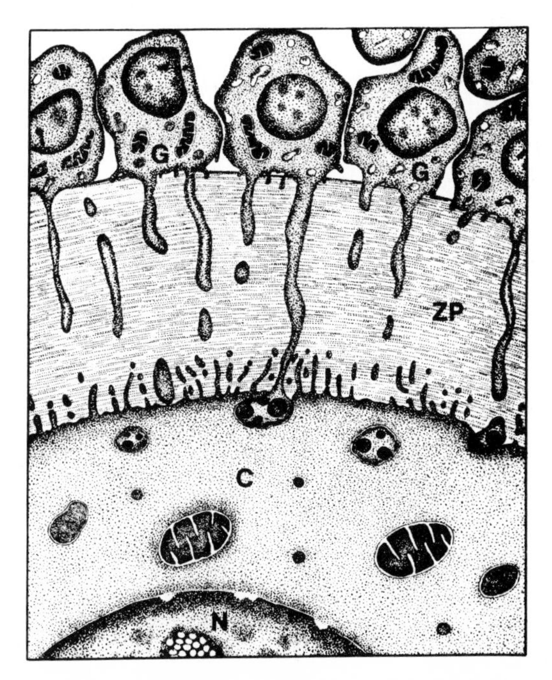

Figure 4.23 Structure of fully formed zona pellucida (ZP) around an oocyte in a graafian follicle. Microvilli arising from the oocyte interdigitate with processes from the granulosa cells (G). These processes penetrate into the cytoplasm of the oocyte (C) and may provide nutrients and maternal protein. N, oocyte nucleus. [*From T. G. Baker, in "Germ Cells and Fertilization," C. R. Austin (ed.), Cambridge, 1972.*]

presumably responding to the same need for programming found in other eggs.

In the mammal, the growth and development of the follicle, as well as subsequent steps leading toward reproduction, are under the control of changing concentrations of several hormones (Fig. 4.24). In the human female, follicular growth is brought about by the combined presence of the anterior pituitary hormones, follicle-stimulating hormone (FSH) and luteinizing hormone (LH), and the hormone from the follicle cells themselves, estrogen. The concentrations of these various hormones are determined by complex, poorly understood feedback loops operating between the hypothalamus, pituitary, and gonad. At about the midpoint in the menstrual cycle, there is an abrupt surge of LH secretion (and a lesser surge of FSH secretion) which acts on the follicle—possibly by activating enzymes that dissolve its surface layer—to cause the sudden release of the oocyte and its surrounding layer of follicle cells (Fig. 4.25). In human females, ovulation occurs approximately 36 hours after the LH peak. It is during this time that the events of maturation (described later) occur. After ovulation, the follicle, now lacking an oocyte, transforms into an endocrine structure, the *corpus luteum*. This structure now secretes considerable quantities of estrogen and progesterone; they act on the hyothalamus, which in turn acts to greatly decrease the secretion of FSH and LH by the pituitary. The hormones of the corpus luteum act on the uterus to prepare it for the ovulated oocyte—now awaiting fertilization at the second meiotic metaphase—which will be implanted in the uterus several days later. If implantation does not occur, some mechanism (at present unknown) causes the corpus luteum to stop hormone production, and part of the uterine wall is sloughed during menstruation. In the absence of activity by the corpus luteum, FSH and LH secretions resume to initiate a new cycle of follicle growth.

The invention of the birth control pill has been a direct outcome of basic research in reproductive physiology. These pills generally contain combinations of estrogen- and progesteronelike substances. The earlier group of pills contained sufficient quantities of hormone to inhibit follicle growth and ovulation by acting on the hypothalamic-mediated release of anterior pituitary hormones. With continued research, it was determined that smaller quantities of hormones could still prevent pregnancy without interfering so drastically with the ovarian cycle. It appears that a variety of targets, including the uterine wall and the cervical mucous, can be altered physiologically in some unknown manner to prevent fertilization and/or implantation.

4. Maturation

As pointed out earlier, the essential activities of oogenesis occur within an oocyte suspended in its first meiotic prophase. When the oocyte has reached full maturity, it is ready to undergo *maturation*, the process which converts the cell into a fertilizable egg. Maturation begins with the breakdown of the germinal vesicle (Fig. 4.26), an event that results in the mixture of materials previously restricted to either the nuclear or cytoplasmic compartment. Germinal-vesicle breakdown (GVBD) is followed by the condensation of the chromatin into compact chromosomes, the formation of the meiotic spindle, and the continuation of meiosis, which previously had been suspended in diplotene. The stage to which meiosis proceeds during maturation depends on the species (see Table 5.1). In most vertebrates, the reinitiation of meiosis continues to the second meiotic metaphase. It is at this point that maturation is arrested and the egg awaits fertilization.

The process of maturation, just as that of ovulation (which occurs in most mammals after the oocyte reaches metaphase II), results from specific hormonal

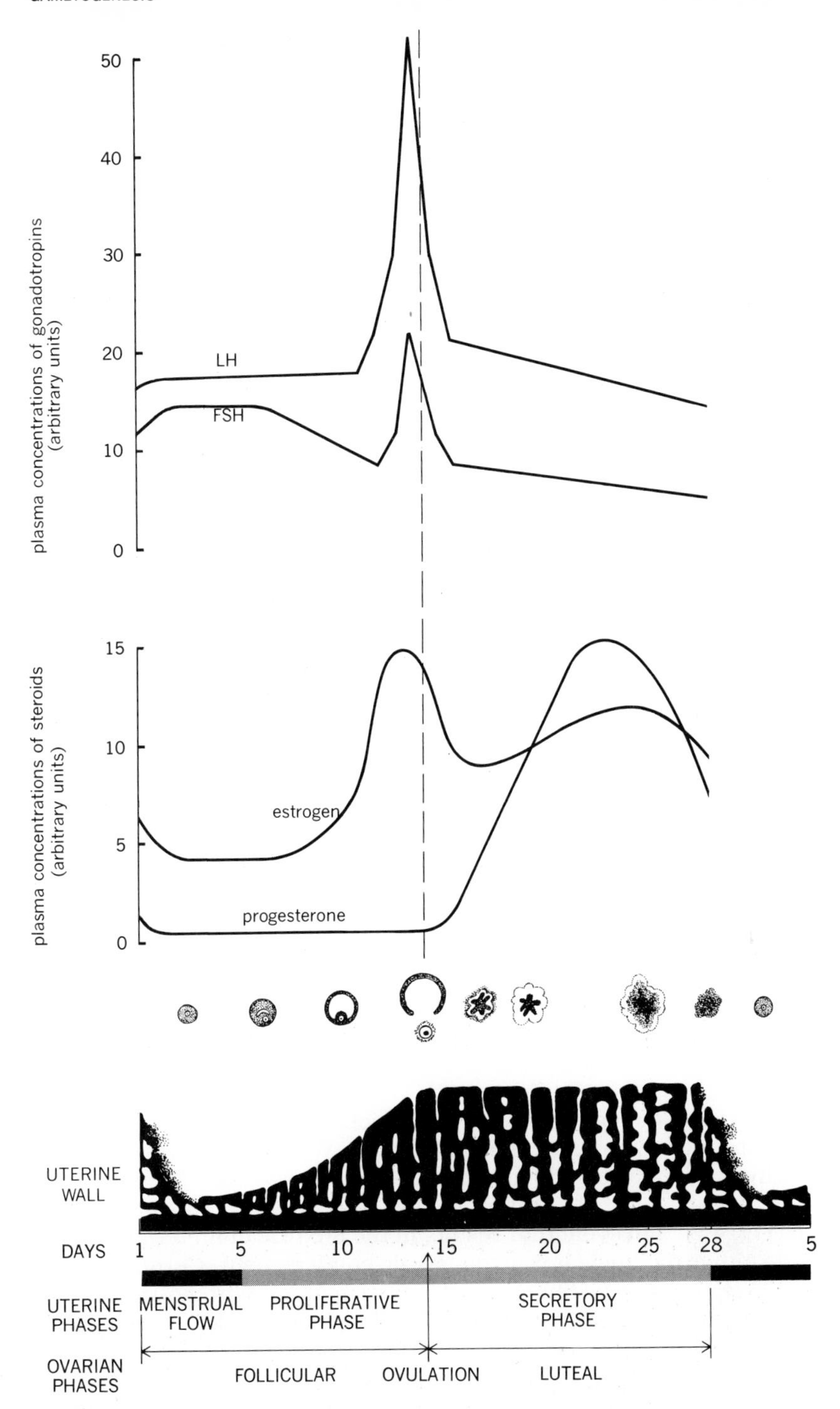

Figure 4.24 Diagram of the ovarian cycle showing levels of the four primary hormones (LH, FSH, estrogen, and progesterone), the condition of the follicle, and the condition of the uterus. [*From A. J. Vander, J. H. Sherman, and D. S. Luciano, "Human Physiology," 2d ed., McGraw-Hill, 1974.*]

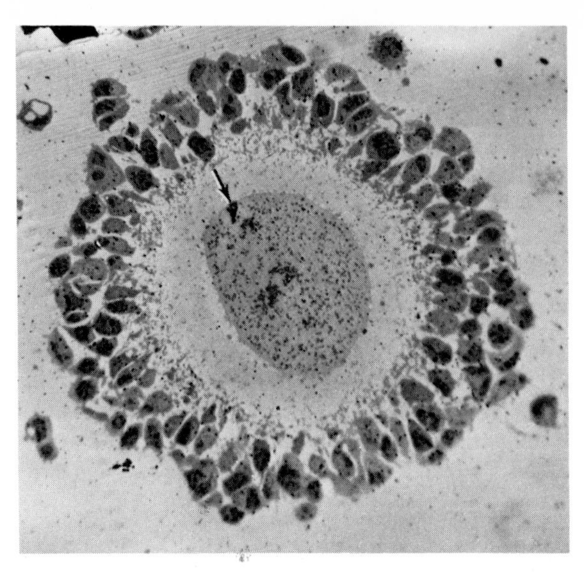

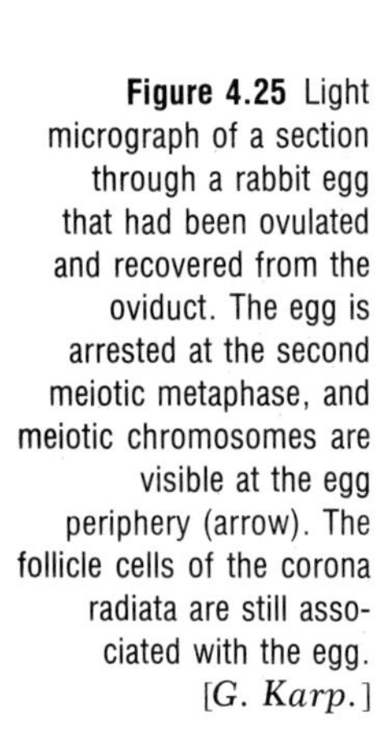

Figure 4.25 Light micrograph of a section through a rabbit egg that had been ovulated and recovered from the oviduct. The egg is arrested at the second meiotic metaphase, and meiotic chromosomes are visible at the egg periphery (arrow). The follicle cells of the corona radiata are still associated with the egg. [*G. Karp.*]

stimulation. The steps that occur prior to and during maturation are best understood in the amphibian, although considerable data on the starfish and the mouse reveal the existence of certain comparable activities. Oocyte maturation in the amphibian is triggered by the release of gonadotropins from the pituitary. The target in this case is the follicle cells that surround the oocyte. These cells respond to the stimulus with the secretion of a steroid hormone, believed to be progesterone, which in turn acts on the oocyte to initiate maturation. Oocytes can be stripped of all follicle cells and caused to proceed to metaphase II simply by exposure to progesterone.

Figure 4.26 Maturation of *Nereis* egg. *a* Early prophase of first maturation division, beginning of breakdown of germinal vesicle (shortly after fertilization), showing nucleoplasm, nucleolus, and chromatin. *b* First maturation division (metaphase). Note spreading of fluid nucleoplasm into surrounding cytoplasm as the nuclear membrane of the germinal vesicle disappears, leaving behind the chromatin and nucleolus; also note the eccentric location of the maturation-division asters and spindle, which results in unequal division of the oocyte to form the first polar body. [*Courtesy of D. P. Costello.*]

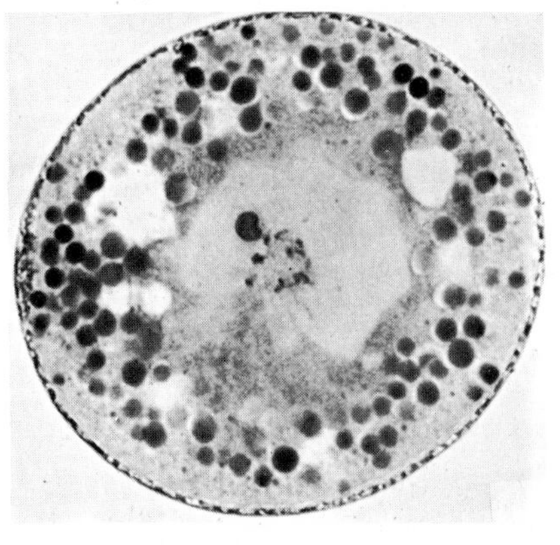

a

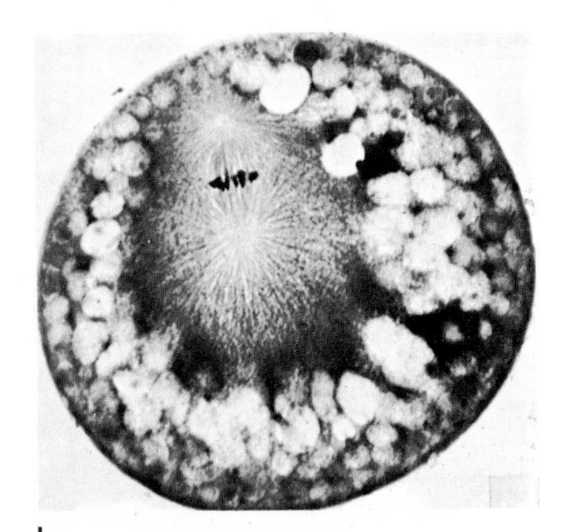

b

A variety of different types of studies indicate that progesterone initiates maturation in amphibians by combination with a hormone receptor present at the external surface of the oocyte plasma membrane. In other words, in this case, the steroid hormone does not combine with a cytoplasmic receptor, as occurs during response by the oviduct to progesterone. This conclusion is based on a number of experimental observations. For example, progesterone injected directly into the oocyte does not induce maturation. In contrast, when oocytes are incubated in a medium containing progesterone that has been linked to a large nonpenetrating polymer, the hormone is still active in stimulating the event. It appears that the interaction between oocyte plasma membrane and progesterone leads to the release of calcium ions within the cell as well as a decrease in the concentration of cyclic AMP. Both these changes are believed to be important in mediating the hormonal response. For example, the introduction of Ca^{2+} by ionophoresis is sufficient to induce maturation in the absence of progesterone. However, the injection of EGTA (which forms a complex with Ca^{2+}) results in the inhibition of the response to the hormone. Similarly, the addition of cyclic AMP to the oocyte medium also is capable of suppressing the response.

Whatever the precise mechanism by which changes in calcium and/or cyclic nucleotide concentration act, the amphibian oocyte responds by the production of certain very important cytoplasmic factors. The best studied of these factors, termed *maturation promoting factor* (MPF), is responsible for the breakdown of the germinal vesicle. If cytoplasm is removed from an oocyte that has been exposed to progesterone for several hours and injected into an untreated oocyte, the latter responds by germinal-vesicle breakdown. Another factor in oocyte cytoplasm is responsible for the arrest of meiosis at metaphase II. This substance, termed the *cytostatic factor,* is also capable of arresting mitotic cells at metaphase as evidenced by this result following injection of oocyte cytoplasm into

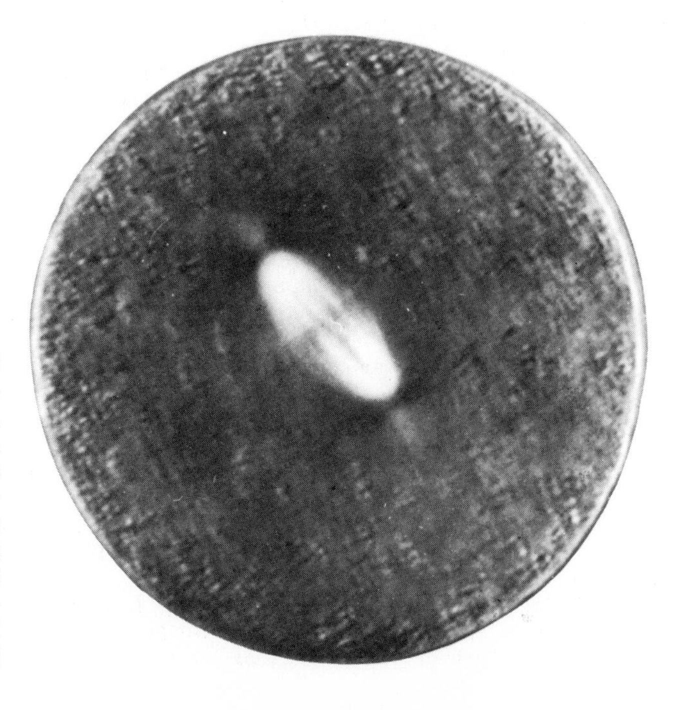

Figure 4.27 An oocyte of *Pectinaria* undergoing meiotic division. The striking birefringence of the metaphase spindle as seen in this polarization micrograph is a result of the large numbers of oriented microtubules. [*Courtesy of G. Sato and S. Inoué.*]

dividing blastomeres. Further discussion of the effects of oocyte cytoplasmic factors can be found in Section 8.2.

The primary importance of the animal egg relates to its large size. Rather than producing four equal-sized cells, the meiotic divisions in the female produce one large cell and two or three small cells, called *polar bodies*. The large egg size is retained by means of two extremely unequal meiotic divisions. They are produced by a meiotic spindle that is greatly displaced toward one side, the animal side, of the oocyte (Fig. 4.27). If the spindle apparatus is caused to move by centrifugation to the center of the oocyte, the resulting divisions can be essentially equal, pointing clearly to the role of the apparatus in determining the relative volumes of the daughter cells. The unequal divisions has the effect of maintaining the necessary large size of the principal oocyte despite the occurrence of the two meiotic divisions.

In conclusion, in this section on oogenesis we have reviewed the main features that relate specifically to the programming and packaging of the egg, so that the embryo can emerge from its organization. There are examples of considerable development in the absence of a sperm, clearly indicating the totality of the information contained within the egg. In one case, that of the anucleolate mutant, we have been able to assay the extent to which the activity of oogenesis provides for the welfare of the developing embryo. The other examples of the impact of oogenesis on development (the direction of shell coiling in a snail and the o mutation in the axolotl) are presented in Chapters 6 and 8.

Readings

AUSTIN, C. R., and R. V. SHORT (eds.), 1972. "Germ Cells and Fertilization," Cambridge University Press.

BACETTI, B. (ed.), 1970. "Comparative Spermatology," Academic.

———, and B. AFZELIUS, 1976. "The Biology of the Sperm Cell," Karger.

BAULIEU, E.-E., et al., 1978. Steroid-Induced Meiotic Division in *Xenopus laevis* Oocytes: Surface and Calcium, *Nature,* **275**:593–598.

BIGGER, J. D., and A. W. SCHUETZ (eds.), 1972. "Oogenesis," University Park Press.

BISHOP, J. O., 1977. Lampbrush Chromosomes Brought Up to Date, *Nature,* **268**:588–590.

BLACKLER, A. W., and C. A. GECKING, 1972. Transmission of Sex Cells of One Species through the Body of a Second Species in the Genus *Xenopus, Develop. Biol.,* **27**: 376–394.

BROWN, D. D., and I. B. DAWID, 1968. Specific Gene Amplification in Oocytes, *Science,* **160**:272–280.

———, and J. B. GURDON, 1964. Absence of Ribosomal RNA Synthesis in the Anucleolate Mutant of *Xenopus laevis. Proc. Nat. Acad. Sci. U.S.,* **51**:139–146.

CALLAN, H. G., 1967. The Organization of Genetic Units in Chromosomes, *J. Cell Sci.,* **2**: 1–7.

CLERMONT, Y., 1972. Kinetics of Spermatogenesis in Mammals: Seminiferous Epithelium Cycle and Spermatogonial Renewal, *Phys. Revs.,* **52**:198–236

———, and C. P. LEBLOND, 1953. Renwal of Spermatogonia in the Rat, *Amer. J. Anat.,* **93**:475—502.

COMINGS, D. E., and T. A. OKADA, 1971. Fine Structure of the Synaptonemal Complex, *Exp. Cell Res.,* **65**:104–116.

DAVIDSON, E. H., 1976. "Gene Activity in Early Development," 2d ed., Academic.

DORRINGTON, J. H., and D. T. ARMSTRONG, 1979. Effects of FSH on Gonadal Function, *Rec. Prog. Hormone Res.,* **35**:301–342.

DUMONT, J. N., 1978. Oogenesis in *Xenopus laevis* (Daudin), *J. Morph.,* **136**:153–180.

———, and A. R. BRUMMETT, 1978. Oogenesis in *Xenopus laevis:* Relationship between Developing Oocyte and Their Investing Follicular Tissues, *J. Morph.,* **155**:73–98.

FAWCETT, D. W., 1979. The Cell Biology of Gametogenesis in the Male, *Persp. Biol. Med.*, **22**:556–573.

———, 1975. The Mammalian Spermatozoa, *Dev. Biol.*, **44**:394–436.

———, 1975. "Ultrastructural Aspects of Gametogenesis," 33rd Symp. Soc. Dev. Biol., Academic.

FORD, P. J., T. MATHIESON, and M. ROSBASH, 1977. Very Long-Lived mRNA in Ovaries of *Xenopus laevis*, *Develop. Biol.*, **57**:417–426.

HAMILTON, D. W., and R. O. GREEP (eds.), 1975. "Handbook of Physiology," vol. 5, American Physiological Society.

HESS, O., and G. F. MEYER, 1968. Genetic Activation of the Y Chromosome in *Drosophila* during Spermatogenesis, *Advan. Genet.*, **14**:171–223.

HOGARTH, P. J., 1978. "Biology of Reproduction," Halsted.

KIEFER, B. I., 1973. "Genetics of Sperm Development in *Drosophila*," 31st Symp. Soc. Dev. Biol., Academic.

KOEHLER, J. K., 1973. Studies on the Structure of the Postnuclear Sheath of Water Buffalo Spermatozoa, *J. Ultras. Res.*, **44**:355–368.

LONGO, F. J., and E. ANDERSON, 1974. Gametogenesis, in J. Lash and J. R. Wittaker (eds.), "Concepts in Development," Sinauer.

MACGREGOR, H. C., 1972. The Nucleolus and Its Genes in Amphibian Oogenesis, *Biol. Rev.*, **47**:177–210.

MASUI, M., and H. J. CLARKE, 1977. Oocyte Maturation, *Int. Rev. Cytol.*, **57**:186–282.

MIDGLEY, A. R., and W. A. SADLER (eds.), 1979. "Ovarian Follicular Development," Raven.

MOSES, M. J., 1968. Synaptonemal Complex, *Ann. Rev. Genet.*, **2**:363–412.

PHILLIPS, D. M., 1974. "Spermiogenesis," Academic.

RAVEN, C. P., 1961. "Oogenesis: The Storage of Developmental Information," Pergamon.

RICHARDS, J. S., 1979. Hormonal Control of Ovarian Follicular Development, *Rec. Prog. Hormone Res.*, **35**:343–373.

ROOSEN-RUNGE, E. C., 1977. "Process of Spermatogenesis in Animals," Cambridge University Press.

SALISBURY, G. W., R. G. HART, and J. R. LODGE, 1977. The Spermatozoon, *Persp. Biol. Med.*, **20**:372–393.

SCHEER, U., W. W. FRANKE, M. F. TRENDELBURG, and H. SPRING, 1976. Classification of Loops of Lampbrush Chromosomes According to the Arrangement of Transcriptional Complexes, *J. Cell Sci.*, **22**:503–519.

SCHUETZ, A. W., 1974. Role of Hormones in Oocyte Maturation, *Biol. Rep.*, **10**:150–185.

SMITH, L. D., 1975*a*. "Germinal Plasm and Primordial Germ Cells," 33rd Symp. Soc. Dev. Biol., Academic.

———, 1975*b*. Molecular Events during Oocyte Maturation, in R. Weber (ed.), "Biochemistry of Animal Development," vol. 3, Academic.

STERN, H., 1977. DNA Synthesis During Microsporogenesis, in L. Bogorad and J. H. Weil (eds.), "Nucleic Acid and Protein Synthesis in Plants," Plenum.

Symposium on Biochemical Actions of Progesterone, 1977. *Ann. N. Y. Acad. Sci.*, **286**:1–449.

Symposium on Biochemistry of Spermatogenesis, 1978. *Fed. Proc.*, **37**:2570.

WALLACE, R. A., and Z. MISULOVIN, 1978. Long-Term Growth and Differentiation of *Xenopus* Oocytes in a Defined Medium, *Proc. Nat. Acad. Sci. U.S.*, **75**:5534–5538.

WASSERMAN, P. M., and W. J. JOSEFOWICZ, 1978. Oocyte Development in the Mouse: An Ultrastructural Comparison of Oocytes Isolated at Various Stages of Growth and Meiotic Competence, *J. Morphol.*, **156**:209–236.

WILSON, E. B., 1925. "The Cell in Development and Heredity," 3d ed., Macmillan.

WISCHNITZER, S., 1976. The Lampbrush Chromosomes: Their Morphological and Physiological Importance, *Endeavour*, **35**:27–31.

ZUCKERMAN, S., and B. J. WEIR (eds.), 1977. "The Ovary," 3 vols., Academic.

Chapter 5 Fertilization

Developmental phenomena have been most intensively studied in the frog and the sea urchin. Frogs offer many advantages: they are vertebrates, abundant, cheap, and easy to keep. They can be injected with hormones and caused to ovulate, and the eggs can be fertilized in vitro and raised synchronously in large numbers. Most frogs breed seasonally, although *Xenopus* can be induced to ovulate year round. Amphibian eggs are large and many intricate surgical experiments can be performed. One of the main difficulties is the impermeability of amphibian embryos to external molecules, which means that isotopes, drugs, etc. generally must be injected directly into the egg.

Of the many possible marine invertebrates, sea urchins have been most widely used because of the ease of procuring large numbers of gametes. Spawning in ripe sea urchins is readily obtained either by injection of a solution of KCl or by stimulation with a weak electric current. If the sea urchin is a female, it is simply turned, oral side up, over a beaker of seawater and eggs come streaming out of the five gonopores. A large sea urchin can spawn many milliliters of packed eggs. If the sea urchin is a male, determined by the white seminal fluid emanating from the gonopores, it is turned over a small dish and great numbers of sperm become available. Eggs are readily fertilized and embryos are raised in large, synchronous cultures. The embryos are transparent, and internal processes can be observed. Unlike the amphibian, the embryos readily take up low-molecular-weight materials and macromolecular metabolism is easily studied.

Generally, one must attempt to find the system that is best suited for the study at hand. Different embryos are obviously adapted to different experiments. If one is, for example, interested in developmental genetics, *Drosophila* is a likely choice, considering all the genetic background available on this one animal. If lampbrush chromosomes are desired, salamander oocytes are a good choice because of the size of these structures in these cells. Once a particular group of animals becomes widely studied, a background of information is built up that makes further study more profitable. The biggest difficulty with this approach is our tendency to generalize findings to all embryos when we have information about only a very few.

Mammals are very difficult to work with. They are hard to keep, their embryos are small and require very special conditions, even for short periods, and only a

limited number of embryos can be obtained. One female is capable of providing only about 50 embryos after special hormone treatment, compared with the millions obtained from a large sea urchin. This makes it very difficult to perform biochemical studies on the early, small stages. It is a fairly safe assumption that if humans were not mammals, we would know next to nothing about mammalian embryonic development. At the molecular and cellular levels, however, great similarities can be observed between the frog and sea urchin egg and that of a human. Large numbers of studies performed on the eggs of lower animals provide concepts of development that can then be tested in a defined way on embryos closer in evolution to ourselves.

1. Strategies of reproduction

Certainly one of the traits most strongly affected by natural selection is the ability of a population to fertilize its eggs. If the number of offspring that survive to reproduce diminishes, the future of that population is in obvious danger. The number of diverse, and often bizarre, reproductive strategies that have evolved illustrates this point. A species usually produces the number of eggs required to ensure its survival. In the case of some mammals, only one egg is produced at a time. To compensate for this drastically low fecundity, the sole egg is nurtured in the womb and protected after birth. On the other extreme are animals such as parasites, many of which produce thousands of eggs per day. The chance for reproductive success is very remote, and selection has favored high fecundity in these groups.

Fertilization requires the union of two cells from two different individuals, and mechanisms have evolved to ensure that the two will be present in the same place at the same time. For many animals, including mammals, internal fertilization clearly satisfies that role. Even here there is room for improvement, and some mammals, including the prolific rabbit, delay the process of ovulation until copulation. This provides the necessary trigger for ovulation and virtually ensures fertilization. In many animals having external fertilization, complex physiological mechanisms have evolved to deliver large quantities of spermatozoa close to numbers of ripe eggs at the right time. The most primitive situation is probably that of many marine invertebrates, including the sea urchin, where communities of sexually mature adults shed (spawn) eggs and sperm freely into the surrounding water. The members usually become ripe in unison under the stimulation of common environmental cues of light and dark, tidal changes, and temperature of the water. To provide for simultaneous spawning, chemicals present in the fluid spawned by members of one sex will often trigger release of the gametes of the opposite sex. Since these animals cannot see, touch, or hear each other, diffusible chemicals must provide the communication.

This phenomenon can readily be verified by a visit to a marine laboratory during the reproductive season, where one animal spawning in a tank will promptly set all the remaining members of the species into action. One of the most dramatic examples of reproductive timing is the palolo worm of Samoa. In these worms the posterior half develops ovaries or testes; this half then breaks off, swims to the surface of the water under the moonlight, and eggs and sperm are shed. Under the influence of photoperiod, tides, and neuroendocrine factors, the exact date when these worms reproduce can be predicted (a particular night in November after the last quarter of the moon).

Another factor to be considered in the timing of reproductive activity between the two sexes is the limited life span of both gametes. Sea urchin sperm, for example, will remain fertile if kept in a concentrated state, in the cold, for more

than a day. Once the sperm cells are diluted, however, their respiratory activity greatly increases and their life span is reduced to minutes or less. Similarly, sea urchin eggs will remain fertile for a few hours, but then rapidly deteriorate. Animals that live in freshwater face more severe problems as a result of the shortened life of spermatozoa in solutions of low salt. In these animals, including fish, amphibians, and invertebrates, sperm cells are delivered directly to the eggs at the moment of laying, if not before. In animals living on land, spermatozoa are generally stored in a physiological medium that maintains their life. In many forms, spermatozoa must be delivered internally via the insertion of a sexual appendage. Yet even in mammals, millions of sperm cells must be ejaculated in order that a sufficient number reach the upper end of the fallopian tube, where one or more ripe eggs may be descending. In mammals, as in other animals, the life of the gamete is very limited. In humans, sperm cells remain fertile for about 2 days, and the egg is fertile for no longer than 24 hours. In rats, it is well established that the incidence of abnormal pregnancies rises sharply with the time elapsed between ovulation and fertilization. In humans, one effect of fertilization of aging eggs is believed to be the production of triploid embryos, which account for approximately 20 percent of abortive pregnancies.

2. Egg envelopes

Eggs are surrounded by a variety of extracellular covers with various names, origins, and functions. In an attempt to categorize these outer covers, we will divide them into structures formed (1) within the ovary, which can result from secretions of the oocyte, surrounding follicle cells, or both, and (2) within the oviduct. During the course of this chapter we will discuss the vitelline layer and jelly coat of the sea urchin, the zona pellucida of the mammal, the chorion of a fish, the vitelline layer of an amphibian, and the chorion of an ascidian. All these are examples of structures formed within the ovary, produced by the egg and/or follicle cells.

The oviduct is responsible for the production of such materials as the jelly in which amphibian eggs are wrapped and the majority of the contents of a chicken egg. In the case of the chicken, the product of the ovary, the ovum, consists of the yolk and a small area of cytoplasm on top of it, into which the sperm will enter and from which the embryo will begin to emerge. Surrounding this large cell is a thin membrane, the vitelline membrane, formed within the ovary. After fertilization, the egg passes through the oviduct and receives a covering of albumin (the egg white), the shell membrane (the thin membrane just beneath the shell), and the shell itself. Other examples of oviduct-derived envelopes include the shell membranes of reptile eggs and the egg cases of mollusks and certain fish.

Whatever functions are served by external coats, jellies, or cell layers, these structures often present barriers to fertilization of the egg by spermatozoa, and special enzymes must be incorporated into the sperm acrosome to facilitate penetration. These *egg-membrane lysins* have been extracted from the sperm of a variety of invertebrates and vertebrates, including humans. In the mammal, when the oocyte is released from its follicle (see Fig. 4.25) at ovulation, it is surrounded by the zona pellucida and several layers of follicle cells. Although the topic is a controversial one, it is believed that the acrosome contains at least two enzymes that allow it to reach the egg. One is a hyaluronidase, which is capable of digesting the material that holds the follicle cells together, thereby allowing the sperm to pass between the cells. The other is a proteolytic enzyme, termed *acrosin,* which has been postulated as the means by which the spermatozoon can burrow through the zona pellucida. This protease appears to be stored in an inactive

form, *proacrosin,* until just prior to sperm activation (see subsequent discussion).

3. State of egg at fertilization

Eggs of different species are typically fertilized at different stages during the maturation process. As previously described, the eggs of vertebrates generally proceed to the second meiotic metaphase before sperm penetration. In the sea urchin, both meiotic divisions are completed before fertilization. At the other extreme there are animals whose oocytes have not begun meiosis at the time of fertilization. In these cases, the sperm nucleus remains in the cytoplasm awaiting the maturation process before the subsequent events can occur. Table 5.1 lists representatives that are fertilized at the various stages.

Table 5.1 Stage of egg maturation at which sperm penetration occurs in various animals

Young primary oocyte	Fully grown primary oocyte	First metaphase	Second metaphase	Female pronucleus
Brachycoelium	*Ascaris*	*Aphryotrocha*	*Amphioxus*	Coelenterates
Dinophilus	*Dicyema*	*Cerebratulus*	Most mammals	Echinoids
Histriobdella	Dog and fox	*Chaetopterus*	*Siredon*	
Otomesostoma	*Grantia*	*Dentalium*		
Peripatopsis	*Myzostoma*	Many insects		
Saccocirrus	*Nereis*	*Pectinaria*		
	Spisula	Ascidians		
	Thalassema			

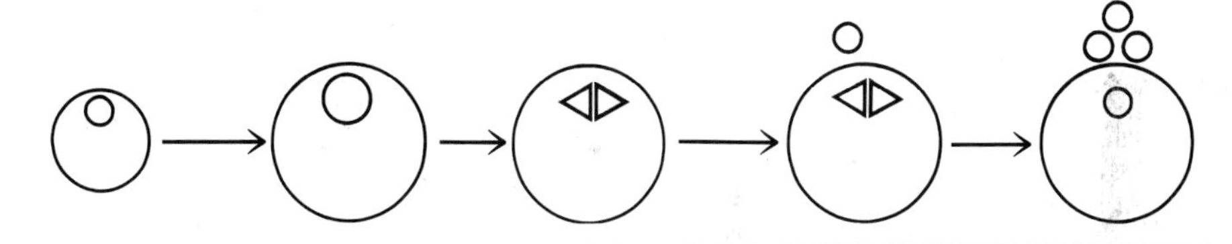

Source: C. R. Austin, "Fertilization," © 1965. Reprinted by permission of Prentice-Hall, Inc. Englewood Cliffs, N.J.

4. Chemotaxis versus trap action

If the eggs of a sea urchin, or any other animal, are placed on a slide and a tiny drop of a sperm suspension is added, in a short time a large number of sperm cells will have collected at the surface of the egg. The appearance of these events under the microscope suggests that the sperm cells have been attracted to the surface of the egg. If spermatozoa are being attracted to the egg, the condition would be termed *chemotaxis,* implying that the sperm cells were responding to some chemical being liberated from the egg or its surroundings. Since the egg is a sphere, this chemical would be expected to diffuse outward in all directions, thereby producing a gradient with its highest concentration at the egg surface and a diminishing concentration at greater distances from the source. If chemotaxis does occur, then the sperm cells (by utilization of some type of receptor system) must be able to respond to tiny differences in the concentration of the chemical to be directed to the egg surface by such a gradient. Another explanation can provide for the accumulation of spermatozoa at the egg surface, that of *trap action.* In the case of trap action, the spermatozoa would swim in a nondirected random manner, but those happening to reach the egg surface would become

trapped and remain. An analogy might be made to a group of people wandering randomly in the dark on a large plot of land that has a small well in the center of the area. By the break of day, one would likely find a number of unfortunates at the bottom of the well, and chemotaxis could not be blamed for their fate.

The best way to determine if spermatozoa are attracted to an egg is to track individuals as they swim. If the sperm cells are moving randomly, they will move away from the egg as often as they move toward it. For many years the conclusion had been that chemotaxis was present in ferns, liverworts, and mosses, but was absent among animals. As a result of a number of more recent investigations, however, several examples of chemotaxis in animals have been uncovered, and many more are likely to be found. In the herring, as is typical of fish, there is a small opening, termed the *micropyle* (see Fig. 5.19), in the thick outer cover, or *chorion*, of the egg. It is through the micropyle that the fertilizing sperm cell must enter, and if spermatozoa in the vicinity of the opening are watched, they are seen to speed up and to be directed through the micropyle. In several coelenterates, such as the hydroid *Campanularia*, the eggs are kept within a vaselike structure, the *gonangium* (Fig. 5.1). If spermatozoa are tracked, they are seen to be directed through the opening of the gonangium and down toward the egg in response to the diffusion of a specific molecule from the egg. In hydroids, these substances are of relatively low molecular weight and possess some species specificity; i.e., spermatozoa of closely related species do not respond to the same degree as the homologous sperm. Other cases of chemotaxis have been found in tunicates and chitins, suggesting its occurrence might be widespread.

5. The response by spermatozoa to the egg surroundings

Regardless of how it reaches the vicinity of an egg, an approaching sperm finds itself in a region capable of exerting great influence over it. Investigations into this subject date back to 1913, when F. R. Lillie found that sea urchin spermatozoa clump together if dropped into a solution of seawater in which eggs of the same species had been shed. This solution, termed *egg water*, contains macromolecules released from the jelly coat of the egg. Lillie called this substance *fer-*

Figure 5.1 The female gonangium of a coelenterate *C. flexuosa*. *a* The ovum (O), waiting to be fertilized, is retained within the body of the gonangial structure. *b* Plotted tracks of *C. flexuosa* sperm attracted to the distal end of a mature female gonangium of *C. flexuosa*. [*From R. L. Miller,* J. Exp. Zool., **162**:*27, 31 (1966).*]

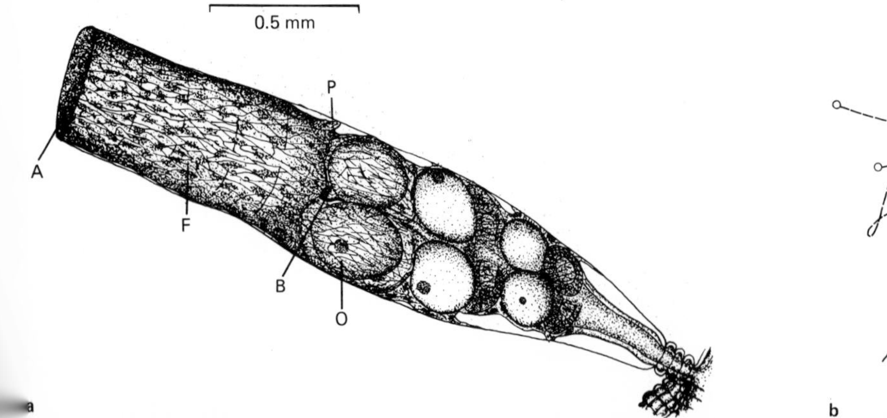

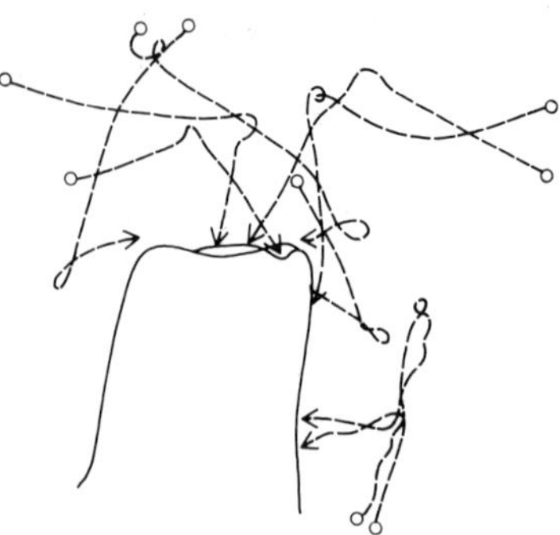

tilizin, and later studies suggested that it accomplished its agglutination of homologous spermatozoa by interaction with a sperm surface protein termed *antifertilizin*. The reaction was thought to occur in a manner analogous to that between antigen and antibody. A great deal of research has been performed on the nature of the fertilizin-antifertilizin reaction and its function in fertilization, although very few firm conclusions have been generated. Functions attributed to fertilizin have included the activation of the sperm, the blockade of foreign sperm, and the attachment of the sperm to the egg. For various reasons (see Collins, 1976, and Metz, 1978), the fertilizin-antifertilizin concept has been largely abandoned and replaced with a more molecular analysis of the early events of gamete interaction, as discussed later.

As a sperm cell approaches an egg, a striking transformation of the anterior region of the sperm head occurs in response to material in the egg environment. This response, termed the *acrosomal reaction*, is invariably found to be a prerequisite for the fertilization of an egg by a given sperm. Early studies of the acrosomal reaction and of early events of fertilization were performed on invertebrates, particularly on the hemichordate *Saccoglossus* and the polychaete *Hydroides*. More recently, these events have been described for several mammals, and many of the basic processes are quite similar. We will first discuss the overall process in *Hydroides* (Figs. 5.2 and 5.3) and then reexamine the events at a more molecular level in the sea urchin and mammal.

In *Hydroides*, the reaction begins when the sperm tip contacts the outer egg envelope. The first response is a fusion of two sperm membranes, the outer cell membrane of the sperm head and the anterior membrane of the acrosomal vesicle (Figs. 5.2*b* and 5.3). Once membrane fusion has occurred, the posterior wall of the acrosomal vesicle is seen to form a number of rodlike projections (Fig. 5.2*b* and *c*). These grow in length (Fig. 5.2*d*) and make the first actual contact with the plasma membrane at the egg surface (Fig. 5.2*e*). Again a process of membrane fusion occurs, but this event involves the acrosomal membrane of the sperm and the plasma membrane of the egg. At this point we can consider the two gametes as one cell, the *zygote*. This latter fusion event provides an open

Figure 5.2 Stages of sperm-egg association (of polychaete *Hydroides*). *a* Unactivated spermatozoon at about the time of initial contact with the egg envelope. *b* Acrosomal reaction is beginning. The plasma membrane of the sperm and the acrosomal membrane have fused to one another to provide an outlet for the contents of the acrosome. *c* Contents of the acrosome (including lysins) are being released and the posterior wall of the acrosomal vesicle is beginning to evert to form acrosomal tubules, which will make the first contact with the egg. *d* Acrosomal tubules are leading sperm penetration through the outer egg covering. *e* Acrosomal tubules are initiating contact with the egg plasma membrane, which forms microvilli in response. *f* Plasma membranes of the sperm and egg have fused to form the zygote. *g–i* Successive stages in the engulfment of the sperm by the fertilization cone formed by the egg in response to sperm contact. [*After A. L. and L. H. Colwin, in C. B. Metz and A. Monroy (eds.), "Fertilization," vol. 1, Academic, 1967.*]

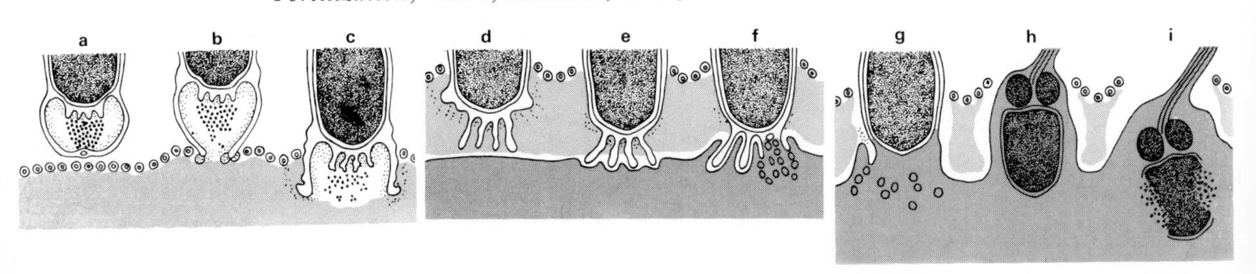

Figure 5.3 *a* Electron micrograph of stage of initial contact of sperm and egg of *Hydroides*. Sperm plasma membrane meets egg envelope and ruptures, permitting the interior of acrosomal vesicle to open to outside. Egg plasma membrane is still separated from sperm cell by bulk of egg envelope. *b* Electron micrograph of contact stage of sperm and egg of *Hydroides*, showing acrosomal tubules indenting egg surface, and the beginning of the rise of the fertilization cone. The two gametes plasma membranes closely confront each other in the region of the interdigitation. [*Courtesy of A. L. and L. H. Colwin.*]

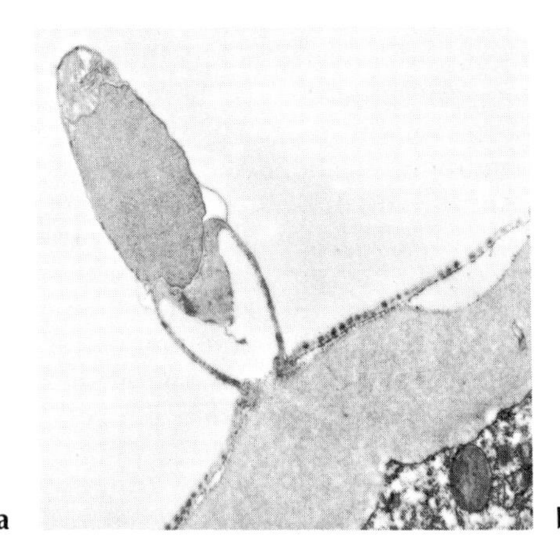

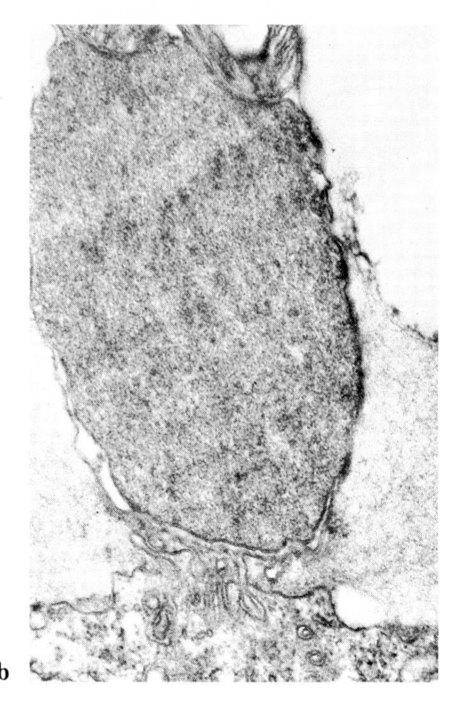

channel to the interior of the sperm head (Figs. 5.2*f* and *g*), and the male nucleus and components of the midpiece gain entrance to the egg cytoplasm.

In the hemichordate *Saccoglossus*, as well as in the river lamprey, a variety of echinoderms, and certain bivalve mollusks, the acrosomal reaction involves the formation of one long acrosomal tubule rather than the several small ones seen in *Hydroides*. The formation of the acrosomal tubule, which occurs in an explosive manner, results from the polymerization of monomeric actin molecules into the polymerized filamentous state (Fig. 5.4). This polymerization reaction is triggered, apparently indirectly, by a component in the egg jelly coat. In the sea ur-

Figure 5.4 *a* Electron micrograph of unreacted sperm showing the potential site for membrane fusion where the acrosomal and plasma membranes are in close apposition. *b* Sperm that has undergone the acrosomal reaction as triggered by egg jelly at pH 8.75. Filamentous material, extending from the nuclear apical fossa, occupies the core of the process. [*Courtesy of G. L. Decker, D. B. Joseph, and W. J. Lennarz.*]

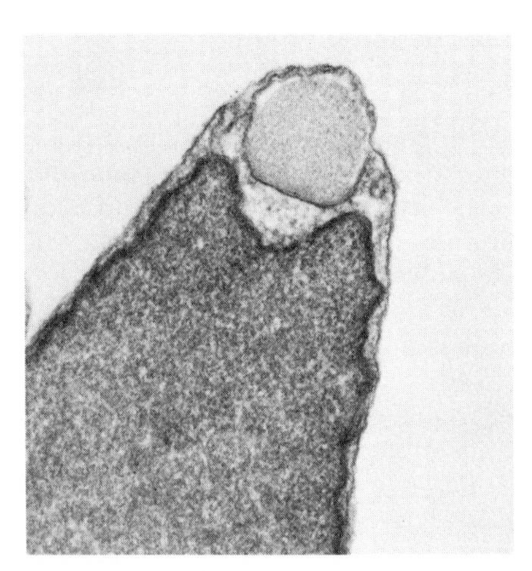

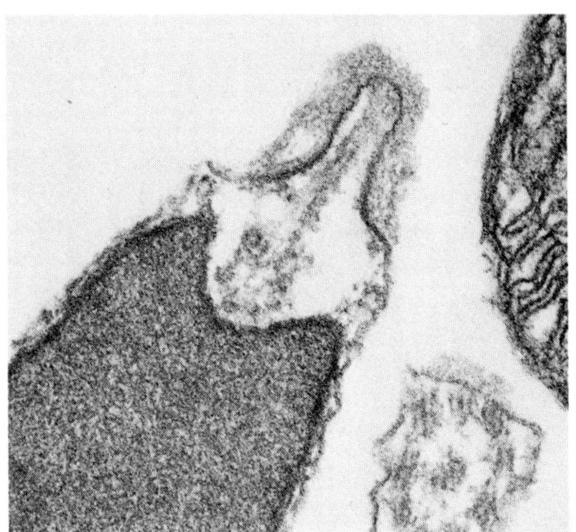

chin, the jelly of the egg consists of two distinct macromolecular components, one a polysaccharide composed almost entirely of polymerized fucose sulfate and the other a sialoprotein, i.e., a glycoprotein containing large amounts of sialic acid (41 to 72 percent of the molecule). The fucose sulfate polysaccharide is believed responsible for the initiation of the acrosomal reaction, presumably as a result of an interaction with the sperm plasma membrane. This surface interaction is coupled to the acrosomal reaction itself via an increased permeability of the sperm plasma membrane toward calcium ions. In fact, the acrosome reaction can be initiated in the absence of jelly material by the introduction of the calcium ionophore A23187, a substance which specifically binds calcium and transports it across membranes. In this case calcium ions are moved from the seawater into the sperm cell across the plasma membrane. The presence of calcium ions in the medium is an absolute requirement for the initiation of the acrosomal reaction by either egg jelly or the ionophore. The need for calcium ions is believed to center on the processes in the spermatozoa that lead to membrane fusion. As a consequence of their lipid bilayer, certain membranes possess the capacity to fuse with one another to become continuous structures. Although the underlying basis for membrane fusion is poorly understood, this phenomenon is commonly found to be the means by which cells release membrane-bound secretions.

Membrane fusion involves (1) the interaction between specialized regions within the opposing membranes, regions that can be visualized in freeze-fracture replicas (see Satir, 1975); (2) the introduction of instability in the lipid bilayers, possibly by the formation of altered phospholipids (lysophospholipids that lack one of the two fatty acid chains typically present in membrane glycerides); and (3) calcium ions. Membranes derived from Golgi vesicles, as in the case of the acrosome, seem to be particularly able to fuse with the plasma membrane. In one recent study on the sperm of *Limulus*, the horseshoe crab, freeze-fracture replicas revealed the presence of corresponding rings of particles in both the plasma membrane and the membrane of the acrosomal vesicle at the site of the upcoming fusion. It is suggested that the corresponding arrays of particles serve to hold the two membranes very close together as well as maintain an instability in the pool of lipids within the circle. It is within the particle-free lipid space that subsequent fusion actually takes place. It is believed that it is the sudden increase in the Ca^{2+} concentration within the sperm head that converts this potentially unstable membrane region into a site of active fusion. In addition to the formation of the acrosomal tubule, the opening of the acrosomal vesicle via fusion of the acrosomal and plasma membranes leads to the release into the medium of various hydrolytic enzymes that had been stored in the acrosomal vesicle. These enzymes are used by spermatozoa in penetration of the egg surface envelopes and possibly as an aid in promoting contact with the egg plasma membrane itself.

Although Ca^{2+} may provide the stimulus for membrane fusion, it is not sufficient to bring about the formation of the elongated acrosomal tubule. Studies of various systems suggest that the polymerization of actin, i.e., the transformation of monomeric G-actin to polymeric F-actin, occurs upon dissociation of a binding protein from actin monomers. It is proposed that the binding protein serves to maintain the actin in its depolymerized state. The dissociation of the actin-binding protein is believed to result from a rise in the internal pH of the sperm head, which, in turn, results from an efflux of protons out of the sperm across its plasma membrane.

Once spermatozoa have been activated by contact with egg jelly material, their respiratory activity is greatly elevated and they become highly motile.

Spermatozoa can be maintained in this activated state for only a short period: if they do not reach the egg surface in a minute or so, their fertilizing capacity is greatly diminished.

In mammals, the acrosomal reaction is only the last alteration in sperm morphology that allows it to penetrate the egg. The spermatozoa that leave the testis proper are at least two steps from being fertile. One maturational step occurs in the tubules of the epididymis, while a second step, termed *capacitation*, occurs within the female reproductive tract. Capacitation, which can occur in vitro in a completely defined medium, involves the removal of some type of inhibiting material from the outer surface of the spermatozoa. Marked changes in the distribution of plasma membrane particles also have been noted after capacitation. One unusual experimental technique for rendering mammalian spermatozoa fertile is to mix them with Sendai virus, which causes them to agglutinate head to head (Fig. 5.5). Sendai virus is a common tool to bring about the fusion of membranes between two cells, and in this case the altered spermatozoa are now capable of penetrating the ovum.

In contrast to invertebrates, the acrosomal reaction of the mammalian spermatozoa involves the fusion of two membranes, the plasma membrane covering the sperm head and the acrosomal membrane, at several sites along the length of the sperm (Fig. 5.6). Consequently, membrane fusion results in the formation of a shroud of vesicles at the tip of the sperm and the early release of the acrosomal contents to digest the intercellular cement of the follicle cells. No acrosomal tubule is formed. By the time the sperm cell reaches the zona pellucida, there is little evidence of any remaining acrosomal material, yet the penetration of the zona is believed to require the activity of a proteolytic "zona lysin." It may be that this enzyme, termed *acrosin*, is present in association with the inner acrosomal membrane, which, after the acrosomal reaction, is present at the sperm surface. Penetration of the zona requires several minutes and leaves the sperm cell in virtual contact with the microvilli of the egg surface.

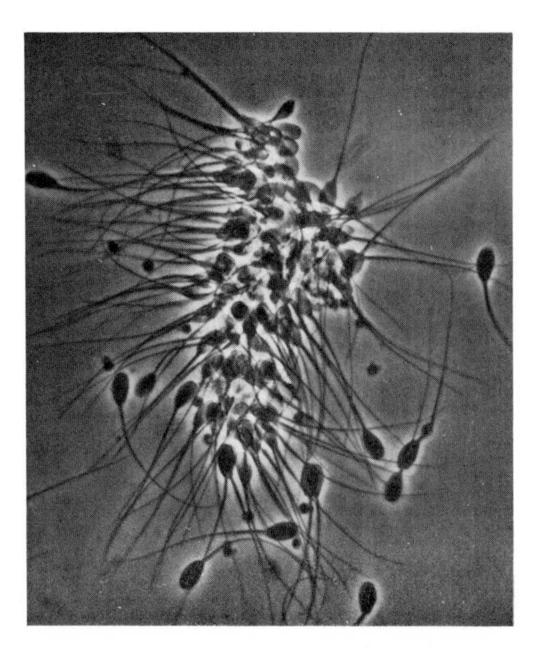

Figure 5.5 Head-to-head agglutination of live epididymal rabbit sperm after being mixed with Sendai virus. Fertilization occurs when rabbit ova are cultured in vitro with epididymal sperm into which Sendai virus is adsorbed. These sperm do not require capacitation in vivo in order to fertilize. [*From R. J. Ericsson, D. A. Buthala, and J. F. Norlund,* Science, **173**:54 *(1971); copyright © American Association for the Advancement of Science.*]

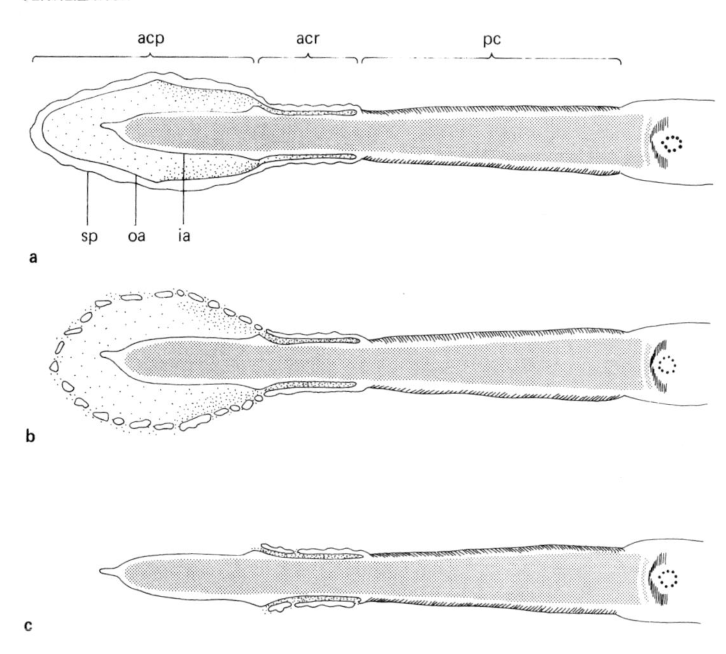

Figure 5.6 Diagrams of acrosomal reaction in capicitated hamster sperm. Before *a*, during *b*, and after *c* acrosomal reaction: acp, acrosomal cap region; acr, acrosomal collar region; ia, inner acrosomal membrane; oa, outer acrosomal membrane; pc, postnuclear cap region; sp, plasma membrane of the spermatozoon. [*From R. Yanagimachi and Y. D. Noda,* Am. J. Anat., **128**:*480 (1979).*]

In contrast to invertebrates, membrane fusion between mammalian gametes does not occur at the anterior tip of the sperm surface, but rather at a more posterior location in the region of the *postnuclear cap* (Fig. 5.7). Presumably those membranes which undergo the fusion process—regardless of which they happen to be—have acquired a special fusion-mediating property. In the sea urchin, the spermatozoa have been found to carry an enzyme (a phospholipase) capable of removing one of the fatty acids of membrane phospholipids. As mentioned earlier, the conversion of phospholipids to lysophospholipids (ones lacking one of the two fatty acids) introduces an instability into the lipid bilayer that has been shown to lead to the fusion among artificial phospholipid bilayers.

In every species that has been studied, the acrosomal reaction of the spermatozoa is a prerequisite for fertilization. Even though a nonreacted sperm cell may adhere to an egg, it will not lead to its fertilization. In the sea urchin, spermatozoa have been injected directly into the egg, bypassing the surface-mediated events. These injected spermatozoa do not activate the egg, but remain intact and mobile within the egg cytoplasm. In this case, at least, membrane fusion is the only means by which gametes can be united.

6. The response by the egg to sperm contact

Although the spermatozoon is a tiny cell relative to the gigantic egg, its imprint at one point on the egg surface releases an activation response within the egg of tremendous magnitude. The nature of this preprogrammed response has come under intensive investigation in recent years, and a great deal of progress toward an understanding of the events of fertilization has been made. The response by the egg to sperm penetration includes a number of diverse activities, some of which can be seen with the aid of a light or electron microscope, others of which require the use of various types of physiological probes. Fortunately for their study, the various events occur in a highly regimented manner, each at its own

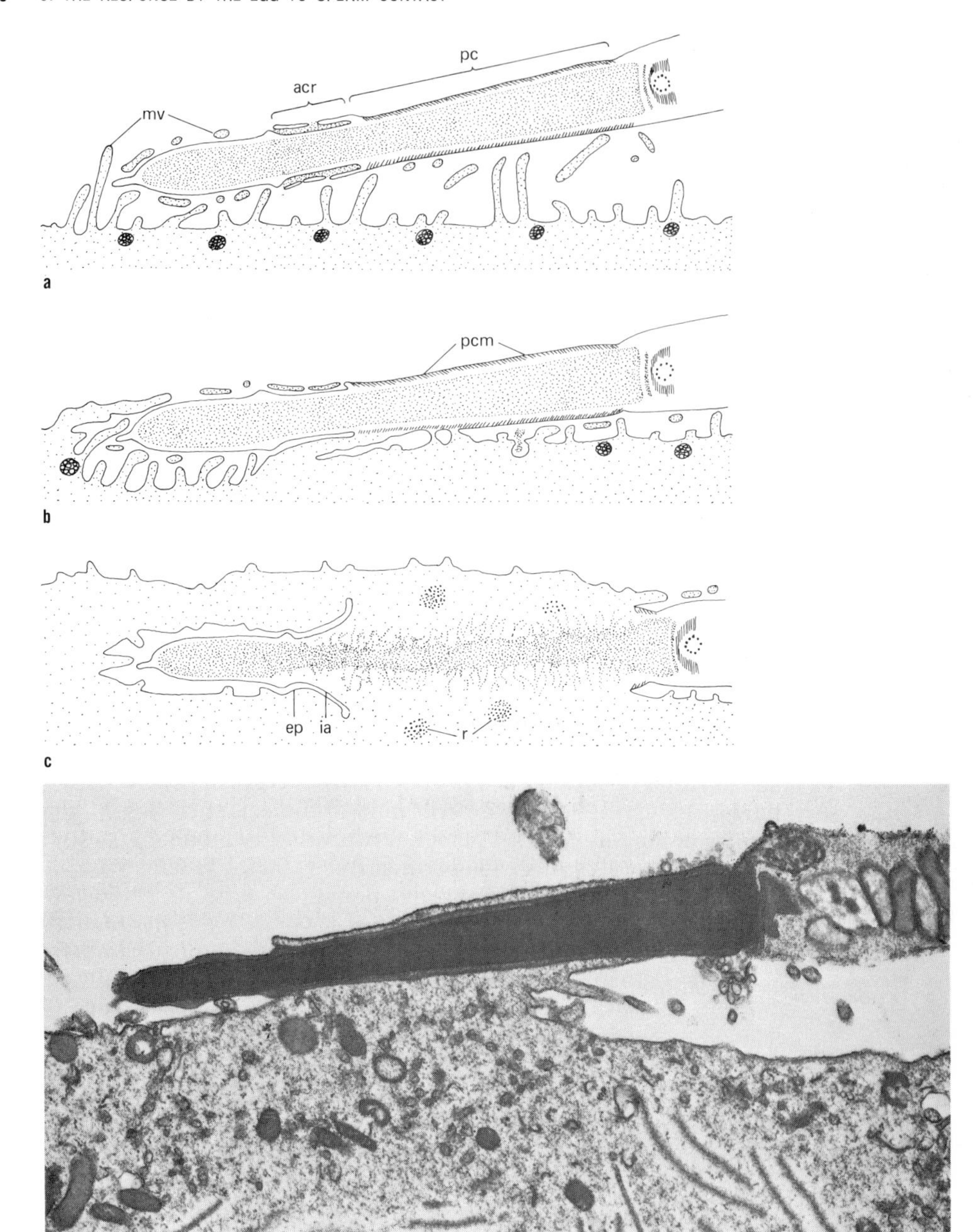

Figure 5.7 *a–c* Diagrams of the incorporation of the hamster spermatozoon into the egg. *a* A spermatozoon is trapped by egg microvilli. *b* Fusion of the egg plasma membrane (microvilli) and the plasma membrane of the postnuclear-cap region of a spermatozoon and simultaneous dislodging of the outer acrosomal membrane of acrosomal collar region of spermatozoon. Cortical granules of the egg have started to break down. *c* Inner acrosomal membrane of the spermatozoon is gradually separated from the sperm head as swelling of the sperm nucleus continues; acr, acrosomal collar region; mv, microvilli of egg; pc, postnuclear-cap region; pcm, postnuclear-cap material; cg, cortical granule; ep, egg plasma membrane; ia, inner acrosomal membrane; r, aggregates of particles. *d* Electron micrograph of a hamster spermatozoon in an early stage of incorporation into the egg. [*From R. Yanagimachi and Y. D. Noda,* Am. J. Anat., **128**:*429 (1970).*]

characteristic time after fertilization.[1] Any discussion of the response by the egg to sperm contact invariably centers around the echinoderm egg, since work on these animals has so dominated the field. In the sea urchin, the events of fertilization are generally divided into two groups: the "early events," which occur within the first minute or so, and the "late events," which begin at approximately 5 minutes and carry the zygote toward its first mitotic division. Some of these events are listed in Fig. 5.8 and will be discussed in the text that follows. One of the primary goals of investigators in this area of research is to understand the relationships between these seemingly independent activities. Which of the events of Fig. 5.8 can be considered primary changes that in turn lead to other secondary changes? In order to make such assignments we will have to try to dissect the postfertilization program and determine cause and effect.

Initiation of the fertilization response occurs when the surfaces of the two gametes make contact with one another. Important advances have been made in recent years in determining the nature of the complementary molecules present at the gamete surfaces which mediate their interaction. A protein has been isolated and purified from the acrosomal granules of sea urchin sperm that appears to constitute the egg-receptor molecule. This protein, termed *bindin,* has been used in the following types of experiments. Antibodies prepared against bindin attach to the membrane of activated sperm at the acrosomal process as well as to the vitelline layer of a fertilized egg at the site to which a sperm is bound. When homologous eggs, i.e., eggs of the same species as the sperm source, are incubated with a preparation of bindin, the eggs agglutinate to form clusters held together by linking bindin molecules. Success also has been achieved in the isolation of the sperm receptor present as part of the vitelline layer of the unfertilized egg surface. Several laboratories have isolated a large glycoprotein from the egg surface that has been used in the following types of experiments. Antibodies prepared against the molecule will bind to the egg surface and block fertilization. The glycoprotein itself will block fertilization by binding to activated sperm, presumably via attachment to sperm-surface bindin. Finally, it has been shown that agarose beads coated with this egg protein serve as a specific attachment site for activated sperm; beads coated with other proteins do not have this capacity.

Experiments with purified gamete receptors indicate that they function in a highly species-specific manner. Bindin will only agglutinate eggs of the same species, and the egg glycoprotein will only bind to sperm of the same species. These observations and others suggest that the receptors of the egg and sperm surface serve as the primary barrier against fertilization of the egg by the sperm of another species and account for the general (although not absolute) lack of cross-fertilization observed between gametes of different species.

An important sidelight to the study of sperm and egg receptors is the potential that this research provides in the development of new types of antifertility procedures. At the time of this writing, numerous laboratories are engaged in research aimed at the prevention of fertilization by the use of interfering antibodies. The best results to date have been obtained with antibodies produced against various egg extracts (as opposed to antibodies against sperm). For example, antibodies directed against antigens of the hamster zona pellucida are not only effective in blocking fertilization in vitro, but they also can be injected into an animal, where they serve to block fertility for several estrous cycles. This type of passive immu-

[1] The term *fertilization,* in this sense, refers to the contact and penetration of an egg by a spermatozoon. In a broader sense, the term refers to the entire sequence of events that lead up to the first mitotic division of the zygote, i.e., the topics of this chapter.

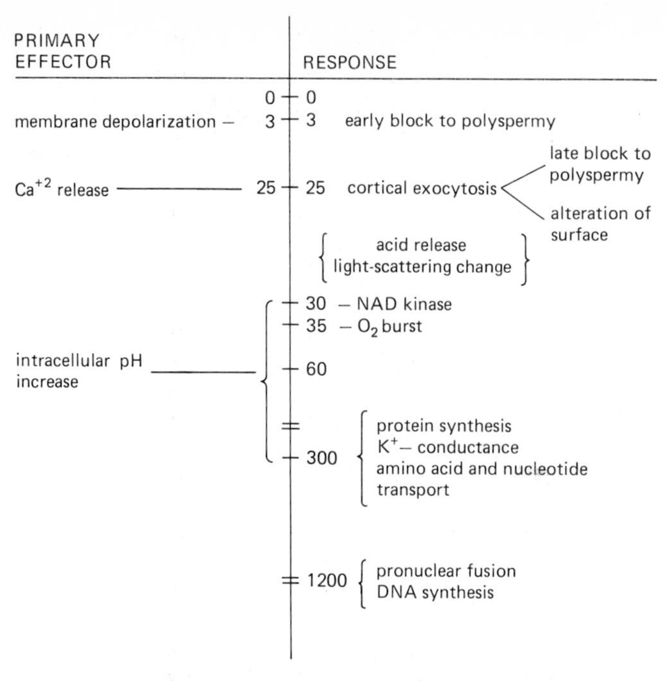

Figure 5.8 The program of changes that occur during fertilization of the egg of the sea urchin *S. purpuratus* at 17°C. There are believed to be three primary ion fluxes (*left side*) that serve to trigger the remaining changes. Time of various events are given in seconds following sperm contact. [*From diagrams of D. Epel.*]

nization, i.e., immunization by the use of antibodies produced in another organism (as in tetanus antitoxin, which is a horse serum), should prove to be safe, effective, and temporary. Most important, it takes advantage of the highly selective interactions between antigen and antibody. Consequently, as in the case of other vaccines, the only site of action of the antibody is the particular target antigen; no other immunological damage would occur.

A. Early events

The first visible response to sperm contact centers on the sperm cell itself. In the sea urchin, once contact between the gametes is made, the fertilizing sperm cell is seen to remain quite active for a period, rotating about its point of attachment. Then, suddenly at about 20 seconds, the fertilizing sperm stops its motion and, within a second or two, begins to be passively engulfed by a protrusion of cytoplasm termed the *fertilization cone* (Fig. 5.9). Although the first visible response occurs at the site of sperm entry, the entire surface of the egg is seen to be rapidly involved in the activities. If one examines the newly fertilized sea urchin egg under special dark-field illumination, a color change is seen to spread around the egg from the point of sperm penetration. Depending on the species and the temperature of the water, this wave begins at about 30 seconds after sperm contact and is completed approximately 20 seconds later. This wave of color change (a reflection of the cortical changes described later) is followed by the elevation of a membrane, the *fertilization envelope,* from the surface of the egg (Fig. 5.10). Membrane elevation also begins at the site of sperm contact, initially as a small

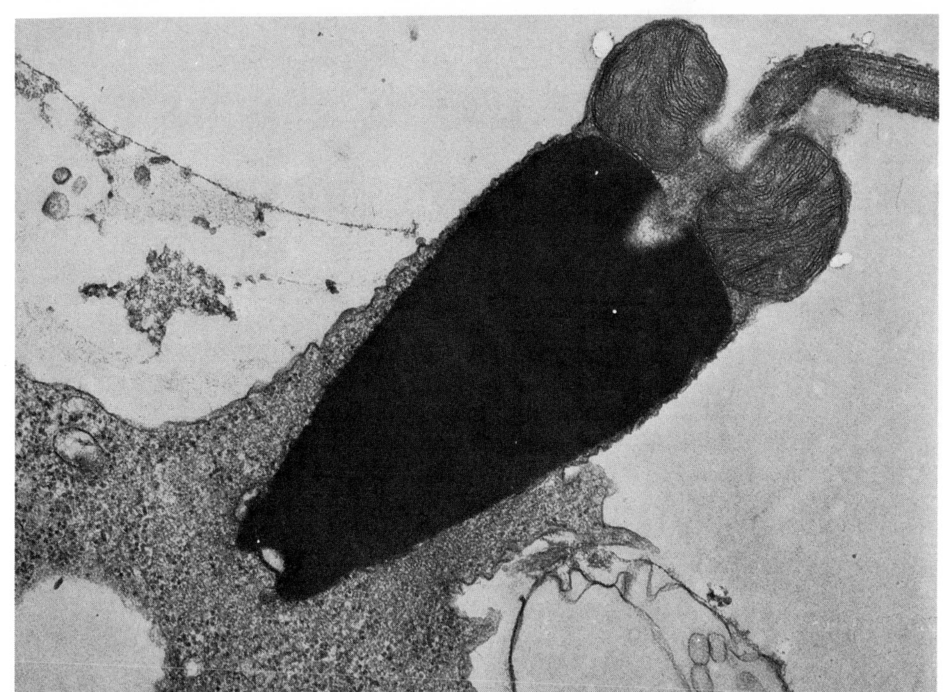

Figure 5.9 Electron micrograph of *Arbacia* sperm being taken into the egg through the fertilization cone. [*Courtesy of E. Anderson.*]

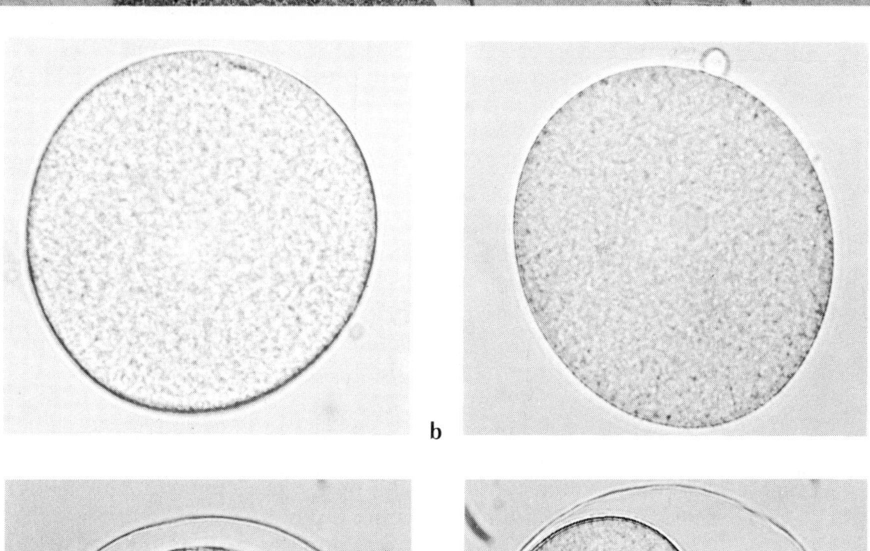

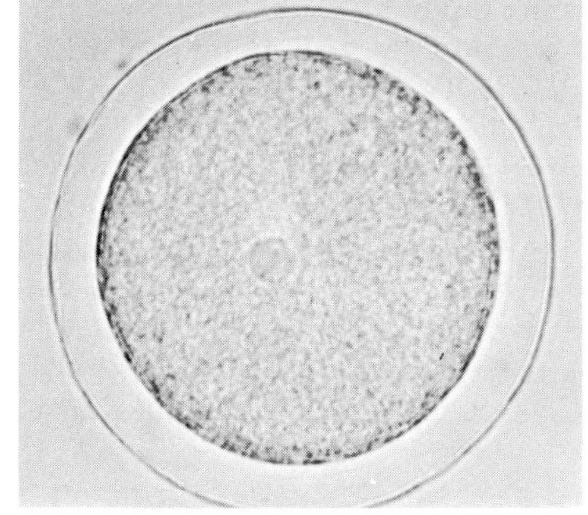

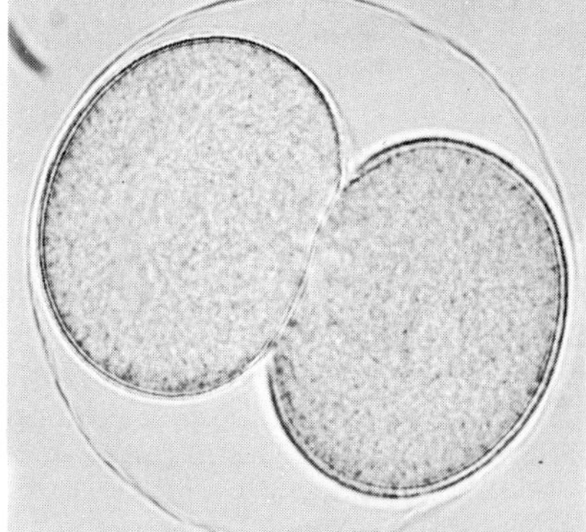

Figure 5.10 Maturation and fertilization in the egg of the sea urchin. *a* Egg showing metaphase of first maturation division. *b* Maturation complete, polar body evident, and pronucleus at center. *c* Fertilization membrane lifted off. *d* First cleavage. [*Courtesy of T. Gustafson.*]

blister which gradually enlarges to encircle the egg (Fig. 5.10*b*). This process is shown in Fig. 5.11*a*, and the completed envelope is shown in Figs. 5.10*c* and 5.14.

The morphological basis for this surface response is well established. The surface of the sea urchin egg represents a complex of components, each of which plays a role in the early response to sperm contact. The surface of the unfertilized sea urchin egg is composed of two distinct layers closely apposed to one another. There is an outer, thinner proteinaceous vitelline layer containing the sperm receptors to which the acrosomal process specifically binds. Beneath the vitelline layer is the plasma membrane, and just beneath the membrane and attached to it there exists a layer of granules, the *cortical granules* (Fig. 5. 11*b*), embedded in a cortical matrix believed to contain considerable amounts of actin present in an unpolymerized state. Within the first minute after fertilization, the morphology of the surface is completely altered. This alteration begins at the point of sperm contact by the fusion of the membranes of the cortical granules with the overlying

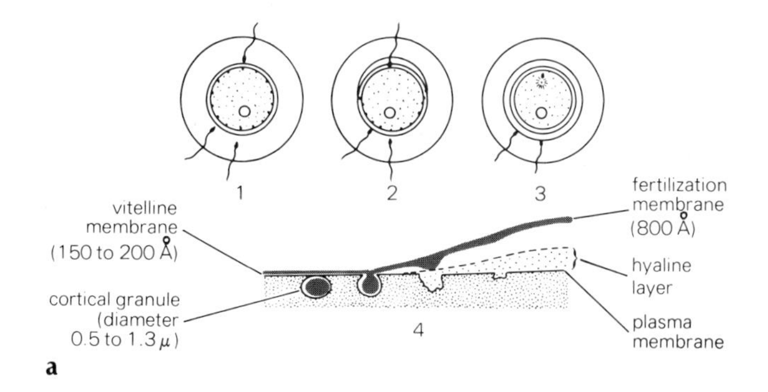

Figure 5.11 *a* Drawings 1 to 3 show the course of events in cortical-granule breakdown and fertilization-membrane elevation in the sea urchin egg. Drawing 4 illustrates the mechanism believed to underlie these events. [*After C. R. Austin, "Fertilization," 1965. Used by permission of Prentice-Hall, Inc., Englewood Cliffs, N. J.*] *b* Freeze-etch replica of an unfertilized egg of the sea urchin egg *Strongylocentrotus purpuratus*. In the upper part of the micrograph, the plasma membrane (pm) is seen in surface view extending between and over the surfaces of the regularly spaced microvilli (mv) that project from the egg surface. In the lower part of the micrograph, the fracture plane passes through the oplasm, outlining the outer contour of a cortical granule (cg). (×64,000) [*Courtesy of W. J. Humphreys.*]

plasma membrane (Figs. 5.11*a* and 5.12). As a result of this fusion, the membrane of the cortical granules becomes included in that covering the surface of the egg. These regions of newly inserted membrane can be seen in the scanning electron microscope as smooth patches interspersed among the original plasma membrane regions, which can be distinguished by the presence of microvilli (Fig. 5.13*a*). The fact that the surface of the egg retains this type of mosaic pattern for a period suggests that the lateral mobility of surface components is severely restricted.

Figure 5.12 Cortical reaction and fertilization-membrane elevation in the sea urchin *Clypeaster japonicus*. *a* Unfertilized egg. *b* Fusion of egg plasma membrane and membrane of cortical granule with beginning of release of granule contents. *c* Adhesion of electron-opaque material to the vitelline membrane now lifted up; complete fusion of this material with the membrane will give rise to the fertilization membrane. Other material of the cortical granules is expelled into the perivitelline space, while other contents remain close to the egg surface to form the hyaline layer. *d* The egg surface upon completion of the events. [*After* Y. *Endo,* Exp. Cell Res., **25:**383 (1961).]

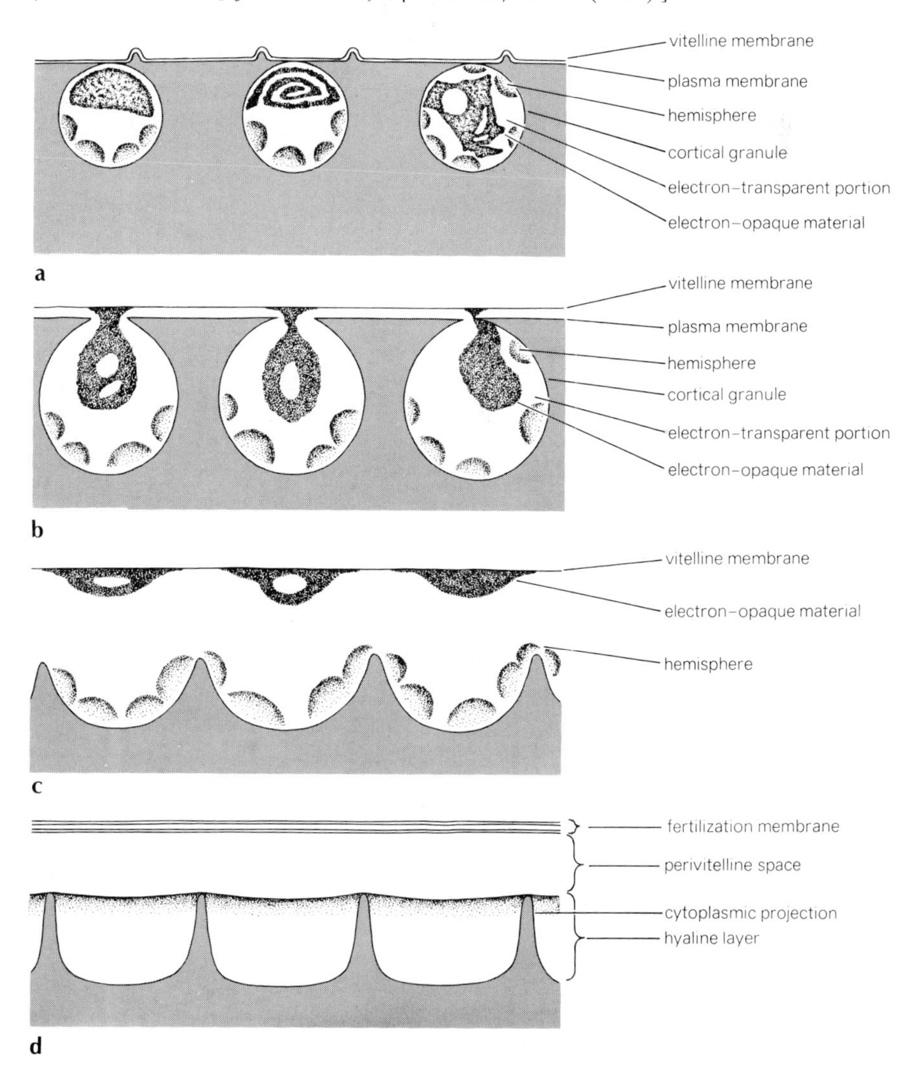

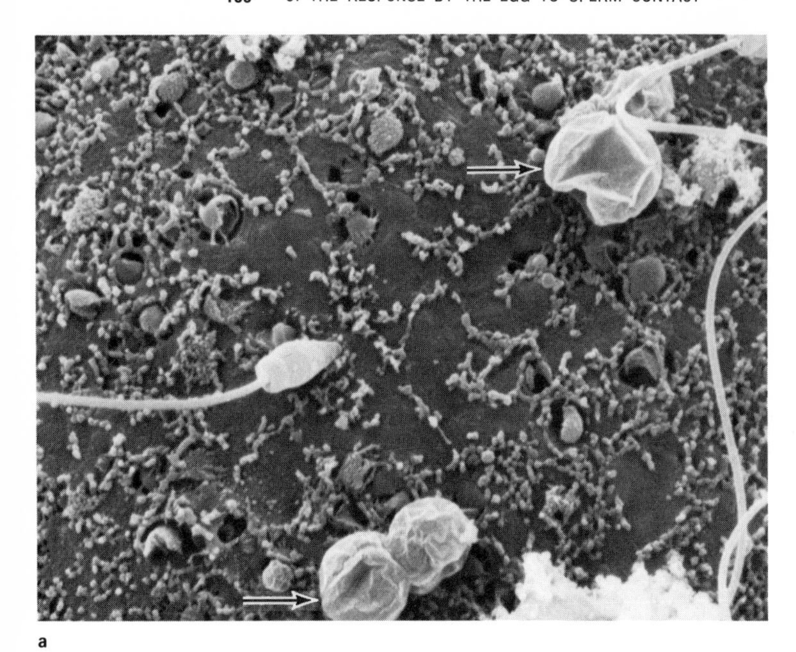

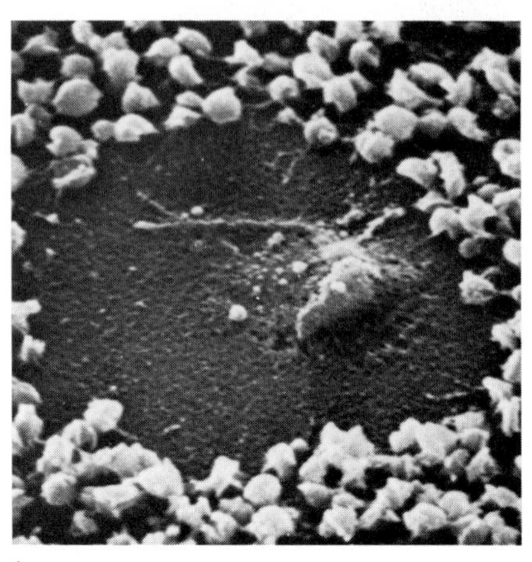

a b

Figure 5.13 *a* The mosaic surface of a sea urchin egg (treated to prevent elevation of the fertilization envelope) fixed 25 seconds after insemination. The fertilizing sperm is surrounded by patches of smooth membrane donated by the cortical granules and patches of microvilli representing the original egg surface membrane. Exocytosis of cortical granules (open pits) is in the process of occurring peripheral to this zone. [*Courtesy of E. M. Eddy and B. M. Shapiro.*] *b* The surface of a sea urchin egg of approximately the same stage of fertilization as that of *a*, but viewed from the *inner* surface of the plasma membrane. This type of preparation is obtained by sticking the outer surface of the egg onto a polylysine-coated plate and tearing away the contents of the egg, leaving only the cortical surface. The spermatozoon is seen to be in the process of penetration through the surface and exists within a patch of cortex that is devoid of cortical granules owing to previous rupture. [*Courtesy of G. Schatten and D. Mazia.*]

Cortical granule–plasma membrane fusion begins (Fig. 5.13*b*) at the site of sperm entry, causing the vitelline layer to locally detach from the plasma membrane and lift away from the egg surface forming a small blister at that point. The reaction of the cortical granules continues around the egg, leaving in its wake an elevated encircling fertilization envelope (Figs. 5.10*c* and 5.14). This wave of cortical-granule breakdown that sweeps around the body of the egg leads to dramatic changes at the egg surface. It was already mentioned that the cortical-granule membrane becomes part of the plasma membrane of the fertilized egg. This is a huge amount of surface (an estimated 18,000 cortical granules containing a total membrane surface greater than that of the original plasma membrane) that must be absorbed. It is believed that much of this extra membrane is taken up into the microvilli, which undergo a marked elongation in the first few minutes after fertilization. The lengthening of the microvilli appears to be aided by the polymerization of much of the monomeric actin present within the cortical cytoplasm (see Fig. 6.8).

So much for the membrane of the cortical granules, what of their internal contents? As cortical granules fuse with the plasma membrane, their contents are released by exocytosis into the space formed between the two previously apposed

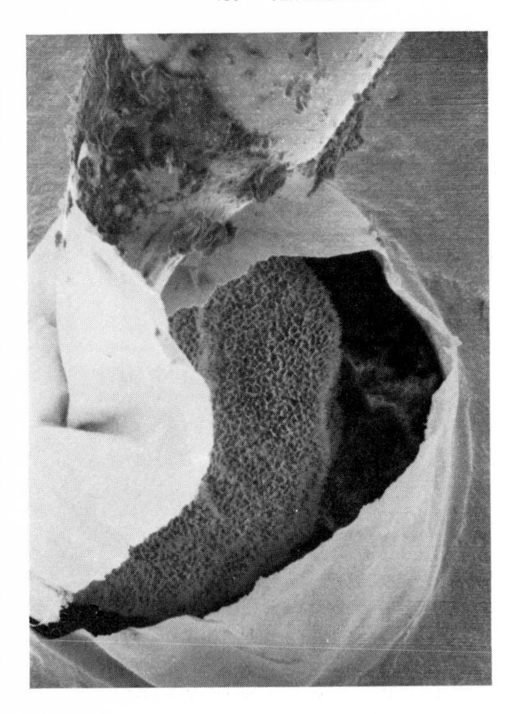

Figure 5.14 Scanning electron micrograph of a sea urchin egg 5 minutes after fertilization. The spatial relationship between the egg surface and the fertilization envelope (which has been ripped open with a needle) is apparent. [*Courtesy of E. M. Eddy and B. M. Shapiro.*]

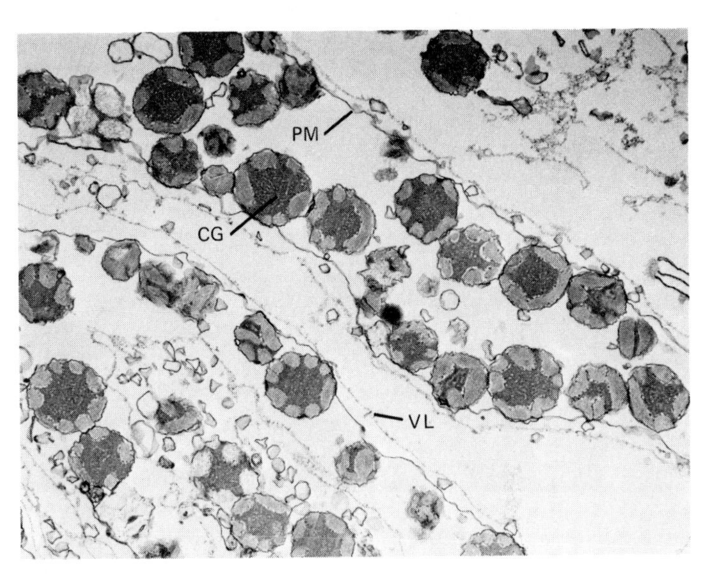

Figure 5.15 Electron micrograph of isolated cortices of an unfertilized sea urchin egg. The close association of the cortical granules (CG) with the plasma membrane (PM) is evident, as is the bipartite nature of the surface layer, which is composed of a plasma membrane and an overlying vitelline layer (VL). Isolated cortices of this type are capable of undergoing an activation response upon addition of sperm. [*Courtesy of N. K. Detering, G. L. Decker, E. Schmell, and W. J. Lennarz.*]

layers. This material extruded into the *perivitelline space* is used in several ways. A crystalline portion of the expelled contents is seen to fuse with the underside of the lifting vitelline layer forming lumps (Fig. 5.12) that soon smooth out to form the final hardened and insoluble fertilization envelope. Another portion of the cortical-granule material is seen to remain close to the egg surface, take up water, and form the *hyaline layer*. The formation of the hyaline layer, which consists primarily of glycosaminoglycan, requires the presence of calcium in the surrounding seawater. The hyaline layer forms an extracellular casing around the embryo to which the early blastomeres firmly attach (discussed further in Section 6.5). Still another contribution from the cortical granules consists of a variety of enzymes that function in the extracellular space (see Section 5.6.C).

Recently, techniques for the preparation of the outer crust, i.e., the *cortex* of the egg, have been developed. These isolated cortices, which are composed primarily of the layer of cortical granules attached to the overlying plasma membrane–vitelline layer complex (Fig. 5.15), remain capable of undergoing the cortical reaction. If spermatozoa are added to a preparation of this type, the sperm cells bind specifically to the outer surface of the vitelline layer and initiate a propagated breakdown of the cortical granules, which leads to the elevation of a fertilization envelope. All components needed for the cortical reaction are present within the cortex itself. The expulsion of cortical granule material at fertilization is found in many groups besides the echinoderms, including fish, amphibians,

and mammals. Many groups of animals, however, lack this phenomenon; even in sea urchins whose cortical-granule breakdown has been inhibited, normal development can result.

Considerable attention has been paid to the cortical reaction because it is the best studied early response during fertilization. However, the fact that it is not essential for normal development suggests that it is not a primary underlying event, but rather results from a more basic activation response. It has long been felt among researchers that the primary event in egg activation consists of some type of invisible, self-propagating, rapid "fertilization wave." The event most analogous to the proposed fertilization wave is the nerve impulse, and inherent in the theory is the concept that the wave would be accompanied by a change in the distribution of ions at the cell surface. Results of numerous studies in recent years have provided strong support for the existence of rapid changes in plasma-membrane voltage during fertilization. It has been found that the plasma membrane of the sea urchin egg undergoes a rapid depolarization beginning as early as 1 to 3 seconds after sperm contact. In one study, for example, the potential was found to change from approximately -60 millivolts (mV) (inside negative) to values as high as $+10$ mV within the first few seconds. This depolarized state is maintained for approximately 1 minute before the potential drops. This event (whose proposed role will be discussed later) results from an influx of Na^+ and represents the most rapid of the fertilization-related changes yet recorded.

The rapid sodium influx just described is not the only ionic flux to occur in the newly fertilized egg. Before discussing these other events, it is necessary to digress and consider the activation of an unfertilized egg without the intervention of a fertilizing sperm. It has been known for many years that various types of treatments applied to a variety of eggs are capable of experimental activation of the egg. If the normal trigger, the sperm, is bypassed, then the experimental treatment must in some way connect with the primary activation event(s), since the entire activation syndrome often results. It might be expected that the analysis of agents capable of causing activation would point toward the nature of the underlying reaction system within the egg. The difficulty with this line of reasoning becomes evident when one considers the great variety of treatments that are capable of providing the needed trigger. These treatments range from physical damage, such as the prick of a needle, to temperature shocks, to probably hundreds of chemicals, to salt solutions of different concentrations, etc. Clearly, the egg represents a preprogrammed system awaiting a stimulus that can be quite nonspecific in nature under experimental conditions.

It was mentioned previously (Section 5.5) that compounds called *ionophores* exist which are capable of transporting specific ions across membranes. The calcium ionophore A23187 has been used successfully in activating every type of egg on which it has been tested. As in the previous cases, these results strongly implicate the calcium ion in the events under study. Significantly, the activation of the echinoderm egg by A23187 is not dependent on the presence of calcium ions in the external medium. For example, eggs placed in artificial seawater that has been prepared without the addition of calcium salts (calcium-free seawater) are still capable of being activated by the ionophore (or fertilized by acrosome-reacted sperm). It is apparent that the ionophore carries out its action by entering the egg and causing the release of calcium ions into the cytoplasm from sites within the egg to which the ions had been bound.

The release of bound intracellular calcium within eggs soon after sperm contact has been revealed in a dramatic manner, initially in the large eggs of a fish and then in the smaller sea urchin egg. There is a protein, termed *aequorin*,

which luminesces when complexed with calcium ions. When a sample of this protein is injected into an egg, the level of luminescence provides a sensitive measure of the concentration of *free* Ca^{2+} within the egg cytoplasm. When eggs of the medaka fish are injected with aequorin and then fertilized, the eggs virtually light up: the glow following fertilization results from a 300-fold increase in calcium ion levels and is sufficiently bright to make a single egg visible in a dark room with the naked eye. Using methods of image intensification, the aequorin luminescence is seen to begin in the region beneath the micropyle and then travel down the egg as a peripheral band of light.[2] This transient "calcium wave" is believed to be self-propagated. The Ca^{2+} levels are initially raised at the site of sperm penetration, and this in turn stimulates the release of calcium from adjacent sites of storage. As the wave passes, the calcium ions are once again sequestered in some type of calcium sink.

Calcium ions have been intensively studied as the major factor in the initiation of exocytosis in cells throughout the animal kingdom. In keeping with this role, the release of calcium upon fertilization has been implicated as the stimulus responsible for initiation of the cortical reaction. The sensitivity of the cortical granules to a rise in calcium levels is best seen under a particular set of in vitro conditions. In this experiment, the bottom of a plastic culture dish is covered with a coating of protamine, a positively charged protein, and sea urchin eggs are then allowed to settle on the dish in the presence of calcium-free medium. After this procedure, one is left with the plasma membrane, inner surface facing up, and the attached cortical granules (Fig. 5.16). However, this "lawn" of cortical granules is maintained as such only in the absence of calcium. If calcium-containing medium is added to the dish, the granules immediately discharge their contents. Alternatively, if the cortical granules in the center of the dish are broken open mechanically, this triggers a chain reaction such that breakdown of granules radiates outward across the entire lawn as an expanding circle of transparency. It is believed that the breakage of the central granules releases sufficient calcium ions to promote the breakdown of adjacent granules; soon the stimulus is propagated across the entire lawn, much like it is propagated through the cortex of the newly fertilized egg.

The first changes, i.e., the depolarization of the membrane and the cortical reaction, are propagated changes at or near the cell surface. Soon, however, the entire egg is drawn into the process; somehow the activation must spread inward toward the center of the egg. There are indications of cytoplasmic activation even within the first minute. For example, a transient burst in oxygen consumption (associated with fatty acid oxidation and/or H_2O_2 formation) occurs within this period, as does the activation of a calcium-dependent enzyme NAD kinase, an enzyme that rapidly converts much of the egg's NAD stores to NADP (see Fig. 5.8). However, the major activation of the egg's metabolism is briefly delayed.

B. Late events

Beginning at approximately 5 minutes after fertilization, a second set of physiological changes can be detected in the sea urchin egg. Included in this group is the activation of a number of membrane transport systems (amino acids, phosphate, nucleosides), a slow hyperpolarization of the membrane (a result of K^+ conductance), an activation of protein synthesis (see Fig. 5.8), and the synthesis of DNA prior to the upcoming mitotic division. An important insight into the

[2] In these fish, the nonyolky cytoplasm is restricted to a thin cortical strip which sits on top of a large yolky mass.

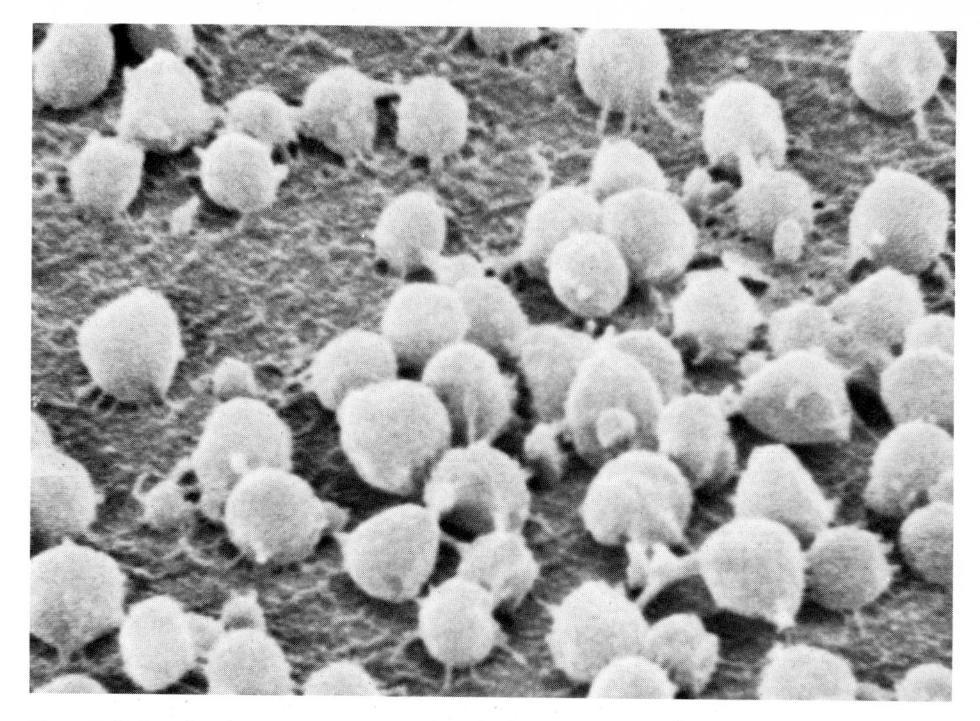

Figure 5.16 Scanning electron micrograph of cortical granules bound to the inner surface of the plasma membrane. A preparation of this type will undergo cortical-granule breakdown upon mechanical damage or the addition of calcium ions. [*Courtesy of V. D. Vacquier.*]

mechanism underlying this set of late changes was obtained when it was found that these events can be triggered in the absence of the early set of changes by treating unfertilized eggs with seawater to which ammonium hydroxide had been added. As a result of such treatment, protein synthesis is activated, as is DNA synthesis, chromosome condensation, and the subsequent splitting of the duplicated chromatids from each other. In other respects, however, the egg remains unfertilized. The addition of sperm to ammonia-treated eggs renders them "fertilized"; they undergo membrane depolarization, the cortical reaction, and the elevation of the fertilization envelope. The findings obtained with ammonia indicate that the late events can be dissociated from the early changes, i.e., that the initial membrane depolarization and subsequent cortical reaction are not essential underlying events leading to the initiation of the later series of changes. The presence of ammonia has somehow short-circuited the system, calling forth only the later changes. Is there any relationship between the ammonia effect and the normal events responsible for the activation of the late changes? There does appear to be a definite relationship, as will be evident from the following findings.

It was noted in earlier studies that sea urchin eggs release acid soon after fertilization. More recent measurements of intracellular pH with the use of microelectrodes have indicated that the pH of the cytoplasm increases approximately 0.4 units (from 6.84 to 7.27 in *Lytechinus pictus*), beginning approximately 1 minute after fertilization. In an independent series of experiments it was noted that the early development of the sea urchin egg is dependent on the presence of sodium ions. If these eggs are fertilized and kept in Na^+-free seawater (choline ions are substituted for sodium ions), neither the late changes nor subsequent de-

velopment occurs. Further analysis indicated that the need for Na^+ occurred very soon after fertilization, from approximately 1 to 5 minutes after sperm contact. For example, eggs could be fertilized in complete seawater and then transferred to Na^+-free seawater at 10 minutes and normal development ensued. The drop in pH and the need for Na^+ appear to be related by the presence in the membrane of an H^+/Na^+ exchange system which becomes activated soon after fertilization. The rise in intracellular pH results from the efflux of hydrogen ions, which are released *in exchange* for the Na^+ taken up from the medium. The situation is strikingly similar to that occurring during the acrosome reaction (Section 5.5). It has been shown that the important factor with respect to activation of the late events is the rise in pH (rather than the influx of Na^+). In fact, this is precisely how ammonia appears to initiate the late events; it raises the pH of the cytoplasm (either directly or indirectly) without requiring exchange with Na^+. The mechanism by which the rising pH initiates the late events remains obscure.

Considering the number of events that occur during the first minutes after fertilization, it is very difficult to assign causal relationships in an attempt to explain the underlying basis of activation. In the past few pages we have concentrated on the ionic fluxes because they are believed to be the most important in initiating subsequent responses. The rapid depolarization that results from the initial Na^+ influx is believed primarily responsible for establishing a rapid block to the penetration of the egg by additional sperm (discussed in the next section). Depolarization by itself does not appear to activate the egg, since the membrane can be experimentally depolarized with electrodes without causing a general activation. The calcium wave is believed directly responsible for the cortical reaction, as well as the other early events. The Na^+/H^+ exchange precipitates the late events. The manner in which these relatively independent sets of activities are normally linked to one another is less clear.

C. Blocks to polyspermy

In a variety of species (including the salamander described later), several spermatozoa routinely enter the egg, although only one male pronucleus participates in the formation of the cleavage nucleus. The penetration of the egg by more than one sperm is termed *polyspermy*. Where physiological polyspermy exists, it is usually found among animals having large, yolk-laden eggs. Polyspermy in its pathological form is incompatible with normal development. If two spermatozoa enter a sea urchin egg, bringing two centrioles for use in division, a quadripolar spindle apparatus generally forms and four cells are formed at first cleavage. If this were all, it might not matter. In such circumstances, however, the chromosomes of the nucleus formed by the fusion of three pronuclei are distributed in a most irregular manner, and no daughter cell has a normal complement. This irregularity is passed on to subsequent cell generations. Considering the extreme sensitivity of normal activities to a proper balance of individual chromosomes, it is not surprising that this condition results in a premature death for these embryos. Most, if not all, species therefore have some sort of structural or physiological mechanism whereby fertilization is restricted to the active participation of a single spermatozoon.

The addition of a moderate amount of sperm in the vicinity of an unfertilized egg is usually followed by the rapid congregation of a number of spermatozoa at the surface of the egg. How is it that of these numbers of spermatozoa, only one (presumably that which arrives first) is able to enter the egg. The earlier work on the sea urchin (see Rothschild, 1954) suggested that the block to polyspermy was a two-phased process. The first phase was postulated to occur rapidly (within a

second or so) and to convey a partial block to further sperm penetration. Recent experiments have provided strong evidence that the membrane depolarization which follows rapidly upon sperm contact results in precisely this type of block to polyspermy. This conclusion has been made on the basis of experiments in which the potential difference across the membrane of the unfertilized sea urchin egg is experimentally manipulated by the use of electrodes. Normally upon fertilization the voltage rises to approximately +5 mV. If the voltage across the membrane of an unfertilized egg is experimentally raised to this value (by application of electric current) before the addition of sperm, the added sperm will attach to the egg, but none will enter; the egg cannot be fertilized. If, however, the potential difference is allowed to drop into the negative voltage range, the eggs immediately become fertilized. Similarly, if the potential differences of egg membranes are held at a negative voltage (subjected to a voltage clamp) in the presence of sperm, the eggs rapidly become polyspermic.

The second block to polyspermy occurs as a result of the cortical reaction and subsequent elevation of the fertilization envelope. Once formation of the fertilization envelope is complete at approximately 1 minute, no additional sperm are capable of entering the egg. If the fertilization envelope is removed after it is fully formed, additional sperm will readily penetrate the egg surface long after it was originally fertilized; the egg is said to be "refertilized." The ability to refertilize sea urchin eggs indicates that the rapid block to polyspermy which results from membrane depolarization is a transient one (as is the depolarization itself). Furthermore, it suggests that sperm receptors (if they are required at all) are not located exclusively on the outer surface of the vitelline layer where sperm normally bind, but are present on the underlying plasma membrane as well.

If the fertilization process is stopped at various stages and the eggs are examined under the scanning electron microscope, it is seen that as the vitelline layer lifts off from the surface of the egg, spermatozoa previously attached to that layer are detached (Fig. 5.17). The basis for sperm detachment resides in the release of proteolytic enzymes from the cortical granules during exocytosis. The proteases released by the cortical granules are inhibited by a substance extracted from soybeans, the soybean trypsin inhibitor (SBTI). If eggs are fertilized in the presence of SBTI, the protease cannot function, the fertilization envelope does not form, and spermatozoa that are able to attach cannot become detached. The result is polyspermy.

A method has been developed to obtain preparations of these enzymes. The procedure involves treatment of unfertilized sea urchin eggs with dithiothreiotol (a reducing agent of disulfides), which essentially eliminates the vitelline layer. If these eggs are fertilized at high egg concentration, the cortical granules rupture, expelling their contents into the medium. This supernatant, termed the *fertilization product*, can be removed and studied. If unfertilized eggs are treated with this fertilization product, spermatozoa cannot attach because their binding sites have been digested; the eggs do not fertilize. Two distinct proteolytic enzymes have been isolated among the products of cortical granule exocytosis. One of these enzymes is specific in its hydrolysis of the sperm binding sites; the second is specific in its hydrolysis of the linkage material which holds the vitelline layer and plasma membrane together (visible in Fig. 5.15) in the unfertilized egg. Digestion of this linker material leads to elevation of the vitelline layer.

Results with other animals suggest that the basic processes just described are widespread. In frogs, the release of cortical granule material also causes an alteration in the nature of the vitelline layer. In *Xenopus*, much of this material passes through the vitelline layer to form a dense precipitate at its outer surface

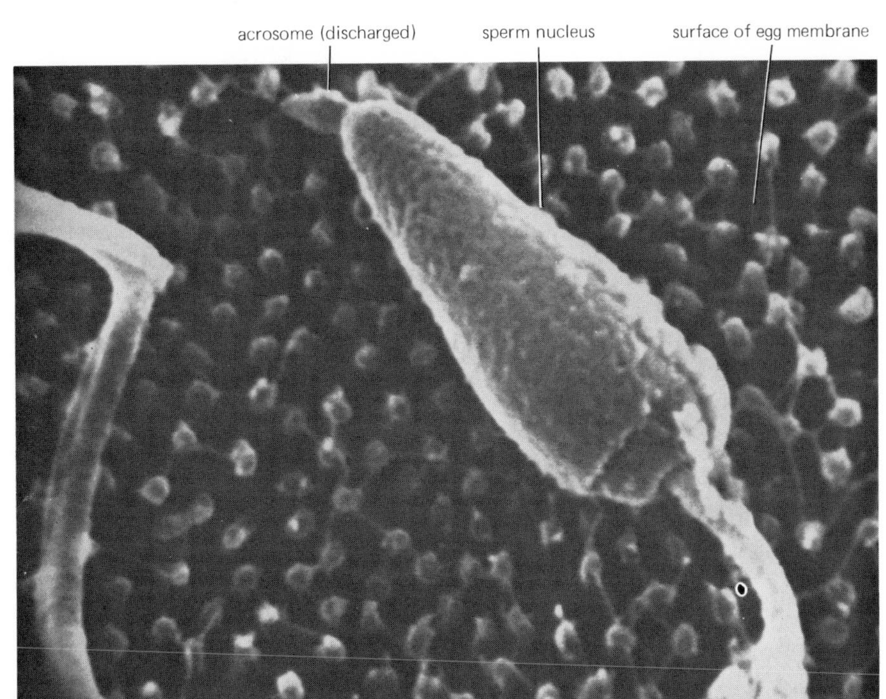

a

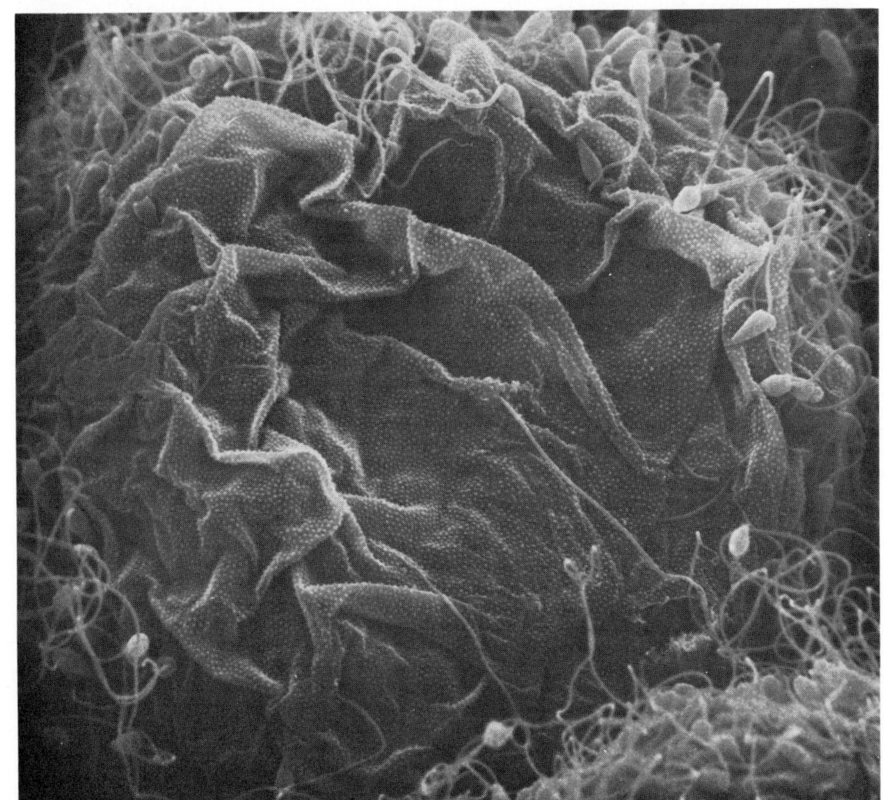

b

Figure 5.17 Scanning electron micrographs of fertilization in the sea urchin *Strongylocentrotus*. *a* The head of this spermatozoon has undergone the acrosomal reaction and is attached to the surface of the egg. The projections of the vitelline layer are believed to reflect the locations of the microvilli projecting from the egg surface. These sites on the vitelline layer contain the sperm receptor glycoprotein. *b* Egg surface shown 30 seconds after fertilization. Vitelline membrane has lifted over a portion of the surface, detaching the sperm in the process. Remaining egg surface has yet to lose attached sperm, whose tails are clearly visible. [*Courtesy of M. Tegner and D. Epel;* Science, **179:***685 (1973); copyright © 1973 American Association for the Advancement of Science.*]

Chapter 6 Polarity, cleavage, and blastulation

1. Polarity

It was pointed out at some length in Chapter 4 that oogenesis produces an egg with the beginnings, at least, of a developmental program. At first glance, the eggs of most animals appear as spherical, rather homogeneous cells without any conspicuous differences at one end or the other. In virtually every case, however, closer examination reveals that at least one major axis of the egg is differentiated in some way. This primary axis is called the *animal-vegetal axis* and marks the *polarity* of the egg. One feature, the site of polar-body formation, will mark the location of the *animal pole*. The egg pronucleus begins its migration from this end of the egg. In those eggs with large amounts of yolk, as in amphibians, the yolk, being more dense than the cytoplasm, settles by gravity to one pole, the *vegetal pole,* opposite the egg chromosomes. In the frog egg, therefore, the axis of egg polarity approximates the axis of gravity, and the egg floats with its animal pole upward. In eggs with relatively little yolk, such as the sea urchin, the yolk is evenly distributed and the manner in which the egg lies is unrelated to its animal-vegetal axis.

Since polarity is generally not evident in the very early stages of oogenesis, i.e., in the oogonia, it is believed to arise during the growth and differentiation of the oocyte, being imposed on the unpolarized germ cell from the outside. In the examination of a number of oocytes, such as those of echinoderms and mollusks, there is a predictable relationship between the position of the oocyte within the ovary and the future animal-vegetal axis. The point at which the oocyte is attached to the ovarian wall, and therefore the point of entrance of supplies to the oocyte from the outside, becomes the vegetal pole. In the ovaries of many animals, including mammals, oocytes are completely surrounded by follicle cells, and there is no obvious relation of any ovarian feature to polarity.

Although polarity is a universal feature of animal eggs, there are examples among certain algae where a differentiated axis is lacking in the egg at the time of fertilization. This is the case in certain seaweeds, such as *Fucus* and *Pelvetia,* in which polarity is expressed morphologically as a visible protrusion at about 12 hours after fertilization. The sequence of events in *Fucus* is shown in the micrographs in Fig. 6.1. As a result of the protrusion at one pole, the egg becomes pear-shaped; it then divides unequally to form two cells that differ in structure, bio-

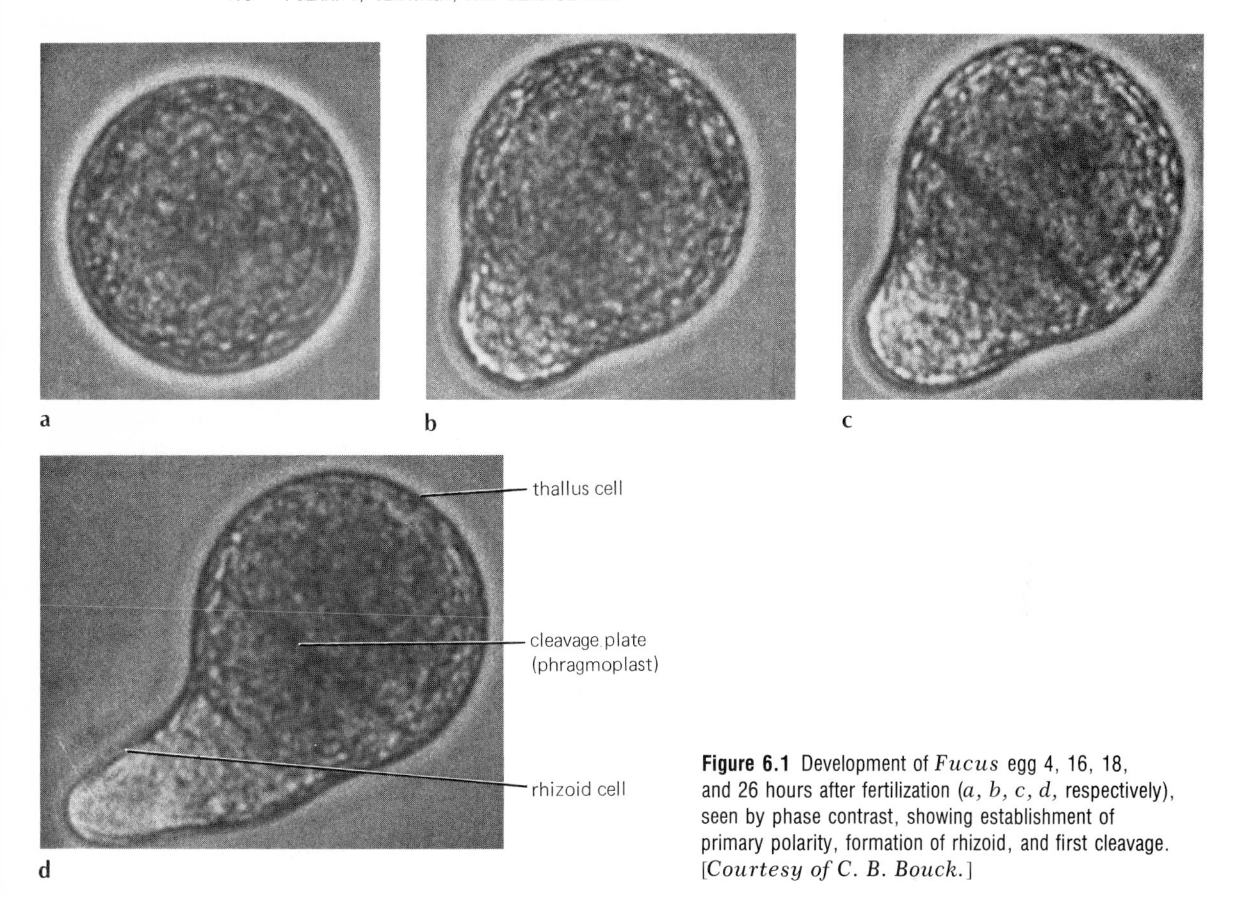

Figure 6.1 Development of *Fucus* egg 4, 16, 18, and 26 hours after fertilization (*a, b, c, d,* respectively), seen by phase contrast, showing establishment of primary polarity, formation of rhizoid, and first cleavage. [*Courtesy of C. B. Bouck.*]

chemical composition, and developmental fate. The initial protrusion remains as part of the smaller rhizoid cell, which will eventually form a holdfast to anchor the plant in the surf. The other cell, the thallus cell, will become the bulk of the plant.

Since *Fucus* begins its development without an inherent polarity, the study of this egg provides information on the types of external stimulation that can affect the internal organization of an egg. Furthermore, it tells us about the manner in which the spatial organization within an embryo can arise. As will be discussed at great length in the following chapter, the topographical organization of the cytoplasmic components within an egg and early embryo is of vital importance in the development of that embryo. The transformation of the homogeneous *Fucus* egg into a polarized structure can be studied as a model system for the regulation of intracellular localization and the formation of biological patterns. It has been known for a long time that a variety of external gradients can determine polarity in *Fucus*. If a light beam is focused on a fertilized *Fucus* egg, the rhizoid will form at a site away from the light source. If eggs are put in an electric current, the rhizoid forms toward the anode. In a group of eggs, the rhizoids of each develop toward the center of the cluster. Regardless of the specific manner in which the environment becomes polarized, this environmental differentiation becomes imposed on the internal contents of the egg. Further examination of these types of experiments brings out another important aspect of this topic. The egg pro-

ceeds through a period where its axis of polarity can be redirected by alterations in the environment. Then a point is reached where further changes in the environment no longer affect the future site of rhizoid formation; the axis is said to be fixed. For example, if the direction of light impinging on an egg is reversed at about 6 to 10 hours postfertilization, the rhizoid forms in keeping with the second light beam. However, if one waits until after 10 hours postfertilization, the reversal of light is no longer capable of reversing the direction of egg polarity, even though overt morphological indications of rhizoid formation are still lacking.

An important insight into the underlying mechanism of polarization in fucoid algae was made when it was discovered that the fertilized eggs drive an electric (ionic) current through themselves. Furthermore, the direction in which the ions move into, through, and out of each egg is related to the direction of the axis of polarity, as indicated by the experiment illustrated in Fig. 6.2. When a number of eggs are placed in small tubes and light is shone in one direction, a current can be measured across these eggs, whose rhizoids will all be aligned. In contrast, if the eggs are simply allowed to develop rhizoids at random locations, no such current is detected. It was proposed that the current detected in the aligned eggs resulted from a flow of positive ions into the egg at the point where the rhizoid forms and out at the opposite end; the return through the medium forms a current loop (Fig. 6.2*c*) estimated at 9×10^{-10} amps per embryo. Further analysis on individual eggs has confirmed and extended the original conclusion.

Several important questions arise in connection with these findings. How is the path of the current loop determined? Is there a causal relationship between the current loop and the fixation of the axis of polarity? Is the current loop directly involved in the events leading to rhizoid formation? Analysis of single eggs indicate that steady ionic currents begin as early as 30 minutes after fertilization. Initially, multiple sites of ion entry and exit may exist over the surface of a single egg, but as time progresses, a single region of ion entry comes to dominate the current patterns (Fig. 6.3). Barring changes in the nature of the environment, this site will become the rhizoid. If one considers that both entry and exit of ions through a cell are mediated by membrane protein channels, then questions concerning this phenomenon center on the spatial organization of the plasma membrane. The change in electrical properties of the egg during postfertilization development are believed to reflect movements of ionic channels within the plane of the membrane. In some manner, gradients (electrical, light, pH, etc.) within the environment appear to be responsible for the directed mobility of membrane

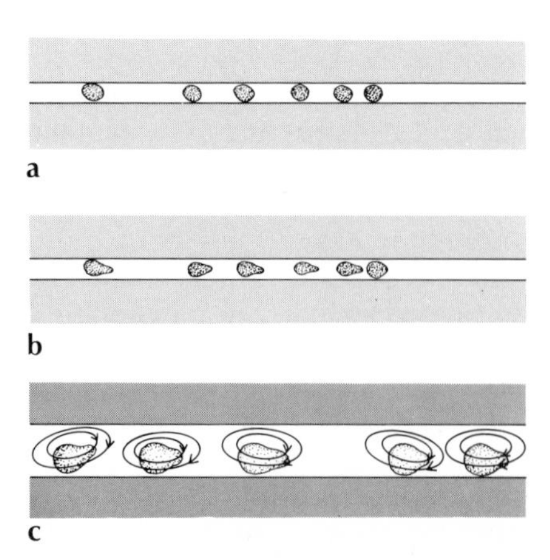

Figure 6.2 Measurement of electric potentials in fertilized *Fucus* eggs. *a* Eggs drawn into glass capillary tube of diameter slightly less than that of the eggs. *b* A day later, all eggs have germinated in the same direction, dividing into a rhizoid and a thallus cell. *c* Suggested scheme of inferred current pattern in a tube. [*After L. Jaffe,* Proc. Nat. Acad. Sci. U.S., **56:***1102* (*1966*).]

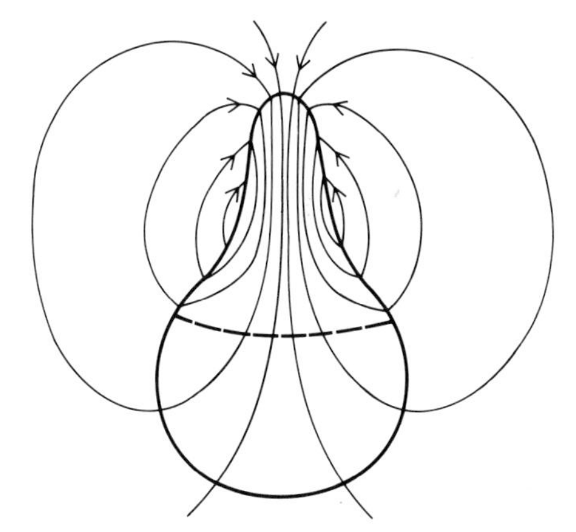

Figure 6.3 The spatial-current pattern inferred from the current-density measurements. [*From R. Nucitelli and L. Jaffe,* J. Cell Biol., **64**:*640 (1975).*]

components. If the direction of the gradients is reversed, the directions of the ionic currents are slowly shifted, but only up to that point in time at which the direction of the axis is irreversibly fixed. It has been proposed that the specific event responsible for the fixation of polarity is the stabilization of the ionic channels within the membrane. Support for this concept has come from studies utilizing cytochalasin B, a substance that results in the disappearance of microfilaments, organelles whose role in anchoring integral membrane proteins was discussed in Section 3.1.A. Cells incubated in cytochalasin B are not capable of fixing their axis of polarity (as indicated by the subsequent ability of directed light to shift the position of rhizoid formation).

As described earlier, the polarity of an egg represents a primary heterogeneity in the distribution of intracellular materials. Therefore, one of the most important aspects in the analysis of polarity concerns the means by which materials can be moved through the egg. In the best-studied cases (such as secretory-granule movement, axonal transport, pigment dispersal, etc.), organized networks of microfilaments and microtubules have been implicated in the directed movements of macromolecules and small organelles. Studies on these algal eggs have raised the possibility of a very different type of localizing force, namely, the electric fields that are generated by the presence of ionic currents. For example, the movement of a positive current into the rhizoid site could serve as a powerful influence for the movement of negatively charged macromolecules or particles through the cytoplasm (or within the membrane) to that site. In fact, the development of the current loops in *Fucus is* followed by the selective migration of materials toward the rhizoid and their eventual accumulation in the rhizoid wall. Whether these events are causally related remains to be demonstrated. In addition to the localizing potential that the ionic current may possess, the fact that calcium ions are an important component of the current adds another dimension to the subject. During the polarization of eggs by unilateral light, approximately 5 times more calcium ions enter the presumptive rhizoid region as enter the opposite side. Considering the role of Ca^{2+} in so many intracellular activities, this influx could have important consequences for events occurring at the rhizoid site. Regardless of the precise mechanism, it would appear that the influence exerted by the environment on the egg is mediated by the plasma membrane, which is not a surprising finding.

In the case of the typical animal oocyte, polarity is established in an environment of much greater complexity than that of the sea in which *Fucus* polarity

arises. Whether similar mechanisms for intracellular localization exist within an organized ovary is totally unknown. Whatever the means of formation, the consequences of polarity are significant. The animal pole will generally form the anterior end of the animal and differentiate into the external covering of the embryo. The vegetal end typically is pushed into the interior of the egg at gastrulation and will differentiate into endodermal and, in some cases, mesodermal structures. As will be discussed later in this chapter, the cleavage planes are formed in relation to the axis of polarity. The first and second planes of the sea urchin and frog, for example, coincide with the animal-vegetal axis; i.e., they are meridional. The third is perpendicular to it; i.e., it is equatorial. In the sea urchin the axis of polarity becomes the axis of radial symmetry during cleavage, while in the frog it comes to lie within the plane of bilateral symmetry.

When one examines the egg of the frog *Rana pipiens*, the polarity is obvious. Yolk is concentrated toward one end; the chromosomes lie at the opposite end, where the polar bodies are given off; and there is a covering of black pigment over the egg in all regions except around the vegetal pole, which is unpigmented. These are the obvious markers of polarity, but they are not necessarily its primary features. The animal-vegetal axis of the frog egg, or of any egg, represents a basic structuring of the egg protoplasm. Many of the visible signs, such as yolk and pigment, are believed to have only secondarily responded to a previous, more basic differentiation. This is clearly illustrated by experiments in which these visible markers of polarity are displaced without disrupting the more fundamental organization of the egg.

One of the simplest and most effective methods of disturbing the visible organization of the egg is to place it in a centrifugal field. Under the influence of centrifugation, those components of the greatest density move fastest and become sedimented against the inner surface of the egg, facing the centrifugal pole (away from the center of the centrifuge rotor). Any materials less dense than the cytoplasmic fluid of the egg, such as oil droplets, float to the opposite end, the centripetal pole. Centrifugation produces an egg whose components are stratified into well-defined layers. Dense particulate materials (pigment granules, yolk granules, mitochondria, etc.) are in layers at one end, and above this is a clear zone containing the fluid cytoplasm in which the nucleus is generally found. This is capped by a collection of oil droplets. The eggs of the polychaete *Nereis*, after centrifugation, are shown in Fig. 6.4.

Early centrifugation studies by T. H. Morgan on the sea urchin embryo indicated that the underlying basis of polarity is resistant to moderate centrifugal forces. In these pioneering experiments, Morgan used the eggs of *Arbacia*, whose animal pole is marked by an indentation in its jelly coat that is easily identified in the presence of india ink. Upon centrifugation, sea urchin eggs do not orient with respect to gravity; i.e., they will settle with their animal-vegetal axis oriented randomly in the centrifugal field. Therefore, they will become stratified in all possible relations to their axis of polarity. If one observes the cleavage pattern in these eggs after centrifugation, the first cleavage furrow is usually perpendicular to the stratification (cuts across all layers), the second is parallel, and the third is at right angles to the first two cleavages. In other words, these three cleavages are now related to the stratification. Since the stratification is independent of the animal-vegetal axis, the first three cleavages also must be independent of the original egg polarity. Presumably the stratification imposes mechanical restraints on the mitotic apparatus and establishes a new cleavage pattern.

When Morgan examined the fourth cleavage, he observed that the micromeres were formed at or near the original vegetal pole (across from the jelly fun-

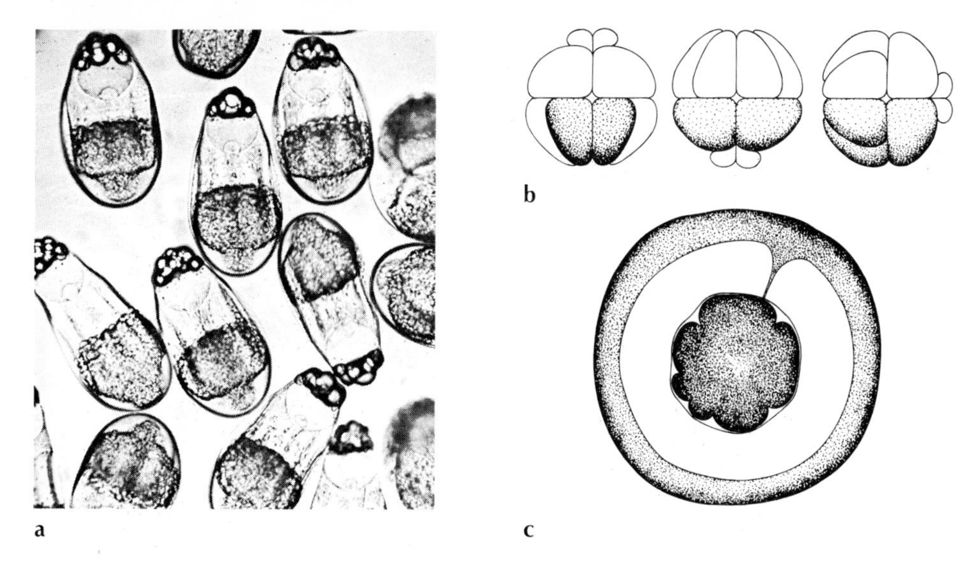

Figure 6.4 *a* Centrifuged unfertilized eggs of *Nereis*, showing zones of stratification, with oil drops at top, hyaline protoplasmic layer containing the nucleus, a zone of yolk spheres, a second hyaline zone, and a zone of jelly-precursor granules. [*Courtesy of D. P. Costello.*] *b* The fourth cleavage of centrifuged *Arbacia* occurs in keeping with the original animal-vegetal axis regardless of the stratification of the egg. The micromeres have formed at the centripetal pole, the centrifugal pole, and at the side, respectively. *c* The method used by Morgan to identify the original animal-vegetal axis. The india ink marks the position of the animal pole opposite which the micromeres are formed. [*After T. H. Morgan, "Experimental Embryology," Columbia, 1927.*]

nel) regardless of how the first three cleavages had occurred (Fig. 6.4*b*). Similarly, gastrulation occurred at the original vegetal pole just as it would have in the uncentrifuged eggs. It is clear from these experiments that the visible components of the egg can be greatly disturbed and cleavage can be affected, but the underlying axis of polarity is resistant to these treatments. In numerous other animals, stratification does not affect even the first cleavages, which continue to occur with respect to the axis of polarity.

2. Bilateral symmetry

In animals with a symmetrical body plan that is roughly bilateral, such as vertebrates, three body axes can be defined: the anteroposterior, the dorsoventral, and the left-right. The anteroposterior axis is typically related to the axis of polarity and as such is established in the ovary. The other two axes become established simultaneously, although in a different manner among various animals. In some species, including the cephalopod mollusks and some insects, the unfertilized egg already possesses these three identifiable axes as a result of oogenesis. In some insects these axes become established within the oviduct. In other cases, the point of fertilization is clearly related to the plane of bilateral symmetry, as in many mollusks, annelids, and amphibians; in other animals, such as the birds, reptiles, and mammals, these axes are not established until the end of cleavage.

This discussion concentrates on those cases in which the plane of bilateral symmetry (the median plane separating right and left halves of the embryo) becomes established at fertilization, as a result of the point of sperm penetration. The history of this subject can be traced back to 1885, when Roux was able to

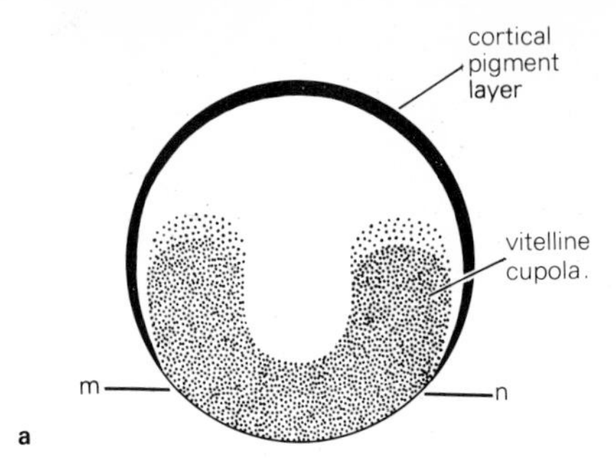

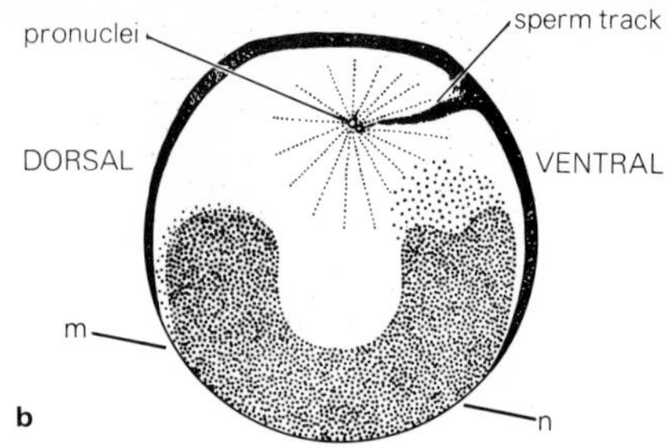

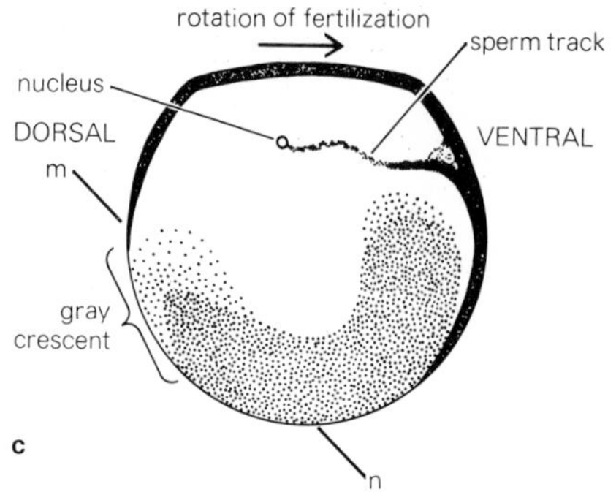

Figure 6.5 *a* Diagrammatic axial section of an unfertilized egg of *Rana fusca*, m–n representing the interior limit of the cortical pigment layer. *b* The reaction of the egg to the spermatozoon when it enters. Note in particular the 10° inclination of m–n, and that the vitelline horn has moved closer to the future dorsal side. [*After P. Ancel and P. Vintemberger*, Biol. Bull., *suppl.*, **31**:377 (*1948*).] *c* An axial section through the center of the gray crescent of *R. fusca* showing the modifications that have occurred in the egg after the rotation of fertilization (*rotation of symmetrization*). The arrow indicates the direction of the rotation of fertilization. The gray crescent, which marks the dorsal side, is formed. Within this dorsal area, only slight pigmentation remains after the pigmented cortical layer has receded. The vitelline horn on the dorsal side has moved nearer the cortex, and the line m–n has inclined through 30°.

fertilize a frog egg at any desired point. He placed a fine silk thread on an egg and let a tiny drop of sperm creep along the thread to contact the egg at one identifiable point. Analysis of eggs fertilized at specific points indicates that the future plane of bilateral symmetry of the animal passes through the point of sperm penetration; in the majority of cases this also coincides with the plane of the first cleavage furrow. The conclusion is therefore reached that the point of sperm penetration establishes the entire bilateral organization of the future adult. Thus, as in the case of polarity, this axis is imposed on the egg by an external factor.

To understand the relationship between fertilization and the development of bilateral symmetry in the amphibian egg, we need to consider a series of events that occurs between the time of fertilization and first cleavage. In the typical case, the unfertilized egg outwardly consists of two regions: a dark animal portion, which extends to below the equator (*m* to *n* of Fig. 6.5*a*), and a whitish vegetal portion. The coloration of the animal region results from a layer of black pigment granules that lie in the cortex just beneath the plasma membrane. The vegetal region of the egg lacks these pigment granules. Internally, a cup-shaped mass of white yolk granules occupies the vegetal region, with the central concavity opening toward the animal pole. The frog egg is only susceptible to sperm penetration on its animal hemisphere. The actual point of sperm penetration is visible externally in some species (urodeles) by the presence of a small depression (Fig. 6.6).

Within an hour or two following fertilization, there appears to occur a rotation of the cortex of the egg relative to the internal endoplasm. In Fig. 6.5 this rotation of the "outer crust" is reflected in the shift of the pigment from *m* to *n* in Fig. 6.5*a* to that shown in Fig. 6.5*c*. Since the site of sperm penetration is marked internally by a track of pigment carried into the endoplasm by the male pronucleus, one can readily determine the spatial relationship of the point of fertilization to the events occurring in the cortex and to the bilateral symmetry of the embryo.

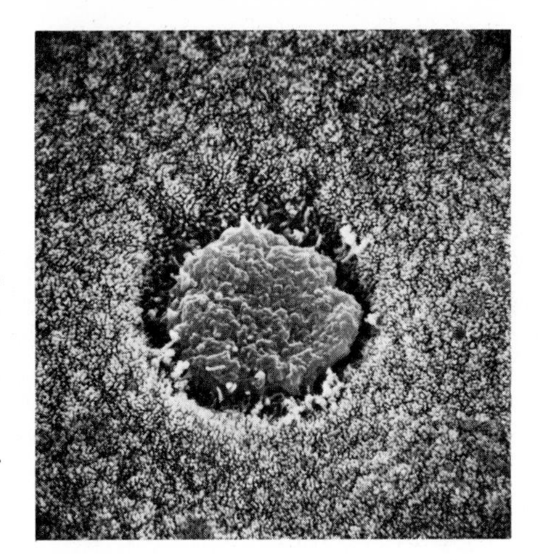

Figure 6.6 Scanning electron micrograph of a frog egg fixed 20 minutes after fertilization showing the surface depression associated with the point of sperm entry. [*Courtesy of R. P. Elinson and M. E. Manes.*]

That side of the egg at which the pigmented cortex lifts toward the animal pole will become the future dorsal surface of the embryo. The upward movement of the pigmented cortex on this side of the egg leaves a relatively unpigmented crescent-shaped section of egg surface near the midline, which is referred to as the *gray crescent.* The midpoint of the gray crescent is located on the meridian 180° from the point of fertilization. The gray crescent is middorsal, the meridian of sperm penetration is midventral, and the plane connecting these two meridians forms the plane of bilateral symmetry, which (in the frog) is usually marked by the first cleavage furrow. Although the presence of a visible cresent on the dorsal midline is not evident in all amphibian species, this area of the embryo is of particular importance in amphibian development, as will be discussed in later chapters. On occasion a frog egg is fertilized by two spermatozoa; in such cases the gray crescent forms on the meridian midway between the two points of sperm penetration (Fig. 6.7). The manner in which the bilateral symmetry of the embryo is determined by the point of sperm entry remains obscure, although some evidence suggests that the sperm aster plays an important role in this process.

Figure 6.7 Equatorial section through dispermic frog egg showing that the gray crescent (*position indicated by thin outline opposite thickest part of outline*) forms at the midpoint between the two points of sperm entry. Broken line indicates plane of symmetry. [*After M. Herlant,* Arch. Biol., **26** *(1911).*]

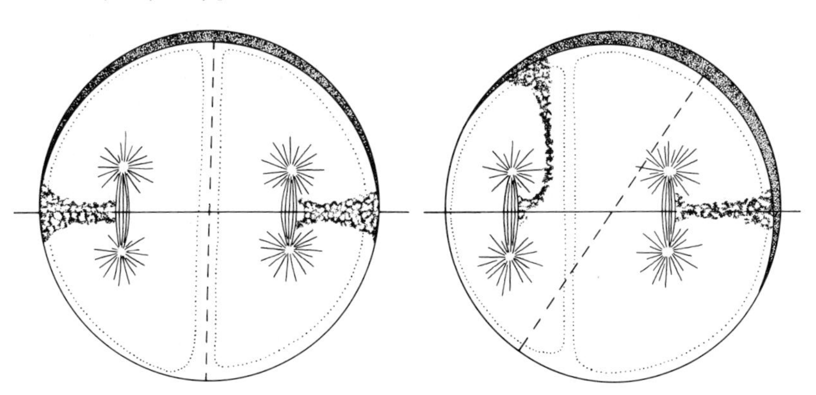

3. The egg cortex

In the fertilized egg of the sea urchin *Arbacia* (after the cortical granules have expelled their contents), there are numerous red pigment granules whose position can be followed. Most of these granules are in the central endoplasm, although they are found in the cortex as well. If these eggs are centrifuged, the granules in the endoplasm move to the centrifugal pole of the egg, while those in the cortex do not. Why should the same type of granule behave so differently in these two locations? The most accepted answer is that the cortex of the sea urchin egg is a region of much higher viscosity; i.e., it is gelated with respect to the more fluid (more solated) endoplasm. As a result of this cortical gel, granules located in that region are restricted from displacement by the increased viscosity of their surroundings.

The gelated material that everyone is most familiar with is the gelatin gel, which differs markedly from protoplasmic gel in the following ways. The protoplasmic gels increase their volume and absorb heat upon gelation, while a gelatin gel melts as it warms. These properties of protoplasmic gels cause them to become solated (made more fluid) by either an increase in hydrostatic pressure or a decrease in temperature, or a combination of both. This means that if a sea urchin egg is placed in a cylinder to which a piston is attached and hydrostatic pressure is applied, the gelated regions of the egg will become solated. The greater the gelated state of the cytoplasm, the greater the pressure required to solate it. The invention of the pressure centrifuge allowed eggs to be subjected to hydrostatic pressure and to be centrifuged simultaneously. Investigators were able to utilize this device to determine that cortical pigment granules as well as endoplasmic granules could readily be displaced by centrifugation as long as the eggs were first solated by the pressure. These findings provide convincing evidence of the gelated nature of the cortex.

Several other approaches to investigation of the nature of the cortex illustrate the variety of techniques that can be brought to bear on one specific problem. One of the great traditions in developmental biology has been the micromanipulative procedure. The tools of the workers are delicate instruments: microneedles, micropipets, fine scissors and forceps, and in many cases, a micromanipulator. The needles and pipets can be attached to the micromanipulator and their operation controlled by a system that translates the coarse movements of the hands into very small movements of the tips of the instruments. Various controls direct the instruments in the various possible directions as the investigator observes their progress through the attached microscope. Inasmuch as the sea urchin egg and the mammalian egg are approximately 100 μm in diameter, the tasks that have been accomplished are impressive.

If a blunt-tipped microneedle is inserted into a sea urchin egg and pushed through to the opposite side, the tip of the needle does not come in contact with the very inner edge of the egg. Instead, when the tip reaches a position of about 3 μm from that edge, the surface begins to bulge and the needle moves no closer to the edge itself. Presumably there is some layer of approximately 3 μm resisting the penetration of the needle tip. If the needle is slowly withdrawn from the cortical layer back into the endoplasm, adhering gelled material can be seen flowing into the more fluid internal cytoplasm. Similarly, if an oil droplet is injected into the sea urchin egg, it never comes in direct contact with the inner edge of the cell membrane; rather it remains a few microns from the edge itself. Even when the cell divides, the advancing tip of the furrow is seen to be separated from the surface of the oil by a gap of several microns and remains at this distance as the droplet is split in two.

Not only has the existence of the cortex been indicated by the preceding techniques, but it also has been isolated in bulk after disruption of eggs in 0.1 *M* $MgCl_2$, which lyses the cells and keeps the cortex gelated. By this technique, large numbers of cells can be broken in a way that releases the internal contents, leaving the plasma membrane and associated cortical cytoplasm as particulate material that can be readily isolated. It was mentioned in the previous chapter that fertilization in the sea urchin was followed by a marked polymerization of cortical actin. The filamentous network of the cortex is evident in the scanning electron micrograph of Fig. 6.8. It is the presence of these microfilaments that is believed to provide the region with its increased viscosity. The sensitivity of microfilaments to depolymerization as the result of high hydrostatic pressure explains the experiments with the pressure centrifuge just described.

Evidence has accumulated that assigns to the cortex a very important role in the storage of developmental information. In the sea urchin, for example, the seat of polarity is capable of withstanding centrifugation, and it appears that the cortex is the only region of the egg for which the same can be said. One can still argue that the information is present in the endoplasm but is simply not connected with the particulate materials being displaced by the centrifuge. Even though it is difficult to conceive of an endoplasmic structure that retains its integrity in the midst of the centrifugal traffic, more direct evidence is needed. The best argument against the requirement of any cytoarchitecture of the endoplasm for polarity is the finding that up to 50 percent of the internal cytoplasm can be indiscriminately removed with a micropipet without disturbing either the polarity or the course of normal development.

4. Cleavage

The first mitotic divisions of each individual comprise the *cleavage* divisions; the cells produced are called *blastomeres*. The period of cleavage extends from fertilization to the formation of a characteristic developmental stage, the *blastula*, and is generally characterized by rapid, successive divisions without intervening periods of growth. In many organisms the rapid cleavage divisions are highly synchronous, occurring throughout all or most of the embryo at precisely the same time. This feature of development is best revealed using time-lapse cinema-

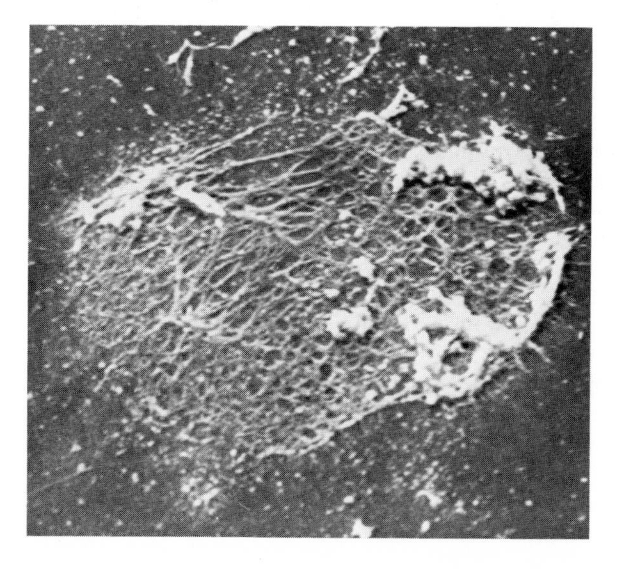

Figure 6.8 After completion of the cortical-granule discharge, the cortex of the sea urchin egg appears as an overlapping network of fibers 0.2 to 0.5 μm thick. [*Courtesy of G. Schatten and D. Mazia.*]

myosin has been found in the cortex of the cleavage furrow, it does not appear to be present as distinct filaments, since there is no evidence of such filaments in the electron microscope. At the same time, however, it is difficult to imagine how actin filaments could function in the absence of some type of myosinlike material. Indirect evidence for the involvement of myosin has come from experiments in which cytokinesis is rapidly blocked following the injection of antimyosin antibodies into cleaving eggs.

Before discussing the mechanism of cytokinesis further, it is worthwhile to consider the nature of the cleavage furrow as it pinches the cell in two. In most cases, the microfilaments first appear within the cortex as an encircling equatorial ring just prior to the furrowing event. It is uncertain as to whether the bulk of microfilaments arise by a process of polymerization in place within the cortex as opposed to being recruited from a preexisting population in other sites within the egg. Nonmuscle cells, in general, appear to possess large pools of actin monomers (the G-actin form) from which filaments (the F-actin form) arise. The manner in which the cell controls the time, location, or orientation of the polymerization is largely unknown.

As the cleavage furrow forms and then deepens, the contractile ring (Fig. 6.11) remains approximately constant in width (about 8 μm in the sea urchin)

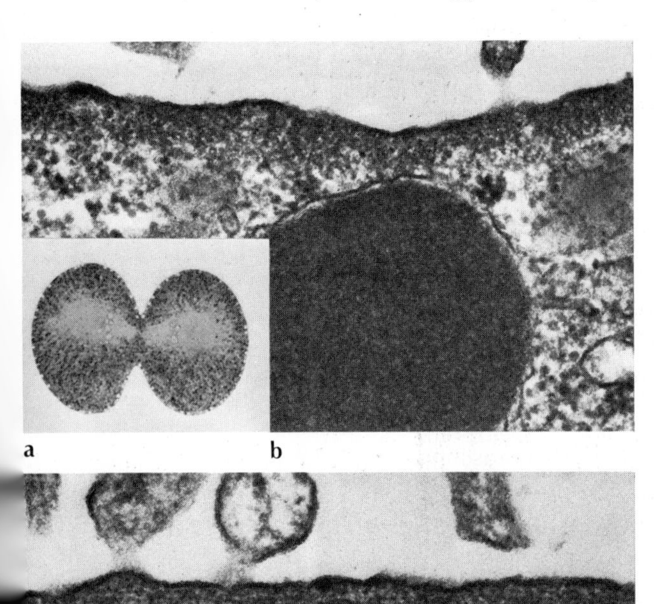

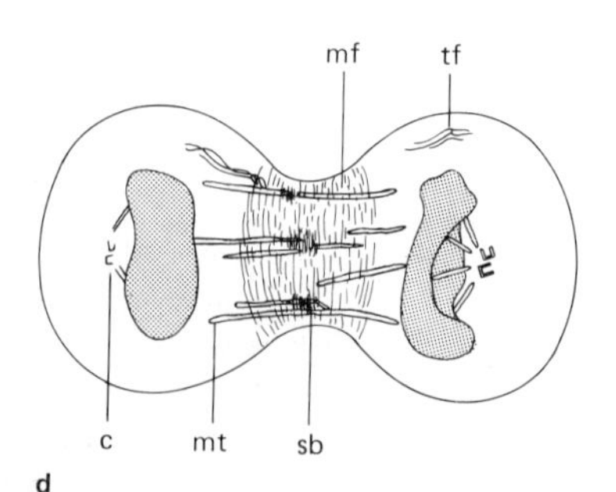

Figure 6.11 Photomicrographs of half-cleaved *Arbacia* eggs. *a* Light micrograph of an egg sectioned parallel to the long axis of the mitotic spindle. *b* Electron micrograph showing the cleavage furrow of a similar egg sectioned in the same plane as *a*. The layer of "dots" beneath the plasma membrane is the contractile ring, whose microfilaments are seen in cross section when cut in this manner. The dark round body is a yolk granule. *c* Electron micrograph of a cleaving egg sectioned perpendicularly to those of *a* and *b*. This section is parallel to the furrow and, therefore, parallel to the contractile-ring microfilaments. As a result, the filaments now appear as filaments and some spindle fibers are cut in cross section. Three mitochondria are seen in addition to part of a yolk granule. [*Courtesy of T. E. Schroeder.*] *d* Diagram of a single cell midway through cleavage showing the organization of the contractile-ring microfilaments (mf). Microtubules (mt) are seen to form stem bodies (sb) in the equatorial plane. c, centrioles; tf, thick filaments in bundles; cross hatching, nuclei. [*From T. E. Schroeder*, Am. Zool., **13:**950 (1973).]

and depth (0.1 to 0.2 μm), although its circumference decreases much like a purse string narrows the diameter of an opening. Since the decreasing diameter of the ring is not accompanied by a corresponding increase in its width or thickness (or in the packing density of the filaments), it has been proposed that a depolymerization of the actin filaments accompanies cytokinesis. However it is accomplished, it is believed that the constriction of the egg and its blastomeres occurs by the sliding of the actin filaments over one another, presumably as a result of their attachment to myosin molecules. Furthermore, it is assumed that the actin filaments are firmly attached to the overlying plasma membrane. Consequently, as the diameter of the contractile ring decreases, tension is generated and the plasma membrane in the furrow region is carried inward. It should be mentioned that there are cases, such as the coelenterates, ctenophores (Fig. 6.12), and cephalopods, where the cleavage furrow advances from only one end of the egg rather than appearing as an encircling ring. In these cases, the zone of microfilaments is initially restricted to a portion of the equatorial surface and then expands laterally as cytokinesis progresses. Regardless of the manner in which the furrow cuts through the egg, the contractile apparatus is a short-lived structure which disappears soon after the event is completed.

The cleavage of a single egg cell into hundreds of blastomeres results in a tremendous increase in total cellular surface area. The formation of two spheres from one results in a 26 percent increase in surface area that must be covered by plasma membrane. In some cases, the surface of the uncleaved egg contains large numbers of microvilli or blebs which, if they were to become flattened out as cleavage proceeds, could provide much of the needed membrane. However, direct evidence for this proposal has not been forthcoming. In other cases, new membrane is inserted directly into the plasma membrane of the cleavage furrow via the fusion of cytoplasmic vesicles or special membranous lamellar structures. The introduction of new membrane into the advancing cleavage furrow is best illustrated by work on the frog egg. A quick examination of a cleaving frog egg reveals a marked difference in pigmentation between the outer surface of the embryo and the membrane surfaces between the blastomeres. The pigment, which is produced during oogenesis, is embedded in the cortical layer of the egg, which is, in turn, firmly adherent to the plasma membrane. Very little of this "old" surface is seen between the blastomeres; the "new" intercellular surface is essentially unpigmented. In the electron microscope, the plasma membrane be-

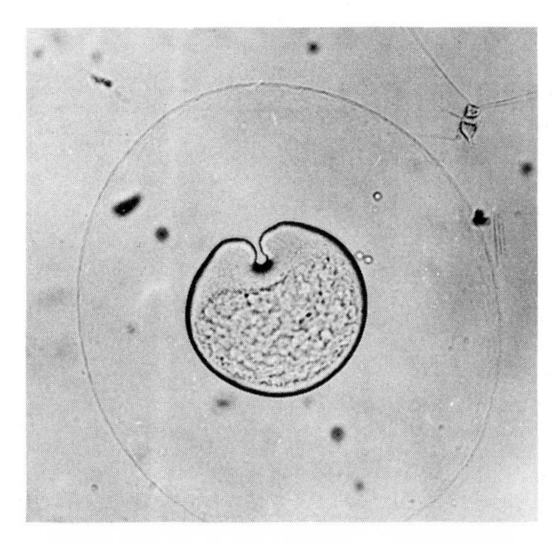

Figure 6.12 Photograph of a ctenophore egg undergoing its first cleavage. The cleavage furrow in this group begins at one pole and moves unidirectionally through the egg. Cleavage in ctenophores is characterized by a striking movement of cytoplasmic materials (an example of ooplasmic segregation of the type discussed in Section 7.4). Note the presence within the egg of two distinct cytoplasmic regions. Material collected in the region beneath the furrow had previously been spread rather uniformly within the cortex of the egg surrounding the yolky endoplasm. Between fertilization and first cleavage, this cortical cytoplasm flows toward one end of the egg and collects as shown. [*Courtesy of G. Freeman.*]

tween the cells has a reduced cell coat, a less distinct trilaminar appearance, and a lower electric resistance, as measured physiologically, than the plasma membrane at the outer surface of the blastomeres. The presence of the new intercellular surface, which is inserted by vesicle fusion, is best revealed by treatment of a cleaving egg with cytochalasin B, a substance that disperses the microfilaments of the contractile ring. Treatment of a cleaving egg with this drug causes a rapid regression of the cleavage furrow, even if the drug is added well after furrowing has begun. A series of micrographs of an egg treated in this manner is shown in Fig. 6.13. The unpigmented band represents newly incorporated cell surface. As time passes, the unpigmented band in this cleavage-arrested *Xenopus* egg grows in width by the addition of more new surface material, material that normally would reside between the forming blastomeres. If one looks closely at the micrographs, four black particles are visible at the border of the pigmented and nonpigmented region. These are four iron oxide particles that were placed as markers on the egg surface prior to the onset of cleavage so that relative shifts in position of the old surface could be followed. In the course of time, the distance between each lateral pair remains constant, but the distance across the unpigmented zone increases greatly.

C. The mitotic apparatus

Cytokinesis is one aspect of cell division. The other event that is almost invariably linked to it is nuclear division or mitosis (Section 3.11), which is accomplished by the mitotic apparatus (see Fig. 6.15). Although these two processes appear to be causally linked, a surprising degree of dissociation is possible. Experiments have shown that the furrowing events can occur after complete removal or disruption of the mitotic apparatus, but only if performed after the onset of metaphase. For example, the entire mitotic apparatus can be sucked out of an

Figure 6.13 Photograph of a *Xenopus* egg undergoing cleavage in the presence of cytochalasin B. The unpigmented zone represents cell surface that normally would have formed in the region between blastomeres, but as a result of the drug-induced furrow regression, the unpigmented zone appears at the egg surface. The black dots are four iron oxide particles that have been placed on the egg surface for subsequent observation. In the course of time, the distance between each lateral pair remains constant, but the distance across the unpigmented zone increases greatly as more and more unpigmented surface is added. [*Courtesy of J. G. Bluemink and S. W. de Laat.*]

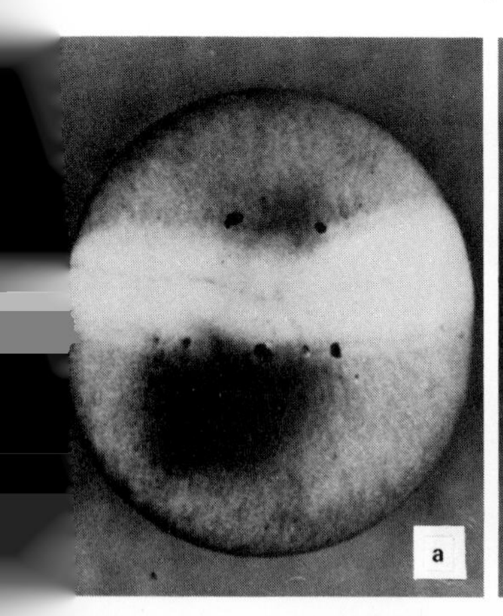

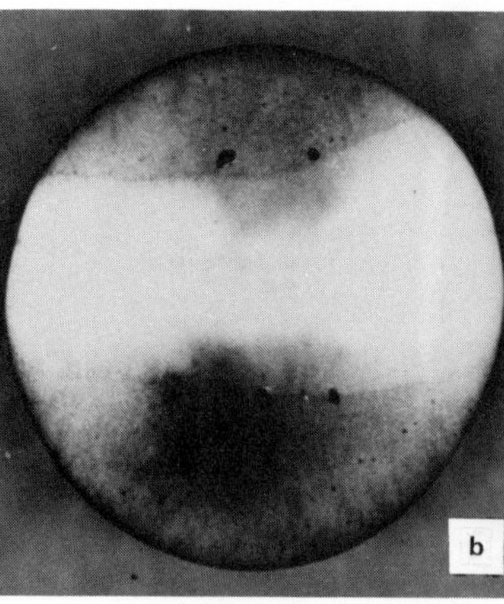

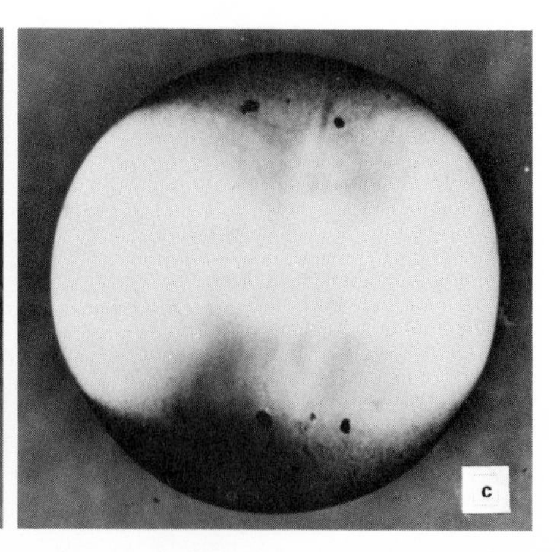

anaphase cell well in advance of furrowing, and cytokinesis will occur in the same place it would have in the undisturbed cell. Figure 6.14 shows the furrow in a cell in which a large oil droplet had been placed at anaphase prior to the onset of cytokinesis. In this case, the oil droplet has completely displaced the mitotic spindle, yet the furrow forms along the same plane it would have had the oil not been injected.

The conclusions emerging from a large-scale investigation of first cleavage implicates the mitotic apparatus, and especially the asters, as the source of the stimulus for cytokinesis. For example, furrowing can be initiated between two asters even in the absence of the spindle and chromosomes. Presumably, some substance(s) radiating from these two sources at each end of the mitotic spindle serves to condition one part of the surface for cytokinesis. Some evidence exists implicating the timed release of calcium ions, once again, as a major component in the triggering of a developmental event. If calcium is injected beneath the surface of an uncleaved amphibian egg, a local furrowing is induced in the surface adjacent to the site of injection. Conversely, agents such as EGTA which complex Ca^{2+} block furrowing. Furthermore, calcium ions appear to be concentrated in the vesicles associated with the mitotic spindle of the cleaving sea urchin egg, and there may be a timed release of this ion from these or some other compartments prior to furrow formation.

The materials needed for the division process, particularly the microtubule proteins of the mitotic spindle, account for a significant percentage of the total egg mass. Analysis of the contents of the unfertilized egg indicates that the bulk of the material is present in a preformed state awaiting only the polymerization process to become incorporated into the structural machinery for division. This is

Figure 6.14 Injection of oil droplet into this sea urchin egg at anaphase has completely disrupted the mitotic apparatus, but has no effect on the position of the ensuing cytokinesis. [*Courtesy of Y. Hiramoto.*]

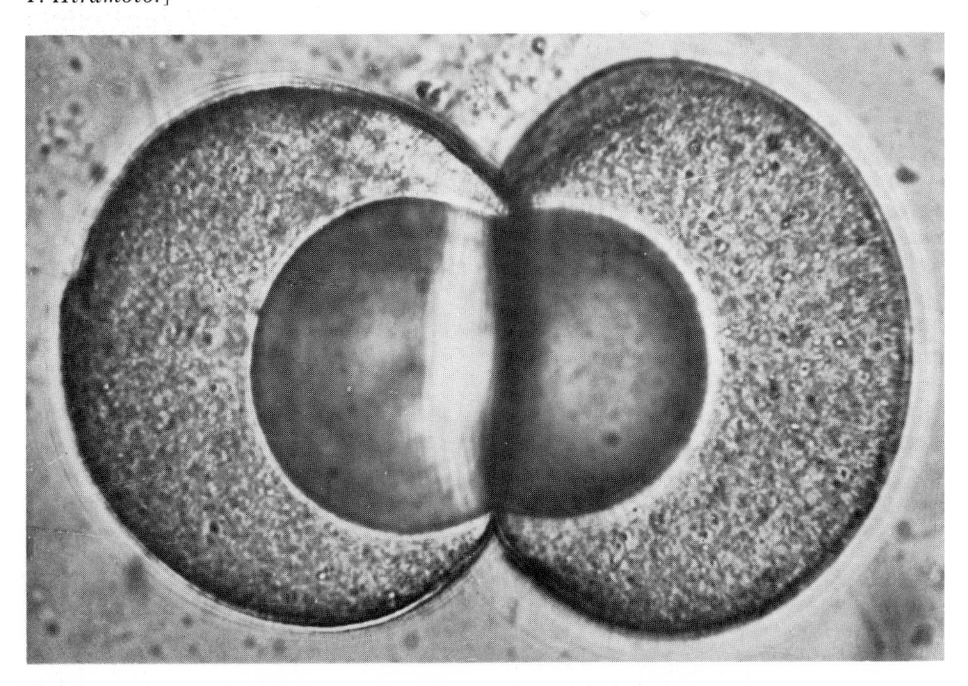

not surprising if one considers the short time available to the egg between the time of fertilization and first cleavage. It is estimated that less than 0.5 percent of the total egg protein is synthesized in this period. The use of preformed material is typical of many types of events within the cell; the precursors are not necessarily synthesized at the last minute, but may exist in a "precursor pool" well in advance of the time they are needed. In the case of the mitotic spindle, the cell is able to determine whether the polymerization or depolymerization reactions are favored at a given location within the cell at a given time.

D. Cleavage pattern

For a given species, the relationship of one cleavage furrow to the next does not occur in a helter-skelter fashion, but rather is a carefully determined genetic trait. Several general patterns of cleavage are found, as well as numerous variations. Before we can begin to describe these patterns, a few general points should be made.

The process of mitosis was described in Section 3.11. The *mitotic apparatus* (or *mitotic spindle*) which brings about the separation of the duplicated chromosomes is shown in Fig. 6.15. The mitotic apparatus consists of a spindle-shaped basket of microtubules that converges at its poles. Each of the polar regions of the apparatus contain a pair of centrioles (disposed at right angles to one another), and these are surrounded by an array of short microtubules, the *astral fibers,* which form a "sunburst" arrangement known as the *aster.* Between the pairs of centrioles stretch the microtubular spindle fibers, some of which engage the chromosomes while others are continuous from pole to pole. The chromosomes lie between the centriole pairs, in most cases arranged along the cell's equatorial plane. The long axis of the mitotic apparatus runs between the asters, and the cleavage furrow will always, under normal conditions, form perpendicular to this axis of the spindle. In other words, the cleavage furrow will lie in the same plane in which the chromosome had previously been situated. It follows, therefore, that the cleavage pattern will ultimately be determined by the arrangement of the mitotic apparatus of the various blastomeres. The orientation of the mitotic apparatus, in turn, depends on the organization of the cytoplasm, which dictates

Figure 6.15 The mitotic spindle of the flatworm *Polycoerus,* showing chromosomal and centriolar orientation at metaphase. *a* Polar view of central spindle showing 34 chromosomes. *b* Centrioles with long axes at right angle to each other. [*After D. P. Costello,* Biol. Bull., **120:***285* (*1961*).]

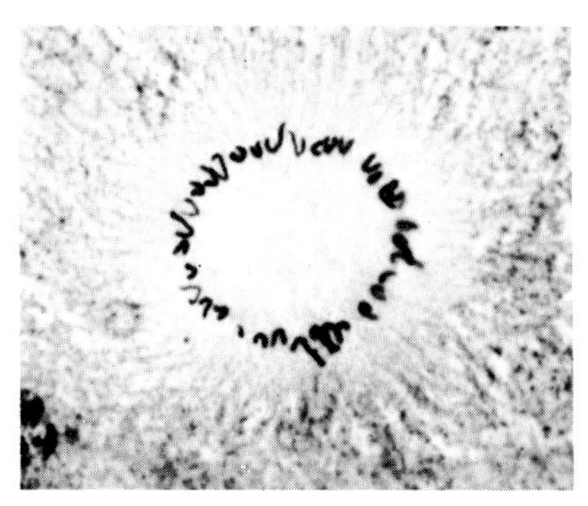

a

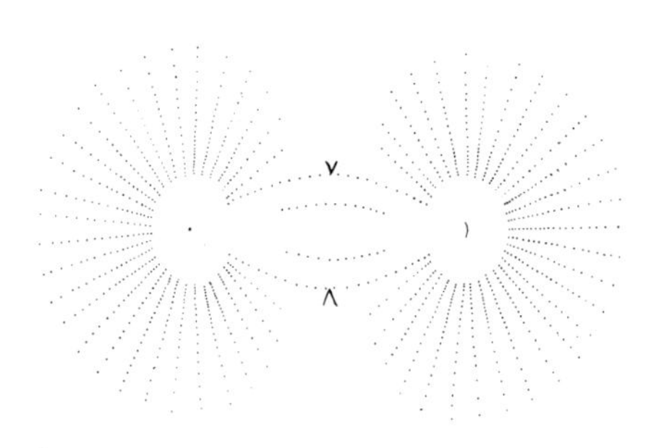

b

where it will form. This, in turn, depends on the cytoplasmic package and program determined by oogenesis. We can push all this back one more level, to the genetic material responsible for the activities of oogenesis, and conclude that the nature of cleavage is determined by genetic determinants acting upon cytoplasmic organization during oogenesis. This is clearly illustrated by cleavage in *Lymnea,* described below.

The cleavage pattern of most animals falls roughly into three categories: radial, bilateral, and spiral. In radial cleavage, the cleavage spindles are oriented with their long axis (centriole to centriole) either perpendicular or parallel to the axis of polarity (Figs. 6.16 and 6.19*d*). In the first two cleavages, the long axis is perpendicular to the animal-vegetal axis, and the furrow is parallel to it. A division along this line is a vertical (or meridional) cleavage. Although the long axes of the spindle of the first two cleavages are both perpendicular to the animal-vegetal axis, they are at right angles to each other, as are the resulting planes of the furrows. The third cleavage, occurring in each of the four existing blastomeres, is generally a horizontal cleavage. In radial cleavage, cells rest directly above one another as they cleave, forming tiers of cells (Fig. 6.9) and producing an embryo that has radial symmetry. Radial symmetry is the symmetry of a cylinder or a sphere or a star. One axis is differentiated along its length, i.e., the animal-vegetal axis, so that one end is different from the other. However, starting at any point on this axis and proceeding out in any direction, one encounters the same structures regardless of the direction. The consequence of radial symmetry is the anonymity of the blastomeres. No blastomere in a given tier of cells can be recognized from any other, since no distinguishing landmarks exist on any one meridian, much as one section of an orange becomes lost following a simple twirl.

In bilateral cleavage, a right and left side become apparent, separated by a plane, the plane of bilateral symmetry. Cleavage activity on one side is mirrored by activity on the other side. Vertebrates cleave in an essentially bilateral manner, as does one class of mollusks, the cephalopods. In most cases, the plane of

Figure 6.16 Development of an idealized radially cleaving egg such as occurs among some echinoderms *a–c* Early cleavage. *d–f* Conversion of morula to blastula. *g* Invagination of blastula to form a gastrula. *h* Evagination from the invaginated enteron to form a mesodermal or coelomic pouch, thereby establishing the three primary germ layers: mesoderm, endoderm, and ectoderm.

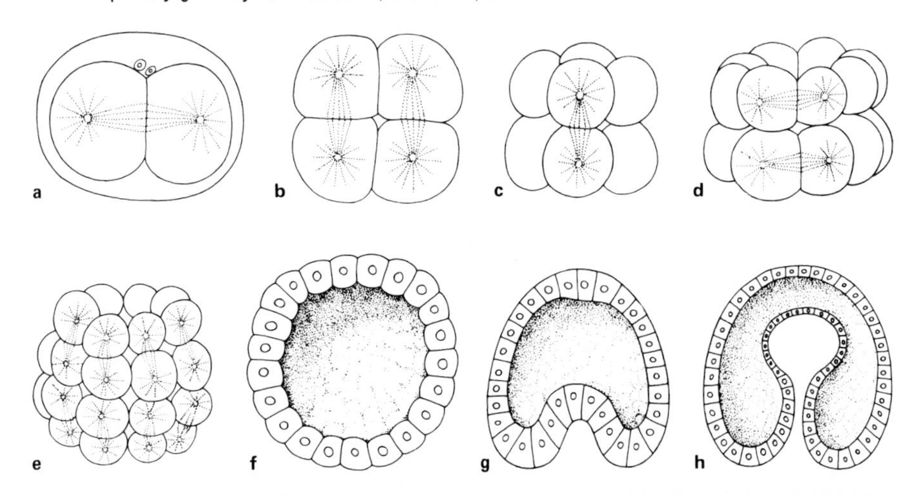

bilateral symmetry is established by the plane of the first cleavage furrow, and cleavage is bilaterally symmetrical.

In spiral cleavage, an entirely different pattern is seen. In contrast with radial cleavage, the cleavage spindles lie at an oblique angle (some angle other than 90 or 180°) to the animal-vegetal axis (Fig. 6.17). The resulting tiers of cells come off in a spiral fashion to the right or the left and come to lie over the furrow of the underlying cell rather than squarely on top of it (Fig. 6.18). The consequences of such cleavage are that each cell is readily identifiable from one minute to the next from embryo to embryo. As a result, many embryologists have painstakingly followed the fate of each cell of many of these embryos and have come to some remarkable conclusions. Animals with spiral cleavage, the spiralia, include members of such diverse groups as annelids, mollusks, turbellarian flatworms, some brachiopods, and echuroids; even nematodes have a type of such cleavage. If one follows the fate of individual cells (a cell lineage study) in the annelids, mollusks, and flatworms, the same cells are seen, and their fate in many cases is identical among these three groups.

Discussion of specific examples involves some cell-lineage terminology. In spiral-cleaving eggs, the first division produces two cells, the AB and the CD cells. After the second division, there are the A, B, C, and D cells. Often there are differences in the sizes of these cells, which allow ready identification. At the third cleavage, two tiers of cells are formed. The animal cells are *micromeres*, and the vegetal tier cells are *macromeres*. Usually, as the name implies, the macromeres are larger; but among spiral eggs this term defines the animal and vegetal cells, and the micromeres may be larger. The A cell gives rise to a macromere, the 1A, and a micromere, the 1a, and so on, for the B, C, and D cells. At the end of the third cleavage, there are four micromeres (the first quartet of micromeres) and four macromeres. At the next division, the 1A produces the macromere 2A and a micromere 2a, while the micromere 1a divides to form two micromeres, $1a^1$
the 1A, and a micromere, the 1a, and so on, for the B, C, and D cells. At the end
cell stage are the 3A, 3a, $2a^1$, $2a^2$, $1a^{11}$, $1a^{12}$, $1a^{21}$, and $1a^{22}$. Figure 6.18 shows a

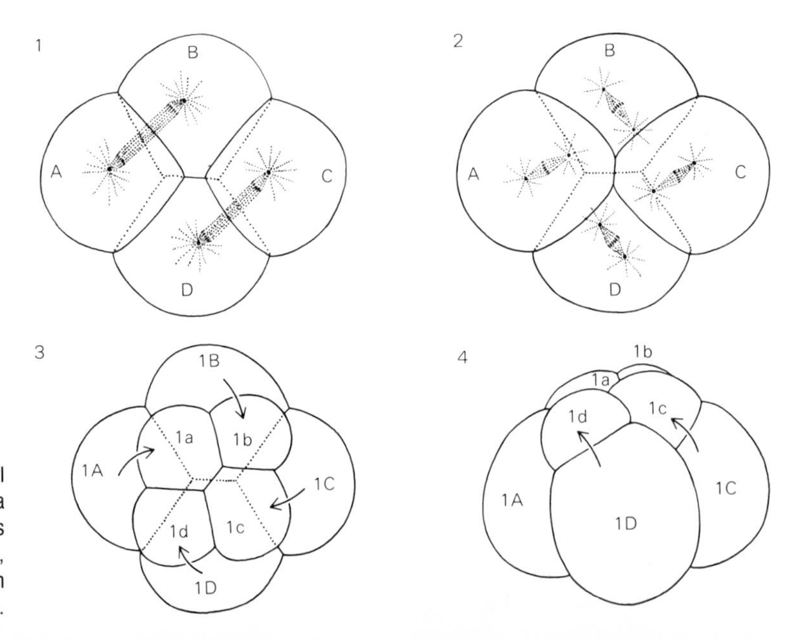

Figure 6.17 Spiral cleavage of egg of a mollusk. Last drawing is viewed from the side, others are seen from the animal pole.

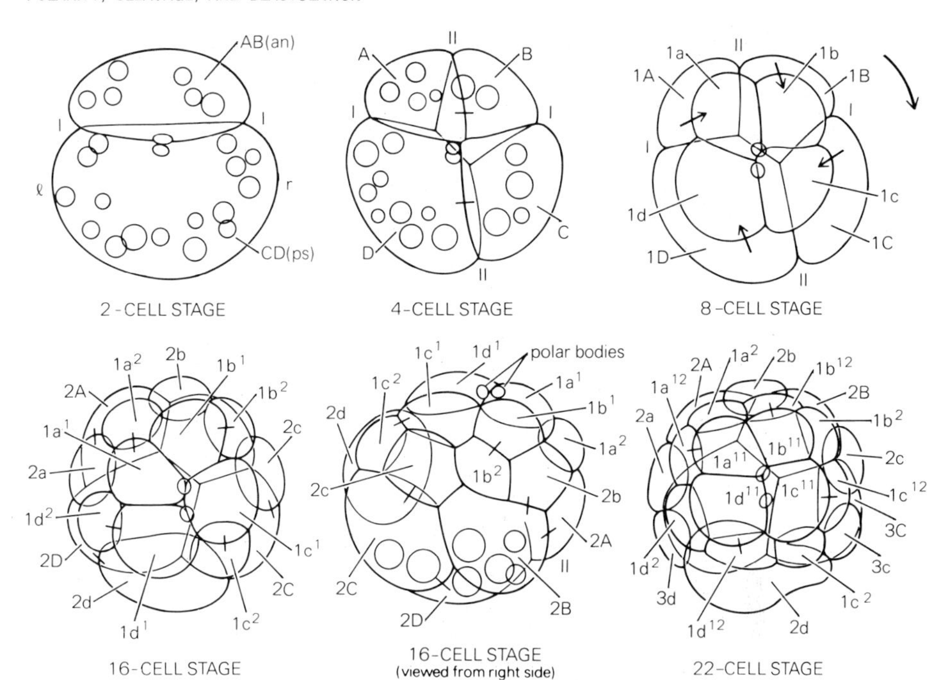

Figure 6.18 Spiral cleavage of the egg of the polychaete *Nereis*. All drawings are polar views except as noted for the 16-cell stage. The circles in the 2-, 4-, and 16-cell stages are oil droplets. [*After D. P. Costello*, J. Elisha Mitchell Sci. Soc., **61**:277 (*1945*).]

Nereis egg part of the way through the fifth cleavage, i.e., some of the cells have divided but others have not. If we had followed the D cell rather than the A cell, we would have encountered cells with a more formative destiny. The large 2d cell (see Fig. 6.18), for example, is responsible for a very large part of the ectoderm of the larva, and the 4d cell is responsible for nearly all the mesodermal structures.

It is a remarkable feature that the same cells form essentially the same structures in many turbellarians, polychaete annelids, and mollusks (the main exception here being the cephalopods). All these groups have been separated from each other for what probably approaches a half billion years; yet their early development, complex as it is, is virtually identical. This clearly illustrates the conservative nature of embryological development. Very little change appears to be tolerated in embryonic stages in species whose adult stages may bear no relation to one another. A similar result is found when the early postgastrulation stages of vertebrate development are compared, although the evolutionary separation time of the various classes is not nearly so great.

Where the location of each cleavage furrow is highly predictable and the blastomeres identifiable, as in the spiralia, cleavage is called *determinate;* radial cleavage, in which the cells become lost as their numbers grow, is called *indeterminate*.

The cytoplasmic organization appears to determine the orientation of cleavage planes. In the spiralia, a cleavage can be a right-handed one, when the spindles are oriented so that the micromeres are formed to the right of the macromere, or a left-handed one. All the mitoses at one cleavage are either right (dextral) or left (sinistral), and the direction alternates at successive cleavage cycles. In other words, if one cleavage is left-handed (counterclockwise as viewed from the animal pole), then the next divisions of the cells will be right-handed (clockwise as viewed from the animal pole).

In snails with coiled shells, the direction of the coiling and therefore the entire viscera can be traced back to the direction of the third cleavage—whether it was dextral or sinistral. Among a population of the snail *Lymnaea*, both directions are found, and a single gene locus is responsible. Dextral coiling is dominant over sinistral. Therefore, one would expect that if a homozygous sinistral female is crossed with a homozygous dextral male, all the offspring would cleave dextrally, since it is dominant. The exact reverse is true, and all offspring cleave sinistrally. The reason for this apparent genetic paradox is that the cytoplasmic organization of the oocyte is produced under the guidance of the sinistral genes of the mother, and the cytoplasm becomes programmed for sinistral cleavage. In this case, the sperm genome has no voice in the entire asymmetry of the individual's internal organs, which is dictated by the early cleavage pattern. In each case, the embryo's cleavage is determined by the genes of its mother rather than by its own, and it is thus one generation behind in this respect.

E. Cleavage of the sea urchin egg

The early stages of sea urchin cleavage are shown in the micrographs in Fig. 6.19 and the diagrams in Fig. 6.20. The first two divisions are vertical and at right angles to each other; the third is horizontal in each of the cells, producing an animal tier and a vegetal tier. In the fourth cleavage, the first indications are seen that different parts of the egg will have different properties. In the animal half, each cell divides with an equal vertical cleavage, producing one tier of eight cells of similar volume, the *mesomeres*. In the vegetal half, each cell undergoes a horizontal cleavage, which is also an unequal cleavage. It produces four small cells at the vegetal pole, the micromeres, and four large cells, the macromeres (Fig. 6.19*d* and *e*).

As in the case of *Lymnaea*, the unequal division at the fourth cleavage points to some underlying cytoplasmic program that will dictate where the mitotic apparatus will lie within each cell and where the cleavage furrow will strike. Not only does the program determine the place of micromere formation, but it determines the time of this cleavage as well. One of the best ways to analyze the underlying mechanism of an event is to dissociate the event under scrutiny from other processes that normally accompany it. For example, if we are interested in what factors determine that an unequal fourth cleavage should occur in sea urchins, we can try to analyze which processes of the embryo can be separated from this unequal cleavage and which cannot. Presumably the latter may be related in a causal way to the unequal cleavage. Numerous treatments of a sea urchin egg are capable of suppressing cleavage divisions. For example, by the use of ultraviolet radiation, shaking, or hypotonic seawater, the first one, two, or three cleavages can be inhibited. In any of these cases (Fig. 6.20*b* and *c*), at the time when the fourth cleavage would normally have occurred (even if no prior cleavage had been allowed), an unequal horizontal cleavage takes place just as if no disruption had been inflicted on the egg. In some way the egg has kept track of the time after fertilization, and this "micromere clock" within the egg has determined that a horizontal, unequal cleavage should occur. In other words, since this cleavage has been dissociated from the three previous cleavages, we can conclude that its formation is not dependent on any earlier cleavage. The elimination of what it is not related to tells us little about what it is dependent on. The question then becomes: Are there any treatments that the egg can be subjected to that will selectively interfere with the micromere cleavage, and if so, what else is coordinately inhibited? We may be able to answer this, but first a digression must be presented.

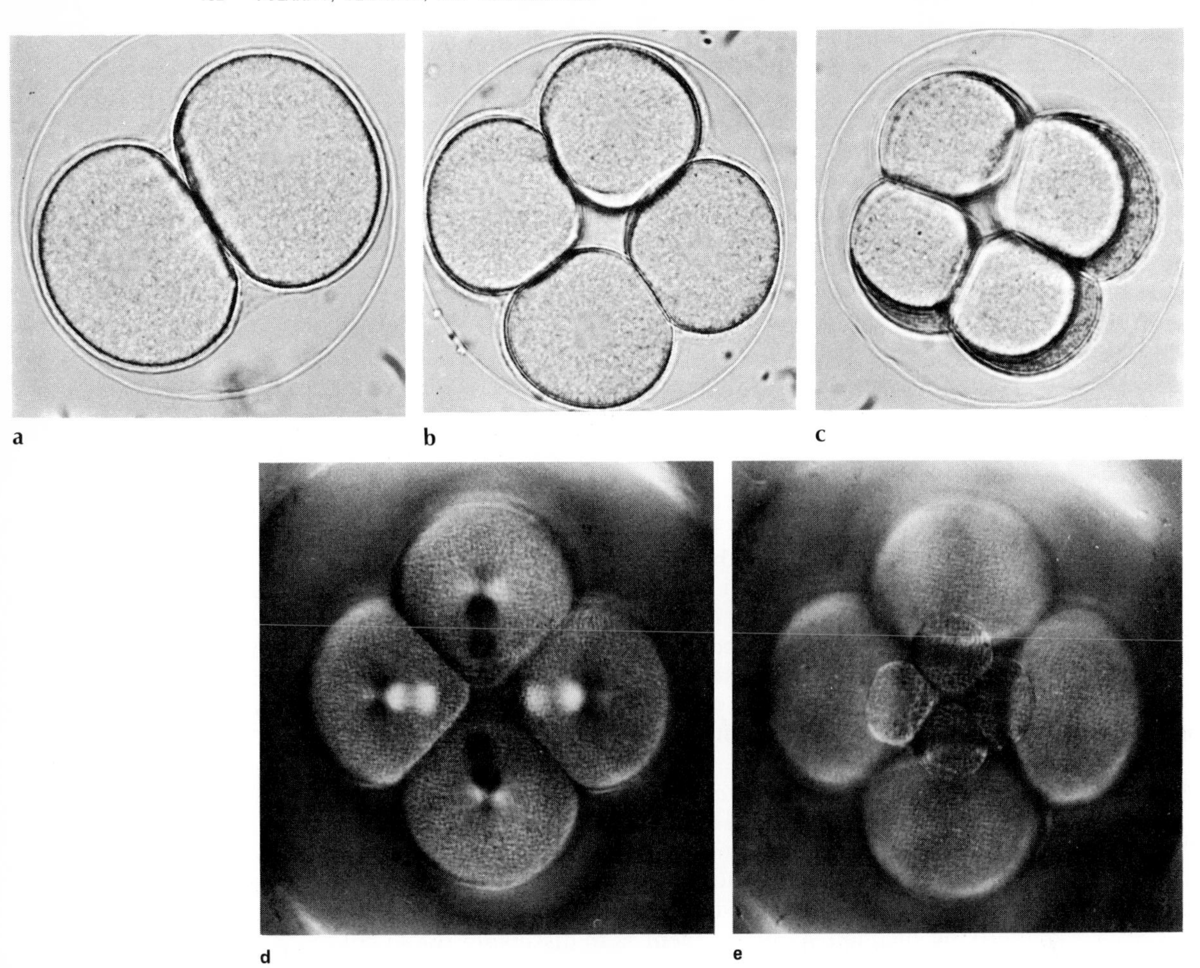

Figure 6.19 *a–c* First three cleavages of sea urchin egg within fertilization membrane, showing hyaline layer, flattening of interface between adjoining cells, and curved free surfaces. [*Courtesy of T. Gustafson.*] *d* A view of the vegetal pole of an echinoderm egg undergoing fourth cleavage as seen in the polarization microscope. The polarization microscope has a special function in the study of cells in that it reveals structures composed of highly oriented components. The mitotic spindle is such a structure since it consists of large numbers of aligned microtubules causing the spindle to appear birefringent, i.e. brighter or darker than the background cytoplasm. In this photograph the four spindles of the dividing macromeres and their perpendicular orientation to one another is evident from their alternating bright and dark appearance. The fourth cleavage in echinoids (as in the sand dollar shown here) is highly unequal and the mitotic spindles are accordingly located much closer to the inner edge of the dividing cells. *e* Polarization micrograph of the embryo shown in *d* following cleavage. The cells have cleaved perpendicular to the spindle axis and have given rise to four micromeres and four macromeres. [*From S. Inoué and D. P. Kiehart in "Cell Reproduction: In Honor of Daniel Mazia," E. R. Dirksen, D. M. Prescott, and C. F. Fox, eds., Academic, 1978.*]

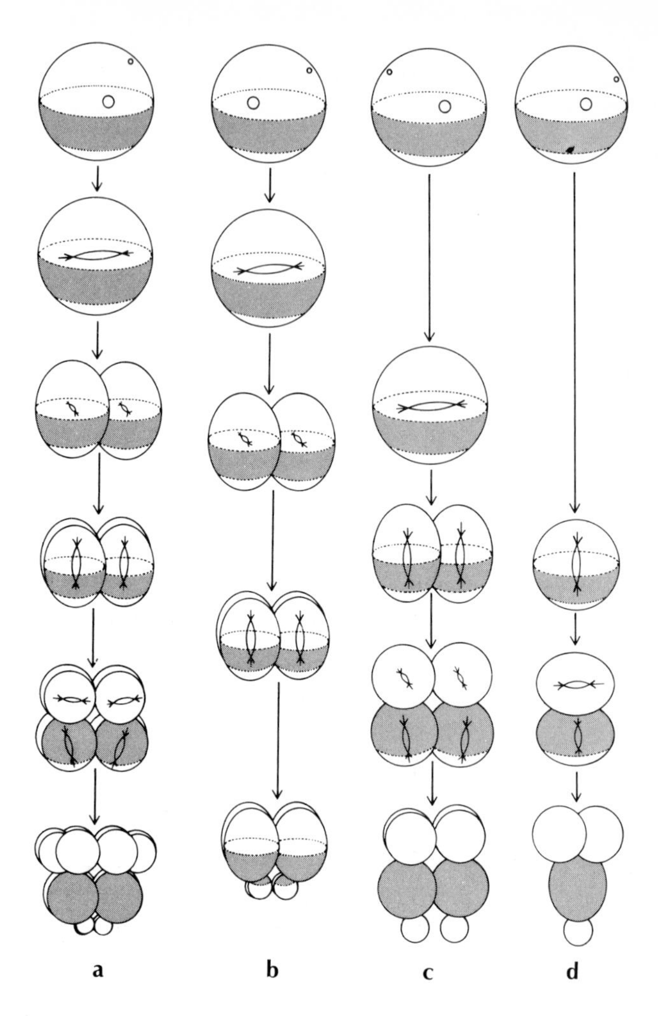

Figure 6.20 *a* Normal cleavage in the sea urchin. *b–c* In these cases, the normal cleavage pattern is altered by treating the eggs with hypotonic seawater or shaking. In both cases, one of the first cleavages has been skipped (a horizontal cleavage is missed in *b* and a vertical cleavage is missed in *c*), but the micromeres are formed at the vegetal pole at the same approximate time as in the control cultures. As a result, the 8-cell embryos of *b* and *c* demonstrate a differentiation that normally occurs at the 16-cell stage. *d* In this case, the fertilized egg was cut in half meridionally and each fragment demonstrates a half-cleavage pattern. [*After S. Horstadius,* Biol. Revs., **14:***139 (1939)*.]

If sea urchin eggs are homogenized in distilled water, all the materials soluble in water will be extracted and those insoluble can be centrifuged to the bottom of the tube. If this pellet, after extraction in water, is then resuspended in 0.6 *M* KCl, a certain group of proteins that were not soluble in water are now solubilized in this salt solution. One of the amino acids that compose proteins, cysteine, has a sulfhydryl (—SH) group, and the amount of SH in the KCl extract is easily determined. It has been found that if the amounts of SH present in the KCl extract are measured, using eggs of different stages of cleavage, there are striking changes that can be closely correlated with the cleavage cycle (Fig. 6.21). The maximum SH levels are found during cleavage and the minimum SH levels during mid-interphase. By the fourth cleavage there have occurred four SH cycles. How do these SH cycles relate to the micromere clock? If sea urchin eggs are irradiated with UV light or treated with the inhibitor 2,4-dinitrophenol (DNP) for a short time, the next cleavage is blocked, but the SH cycle continues. At the time when untreated eggs undergo their fourth cleavage, the treated eggs undergo their third cleavage (although their fourth SH cycle) and cleave unequally to produce micromeres at the vegetal pole in the eight-cell stage.

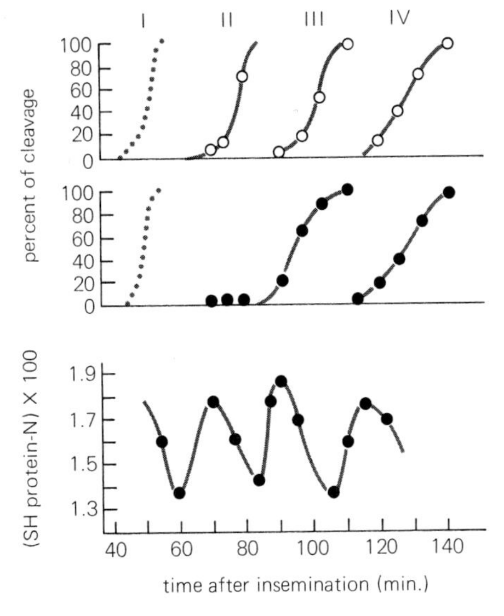

Figure 6.21 Lower curve illustrates the variation in SH content of extracted protein with time after fertilization. Maximum SH content coincides with midmitotic period through each of the first four cleavages. The upper two curves show the percentages of cells that have actually undergone division at each cleavage period. The very top graph illustrates a control culture undergoing the first four cleavages. The middle graph illustrates the effect of a 1-minute exposure to ultraviolet radiation; the second cleavage is skipped, and the eggs wait until the time the controls are undergoing their third cleavage before they undergo their second. The second SH cycle continues unchanged in these irradiated eggs, even though the second cleavage has not occurred. At the time the controls undergo their fourth cleavage, the irradiated eggs undergo their third cleavage, which is now producing unequal micromeres at a premature cleavage stage. [*After M. Ikeda,* Exp. Cell Res., **40**:282 (1965).]

If cleaving sea urchin eggs are placed in seawater containing a bit of ether (0.6 volume percent), the SH cycle is blocked and the level of KCl-soluble SH remains constant until the eggs are removed. Even though the SH cycle is suppressed, nuclear division can continue. If these ether-treated eggs are returned to normal seawater and their cleavage is analyzed, it is found that the fourth cleavage (the third SH cycle) produces essentially an equal cleavage and no micromeres are formed, just as if it had been the third cleavage.

Many biological phenomena can be shown to occur with a predictable time interval, suggesting that mechanisms exist that are capable of measuring periods of time. Very little is known about these mechanisms—whether they relate to annual migrations of animals, monthly variations in reproductive activity, daily cycles of feeding behavior, or, in this case, rhythmic increases in cell number. The implications from the present study of micromere formation are that SH cycles keep track of cleavage number in anticipation of the fourth unequal cleavage. These two events cannot be dissociated from each other, in contrast with the dissociation that has been achieved between micromere formation and the number of nuclear divisions or cytokineses. Even though answers such as this one, if the interpretation of these results is correct, serve to push back the questions to a more basic level (i.e., what causes the cycles in SH level), they still serve to explain in a very real sense how complex developmental processes are controlled.

F. Alterations of cleavage pattern

Even though much of the program for the cleavage pattern is contained within the cytoplasmic organization of the egg, mechanical factors can be demonstrated to play a role as well. For example, if a fertilized sea urchin egg is compressed between two coverslips so that its top and bottom become flattened, the spindles are forced to lie with their long axes parallel to the coverslips. In such a condition, the ensuing cleavage furrow will be perpendicular to the planes of the coverslips. As long as the compression is maintained, the cleavages will continue to be vertical, producing a single tier with an increasing number of cells. If the compression is removed, even as late as after the sixth vertical cleavage, normal development can result. This experiment indicates that the manner in which the cytoplasm is cleaved is not of utmost importance to the future development of the embryo, although this does not necessarily pertain to all embryos.

Another type of experiment also suggests that certain developmental processes, at least, can continue in the absence of the typical cleavage pattern. When unfertilized eggs of the polychaete *Chaetopterus* are treated with a solution of potassium chloride, a remarkable response is seen. Initially, the egg becomes activated and completes its meiotic divisions. Chromosomes are then replicated in anticipation of first cleavage, but the egg is not split and it remains unicellular. One might expect that blocking cleavage would leave the egg in an arrested state, but instead, striking changes occur whose parallel can be found in the normally developing embryo (Fig. 6.22). Various movements occur within the cell, and the nonyolky cytoplasm moves to surround the yolky mass in a manner similar to that which occurs during gastrulation. Meanwhile, the cell elongates to resemble the normal trochophore larva, and the surface of the cell becomes ciliated and motile. The extent to which this process parallels normal development can be debated, but it appears that some developmental events can proceed independently of the parceling out of the cytoplasm. Again, it is apparent that some of the events occurring during development can be dissociated from one another, presumably reflecting their different underlying mechanisms.

Figure 6.22 Differentiation without cleavage in the egg of the polychaete *Chaetopterus*. *a* Normal larva. *b* Unfertilized egg after treatment with dilute solution of potassium chloride in seawater for 24 hours. [*After F. R. Lillie,* Arch. Entwickl.-mech., **14**:477 (*1902*).]

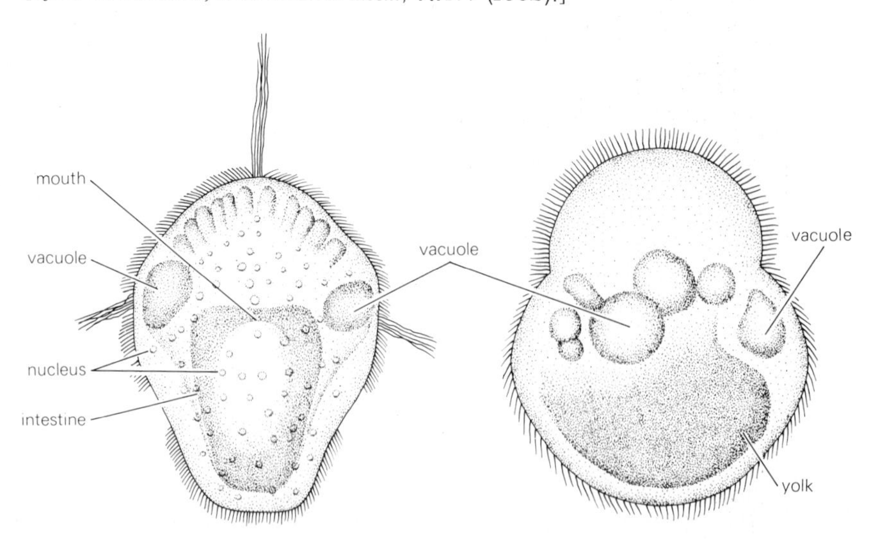

5. Blastulation

Although the basic mechanisms involved in cleavage appear to be very similar in all animals, the patterns that are formed by the cleaving cells are dramatically dissimilar. The latter part of cleavage is marked by a decrease in the rate of cell division and the formation of the *blastula*. The blastula is generally a brief stage, awaiting a much more dramatic rearrangement of cells during gastrulation. The process of blastulation is so different among different animals that generalizations cannot be made. Most blastulas are characterized by an internal, fluid-filled cavity, the *blastocoel*, although a variety of animals possess a blastula completely filled with cells, a *stereoblastula*.

One of the better-studied examples of blastulation is in the sea urchin, an embryo with an extremely spacious blastocoel, as is common among embryos developing from isolecithal eggs. In the sea urchin a cavity appears at a very early stage in cleavage (Fig. 6.19); this cavity continues to enlarge as the cells grow in number and decrease in thickness to form a thinner and thinner outer wall (Fig. 6.23). As each blastomere divides into two cells, one would expect that the large

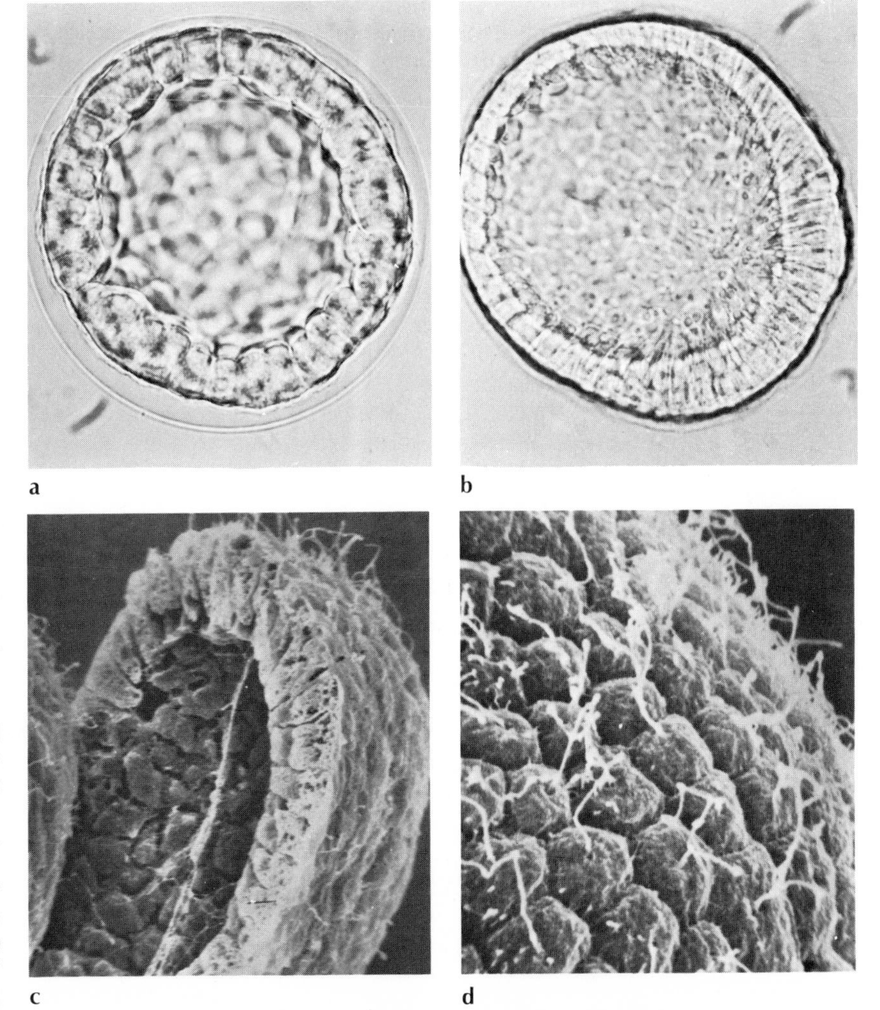

Figure 6.23 Early *a* and late *b* blastula of a sea urchin, showing blastocoel, single-cell-thick layer of the blastula wall, and external noncellular hyaline layer. [*Courtesy of T. Gustafson.*] *c*, *d* Scanning electron micrographs of hatched sea urchin blastula. *c* Bisected blastula showing blastula wall and blastocoel. *d* Small area of external surface, each cell bearing a single cilium. [*Courtesy of W. J. Humphreys.*]

spherical egg would become chopped up into a mass of smaller cells, packed into the original volume; the formation of this entirely different structure is difficult to explain and presents formidable engineering problems. Why do the cells remain at the outer wall and allow the central region to become cavernous? Two theories have been presented to explain this formation. One of the basic premises of both theories is this: The reason the cells form an outer wall rather than a solid mass is that each cell is attached to the hyaline layer which tightly surrounds the entire embryo. In Dan's theory, the blastocoel forms and enlarges as a result of osmotic pressure of the blastocoelic fluid. This fluid contains a considerable quantity of macromolecules and might exert osmotic pressure via the osmotic uptake of water from the environment. If water were to move into the blastocoel, it could exert pressure on the cells attached to the hyaline layer and force them outward.

In the theory of Gustafson and Wolpert, blastocoel formation and expansion result from the manner in which the cells are packed. Just before each cell divides, the surface tension of the outer cell layer increases and the cells round up, divide, and produce two rounded cells. When one spherical cell divides into two spherical cells, the combined diameter of the two daughter cells exceeds that of the original cell. As long as the dividing cells remain in one layer and their outer surface remains attached to the hyaline layer, the layer of cells must inevitably push itself outward as a result of the increasing diameters of the cells. It is the radial pattern of cleavage that keeps the cells side by side rather than forming layers of cells. The rounding up of cells prior to their division suggests that changes occur in the surface tension of the membrane and in the adhesiveness of one cell to its neighbors. The importance of these factors is illustrated in Fig. 6.24. Transitions between *a*, *b*, and *c* are possible simply by changes in surface

Figure 6.24 Possible configurations of the cross section of a tube comprising 24 cells in a single layer attached to a supporting membrane (H), with no change in cell volume. *a* Considerable contact between cells and no cavity. *b* Moderate contact, and a cavity is present. *c* Point contacts between the rounded-up cells. *d* Considerable contact with the supporting membrane. [*After T. Gustafson and L. Wolpert,* Biol. Revs., **42:**458 (1967).]

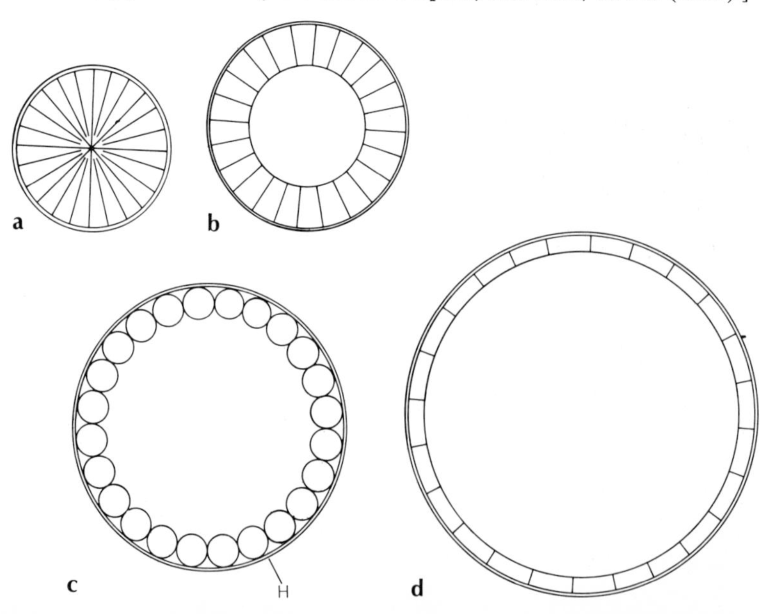

tension and cell-cell adhesiveness. Case *d* would require additional factors. The view of Wolpert and Gustafson, therefore, seems to explain echinoderm blastulation simply by changes in the mechanical properties of the cells and geometrical packing of an increasing number of cells, produced by radial cleavages, that remain attached to an elastic hyaline layer.

Blastocoel formation in the oligolecithal egg of *Amphioxus* (Figs. 6.9 and 6.25*a*) resembles that in the sea urchin. In embryos containing more yolk, blastula formation must be governed by different physical forces.

Not only is the amphibian egg larger than the invertebrate kinds just discussed, but it contains proportionately more yolk. If a frog egg is centrifuged so that all the yolk is compacted toward one side, the yolk is seen to occupy about one-half the egg interior. Normally it is distributed as a gradient, so that the ratio of yolk to cytoplasm is least toward the animal pole and greatest toward the vegetal pole. As a result, the egg nucleus resides considerably above the equator. The relative concentration of yolk in the vegetal half of the egg and the location of the nucleus in the animal half have important consequences with regard to cleavage, blastula formation, and gastrulation (Fig. 6.25).

The first two cleavages are both in the plane of the polar axis at right angles to each other. The mitotic apparatus, however, forms in conjunction with the eccentrically placed nucleus, so that the cleavage furrows first appear near the animal pole and progressively extend toward the opposite pole. The third, or horizontal, cleavage is also influenced by the position of the nucleus in each of the first

Figure 6.25 Schematic sectional diagrams comparing the blastulas of *Amphioxus*, frog, and chick. [*Modified from A. F. Huettner, "Fundamentals of Comparative Embryology of the Vertebrates." By permission of The Macmillan Company, New York, 1949.*]

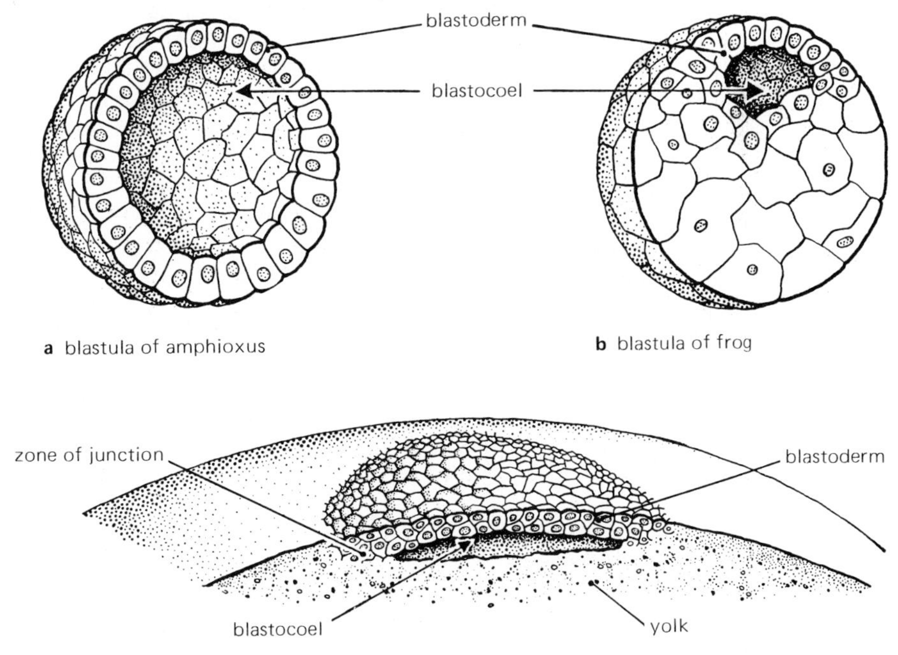

four blastomeres; it divides each cell into a relatively small animal cell, or micromere, containing relatively little yolk and a relatively large, yolky, vegetal macromere (Fig. 6.9*b*). By so doing, it sets up a differential. From this time on, the micromeres divide relatively rapidly, free from the burden of cleaving through a yolk-dense region; the macromeres divide very slowly, since the cytoplasm is now loaded with yolk throughout. Yolk granules or platelets have the double effect of conferring physical inertia on the dividing mass and diluting the active cytoplasm with metabolically inactive inclusions.

As cleavage proceeds, a stage is reached that consists of an upper, animal region and a lower, vegetal region. The animal region is composed of numerous, small, rapidly and synchronously dividing cells that cover a comparatively small blastocoel. The vegetal region is at least half the whole embryo and consists of relatively large, slowly dividing cells. In most amphibians—both frogs and salamanders—the cells constituting the roof of the blastocoel at first form an epithelium one cell thick; soon they establish a layer two cells thick, presumably because their rate of multiplication exceeds the capacity of the blastula to expand. In *Xenopus*, however, the blastula roof remains as a single layer. In meroblastic cleavage (Fig. 6.25), the yolk remains uncleaved; the thin, cleaving disc of cytoplasm becomes slightly elevated above the yolk to form a blastocoelic cavity.

In mammals an entirely different means of blastulation is seen. In the isolecithal sea urchin egg, the blastocoel can be found as soon as cleavage begins, since the cells remain in one peripheral layer. In the isolecithal mammalian egg, cleavage divisions are not radial and the embryos become a solid ball of cells, some of which are completely surrounded by other blastomeres. This solid intermediate state is the *morula,* and blastula formation results from a complete rearrangement of the blastomeres. The process, termed *cavitation,* is characterized by the uptake of fluid between the cells and therefore the formation of an internal cavity. As the cells become separated from each other, the embryo expands greatly in volume and a layer of cells forms around a central cavity (Fig. 6.9). Eventually the outer layer becomes thinner and more and more flattened; it is called the *trophectoderm*. During this process of cavitation, not all the cells are incorporated into the thin, outer trophectoderm layer. Instead, some of the cells collect in a clump (the inner cell mass) and are found within the blastocoel at one edge in contact with the trophectoderm. In mammals this stage is called a *blastocyst,* but it corresponds to the blastula of other animals. The formation and development of the mammalian egg are described in more detail in Chapter 7.

Although some type of cleavage process is a prerequisite in the conversion of a single-celled egg into a multicellular embryo, the manner in which this takes place illustrates a remarkable diversity among animal groups. As a result of cleavage, the swollen cytoplasmic/nuclear volume ratio is corrected; the embryo is provided with a sufficient number of cells with which to begin the business of reorganization into a complex, multilayered structure. Although the process of cleavage is essential, the precise way it occurs is not always of great importance; in numerous cases the disruption of the normal series of cleavage furrows does not interfere with the embryogenesis of that individual. Cleavage is accomplished by a series of mitotic divisions whose study has provided insights into the mitotic process itself. The establishment of the contractile ring, as manifested in the contractile properties of its cytoplasmic filaments, has provided answers to questions that have been posed for many decades. In the following chapter we will return to the cleavage period with an eye to the informational organization of the early embryo.

Readings

ARNOLD, J. M., 1976. Cytokinesis in Animal Cells: New Answers to Old Questions, in G. Poste and G. L. Nicolson (eds.), "The Cell Surface in Animal Embryogenesis," Elsevier.

———, 1969. Cleavage Furrow Formation in a Telolecithal Egg (*Loligo pealii*). I. Filaments in Early Furrow Formation, *J. Cell Biol.*, **41**:894–904.

BELLAIRS, R., F. W. LORENZ, and T. DUNLAP, 1978. Cleavage in the Chick Embryo, *J. Embryol. Exp. Morphol.*, **43**:55–69.

BENNET, M. V. L., and J. P. TRINKAUS, 1970. Electrical Coupling between Cells by Way of Extracellular Space and Specialized Junctions, *J. Cell Biol.*, **44**:592–610.

BOYCOTT, A. E., C. DIVER, S. L. GARSTANG, and F. M. TURNER, 1930. The Inheritance of Sinistrality in *Limnaea peregra* (Mollusca, Pulmonata), *Phil. Trans. Roy. Soc. London*, **B219**:51–131.

CLAVERT, J., 1962. Symmetrization of the Egg of Vertebrates, *Advan. Morphog.*, **2**:27–60.

COSTELLO, D. P., 1955. Cleavage, Blastulation, and Gastrulation, in B. H. Willier, P. Weiss, and V. Hamburger (eds.), "Analysis of Development," Saunders.

DAN, K., 1960. Cytoembryology of Echinoderm and Amphibia, *Int. Rev. Cytol.*, **9**:321–367.

———, and M. IKEDA, 1971. On the System Controlling the Time of Micromere Formation in Sea Urchin Embryos, *Development, Growth, and Differentiation*, **13**:285–301.

ELINSON, R. P., and M. E. MANES, 1978. Morphology of the Site of Sperm Entry on the Frog Egg, *Develop. Biol.*, **63**:67–75.

FREEMAN, G., 1976. The Effects of Altering the Position of Cleavage Planes on the Process of Localization of Developmental Potential in Ctenophores, *Develop. Biol.*, **51**: 332–337.

FUJIWARA, K., M. E. PORTER, and T. D. POLLARD, 1978. Alpha-Actinin Localization in the Cleavage Furrow during Cytokinesis, *J. Cell Biol.*, **79**:268–275.

GUSTAFSON, T., and L. WOLPERT, 1967. Cellular Movement and Contact in Sea Urchin Morphogenesis, *Biol. Rev.*,**42**:442–498.

HARVEY, E. B., 1956. "The American *Arbacia* and Other Sea Urchins," Princeton University Press.

HIRAMOTO, Y., 1970. Rheological Properties of Sea Urchin Eggs, *Biorheology*, **6**:201–234.

———, 1965. Further Studies on Cell Division Without Mitotic Apparatus in Sea Urchin Eggs, *J. Cell Biol.*, **25**:161–167.

LANDSTROM, U., and S. LOVTRUP, 1975. On the Determination of the Dorso-Ventral Polarity in *Xenopus laevis* Embryos, *J. Embryol. Exp. Morphol.*, **33**:879–895.

MARSLAND, D., 1956. Protoplasmic Contractility in Relation to Gel Structure: Temperature-Pressure Experiments on Cytokinesis and Amoeboid Movement, *Int. Rev. Cytol.*, **5**:199–227.

———, and J. V. LANDAU, 1954. The Mechanics of Cytokinesis: Temperature-Pressure Studies on the Cortical Gel System in Various Marine Eggs, *J. Exp. Zool.*,**125**:507–539.

MIKI-NOUMURA, T., and F. OOSAWA, 1969. An Actin-like Protein of the Sea Urchin Eggs. I. Its Interaction with Myosin from Rabbit Striated Muscle, *Exp. Cell Res.*,**56**:224–232.

MITCHISON, J. M., and M. M. SWANN, 1954. The Mechanical Properties of the Cell Surface, *J. Exp. Biol.*, **31**:443–472.

MORGAN, T. H., 1927. "Experimental Embryology," Columbia University Press.

NIEUWKOOP, P. D., 1977. Origin and Establishment of Embryonic Polar Axes in Amphibian Development, *Curr. Topics Dev. Biol.*, **11**:115–132.

NUCITELLI, R., 1978. Ooplasmic Segregation and Secretion in the Pelvetia Egg is Accompanied by a Membrane-Generated Electrical Current, *Develop. Biol.*, **62**:13–26.

QUATRANO, R. S., 1978. Development of Cell Polarity, *Ann. Rev. Plant Physiol.*, **29**: 487–510.

RAPPAPORT, R., 1975. Establishment and Organization of the Cleavage Mechanism, in S. Inoué and R. E. Stephens (eds.), "Molecules and Cell Movement," Raven.

———, 1974. Cleavage, in J. Lash and J. R. Whittaker (eds.), "Concepts of Development," Sinauer.

ROBINSON, K. R., and L. F. JAFFE, 1975. Polarizing Fucoid Eggs Drive a Calcium Current Through Themselves, *Science*, **187**:70–72.

SAKAI, H., 1978. The Isolated Mitotic Apparatus and Chromosome Motion, *Int. Rev. Cytol.*, **55**:23–48.

SCHROEDER, T. E., 1975. Dynamics of the Contractile Ring, in S. Inoué and R. E. Stephens (eds.), "Molecules and Cell Movement," Raven.

———, 1972. The Contractile Ring. II. Determining its Brief Existence, Volumetric Changes and Vital Role in Cleaving *Arabacia* Eggs, *J. Cell Biol.*, **53**:419–434.

SHERIDAN, J. D., 1976. Cell Coupling and Cell Communication, in G. Poste and G. L. Nicolson (eds.), "The Cell Surface in Animal Embryogenesis," Elsevier.

SZOLLOSI, D., 1970. Cortical Cytoplasmic Filaments of Cleaving Eggs: A Structural Element Corresponding to the Contractile Ring, *J. Cell Biol.*, **44**:192–209.

WILSON, E. B., 1925. "The Cell in Development and Heredity," 3d ed., Macmillan.

Chapter 7 Early development: experimental analyses

1. Epigenesis and preformation

One of the most profound questions of concern to biologists for several hundred years has been how the complex structure of an embryo and subsequent adult can develop from what appears to be the simple, unordered structure of the fertilized egg. As is most spectacularly conveyed by time-lapse cinematography, this simple, large cell becomes chopped into many cells and undergoes dramatic rearrangement; soon complex structures begin to appear here and there as the entire embryo continues to be shaped according to some preexisting plan of construction. Each step along the way is predictable; little is left to chance.

Historically, two theories were called upon to explain the events of development. One theory, that of *preformation*, simply denied that development results in an increase in structural complexity. The early preformationists of the seventeenth and eighteenth centuries held that a tiny but complete adult is present in each egg and development simply results in its growth and emergence.

As biologists began to look more closely at the structure of the egg, no such preformed organization could be found; the theory of preformation became more untenable and was abandoned. The theory opposing preformation was referred to as *epigenesis*, a product of the eighteenth and nineteenth centuries. Epigenesis suggests that the structure of the embryo is able to emerge from a formless, protoplasmic mass. The foremost proponent of this theory, E. Wolff, closely watched the development of plants and animals and saw from these studies that the particular organs developed in a gradual manner from more generalized embryonic parts.

By the end of the nineteenth century, experimental embryology had found its beginnings, and each event demanded an explanation in a mechanistic fashion. One way out of the dilemma, in keeping with the newly emerging concepts, was to suggest that there was a type of preformation in the egg protoplasm that fore-

shadowed events to come; this preformation was not a morphological one, but a molecular, biochemical one.

The current explanation for the emergence of the embryo from the egg is in a real sense a composite of these two opposing concepts, epigenesis and preformation. The information for total development is present in the egg in an unrecognizable form partially in the organization of the egg cytoplasm, but primarily in the genetic inheritance passed on to the egg by meiosis. If all the information for all processes of the organism is somehow present in the nucleotide sequence within the DNA, then the overall problem of development becomes one of information retrieval from this information storage bank. As discussed in Chapter 2, the concept of sequential gene activation requires only that the proper initiating events occur. Once the first genes are expressed, then presumably part of that expression entails the activation of the next set of genes, and so on, in an ordered and predictable fashion, until the program for development has run its course. The information is therefore present in a coded, preformed state, but the morphology of the embryo arises epigenetically from what is primarily unordered protoplasm under the direction of the genetic material. The developmental biologist is left with the assignment of describing the multitude of steps that occur and of unraveling their underlying mechanisms.

We will return to the role of the genetic material in early development in the following chapter. First we will need to examine the organization of the cytoplasm, for it is there that the initial modulators of genetic activity appear to reside. As discussed in Chapter 4, the very early stages of development, those in which the broadest features of the embryonic body plan are laid down, spring from a program constructed within the cytoplasm of the egg during oogenesis. To a large degree the questions to be addressed in this chapter are ones concerning the spatial organization within the cytoplasm of the egg and early embryo. As we will see, certain of the most basic features of various animal embryos can be traced directly to patterns of cytoplasmic components present within the early developing stages. In addition, the heterogeneity present within the egg cytoplasm is believed to be of great importance in initiating the process of differential gene expression ultimately responsible for embryonic differentiation. Specific genetic modulators (activators and/or repressors) might be expected to be present in some defined spatial order within the egg cytoplasm. With the production of large numbers of nuclei by mitosis during cleavage, the effects of localized cytoplasmic influences would be brought to bear on specific nuclei that come to lie in a given region of the total egg compartment. Once various nuclei are exposed to different cytoplasmic influences, they will respond in a variable manner to produce different genetic transcripts that ultimately should lead the various cells along diverse paths.

During the course of this chapter we will deal with the early stages of a wide variety of different embryos. We will focus our attention on the manner in which the cytoplasmic program is expressed in these embryos, particularly as revealed by the experimental, manipulative approach. In the following chapter, we will return to some of these same embryos and stages and consider some of the biochemical studies on the analysis of gene expression.

2. Echinoderms

Development begins with one cell whose cytoplasm becomes divided into many cells, each of which has a specific destiny, to give rise eventually to cells that have specialized differentiated properties. One of the basic questions in develop-

mental biology, and one of the first to be asked, is whether or not a cell can be made to develop into tissues other than those it would normally form.

The history of this question goes back to 1888, when Wilhelm Roux killed one of the first two cells of a frog egg with a hot needle.[1] The embryo that resulted from the living member of this pair of cells resembled a half-embryo. The interpretation seems clear enough: each cell produced during cleavage will form the same structures it normally produces, regardless of what operations are performed on the embryo. A few years later (1892), Hans Driesch, working with the sea urchin, obtained opposite results.

Driesch separated the first two cells of the sea urchin embryo by vigorously shaking the embryos for several minutes until a few fertilization membranes had burst and the first two cells or four cells separated. He found that each of the first two or four cells could, in isolation, give rise to complete and normal larvae, although of reduced size (Fig. 7.1*a*). These results contradict those of Roux, and subsequent reinvestigations of the fate of the first two cells of a frog embryo supported the findings of Driesch. The original experiment had indicated each cell in the frog could form only a half-embryo because the dead cell remained as an influence on the activities of the living member. It seems that even though the cell was dead and therefore unable to differentiate, the living cell recognized its presence and was thereby limited in its own differentiation. Two-cell separation in *Amphioxus* also results in twins (Fig. 7.1*b*).

The striking regulation of blastomeres into whole embryos led Driesch to describe the egg as a *harmonious equipotential system*—equipotential because a part had a potency to form the whole, and harmonious because the parts work together so well. He defined the term *prospective significance* (or *prospective fate*) as the fate of an embryonic cell under normal conditions of development, and he showed experimentally that the *prospective potency* of an embryonic cell is much greater than its prospective significance. He stated that *the fate of a cell is a function of its position in the whole.* In fact he not only emphasized the epigenetic and positional aspects of early development, but also spoke of both chemical and contact induction processes between cells and tissues, of polarity of the egg as an arrangement of polarized constituents, and of developmental controls mediated by nucleus and enzymes. The two foundation publications by Roux (1888) and Driesch (1892) can be read in English translation in *Foundations of Experimental Embryology*, edited by Willier and Oppenheimer (1964).

These remarkable observations rank among the most important in the science of biology. How can a cell that is to form a part of an embryo suddenly form additional parts not in its normal repertoire? How can we view, on the one hand, the processes of development in a mechanistic fashion and, on the other hand, conceive that an embryo can repair its own damage? What machine do we know that can be cut in half and both halves continue to function as if nothing had happened? How can a part have a sense of the whole? These are basic questions, and we cannot yet provide answers.

In sea urchin development, the third cleavage is the first horizontal cleavage and therefore the first time cells are found whose cytoplasm does not extend from the animal to the vegetal pole. This horizontal cleavage is seen to have a profound effect on the ability of the blastomeres to develop in isolation; they are no longer capable of differentiation in a totipotent manner.

[1] Actually a similar, but less well known experiment was reported in 1887 by Laurent Chabry working on an ascidian. Like Roux, he found that the destruction of blastomeres resulted in the loss of larval structure, but unlike Roux, his conclusions on ascidians were actually shown to be correct (see Section 7.4).

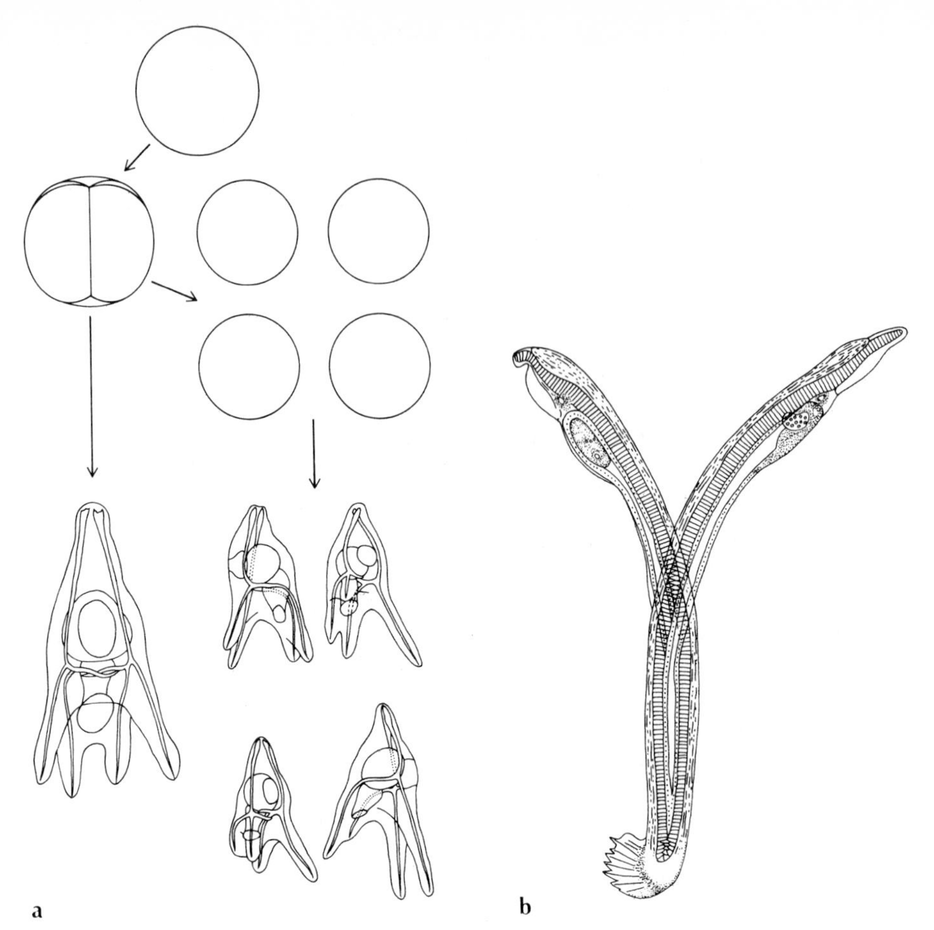

Figure 7.1 *a* Development of isolated blastomeres of the four-cell stage of the sea urchin. Each develops into a pluteus larva of normal proportions but reduced size. *b* Double monster of *Amphioxus*, produced by mechanical disarrangement and partial separation of the blastomeres of the two-cell stage. [*After E. G. Conklin,* J. Exp. Zool., **64**:*373 (1933).*]

The basis of this change in potency is clearly related to the separation of animal from vegetal materials. If the eight-cell stage is severed into two halves by a meridional operation, each group of four cells gives rise to an intact larva. In contrast, if the eight-cell stage is separated into two halves by a horizontal operation, neither the animal half nor the vegetal half can produce a normal embryo. The animal half differentiates into a hollow ball of cells with very long cilia; this abnormal embryo, called a *dauerblastula,* is devoid of endodermal and mesodermal differentiations. The vegetal half develops more normally but is characterized by a lack of arms, a more ovoid body shape, and the formation of an enlarged gut that is often evaginated outward (exogastrulated) rather than pushed within the embryo. The differentiations of the vegetal half reflect an exaggeration of endodermal differentiation at the expense of ectodermal structures. We do not have to wait, however, until the third cleavage to obtain embryos with these abnormalities. The unfertilized egg can be cut horizontally and each half fertilized. The animal and vegetal halves then develop as described for the eight-cell operation. An animal-vegetal differential with regard to prospective potency apparently already exists by the end of oogenesis.

The question that must be analyzed is how these two parts, the animal and vegetal halves, have become so different in their abilities to differentiate in isola-

tion. A theory of gradients based on the work of Runnström and Hörstadius seems to explain this and many other results. According to this theory, the direction in which a given cell of a sea urchin embryo will differentiate is under the control of two influences acting coordinately throughout the embryo. One of these influences (activities or substances of unknown nature) is at its greatest concentration at the animal pole; the other is at its greatest level at the vegetal pole. Each influence spreads out from its respective pole, decreasing until it has essentially disappeared in the region of the opposite pole. Since each diminishes in strength, each is considered to occur as a gradient, and together they form a *double gradient*.

In this theory, the nature of differentiation of each part of the embryo is based on the relative levels of each member of the double gradient. The concept is expressed diagrammatically in Fig. 7.2. In the lower row, each factor is represented by a triangle: the broadest part represents the highest concentration, and the narrow part, the lowest concentration. In Fig. 7.2*e*, the normal condition is shown, in which both animal and vegetal influences are balanced and development is normal. If the embryo is placed under conditions that reduce the vegetal influence further and further (through an increasing imbalance), the embryo develops more and more in an animal-half direction; i.e., it becomes animalized. If the reverse occurs and the embryo's vegetal influence predominates (Fig. 7.2*f* and *g*), the embryo develops as a vegetal half would and shows an enlarged gut that may exogastrulate; the embryo has been vegetalized. The differentiation of each region is believed to be directed by the specific animal/vegetal ratios of that region. Where the animal influence predominates, ectodermal structures are formed. Normally this is only near the animal pole, but if the vegetal influence is somehow depressed or the animal influence is increased, ectodermal differentiations will occur in a greater part of the embryo (Fig. 7.2*a*). The reverse is true for the other gradient system. Techniques to alter the gradients are described later.

Physiological gradients have been invoked for a long time to explain a wide variety of developmental phenomena, including embryonic induction and nerve

Figure 7.2 Sea urchin larvae of about the same age but varying in the animal and vegetal tendencies. *e* Normal gastrula. *a–d* Larvae animalized to varying extents. *f–g* Larvae vegetalized to different extents. The hypothetical double-gradient system corresponding to each larva is indicated. Animalization can be brought about by a reduction of the amount of vegetal material, by treatment with various chemical agents such as SCN^- before fertilization, or by treatment with *o*-iodosobenzoic acid or 2,5-thiomethylcytosine during cleavage stages. Vegetalization can be brought about by lithium or chloramphenicol treatment during cleavage. [*After T. Gustafson, in "The Biochemistry of Animal Development,"* R. Weber (ed.), *Academic, 1965.*]

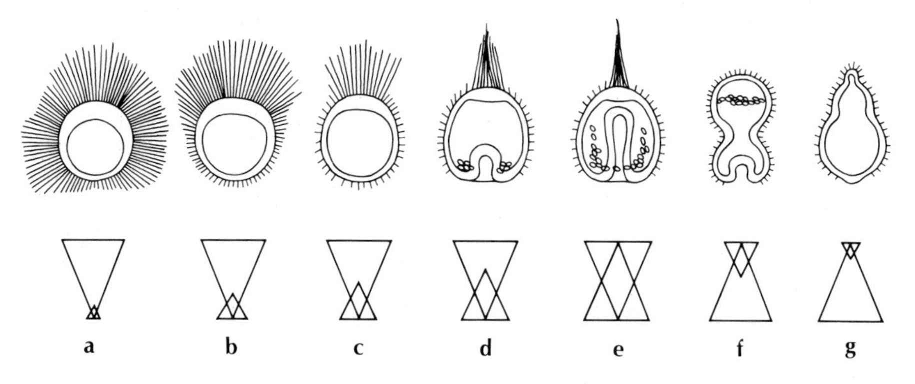

growth and regeneration. The underlying mechanism in a gradient hypothesis is that each part of the whole has a unique identity, a certain level of whatever is in the gradient. In a double gradient the important determinant is the relative value of each influence rather than its absolute amount. Runnström first proposed a double-gradient mechanism underlying sea urchin development; Hörstadius, in a series of elegant microsurgical experiments (reviewed in 1939 and in 1973), is primarily responsible for providing evidence in its favor.

Before describing the experiments of Hörstadius, one must become familiar with the different cells of the early sea urchin and their prospective fates (Fig. 7.3). At the 16-cell stage, there are 8 mesomeres, 4 macromeres, and 4 micromeres (8 + 4 + 4 in the terminology of Hörstadius). At the 32-cell stage, the animal half is composed of two rings of cells (8 per ring), called an_1 and an_2. By the 64-cell stage, the macromeres have divided into two rings of 8 cells each, veg_1 and veg_2. The prospective fate of an_1, an_2, and veg_1 is the outer ectodermal covering of the embryo. Veg_2 will form the gut. The micromeres will differentiate into the skeleton. Hörstadius has isolated the various cells and clusters of cells of the cleaving embryo and recombined them in virtually every possible combination in an analysis of their prospective potencies.

If the double-gradient theory is correct, differentiations should be predictable on the basis of the relative animal-to-vegetal influence of the part in question. An isolated animal half (8 + 0 + 0) differentiates in isolation into a hollow ciliated ball. If, however, four micromeres are added (8 + 0 + 4), a pluteus larva can develop; the micromeres provide the required vegetal influence to balance the animal half and restore the animal/vegetal ratios compatible with normal development. The isolated macromeres (0 + 4 + 0) represent a piece of the middle of both gradients; and although they lack the extreme values of each influence, they retain the opposing concentrations. In isolation, the macromeres develop into a recognizable pluteus larva. It appears that the 4 macromeres are not required for the

Figure 7.3 Development of sea urchin *Paracentrotus lividus* egg. *a* Uncleaved egg. *b* 4-cell stage. *c* 8-cell stage. *d* 16-cell stage formed by equal vertical cleavages in animal half and unequal horizontal cleavages in vegetal half. *e* 32-cell stage with two tiers of blastomeres (an_1 and an_2). *f* 64-cell stage. Macromere descendants form two tiers of blastomeres (veg_1 and veg_2). *g* Early blastula. *h* Later blastula, with apical organ, before formation of primary mesenchyme. *i* Migration of primary mesenchyme. *j* Gastrula: secondary mesenchyme shown at tip of archenteron and two triradiate spicules formed to the sides of it. [*After S. Hörstadius,* Biol. Revs., **14:***139* (*1939*).]

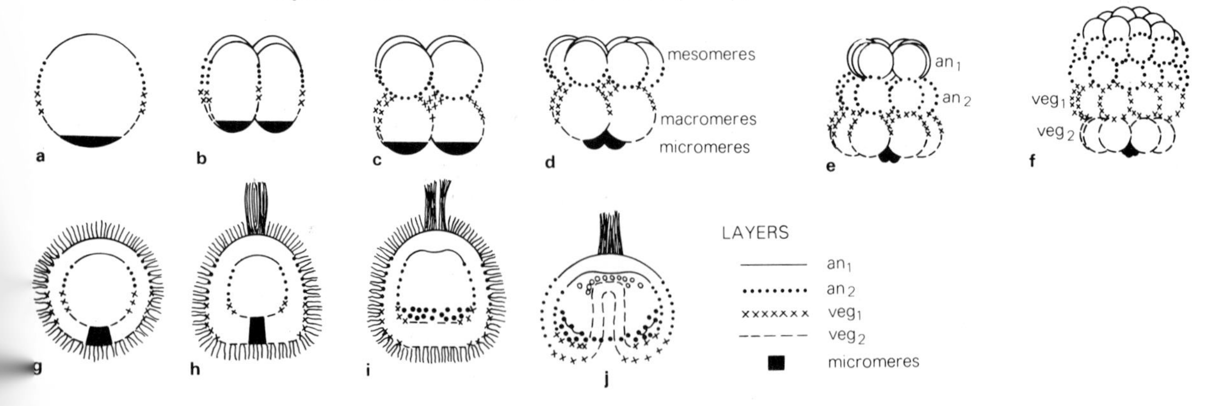

near-normal differentiation of the remaining 12 cells and vice versa. The animal half (eight mesomeres) greatly dominates one macromere (8 + 1 + 0), but if two mesomeres are combined with one macromere (2 + 1 + 0), a pluteuslike larva can form. An entire embryo can be shifted in its differentiation if its animal-vegetal balance is altered; this can be accomplished by the addition of 20 extra micromeres to an intact embryo (8 + 4 + 24). Under the influence of the greatly increased vegetal activity, the whole embryo develops as if it were a vegetal half.

These relative animal and vegetal activities can be essentially titrated, one against the other. If the layers of the 64-cell embryo are separated, their differentiation in isolation or after recombination with one, two, or three micromeres can be determined (Fig. 7.4). In isolation the veg_2 layer shows the most balanced differentiation. As micromeres are added to the veg_2 layer, a more and more vegetalized embryo will develop. In contrast, the an_1 layer can produce, in isolation, only the hollow ball of ciliated cells. Addition of micromeres to the an_1 layer begins to restore the necessary balance, and after four micromeres are added, a rather normal pluteus larva can be produced.

A chemical approach to the analysis of sea urchin development has gone hand in hand with the operative approach from the start. As early as 1892, Herbst discovered that lithium ions (from LiCl) added to seawater caused an increase in the amount of endoderm and led to exogastrulation (Fig. 7.5). Since it causes an entire embryo to develop as if it were a vegetal half, lithium is said to be a vegetalizing agent and to have vegetalized the embryo. Other agents, particularly thiocyanate (SCN^-) and trypsin, are capable of animalizing an embryo so that it will develop as an animal half. Superficially, it seems that lithium and micromeres have the same vegetalizing effect on the embryo. In fact, experiments have indicated that alterations in differentiation that result from surgical disruption of the normal gradients can be balanced by a chemical disruption acting in the opposite

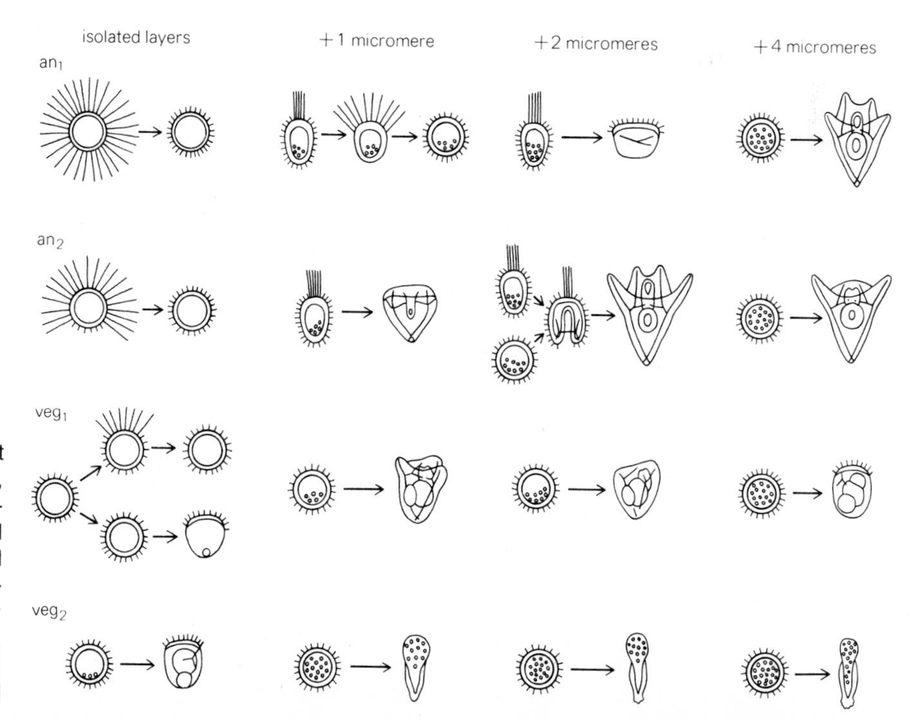

Figure 7.4 Development of the layers an_1, an_2, veg_1, and veg_2 in isolation (*left column*) and with one, two, and four micromeres implanted. [*After S. Hörstadius*, Biol. Revs., **14:***132 (1939)*.]

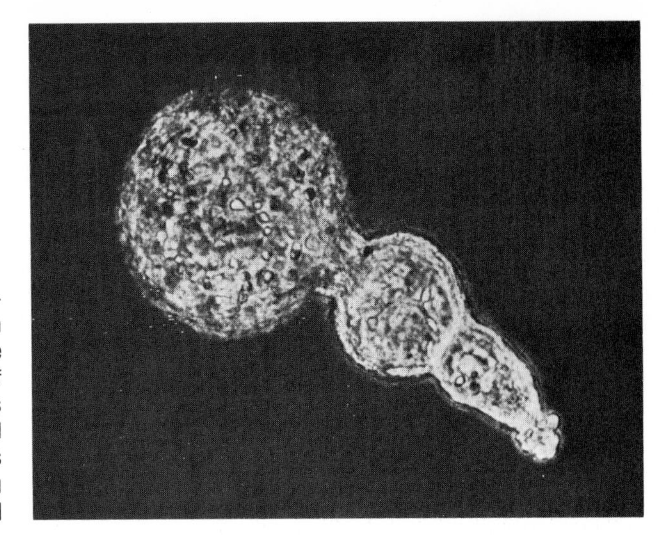

Figure 7.5 Exogastrulation of the sea urchin *Lytechinus,* due to the presence of lithium. Large bulb is a thin-walled ectodermal sac with the regions of the gut extending outward. [*G. Karp.*]

manner. For example, treatment of an animal half with lithium will cause it to differentiate along nearly normal lines. This experiment indicates that the chemical treatments somehow interact specifically with the proposed double gradients and that study of their effects might provide information on the nature of the gradients themselves.

If gradients exist, there should be some way to demonstrate their presence. This appears to have been accomplished by the use of dyes that are capable of accepting electrons from the oxidative metabolism of the embryo in the absence of oxygen. For example, if an embryo is placed in seawater with the dye Janus green and the drop of stained embryos is sealed from the environment, the oxygen will eventually be exhausted as a means to accept electrons from the oxidative phosphorylation chain. When this happens, the Janus green will become the electron acceptor and will be reduced in the process. The dye is colorless in the reduced state; therefore those regions of the embryo undergoing the highest level of oxidative metabolism should become colorless before less active parts. Under these conditions, the vegetal pole loses its stain first, and the reaction spreads in an animal direction. As this reaction is proceeding, a second point of destaining at the animal pole spreads vegetally.

These results suggest that a correlation exists between a double gradient of metabolism and the double gradient of differentiation discussed earlier. The causal link between these two phenomena comes from the analysis of the metabolic gradients under conditions that are known to affect the morphogenetic gradients. The two appear to be linked. For example, lithium treatment abolishes the center of metabolic activity at the animal pole without affecting that of the vegetal region. Animalizing agents have the opposite effect and leave the embryo in possession of an intact animal gradient but lacking a counterbalance from the opposite pole. In each case, the ratios of the two influences are greatly disturbed, and the expected abnormal differentiation results.

Micromeres, when added to the side of an intact embryo, initiate a new vegetal gradient at that point; the morphological outcome is the invagination of a secondary archenteron from that region. Addition of extra cells is an *implantation* experiment. Micromeres are said to have "induced" the formation of a secondary

gut. The term *induction* specifies that the micromeres, in this case, have caused the surrounding tissue to differentiate in a way it would not have without this external influence. If the metabolic gradients of embryos are analyzed after implantation of micromeres, a third center of metabolic activity—in addition to that of the animal and vegetal poles—is seen within the added cells. Taken together, these results lend strong support for the double-gradient hypothesis of sea urchin development, the developmental system most extensively analyzed to date. Since the development of the sea urchin embryo is resistant to centrifugation and the removal of much of the endoplasm (Section 6.3), the cortex has been implicated as the site in which the controlling influences reside.

The foregoing analyses of the development of the sea urchin egg relate to two phases: (1) the early establishment of three primary germ layers represented by an outer ectodermal layer, an inner endodermal tubular layer confluent at each end with the ectoderm, and an intermediate mesoderm component; (2) the development of a specific type of larva capable of maintaining itself for a long period as a swimming, feeding, and growing planktonic organism.

Neither the symmetries nor the structures of the juvenile-adult type of organism—a radially symmetrical, crawling, and browsing seafloor creature of great complexity—appear to be anticipated during the development of the egg to the pluteus stage. Only after a period of growth as a pluteus larva does the profound metamorphosis of the adult pattern begin (Fig. 7.6). There is thus a dual quality to development: (1) the development of the zygote as a specially programmed cell to become a particular type of transient larva, and (2) the development of the zygote as a totipotent cell capable of proliferation eventually sufficient to initiate the parental organization. Comparable dual systems are seen in the development of amphibians and ascidians.

Figure 7.6 Photographs of live stages of developing sea urchin *Echinus esculentus*. *a* Advanced 17-week-old bilaterally symmetrical pluteus larva with eight arms, ciliated epaulettes, and larval skeleton. *b* Young sea urchin soon after metamorphosis (7 weeks old, from fertilization) showing radial symmetry spines and tube feet. [*Courtesy of D. P. Wilson.*]

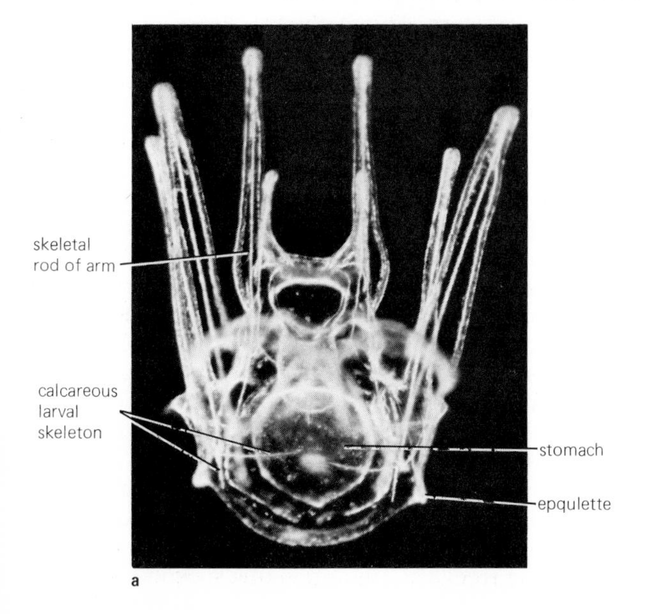

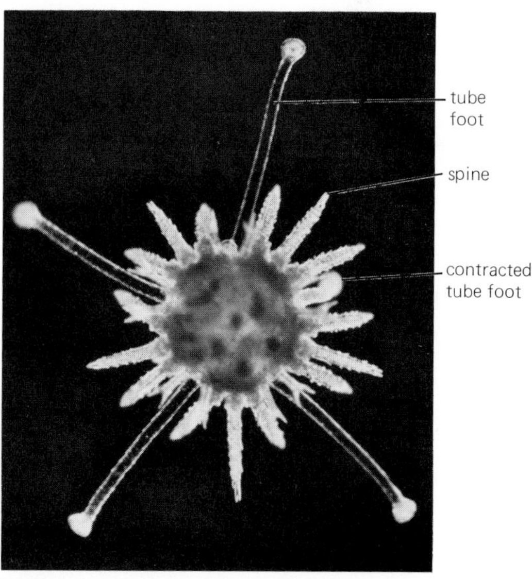

3. Amphibians

In the previous section we mentioned the classic experiment of Wilhelm Roux that erroneously suggested that each of the first two blastomeres of a frog could form but half an embryo. Further experiments with isolated blastomeres indicated that the situation in amphibians was more complex. Unlike the early sea urchin embryo, which is radially symmetrical, the amphibian embryo reveals its bilateral symmetry even before the first cleavage, initially by the formation of the gray crescent (Section 6.2). In some cases the plane of the first cleavage furrow coincides with the plane of bilateral symmetry, while in others it bears no relationship to the future body plan. Experiments in which the first two blastomeres of the urodele *Triturus* were separated indicated that the potential for future development of one of these half-embryos depended on the manner in which the cleavage plane had divided the egg. If the furrow had divided the egg into right and left halves, i.e., had split the egg near the future axis of symmetry, then each cell was capable of regulation and could form a normal diminutive embryo in isolation. However, if the cleavage plane happened to split the egg into dorsal and ventral blastomeres, these two cells had very different fates when separated. Whereas the dorsal blastomere went on to develop into an essentially normal embryo, the developmental capacity of the ventral cell was much more restricted; it formed a "belly piece" lacking in axial structures, such as notochord and nervous system. The ventral half of a two-cell stage does, however, make an attempt to gastrulate, and some mesoderm can move into the interior and differentiate into kidney tubules and blood cells. Similarly, dorsally located blastomeres of the four-cell stage are capable of developing in a largely normal manner in isolation, while ventral blastomeres are not.

The results of these early experiments stimulated a large-scale effort to determine the basis for the importance of the egg's dorsal region. Two experimental approaches have provided further evidence that the gray-crescent region gains special properties soon after fertilization. In an earlier series of experiments (see Curtis, 1962), gray-crescent cortex was removed from very early embryos, while in more recent experiments (see Malacinski et al., 1975), early embryos were subjected to localized ultraviolet radiation. In the following discussion, we will concentrate on the latter experiments, which are less controversial, since they involve less surgical manipulation (which can cause severe abnormalities). Several reports have indicated that ultraviolet radiation of the vegetal hemisphere, but not the animal hemisphere, of frog eggs results in a marked effect on later development. When the vegetal hemisphere of fertilized, uncleaved eggs are irradiated, they typically develop into embryos lacking head structures (acephalic) or nervous tissue in general (aneural). Attempts to localize the UV-sensitive site of the egg suggest that the region just beneath the pigmented cap (the marginal zone) on the dorsal side serves as the primary target. As discussed in Chapter 10, this region of the egg, which includes the gray crescent, will later become the dorsal lip of the blastopore, i.e., the site of initiation of gastrulation. Accompanying nuclear transplantation experiments have indicated that the radiation effect is not occurring as a result of damage to nuclear material of the zygote. Surprisingly, the newly fertilized frog egg is only sensitive to UV radiation for a period up to about 90 minutes after fertilization, a time before the definitive appearance of the gray crescent. Since UV radiation is only capable of penetrating the outer 5 to 10 μm of the egg, the change in sensitivity suggests that an important UV-sensitive component(s) shifts from the peripheral region of the frog egg to a more internal location during the second hour after fertilization. The damage inflicted on the egg by UV radiation can be largely corrected by (1) injection of marginal-zone cytoplasm from nonirradiated eggs or (2) the replacement of the dorsal lip of the

early gastrula (that forms from the irradiated egg) with the dorsal lip of a nonirradiated embryo.

The results of the preceding experiments suggest that some type of cytoplasmic material (likely nucleic acid), seemingly of particular importance in the future development of neural structures, is present in the egg and localized in the dorsal vegetal area of the early embryo. Other observations of long-standing have revealed the presence and localization of an entirely different determinant within the amphibian egg, one having special significance for germ-cell differentiation. It was mentioned in Section 4.3.A.ii that the primorial germ cells arise within the endoderm of the early embryo. In fact, elements of the cytoplasm of the primordial germ cells can be traced all the way back to the unfertilized egg. If the vegetal pole of an unfertilized egg, fertilized egg, or two-cell stage is subjected to UV irradiation, the embryo that develops will lack germ cells and, consequently, will be sterile. Electron microscopic examination of the cytoplasm just beneath the vegetal-pole cortex (the *germ plasm*) reveals the presence of a particular type of cell inclusion referred to as a *polar granule* (Fig. 7.7). The presence of these particulate bodies is closely correlated with the origin of the primordial germ cells into which they become segregated as cleavage progresses. For example, the UV-sensitive period (with respect to producing sterility) in *Xenopus* extends only through the very early cleavage period. By the 8- to 16-cell stage, *Xenopus* embryos are no longer affected by vegetal-pole irradiation, and cytological examination indicates that the polar granules are now found further into the interior of the embryo, out of range of UV damage. The presence of germ-cell determinants as part of the cytoplasmic inheritance of the egg is a very common feature of animal development. Furthermore, these determinants, regardless of their source, typically appear similar to that of the polar granules shown in Fig. 7.7.

Figure 7.7 Electron micrograph of a portion of the vegetal region of an unfertilized frog egg. Numerous germinal granules are present within a mitochondrial cluster lying below the cortical granules. GG, germinal granule; R, ribosomes, G, glycogen; L, lipid; M, mitochondrion. [*Courtesy of M. A. Williams and L. D. Smith.*]

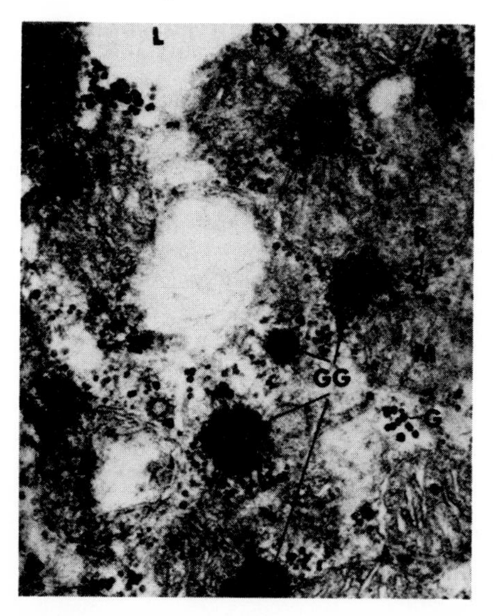

4. Ascidians

We have seen that upon isolation of the first two or four cells of the sea urchin embryo, each is totipotent; development is regulative. If a similar operation is performed on a mollusk, annelid, ascidian, or one of several other groups, the result is different. To a lesser or greater degree the capability for differentiation of isolated cells of these embryos is restricted; i.e., their prospective potency may be little more than their prospective fate. The development of such embryos is termed *mosaic* (discussed in detail in Section 7.6).

An egg of this type is that of the ascidian *Styela,* in which a variety of distinct cytoplasmic regions are demarcated. The zygote develops into a tadpolelike larva (Fig. 7.8), which subsequently metamorphoses into a sessile ascidian. In the unfertilized egg, three distinct regions can be seen: a peripheral layer in which is embedded yellow lipid material in association with mitochondria, gray yolk material, and a germinal vesicle (Fig. 7.9*a*). Immediately after fertilization, there is a dramatic shift in the zygote contents (Fig. 7.9*b*) that can be noted by the movement of the visible cytoplasmic materials (shown in the classical cell lineage study by E. G. Conklin, 1905). The germinal vesicle breaks down and its contents flow to the animal pole. The yellow plasm flows to the vegetal pole and forms a zone at that location. Above the yellow vegetal plasm is a thin clear zone, above which is the gray yolk that fills most of the cell (Fig. 7.9*c*). The egg contents do not remain in this position but are rearranged once more (Fig. 7.9*d–g*). At the future anterior side a gray crescent is formed. On the future posterior side two crescents have been described, a yellow crescent beneath the equator and a clear crescent above it. Most of the yolk becomes shifted to the anterior half of the zygote. The bilateral organization of the embryo is clearly defined by these visible regions.

The first cleavage plane always divides the egg in the plane of bilateral symmetry. As successive cleavages proceed, the material of the yellow crescent becomes segregated in two of the first eight blastomeres; these continue to divide to form a total of 36 cells, the future muscle cells of the tadpole larva. Similarly, the material in the anterior crescent becomes segregated as the 38 to 40 cells that differentiate as notochord cells. The gray yolky protoplasm forms the endoderm. The segregation of mitochondria into the larval muscle cells of *Ciona* is shown in Fig. 7.10.

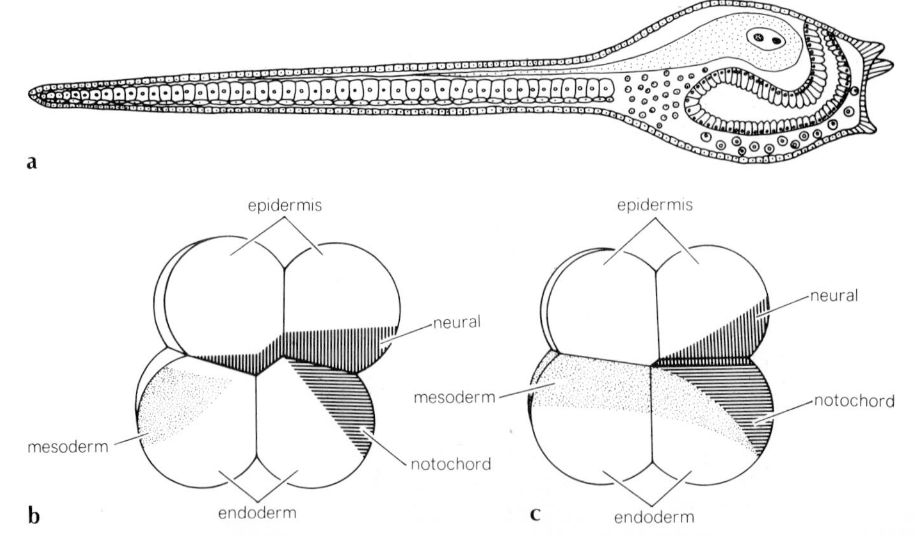

Figure 7.8 *a* Fully developed ascidian tadpole larva possessing notochord, dorsal nerve cord, and pharyngeal gill slits. This motile larva undergoes metamorphosis to form the sessile adult, which lacks the chordate characteristics of a notochord and dorsal nerve cord. *b* Fate map of eight-cell stage of ascidian embryo. *c* Fate map of eight-cell stage of *Amphioxus,* for comparison.

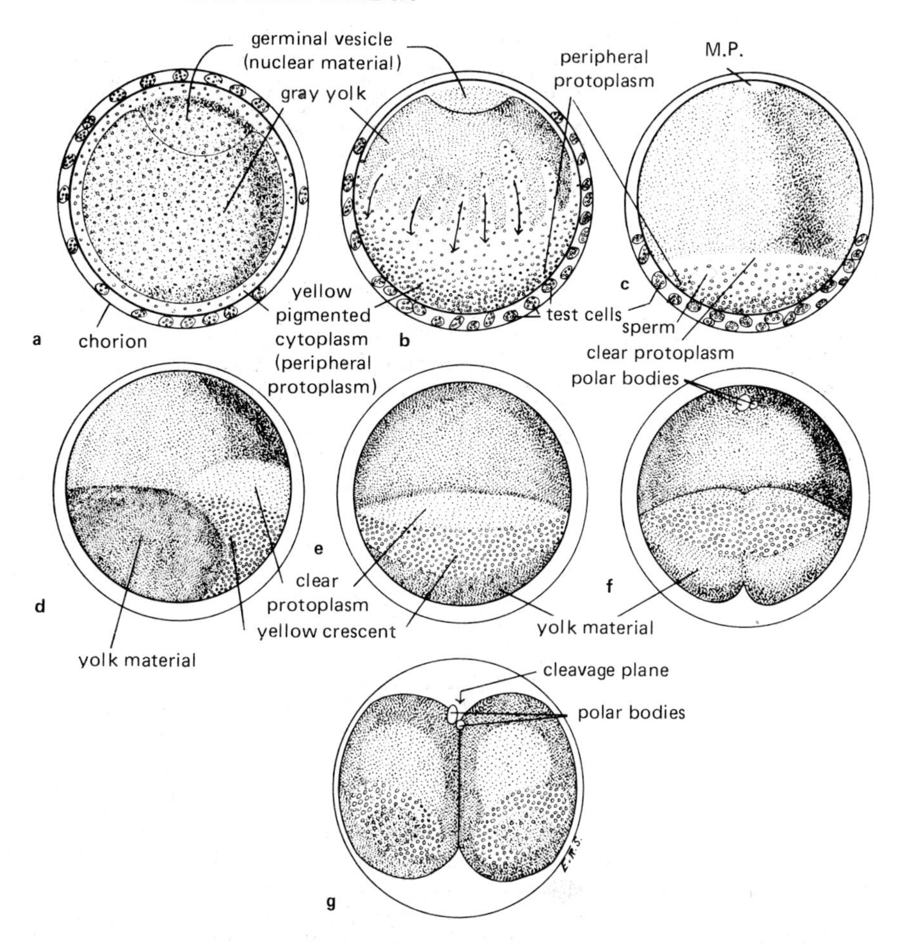

Figure 7.9 Movement of cytoplasmic materials (ooplasmic segregation) in the egg of the ascidian *Styela* following fertilization. *a* Unfertilized egg after germinal-vesicle breakdown. *b* Egg 5 minutes after fertilization. Peripheral cytoplasm streams toward the vegetal pole. The clear cytoplasm (previously from the germinal vesicle) also flows vegetally leaving the gray yolk concentrated in the animal hemisphere. *c* Side view of egg following cytoplasmic movements. *d* Side view of egg showing the yellow and clear crescents on the future posterior side. *d–f* Views of the posterior side of the egg at successive stages. The posterior crescents are bisected equally by the cleavage furrow, which lies in the plane of the median axis of the embryo. [*After E. G. Conklin*, J. Acad. Sci. (*Philadelphia*), **13**:*1* (*1905*).]

If the first two cells of an ascidian are isolated, each develops into a half-embryo. At the four-cell stage the anterior half can be separated from the posterior half; each develops in isolation into the same structures it would form in the intact embryo. In this latter case, for example, the anterior blastomeres form ectoderm, nervous system, notochord, and endoderm, but not muscle or other mesodermal tissue. The prospective potency and the prospective fate are identical. If pairs of blastomeres are isolated at the eight-cell stage, the anterior vegetal pair, posterior animal pair, and posterior vegetal pair differentiate in keeping with their prospective fates. The anterior animal pair, however, does not differentiate into neural tissue as expected and requires the presence of adjacent anterior vegetal blastomeres in order to differentiate. We will return to these results in Section 7.6.

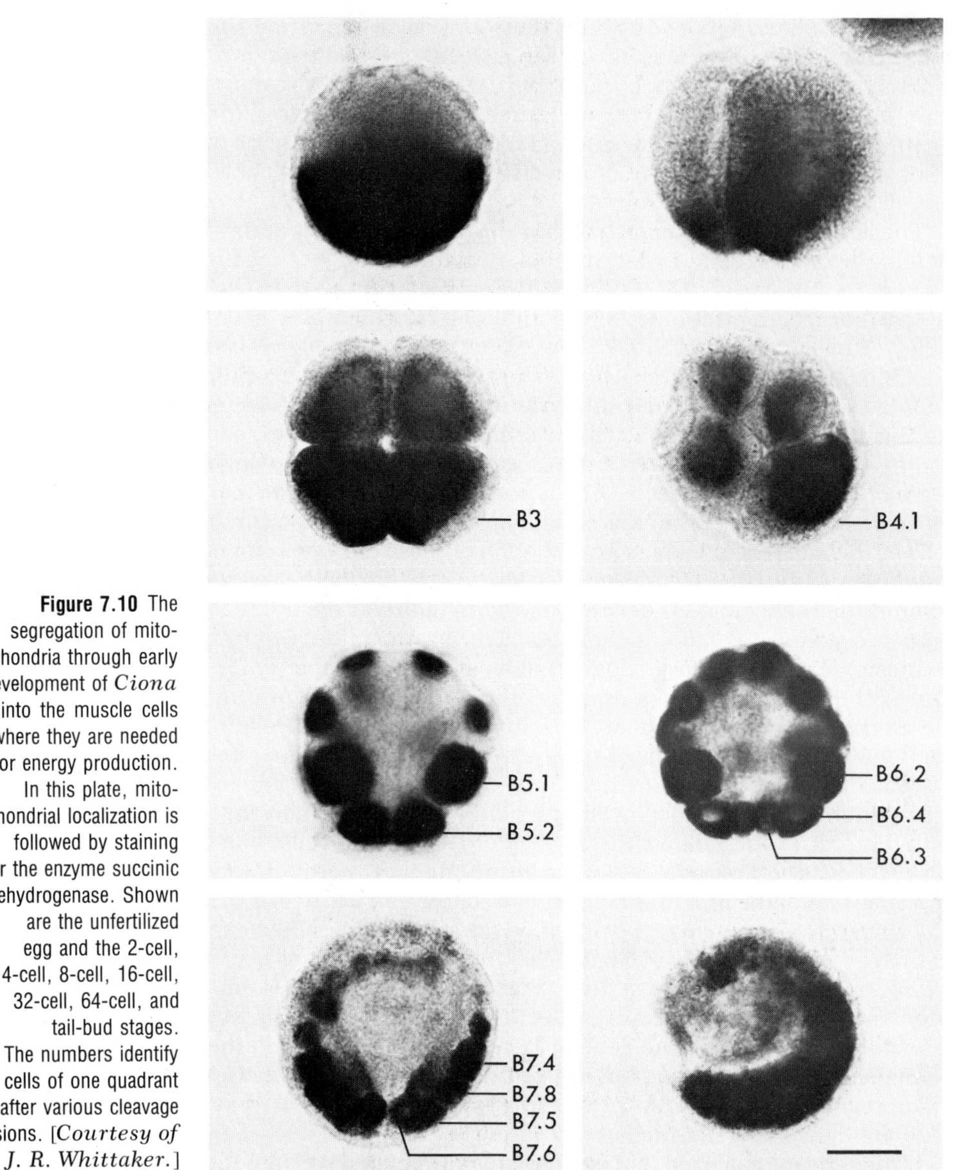

Figure 7.10 The segregation of mitochondria through early development of *Ciona* and into the muscle cells where they are needed for energy production. In this plate, mitochondrial localization is followed by staining for the enzyme succinic dehydrogenase. Shown are the unfertilized egg and the 2-cell, 4-cell, 8-cell, 16-cell, 32-cell, 64-cell, and tail-bud stages. The numbers identify cells of one quadrant after various cleavage divisions. [*Courtesy of J. R. Whittaker.*]

The observation that different parts of the egg contain visibly different materials seems to be correlated with the restricted nature of their prospective potency. Are the various cytoplasmic components directly responsible for the differentiation of the cells in which they become segregated?

Conklin later established the causal relationship by centrifugation experiments. If *Styela* eggs are centrifuged after fertilization, the visible cytoplasmic regions (the yellow cytoplasm, the gray yolk, and the clear plasm) can be displaced in various ways. Under these circumstances, development is always abnormal. He found that in ascidian embryos it is possible to recognize the endoderm, chorda, neural plate, and muscle cells by their colors, shapes, sizes, and

histological characters even when these tissues are far from their normal positions. When egg substances have been dislocated by centrifuging, it is easy to see that the larval parts to which they typically give rise are also dislocated. Thus larvae may be turned inside out, with the endoderm, muscles, and chorda on the outside and the ectoderm, neural-plate cells, and sense organs on the inside. In other words, each type of cytoplasm differentiates into its expected tissue regardless of its position in the embryo.

These results have suggested that the ascidian egg contains morphogenetic factors (or morphogenetic determinants) that become localized into different regions of the embryo (Fig. 7.9). Each of these factors is in some way responsible for the path of differentiation taken by that cell in which they become localized. A series of experiments with *Ciona* eggs employing the cleavage-inhibiting drug cytochalasin B (acting on microfilaments) has established a relationship between cytoplasmic segregation and the differentiation of tissue-specific proteins. During the development of *Ciona*, three specific enzymes appear, each within a particular type of differentiated larval cell. The enzyme acetylcholinesterase appears only within the muscle cells, tyrosinase only within the pair of large black pigment cells of the brain, and alkaline phosphatase only within the endodermal cells of the gut. The presence and location of these enzymes can be demonstrated histochemically, i.e., by staining procedures specific for the products of one or another of these enzymes. If developing *Ciona* embryos are treated with cytochalasin B at a given stage, no further cleavages occur (mitosis continues without cytokinesis) and the morphological development of the embryo is arrested. However, if one waits an appropriate number of hours, the presence of all three of these enzymes can be demonstrated. In other words, the biochemical differentiation leading to the synthesis of these three enzymes continues despite the lack of morphological progress. Furthermore, each enzyme appears only within those cells which would have led to the particular tissue in which that enzyme is normally found. If one follows the lineage of the muscle cells, one can trace them back to both the first two cells (the forerunners of the right and left half of the larva), to two of the first four cells (the posterior pair), to two of the first eight cells (the vegetal posterior pair), and so on in a highly specified manner. If *Ciona* cleavage is arrested at various stages and the locations of acetylcholinesterase are determined several hours later (Fig. 7.11), the enzyme is found to be synthesized specifically within those cells of the muscle lineage. It would appear that determinants related to the development of this enzyme (and the other two) are localized in the zygote and segregated during cleavage into appropriate future tissue regions. The basis for the localized expressions of these particular polypeptides will be considered further in Chapter 8.

It must be emphasized in a discussion of ascidian development that the precociously differentiating cells and tissues relate to the development of the special larval structures that constitute the distinctive chordate character, namely, notochord, lateral bands of locomotory tail muscle, dorsal tubular nerve cord, and anterior brain vesicle with sense organs. None of these survives metamorphosis to become part of the adult ascidian. At the time of fertilization the ascidian egg is essentially preprogrammed to develop into a tadpole larva. Although the special properties of the egg appear to be related to this production, not all the egg is directed to the process, for some features of the permanent organism slowly develop during the rapid formation of the tadpole structure. When the period of tadpole larval activity comes to an end, however, and larval tissue is resorbed, the residual tissues rapidly complete their development to become a small but functional ascidian with beating heart, active gills, functioning gut, and so on.

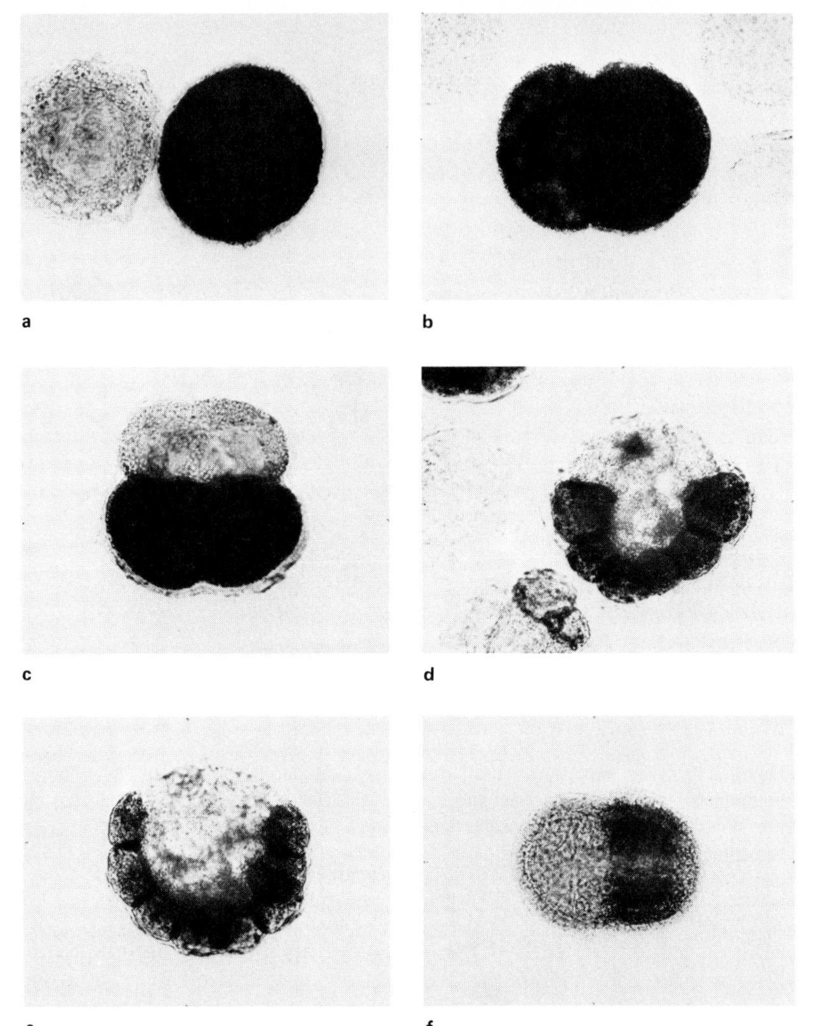

Figure 7.11 Histochemical localization of acetylcholinesterase at 15 to 16 hours after fertilization in *Ciona* embryos that had been cleavage-arrested by addition of cytochalasin B at the *a* 1-cell, *b* 2-cell, *c* 4-cell, *d* 16-cell, and *e* 64-cell stages. *f* Localization in a 9-hour control embryo. [*Courtesy of J. R. Whittaker.*]

There are many examples among ascidians where the tadpole organization has been entirely eliminated in the development of small, virtually yolkless eggs of animals living on the open seafloor, where the functional value of the tadpole larva is minimal. In these cases, the egg size, cleavage rate, time of gastrulation, time of hatching, and course of posthatching development remain the same as in closely related species that form a tadpole during the period between gastrulation and hatching. In the case of one species that produces an anural larva (no tadpole stage), the enzyme acetylcholinesterase is still produced as a vestigial biochemical product within the muscle-cell lineage, even though these muscle cells are entirely nonfunctional. The morphogenetic determinant for this enzyme continues to be segregated.

In ascidians, the visible components cannot be dissociated from the morphogenetic factors. This is not the case in many other forms. In many experiments it seems that the visible components reflect a more basic differentiation of the cytoplasm, one that remains in operation even after the visible components have been displaced. An example of this is described in the next section.

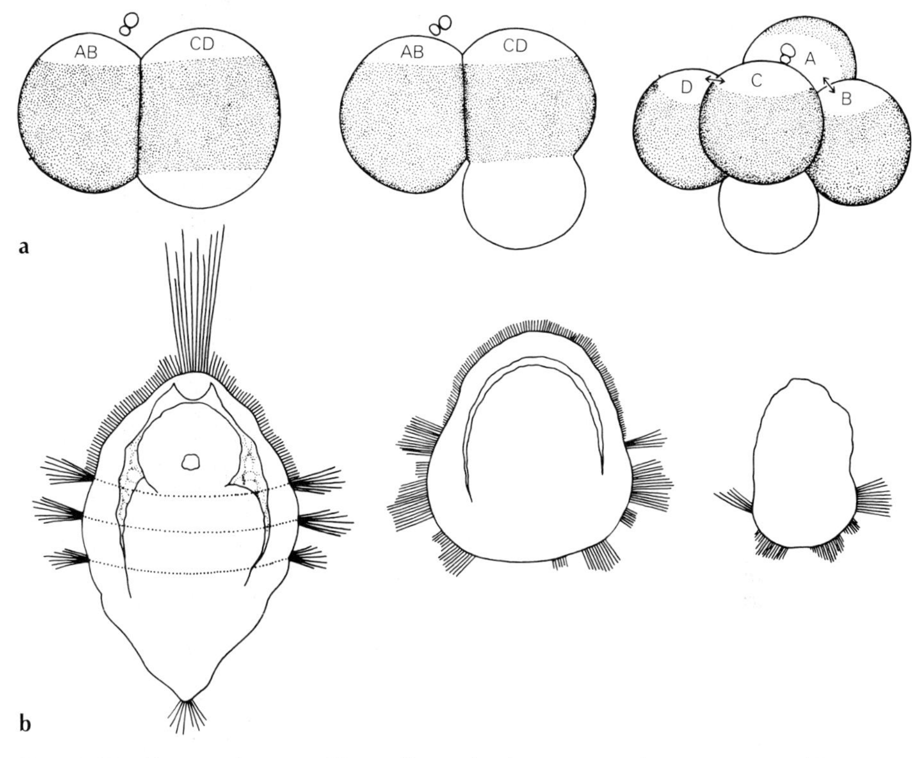

Figure 7.12 *a* First two cleavages of *Dentalium*, showing more precocious appearance and larger proportions of the first and second polar lobes (associated with blastomere CD and D). *b* Normal trochophore of *Dentalium*, together with larva of same age developing after removal of polar lobe (posttrochal region and apical tuft missing), and defective larva developing from one of small A, B, or C blastomere not containing the yolk lobe. [*After E. B. Wilson, 1892.*]

5. Spiralia

Many eggs, particularly those of the spirally cleaving annelids and mollusks, are characterized by visible plasms that become shifted in various ways during the early postfertilization stages. This movement is termed *ooplasmic segregation.*[2] In a number of mollusks, such as *Dentalium* and *Ilynassa*, a specialized cytoplasmic component called the *polar lobe* has evolved as an apparent mechanism for the segregation of morphogenetic factors. The polar lobe appears as a large protrusion at the vegetal pole prior to first cleavage (Figs. 7.12*a* and 7.13*a*). The lobe forms by the constricting action of a band of microfilaments. A large number of surface projections give the lobe a strikingly convoluted appearance in the scanning electron microscope (Fig. 7.13*b*). In *Ilynassa* the first lobe disappears and a second one soon forms (Fig. 7.12*a*). The first cleavage furrow splits the egg into an AB cell and a CD cell, with the polar lobe attached by a cytoplasmic bridge to the CD cell. The appearance, however, is of three cells (Fig. 7.13*c*), and the stage is called the *trefoil.* After cleavage the entire content of the polar lobe goes into the CD cell. A third polar lobe reforms before the second cleavage, and the process is repeated, with all the polar-lobe material finding its way into the D cell of the four-cell stage.

[2] A particularly clear example of ooplasmic segregation occurs among ctenophores, which are not discussed in this text. The segregation of cytoplasmic contents is illustrated in the photomicrograph of a living egg in Fig. 6.12.

The role of the polar lobe and the value of each of the blastomeres has been analyzed in a series of experiments where parts of the early molluskan embryo were removed. Experiments of this type were originally performed on the scaphopod *Dentalium* (Fig. 7.12*b*) and more recently on the gastropod *Ilynassa*. An experiment that determines the effect of cell removal on an embryo is referred to as a *defect experiment*. If the missing cell contains some essential factor for the development of a structure or structures, it might be expected that its loss would leave the embryo unable to form that tissue. For example, removal of the first polar lobe of *Ilynassa* results in the embryo's failure to develop an anteroposterior axis; it also lacks a velum, foot, shell, heart, intestine, and eyes. A normal veliger larva is shown in Fig. 7.14. It would appear that the polar lobe contains morphogenetic determinants that are responsible for the differentiation of those structures formed by the cells receiving the materials and missing in the lobeless embryo. To determine when possible morphogenetic determinants become segregated, the D macromere was removed at different cleavages in different

Figure 7.13 *a* Diagram of the changes that occur in the shape of the fertilized *Ilynassa* egg before and during first cleavage. By 1½ hours after being laid, a prominent polar lobe is visible at the vegetal region of the egg. The pronuclei fuse at the time of resorption of the lobe. Approximately 40 minutes later, at the time of spindle formation, another polar lobe appears in the same region. As the polar lobe becomes constricted from the body of the egg by the action of a band of microfilaments, the first cleavage furrow appears at the pole opposite the lobe. As the cleavage furrow deepens, the connection of the polar lobe with the blastomeres becomes very tenuous (trefoil, last stage shown). Following cleavage, the lobe is resorbed by the CD cell. [*From G. W. Conrad, D. C. Williams, F. R. Turner, K. M. Newrock, and R. A. Raff,* J. Cell Biol., **59**:229 (1973).] *b* Scanning electron micrograph of an uncleaved egg of the mollusk *Buccinum*. Only the polar lobe is covered with ridges. [*Courtesy of M. R. Dohmen and J. C. A. van der Mey.*] *c* Scanning electron micrograph of an *Ilynassa* egg at the trefoil stage. The cleavage furrow (CF) and polar-lobe constriction (PLC) are indicated. [*Courtesy of G. W. Conrad.*]

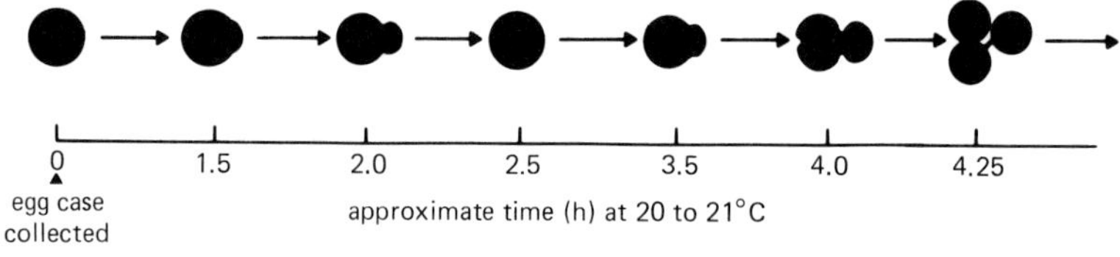

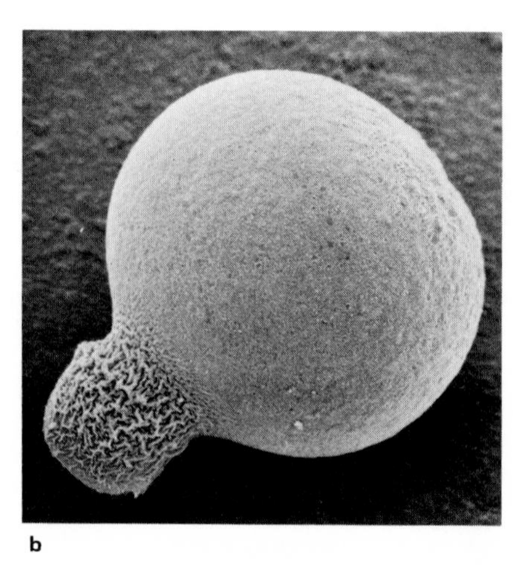

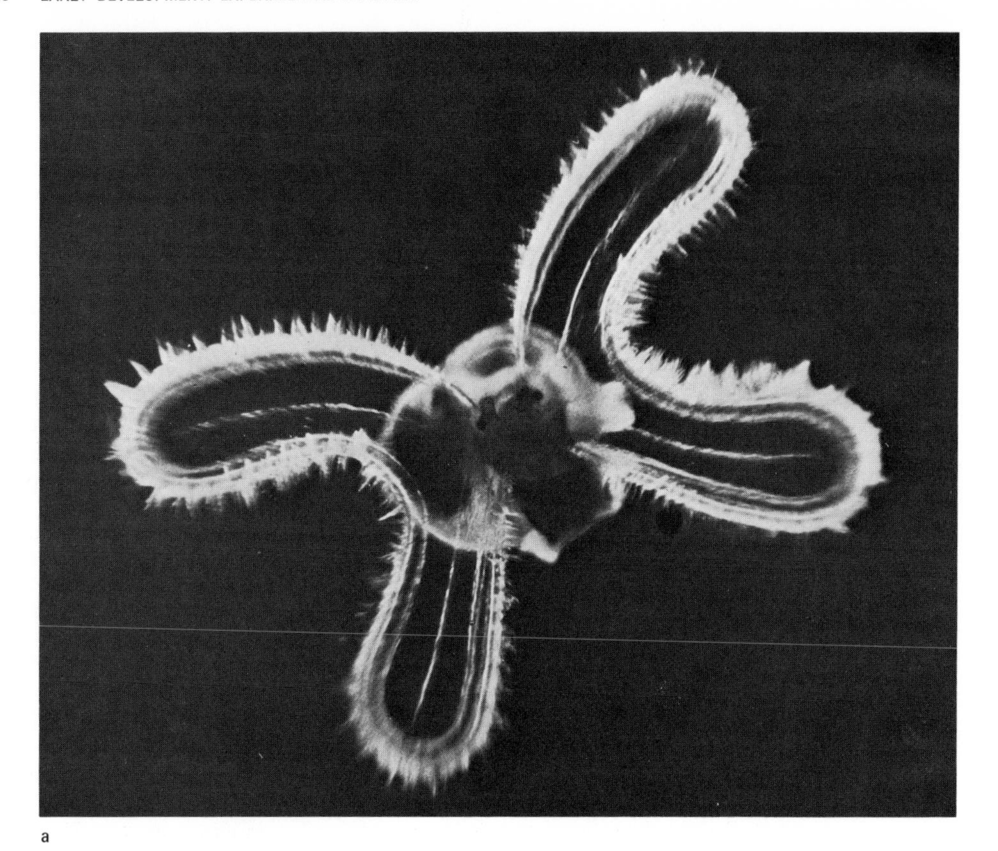

a

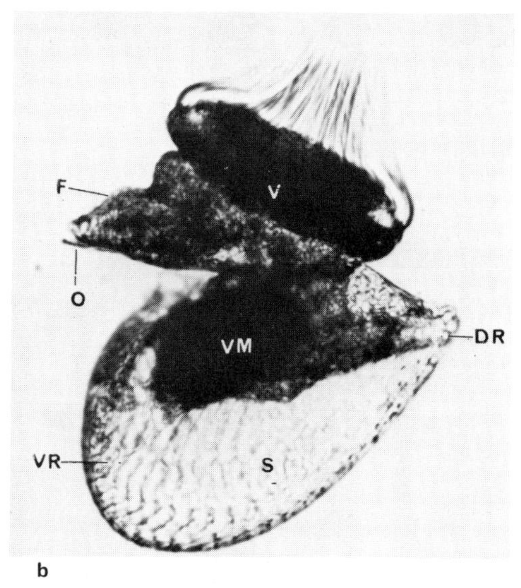

b

Figure 7.14 *a* The planktonic veliger larva of a marine gastropod mollusk, *Nassarius incrustus,* swimming with fully extended velum. The velum is a transient larval structure that is resorbed during final metamorphosis. The essential structure of the adult-type organization is already present in the compact body within the velar larva; included, for example, are the eyes, siphon, and shell gland. [*Courtesy of D. P. Wilson.*] *b* Veliger of another gastropod mollusk, the limpet *Acmaea scutum.* If the larva is agitated, the velum and foot are retracted into the shell with the opening closed by the operculum. VR, ventral retractor muscle; DR, dorsal retractor muscle, S, shell; F, foot; O, operculum; V, velum; VM, visceral mass. [*G. Karp.*]

embryos. Removal of the D cell (leaves ABC) or removal of the 1D (leaves ABC descendants + 1d) or removal of the 2D (leaves ABC descendants + 1d + 2d) has essentially the same effect as removal of the polar lobe at the trefoil stage. If one waits until after the fourth cleavage and removes the 3D (leaves ABC descendants + 1d + 2d + 3d), the remaining embryo will now form velum, eyes, shell, and foot, but it still lacks the heart and intestine. After the fifth cleavage, the 4D can be removed, but the remaining embryo can form as an essentially complete larva; the 4D has no morphogenetic role. The conclusion from these experiments is that the factors of the lobe of *Ilynassa* become segregated in two main steps, one upon the formation of the 3d cell and the other upon the formation of the 4d cell.

A most important question concerns the nature of the morphogenetic determinants being segregated, although at the present time there exists very little relevant data. Studies on centrifuged eggs suggest that the information in the polar lobe is not displaced by relatively low centrifugal forces. If the egg of *Ilynassa* is held in an inverted manner and centrifuged, all the visible components of the vegetal endoplasm will be driven to the opposite pole. In this case, the polar lobe which forms at the original vegetal pole has an entirely different composition; yet its effect on development (as determined by defect experiments) is unchanged. This type of study has implicated the cortex (including the plasma membrane itself) as the site of information storage. However, in a few mollusks, the polar cytoplasm is seen to contain distinct cell inclusions not found in the remainder of the egg. In *Bithynia*, for example, the small polar lobe contains a cup-shaped structure termed the *vegetal body* (Fig. 7.15), which is composed of a mass of small vesicles containing electron-dense material. This aggregate of vesicles, which appears to contain RNA, is not displaced by centrifugation, suggesting that it is firmly attached to the cortex. The relationship between these vesicles and the fate of the cells which contain them remains obscure. The effect of polar-lobe materials on transcriptional and translational activities will be discussed in Section 8.3.B. It should be kept in mind that most spiralian embryos develop without the formation of any type of polar lobe, yet the results of cell-lineage studies and defect experiments indicate that the same types of segregation patterns exist throughout the group.

Figure 7.15 Section of a *Bithynia* egg at first cleavage showing the small polar lobe, with the vegetal body indicated by the arrow. [*Courtesy of M. R. Dohmen and N. H. Verdonk.*]

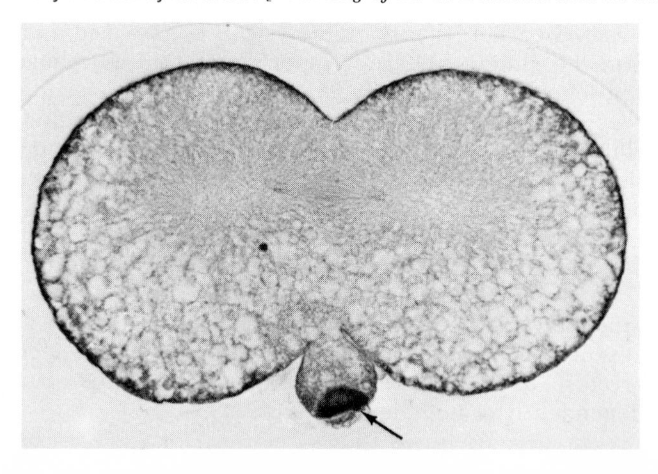

6. Mosaic versus regulative development

The preceding discussion of the sea urchin, frog, ascidian, and snail has led into a theoretical topic in which there is much difference of opinion. Before continuing with other cases (squid, insect, and mammal) in which other principles of development will be illustrated, it is worthwhile to reexamine the previous examples in an attempt to integrate some of the information. In the case of the ascidian and snail, the prospect was considered that development may be controlled by the segregation of morphogenetic determinants into different blastomeres. The development of embryos with such determinants—assuming this concept is valid—is *mosaic*, a term that implies that the egg is a patchwork of materials with morphogenetic influence. The fate of the cells that contain these factors is fixed as a result of their presence; i.e., their prospective potency is limited to their prospective fate (Section 7.2). The term for this type of commitment is *determination*. The degree to which a cell or group of cells is determined can be difficult to ascertain. To assay for determination one must subject the cells in question to a variety of alternate environments in an attempt to influence the direction of differentiation. If the cell(s), upon isolation or after recombination with other types of cells, can still differentiate only into the structures that it would normally form, it is said to be fully determined. It should be kept in mind that one can never be certain a group of cells could not be directed along some other path of differentiation if only the proper environment were provided. Since we do not understand the molecular basis of determination, we have to define it in terms of various experimental procedures. The determination of a cell or region of an embryo is usually accompanied by a gain in independence from its neighbors. This autonomy is reflected in the ability of the cells to differentiate when cultured in isolation, a process termed *self-differentiation*. Embryos referred to as mosaic are characterized by the early determination of their cells. Cells isolated from a mosaic embryo are typically capable of self-differentiation, and the remaining parts of the embryo are deficient in the structures normally formed by the missing cells. For example, if one of the first micromeres of a mollusk is isolated, it will divide twice and form four cilia-bearing trochoblasts. This is exactly its fate in the undisturbed embryo.

At the opposite extreme from the more highly mosaic embryos are the embryos whose cells are not determined until considerably later stages, such as the echinoderm and the vertebrate. These embryos are capable of adjusting, or regulating, to damage imposed on them; they are classified as *regulative embryos*. In the sea urchin, for example, one can wait until the blastula stage and cut the embryo into two meridional halves, and each will form an entire larva. In another example discussed earlier, the macromeres can be removed and the remaining cells can form the pluteus. There is no evidence that morphogenetic determinants become segregated in macromeres or in any other specific cell. The prospective potencies of the other 12 cells (at the 16-cell stage) include the formation of the gut, which is normally the prospective fate of the deleted macromeres. The other cells have regulated their activities to differentiate into structures they would not normally become.

Regulative development is characterized by a greater interaction of the embryonic parts. Two sea urchin embryos can be fused together, and one giant larva will form. Each of the original embryos is now forming only one-half an embryo rather than the total product; this is another form of regulation. Each egg recognizes the presence of the other and adjusts its development accordingly. If two mosaic embryos are fused, each will continue to form one member of twins. Each cell acts independently of its neighbors.

Another way that intercellular interaction manifests itself is in inductive contacts, i.e., contacts where one group of cells stimulates another group of cells to differentiate as it could not in isolation. Mosaic embryos tend to develop with a minimum of induction. That is, tissues do not require specific contacts with other cells to differentiate; this is shown in their ability to self-differentiate.

In actual practice the distinction between these two types of development is difficult to make; the view is generally accepted that the differences are relative ones and there is a continuous spectrum between the extremes. For example, we have pointed out that even in the highly mosaic ascidian, inductive interactions are required for normal development. We could have pointed out the same fact in our example of the mosaic development of *Ilynassa*, since the ABC + 1d + 2d embryo lacks a foot. Normally the foot is derived from the 2d cell, as revealed by cell-lineage studies; this suggests that its failure to develop in embryos from which only the 2D is removed is due to a missing inductive influence from the 2D. Conversely, the *Ilynassa* embryo is capable of some regulation, as illustrated in the following deletion experiments. Removal of the 2c and 3c micromeres interferes with the formation of the right statocyst and heart, respectively. However, removal of the C blastomere at the four-cell stage (the cell which gives rise to the 2c and 3c) does not block the formation of these two structures in a substantial percentage of larvae.

The development of the amphibian illustrates the difficulty in assigning an embryo to a mosaic or regulative position. If left and right blastomeres are separated at the two-cell stage, each can form an entire embryo; this indicates regulation. However, if a ventral and dorsal pair are separated, only the latter can continue to form even the semblance of a normal embryo. This difference appears directly related to the presence or absence of the dorsal marginal region (possibly the gray crescent) of the embryo. The UV experiments suggest the localization of some morphogenetic determinant(s); the egg is a mosaic.[3] Similarly, in the frog egg there is extensive evidence for the localization of material that is needed for primordial germ-cell differentiation at a much later stage. In nearly every other way the frog embryo is highly regulative; it is only after gastrulation that the other parts are determined and are capable of self-differentiation.

In the frog and sea urchin, as well as many other embryos, stepwise changes in determination can be demonstrated as development proceeds. With increasing age of development, the potencies of groups of cells become restricted and the capacities of these groups for self-differentiation increases. The early regulative stages give way to more mosaic later stages. For example, the isolated animal half of a 16-cell sea urchin embryo forms a hollow ciliated ball. If this same animal half is isolated at the blastula stage, it is capable of forming a mouthlike structure, the apical tuft of cilia, and so on, which are structures this region would normally develop. Presumably the same types of determinative events that occur by the blastula and gastrula stages in the frog and sea urchin occur at a much earlier stage in the snail and ascidian. In fact, even in the most strictly mosaic eggs an early change in commitment may be demonstrated. For example, if *unfertilized* eggs of *Ascidia* are fragmented into equal pieces by equatorial, meridional, or oblique sections and all fragments are then fertilized, each pair gives rise to twin tadpoles, although they are proportionately small. In contrast, if the first two blastomeres are separated, each develops into a highly defective larva.

[3] It should be noted that there are other types of experiments which argue against the localization of morphogenetic determinants in the dorsal region. The reader can consult the paper by Ulf Landstrom and Soren Lovtrup, 1975, *J. Emb. Exp. Morph.* **33**:879, for an example.

7. The cephalopod egg: cortex and pattern

There is little doubt that cephalopods, e.g., squid and octopus, evolved from a more primitive group of mollusks and at one time shared the spiralian inheritance. Adaptive changes, however, not only brought the cephalopods to the highest level of invertebrate evolution, but resulted in the production of relatively large and yolky eggs. Development is direct, with no trace of either spiral cleavage or trochophore or veliger larva (Fig. 7.16).

Figure 7.16 Direct development of the squid egg from early morphogenesis to newly hatched stage. *a* Yolk mass of egg nearly enclosed by epibolic spread of cellular layer. The mantle and the pair of eyes are already observable as delimited areas in the cellular layer of the animal hemisphere. *b* Later stage showing primordia of much of the primary organization, i.e., shell gland, mantle, funnel, eyes, and arms. Much of the yolk vegetal hemisphere, now fully enclosed, will remain as the external yolk sac. *c* Progressive development of primordia already present. *d* Newly hatched stage with small external yolk sac still present. [*Courtesy of J. Arnold.*]

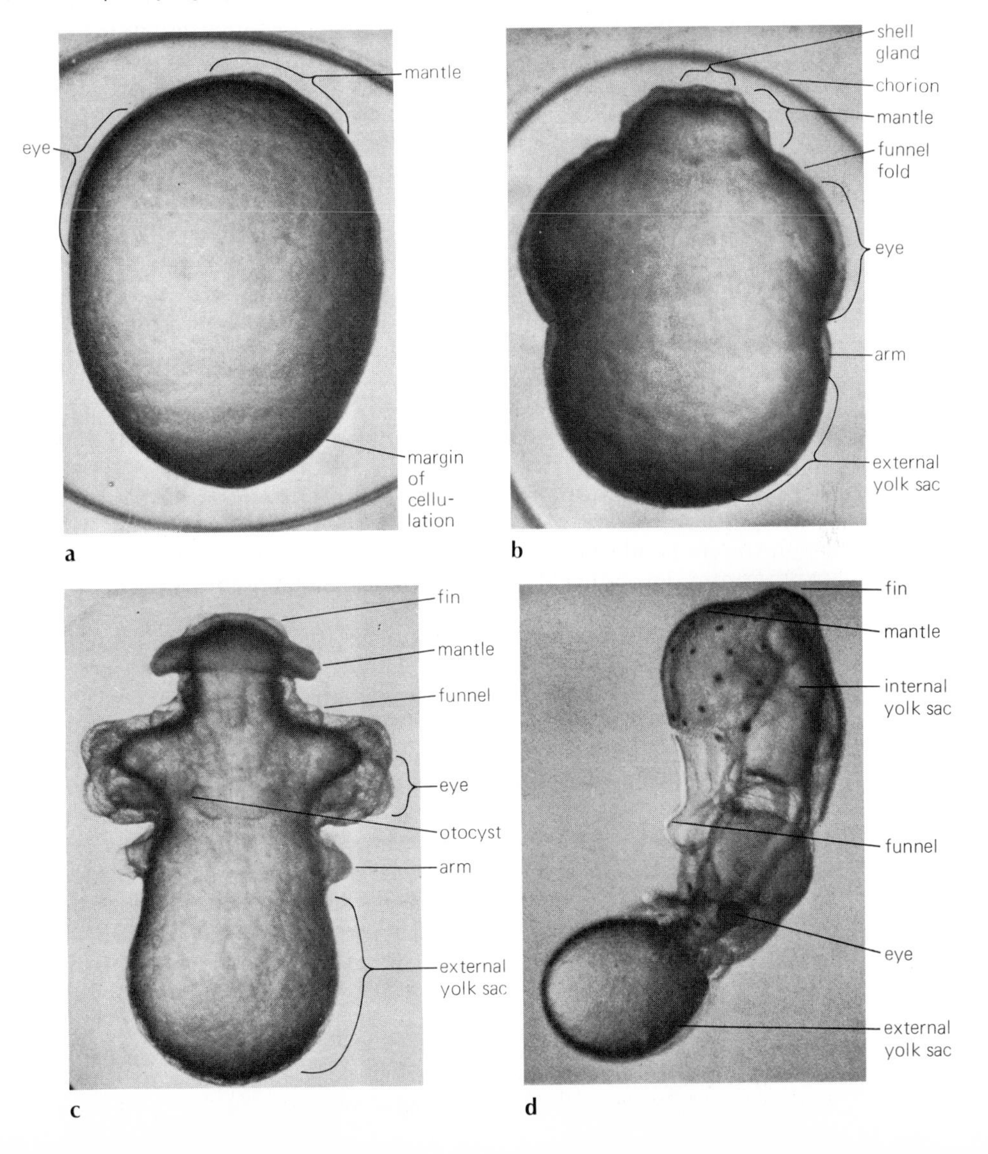

The cleaving squid egg consists essentially of three parts: a yolky central mass, a disc of cytoplasm at the animal pole in which meroblastic cleavage occurs, and a thin layer of cortical cytoplasm continuous with the blastodisc but surrounding the yolk mass. Although the beginnings of embryonic development occur in a restricted section of the egg surface, the forming blastoderm expands by peripheral mitosis toward the vegetal pole so that more and more of the thin cortical cytoplasm surrounding the yolk becomes incorporated into the developing embryo. The process is analogous to that of epiboly, which occurs in fishes (see Section 10.2.A). As the margin of the blastoderm expands over the yolk, the original egg surface becomes cellulated but not displaced. Consequently, specific locations within the original cortex remain relatively constant during development.

In this and the previous chapter, the importance of the egg cortex as a site of developmental information has been stressed. Since the squid egg consists largely of a cortex that serves as the forerunner in space of the developing embryo, it serves as an excellent system for the analysis of questions of cortical function. In this respect, it has been found that microbeam irradiation of small regions of the egg surface during early cleavage leads to the formation of localized defects during later periods of organogenesis. For example, irradiation of the future eye region leads, in many cases, to the formation of an embryo totally lacking an eye on the irradiated side. Similar results are found when small regions of the surface are tied off by ligation. Although the number of experiments on cephalopods is small, it would appear that a two-dimensional pattern exists within the cortex of the uncleaved egg (either within the thin layer of cytoplasm or associated with the plasma membrane itself) that determines the direction in which that part of the future blastoderm will differentiate.

8. Insects

The very early stages of insect development were described in Chapter 6. The zygote nucleus resides within the egg interior, where it undergoes numerous mitotic divisions (eight in *Drosophila*) to form a single-celled multinucleate cleavage stage (Fig. 7.17). The daughter nuclei (termed *energids*) then migrate to the egg periphery, where they continue their mitotic divisions. Those nuclei which happen to reach the posterior pole of the egg become surrounded by plasma membrane to form *pole cells*, the forerunners of the germ cells, and are segregated somewhat from the remainder of the egg (see Fig. 7.26). Although the pole cells are the first to form, the remainder of the nuclei residing in the cortical cytoplasm are soon separated by the formation of cell boundaries. The egg has now reached the *cellular blastoderm* stage (approximately 3 hours at 25°C forming about 5,000 cells in *Drosophila*), in which virtually all cells of the embryo exist as part of a single-layered, two-dimensional crust. The cells of the blastoderm are not uniformly distributed over the surface of the embryo, but rather there is a marked concentration in one region on the ventral surface, a region which ultimately will reside in the prothorax of the organism to be. This region, termed the *germ anlage,* is transformed after gastrulation into the *germ band* (Fig. 7.18), which continues to differentiate anteriorly and posteriorly from this site. The remainder of the cells of the blastoderm are destined to form the extraembryonic covers, the amnion and serosa.

The eggs of insects vary greatly in their regulatory powers, not only among major groups, but within certain insect orders. Some eggs, such as those of the dragonfly, possess great regulatory abilities, being able to form a complete embryo from a fraction of the original egg structure (Fig. 7.19). In contrast, the eggs

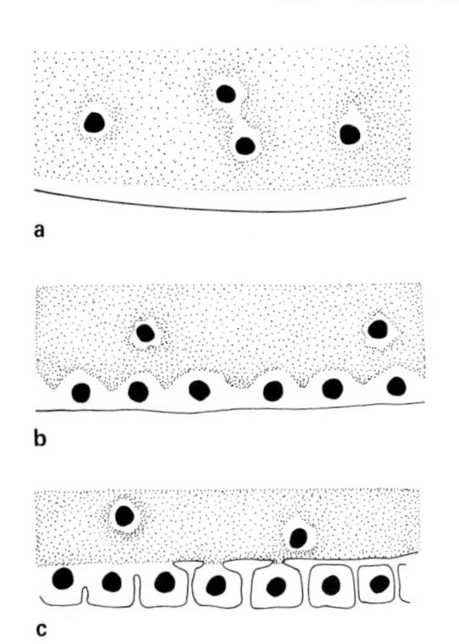

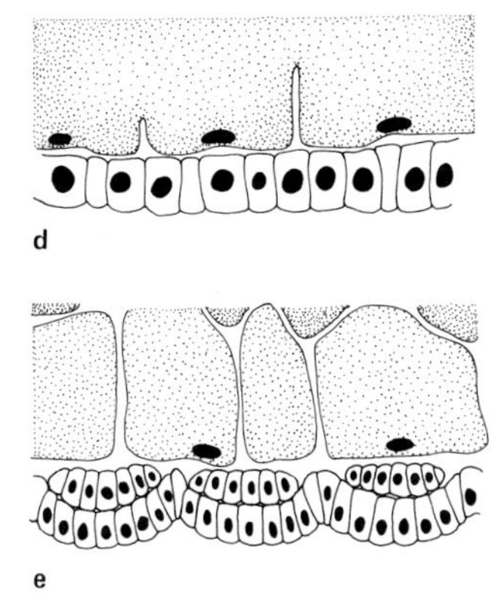

Figure 7.17 Diagram of early development of an insect egg. *a* Multiplication of nuclei in the yolky region. *b* Migration of the nuclei to the egg cortex. *c* The formation of cells (from left to right) by infoldings of the plasma membrane. *d* The superficial epithelium (blastoderm) has formed and the yolk system is being divided into large yolk cells. *e* The superficial epithelium has become two-layered by gastrulation and is being subdivided by segment borders. [*From K. Sander, "Cell Patterning," Ciba Symp. No. 29, 1975.*]

Figure 7.18 Early development of the dragonfly *Platycnemis*, showing the first three divisions of the nucleus, the spreading of the daughter nuclei through the egg cytoplasm (*the two sister nuclei from a division are joined by a dotted line*), the multiplication of the nuclei to form a blastoderm, and their aggregation to produce the germ band. [*After F. Seidel,* Arch. Entwickl.-mech., **119:***322 (1929).*]

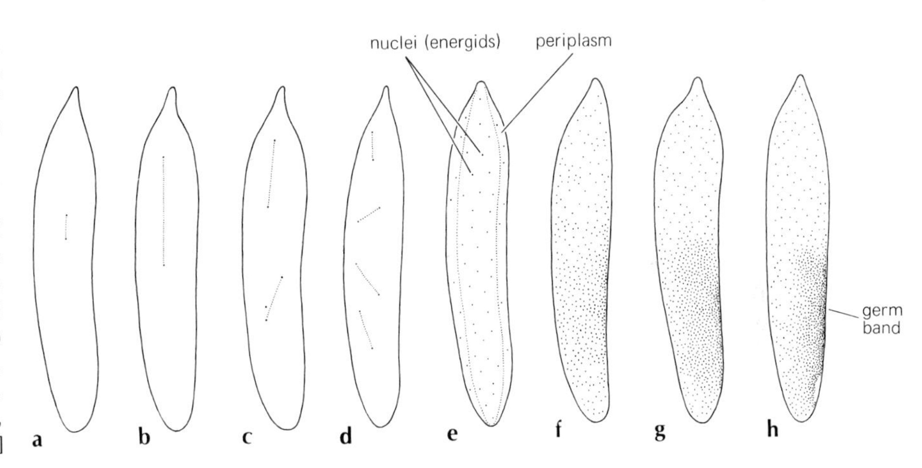

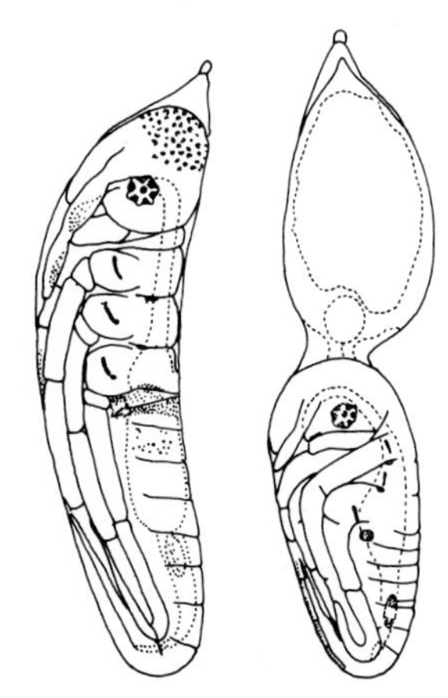

Figure 7.19 Regulation in the insect egg. *a* Normal embryo of the dragonfly *Platycnemis penipes,* seen from the left side, and dwarf embryo, obtained by partial constriction of the egg at the four-nucleus stage. The dwarf is normally proportioned and developed, and its organs have arisen from regions the presumptive fates of which were quite different; their fates were therefore not irreversibly determined at the stage operated upon, and regulation has been possible. [*After F. Seidel,* Arch. Entwickl.-mech., **119:***322 (1929).*]

of higher Diptera and Lepidoptera (holometabolous insects that undergo drastic, complete metamorphosis) are severely limited in their ability to regulate following various types of experimental damage. Although it was believed for a period that many insect eggs were organized in a strict mosaic manner at the time of ovoposition (egg-laying), it is now generally believed that, with the exception of the germ cells, the determination of the parts occurs in a stepwise manner as discussed later. To whatever degree the newly laid egg has become differentiated during oogenesis, this differentiation resides within the outer layer of cortical cytoplasm, since the nuclei at the cellular blastoderm stage, and in later stages, have been shown to be equivalent and totipotent (see Chapter 8).

Results from various types of experiments indicate that at least by the cellular blastoderm stage, the anterior and posterior portions of the embryo have distinctly different developmental potential. This can be demonstrated in the following experiment. Cellular blastoderms were cut into anterior and posterior halves, and then each fragment was mixed with a whole blastoderm from a genetically distinct strain. The preparations were then dissociated to form a suspension of intermixed single cells which were then aggregated by centrifugation. These mixed aggregates were placed into culture to determine their developmental potential. Although the cells derived from the whole embryos were found within adult structures from all body regions, those from anterior halves gave rise only to structures of the head and thorax, while those from posterior halves formed only thoracic and abdominal structures. These results indicate that cells of one half of the blastoderm are already committed to form only structures characteristic of that half and cannot be influenced by the presence of cells from other sections of a whole embryo.

A different type of experiment performed on several different insects seems to reveal the stepwise nature of the initial determinative events. In these experiments, young embryos are fragmented with a blunt razor blade without severing the surrounding vitelline membrane, and the fate of the fragments are followed. When bisection is performed on the very early nuclear multiplication stages, each half forms a highly deficient portion of the embryo. In the anterior half, the most anterior region becomes well differentiated, while the potential for differentiation drops off in the regions closer to the site of ligation. Similarly, it is the most posterior section of the posterior half that is capable of expressing its prospective fate. It would appear that the extreme anterior and posterior ends of the newly laid egg have achieved a more advanced state of determination than the intermediate section of the egg. The longer one waits before ligation, the more complete the differentiation of each half and the less the "gap" with respect to differentiation of the middle portion of the embryo. If the ligation is performed at some point after the formation of the cellular blastoderm, each part self-differentiates according to its prospective fate, and together the two fragments "add up" to the complete embryo. Findings of this type have led to the proposal that development of the basic body plan of insects is governed by the activity of some type of double gradient, much like that formulated for the sea urchin embryo (Section 7.2). Further support for this concept comes from a series of experiments performed on another dipteran, a chironomid midge. It has been found in these studies that damage inflicted locally at the anterior pole of the cleaving embryo can lead to the formation of an abdomen in the anterior half of the embryo. In extreme cases, the embryo that forms after such localized damage contains two abdomens organized in a mirror-image arrangement and totally lacks any head or thoracic structures (Fig. 7.20). Although the basis for the reprogramming of the anterior half is unclear, it may be that damage to the anterior tip destroys some type of

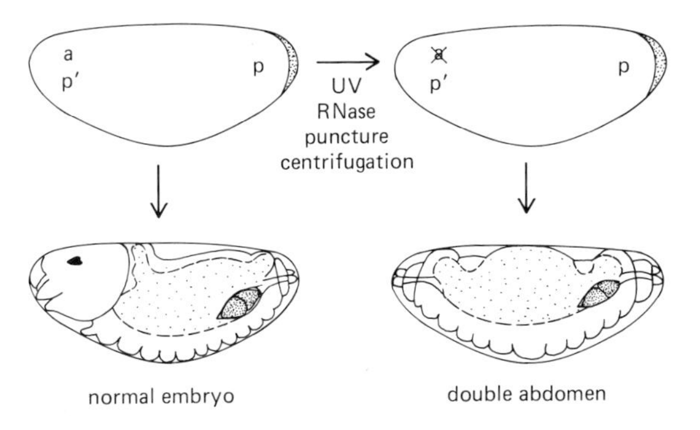

Figure 7.20 Diagrammatic representation of double-abdomen induction in *Smittia* by various types of experimental interference, all of which appear to inactivate or displace anterior determinants designated as a. These are thought to cooperate with other factors (p') in the anterior half of the embryo so as to allow the formation of head and thorax. Upon inactivation of a, p' is assumed to cause abdomen formation in the anterior half, although germ cells (*shaded*) are lacking. [*From K. Kalthoff, "37th Symposium of the Society for Developmental Biology," Academic, 1979.*]

localized factor required for anterior differentiation. In the normal embryo, this factor and a different one from the posterior pole would diffuse across the egg leading to the progressive determination of a greater and greater portion of the egg contents. Once the cellular blastoderm stage has been reached, local destruction of the anterior pole leads only to the formation of head defects: the basic body plan of the embryo has been established.

The development of an abdomen in the anterior half of the midge embryo has a counterpart in the genetic mutant of *Drosophila* known as *bicaudal*. One of the great values of using mutants is the information they provide about normal genetic functions. In the absence of the mutation there is no distinguishing feature to tell us of the existence of such a gene. It is only when the particular gene product is missing or defective that its role can be appreciated. In the case of bicaudal, it would appear that some genetic locus is responsible for the production of a substance that is somehow required for the anterior half of the egg to develop into head and thoracic structures. In the absence of this substance, an abdomen is formed in their place. Further analysis of bicaudal indicates that it is a maternal-effect mutant, i.e., one that appears in the offspring of the affected female rather than in her own phenotype and is independent of the genotype of the fertilizing sperm. The situation is analogous to that of shell coiling in *Limnaea* discussed in Section 6.4.D. It is assumed that some gene product formed during oogenesis in the normal fly (probably by adjacent nurse cells) becomes localized in the anterior region of the egg and is utilized in the development of anterior structures.

A. Imaginal discs

During the embryonic development of holometabolous insects such as *Drosophila*, a remarkable segregation takes place. One population of embryonic cells become determined for forming the larval body, i.e., the cells differentiate so that at the time of hatching the organs and systems of the first instar larva (the stage before the first molt) are ready to function. A second system is set apart within this primary embryonic one in the form of small isolated territories, the *imaginal discs* (Fig. 7.21), which individually undergo some growth during larval existence but do not otherwise proceed with further development until metamorphosis, i.e., the transformation of the third larval instar into the adult (imago). The

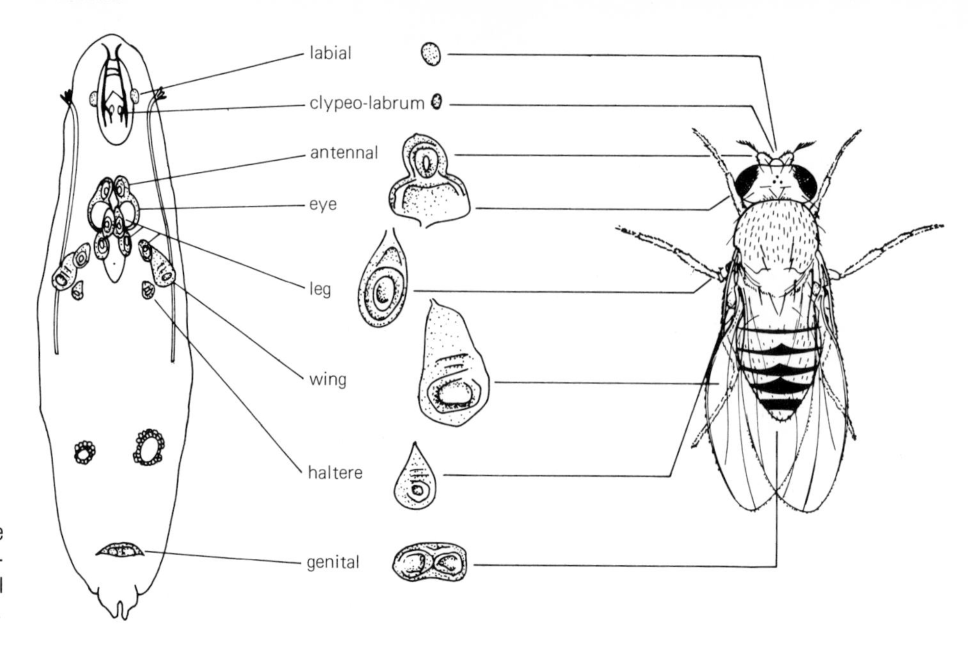

Figure 7.21 The locations and developmental fates of imaginal discs of *Drosophila*.

insect egg is, in essence, a carrier of two embryos. Differentiation is postponed in the second system, but cell division is not; for the discs which are so small in the embryo that they are hardly recognizable, each consists of thousands of cells when they reach their final size at the beginning of metamorphosis.

At metamorphosis most of the larval organs break down within the pupal case and their cells disintegrate. Simultaneously, the cells of the imaginal discs, which were previously homogeneous in appearance, begin to differentiate into a variety of specific adult cell types. As a result, the cells of each disc collectively form a structure characteristic of the location of the disc on the body wall. Thus three pairs of anterior discs form the head (with its special mouth parts and sense organs), a pair of discs forms the dorsal prothorax, three pairs of ventral thoracic discs give rise to the six legs, two pairs of more dorsal thoracic discs form the wings and specialized wing rudiments (the halteres), and a single posterior bilobed disc gives rise to the genital structure. The adult abdomen forms from imaginal cells that are not present as discs but rather as nests of abdominal histoblasts embedded within the larval epidermis.

Drosophila has four major developmental stages, as do *Cecropia* (see Fig. 19.8) and other holometabolous insects. Embryonic development persists for about 22 hours; then the larva hatches, grows for about 4 days, forms a pupa, and in 4 days undergoes metamorphosis to an adult having a life span of several weeks. In *Drosophila*, during the prepupal period, the imaginal discs are converted in a matter of hours into the basic form of the adult insect. During larval life, the discs are present as an invaginated single-layered epithelium which is concentrically folded and surrounded by a noncellular envelope (Fig. 7.22). In response to increasing levels of the steroid hormone ecdysone at the time of metamorphosis (see Chapter 19), the discs undergo a remarkable transformation. They rapidly evaginate and begin to synthesize the various parts of the adult cuticle. Each disc develops into a structural complex that connects with neighboring complexes, and together they construct the complete adult.

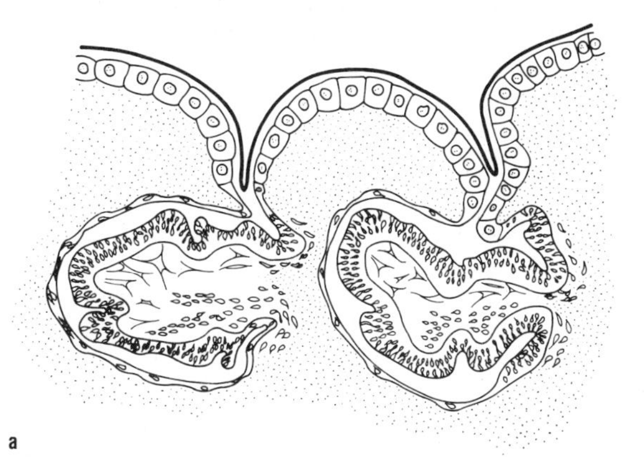

a

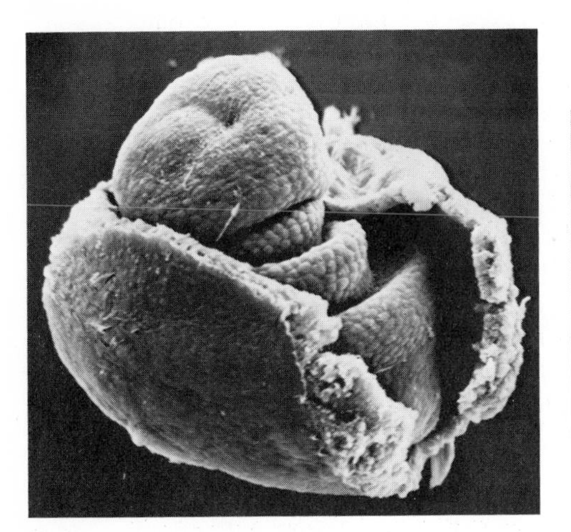

b

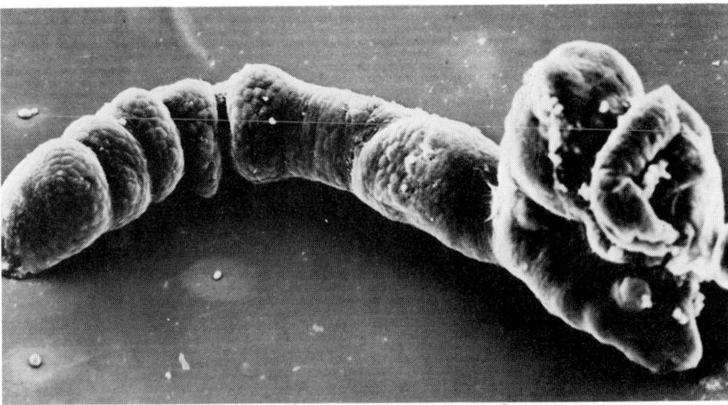

c

Figure 7.22 *a* Longitudinal section of invaginated imaginal discs of the wings of an ant. The discs evaginate during pupation (metamorphosis). [*After V. Wigglesworth, "The Physiology of Insect Metamorphosis," Cambridge, 1954.*] Scanning electron micrographs of an unevaginated *b* and fully evaginated *c* leg disc. [*Courtesy of D. Fristrom.*]

B. Determination of imaginal-disc cells

In order to assess the state of determination of imaginal disc tissue prior to its activation at metamorphosis, an entire disc, or a disc fragment, can be removed from a larva and placed under conditions that will reveal the extent of its commitment to form a particular type of structure. In most experiments, imaginal discs are cut out of an advanced larva with a fine needle and then implanted by means of a micropipet into the body cavity of another larva, where it is essentially cultured in vivo. During their stay in the host larva, the disc cells can multiply, but they remain undifferentiated until the time arrives for the host to begin its own transformation into an adult. Until then, both host and graft tissue are exposed to a high level of juvenile hormone secreted by the corpora allata (Section 19.10), which has the combined effect of sustaining the larval organization and inhibiting disc differentiation. When metamorphosis begins, the level of this hormone drops as the concentration of ecdysone increases and disc differentiation (both graft and host) is allowed to proceed.

When an imaginal disc from a late third-stage larva is implanted into a host larva, it invariably differentiates in keeping with its prospective fate (Fig. 7.23). The inventory of differentiated structures formed by the disc corresponds exactly to that which is formed in the intact animal. Conversely, the animal from which the disc is removed is invariably lacking the structure in question. Not only is the

whole disc of the late third-instar larva a rigorously determined primordium, so too are its component parts. This can be demonstrated by the implantation of defined fragments of a late third-instar disc into a host larva *of the same age;* it gives rise to the same set of cuticular structures predicted from the fate map of the disc (such as that of Fig. 7.23). A leg disc, for instance, contains cell subdistricts that develop into claws, tarsal parts, tibia, femur, trochanter, coxa, and adjoining parts of the thorax, each such subdistrict having bristles and hairs in specific numbers and patterns. Similarly, the genital disc (male) is a mosaic of subdistricts (Fig. 7.23) more or less determined as prospective sperm pump, penis, anal plates, and hindgut, and each of these in turn is a mosaic of still smaller subdistricts. In fact, the state of determination appears to extend to each individual cell of the late third-instar larval disc. This conclusion is drawn from

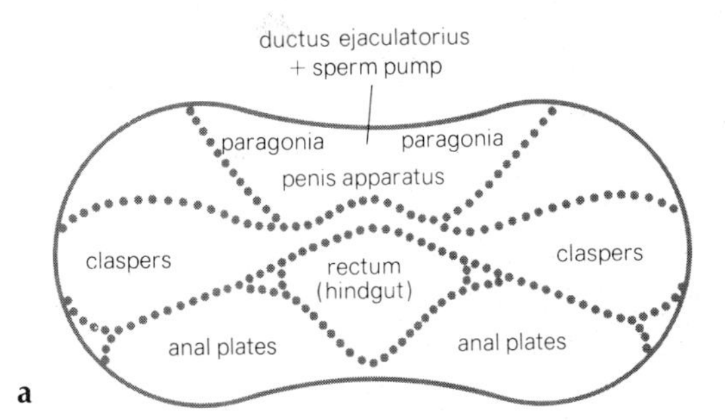

Figure 7.23 *a* A map of the anlagen (presumptive areas) in a male genital disc of *Drosophila melanogaster*. *b* Result of a fragmentation experiment; below, the metamorphosed structure obtained by the anterior half (*left*) and the posterior half (*right*). *c* Result of a localized irradiation indicated by hatching; below, the metamorphosed implant in which no anal plates are differentiated. [*After E. Hadorn,* Brookhaven Symp. Biol., **18:***152* (*1965*).]

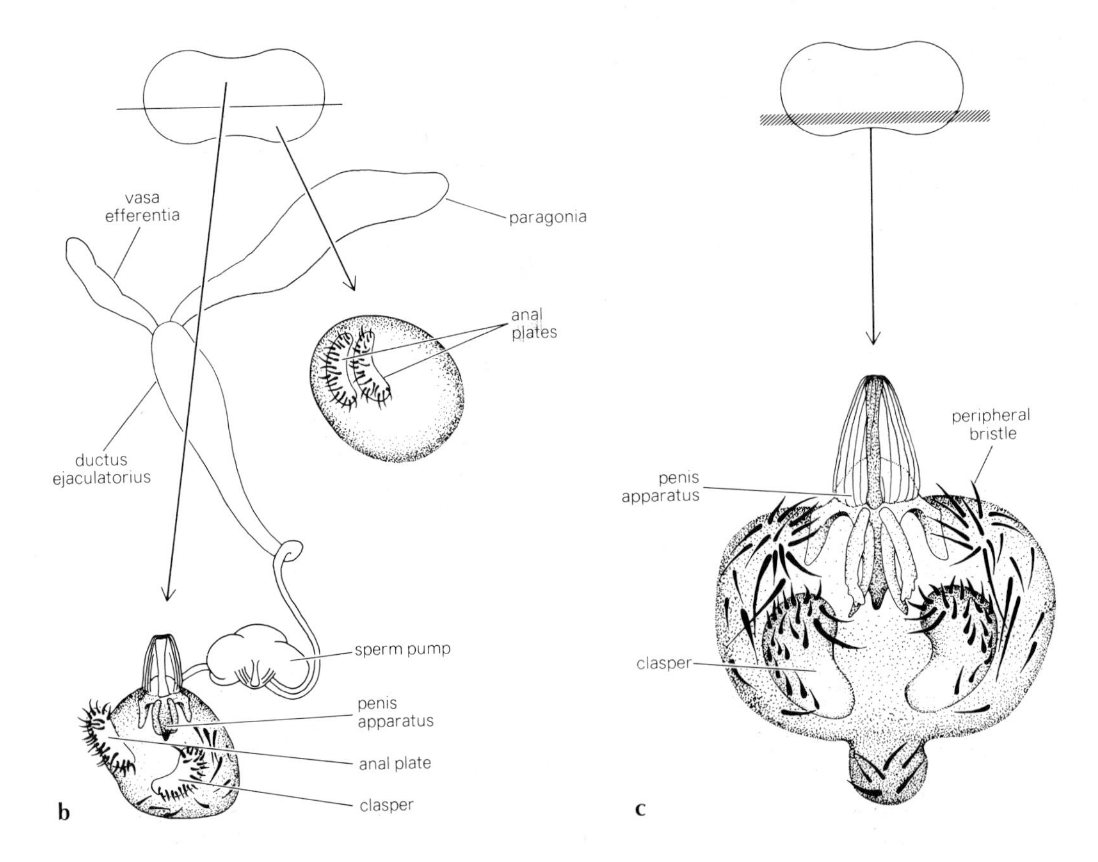

experiments in which genetically marked discs are dissociated into single cells and allowed to reaggregate. If cells from a wing disc and a genital disc are mixed in this manner, an integrated mosaic structure does not form. Rather, the two types of cells separate from each other by self-association and proceed to differentiate into genital and wing structures side by side. Occasionally, a single cell or a very small group of cells from a genital disc was trapped in a large area of wing-disc cells and forced to metamorphose there. When this occurred, the cells differentiated autonomously into characteristic genital bristles. The surrounding wing-disc cells were unable to alter the commitment of the isolated genital cell(s).

The degree to which a fragment of a disc is able to regulate and differentiate into parts other than those of its prospective fate seems to depend on the stage at which the fragment is removed and the time between removal and differentiation. For example, the fragments of discs from the early third-stage larvae are capable of greater regulation than those from late third-stage larvae with respect to forming other structures derived from the *same* disc. However, the ability of a fragment to reorganize itself and give rise to an essentially complete structure appears dependent on the cells having multiplied in the host larva before differentiation. If a fragment of a wing disc is caused to metamorphose immediately, it will form only those structures it would have formed in the original donor. In contrast, if the fragment is allowed to grow for a period, it can be caused to form an entire wing. The commitment to form a particular part of the wing is less binding than the commitment to form "wing" in general.

C. Cell-lineage studies

The overall results of experiments in which parts of the early *Drosophila* embryo are subjected to various types of damage suggest that specific regions of the adult are laid out at about the time of blastoderm formation, i.e., the time the multinucleate cleavage stage is divided simultaneously into several thousand cells. In order to follow the lineage of these cells, special techniques have been developed which take advantage of the large variety of genetic mutants available in this organism. In the most widely used technique, the genotype of randomly positioned cells is altered in such a way that this cell and all its progeny can be recognized simply by examining the surface of the adult. In order to produce a *genetic mosaic* of this type, i.e., an organism containing cells of more than one genotype, the embryo is treated with x-rays, which induce a process termed *mitotic recombination* in a very small percentage of the irradiated cells. For example, an individual that is heterozygous for traits affecting bristle color and bristle appearance ($y\ sn^+/y^+\ sn$) will have a wild-type phenotype. If one of the cells of this embryo is struck by x-rays and induced to undergo crossing-over during one of its mitotic divisions, then it may produce daughter cells that are homozygous for one or the other of these traits ($y\ sn^+/y\ sn^+$ and $y^+\ sn/y^+\ sn$). As each of these cells divides it will generate a clone of cells which, after metamorphosis, can be seen as a patch of *yellow* or *singed* bristles surrounded by a background of wild-type bristles. An example of this technique utilizing eye-color markers is shown in Fig. 7.24.

The analysis of clones of cells formed by mitotic recombination has led to the study of development at the cellular level. It has provided important information on a number of basic questions on the development of these insects. For example:

The size of the patch (i.e., clone) is dependent on the number of cell divisions that occurred between the time of clone induction and that of imaginal differentiation (when cell division ceases). The later the stage irradiated, the smaller the area of the cuticular surface covered by the clone.

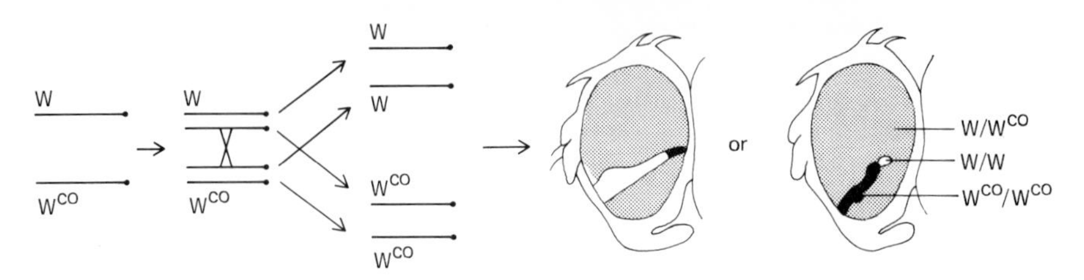

Figure 7.24 Somatic recombination in the X chromosome of *Drosophila* and the resulting twin mosaic spots in the eye. [*After H. J. Becker,* Verh. Deutsch Zool. Ges. (*1956*).]

The ratio of the total number of cells in the mature disc to the average size of the clones induced at time t gives the number of cells present in the disc (or disc primordium) at time t. For example, if an embryo is irradiated at the blastoderm stage and a clone is induced that covers one-twentieth of the wing surface, one can estimate that approximately 20 cells present at the blastoderm stage are destined to give rise to the tissues of the wing disc. The 20 or so cells of this group are not closely related to each other; i.e., they are not descendants of a particular cleavage-stage nucleus. Rather they represent those cells which formed with nuclei that happened to migrate to a particular peripheral region of the egg.

The shape of the patch tells the observer about the morphogenetic processes operating during development. In the leg, for example, a patch of marked cells tends to extend as a thin strip (a few cells wide and over 100 cells long) lengthwise along the appendage. This indicates that there is a predominantly radial arrangement of daughter cells in the developing disc, which may be one of the important factors determining its structure. In the eye, for example, clones in the anterior region are larger than those in the posterior region (Fig. 7.24), indicating the presence of regional growth control among the cells of a disc.

The most important information obtained from the analysis of genetic mosaics relates to the lineages of embryonic cells and the degree to which their developmental capacity may be restricted at various stages of development. The results of studies that bear on these aspects will be considered at length in the remainder of this section.

When the embryos of the milkweed bug *Oncopeltus* are irradiated at the multinucleate cleavage stage, some of the larvae show marked clones that are split into patches of hundreds or thousands of cells scattered over *several* abdominal segments (Fig. 7.25*a*) and often on the head and thorax as well. It would appear that in these cases the nucleus induced to undergo mitotic recombination gave rise to daughter nuclei which migrated into different regions of the blastoderm. In contrast, when embryos are not irradiated until the end of the cellular blastoderm stage, the marked clones that appear are always restricted to a single segment. When the cells reach the border between segments, they run along the boundary line for a varying distance, but they do not cross into the tissue of the neighboring segment (Fig. 7.25*b*). These results indicate that the prospective fate of each blastoderm cell is restricted to the formation of those structures present within a given segment.

An important question that can be answered by clonal-analysis techniques is whether or not an individual, genetically marked blastoderm cell can give rise to

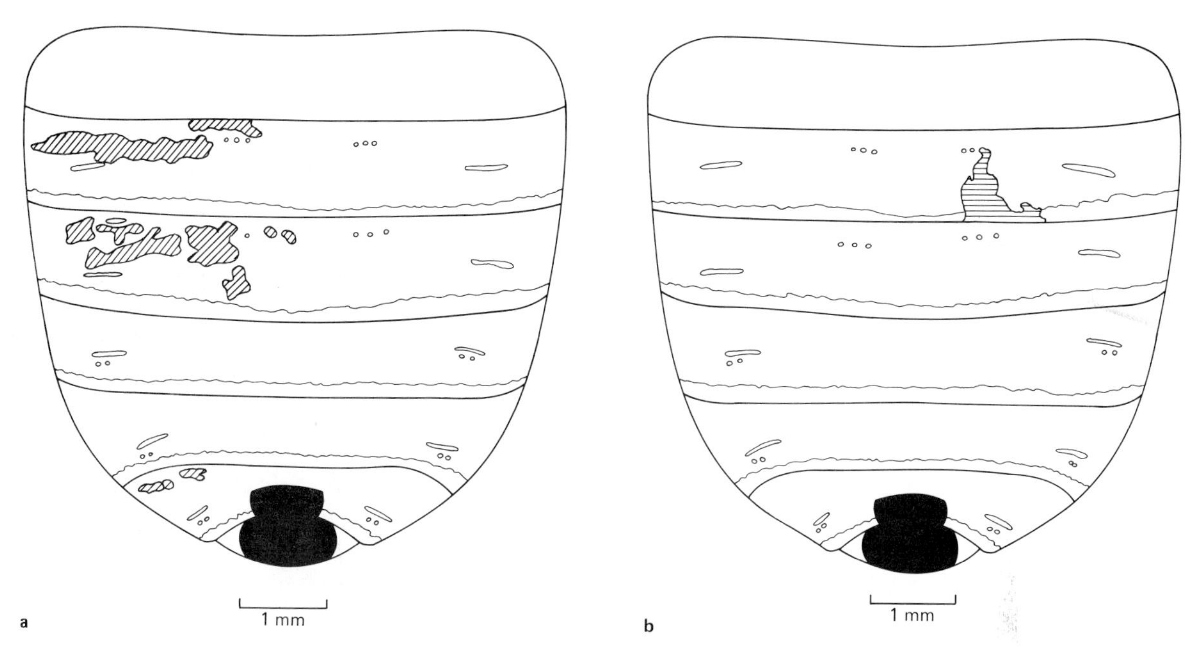

Figure 7.25 Clones (*shaded regions*) produced by x-irradiation of *Oncopeltus* embryos as eventually seen on the abdomen of a fifth-stage larva. *a* When irradiation is performed during late cleavage, the clone extends to three different abdominal segments. *b* When irradiation is performed at the end of the blastoderm stage, the clone is confined to a single segment and respects the intersegmental boundary. [*After P. A. Lawrence,* J. Emb. Exp. Morph., **30:***681* (*1973*).]

structures derived from more than one imaginal disc. When embryos are irradiated at the beginning of the cellular blastoderm stage, most of the marked clones generated are restricted to a given set of disc structures. However, there are notable exceptions. In one study, 7 of 31 clones covering the appropriate area were found to overlap structures derived from the wing and second leg disc. This finding clearly indicates that the founder cells of these various clones were not fully determined at the blastoderm stage, i.e., the time of irradiation. Rather there were at least two distinct developmental pathways still open to the cell's progeny, even though both led to structures of the same (mesothoracic) segment. In contrast, when embryos of 7 and 10 hours were irradiated, all the marked clones of the thoracic region were restricted to a single disc. However, under these latter experimental conditions, clones were still found to extend into both eye and antennal structures: a commitment with respect to the eye versus the antennal disc had not been made by 10 hours of development. These results are consistent with the concept of stepwise determination discussed in Section 7.6. By 10 hours of development, the cells in question are committed to the formation of head structures as opposed, for example, to the formation of abdominal structures, yet they are still capable of giving rise to cells of more than one head disc. In contrast, fragmentation experiments on discs of late third-stage larvae indicate that by this stage all cells are determined to form structures derived from a single disc (discussed in Section 7.8.B).

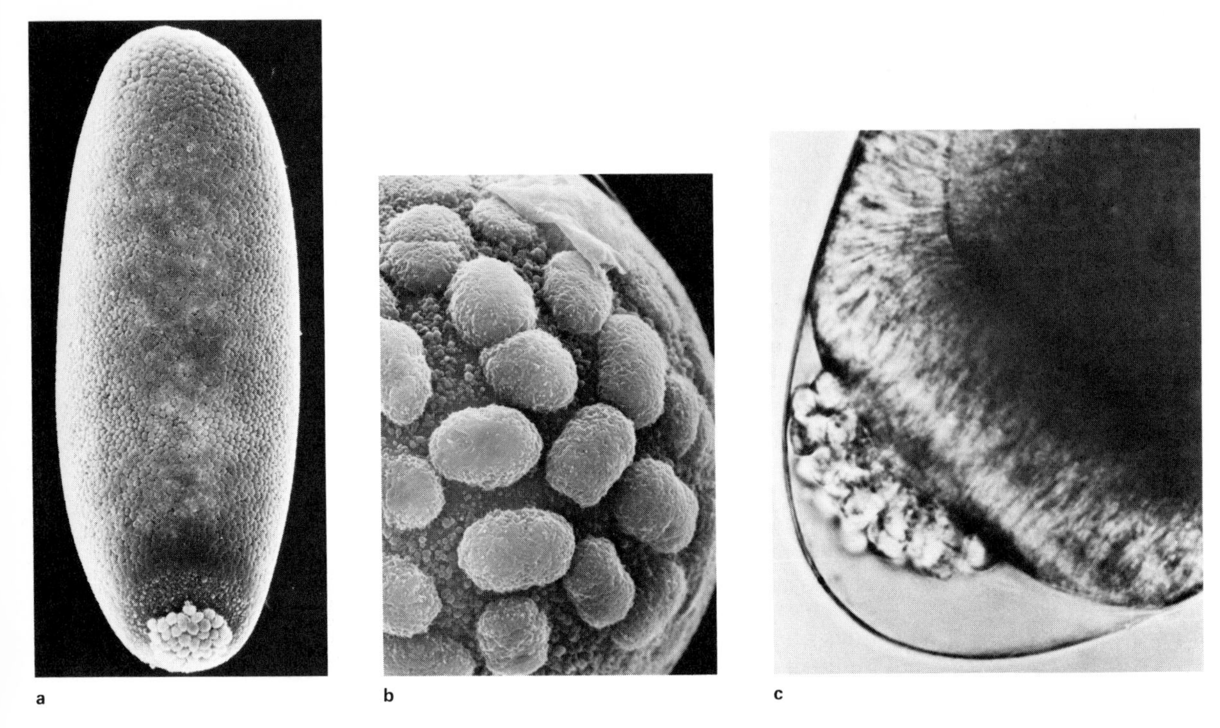

Figure 7.26 Pole cells of *Drosophila*. *a, b* Scanning electron micrographs showing the distribution of pole cells at the posterior end of the early gastrula stage. Note the apparent division planes in several of the pole cells of *b*. [*Courtesy of F. R. Turner and A. P. Mahowald.*] *c* Posterior region of a living embryo in the phase-contrast microscope. [*Courtesy of M. Bownes.*]

D. Germ-cell determinants

A striking exception to the occurrence of stepwise determination is seen in the development of the germ cells. The pole cells (Fig. 7.26), which form at the posterior tip of the egg, owe their special properties to the cytoplasmic material present in that region of the egg. As in amphibians, the presence in the egg of polar granules containing germ-cell determinants seems indisputable. This has been clearly demonstrated in a remarkable experimental study. Cytoplasm from the posterior tip of the egg can be removed with a micropipet and injected into the anterior portion of a second egg. When the cleavage nuclei reach this foreign cytoplasm, they become incorporated into pole cells just as those reaching the posterior pole of the same egg. Although the anterior pole cells cannot develop into germ cells in this abnormal position of the embryo, they can be transplanted to the posterior region of a third egg (having a different recognizable genotype), where they will proceed to produce viable gametes in the adult. One can therefore conclude that the injected *germ plasm* was able to act in an autonomous manner and influence the nuclei it came into contact with so that the cells that formed could develop into bona fide germ cells. Recent successes in the isolation and purification of insect egg polar granules have provided a rare opportunity to learn more about the macromolecular nature of the germ-cell determinant(s). It has been found that in addition to RNA, the polar granules contain large amounts of a single protein, one with a preponderance of basic amino acid residues. It may

be that this protein serves as a scaffolding to which specific mRNAs can bind, thereby promoting the localization of maternal information required in germ-cell formation.

Since the germ plasm is already localized in the unfertilized egg, it must be laid down during oogenesis. Confirmation of this belief comes from studies of a particular mutant termed *grandchildless*. Homozygous (*gs/gs*) females appear perfectly normal, but their offspring lack germ cells and are thus sterile, hence the name. It would appear that this maternal-effect mutation leads to the production of eggs lacking some type of substance that is essential for the formation of germ cells. As in the case of amphibians, a similar type of sterility can be achieved by irradiation of the posterior pole of the egg with ultraviolet light. In these latter experiments, the eggs can be rescued, i.e., made fertile, by injection of polar cytoplasm from nonirradiated eggs, once again confirming the presence of a UV-sensitive cytoplasmic determinant. Taken as a whole, the studies on germ plasm have provided the best (and only clear-cut) evidence for the existence of localized oogenetic factors.

E. The compartment hypothesis

The analysis of genetically marked clones has led to the concept that the insect body is made up of specific regions, termed *compartments*, which have both a developmental and genetic basis. A particular compartment is a unit defined by its boundary lines and cell lineage. The cells of each compartment are derived from a small group of neighboring cells that was functionally isolated at a much earlier stage of development. It is proposed that some determinative event occurs simultaneously within this group of cells such that the developmental capacity of progeny cells is appropriately restricted. In this sense, the compartment can be thought of as a unit of determination. Since the cells are derived from a *group* of founder cells (which are not themselves clonally related but only neighbors) rather than a single founder cell, they are referred to as a *polyclone*. As the cells of a polyclone proliferate, they are restricted in their range to developing within a set of boundary lines; they cannot cross the border into another compartment. The parameters of a given compartment are defined in both space and time. As the time of development increases, the descendants of one polyclone can become divided into daughter polyclones, each destined to form a smaller compartment having narrower boundary lines than that formed by the primordial polyclone of the blastoderm stage. With each subdivision, a population of homogeneous cells is split into two or more subpopulations having different restrictions, i.e., having different developmental potential. It is assumed in the hypothesis that the same compartments, i.e., ones having the same boundaries, are present in all members of the species. The formation of successive compartments is therefore a genetically programmed event.

The formation of compartment boundaries and their subsequent subdivision into smaller compartments can be illustrated using the mesothorax, the segment bearing the wings and second legs. If mitotic recombination is induced at the cellular blastoderm stage, the marked clones of the mesothorax can extend between the leg and the wing. One can conclude that the commitment to form leg versus wing structures has not been made by the cells of the blastoderm stage. However, none of the clones extends between the anterior and posterior portion of the wing or between the anterior and posterior portion of the second leg. It would appear that the whole mesothoracic polyclone is subdivided at a very early stage into separate polyclones that define anterior and posterior compartments of this segment. Cells of each polyclone will eventually populate both leg and wing

discs and contribute to the anterior *or* posterior portion of both structures. In later stages of development, the polyclones of the anterior and posterior compartments are subdivided into smaller polyclones so as to form dorsal, ventral, wing-blade, and thoracic parts. The proposed sequence of events in the compartmentalization of the mesothorax is shown in Fig. 7.27.

If we reexamine the subdivision of the mesothoracic segment into anterior and posterior compartments, we can illustrate certain other features of the compartment hypothesis. Figure 7.28*a* shows the boundary line between these two compartments of the wing. Since each compartment is populated by the descendants of a polyclone, a given marked clone normally occupies only a fraction of the entire compartment area. In order to best reveal the boundaries of a compartment, a special strain of flies is utilized. There is a dominant mutation termed *Minute* in which a heterozygous larva grows very slowly but hatches eventually as a normal-sized adult. If such a heterozygote (M^+/M) is treated with x-rays, any wild-type cell formed after mitotic recombination will have a greatly elevated mitotic rate and will overgrow the heterozygous tissue. When this occurs within either polyclone of the wing at the blastoderm stage, that cell proliferates, but its range is still restricted by the boundary line of the compartment (referred to as the *line of clonal restriction*). Consequently, the wild-type cells fill out most of the compartment (Fig. 7.28*b*), revealing its borders in a clear-cut manner. Moreover, these results indicate that given the opportunity, a cell can divide a greater number of times than normal, thereby forming a greater part of the animal. Even though the fate of the entire polyclone is sealed, that of the various cells within the group has not been irreversibly defined. It should be noted that the line of clonal restriction which cuts across the wing does not correspond to any obvious morphological or physiological subdivision. Rather it relates only to a position on the insect and appears only to have developmental significance. If the techniques for analysis of cell lineage had not been developed, this type of boundary line would not have been detected. We can conceive the body of an insect to be analo-

Figure 7.27 Diagram of compartmentalization of the mesothoracic cuticular structures. The first step (1) is the segregation of anterior and posterior polyclones, and this occurs probably at the blastoderm stage. The second step (2) consists of the separation of the dorsal part of the segment (wing disc) and the ventral part (leg disc), which occurs before 10 hours of development. The wing disc becomes subsequently divided (3) into dorsal and ventral compartments, which are divided (4) into thorax and wing blade. The anterior portion of the leg disc may become subdivided as shown. [*From G. Morata and P. A. Lawrence,* Nature, **265**:*215 (1977).*]

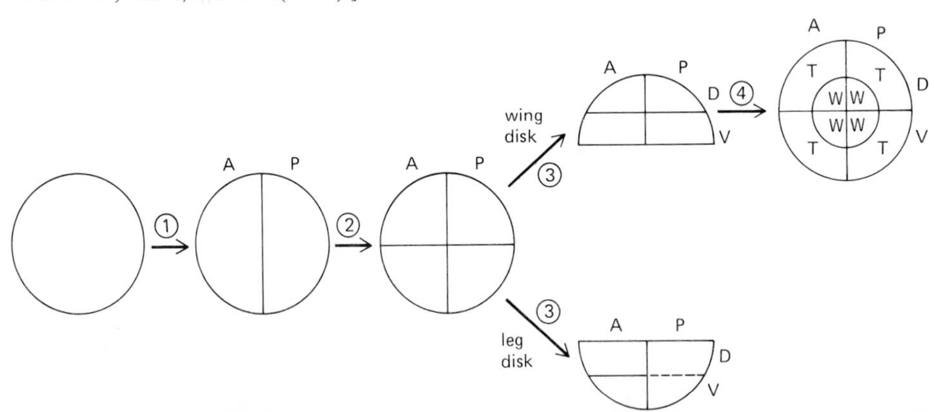

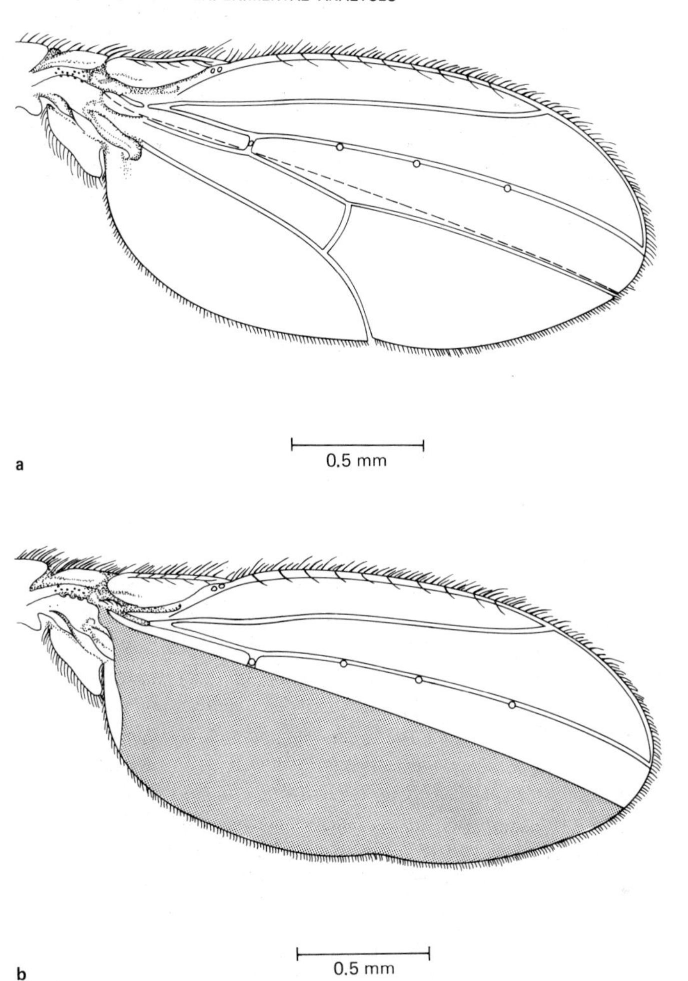

Figure 7.28 *a* The wing of *Drosophila,* to show the position of the antero-posterior compartment border (*dotted line*). *b* Distribution of the cells of a wild-type clone (*shaded portion*) that has overgrown the minute background. Note how the clone, which extends to both dorsal and ventral surfaces, respects the compartment border. [*From F. C. Crick and P. A. Lawrence,* Science, **189**:*340 (1975); copyright © 1975 by the American Association for the Advancement of Science.*]

gous to a jigsaw puzzle in which each piece is represented by a compartment. As in the case of a puzzle, the pieces do not necessarily correspond to distinct structures in the overall picture, yet their borderlines are identically placed in each individual. This concept has important consequences for the type of determinative events occurring during insect development. Most important, cells of a given polyclone would become determined to form a particular area of the insect rather than a particular type of cell. For example, in the case just discussed, a cell becomes determined to form part of the anterior or posterior portion of the mesothorax, although it can still form tissue of either leg or wing.

The existence of distinct lines of clonal restriction suggests that the cells of one compartment are not able to freely intermingle with cells of a neighboring

area. If cells of one polyclone are not able to cross into a region occupied by cells of another polyclone, one might expect this difference to be mediated by properties of their cell surfaces. As mentioned earlier, cells from two different imaginal discs can be mixed, but they do not remain associated with one another. Rather they sort out into aggregates containing one or the other cell type. Similar types of cell-surface differences are believed to distinguish the cells of different compartments, although the evidence for this is less clear.

If the subdivision of a polyclone is accompanied by a change in cell affinities, then some type of genetic function must exist to control these as well as other properties that distinguish neighboring compartments. It has been proposed that the development of each compartment is under autonomous genetic control in that the commitment by a polyclone to develop along a certian pathway requires the expression of a particular "selector" gene. The product of the selector gene would then directly or indirectly lead to the determination of those cells. The subdivision of that compartment into smaller ones at a later stage would result from the activation of new selector genes in each new polyclone. In each case, one of the consequences would be the acquisition of cell-surface properties restricting the movement and interaction of the descendants of the polyclone.

If such genes exist, one would expect to find mutants whose altered phenotype reflected some disturbance in the selection of the appropriate developmental pathway. A variety of mutants have been isolated which appear to satisfy this expectation. Among these mutants is a group termed *homeotic* (or *homoeotic*) *mutants* in which one body part is replaced by another. For example, in the homeotic mutant *antennapedia,* various parts of the antenna are transformed into mesothoracic leg structures. A particularly well-studied mutant is *bithorax,* in which the anterior half of the haltere is transformed into the anterior half of the wing. The presence of this mutation suggests that the normal product of this gene is involved in the determination of the cells of the anterior haltere. If the normal gene product is absent, this developmental pathway is not selected and the tissue differentiates into wing. Further studies have provided information on the times during development at which particular homeotic genes are expressed. In these studies, heterozygotes for bithorax are subjected to x-rays at various stages of development. It has been found that homozygous bithorax cells generated in the anterior haltere region as late as the third-stage larva are still capable of differentiation as wing cells. Even though these cells had existed as heterozygotes through the bulk of the larval period, they were not irreversibly determined to form haltere. Presumably, continuous expression of the normal gene is needed almost to the time of metamorphosis in order to form the haltere.

In certain cases, the homeotic mutations affect more than a single disc. For example, in the bithorax mutant, the anterior portion of the haltere is not the only part affected, the anterior part of the metathoracic leg is also replaced by anterior parts of the leg of the mesothoracic (wing) segment. It is apparent that the mutation is not specific for a type of structure, since both haltere and leg parts are transformed. Rather, this mutation appears to involve a particular location, namely, the anterior portion of the metathoracic segment, which is replaced by structures normally found only in the anterior portion of the mesothoracic segment. Analysis of a number of these homeotic mutants suggests that the replacements are ones involving specific compartments, such as the anterior compartment of the wing described earlier. Regardless of the validity of the compartment hypothesis, it is hoped that the analysis of these mutants will provide an insight into the genetic control of morphogenesis, one that is not available from studies on any other organism.

F. Transdetermination

In a previous section, experiments were described in which an imaginal disc or disc fragment was implanted into a larval host, where it received full exposure to the metamorphic hormone during pupation. In a different type of experiment, pieces of imaginal discs have been inserted into the abdomens of adult flies. In the larval host, the disc cells stop dividing and begin to differentiate as soon as the host pupation begins. In the adult carrier, they continue to grow and divide indefinitely, as long as a sample of proliferating cells is transferred to a new adult abdomen every 2 weeks, i.e., as long as they are regularly subcultured as in typical tissue-culture procedure. By this means, the cultured disc cells live for years, from fly to fly, without undergoing any observable differentiation. Since cell populations cultured in this way have been maintained for several years, passing through hundreds of adult fly generations, there has been ample time and opportunity to check on cell potentialities.

Experiments utilizing adult flies as culture vessels began with a male genital disc (Fig. 7.23). Half of such a disc, the stem piece, was implanted in the abdomen of an adult fly; the other half, the test piece, was implanted into a larva, where it could differentiate during pupation, thereby revealing its potential (Fig. 7.29). After 2 weeks, the stem piece was recovered and divided into two; one piece was again inserted into an adult abdomen as the stem piece, and the other into a larva as the test piece to see if it still differentiated as before. At first the test pieces continued to differentiate into typical genital structures. The disc cells were able to maintain their determined state even though the act of differentiation was postponed for months or even years. Since the cells were able to maintain this state over a course of many cell divisions, some mechanism must exist by which the determined condition is passed from parent to daughter cells. As the number of transfers continued, test pieces, here and there, began to differentiate

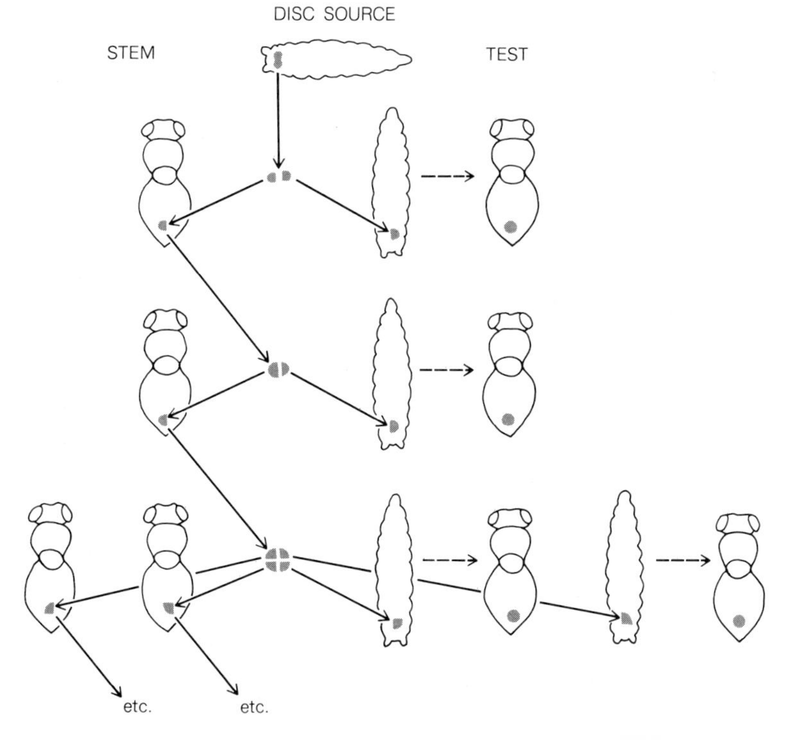

Figure 7.29 Method used for permanent cultures in vivo. Stem-line fragments grow but remain undifferentiated; test fragments pass metamorphosis and develop into adult structures. [*After E. Hadorn*, Brookhaven Symp. Biol., **18**:*154* (*1965*).]

into leg parts or antennas, switching from the original determined path to another, a change that has been termed *transdetermination* (Fig. 7.30). Evidence that transdetermination is not due to somatic mutation is provided by the discovery that it occurs simultaneously in groups of contiguous cells not clonally related and is too frequent to be due to mutation.

Transdetermination can proceed from state to state. In a particular transfer series, for instance, genital disc cells gave rise to head and leg structures in the eighth transfer generation, or after about 4 months. After the thirteenth transfer, they gave rise to wings. After nineteen transfers, a thoracic type of structure appeared. There appears to be a given transdetermination sequence (Fig. 7.31); genital disc cells may give rise to head or leg cells, which may give rise to wing cells, which may give rise to thoracic cells, but apparently the genital cells cannot transform directly into wing cells. Haltere discs transdetermine directly to wing, but not to leg or genitalia. Leg discs transdetermine directly to wing, but not to haltere. In some sense, haltere is closer to wing than to leg or genitalia. Both haltere and leg discs can transdetermine to wing, but not to each other. Moreover, the transdetermination may become stabilized at any stage and need not proceed to a specific end point. Each type of transdetermination occurs with a characteristic probability per transfer generation from adult to adult. In each case, the new state of determination is clonally heritable in the transdetermined tissue line, indicating that it too is a self-maintained condition. The fact that insect cells can be maintained in a determined undifferentiated state for long periods in vitro makes them one of the best systems for analysis of the underlying molecular basis of determination.

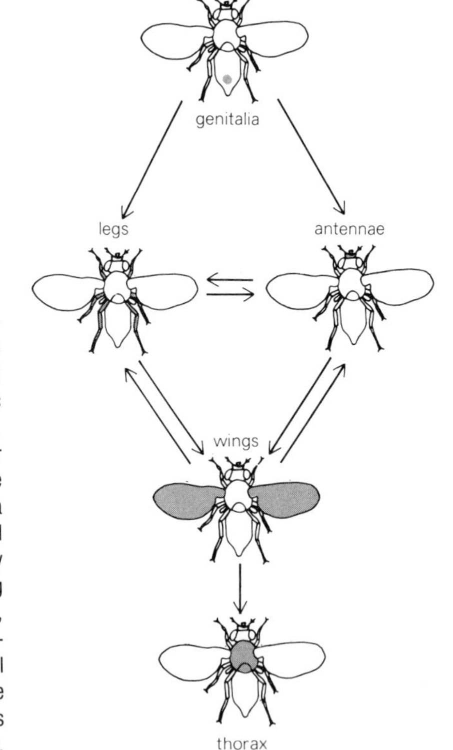

Figure 7.30 Transdetermination sequence undergone by several kinds of imaginal disc cells is shown by arrows. Genital cells, for example, may change into leg or antenna cells, whereas leg and antenna cells may become labial or wing cells. In most instances, the final transdetermination is from wing cell to thorax cell; the hange to thorax appears to be irreversible.

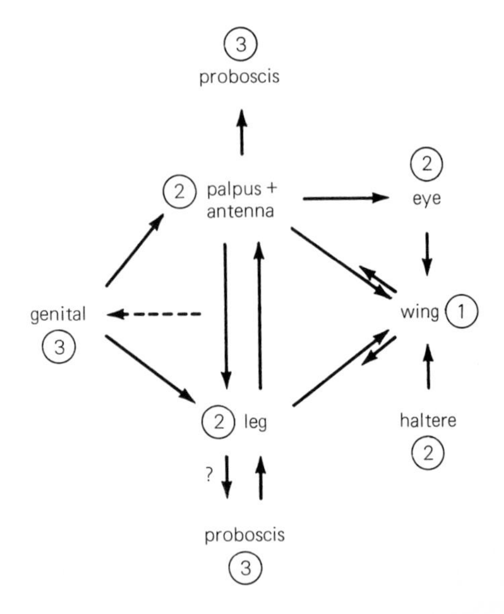

Figure 7.31 Flow-pathway diagram of transdetermination. [*After W. Gehring, in "The Stability of the Differentiated State," W. Beerman, J. Reinart, and H. Ursprung (eds.), Springer-Verlag, 1968.*]

9. Mammals

Mammalian eggs and embryos, though awesome in their developmental potential, are unassuming in appearance and very difficult to work with. By vertebrate standards, mammalian eggs are small (approximately 100 μm in diameter), essentially yolkless, and most unusual in that they develop entirely within the female reproductive tract. It is this latter feature which gives great benefit to the embryo and considerable frustration to the developmental biologist. Some of the major difficulties in the study of mammalian development include:

The small number of eggs that are ovulated in a given reproductive cycle. With the use of hormones (Section 4.3.B), increased numbers of eggs are brought to maturity and ovulated on a given day. *Superovulated* females generally provide 50 or so eggs (or embryos if mated), which, when removed from the oviduct and pooled with those from other animals, allows one to carry out considerable biochemical analysis.

The small size of the mammalian egg. In recent years a variety of microsurgical techniques have been worked out by which mammalian embryos can be dissected into their parts, or fused with one another, or dissociated into single cells which can then be injected into another embryo. With the use of these manipulations, some indications have been obtained concerning the fates of embryonic cells and the degree to which they are determined at various stages.

The development of the embryo within the body of the mother. Attempts to overcome this difficulty have led investigators to devise in vitro culture conditions that can support embryonic development. Although a great deal more work is required in this area, conditions have been defined which will allow eggs to be fertilized in vitro and raised through the early stages of development in culture. Similarly, later-stage embryos can be removed from their attachment to the uterus and cultured for several days during the period in which organ formation (*organogenesis*) occurs. Development in these cases can continue in a normal manner to a remarkable degree (see Fig. 7.41).

We will return to these matters in later sections.

A. Early development

Mammalian development is typically divided into two major periods. During the *preimplantation period,* the fertilized egg develops into a microscopic blastocyst which becomes firmly attached to the uterine wall. It is only after this point, in the *postimplantation period,* that the embryonic axis is established and organogenesis occurs. It is during the embryonic stage (through the eighth week in humans) that the organ systems are laid down and the placenta is constructed. The fetal stage which follows (ninth week to term) is characterized primarily by growth. In the following discussion we will be largely concerned with the events that occur prior to and during implantation, primarily as observed in the best-studied mammalian embryo, that of the mouse. The timetable for mouse development is given in Table 7.1. It should be kept in mind that developmental events (particularly those involving implantation) in other mammals may be quite different. We will pick up the story once again in Chapter 10, when we discuss gastrulation.

At the time of ovulation (Section 4.3.B), one or a few eggs are released from the ovary and swept into the *fimbria,* the funnel-shaped opening of the oviduct, by the activity of the cilia lining its wall. Once within the oviduct (Fig. 7.32), the egg awaits its chance encounter with a sperm. Once fertilized, the events of

Table 7.1 Developmental Timetable of Mouse

Days of gestation	Stage
1	One- to two-cell, in uppermost part of oviduct
2	Two- to sixteen-cell, in transit to uterus
3	Morula (solid ball of cells), in upper part of uterus
4	Free blastocyst in uterus, shedding of zona pellucida
4½	Beginning of implantation
5	Inner cell mass elongating, primitive streak evident, and proamniotic cavity
6	Implantation complete, extraembryonic parts developing
7	Ectoplacental cone, amniotic folds, primitive streak, heart and pericardium forming, head process evident
7¼	Early neurula, neural plate, chorioamniotic stalk, allantoic stalk beginning, inner cell mass with three cavities, somites beginning to differentiate, foregut present
8½	Neural tube formed, embryo established
9–19	Growth of fetus
19–20	Birth

Source: C. R. Austin, "Fertilization," © 1965. Reprinted by permission of Prentice-Hall, Inc. Englewood Cliffs, N.J.

meiosis are completed and those of cleavage begin, at first very slowly. First cleavage of a mammalian zygote typically occurs about 24 hours after fertilization, the second and third cleavages being spaced about 10 to 12 hours apart. Since there are no distinguishing features in the embryo during these stages (Fig. 7.33), the individual blastomeres cannot be identified. It is at the 8- to 16-cell stage that the first obvious morphogenetic change occurs. Prior to this point the blastomeres appear generally rounded and not packed very closely together. Following the third cleavage in the mouse, a process termed *compaction* occurs in

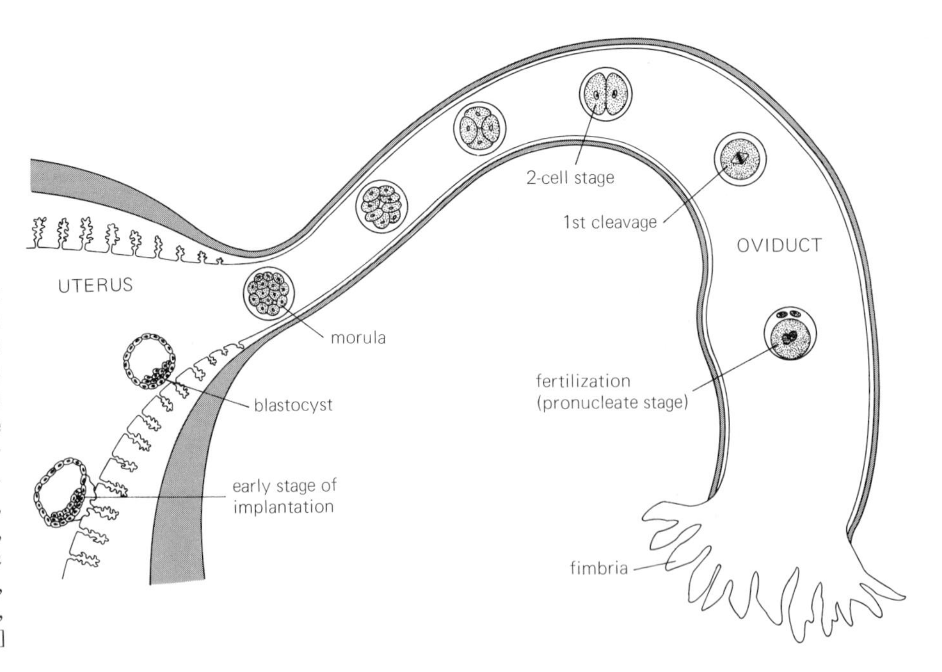

Figure 7.32 Development of the human embryo in the reproductive tract, from fertilization in the oviduct to implantation in the uterus. [*After H. Tuchmann-Duplessis, G. David, and P. Haegel, "Illustrated Human Embryology," vol. 1, Springer-Verlag, 1972.*]

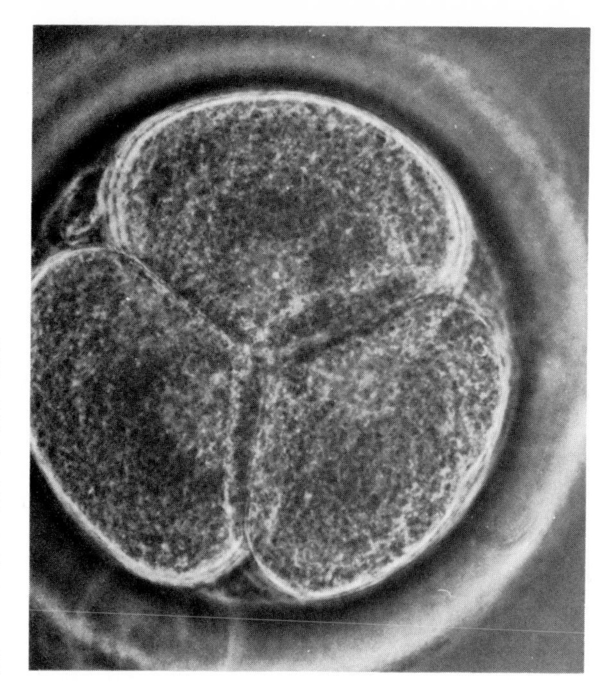

Figure 7.33 Cleaving egg of rabbit fertilized and cultured in vitro, about 0.15 mm in diameter. [*From R. J. Ericsson, D. A. Buthala, and J. F. Norlund,* Science, **173**:*54 (1971); copyright © American Association for the Advancement of Science.*]

which the cells become pressed against one another along their lateral surfaces causing the embryo to appear almost as a single, highly irregular-shaped cell (Fig. 7.34). Scanning and transmission electron micrographs reveal that compaction is accompanied by major changes in the cell surfaces of the blastomeres. The microvilli, for example, become very unevenly distributed. Some microvilli are seen in the basal region between the cells prior to their becoming pressed

Figure 7.34 Scanning electron micrographs of *a* uncompacted and *b* compacted eight-cell mouse embryos. Note the change in cell shape and maximization of cell-cell contact. The polar bodies do not participate in compaction. [*Courtesy of T. Ducibella, T. Ukena, M. Karnovsky, and E. Anderson.*]

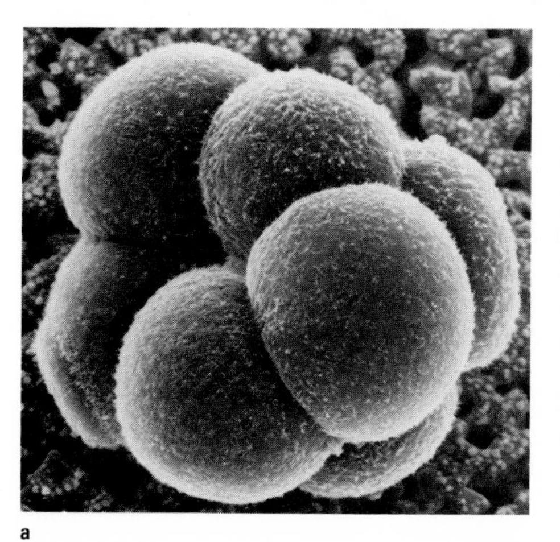

a

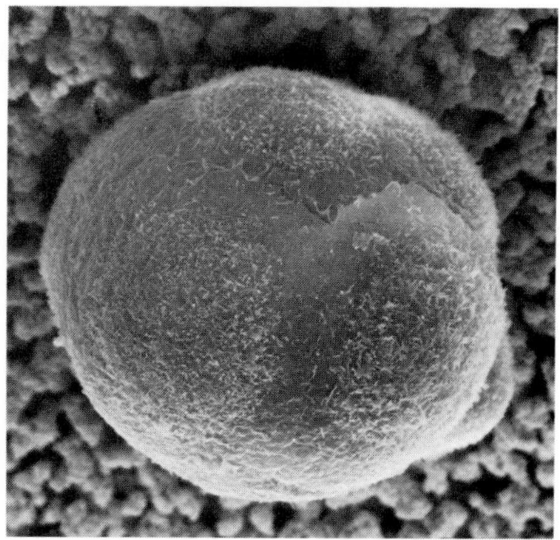

b

against one another. Since microvilli are known to be dynamic structures containing bundles of contractile microfilaments, they may function in bringing together the surfaces of adjoining cells. The remainder of the microvilli are localized on the outer (apical) surface of the blastomeres, giving each cell a distinct polarity. The apposition of adjoining surfaces leads to the formation, over a period of time, of desmosomes, gap junctions, and tight junctions between the cells of the embryo (Fig. 7.35), which are believed to have important structural and developmental roles. During this period of early development, the cleaving embryos are moving down the oviduct by the action of the cilia and/or musculature of the wall.

As cleavage progresses, a ball of cells termed the *morula* is formed. It is at about this stage, 16 to 32 cells in number, that the embryo passes into the uterus. Among the cells of the morula, a very few occupy completely internal positions (3 to 5 at the 32-cell stage), the remainder forming a layer of cells at the outer surface of the embryo. Tight junctions (*zonula occludens*) form at the apical surfaces between the cells of the outer layer, thereby sealing off the extracellular space within the embryo from that of the outside uterine environment. Soon a major reorganization occurs among the cells of the embryo. Fluid is secreted into the spaces between the cells, and the embryo grows rapidly in volume forming a blastocyst. By the early blastocyst stage (3½ days *post coitum* in the mouse), two

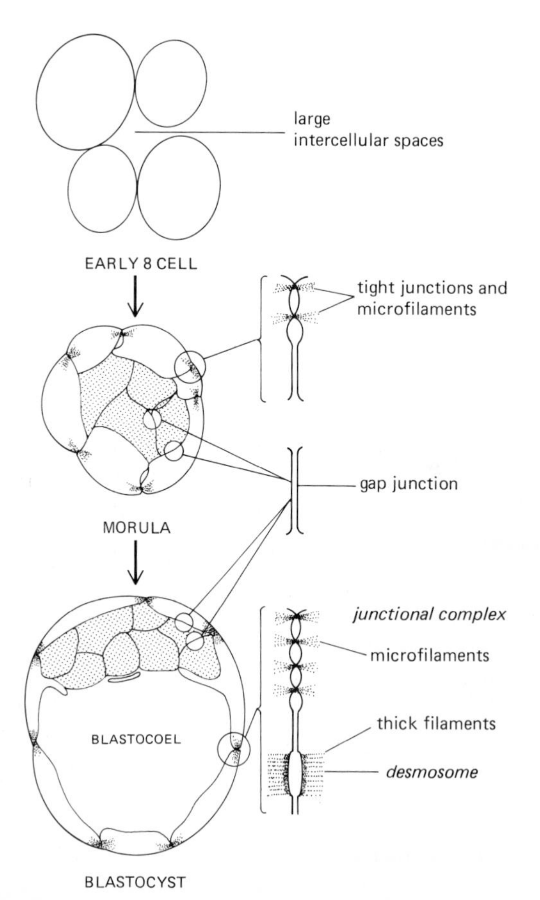

Figure 7.35 Schematic summary of the development of intercellular junctions in the preimplantation mouse embryo. [*From T. Ducibella, in "Development in Mammals," vol. 1, M. H. Johnson (ed.), Elsevier, 1977.*]

distinct populations of cells can be distinguished (Fig. 7.35). One group of cells is present as an outer single-celled layer termed the *trophectoderm* (or *trophoblast*), which encloses an increasingly spacious fluid-filled blastocoel.[4] As discussed later, the trophectoderm is concerned with the establishment of an intimate relationship between the mother and embryo. At one side of the blastocoel is a cluster of cells pressed up against the inner surface of the trophectoderm. This cluster is referred to as the *inner cell mass* (*ICM*), which, as we will see, ultimately gives rise to the entire structure of the embryo. The mouse blastocyst of 3½ days contains about 64 cells, three-quarters of which make up the trophectoderm. The cells of the trophectoderm and ICM can be distinguished at this stage by several physiological and biochemical criteria; i.e., they have become somewhat differentiated from one another. For example, trophectoderm cells are phagocytotic, are able to initiate implantation, and are capable of accumulating blastocoelic fluid, while those of the ICM lack these properties. Most important, these two types of cells have been shown to be synthesizing somewhat different populations of proteins. Two-dimensional gel electrophoresis indicates that cells of the ICM synthesize a number of proteins not detectable among the products of the trophectoderm, and vice versa.

Now that we have reached a stage in the development of the mammal in which two distinct populations of cells have been formed, the obvious question concerns the role of each cell population in the future development of the animal. What is the prospective fate of each cell type? In the snail (Section 6.4.D), cells could be identified and followed as a result of the determinate spiral cleavage pattern; in the ascidian, the segregation of cytoplasmic materials proved useful in cell-lineage studies; in the insect, recognizable genetic variants were induced randomly by radiation. Much of the early work on the analysis of mammalian development was accomplished by examination of serial sections of embryos at steadily increasing stages. This approach, however, has inherent limitations. One can only look at "time-frozen" sections of entirely different specimens; there is no way to follow the progress of a given cell or group of cells over a period of time.

During the 1960s, researchers began to devise techniques for handling and manipulating living mammalian embryos. Preimplantation embryos of the desired stage can be flushed from the oviducts of mated animals, the zona pellucida removed, and the naked embryo treated in various ways prior to its reinsertion into the uterus of a hormonally prepared (pseudopregnant) female in which it will implant. We will consider the use of these techniques in the study of regulation and determination in a later section. In the present discussion we will simply point out the manner in which the fate of the two cell types have been followed and then proceed with a description of early mouse development. In order to follow the prospective fate of a cell or group of cells, two requirements must be met. First, the cells must possess some recognizable feature by which they and their progeny can be distinguished. Second, the spatial relationship of the cells to their neighbors should be disturbed as little as possible. In the case of mammals, these criteria are best met by the construction of genetic mosaics in which the ICM is derived from one source and the trophectoderm from another. In order to construct this type of chimera, blastocysts from two mouse strains are dissected into their ICM and trophectoderm components and the ICM of one is inserted into the empty blastocoel of the other. The embryo and its associated extraembryonic

[4] In the following discussion we will use the term *trophectoderm* for the outer cellular layer of the blastocyst and reserve the term *trophoblast* for the postimplantation derivatives of this layer.

membranes can be recovered after implantation and surveyed for the distribution of the two types of cells. In a related approach, rat-mice chimeras have been produced, and the localization of each species' cells can be determined by treatment of sections with fluorescent antimouse or antirat antiserum. The prospective fates of the cells discussed in the following paragraphs (and shown in Fig. 7.39) have generally been revealed by these means.

As the blastocyst expands, important changes occur within the two major cell populations. During the fifth day, the cells of the trophectoderm that line the ICM become distinguished from those in direct contact with the blastocoel. The cells lining the blastocoel, the mural trophectoderm cells, become larger and larger but do not undergo cell division. Determination of the DNA content of these cells (referred to as *giant cells*) indicates that replication is occurring in the absence of mitosis; the cells are becoming increasingly polyploid, some containing over 1,000 times the normal DNA content. The mural trophectoderm is a highly phagocytotic cell layer that plays an important role in the invasion of the uterine lining at implantation. In contrast, the cells of the *polar trophectoderm*, that portion in contact with the ICM (Fig. 7.36), continue to proliferate, forming a conical mass of cells termed the *ectoplacental cone*. The ectoplacental cone, and a population of secondary giant cells derived from it at its periphery, will be of future importance in establishing the trophoblastic connection with the maternal blood supply. In addition to its contribution to the ectoplacental cone, the polar trophectoderm is also believed to give rise to the extraembryonic ectoderm (Fig. 7.36*b*). This mass of cells eventually forms the ectodermal layer of an extraembryonic structure, the chorion, which in mammals is also incorporated into the fetal vasculature of the placenta. As the extraembryonic ectoderm grows in mass, it pushes further into the blastocoel displacing the ICM toward the interior (Fig. 7.36*c*).

As the trophectoderm is undergoing its differentiation into mural and polar regions, the inner cell mass is also diverging along different pathways. Initially, a layer of cells is delaminated (split) from the bulk of the ICM on its blastocoelic surface. These cells, referred to as the *primitive endoderm*, grow down along the inner surface of the mural trophectoderm (Fig. 7.36*b*), converting it into a double-layered structure. The proximal and distal segments of the primitive endoderm give rise to the endodermal layers of the extraembryonic amnion, allantois, and yolk sac (structures discussed in Chapter 12). The remainder of the

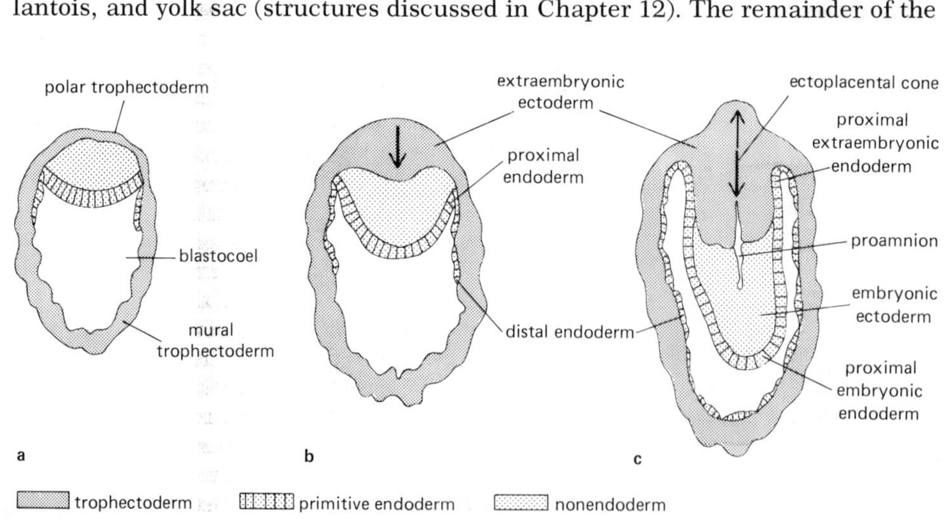

Figure 7.36 Diagrammatic representation of the changes in cellular organization from the blastocyst to the egg-cylinder stage in the mouse showing the proposed origin of embryonic and extraembryonic parts. *a* Blastocyst (4½-day-old) before local proliferation of polar trophectoderm has begun. *b* Local proliferation results in accumulation of polar cells at the mesometrial (upper) pole of the blastocyst, which leads to invagination of the ICM and future extraembryonic ectoderm into the blastocoel. The latter pushes inside the sheath of primitive endoderm, which is anchored mesometrially to the adjacent mural trophectoderm cells. Further proliferation of the polar trophectoderm leads to the mesometrial outgrowth of the ectoplacental cone into the space made available by the degeneration of the uterine epithelium. [*From R. Gardner and V. E. Papaioannou, in "Early Development in Mammals, M. Balls and A. E. Wild (eds.), Cambridge, 1975.*]

cells of the ICM are referred to as the *embryonic ectoderm,* or *primitive ectoderm* (Fig. 7.36*c*). This is a particularly unfortunate choice of terms, since it is from this group of cells that the entire structure of the fetus (ectodermal, mesodermal, and endodermal derivatives as well as the germ cells) is believed to emerge. As shown in Fig. 7.36*c*, the primitive ectoderm is situated between the extraembryonic ectoderm and the primitive endoderm. As the mass of extraembryonic ectoderm grows down into the blastocoel, so too does the mass of primitive ectoderm. Together the extraembryonic and primitive (or embryonic) ectoderm form an elongated structure, referred to as the *egg cylinder* (Fig. 7.37), that is suspended in the blastocoel. A proposed scheme for the fate of the various cells of the egg cylinder is given in Fig. 7.38.

Figure 7.37 Diagram *a* and photomicrograph *b* of a sagittal section of a mouse egg cylinder. A constriction in the middle of the cylinder marks the boundary between embryonic and extraembryonic ectoderm. [a *From G. D. Snell and L. D. Stevens in "Biology of the Laboratory Mouse," E. L. Green (ed.), McGraw-Hill, 1966;* b *Courtesy of J. Rossant and V. E. Papaioannou.*]

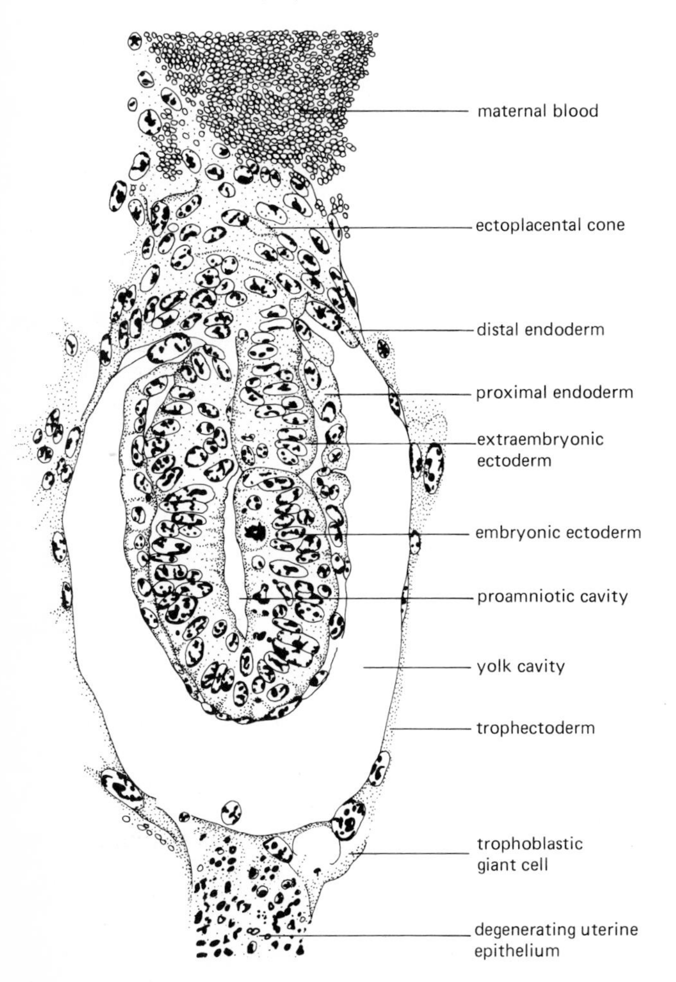

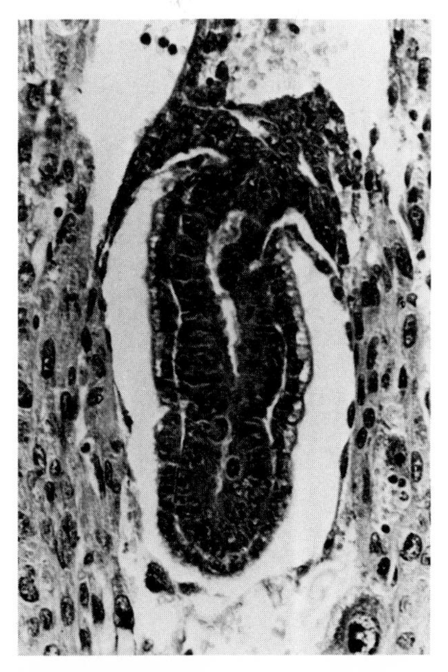

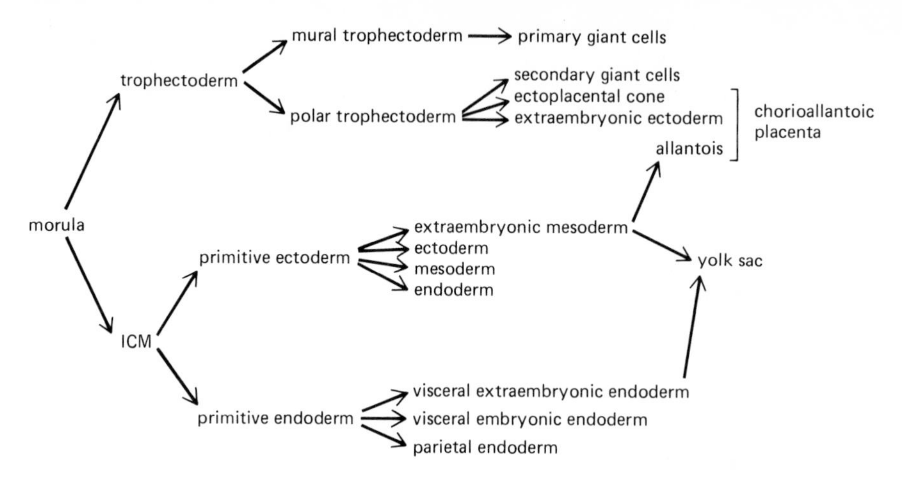

Figure 7.38 Proposed plan of cell-lineage relationships in early mouse embryogenesis. [*From the work of R. Gardner, J. Rossant, V. E. Papaioannou, and others.*]

B. Implantation

The process of implantation can be divided into several stages occurring over a period of a few days. The first indication that the blastocyst is preparing for the upcoming attachment to the uterine wall is the escape of the embryo from the zona pellucida, an event referred to as *hatching.* The hatching process is accomplished by the regular pulsating movements of the blastocysts. Although blastocyst hatching can occur outside the uterine environment, i.e., in ectopic sites (nonuterine regions of the body) or in vitro, there is evidence that the uterus provides some substance that aids in the escape event. Once freed from the zona, the blastocyst enters an *apposition phase,* where it aligns and orients itself in preparation for the upcoming attachment. During the *adhesion phase* of implantation, the blastocyst becomes firmly attached to the uterine epithelium by the mural trophoblast cells at the ebembryonic pole. As a result of the action by giant cells of the mural trophoblast, the *invasive phase* takes place, during which the blastocyst may penetrate deeply into the uterine wall. As this occurs, the trophoblast cells of the ectoplacental cone, originally derived from the polar trophectoderm, also become invasive giant cells and eventually penetrate the maternal blood spaces. In the end, the maternally circulating blood comes into intimate contact with the fetal capillary system, and exchange of substances can occur. The uterus is not simply a passive partner in the implantation event. Beginning approximately at the time of attachment, the connective tissue of the uterine stroma rapidly proliferates, displacing the uterine glands and forming a casing, termed the *decidua,* around the embryo. The degree of invasiveness required to bring about maternal-fetal exchange varies greatly among species. In humans and mice, for example, tissue penetration is much greater than that occurring in the horse or pig. In these latter species, the blastocyst does not develop invasive properties but simply lies in intimate apposition to the uterine epithelium. The general course of events in the implantation of the human embryo is shown in Fig. 7.39.

The attachment of the blastocyst to the uterine wall must be mediated, to a large degree, by events occurring in the apposing membranes of the trophectoderm and uterine cells, since expected changes in the properties of the trophectoderm surface can be detected prior to implantation. Relative to earlier stages, the

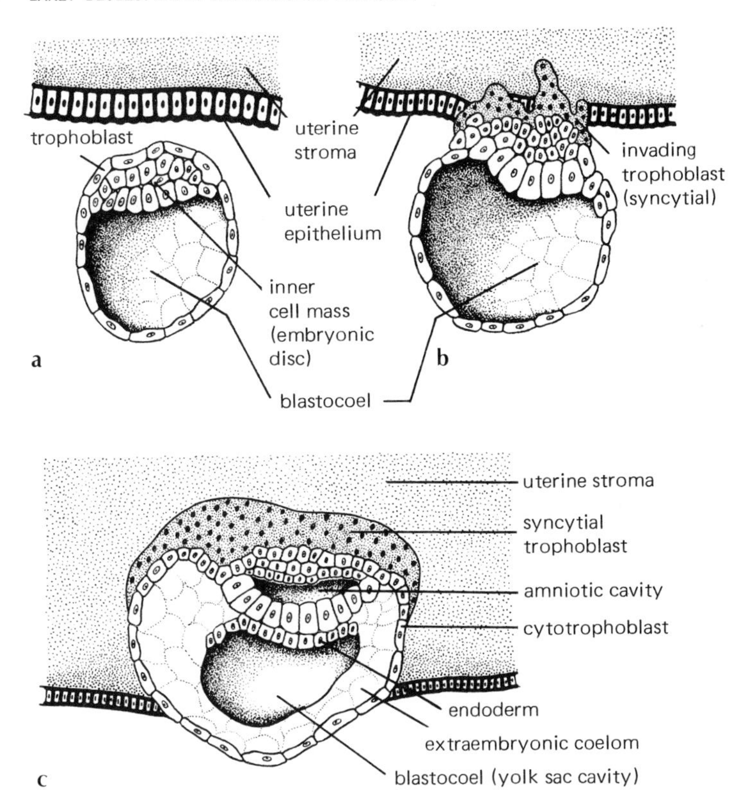

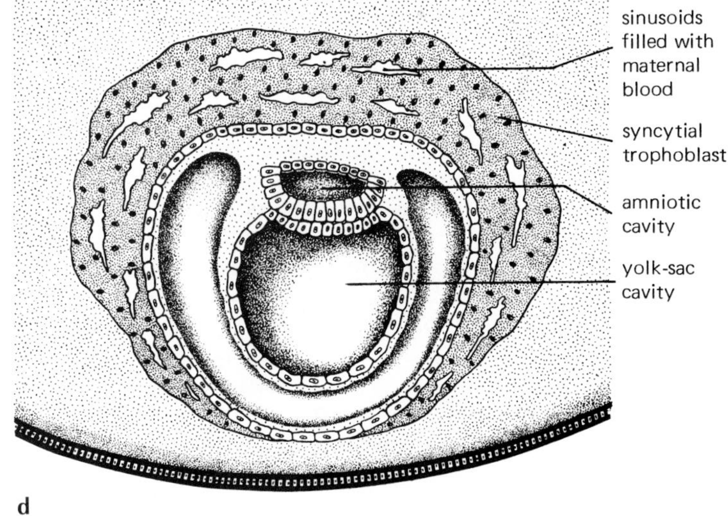

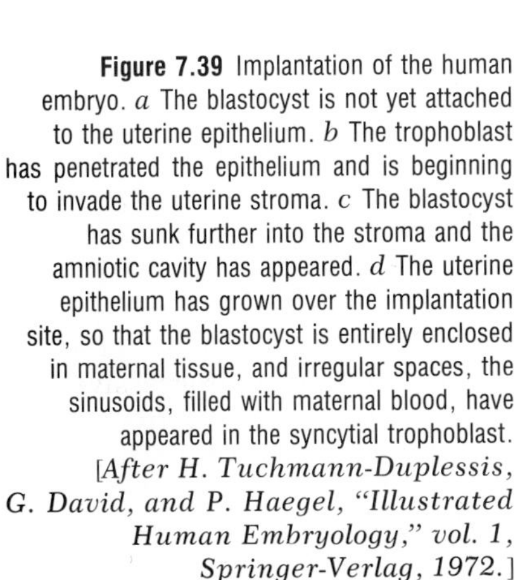

Figure 7.39 Implantation of the human embryo. *a* The blastocyst is not yet attached to the uterine epithelium. *b* The trophoblast has penetrated the epithelium and is beginning to invade the uterine stroma. *c* The blastocyst has sunk further into the stroma and the amniotic cavity has appeared. *d* The uterine epithelium has grown over the implantation site, so that the blastocyst is entirely enclosed in maternal tissue, and irregular spaces, the sinusoids, filled with maternal blood, have appeared in the syncytial trophoblast. [*After H. Tuchmann-Duplessis, G. David, and P. Haegel, "Illustrated Human Embryology," vol. 1, Springer-Verlag, 1972.*]

surface of the mouse blastocyst binds less concanavalin A, is less agglutinable by concanavalin A, and appears to have a reduced negative charge as revealed by the decreased binding of colloidal iron particles. There is also an apparent disappearance of histocompatability antigens, a change that may cause the embryo to present an antigenically neutral outer surface to the maternal circulation.

Much of what we know about implantation has come from studies conducted in vitro. Blastocysts are capable of hatching in culture and attaching to collagen coated plastic surfaces or to strips of uterus. Once attached, the trophoblast begins an *outgrowth* activity that resembles its invasive behavior in vivo. Although the outgrowth of the trophoblast into the substrate appears to be primarily a mechanical process, associated phagocytotic or hydrolytic functions may be operating as well. When implantation occurs in extrauterine sites such as the kidney, spleen, or testis, the period of invasion, which becomes extremely destructive, is much extended. In some locations, such as abdominal mesentery (particularly in rabbits), abnormally sited implantations of an entire litter may develop to term, although delivery of the young may be impossible and fetal death follows. The uterus is prepared for implantation by the action of the ovarian hormones progesterone and estrogen (Section 4.3.B). When the uterine endometrium is in an appropriately sensitized condition, a decidual reaction can be induced not only by the blastocyst as a stimulus, but by an artificial stimulus such as the scratching of a needle.

C. Culture of eggs and embryos

The difficulties that arise in the study of an embryo that develops within its mother are obvious. In order to circumvent these problems and learn more about the conditions favoring embryonic and fetal growth, investigators have sought to devise in vitro cultures capable of supporting mammalian development. It is not surprising that the preimplantation embryo is more amenable to in vitro development then postimplantation embryos. Oocytes can be removed from the ovary, caused to mature in vitro, fertilized, and raised to the blastocyst, which, if allowed to implant in a pseudopregnant animal, will develop to term. These same results can be achieved with human eggs, as attested to by the birth of the first "test-tube" baby in 1978. The most difficult period to traverse in culture is that of implantation. If cultured blastocysts are allowed access to a collagenous mat or a strip of uterine tissue, implantation of a sort can occur, and a small percentage of the embryos can proceed to an early somite stage (can contain a beating heart, neural tube, etc.), but the morphogenetic processes tend to produce a highly disorganized embryo. Whereas development to the blastocyst stage can occur in a completely defined medium, i.e., one in which all the components are known, development beyond the blastocyst requires the addition of a much more complex (serum-containing) medium.

Striking success has been made in recent years in the in vitro development of postimplantation embryos. Mouse embryos, for example, can be removed at the egg cylinder stage (Fig. 7.37) and raised in culture for a period of 4 to 5 days, during which they develop at near normal rates to approximately the 40-somite stage. The striking progress made by these embryos during this period is shown in Fig. 7.40. Embryos explanted at later and later stages develop for progressively shorter periods. The most advanced stage attainable in culture is a fetus with limbs and developing digits (approximately 15 days in the rat). Development beyond the early fetal stages requires support from the placenta. Although the yolk sac grows well in culture and takes on a placental function, no in vitro system has yet been able to substitute for the more complex allantoic placenta.

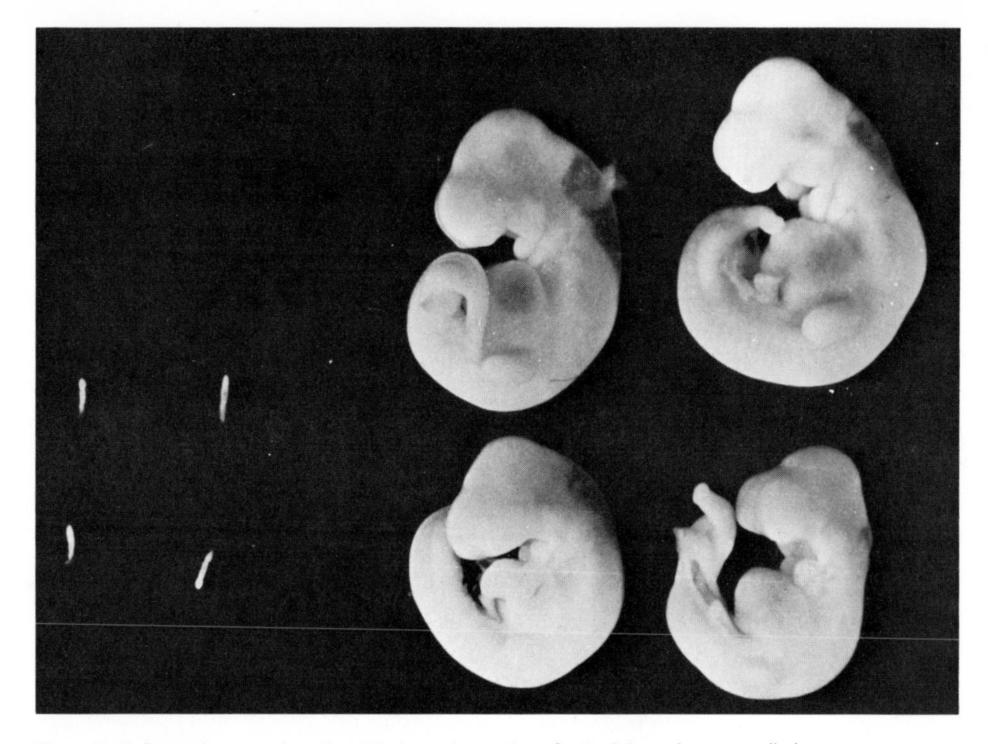

Figure 7.40 Rat embryos explanted at 7½ days of gestation. On the left are four egg cylinders as explanted. On the right are four embryos explanted at 7½ days and cultured for 120 hours. [*Courtesy of S. K. L. Buckley, C. E. Steele, and D. A. T. New.*]

D. Regulation and determination

In Section 7.9.A we described a variety of experimental procedures that have led to the tentative assignment of the prospective fates of the cells of the mammalian blastocyst. We will now return to these types of micromanipulative studies to pursue related questions concerning the developmental potential of the cells of an early mammalian embryo. The ability of a cleaving rodent embryo to undergo regulation is readily demonstrated. In early studies, needles were used to destroy blastomeres of two-, four-, and eight-cell embryos without apparent effect on the ability of the remaining cells to give rise to a normal animal (Fig. 7.41). In one experiment, a single blastomere of an eight-cell rabbit embryo was reportedly able to develop into a complete adult.

With the development of procedures to remove the zona pellucida, a greater variety of manipulative experiments could be carried out. Surfaces of embryonic cells from which the zona has been removed are sticky in nature, and aggregates between cleavage-stage embryos are readily obtainable. Aggregates of two or more cleaving embryos are capable of developing to the blastocyst stage in vitro and then proceeding to term when transferred to the uterus of a hormonally prepared foster mother. The ability of two or more embryos to form a single animal attests to their regulatory capacity (Section 7.6). Not only is the newborn morphologically normal, it also has normal size; size regulation is believed to occur at about the time of implantation.

Since aggregates are readily formed between mice of different genetic stock, intraspecific chimeras of any type of mouse can be constructed (Fig. 7.42). Genetic mosaics produced in this manner have four (or more) parents and are re-

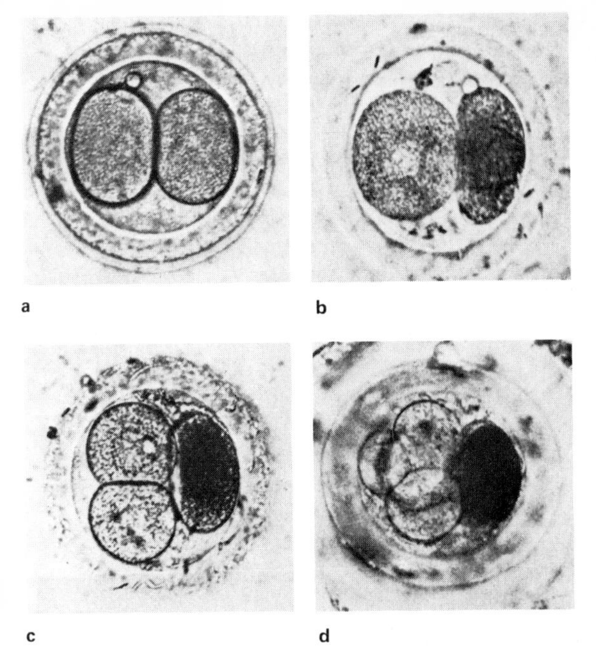

Figure 7.41 Regulation in the rabbit embryo. *a* Normal two-cell stage. *b* Two-cell stage after cell on right has been killed with a needle. *c, d* Second and third cleavages of the remainder of the embryo. *e* Genetically dissimilar parent and offspring that had developed after transfer of an embryo treated as shown to the uterus of a foster mother. [*From F. Seidel,* Naturwissenschaften, **39:***355 (1952).*]

ferred to as *allophenic* mice. Such mice have proved invaluable in the analysis of several different developmental questions. Fusion of cleaving embryos can give rise to normal, though large, blastocysts, even when the fused embryos are at different stages. For example, fused eight- and four-cell stages (12 hours out of phase) and eight-cell and late morula stages both produce normal blastocysts.

The results of a wide variety of experiments suggest that at least some blastomeres of cleaving mouse, rat, or rabbit embryos retain their totipotency until at

Figure 7.42 Experimental procedures for producing allophenic mice from aggregated embryos. Cleavage stages are denuded of zona pellucida and are fused together. The combined embryos are then introduced into the uterus of a foster mother, where they implant and develop. [*After B. Mintz,* Proc. Natl. Acad. Sci. U.S., **58:***345 (1967).*]

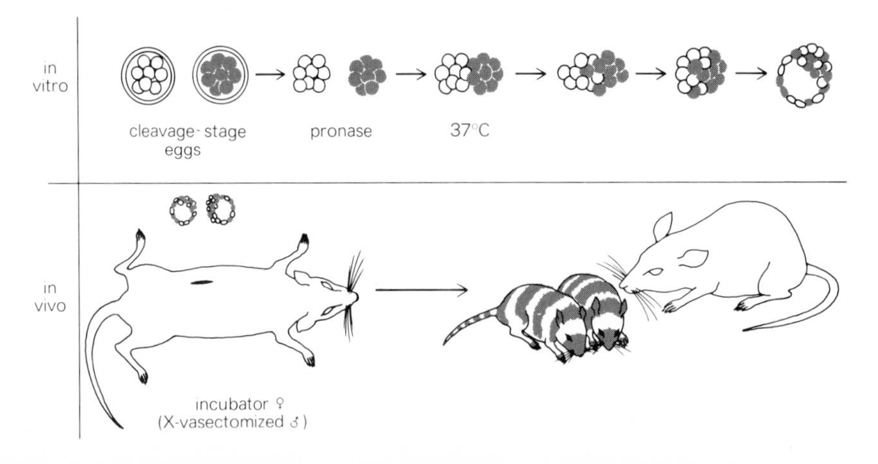

least the eight-cell stage. Although individual blastomeres of a four- or eight-cell embryo do not develop well in isolation, their developmental potential can be revealed by combining them with "carrier" embryos of a different genetic background and following the distribution of their progeny cells in the tissue of the embryo and fetus. One experiment of this type is illustrated in Fig. 7.43. In this case, four-cell embryos were dissociated into single cells and each cell was allowed to divide once before each of its daughter cells was combined with four embryonic cells of a different mouse strain. These genetic mosaics were allowed to develop to the blastocyst stage; then they were implanted, two at a time, into a set of four foster mothers. When these embryos were recovered on the tenth day of pregnancy it was found that cells of the donor genotype were present in the embryo, extraembryonic membranes, and trophoblastic portion of the placenta. No developmental restriction is apparent among the cells of the eight-cell stage.

When do the cells of the mouse embryo become committed to different developmental pathways? Results of experiments involving isolation, transplantation, and rearrangement of cells indicate that the first major determinative event, one that irreversibly separates the pathways of the ICM and trophectoderm, occurs rather late in development. Although the time at which this is estimated to occur may vary somewhat among different experimental procedures, it is generally placed at about 3½ days post coitum. This is the time at which the fully expanded blastocyst has formed and not long before each cell type differentiates into its own distinct derivatives: ICM into primitive endoderm and primitive ectoderm and trophectoderm into mural and polar regions. In the remainder of this section we will consider some of the experiments leading to this conclusion.

One test of determination applied in Section 7.6 is the ability of the cells in question to differentiate according to their prospective fates when kept in isolation. Although isolated portions of mammalian embryos are generally unable to undergo extensive development on their own, they can provide some indication of their potential. The isolated trophectoderm of the 3½-day blastocyst exists as a

Figure 7.43 Protocol for experiment demonstrating the totipotent nature of early blastomeres in mammalian development. Step 1: The donor type embryo is obtained at the four-cell stage and its zona is removed. Steps 2 and 3: The blastomeres are combined after one division with the carrier-type blastomeres. Step 4: The composites are cultured to the blastocyst stage and transferred as a group to the uterus of a pseudo-pregnant recipient. Donor and host differ genetically with respect to glucose phosphate isomerase and coat color. [*From S. J. Kelly, in "Early Development of Mammals," M. Balls and A. E. Wild (eds.), Cambridge, 1975.*]

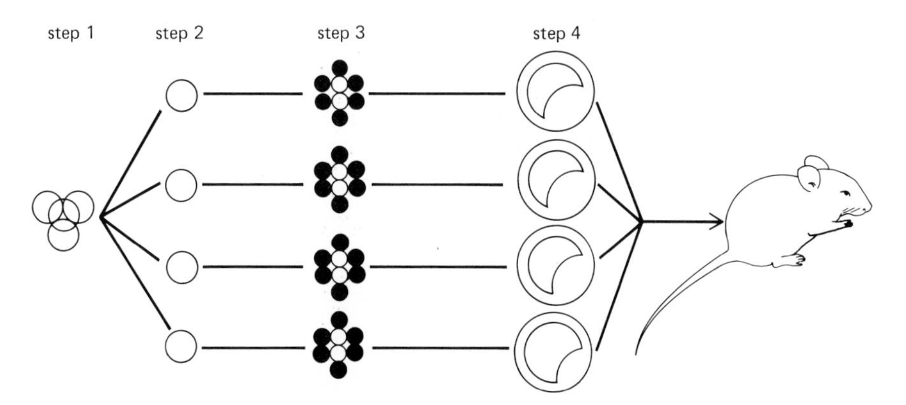

vesicular structure that is capable of implanting and differentiating into invasive giant cells, which is its normal fate. However, these cells do not proceed to form the differentiated tissues of the placenta. Moreover, they show no indication of forming any of the tissues normally derived from the ICM. Whereas isolated trophectoderm differentiates more extensively in utero than in vitro, the developmental potential of ICM isolates is best expressed in culture. When the ICM from an embryo as late as 3⅓ to 3½ days is removed and placed in culture, a remarkable transformation occurs. Within a few days, a blastocystlike structure has been regenerated (Fig. 7.44). Since the trophectoderm cells of the ICM-derived blastocyst proceed to differentiate into sheets of giant cells, one can conclude that the cells of the ICM are capable of reversing the direction in which they develop. This finding is particularly surprising when one considers that the ICM and trophectoderm of the 3½-day blastocyst have considerably different properties (Section 7.9.A); i.e., they have already undergone some degree of differentiation. These results suggest that the initial steps of differentiation of these cells precedes their full determination.

Alternatively, a test for determination can be made by transplanting cells to various locations to see if they yield reproducible patterns of differentiation regardless of their surroundings. This type of experiment is best accomplished by injection of genetically marked cells from one blastocyst into another. When trophectoderm cells from a 3½-day blastocyst are injected into the blastocoel of another embryo, they do not integrate with cells of the host ICM or contribute to its derivatives. Similarly, when cells of the ICM are injected into a genetically distinct blastocyst, descendants of the donor cells are restricted to those tissues normally derived from the ICM. Donor ICM cells never give rise to ectoplacental cone or giant cells, both of which derive from the trophectoderm. Although a given ICM cell cannot be made to form trophectoderm structures, it remains able to form any variety of ICM derivatives. For example, single injected ICM cells are able to develop into both embryonic and extraembryonic structures. The results of these types of studies once again point to the progressive determination of embryonic cells. In the mammal, ICM versus trophectoderm determination in the 3½-day blastocyst is followed by successive steps that further restrict their developmental potential. Determination of parts *within* the embryo or extraembryonic

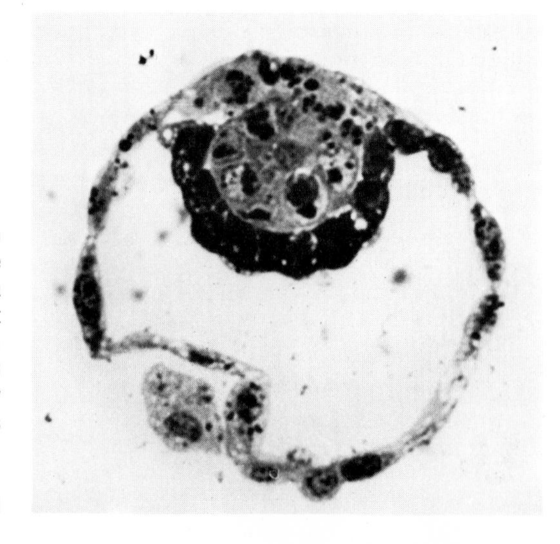

Figure 7.44 Section through a structure closely resembling a normal late blastocyst that had developed from the isolated inner cell mass of an early blastocyst after 3 days in culture. [*Courtesy of B. Hogan and R. Tilly.*]

membranes does not occur until a later blastocyst stage. It is for this reason that additional ICM cells can be added to the embryo without upsetting its development. Conversely, cells of the 3½-day ICM can still be removed from the blastocyst without affecting development; other cells can take over the function of those which have been lost. The results of an experiment in which cells (radioactively labeled in this case) have been injected from one embryo to another is shown in Fig. 7.45.

In addition to providing insights into the state of determination of cells, the formation of chimeras has led to speculation concerning the number of cells present in the blastocyst that are actually involved in the formation of the embryo as opposed to the extraembryonic tissues. The fact that a single injected ICM cell is able to colonize large parts of the developing embryo and fetus indicates that very few cells in the host are truly destined to become embryonic, for otherwise a single additional cell could not contribute to so much of the future mouse. When single injected ICM cells do lead to the formation of chimeras, approximately 30 percent of the mouse is typically derived from the injected cell. Based on this percentage, one would conclude that three cells of the 3½-day ICM are utilized in embryo formation. Moreover, when mice are formed from the fusion of two cleavage-stage embryos of different genetic strains (allophenic mice), approximately 75 percent of the double embryos yield chimeric adults. If the cells of each of the two combined embryos have an equal chance to contribute to the chimeric adult, then 75 percent of the double embryos would be expected to form chimeric adults if three cells were set aside for embryo formation. The remaining 25 percent would, by chance, contain cells derived exclusively from one or the other parental strain. More recently, chimeric mice have been formed by aggregation of three genetically distinct cleavage-stage embryos. A small percentage of these

Figure 7.45 Transplanted mouse blastomeres. *a–b* Recipient blastocysts with labeled cells 40 hours after transplantation. *c* Litter of five fetuses typical of chimeras produced when embryonic cells are transplanted into preimplanted mouse blastocysts. At birth and during postnatal development, chimerism is expressed in coat color, eye pigmentation, and germ cells. [*Courtesy of L. A. Moustafa.*]

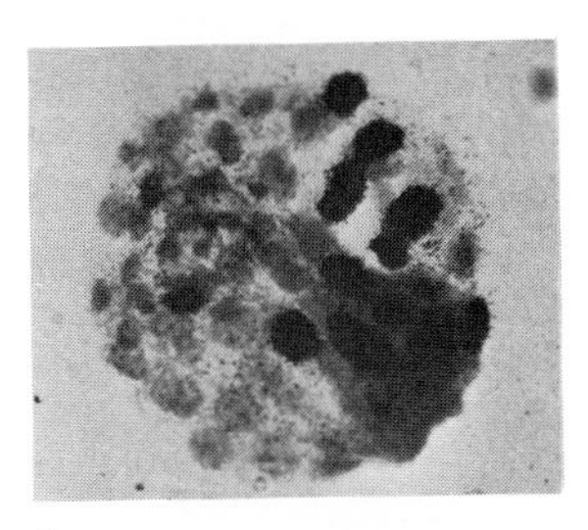

a

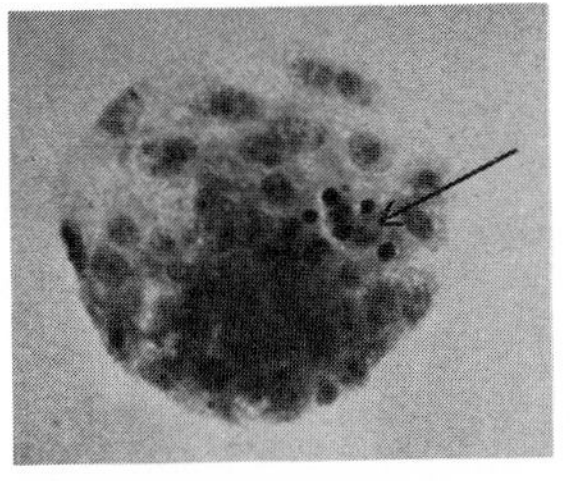

b

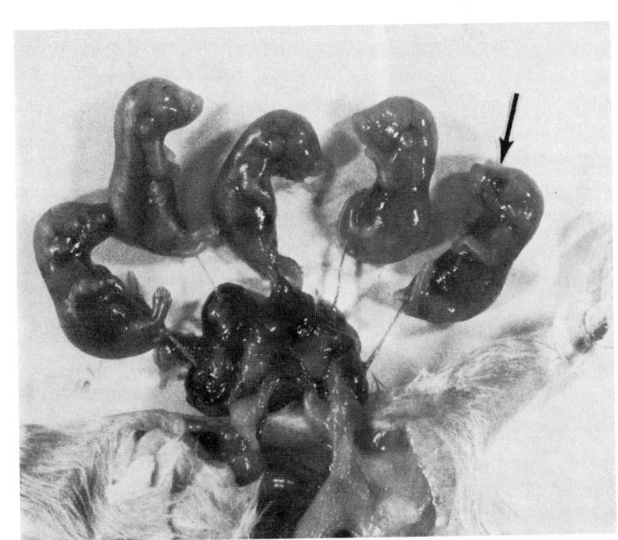

c

triple embryos grew into triple chimeras, i.e., mice with cells of all three genotypes. Clearly, some stage exists at which at least three (and likely only three) cells of a clonally unrelated nature are set aside to produce the embryo and future adult. It will be interesting to discover how great a variety of offspring are produced by triple-chimeric mice, for this will tell us about the formation of primordial germ cells.

i. Inside-outside concept. It is apparent from the previous discussion that the most important morphogenetic event in the early development of mammals is the separation of cells into ICM and trophectoderm portions of the blastocyst. Ultimately this decision may determine whether or not a cell will contribute to the embryo or be discarded with the placenta at birth. On what basis is a cell directed along one or the other of these two pathways? At the present time, the most widely accepted theory states that the decision is made on the basis of the position of a cell within the cleavage-stage embryo. Those cells which happen to lie in an exterior position throughout cleavage become trophectoderm, and those which reside in the interior during this same period become ICM. There are several predictions that derive from this proposal. First, one should be able to demonstrate a consistent relationship between the presence of a cell in an inner or outer position at the morula stage and its likelihood of becoming ICM or trophectoderm. Second, one should be able to modify the fate of a cell by altering its position within the early embryo. Results of studies designed to test these predictions have provided support for the "inside-outside" concept.

With regard to the first of these predictions, it has been shown that cells located in the outer regions of the morula do show a marked tendency to become trophectoderm, while those present within the interior show a strong likelihood to become part of the ICM. This has best been shown by the injection of tiny silicone oil droplets (Fig. 7.46) into different positions within the uncleaved egg. Once positioned, these oil droplets remain in place, allowing one to follow the fate of different cytoplasmic regions. When the oil droplet is placed in a peripheral region of the egg, it becomes localized in the outer cells of the morula and subsequently in the trophectoderm. In contrast, if the droplet is placed in the central region of the egg, it tends to become localized in the internal cells of the morula, from which it passes into the ICM. Results of these studies indicate that the cellularization of the mouse egg is accomplished with a minimum of disturbance to its cytoplasmic organization. Peripheral egg cytoplasm is passed to the outer cells and then to the trophectoderm; central cytoplasm is passed to the interior cells and then to the ICM. It is believed that cells in an internal position in the morula are exposed to different environmental influences than those residing in an external position, and it is this difference in microenvironment that leads to their different prospective fates. An event believed to be of particular importance in this regard is the establishment of tight junctions between the outer cells of the early morula. Once assembled, these occluding junctions form an effective seal between the outer cells, thereby ensuring that the extracellular space within the embryo will be different from that of its surroundings.

With regard to the second prediction, that concerned with the effect of blastomere arrangement on cell fate, the following experiments have proved the most enlightening:

1. If the blastomeres of one embryo are disaggregated and placed around the periphery of another intact embryo, the outer cells are almost always found

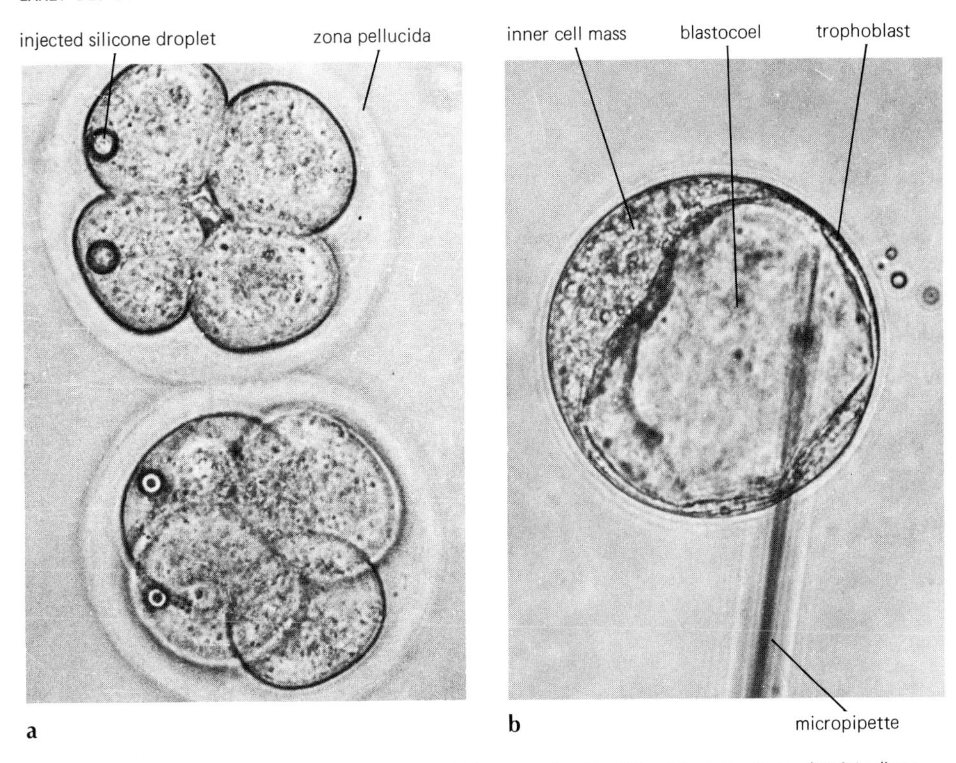

Figure 7.46 *a* Four-cell stages, each enclosed in thick zona pellucida. Microinjected silicone droplets (here seen in two cells in each case) are employed as cell tracers in subsequent development. *b* Blastocyst showing differentiation into inner cell mass and trophoblast, with large blastocoel. Hatching from the zona has already occurred. The end of a micropipet used for injection of droplets or of cells into the developing system. [*Courtesy of I. B. Wilson.*]

in the trophectoderm layer of the blastocyst that develops 2 days later. Under normal circumstances, these blastomeres, which represent an entire embryo, would have formed both ICM and trophectoderm. Their outer position has shifted their prospective fate in the direction of trophectoderm differentiation. Figure 7.47 shows a case in which an intact embryo is surrounded by eight disaggregated blastomeres. In this example, cells of the two embryos are distinguished by genetic differences (electrophoretic mobility of glucose phosphate isomerase). In other cases, cells of one embryo are distinguished by the use of tritiated thymidine.

2. If a whole morula is labeled with tritiated thymidine and is surrounded on all its surfaces by a total of 15 unlabeled morulas, then at least in some cases, labeled cells are found only in the ICM of the large blastocysts that develop. In these cases, none of the cells of the central embryo form trophectoderm, whereas normally approximately 75 percent would have differentiated in this direction. Surprisingly, these giant blastocysts can be transferred to pseudopregnant recipients, where they develop into apparently normal 13-day embryos (Fig. 7.47). None of these composite embryos come to term; they are resorbed by the sixteenth day of pregnancy.
3. If the central cells are isolated from these composites after 2 days, it is found that they have irreversibly lost their ability to form trophectoderm

———, 1971. The Design of the Mouse Blastocyst, Symposium, *Soc. Exp. Biol.*, **25**:371–378.

HILLMAN, N., M. I. SHERMAN, and C. F. GRAHAM, 1972. The Effect of Spatial Arrangement on Cell Determination during Mouse Development, *J. Embryol. Exp. Morphol.*, **28**:263–278.

HOGAN, B., and R. TILLY, 1978. In Vitro Development of Inner Cell Masses Isolated Immunosurgically from Mouse Blastocysts, *J. Embryol. Exp. Morphol.*, **45**:107–121.

HSU, Y.-C., 1979. In Vitro Development of Individually Cultured Whole Mouse Embryos from Blastocyst to the Early Somite Stage, *Develop. Biol.*, **68**:453–461.

JOHNSON, M. H., 1977. "Development in Mammals," 3 vols, Elsevier.

KELLY, S. J., 1977. Studies of the Developmental Potential of 4- and 8-Cell Stage Mouse Blastomeres, *J. Exp. Zool.*, **200**:365–376.

MARKERT, C. L., and R. M. PETTERS, 1979. Manufactured Hexaparental Mice Are Derived from 3 Embryonic Cells, *Science*, **202**:56–58.

MCLAREN, A., 1979. Early Events in Mammalian Embryogenesis, in J. D. Ebert and T. S. Okada (eds.), "Mechanisms of Cell Change," Wiley.

MINTZ, B., 1967. Mammalian Embryo Culture, in F. H. Wilt and N. K. Wessells (eds.), "Methods in Developmental Biology," Crowell.

MOUSTAFA, L. A., and R. L. BRINSTER, 1972. The Fate of Transplanted Cells in Mouse Blastocysts in vitro, *J. Exp. Zool.*, **181**:181–192, 193–202.

NEW, D. A. T., 1978. Whole Embryo Culture and the Study of Mammalian Embryos during Organogenesis, *Biol. Revs.*, **53**:81–122.

NEWMAN, H. H., 1917. "The Biology of Twins," University of Chicago Press.

PATTERSON, J. T., 1913. Polyembryonic Development in *Tatusia novemcineta*, *J. Morphol.*, **24**:599–684.

SCHLAFKE, S., and A. C. ENDERS, 1975. Cellular Basis of Interaction between Trophoblast and Uterus at Implantation, *Biol. Rep.*, **12**:41–65.

SHERMAN, M. I., 1979. Developmental Biochemistry of Preimplantation Mammalian Embryos, *Ann. Rev. Biochem.*, **48**:443–470.

———(ed.), 1977. "Concepts in Embryogenesis," MIT Press.

———, and L. R. WUDL, 1976. The Implanting Mouse Blastocyst, in G. Poste and G. L. Nicolson (eds.), "The Cell Surface in Animal Embryogenesis," Elsevier.

STERN, S. M., and I. B. WILSON, 1972. Experimental Studies on the Organization of the Preimplantation Mouse Embryo, *J. Embryol. Exp. Morphol.*, **28**:247–261.

STORRS, E. E., and R. J. WILLIAMS, 1968. A Study of Monozygous Quadruplet Armadillos in Relation to Mammalian Genetics, *Proc. Nat. Acad. Sci. U.S.*, **60**:910–914.

TARKOWSKI, A. K., 1975. "Induced Mammalian Parthenogenesis and Early Development," 33rd Symp. Soc. Dev. Biol., Academic.

———, and J. Wroblewska, 1967. Development of Blastomeres of Mouse Eggs Isolated at the 4- and 8-celled Stage, *J. Embryol. Exp. Morphol.*, **18**:155–180.

VAN BLERKOM, J., S. C. BARTON, and M. H. JOHNSON, 1976. Molecular Differentiation in the Preimplantation Mouse Embryo, *Nature*, **259**:319–326.

WILSON, I. B., E. BOLTON, and R. H. CUTLER, 1972. Preimplantation Differentiation in the Mouse Egg as Revealed by Microinjection of Vital Markers. *J. Embryol. Exp. Morphol.*, **27**:467–479.

Chapter 8 The molecular biology of early development

1. Equivalence of nuclei

The assumption is made throughout this book that all cells contain all the genes and that differentiation results from selective gene expression in each type of cell. In other words, each cell contains all the genetic information but utilizes only that fraction which is relevant to its function. This view of differentiation has not always been held. In an earlier hypothesis, August Weismann proposed that the genetic determinants, later shown to be the chromosomes, were parceled out in some manner that determined the ultimate path of differentiation that a given cell might take. In other words, cells would achieve a particular specialized state by retaining those parts of the chromosomes required for that condition and eliminating those not needed. In this theory, special consideration must be given to the germ cells, for they are responsible for producing a complete individual in the next generation. Germ cells must retain all genes. The loss of genetic information by somatic cells is a reasonable explanation for the phenomenon of differentiation, but one that has virtually no experimental support. With the exception of a few insects and nematodes (discussed in the following section), the nuclei of organisms are believed to contain equivalent and complete genetic complements.[1]

A. Chromosome loss

Examination of the very early stages of development in the nematode *Ascaris* revealed at the end of the nineteenth century that pieces of chromosomes were being lost as the egg underwent cleavage (Fig. 8.1). After the first cleavage (which is unusual in being horizontal in *Ascaris*), the chromosomes in one of the two cells undergo fragmentation and certain of the fragments are not passed on to the daughter cells at second cleavage. In the other cell, the one that leads to the formation of germ cells, no fragmentation occurs and each daughter cell receives a total set of chromosomes. Chromosome loss, termed *chromosome dimi-*

[1] It should be noted that major rearrangements of DNA sequences do occur during the development of the immune system. There is also evidence in these cells for the expansion of the number of genes that code for antibodies. These topics are discussed in detail in Chapter 17.

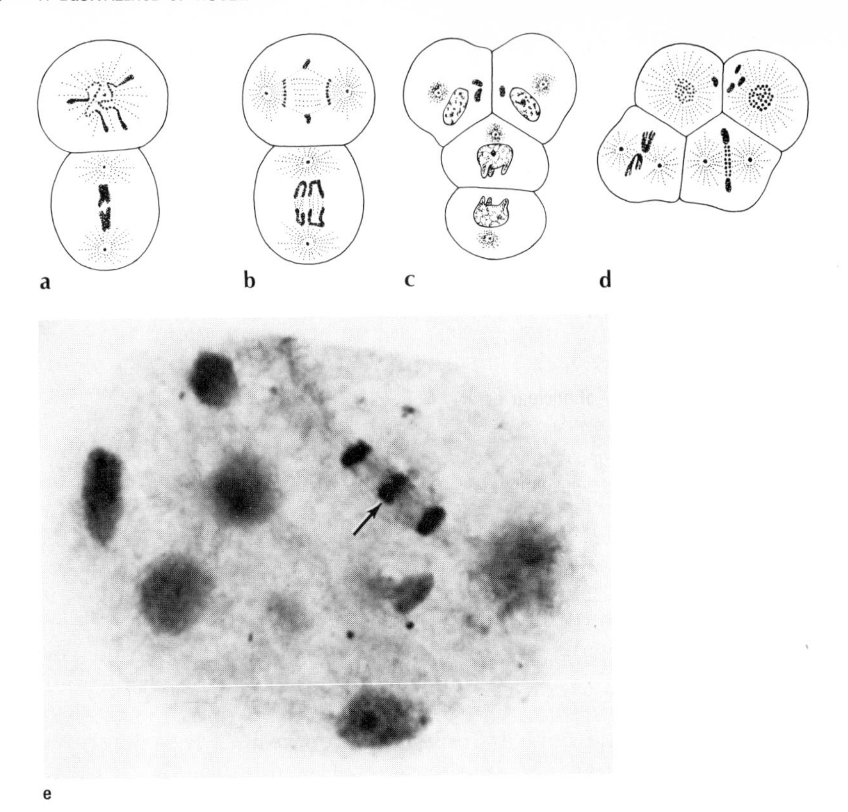

Figure 8.1 *Ascaris*. Development of an egg from the two-cell stage to the end of the four-cell stage. *a* Two-cell stage, upper cell undergoing chromosome diminution, lower cell showing normal chromosome behavior. *b* Two-cell stage, slightly later than in *a*, upper cell in side view. The ends of the diminished chromosomes remain in the center of the cell between the separating chromosome fragments. *c* Four-cell stage, resting nuclei. In the upper two cells, degenerating chromosome ends lie outside the nuclei. In the lower two cells, the undiminished chromosomes are recognizable by the irregular shape of the nuclei. *d* Four-cell stage, nuclei dividing. Upper cells show the earlier diminished chromosomes; lower right cell undergoes diminution; lower cells show normal chromosome behavior. The lower left cell corresponds to the lowermost cell of *c*, which has changed its position. [*After T. Boveri, 1899.*] *e* Photograph of a section through a cleaving *Ascaris* egg showing chromosomal fragments (*arrow*) left behind at anaphase after spindle-attached chromosomes have moved toward the poles. [*Courtesy of P. Goldstein.*]

nution, continues in the succeeding several divisions, only those blastomeres of the germ-cell lineage being spared.

Attempts to learn of the nature of those DNA sequences lost during chromosome diminution have led to contradictory reports. In one study it was reported that only highly repeated sequences (*satellite DNA*) were lost, while in other cases it has been reported that a large fraction of the eliminated DNA consisted of unique sequences. If the latter findings are confirmed, they would indicate that an enormous amount of genetic information is lost to somatic cells. Chromosome elimination also occurs in the somatic cell lines of certain insects. In two species of midge, induced chromosomal elimination in germ-line cells results in sterility in otherwise normal adults. It would appear that at least some genetic information qualitatively different from that of the retained portion is contained in the eliminated chromatin, and this information is indispensable for normal gametogenesis.

B. Transplantation of nuclei

The most direct approach to the question of nuclear equivalence has utilized the nuclear-transplantation experiment. These studies once again illustrate the potential for experimental manipulation in the study of development—if the biologist can clearly formulate the question, plan the necessary experiment, and overcome the technical difficulties inherent in any new probe at the subcellular level. The question is: Do irreversible (stable) changes in the genetic content of the nuclei occur during development? The experimental plan is to remove nuclei from cells of progressively older embryos, insert these nuclei into uncleaved eggs whose own nuclei have been removed, and determine if these later-stage nuclei are capable of supporting the development of the recipient eggs.

The first successful technique for embryonic nuclear transplantation was developed by Robert Briggs and Thomas King using the frog *Rana pipiens*. In these studies, the chromosomes of the recipient egg are removed from the animal region just after the egg is experimentally activated with a needle (Fig. 8.2). A substitute donor nucleus is obtained by using a micropipet having a diameter less than that of the donor cell so that the cell is ruptured during the process of nuclear uptake. The isolated nucleus, together with a protective coat of surrounding cytoplasm, is injected into the cytoplasm of the already activated egg. Development proceeds, to whatever extent, under the direction of the newly acquired genetic material. Since the early work on *Rana*, a second amphibian system, that of *Xenopus*, has been utilized to probe the same question. There appear to be certain differences between these two series of investigations and some difference in interpretation. We will begin with a discussion of the results on *Rana* and then consider the work on *Xenopus*.

The goal of the nuclear-transplantation studies is to determine if the nuclei remain equivalent to one another despite the alterations in the differentiating cytoplasm that surrounds them. To this end, nuclei of later and later stages are tested for their ability to support the development of the recipient enucleated egg. It must be kept in mind that even if nuclei do become irreversibly altered so that they can no longer support total development, this does not mean that they have undergone gene loss. Other types of genetic alterations that cannot be reversed and that fall short of genetic deletion are not only possible, but also more likely.

When a nucleus from an early stage, such as a blastula or early gastrula, is transplanted, a high percentage of the transplants develop into normal tadpole larvae. The interpretation is clear: No irreversible nuclear changes have occurred during this early period; the nuclei remain equivalent and totipotent. This

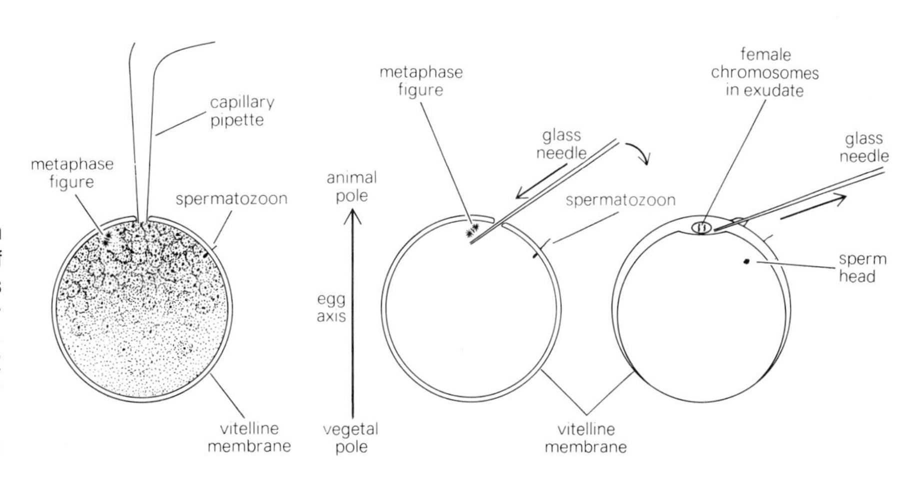

Figure 8.2 Enucleation of frog egg by means of capillary pipet or glass needle. [*After R. Rugh, "Experimental Embryology," p. 180, Burgess, 1962.*]

is a very important conclusion. By the early gastrula stage, thousands of cells have been formed without the occurrence of stable nuclear alterations. When the nuclei are taken from the late gastrula or a later stage, the morphological development becomes restricted. As nuclei are taken from increasingly later-stage *Rana* embryos, the resulting transplants are progressively more abnormal, and the percentage that develop into normal larvae is essentially zero by the neurula and tail-bud stages. Even the percentage of transplants that reach the blastula stage becomes severely curtailed (65 percent from late gastrula, 33 percent from neurula, and 17 percent from tail-bud stages when endodermal nuclei are donors). Moreover, it has been shown using the technique of *serial transplantation* that nuclear restrictions are stable. In a serial transplant (Fig. 8.3), the donor nucleus is injected into the egg, which is then allowed to develop to the blastula stage; nuclei are then removed and injected into new enucleated eggs, and so forth. In this way nuclei can undergo many more divisions than they normally would if they were simply allowed to become part of a deficient larva that is arrested in its cell divisions. It has been shown in a number of studies that nuclei passed through a series of transplants retain their specific deficiencies, deficiencies that are revealed by allowing the recipient egg to express its developmental potential. The opportunity to undergo repeated divisions in egg cytoplasm does not seem to be of any value in reversing the nuclear restrictions in these experiments on *Rana*.

There are several possible explanations to account for the limited ability of older-stage nuclei to support development: (1) that as nuclei are taken from older stages, there is an increased risk of damage to the nuclei by the transplantation procedure; (2) that the nuclei are becoming differentiated in a stable, irreversible manner as development proceeds and they are no longer equivalent or totipotent; and (3) that the nuclei are not irreversibly altered, but as they get older they are less able to manage in egg cytoplasm, which becomes increasingly more distant from their own cytoplasmic state.

The first of these explanations can be ruled out: endodermal nuclei of the tail bud are no more sensitive than ectodermal nuclei of the early gastrula, yet the

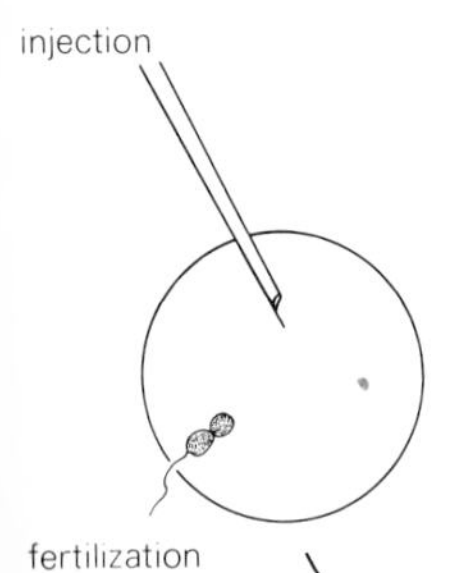

Figure 8.3 Serial transplantation of nuclei, from an embryo with nucleus derived from an egg arrested at the blastula stage. In each generation, the development of the egg is arrested at the blastula stage, indicating that a permanent change has occurred, but it is not established whether the changes have occurred before or after transplantation. [*After C. L. Markert, "The Nature of Biological Diversity," p. 110, McGraw-Hill, 1963.*]

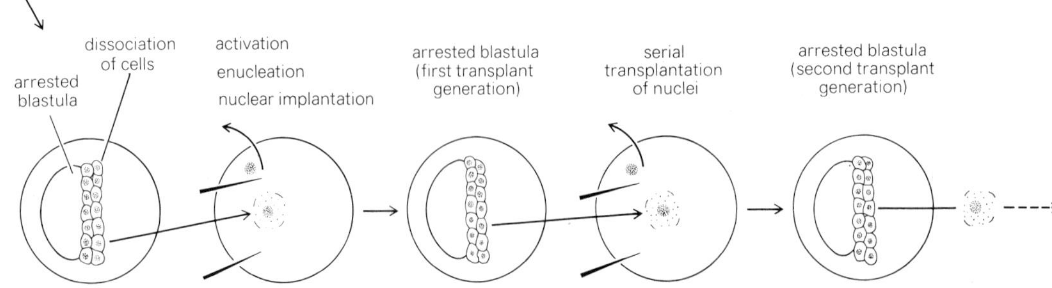

latter remain totipotent. The difference between the second and third possibilities relates to the time that the nuclear changes occur. In the second explanation the changes occur before transplantation as a result of developmental events, while in the third explanation they result *after* transplantation as a result of nuclear-cytoplasmic incompatibility. There is no doubt that the cytoplasm of the egg is very damaging, since gross chromosomal aberrations are often visible. It is believed that essentially all nuclei from late embryonic stages sustain chromosomal damage, whether it is obvious under the microscope or not. This is not surprising; a nucleus from a late stage undergoes division, and therefore replication, in a much more leisurely fashion than it would in a newly fertilized egg with a cleavage regime demanding the most rapid replication. In the first cleavage cycles, the later-stage donor nucleus seemingly cannot replicate fast enough, and it receives chromosomal damage. Even though damage is sustained after transplantation, the question still remains as to whether or not developmental changes also had occurred before the operation. The changes could be irreversible, such as gene loss or irreparable genetic repression, or they could be reversible ones that would require the proper environment for return to the original state. This environment is not provided by uncleaved egg cytoplasm. It is very difficult to distinguish with certainty between these possibilities. There is, however, one type of transplantation experiment that serves as a type of control for the others. We know that germ cells must neither lose genetic information nor irreversibly repress it, since it must be passed on in a functional state to the next generation. Do transplantation experiments using nuclei of germ cells fare better than those of somatic cells? The answer is that they do not. Nuclei taken from adult *Rana* spermatogonia develop relatively well following transplantation into enucleated eggs, but they demonstrate the same types of chromosomal damage and limited developmental potential as do somatic-cell nuclei. We can assume that this genetically totipotent nucleus must be receiving some type of damage in the process of being thrust into enucleated cytoplasm.

Although nuclei from late developmental stages of *Rana* show severe restrictions after transplantation, at least one type of adult somatic nucleus is able to support development to an advanced embryonic stage. These nuclei are derived from a frog tumor, a renal adenocarcinoma (Fig. 8.4). The larvae that develop from these transplants (7 of 143 became swimming tadpoles) contain a wide variety of tissues (Fig. 8.4*b*), the genetic information for which must have been contained in the nucleus of the adult tumor cell. Unfortunately, the fact that the nucleus is derived from an abnormal cell of unknown origin, and that the larvae die before feeding, makes the finding subject to interpretation.

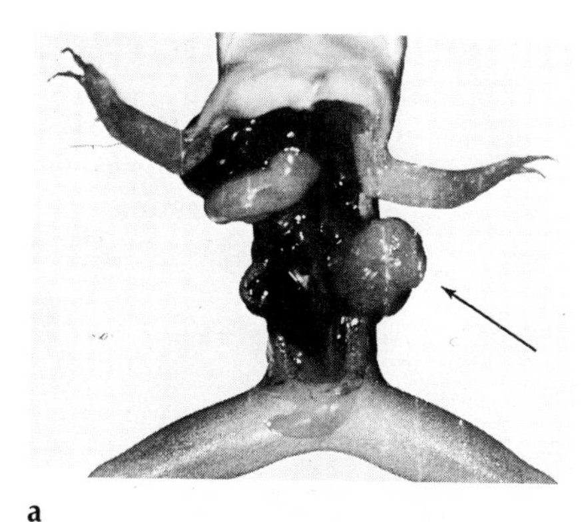

a

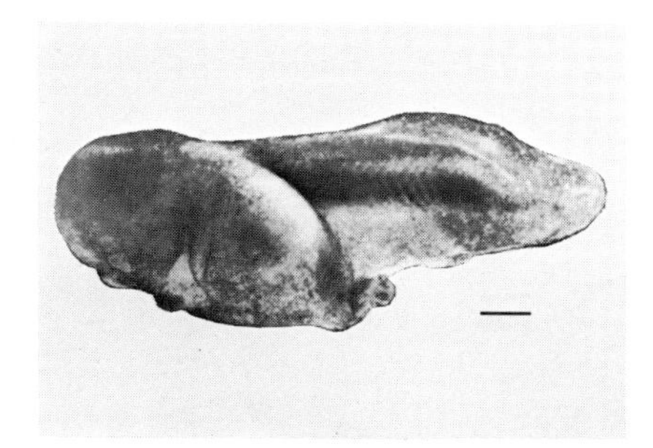

b

Figure 8.4 *a* Left renal tumor (*arrow*) of recently metamorphosed triploid frog. *b* Tadpole developed from egg with transplanted triploid tumor nucleus, with well-formed head, body, and tail (scale = 1 mm). [*Courtesy of R. G. McKinnell.*]

The other main series of nuclear-transplantation studies, using *Xenopus* rather than *Rana*, has been carried out by John Gurdon and coworkers. As in the case of *Rana*, when *Xenopus* cells differentiate, their nuclei are less able to support normal development after transplantation. In the studies of embryonic *Xenopus* donor nuclei, Gurdon has concentrated on the larval intestinal epithelial cells of the tadpole. Because of their large size, it is relatively easy to obtain undamaged nuclei from these cells. In addition, they represent a differentiated cell type, shown by a striated border and the capability of food absorption. In *Xenopus*, the egg chromosomes cannot be removed mechanically and instead are destroyed by irradiation. The procedure used in these experiments is shown in Fig. 8.5. In these experiments, the majority of endodermal nuclei promote very little development, although approximately 20 percent produce tadpoles with functional nerve and muscle cells. In a few cases, transplanted nuclei were found to promote the development of normal, fertile adults. To be certain that the transplanted nucleus is indeed providing the chromosomes, the donor nucleus carries an identifiable genetic marker. Donor nuclei are heterozygous for the nucleolar deletion and thus have only one nucleolus. The recipient egg that is irradiated contains the normal chromosome set.

The point of greatest controversy in these experiments has centered on the very small percentage of transplantations (2 percent) that lead to completely normal development. It has been argued that the nuclei that support development might reside in less specialized cells, ones that were not fully differentiated. Gurdon has argued that as long as any of these nuclei can be shown to be totipotent, one can conclude there are no irreversible changes along the path to the differentiated state. The transplants that do not produce normal development can simply be attributed to damage to the nucleus during or after transfer. Even in these cases, the deficient larvae show a wide series of differentiations other than those of the cells from which the nucleus was taken (an intestinal type). They must, therefore, contain genetic material for a great variety of differentiated pathways.

In recent years, the most important questions raised by nuclear-transplantation experiments appear to have been answered by the transfer of nuclei from differentiated adult cells into enucleated eggs. It has been found, for example, that nuclei taken from cultured adult skin cells will promote the development of an enucleated egg to an advanced embryonic stage. In this case, the cells were ones that had migrated out of an explant of adult frog skin. Their state of differentiation was demonstrated by their possession of keratin, a protein that appears in skin cells during the later steps of differentiation. Embryos that develop with these nuclei may reach a stage containing functional muscle and nerve, a heartbeat and blood circulation, eyes with lenses, and other types of differentiated cells. Similar results have been obtained using nuclei from fully differentiated adult frog lymphocytes. In either case, many types of cells are formed other than the type from which the nucleus was taken. At the same time, none of these embryos ever develop to the feeding larval stage, and therefore the question of full reversibility still remains open. Anyone interested in pursuing the subject of nuclear transplantation should refer to a recent volume devoted entirely to the subject (*Int. Rev. Cyt.* Suppl. 9, 1979).

Although amphibians have been the most exploited animals for nuclear-transplantation studies, they have not been the only source. Similar types of experiments with insect embryos also have indicated that nuclei, at least to the gastrula stage, remain capable of supporting the development of enucleated eggs. Experiments with mammalian eggs, whose volume is 1,000 times less than a frog's egg, have only recently met with much success. It has now been reported

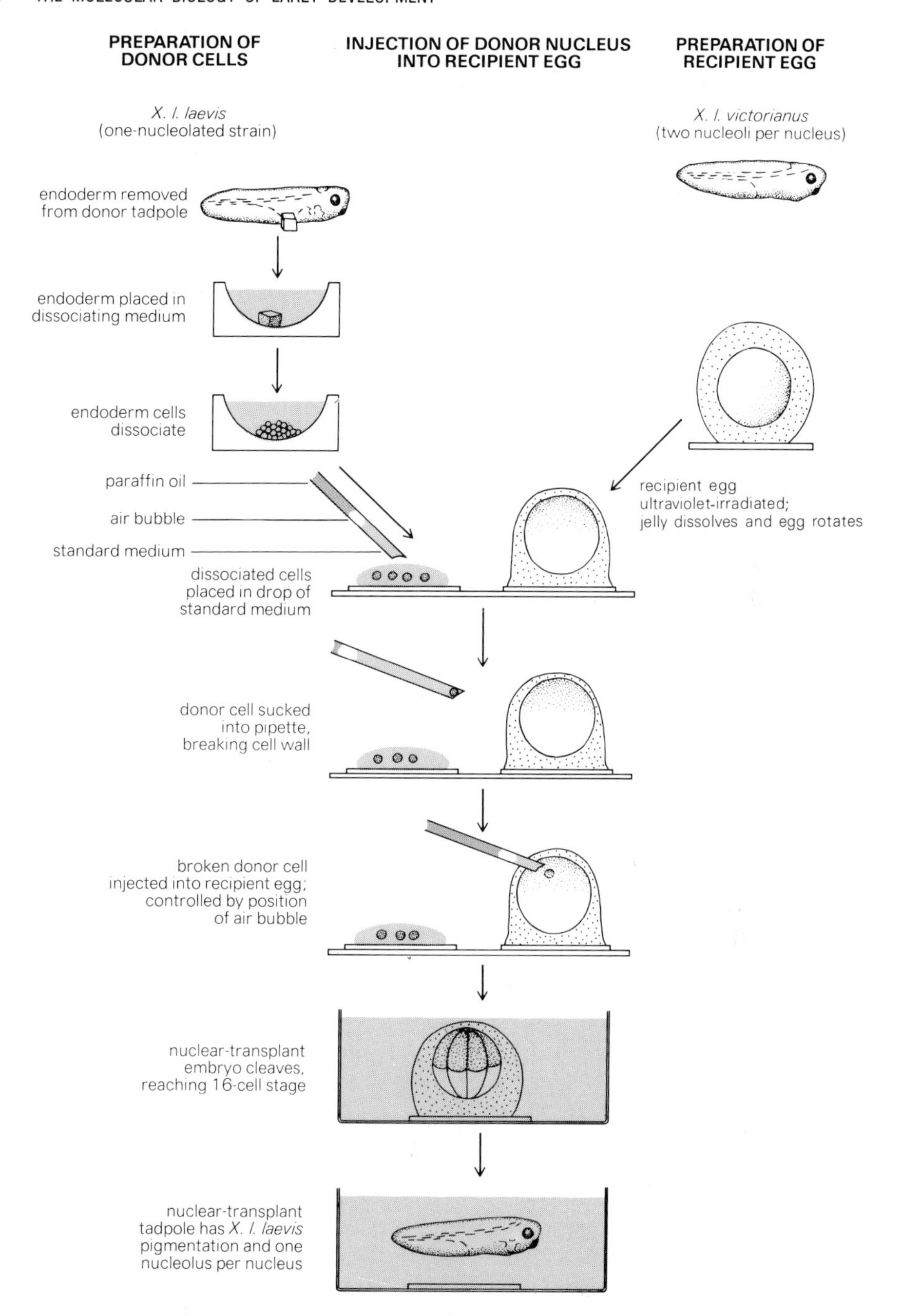
PREPARATION OF DONOR CELLS
INJECTION OF DONOR NUCLEUS INTO RECIPIENT EGG
PREPARATION OF RECIPIENT EGG
X. l. laevis
(one-nucleolated strain)
X. l. victorianus
(two nucleoli per nucleus)
endoderm removed from donor tadpole
endoderm placed in dissociating medium
endoderm cells dissociate
paraffin oil
air bubble
standard medium
recipient egg ultraviolet-irradiated; jelly dissolves and egg rotates
dissociated cells placed in drop of standard medium
donor cell sucked into pipette, breaking cell wall
broken donor cell injected into recipient egg; controlled by position of air bubble
nuclear-transplant embryo cleaves, reaching 16-cell stage
nuclear-transplant tadpole has X. l. laevis pigmentation and one nucleolus per nucleus

Figure 8.5 The principal stages involved in transplanting nuclei in *Xenopus laevis*. Donor nuclei are taken from a strain of the subspecies *Xenopus laevis laevis*, which has only one nucleolus in each nucleus. The young tadpoles of this subspecies have many pigment cells on their bodies. Recipient eggs are taken from the subspecies *Xenopus laevis victorianus*, in which nuclei of most diploid cells have two nucleoli and in which the young tadpoles have no body pigment. The nuclear-transplant tadpoles have the characteristics of the nuclear parent, not of the cytoplasmic one, showing that their nuclei are derived from the transplanted nucleus. [*After J. B. Gurdon*, Endeavour, **25:***96 (1966)*.]

that nuclei removed from inner mass cells of mice embryos are able to support development to fetal stages. The results from nuclear-transplantation experiments indicate that nuclei from differentiated cells retain the genetic information for the formation of many different cell types. This same type of conclusion has been drawn from a very different type of experiment, one involving the hybridization of specific mRNA sequences to DNA extracted from different adult tissues. If differentiated cells of one type retain the DNA sequences needed for the formation of other cell types, their presence should be revealed by hybridization experiments. For example, DNA extracted from adult liver cells should contain genes for globin synthesis, even though this protein is produced only by cells of the erythrocyte line. The presence of globin DNA sequences within the duck liver cell genome is shown in Fig. 8.23.

2. Cytoplasmic control of nuclear activity

The nuclear-transplantation studies on *Xenopus* have provided information far beyond the question of nuclear equivalence. They have provided a valuable window through which one can observe the cytoplasmic control over nuclear activities. In the previous section we discussed one type of recipient cell into which *Xenopus* nuclei have been transplanted, the activated irradiated egg. The synthetic state characteristic of the early postfertilization stage is intensive DNA synthesis and negligible RNA synthesis (including the complete absence of rRNA and tRNA synthesis, which is not activated until much later). Nuclei transplanted into the activated egg undergo a predictable pattern of activities. These nuclei assume the synthetic activity characteristic of the nucleus they replace, regardless of their properties before transplantation. They become reprogrammed. For example, a nucleus at the neurula stage is synthesizing all kinds of RNA but is not likely to be making DNA. Within an hour after transplantation, the synthesis of RNA has been suppressed and DNA synthesis has been activated to the typical early cleavage level. If the synthetic activity of the descendants of the transplanted nucleus is followed, we find that each type of RNA synthesis is sequentially reactivated at the same stage it occurs during normal development. Transfer RNA synthesis becomes reactivated at the blastula stage, and rRNA begins to be synthesized at gastrulation. Such experiments strongly suggest that the synthesis of the various RNA classes is controlled independently by different cytoplasmic components.

The changing synthetic pattern after transplantation is accompanied by striking morphological changes in the donor nucleus that similarly reflect the expected state of an early cleavage nucleus. For example, the nucleus swells to the volume expected for that stage, an increase of over thirtyfold. The nucleolus of the transplanted nucleus disappears until its re-formation within the nuclei of the gastrula. The conditioning effect on nuclei by egg cytoplasm can be extended to injected nuclei from nonembryonic sources. If brain nuclei from adult *Xenopus* are injected into activated egg cytoplasm, a marked swelling of these nuclei is found, together with the activation of DNA synthesis (see Fig. 8.7*c* and *d*) in nuclei that had stopped making it before their differentiation into neurons (years earlier). Swelling appears to be a prerequisite for DNA synthesis; those nuclei which remain unswollen for whatever reason do not synthesize DNA. The activation of DNA synthesis is not restricted to *Xenopus* nuclei, or even to nuclei from amphibian cells. If mouse liver nuclei are injected into amphibian eggs, they become stimulated to synthesize DNA just as if they had been taken from a *Xenopus* neurula cell. The replication signal would appear to have little species specificity.

What is there in egg cytoplasm that produces these rapid changes in injected nuclei? The answer is not known, but an interesting series of experiments indicates that proteins from the egg cytoplasm rapidly enter the nucleus as it swells. The evidence comes from injection of nuclei into eggs that had been prelabeled with [^{3}H]amino acids. The labeled amino acids are injected into the female body cavity at the same time hormones are administered that cause maturation, ovulation, and release of eggs (shed 12 to 16 hours after hormone treatment). When brain nuclei are injected into these prelabeled eggs, labeled protein is found in the nucleus in 10 minutes, which is prior to swelling or DNA synthesis. As the nuclei swell, the labeled protein accumulates within the nucleus to a concentration greater than that of the cytoplasm. It has been proposed that entering proteins become concentrated within the nucleus as a result of their specific attachment to the chromosomes. The most swollen nuclei accumulate the greatest concentration of labeled protein, and in turn these nuclei synthesize the most DNA.

After it was demonstrated that nuclei would begin replication when transferred to egg cytoplasm, it was shown that purified DNA also would be replicated upon injection into the cytoplasm of an egg. Egg cytoplasm appears to contain all the components necessary for the entire replication process, a synthetic activity known to require a wide variety of enzymes and protein factors. It has been demonstrated that at least some of these components are stored in large quantities in the egg for use by the cleavage-stage nuclei during the rapid advance of the embryo to the blastula stage.

Whereas the cytoplasm of the activated egg leads to a stimulation of DNA synthesis and a cessation of RNA synthesis in a transplanted nucleus, the cytoplasm of a growing oocyte induces a reversed synthetic pattern. We will briefly consider the effect on a blastula nucleus of its transplantation into the cytoplasm of a growing oocyte. Prior to transplantation, the oocyte is making large quantities of RNA but has already replicated its DNA. In contrast, the blastula nucleus is still making DNA at a fairly high rate but has not yet begun to synthesize large amounts of RNA. Within a few hours after injection into oocyte cytoplasm, a number of changes begin. The swelling of blastula nuclei in eggs versus oocytes is shown in Fig. 8.6. Although nuclei swell considerably in egg cytoplasm, they enlarge many times more in oocyte cytoplasm, as would be expected if the transplanted nucleus is to mimic the large germinal vesicle. While the morphological changes are occurring, the blastula nuclei shut down their synthesis of DNA and are found to be actively manufacturing RNA (Fig. 8.7*a*). If blastula nuclei had

Figure 8.6 Blastula nuclei swell following their injection into *a* an egg or *b* an oocyte. The nucleus in the oocyte enlarges 200 times. *c* Blastula nuclei of normal dimensions. [*Courtesy of J. B. Gurdon.*]

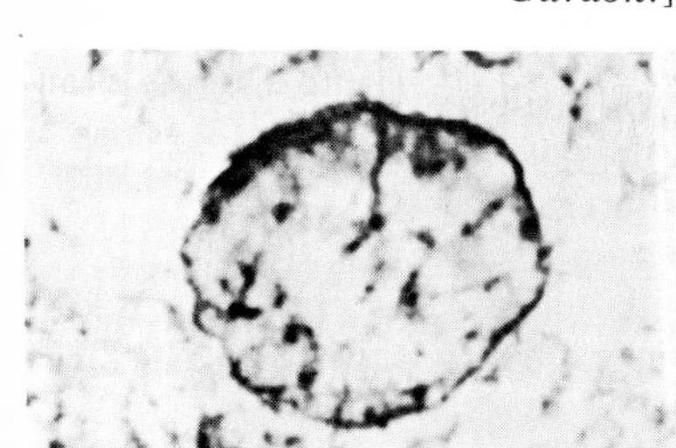

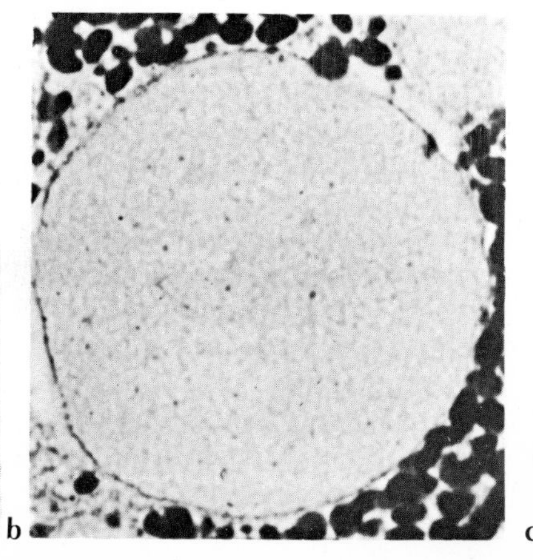

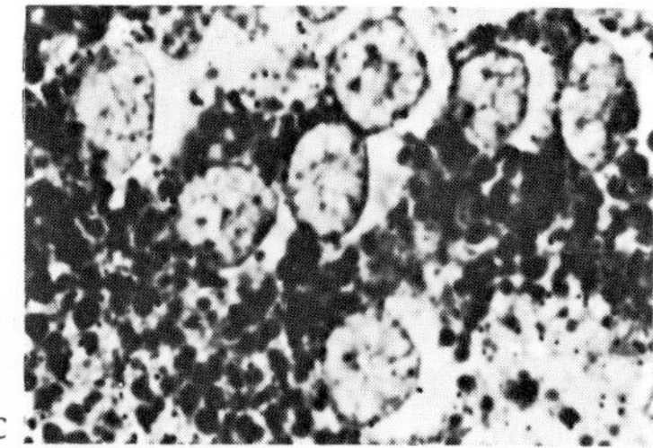

instead been injected into the activated egg, no RNA synthesis would have been initiated (Fig. 8.7*b*). The conditions for DNA synthesis are reversed (Fig. 8.7*c* and *d*). Oocyte cytoplasm does not activate DNA synthesis, while egg cytoplasm does. These same results are observed following the injection of purified DNA into the two types of cells. The DNA is replicated in egg, but not oocyte, cytoplasm. There are two possible explanations to account for the lack of replication in oocyte cytoplasm: the absence of one or more necessary factors and/or the presence of an inhibitor. Evidence for both these explanations has been obtained. Evidence

Figure 8.7 *a–b* Activation of RNA synthesis in transplanted nuclei. Midblastula nuclei, which synthesizc DNA but very little RNA, are injected into *a* an oocyte and *b* an egg, and RNA synthesis is followed autoradiographically by monitoring the incorporation of ^{3}H-precursors. Transcription is activated in the oocyte (presence of black silver grains) but not in the egg, in keeping with the synthetic activities of the host-cell environment. *c–d* Activation of DNA synthesis in transplanted nuclei. Frog-brain nuclei, which synthesize RNA but not DNA, are injected into *c* an oocyte or *d* an egg, and DNA synthesis in the donor nuclei is monitored autoradiographically. Replication is activated in the egg but not the oocyte, once again as directed by the cytoplasm of the host cell. [*Courtesy of J. B. Gurdon.*]

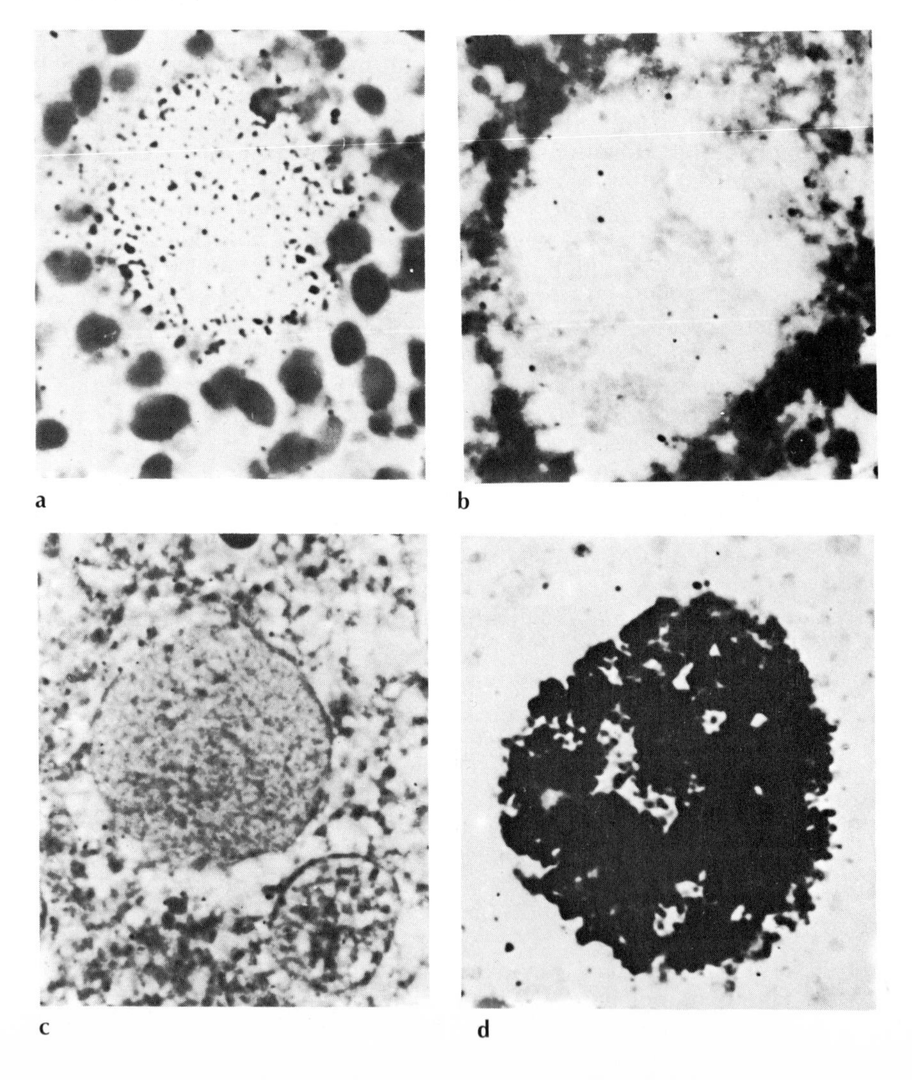

also has been gathered for the presence within oocyte cytoplasm of substances that stimulate RNA synthesis. One of these activities appears specifically to activate RNA polymerase I, which is responsible for synthesis of ribosomal RNA.

It is obvious from the previous discussion that oocyte cytoplasm promotes the general activation of RNA synthesis, but this tells us little about the specific nature of the transcripts being produced. Certainly one of the foremost questions in molecular biology concerns the basis of selective gene expression. As will be apparent in the remainder of this section, the amphibian oocyte holds great promise for research on this question. It was mentioned earlier that adult cells are able to respond to signals present in egg or oocyte cytoplasm. For example, when nuclei from cultured *Xenopus* kidney cells are injected into oocytes, they enlarge greatly, their chromatin disperses, and they synthesize RNA. The question of importance concerns the nature of the transcripts. Are they similar to those made by a kidney cell, i.e., do the nuclei continue to synthesize the same species of RNA produced prior to transfer, or are the newly synthesized transcripts of the oocyte variety? In order to answer this question, a great deal must be known about the synthetic activities of the cells involved. The best available technique for the analysis of large numbers of proteins is two-dimensional electrophoresis (Section 21.8). In this technique, proteins are separated in one direction on a flat plate according to their molecular weight and in a second direction according to their isoelectric point. Hundreds of proteins can be separated in this manner. Since the spectrum of proteins being synthesized in a cell reflects the population of mRNAs it contains, this protein-fractionation procedure allows one to compare the transcriptive activities of nuclei in different cytoplasmic environments. As might be expected, the pattern of proteins synthesized by *Xenopus* oocytes is quite different from that of *Xenopus* cultured kidney cells. Although there is much overlap, each synthesizes a number of proteins absent from the profile of the other cell type. Therefore, it should be possible to determine the extent to which the synthetic program of a transplanted kidney nucleus is reprogrammed by factors in oocyte cytoplasm. There is one further catch, however. How can one tell whether the proteins synthesized in the oocyte following transplantation are being made using existing oocyte mRNAs as opposed to mRNAs synthesized by the newly transferred kidney nucleus? The difficulty and its solution are similar to that encountered in Section 4.3.A.ii; we have to introduce another species into the experimental picture. In this case, the *Xenopus* kidney nucleus is injected into the cytoplasm of an oocyte of *Pleurodeles*, another amphibian. In this experiment it is hoped that the two species are (1) close enough relatives that the cytoplasmic factors of *Pleurodeles* will activate oocyte-specific genes in the transplanted *Xenopus* nucleus and (2) distant enough relatives that the proteins synthesized by *Pleurodeles* oocytes can be distinguished from those of *Xenopus* oocytes. As it turns out, there are several proteins synthesized by *Xenopus* oocytes that migrate differently from any of the proteins made by either *Pleurodeles* oocytes or *Xenopus* cultured kidney cells. It has recently been found that several of these proteins (*Xenopus* oocyte-specific) appear in the *Pleurodeles* oocytes after transplantation of the *Xenopus* kidney nucleus. Moreover, the recipient oocyte does not synthesize any of the eight detectable *Xenopus* kidney-specific proteins. These results strongly suggest that oocyte cytoplasm (1) contains factors capable of the selective activation of oocyte-specific genes that were previously inactive in the kidney cell and (2) lacks factors necessary to support kidney-specific transcription.

The amphibian oocyte has proved to be an excellent system for the study of a variety of molecular-level questions. Its tolerance to micromanipulation and its large size make it suitable for the injection of nuclei as well as nanogram quanti-

ties of nucleic acids or other substances. Messenger RNAs can be injected to determine their translation activity; nuclei and DNA can be injected to determine their transcriptive activity; and regulatory factors can be injected along with the mRNA, DNA, or nuclei to test their effects on the synthetic activity of the oocyte. In many ways these types of assays resemble those normally carried out by cell-free systems (see Chapter 21), but the oocyte is much more efficient and much longer-lived. An injected mRNA, for example, can be translated thousands of times over a period of several weeks. A new line of inquiry has emerged with the finding that purified DNA can be injected into the germinal vesicle of an oocyte where it will be faithfully transcribed. For example, the injection of SV40 viral DNA is followed by the appearance of known viral proteins in the cytoplasm. This finding indicates that normal transcription and/or processing is occurring such that the proper mRNAs are being provided for translation. It is hoped that this system can be utilized to learn more about the regulatory factors involved in these activities. It might be possible, for example, to inject naked DNA into the oocyte and then "fish out" associated regulatory proteins by recovering the DNA after a period in the oocyte. Alternatively, one could inject DNAs of different structure to determine the effects of various parts of the sequence on their transcriptional activity. Not only are injected DNAs transcribed in the oocyte, but they are converted into authentic chromatin by combination with histones and other proteins. It is estimated that 1,000 times as much DNA can be converted to chromatin in an oocyte than is present in the normal oocyte's own nucleus. This illustrates, once again, the storehouse of materials present in an amphibian oocyte that are normally utilized during early development.

The experiments with *Xenopus* oocytes give us a look at one of the most basic processes in cell biology, the cytoplasmic control of nuclear activity. Another series of experiments relates directly to this question, although the cells employed are not usually embryonic. The technique, termed *cell fusion,* involves the association of two cells, the dissolution of the membranes separating the cells, and the subsequent fusion of their cytoplasms. In most cases, fusion is brought about by the addition of an inactivated (therefore noninfective) virus, which attaches to the plasma membranes and makes the cells sticky. For a period of time, fused cells contain two or more nuclei; they are *multinucleate.* Cells derived from different types of tissue, or even from different species of animals, are still capable of fusing and forming multinucleate cells (they are called *heterokaryons* if they are from two species). Various fates can befall these hybrid cells, one of which is the formation of a population of healthy dividing cells with mixtures of both types of chromosomes within each nucleus. For two sets of chromosomes to fuse within one nucleus, an intervening mitosis must occur. As the nuclei within the fused cell prepare for mitosis, the nuclear membranes break down, the chromatids separate into daughter cells, and a nuclear membrane re-forms in each cell, enclosing a mixed population of chromosomes. Cell fusion has been an invaluable tool in the study of several aspects of cell function, including the nature of the malignant cell (Chapter 18) and human cell genetics.

We will briefly consider one case of cell fusion from the standpoint of reprogramming nuclear activity. (Other examples can be found in the books by H. Harris and by N. R. Ringertz and R. E. Savage listed at the end of the chapter.) The chicken erythrocyte is a circulating red blood cell, the terminal stage in a series of differentiation steps. The nucleus of this cell is synthetically inactive: it makes no DNA and no RNA and its chromatin is tightly condensed. The human HeLa cell is a rapidly dividing, active malignant cell that makes RNA continuously and DNA during its periodic S phases. When these two cells are fused, there is a striking series of changes in the erythrocyte nucleus, including a twen-

tyfold to thirtyfold volume increase, a dispersal of the condensed chromatin into a more synthetic state, the accumulation of human nuclear proteins, the appearance of a nucleolus, and the reactivation of both DNA and RNA synthesis (Fig. 8.8). The chicken erythrocyte nucleus, dormant in its own cytoplasm, can be rapidly reprogrammed under cytoplasmic influence to a full synthetic activity. Fusions of various combinations of cells indicate that (1) cytoplasmic factors act to stimulate nuclear synthetic activities rather than inhibiting them, (2) activation of RNA and DNA synthesis is controlled independently, and (3) the signals generally lack species specificity.

In summary, we have shown in this discussion that nuclei retain their full genetic complement and that their expression is dictated by their cytoplasmic environment. These two basic findings rest at the foundation of our present concept of the utilization of genetic information during embryonic development. Nuclei formed during cleavage come to reside in different regions of egg cytoplasm and are therefore influenced in different ways. As a consequence of the differential expression of embryonic nuclei, new types of localized cytoplasmic influences are generated that exert a local effect on those nuclei in that region, and so forth

Figure 8.8 The reactivation of the erythrocyte nucleus. *a* A heterokaryon containing one HeLa nucleus and one hen erythrocyte nucleus, on the first day after fusion. The nuclear bodies are visible in the erythrocyte nucleus. *b* A heterokaryon containing one HeLa nucleus and one hen erythrocyte nucleus which has now begun to enlarge. The nuclear bodies are more diffuse and stain less deeply. *c* A heterokaryon containing one HeLa nucleus and one hen erythrocyte nucleus at a later stage of reactivation. The nucleus is now very much larger and the nuclear bodies are not visible. *d* Autoradiograph of a heterokaryon from a 24-hour culture exposed for 20 minutes to tritiated uridine. The autoradiograph was developed after exposure for 3 days. The cell contains four HeLa nuclei and four hen erythrocyte nuclei. All the nuclei are synthesizing RNA. [*Courtesy of H. Harris.*]

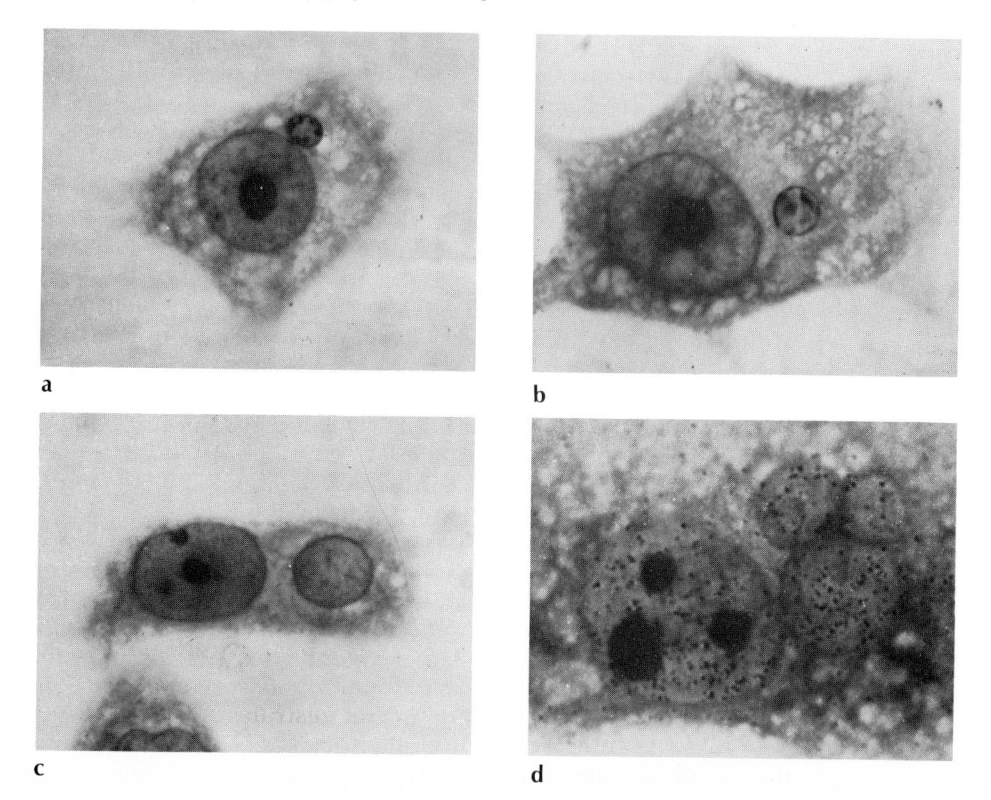

throughout the course of embryonic development. Together the temporal and spatial restrictions placed on the reciprocal interactions between nuclei and cytoplasm lead eventually to the formation of highly specialized body parts and hundreds of different types of cells, each possessing nuclei of equivalent information content. With this general picture in mind, we can turn now to studies that attempt to monitor gene expression during development.

3. Gene expression during early development

In previous chapters we examined a number of aspects of early development in a variety of organisms. In this chapter we will return for the last time to the period before gastrulation, concentrating here on the molecular aspects of gene expression. A large-scale effort has been mounted in this area in recent years, and much has been learned, primarily with regard to development in the sea urchin. Although we are in possession of considerable data concerning the types of molecules synthesized at different stages of development, we know very little about the manner in which their synthesis is controlled. Consequently, this chapter will deal with descriptive aspects of molecular activities rather than a mechanistic analysis. Anyone interested in exploring this subject in a comprehensive manner should consult the monograph by Eric Davidson (1976).

It was pointed out in Section 4.3.A that the egg contains a considerable store of RNA as a result of the synthetic activities of the oocyte. This RNA can be extracted from a batch of unfertilized eggs and analyzed in a variety of ways. For example, one can estimate the variety (or diversity) of the transcripts using molecular hybridization analysis. The RNA present in unfertilized sea urchin eggs is capable of hybridizing to approximately 3 percent of the single-copy DNA sequences (therefore, 6 percent of the single-copy genome, assuming that RNA molecules are transcribed only from one strand of the DNA duplex). The collection of sequences represented by this population of RNA is referred to as the *single-copy sequence set* of the unfertilized egg. It is equivalent to 3.7×10^7 nucleotide pairs of DNA, which, in turn, is sufficient to code for more than 20,000 polypeptides. A similar diversity of transcripts is present in unfertilized *Urechis* and *Xenopus* eggs.

A. Embryonic transcription

We can conclude from the preceding results that development of the egg begins with a legacy of template information, but when does the embryonic genome begin to be expressed? There is a variety of related questions to be considered here. When does RNA synthesis begin in the embryo? What types of RNA are made and at what stages by which cells? When does the RNA made by the embryo become necessary for continued normal development? Although these questions are obviously related, each is distinct, each requires a different analytical approach, and each provides a different answer. As might be expected, the answers to all these questions depend entirely on the animal chosen for investigation. Rather than try to present a comparative discussion, we will concentrate on the sea urchin and bring in a few other animals (particularly the frog and mouse) to illustrate the variety of approaches that have been taken.

When does transcription begin, and what types of RNA are made? Measurements of RNA synthesis (see Section 21.10) are carried out with the use of radioactively labeled RNA precursors. Although the patterns of transcription are quite variable among diverse groups, all animals investigated have been found to be synthesizing RNA prior to the time of gastrulation. In *Xenopus* (Fig. 8.9), the level of RNA synthesis remains low during the cleavage period, although the production of heterogeneous nuclear RNA (hnRNA) has been detected. As embryos

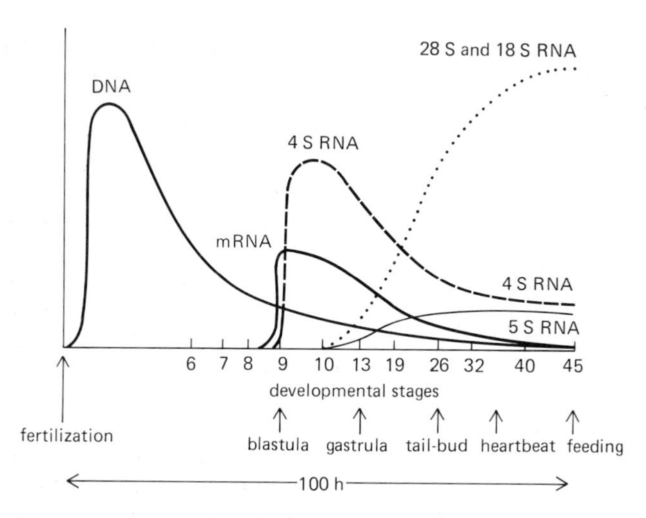

Figure 8.9 Diagram of the relative rates of nucleic-acid synthesis during amphibian development. Rates of synthesis are indicated very approximately on a low-to-high scale. [*From J. B. Gurdon, in "Control of Gene Expression in Animal Development," Harvard, 1974.*]

reach the blastula stage, a rapid activation of hnRNA synthesis (and presumably its mRNA derivatives) occurs, followed closely by the synthesis of significant quantities of tRNA for the first time. At about the beginning of gastrulation, ribosomal RNA synthesis begins, following the appearance of definitive nucleoli in which it is synthesized. The synthesis of rRNA in the embryo occurs without the benefit of rDNA amplification, as occurred during oogenesis. Since the synthesis of different types of RNA is activated at different times, each is presumably controlled independently. We know from the study of anucleolate mutants (Section 4.3.A.i) that in the absence of rRNA synthesis, development can proceed well past gastrulation; this indicates that rRNA synthesis begins well before the appearance of new rRNA molecules is required. In *Xenopus,* the pregastrular activation of RNA snythesis occurs at different times in different parts of the embryo, suggesting that activation signals do not act across the entire embryo.

The pattern of transcription in the sea urchin is quite different in some respects from that of *Xenopus.* RNA synthesis is activated in earnest in sea urchins at about the time of the third or fourth cleavage. Unlike *Xenopus,* the early cleavage nuclei of the sea urchin are very active in the synthesis of hnRNA and mRNA, although the scarcity of nuclei causes the overall level per embryo to be relatively low. Considerable controversy exists over the time at which ribosomal RNA synthesis begins. Earlier studies suggested that it was not until gastrulation that embryonic rRNA was produced, and even at this stage its rate of synthesis was very low. More recent studies suggest that embryos of the blastula or earlier stages are engaged in rRNA synthesis, but the amount made per nucleus is so low it becomes difficult to detect against the background of other labeled species. Transfer RNA synthesis appears to begin at about the blastula stage.

The period leading up to the formation of the blastula is marked by a progressive increase in the level of embryonic RNA synthesis (although an actual decrease per nucleus). Following blastulation the levels remain roughly constant through the remainder of development. The general pattern of incorporation of [^{3}H]uridine into messenger RNA during sea urchin development is shown in Fig.

8.10. In this figure, labeled polysomal RNA is divided into three distinct classes: poly(A)-containing mRNA, nonadenylated nonhistone mRNA, and histone mRNA, which represents a sizable, well-studied fraction. During cleavage, approximately 50 percent of the labeled polysomal RNA codes for histones, approximately 30 percent represents mRNA lacking the 3′ poly(A) tail, while the remaining 10 percent is polyadenylated. Eggs and embryos are unusual among cells in having such a large percentage of nonadenylated message. The reason for the preponderance of poly(A)$^-$ mRNA is not known, although its potential role in mRNA turnover is under investigation. As development proceeds, the synthesis of histone mRNA drops sharply and the synthesis of poly(A)-containing species rises proportionately (Fig. 8.10).

This picture of early mRNA synthesis and delayed rRNA synthesis is not universally found. For example, in both the nematode *Ascaris* and the mouse, rRNA synthesis begins soon after fertilization. Presumably the stage at which a particular RNA is made, whether it is rRNA or a particular mRNA, is related in some way to the activities of that embryo. We have described in detail the packaging of the amphibian egg with ribosomes; as a result, the embryo is able to delay its own ribosome production until gastrulation or much later. However, a mammalian embryo begins its life with a scarcity of cytoplasmic ribosomes, and this deficiency is reflected in the activation of rRNA synthesis at an earlier stage of development. Consequently, the cells of a mammalian blastocyst have a much greater content of ribosomes than the cells of the morula.

B. Preformed RNA versus embryonic transcription

The embryo begins as a zygote and rapidly becomes transformed into a multicellular form whose activities carry it toward an increasingly complex morphology. From what we know of the role of proteins in all cellular activities, we would predict that these changes require a continual input of newly synthesized proteins.

Figure 8.10 The relative amounts of three classes of newly synthesized mRNA in free polyribosomes during the embryonic development of the sea urchin. The three classes represent poly(A)$^+$ mRNA (●), nonhistone poly(A)$^-$ mRNA (○), and histone poly(A)$^-$ mRNA (▲). [*From M. Nemer, Cell,* **6**:*564 (1975).*]

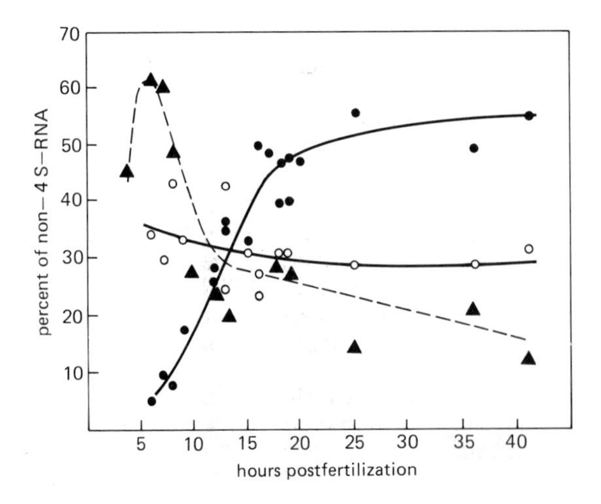

This is readily confirmed by the use of substances that inhibit protein synthesis. Inhibitors such as puromycin or cycloheximide bring all the outward signs of development to a rapid halt. The need for new proteins is clearly established, but on which RNA templates, maternal or embryonic, are they being translated? In other words, to what extent do the preformed, maternally derived mRNAs direct development as opposed to those made by the embryo after fertilization? In a broader sense, this question brings us to the heart of the matter of translational versus transcriptional control of early development.

An early approach to distinguishing between maternal and embryonic sources of developmental information utilized interspecies hybrids. In a normal embryo there is no way to identify a period when the embryonic genes have been active, since we cannot tell which events in development involve maternal templates and which utilize embryonic templates. Once a foreign sperm has entered an egg, the effect of a set of foreign chromosomes can be observed. Developing embryos of even the most closely related species show observable differences. As soon as development deviates, either morphologically or biochemically, from that of the maternal species toward that of the paternal species, we know that paternal embryonic genes have been transcribed and translated.

Studies of interspecies hybrids in frogs and sea urchins indicate that early development depends heavily on maternal mRNAs. In the sea urchin it is generally found that the first *morphological* evidence of paternal gene involvement does not occur until gastrulation or beyond. In amphibians the paternal characteristics generally appear only after gastrulation is completed. In ascidians the development of hybrids proceeds in a strictly maternal manner throughout the formation of the complex ascidian tadpole, even in hybrids where the maternal chromosomes of the egg had been removed and all genes in the embryo are derived from the sperm.

Morphological analysis of hybrids calls for numerous assumptions and tells us only about RNAs that result in visible or measurable effects. It is entirely possible that many RNAs made by the embryo are being translated at early stages but are not being detected in the hybrids as a paternal trait. Since the embryonic genome is active soon after fertilization, it seems quite suspect that it is not until gastrulation that its effects are seen. In fact, other analyses of a more molecular nature (e.g., that of Fig. 8.17) indicate that RNAs made by the hybrid embryo are translated and that the resulting proteins do have an important role in early development.

Several lines of evidence point to the necessity of embryonic transcription if development is to proceed to any significant extent. One approach is to abolish embryonic transcription and let the embryo show whether it needs this transcription by revealing how far it can develop in its absence. Several techniques can be used for this type of destructive analysis. For example, one can produce an activated frog egg that lacks functional chromosomes. Irradiation of a sperm destroys its chromosomes but does not affect its fertilizing capacity. The paternal chromosomes are irreversibly damaged and do not participate in any subsequent events. Once the egg is activated by such a sperm, the egg chromosomes move very close to the animal pole surface to complete meiosis; when this occurs, they are removed with a needle. Such an enucleated frog egg, which lacks chromosomal DNA to be transcribed, cleaves and develops into a seminormal blastula, at which time development is arrested. There is no indication of gastrulation.

A less drastic means to abolish transcription and confirm the results of physical enucleation is to inject the amphibian egg with actinomycin D and achieve a "chemical enucleation." When this is done, development proceeds to the blastula

stage and stops there. However, we cannot be certain in either the physical or chemical enucleation experiments that development of the embryo stops as a result of the lack of RNA synthesis rather than by an unrelated side effect. From both experiments, however, we can conclude that development in the frog can proceed *at least* to the blastula stage in the absence of embryonic RNA synthesis.

It is not surprising that the frog embryo cannot gastrulate in the absence of RNA synthesis. As previously described, gastrulation is foreshadowed by a general activation of transcription, which presumably provides the embryo with informational RNAs necessary for the complex morphogenetic events that follow. The development of the sea urchin following enucleation (chemical or physical) is very similar to that of the frog; development proceeds to the blastula stage and stops (Fig. 8.11).

In order to decide how far an embryo can progress before it must call on its own genes to continue its development, we need a less damaging approach than those just described. The most suitable genetic damage to utilize in the study of gene function is the genetic mutation. Most mutations involve only a very small part of a single gene. We can therefore assume that any observable deficiency in development is a result of the absence of one correct RNA that cannot be made by the embryo.

As one might expect, the analysis of development employing genetic mutations has reached its most sophisticated level in studies on *Drosophila*. Early studies had indicated that fruit flies beginning life with missing pieces of chromosomes could expect to undergo developmental arrest during the nuclear cleavage or early blastoderm stage (Section 7.8). Presumably certain genes are required (and missing in the deleted chromosome fragments) to support development even during the first few hours. Another group of studies has taken advantage of a special class of mutations, the *temperature-sensitive (ts) mutations*, to determine the time during development at which specific genes are required. Flies carrying a temperature-sensitive mutation will appear perfectly normal if raised at a lower temperature (e.g., 17°C, termed the *permissive temperature*), but they will show

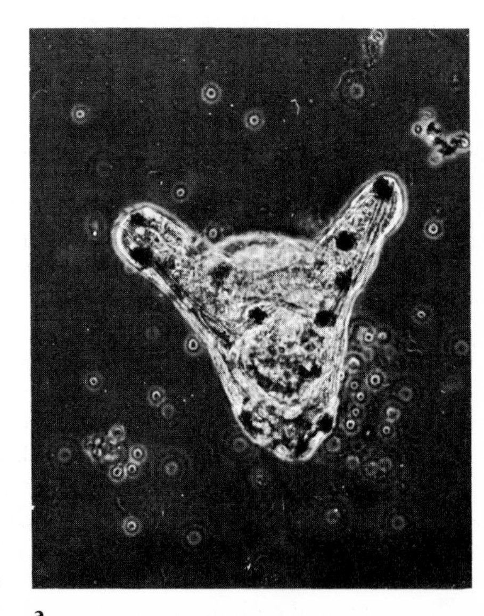

a

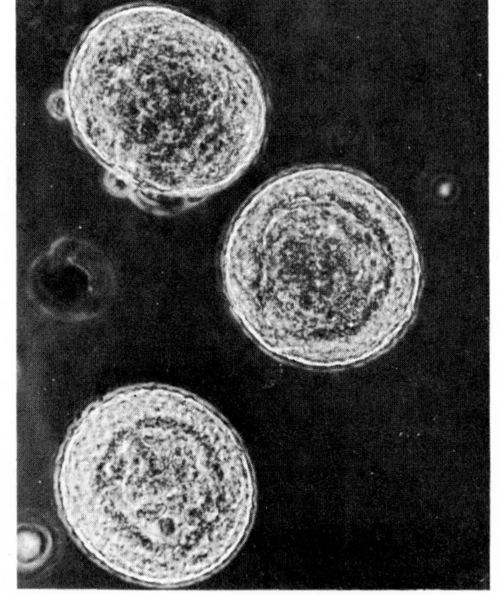

b

Figure 8.11 Effect of actinomycin D on sea urchin development. *a* Normal *Arbacia punctulata* pluteus larva. *b* Arrested *Arbacia* blastulas after continuous treatment with actinomycin D. [*Courtesy of C. H. Ellis.*]

the mutant phenotype, which is usually death, if raised at a higher temperature (e.g., 29°C, termed the *restrictive temperature*). Why should the mutant phenotype show up only at the elevated temperature? The answer lies in the physical chemistry of proteins. Increased temperature places certain stresses on protein molecules that tend to disrupt their three-dimensional structure and thus their function. In the normal (wild-type) protein these elevated temperatures are not sufficiently destructive to alter function. The mutant protein is more sensitive to disruption because of its altered amino acid sequence; when the temperature is raised, the protein cannot function and the mutant phenotype results.

Gene function can be described at several successive steps: the period during which (1) the gene is transcribed, (2) the RNA is processed and transported to the cytoplasm, (3) the mRNA is translated, (4) the protein functions, and (5) a visible effect is observed. No set time separates these steps; rather it depends on the particular gene and its cytoplasmic environment. Analysis of *ts* mutants tells us about the last two steps. The period during which the protein is functioning is the *temperature-sensitive period* (*TSP*), which is determined by raising and lowering the temperature (Fig. 8.12). If the fly is living at the elevated temperature during a significant part of the TSP of its mutant gene, it will show the mutant phenotype (death, a missing structure, a different morphology). The time between the TSP and the stage at which the mutant phenotype is observed can vary from minutes to hours or even longer. Some genes act early in development and others act late; certain genes act one time, others twice, still others several times; some act in an on-off manner, while others act continually. There are even established differences in the same gene in different sexes. The studies on *ts* mutants in *Drosophila* have provided considerable information on the timing of gene function and have corroborated what would have been predicted from other types of studies.

Genetic modifications also have been used in the study of mammalian development. In the mouse a mutant with an extremely early effect on development has been isolated. When this mutant, known as t^{12}, is present in a homozygous condition, it produces effects on development as early as the two-cell stage. Development is arrested at the morula, a few cleavages later. The defect manifests

Figure 8.12 The temperature-sensitive periods for a gene (shi^{ts1}). Elevation of the temperature during any of the periods blocked out in the figure causes the death of the organism. It appears that this gene is active in an on-and-off manner throughout early development. [*After D. T. Suzuki, in "Concepts in Development," J. Lash and J. R. Whittaker (eds.), Sinauer, 1974.*]

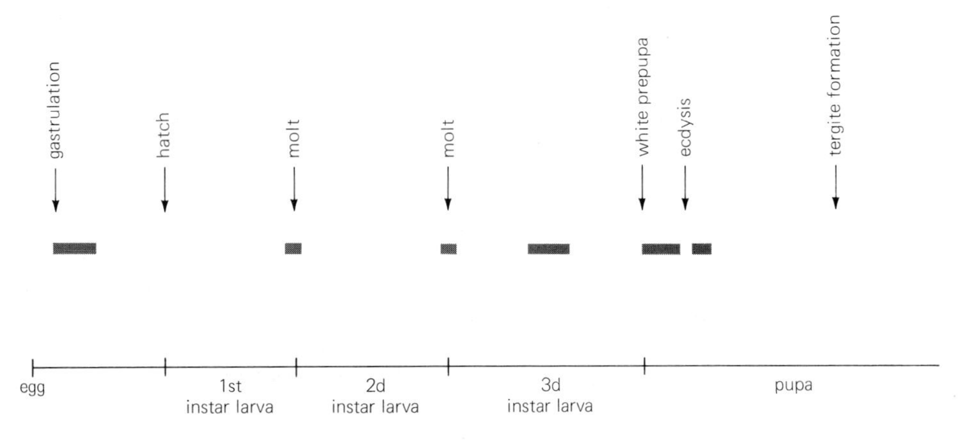

itself in several abnormalities, but it is difficult to be sure which of the alterations is the primary genetic defect. Other types of studies also point to a role for the embryonic genome at a very early stage. For example, when mice of different genetic strains are mated, paternal characteristics appear by early stages of development, often during cleavage. The paternal strain of the enzyme β-glucuronidase is detectable by the eight-cell stage. Similarly, mammalian embryos are particularly sensitive to inhibitors of RNA synthesis. One of the most potent and specific inhibitors of mRNA synthesis is the mushroom toxin α-amanitin, which acts by interference with eukaryotic polymerase II, the enzyme responsible for hnRNA and mRNA synthesis. Unlike the sea urchin or frog, fertilized mammalian eggs are capable of only a couple of divisions when kept in α-amanitin. Whether the much slower cleavage schedule of the mammalian egg is a factor in its dependence on newly synthesized RNA is unclear.

One of the most important, and most neglected, aspects of the molecular biology of development concerns the manner in which developmental information is localized in particular regions of the egg and embryo. Most biochemical procedures begin with the homogenization of a batch of eggs or embryos followed by the analysis of a particular subcomponent. This approach, while extremely informative, tells us little about the spatial organization of molecular information. The concept of localized morphogenetic determinants was discussed at length in Chapter 7. Morphogenetic factors are generally envisioned as substances that act to modify the transcriptional and/or translational activities of the cells into which they are segregated. For example, the appearance of tyrosinase in the pigment-cell line and acetylcholinesterase in the muscle-cell line of ascidian embryos (Section 7.4) was attributed to the segregation of cytoplasmic determinants into different blastomeres. When these embryos are incubated in actinomycin D (or other transcriptional inhibitors) prior to enzyme appearance, the production of both enzymes is blocked. This result suggests that the enzyme-determining factors being segregated act at the transcriptional level to promote the synthesis of mRNAs for these enzymes. Other studies on these same embryos suggest that maternal factors influencing translation (rather than transcription) also are segregated during the cleavage period. The enzyme alkaline phosphatase appears within the intestinal cells of the ascidian embryo (Section 7.4) regardless of the presence of actinomycin D. Since the enzyme is synthesized in the absence of embryonic RNA synthesis, we can conclude that the mRNA for this enzyme was present in the egg at the time of fertilization. Since the message is translated only within the cells of the gut, we can further conclude that this maternal mRNA (or a factor responsible for its selective translation) is segregated into the appropriate blastomeres during cleavage.

This type of segregation of developmental information is precisely that expected on the basis of the mosaic theory discussed in Chapter 7, although there is no reason at present to believe that similar types of localization do not occur among more regulative embryos. In fact, some evidence does indicate that qualitative differences exist between the populations of mRNAs present in the micromeres, macromeres, and mesomeres of the 16-cell sea urchin embryo.

C. Protein synthesis during early sea urchin development

As discussed in Chapter 5, there is a marked activation of protein synthesis in the sea urchin beginning a few minutes after fertilization (Fig. 8.13). It was indicated in Section 5.6.B that the Na^+/H^+ exchange, and the accompanying increase in intracellular pH, appears to be the direct trigger for the initiation of the late events, one of which is the activation of protein synthesis. Recent experiments

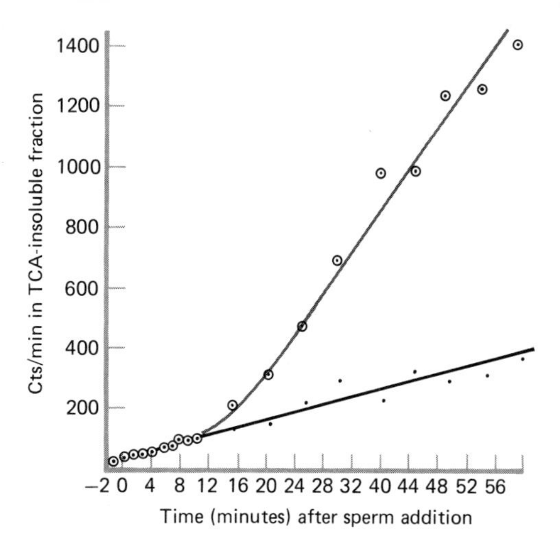

Figure 8.13 The cumulative incorporation of [^{14}C]leucine by unfertilized (*dots*) and fertilized (*circles*) eggs of the sea urchin. The 0 time marks the point of fertilization, which is followed, after a brief lag, by a marked elevation in the rate of protein synthesis. [*From D. Epel,* Proc. Natl. Acad. U.S., **57**:*901 (1967)*.]

have indicated a striking relationship between the pH of the egg and its level of protein synthesis. If unfertilized eggs are incubated in seawater containing ammonia, the cytoplasmic pH rises, as does the translational activity. If the ammonia is washed out of the eggs and sodium acetate is added, thereby lowering the pH of the egg cytoplasm, the rate of protein synthesis falls well below the typical value for the unfertilized egg. It is estimated that a rise (or fall) of 0.1 pH unit results in a threefold rise (or fall) in protein synthesis. The dramatic effect of a small decrease in cytoplasmic pH on the metabolism of a *fertilized* egg is shown in Fig. 8.14. In this case, sodium acetate is added at 30 minutes postfertilization and is seen to cause a striking drop in the rate of protein synthesis back to essentially the same level found in the unfertilized egg. If the pH is allowed to rise once again, the translational activity of the cleaving egg returns to normal levels (Fig. 8.14). The manner in which these pH changes affect protein synthesis is totally obscure. The fact that there is a lag of several minutes between the pH increase and the change in synthesis suggests that the effect is an indirect one.

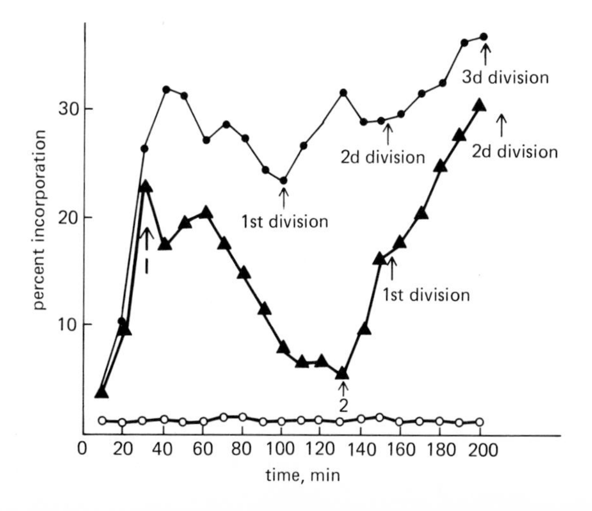

Figure 8.14 The effects of pH change (brought about by treatment with sodium acetate) on the rates of protein synthesis: unfertilized control (○); fertilized control (●); fertilized with sodium acetate treatment as described below (▲). Unfertilized eggs were fertilized in normal seawater at time = 0. At 30 minutes after fertilization, one batch of eggs (▲) was resuspended in seawater containing 10 *mM* sodium acetate, pH 6.5 (*arrow 1*). At 130 minutes, this same batch of eggs was washed twice in pH 8 seawater and resuspended in normal sea water (*arrow 2*). Samples also were scored every 15 minutes for mitotic activities and cell division. [*From J. L. Grainger, M. M. Winkler, S. S. Shen, and R. A. Steinhardt,* Dev. Biol., **68**:*401 (1979)*.]

The rate of protein synthesis climbs steeply through the cleavage period and levels off at about the blastula stage. It is calculated (see Section 21.10) that the absolute rate of protein synthesis in the gastrula of *Strongylocentrotus* is 0.92 nanograms per hour (ng/h), which is a 113-fold increase over the level of the unfertilized egg. The translation of mRNA occurs on polyribosomes that are best analyzed by sucrose-density-gradient centrifugation (see Section 21.2), which serves to separate polyribosomes on the basis of the number of attached ribosomes. The greater the number of ribosomes attached to an mRNA strand, the faster it will move through the gradient and therefore the closer it will be to the bottom of the tube at the end of the centrifugation. Sucrose density gradients of unfertilized egg cytoplasm indicate the virtual absence of polyribosomes; it is estimated that only 0.75 percent of the ribosomes of the egg are present in polyribosomes. The lack of translational activity is certainly not due to the lack of translatable mRNA, whose presence is readily demonstrated in vitro, nor is it due to the presence of nonfunctional ribosomes; ribosomes prepared from unfertilized eggs are as capable of supporting protein synthesis in a cell-free system as ones taken from fertilized eggs. Rather it appears that the ribosomes and mRNA of the unfertilized egg are not capable of association with one another in the manner required to initiate polypeptide formation. Moreover, the responsibility for this lack of interaction appears to reside with the mRNA, which is somehow present in the unfertilized egg in a "masked" condition.

When changes in the rate of protein synthesis occur in the absence of comparable changes in mRNA concentration, as is seen in the response to fertilization in the sea urchin, regulation is said to be operating at the *translational level.* The increased availability of mRNA for recruitment into polyribosomes is one means by which translational-level control can operate. Messenger RNAs of animal cells are not found in a naked state, but instead exist in association with proteins to form messenger ribonucleoprotein (mRNP) particles. The mRNP particles of most cells are readily translated, both in vivo and in vitro. However, when the mRNP particles of unfertilized sea urchin eggs are isolated under conditions that preserve their stability, they are not translatable in cell-free systems. In contrast, mRNP particles prepared in a similar manner from fertilized eggs can be translated in vitro. Presumably some alteration has occurred in the structure of the particle following fertilization which renders it available for interaction with ribosomes.

Figure 8.15 shows the polyribosome profiles of sucrose gradients prepared at various times after fertilization. There are virtually no detectable polyribosomes in the unfertilized egg (not shown in Fig. 8.15), but the distribution of ribosomes in the gradient changes slowly after fertilization as two distinct classes of polyribosomes appear. By 2 hours after fertilization, approximately 20 percent of the ribosomes are now in polyribosomes, primarily found in a heavier class seen as a major peak toward the bottom of the gradient. These are the heavier r-polyribosomes (r denoting rapid sedimentation), which contain an average of 23 attached ribosomes and have an average molecular weight of nascent (being synthesized) protein of 57,000 daltons. By 6 hours after fertilization, a second, lighter class of polyribosomes is clearly visible. These are the s-polyribosomes (s denotes slow sedimentation), which have an average number of 9 ribosomes and an average nascent protein molecular weight of 9,300 daltons.

To determine which of the polyribosomes are most active in the incorporation of amino acids into protein, the embryos are given radioactively labeled amino acids just prior to homogenization. These labels will be taken up and incorporated into the growing polypeptide chains that remain attached to the polyribo-

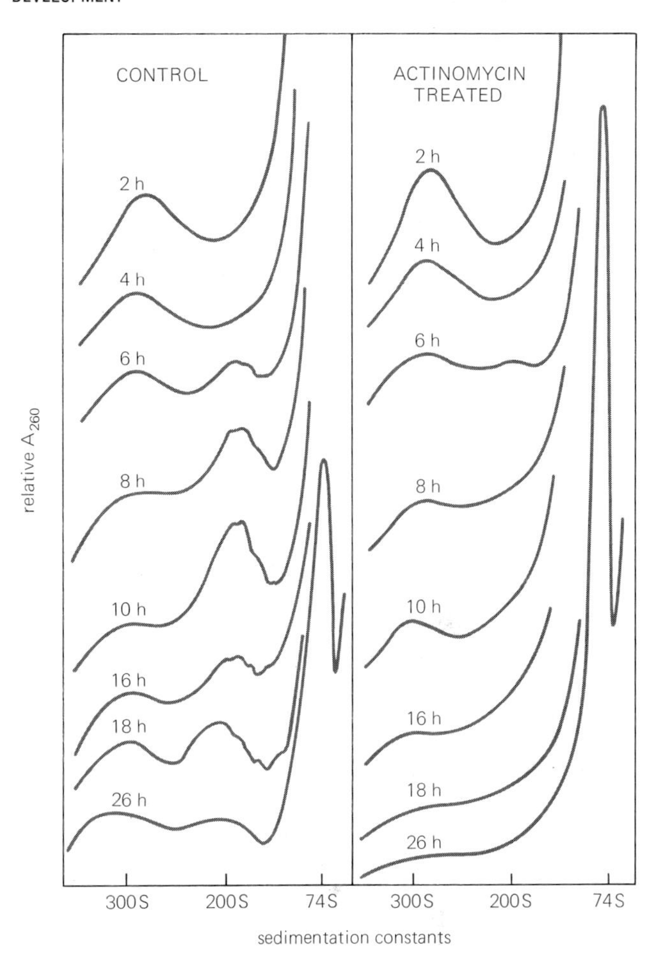

Figure 8.15 Sedimentation profiles of polyribosomes from sea urchin embryos at various times after fertilization in the absence (*left*) and presence (*right*) of actinomycin D, which blocks the synthesis of RNA by the embryo. The first polyribosome peak to appear after fertilization (none are present in the unfertilized egg) occurs toward the bottom of the tube (*far left of each gradient is the bottom*). This peak represents the r-polyribosomes that sediment rapidly. The peak appears equally as well in the presence of actinomycin D, indicating that it forms as a result of the recruitment of preformed maternal mRNA and is not dependent on newly synthesized embryonic messages. By 6 hours after fertilization, the control embryos are showing a second peak, toward the top of the centrifuge tube. This peak represents the smaller, slowly sedimenting s-polyribosomes that increase in amount during the next hours. These are the histone-synthesizing polyribosomes and their accumulation is sensitive to actinomycin D (compare the 10-hour control and drug-treated). [*After A. A. Infante and M. Nemer,* Proc. Nat. Acad. Sci. U.S., **58:***681* (*1967*).]

somes within the gradient. The distribution of radioactivity incorporated into nascent chains is shown in Fig. 8.16*a*. The results suggest that all the polyribosomes are engaged in protein synthesis to an equivalent degree. The smaller polyribosomes have less label, but this is expected because they are making smaller proteins; therefore, each ribosome will have smaller labeled nascent chains.

The use of labeled amino acids tells us about protein synthesis but nothing about the nature of the mRNA threads by which the ribosomes are held together. How can we distinguish polyribosomes with preformed mRNA from those with embryonic mRNA. There are several ways. One approach is to provide the embryos with radioactively labeled RNA precursors, such as [^{3}H]uridine, prior to homogenization. This label will be incorporated into RNA in the nucleus, some of which will enter the cytoplasm as mRNA and become rapidly attached to ribosomes. Since the synthesis of tRNA and rRNA does not occur during the early stages of sea urchin development, we can be certain that all the label in the polyribosomes is in mRNA. Polyribosomes containing preformed mRNA will remain unlabeled, while those with embryonic mRNA, which was synthesized during the time the label was available, will be radioactive. In this experiment (see Fig. 8.16*b*), the label occurs in all sizes of polyribosomes, but the lighter class (s-

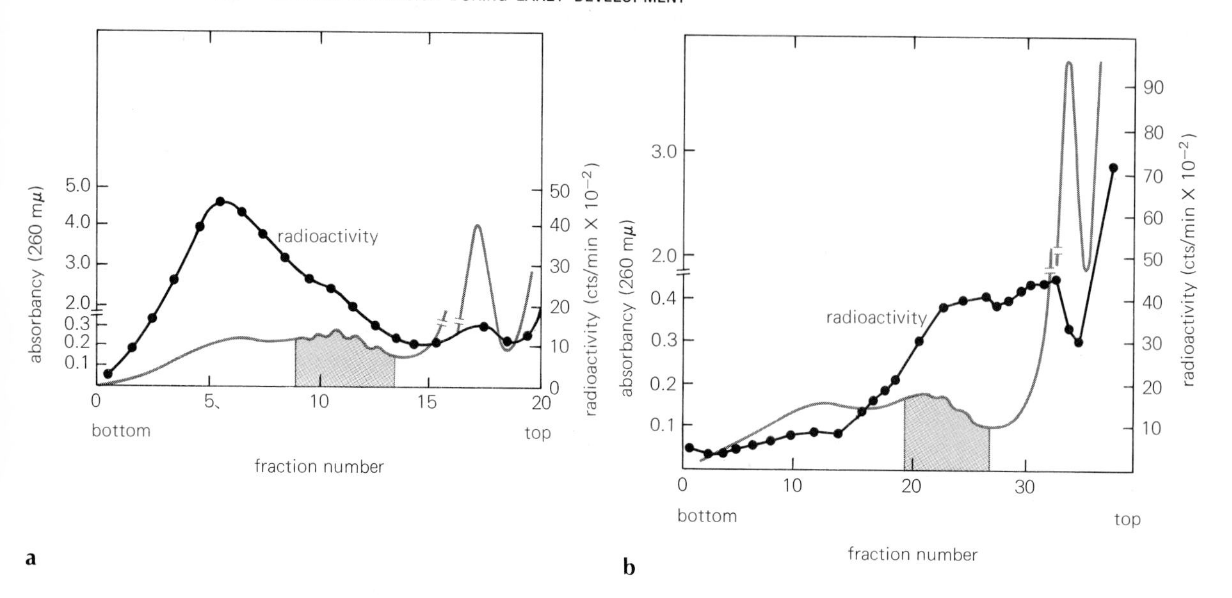

Figure 8.16 Sucrose-density-gradient profiles showing polyribosomes of the sea urchin during cleavage. Embryos were incubated in radioactive precursors as described below, washed, homogenized, and the homogenate centrifuged at 15,000 *g* to remove cell organelles. The supernatant containing the ribosomes and polyribosomes was then centrifuged through the gradient. In both these curves the solid line represents the absorbance of each fraction from the centrifuge tube. Since absorbance reflects amount of material present, this curve indicates the amount of ribosomes present throughout the gradient. The peak on the right-hand side of the graph (*close to the top of the gradient*) indicates the monoribosome peak, i.e., those not present in polyribosomes. The hatched part of the curve shows the region of the s-polyribosomes. The area to the left of the shaded part is the region containing the r-polyribosomes. The dotted line in each graph shows the radioactivity present in each of the fractions of the gradient.

a Embryos were incubated in a [^{3}H]amino acid for 10 minutes before homogenization. The dotted line, therefore, reveals the newly forming protein chains that remain attached to the polyribosomes as they sediment through the sucrose. The r-polyribosomes are seen to incorporate much more radioactivity than the s-polyribosomes. If, however, the incorporation is corrected for the smaller size of the s-polyribosomes, all the different-sized polyribosomes are found to be equally active in protein synthesis.

b The embryos were incubated in [^{3}H]uridine for 90 minutes prior to homogenization. During this period the [^{3}H]uridine is incorporated into RNA, which leaves the nucleus and becomes recruited into polyribosomes. Only those polyribosomes using RNA synthesized by the embryo will be labeled. Polyribosomes made up with preformed, maternal RNA will contribute to the absorbance (*line without dots*) of the fractions in which they occur, but they will not contribute to the radioactivity (*dotted line*) of those fractions. The curve of the radioactivity of *b* indicates that the s-polyribosomes are composed of newly synthesized embryonic RNA to a much greater extent than r-polyribosomes. [*After L. H. Kedes and P. R. Gross,* J. Mol. Biol., **42:**559 (1969).]

polyribosomes) is preferentially labeled. Since the labeled mRNA is found throughout the gradient, we can conclude that the embryo is synthesizing a wide variety of sizes of mRNA and thus is providing templates for many different proteins.

A radioactive label can always be employed to examine a substance, such as embryonic mRNA, that has just been synthesized; it is more difficult to examine substances that are already present, such as preformed, maternal mRNA. One method is to block the formation of new RNA (using actinomycin D) and then follow the polyribosome profiles after fertilization. If polyribosomes appear in the

presence of actinomycin D (and thus in the absence of embryonic RNA synthesis), we know they are forming with maternal mRNA as a template. The right-hand column of Fig. 8.15 shows the polyribosome profiles at corresponding times after fertilization in embryos that had been raised in actinomycin D from before the time of fertilization. The r-polyribosome peak at 2 and 4 hours forms similarly to the controls. These polyribosomes are using maternal mRNA as a template. The most significant effect of the drug is to greatly decrease the formation of the s-polyribosome peak (Fig. 8.15), indicating its dependence, to a larger extent, on newly synthesized mRNA. Neither class of polyribosomes is using only one or the other class of mRNA; each is a mixed group.

These results tell us something of the distribution of preformed versus embryonic information and how it is used, but little of the nature of the specific proteins involved. The s-polyribosomes appear to be totally responsible for the synthesis of histones. These small, highly basic proteins are found closely associated with DNA as part of the chromosomes. One would expect proteins with this property to increase in proportion to increasing DNA levels. Cleavage is a time of intense DNA synthesis, and unless there is a large pool of preformed histone (which is the case in amphibians but not in sea urchins), the synthesis of new histones must keep pace if chromosomes are to be produced. The appearance of the s-polyribosomes as a major peak by 6 hours, the rise of the peak to contain about 35 percent of the embryo's ribosomes by the 10-hour early blastula, and its relative decline after 20 hours reflect the relative need for chromosomal proteins at each of these times. Histones are not necessarily of greater importance for development than other proteins, but they have received more attention because of special properties that make them more amenable to biochemical analysis.

If the histones are extracted from a sea urchin embryo and are fractionated by electrophoresis (see Section 21.8), five distinct groups are found (see Fig. 8.19). These five classes of histones differ from each other in molecular weight and amino acid composition. If labeled amino acids are given to the embryo prior to homogenization and histone preparation, all five classes become labeled; this indicates that all these proteins are being made by the embryo rather than simply being selected from a pool of histones already present at fertilization (which does occur to a large degree in amphibian embryos). Are they being translated on preformed or embryonic mRNA? To answer this question we can examine the effect of actinomycin D on histone production. Since, as already indicated, the s-polyribosomes are significantly reduced in the presence of actinomycin D and these are the ones making histones, we would expect to find an effect. In fact, there is a considerable reduction in histone synthesis. However, all five classes of histone continue to be produced in the absence of embryonic RNA synthesis. This clearly indicates that preformed mRNA for all classes of histone is present in the unfertilized egg. This can be directly confirmed by extracting the RNA at this stage and using it to direct the synthesis of all classes of histone in a cell-free protein-synthesizing system. It is estimated that 4 to 8 percent of the maternal mRNA stockpile codes for histones. The involvement of the embryonic genome in the synthesis of histone mRNA is also readily demonstrated. For example, the appearance in sea urchin hybrids of paternal histone species by the blastula stage (Fig. 8.17) reveals the previous expression of the embryonic genome. In a more direct manner, the mRNA of the histone-synthesizing s-polyribosomes can be extracted and analyzed. If the purified RNA of the light polyribosomes is run on a sucrose gradient (Fig. 8.18), a peak of RNA sedimenting at about 9 S dominates the profile. The 9 S peak contains the histone mRNA of the embryo and can be fractionated

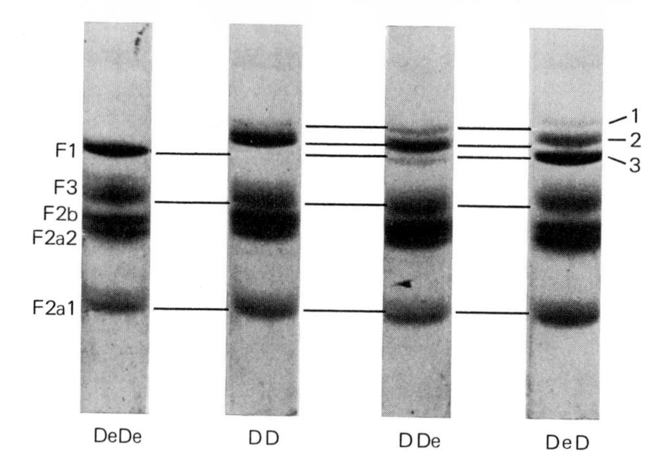

Figure 8.17 Patterns of histones after electrophoresis in acrylamide gels. The five classes of histones are labeled F1, F3, F2b, F2a2, and F2a1. The four gels represent the patterns of histones from the hatched blastula stage of sand dollar embryos (DeDe), sea urchin embryos (DD), and the hybrids having a sea urchin egg fertilized by a sand dollar sperm (DDe), and vice versa (DeD). Sand dollar embryos give one band of F1 class histone, while sea urchin embryos give two bands, both of which migrate differently than the sand dollar band. In both hybrids, all three bands are present by the hatched blastula stage, indicating the activity of the paternal chromosomes prior to this stage. [*Courtesy of D. Easton.*]

in further steps into subgroups that code for each of the histone classes. It is estimated that up to 60 percent of the mRNA synthesized by the sea urchin during cleavage codes for histone and that by the blastula stage, histone mRNA synthesized by the embryo accounts for approximately two-thirds of the histone being produced.

Whereas the earlier studies on histones established the existence of five distinct classes (H1, H2a, H2b, H3, and H4), more recent analyses using different types of electrophoretic techniques have revealed that each of the classes is com-

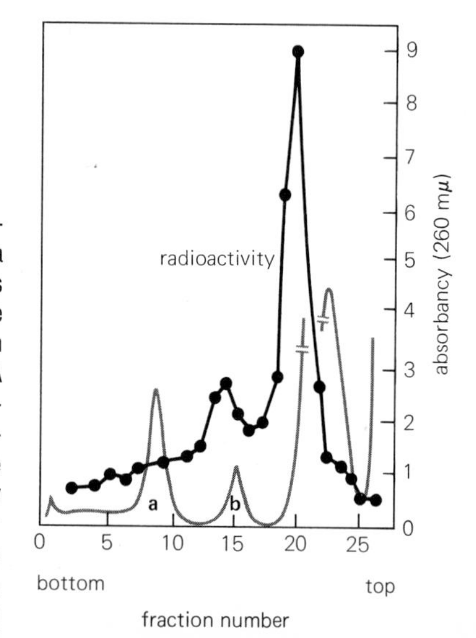

Figure 8.18 Sucrose-density-gradient profile of the RNA extracted from the s-polyribosomes. Embryos were incubated in [^{3}H]uridine for 90 minutes, homogenized, and a polyribosome gradient prepared as described in the legend of Fig. 8.16. Those fractions containing the s-polyribosomes (*hatched area of Fig. 8.16*) were taken and the RNA was extracted from these polyribosomes. Several types of RNA are extracted in this procedure: the two species of rRNA, the tRNAs, and the mRNA. Only the mRNA that had been synthesized during the 90-minute incubation in [^{3}H]uridine will be radioactively labeled. In this graph, the solid line is a measure of the amount of RNA (determined by absorbance) in each part of the gradient; the two peaks (a and b) are the two ribosomal RNAs, neither of which is labeled. The dotted line shows the radioactivity of the various parts of the gradient and the profile is dominated by one large peak near the top of the tube. This peak contains the mRNAs for all the histones, as can be demonstrated by using this RNA as a template for histone synthesis in a cell-free system. [*After L. H. Kedes and P. R. Gross,* J. Mol. Biol., **42**:*559 (1969).*]

posed of more than a single species of subtype (Fig. 8.19). Moreover, the synthesis of various members of each of the classes undergoes predictable changes (Fig. 8.20) during the course of early sea urchin development in a tightly regulated manner. In the following discussion, we will consider only the H1, H2A, and H2B classes, which have been best studied. During the first two cleavages, a single subtype of each of the H1, H2A, and H2B classes is synthesized and incorporated into the chromatin of the first few blastomeres. These histones are referred to as CS-H1 (cleavage-stage H1), CS-H2A, and CS-H2B subtypes. During the third cleavage cycle, the synthesis of a new species, the α-subtype, of each histone class begins, and these new histones are soon found within the newly forming chromatin. Finally, at about the mesenchyme blastula stage, when cell division has greatly slowed, a second shift occurs and the synthesis of a new set of subtypes begins. In this case, more than one new species of each class appears; there is β- and γ-H1; β-, γ-, and δ-H2A; and γ- and δ-H2B. All these histones are distinct species as opposed to being simply modifications of preexisting forms. This conclusion is based on amino acid sequence analysis, which indicates that these histones contain a different primary structure, and on DNA-RNA hybridization, which indicates that they are synthesized on different mRNA templates. It is interesting to note that even though the cleavage-stage histones are likely synthesized only for a fleeting period during very early development in the entire life cycle of the animal, the vast majority of the several hundred histone genes in the sea urchin genome code for these species. The demand for histones during these rapid divisions is so great that a multiplicity of DNA templates must be utilized.

The stage-specific shifts in the synthesis of histone subtypes provides an excellent system to study the roles of transcriptional and translational control mechanisms in early development. When the template activity of purified unfertilized

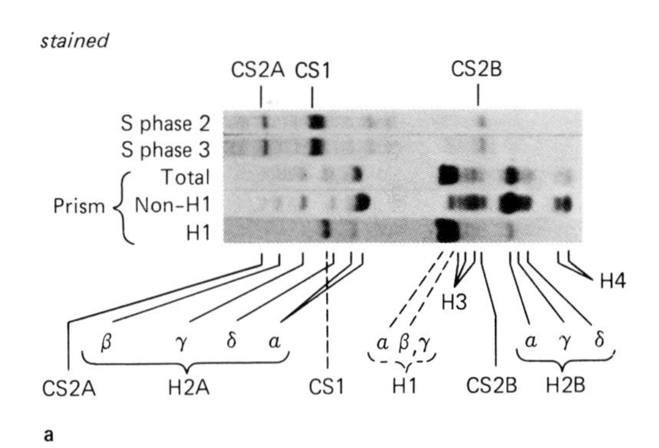

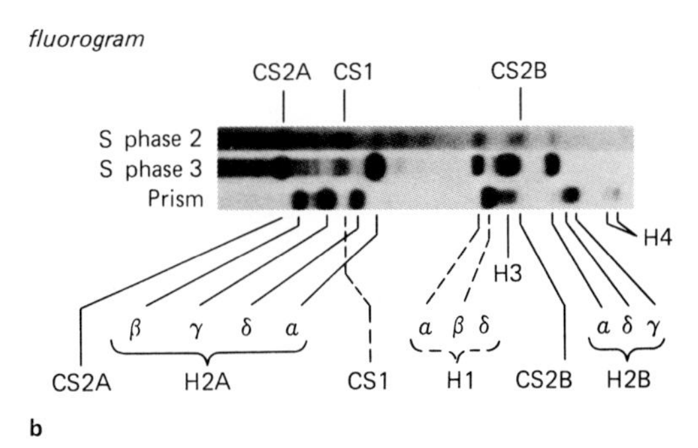

Figure 8.19 Acid-extracted histones of cleavage-stage sea urchin embryos displayed on triton-acid-urea gels. Proteins were labeled with [^{3}H]leucine during the second or third S phase, and the histones were extracted from chromatin at the following interphase. Labeled prism-stage markers are shown for reference. *a* Coomassie blue-stained gel showing the relative amounts of each histone species. *b* Fluorograph showing radioactively labeled species in the same gel. [*From K. M. Newrock, R. Alfageme, R. V. Nardi, and L. H. Cohen,* Cold Spring Harbor Symp. Quant. Biol., **42**:422 (*1977*).]

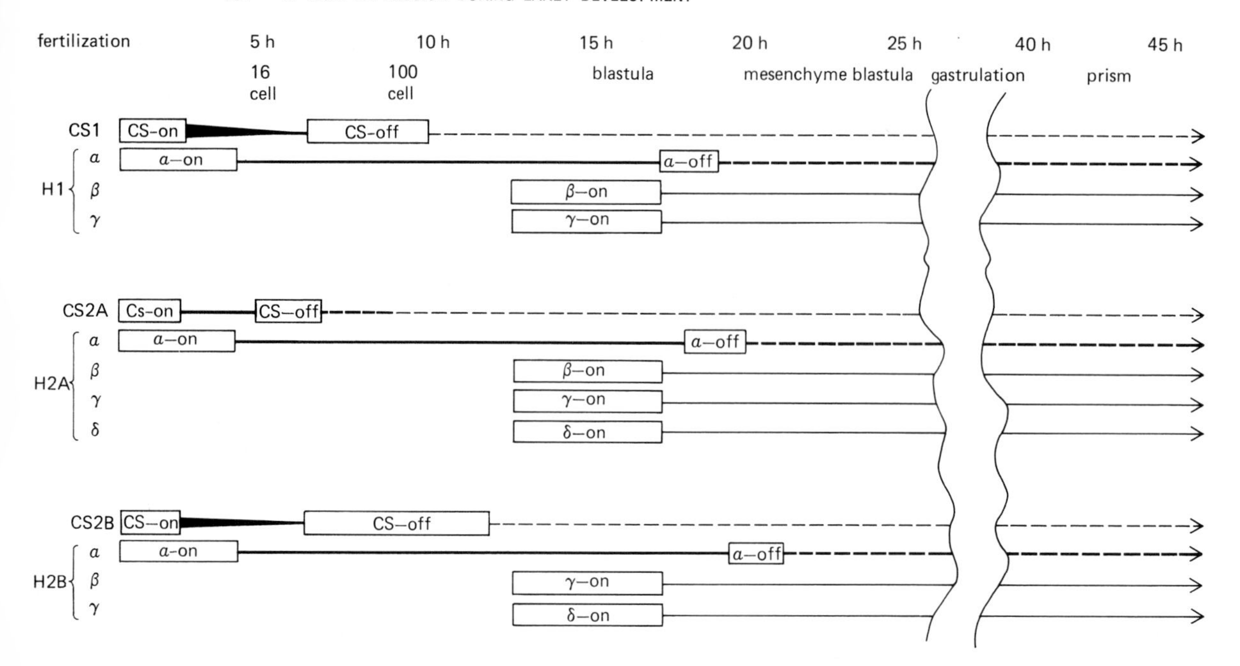

Figure 8.20 Schedule of synthesis and incorporation into chromatin of subtypes of H1, H2A, and H2B during the early development of the sea urchin *S. purpuratus*, as detected by incorporation of [^{3}H]leucine. Heavy line indicates an abundant component; narrow line indicates a minor component; solid line shows the period during which a component is synthesized; dashed line shows the period during which a protein is no longer synthesized, but is still maintained in the chromatin. Switches in the synthesis of components are indicated by the rectangles. The length of a rectangle shows the imprecision with which the time of the switch has been determined. Thus it is not intended to imply that the on switches of β, γ, and δ subtypes are necessarily simultaneous. [*From K. M. Newrock, C. R. Alfageme, R. V. Nardi, and L. H. Cohen,* Cold Spring Harbor Symp. Quant. Biol., **42:**427 (1977).]

egg RNA is assayed in a cell-free system, all the cleavage-stage (CS) subtypes and α-subtypes are found among the products. Therefore, the mRNAs for both groups of histones are present in the maternal mRNA population. In contrast, none of the proteins that normally appear at the mesenchyme blastula stage are found to be synthesized at the direction of the unfertilized egg mRNA. Further insight into these events is obtained by raising the embryos in the presence of actinomycin D. In the absence of RNA synthesis, the CS subtypes begin to be produced very soon after fertilization, as in the control embryos. Similarly, as in the controls, the synthesis of the α-histones begins at a specified time after fertilization. Unlike the controls, the synthesis of the CS histones does not stop in the actinomycin-treated eggs, nor is the synthesis of the β, γ, and δ species ever detected in the presence of the drug.

Several other specific proteins also have been examined to determine whether or not they are translated on preformed mRNAs. These include the microtubule proteins, which are synthesized at all times after fertilization; the enzyme ribonucleotide reductase, which first appears 1 hour after fertilization; and the hatching enzyme, which is made by the midblastula to destroy the fertilization envelope and release the embryo from this container (Section 5.6.A). Some evidence indicates that all three of these proteins are made on preformed mRNAs. The microtubule proteins are made in enucleated halves of eggs that have been

artificially activated; the hatching enzyme is strictly maternal in interspecies hybrids; and the ribonucleotide reductase appears in actinomycin D–inhibited embryos.

The ability of embryos to operate on preformed mRNAs is a remarkable biochemical feat. Unfortunately, the mechanisms by which translational-level control (Section 2.6.B) is exerted are poorly understood. We conceive of the egg as being filled with a massive number of messengers, yet it is only after fertilization that most of them are being translated. Are particular mRNAs recruited into polyribosomes at particular times, i.e., is there qualitative control over translation? Or, are the mRNAs utilized after fertilization simply a cross section of those present in the cytoplasm? The answer to this very important question has not been clearly obtained. When specific proteins are examined, there appears to be good evidence for the programmed release of stored mRNAs into polyribosomes. Ribonucleotide reductase, hatching enzyme, and α-histones all appear to be selectively synthesized at particular times on preformed mRNA templates. Similarly, the appearance of certain of these proteins in actinomycin D–treated embryos argues for the existence of qualitative control mechanisms. Other types of studies, however, suggest that translational control is primarily quantitative in nature. For example, DNA-RNA hybridization analyses indicate that the mRNAs present in the polyribosomes during cleavage are the same species as the nonpolysomal cytoplasmic mRNAs. In these studies (which focus on the overall population of molecules), it is argued that the nonpolysomal mRNAs simply represent an excess supply of those being translated. If qualitative translation control was *widespread*, one would expect the two populations of mRNA to be quite different. One also would expect the variety of proteins made after fertilization to be quite different from those made in the unfertilized egg. This does not appear to be the case. As described in Section 21.8, two-dimensional electrophoresis is the most powerful technique available for separating large numbers of proteins. When labeled proteins synthesized by unfertilized and fertilized eggs are compared using this technique, the patterns of approximately 400 labeled spots (histones excluded) from the two stages are essentially identical. It is not until the early gastrula stage that extensive changes appear in the patterns, although even at this stage the majority of spots are ones present at earlier stages.

Although translational-control mechanisms operate during early sea urchin development, they are even more obvious during the comparable period in the amphibian. As with the sea urchin, cleavage in the amphibian represents a time of extremely short cell cycles and rapid cell division. However, the egg of the amphibian has over 1,000 times the volume of the tiny sea urchin gamete. What effect could gene expression from one or a few nuclei have on the synthetic events in an egg of such magnitude? The answer would seem to be very little. It was mentioned earlier that RNA synthesis during amphibian cleavage is very low (even on a per-nucleus basis). Unlike the sea urchin, which produces a large number of new mRNA templates during cleavage (particularly for histones), the amphibian egg hardly seems to make an attempt at transcriptional-level control. This is best illustrated for the histones, which are needed in tremendous quantities during cleavage in order to construct the chromatin for the approximately 20,000 nuclei that appear in a matter of a few hours. Unlike the sea urchin, there is no evidence of the synthesis of histone mRNAs during cleavage. Instead, the amphibian egg draws on its stores, which include both histones and histone mRNAs that had been made by the oocyte prior to fertilization. Protein synthesis is high during amphibian cleavage, and it operates almost exclusively on maternal mRNA templates, i.e., by translational-level control. Furthermore, transla-

tional control appears to operate in a selective, i.e., qualitative, manner. If one examines the types of histones being synthesized during the periods of late oogenesis, maturation, and early development, one finds considerable variations in the ratio of synthesis of H1 to those of the other four classes. During the latter stages of oogenesis, all histone species are made at roughly equal proportions. However, during oocyte maturation and embryonic cleavage, H1 synthesis is very low, returning to normal levels by the late blastula stage. There clearly occurs a change in the relative synthesis of this particular protein. However, no corresponding changes occur in the level of mRNA. If one extracts the RNA from oocytes, eggs, and cleaving embryos and uses it to direct the synthesis of proteins in a cell-free system, the same relative amounts of each histone class are produced by the mRNA preparations of all stages. Despite the fact that mRNA levels remain constant, the cell appears able to turn on and off the use of the H1 mRNA template, an excellent example of translational-level control.

Whereas the early development of the sea urchin is characterized by the lack of appearance of new proteins, that of the mouse (and rabbit) is accompanied by major shifts in the species of proteins being synthesized. In one study in which approximately 600 labeled spots could be detected in autoradiographs made from the gels, 36 proteins were found to be expressed during only a portion of the period of early mouse development (to the blastocyst stage). Of these 36 *stage-specific proteins*, 18 represented proteins whose synthesis ceased during early development, and the other 18 were proteins whose synthesis (in detectable amounts) began at some point after fertilization (Fig. 8.21). For all but 2 of these 36 proteins, cessation or initiation of synthesis occurred prior to the third cleavage. Once the four-cell stage is reached, very few additional changes occur in the synthetic pattern despite the prominent morphogenetic activities involved in blastocyst formation. In keeping with the early involvement of the mammalian genome (Section 8.3.B), these changing translational profiles presumably reflect the input of new species of RNA, i.e., transcriptional control, rather than the unfolding of an extensive maternal program, i.e., translational control. Further studies with embryos whose transcriptional activities are blocked, preferably with α-amanitin, should answer this important question.

Can the major differences among these groups be summarized? The risks of misinterpretation are great, but the following generalities may have some validity. At one extreme, there is the amphibian egg with its large size and rapid divisions. Early development of these eggs is controlled almost exclusively at the translational level. At the other extreme is the mammal with its relatively small egg and very slow early development. These embryos appear to operate to a considerable degree at the transcriptional level. Between these two types of animal lies the sea urchin, having an egg of small size that undergoes rapid early development. In this case, the importance of both transcriptional and translational control mechanisms are readily demonstrated.

D. Sequence analysis of sea urchin embryonic transcripts

Over the past several years a series of studies, primarily in the laboratory of Roy Britten and Eric Davidson, has provided an extensive profile of the similarities and differences among sequences present in cytoplasmic and nuclear RNAs at different stages of sea urchin development and in different tissues of the adult. In many ways these studies have raised more questions than they have answered and have made us realize that a simplistic picture of the role of gene expression in embryonic development is not likely to be a correct one. In this section we will be discussing the sequence complexity of two populations of RNA, the mRNAs

a

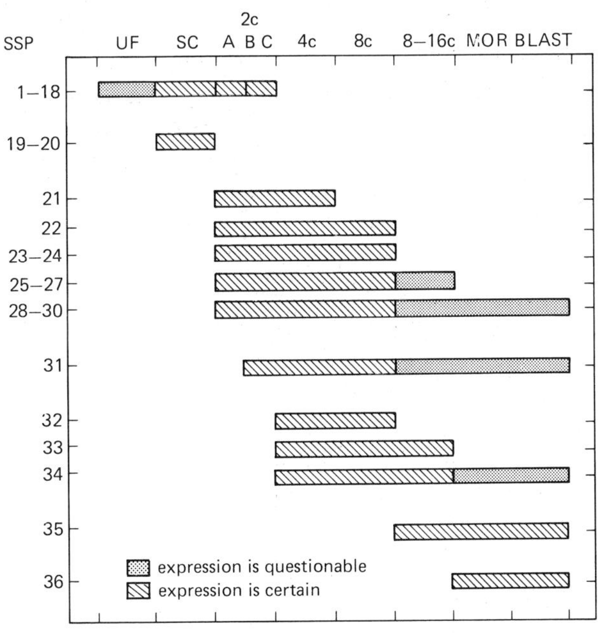

b

Figure 8.21 *a* Pattern of radioactive spots obtained from an autoradiogram after two-dimensional gel electrophoresis of a lysate prepared from over 100 [^{35}S]methionine-labeled single-cell zygotes. Isoelectric focusing was carried out in the horizontal direction (*basic end on left*), and SDS gel electrophoresis was carried out in the vertical direction with the origin at the top. Upward-pointing arrows identify stage-specific proteins (SSPs) that are not found in gels of earlier embryos. Downward-pointing arrows identify SSPs that are decreasing according to the key in *b*. *b* Map of the time during which the expression of 36 SSPs occurs as defined in two-dimensional gels. Stages are given along the top. The two-cell stage has been split into three protein-synthetic periods labeled *A*, *B*, and *C*, during which the first 18 SSPs were differentially decreasing in their rates of synthesis. Similarly, the initial appearance of SSP 31 appears to subdivide the two-cell stage. [*From J. Levinson, P. Goodfellow, M. Vadeboncoeur, and H. McDevitt,* Proc. Natl. Acad. Sci. U.S., **75**:*3332* (*1978*).]

that are isolated from cytoplasmic polyribosomes and the nuclear RNAs (hnRNAs) that are prepared from isolated nuclei. Analysis of the mRNA population provides us with an overall look at the variety of proteins being synthesized in the embryo (or adult tissue). However, analysis of the nuclear RNA population provides us with an overall look at the variety of DNA sequences being transcribed in the embryo (or adult tissue) without regard to their ultimate function in the cell. The experimentation that led to certain of the following observations is provided in detail in the appendix to this chapter. The results of these studies will be summarized in this section.

The mRNAs of sea urchin cells can be divided into two classes based on their frequency, i.e., the number of copies per cell, as measured by their rates of reaction with complementary DNA sequences. There is a *complex class* that comprises approximately 90 percent of the diversity but only about 5 to 10 percent of the actual mass of mRNA. It is estimated that at the gastrula stage, members of the complex message class are present at only one or a few copies per cell (assuming that all cells of the embryo contain the same population of mRNAs). The other group, referred to as the *prevalent* (or *abundant*) *class* comprises 90 to 95 percent of the mRNA bulk but represents only a few hundred different species. The prevalent mRNAs are present at a concentration at least 100 times that of the complex species. The diversity of the two mRNA classes is revealed in different ways. Most DNA–RNA hybridization experiments focus on the complex class, while the diversity of the prevalent messages is best analyzed by techniques, such as two-dimensional electrophoresis, designed to fractionate newly synthesized proteins.

Although morphological complexity increases with development, the variety of mRNA sequences actually decreases with advancing embryonic age (Fig. 8.22). The unfertilized egg is estimated to contain a population of maternal

Figure 8.22 Diagram illustrating the loss of mRNA sequences during early development in the sea urchin. In the series of experiments whose results are summarized here, RNAs from various stages were hybridized in excess with labeled single-copy DNA representing a particular fraction of the genome, namely, that fraction complementary to the total RNA of the unfertilized egg. To obtain this [^{3}H]DNA fraction, termed *oDNA,* the total single-copy [^{3}H]DNA fraction was incubated with unfertilized-egg RNA in excess and the [^{3}H]DNA of the hybrid subsequently isolated and recycled for further purification. The [^{3}H]oDNA was then incubated with excess amounts of (1) total-ovary RNA (○), (2) unfertilized-egg RNA or total cytoplasmic RNAs from embryos of different stages (●), or (3) mRNAs from embryos of different stages (■). The left ordinate represents the saturating hybridization value for each RNA preparation incubated with [^{3}H]oDNA, plotted as the percentage of the reaction of [^{3}H]oDNA bound with RNA of unfertilized eggs (assigned a value of 100 percent). The right ordinate represents the corresponding RNA complexity. [*From B. R. Hough-Evans, B. J. Wold, S. G. Ernst, R. J. Britten, and E. H. Davidson,* Dev. Biol., **60**:272 *(1977).*]

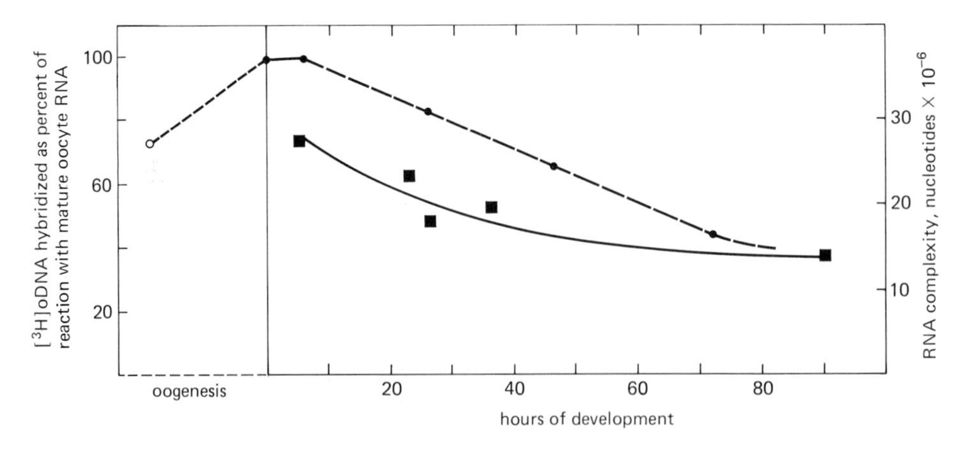

mRNA that is synthesized from approximately 6 percent of the single-copy genome (equivalent to 3.7×10^7 base pairs of DNA).[1] Comparable values for the blastula, gastrula, and pluteus are 4.5, 2.8, and 2.2 percent, respectively. Adult cells, such as that of the intestine, contain an even more restricted population of polysomal mRNAs, one equivalent to approximately 1 percent of the single-copy genome. The reason for the decreasing mRNA complexity is not at all apparent; there is no reason to believe that the events occurring prior to blastula formation require a greater variety of newly synthesized proteins than the events of gastrulation, or pluteus formation, or metamorphosis, or the maintenance of complex adult tissues.

Further analysis by DNA-RNA hybridization indicates not only that the later stages have fewer mRNA species than the unfertilized egg, but also that the species they do have appear to be simply a subset of the maternal sequences. For example, when the sequences present among the gastrula mRNAs are compared with those of the unfertilized egg (see Fig. 8.28), the entire gastrula single-copy sequence set is found to be contained at the earlier stage. Similarly, all the sequences in the smaller population of pluteus mRNAs are already present by the gastrula stage. We can conclude that mRNA sequences present at earlier stages are lost from the embryo, but that new sequences are not gained in detectable amounts. There is an important qualification to these results. Since the prevalent class of mRNAs represents such a small fraction of the diversity, one cannot say with assurance that this decrease in message complexity applies to the more abundant mRNAs as well. It is possible that such changes are escaping detection. The fact that some new proteins appear by the gastrula stage when studied by two-dimensional electrophoresis suggests that at least a few new stage-specific prevalent mRNAs appear in the polyribosomes as development proceeds.

These findings have come as some surprise. It has been assumed for many years that development is accompanied by the selective activation of new sets of genes, a phenomenon that should result in the appearance of many new species of mRNA. Instead we seem to find the opposite situation (at least with respect to the complex class of messages). In one recent theory (see Caplan and Ordahl, 1978) it is proposed that the decreasing variety of mRNAs is correlated with the increasing restriction in developmental potential. We have pointed out in previous chapters that determination occurs in a stepwise manner gradually limiting the prospective potency of an embryonic cell. In this latter theory it is suggested that embryonic determination is accompanied by the irreversible repression of DNA sequences, thereby leading to a restricted transcriptional potential. However, as we will see later, the shrinking population of mRNAs may not be reflecting a comparable shrinkage in transcriptional diversity.

Even though virtually all the mRNA sequences of the gastrula are found within the unfertilized egg, the gastrula mRNAs do not represent a group of molecules that are simply holdovers from the maternal population. Various kinetic measurements indicate that the half-life of sea urchin embryonic mRNAs (both

[1] In this discussion, all the RNAs utilized in the hybridization experiments are derived from either isolated polyribosomes or isolated nuclei. They are termed *mRNA* or *nuclear RNA*, respectively. The RNA of the unfertilized egg, i.e., the maternal RNA, is the single exception to this statement; it represents the total RNA of the egg cell. Since there are very few polyribosomes and only a single small pronucleus in the unfertilized egg, it would be virtually impossible to isolate mRNA or nuclear RNA populations even if this were desirable. Since we know that at least 73 percent of the maternal RNA (ignoring the rRNA and tRNA) does find its way into polyribosomes during cleavage, most of it is clearly *maternal mRNA*. Although it may be somewhat of an oversimplification, we will consider the heterogeneous RNA of the unfertilized egg to be wholly messenger in nature. An argument that supports this conclusion based largely on kinetic data can be found in G. Galau et al. (1976).

prevalent and complex classes) is approximately 5 to 6 hours. Therefore, the population of mRNAs in the polyribosomes of the gastrula are, for the most part, molecules that have been recently synthesized. Beginning during early cleavage, the mRNAs of the embryo are gradually destroyed. Some of the species of the maternal sequence set, such as those coding for histones, are resynthesized while others disappear without being replaced. It is this latter group of messages that accounts for the decreasing sequence complexity of the mRNA population. This type of turnover illustrates the dynamic steady-state nature of cellular activity, in which the numbers of molecules present at any given time represent a balance between synthesis and degradation.

The sequence properties of the nuclear RNAs are very different from those of the messenger RNAs. The nuclear RNAs are transcribed from a much greater percentage of the single-copy genome. It is estimated that the single-copy sequence set of nuclear RNAs from either embryos or adult tissues is equivalent to approximately 1.8×10^8 base pairs of DNA, or 34 percent of the single-copy genome.

When the nature of the sequences present in various nuclear RNA and mRNA populations is compared, a surprising result emerges. The nuclear RNAs, *regardless of their source,* appear to contain all the mRNA sequences that have been detected. For example, essentially all the mRNA sequences of the blastula appear to be present within the nuclear RNA of intestinal cells, even though the mRNA population of this adult organ contains only about 20 percent of the sequence complexity of the blastula mRNAs. Similarly, the single-copy sequence sets of the blastula and gastrula nuclear RNA are both equivalent to approximately one-third of the nonrepeated DNA, and they are indistinguishable from one another, even though approximately 45 percent of the mRNAs of the blastula are no longer present in the pluteus. It would appear that the entire set of structural genes that codes for embryonic mRNAs in the sea urchin is transcribed in all cells regardless of whether or not the corresponding mRNAs actually make their way into cytoplasmic polyribosomes.

Based on what we know of gene expression, we have come to regard the process of differentiation as being primarily under transcriptional control, with a bit of translational control here or there. However, if transcriptional control is of overriding importance, and if mRNAs are derived from nuclear RNA precursors, then we should see equivalent changes in nuclear RNA and mRNA populations. Instead it appears that the mRNA sequence set is regulated in the absence of equivalent regulation in the nuclear sequence set. Why, for example, should the cells of the intestine transcribe structural genes required for gastrulation, particularly when these sequences never seem to leave the nucleus? There are explanations to account for these findings, although it remains to be seen if they have any validity. At one extreme, it is possible that developmental events are not regulated at the transcriptional level at all, but rather are under posttranscriptional control. One might refer to this as the "extreme processing model." All cells would synthesize the same all-encompassing set of nuclear RNAs from which specific sets of mRNAs would be carved. In this model, the selection of polyribosomal mRNAs would depend on differential processing steps occurring in different types of cells. At the other extreme, there is an explanation compatible with the data that allows us to salvage a model of transcriptional control. It is possible that there are two classes of nuclear RNA molecules and that the synthesis of each is regulated independently. One class would contain mRNA precursors, i.e., molecules from which polyribosomal mRNAs specific for that cell would be derived. This group of molecules would remain under transcriptional control and

its sequence diversity would vary among different types of cells. The other (more diverse) class would contain RNAs of unknown function whose sequence diversity would be essentially constant in all cells. Hopefully answers to this and other questions posed in this section will be forthcoming.

E. Preformed proteins

In Section 8.3.D, the relative roles of preformed and embryonic RNAs were discussed. The embryo begins development with another preformed macromolecular store, i.e., the proteins synthesized during oogenesis and inherited by the embryo for use after fertilization. Some of these proteins are presumed to be morphogenetic substances involved in the activation of the embryonic genome. Information on this subject is relatively scarce, but there is one well-studied example uncovered in vertebrates, in this case the Mexican axolotl. The axolotl is the most genetically defined amphibian available, one in which a wide variety of mutants has been isolated (see the symposium in the *American Zoologist,* **18**). One of these, the *o* mutation (*o* for ova deficient), is a maternal effect mutation. Females homozygous for this gene produce eggs that invariably arrest at gastrulation, regardless of the genetic composition of the fertilizing sperm. As in the *Drosophila* examples (Section 8.8.D), eggs of homozygous mutant females contain some type of cytoplasmic deficiency that leads to an early developmental arrest. This deficiency can be corrected by injecting eggs of *o/o* females with cytoplasm from normal eggs. The corrective component is found within the germinal vesicle of large oocytes, i.e., within the nucleoplasm, or in the egg cytoplasm following germinal-vesicle breakdown.

The substance capable of correcting the *o* deficiency, presumably the product of the normal *o* allele, appears to be a protein that is synthesized during the latter stages of oogenesis and functions in the genetic activation that occurs in amphibian embryos prior to gastrulation. Autoradiographic studies of RNA synthesis in the axolotl embryo indicate that the level of transcription is very low up to about the midblastula stage, when a sweeping activation of RNA synthesis occurs. In contrast, the eggs of *o/o* females completely fail to undergo this activation; unless rescued by the injection of normal cytoplasm, these deficient embryos continue to synthesize little or no RNA. The *o* substance can be presumed to be some type of regulatory substance that acts, directly or indirectly, on the embryonic nuclei leading to the synthesis of RNAs required for gastrulation.

Additional information on the *o* gene has been obtained from nuclear-transplantation experiments. It has been found that nuclei transplanted from *normal* middle to late blastulas into enucleated recipient *mutant* eggs are capable of supporting normal development. We can conclude that the *o* substance of the normal egg has already acted by the mid to late blastula stage in such a way that the nuclei have no further need for o^+ cytoplasm in order to support development. We can further conclude that the nuclear alteration induced by the *o* substance is a stable, inheritable change, i.e., one that is passed on from nucleus to nucleus at mitosis. If this were not the case, a single "activated" nucleus from a normal embryo could not support the development of a mutant egg into a larva containing thousands of nuclei. This result is somewhat surprising considering the ability of egg cytoplasm to reprogram transplanted nuclei (Section 8.2). In this case, as in *Xenopus,* the late blastula nucleus stops making RNA and reverts to the synthetic state characteristic of a newly fertilized egg. However, despite the lack of RNA synthesis, we know that the nuclear alteration must be stable in some way; otherwise this nucleus could not lead to the formation of a normal larva in the further absence of the *o* substance.

In contrast to these results, if a normal *early* blastula nucleus is transplanted into an enucleated mutant egg, its development arrests at gastrulation, just as it would have had it not received the transplant. In this latter case, the normal nucleus had not undergone alteration in its own cytoplasm prior to transplantation and, consequently, was of no value to the deficient egg. Transplantation experiments in the opposite direction, i.e., ones in which the nucleus from a mutant embryo is placed into the cytoplasm of a normal egg, also have been performed with interesting results. When nuclei from a mutant early blastula are transplanted, the recipient develops normally, which indicates the ability of the mutant nuclei to respond to normal cytoplasm. However, if nuclei from a mutant late blastula are transplanted, the recipient arrests at gastrulation. It would appear that once the nucleus passes the midblastula stage without undergoing normal activation, it receives some type of irreversible damage that prevents it from responding to the *o* substance in the recipient egg and supporting normal development.

4. Appendix

A. Nucleic acid hybridization

Results obtained by nucleic acid hybridization (either DNA to DNA or DNA to RNA) are discussed in several places in this book. The technique is complex, and so many variations can be used that a discussion of these methods rapidly moves beyond the limits of an introductory textbook. However, the technique is the only one available to analyze populations of nucleic acids from the standpoint of their nucleotide sequences, from which their entire informational content is derived. Without some understanding of the potential that the technique offers in the study of nucleic acids, one cannot appreciate our current knowledge of the utilization of molecular information during development or any other event.

The underlying principle of all hybridization experiments is that single-stranded nucleic acids of complementary sequence will hybridize to one another to form a double-stranded molecule held together by hydrogen bonds. Hybridization can occur between two single-stranded DNA molecules or between a single-stranded DNA molecule and an RNA molecule (see Section 2.2 for a general discussion).

Certain preparatory steps are required in all hybridization experiments. The nucleic acids to be used must first be purified. DNA must be converted to a usable form, which requires denaturing the DNA duplex into a single-stranded form and usually breaking up the DNA into uniform-sized fragments. The nucleic acids to be hybridized must be mixed together at sufficient concentration and given sufficient time of incubation for the complements to find one another. The temperature and the ionic composition of the medium are of critical importance in determining how closely two molecules must be complementary to one another before they will reanneal. The higher the temperature and the lower the salt, the closer the two strands must complement each other before they will remain attached; i.e., the percentage of possible hydrogen bonds formed between the two strands must be greater. In most experiments, two very different populations of nucleic acids are brought together to hybridize. In these cases, one population is radioactively labeled and present at very low concentration (the *tracer molecules*), while the other population is unlabeled and present at far greater concentration (the *driver molecules*). After the desired incubation period, the hybridized nucleic acids—whether they are DNA–DNA hybrids or DNA–RNA hybrids—must be distinguished from the nonhybridized fraction, so that the percentage of radioactivity in each can be determined. In most cases, this separation of hybrid from nonhybrid is carried out by hydroxyapatite chromatography (discussed

later), although in other cases, the determination is made using an enzyme, such as S1 nuclease, which selectively degrades single-stranded DNA but not DNA present in a hybrid state.

The reannealing curve in Fig. 2.3 illustrates the basic hybridization procedure. Double-stranded DNA is converted to single-stranded molecules by heat, and the solution is cooled to a temperature (for example, 60°C) at which reannealing can proceed. The longer the period of time (or the greater the Cot, see Fig. 2.3), the greater the percentage of DNA in a double-stranded state. Those sequences present at the greatest concentration reanneal more rapidly. In order to plot a reannealing curve, some method must be used to determine the percentage of the original DNA that has reannealed at various times of incubation. Reannealed, double-stranded DNA has different properties from nonreannealed, single-stranded DNA, and this difference is used for their separation. It has been found that a particular type of calcium phosphate, called *hydroxyapatite,* is capable of binding DNA. The hydroxyapatite crystals are packed into a column through which solutions of DNA can be passed. At low phosphate concentrations (for example, 0.03 *M*), both single- and double-stranded DNA binds to the hydroxyapatite. At intermediate phosphate concentrations (for example, 0.12 *M*), only double-stranded DNA binds, while single-stranded DNA passes through and can be collected. At high phosphate concentration (for example, 0.5 *M*), neither DNA will bind. To determine the percentage of DNA that has reannealed after a given period of time, the solution is passed through the column in 0.12 *M* phosphate. The double-stranded molecules are bound, while the single-stranded DNA comes through and is measured. The phosphate concentration in the column is then raised to 0.5 *M*, causing the bound DNA to be eluted for measurement. The reannealing curve in Fig. 2.3 is obtained in this manner. The rate at which a given DNA sequence reanneals reflects its concentration in the solution and thus its frequency within the genome. Reannealing curves obtained from eukaryotic DNA have a complex shape reflecting the variation in frequency at which different sequences are present within the genome.

QUESTION: Are mRNAs transcribed from repetitive or single-copy (nonrepeated) DNA sequences? Before one can begin to ask what type of DNA sequence is responsible for the coding of a specific protein, one must be able to obtain large quantities of the nucleic acid that carries the code for that polypeptide. This generally requires that the synthetic activities of a cell are dominated by the production of one particular protein species. One such cell that fits this requirement is the hemoglobin-producing erythroblast. From the polysomes of this cell, a small quantity of hemoglobin mRNA can be isolated. To convert this small quantity of unlabeled RNA into a larger quantity of highly radioactive DNA with the code for hemoglobin, we take advantage of a special enzyme that can be prepared from certain cancer-causing viruses. This enzyme, *reverse transcriptase,* uses RNA as a template to synthesize a complementary copy of DNA, hence the name.

Once the specific labeled DNA (a *cDNA*) is available, a small amount of it is mixed with a large excess-of-total unlabeled DNA; the mixture is heated to separate the strands of the double helix and is then allowed to hybridize. The DNA reanneals to form a curve of the type shown in Fig. 8.23. In this case, however, one sequence of all possible sequences is represented by a few radioactive molecules. If this sequence is identical with a sequence in the repetitive DNA, it should be found in the rapidly reannealing fraction (left part of the curve). If it is identical with a sequence in the single-copy fraction, radioactivity should appear in the hybrid fraction very slowly (right part of the curve). Analysis of several

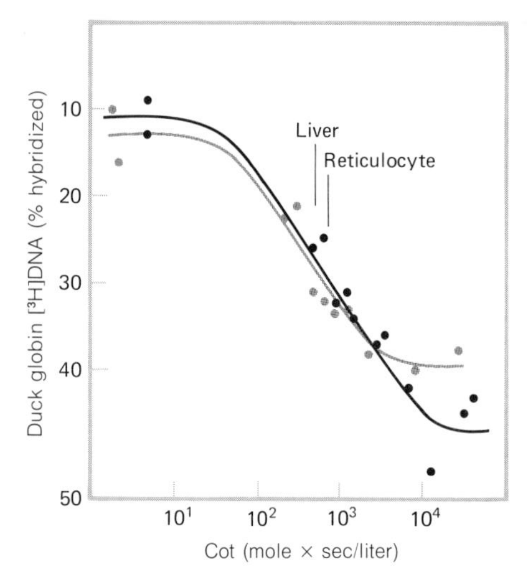

Figure 8.23 Kinetics of hybridization of single-stranded [^{3}H]duck-globin cDNA with a large excess of unlabeled DNA extracted from either globin-producing cells (duck reticulocytes) or nonproducing cells (duck liver). In both cases, the labeled sequences hybridize with kinetics which suggest that their complements within the unlabeled DNA fraction are present in one or, at most, a few copies. Consequently, these results provide no evidence for the amplification of the globin sequence in reticulocyte DNA or the deletion of the globin sequence in liver DNA. [*From S. Packman, H. Aviv, J. Ross, and P. Leder,* Biochem. Biophys. Res. Comm., **49:***816 (1972).*]

mRNAs, including that for the polypeptide chains of hemoglobin, suggests that they are transcribed from genes whose sequence is not repeated in the haploid genome. Messenger RNAs for the histones, however, hybridize with DNA sequences that are repeated up to a few hundred times (depending on the species) and thus are assumed to be transcribed from moderately repeated genes.

QUESTION: How great a variety of mRNAs and/or nuclear RNAs are present at different stages of development in the sea urchin? Because experiments with repetitive and single-copy DNA are handled separately, the question becomes: How great is the variety of RNAs transcribed from each DNA fraction? We will begin with an examination of the diversity of transcripts complementary to single-copy DNA sequences.

Before one can hybridize RNA to single-copy DNA, this fraction of the total DNA must be isolated in a pure state uncontaminated by the presence of repeated sequences. This is accomplished by allowing the total population of DNA fragments to reanneal long enough for all the repetitive sequences to form hybrids, but with most of the single-copy population still present in single-stranded fragments. When the reaction has proceeded to this point, the reaction mixture is passed through a column of hydroxyapatite. The double-stranded (reannealed, repetitive) DNA sequences bind to the column, while a nearly pure fraction of single-copy sequences is collected as it drips through. The single-copy DNA can be further purified by an additional reannealing cycle, and its purity can be tested by allowing it to hybridize to excess total DNA.

In the type of experiment shown in Fig. 8.24, the purified single-copy DNA is combined with a great excess of the RNA to be tested and the two nucleic acids are allowed to hybridize. To give the investigator a simple means of following the tracer DNA sequences, radioactively labeled DNA is used. In earlier studies, in vivo labeled DNA was obtained by incubating embryos in [^{3}H]thymidine for its incorporation into DNA during replication. In more recent studies, the DNA is labeled to a much higher specific activity (counts per minute per microgram of DNA) by an in vitro procedure termed *nick translation*. In this technique, DNase I is used to introduce single-stranded nicks, which then serve as initiation sites for bacterial DNA polymerase I. Once the DNA and RNA are mixed, the labeled

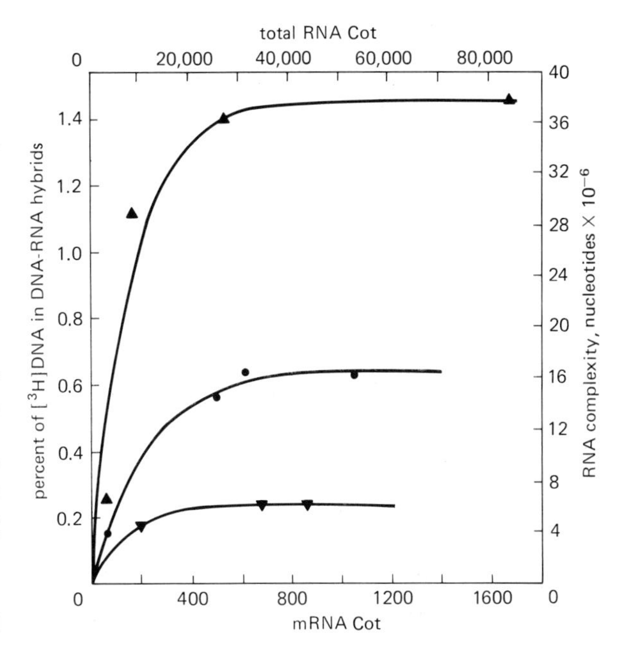

Figure 8.24 Determination of the sequence complexities of three mRNA preparations as measured by their hybridization in excess to single-copy [^{3}H]DNA. Intestine mRNA (▼), gastrula mRNA (●), total-unfertilized-egg RNA (▲). As development proceeds (summarized in Fig. 8.22), there is a decreasing ability of mRNA preparations to hybridize to single-copy DNA sequences. This decrease reflects the progressive loss of mRNA sequences with development. [*From G. A. Galau, W. H. Klein, M. M. Davis, B. J. Wold, R. J. Britten, and E. H. Davidson,* Cell, **7**:*490 (1976).*]

DNA sequences can be divided into two groups, those to which the RNAs can bind and those to which they cannot. Presumably those DNA sequences which form DNA–RNA hybrids were the ones that were active in the synthesis of this species of RNA. Even though there is only one copy of each DNA sequence per haploid amount of DNA and some of the RNAs may be quite rare, the RNA is in such great excess that there are sufficient collisions to drive the reaction essentially to completion; all DNA sequences that have complements will hybridize. The course of hybridization of single-copy DNA with various RNA preparations is shown in Fig. 8.24. In most cases, the DNA sequences are present at such a low concentration that there is very little DNA–DNA self-reaction. However, if the level of DNA–RNA hybridization also is relatively low, as in the case of the reaction with intestinal mRNA (Fig. 8.24), then the DNA–DNA hybrid content of the overall hybrid fraction must be determined. This measurement is made utilizing a ribonuclease that destroys the RNA in a DNA–RNA hybrid, thereby converting this hybrid DNA to the single-stranded state. The double-stranded DNA that remains is measured and substracted from the total hybrid (DNA–DNA and DNA–RNA) radioactivity. Since only one of the two strands (the sense strand) of the DNA is transcribed, only one is complementary to the RNA produced. Therefore, the saturation values shown in Fig. 8.24 can be doubled to arrive at the percentage of the single-copy genome responsible for the synthesis of a given RNA population. In some experiments, the saturation values also must be corrected for a fraction of the DNA sequences that becomes too small to hybridize to an RNA regardless of its complementarity.

The curves shown in Fig. 8.24 follow the course of hybridization to single-copy DNA of mRNA populations isolated from unfertilized eggs, gastrulas, or adult intestinal cells. In these experiments, total RNA from unfertilized eggs is used (see footnote, Section 8.3.D), while polyribosomal RNA from other stages is used. Approximately 4 percent of the RNA of the polyribosomes is mRNA. It is apparent

that as development of the sea urchin proceeds, the complexity of the mRNA populations (as indicated by the percentage of the DNA driven into hybrids) sharply decreases. The rate at which hybridization occurs provides information on the concentration of the mRNA sequences. If all RNA sequences were present at equal concentration, a simple shaped curve with a single rate constant would be obtained as in the reannealing of single-copy DNA, which is composed of sequences present at equal concentration. However, in the case of reaction with polyribosomal RNA, the kinetics of hybridization indicate that over 90 percent of the mRNA is accounting for very little hybridization; these sequences are few in number, but each is present at high concentration. This is the prevalent class of mRNA (Section 8.3.C). Kinetic analysis indicates that the reaction is being driven by a complex class of mRNA containing over 90 percent of the mRNA diversity but representing less than 10 percent of the mRNA mass. When similar types of experiments are carried out using nuclear RNA rather than polysomal RNA, much higher levels of hybridization are found (Fig. 8.25). Furthermore, it makes little difference whether the nuclear RNA is taken from blastulas, plutei, or adult tissues; it hybridizes to approximately the same degree.

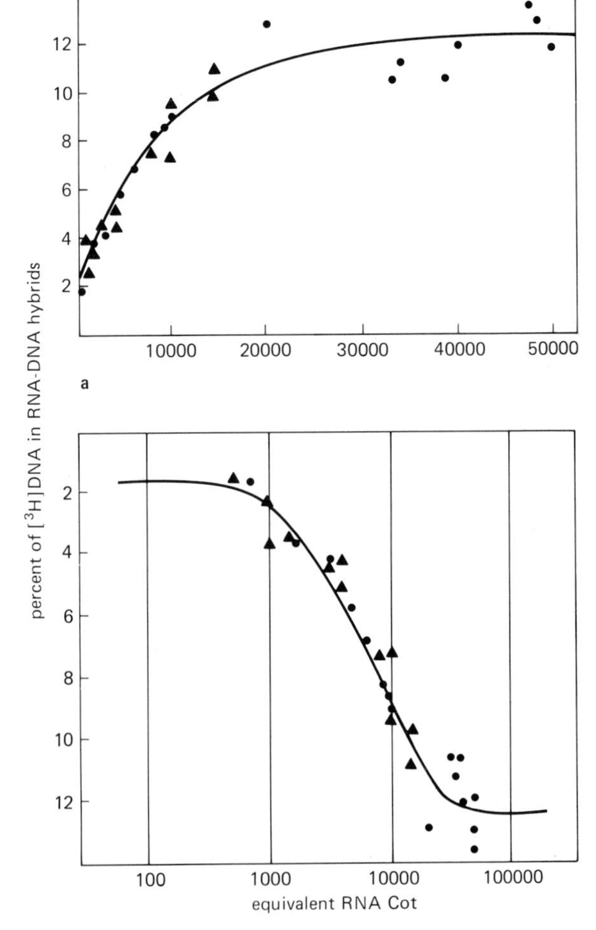

Figure 8.25 Determination of the sequence complexity of a nuclear RNA preparation as measured by its ability to hybridize in excess to single-copy [^{3}H]DNA. The two curves simply represent two methods of presenting the same data. The two types of symbols represent two different preparations of gastrula nuclear RNA. It is evident that gastrula nuclear RNA is capable of hybridizing to a much greater percentage of the single copy DNA sequences (12.4 percent) than is gastrula mRNA (0.6 percent, from previous figure). Nuclear RNA taken from various embryonic stages or adult tissues has a similar hybridization capacity; its sequence complexity is at least 1.74×10^8 nucleotides. Calculations of the rate of hybridization indicate that 2.5 to 3.1 percent of the nuclear-RNA preparation is actually driving the reaction, i.e., accounts for the vast complexity of the sequences. The remainder of the RNA represents sequences present at much greater abundance. These abundant sequences would include ribosomal RNA, mRNA contaminants, and some abundant nuclear species. This contamination does not affect the complexity measurement, but simply reduces the measured driver concentration. [*From B. R. Hough, M. J. Smith, R. J. Britten, and E. H. Davidson,* Cell, **5**:293 *(1975)*.]

Up to this point we have restricted the discussion to RNA sequences transcribed from nonrepeated DNA sequences. The repeated fraction of the genome also is represented among embryonic RNAs, particularly those found in the nucleus. Analysis of nuclear RNAs indicates that individual molecules contain portions transcribed from both repeated and nonrepeated DNA sequences. This is not unexpected considering the large size of the hnRNA molecules and the interspersed organization of the genome (Section 2.2). Further analysis suggests that most, if not all, of the repeated sequence families are represented among the nuclear RNAs. Unfortunately, the role of most of the repeated DNA sequences remains obscure.

Experiments with repetitive DNA sequences involve a considerably different technology. The results of a large number of earlier experiments are summarized in Eric Davidson's, "Gene Activity in Early Development," 2d ed. (1976). In this section we will briefly describe a more recent approach to determining the degree to which repetitive DNA sequences are represented in RNA populations. The actual hybridization reaction is similar to that just described. A large excess of RNA is incubated with a radioactively labeled tracer DNA. In this case, however, the labeled DNA consists of homogeneous pieces of DNA, i.e., strands of DNA containing the exact same repetitive DNA sequence. In order to obtain a homogeneous population of DNA molecules, sea urchin DNA was fragmented, combined with bacterial plasmid DNA, and cloned within bacterial cells. Using this procedure (described in R. H. Scheller et al., *Science,* **196:**197, 1977), nine different DNA fragments have been isolated, each consisting of a different repetitive sequence. As a group, these sequences are found at a reiteration frequency that ranges from approximately 20 to 1,000 times per haploid genome. The hybridization of nuclear RNA populations from blastulas and adult intestinal nuclei to three of these cloned tracer DNAs is shown in Fig. 8.26. Each section of the figure shows the hybridization to a different repetitive sequence.

In the experiments shown in Fig. 8.26, it is necessary not only to dissociate the two strands of the duplex prior to hybridization, but also to separate these strands from one another, a feat accomplished by an electrophoretic technique. It is essential that the incubation mixture contain a purified preparation of only one of the two strands of the cloned tracer; otherwise the DNA would self-anneal. Another difference in the conditions from that just described concerns the temperature and ionic strength utilized for incubation. Hybridizations with single-copy DNA are typically carried out at 60°C and 0.12 M phosphate buffer (or higher). The reactions shown in Fig. 8.26 were carried out at 0.5 M phosphate and 55°C, which reduces the stringency of the reaction, i.e., allows molecules with a higher degree of mismatching to bind to one another. As mentioned in Section 2.2, repetitive sequences exist in families whose members show some variation in nucleotide sequence. By lowering the stringency of hybridization, all members of a family can hybridize to a single representative DNA sequence.

Results with cloned repetitive DNA sequences tentatively suggest (1) that each and every family of repeated sequences is represented within the nuclear RNAs of all embryonic stages and all adult tissues, but (2) that the frequency with which a given sequence is represented varies markedly from stage to stage or tissue to tissue. This is clearly seen in the different rates of hybridization of blastula and intestinal-cell nuclear RNAs with a given repeated sequence. The existence of tissue-specific quantitative differences in the representation of repetitive sequences argues that their synthesis is under close regulation. Other experiments (not illustrated) indicate that both strands of a given repetitive sequence react with a preparation of nuclear RNA; i.e., both strands are

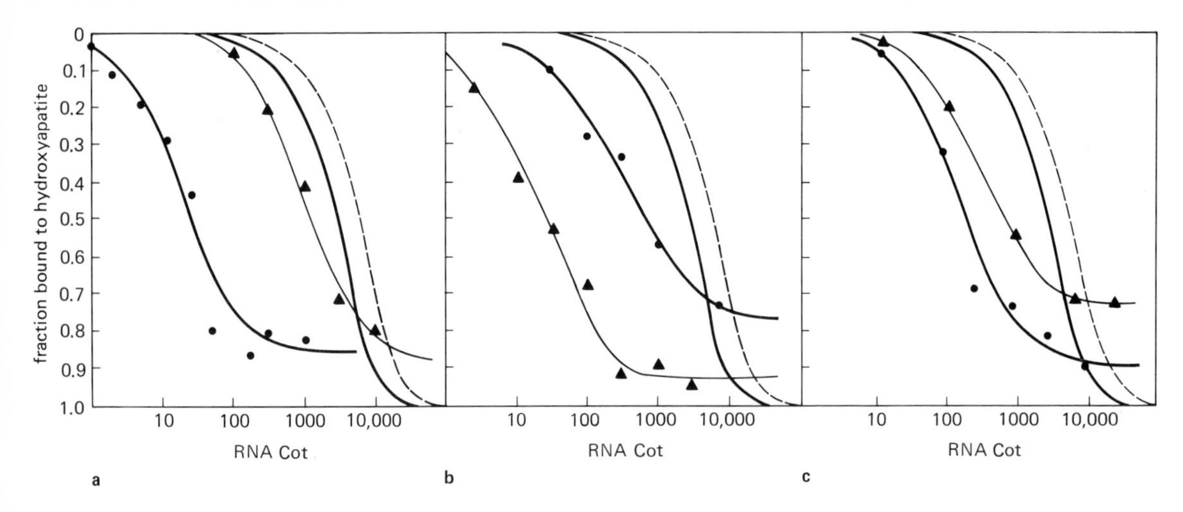

Figure 8.26 Each panel displays the kinetics of reaction between a labeled, strand-separated, cloned, repeated DNA sequence and excess nuclear RNA from intestine (●) and gastrula (▲). In each panel, the two right-hand curves are kinetic standards presented for purposes of comparison. (The right-hand solid line indicates the kinetics of the reaction of a *single-copy* [^{3}H]DNA tracer with excess intestine nuclear RNA under the same conditions, and the right-hand dashed line shows the reaction of the *single-copy* [^{3}H]DNA tracer with excess gastrula nuclear RNA.) Each of the cloned repeated sequence tracers reacts at a particular rate with each nuclear RNA. Comparison of the panels indicates that the set of repetitive sequences highly represented in intestine nuclear RNA is different from that highly represented in gastrula nuclear RNA. The quantitative patterns of repititive-sequence representation in RNA are specific to each cell type. [*From R. S. Scheller, F. D. Constantini, M. R. Kozlowski, R. J. Britten, and E. H. Davidson,* Cell, **15:***189 (1978).*]

transcribed. Since the same sequence is present in many places in the genome, this does not necessarily mean that both strands at a given site are transcribed (termed *symmetrical transcription*). Whereas nuclear RNAs react with these cloned tracers, polysomal RNAs do not, indicating the absence of these repetitive sequences in mRNAs. Interestingly, unfertilized egg total RNA does react with all nine cloned tracers, which raises the possibility that these repetitive sequences may be present in maternal RNA and subsequently removed after fertilization prior to assembly on polyribosomes.

QUESTION: To what extent are populations of mRNA and/or nuclear RNAs present at different embryonic stages or among different adult tissues similar to one another? This question, unlike the previous ones, asks for a comparison between two or more populations of RNA. In the following paragraphs we will briefly consider a few experiments that have been carried out to answer this type of question.

It is apparent from Fig. 8.24 that mRNA complexity decreases as development proceeds, but this type of experiment does not inform us about the nature of the mRNA populations. We will begin with the question: To what extent are the mRNAs of the gastrula similar to the RNAs of the unfertilized egg? This question has been answered by isolating those *DNA* sequences which are complementary to gastrula mRNA and then using this particular set of DNA sequences in additional hybridization experiments with unfertilized egg RNA. In order to carry out this experiment, single-stranded, single-copy [^{3}H]DNA is incubated

with gastrula mRNA and the DNA–RNA hybrids are separated from the unhybridized DNA on hydroxyapatite. Approximately 1.1 percent of the single-copy DNA hybridizes with the gastrula messages; this DNA represents the *gastrula mDNA*. The remainder of the DNA, which consists of those sequences absent from the gastrula mRNA, is referred to as *gastrula null mDNA*. The RNA is digested from the mDNA-RNA hybrid and the DNA fragments of both the mDNA and the null mDNA are further purified by another round of hybridization with gastrula mRNA. The results of further hybridization reactions with gastrula mDNA and null mDNA are shown in Fig. 8.27*a*. When gastrula mDNA is tested against gastrula mRNA, 57 percent of the mDNA fragments (equivalent to 1.7×10^7 nucleo-

Figure 8.27 *a* Hybridization of labeled gastrula mDNA and null mDNA with four different preparations of gastrula mRNA. The gastrula mDNA reacts with gastrula mRNA at a rate similar to that using total single-copy DNA (Fig. 8.24), indicating that the same complex-class mRNAs that react with unfractionated single-copy [^{3}H]DNA react with the [^{3}H]mDNA. These results indicate that the DNA sequences complementary to gastrula mRNA have been purified 46-fold compared with total single-copy [^{3}H]DNA. The lower figure shows the absence of detectable reaction between the null mDNA and the gastrula mRNA, indicating the successful removal of DNA sequences whose transcripts are present within gastrula polyribosomes. *b* Hybridization of labeled gastrula mDNA and null mDNA with total-unfertilized-egg RNA (▲) or gastrula mRNA (*thinner line showing same results as in* a). It appears that gastrula mDNA reacts as well with unfertilized-egg RNA as it does with gastrula mRNA, indicating that essentially all the mRNA sequences of the gastrula polysomes are already present in the maternal message population of the unfertilized egg; there is no appearance of large numbers of new mRNA sequences during early development. In contrast, the comparison between unfertilized-egg RNA and gastrula mRNA to bind to gastrula null mDNA is striking, indicating that a large variety of cytoplasmic RNA sequences are lost between the time of fertilization and gastrulation. In fact, a greater complexity of sequences is lost (20×10^6 nucleotides) than is still present (17×10^6 nucleotides) in the gastrula mRNA. [*From G. A. Galau et al.*, Cell, **7**:*496* (*1976*).]

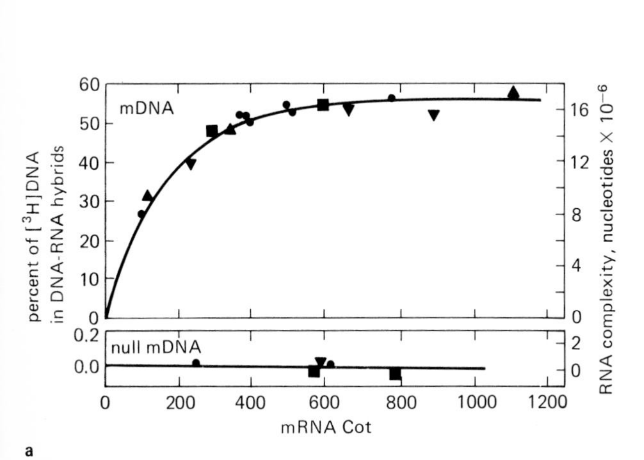

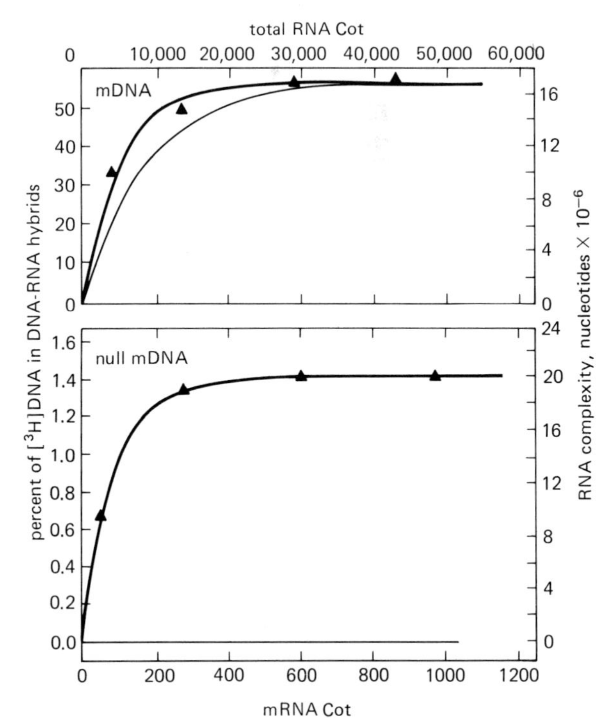

tides) form hybrids; after corrections are made for DNA fragments that are no longer reactable, this represents a 46-fold purification over a preparation of *total* single-copy DNA. When gastrula mDNA is incubated with a large excess of unfertilized egg RNA (Fig. 8.27*b*), it hybridizes to approximately the same extent as with gastrula mRNA. This finding suggests that the entire population of mRNA sequences (the single-copy sequence set) of the gastrula is already present in the egg at the time of fertilization. Approximately 45 percent of the unfertilized egg RNA sequences (equivalent 1.7×10^7 nucleotides) reacts with gastrula mDNA. As expected, the remainder of the RNA sequences of the unfertilized egg (55 percent, or 2.0×10^7 nucleotides) reacts with gastrula null mDNA. The results of a wide variety of experiments with gastrula mDNA and null mDNA are shown in Fig. 8.28. The overall drop in complexity of mRNA populations with development is seen in the drop in overall height of the bars, and the similarity of mRNA populations with those of the gastrula is seen by comparing the darkened part of each bar. There does appear to be a set of mRNA sequences that is present in all cells. This set presumably corresponds to the minimum variety of messages involved in maintaining the life of sea urchin cells. The structural genes responsible for the production of these ubiquitous mRNAs are referred to as *housekeeping genes*. The maximum complexity of the housekeeping sequence set is 0.3 percent of the single-copy DNA, represented essentially by the darkened portion of the bar for intestinal mRNA. This set corresponds to approximately 1,000 to 1,500 genes.

The results just described indicate that each embryonic stage or adult tissue contains a different, although partially overlapping set of structural gene transcripts, i.e., mRNAs. A question of related interest concerns the similarity between various messenger RNA and *nuclear* RNA populations. In one experi-

Figure 8.28 Sets of structural genes active in sea urchin embryos and adult tissues. The shaded portion of each bar indicates the amount of single-copy sequence shared between gastrula mRNA and the RNA preparation listed along the abscissa. The unshaded portions show the amount of single-copy sequence present in the various RNAs studied but absent from gastrula mRNA. Dashed lines indicate the maximum amount of null mDNA that could have been present and escaped detection, in terms of complexity, for cases where no apparent null mDNA reaction was observed. The total complexity of each RNA is indicated by the overall height of each bar. Complexity is calibrated in three ways along the three ordinates shown. From left to right, these are in nucleotides of single-copy sequence, as percentage of gastrula mRNA complexity, as percentage of total single-copy sequence. [*From G. A. Galau et al.*, Cell, **7**:*499* (*1976*).]

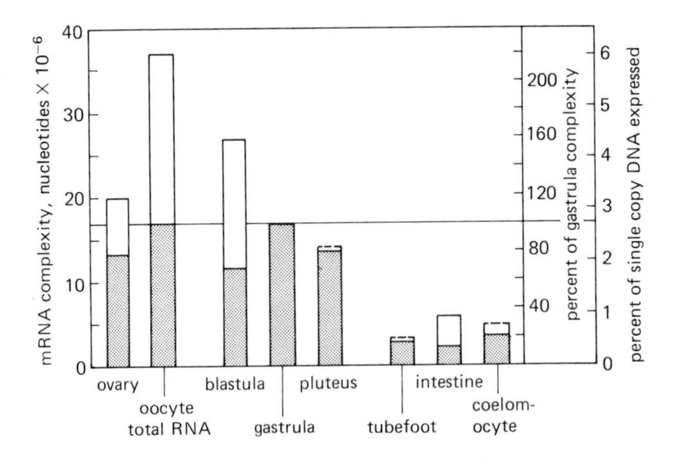

ment, total single-copy [^{3}H]DNA was incubated with an excess of blastula mRNA, and the complementary DNA fragments, i.e., blastula mDNA, were isolated. When tested against blastula mRNA, 78 percent of the reactable mDNA is driven into hybrids. This indicates that the blastula mDNA represents a 37-fold purification of blastula mRNA sequences as compared with total single-copy DNA. The blastula mDNA was then incubated with various nuclear RNA preparations (Fig. 8.29) to determine the extent to which the sequences present in blastula polyribosomal mRNA are represented in nuclear RNA populations. The curves in Fig. 8.29 indicate that virtually all the blastula mDNA that can react with blastula mRNA (78 percent of the fragments) is driven into the hybrid state by each of three different nuclear RNAs. It would appear based on these findings that cells, such as those of the adult intestine, transcribe structural genes regardless of whether or not the corresponding mRNA ever appears in the cytoplasmic polyribosomes. It should be kept in mind that the results of these types of hybridization reactions reflect similarities among the complex class of mRNA sequences (Section 8.3.C); it remains possible that the prevalent blastula messages, whose importance in protein synthesis is predominant, are not represented in the nuclear RNAs of adult cells.

Finally, we can ask the question: How similar are nuclear RNAs from different embryonic stages or adult tissues? At the time of this writing, two such studies with sea urchin nuclear RNAs had been carried out, one comparing populations present at the blastula and pluteus stages (Kleene and Humphreys, 1977),

Figure 8.29 Hybridization of blastula mDNA with nuclear RNAs from intestine (▲—▲), coelomocytes (●– –●), and gastrula stage embryos (■------■). Virtually all the blastula mDNA that can react with blastula mRNA is driven into hybrid by each of the nuclear RNAs. Thus at least 90 percent of the blastula structural gene set, i.e., the DNA sequences that code for mRNAs of the blastula, is transcribed in the nuclei of three very different cell populations, representing both embryonic and adult tissues. [*From B. J. Wold, W. H. Klein, B. Hough-Evans, R. J. Britten, and E. H. Davidson,* Cell, **14**:*945 (1978).*]

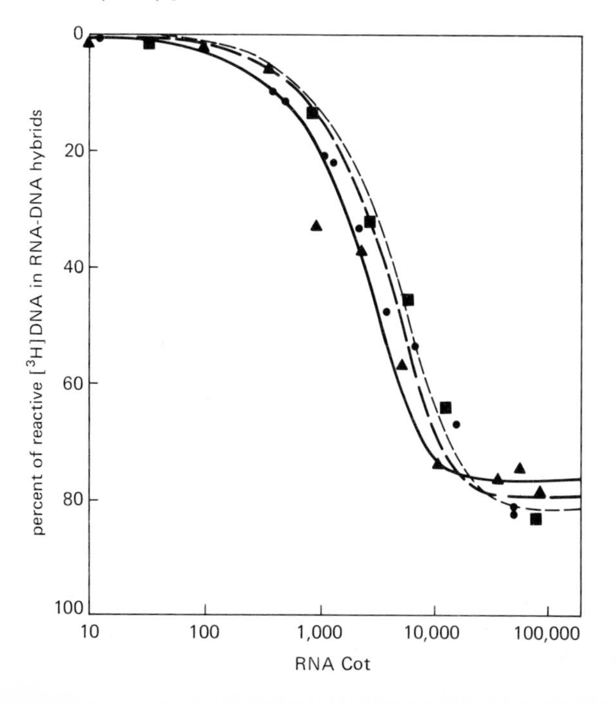

and the other comparing populations present at the gastrula stage to that of adult intestinal cells (S. G. Ernst, R. G. Britten, and E. H. Davidson, 1979). The two studies use different types of hybridization procedures and come to similar, but not identical, conclusions. We will begin with the earlier study. In the previous examples, comparisons were made by isolating DNA sequences complementary to one group of RNAs and its subsequent hybridization to a different RNA preparation. In the present experiment, the comparison is made in a different way, one in which the two RNA populations are mixed prior to incubation with single-copy DNA. If the two nuclear RNA populations being compared contain totally different sets of sequences, they should hybridize to twice the variety of DNA sequences. Since the RNA is in great excess, all complementary DNA fragments will still be driven into hybrids. If the two populations are identical, there will be no added hybridization over that level observed with either RNA preparation incubated alone. No increased hybridization was found in this type of *additive* experiment, suggesting that the two RNA preparations are essentially identical. It is possible, however, in this type of experiment, for significant differences between populations to be overlooked. In the more recent study, single-copy [^{3}H]DNA was hybridized to gastrula nuclear RNA, and the DNA that *failed* to bind to the RNA was isolated. This fraction, consisting of sequences absent from gastrula nRNA, is referred to as the *gastrula null nDNA*. When the labeled gastrula null nDNA is incubated with excess nuclear RNA prepared from intestinal cells, a significant level of hybridization is found to occur. It was estimated that approximately 3.6 percent of the sequences in the gastrula null nDNA that are absent in gastrula nuclear RNA are present in intestinal nuclear RNA. Although this represents a relatively small percentage, it consists of a considerable amount of information, approximately 3.5×10^7 nucleotides, or approximately 20 percent of the total intestinal nRNA complexity. Based on the work previously discussed, it is believed that this subset of nuclear RNA sequences is not coded by structural genes, i.e., ones coding for proteins. Rather it is believed that the differences in nuclear RNA populations have to do with intranuclear function. Further experiments will be needed to decide this point. Regardless of the existence of significant differences in nuclear RNA populations, it is apparent that mRNA populations are much more dissimilar and, therefore, that changes in nuclear and messenger RNAs need not occur in parallel fashion.

Readings

Ballantine, J. M., H. R. Woodland, and E. A. Sturgess, 1979. Changes in Protein Synthesis during the development of *Xenopus laevis*, *J. Embryol. Exp. Morphol.*, **51**:137–153.

Bennett, D., 1975. The T-Locus of the Mouse, *Cell*, **6**:441–454.

Brandhorst, B. P., 1976. Two-Dimensional Gel Patterns of Protein Synthesis before and after Fertilization in Sea Urchin Eggs, *Develop. Biol.*, **52**:310–317.

———, D. P. S. Verma, and D. Fromson, 1979. Polyadenylated and Nonadenylated Messenger RNA Fractions from Sea Urchin Embryos Code for the Same Abundant Proteins, *Develop. Biol.*, **71**:128–141.

Briggs, R., 1973. "Developmental Genetics of the Axolotl," 31st. Symp. Soc. Dev. Biol., Academic.

———, and T. J. King, 1952. Transplantation of Living Nuclei from Blastula Cells into Enucleated Frog Eggs, *Proc. Nat. Acad. Sci. U.S.*, **38**:455–463.

Brothers, A. J., 1976. Stable Nuclear Activation Dependent on a Protein Synthesized during Oogenesis, *Nature*, **260**:112–115.

Caplan, A. I., and C. P. Ordahl, 1978. Irreversible Gene Repression Model for Control of Development, *Science* **201**:120–130.

CARROLL, A. G., and H. OZAKI, 1979. Changes in the Histones of the Sea Urchin *Strongylocentrotus purpuratus*, *Exp. Cell Res.*, **119**:307–315.

COLLIER, J. R., and MCCARTHY, M. E., 1979. Protein Synthesis in the Polar Lobe of the *Ilynassa* Egg, *Amer. Zool.*, **19**:950.

DANIELLI, J. F., and M. A. DIBERARDINO (eds.), 1979. "Nuclear Transplantation," *Int. Rev. Cytol. (suppl.)*, **9**.

DAVIDSON, E. H., 1976. "Gene Activity in Early Development," 2d ed., Academic.

DAVIDSON, R. L., 1973. "Control of the Differentiated State in Somatic Cell Hybrids," 31st Symp. Soc. Dev. Biol., Academic.

DEROBERTIS, E. M., J. B. GURDON, G. A. PARTINGTON, J. E. MIRTZ, and R. A. LASKEY, 1977. Injected Amphibian Oocytes: A Living Test Tube for the Study of Eukaryotic Gene Transcription, *Biochem. Soc. Symp.*, **42**:181–191.

DEROBERTIS, E. M., R. F. LONGTHORNE, and J. B. GURDON, 1978. Intracellular Migration of Nuclear Proteins in *Xenopus* Oocytes, *Nature*, **272**:254–256.

DIBERARDINO, M. A., and N. HOFFNER, 1970. Origin of the Chromosomal Abnormalities in Nuclear Transplants, A Reevaluation of Nuclear Differentiation and Nuclear Equivalence in Amphibians, *Develop. Biol.*, **23**:185–209.

DOLECKI, D. J., R. F. DUNCAN, and T. HUMPHREYS, 1977. Complete Turnover of Poly A on Maternal mRNA of Sea Urchin Embryos, *Cell*, **11**:339–344.

DUPASQUIER, L., and M. R. WABL, 1977. Transplantion of Nuclei from Lymphocytes of Adult Frogs into Enucleated Eggs. Special Focus on Technical Parameters, *Differentiation*, **8**:9–29.

EPHRUSSI, B., 1972. "Hybridization of Somatic Cells," Oxford University Press.

ERNST, S. G., R. J. BRITTEN, and E. H. DAVIDSON, 1979. Distinct Single-Copy Sequence Sets in Sea Urchin Nuclear RNAs, *Proc. Nat. Acad. Sci. U.S.*, **76**:2209–2212.

FANKHAUSER, G., 1955. The Role of Nucleus and Cytoplasm in Development, in B. H. Willier, R. Weiss, and V. Hamburger (eds.), "Analysis of Development," Saunders.

GALAU, G. A., et al., 1976. Structural Gene Sets Active in Embryos and Adult Tissues of the Sea Urchin, *Cell*, **7**:487–505.

GRAINGER, J. L., M. M. WINKLER, S. S. SHEN, and R. A. STEINHARDT, 1979. Intracellular pH Controls Protein Synthetic Rate in the Sea Urchin Egg and Early Embryo, *Develop. Biol.*, **68**:396–406.

GROSS, P. R., and COUSINEAU, G. H., 1963. Macromolecule Synthesis and the Influence of Actinomycin D on Early Development, *Exp. Cell Res.*, **33**:368–395.

GURDON, J. B., 1974*a*. "The Control of Gene Expression in Animal Development," Harvard University Press.

———, 1974*b*. Molecular Biology in a Living Cell, *Nature*, **248**:772–776.

———, 1968. Transplanted Nuclei and Cell Differentiation, *Sci. Amer.*, **219**:24 (December).

———, R. A. LASKEY, and O. R. REEVES, 1975. The Developmental Capacity of Nuclei Transplanted from Keratinized Skin Cells of Adult Frogs, *J. Embryol. Exp. Morphol.*, **34**:93–112.

———, E. M. DEROBERTIS, and G. PARTINGTON, 1976. Injected Nuclei in Frog Oocytes Provides a Living System for Study of Transcriptional Control, *Nature*, **260**:116–120.

HARRIS, H., 1974. "Nucleus and Cytoplasm," 3d ed., Oxford University Press.

HOTTA, Y., and S. BENZER, 1973. "Mapping of Behavior in *Drosophila* Mosaics," 31st Symp. Soc. Dev. Biol., Academic.

HUMPHREYS, T., 1969. Efficiency of Translation of Messenger RNA before and after Fertilization in Sea Urchins, *Develop. Biol.*, **20**:435–458.

ILLMENSEE, K., 1973. The Potentialities of Transplanted Early Gastrula Nuclei of *Drosophila melanogaster*. *Roux Arch. Entwickl.*, **171**:331–343.

INFANTE, A. A., and M. NEMER, 1967. Accumulation of Newly Synthesized RNA Templates in a Unique Class of Polyribosomes during Embryogenesis, *Proc. Nat. Acad. Sci. U.S.*, **58**:681–688.

JEFFERY, W. R., and D. G. CAPCO, 1978. Differential Accumulation and Localization of Maternal Poly(A)-Containing RNA during Early Development of the Ascidian, *Styela*, *Develop. Biol.*, **67**:152–166.

Jenkins, N. A., J. F. Kaumeyer, E. M. Young, and R. A. Raff, 1978. A Test for Masked Message: The Template Activity of Messenger Ribonucleoprotein Particles Isolated from Sea Urchin Eggs, *Develop. Biol.*, **63**:279–298.

Kedes, L. H., 1979. Histone Genes and Histone Messengers, *Ann. Rev. Biochem.*, **48**:837–870.

———, and P. R. Gross, 1969. Synthesis and Function of Messenger RNA during Early Embryonic Development, *J. Mol. Biol.*, **42**:559–575.

Kleene, K., and T. Humphreys, 1977. Similarity of hnRNA Sequences in Blastula and Pluteus Stage Sea Urchin Embryos, *Cell*, **12**:143–155.

Kunkel, N. S., and E. S. Weinberg, 1978. Histone Gene Transcripts in Cleavage and Mesenchyme Blastula Embryos of the Sea Urchin, *S. purpuratus*, *Cell*, **14**:313–326.

Levinson, J., P. Goodfellow, M. Vadeboncouer, and H. McDevitt, 1978. Identification of Stage-Specific Polypeptides Synthesized during Murine Preimplantation Development, *Proc. Nat. Acad. Sci. U.S.*, **75**:3332–3336.

Manes, C., 1975. "Genetic and Biochemical Activities in Preimplantation Embryos," 33rd Symp. Soc. Dev. Biol., Academic.

McKinnell, R. G., R. A. Diggins, and D. D. Labat, 1969. Transplantation of Pluripotential Nuclei from Triploid Frog Tumors, *Science*, **165**:394–396.

Merriam, R. W., and R. J. Hall, 1976. The Germinal Vesicle Nucleus of *Xenopus laevis* Oocytes as a Selective Storage Receptacle for Proteins, *J. Cell Biol.*, **69**:659–668.

Mescher, A., and T. Humphreys, 1974. Activation of Maternal mRNA in the Absence of Poly(A) Formation in Fertilized Sea Urchin Eggs, *Nature*, **249**:138–139.

Newrock, K. M., C. R. Alfageme, R. V. Nardi, and L. H. Cohen, 1977. Histone Changes during Chromatin Remodeling in Embryogenesis, *Cold Spring Harbor Symp. Quant. Biol.*, **42**:421–431.

Peterson, J. A., and M. C. Weiss, 1972. Expression of Differentiated Functions in Hepatoma Cell Hybrids: Induction of Mouse Albumin Production in Rat Hepatoma–Mouse Fibroblast Hybrids, *Proc. Nat. Acad. Sci. U.S.*, **69**:571–575.

Ringertz, N. R., and R. E. Savage, 1977. "Cell Hybrids," Academic.

———, U. Krondahl, and J. R. Coleman, 1978. Reconstitution of Cells by Fusion of Cell Fragments, *Exp. Cell Res.*, **113**:233–246.

Rodgers, W. H., and P. R. Gross, 1978. Inhomogeneous Distribution of Egg RNA Sequences in the Early Embryo, *Cell*, **14**:279–288.

Ruderman, J. V., and P. R. Gross, 1974. Histones and Histone Synthesis in Sea Urchin Development, *Develop. Biol.*, **36**:286–298.

———, H. R. Woodland, and E. A. Sturgess, 1979. Modulation of Histone Messenger RNA during the Early Development of *Xenopus laevis*, *Develop. Biol.*, **71**:71–82.

Schultz, G. A., and R. B. Church, 1975. Transcriptional Patterns in Early Mammalian Development, in R. Weber (ed.), "The Biochemistry of Animal Development," Academic.

Senger, D. R., R. J. Arceci, and P. R. Gross, 1978. Histones of Sea Urchin Embryos, *Develop. Biol.*, **65**:416–425.

Slater, D. W., and S. Spiegelman, 1966. An Estimation of Genetic Messages in the Unfertilized Echinoid Egg, *Proc. Nat. Acad. Sci. U.S.*, **56**:164–170.

Subtelny, S., 1975. Nucleocytoplasmic Interactions in Development of Amphibian Hybrids, *Int. Rev. Cytol.*, **39**:35–88.

Suzuki, D. T., 1974. Developmental Genetics, in J. Lash and J. R. Whittaker (eds.), "Concepts in Development," Sinauer.

Symposium on Spiralian Development, 1976. *Amer. Zool.*, **16**(3).

Van Blerkom, J., and R. W. McGaughey, 1978. Molecular Differentiation of the Rabbit Ovum, *Develop. Biol.*, **63**:151–164.

Whiteley, H. R., and A. H. Whiteley, 1975. Changing Populations of Reiterated DNA Transcripts during Early Echinoderm Development. *Curr. Topics Dev. Biol.*, **9**:39–88.

Wilt, F. H., 1977. The Dynamics of Maternal Poly(A)-Containing mRNA in Fertilized Sea Urchin Eggs, *Cell*, **11**:673–681.

———, R. E. Maxson, and H. R. Woodland, 1979. Gene Activity in Early Development:

Oogenetic and Zygotic Contributions to mRNA, in J. D. Ebert and T. S. Okada (eds.), "Mechanisms of Cell Change," Wiley.

WOLD, B. J., W. H. KLEIN, B. R. HOUGH-EVANS, R. J. BRITTEN, and E. H. DAVIDSON, 1978. Sea Urchin Embryo mRNA Sequences Expressed in Nuclear RNA of Adult Tissues, *Cell,* **14:**941–950.

WUHL, L., and V. CHAPMAN, 1976. The Expression of β-Glucuronidase during Preimplantation Development of Mouse Embryos, *Develop. Biol.,* **48:**104–109.

Chapter 9 Morphogenesis: examples in lower chordates

1. The development of *Amphioxus*
2. The development of the ascidian tadpole larva

Morphogenesis is the progressive attainment of bodily forms and structures during development. It has been called the quintessential problem of the developmental biologist. The phenomenon is the most easily observed, the most easily described, and probably the least understood process of development. In spite of recent advances in the understanding of the behavior of macromolecules and consequent reexamination of the classic problems of development, the problems remain. The questions persist because the new analytical methods—which involve extraction, purification, degradation, and identification of the macromolecular constituents of living matter—are intrinsically destructive.

The very essence of development is continuous change, and it is difficult to come to grips with it. It has to be seen to be appreciated. One can easily see great changes from day to day in the development of the comparatively large eggs of fish, frogs, and salamanders, but only stop-action speeded-up films can properly portray the steady, continuous process of change. Many such films are now available for presentation and study.

The full impact of the living developmental process in its totality, however, is best experienced by observing the development of the very small translucent eggs of certain marine animals. In particular, the holoblastic cleavage and development of echinoderms (starfish, sea urchin, sand dollar) and of protochordates (ascidian, *Amphioxus*) offer an ongoing spectacle of great beauty and wonder within a matter of hours, with perhaps one sleepless night for the observer. Depending on the temperature, a sea urchin egg fertilized early one morning may develop into a pluteus larva by late the following day, with changes observable hour by hour. Ascidian eggs become swimming tadpole larvae (see Fig. 7.8) during the same time, the smallest doing so between dawn and dusk. These forms are readily obtained and studied at any of the marine biological laboratories scattered around the world, and even in inland laboratories during this age of jet delivery. *Amphioxus,* however, the very prototype of chordate development and therefore of vertebrate development that has grown out of it, is less amenable, since it spawns only during a few days in the year and is not commonly abundant. Nevertheless, its development has been well described, most recently by E. G. Conklin.

In *Amphioxus* and the other forms just mentioned, the egg is about 0.1 mm in diameter. At a given temperature, all such eggs develop at much the same rate and undergo cleavage in the early stages about every half hour. Cleavage proceeds either until cells are about the same small size as the somatic cells of the species or, in certain parts, until cell division is replaced by precocious cell dif-

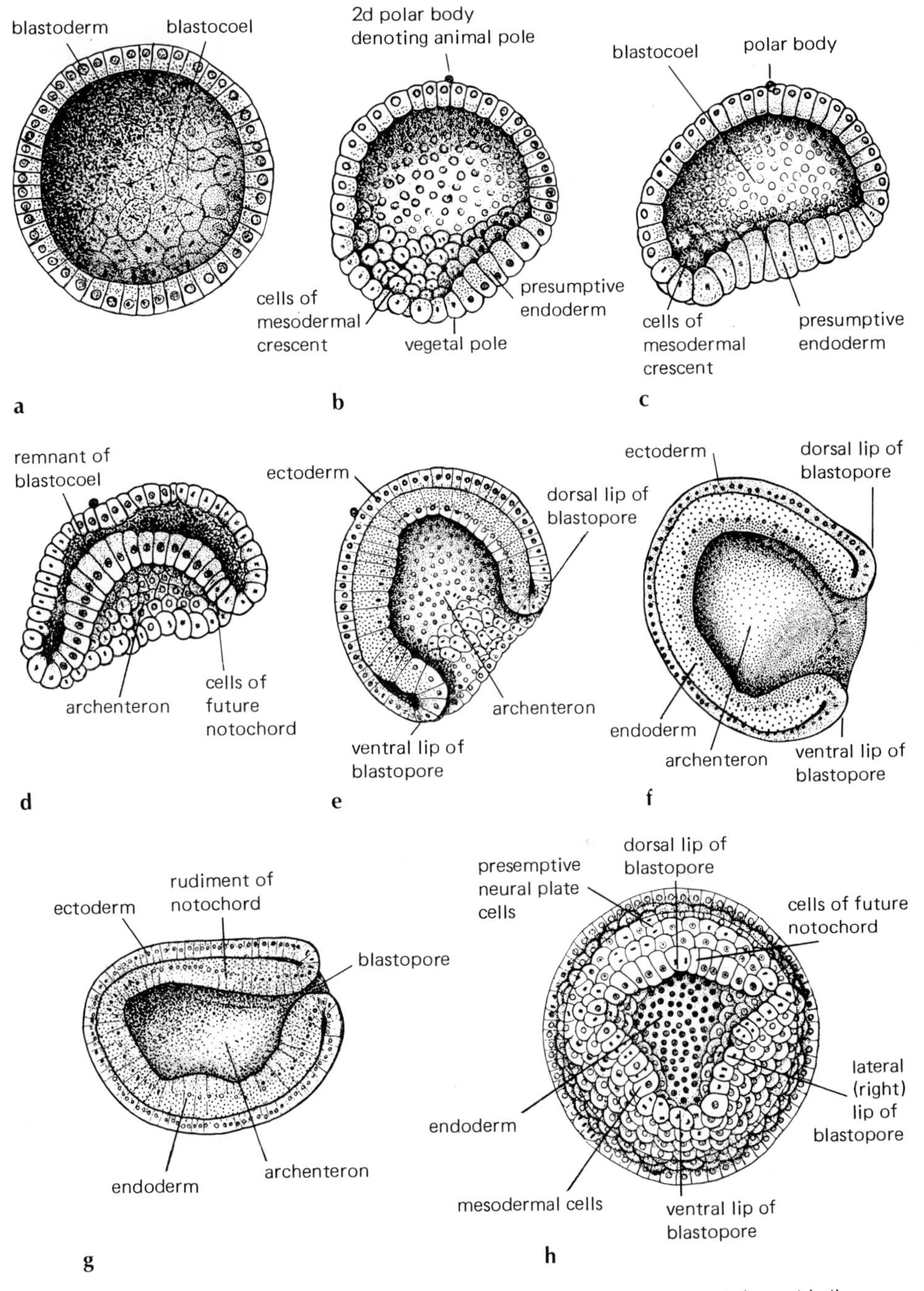

Figure 9.1 Stages of gastrulation of *Amphioxus*. The embryos in a–g are represented as cut in the median plane. a Blastula. b, c Beginning of invagination. d Invagination advanced, the embryo attaining the structure of a double-walled cup with a broad opening to the exterior. e, f Constriction of the blastopore. g Completed gastrula. h Middle gastrula, whole, viewed from side of blastopore. [*From* E. G. Conklin, J. Morph., **54**:69 (1932).]

ferentiation. In most cases, zygotes of this type divide to form several thousand cells. Cleavage rate is rapid at first but steadily slows to a standstill, as though approaching a state of equilibrium. During and immediately following this period, the community of cells must have attained a morphological structure and a cellular difference capable of serving as a self-supporting functional organism. *What happens when* during this whole process is the focus for students of morphogenesis. The transformation of the zygote into a blastula, together with a molecular approach to the process of gastrulation, has already been described. Here we are concerned with the events that follow.

Gastrulation is the most critical phase of early embryonic development. During the development of small holoblastically cleaving embryos, the single-layered hollow sphere of cells, the blastula, infolds to form two and then three layers of cells that constitute the gastrula. The outermost layer becomes the *ectoderm*, the innermost layer the *endoderm*, and the intermediate layer the *mesoderm*. (Similarly, although by ingrowing rather than by infolding, the single-layered blastodisc of meroblastically cleaving embryos become a three-layered or trilaminar structure.) Each of the three germ layers thus established continues to develop morphologically and histologically along its own special path. Gastrulation is the whole event during which the three layers are established. It can be approached from several standpoints, or stated differently, it raises several very different problems. What are the means, in cellular terms, by which the various morphogenetic movements are effected? What instigates such movements? What is accomplished by these movements? How is the timing controlled? Questions of this sort apply to morphogenetic phenomena generally, not only to gastrulation.

1. The development of *Amphioxus*

Cleavage in *Amphioxus* results in a blastula that consists of a single layer of columnar cells surrounding a large blastocoel (Fig. 9.1). The greater the number of cleavages before gastrulation begins, the smaller the constituent cells. Gastrulation in *Amphioxus* occurs when about 800 cells have been formed, between the ninth and tenth cleavages. It begins as a flattening of the blastula wall at the vegetal pole, followed by invagination of the whole vegetal half of the blastula so that double-walled cup-shaped gastrula forms. As a result, the blastocoel becomes completely obliterated. The new cavity is the *archenteron;* the opening is the *blastopore*. As cell division continues, the gastrula elongates and the blastopore becomes progressively smaller as its dorsal, lateral, and ventral lips converge.

During this process, the cells derived from the chordal crescent and those derived from the mesodermal crescent of the zygote, at first widely separated from one another, are brought into juxtaposition. As a result, the presumptive notochord is now flanked by presumptive mesoderm. At the same time, the presumptive neural material that initially lay anterior to the chordal crescent is in effect stretched over the presumptive notochord as the neural plate.

There is no evidence that the constituent cells in different territories of the *Amphioxus* gastrula or the neurula are dividing at different rates. In other words, neither the process of gastrulation nor that of neurulation can be explained, with validity, as the result of differentials in cell division rates between one territory and another—e.g., between prospective neural plate and prospective epidermis. At the same time, it may be highly significant that cell division is out of phase in the various recognizable and distinctive territories of the developing embryo (Fig. 9.2). This suggests that in each prospective territory the cells are at least beginning to take a special, presumably biochemical, path that is distinct. With contin-

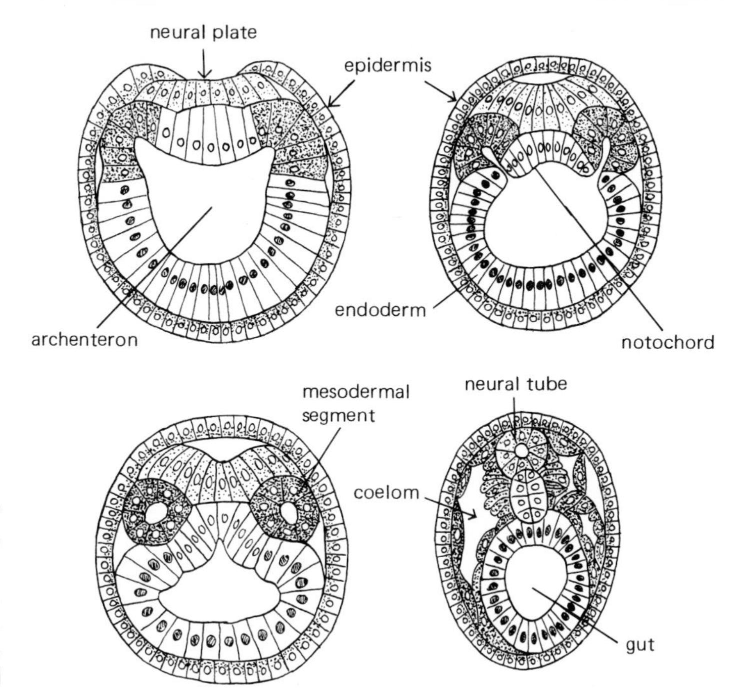

Figure 9.2 Formation of the primary organ rudiments in *Amphioxus*. Transverse sections. [*After Hatchek, from E. Korschelt, 1936.*]

uing elongation and cell division, the flattened neural plate rolls up on itself to form a neural tube, open in front as a *neuropore*. Simultaneously, ectodermal folds along its side stretch over the neural tissue to fuse together as a continuous sheet (Fig. 9.3). This phase of development is known as *neurulation* and the embryo is termed a *neurula*.

Meanwhile, further development has occurred within the embryo. When fully formed, the archenteron develops ridgelike outgrowths as a result of outpushing, or evagination, from the middorsal and from each dorsolateral region. These slowly constrict from the archenteron proper to become the rudimentary notochord and the lateral mesoderm tissue, respectively (Fig. 9.3). The notochord separates as a continuous middorsal rod immediately below the developing neural tube. The mesoderm on each side separates from the archenteron as a series of pocketlike evaginations, each of which becomes a mesodermal segment, or *somite*, with its own cavity. The series does not arise simultaneously throughout its length; the outpocketings appear at first anteriorly and then in quick succession toward the posterior end of the embryo. At a particular moment, before completion of the process, there is accordingly a *gradient* with regard to the stage of formation of mesodermal pouches along the anteroposterior axis, i.e., an axial gradient. The remainder of the archenteron, after outpocketing, persists as the endoderm proper, from which most of the digestive system derives. A large evagination from its ventral side becomes the liver diverticulum. An invagination from the anterior end of the ectoderm of the embryo extends posteriorly to fuse with the anterior end of the archenteron, thus forming the mouth. In an apparently simple and direct manner, a spherical epithelial envelope one cell thick transforms, by means of stretchings and foldings, into an elongate chordate orga-

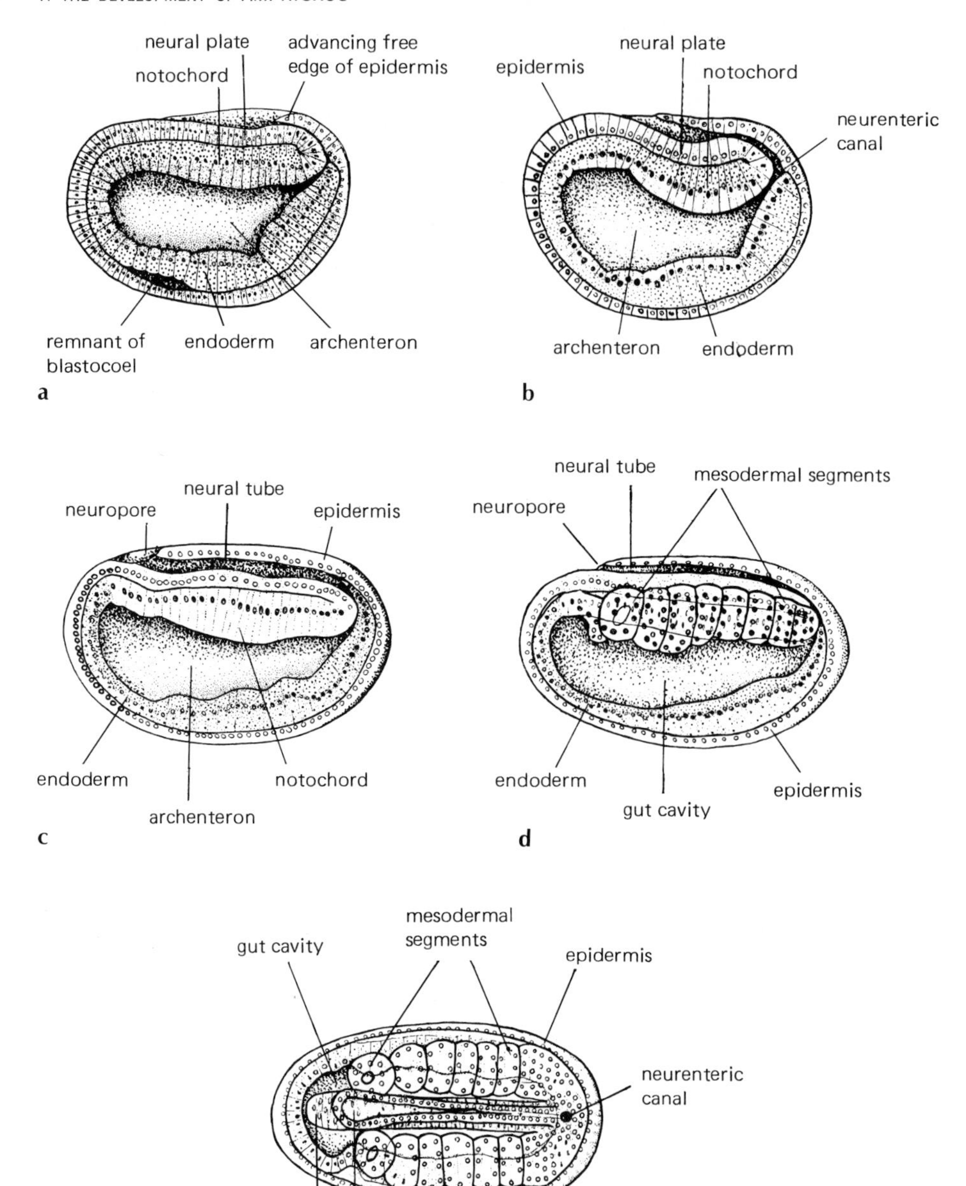

Figure 9.3 Stages of neurulation of *Amphioxus*. In *a*, *b*, and *c*, the embryos are represented as being cut in the median plane. *a* Earliest stage of neurulation. *c* Almost completed neurula. *d* Slightly later stage than *c* but cut paramedially so that the right row of mesodermal segment can be seen. *e* Completed neurula whole, seen from the dorsal side. The transparency of the embryo allows one to see at the same time the various parts superimposed over one another (neural tube, notochord, mesodermal segments, and gut). [*From E. G. Conklin,* J. Morph., **54:**69 (1932).]

nism. It is complete, with a digestive tube open at both ends, with a supportive notochord and a tubular nerve cord, and with a series of mesodermal somites on each side between gut and epidermis.

2. The development of the ascidian tadpole larva

In contrast with the preceding, although essentially similar in pattern, the embryo of the ascidian undergoes gastrulation between the sixth and seventh cleavages; i.e., it begins and completes the process between the 64- and 128-cell stages (Figs. 9.4 and 9.5). The overall effect, however, is the same as in *Amphioxus,* namely, to bring the chordal and mesodermal territories together as prospective tail, while neural plate dorsal to the notochord transforms into a neural tube. The whole development closely resembles that of *Amphioxus,* except that the morphogenetic process appears to be accelerated, or condensed, relative to the course of cleavage. Fewer cells are present at each morphogenetic stage. It should be noted, however, that in the ascidian the mesoderm does not subdivide to form segments nor does it develop a coelom (Figs. 9.4 and 9.5).

The first major morphogenetic event in the development of most animal embryos is gastrulation. The sweeping transformation of a blastula into a gastrula is

Figure 9.4 Neurula stage of development of ascidian. *a* Dorsal view (*left*) and side view (*right*) of neurula of *Styela,* and cross section of open-neural-plate stage of younger embryo. [*After E. G. Conklin, 1905.*] *b, c* Early and late neurula of *Clavelina* from side, showing neuropore, neural tube, notochord (*stippled*), and superimposed tail mesoderm and mesenchyme. [*After Van Benedin and Julin,* Arch. Biol., **2:**317 (*1884*).]

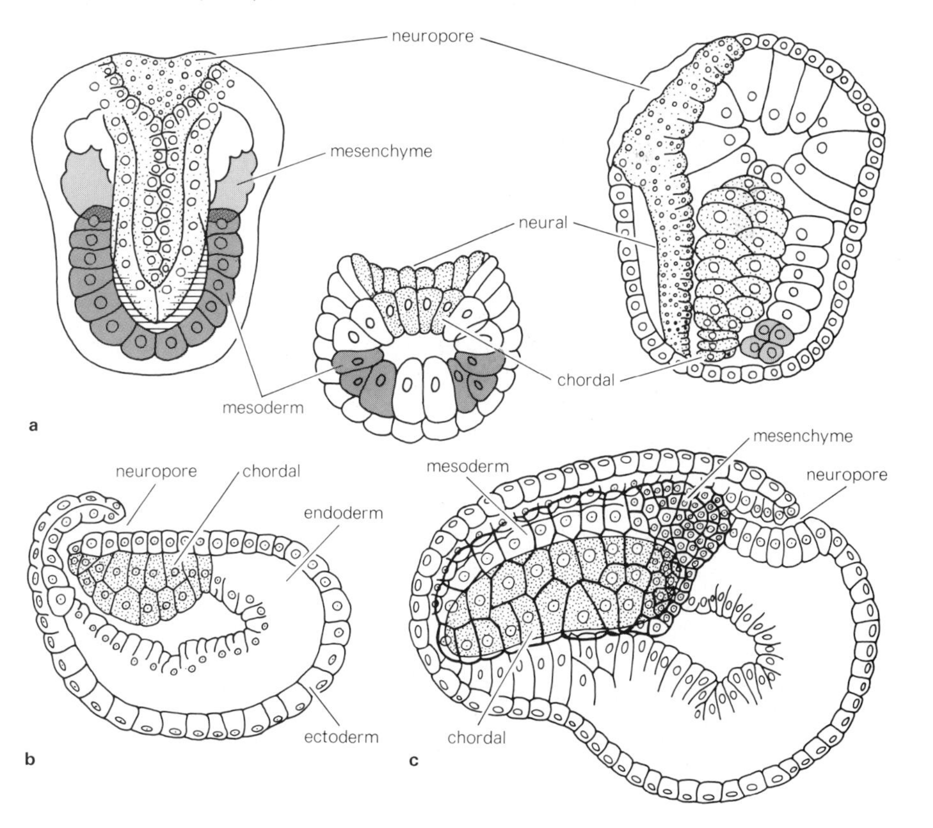

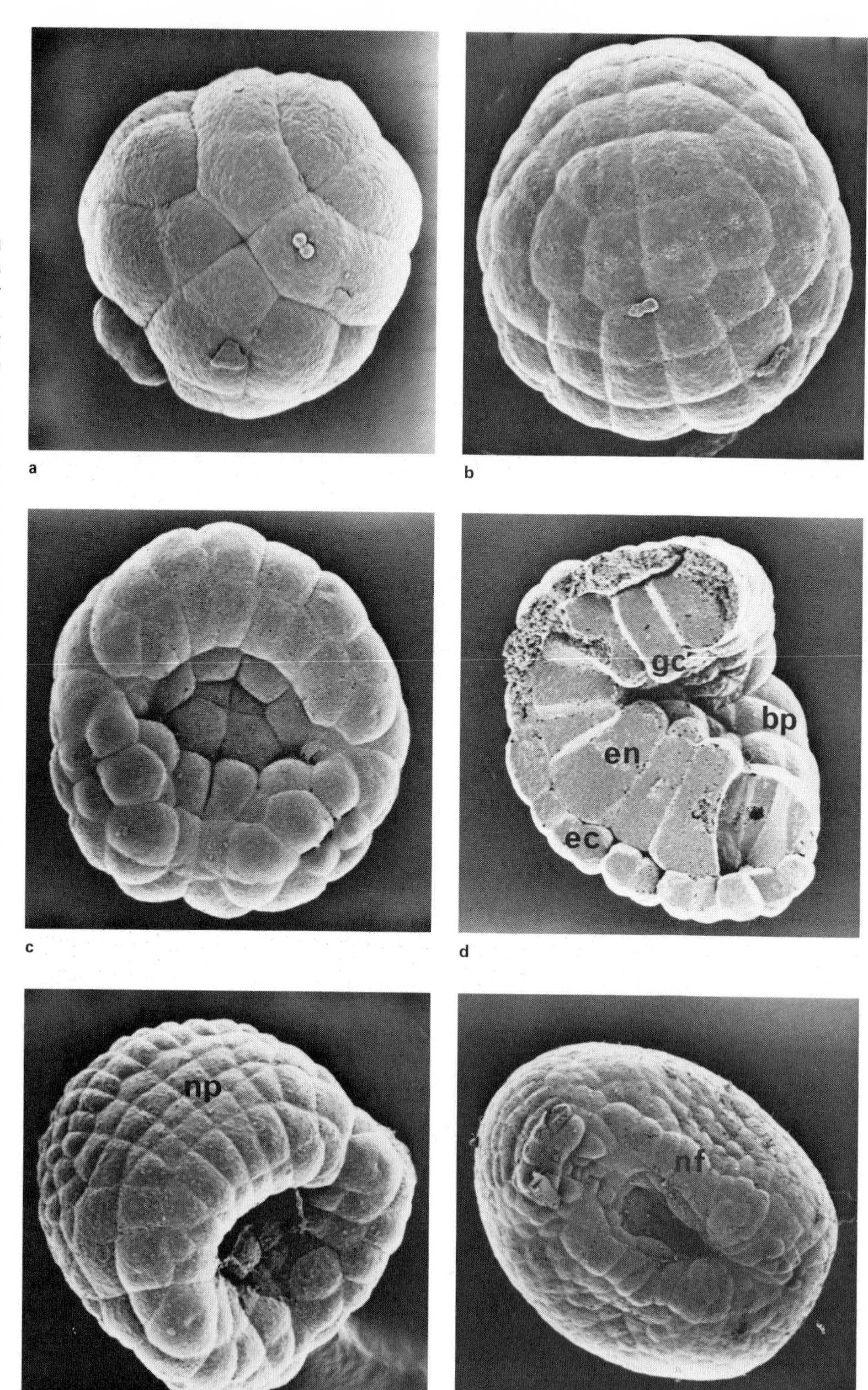

Figure 9.5 Scanning electron micrographs showing the cellular morphology and architecture during the early morphogenesis of the ascidian *Halocynthia roretzi*. *a* Thirty-two-cell stage viewed from the animal pole. The two small spheres are polar bodies. The plane of bilateral symmetry is evident. *b* Early gastrula stage viewed from the ventral (animal) pole. The ectoderm cells have divided seven times. *c* Early gastrula stage viewed from the dorsal pole showing the beginnings of invagination. *d* Dissected midgastrula stage along the midsagittal plane. By further enfolding movement of the ectoderm cells (ec), the embryonic gut, or gastrocoel (gc), is formed: bp, blastopore; en, endoderm. *e* Late-gastrula stage viewed om the dorsal side. The blastopore has shifted from the mid-orsal portion of the embryo to the posterior ortion. The neural-plate lls (np) are distinguishable. *f* Middle-neurula stage viewed from the lorsal pole. Neural-tube formation takes place m the posterior end of embryo to the anterior nd. At the anterior end of the neural fold (nf), cells are heaped in a mass. [*Courtesy of N. Satoh.*]

essential to further development and presents a challenge at several levels. The process is described in the following chapter in terms of cell and tissue movements and mechanics. Apart from the molecular control of the event, initiated or at least activated following fertilization, the primary question is: Are such movements independently prepared for in earlier stages, or are they the inevitable response to processes of cell differentiation already occurring? Are the gastrular changes responses to differentiation, or are they responsible for differentiation?

In this connection, a comparison of the development of three lower chordates is informative. These are *Amphioxus* and the tunicates *Styela* (an ascidian tunicate) and *Oikopleura* (a larvacean tunicate). All three have translucent eggs of roughly the same size (about 120 μm in diameter); all develop the basic chordate organization including notochord, dorsal tubular nerve cord, lateral tail musculature, and gill slits (Fig. 9.6). In all three, at or before fertilization, a flowing cytoplasm establishes crescentic zones in relation to the egg cortex, imposing bilateralism on the egg. Crescentic zoning, bilateral cleavage pattern, and the association of the anterior crescent with chordaneural tissue and the posterior crescent with mesoderm (muscle) indicate a fundamental relationship between the three.

In each of the three forms, cell division within the crescentic zones ceases as histological differentiation becomes visible. In all three there is a close correlation between the rates of differentiation of notochordal and mesodermal tissues. There is also a correlation between this and the time of onset (in terms of cleavage) of gastrulation. *Amphioxus* gastrulation occurs between the ninth and tenth cleavages, in *Styela* between the sixth and seventh, and in *Oikopleura* between the fifth and sixth. Correspondingly, the number of notochordal and muscle cells in the *Amphioxus* embryo is 300 to 400, representing from eight to nine divisions of the initial territory; in *Styela* there are about 40 cells in each, representing

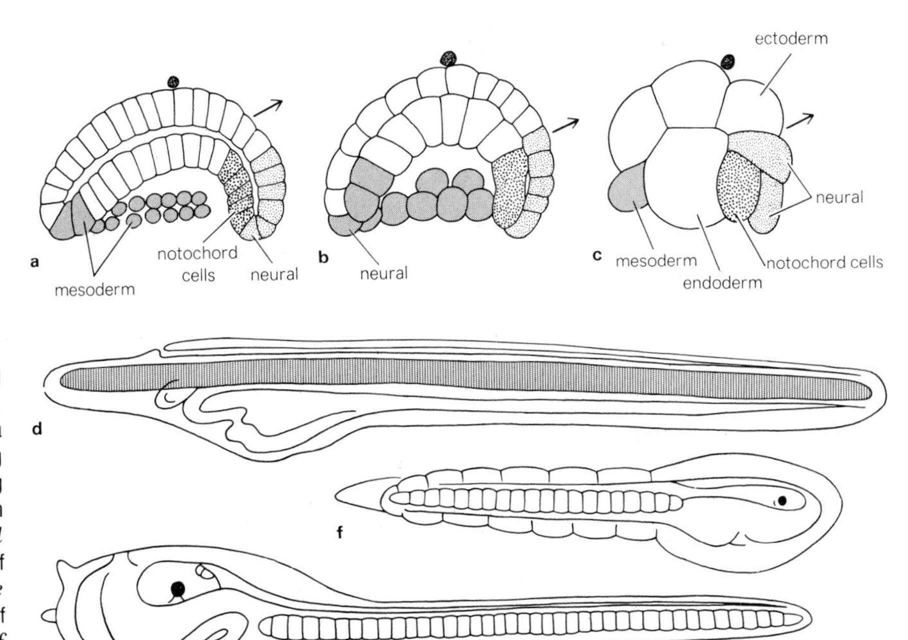

Figure 9.6 Gastrulas and larvae of *Amphioxus*, an ascidian, and a larvacean tunicate showing effect of varying ratios of differentiation and cleavage rates. *a*, *d* Gastrula and larva of *Amphioxus*. *b*, *e* Gastrula and larva of ascidian *Ciona*, *c*, *f* Gastrula and larva of *Appendicularia*.

from five to six divisions; and in *Oikopleura* the number of each kind of cell is 20, or from four to five divisions. In each form the size of the cells produced is inversely proportional to their number. Differentiation relative to cleavage rate is accelerated in ascidians compared with *Amphioxus,* and in *Oikopleura* compared with ascidians. Thus cell division in the crescent-derived tissues ceases three divisions sooner in the ascidian than in *Amphioxus* and one division sooner in *Oikopleura* than in the ascidian. The onset of gastrulation is advanced to exactly the same degree. This correlation strongly suggests gastrulation is a response to the already advancing differentiation of the cleavage embryo into sharply delineated histogenetic territories.

In spite of the minute scale of these overall events, we are confronted with the primary problem of development—no less in *Amphioxus* and ascidian than in mice and humans—namely, the emergence and nature of pattern or organization. The great advances in molecular biology at the level of the individual cell have yet to illuminate this fundamental phenomenon of multicellular development in any truly satisfactory way.

Throughout development we find an interplay between cellular differentiation and multicellular morphogenesis. We also find two opposite points of view concerning the relative role of individual cell character and that of possible supervising agencies. These viewpoints may be briefly stated as follows: (1) cells and tissues tend to acquire locations according to the character of their individual or collective differentiation, and (2) cells and tissues tend to differentiate according to their location in the organized system. Both concepts are prominent in contemporary developmental biology, although both were formulated by Driesch at about the turn of the century. We do not necessarily have to choose between two such apparently conflicting concepts. It is common experience for biologists to find that when a question is posed to nature as a choice between one interpretation and another, the response is that both are valid in varying degree in varying circumstances. This is true of the concepts we have just stated, which implies that we are faced with an intriguing but difficult intellectual problem.

Readings

Conklin, E. G., 1933. The Development of Isolated and Partially Separated Blastomeres of *Amphioxus, J. Exp. Zool.,* **64:**303–375.

———, 1932. The Embryology of *Amphioxus, J. Morphol.,* **54:**69–119.

———, 1931. The Development of Centrifuged Eggs of Ascidians, *J. Exp. Zool.,* **60:**1–80.

Johnson, K. E., 1974. Gastrulation and Cell Interactions, in J. Lash and J. R. Whittaker (eds.), "Concepts of Development," Saunders.

Reverberi, G., 1961. The Embryology of Ascidians, *Advan. Morphog.,* **1:**55–103.

Satoh, N., 1978. Cellular Morphology and Architecture during Early Morphogenesis of the Ascidian Egg: An SEM Study, *Biol. Bull.,* **155:**608–614.

Trinkaus, J. P., 1965. Mechanisms of Morphogenetic Movements, in R. L. DeHaan and H. Ursprung (eds.), "Organogenesis," Holt, Rinehart and Winston.

Weiss, P., 1950. Some Perspectives in the Field of Morphogenesis, *Quart. Rev. Biol.,* **25:**177–198.

Wilde, C. E., 1974. Time Flow in Differentiation and Morphogenesis, in J. Lash and J. R. Whittaker (eds.), "Concepts of Development," Saunders.

Chapter 10 Gastrulation

Gastrulation is the first and most crucial step in the transformation of the cleaving egg into some semblance of the embryo-to-be. Gastrulation is characterized by an extensive series of coordinated morphogenetic movements by which regions of the blastula are displaced into radically different locations. At the completion of this rearrangement, the primitive body plan of the organism has become established and the various organ-forming areas are in their proper positions awaiting the steps that will convert them into differentiated tissues.

One of the most important concepts generated by the early comparative embryologists was the division of the early embryo into three parts, or *germ layers*, the *ectoderm*, *mesoderm*, and *endoderm*. In its initial form, the concept of embryonic germ layers was deeply rooted in speculative theories of the evolution of multicellular organisms and their phylogenetic divergence. Over the years, the evolutionary aspects of the proposal have been largely abandoned, but the embryological aspects have been retained. In essence, the germ-layer theory points out the basic organization of the embryo (and subsequent adult) as being composed of three layers, an inner, middle, and outer layer. The inner layer is represented by the digestive tract and its derivatives, the outer layer by the skin and nervous system, and the middle layer by the remaining body components. The cells of the endoderm, ectoderm, and mesoderm give rise to these three parts of the body, respectively. Since all animals are organized in a similar manner (at least in having a central gut and an outer covering), the terms *ectoderm*, *mesoderm*, and *endoderm* can be applied universally throughout the animal kingdom. Keep in mind that no evolutionary relationships are meant by the use of these terms in this text; i.e., the ectodermal derivatives of flatworms or mollusks are not homologous to the ectodermal derivatives of frogs and mammals.

The concept of germ layers is raised in this discussion because it is during gastrulation that the embryonic gut cavity forms and the mesoderm is caused to slide into its final position between the inner endodermal and outer ectodermal layers, leaving ectoderm as the sole occupant of the embryonic surface. By the time gastrulation is completed, the three germ layers have become arranged in their final "concentric" organization. As a morphogenetic process, gastrulation takes place in markedly different ways in different types of developing embryos. It is therefore preferable to describe and analyze the events separately as they occur in small

holoblastic eggs (those of echinoderms and *Amphioxus*, Figs. 10.1 and 9.1*c* and *d*), in larger relatively yolky holoblastic eggs (of amphibians especially), and in meroblastic eggs (particularly teleosts and birds).

Even though the events of gastrulation appear superficially to be very different among diverse groups, the types of underlying morphogenetic activities can be remarkably similar. Several themes will recur throughout the discussion of gastrulation. In all cases there is extensive movement, often involving cells of an epithelium (an unbroken cohesive cell layer), as well as the more independent types of motility displayed by cells of a mesenchyme (a cluster of loosely orga-

Figure 10.1 *a* Photomicrographs of early sea urchin gastrulas. On the left is an embryo just prior to the onset of invagination. The vegetal plate has become flattened, and primary mesenchyme cells have been released into the blastocoel prior to beginning their migration. On the right is an embryo in which the vegetal plate has pushed in to begin the formation of the archenteron. Primary mesenchymal cells are beginning to move up the wall of the embryo. [*Courtesy of T. Gustafson and L. Wolpert.*] *b* Diagrams of two views of an advanced gastrula showing the distribution of primary and secondary mesenchyme cells. [*From T. Gustafson,* Exp. Cell Res., **32**:*563* (*1963*).]

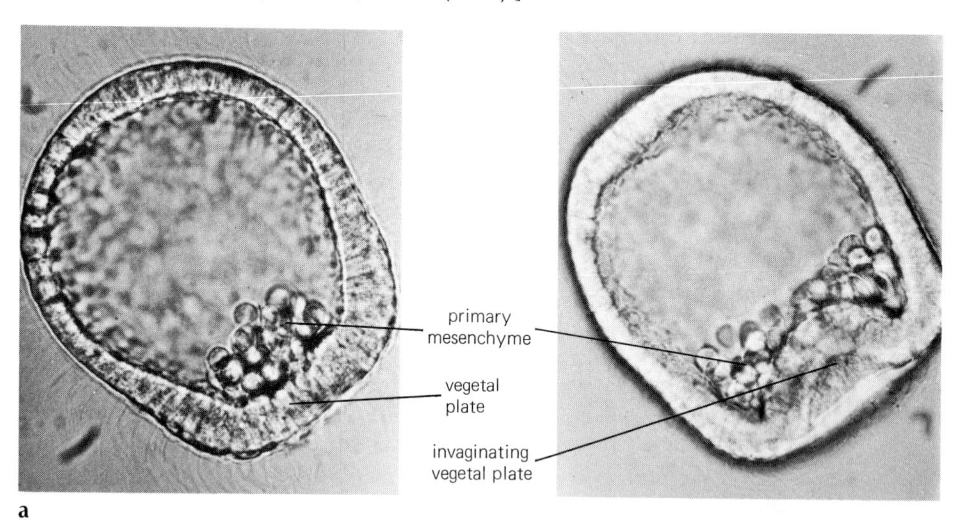

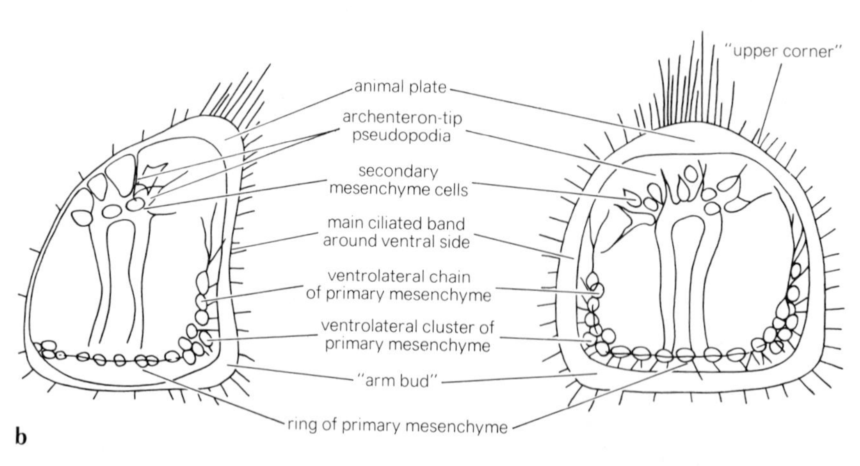

nized cells).[1] As we will see, the epithelial versus mesenchymal organization of cells is not a fixed property; there are many examples during gastrulation where epithelial layers become dispersed as mesenchymal cells and vice versa. As these events occur, there are often striking changes in the shapes of cells. Changes at the outer surfaces of cells are also important during morphogenesis. In this latter regard, changes in the adhesive properties of cell membranes as well as the composition of extracellular materials have been shown to be of importance. The role of these cellular activities will be apparent in the following pages.

1. Gastrulation in embryos which undergo holoblastic cleavage

A. Gastrulation in echinoderms

Gastrulation in small holoblastically cleaving embryos has been studied primarily in echinoderms, whose transparency makes internal events easy to follow. Gastrulation in echinoderms involves two simultaneous but relatively independent processes, the invagination to form the archenteron and the migration of the primary mesenchyme cells. The first indication of approaching gastrulation is seen when the vegetal region of the swimming blastula becomes noticeably flattened, forming the *vegetal plate*. Soon cells are seen to be released from the vegetal plate into the blastocoel, where they remain for a brief period before they begin to migrate. These cells constitute the *primary mesenchyme*, and we will return to them later.

As the primary mesenchyme cells collect at the vegetal region of the blastocoel, an indentation appears in the center of the vegetal plate (Fig. 10.2*a*) marking the onset of invagination.[2] Many attempts have been made in this century to explain how invagination is accomplished. The early interpretations were based on the following assumption: the forces that bring about invagination result from the pressure of dividing cells acting in a plane of the polar axis so that the cells of one hemisphere are turned in. This concept was tested in a series of direct operational experiments on echinoderm embryos, principally by cutting away parts of the embryo above the vegetal plate. The surprising result of the excision of the vegetal plate is the continuation of the process of invagination in the isolated plate. The inpushing deepens, and the rim of the isolated plate rolls up and closes over the central invagination to form a relatively small gastrula (Fig. 10.2*b*). Experiments such as these show that the forces producing invagination act in the plane of the vegetal plate and do not require the participation of other parts of the embryo. Moreover, if a radial cut is made from the periphery to the center of the isolated plate, the cut edges spring apart, suggesting that the whole plate is normally under tension.

Since the inpocketing of the vegetal plate in sea urchins has not been the subject of extensive analysis, we know very little about the underlying mechanism for the event. However, there have been studies of numerous other events in which an epithelial layer, such as the wall of the sea urchin embryo, undergoes some type of folding process. Where it has been studied (most thoroughly in the case of neurulation, Section 11.1.A), the infolding appears to be driven by a change in shape of the cells of the epithelium. The cells of the embryonic wall are tall columnar cells whose sides are roughly parallel to one another and pressed

[1] It should be noted that the terms *epithelium* and *mesenchyme* refer to the organization of cells, not to their origin from one or another germ layer. Epithelia can form from ectoderm (epidermis, neural tube), mesoderm (kidney), or endoderm (gut wall), as can masses of mesenchymal cells (neural crest, dermis, and liver mesenchyme form from ectoderm, mesoderm, and endoderm, respectively).

[2] The term *invagination* is used here to refer to an inpocketing of an unbroken sheet of tissue as occurs during gastrulation in *Amphioxus* and sea urchins. As we will see in later sections of this chapter, the process of gastrulation in meroblastic eggs is very different.

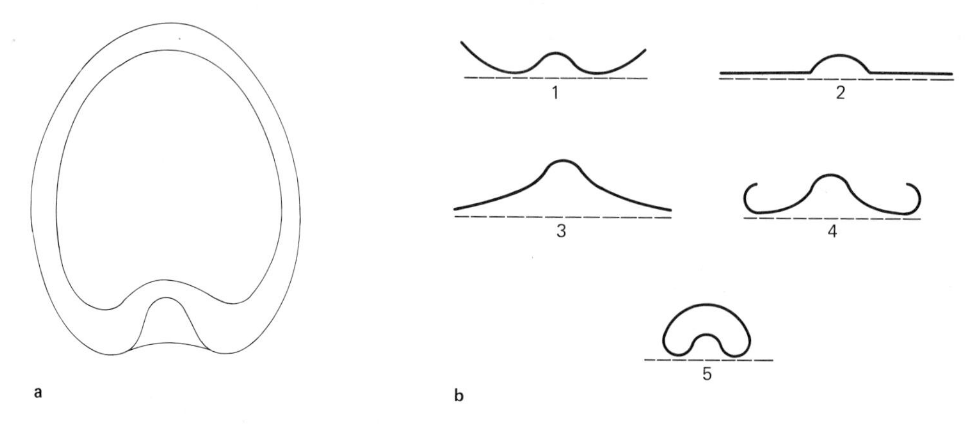

Figure 10.2 *a* Outline of a gastrula at an early stage of invagination. *b* Horizontal view of the positions assumed successively by the isolated gastral plate of the starfish *Patiria* during the first hour following excision. [*After A. R. Moore,* J. Exp. Zool., **87**:*101* (*1941*).]

rather closely against the lateral surfaces of their neighbors. Cells of an epithelial layer generally exhibit marked differences at their inner and outer edges, which is not surprising considering the very different nature of the environment at these two sites. In the case of the sea urchin embryo, one edge faces the external seawater, the other faces the blastocoel. Consider what would happen if the outer surface of the epithelium were to decrease in surface area as a result of some type of constrictive activity (Fig. 10.3). As long as the cells remained elongated and adherent to their neighbors, the layer would inevitably buckle in the manner shown in Fig. 10.3. Electron microscopic analysis of these types of events invariably reveals the presence of a dense band of microfilaments at the constricted end of the cell and arrays of microtubules aligned parallel to the cell's long axis. We can presume that the contractile activity of this band of microfilaments results in a change in the shape of the cells from columnar to wedge-shaped (pyramidal), thereby providing the tension in the layer that results in its folding response. As this occurs, the elongated shape of the cells is maintained by the skeletal function of the microtubules.

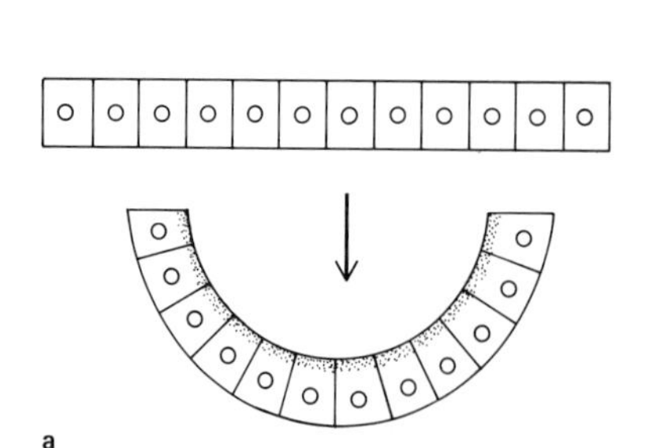

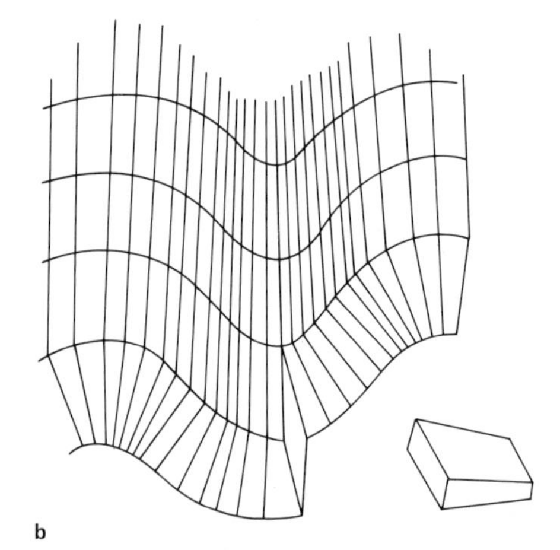

Figure 10.3 *a* Constriction of one edge of the cells of an epithelium leads to the folding of the entire cell layer, as shown in *b*.

Although a few surgical studies have been performed on the small sea urchin embryo, most of what we know about their gastrulation comes from the analysis of time-lapse movies made of single embryos trapped in tiny chambers within a piece of nylon net (Fig. 10.4*a*). The films show the invagination of the archenteron occurs in two phases (Fig. 10.4*b*). The first phase of invagination, which results from forces existing within the vegetal plate itself, takes the tip of the invaginating tube about one-third of the way through the blastocoel. There is a pause, followed by vigorous pseudopodial activity at the tip of the archenteron. These pseudopodia reach out to attach to the inner surface of the gastrula wall, and their subsequent contraction is believed to help pull the archenteric tip forward. As the archenteron approaches the wall of the animal region, the pseudopodial-forming cells become disengaged from the archenteron and form a crowd of *secondary mesenchyme cells* that spread over the inner surface of the adjacent ectoderm. Many of these cells remain attached to the archenteron by pseudopodial bridges that are soon supplemented by new bridges formed by pseudopodia extended between the two sites. As the tip of the archenteron approaches the ectodermal wall it is seen to bend toward the ventral surface below the animal plate, presumably as a result of stronger pseudopodial connections from that direction. With further development, the archenteron buds off a large coelomic sac (which later constricts to form a left and right sac) from near its tip. Following this an ingrowth from the oral ectoderm forms the stomodeum, which fuses with the tip of the archenteron to form a continuous digestive tube through the interior of the larva to the open blastopore, which becomes the anus.

Now that we have traced the general course of events taken by the invaginating archenteron, we can return to an earlier stage and follow the activities of the primary mesenchyme. The cells of the primary mesenchyme are derived from cells of the vegetal plate via a striking series of changes of shape. As part of the wall, the precursors of the mesenchyme are columnar cells (Fig. 10.5*a*) that ad-

Figure 10.4 *a* Photomicrograph of a sea urchin embryo trapped within a chamber of a nylon net. Once the embryo is immobilized in this manner, developmental events can be followed by time-lapse cinematography. [*Courtesy of T. Gustafson.*] *b* The progress of invagination as studied by this cinematic technique. The abscissa shows the time scale, in film frames. The time between frames is 20 seconds. The arrow indicates the time when pseudopodia were formed. [*After T. Gustafson and L. Wolpert,* Exp. Cell Res., **22**:*443* (*1961*).]

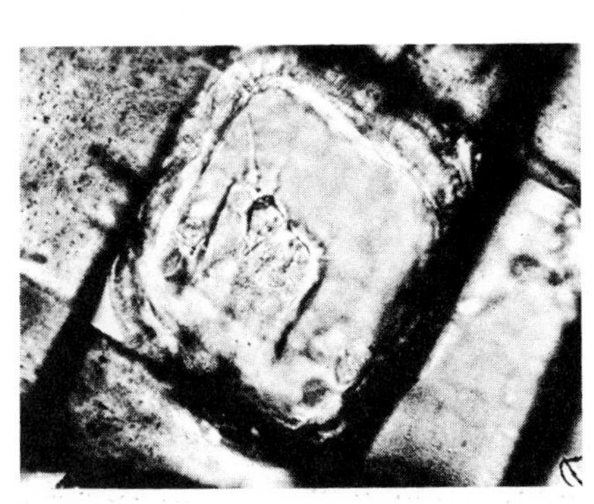
a

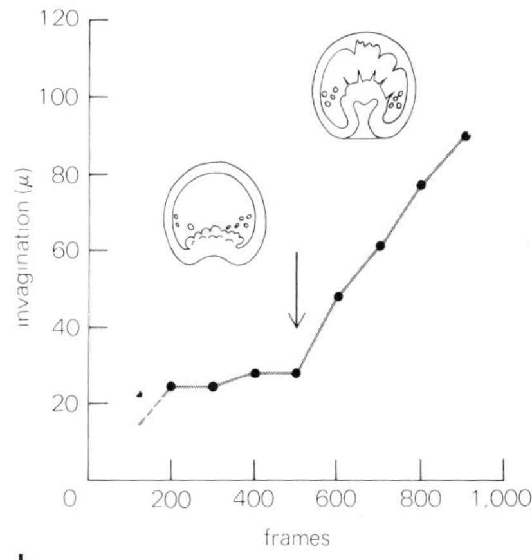

b

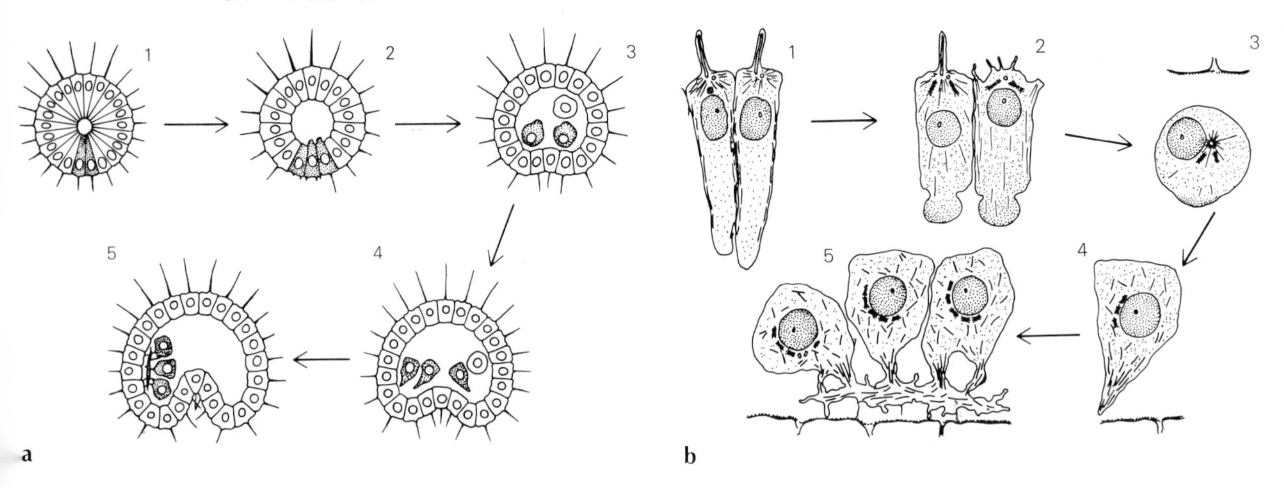

Figure 10.5 *a* Changes in cell shape of primary mesenchyme cells (*shaded*) during their formation and differentiation. In the early blastula, (1) the cells have a long conical shape. As blastulation proceeds, these cells shorten (2) and undergo pulsatory movements that produce small lobes on their basal surface. Shortly thereafter these cells migrate into the blastocoel, where they take on a nearly spherical shape (3) and then send out pseudopodia (4) with which they migrate. At about this time, the archenteron begins to indent. The pseudopodia of the primary mesenchyme cells come into contact and fuse, thus forming a common cable of cytoplasm that is specifically oriented within the blastocoel (5). It is within this cable that the calcareous skeleton of the pluteus is deposited. *b* The distribution of microtubules at each stage in the formation and differentiation of the primary mesenchyme (*stages correspond to those in* a). In every case, the microtubules are oriented parallel to the direction of cell asymmetry. When the cell is spherical (3), they radiate from a central spot. [*From J. R. Gibbins, L. G. Tilney, and K. R. Porter,* J. Cell Biol., **41**:*214 (1969)*.]

here strongly to their neighbors and possess numerous microtubules oriented parallel to the long axis of the cell (Fig. 10.5*b*). At the late blastula stage, these cells lose their single cilium and begin to show pulsating activity at their inner end (indicated as a knob on the cells of part 2, Fig. 10.5*b*). The release of the cells into the blastocoel is believed to be foreshadowed by a decrease in adhesiveness to their neighbors and to the outer hyaline layer (Section 5.6.A). When first released, the primary mesenchyme cells are essentially spherical, and the microtubules are confined to one region of the cell, where they converge from all directions onto a centriole (part 3, Fig. 10.5*b*). Long thin pseudopodia (typically 0.5 μm in diameter and up to 30 μm long), often termed *filopodia*, soon appear at the surface of these cells providing them with a locomotory capacity. Using their pseudopodia, the primary mesenchyme cells begin to migrate over the inner surface of the ectodermal wall in an exploratory manner. Careful analysis of the paths taken by individual cells suggests that their migratory activity is random in nature, yet after a period of time the cells have taken up a predetermined arrangement within the embryo. If one watches these cells as they migrate, the pseudopodia appear to "feel" their way along the inner surface of the wall, making contacts, pulling themselves forward, then breaking the contacts and moving on. Presumably the stronger (or more stable) the contact between pseudopod and ectoderm, the more likely it is that the mesenchyme cell will remain attached to that site. If the inner surface of the ectoderm possesses topographical differences in its adhesiveness toward mesenchyme cells, then these differences could form the basis for the eventual pattern of the mesenchyme cells. Viewed in this way,

the ectoderm serves as a template over which mesenchyme cells migrate and subsequently become distributed in some predetermined way as a result of selective adhesion. Specifically, the mesenchyme cells ultimately form a ring around the lower ectoderm in a plane that is roughly parallel to the vegetal plate. From this ring, two branches emerge that extend along the ventral surface in an animal direction (see Fig. 10.8).

The ordered congregation of the mesenchyme cells is accompanied by a progressive fusion of their pseudopodia to form a thick cytoplasmic cable (Fig. 10.6) from which numerous thin pseudopodia extend to explore the ectodermal wall. The pseudopodia of the primary mesenchyme cells appear to be particularly rigid and capable of generating considerable tensile (pulling) forces. As expected, the pseudopodia contain large numbers of microtubules oriented parallel to the long axis of the extension. The syncytial cables also have a distinct microtubular "cytoskeleton." Treatments that affect microtubule morphology are found to have a severe effect on mesenchyme cell activity, indicating the importance of these organelles. When colchicine or hydrostatic pressure (which causes microtubules to disassemble) are applied to gastrulas, the microtubules disappear and the mesenchyme cells quickly lose their pseudopodia and their migratory ability. When D_2O (which tends to stabilize microtubules) is applied, the microtubules become "frozen" in place. Since disassembly and reassembly is a prerequisite for cell motility, this stabilization is as immobilizing as the previous treatments.

More recently, attention has been focused on the surfaces of migratory cells as well as on their cytoskeletal organelles. The movement of the primary mesenchyme cells depends on the synthesis of sulfated glycosaminoglycans destined for the cell surface. If the production of these materials is inhibited, the primary mesenchyme cells appear in the blastocoel but do not migrate (Fig. 10.7). In the

Figure 10.6 *a* Light micrograph of the primary mesenchyme cells within the blastocoel of a living sea urchin gastrula. As the cells move up the wall, their filopodia fuse to form a cablelike complex. [*Courtesy of T. Gustafson.*] *b* Attachment of primary mesenchyme pseudopods to the ectoderm. [*From T. Gustafson,* Exp. Cell Res., **32**:*576 (1973).*]

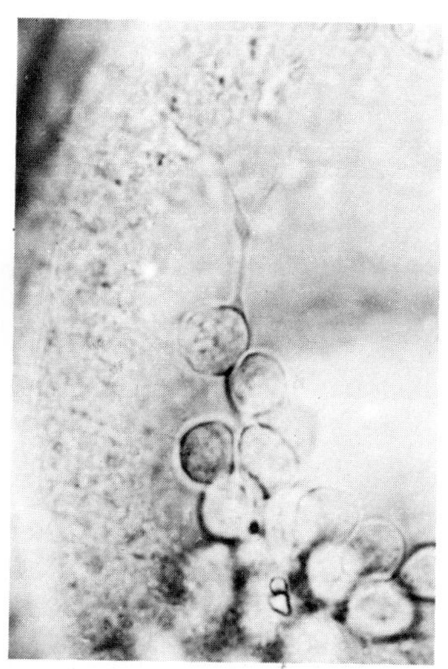

a

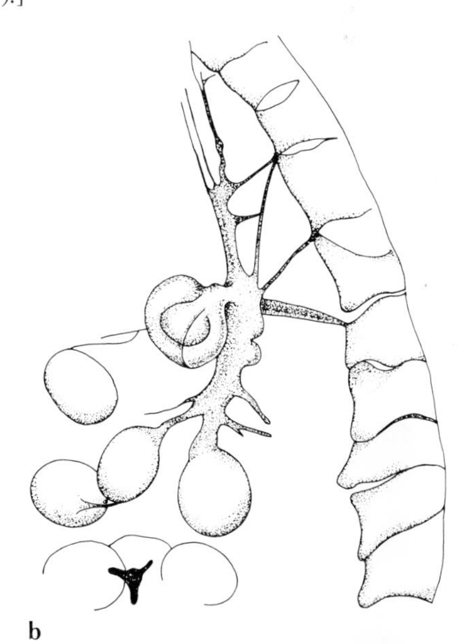

b

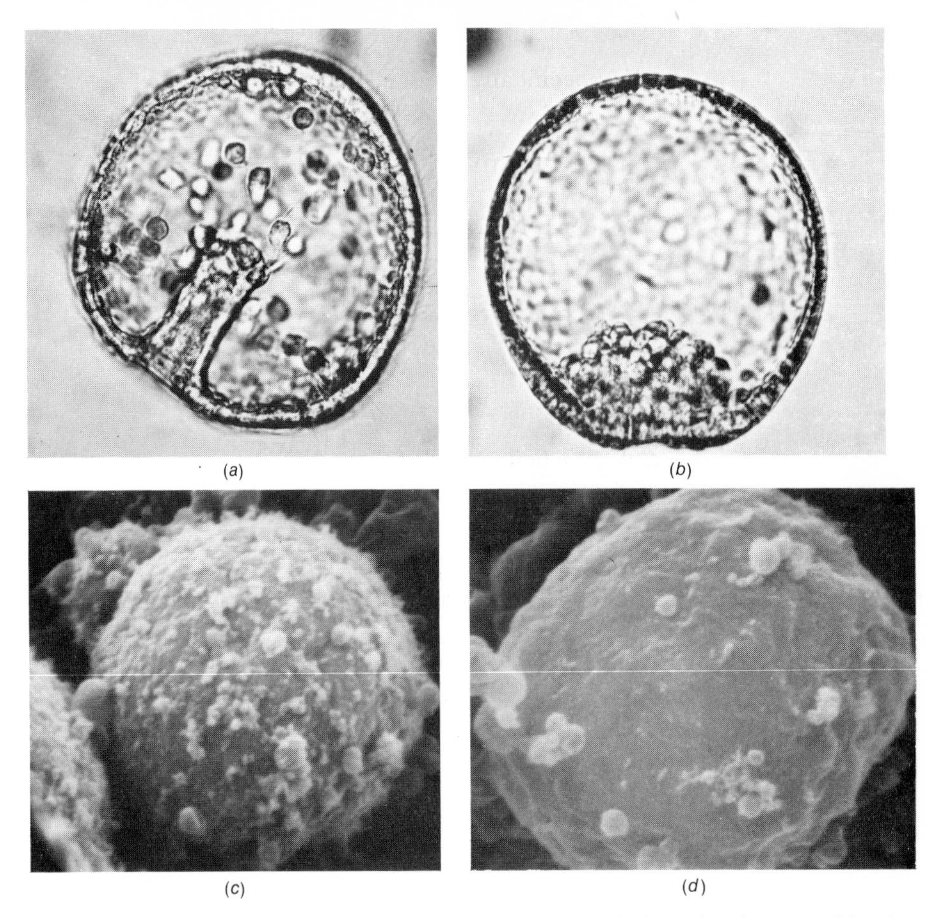

Figure 10.7 The effect of inhibition of sulfated glycosaminoglycan synthesis on the migratory activity of primary mesenchyme cells. *a* A normal gastrula raised in complete seawater. *b* An embryo of the same age as that of *a* that had developed in sulfate-free seawater. Primary mesenchyme cells collect in the blastocoel without migrating. *c*, *d* Scanning electron micrographs of primary mesenchyme cells from an embryo raised in normal (*c*) and sulfate-free (*d*) seawater. [*Courtesy of G. C. Karp and M. Solursh.*]

case depicted in Fig. 10.7*b*, inhibition of synthesis is brought about by raising the embryos in seawater lacking sulfate ions, sulfate being an integral component of these polyanionic macromolecules. Scanning electron microscopic examination (Fig. 10.7*d*) of the surface of the primary mesenchyme cells of embryos raised in sulfate-free seawater reveals a relative deficiency of extracellular materials as compared with controls. It is presumed that the immobility of the mesenchyme cells is a consequence of their altered surface morphology, which, in turn, reflects the relative absence of newly synthesized sulfated glycosaminoglycans.

Once the primary mesenchyme cells attain their proper position, they begin to carry out their major activity, the formation of the larval skeleton. The crystallization of calcium carbonate, which occurs inside the cytoplasm within an organic matrix, begins at the sites where the ventral branches meet the mesenchymal ring. Initially, a triradiate spicule is formed at each branch (Fig. 10.6), and each spicule then grows within the pseudopodial cables to form a relatively complex skeletal rod. The growth of the skeleton into the arm buds is believed to be

the primary cause of arm formation during the development of the pluteus larva (Fig. 10.8).

B. Gastrulation in amphibians

The eggs of amphibians, certain fish, and the jawless vertebrate, the lamprey—all of which are laid in freshwater—represent what may be called the *primitive vertebrate egg.* Such eggs are much larger than those of the sea urchin, the ascidian, or *Amphioxus*, having a diameter of 1 to 3 mm. Moreover, these eggs contain proportionately more yolk, which affects their manner of cleavage (Section 6.4.A), blastulation (Section 6.5), and gastrulation. Most important, the relative size of the blastocoel is reduced and the wholesale invagination of the vegetal wall of the embryo, as occurs in sea urchins, is impossible. Instead, the cells of the surface must arrive in the interior by more devious means.

Three main regions can be distinguished in the amphibian blastula—an upper part around the animal pole, a lower part around the vegetal pole, and an intermediate or marginal zone between the others that extends around the equator. In many amphibian species the animal region is most deeply pigmented, while the marginal zone, also pigmented, is typically grayish in contrast with the white vegetal region. The three regions approximately represent the future three primary germ layers, and by the end of gastrulation, the cells originally occupying the marginal and vegetal zones on the surface of the egg have become enclosed within the embryo's interior.

One of the first approaches to the study of gastrulation was taken by W. Vogt, who developed a technique to label embryonic cells. Small pieces of agar were impregnated with various kinds of vital dyes, i.e., dyes that do not injure the living cells, particularly neutral red and Nile blue sulfate, and the agar was pressed

Figure 10.8 Late gastrula *a* (prism larva) and *b* pluteus larva showing extension and subdivision of invaginated digestive tube, its union with the stomodaeal invagination, and the relation of tips of larval skeleton to location of outgrowth of pluteus arms. [*Courtesy of T. Gustafson.*]

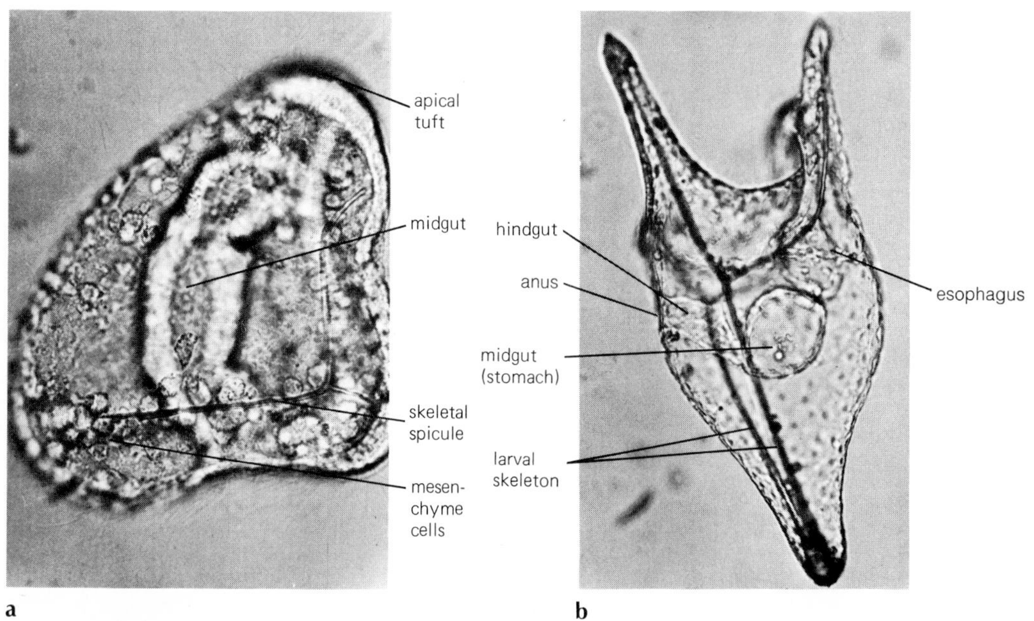

against the surface of the blastula (Fig. 10.9*a*). Subsequent shape changes and migrations of the stained patches of the blastula were then followed, revealing how extensively the parts of the blastula became rearranged. As a result of his studies, Vogt was able to construct a *fate map* (Fig. 10.9*b*), which showed the future destiny of each region of the blastula in the embryo-to-be. It is important to keep in mind that this type of fate map tells us nothing about the developmental potentialities (prospective potency) of the regions of the blastula, but only about their prospective fates.

Although a great deal of our knowledge of the morphogenetic movements of gastrulation are based on the Vogt approach, there is an important drawback that has to be considered: it limits investigation to cells of the blastula surface [see Keller (1976) for a critical discussion]. It is convenient to portray the parts of an embryo as a two-dimensional projection on the surface of an earlier stage (as in Fig. 10.9*b*), but in so doing one ignores the fate of the vast majority of the cells that lie at some depth beneath the outer layer.

i. The gastrulation process. The first indication of the onset of gastrulation in amphibian development is the appearance of a depression or groove on the dorsal midline at the border between the vegetal region and the marginal zone. The fold above the groove represents the upper, or dorsal, lip of the developing blastopore and is derived directly from the gray crescent. The formation of the dorsal lip marks the initiation of a process that underlies the entire phenomenon of am-

Figure 10.9 *a* Mapping of the movement of various areas of the salamander egg and embryo by means of vital-dye staining devised by Vogt [Arch. Entwickl.-mech., **106** (*1925*)]. The egg is held against pieces of agar stained with Nile blue or neutral red. *b* Fate map of a urodele embryo. Prospective fates of cells present on the surface of the embryo at the beginning of gastrulation. A, animal; V, vegetal; NE, neural ectoderm; EE, epidermal ectoderm; N, notochord; S, somites; PE, pharyngeal endoderm; V, vegetal endoderm; LPM, lateral plate mesoderm; TM, trunk mesoderm. [*After W. Vogt,* Roux Arch. Entwickl., **120:***392* (*1929*).]

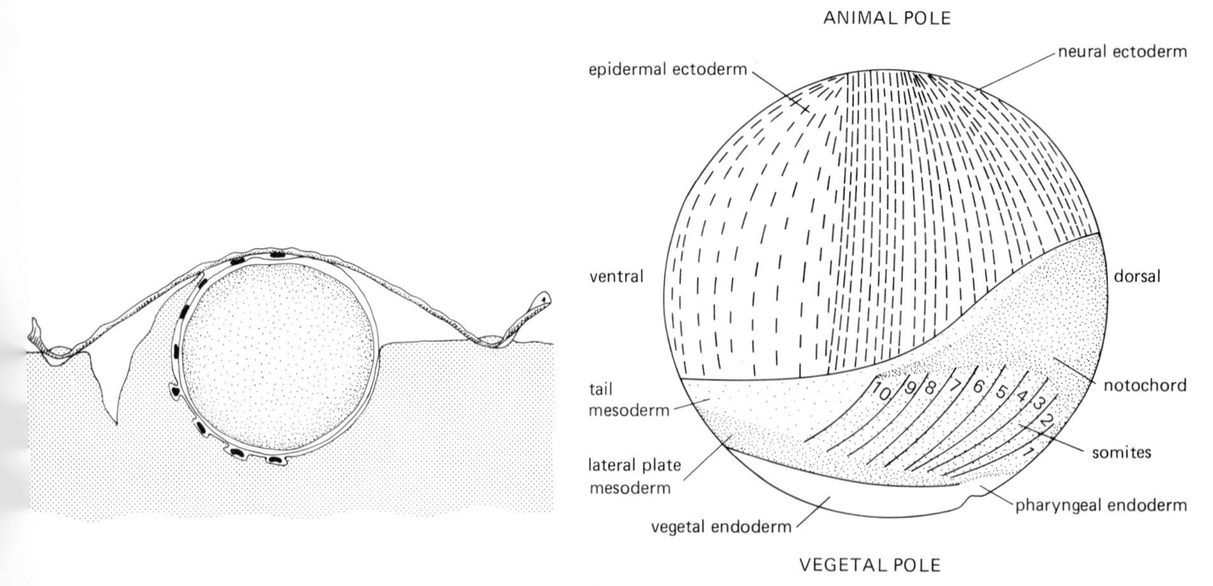

phibian gastrulation, namely, the inrolling, or *involution,* of cells. Cells at the rim of the blastopore undergo a change in shape (described later) that causes them to sink beneath the surface. Involuting cells are replaced by new cells that are moved from their original position on the surface of the embryo to the blastoporal lip. As a result, there is a flowing or streaming movement of an entire sheet of cells toward and around the blastoporal lip, as shown by the elongation of dye spots that were initially round in shape. The lip of the blastopore is therefore a continually changing structure.

Once inside the embryo, the involuted cells move away from the blastopore and deeper into the interior, forming the walls of an increasingly spacious cavity, the archenteron. The archenteric tube remains open to the outside via the blastopore. In the early stages of gastrulation, the tissue of the dorsal lip is composed of endodermal cells destined to become the foregut (prospective pharyngeal endoderm). These endodermal cells of the marginal zone (Fig. 10.10) are less yolky and more motile than their endodermal counterparts in the vegetal region. The

Figure 10.10 Gastrulation and the establishment of the germ layers, as seen in sagittal sections. [*From B. M. Patten and B. M. Carlson, "Foundations of Embryology," 3d ed., McGraw-Hill, 1974, based on W. Vogt, 1929.*]

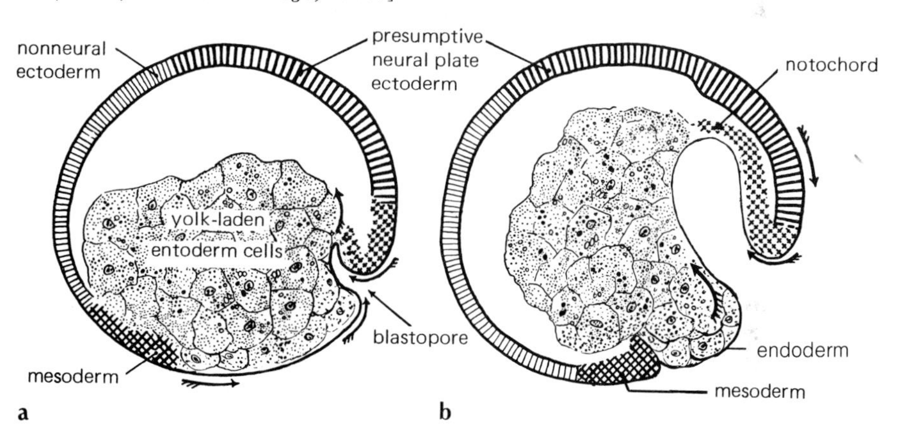

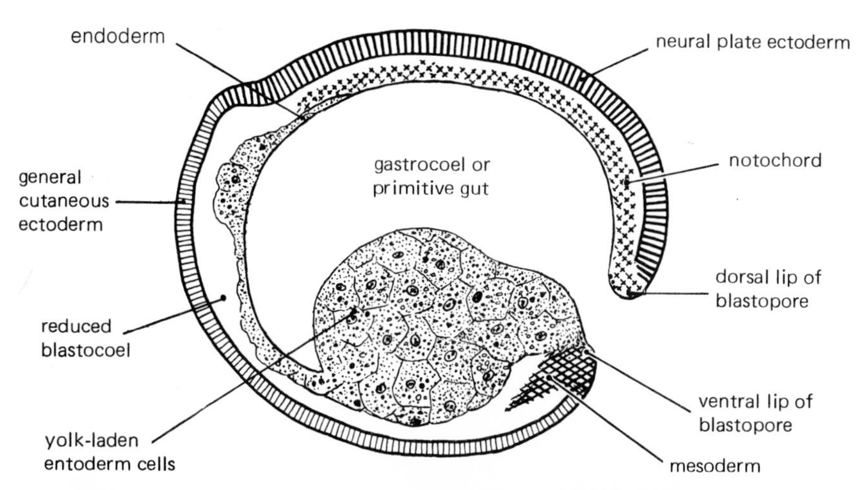

pharyngeal endoderm is followed over the dorsal lip by prospective mesodermal cells of the future head (prechordal mesoderm) and notochord (chordamesoderm). As gastrulation proceeds, increasingly more posterior internal structures are found undergoing involution at the dorsal lip (Fig. 10.10). As the process of involution continues, the archenteron lengthens, somewhat in the manner of the lengthening of the archenteric tube in the sea urchin gastrula, and the blastocoel is gradually obliterated. Eventually, the blind end of the archenteron reaches the inner surface of the ectoderm opposite the blastopore, where later a small invagination of the ectoderm fuses with the pharyngeal endoderm. The ectodermal invagination forms the mouth cavity, or stomodeum; the adjacent endoderm becomes the pharynx.

The sequence of events occurring at the dorsal lip during gastrulation is best visualized by following the shift in position of circular dye spots placed at strategic locations on the dorsal median of a late blastula from the animal to the vegetal pole (Fig. 10.11). The positions and shapes of these dye spots in the early tail bud (following gastrulation and neurulation) is shown in Fig. 10.11. It is apparent that the spots of the vegetal region (spots 9, 10, and 11) become the floor of the archenteron, spot 7 finds its way to become pharyngeal tissue, spot 6 becomes prechordal mesoderm, and spot 5 becomes notochord. The dye spot in position 4 marks the last of the mesoderm to roll over the dorsal lip, becoming part of the posterior portion of the embryo. The spots in positions 1, 2, and 3 remain at the dorsal surface of the gastrula, but then sink into the interior during neurulation to form a hollow tube, the dorsal nerve cord. The marked elongation of most of the dye spots in all three germ layers reflects the extension of embryonic tissues in the anteroposterior axis during gastrulation and neurulation.

At first the blastopore is just a short indentation on the dorsal surface, but as the inward movement of tissue continues, the two ends of the blastoporal groove extend horizontally around the embryo (Fig. 10.12*a*). The extension of the groove results from the involution of lateral mesoderm in an increasingly greater arc, forming what are termed the *lateral lips* of the blastopore (Fig. 10.14*a*). Once inside, the mesodermal cells move anteriorly as a loosely organized sheet termed the *mesodermal mantle* (Fig. 10.13). There is also a marked tendency on the part of the mesodermal cells to become concentrated (converge) toward the dorsal midline (Fig. 10.14*a* and *b*). As indicated in Fig. 10.13, the mesodermal mantle

Figure 10.11 *a* Lateral view of a late blastula indicating the positions along the median at which 11 different dye spots might be placed. *b* Embryo at the early tail-bud stage showing the location to which each of the 11 dye spots would subsequently have moved. [*After W. Vogt,* Arch. Roux Entwickl., **120:***399 (1929).*]

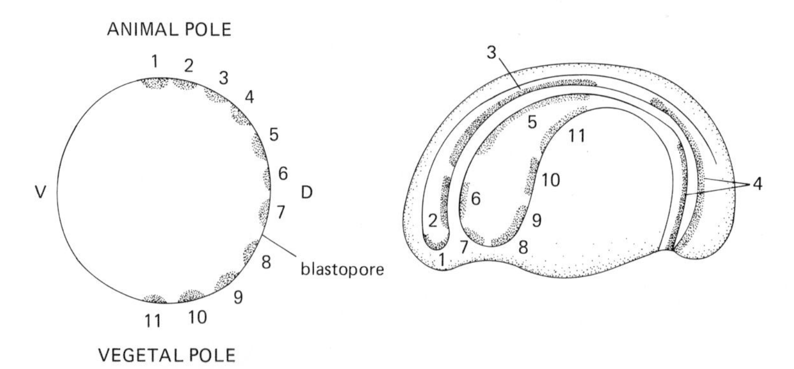

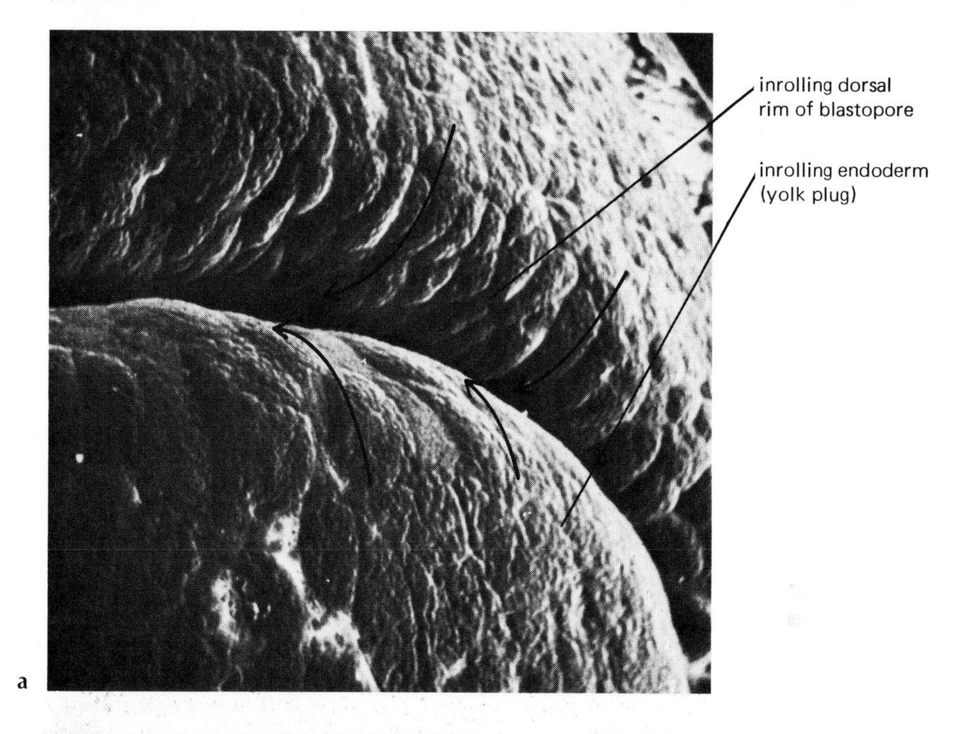

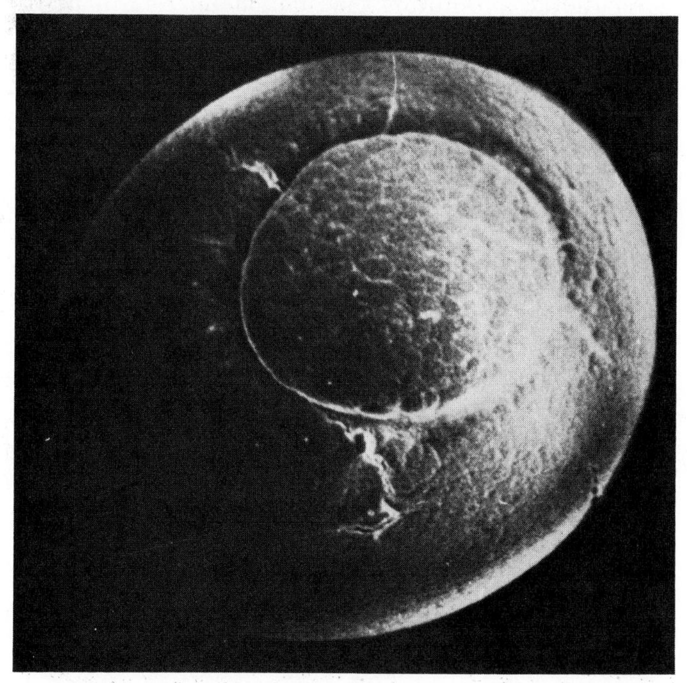

Figure 10.12 Scanning electron micrographs of late gastrulation (yolk-plug stage) of the African clawed toad *Xenopus* from the vegetal pole. *a* Inrolling of the blastopore rim around the diminishing margin of the yolk plug. *b* Epibolic enclosure of the large, yolky vegetal cells by the pigmented ectodermal-cell layer. [*Courtesy of D. Tarin.*]

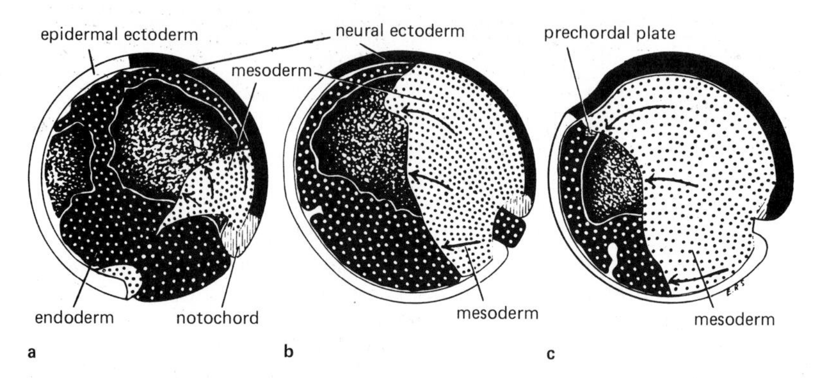

Figure 10.13 Anterior movement of the mesoderm from the blastoporal-lip area after involution in the urodele *Pleurodeles*. The progressive inward migration of the mesodermal mantle is indicated by the white area stippled with coarse dots. *a* Early gastrula. *b* Late gastrula. *c* Early neurula. [*From O. E. Nelsen, "Comparative Embryology of the Vertebrates," Blakiston, 1953.*]

does not reach into the most anterior region of the embryo during the stage of gastrulation. The extension of the blastoporal lips around the embryo continues until the ends meet on the ventral side, forming the ventral lip opposite the starting point. The formation of a ring of involuting tissue occurs by surrounding the bulk of the large, yolk-laden cells of the vegetal hemisphere. As gastrulation proceeds, the ring moves slowly in a vegetal direction sweeping down over the yolky endoderm and enclosing it within the interior.

During this latter process when the circular lips of the blastopore are moving to enclose the yolky endoderm, the upper ectodermal layer of cells extends progressively over the lower vegetal region. This process, called *epiboly*, is particularly apparent in frog embryos, where animal-pole cells are deeply pigmented and vegetal cells are white. The cells of the surface ectoderm form a tightly packed sheet that does in fact expand in surface area as it replaces the regions lost to the interior. The expansion of the sheet is believed to result from an active flattening on the part of the component cells. Eventually, the original sickle-

Figure 10.14 *a* Overall movements occurring during gastrulation in the amphibian. *b* Germ-layer relationships in the late gastrula of the frog as seen from the dorsal side. The neural plate is not indicated. [*From R. Rugh, "The Frog," McGraw-Hill, 1951.*]

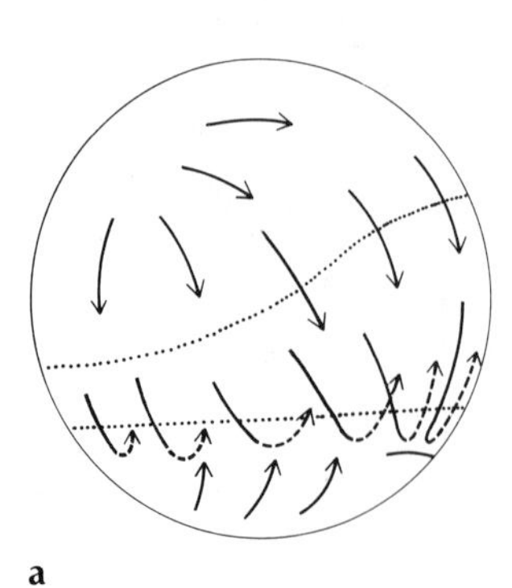

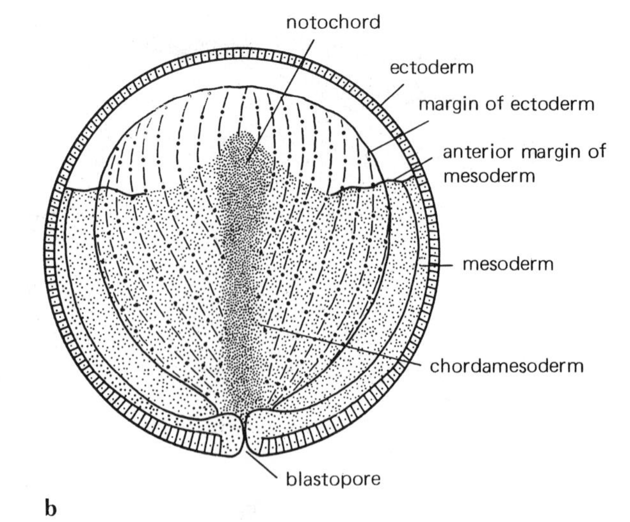

shaped blastopore becomes first an open and then a closed circle of steadily diminishing size as it sweeps vegetally to enclose the endoderm. Finally, a *yolk plug* (Fig. 10.12*b*) surrounded by dark ectoderm is all that remains at the surface of the mass of vegetal cells. At last only a narrow vertical groove denotes the blastopore and the termination of gastrulation. At this time, the ectoderm alone covers the surface of the embryo.

The apparent epibolic growth of an expanding pigmented ectoderm over a seemingly passive unpigmented vegetal region is, to some extent, an illusion. The vegetal region is also dynamically involved, even though the constituent cells are comparatively large, few, and yolky. Simultaneous with the downward- and inward-flowing stream of presumptive mesoderm cells around the extending rim of the blastopore, the presumptive endoderm streams upward and inward in a similar but grosser manner, the upper and lower streams rolling in together in a marvelously integrated motion (Fig. 10.14*a*). The full impact and beauty of this event can be conveyed only by time-lapse cinematography, for in nature the process is too slow to be visually appreciated and neither words nor still pictures can adequately portray the reality.

The diagrams in Fig. 10.14 and the description just provided are based largely on the early studies of urodele amphibians employing vital dyes. In these studies, the cells of the prospective chordamesoderm map *on the surface* of the blastula in the marginal zone above the pharyngeal endoderm. Consequently, the roof of the early archenteron consists of prechordal and notochordal mesoderm that follows the pharyngeal endoderm over the dorsal lip into the interior (Fig. 10.10). In these embryos it is believed that the endoderm of the floor of the archenteron secondarily sweeps up its wall to fuse on the dorsal surface, thereby underlying the original mesodermal roof and enclosing the future gut in endoderm. Results of dye-mapping studies on the anuran *Xenopus* provide a considerably different picture of events. In this case it appears that the surface of the blastula in the marginal zone consists entirely of prospective endodermal cells, which, following involution, form a continuous endodermal lining for the archenteron. In *Xenopus*, all the prospective mesodermal cells appear to be located in the deeper regions beneath the embryo surface. Whether these important differences in regional fates reflect actual species differences or experimental error remains to be determined.

ii. Integration of gastrulation. Gastrulation as it occurs in amphibians consists of several distinct types of activities that are integrated into an overall morphogenetic process in which each and every part of the embryo participates. Gastrulation begins with the formation of a dorsal depression within the prospective endoderm. Sections cut through this region indicate that the initial sinking of cells into the blastoporal pit is accompanied by a striking change in their shape. The cells that form the dorsal lip are referred to as *bottle* or *flask* cells (Fig. 10.15*a* and *b*); they contain a long narrow neck at their outer (apical) end and a rounded shape at their inner end. Since their discovery by early embryologists, flask cells have been thought to be largely responsible for the involution occurring at the dorsal lip. Although their importance in gastrulation has been questioned, at the very least they do appear responsible for the initiation of the process. It has been proposed in various ways that the inward sinking of the bottle cells exerts a pull on the adjoining cells causing them to be drawn into and over the dorsal lip. This concept presupposes that the bottle cells are capable of (1) generating and (2) transmitting such a force. The two ends of the bottle cells are believed to be engaged in very different activities. The constriction of the neck of the cell at its

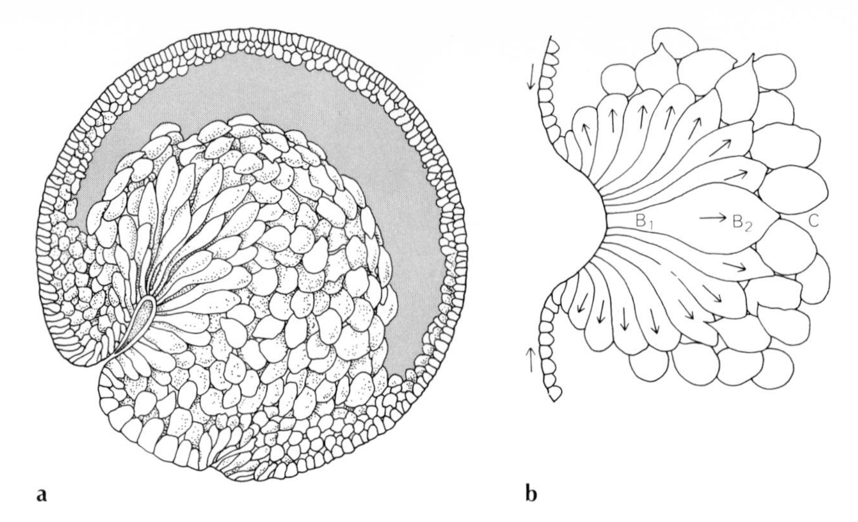

Figure 10.15 *a* Slightly schematized section through an advanced gastrula showing active extension of the bottle (flask) cells that occurs during involution. [*After J. Holtfreter*, J. Exp. Zool., **94**:*274* (*1943*).] *b* Bottle cells extending from the blastopore.

apical end results from the contractile activity of a band of microfilaments (see Fig. 11.7) in a manner analogous to the constriction of the cleavage furrow during cytokinesis (Section 6.4.B). The coordinated contraction of the group of cells in the region of the blastopore could easily account for the formation of the initial depression. The withdrawal of the bottle cells into the deep interior would seem to require a directed penetration of the surrounding tissue by the rounded inner end of the bottle cells (Fig. 10.15*b*). Although there is little morphological evidence of motile activity (such as pseudopodia or lamellipodia) at the bulbous end of these extended cells, they clearly recede deeper into the interior. The role of the bottle cells in pulling the adjoining surface cells into the blastopore is much less clear. It was proposed in earlier studies that the pull of the bottle cells was transmitted to neighboring cells via an elastic surface coat. This coat was believed to be tightly adherent to the outer surface of the cells, linking them together. Although the outer surfaces of the cells of the embryo do have a distinctly different, shiny appearance, no such surface coat is seen in the electron microscope. Since the cells of the surface layer are tightly packed and joined by intercellular junctions, the concept of involution just described does not depend on a distinct extracellular layer to coordinate intercellular activities. Moreover, other types of observations also are inconsistent with a mechanism whereby cells are pulled into the interior.

One of the difficulties in the study of amphibian gastrulation is the inability to directly observe the activities occurring within the embryo. In recent years, techniques have been developed whereby amphibian embryos can be cracked open without interfering greatly with the internal migratory events. A scanning electron micrograph of this type of preparation is shown in Fig. 10.16. Unlike the cells at the surface of the embryo, which move toward the lip of the blastopore as a tightly packed sheet, the cells of the pharyngeal endoderm and chordamesoderm migrate within the embryo in a relatively independent manner. Each cell moving along the roof of the archenteron appears to have its own locomotory or-

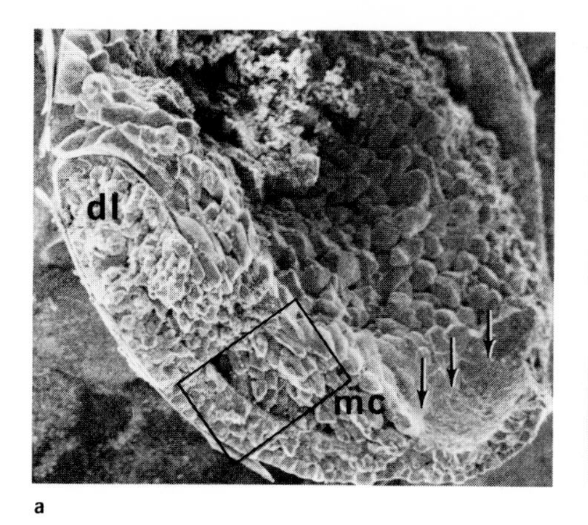

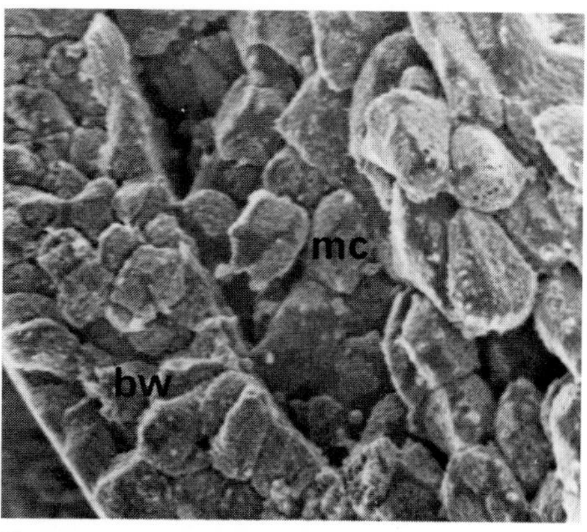

a b

Figure 10.16 Scanning electron micrographs of a bullfrog gastrula. *a* Low-magnification micrograph of a gastrula fractured in the midsagittal plane. Dorsal lip (dl) and migrating cells (mc) are visible. Frontal edge of the migrating cells is shown by arrows. *b* Higher magnification of the area shown by the rectangle in *a*. Blastocoel wall (bw) and migrating cells (mc) are shown. [*Courtesy of N. Nakatsuji.*]

ganelles, such as pseudopodia and lamellipodia, and each is not closely attached to its neighbors as it migrates along the overlying ectodermal substrate.

As in the case of the sea urchin, the migration of the involuting cells appears to be correlated with changes at their surface. In one study, which used an assay in which cells were placed in an electric field and their mobility measured, a striking correlation was found between the surface charge of the cell and its morphogenetic activity. When the electrophoretic mobility of cells from various regions of the blastula and gastrula were determined, a rapid increase in surface negativity was found to occur in the cells of the dorsal lip at the onset of gastrulation; no other region was so affected. Similarly, as cells moved into the dorsal lip, their surface charges increased. In other studies it was shown that the cells of the dorsal lip become very active synthesizers of glycosaminoglycans and glycoproteins. Once again, the synthesis of these extracellular carbohydrate-bearing molecules is correlated with cellular migratory activity.

As the cells of the endoderm and mesoderm migrate into and over the lips of the blastopore, their positions on the surface of the embryo are replaced by ectoderm cells. The increased coverage on the part of the ectoderm, i.e., epiboly, is accomplished by the actual expansion or flattening of individual cells rather than a recruitment of underlying cells into the superficial layer. We will consider the process of epiboly further in connection with gastrulation in fish and chick embryos, where it is better studied.

It is apparent from the foregoing discussion that gastrulation in amphibians encompasses several types of activities that must be coordinated if development is to proceed in a harmonious manner. Each type of morphogenetic activity, whether it is the expansion of a sheet, the involution of cells, or the movement of mesenchymal cells, results from mechanicochemical processes occurring within each of the component cells. At the same time, however, some type of supracellular agency must be at work so that groups of cells can act in a coherent manner. An important question concerns the degree to which embryonic regions are capable of independent morphogenetic movements. Can an isolated portion of an embryo carry out its morphogenetic activities in a normal manner, or do these events require the presence of the whole? An early insight into the capability of

the parts was demonstrated by the classical experiments of Hans Spemann beginning near the turn of the century. In one series of experiments, parts of one gastrula were transplanted to a neutral region of a host embryo to test its ability to self-differentiate (discussed at length in Section 11.3). When the dorsal lip was transferred from one embryo to another, a second archenteron developed at the site of the graft. It would appear that the dorsal-lip region is capable of acting in an independent manner to initiate the formation of a second site of invagination.

In an extensive series of experiments in the 1930s and 1940s, most notably by Johannes Holtfreter, the morphogenetic potential of various parts of the amphibian embryo was assessed. It was found, for example, that the spreading of the superficial layer of ectoderm that occurs during epiboly is readily demonstrated in vitro. A flap of ectoderm isolated from any region of the gastrula will expand outward in all directions. It is presumed that in the embryo the expanding ectoderm replaces the involuting mesoderm simply by following the direction of least resistance. In another series of experiments, aggregates of ectoderm, mesoderm, or endoderm were stained with Nile blue sulfate and placed on the surface of a thick layer of endoderm (Fig. 10.17). In the experiment depicted in Fig. 10.17*a*, the stained graft is derived from the blastoporal lip of an early gastrula and therefore contains prospective pharyngeal endoderm in the process of folding inward. When first isolated, the lip tissue contracts to form a solid ball. When placed on the endodermal substratum, the graft immediately adheres to the underlying layer of cells and then sinks as a whole into the host tissue. As it moves in, the graft loses its spherical shape and the individual cells (detectable by their blue stain) stretch out to form the typical bottle shape and squeeze themselves in between the unstained host tissue. The penetration process continues until the original blastoporal groove is reestablished. A very different result is obtained when the graft cells are taken from a different region of the gastrula. When a group of flattened prospective endodermal cells taken from the lining of the *blastocoel* are grafted to the same underlying endodermal substratum (Fig. 10.17*b*), the graft cells sink into the host tissue, but in contrast to a graft derived from the blastoporal lip, these cells do not form the highly attenuated bottle shape. Instead, the stained cells are simply incorporated into the lower layer without the formation of a blastopore.

Other types of combinations have revealed the importance of selective affinities of cells for one another. Up to the neurula stage, cells of different germ layers will readily unite with each other, although the adhesion between similar cells is stronger than between cells of different germ layers. For example, a ball of ectoderm or mesoderm that has been incorporated into an endodermal layer can still be lifted out of the host tissue. With the progress of differentiation, the alienation between cells of different germ layers increases, for after 2 or 3 days an imbedded piece of ectoderm appears once again at the surface of the endodermal substrate. An example of the alienation of germ layers is shown in Fig. 10.17*c*. Here a stained ball of endoderm (derived from the lining of the blastocoel, not the lips of the blastopore) partly covered by ectoderm was placed on a thick endodermal layer. The endodermal portion of the graft was able to penetrate into the host substratum, while the ectoderm began spreading over the surface. Ectoderm and endoderm have sufficiently incompatible surfaces that these two germ layers do not adhere to each other. However, in mixtures of ectoderm, mesoderm, and endoderm, the mesoderm adheres to both ectoderm and endoderm, thereby uniting two nonadherent layers. It is apparent that the complex movements that occur during gastrulation require that germ layers respond to contact with each other in appropriate ways. Mutual adhesion of cell surfaces is a highly selective event

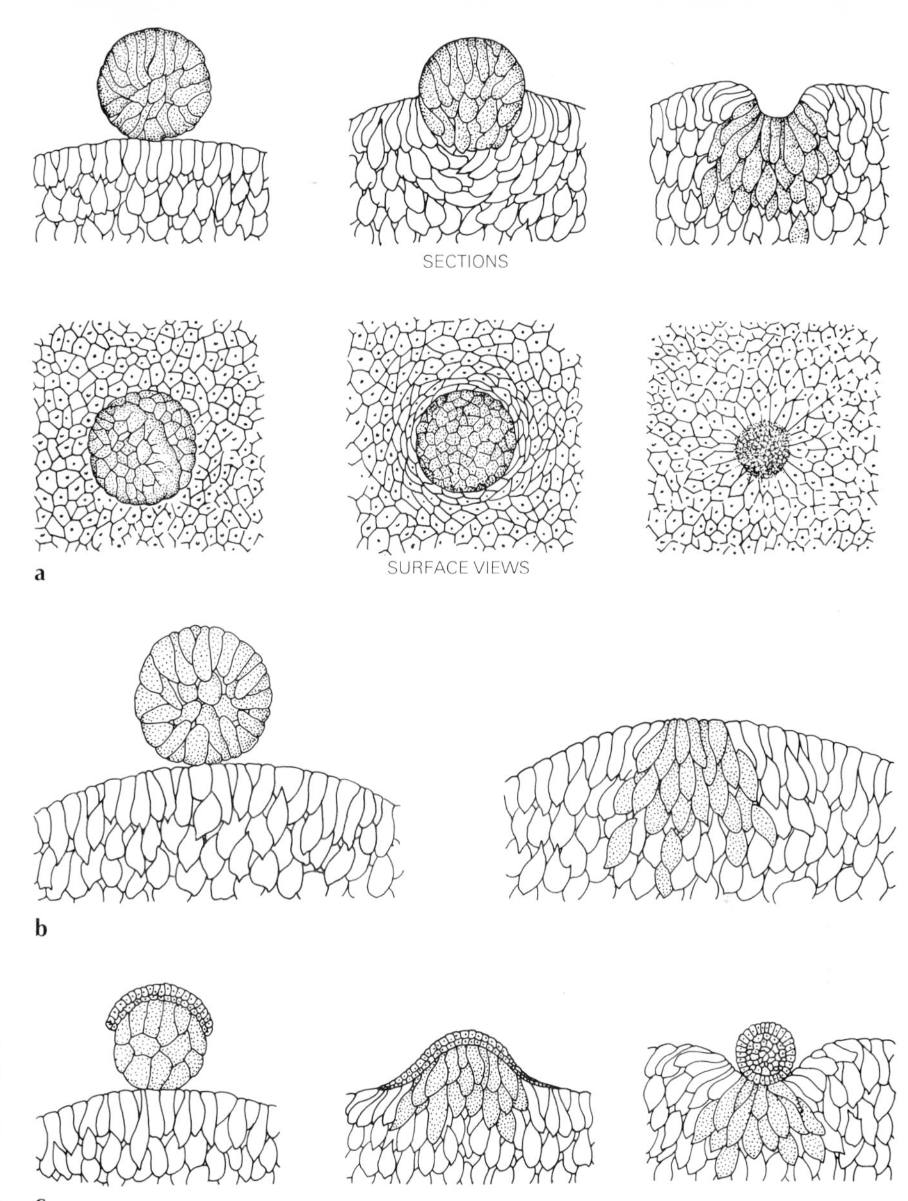

Figure 10.17 Experiments utilizing isolated fragments of amphibian gastrulas as described in the text. [*From J. Holtfreter,* J. Exp. Zool., **95:***171* (*1944*).]

and involves much more than mere differences in stickiness. The topic of selective cell adhesion is taken up in more detail in Chapter 16.

iii. Cell division and morphogenesis. The role of cell division appears to be mainly permissive rather than directive, in much the same way that the number of bricks available for building limits what may be built. Nevertheless, morphogenetic processes may proceed more or less normally in spite of notable departure from normal cell numbers. Thus in salamander larvae with nuclei that are haploid, diploid (normal), and pentaploid, respectively, cell size in various tissues

varies directly with the degree of ploidy. Although the overall size of the eggs, embryos, and larvae is essentially the same in all and constituent organs in the larvae are similar in size and form, the cellular makeup of tissues and organs is numerically very different. Larger and fewer cells are produced as the degree of ploidy increases. For example, cross sections of kidney tubules of haploid, diploid, and pentaploid larvae show eleven, six, and three cells, respectively; yet all tubules have the same diameter and wall thickness (Fig. 10.18). A similar situation is seen in sections of the lens epithelium of corresponding larvae. In all these cases, normal morphogenesis occurs.

Further evidence that embryonic organization is notably independent of cell number comes from experiments involving suppression of mitosis. When early blastulas of *Xenopus* are exposed to the drugs colcemid or mitomycin C, the mitotic cycle of the cells is blocked, with the result that normal increase in cell number is prevented from that time on. Such blockage is complete either immediately or after one further cell division. If blockage is begun as late as the stage of dorsal-lip formation, essentially normal morphogenesis ensues up to tail-bud stages, including differentiation of ectodermal structures, notochord, and twitching somitic muscle. Blockage induced at earlier stages results in developmental arrest, possibly because the smaller number of cells present is too small to permit the various morphogenetic tissue movements to take place.

2. Gastrulation in embryos which undergo meroblastic cleavage

A. Teleosts

The cleavage differential seen along the animal-vegetal axis in the amphibian egg is carried to an extreme in the eggs of bony fishes and birds and reptiles. These groups are believed to have evolved from primitive freshwater fish with amphibian-type eggs and embryos. In the teleost egg, following fertilization, yolk and cytoplasm are distinctly segregated. The core of the egg is composed of a fluid yolk sphere upon which sits a polar cap of cytoplasm (the *blastodisc*) that contains the single egg nucleus. The cytoplasm of the blastodisc is continuous with a thin strip of cortical cytoplasm, termed the *yolk cytoplasmic layer* (*YCL*),

Figure 10.18 Maintenance of normal structure in heteroploid tissues. *a* Cross sections of pronephric tubules from a haploid larva (35 days old), from a diploid, and from a pentaploid. Size of tubules and diameter of wall remain approximately the same, in spite of differences in cell size, through changes in cell shape. *b* (*Left*) Small portions of lens epithelium of the same haploid, diploid, and pentaploid larvae. Thickness of epithelium remains nearly constant. (*Right*) Corresponding portions of lens epithelium of an older diploid larva and of a pentaploid of the same age and stage. Cells and nuclei of the pentaploid again flattened to about same diameter as in the diploid. [*After G. Fankhauser*, Quart. Rev. Biol., **20**:20 (*1945*).]

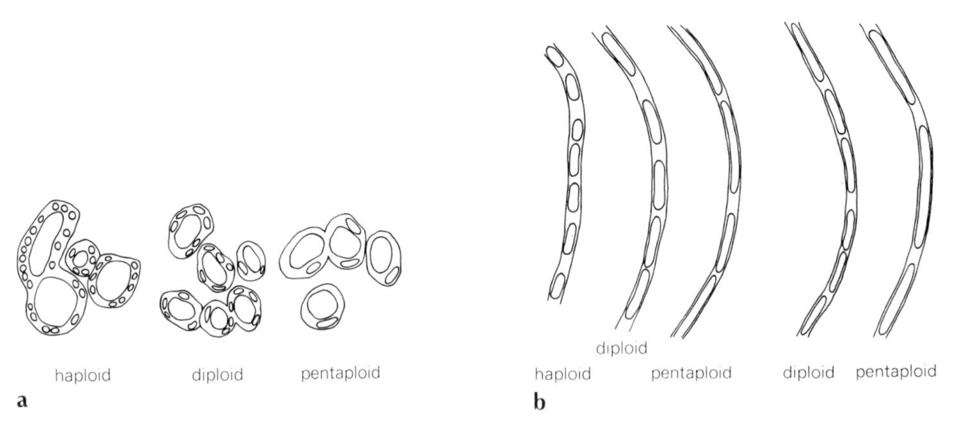

that completely surrounds the yolk mass. Cleavage of the fish egg is partial, i.e., meroblastic, being confined to the blastodisc. As cleavage progresses, several distinct regions of the blastodisc are formed (Fig. 10.19). There is an outer cellular epithelial layer, termed the *enveloping layer* (*EVL*), and an inner noncellular layer, termed the *yolk syncytial layer* (*YSL*), or *periblast*. The YSL sits directly on the polar end of the yolk mass; it contains nuclei formed from mitotic divisions of the original zygote nucleus, but cleavage furrows do not form between the cells, thereby producing a multinucleated (syncytial) layer. The YSL is continuous with the YCL surrounding the bulk of the yolk. In between the EVL and YSL lies a dense population of individual cells, referred to as *deep cells*. The cellular part of the egg (the EVL and deep cells) is termed the *blastoderm*.

Gastrulation in bony fishes is characterized by a remarkable process of epiboly by which the blastoderm and underlying YSL sweep down to the vegetal pole to completely enclose the central yolk. The progress of these events is indicated in Fig. 10.20. Although both YSL and blastoderm are stretched over the yolk mass, it is the YSL that leads the way and apparently supplies the active force. The blastoderm is attached to the underlying YSL at its margin. If this marginal adhesion is cut, the blastoderm immediately retracts. If the blastoderm is completely re-

Figure 10.19 *a* Diagram of an early gastrula of the teleost *Fundulus*. EL, enveloping layer; DB, deep blastomeres; P, periblast; SC, segmentation cavity; PN, periblast nucleus; YCL, yolk cytoplasmic layer; Y, yolk. [*From T. L. Lentz and J. P. Trinkaus,* J. Cell Biol., **32**:*124* (*1967*).] *b* Diagram showing the relationship of the enveloping layer (EVL), deep cells, internal-yolk syncytial layer (I-YSL), external-yolk syncytial layer (E-YSL), and yolk cytoplasmic layer (YCL) in *Fundulus*. [*From T. Betchaku and J. P. Trinkaus,* J. Exp. Zool., **206**:*385* (*1978*).]

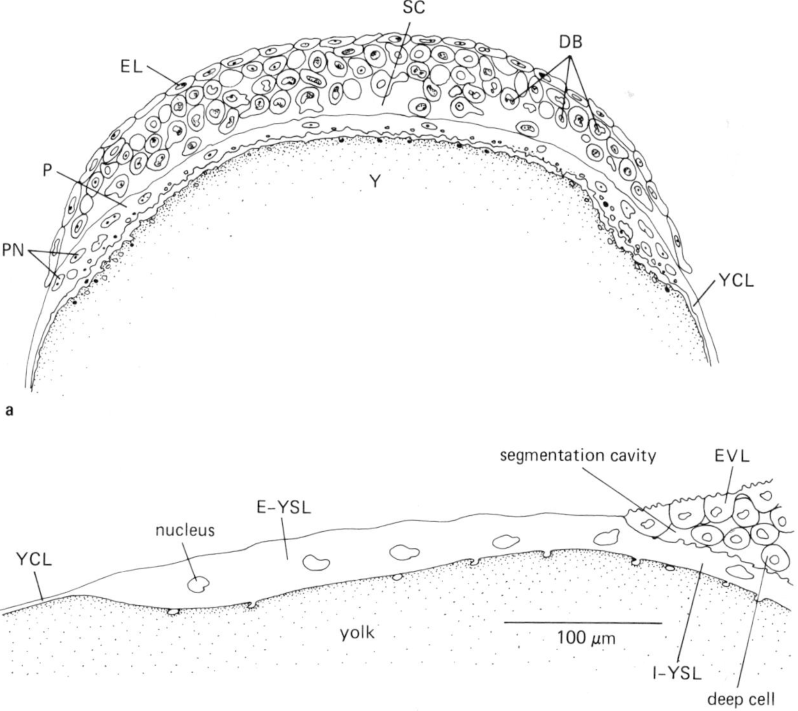

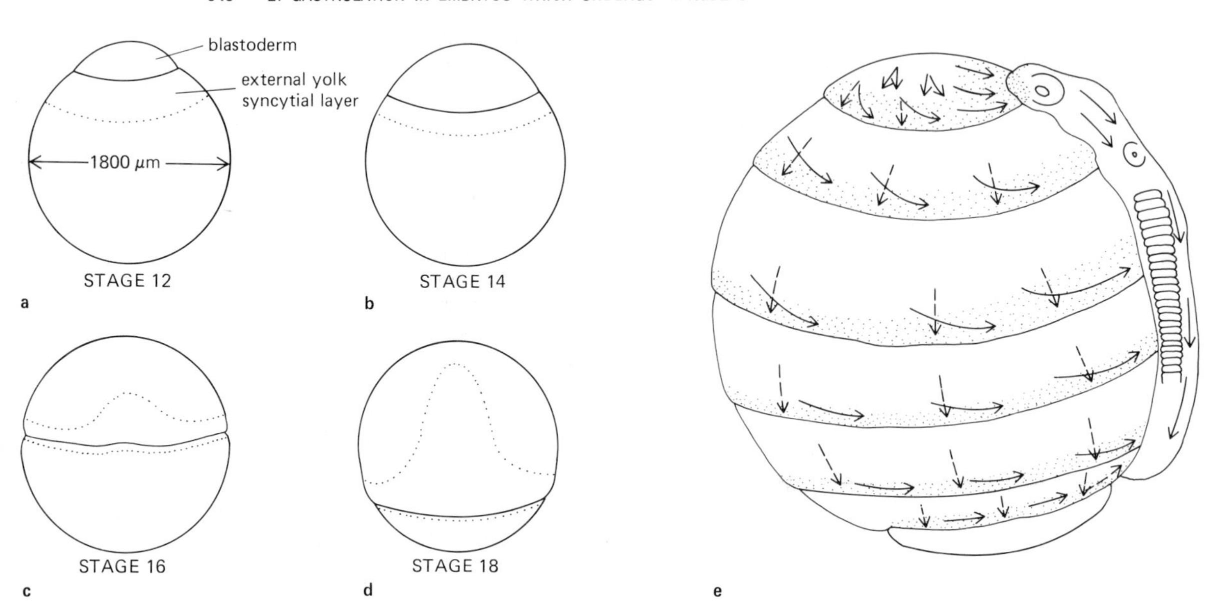

Figure 10.20 *a–d* Outline drawings of whole eggs showing the extent of the blastoderm and yolk syncytial layer for each stage. The germ ring and embryonic shield are represented in outline on the blastoderms of stages 16 and 18. [*From T. Betchaku and J. P. Trinkaus,* J. Exp. Zool., **206**:*385* (*1978*).] *e* Diagram of a later stage in embryo formation. Cells of the outer epiblast accomplish the completion of the yolk sac by epiboly (*dashed arrows*), while underlying cells descend with the germ ring but continually converge toward the body axis forming somites and lateral plate. [*From W. W. Ballard,* J. Exp. Zool., **184**:*44* (*1973*).]

moved, the YSL continues to carry out the entire process of epiboly by itself. In fact, in the absence of the blastoderm, the YSL sweeps over the yolk in a more rapid than normal manner, suggesting that the blastoderm slows down the progress of the underlying YSL. It would appear that the blastoderm is carried along passively in piggy back style during gastrulation. Both the EVL and YSL have been shown to contain extensive networks of microfilaments, and both layers are under considerable tension, as if they are being stretched over the yolk. If, for example, either layer is isolated, it becomes rapidly contracted to form a dense ball. Although the contractility of the EVL and YSL are believed very important in the process of epiboly, the manner of this involvement remains obscure.

Although epiboly in fish is accompanied by mitosis and cell division in the blastoderm (and mitosis in the YSL syncytium), these events can be blocked without disruptive effects on the spreading process. We can conclude therefore that epiboly occurs primarily as a result of the extension of cells rather than their multiplication. With the continuation of epiboly, the thickness of both the blastoderm and YSL decreases as their cytoplasm is made to cover a greater and greater surface.

As the margin of the YSL moves down over the yolk, the form of the embryo proper begins to take shape within a portion of the blastoderm, termed the *embryonic shield* (Fig. 10.20*d*). The embryonic shield, and the embryo contained within, forms primarily from a convergence of the deep cells toward the dorsal region of the egg. Contrary to earlier findings, there appears to be no movement of cells from the overlying EVL to the deeper regions of the enbryo. Rather, the

EVL contains only prospective ectoderm and is involved only in formation of the epidermis, the outer covering of the animal. The deep cells are believed responsible for the formation of the mesodermal and endodermal organs as well as the embryonic nervous system. At about the late blastula stage, the rounded deep cells begin to form blebs that elongate and take the form of fingerlike processes referred to as *lobopodia* (Fig. 10.19*a*). The lobopodia make attachments and pull the rounded cell body forward. Gradually the deep cells converge toward the dorsal side (indicated by the lateral arrows of Fig. 10.20*e*) and somehow become arranged to form the complex tissues of the dorsal axis; there is no evidence of any type of involution or archenteron formation. As the embryo forms, the YSL and the remainder of the EVL develop into the yolk sac, which functions to make the nutrients taken from the yolk available to the growing embryo.

Although most fish eggs develop essentially in the manner just described, one group of these animals, the annual fish, shows certain striking differences. This group of fish live only in temporary water holes or swamps in parts of South America and Africa. When their habitat dries up, as it does in yearly climatic cycles, the adults and juveniles die off, but their presence continues in the form of eggs buried in the mud. When the rainy season returns and the ponds refill, the eggs hatch and the larvae rapidly grow to maturity and reproduce. Their unusual life cycle is accompanied by very unusual developmental characteristics. First, after cleavage produces a certain number of blastomeres, a striking event occurs: the embryo-forming cells enter a stage of complete dispersion and become scattered around the egg, after which they reaggregate before embryogenesis can begin. Second, annual fish eggs are capable of undergoing *diapause*, or developmental arrest, at one or more distinct embryonic stages, depending on environmental conditions.

Stages in the early development of *Austrofundulus*, an annual fish, are shown in Fig. 10.21. Two populations of blastomeres are formed during cleavage, and they segregate at the blastula stage in the typical manner of teleostean development (Fig. 10.21*a*). By the early blastula stage there are flattened periblast cells adjacent to the yolk and continuous with a surface-enveloping layer, and there is a compact mass of ameboid, more or less spherical, blastomeres (Fig. 10.21*b*). As epiboly commences, the ameboid blastomeres consolidate into a mass and then migrate away from that mass as individual cells. They move through the space formed between the enveloping layer and the periblast as they advance over the yolk. When epiboly is completed, the ameboid blastomeres of the 3-day egg, which will later form the embryo, are completely dispersed (Fig. 10.21*d*). At this stage, no embryonic shield is present; this would be an obvious impossibility.

During the late dispersed phase, cell contacts increase in number and duration until, after a few days, these cells come together to form a definitive aggregate within which embryogenesis occurs (Fig. 10.21*e*). In 10 days, a solid ridge has formed in the aggregate, while development of a typical teleostean embryo follows (Fig. 10.21*f*). The development of five other genera of annual fishes appears to be essentially similar. The reaggregation may therefore be considered identical to the embryonic shield of nonannual teleosts. The striking differences in the manner in which gastrulation occurs among various fishes, as well as between fish and amphibians, illustrate that contrary to popular belief, early development need not be evolutionarily conservative. Hundreds of millions of years have separated various vertebrate groups from each other, and even though there are obvious basic similarities in the development of vertebrates, there also are basic differences.

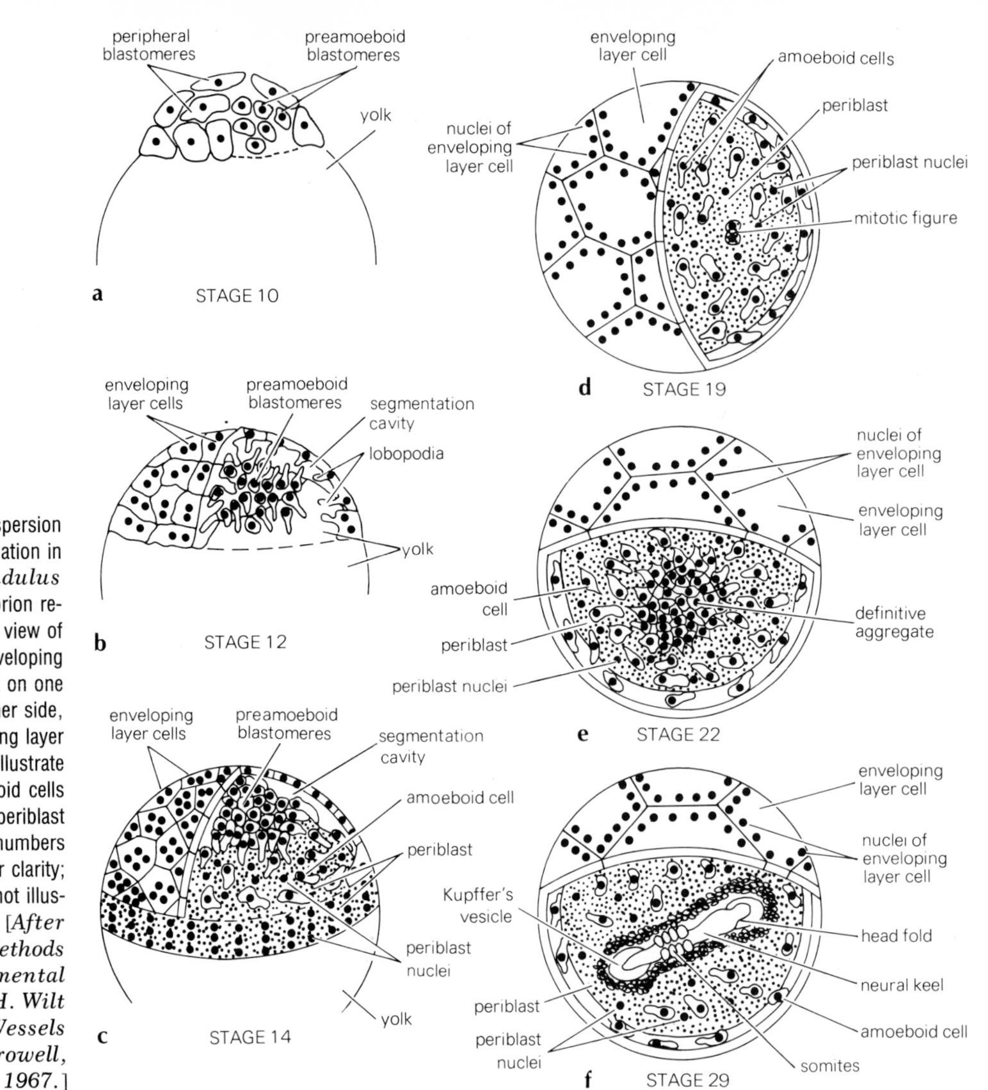

Figure 10.21 Dispersion and reaggregation in *Austrofundulus myersi*. Chorion removed; surface view of the egg with enveloping layer intact on one side; on the other side, the enveloping layer is removed to illustrate underlying amoeboid cells and periblast; periblast stippled; cell numbers reduced for clarity; oil droplets not illustrated. [*After J. Wourms, "Methods in Developmental Biology," F. H. Wilt and N. K. Wessels (eds.), Crowell, 1967.*]

3. Gastrulation in amniotes

A. Birds and reptiles

As in the teleost, cleavage in the meroblastic eggs of birds and reptiles is initially confined to a disc of cytoplasm, i.e., a blastodisc, at the animal pole of the egg. Cleavage furrows continue to divide the disk superficially, leaving the cells still continuous with undivided yolk below (see Fig. 6.10). Then horizontal furrows cut away the superficial cells, except at the disk margin, leaving a space, the blastocoel, above the yolk (see Fig. 6.25). The blastodisc is thereby converted into the blastoderm, which becomes divided into three or four cellular layers by horizontal divisions.

The central cells overlying the blastocoel constitute the clear *area pellucida*, which first appears as a circular region at the egg surface. The marginal cells of the blastoderm which surround the area pellucida and adjoin the yolk are darker in appearance and form the *area opaca*. The outermost marginal cells that con-

nect the blastoderm with undivided cytoplasm of the yolk surface (the *periblast*) form the *germ wall*. These early stages of development occur while the fertilized egg is still passing down the oviduct of the hen. At the time of laying, about 60,000 cells are present in the chick blastoderm. An important event that occurs at about the time of laying is the formation within the blastoderm of two distinct layers, an upper *epiblast* and a lower, thinner *hypoblast*. The manner in which these two layers form is still not completely clear. It is generally believed that the hypoblast arises from two semiindependent processes: the movement of cells vertically from the epiblast down to the lower level (a process termed *delamination*, or *polyinvagination*) and the migration of cells horizontally in an anterior direction out from the posterior marginal zone of the area pellucida (i.e., the *germ wall*). Cells from these two sites are believed to merge to form a uniform layer beneath the epiblast.

In the chick, the two aspects of gastrulation, namely, the advance of the blastodermal margin around the yolk mass and the establishment of a visible embryonic axis, are regionally separate processes. The growth of the blastoderm margin by epiboly occurs in a manner similar to that in teleosts, although in the chick it is concerned exclusively with enclosure of the yolk within the yolk sac and not with the formation of the embryo. In contrast to the teleosts, the embryo-forming area in the chick lies within the margins of the initial blastoderm. In the chick, the expansion of the blastoderm is accomplished by the active movement of the margin of the epiblast along the inner surface of the overlying vitelline membrane (Fig. 10.22). The vitelline membrane is an extracellular fibrous structure that is formed in the oviduct and adheres tightly to the outer surface of the egg. Consequently, the enclosure of the yolk, which takes about 4 days to com-

Figure 10.22 Scanning electron micrograph of a typical region of the ventral face of the edge of a 24-hour (stages 7 to 9) epiblast (epi) seen on the fibrous vitelline membrane (vm). Visible on the left is a broad lamellipodium (lam) with small filopodia along its edge. Toward the right is an area where filopodia are more prominent (fil). Epiblast cells just behind the edge are very tightly apposed. Cell boundaries are virtually indistinguishable. Yolk is present in the lower right quadrant of the figure. [*Courtesy of E. A. G. Chernoff and J. Overton.*]

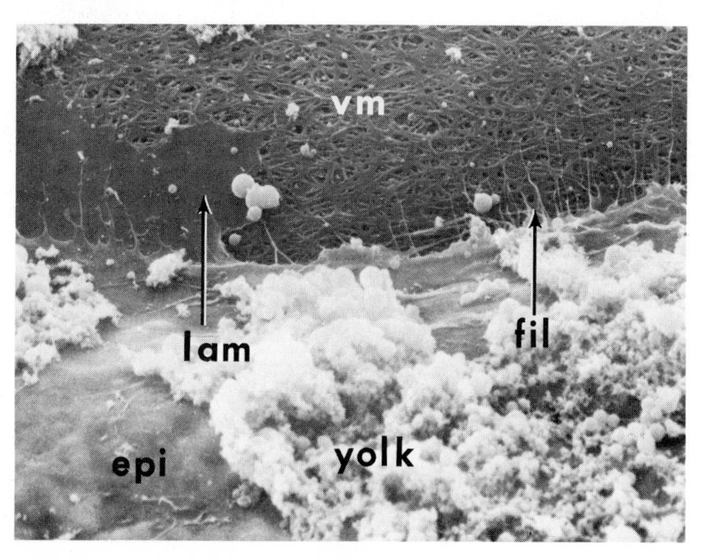

plete, occurs when the edge of the blastoderm squeezes through the thin space between the yolk and overlying vitelline membrane clinging to the undersurface of the latter structure. As in the case of the teleost, the blastoderm is under considerable tension as it grows to cover a greatly expanded surface area.

The formation of the embryo in birds, reptiles, and mammals centers around the establishment within the area pellucida of a structure referred to as the *primitive streak.* A foreshadowing of the primitive streak first appears as a thickening in the midline at the posterior region of the area pellucida (Fig. 10.23*a*). This thickening results from the movement of cells located within the posterior portion of the epiblast toward the midline. This thickening in the epiblast spreads anteriorly, first over a rather broad area, then gradually constricting to form a narrow streak (Fig. 10.23*b*). The elongated axis of the primitive streak marks the anteroposterior axis of the future embryo.

Figure 10.23 Surface-view drawings of photographs of the developing primitive streak. *a* Initial streak: a short, conical thickening at the posterior end of the blastoderm. *b* Intermediate streak: the thickened streak area approaches the center of the area pellucida. *c* Definitive streak: the primitive groove, primitive pit, primitive fold, and Hensen's node are present. *d* Head-process stage: the notochord or head process is visible as an area of condensed mesoderm extending anteriorly from Hensen's node. The proamnion is indicated in the front portion of the area pellucida; the head fold is not yet present. [*From O. E. Nelsen, "Comparative Embryology of the Vertebrates," Blakiston, 1953; after V. Hamburger and H. L. Hamilton,* J. Morph., **88**:*49* (*1951*).]

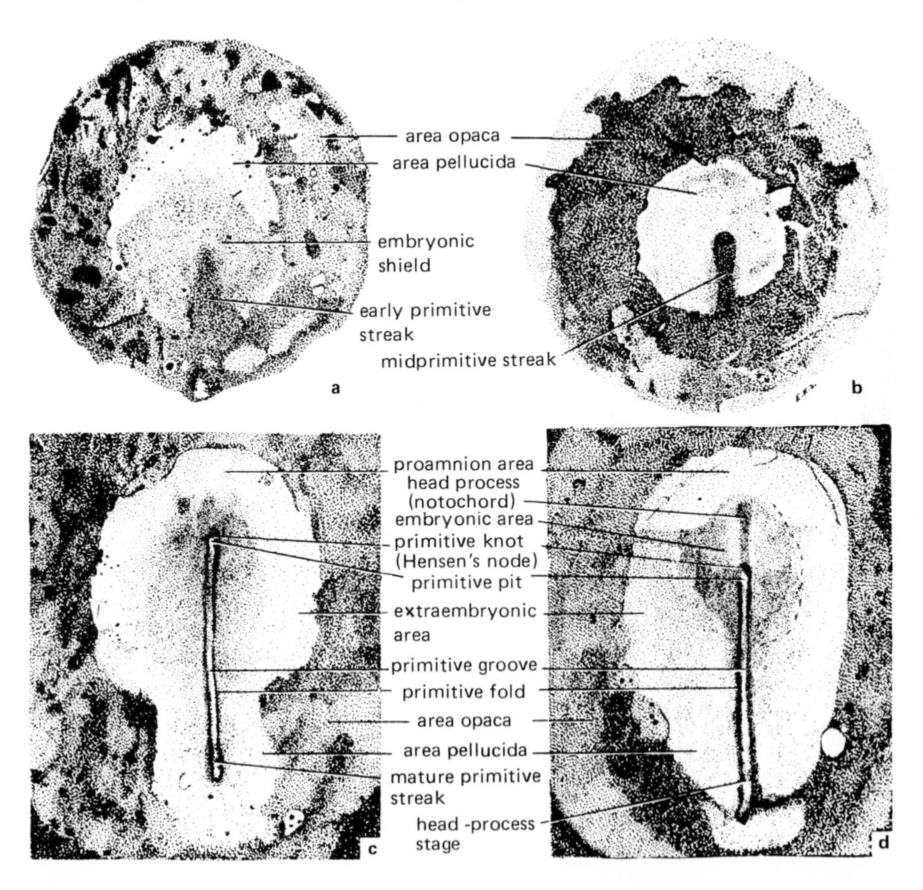

It often has been suggested that streak elongation is due to an incorporation of new material added to one end or the other. Markers added to the anterior and the posterior tip, however, remain closely associated with these two ends of the streak. Streak lengthening therefore results primarily from a very active stretching of the area in which the streak has appeared. The lengthening is due to an active pushing forward and backward, which would explain how the circular area pellucida gradually becomes pear-shaped (Fig. 10.23*d*) and why an early primitive streak already possesses bipolarity, since both halves of the younger streak are remodeling in opposite directions.

Gastrulation in higher vertebrates is closely related to the whole story of the primitive streak. The morphogenetic movements leading to the formation of the primitive streak begin as soon as the incubation temperature rises to the normal level (38.5°C). In order to follow the movement of cells from various regions of the blastoderm, several types of marking procedures have been employed. In earlier studies, vital dyes or carbon particles were applied externally to a patch of cells and the embryo was allowed to proceed with development so that the course of the marked cells could be followed. In some studies, embryos were labeled in ovo, i.e., without removal of the embryo from the egg, while in other studies, the blastoderm was removed from the yolk and cultured in vitro on an artificial substrate while gastrulation proceeded.

Unfortunately, these types of marking procedures have a definite drawback: one can never be certain that the dye or carbon particles are remaining associated only with those cells initially marked. To overcome this uncertainty, investigators have come to rely on procedures by which cells can be distinguished by nondiffusible internal markers. In one series of studies, the DNA of the cells being followed is labeled by exposure to [^{3}H]thymidine for several hours. Since cells in all regions of the embryo are undergoing replication and division during these early stages, a large percentage of the cells of the embryo become labeled. After incubation with the radioactive DNA precursor, a small piece of the labeled embryo is removed and transplanted to the corresponding site of an unlabeled embryo (after excision of that region of the recipient prior to transplantation). Once the labeled graft is in place, the wound rapidly heals and labeled cells move out into the unlabeled regions of the host blastoderm. The position of the labeled cells can be determined at various times after transplantation by autoradiography. Finally, similar types of experiments have been carried by formation of Japanese quail-chick chimeras. In these latter studies, parts of a blastoderm from one species are transplanted to a corresponding region in the blastoderm of the other species. Since the cells of the two birds are readily distinguished (the Japanese quail cells have heterochromatic bodies in their nuclei), the movements of donor cells can be followed.

Detailed marking studies indicate that the primitive streak forms as a result of the convergence of epiblast cells to the dorsal midline of the blastoderm. As the definitive streak is formed, a groove (the *primitive groove*) appears along its length, and it is toward this groove that cells migrate and via this groove that cells reach the interior. The thickening of the blastoderm that constitutes the primitive streak has been likened to the margin of the amphibian blastopore, here fused as a single structure. The thickened anterior end of the streak, known as the *primitive knot,* or *Hensen's node* (Fig. 10.23*c* and *d*), would correspond to the dorsal lip. The *primitive ridges,* which form the edges of the groove along its length, would correspond to the lateral lips. As in the amphibian, the edges of the groove are continually changing structures comprised of cells in the process of leaving the surface layer (epiblast) and moving into the interior. Results from

various studies suggest that all of the ectodermal, mesodermal, and endodermal cells *of the embryo* are originally derived from cells located in the epiblast. Of this collection, the endodermal and mesodermal cells migrate through the streak into the lower layers (Fig. 10.24) leaving the ectodermal cells as the sole residents of the definitive epiblast layer. Although both prospective endodermal and mesodermal cells migrate through the primitive streak, these two groups of cells migrate to different regions of the embryo. The endodermal cells move into the streak and take up occupancy within the dorsal regions of the hypoblast, gradually displacing the original hypoblast cells into more distal regions where they

Figure 10.24 Grastulation in chick embryo. Portion of 24-hour embryo (3- to 4-somite stage) showing embryo with head fold, open neural groove and neural folds, somites, the node, and primitive streak. The blastoderm has been transected about midway along the length of the primitive streak to show the cellular constitution of the area pellucida and area opaca. Inset, an enlargement of the primitive streak showing the cells of the epiblast moving down and migrating laterally as the mesoblast. The hypoblast is seen as a thin underlying epithelium. Compare with the invaginative dorsal-lip process of the amphibian embryo seen in Fig. 10.15. Following immigration, the epiblast, mesoblast, and hypoblast correspond to the three primary germ layers, ectoderm, mesoderm, and endoderm. [*From E. D. Hay, "Epithelial-Mesenchymal Interactions," Williams and Wilkins, 1968.*]

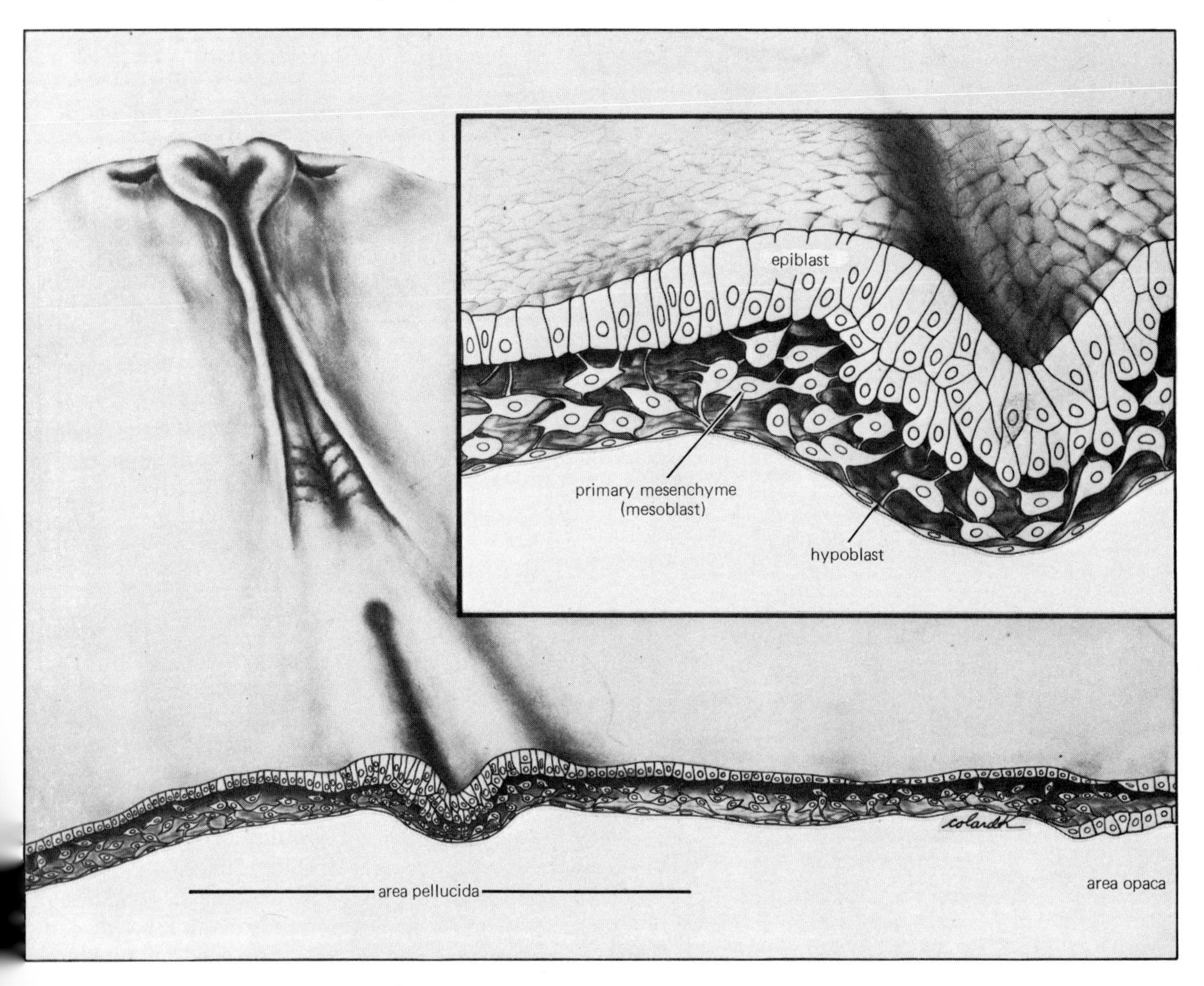

participate in the formation of extraembryonic membranes. In contrast, the mesodermal cells that migrate through the streak move out as a middle layer (Figs. 10.24 and 10.25) which forms between the epiblast and hypoblast.

Unlike gastrulation in amphibians, where cells move toward the blastopore and over its lips as a relatively cohesive sheet, the cells in the chick gastrula move as a loosely connected network, i.e., a *mesenchyme*. Although the cells of the mesenchyme are largely independent of one another, they all move in a

Figure 10.25 *a* Schematic cross section of the chick embryo in the primitive-streak stage showing cells of the epiblast entering the streak and turning in between the layers to form the mesoblast. [*From B. M. Patten and B. M. Carlson, "Foundations of Embryology," 3rd ed., McGraw-Hill, 1974.*]
b, c Autoradiographs demonstrating the actual movements of presumptive mesodermal cells. A graft of [^{3}H]thymidine-labeled epiblast from a donor embryo was placed into the epiblast of an unlabeled recipient at a point between the two arrows. After a period of time, the embryo was fixed and an autoradiograph prepared (*b*), revealing the distance that the cells had moved within the epiblast. If an embryo with this type of graft is allowed to progress to a stage of organogenesis (*c*), autoradiographs show the labeled cells scattered through the mesodermal organs. Note that labeled cells are found in tissue on both sides of the midline, even though the original graft had been placed at a discrete spot on one side of the embryo. [*Courtesy of G. C. Rosenquist.*]

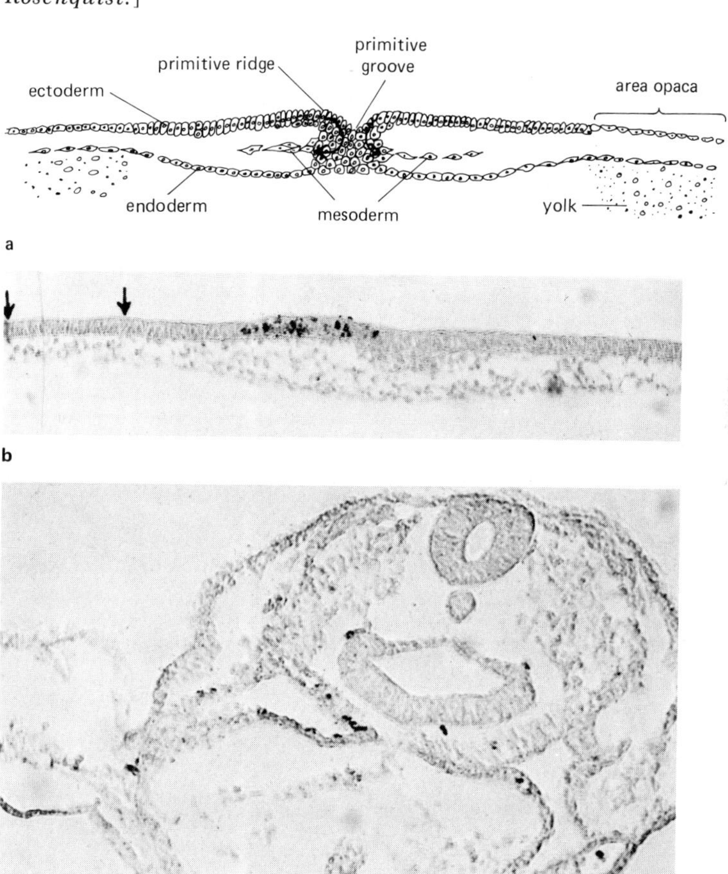

directed manner toward predetermined sites (Fig. 10.25*c*). Once they arrive, the mesenchyme cells often regroup and form a fully cohesive epithelial layer. For example, the notochord, somites, and lateral plate all represent epithelial structures formed following the aggregation of loosely organized mesenchymal cells.

Even though there is considerable extracellular space within the mesenchyme, cells are seen to be connected to one another at scattered points on their surface, as well as to the substrate cells over which they migrate. Electron micrographs of these contacts indicate that they represent sites of small but organized intercellular junctions. Freeze-fracture analysis suggests that the contacts consist of small gap junctions and short sections of tight junctions. It may be that these rudimentary junctions represent remnants of more extensive junctions that previously existed in the epithelium of the epiblast. In addition to making contacts with one another, as these migrating mesenchymal cells move away from the primitive streak, they send out filopodia that make transient contacts with an extracellular basal lamina that lines both the epiblast and hypoblast layers. It has been proposed that contacts with the surfaces lining the cavity through which the cells move may be important in guiding them in their migration.

As indicated elsewhere in discussions of other morphogenetic events, considerable insight into a process can be gained by analysis of the shapes of the cells involved. Although there is always some danger of the formation of artifacts, examination of preparations in the scanning electron microscope provides one of the best techniques for the study of cell shape. As shown in Fig. 10.26, there are striking shape changes in cells as they move through the primitive streak. As the

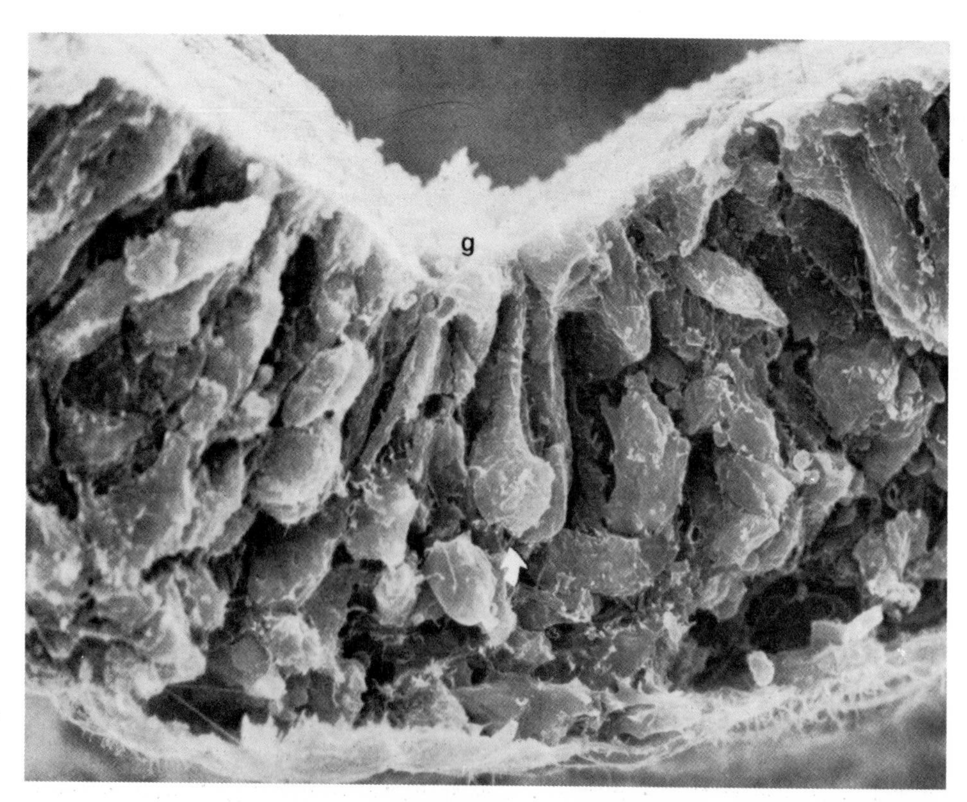

Figure 10.26 Scanning lectron micrograph of a chick embryo fractured ross the primitive streak. [*Courtesy of . Solursh and J. P. Revel.*]

cuboidal cells of the epiblast move into the streak, they become elongated and then flask-shaped. As in the lips of the amphibian blastopore, the apical end of the flask-shaped cells become constricted, forming a long thin neck that tapers into a basal rounded portion at the other end of the cell. As the apical end of the epiblast cell is released from its point of contact at the surface of the primitive streak, the whole cell becomes shortened and rounded and advances into the stream of migrating mesenchymal cells. Once in the space between the epiblast and hypoblast, the mesenchymal cells become flattened and send out locomotory processes (lamellipodia and filopodia), by which they move in typical single-celled fashion.

As in the gastrulation of other animals, the synthesis of extracellular materials is believed to be of particular importance in the migratory activities of embryonic chick cells. Gastrulation in the chick is a time of considerable synthesis of extracellular materials, particularly hyaluronic acid, a nonsulfated glycosaminoglycan. In this case, hyaluronic acid accumulates in the jellylike space between the epiblast and hypoblast and becomes particularly concentrated at the surface of the mesenchymal cells. One of the properties of this material is its ability to take up large quantities of water and swell; its hydrated volume is approximately 10^3 times that of the anhydrous state. There is evidence that the hyaluronic acid secreted by cells in the region of the primitive streak may be of particular importance in maintaining the extensive extracellular spaces in the mesenchyme. When this glycosaminoglycan is digested by the injection of hyaluronidase into the embryo-forming region, the space through which the mesenchymal cells are migrating is greatly decreased and the mesenchymal cells tend to become clumped together. It may be that the synthesis of hyaluronic acid at the beginning of gastrulation plays a causative role in the conversion of cells from a cohesive epithelium into a loosely organized mesenchyme. Conversely, the normal destruction of this same extracellular material toward the end of gastrulation may be important in promoting the aggregation of mesenchymal cells back into epithelial structures, such as the notochord and somites. In addition to the amorphous glycosaminoglycans, the extracellular space between the epiblast and hypoblast also contains the fibrous proteins collagen and fibronectin. These materials also have been implicated in the movement of mesenchymal cells, possibly by providing such cells with tactile directions for orientation and locomotion.

Since virtually the entire chick embryo arises from cells originally located within the epiblast (the primordial germ cells are one exception, Section 12.4.B), a fate map of this layer (Fig. 10.27) made before gastrulation contains all the prospective organ-forming regions. The first cells to migrate into the primitive streak comprise the prospective foregut. These endodermal cells are located directly around Hensen's node and enter the primitive streak at its anterior end, taking their place in the hypoblast. There is no formation of an archenteron in the chick. The endodermal cells that have moved through the primitive streak become spread along the midline within the hypoblast. At a later stage, the embryo is undercut by ventral folds that lead to the formation of the digestive tract (Section 12.2.D).

The cells of the prospective foregut (the pharyngeal endoderm) are followed by prechordal mesoderm and chordamesoderm, which move down into the streak and then *anteriorly* in the midline between the overlying epiblast and underlying hypoblast. These chordamesoderm cells, which will differentiate into the anterior portion of the notochord, extend from a point beneath Hensen's node anteriorly as the *head process* (Figs. 10.23*d* and 10.28). At later stages, cells of the

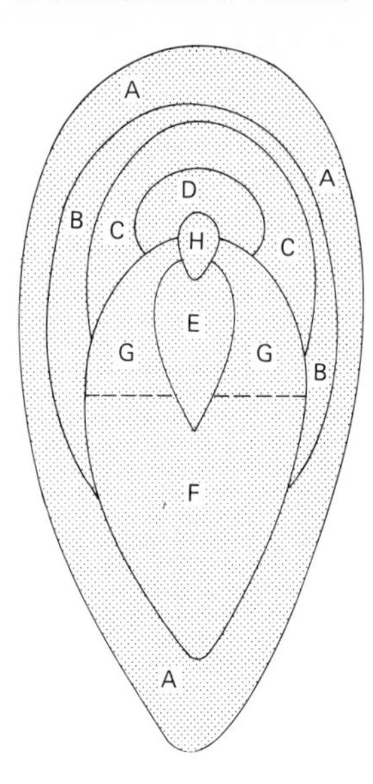

Figure 10.27 Fate map of a chick embryo showing the prospective fates of the various regions of the epiblast of the blastoderm at a stage prior to the onset of gastrulation. A, presumptive extraembryonic ectoderm; B, presumptive epidermal ectoderm; C, presumptive dorsal neural tube ectoderm; D, presumptive ventral neural tube ectoderm; E, presumptive endoderm; F, presumptive lateral plate mesoderm; G, presumptive paraxial mesoderm; H, presumptive notochord. [*From G. C. Rosenquist,* Contr. Embryol., **38:**77 *(1966).*]

presumptive somites and lateral plate migrate through the streak and take up residency at the sides of the head process. After a period of several hours, migration of cells through the anterior portion of the primitive streak appears to have run its course and a new type of activity, the regression or shortening of the primitive streak, begins. In this latter phase of gastrulation, the shortening of the primitive streak is accomplished by the movement of Hensen's node from its initial position near the center of the area pellucida in a posterior direction along the streak. As the streak shortens, endodermal cells migrate into the hypoblast to form the remainder of the digestive tract, cells of the chordamesoderm migrate through the streak to form the notochord, and cells of the paraxial mesoderm migrate through the streak to the sides of the notochordal process to form the somites and lateral plate. In other words, as the anterior tip of the streak moves posteriorly down the midline, it leaves the entire dorsal axis of the embryo in its wake. The further the node progresses, the more posterior is the section of the embryo being laid down. As the streak closes behind the regressing node, the epiblast (which now contains only ectoderm) in the dorsal midline becomes transformed into the embryo's central nervous system. This process is discussed further in the next chapter.

Figure 10.28 Diagram of a median longitudinal section of an 18-hour chick embryo showing the notochord and primitive streak. [*From B. M. Patten and B. M. Carlson, "Foundations of Embryology," 3rd ed., McGraw-Hill, 1974.*]

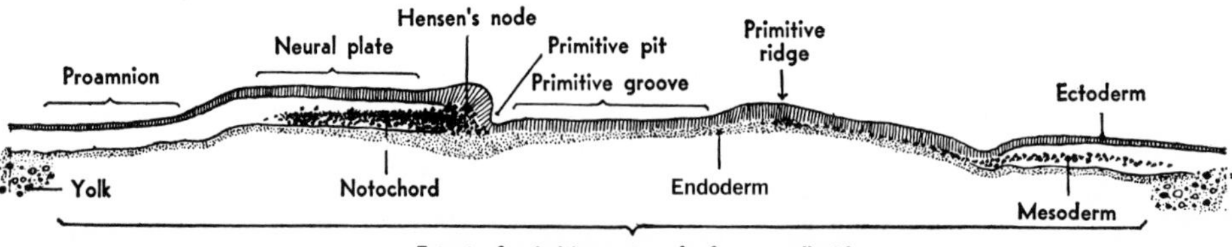

B. Mammals

In our discussion of mammalian development in Chapter 7, we followed the progress of the mouse embryo up to the start of gastrulation. In the diagram of a mouse embryo shown in Fig. 7.36, the inner cell mass had formed two main layers (Section 7.9.A), the primitive ectoderm and the primitive endoderm. This latter tissue had split off and become extended as a single-celled layer within the trophectoderm that enclosed a fluid-filled space. Although there is no yolk mass in the mammalian egg, this layer of cells is believed to be homologous to that portion of the blastoderm which grows out over the yolk in the bird. It is therefore referred to as the *yolk sac*. At its dorsal roof, the primitive endoderm is closely applied to the overlying layer of primitive ectoderm. These two layers (Fig. 10.29) form the *blastodisc* of the mammalian embryo and are homologous to the epiblast and hypoblast of the chick. As in the chick, the upper layer (primitive ectoderm, or epiblast) is believed to contain all the prospective embryonic cells. This layer in the mammal becomes greatly thickened as its cells become columnar, and the primitive streak appears toward its posterior edge. As in the chick, the prospective mesodermal and endodermal cells migrate through the primitive streak and take their place beneath the epiblast.

Unfortunately, we have very little experimental evidence concerning the activities of the embryonic cells during mammalian gastrulation. Unlike the chick embryo, that of the mammal is very small and highly inaccessible. The types of marking procedures discussed in connection with the analysis of chick gastrulation cannot be applied readily to the study of mammalian embryos. Consequently, most of what we know of mammalian gastrulation comes from studies of serial sections of embryos obtained from experimental animals, although some information has been obtained recently from experiments in which genetically

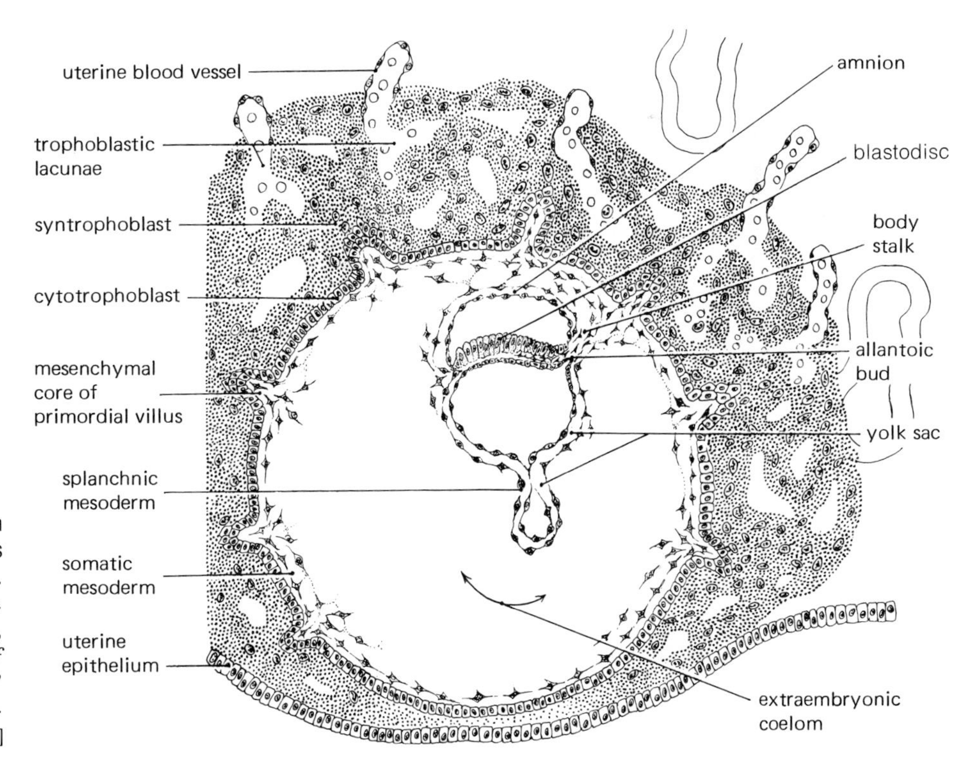

Figure 10.29 Human embryo of about 13 days fertilization age. [*From B. M. Patten and B. M. Carlson, "Foundations of Embryology," 3rd ed., McGraw-Hill 1974.*]

labeled cells of one layer are injected into recipient embryos. At the present time we have to assume, based primarily on morphological similarities, that gastrulation in mammals is basically similar in nature to that in the chick.

C. Multiple embryo formation

The potentiality and lability of the chick blastoderm preceding the establishment of the primitive streak have been demonstrated, as in fish, both by environmental conditioning and by operative procedures. Because development of a bird egg normally proceeds at a high incubation temperature, the system is comparatively sensitive to temperature change. Cooling the developing egg of a chick at the critical pre-primitive-streak stage retards or arrests the developmental process. Such cooling may be followed by a simple physical change such as cytoplasmic gelation; in any case, it is reversible. Following the return to normal incubation temperature, two or three primitive streaks commonly appear on the blastoderm in place of the one that would have developed normally. Consequently, two or three well-formed embryos develop. However, since they are associated with the same yolk sac, they become abdominally conjoined at hatching, as in fish, where partial or complete twinning also is induced by critical temperature changes during a particular phase of blastodermic development (Fig. 10.30). Operative procedures in chicks offer a wider range of experiments:

1. If the early blastoderm is cut into two pieces while it is still on top of intact yolk, twins develop (Fig. 10.31).
2. If the blastoderm is cut into four pieces by two crosscuts and grown on a culture medium, each piece develops a well-formed embryo with its head toward the original center of the blastoderm (Fig. 10.32).
3. Conversely, if three blastoderms, each having had its posterolateral parts cut away, are joined so that the three prospective organization centers unite in the middle of the fused disk, three embryos form, each with its head toward the middle; i.e., the original blastodermal polarity is reversed.

In all such cases, embryos are formed from cell populations that otherwise would have contributed either to the formation of extraembryonic tissues or to entirely different parts of the embryo. The blastoderm is evidently a two-layered disk of tissue that is essentially undetermined with regard to specific embryonic axes until after considerable growth or expansion.

Figure 10.30 Twins (*a*) and triplets (*b*) in teleost development.

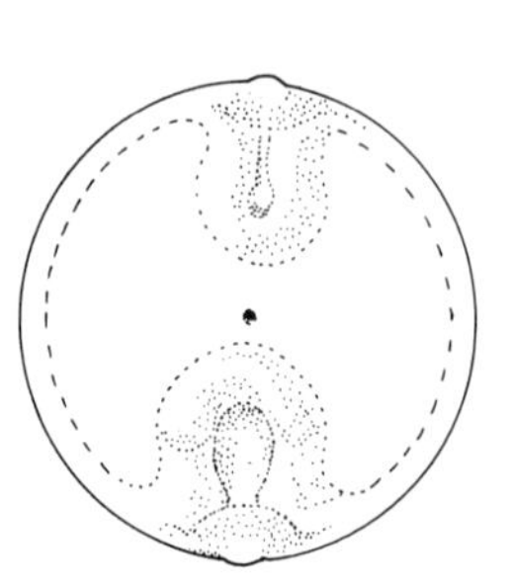

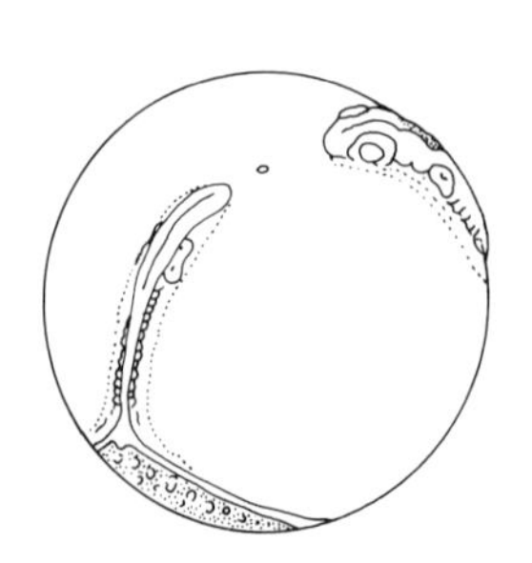

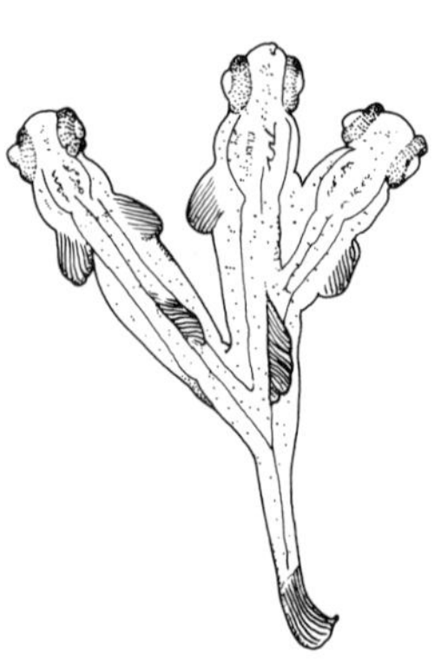

a

b

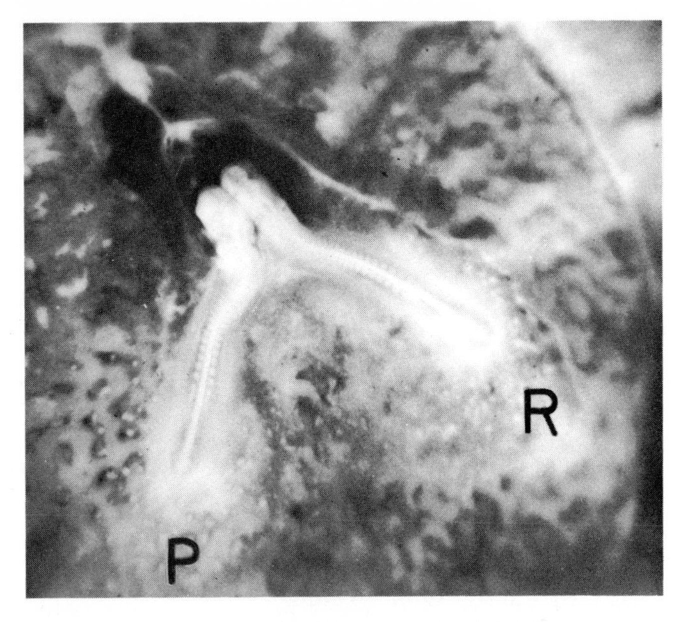

Figure 10.31 Chick twins produced by cutting unincubated blastoderms. [*Courtesy of N. T. Spratt, Jr.*]

Figure 10.32 Potentiality and polarity in chick blastoderm. Four individuals resulting from division of blastoderm into four segments. [*After N. T. Spratt and H. Haas,* J. Exp. Zool., **145:***113 (1960)*.]

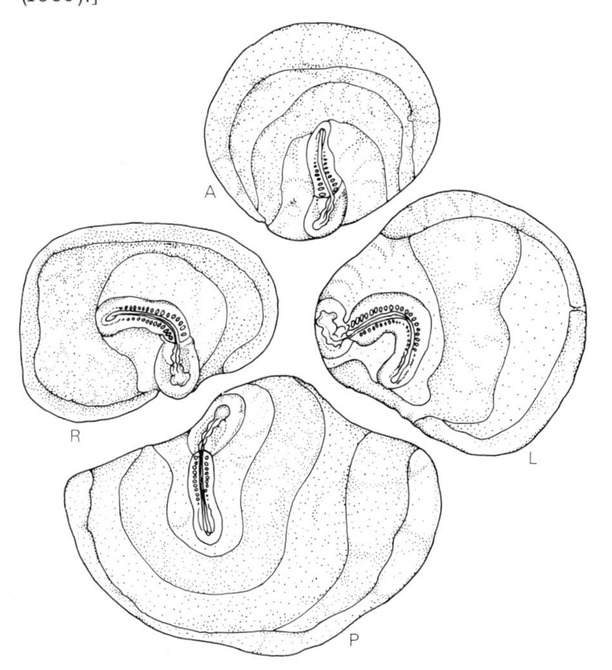

D. Twinning in mammals

Twins can be *monozygotic* (one-egg, identical) or *dizygotic* (two-egg, fraternal). Dizygotic twins are formed when two eggs are shed in a single ovulation period and are fertilized by two separate spermatozoa. The resulting young resemble each other genetically no more than do any other brothers and sisters. Injection of gonadotrophic hormones can induce multiple ovulations and hence promote dizygote twinning. This technique finds practical application in cattle breeding, while in women multiple births are an undesirable side effect of the use of gonadotrophins to combat sterility.

The production of two or three embryos from a single blastoderm is not infrequent in fish, reptiles, and birds, but the offspring are fated to an early death. In all these, the two or more embryos developing from the single blastoderm inevitably share one and the same yolk sac. When the yolk is finally resorbed, the embryos become united at their abdominal surfaces as Siamese twins. In mammals the situation is different because the yolk sac is a small remnant, empty of yolk, and plays no part in later development. Mammalian identical (monozygotic) twins are united only in that they share a common placenta, although each has its own umbilical cord and amnion. Inasmuch as the cords are in any case broken at birth, the offspring are set free from one another.

True twinning cannot be studied in most mammals because the event itself cannot be anticipated and is known only after the critical process has occurred. Various species of the armadillo, however, produce identical twins as the normal routine; the offspring number two, four, and commonly, eight, according to the species.

The nine-banded Texas armadillo *Dasypus novemcinctus* produces four monozygous offspring routinely. The study of normal development in this species is inevitably the study of quadruplet formation. The basic studies are by Patterson (1913). These are discussed at length in "The Biology of Twins" (Newman, 1917) and were recently reexamined by Storrs and Williams with regard to biochemical and anatomic variation within sets of quadruplets. Analysis of this development strongly confirms the conclusion that the mammalian egg is completely labile until well after implantation of blastocyst in the number of embryos that can be initiated.

In nontwinning mammals, the embryo first appears as a primitive streak in a thickened area, or apical pole, of ectoderm, forming the floor of the vesicle and thereby establishing bilateralism (the embryonic axis). In the armadillo, after a period of growth of the blastocyst, two embryonic ectodermal thickenings appear. These are located at two ends of the same bilateral axis, each with its head toward the apical pole (Fig. 10.33). They are the primordia of the two *primary embryos*. According to Newman, the original bilateralism of the vesicle, which determines this primary axis, has been secondarily imposed by the preexisting bilateralism of the uterus; no other explanation of the coincidence between uterine and embryonic axes is apparent.

Secondary embryos first appear as shorter secondary outgrowths of the ectodermal plate on an axis at right angles to the primary axis after the primary embryonic areas are already identifiable (Fig. 10.33). In another species (*Dasypus hybridus*), however, the typical number of monozygous embryos produced ranges from seven to twelve, with a preponderance of eight. In this species, one or two tertiary axes probably form across the blastodermic ectoderm between the primary and secondary axes.

It is evident that in the mammal—as in fish, reptiles, and birds—there is no preprogramming of the egg for a specific development of an embryo and no defin-

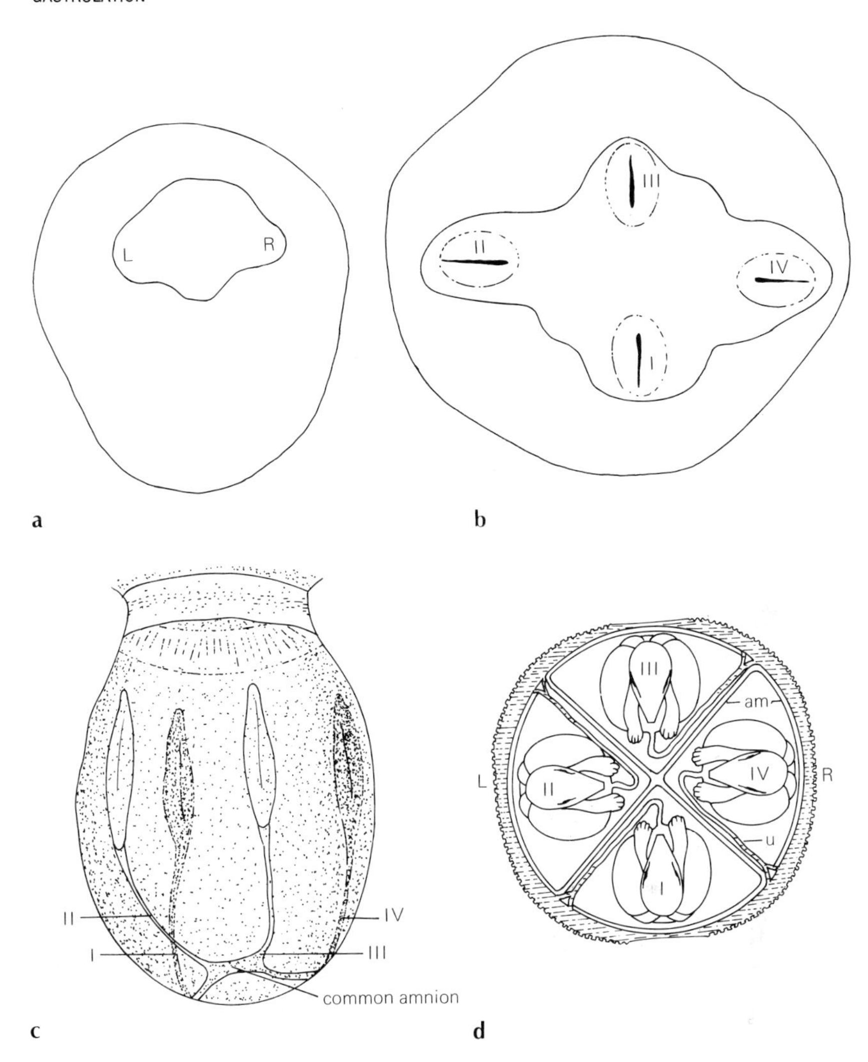

Figure 10.33 Quadruplet production in the armadillo. *a* Blastocyst with outline of ectodermal vesicle developed from an inner cell mass. One axis of the vesicle is longer than the other, and a left and right "primary bud" form at the ends of the longer axis. *b* Later stage showing four embryos at late-primitive-streak stage, resulting from two "secondary buds" forming between the two primary buds, at the ends of the shorter axis; all four are oriented toward the center of the complex or common amnion. *c* Later stage showing that the primary embryos (II and IV) are in advance of the secondary embryos (I and III). *d* Full-term quadruplets seen from one pole, occupying four quadrants with partitions.

itive determination; any part of the ectodermal blastocyst disk can establish an embryonic field capable of self-directed development into a typical embryo.

Quadruplets derived from a single zygote presumably have identical genes. Nevertheless, such quadruplets are only approximately identical, and recent biochemical analyses have shown that a remarkable degree of variation actually exists.

It was observed that enormous physiological variations exist within very closely inbred rats, fowl, etc. The question arose whether unknown factors, aside

from the gene pool itself, also control the intricate process of differentiation, particularly *the extent to which each of the numerous types of differentiated cells proliferates*. During the production of the four primordial buds in the armadillo in two stages, small variations in initial size and cell constitution (apart from chromosomal content) are inevitable. These could in turn lead to varying differences in the sizes of organs and in biochemical parameters related to these varying organ sizes. Analyses show that differences are always present and are detectable for 20 parameters measured. It was concluded that such differences are due to inescapable, accidental differences in the makeup of the cytoplasms of the four primordial buds. How many of each kind of cell and how much of each tissue will be produced during development are of paramount importance, because they may govern the most fundamental and important characters of the organism. Much of human variability, accordingly, may be the result of circumstantial quantitative differences arising during the initial phase of embryo establishment, in addition to those from genetic variability.

Readings

BAKER, P. C., 1965. Fine Structure and Morphogenetic Movements in the Gastrula of the Tree frog *Hyla regilla*, *J. Cell Biol.*, **24**:95–116.

BALLARD, W. W., 1976. Problems of Gastrulation: Real and Verbal, *Bioscience*, **26**:36–39.

———, 1973. Morphogenetic Movements in *Salmo gairdneri* Richardson, *J. Exp. Zool.*, **184**:27–48.

BELLAIRS, R., 1971. "Developmental Processes in Higher Vertebrates," University of Miami Press.

———, 1969. Experimental Twinning and Multiple Monsters in Chick Embryos, in E. Bertelli (ed.), "Teratology," Excerpta Medica.

BETCHAKU, T. B., and J. P. TRINKAUS, 1978. Contact Relations, Surface Activities, and Cortical Microfilaments of Marginal Cells of the Enveloping Layer and of the Yolk Syncytial and Yolk Cytoplasmic Layers of *Fundulus* before and during Epiboly, *J. Exp. Zool.*, **206**:381–426.

BRUMMETT, A. R., 1968–1969. Deletion-Transplantation Experiments on Embryos of *Fundulus heteroclitus*. I. The Posterior Embryonic Shield, *J. Exp. Zool.*, **169**:215–253. II. The Anterior Embryonic Shield, *J. Exp. Zool.*, **172**:443–463.

CHILD, C. M., 1940. Lithium and Echinoderm Exogastrulation: With a Review of the Physiological-Gradient Concept, *Physiol. Rev.*, **13**:4–41.

FISHER, M., and M. SOLURSH, 1977. Glycosaminoglycans Localization and Role in Maintenance of Tissue Spaces in the Early Chick Embryo, *J. Embryol. Exp. Morphol.*, **42**:195–207.

FONTAINE, J., and N. M. LEDOUARIN, 1977. Analysis of Endoderm Formation in the Avian Blastoderm by the Use of Quail-Chick Chimeras, *J. Embryol. Exp. Morphol.*, **41**:209–222.

GIBBINS, I. R., L. G. TILNEY, and K. R. PORTER, 1968. Microtubules in the Formation and Development of the Primary Mesenchyme of *Arbacia punctulata*, *Develop. Biol.*, **18**:523–539.

GUSTAFSON, T., and L. WOLPERT, 1967. Cellular Movement and Contact in Sea Urchin Morphogenesis, *Biol. Rev.*, **42**:442–498.

HAMBURGER, V., 1960. "A Manual of Experimental Embryology," University of Chicago Press.

HOLTFRETER, J., 1944. A Study in the Mechanics of Gastrulation. II. *J. Exp. Zool.*, **95**:171–212.

———, 1943*a*. Properties and Functions of the Surface Coat in Amphibian Embryos, *J. Exp. Zool.*, **93**:251–323.

———, 1943*b*. A Study in the Mechanics of Gastrulation. I. *J. Exp. Zool.*, **84**:261–318.

———, and V. HAMBURGER, 1955. Amphibians, in B. J. Willier, P. Weiss, and V. Hamburger (eds.), "Analysis of Development," Saunders.

HÖRSTADIUS, S., 1973. "Experimental Embryology of Echinoderms," Oxford University Press.

JOHNSON, K. E., 1977. Extracellular Matrix Synthesis in Blastula and Gastrula Stages of Normal and Hybrid Frog Embryos, *J. Cell Sci.*, **25**:313–354.

KARP, G. C., and M. SOLURSH, 1974. Acid Mucopolysaccharide Metabolism, the Cell Surface, and Primary Mesenchyme Cell Activity in the Sea Urchin Embryo, *Develop. Biol.*, **41**:110–123.

KELLER, R. E., 1976. Vital Dye Mapping of the Gastrula and Neurula of *Xenopus laevis*, *Develop. Biol.*, **51**:118–137.

MOORE, A. R., 1941. On the Mechanics of Gastrulation in *Dendraster excentricus*, *J. Exp. Zool.*, **87**:101–111.

NAKATSUJI, N., 1976. Studies on the Gastrulation of Amphibian Embryos: Ultrastructure of the Migrating Cells of Anurans, *Roux Arch. Entwickl.*, **180**:229–240.

NICOLET, G., 1971. Avian Gastrulation, *Advan. Morphog.*, **9**:231–262.

OPPENHEIMER, J. M., 1972. Regulation of Partially Dissociated and Reaggregated Portions of *Fundulus* Embryonic Shields, *J. Exp. Zool.*, **179**:63–80.

———, 1940. The Non-Specificity of the Germ Layers, *Quart. Rev. Biol.*, **15**:1–27.

SPRATT, N. J., and H. HAAS, 1960. Integrative Mechanisms in the Development of the Chick Embryo. I. *J. Exp. Zool.*, **145**:97–137.

TRELSTAD, R. L., E. HAY, and J. P. REVEL, 1967. Cell Contact during Early Morphogenesis in the Chick Embryo, *Develop. Biol.*, **16**:78–106.

TRINKAUS, J. P., 1976. On the Mechanism of Metazoan Cell Movements, in G. Poste and G. L. Nicolson (eds.), "The Cell Surface in Animal Embryogenesis," Elsevier.

———, 1969. "Cells into Organs," Prentice-Hall.

WOURMS, J. P., 1972. Developmental Biology of Annual Fishes. II. Naturally Occurring Dispersion and Reaggregation of Blastomeres during the Development of Annual Fish Eggs, *J. Exp. Zool.*, **182**:160–200.

———, 1967. Annual Fishes, in F. H. Wilt and N. K. Wessells (eds.), "Methods in Developmental Biology," Crowell.

Chapter 11 Neurulation and primary induction

Toward the end of gastrulation in amphibian embryos, when the yolk plug is disappearing and the blastopore is becoming a dorsoventral groove, the embryo is still almost spherical. At this time, faint streaks appear on the dorsal side, extending from the dorsal edge of the blastopore to the anterior end of the embryo (Fig. 11.1). They are the first external signs of the *neural plate*. A median streak represents the future neural groove; the marginal streaks represent the neural fold of each side. *Neurulation* is primarily the process whereby a flattened neural plate forms and transforms into a hollow *neural tube*.

Neurulation continues until the whole neural tube, representing the brain and the spinal cord and their outgrowths, has been entirely segregated from the overlying surface ectoderm. In amphibians and other lower vertebrates, or anamniotes, the enclosure of the neural tube is completed at much the same time throughout its length. During the whole process, the stage is known as the *neurula*. In land vertebrates of all classes, the process of neurulation occurs in a progressive manner from the anterior to the posterior part of the embryo. For example, in the embryo in Fig. 11.2, neural-tube formation has been completed anteriorly, stretching as far back as the region of the third somite. At this stage the embryo contains a well-infolded brain, the trunk region has a fused neural tube anteriorly and an open neural plate posteriorly, and the most posterior part of the embryo is still undergoing gastrulation. Therefore, although the neurula stage is precisely definable in amphibian development, *neurula* and *neurulation* are terms that apply in a temporal succession to the stage and process seen along the embryonic axis. They are associated with an axial gradient with regard to time and stage of axial development.

Neurulation, although specifically the process of neural-tube formation, is also the phase of development during which the mesoderm and endoderm undergo primary differentiation into particular structural entities, as already noted in *Amphioxus* (Section 9.1). During neural-plate formation, the mesoderm, which is already separated from the endoderm of the archenteric roof, completes its extension into the midventral portion of the embryo (Fig. 11.3) and becomes demarcated into five strips of tissue representing the median dorsal notochord and, on each side, the band of somites and lateral plates (Fig. 11.3). Each tissue

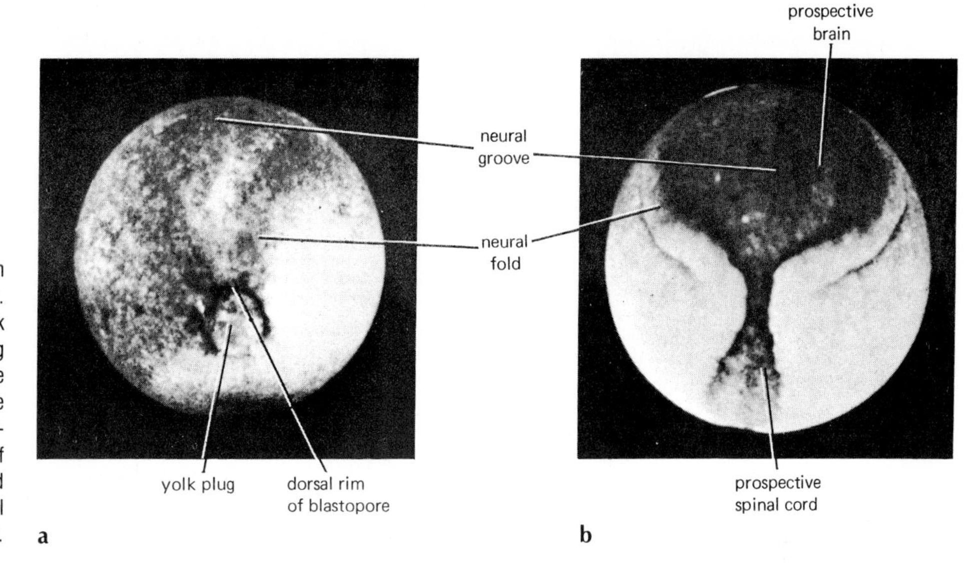

Figure 11.1 Neurulation in the salamander. *a* Early stage, with yolk plug still evident, showing shallow neural groove and neural folds. *b* Middle stage, showing great difference in width of prospective brain and prospective spinal cord.

Figure 11.2 Neurulation in the reptile embryo. Pictured is a four-somite turtle embryo, showing head fold and developing brain, neural tube open posteriorly as a neural groove, four pairs of mesodermal somites with lateral mesodermal plate extending posteriorly on each side of the neural groove, and primitive node.

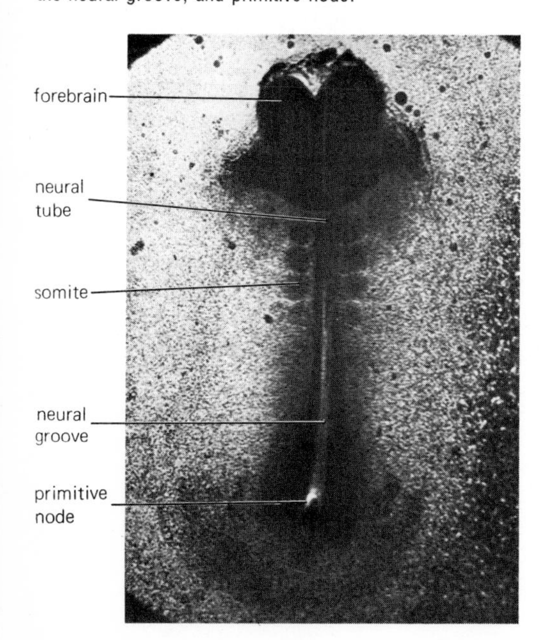

Figure 11.3 Three stages in urodele neurulation in transverse section, showing open neural plate, neural groove with neural folds, and closed neural tube covered by epidermis, respectively. The notochord, mesoderm, and endoderm are also seen in all three stages, in progressive differentiation.

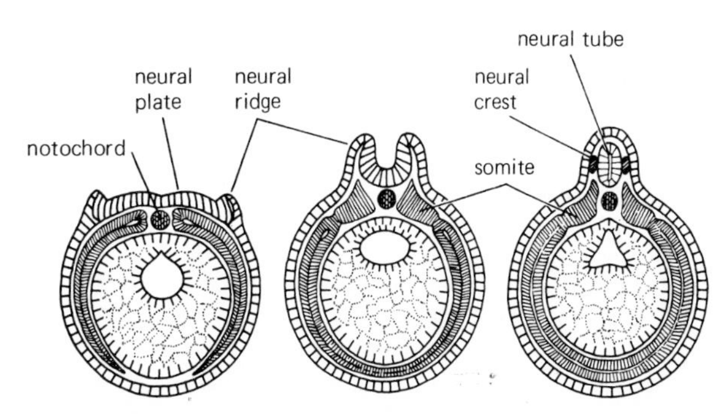

of the neurula thus constituted proceeds with its own characteristic patterns and processes of morphogenesis. These several phenomena, which are integrated into the development of the embryo as a whole, are discussed at length in Chapter 12.

1. The process: neurulation

In amphibians the neural tube forms from as much as 50 percent of the ectoderm, that portion which covers the dorsal hemisphere of the embryo. In this group especially, the topographic aspects of neural-tube formation have been thoroughly studied by means of vital dyes and tissue transplantation and are well understood. It is easy to see what happens because it occurs at the surface of the embryo; it is less clear *how* it happens.

In urodeles, which have been well studied, the presumptive neural ectoderm of the late *gastrula* exists as a single-celled epithelium comprised of cells that are slightly columnar in shape. From this stage throughout the entire process of neural-tube formation, the neural epithelium remains one cell layer thick. The morphogenetic events of neurulation occur as a result of a change in morphology of this single sheet of cells. The movements that occur in this cell layer have been closely followed. In one set of experiments, cells were marked with vital dyes, while in other studies, the shifts in position of naturally pigmented cells have been observed. In either case, a striking rearrangement of the surface cells is revealed (Fig. 11.4). Whereas the cells of the presumptive nervous system cover essentially the entire dorsal hemisphere of the late gastrula, their composite surface area is much more restricted by the time the embryo has reached the early neurula stage. The dorsal surface first loses its rounded shape and flattens into a

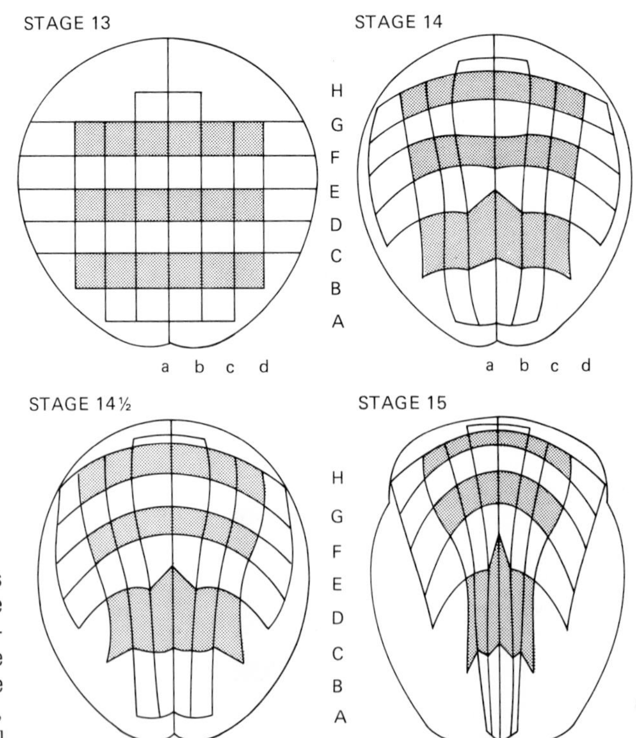

Figure 11.4 Pigmented cells at the intersections of the lines of a grid on stage-13 embryos were traced by time-lapse cinematography through subsequent stages shown. The transformation of the coordinate grid is shown at each stage reflecting the movements of these cells at the surface of the embryo. [*From B. Burnside and A. G. Jacobson,* Dev. Biol., **18**:*543* (*1968*).]

disk that subsequently shrinks to assume the shape of a keyhole. The wider anterior portion will develop into the brain, and the narrow posterior portion will form the spinal cord. Careful analysis of the movements of individual cells in the neurectoderm of the newt *Taricha* has indicated that the change in shape of the neural epithelium prior to formation of the neural tube occurs as a result of two independent activities, a change in shape of the neural epithelial cells and a change in shape of the underlying notochord.

If one follows the cells of the neural epithelium from the late gastrula stage to the early neurula stage, they are found to become markedly elongated, thereby forming the thickened *neural plate.* Since this increase in cell height is not accompanied by a corresponding increase in cell volume, the area of *embryo surface* covered by these cells decreases (Fig. 11.5). It is this decrease in the outer surface area of the cells that is primarily responsible for the shrinkage of the neurectoderm toward the dorsal midline. Furthermore, mapping studies demonstrate that the change in surface area is directly related to the displacement of cells observed during neural-plate formation (Fig. 11.4).

Other types of analyses indicate that the change in height appears to be programmed into these cells well before their actual elongation. For example, the neural epithelium can be surgically removed from the embryo prior to its thickening and cultured as an explant in vitro. When this is done, the cells of the explant undergo elongation and the isolated neural plate undergoes shrinkage. The capacity to elongate at a particular stage of development is clearly programmed into the cells of the neurectoderm. That this capacity is an intrinsic property of the cells themselves is indicated by the fact that single, isolated urodele neural-plate cells retain their columnar shape and continue to elongate in culture.

Figure 11.5 Comparison of numbers of paraxial microtubules counted in cross sections of neural-plate cells at two stages of development. When cells from equivalent regions of the neural plate are compared before and after elongation (stages 13 and 15), one finds a significant decrease in average numbers of microtubules per cell as the cell elongates. Nevertheless, there is a conservation of total cumulative length of microtubules (number of microtubules per cell times cell height). Counts of microtubules in cells from two regions of the same neural plate which have elongated to a different extent do not differ significantly. [*From B. Burnside,* Am. Zool., **13:***1001* (*1973*).]

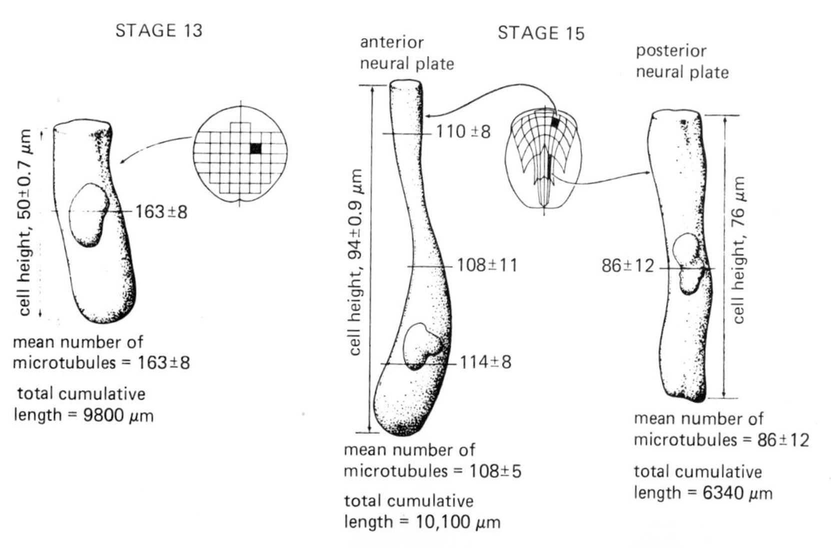

Although the isolated neural epithelium undergoes shrinkage in culture, it does not assume the characteristic keyhole shape. Some other factor besides cell elongation appears to be responsible for the morphogenetic process. Further analysis has indicated that it is the movement of the underlying notochord, which becomes rod-shaped and elongated in the anteroposterior axis, that aids in the morphogenesis of the neural plate. For example, if the neural epithelium is explanted together with the notochord, the characteristic keyhole-shaped structure is formed in culture. Similarly, if the notochord is removed from the intact embryo by cutting it away from below, the neural plate does not undergo its normal morphogenesis. Since the notochord is firmly attached to the median portion of the neural plate, the movements of the two layers become synchronized. As the notochord elongates, it distorts the shape of the overlying sheet by pulling middorsal cells forward into the expanding anterior region, further shrinking the posterior portion, which will ultimately develop into the narrow spinal cord.

Once the thickening of the urodele neural plate is complete, the edges of the plate rise as ridges (Fig. 11.6). As they continue to do so, they become the neural folds. The neural plate between them tends to narrow, forming a wide neural groove (Fig. 11.6). At the same time, the embryo begins to lengthen along its anteroposterior axis. The neural folds then meet in the dorsal midline and fuse. For a time, the folds are still separated anteriorly. Consequently, there is a temporary anterior opening into the tube termed the *neuropore*. Meanwhile, the lateral ectoderm (presumptive epidermis) of each side, which has been spreading dorsally as the edges of the neural folds come together, also meets and fuses at the middorsal line. The neural tube is thereby cut off and lies beneath the completed epidermal layer. As the neural folds fuse middorsally and separate from the covering epidermis, a mass of cells called the *neural crest* migrates out of the dorsolateral region of the neural tube into the surrounding tissue. The cells of the neural crest

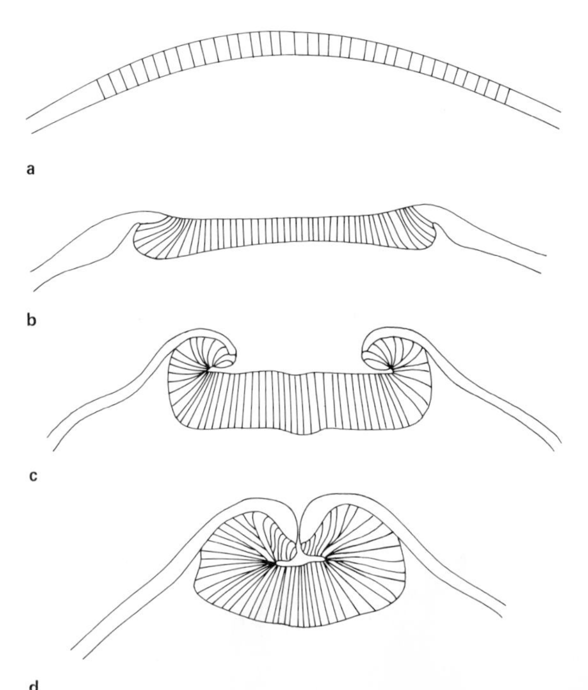

Figure 11.6 Diagrammatic cross sections of the neural anlage of a generalized urodele. *a* Stage-13 embryo (pre-neural plate). *b* Stage-14 embryo (neural plate). *c* Stage-16 embryo (neural folds have begun to move mediad). *d* Stage-19 embryo (neural folds have just met). [*From P. Karfunkel,* Int. Rev. Cytol., **38**:*248* (*1974*).]

are derived from the transition zone between the neural plate proper and adjacent presumptive epidermis. These migratory neural-crest cells will differentiate into a great variety of tissues (discussed in Section 11.2).

The process of neural-tube formation is roughly the same in all vertebrates (with the exception of teleosts), although important species differences do exist. For example, the neural epithelium of *Xenopus* is composed of more than one layer of cells, and the formation of the neural folds occurs at the same time as the cells are elongating. The cell-shape changes associated with neurulation are principally those of the deep layer. Moreover, in the comparatively large embryo of the salamander, neurulation takes about 2 days, whereas in the smaller embryo of *Xenopus*, it is remarkably rapid and is complete within about 3 hours after the neural plate begins to infold. In the chick the neural epithelium is composed of a single layer of very tall columnar cells (ones whose heights are approximately 10 times their widths). The chick neural plate begins to fold at its midline, thereby forming a neural groove with relatively straight sides. As neurulation proceeds, the neural folds become increasingly concave until they meet overhead and fuse. The mode of formation of a hollow tube in fish, however, is very different. The neural tube forms as a solid rod of cells that separates from the ectoderm and subsequently somehow acquires a hollow interior. This suggests that the common method of plate and tube formation characteristic of most vertebrates is not essential to neural differentiation as such, but represents one of possibly several means by which one kind of tissue can separate from another.

A. The mechanism of neurulation

Attempts to analyze the mechanism of neurulation, as in the study of most problems of development, go back to the nineteenth century. In 1874, W. His suggested that form changes in the cells, and eventually in the neural plate as a whole, resulted from growth pressure (tissue expansion) owing to excessive mitosis in the neural ectoderm within a confined area. More than half a century later it was shown that while there is a 23 percent increase in the number of cells present in the whole ectoderm during neurulation, there is no increase in its total volume; cells are simply becoming smaller. Further, His showed that there are no detectable differences in either cell number or cell volume between the neural and the nonneural ectoderm. Differences in cell shape are another matter.

The process of neurulation, as studied primarily in birds and amphibians, is seemingly accomplished by two distinct changes in cell shape: the elongation of the neurectoderm cells to form a tall columnar epithelium and their subsequent apical constriction, which causes them to become flask-shaped. As described earlier, elongation leads to the formation of the thickened neural plate, while apical constriction leads to its conversion into the neural tube (Fig. 11.7). We will consider these two steps in succession.

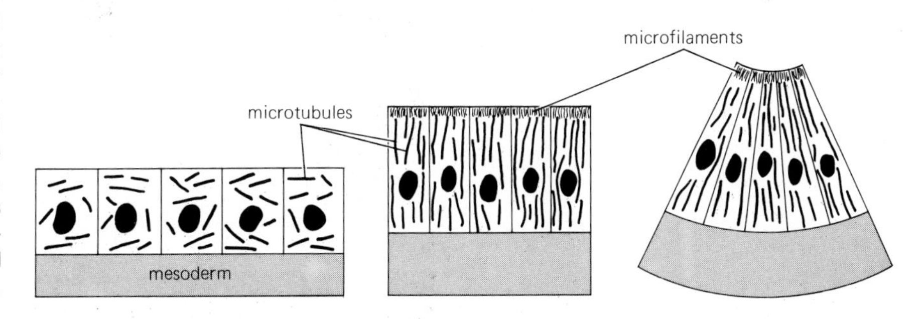

Figure 11.7 Diagrammatic illustration of the overall changes in cell shape that occur during neurulation. Previously random microtubules become oriented parallel to the long axis of the cells as they elongate, and apical microfilaments bring about the constriction of each cell's apical end.

The elongation of the neurectodermal cells has been closely correlated with changes in the microtubular skeleton of the component cells (Figs. 11.5 and 11.8). In the salamander, the entire ectoderm of the late gastrula consists of a simple low-columnar epithelium in which the cells of the neural portion are indistinguishable from those of the epidermal portion. The microtubules are scattered throughout the cytoplasm of both types of cells in an apparently random

Figure 11.8 *a* Two possible routes of ectodermal differentiation; the tall columnar cells of the neural tube and the flattened cells of the epidermis. *b* Orientation of microtubules and microfilaments in elongating neural-plate cells. The microtubules (pmt) are aligned parallel to the axis of the cell, while the microfilaments (mf) are arranged in a circumferential bundle around the apex in purse-string fashion. *c* A flattened epidermal cell. Microtubules are randomly oriented, but the microfilaments (f) are arranged in discrete bundles, often seen in continuity with desmosomal fibers (see Fig. 3.4) and are thought to span the cell from desmosome to desmosome (d). [*After B. Burnside,* Develop. Biol., **26**:*434* (*1971*).]

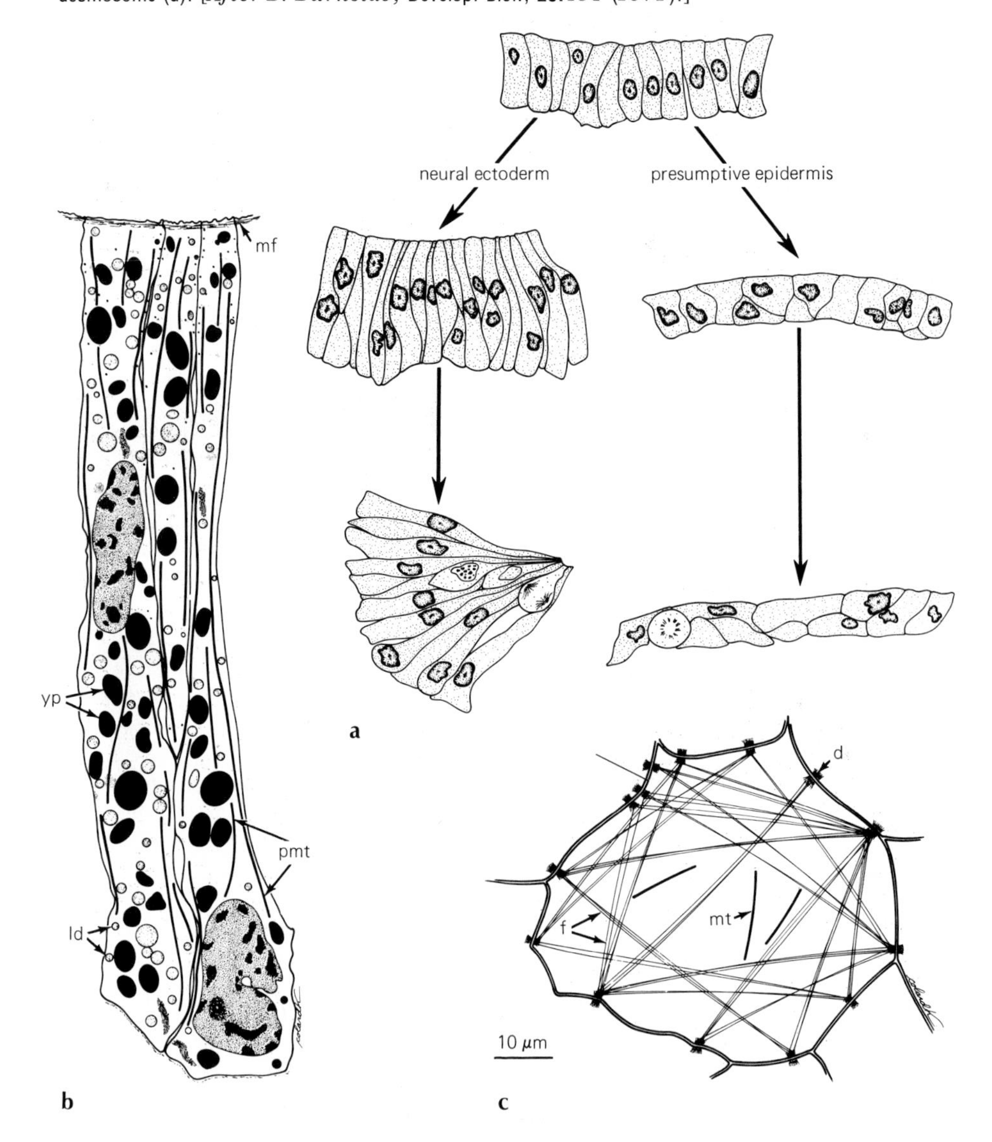

manner. As neurulation begins, the bulk of the microtubules of the neural epithelium become increasingly oriented with their long axes parallel to the axis of cell elongation. These microtubules, which are arranged in vertical groups, are referred to as the *paraxial microtubules* (Fig. 11.8). If we assume that microtubules do play an *active* role in the elongation process, there remain several means by which they might act. It is possible, for example, that their extension at the basal and/or apical ends pushes against the edge of the cell causing it to become longer and thinner. This type of pushing action could occur as a result of the addition of subunits to the ends of the microtubules, i.e., by growth. Alternatively, this same type of pushing action could be generated by causing overlapping microtubules to slide across one another in a manner similar to that believed to occur during muscle contraction or ciliary movement. As a third hypothesis, microtubules might bring about cell elongation indirectly by actively transporting cytoplasmic elements toward the basal end of the cell. Microtubules have been implicated in the movements of materials through a variety of different cells, and this type of directed transport might be expected to apply pressure at one end of the cell causing the cell's extension in that region. As the cell elongates, new subunits could be added to the ends of the microtubules, thereby stabilizing the extension process. Each of these hypotheses leads to certain predictions that should be verifiable, although at the present time a definitive conclusion cannot be drawn.

Following elongation, the cells of the neural epithelium become constricted at their apical ends (free outer surface) causing the cells to become flask-shaped and the layer of cells to curve inward (Figs. 11.7 and 11.8*a*). This second change in cell shape is brought about by the action of microfilaments, which become arranged in a bundle that encircles the apex of each cell in a purse-string fashion just below its free surface. As neurulation proceeds, the thickness of the apical microfilament bundle increases as the circumference of the apical region is reduced. Furthermore, the microfilament bundle of each cell is seen to be connected to the desmosomes located at the surface of the cell near its apical edge. Since microfilaments of adjoining cells attach to both sides of the same desmosome, an interconnected network of filaments is formed which covers the entire neural plate. This extensive interconnection might provide the means (as in the case of the Z-band attachments of actin filaments in muscle fibers) whereby movement of the entire plate is coordinated.

The active role of both microtubules and microfilaments in neurulation is supported by experiments exposing neurulas of *Xenopus* and the chick to a number of agents known to interfere with the assembly or maintenance of these structures. In *Xenopus*, vinblastine has been used to disrupt microtubules, thus preventing neurulation from proceeding as well as reversing changes in cell shape associated with neurulation that may have taken place already. At the least, microtubules appear to be necessary to maintain the elongated shape. In the chick, elongated neurula cells round up when treated with colchicine and lose their constricted state when exposed to cytochalasin B. Both treatments prevent neurulation from proceeding. In the presence of cytochalasin, the neural folds of the chick embryo come away from each other, eventually falling back and flattening, although the neurula cells remain wedge-shaped. Microtubules are still present in these cells, but microfilaments are mostly lacking.

Not all investigations of neurulation have focused on the intracellular components. Materials of the extracellular matrix as well as the cell surface itself have also been implicated in the process, although their roles are less well understood. The most clear-cut involvement of the cell surface during neurulation occurs during closure of the neural tube in mammals. A light and electron micrograph

of a mammalian embryo at the open-neural-fold stage are shown in Fig. 11.9. As the edges of the neural folds approach one another, the surfaces of the cells at the upper edge of the folds (the area of ultimate contact) are seen to be thrown into extensive projections (Fig. 11.10). These projections, which include long thin extensions (filopodia) as well as broad flat ones (lamellipodia), appear to contact one another across the intervening space and pull the edges together. In addition to the cellular processes, considerable amounts of extracellular material appear at the edge of the uplifted neural folds. It has been suggested that this material provides spatial orientation and guidance for the folds as they converge as well as promoting a stable adhesion between opposing surfaces (Fig. 11.11). The importance of this material (presumably glycoproteins and/or glycosaminoglycans) is demonstrated by experiments utilizing concanavalin A, a lectin that binds to exposed α-glucosyl and α-mannosyl residues of glycoproteins. Chick embryos incubated in the presence of this lectin do not undergo neural-tube closure. It would appear that the carbohydrate portions of these surface macromolecules are important in neural-fold activity.

Before leaving the topic of neural-tube formation, some mention should be made of the other cells of the embryo surface. While the cells of the neural and

Figure 11.9
a Human embryo at 8-somite stage (probable age 18 days from fertilization) within amnion, showing large neural folds anteriorly, somites in trunk region, and primitive streak posteriorly. [*From B. M. Patten and B. M. Carlson, "Foundations of Embryology," 3rd ed., McGraw-Hill, 1974.*]

b Scanning electron micrograph of mammalian embryo, removed from amnion. Neural development in the hamster at open neural-fold stage (7¾-day embryo), showing early optic evagination internally from each side of the open brain cavity. [*Courtesy of R. E. Waterman.*]

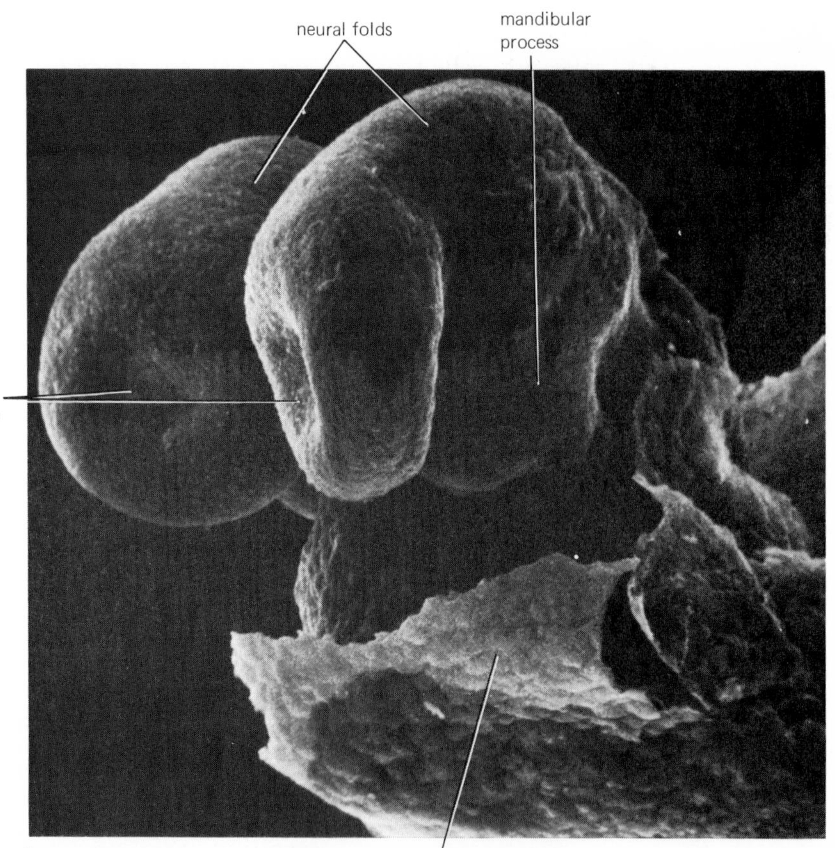

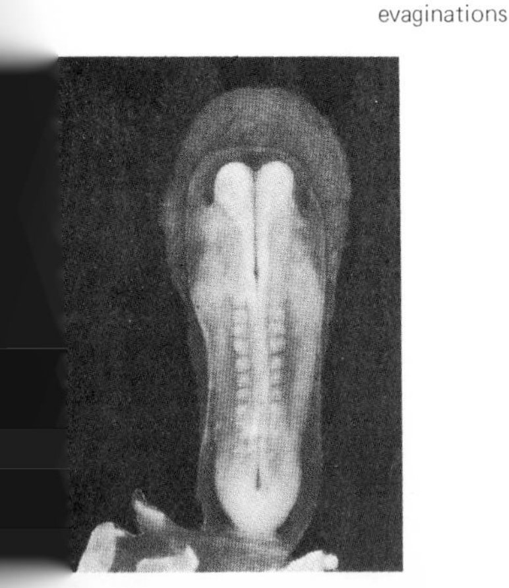

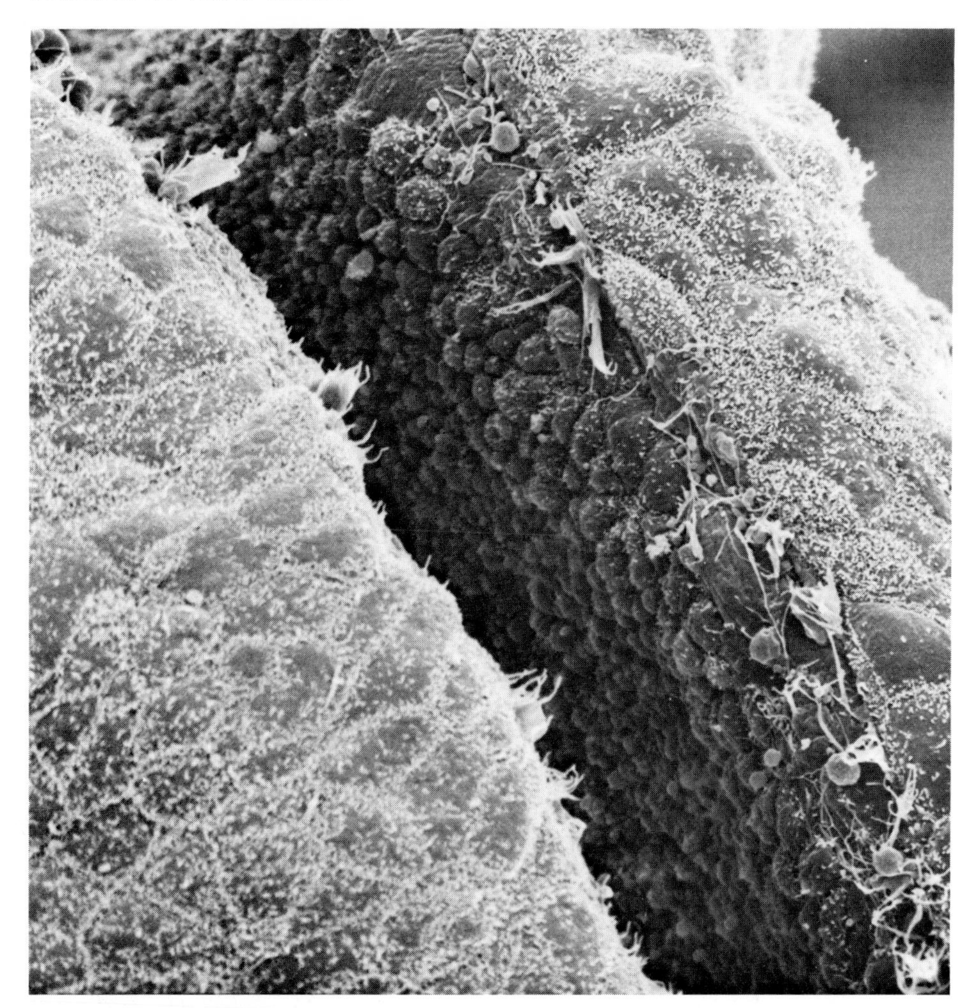

Figure 11.10 Scanning electron micrograph of the neural folds of the mouse prior to initial contact and fusion. A narrow zone of alteration characterized by flattened cells and membranous ruffles is seen between the cells of the surface ectoderm and the rounded cell apices of the neural ectoderm. [*Courtesy of R. E. Waterman.*]

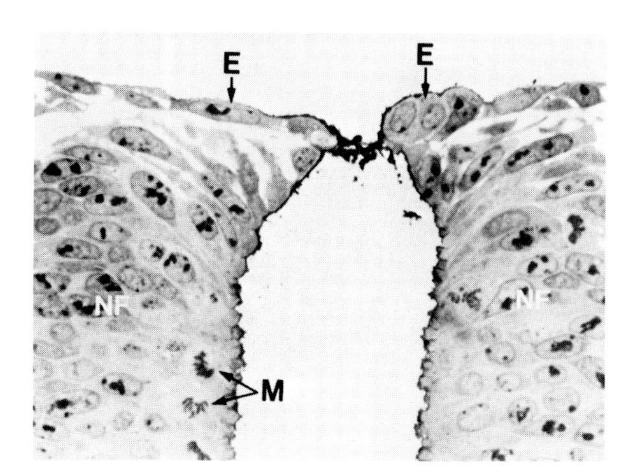

Figure 11.11 Cross section through the hindbrain of a 10-day mouse embryo just before the neural folds contact each other. Considerable amounts of extracellular material, as revealed by ruthenium red staining, are evident at the sites of closure. E, ectoderm; M, mitotic figures; NF, neural fold. [*Courtesy of T. W. Sadler.*]

nonneural ectoderm are indistinguishable in the late gastrula, they become dramatically different in the neurula. While the presumptive neural cells are elongating, the presumptive epidermal cells are gradually flattening to form a squamous epithelium (Fig. 11.8). Unlike the neural cells, the microtubules of the epidermal cells are randomly arranged and the 50- to 70-A microfilaments (ones containing actin) gradually disappear, being replaced by thicker filaments (80 to 100 A) whose function is unclear.

2. Neural crest

The neural crest forms from cells that were originally present as strips located along the lateral edges of the neural epithelium, i.e., in those regions at each side of the neural plate bordering the presumptive epidermis. At about the time the neural folds contact one another during the final stage of neural-tube closure, the neural crests are squeezed from the epithelium, where they form a mass of loosely aggregated cells at each side of the dorsal midline (Fig. 11.12). Those cells which emerge from the expanded neural tube of the brain region form the cranial neural crest (Figs. 11.12*a* and *b*); those from the region of the presumptive spinal cord form the trunk neural crest. Cell migration begins almost immediately after the aggregations are evident. These migrating cells appear in scanning electron

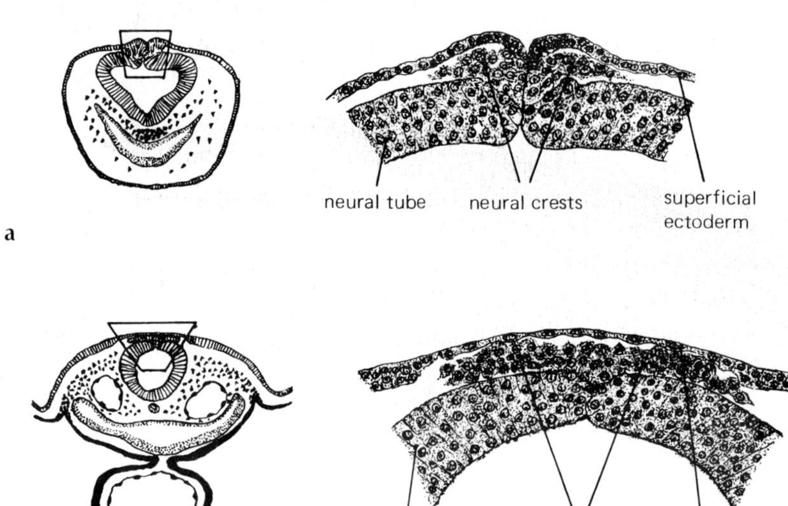

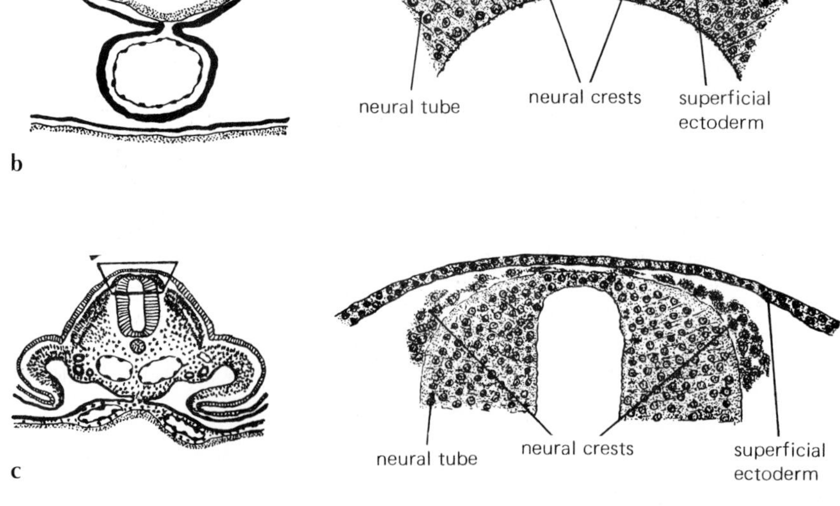

Figure 11.12 Transverse sections to show the origin of neural-crest cells. The location of the area drawn is indicated on the small sketch to the left of each drawing. *a* Anterior rhombencephalic region of 30-hour chick. *b* Posterior rhombencephalic region of 36-hour chick. *c* Mid-dorsal region of cord n 55-hour chick. [*From B. M. Patten and B. M. Carlson, "Foundations of Embryology," 3rd ed., McGraw-Hill, 1974.*]

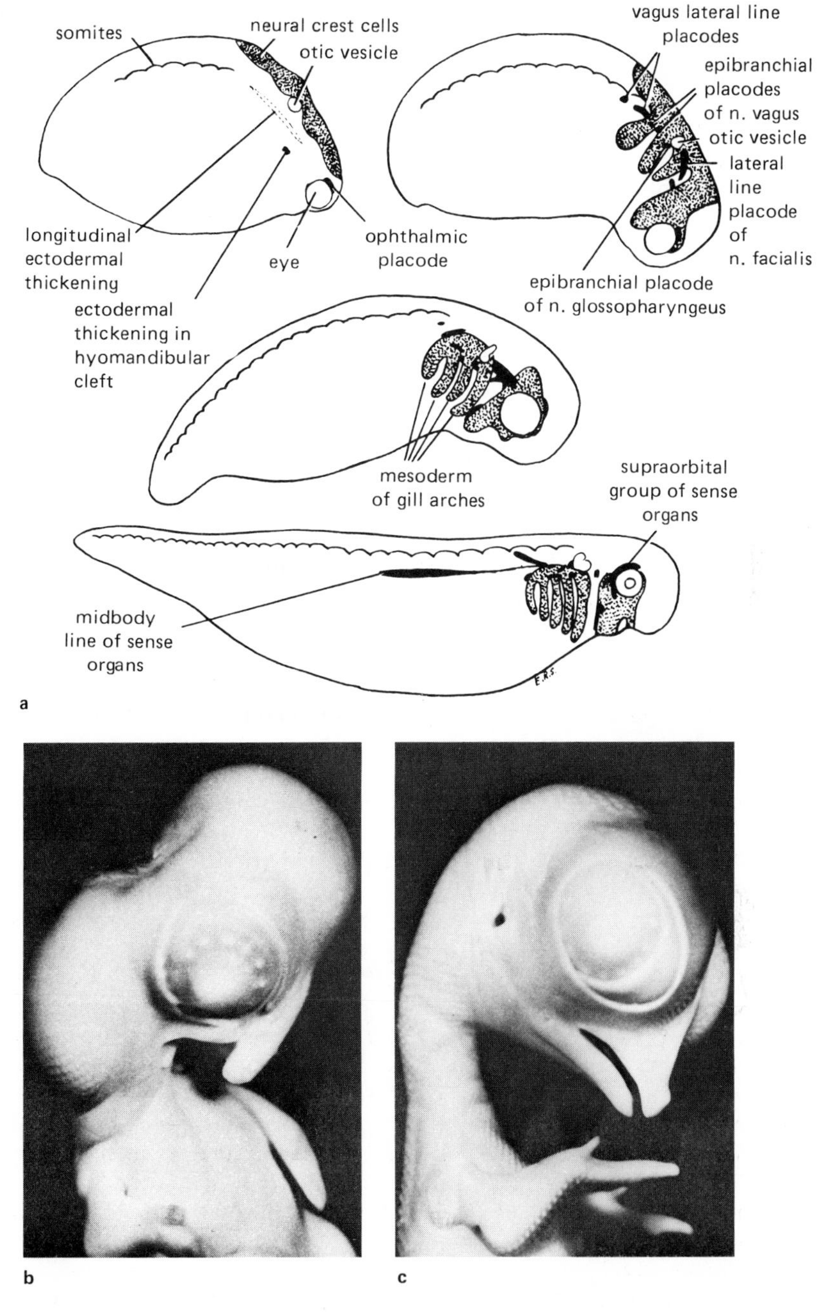

Figure 11.13 *a* Movement of the cranial neural-crest cells in the urodele *Ambystoma*. Neural-crest cells are stippled with the placodes of special lateral sense organs and cranial-nerve ganglia being shown in black. [*From O. E. Nelsen, "Comparative Embryology of the Vertebrates," Blakiston, 1953.*] *b, c* The effect on the morphogenesis of the head and neck of removal of the cephalic neural primordium. *b* The mesencephalic and rhombencephalic primordia of this 8.5-day embryo were removed before the beginning of neural-crest-cell migration. Note the atrophy of the superior part of the beak, of the lids and interocular region, and of the ventral part of the neck. The inferior part of the beak, the mandible, and tongue are missing. *c* Control embryo at the same age. [*Courtesy of N. le Douarin.*]

micrographs as individual flattened cells that are stretched along an axis parallel to the direction of movement. The cells are seen to possess broad lamellipodia as well as slender elongated filopodial processes that make contact with the surfaces of neighboring cells. Like the migratory cells of the gastrula, the cells of the neural crest move into a matrix that is rich in glycosaminoglycans (primarily hyaluronic acid, but also some sulfated material). As in the gastrula, these extracellular mucopolysaccharides probably serve to swell the volume of the extracellular space to facilitate migration and provide a substrate upon which the cells can move.

From their modest beginning as a minor component (in a numerical sense) of the neurectoderm, the cells of the neural crest spread throughout the body and differentiate into a wide variety of strikingly different structures. One group, which migrates only a short distance, becomes localized in segmentally arranged clusters alongside the neural tube. This group forms the ganglia of cranial sensory nerves anteriorly and the dorsal root ganglia of spinal nerves more posteriorly. Some become ganglionic cells of the adrenal medulla and autonomic nervous system. Others become the nonneural nerve-sheath cells (Schwann cells). A large group of crest cells disperse widely beneath the ectoderm to form pigment cells in various locations, often in striking patterns. All the pigment cells of higher vertebrates except those of the brain and retina are derived from the neural crest. A large number of cephalic neural-crest cells (Fig. 11.13) assume a mesenchymal character. These ectomesenchymal cells (of ectodermal origin) contribute to the connective tissue of the head and most notably supply the cartilage of the visceral skeleton (mandibular, hyoid, and gill arches) together with part of the skull. The neural crest does not contribute to the axial or appendicular skeleton (trunk vertebrae and limb bones). The only established mesenchymal derivative of the trunk crest is the dorsal fin of amphibians. A general account and analysis of this remarkable diversity (Table 11.1) can be found in the review by Weston (1970).

The role of the neural crest in the formation of these various structures has been determined over the years using various techniques. Most of the assignments were originally made on the basis of early studies in which the neural crest (or neural folds before crest release) in a particular region of the embryo was removed (a defect experiment) and the corresponding deficiency noted as in Fig. 11.13*b*. However, there are several difficulties in interpreting this type of experiment. For example, the lack of formation of a particular structure is not conclusive evidence of a neural-crest origin; the neural crest may be required in an inductive, i.e., interactive, capacity. Conversely, the formation of a particular structure in the absence of regional neural crest is not conclusive evidence of its non-neural-crest origin. It remains possible that the structure is normally derived from neural crest, but in its absence other cells can regulate and make up the deficiency. In this latter case, it is also possible that neural-crest cells from some distant location are recruited into the deficient region to form the necessary structure.

Alternatively, questions concerning the prospective fates of neural-crest cells can be approached using some type of marker technique by which the cells in question can be followed to their final destination. The same techniques (vital dyes, [^{3}H]thymidine-labeled grafts, quail-chick chimeras) discussed in Section 10.3.A in connection with gastrulation have been employed to follow the migratory pathways of neural-crest cells and learn of their contribution to embryonic structure. In one series of experiments, portions of the neural tube at the time of closure were transplanted from [^{3}H]thymidine-labeled embryos to the same site, i.e., an orthotopic graft (Fig. 11.14), in unlabeled host embryos whose cor-

Table 11.1 Summary of Normal Neural-Crest Fates

Pigment cells	Nervous system		Skeletal and connective tissue
	Sensory	Autonomic	
	TRUNK CREST (INCLUDING CERVICAL CREST)		
1. Melanophores 2. Xanthophores (erythrophores) 3. Iridophores (guanophores) in dermis, epidermis, and epidermal derivatives	1. Spinal ganglia 2. Some contribution to vagal (X) root ganglia	3. Sympathetic Superior cervical ganglion Prevertebral ganglia Paravertebral ganglia Adrenal medulla 4. Parasympathetic Remak's ganglion Pelvic plexus Visceral and enteric ganglia	Mesenchyme of dorsal fin in Amphibia
	5. Some supportive cells Glia (oligodendroglia) Schwann sheath cells Some contribution to meninges		
	CRANIAL CREST		
Small, belated contribution	1. Trigeminal (V)* 2. Facial (VII) root 3. Glossopharyngeal (IX) root (superior ganglia) 4. Vagal (X) root (jugular ganglia)	5. Parasympathetic ganglia Ciliary Ethmoidal Sphenopalatine Submandibular Intrinsic ganglia of viscera	1. Visceral cartilages (except basibranchial 2) 2. Trabeculae craniae (ant.) 3. Contributes cells to posttrabeculae, basal plate, parachordal cartilages 4. Odontoblasts 5. Head mesenchyme (membrane bones)
	6. Supportive cells*		

Source: J. A. Weston, 1970, *Adv. Morphogen,* **8**:74.

* Also receives contribution from placodes in lateral ectoderm. Trunk ganglia of nerves VII, IX, and X derived from ectodermal epibranchial placodes. Some supporting cells of these ganglia presumably also derived from placodal ectoderm.

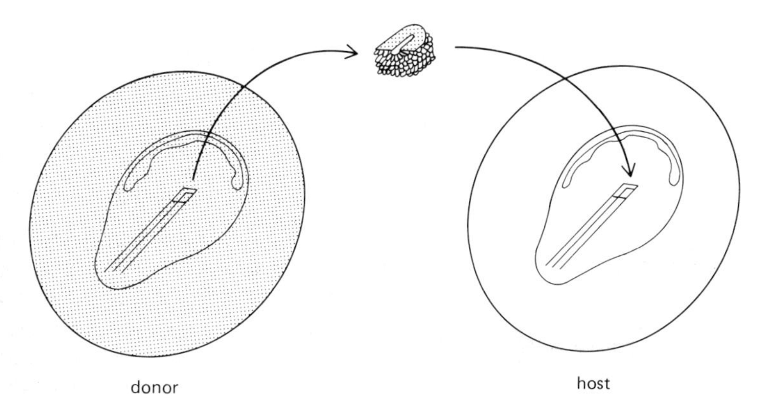

Figure 11.14 Diagrammatic representation of an exchange experiment in which a portion of neural tube from a [^{3}H]thymidine-labeled embryo is transplanted orthotopically to an unlabeled embryo. [*After diagram by G. Nicolet.*]

responding region had been previously excised. Autoradiographs prepared from sections of these host embryos indicated that the neural-crest cells of the trunk migrate in two fairly well-defined streams. One stream moves ventrally into the mesenchyme between the neural tube and the somites. Initially these cells spread out uniformly along the length of the body, but they soon move in a convergent manner toward the segmentally arranged blocks of somite mesoderm. Some cells enter the somite mesoderm, where they form the compact dorsal-root ganglia. Others proceed past the somite mesoderm toward the dorsal aorta, where they form the chain of sympathetic ganglia as well as the adrenal medulla at the posterior thoracic level. The concentration of cells to form the dorsal-root and sympathetic ganglia is strongly dependent on the presence of the segmental somites.

The other stream of cells moves dorsolaterally into the surface layers of the embryo, where it provides the forerunners of the pigment cells. The precise pathway these cells take, whether through the dermis (a mesodermal derivative) or the epidermis (an ectodermal derivative), remains unclear. In all cases, the cells follow definite pathways and, for causes yet unknown, accumulate rapidly at precise destinations. The localization of neural-crest cells does not arise as a result of random migration followed by selective survival at appropriate locations.

The tritium-marking experiments show that neural-crest cells migrate to their particular target locations without getting lost on the way. The mechanism by which these cells find their way through the embryo remains unknown. The initial direction of migration of the ventral stream appears to be dependent on the neural tube. Thus when a neural tube with associated neural crest is inverted, the crest cells continue to migrate but move dorsally instead of ventrally, i.e., in an inverse direction, retaining their orientation relative to the tube itself. It has been suggested that the basal lamina, an extracellular structure rich in collagen and glycosaminoglycans, may be responsible for the initial direction of movement. Once the cells leave the vicinity of the neural tube, they must respond to a complex series of environmental signals that guide them along defined pathways to their final destination. Unfortunately, we know virtually nothing about the nature of these cues.

Questions concerning embryonic determination are of particular interest in the study of the neural crest because these cells, as a group, form such a great variety of structures. To what extent are individual neural-crest cells determined at the time of release from the neural tube? Whereas questions of prospective fate are answered by orthotopic transplantation, questions of prospective potency require heterotopic transplants in which pieces of neural tubes from one axial level of the donor embryo are transferred to a different level in the host. If the donor cells differentiate in the same manner as the cells they have replaced, one can conclude that the donor population has the potency to form all neural-crest–derived structures of this other axial level. In order to follow the fate of the graft cells within the host tissue, the neural tube of the donor embryo must be marked in some recognizable way. The use of [^{3}H]thymidine-labeled donor tissue is not suitable in this type of experiment because the extensive proliferation of the crest cells prior to their final differentiation serves to dilute the isotope within the cell population to a great degree. Since the transfer of quail tissue into chick embryos provides the graft tissue with a permanent marker regardless of cell division, this technique has been widely used in this type of experiment.

Before we can begin to discuss the results of heterotopic transplants, we need to know which neural-crest derivatives are normally formed at the different axial levels. Figure 11.15 gives this information for the neural derivatives of the trunk crest. The autonomic system has served as the focus in these transplantation ex-

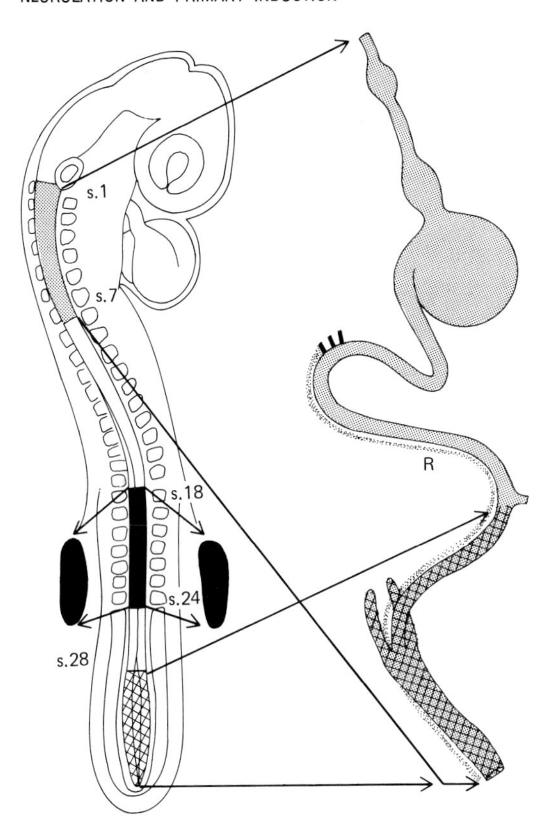

Figure 11.15 Diagram showing the various levels of the embryonic neural axis from which cells of the neural crest arise. The neuroblasts arising from the anterior level (between the levels of somites 1 and 7) colonize the whole gut. Those which come from the posterior level located behind somite 28 contribute only to the formation of the ganglia of the postumbilical gut. The neural crest of the cervical and dorsal regions (from 8 to 28 somites) does not participate in the formation of enteric ganglia but gives rise to adrenergic orthosympathetic neurons and to adrenal medulla cells, which come from the precise level of somites 18 to 24. R, nerve of Remak originating from the lumbrosacral level of the neural axis (behind somite 28). [*From diagrams of N. le Douarin and M.-A. M. Teillet.*]

periments because the cells of its two major divisions, the sympathetic and parasympathetic, (1) are derived from different regions of the neural tube, and (2) once differentiated, can be distinguished from one another on the basis of the neurotransmitter they produce. For example, the neural crest of the vagal region of the chick neural tube (somites 1 through 5) give rise only to parasympathetic ganglia, which contain cholinergic neurons, i.e., neurons that produce acetylcholine for neurotransmission. As indicated in Fig. 11.15, the crest cells from this axial region form the parasympathetic ganglia that populate the embryonic gut. In contrast, the neural-crest cells that migrate from the thoracic region of the chick neural tube (somites 8 through 28) differentiate only into adrenergic neurons, i.e., neurons that produce catecholamines such as adrenalin and dopamine. The crest cells from this axial level form sympathetic ganglia, which use catecholamines for neurotransmission, and the adrenal medulla, which produces large amounts of adrenalin.

When a portion of quail neural tube is taken from the vagal level and transplanted to the thoracic region of a chick recipient, the quail cells move toward the somites of their host and form the sympathetic ganglia characteristic of the thoracic neural crest. Conversely, if thoracic-crest cells from a quail are transplanted to the vagal area of a chick, the quail cells move to the gut, where they differentiate into parasympathetic neurons. Results of a variety of such experiments indicate that crest cells from one trunk region can substitute for those of any other trunk level. The commitment to differentiate as cholinergic or adrenergic neurons is not made prior to the onset of migration.

Similar experiments with cranial neural-crest cells indicate that cells from one cranial level can form the normal derivatives of other cranial levels. For example, the crest cells from the mesencephalon region normally give rise to the rostral parts of the skull. If this portion of the neural tube is transplanted to the more posterior metencephalic level, the crest cells from the graft form the tissue of the lower jaw, the normal metencephalic derivative.

It would appear from these types of experiments that neural-crest cells from one axial level are capable of following a very different pathway of migration and differentiation into structures other than those they would have formed in their original location. At the time of their release from the neural tube, the cells of the neural crest must be pluripotent and capable of responding to a variety of environmental influences. Presumably the environment in which the crest cells find themselves is responsible for leading them to their final destination. Although the environment through which they migrate may influence their developmental potential, the evidence suggests that it is the final destination that causes them to differentiate in a particular way. In at least some cases, the migratory phase of crest-cell development can be short-circuited without hampering the ability of the cells to differentiate. For example, if a piece of thoracic neural tube, which contains presumptive sympathetic neurons, is cultured in the presence of gut tissue, the neural-crest cells invade the gut and differentiate into parasympathetic neurons. The decision to differentiate into adrenergic or cholinergic neurons depends primarily on the site in which they ultimately settle.

Although trunk-crest cells are interchangeable, as are cranial-crest cells, certain difficulties arise when cranial-crest cells are heterotopically transplanted into a trunk region or vice versa. When cranial neural tube is replaced with trunk neural tube, the recipient develops with severe deficiencies in the facial and visceral arch structures. In the normal embryo, the chondrification of cranial-crest cells occurs following their interaction with pharyngeal endoderm. It seems that cells of the trunk neural crest, which normally do not contribute to skeletal tissue, are unable to do so even when they migrate into a cartilage-promoting cranial environment. Conversely, when neural-tube cells from a cranial level are transplanted into a trunk region, some of the crest cells of the graft become chondrified and contribute to the formation of trunk vertebrae. These results indicate that populations of trunk- and cranial-crest cells are not equivalent, at least with respect to their potency to form skeletal tissues. The cells of the cranial crest appear somewhat predetermined to form cartilage. Although they cannot self-differentiate into cartilage cells when cultured in isolation (they require the presence of foregut endoderm), they are capable of differentiating in this manner when present in an environment in which the axial skeleton is forming. Trunk-crest cells, however, do not seem capable of chondrification in any region of the body. The role of neural-crest cells in the formation of pigmentation patterns is discussed in Section 16.2.C.

3. The cause: neural induction

During the first quarter of this century, a remarkable series of experiments was performed, primarily by Hans Spemann and his coworkers, which provided a new insight into the types of interactions that occur during early development. The point was made in Section 7.3 that, to a large degree, the parts of an amphibian embryo become determined relatively late in development. Questions of determination are generally answered by heterotopic transplantation experiments in which a portion of a donor embryo is transferred to a different region of a host embryo in an attempt to influence its differentiation by exposure to a new environment. If a piece of presumptive neural ectoderm (dorsal animal region in the fate map in Fig. 10.9*b*) is removed from an *early* newt gastrula and transplanted to a site on the ventral animal surface (presumptive epidermis) of another embryo of comparable age, the recipient of this graft develops in a perfectly normal manner (Fig. 11.16*a* and *d*). The piece of transplanted ectoderm, which

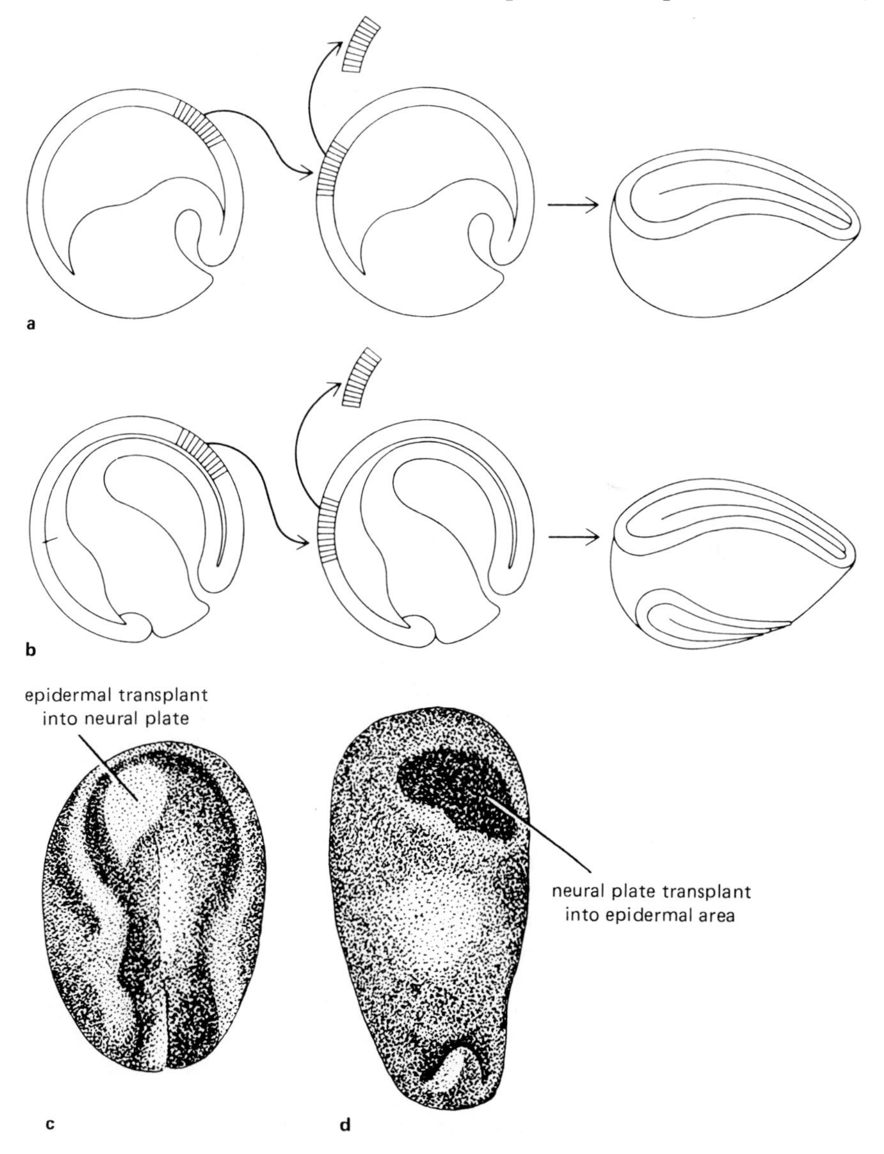

Figure 11.16 Removal and substitution of regions of neural plate at different stages of development. *a* Piece of presumptive neural plate removed from an early newt gastrula and grafted onto the ventral side of another of the same age. The graft develops in accordance with its new surroundings. *b* A corresponding piece removed from a late gastrula develops in the new surroundings according to its original fate. [*After L. Saxén and S. Toivonen, 1962.*] *c*, *d* Examples of transplants made at the early gastrula stage between urodele embryos of different pigmentation. Presumptive epidermal cells transplanted to the neural-plate region become neural tissue (*c*) and presumptive neural tissue becomes epidermis (*d*) when placed in a region of presumptive skin. [*From H. Spemann and H. Mangold,* Roux Arch. Entwickl., **100:***618* (*1924*).]

would have developed into neural tissue if left alone, now develops along a very different pathway and becomes skin in keeping with its newfound surroundings. We can conclude from this type of experiment that the cells of the presumptive neurectoderm of the early gastrula are not yet fully determined; i.e., their prospective potency as determined by transplantation is greater than their prospective fate. Similarly, when a piece of presumptive epidermis of an early gastrula is transplanted into prospective neural region, it develops into neural tissue, as do its surroundings (Fig. 11.16*c*). A wide variety of transplantation experiments indicate that the prospective potencies of the presumptive neural and epidermal regions are similar to one another and far-reaching. If, for example, either of these ectodermal regions is transplanted into the dorsal marginal zone of another early gastrula, it develops into part of the chordamesoderm and foregut, structures normally derived from mesodermal and endodermal cells.

When these same type of grafting experiments are carried out on embryos that have reached the late gastrula stage, very different results are obtained. Presumptive neural ectoderm differentiates into nervous tissue regardless of where it is moved (Fig. 11.16*b*), while presumptive epidermal ectoderm differentiates into skin regardless of the activities of surrounding cells. Although the cells of the neural and epidermal ectoderm of the late gastrula appear identical by morphological criteria, their prospective potencies are now quite different from each other and very restricted.

We can now return to the early gastrula stage and consider the result obtained following transplantation of a portion of the dorsal marginal zone. It should be recalled that this part of the early gastrula has a special origin, being derived from the region of the gray crescent, and a special fate, developing into the dorsal lip of the blastopore. When a portion of the dorsal marginal zone of a late blastula or early gastrula is transplanted to a ventral site in another embryo, it does not develop into skin as did the prospective neurectoderm (Fig. 11.17). Rather, it shows a strong tendency to sink beneath the surface (either by invagination or by being covered with ectoderm) and leads to the development of a remarkably well-formed secondary embryonic axis complete with notochord, somites, pronephric kidney tubules, intestinal tract, nervous system, sense organs, and gut cavity (Fig. 11.17*a*). All these various structures are organized into a harmonious embryonic structure. The results of this transplantation experiment were difficult to understand; for some reason the secondary embryo contained more structures than one would expect from the component parts. The prospective fate of the dorsal marginal zone of an early gastrula is primarily notochord and somite. If this part of the embryo is already fully determined at this stage, one would expect it to undergo self-differentiation following transplantation and form these mesodermal structures. Similarly, one would have expected the overlying ventral ectoderm (which had not been surgically altered) to differentiate into epidermal tissue in keeping with its prospective fate. There is no obvious reason why nervous tissue should form in the region of the graft unless the transplanted dorsal lip contains an influence that affects other regions with which it comes into contact, causing the cells to differentiate along a pathway other than the one they would normally take.

Before any firm conclusions can be made from this transplantation experiment, it is necessary to determine the origin of the various tissues of the secondary embryo. The question of critical importance is whether all the components of the secondary axis are differentiating from the donor tissue or whether some of the host tissue is participating in the composite by differentiating into structures other than those of its prospective fate. In the latter case, we can conclude that

a

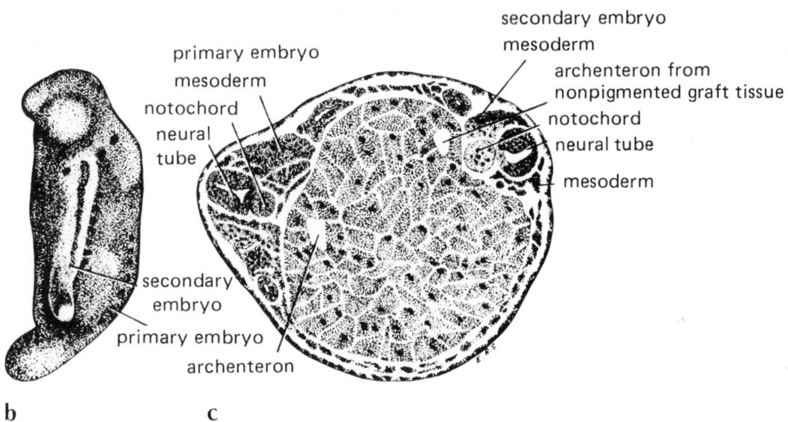

b c

Figure 11.17 *a* Dorsal blastopore lip of an early newt gastrula, transplanted into the blastocoel of another, results in the formation of a complete secondary embryo. [*Courtesy of L. Saxén and S. Toivonen.*] *b*, *c* Results of the original Spemann-Mangold experiment that demonstrated the capacity of the dorsal lip to induce a neural tube in the ventrolateral ectoderm of the host. The nonpigmented graft cells of *T. cristatus* are distinguishable from the pigmented cells of the *T. taeniatus* host. [*From H. Spemann and H. Mangold,* Roux Arch. Entwickl., **100:***618* (*1924*).]

the dorsal-lip graft is somehow responsible for *inducing* a shift in the direction of differentiation taken by host tissue. The crucial experiment, which was reported in 1924, was performed by Hilde Mangold, an associate of Hans Spemann (found in English translation in "Foundations of Modern Embryology," edited by B. H. Willier and J. M. Oppenheimer, 1974). In order to follow the differentiation of graft and host tissues, a dorsal lip from an early gastrula of a nonpigmented species of newt (*Triturus cristatus*) was transplanted into the ventral ectoderm of an early gastrula of a pigmented species (*Triturus taeniatus*) (Fig. 11.17*b* and *c*). An experiment in which tissue is grafted between animals of a different species of the same genus is termed a *heteroplastic* transplantation. Analysis of the

resulting chimera indicated that both the transplanted dorsal lip and the tissue of the host species contributed to the formation of the secondary embryo. The graft formed most of the chordamesoderm and the host supplied essentially all the nervous system (Fig. 11.17*c*). Furthermore, the tissues derived from the two sources were totally integrated within the composite. This experiment clearly showed that presumptive chordamesoderm can induce neural differentiation in prospective epidermal tissue. In addition, it showed that the dorsal-lip territory of the blastopore has a high capacity for self-differentiation and serves also as an organization center, although the nature of the organizing influence or agent still remains elusive.

If a grafted dorsal lip is capable of inducing neurulation in adjacent host ectoderm, we can presume that this same portion of the embryo is responsible for providing a necessary inductive influence in the normal embryo, one that leads to formation of the nervous system. As a result of the early fate-map studies on urodeles, we know that the dorsal marginal zone of the blastula sweeps into the interior, where it becomes concentrated as an elongated strip of chordamesoderm just beneath the presumptive neural ectoderm. The closeness of this association is indicated by the role of the notochord in determining the shape of the overlying neural plate. We conceive of the early embryonic ectoderm as a neutral layer of cells directed along one or another path of differentiation by interaction with adjacent tissue. The need for this type of interaction clearly illustrates the epigenetic nature of development (Section 7.1). In the normal embryo, the inductive process that leads to neurulation evidently occurs during the time when the presumptive chordamesoderm moves inward and forward within the embryo forming the archenteric roof. The dorsal lip of the blastopore, and the chordamesoderm that it forms, is often referred to as the *primary organizer* of the embryo, a term that reflects its central role in the construction of the dorsal axis.

The role of the primary organizer in normal development is nicely illustrated by examination of a particular type of highly abnormal development. Treatment of urodele embryos at the onset of gastrulation with hypertonic medium results in a striking morphogenetic process termed *exogastrulation* (Fig. 11.18). Rather

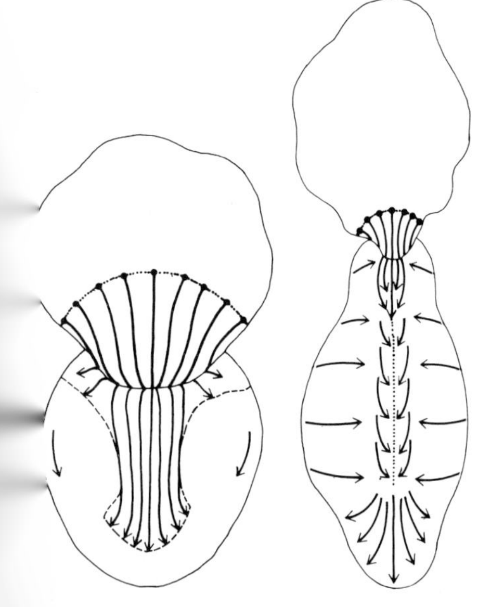

Figure 11.18 Exogastrulation. *a* Mass movements of tissue during exogastrulation in amphibian embryo. Ectoderm above, mesendoderm below. Compare with normal-tissue movement during gastrulation shown in Fig. 10.10. *b* Main differentiations in an exogastrulated embryo. [*After J. Holtfreter,* Roux Arch. Entwickl., **129:***669* (*1933*).]

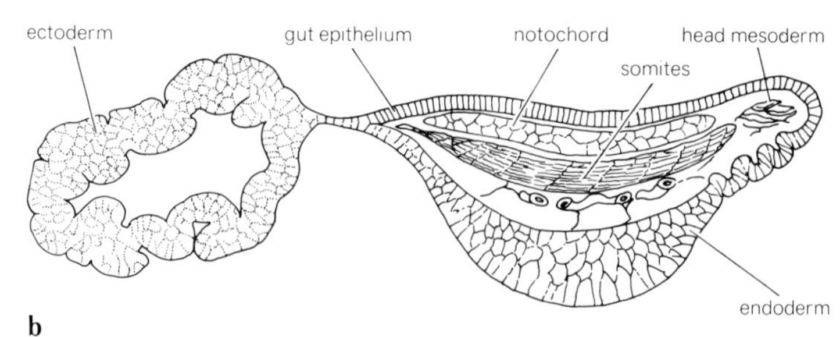

than moving into the interior of the embryo, the complex of endoderm and mesoderm moves in the opposite direction, away from the ectoderm, which survives as an empty sac attached to the posterior endoderm by only a narrow stalk (Fig. 11.18*b*). Even in its exogastrulated position, the marginal zone proceeds with its characteristic movements of dorsal convergence and anteroposterior elongation; at the same time it becomes embedded in the mass of endoderm. The displaced marginal zone develops into an axial notochord associated with somites and pronephric (renal) rudiments, together with various other mesodermal structures, including a beating heart. Similarly, the endodermal tissue, which in this case resides on the outside of the mesoderm rather than at its interior, proceeds to differentiate into an inverted, but essentially complete digestive tract. The ectoderm, however, is much more limited in its capacity to differentiate. Although the ectodermal cells of the exogastrula do form a type of ciliated epithelium (approximately one out of ten cells become ciliated as in the normal *larval* epidermis) with mucous-secreting cells, there is absolutely no evidence of neural differentiation. The differentiation of ectoderm in a neural direction requires that it come into close association with the chordamesoderm. The interaction between the roof of the archenteron and the overlying neurectoderm is referred to as *primary induction*, a term that (1) indicates that it is the first of many inductive interactions occurring during vertebrate development and (2) reflects that it is of primary importance in the establishment of the craniocaudal axis of the embryo.

In a broad sense, there are two types of influences that inductor tissue might exert upon cells in a responding tissue. In one case, the inductor may provide the responding tissue with some type of information that causes it to differentiate in one direction as opposed to another. In this type of interaction, the inductive tissue is said to be exerting an *instructive influence*, which leads to a restriction in the prospective potency of the responding cells. Instructive influences, therefore, are ones involved in the determinative processes that limit the choices of differentiation available to cells. In the other type of interaction, the inductor exerts a *permissive influence* on the responding tissue, which allows it to differentiate along a predetermined direction. In this latter type of induction, the inductor tissue serves as a stimulus for the differentiation of the responding cells by providing some type of required ingredient. Since the ectoderm of the early gastrula can be caused to differentiate in a variety of ways when combined with different types of inductor tissues or transplanted into different regions of the embryo, it would appear that primary induction is an example of an instructive interaction. Presumably the influence of associated chordamesoderm directs the multipotent ectoderm to differentiate in a neural direction. As development proceeds, the potencies of embryonic regions become limited and interactive events shift from ones that are predominantly instructive to those which are predominantly permissive in nature (see Section 15.8 for further discussion).

One of the most important questions in the study of primary induction concerns the manner in which the stimulus is transmitted from the chordamesoderm to the overlying ectoderm. Does the inductive agent(s) pass by free diffusion across a space between the two tissue layers, or is the stimulus transferred as a result of some process requiring cell-to-cell contact? In the latter case, one would presume that the plasma membrane becomes involved in mediating the transfer of information. One approach, illustrated in Fig. 11.19, has been particularly useful in distinguishing between inductive interactions based on free diffusion and those requiring cell-to-cell contact. In this technique, small pieces of inductor and reactive tissues are cultured together but separated from one another by a

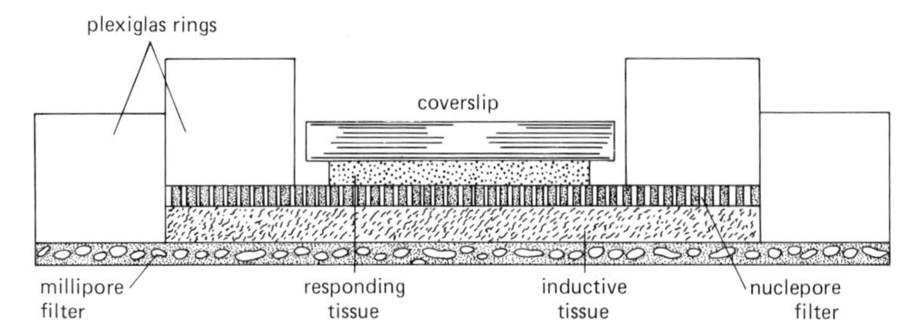

Figure 11.19 System used in transfilter assay for inductive interactions. [*From S. Toivonen, D. Tarin, L. Saxén, P. J. Tarin, and J. Wartiovaara,* Differentiation, **4**:2 (1975).]

nitrocellulose membrane filter of defined thickness containing pores of specified diameter. If the reactive tissue responds to the presence of the inductor by differentiating in a manner unlike that which occurs in the absence of the inductor, one can conclude that the inductive influence has been able to pass through the filter via its pores and reach the responding tissue.

When uncommitted *Triturus* ectoderm is cultured for 24 hours across a membrane filter (25 μm thick, 0.8 μm pore diameter) from a piece of dorsal-lip tissue, the ectoderm becomes determined to differentiate into neural structures. If the ectoderm is removed from the filter and cultured in isolation for an additional 10-day period, the explant proceeds to differentiate into forebrain (and occasional hindbrain) structures. If the piece of ectoderm is cultured on the filter in the absence of dorsal-lip tissue, it fails to differentiate into neural tissue, but rather forms unspecialized epidermis. It would seem apparent from this type of experimental finding that primary induction, at least under in vitro conditions, must result from the free diffusion of an inductor substance across the extracellular space between the cells. This point, however, has been the focus of a great deal of controversy. It has been suggested, with good reason, that these results do not rule out the possibility that long thin processes might project from either or both of the tissues into the filter through the pores. The presence of such processes could lead to an inductive interaction as a result of direct intercellular contact. In order to settle this question, membrane filters are examined with the electron microscope following the period of culture. In the type of experiment just described, no evidence of cellular processes was seen within the filter, but the situation was complicated by the fact that the membrane filters used in earlier experiments contained highly irregular channels making observations very difficult. In more recent experiments, membrane filters with straight-across channels have been utilized, and once again, no evidence of cellular processes have been observed. Based on a number of studies it appears that primary induction, as expressed under transfilter conditions, does not depend on cell-to-cell contact, but rather occurs via the diffusion of an inducing substance(s). It should be noted that certain other types of inductive interactions do seem to be mediated by cell contact (Section 12.4.A).

A. Regional inductive influences

From the time of its initial appearance as a thickened neural plate, the prospective nervous system is characterized by a marked anteroposterior polarity. This polarity along the main body axis is accentuated following neurulation, when the broad anterior portion of the neural tube becomes demarcated into several distinct brain regions (Section 14.4.A). In a broad sense at least, the regionalization

of the central nervous system is determined by the regional properties of the underlying chordamesoderm. This can be shown in several ways. One of the early experiments of Spemann consisted of rotating the animal half of an early gastrula by 180° relative to the vegetal half. Since the nervous system develops from the material of the animal half, while the inductive influences emanate from material present in the vegetal half prior to involution, this type of experiment provides an ideal basis for determining whether the polarity of the nervous system resides in the inductive chordamesoderm or the responding neural ectoderm. It was found that embryos resulting from these rotated gastrulas developed in a normal manner, with the polarities of their nervous systems corresponding to that of the vegetal portions of the eggs.

Further evidence that the regionalization of the nervous system is dependent on the regional character of the underlying chordamesoderm is derived from experiments in which the inductive properties of successive regions of the archenteric roof are determined. Several types of experimental procedures have been devised over the years for determining the inductive powers of small pieces of tissue. A favored approach in earlier studies involved the implantation of a piece of test material directly into the blastocoel of a young gastrula. In this implantation technique, a small window was cut in the relatively thin roof of the blastocoel and the tissue was implanted beneath the prospective ventral epidermis, far from the activities of the embryo's own primary organizer. When successive sections of the archenteric roof of early neurulas were implanted in this manner, it was found that anterior roof tissue induced head organs and forebrain parts, that midroof tissue induced primarily hindbrain and ear structures, and that posterior roof tissue induced spinal cord and tail structures. Furthermore, it was found that one did not have to wait until the neurula stage to demonstrate the regional character of the primary inducing system. Similar results were obtained by implantation of the dorsal lip of the blastopore from successively older gastrulas into the blastocoel of an early gastrula recipient (Fig. 11.20). The dorsal lip from an early gastrula, which will become the anterior part of the archenteric roof, acts as a potent head organizer, while the same region taken from a late gastrula acts as an organizer of trunk and tail structures.

Surgical experiments of the type just described on whole embryos are very informative, but they are also open to certain difficulties of interpretation. The embryo as a whole forms a complex system in which to test the inductive powers of small pieces of tissue. Consequently, it becomes difficult to rule out the involvement of other parts of the embryo in the events occurring at the site of the graft. Once experiments on the intact embryo had been performed and conclusions drawn, it became important to test those conclusions in an independent way under more simplified in vitro conditions. Amphibian tissue had been a favorite for in vitro culture ever since its initial use by Ross Harrison in 1907 (see Section 14.4.C). The use of in vitro culture in the analysis of primary induction was made possible by the development of a simple balanced saline solution (Holtfreter's solution) in which embryonic amphibian cells would remain healthy and differentiate. Cells of amphibian embryos are particularly well-suited for these types of studies because they contain sufficient nutritive materials to support themselves in saline solutions without the need for adding complex nutrients and growth factors, which are required for the culture of cells from higher vertebrates. Most important, the use of these simpler culture media eliminates the possibility that substances from the medium are playing a role in the inductive process being studied.

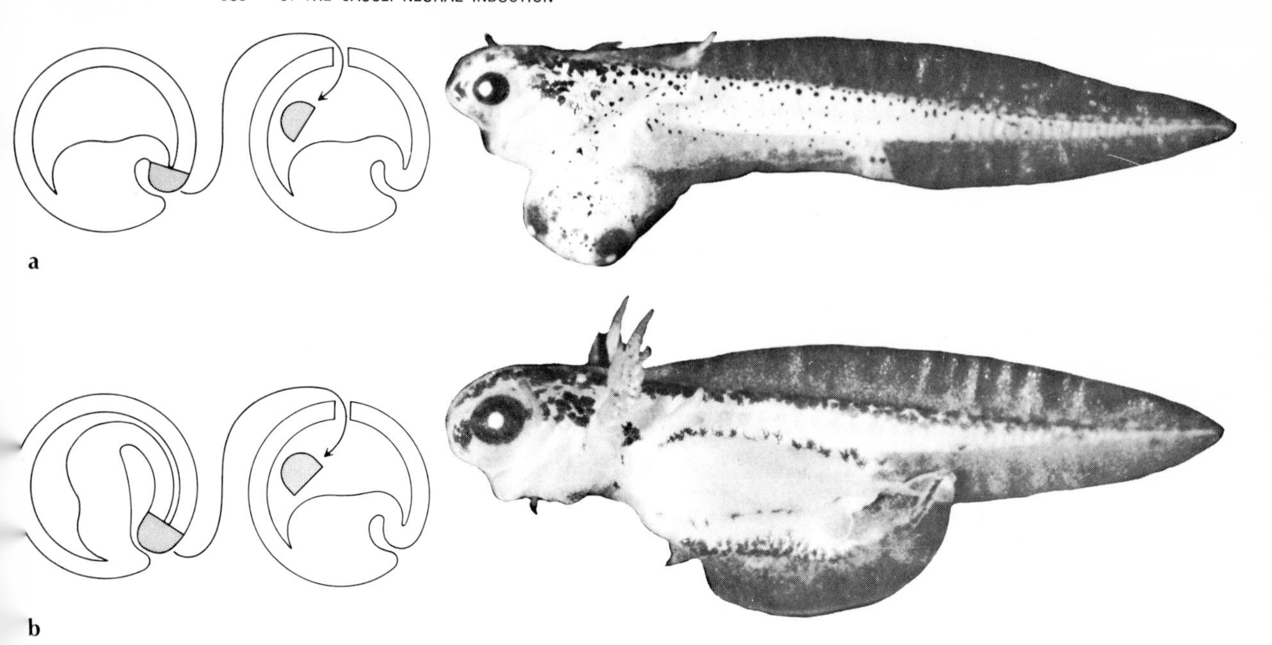

Figure 11.20 *a* Implantation of the margin of the blastopore lip from an early gastrula (head inductor) into another of the same age causes development of a secondary head (*above*). *b* The margin of the blastopore lip of a late gastrula (trunk inductor) induces a secondary trunk and tail when implanted in an early gastrula [*Courtesy of L. Saxén and S. Toivonen.*]

The ability of amphibian embryonic tissue to differentiate in isolation is readily demonstrated by following the fate of an explant taken from the dorsal marginal zone of an early gastrula. As expected from its prospective fate, this tissue differentiates into well-formed notochordal and somite structures. In addition, it generally forms neural and epidermal structures, which are outside its prospective fate. In contrast, explants of presumptive neural ectoderm (or presumptive epidermal ectoderm) are incapable of forming typical tissues and instead form only unspecialized epidermis, as was the case in the exogastrula. The failure of these explants to differentiate cannot be ascribed to inadequacies of the culture medium, because comparable explants from later-stage neurula are fully capable of differentiating into neural tissue (or epidermis if taken from a ventral site).

The failure of uncommitted ectoderm to differentiate into neural tissue in isolation has allowed investigators to test the effects of various types of inductor materials under in vitro conditions. In these experiments, the tissue serving as inductor is generally derived from a species whose cells are visibly distinct from the cells comprising the ectoderm. In earlier experiments, the inductive and reactive tissue fragments were taken from embryos of different genera, a condition referred to as a *xenoplastic* combination. Typically, the inductor tissue is explanted into a folded piece of ectoderm. The edges of the ectoderm then fuse around the inductor tissue enclosing it like a wrapper, thereby forming what is referred to as a "sandwich." When the dorsal lip of an early gastrula is combined with a piece of uncommitted ectoderm in a sandwich culture, a considerable variety of head organs are found among the differentiated tissues of the ectoderm. These include brain, eyes, and mouth structures. The material of the dorsal lip

develops primarily into notochord and somite, as expected, although foregut and neural tissues are also seen among the derivatives of the presumptive chordamesoderm. The formation of ectodermal and endodermal tissues by a fragment of presumptive mesoderm indicates, as in the previous experiments, that presumptive chordamesoderm does not simply self-differentiate to form those mesodermal structures of its prospective fate; it has a much wider potency. Even though the cells of the dorsal lip can be thought of as determined (to the extent that they do not need the influence of other tissues and can self-differentiate in vitro), they still retain the ability to regulate and form other types of structures. The general basis for this type of situation will be considered at greater length in the discussion of morphogenetic *fields* in Section 13.1. Regardless of the basis of the ability of presumptive chordamesoderm to differentiate into nonmesodermal structures, it suggests that within the embryo there are influences from the endodermal and ectodermal parts that act to inhibit the differentiation of the presumptive chordamesoderm along these other lines.

In contrast to the results using the dorsal lip of an early gastrula, sandwiches containing posterior portions of the archenteric roof develop primarily into spinal cord and taillike structures complete with hindgut. The results of these and other experiments clearly indicate that the organizing component of the dorsal mesoderm is itself differentiated along the anteroposterior axis; different regions induce different ectodermal structures. Ideally one would like to demonstrate a one-to-one relationship between a particular region of the nervous system (and associated sense organs) and the inducing capacity of the underlying sector of chordamesoderm. For various reasons, this type of demonstration has not proved feasible. However, the results of a large body of experimental evidence supports the concept of regionalization within the primary organizer and, more specifically, the presence of three distinct induction tendencies. The anterior section of the archenteric roof is an *archencephalic* inductor, i.e., one that tends to induce forebrain, nose, eyes, and suckers. The middle section of the roof is a *deuterencephalic* inductor, i.e., one that tends to induce hindbrain and ear. Finally, the posterior section of roof is a *spinocaudal* inductor, i.e., one that tends to induce spinal cord and tail structures.

B. Induction in the chick: the node

The inductive processes in chick and presumably other amniote embryos appear to be essentially the same as those in amphibians and other anamniotes. If the anterior end of the primitive streak (*Hensen's node*) is extirpated, stripped of all adhering endoderm, and implanted beneath the ectoderm at the side of the area pellucida, it induces a secondary embryonic axis consisting of neural tube, notochord, and somites (Fig. 11.21). This ectoderm does not contribute to the neural tube of the normal embryo. A similar experiment with the same result has been performed in mammalian (rabbit) embryos. As in the case of experiments on amphibian embryos, however, many tissues and substances have been found to induce axial embryonic structure when implanted beneath responsive ectoderm (discussed at length in Section 11.4). Nevertheless, in both cases the power of inducing neural tissue is greater in the anterior end of the normal inductor, i.e., the dorsal lip of the blastopore of the amphibian and the node of the primitive streak of the chick.

In amphibians the dorsal lip of the blastopore persists as such throughout the process of gastrulation, in spite of the diminishing size of the blastopore. It represents a location where chordamesodermal material continues to roll in and to move anteriorly beneath the ectoderm. As it does so, it changes in inductive ca-

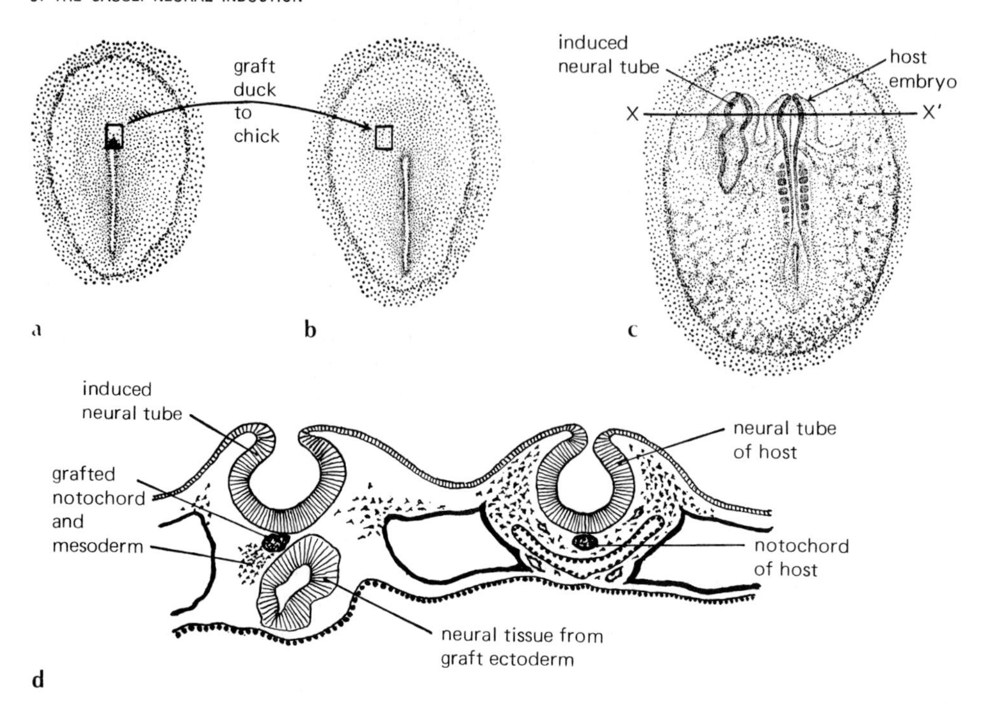

Figure 11.21 Semischematic drawings showing the induction of an accessory neural tube as a result of grafting notochordal tissue from a duck donor into a chick host. *a* Duck embryo showing the location from which the graft was taken. *b* Chick host showing the location where the graft was implanted. *c* Embryo cultivated for 31½ hours after implanting of the graft, showing the location of the induced accessory neural tube. *d* Section at level of the line *X-X′* in *c*. [*From B. M. Patten and B. M. Carlson, "Foundations of Embryology," 3rd ed., McGraw-Hill, 1974.*]

pacity, inducing a head when taken from an early gastrula and inducing a tail when taken from a late gastrula (Fig. 11.20). In the chick, as described in the preceding chapter, the node progressively regresses posteriorly down the primitive streak. That is, as the streak shortens, the node persists as a center of activity comparable with the amphibian dorsal lip.

The success of operative experiments of the kind we have been discussing has depended greatly on technical progress. An amphibian embryo that has received a graft will survive—if not otherwise injured and under suitable conditions—even beyond the stage of hatching. In the beginning, however, it took many years for Spemann to perfect the technical operational skill of grafting fragments from the dorsal lip of one embryo to the flank of another, and to train his students accordingly. In chick embryos, the problem has been even greater because such experiments are virtually impossible in the intact egg and chick embryos removed from the egg rarely live for more than a day. New methods have therefore had to be developed. One of these combines ectoderm and mesoderm from different levels of the chick blastoderm and grows them as sandwiches within the coelomic cavity of older embryos. Another overcomes the difficulty of working within the egg by taking grafts of Hensen's node from full-length primitive-streak blastoderms. The grafts are inserted beneath the ectoderm of blastoderms of the same age lying in situ on the yolk.

In both sets of experiments, forebrain, midbrain, and hindbrain were induced. From these and other experiments, the conclusion was reached that the node is initially a head organizer, but as it retreats it becomes successively a trunk and then a tail organizer. (Compare with amphibian blastopore in Fig. 11.20.) In other words, the differences in the regional character of the induced neural tissue are to a large extent dependent on the position of the node.

Chordamesoderm for all regions derives entirely from the node; i.e., it was part of the node at an earlier stage. Moreover, both the node and the anterior end of the streak possess a certain autonomy. If the node and anterior streak are cultured either in vitro or on the chorioallantoic membrane, the node will regress down the streak and an embryonic axis will differentiate. This whole process, however, is very poorly understood. Yet it holds the key to so much that is characteristic of the development of the higher vertebrates. It is frustrating that the process and consequences of node activity are phenomena normally hidden from direct observation and analysis by confinement within a shelled egg or an inaccessible womb.

C. Competence

Competence is defined as the physiological state or capacity of a tissue that permits it to react in a morphogenetically specific way to determinative stimuli. Whatever it may be, competence is always related to particular stimuli and particular corresponding responses. With regard to primary induction, therefore, we may speak of neural differentiation as a *primary competence* of the ectoderm.

It has long been known that ectoderm of amphibian embryos transplanted from various developmental stages from blastula to early neurula gradually loses neural competence. As the ectoderm ages, it gradually loses its capacity to respond to the inductive stimulus of the chordamesoderm by becoming neural tissue. Similarly, dorsal-lip tissue transplanted under ventral ectoderm at various stages of development shows that neural competence of nonneural ectoderm is not gained until just before gastrulation, then drops sharply near the end of gastrulation and is lost by the onset of neurulation. Isolated ectoderm unexposed to neural induction and ectoderm exposed to a neural inducer at too late a stage differentiate into unspecialized epidermis.

The decrease in neural competence with aging of the tissue has been tested by isolating fragments of gastrula ectoderm, maintaining them in isolation for various lengths of time, and then transplanting them into different locations in a neurula, where various strong inductors are present. Competence to form brain structures markedly decreases at an age corresponding to the late gastrula stage of control embryos. Ectoderm corresponding to the late neurula stage is completely without neural competence; yet the two neural-crest derivatives, namely, mesenchyme and pigment cells, can still be evoked in ectoderm of the tail-bud stage. Older ectoderm, already differentiating as epidermis, is entirely without competence to do anything but proceed toward its intrinsic epidermal destiny.

At the same time, the aging of the ectoderm does not merely restrict its neural competence; it also can bring about new responsiveness not present before. This is known as *secondary competence.* Late neurula ectoderm, no longer convertible into neural tissue, becomes competent to respond to other inductors. Under the influence of optic vesicle, hindbrain, and forebrain, respectively, it differentiates into lens, ear vesicles, and nasal pits (structures normally formed by head ectoderm after neurulation is complete) during postneurula stages of development. Successive states of competence appear synchronously with the succession of primary and secondary inductors.

Transplantation of tissues between different genera and even orders of amphibians demonstrates the complexity of the interacting system of inductors, fields, and in this procedure, genetic competence. These experiments exploit certain morphological differences between various amphibian larvae. Most salamander larvae possess a pair of rodlike temporary lateral head structures known as *balancers*, each consisting of a core of cartilage covered by epidermis. Anuran tadpole larvae have no such structure but do possess a pair of purely epidermal adhesive suckers below the mouth. If competent frog ectoderm, from a site remote from the ventral suckers, is grafted to a site on a salamander embryo corresponding to the sucker location of the frog, an epidermal sucker develops. Conversely, if competent salamander ectoderm, from a site remote from the balancer location, is grafted to a site on a frog embryo corresponding to the balancer location of the salamander, a balancer develops. In each case the transplanted tissue responds to *a position* on the host embryo and develops the structure appropriate to that position as though the host were of the same species as itself, although in each case the host cannot develop such a structure. It is as though the transplant says, "I recognize my new position, but I must respond in my own way."

The guidelines are recognized by implanted tissue even from a different order, but the response is dictated by the specific genomic constitution of the explant. In other words, unstructured tissue recognizes position in the system; for example, it forms a balancer or not, or a ventral sucker or not, depending on whether the implant is urodele or anuran. Position, implying local specific circumstances, is recognized, but the positional properties as such do not carry the precise instructions for the formation of any particular structure. They merely enable a group of uncommitted cells to self-organize, or self-assemble, into one particular pattern of the many permitted by their genomic constitution.

4. The problem: chemical nature of inducing agents

The discovery of the primary inductor quality of the dorsal lip of the amphibian blastopore, the "organizer" of Spemann, led to concerted attempts by teams of embryologists and biochemists throughout the 1930s to identify its chemical nature. It was soon shown that living inductor tissue was not required to initiate ectodermal differentiation; the cells could be heated, frozen, or treated with alcohol or other chemicals. Although these treatments did not destroy the ability of primary inductor tissues to cause uncommitted ectoderm to differentiate into generalized neural tissue, it did abolish the ability of these treated tissues to induce the formation of specific brain structures or mesodermal derivatives. Furthermore, extreme heat (140°C) was also successful in destroying the neuralizing power of the natural inductors. Based on these types of observations, it was thought that two distinct kinds of chemical substances were involved in primary induction, one being a relatively labile mesodermalizing agent and the other a more stable neuralizing factor. These findings served to launch a major investigation of the chemical inductors present in these tissues; sterols, proteins, and nucleoproteins were all implicated in various studies. The massive effort faded before the indigestible discovery that whereas the only live tissue that could induce primary embryonic structures was dorsal-lip material, almost any tissue from the whole animal kingdom could do so if killed by treatment with heat or alcohol (Fig. 11.22). Most important, heat-killed *ectoderm* from an early gastrula served as a potent inducer of neural differentiation when tested with uncommitted ectoderm in a sandwich culture or a blastocoel implant (Section 11.3.A). It was concluded, therefore, that the neuralizing substance was present in cells throughout the embryo, but it existed in some type of masked or inactivated state.

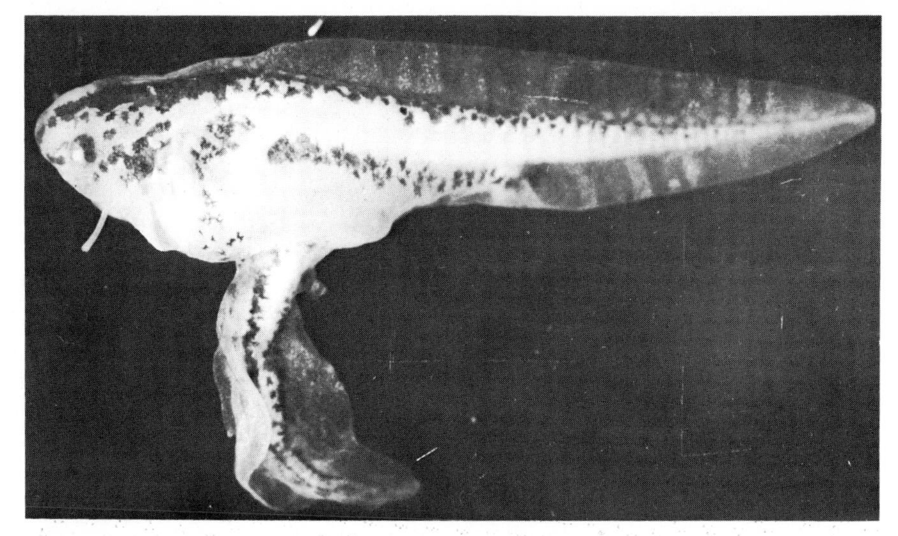

Figure 11.22 Implantation of a piece of alcohol-treated guinea pig kidney tissue into the blastocoel of a gastrula induces tail outgrowth with myotomes, notochord, and spinal cord. [*Courtesy of L. Saxén and S. Toivonen.*]

During normal development, the influence of this agent is released only in the presumptive neural ectoderm as a result of its interaction with the underlying archenteric roof. The inductor becomes an *evocator;* i.e., it does not induce a pattern or a particular differentiation, but rather evokes or triggers a particular response among the many that a cell or a sheet of cells may have in its repertoire. Responses are specific and determined by the substance(s) released in the responding tissue, which subsequently act to alter the course of that cell's differentiation. Evocators of a given response may be widely diverse and may not be chemically related in any significant way to the event called forth, even though different responses may be made to closely related stimuli.

Once it was discovered that evocators were present in a wide variety of unrelated cells, it was shown that inorganic agents such as iodine or kaolin were found to be effective. In fact, no external chemical agent was necessary to evoke the process of neuralization; it could occur in isolated pieces of ectoderm that had been subjected to various kinds of harsh (but not lethal) conditions. All that remained of the early grand concept of a master-chemical embryonic organizer was Holtfreter's concept of *sublethal cytolysis,* namely, the idea that reversible cell injury liberates the neural inductor. As a result of these and other findings, attention turned from the inducing tissue to the tissues being organized.

In most cases, ectoderm cultured with a piece of heat-killed tissue prepared from some exotic source differentiates into neural tissue of a very general nature. The evocator serves to activate the responding cells to undergo cytological differentiation (cytodifferentiation, Chapter 15), but the structure that forms bears little resemblance to any specific brain region. In other words, the ectoderm has been neuralized but lacks the necessary stimulus to gain a regional character. In some cases, however, it was found that specific foreign adult tissues, when combined with uncommitted ectoderm and implanted into the blastocoel of a host gastrula, induced a morphogenetic response in the ectoderm of comparable proportion to that obtained by combination with living chordamesoderm. One of the best artificial inductors proved to be boiled mouse kidney, which induced the formation of a structure possessing all the characteristics of a normal complex brain. It exhibited anteroposterior polarity and bilateral symmetry and consisted of the several regions (forebrain, midbrain, and hindbrain) into which the normal

brain is divided. Eyes and ears were also present, presumably derived as a result of secondary inductions by the forebrain and hindbrain, respectively (Chapter 14). It would appear that the dead foreign tissue is acting as a head organizer in the same way as a piece of the anterior section of the archenteric roof.

In order to appreciate this remarkable finding, it is worthwhile considering the nature of the two components, the dead kidney tissue and the living ectoderm, and their interaction. At best, the piece of dead inductor tissue can serve only as a source of chemical inducing substances. These chemicals, whatever their nature or relation to the normal inductors, must be diffusing out of the dead tissue uniformly in all directions, thereby striking the nearby ectoderm cells in an indiscriminate manner. Since all the ectodermal cells are receiving the same information, i.e., the same evocators, one would expect the ectodermal population to respond in a homogeneous manner. The best one could optimistically hope for is the conversion of the ectoderm into a uniform layer of neural cells, as is usually the case in such experiments. Instead, the uncommitted ectoderm responded by forming a highly structured organ system. How can this high degree of anatomical order and complexity emerge as a result of exposure to a nonpatterned shower of foreign inducing agents? Since we cannot ascribe the information as to position or pattern to the inducing tissue, we must, once again, turn to the responding tissue. In this case, however, the responding tissue is an uncommitted, developmentally flexible, homogeneous-appearing layer of cells. There is no hint of the presence of an "invisible" blueprint or program that would endow this layer with the capabilities for such complex spatially defined differentiation following exposure to an inductive stimulus. Furthermore, if the program for brain formation exists within the ectoderm, so too must the program for formation of other structures, since this layer of undetermined ectoderm can regulate and differentiate in any direction when exposed to the appropriate environment.

As in the experiment of Driesch (Section 7.2), we find once again that the parts of the egg are endowed with the same seemingly mystical powers as the whole egg itself. Both the parts and the whole of the egg or embryo contain some type of blueprint for the construction of a highly complex, patterned structure; yet each can be subdivided into more parts, which can, in turn, give rise to the whole structure. Developmental biologists have long recognized the existence of these properties and have used the term *morphogenetic field* to refer to them. A field is a region of an embryo in which a particular structure will arise. Although the cells of the field may become determined to form that structure, i.e., they can self-differentiate in a foreign environment, the field as a whole retains a developmental flexibility, i.e., an ability to regulate. Rather than digress further into this topic, which would take us far from the discussion of primary induction at hand, we will postpone its discussion until Chapter 13 and return to the nature of primary inducing agents.

The analysis of chemicals capable of evoking preprogrammed responses is less dramatic than the search for inducing molecules that contain the information upon which the response is based. However, the determination of the nature of the chemicals that operate in vivo in carrying inductive messages from one tissue to another remains a major task in developmental biology. The search for the inducers has followed several avenues.

One of the problems apparent from the earlier work was the need to be certain that the response observed was a direct consequence of chemicals being released by the supposed inductor tissue rather than a nonspecific response to sublethal cellular damage. Toward this end, in vitro culture conditions were developed that would maintain small pieces of tissues in a healthy state for long periods to allow

differentiation to take place. In one of the key studies in this area (see Niu and Twitty, 1953) a small piece of embryonic amphibian mesoderm was cultured in a hanging drop for approximately 1 week. After this period, a small piece of early gastrula ectoderm was introduced into the drop (either in the presence of the mesoderm or after its removal), and its fate was followed for the next several days. The result of this procedure is the differentiation of the ectoderm into branched pigment cells and neurons. Ectoderm placed into similar drops in which no previous inducing tissue has been growing does not differentiate. It appears that embryonic mesodermal tissue has released inducing substances into the drop that are then capable of stimulating undetermined ectoderm to differentiate, even after the mesoderm has been removed. No contact between the two tissues is required. The media in the drop has been "conditioned" by the inducing tissues. If the mesoderm is allowed to remain in the drop for a longer period (14 to 18 days) before the introduction of the ectoderm, muscle cells are also found among ectodermal derivatives. This latter observation suggests that the inductive specificity of the chemicals in the drop changes with the age of the inducing tissue.

Studies such as this provide information about the inductive process but little about the chemical nature of the substances involved. Chemical analyses cannot be performed on tiny pieces of tissue or small conditioned drops. Another branch of research on this topic has taken advantage of the findings that a variety of tissues, both embryonic and adult, are capable of causing undetermined ectoderm to differentiate along predictable paths. Guinea pig liver tissue, for example, acts as an archencephalic inducer (nose, eyes, forebrain formed), while guinea pig kidney is a spinocaudal inducer causing a predominant formation of trunk and tail structures. Two sources have been found of a mesodermal inducer that causes the formation of muscle, notochord, and kidney tubules, i.e., mesodermal structures. These latter substances are extracted from either 9- to 11-day chick embryos or from guinea pig bone marrow. The advantage of the study of these nonembryonic (heterologous) inducing tissues is the opportunity to obtain large amounts of material with the capability of purifying and characterizing the active substances. The disadvantage is the uncertainty as to whether there is any meaningful relationship between these substances and those operating normally during embryonic induction.

Analysis of the chemical nature of heterologous inductors has been a matter of some controversy. When inductor substances are purified from adult tissues, the active ingredients are usually found to be proteins or ribonucleoproteins. In cases where ribonucleoproteins are implicated as the inducing substance, treatment with proteolytic enzymes destroys the activity, while treatment of the purified fraction with ribonuclease has no effect. Although some workers hold to the concept of primary induction as an RNA-mediated event, it is generally believed to occur via the transfer of protein molecules from the inductor tissue. A very different view of the underlying basis for primary induction is found in the proposal of Lester and Lucena Barth, where it is suggested that induction may be initiated by release of ions from bound form, representing a change in ratio between bound to free ions within the cells of the early gastrula. Experimentally, induction of nerve and pigment cells in small aggregates of prospective (presumptive) epidermis of the frog gastrula has been found to depend on the concentration of the sodium ion, that is, in the absence of other tissue; also, normal induction of nerve and pigment cells by mesoderm in small explants from the dorsal-lip and lateral marginal zones of the early gastrula is dependent on the external concentration of sodium. Nerve and pigment cells are induced when the culture medium contains

0.088 *M* NaCl; at 0.044 *M* NaCl, the mesoderm differentiates into muscle and mesenchyme, but nerve and pigment cell induction does not occur. The interpretation is that normal embryonic induction depends on an endogenous source of ions and that an intracellular release of such ions occurs during late gastrulation. Considering the recent demonstration of the widespread role of ion fluxes in cell and developmental activity, the possibility of their involvement in primary induction should not be ignored.

Despite the major uncertainty as to the relevance of studies utilizing heterologous inductor tissues, investigators have based much of their thinking concerning the underlying basis of primary induction on results obtained with these materials. As discussed earlier, substances isolated from various adult tissues are capable of inducing specific, extensive differentiations. Moreover, when tested in combination, heterologous inductors are capable of provoking the differentiation of structures that do not appear when either inductor is tested alone. For example, combinations of guinea pig liver (an archencephalic inductor) and guinea pig bone marrow (a mesodermalizing inductor) are together (Fig. 11.23) able to induce a strikingly complete embryonic axis. Presumably certain cells of the ectoderm are responding to the presence of both substances by differentiating in a direction unlike that when either substance is present alone. This type of finding (and others) has led to the concept that primary induction within the embryo is mediated by the action of two diffusible substances that act in a synergistic manner to bring out the formation of the embryonic nervous system.

During the 1950s, two models of primary induction were formulated based on the presence of a pair of inductive agents. In one model, that of Lauri Saxen and Sulo Toivonen, the two substances were present in the amphibian embryo in a graded form, i.e., as a double gradient, and it was the relative concentrations of the two influences in the ectoderm that determined the specific direction of differentiation taken by each part. In this proposal, the neuralizing agent is present at its highest concentration along the entire dorsal midline of the embryo from its anterior to its posterior end (Fig. 11.24), while its concentration decreases at increasing distance in the ventral direction. In contrast, the second agent, a meso-

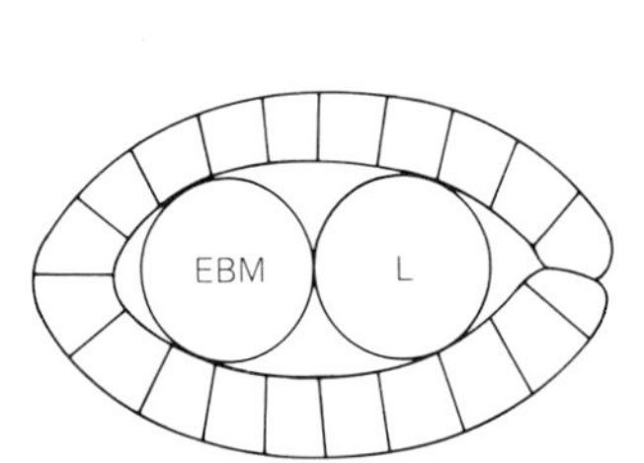

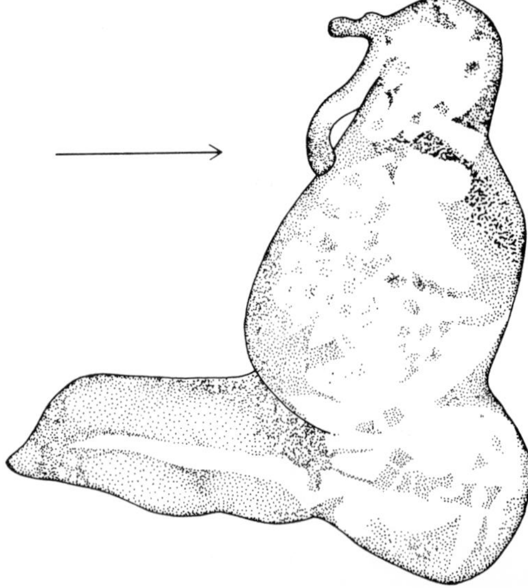

Figure 11.23 The effect of two heterologous inductors on the undetermined ectoderm of an amphibian gastrula. (*Left*) Scheme of the experiment: both inductors are wrapped in a sandwich of ectoderm. (*Right*) An embryolike complex has developed. EBM extract of bone marrow; L, piece of guinea pig liver. [*After S. Toivonen and L. Saxén, 1955.*]

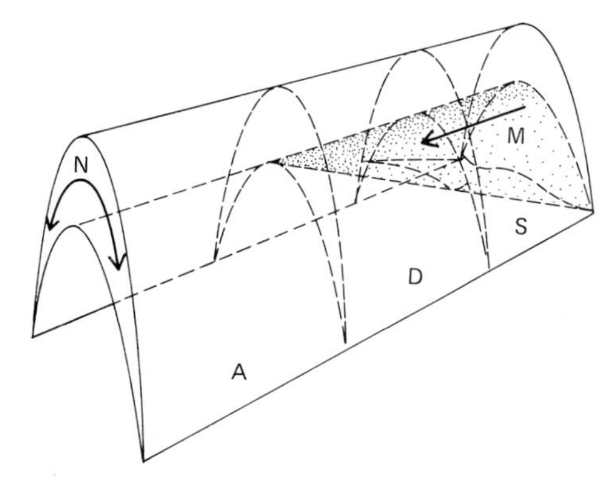

Figure 11.24 Proposed distribution of neuralizing (N) and mesodermalizing (M) activities within the chordamesodermal mantle of the amphibian embryo. A, archencephalic region; D, deuterencephalic region; S, spinocaudal region. [*From S. Toivonen, L. Saxén, and T. Vainio,* Roux Arch. Entwickl., **154**:*296* (*1963*).]

dermalizing factor, has its highest concentration in the dorsal posterior corner of the embryo and a decreasing concentration in both a ventral and anterior direction. In this model, the anterior region of the embryo is subjected solely to the influence of the neuralizing agent and consequently differentiates into forebrain. The mesodermalizing agent makes itself felt only in regions posterior to the forebrain. As the ratio of neuralizing to mesodermalizing influence decreases, the nature of the neurectodermal differentiation takes on a more posterior character, progressing from midbrain to hindbrain and finally to spinal cord.

In the second model, that proposed by P. D. Nieuwkoop, the inductive activities spread during gastrulation through the overlying ectoderm in a caudocranial (posteroanterior) direction as two successive waves. It is proposed that the first wave, which travels ahead and farther than the second, serves to activate or neuralize the overlying ectoderm such that it proceeds to differentiate in a neural direction, i.e., undergo neural-tube formation. It is proposed that the second wave involves the diffusion of a transforming agent that determines the regional character of the various parts of the previously neuralized ectoderm, thereby converting the layer into an organized central nervous system.

Unfortunately, very little additional information has come to light in recent years concerning these matters, and the validity of these and other generalized theories remains open to question. For various technical reasons, investigators have chosen to concentrate more heavily on later secondary inductive steps than on the primary organizing influence occurring during gastrulation. While providing considerable information on the development of a variety of specific organs, this trend had left us very much in the dark on the underlying basis for the first and most important inductive interaction.

Readings

Barth, L. G., and L. J. Barth, 1974. Ionic Regulation of Embryonic Induction of Cell Differential in *Rana pipiens, Develop. Biol.,* **39**:1–23.

———, 1967. Competence and Sequential Induction in Presumptive Epidermis of Normal and Hybrid Frog Gastrulae, *Physiol. Zool.,* **40**:97–103.

Burnside, B., 1973. Microtubules and Microfilaments in Amphibian Neurulation, *Amer. Zool.,* **13**:989–1006.

———, 1971. Microtubules and Microfilaments in Newt Neurulation, *Develop. Biol.,* **26**:416–441.

Cohen, A. M., 1972. Factors Directing the Expression of Sympathetic Nerve Tracts in Cells of Neural Crest Origin, *J. Exp. Zool.,* **179**:167–182.

———, and I. R. Konigsberg, 1975. A Clonal Approach to the Problems of Neural Crest Determination, *Develop. Biol.*, **46**:262–280.

Coulombre, A. J., M. C. Johnston, and J. A. Weston, 1974. Conference on Neural Crest in Normal and Abnormal Embryogenesis, *Develop. Biol.*, **36**:f1–f5.

Gallera, J., 1971. Primary Induction in Birds, *Adv. Morphol.*, **9**:149–180.

Gordon, R. and A. G. Jacobson, 1978. The Shaping of Tissues in Embryos, *Sci. Amer.* (June) **238**:106.

Hay, E. D., 1978. Embryonic Induction and Tissue Interaction During Morphogenesis, in J. W. Littlefield and J. de Grouchy (eds.), "Birth Defects," Excerpta Medica.

Huxley, J. S., and G. R. de Beer, 1934. "The Elements of Experimental Embryology," Cambridge University Press.

Holtfreter, J., 1968. Mesenchyme and Epithelia in Inductive and Morphogenetic Processes, in Fleischmajer (ed.), "Epithelial-Mesenchymal Interactions," Williams and Wilkins.

———, and V. Hamburger, 1955. Embryogenesis: Amphibians, in B. H. Willier, P. A. Weiss, and V. Hamburger (eds.), "Analysis of Development," Saunders.

Jacobson, A. G., 1976. Nature and Origin of Patterns of Changes in Cell Shape in Embryos, *J. Supra. Struct.*, **5**:371–380.

Jacobson, C. O. (ed.), 1979. Formshaping Movements in Neurogenesis, *Zoon*, **6**.

Jacobson, M., 1978. "Developmental Neurobiology," 2d ed., Plenum.

Kallen, B., 1965. Early Morphogenesis and Pattern Formation in the Central Nervous System, in R. L. DeHaan and H. Ursprung (eds.), "Organogenesis," Holt, Rinehart and Winston.

Karfunkel, P., 1974. The Mechanisms of Neural Tube Formation, *Int. Rev. Cytol.*, **38**:245–271.

LeDouarin, N. M., D. Renaud, M. A. Teillet, and G. H. LeDouarin, 1975. Cholinergic Differentiation of Presumptive Adrenergic Neuroblasts in Interspecific Chimeras After Heterotopic Transplantation, *Proc. Nat. Acad. Sci. U.S.*, **72**:728–732.

LeDouarin, N. M., M. A. Teillet, and C. LeLièvre, 1977. Influence of the Tissue Environment on the Differentiation of Neural Crest Cells, in J. W. Lash and M. M. Burger (eds.), "Cell and Tissue Interactions," Raven.

Marx, J. L., 1979. New Information about the Development of the Autonomic Nervous System, *Science*, **206**:434–437.

Needham, J., 1942. "Biochemistry and Morphogenesis," Cambridge University Press.

Nelsen, O. E., 1953. "Comparative Embryology of the Vertebrates," McGraw-Hill.

Nieuwkoop, P. D., 1973. The "Organization Center" of the Amphibian Embryo: Its Origin, Spatial Organization, and Morphogenetic Action, *Advan. Morphog.*, **10**:1–39.

Niu, M. C., and V. C. Twitty, 1953. The Differentiation of Gastrula Ectoderm in Medium Conditioned by Axial Mesoderm, *Proc. Nat. Acad. Sci. U.S.*, **39**:985–989.

Noden, D. M., 1978*a*. Interactions Directing the Migration and Cytodifferentiation of Avian Neural Crest Cells, in D. R. Garrod (ed.), "Specificity of Embryological Interactions," Chapman and Hall.

———, 1978*b*. The Control of Avian Cephalic Neural Crest Cytodifferentiation, *Develop. Biol.*, **67**:296–329.

Revel, J. P., and S. S. Brown, 1975. Cell Junctions in Development with Particular Reference to the Neural Tube, *Cold Spring Harbor Symp. Quant. Biol.*, **40**:443–455.

Saxén, L., and S. Toivonen, 1962. "Primary Embryonic Induction," Prentice-Hall.

Spemann, H., 1938. "Embryonic Development and Induction," Yale University Press (reissued 1968).

———, and H. Mangold, 1924. Induction of Embryonic Primordia by Implantation of Organizers from a Different Species, in English translation, in B. H. Willier and J. M. Oppenheimer (eds.), "Foundations of Experimental Embryology," 1964, Prentice-Hall.

Tarin, D., 1971. Histological Features of Neural Induction in *Xenopus laevis*, *J. Embryol. Exp. Morphol.*, **26**:543–570.

Tiedemann, H., 1967. Biochemical Aspects of Primary Induction and Determination, in R. Weber (ed.) "The Biochemistry of Animal Development," vol. 2, Academic.

Toivonen, S., D. Tarin, L. Saxén, P. J. Tarin, and J. Wartiovaara, 1975. Transfilter Studies on Neural Induction in the Newt, *Differentiation,* **4**:1–7.

Toivonen, S., D. Tarin, and L. Saxén, 1976. The Transmission of Morphogenetic Signals from Amphibian Mesoderm to Ectoderm in Primary Induction, *Differentiation,* **5**:49–55.

Waterman, R. E., 1976. Topographical Changes along the Neural Fold Associated with Neurulation in the Hamster and Mouse, *Amer. J. Anat.,* **146**:151–171.

Wessels, N. K., 1977. "Tissue Interactions and Development," Benjamin.

Weston, J. A., 1970. The Migration and Differentiation of Neural Crest Cells, *Advan. Morphog.,* **8**:41–110.

Yamada, T., 1962. The Inductive Phenomenon as a Tool for Understanding the Basic Mechanism of Differentiation, *J. Cell. Comp. Physiol. (suppl.)*, **60**:49–64.

Chapter 12 Organogenesis

In the following sections of this chapter we will take a brief look at the manner in which the major organs and organ systems of vertebrates develop, i.e., the topic of *organogenesis*. For the most part, we will confine ourselves to a brief description of events, with relatively little concern for underlying mechanisms. A considerable body of experimental study does exist in the area of organogenesis, most of it concerned with the identification of the various inducing tissues involved in the development of the major organs, and much of it remaining highly speculative and controversial. The reader can find these topics discussed at much greater length in the books by Willier, Weiss, and Hamburger (1955) and Balinsky (1976). The organogenesis of the limb, sense organs, and central nervous system are treated separately in following chapters.

1. Visualization of the embryo

It is difficult to visualize the entire ongoing development, even during gastrulation and neurulation, because we are confronted during each phase with a three-dimensional organization of cells and tissues that changes continuously. Postneurula development becomes increasingly complex, and four-dimensional visualization (three spatial dimensions plus time) calls for both concentration and practice. Cinematography is a great aid. For example, in films on salamanders and fish, weeks of continuous development, from egg to feeding larva, are condensed to 40 minutes. A brief description of the external form of early amphibian development must suffice here. More recently, the scanning electron microscope has revealed with striking clarity the topographical changes that occur within the interior of the embryo. The pair of low-power scanning electron micrographs in Fig. 12.1 expose much of the organization of the early chick. The reader can refer back to these micrographs in the discussions of organ formation which follow in this chapter. It should be kept in mind that developmental changes are all part of a continuing process. Dividing this process into time or structure stages such as blastula, gastrula, neurula, and postneurula is arbitrary and more a convenience than a reflection of the evident quality of flow of the entire phenomenon.

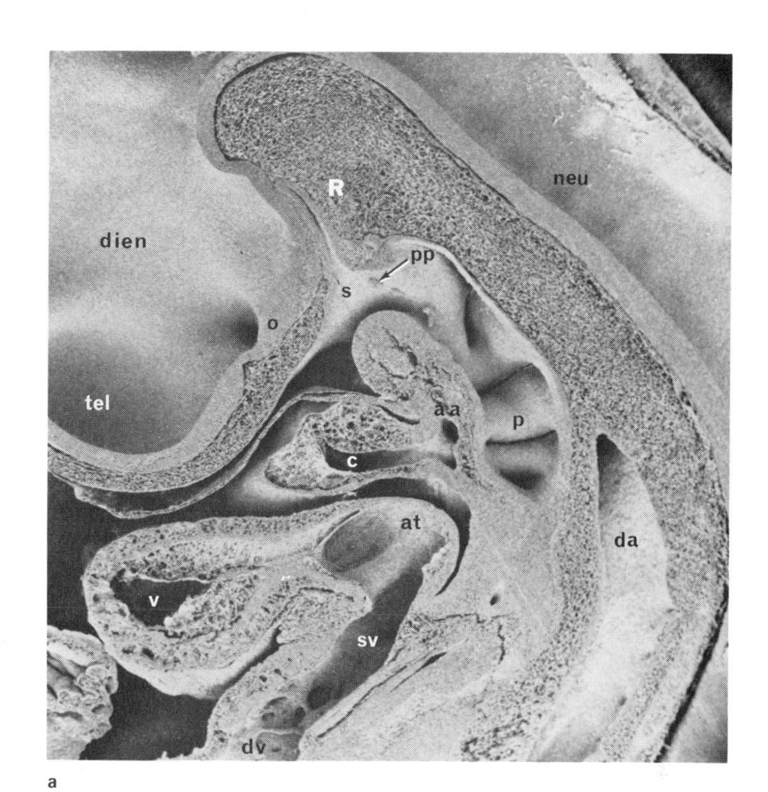

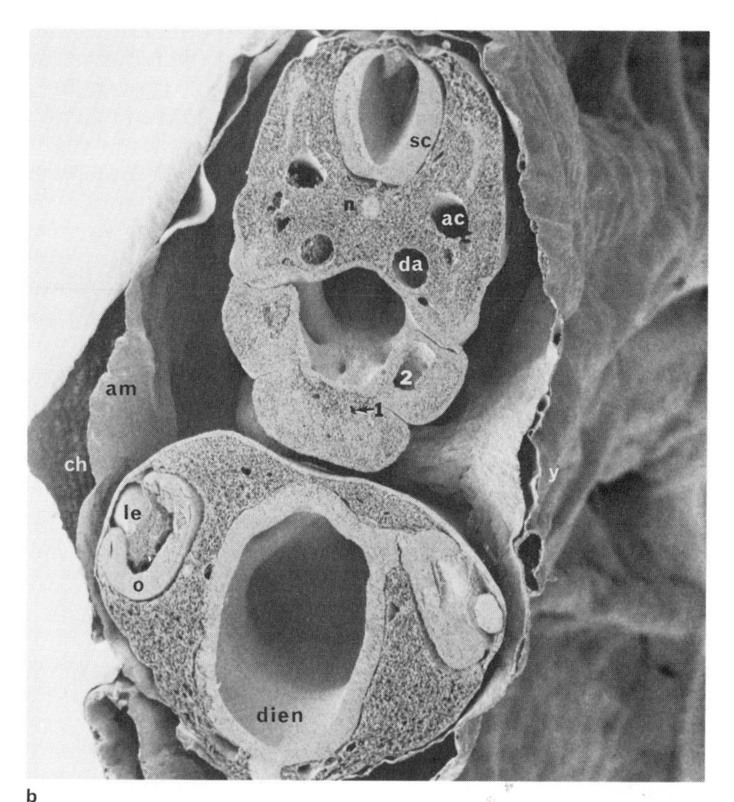

Figure 12.1 Scanning electron micrographs of sectioned 72-hour chick embryos showing the three-dimensional organization of many of the structures discussed throughout this chapter. *a* The right half of the head and pharynx region. The prosencephalon is subdivided into telencephalon (tel) and diencephalon (dien), from which arises the stalk of the optic vesicle (o). Most of the mesencephalon and metencephalon are out of view at the top of the picture. The myelencephalon can be distinguished by the presence of neuromeres (neu) in its ventral wall. Parts of the digestive system include the stomodeum (s) with Rathke's pouch (R) as a dorsal diverticulum and the pharynx with pharyngeal pouches (p) 1 to 4. The separation between pharynx and stomodeum is demarcated by remnants of the pharyngeal plate (pp). The circulatory system is represented by the ductus venosus (dv), which has been invaded by cords of liver tissue, the sinus venosus (sv), the atrium (at), the ventricle (v), the conus (c), the base of aortic arches 3 and 4 (aa), and the dorsal aorta (da). *b* Frontal view of the floor of the pharynx. The plane of section cuts across the diencephalon (dien), eyes (o, eyecup; le, lens), aortic arches 1 and 2, the paired dorsal aortae (da), anterior cardinal veins (ac), spinal cord (sc), and notochord (n). The depression in the floor of the pharynx is the thyroid primordium, and the columns of tissue on each side of the pharynx are the pharyngeal arches 2 and 3. The yolk sac (y), chorion (ch), and amnion (am) surround the embryo. [*Courtesy of P. B. Armstrong and D. Parento.*]

Perhaps the most obvious change, beginning with the onset of neurulation and extending long into postneurula development, is the elongation of the embryo. The amphibian neurula, at first almost spherical, elongates along its anteroposterior axis (Fig. 12.2). With further development, the anterior region of the embryo protrudes as the head, while a small tail bud forms at the posterior end. During this period, various internal developments are visible as external prominences, in particular the bulging of the optic vesicle and gill rudiment on each side of the head (Fig. 12.2).

The harmonious flow or progression of developmental change, however, can be observed only in a superficial way. The related internal events cannot be observed directly. They require disassembly and reassembly to be understood even in terms of anatomic description, as though one were taking a watch or engine apart and putting it together again to know what has to be explained. A three-dimensional picture of a complex embryo is usually obtained by cutting the embryo into a series of two-dimensional sections, studying them individually, and putting

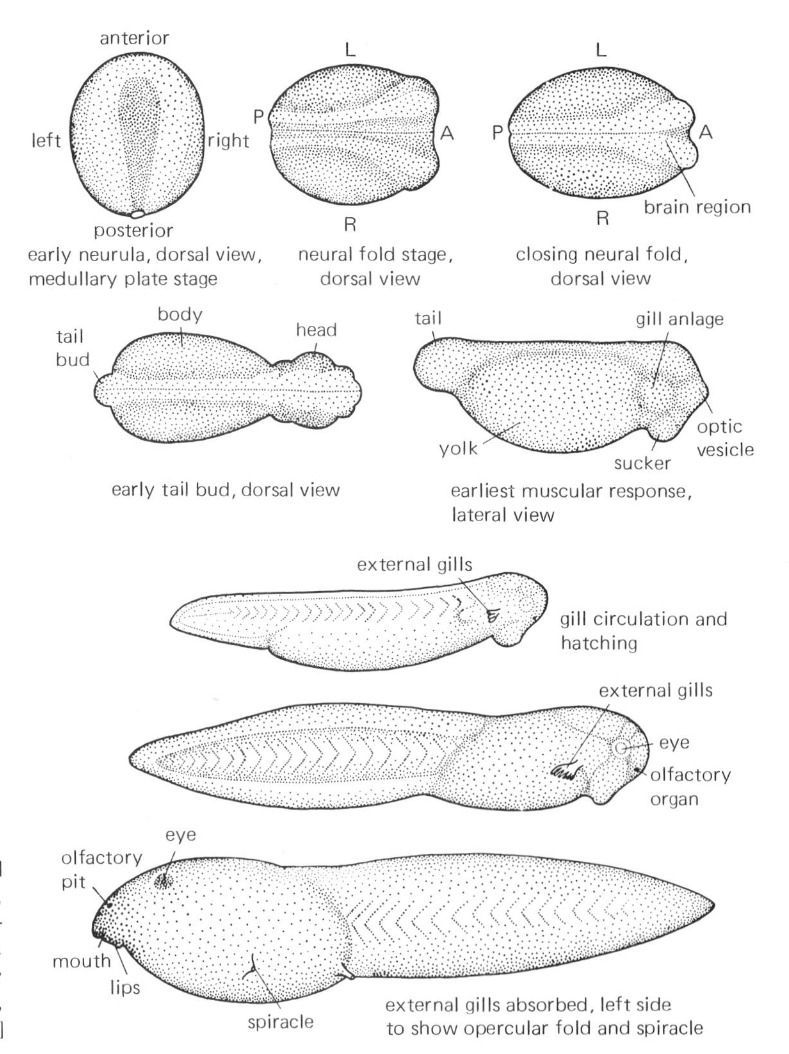

Figure 12.2 Postneural development of frog, from neurula to second-stage tadpole. [*From R. Rugh, "The Frog," McGraw-Hill, 1953.*]

them together again mentally or graphically. Embryos may be sliced in any one of the three spatial dimensions to yield a series of frontal, transverse, or sagittal sections (Fig. 12.3). This tedious process becomes more so as embryos are examined at successively later stages. The problem is not conceptual but is simply the need to visualize happenings not directly observable. We have already had a comparable example in the vital-staining experiments with eggs and blastulas to discern various cell shiftings otherwise difficult to detect.

The primary differentiation of the embryo into ectodermal, mesodermal, and endodermal regions represents fieldlike distributions of corresponding territorial tendencies already existing in the blastula. The differentiation tendencies for notochord, originally distributed over a wider area than the presumptive rudiment, are gradually restricted to the later rudiments. Somite mesoderm develops on both sides of the notochord. The presumptive nephric rudiments appear lateral to the somitic mesoderm, and lateral and ventral portions of the mesodermal mantle give rise to lateral plate and, in the chick, to blood islands, respectively. Each of these structures is believed to form within a certain range of intensity of the influence from the notochordal rudiment. Several of these structural components again show a differentiation into smaller units.

Altogether the developmental process clearly exhibits progression from the general to the particular (*epigenesis*). This is seen in the overall appearance of advanced embryos of higher vertebrates, namely, reptiles, birds, and mammals (Fig. 12.4). They are remarkably alike even though they become widely different in final form.

2. The postneurula embryo

A. Lateral-plate mesoderm

As described in Chapter 10, one of the major processes to occur during gastrulation is the movement into the interior of the embryo of a wall of mesodermal cells (referred to as the *mesodermal mantle*), which slips into the space between the overlying ectoderm and underlying endoderm. Dorsolaterally this mesoderm forms blocks of tissue that constitute the somites (Section 12.3.A). In amphibians, the adjoining portion of the mesodermal mantle lying ventrolateral to the somites is termed the *lateral plate*. The mesoderm of the lateral plate remains as a thin unsegmented sheet that is not uniformly spread throughout the embryo. Owing

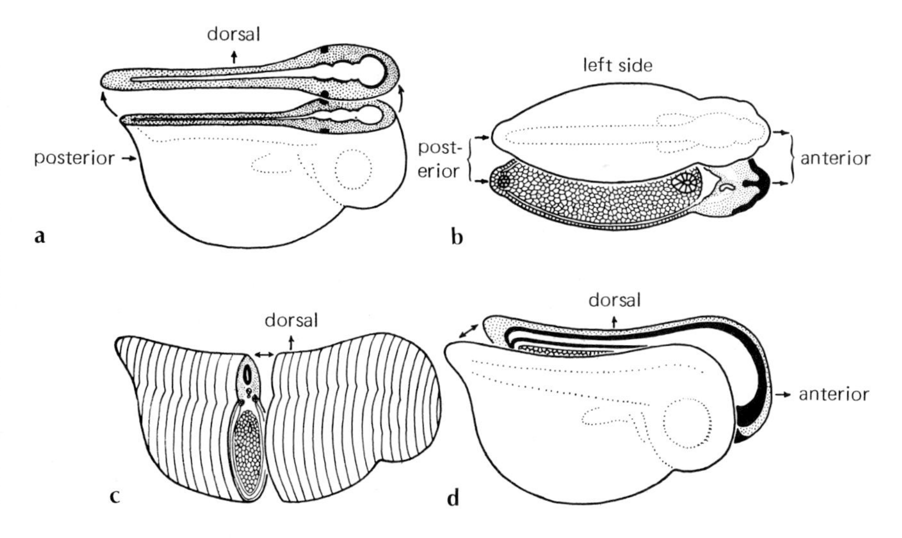

Figure 12.3 Planes in which the embryo may be cut or sectioned, showing the shape and internal structure of the early postneurula stage of the frog embryo: *a* lateral view of frontal section; *b* dorsal view of frontal section; *c* lateral view of transverse serial section; *d* lateral view of sagittal section. [*From R. Rugh, "The Frog," McGraw-Hill, 1953.*]

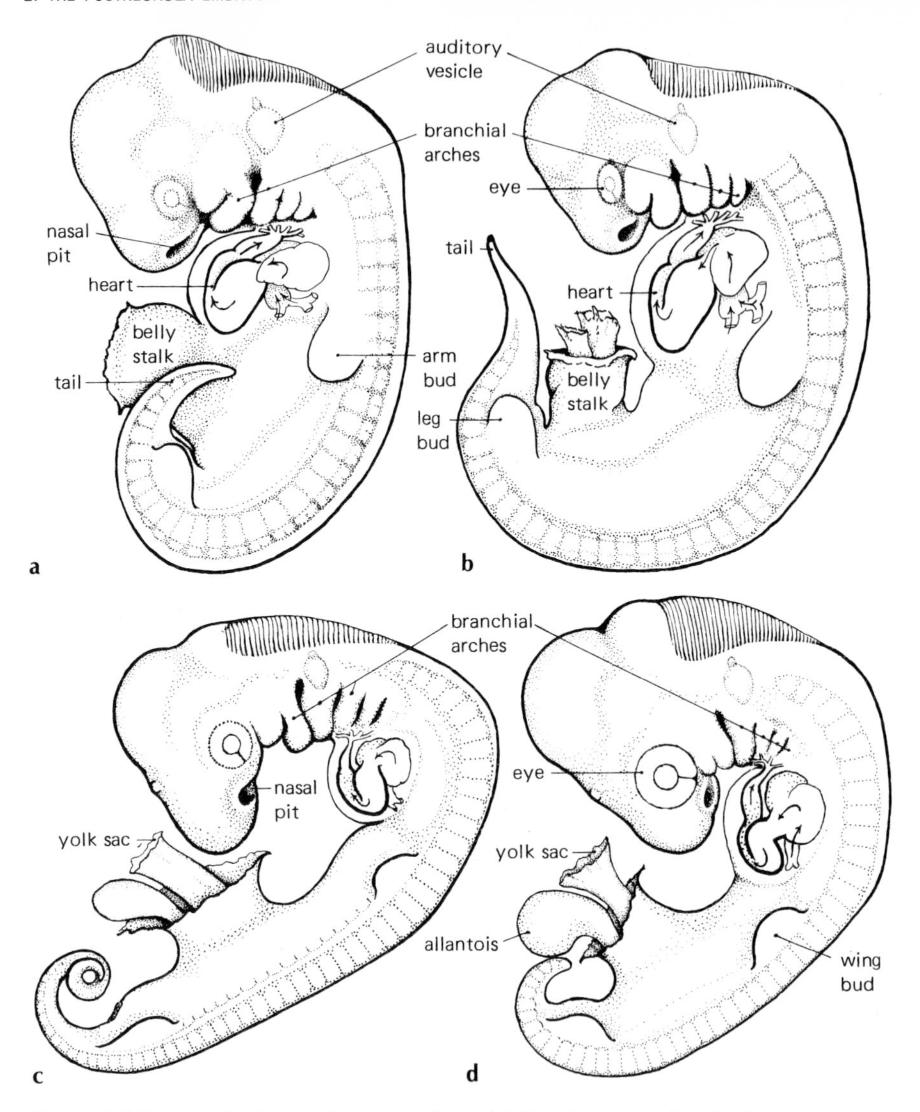

Figure 12.4 Embryos of *a* human, *b* pig, *c* reptile, and *d* bird at corresponding developmental stages. The striking resemblance of the embryos to each other indicates the fundamental similarity of the processes involved in their development. [*From W. Patten, "Evolution," Dartmouth College Press, 1922.*]

to the greater advance of the mesoderm on the dorsal surface of the blastopore and the position of the blastopore on the posterior end of the embryo, the mesodermal mantle of the early neurula does not extend into the ventral region of the anterior end of the embryo. As development in amphibians proceeds, the lateral plate continues to extend in a ventral direction on each side of the embryo (see Fig. 11.3) until they fuse in the ventral midline. As the lateral plates extend around the embryo, each becomes split into an outer *somatic* (or parietal) layer adjacent to the ectoderm and an inner *splanchnic* (or visceral) layer close to the endoderm (see Fig. 12.15). The cavity between the two layers is the body cavity, or *coelom*. Since the lateral-plate mesoderm exists as a continuous unsegmented

sheet, its splitting into two layers forms a continuous coelomic cavity that runs uninterruptedly across most of the anteroposterior axis. At later stages, folds of tissue become inserted across the original coelom, dividing it into a pair of spaces surrounding the lungs, the *pleural cavities*, a space surrounding the heart, the *pericardial cavity*, and a space surrounding the remainder of the viscera, the *peritoneal cavity*. The epithelial linings of these cavities, which are derived from the lateral plate, are the *pleura*, *pericardium*, and *peritoneum*, respectively.

The situation is essentially the same in the chick embryo despite the development of bird embryos on the virtually flat upper surface of the yolk mass; this is in contrast to the frog embryo, which is able to enclose its relatively much smaller, yolky region. Figure 12.5 shows a comparison of equivalent stages of frog and chick, including the procedure for removing a chick embryo from its underlying yolk. In the chick (and mammal), the cavity of the lateral plate extends in a lateral direction far beyond the limits of the embryo itself. Consequently, this space is composed of a medial embryonic coelom that is continuous laterally with an extraembryonic coelom (Fig. 12.15*d*). It is not until a later stage, when the embryo is lifted from the surface of the yolk by the formation of a ventral body wall, that the embryonic and extraembryonic portions of the coelom become separated.

B. The gut

The major changes that occur during the early stages of amphibian development, as evident in a series of midsaggital sections, are shown in Fig. 12.6. In amphibians the alimentary canal differentiates directly from the primitive gut or archen-

Figure 12.5 Comparison of chick and frog embryo in transverse section. *a* Diagram showing how the usual method of removing chick embryos from the yolk in order to prepare them for microscopic study makes the sections appear as if the primitive gut had no ventral boundary. *b* and *c* show how removing a chick from the yolk and pulling its edges together ventrally facilitates comparisons with forms that do not develop with their growing bodies spread out on a large mass of yolk. [*From B. M. Patten and B. M. Carlson, "Foundations of Embryology," 3rd ed., McGraw-Hill, 1974.*]

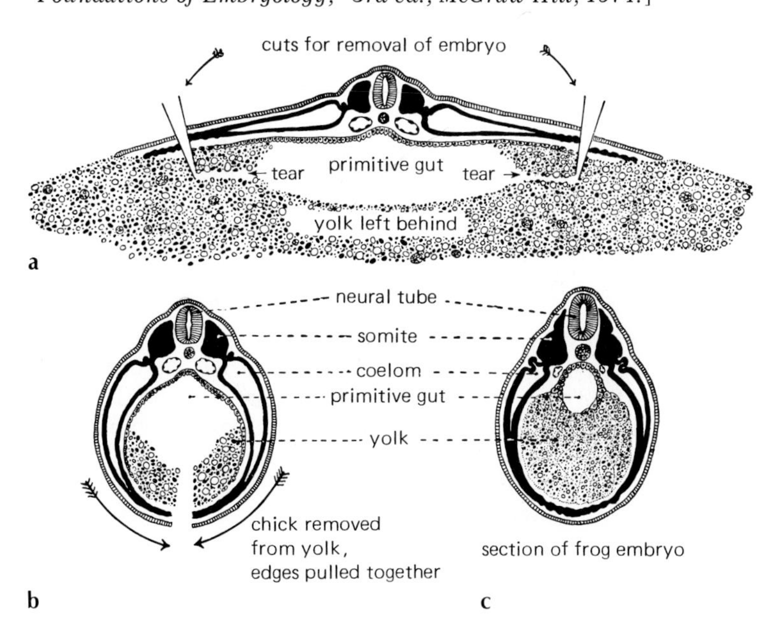

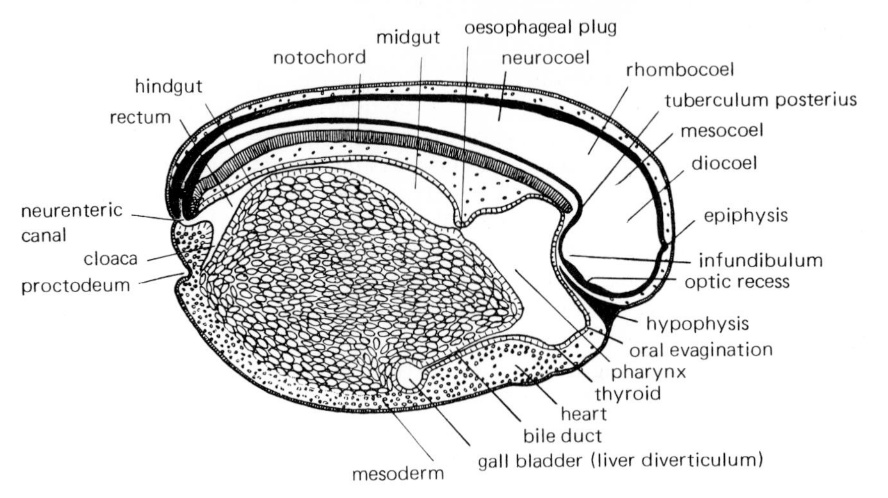

a

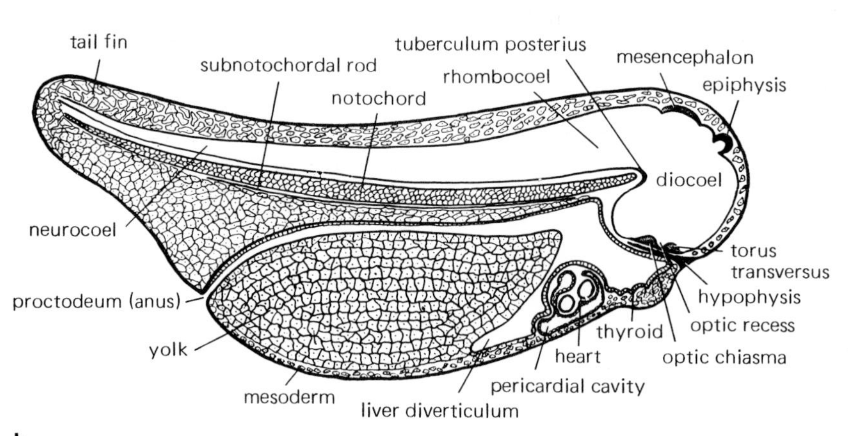

b

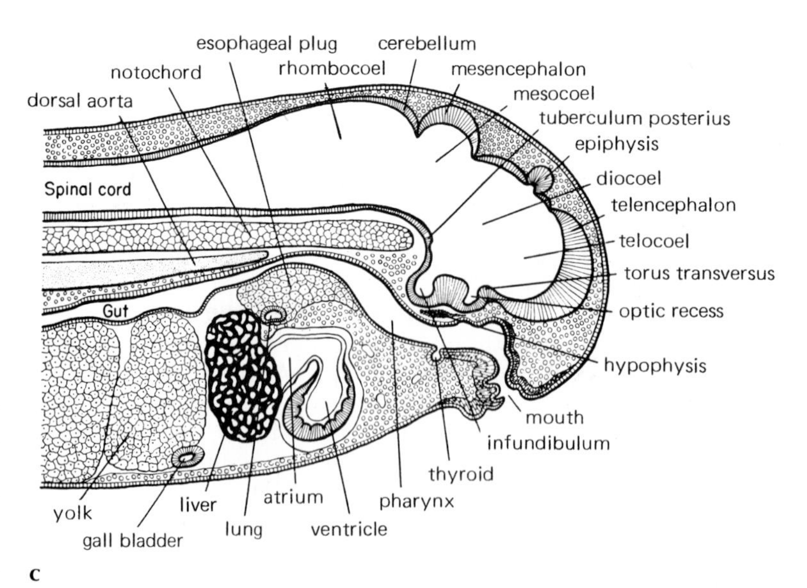

c

Figure 12.6 *a* Longi-[t]udinal section of a 3-mm frog embryo. *b* Longitudinal section of 5-mm [f]rog, showing differentiation of brain and [g]ut with primary regions. *c* Development of the brain and anterior [s]tructures in 11-mm frog tadpole. [*From R. Rugh, "The [F]rog," McGraw-Hill, 1953.*]

teron that forms during gastrulation. At the anterior end of the elongating neurula, the adjacent anterior ectoderm on the ventral surface invaginates as the *stomodeum* and fuses with the anterior, blind end of the archenteron to form an *oral plate* having an inner endodermal and an outer ectodermal lining. Similarly, a *proctodeal* invagination forms at the site of closure of the ventral part of the blastopore, and this ectodermal pocket fuses with the endodermal wall near the posterior end of the archenteron. A continuous tube, eventually open at both ends after rupture of the oral and anal plates, is thus formed. The stomodeum and proctodeum, which are both ectodermal inpocketings, give rise to the linings of the mouth and rectum, respectively. Between these outer compartments stretches the endodermally derived lining of the foregut, midgut, and hindgut. Initially the gut exists as a simple, straight-walled tube. Soon, however, regional differentiation converts the tract into the typical alimentary canal with its dilated stomach, coiled intestine, etc., each with its own distinct histologic character.

The formation of the digestive tract in birds and mammals is accomplished by a very different type of process. The development of the gut in amniotes is independent of the events of gastrulation, during which the mesodermal and endodermal cells immigrate to the interior. In the chick, following gastrulation there exists a ventral layer of endodermal cells resting on the surface of the yolk mass. With the splitting of the lateral-plate mesoderm during coelom formation, the inner splanchnic layer becomes associated with the underlying endodermal layer, together forming the double-layered splanchnopleure, which grows ventrally around the yolk forming the yolk sac. While in the early stages the chick embryo rests directly on the yolk (Fig. 12.7*a*), a fold in the splanchnopleure arises and undercuts the embryo, thereby forming a ventral body wall and lifting it off the yolk (Fig. 12.7*b*). It is this infolding that is responsible for the formation of a floor of the gut and separation of the gut cavity from that of the yolk sac. This ventral fold first occurs beneath the anterior portion of the embryo and forms the subcephalic pocket (Fig. 12.7*c*), which elevates the head to a position over the yolk. The splanchnopleure, which serves to undercut the embryo, is composed of an inner layer of endoderm, which forms the epithelial lining of the digestive tract, and an outer layer of investing mesoderm, which will give rise to the smooth visceral muscle and connective-tissue cells associated with the gut. The portion of the gut formed by the subcephalic pocket is defined as the foregut. Subsequent to the infolding in the head region, a fold develops beneath the posterior end of the embryo lifting the tail, with its enclosed hindgut, off of the yolk. The central part of the embryo, which contains the midgut, remains in open communication with the yolk via the *yolk stalk* (Fig. 12.7*d*). As the head and tail folds extend medially beneath the embryo, the midgut shrinks and the foregut and hindgut increase in length. It is at the yolk stalk, which continues to shrink in diameter, that the splanchnopleure of the embryonic gut meets the extraembryonic splanchnopleure of the yolk sac; this is also where the extraembryonic coelom remains continuous with the embryonic body cavity.

The differentiation of the embryonic gut encompasses the formation of a great variety of structures of both digestive and nondigestive function. It is the foregut from which the greatest variety of derivatives are formed. In all vertebrates, regardless of whether the foregut forms directly from the anterior end of the archenteron or secondarily as a result of being undercut by splanchnopleure, the walls of the *pharynx* (the anterior portion of the foregut) are extended laterally as a series of pouches that grow out toward the ectoderm (Figs. 12.1*a* and *b* and 12.8). These *pharyngeal* (or *branchial*) *pouches* form sequentially in an anteroposterior direction beginning just behind the region of the mouth. As each pouch

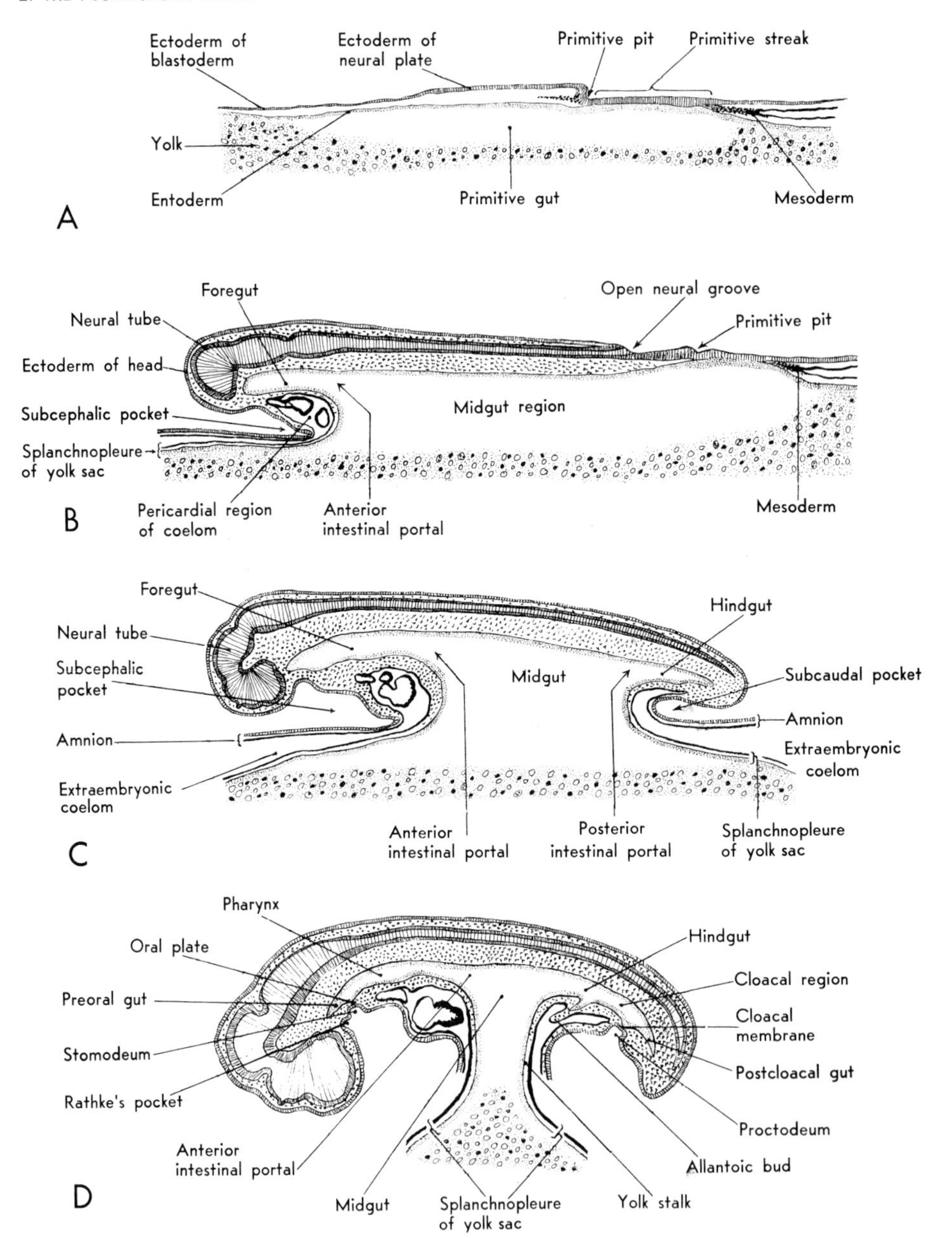

Figure 12.7 Schematic longitudinal section diagrams of the chick showing four stages in the formation of the gut tract. The embryos are represented as unaffected by torsion. *a* Toward the end of the first day of incubation, no regional differentiation of the primitive gut is as yet apparent. *b* Toward the end of the second day, the foregut is established. *c* Chick of about 2½ days; the foregut, midgut, and hindgut are established. *d* Chick of about 3½ days; the foregut and hindgut have increased in length at the expense of the midgut; the yolk stalk has formed. [*From B. M. Patten and B. M. Carlson, "Foundations of Embryology," 3rd ed., 1974.*]

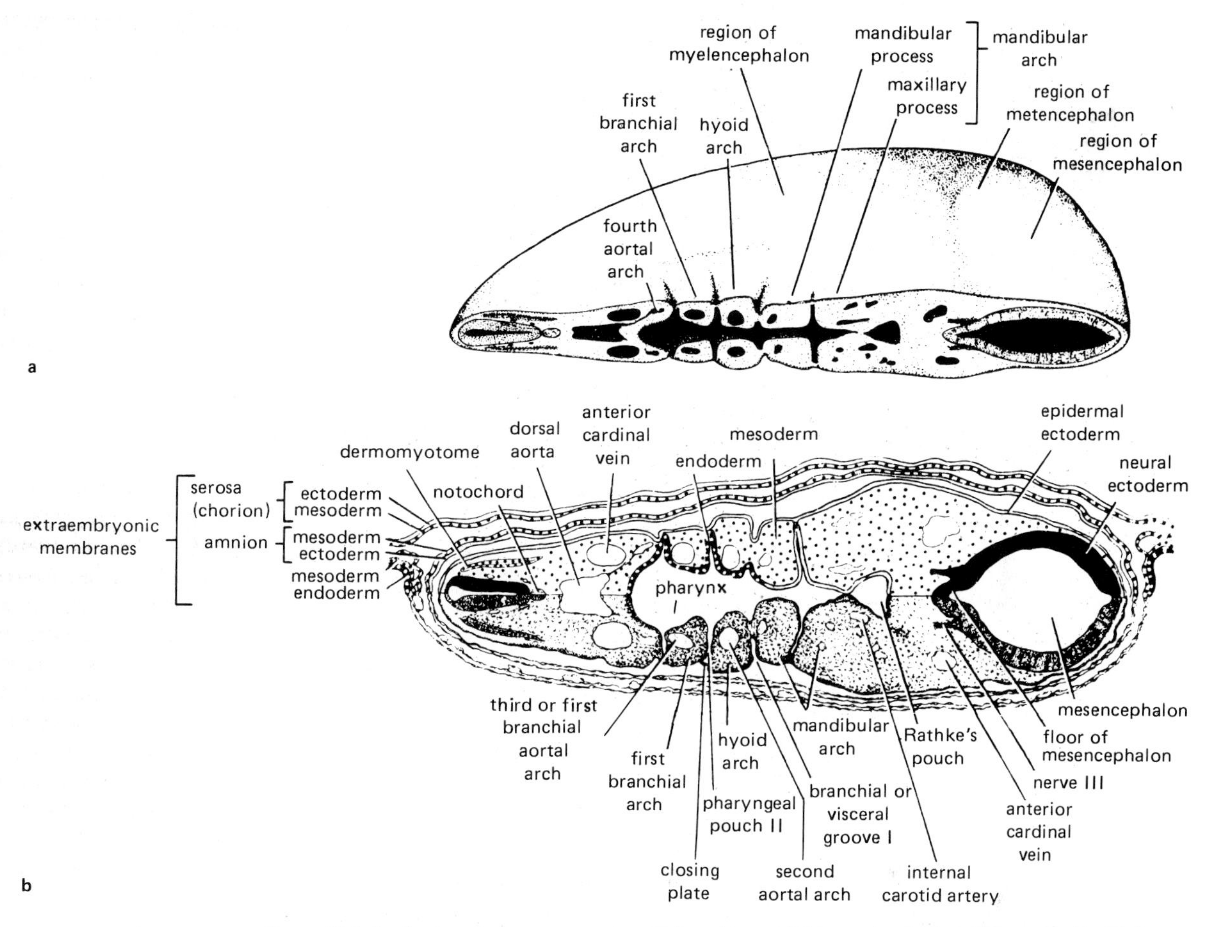

Figure 12.8 Sections and stereograms of a chick embryo of 72 hours incubation. [*From O. E. Nelsen, "Comparative Embryology of the Vertebrates," McGraw-Hill, 1953.*]

grows toward the body wall, a corresponding invagination of the adjoining ectoderm, termed a *branchial furrow,* forms and grows in to meet the outer wall of the pouch. The endodermal wall of the pharyngeal pouch and ectodermal wall of the branchial furrow fuse to form a bilayered *branchial membrane.*

The development of the pharyngeal pouches is closely tied to the migration and differentiation of the cells of the cranial neural crest (Section 11.2). These cells migrate out of the anterior portion of the neural tube in a ventrolateral stream between the pharyngeal pouches and settle in the region adjoining the ventral and lateral walls of the pouches. These cells give rise to a series of *branchial* (or *visceral*) *arches.* In the presence of the inductive influence of the pharyngeal endoderm, the cells of the neural crest initiate chondrogenesis, thereby turning the branchial arches into cartilaginous structures. The first arch, which forms in front of the first pouch, is termed the *mandibular* arch, and the second arch, which forms behind the first pouch, is termed the *hyoid arch.* This first two pairs of branchial arches give rise to parts of the skull, jaws, and middle ear (Table 11.1). The remaining branchial arches are enumerated by Roman numerals and give rise to a variety of structures depending on the type of vertebrate.

The developmental fate of the branchial arches has been widely studied as an example of a particular type of ancestral structure that is modified in various ways over evolutionary time to suit the needs of very different types of animals. The initial steps in the formation of the branchial arches and their associated pouches and furrows is very similar in all groups, but once formed, they develop in a variety of different ways. In the aquatic lower vertebrates, such as the fish and amphibian larvae, the branchial apparatus is primarily involved in the formation of the respiratory apparatus. In these organisms, the branchial membrane ruptures, thereby forming a continuous channel between the pharyngeal cavity and the external aquatic environment. In order to accomplish the task of oxygen uptake and carbon dioxide release, the walls of the pharyngeal clefts develop gill filaments in connection with the capillary beds of the aortic arches that traverse the branchial region. In higher vertebrates, the branchial apparatus no longer functions in a respiratory capacity but has been channeled into the formation of very different structures. In addition to giving rise to the jaw and selected skull bones, the three bones of the middle ear are derived from the anterior arches. The third and fourth pharyngeal pouches are involved in the formation of a pair of important glands, the parathyroids and the thymus. The former is an endocrine gland involved primarily in the control of calcium levels, and the latter, following its invasion by blood-borne cells, becomes an important lymphoid tissue (Chapter 17).

As the pharyngeal pouches are forming and differentiating, other events are occurring elsewhere in the pharynx (Figs. 12.6 and 12.9). From its ventral wall at the level of the second pharyngeal pouch, a mass of endodermal cells breaks away from the epithelium and migrates into the neck region, where it develops into the tissue of the thyroid gland. Elsewhere on the floor of the pharynx, a tube-like evagination of the epithelium forms near the last pharyngeal pouch. This tunneling activity represents the formation of the *laryngotracheal groove* (Fig. 12.9), the forerunner of the embryonic respiratory tract. This tube grows in a caudal direction ventral and parallel to the esophagus and forms the epithelial lining of the larynx and trachea and eventually the remainder of the respiratory system. In response to the presence of investing mesenchyme, the tip of the laryngotracheal bud bifurcates to form the rudiments of the bronchi. This interaction between an endodermal epithelium and a mesodermal mesenchyme occurs in a wide variety of structures throughout the digestive and respiratory systems. A detailed analysis of this type of reciprocal interaction is presented in Chapter 15 in connection with the differentiation of the pancreas. While the epithelium of the respiratory tract is derived from the endoderm, muscle, connective tissue, and cartilage differentiate from the mesodermal cells. Further growth and continued branching converts the paired bronchial buds into a tree of thinner and thinner bronchioles of the developing lung. These bronchioles ultimately terminate in tiny air sacs, or *alveoli*, whose epithelial linings become very thin in order to facilitate gas exchange. The opening of the respiratory tract into the pharynx is the *glottis*.

In more posterior regions of the foregut, the formation of specialized organs of digestion begins (Figs. 12.6 and 12.9). The foregut itself differentiates into regions identified as esophagus, stomach, and duodenum. Just behind the stomach, an evagination appears on the floor of the gut which foreshadows the development of the massive liver complex. As with the other digestive organs, the differentiation of the endodermal derivatives, in this case the cords of hepatocytes, requires an inductive interaction with the surrounding mesenchymal cells. The proximal portion of the liver diverticulum develops into the bile duct

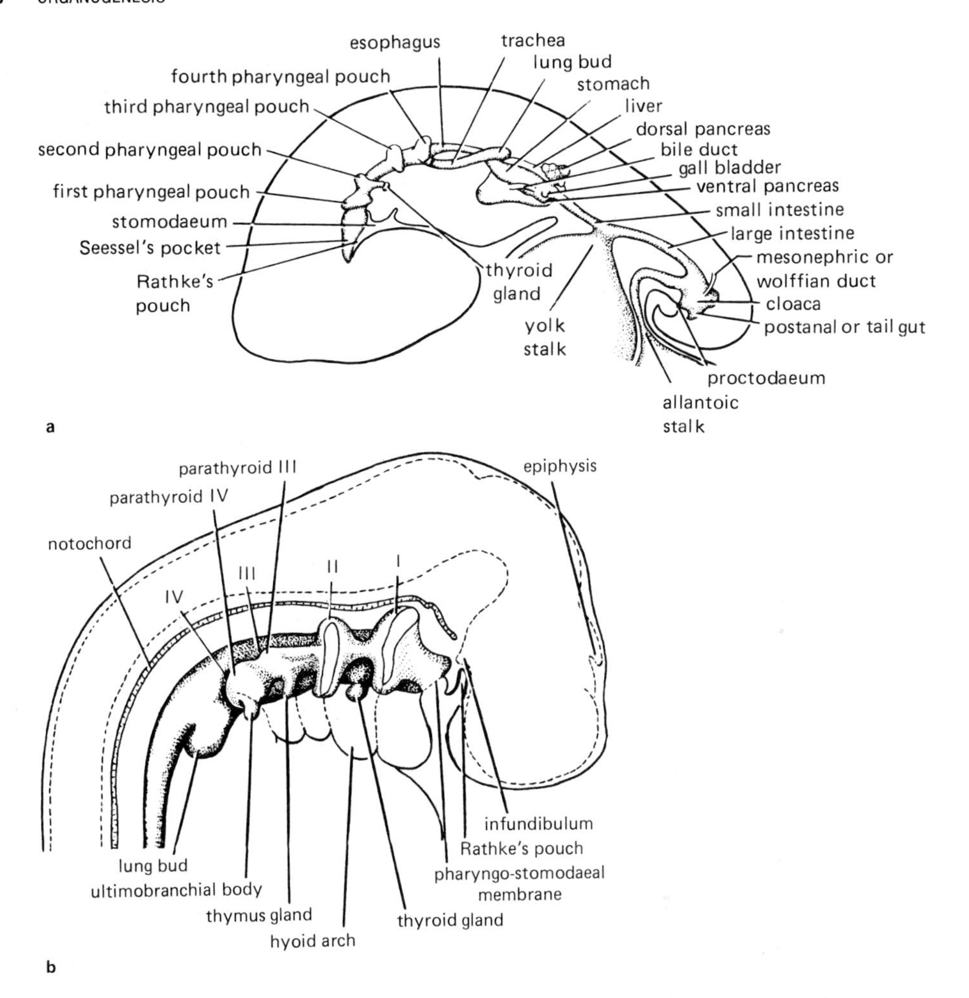

Figure 12.9 Generalized diagrams showing many of the structures derived from the gut in *a* the chick and *b* the human embryo. [*From O. E. Nelsen, "Comparative Embryology of the Vertebrates," McGraw-Hill, 1953.*]

and gall bladder. The pancreas develops from two or three buds of the intestine at the same approximate level as the liver diverticulum. These separate buds fuse into a single glandular mass which, in combination with surrounding mesenchyme, develops both exocrine and endocrine differentiations. In contrast to the foregut, the hindgut remains relatively simple in construction. In birds, the allantois (Section 12.6) arises as a diverticulum from this region of the gut, as does the bursa of Fabricius, an important avian lymphoid organ (Chapter 18).

At least in some degree the anteroposterior differentiation of the endoderm depends on external influences. Endoderm taken from the foregut, destined to become pharynx, differentiates as intestine if experimentally surrounded by posterior ectomesodermal tissue. Midgut endoderm grafted into the anterior ectomesoderm instead produces pharynx, together with other structures. Liver tissue rarely develops in other than its normal location. It seems probable that regional differentiations of the endoderm, as in the nervous system, depends on the mesoderm.

C. The heart

The heart is the first organ to become functional during embryonic development. Whatever the final structure of the heart may be, the organ consists of three basic components: (1) the innermost lining, or *endocardium,* which is continuous with the endothelial lining of the blood vessels; (2) the *myocardium,* or heart muscle; and (3) the *epicardium,* a collagenous connective-tissue layer enclosing the myocardium. In all vertebrates, the heart begins to function when it is little more than a straight tubular structure.

Although unpaired in the adult, the embryonic heart rudiment forms from two distinct heart-forming regions widely separated on either side of the body. The application of vital dyes to urodeles at the neurula stage indicates that the prospective heart cells are present dorsolaterally on each side of the embryo adjacent to the hindbrain (Fig. 12.10). These heart-forming cells are located at the leading edge of the lateral-plate mesoderm, which at this early stage has not penetrated into the anteroventral region of the embryo (see Fig. 10.13). As the free edge of the mesodermal mantle moves across the amphibian embryo (see Fig. 10.14*b*), the paired heart primordia move toward each other at the ventral midline in the region of the pharynx. As they converge, the leading edges of the lateral plate become markedly thickened as a result of cell proliferation. Some of the cells detach from the lateral plate and collect as free mesenchymal cells beneath the pharyngeal endoderm (Fig. 12.11). These mesenchymal cells are initially present as a longitudinal strand but soon become rearranged to form a thin-walled hollow tube, the forerunner of the endocardial layer of the heart. The lumen of this primitive endocardial tube will ultimately become the internal chambers of the heart. Subsequent to the formation of the endocardium, the thickened right and left edges of the mesodermal mantle come together at the ventral midline to form a trough of cells that soon envelopes the endocardial tube. This outer layer of splanchnic mesoderm represents the epimyocardium, the forerunner of the myocardial and epicardial layers of the heart. In birds and mammals, heart development is essentially similar, but as a result of the flattening of the amniote embryo compared with lower vertebrates, the developing heart not only is more amenable to experimentation, but it also shows its bilateral origin more clearly. As in amphibia, the heart primordia develop from thickenings of the splanchnic mesoderm. Mesenchymal cells detach from the mesodermal wall and form two endocardial tubes, one on each side of the midline (Fig. 12.12*a*). Each of these tubes is surrounded laterally by the thickened wall of the splanchnic mesoderm, which will eventually give rise to the myocardial and epicardial heart layers. As the embryo is completed ventrally by the folding processes described in Section

Figure 12.10 Location of presumptive-heart–forming material (*broken line*) at the neurula (*a*) and tail-bud (*b*) stages in the urodele *Ambystoma.* [*From W. H. Copenhaver, in "Analysis of Development," B. H. Willier, P. A. Weiss, and V. Hamburger (eds.), Saunders, 1955.*]

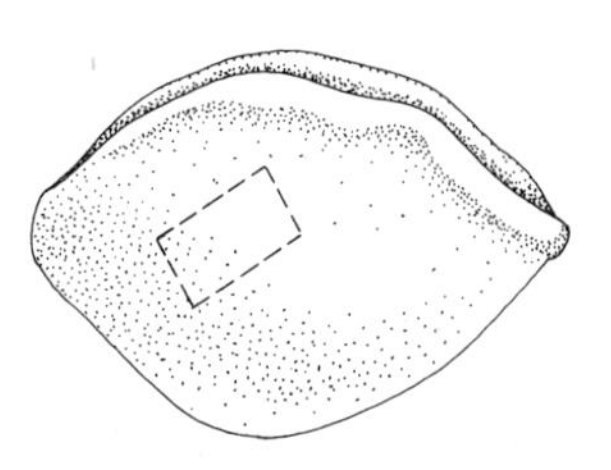

a

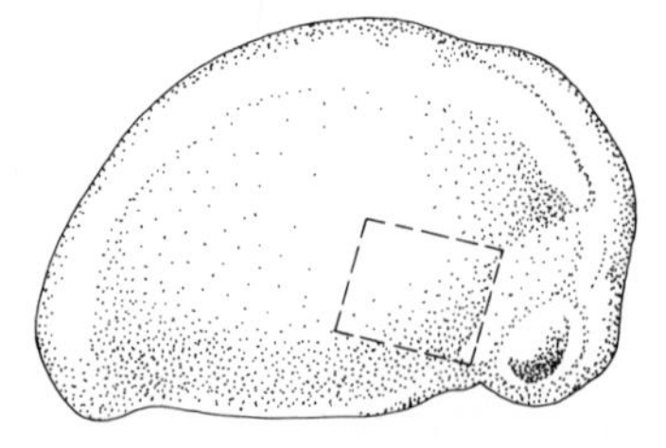

b

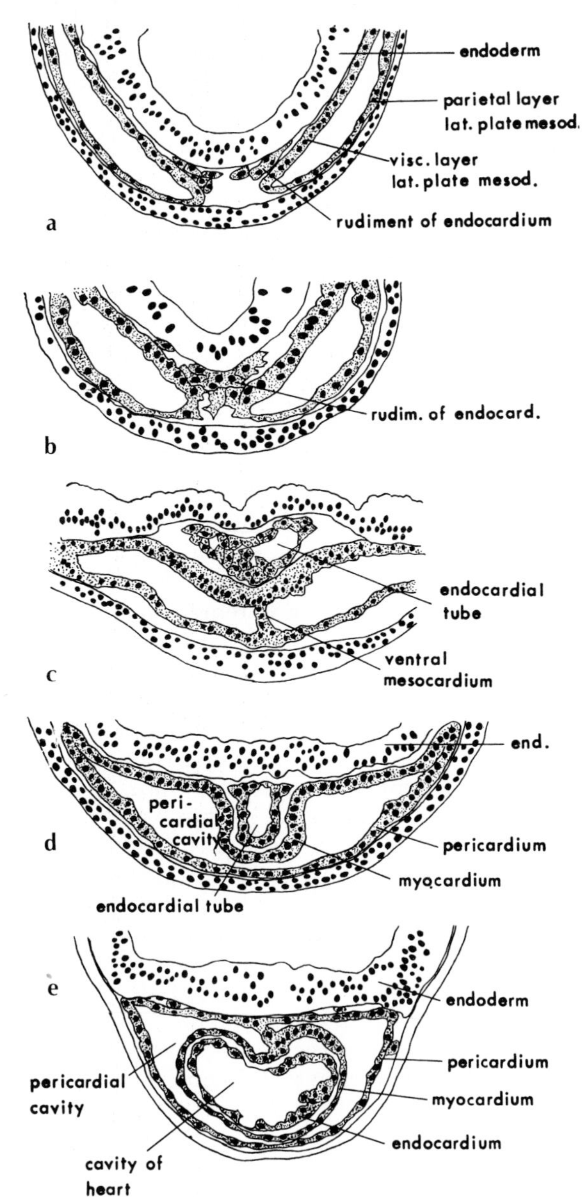

Figure 12.11 Development of the heart in amphibian embryos. [*From B. I. Balinsky, "Introduction to Embryology," 4th ed., Saunders, 1976.*]

12.2.B, the paired primordia of the heart are brought together in the midline and become fused (Fig. 12.12*b* and *c*). The two endocardial tubes fuse in the heart-forming region to become a single hollow channel (Fig. 12.12*d*), but anteriorly and posteriorly they remain paired as the *ventral aortae* and *omphalomesenteric (vitelline) veins*, respectively. As the endocardial tubes are fusing, the two epi-myocardial rudiments envelope the inner conduit to form the outer layers of the embryonic heart, and the right and left coelomic spaces unite to form the pericardial cavity. For the heart and other ventral organs and tissues, this process is best studied by careful examination of a series of transverse sections.

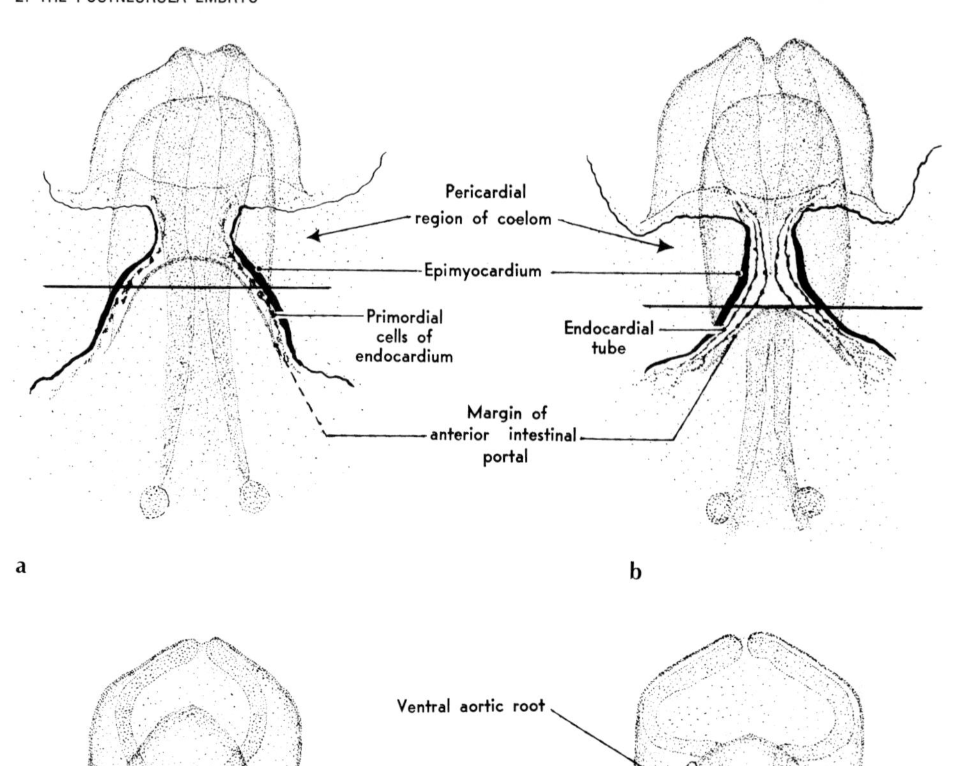

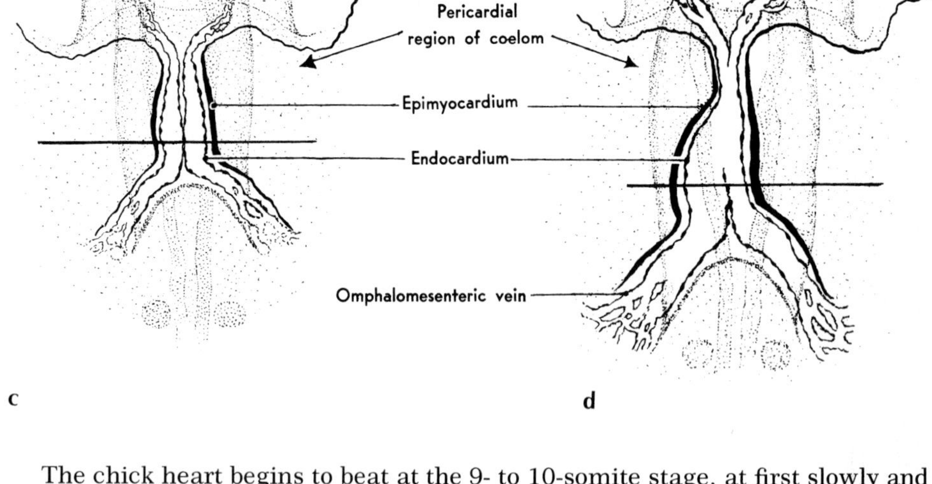

Figure 12.12 Ventral-view diagrams to show the origin and subsequent fusion of the paired primordia of the heart. Horizontal lines indicate successive stages of left-right proximation of endocardium and epimyocardium. *a* Chick of 25 hours. *b* Chick of 27 hours. *c* Chick of 28 hours. *d* Chick of 29 hours. [*From B. M. Patten and B. M. Carlson, "Foundations of Embryology," 3rd ed., McGraw-Hill, 1974.*]

The chick heart begins to beat at the 9- to 10-somite stage, at first slowly and then rapidly. At this time, the heart begins to loop outward toward one side. This folding of the heart into an S-shaped tube appears to be due in part to its growth in length between two fixed points within the less rapidly growing epicardium and in part to greater growth along one side of the tube than the other. Changes in cell shape also may play a prominent role in the looping events. In its early stages the heart tube is composed of four main regions: from intake to output, the sinus venosus, atrium, ventricle, and truncus arteriosus (see Figs. 12.1*a* and 12.14*b*). It is only at later stages that the division of the heart into its various chambers is seen to occur.

The circulation of blood in the chick can be appropriately divided into an embryonic and an extraembryonic circuit, each passing into and out of the heart (Fig. 12.13). Blood enters the sinus venosus from the yolk sac via the omphalomesenteric veins, which collect freshly oxygenated and nutrient-rich blood from the smaller vitelline veins that form in the mesoderm above the yolk. Joining with the blood from the yolk sac is the deoxygenated, nutrient-depleted blood from the embryo proper, which enters the sinus venosus via the common cardinal veins. The mixed blood is then pumped through the heart and out the paired ventral aortae. From the anteroventral region of the embryo, the blood flows around the pharynx in the aortic arches and into the dorsal aorta, from whence it reaches the remainder of the body as well as the yolk sac. The flow of blood into the capillary beds of the yolk sac occurs via the omphalomesenteric arteries. Once in the yolk sac, the blood picks up nutrients that have been solubilized by the endodermal cells of the splanchnopleure.

The heart develops from a combination of a left and right rudiment (Fig. 12.12), but these rudiments are not fully determined. Early studies showed that if the two parts of the early chick heart are prevented from fusing by the insertion of a barrier, two hearts are formed. Similarly, if a lateral half is removed, the remaining half becomes a whole heart. Further, if one rudiment is surgically split

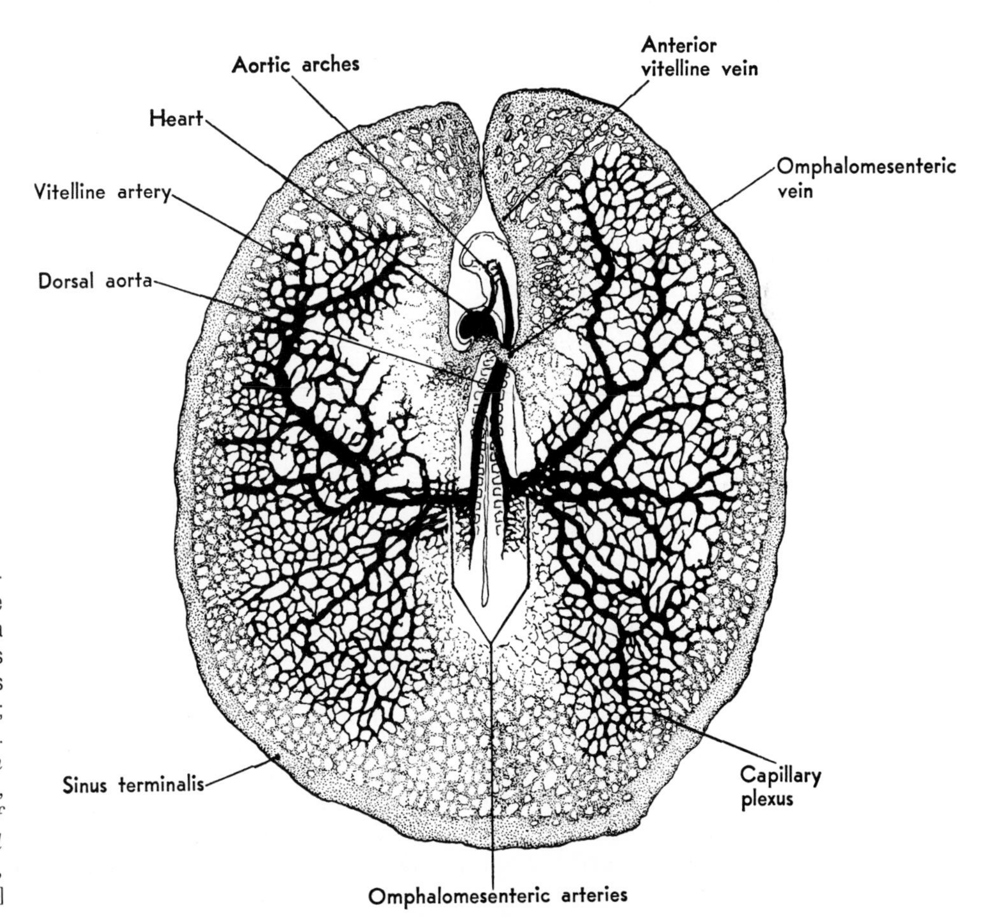

Figure 12.13 Diagrammatic ventral view of the vitelline circulation in a chick of about 44 hours incubation. The arteries are shown in solid black; the veins are stippled. [*From B. M. Patten and B. M. Carlson, "Foundations of Embryology," 3rd ed., McGraw-Hill, 1974.*]

lengthwise, each part becomes a whole heart complete with its own circulatory vessels. Almost any part or any combination of parts of the heart-rudiment stage can form a functioning heart. An anterior half rudiment can do so, and so can combinations of two anterior half rudiments or two posterior half rudiments. This is true also for the mammal.

Although the heart develops from a pair of rudiments uniting in the midventral line of the body, it is an asymmetrical organ with regard to its left and right sides. If the rudiments are prevented from fusing, so that two hearts develop, the left heart develops with the normal asymmetry, whereas the right heart has a reversed asymmetry, a mirror image. This condition is known as *situs inversus*. A similar situation with regard to the left-right asymmetry of the whole abdominal region, i.e., heart, liver, and gut, is often seen in human identical twins.

3. Segmentation and embryonic elongation

One of the most outstanding features of vertebrate organization is the segmentation of the dorsal region of the body into a series of more or less repeating parts, in both neural and mesodermal tissue. During neurulation, the lateral mesodermal sheets extend laterally and ventrally to fuse with each other in the midventral line of the embryo. Simultaneously, the most dorsal portion of the mesoderm, at each side of the notochord, undergoes successive constrictions beginning anteriorly and progressing posteriorly to form a series of segmental blocks, the *somites* (Fig. 12.14).

Mesodermal segmentation, in conformance with accompanying neural segmentation, is also characteristic of two invertebrate phyla, the annelids and the arthropods. The unique feature in vertebrates is that the segmentation process affects only the dorsolateral band of mesoderm at each side of the notochord. The lateral-plate mesoderm on both sides of the body remains unsegmented even though it is still continuous with the segmenting tissue.

Segmentation of a neuromuscular system into a number of units acting in series appears to be a much more effective locomotor mechanism than a single extended muscle sheet system. The positive value of segmentation of certain portions of the body seems clear enough. In vertebrates the unique association of segmented and nonsegmented tissue throughout the length of the body combines the advantages of both conditions. The locomotory apparatus is a highly differentiated segmental system, while the sustaining systems concerned with digestion, circulation, excretion, and reproduction are primarily those of the nonsegmented body. The renal system, deriving from the band of intermediate mesoderm between dorsal and ventral mesoderm, is characterized by both segmental and nonsegmental features.

A. Somites

Somites are first formed as almost solid masses of cells derived from the dorsal mesoderm. Progressive segregation of dorsolateral mesoderm into blocks of somitic tissue is shown in Fig. 12.14. From the first, the cells constituting a somite show a radial arrangement. A central cavity appears and the radial orientation of the outer zone of cells becomes more definite. The central cavity is filled by a core of irregularly arranged cells shown by autoradiographic studies to arise from the somitic wall. With further development, the somites become flattened and stretched dorsoventrally and markedly more thickened in their inner wall. Three distinct regions of each somite become apparent, each having a very different prospective fate. The ventromesial part of the somite loses its originally precise boundaries and merges with the central core. This combined aggregation of

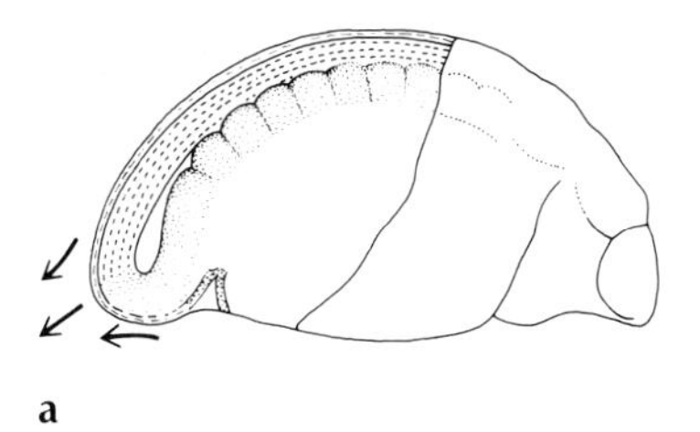

a

b

Figure 12.14 *a* Side view of postneurula of urodele showing somite formation (ectoderm removed in posterior half of embryo) and outgrowth of tail bud, mainly by tissue extension. *b* Chick embryo of 29 somites (about 55 hours incubation) showing caudal fold overlying tail bud, together with differentiating brain, sense organs, heart, and yolk blood vessels. Note that anterior portion of embryo is twisted so that left side of head faces the yolk and only the right side is seen. [*From B. M. Patten and B. M. Carlson, "Foundations of Embryology," 3rd ed., McGraw-Hill, 1974.*]

loosely connected mesenchymal cells, known as the *sclerotome,* loses its original segmented character and extends toward the notochord as a sheet from the somites of each side. The cells of the sclerotome surround the spinal cord and notochord and become resegregated into blocks of discontinuous tissue, blocks that alternate with the spinal ganglia and associated spinal nerves. Under the influence of the spinal cord and/or notochord, chondrification begins at several sites within each block of sclerotome cells. The cartilaginous elements of each block fuse to form a composite *vertebra.* In the thoracic region, extensions of the vertebrae grow out to form the *costal processes,* i.e., ribs. Four stages in somite differentiation are shown in Fig. 12.15.

During the formation of the sclerotome, the outer zone of the somite retains its firm outline and its more or less epithelial character. Part of this zone lies parallel to the epidermal ectoderm and is known as the *dermatome.* It contributes to the dermis (the connective-tissue layer of the skin), although this layer is known to receive cells from the somatic mesoderm generally.

The dorsal portion of the inner wall of the somite becomes the *myotome,* which develops into a muscle segment. In so doing, the cells of the myotome spread ventrally and, together with cells of the lateral plate, form a block of muscle tissue along the side of the embryo. Taken as a whole, the segmented myotomes give rise to a wall of muscle tissue along each side of the embryo, with each of the component segments separated from its neighbors by a thin connective-tissue layer. This organization persists in fish and amphibian larvae, whose aquatic lifestyle is promoted by having segmented bands of muscle capable of moving the vertebral column in a sinusoidal type of locomotion. In higher vertebrates, with their dependence on paired limbs for locomotion, this arrangement is largely lost as the muscle tissue becomes organized into distinct muscle groups. For example, the somites of the head contribute anteriorly to the tongue and muscles of the eye and posteriorly to the muscles of the neck. One remnant of the original segmented nature of the musculature can be seen in the series of intercostal muscles that develops between the ribs.

There are other questions apart from the intriguing ones about the presumptive fate of the various regions of a differentiating somite. The nature of the underlying process that results in the initial segmentations and the nature of the agency that determines how many pairs of somites will be produced remain unknown. Yet there are also minor problems. The formation of more or less repeating parts in an organism is a morphologic event of very great value that has been exploited to its fullest extent in several animal phyla and in plants as well. The answers to these two questions—i.e., what determines the formation of segments and what determines the number produced—almost certainly are closely related. The problem has long been recognized, but there are but few facts to go on. Somites on each side are formed one by one in a posterior direction, as though constricted successively from the front end of the unsegmented mesodermal band on each side of the notochord. Successively formed somites in a series are approximately the same size at the time of demarcation. The interval between the recognizable initiation of successive somites is also approximately the same throughout the succession. The existence of some kind of counting mechanism has been proposed to explain the precision in number produced in a particular species. The number of somite pairs produced is species-specific. This feature is associated, for instance, with the relatively small number of vertebrae characteristic of neck and trunk of amphibians, birds, and mammals and with the large number typical of snakes. The vertebrate embryo, however, is far from being the most conducive to an experimental approach to the segmentation phenomenon.

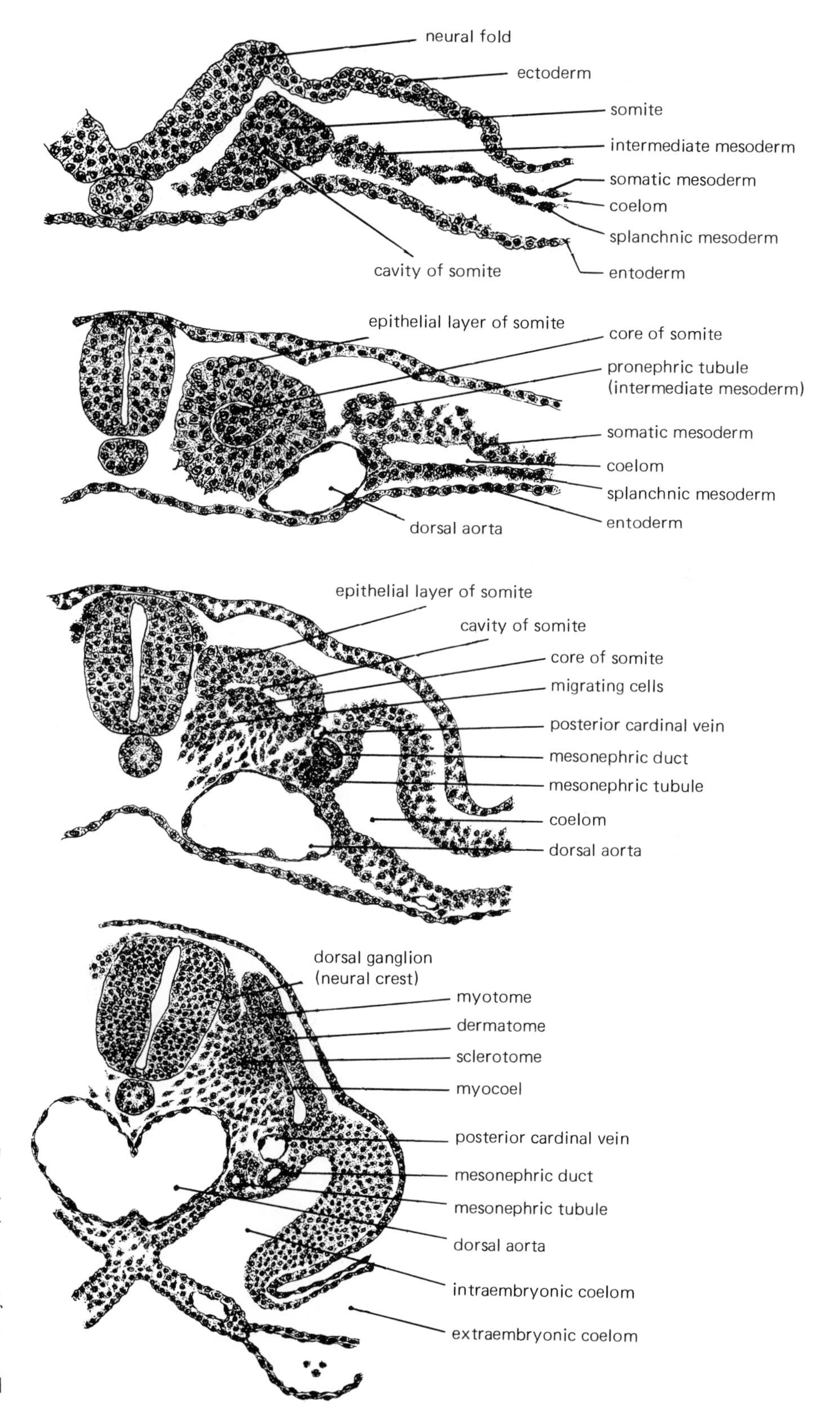

Figure 12.15 Transverse sections of *a* 4-somite, *b* 12-somite, *c* 30-somite, and *d* 33-somite chick embryos. [*From B. M. Patten and B. M. Carlson, "Foundations of Embryology," 3rd ed., McGraw-Hill, 1974.*]

Mesodermal segmentation is much more amenable to observation and analysis in annelids, especially in polychaetes. An understanding of the process, which is similar in principle to that of vertebrates, is more likely to come from studies of this material, particularly in the case of regenerative development. Recent speculation on the control of somite formation in amphibians and chicks can be found in the papers by Jonathan Cooke and by Ruth Bellairs and P. A. Portch in the symposium entitled "Vertebrate Limb and Somite Morphogenesis" (1977).

B. The trunk

Throughout the trunk and tail regions of vertebrates, segmentation clearly involves the spinal cord in addition to the mesodermal segmentation. Each pair of somitic segments and their derivatives is supplied by a corresponding pair of spinal nerves and ganglia. The question is whether the mesodermal segmentation induces the neural segmentation, or vice versa, or even whether both are induced by some external embryonic property. The question, however, is readily answered.

The neural crest is at first a continuous column of cells along each outer side of the closing neural folds. From the start the neural-crest cells appear to have an intrinsic tendency to break up into small groups. The segmental localizations of these groups, which give rise to the segmental sensory ganglia and nerves, are determined by the segmental arrangement of the myotomes. Experimental removal or disarrangement of adjacent myotomes abolishes or correspondingly disarranges the segmental array of the ganglia.

C. The head

A comparable relationship between mesoderm and neural components of the trunk almost certainly prevails during the development of the head. The situation, however, is complicated and confused by the varying degrees of fusion of the embryonic segments that together constitute the integrated entity we call a head. This developmental and evolutionary phenomenon, seen in vertebrates, annelids, and arthropods alike, presents a major subject for experimental analysis. Little has been done in this connection, although the regional patterning of brain and sense organs has received considerable attention (see Chapter 14).

D. Notochord and embryonic elongation

The embryo continually lengthens during neurula and postneurula stages. In amphibians the lengthening of the notochord is considered the chief motivative force underlying the accompanying lengthening of the neural plate and dorsolateral mesoderm. This conclusion is based on the following considerations: (1) isolated parts of the neural system fail to elongate; (2) notochordal rudiments stretch and form elongated rods even if cultivated in vitro; and (3) the notochord accordingly changes its shape actively, whereas the neural plate or tube is pulled in length by the adjacent notochord.

Extension of the notochord is effected by the enlargement of the notochordal cells, through vacuolation, within a confining sheath enclosing the notochordal rod as a whole. In the protochordates (*Amphioxus* and ascidians), this process and circumstance causes the embryonic chordal cells to interdigitate until a column of single cells is attained. The thrust of the collective enlargement of the chordal tissue is accordingly linear, particularly toward the posterior end of the embryo. The situation is similar in the lower vertebrates; however, because of the greater size of the egg, embryo, and notochord, a number of vacuolating and vacuolated cells rather than a single-cell layer are present in a cross section of

the notochord. The same extension mechanism holds for the embryos and larvae of amphibians. In amniote embryos, where a notochord is formed but its cells do not enlarge, body lengthening must take place through other means, such as by polarized cell proliferations or by tissue stretching that changes cell shape.

E. The tail

The preceding considerations should be kept in mind in an examination of tail formation, especially in amphibian development. Whatever the mechanical basis of stretching may be, the embryo as a whole undergoes considerable stretching, particularly in its posterior half. Coinciding with this, a remarkable transformation of the posterior part of the neural plate and tube takes place; they elongate to a greater extent than the ventral part of the embryo. Since the hindmost end of the neural tube is attached to the blastopore, the neural tube becomes bent over and forward, whereby the apex of the bend becomes the tip of the *tail bud* (Fig. 12.14*a*). The neural tube is present in the tail bud as one tube above another conjoined at the bud tip. These two limbs of the neural tube have very different developmental fates. The dorsal part continues into the growing tail as presumptive spinal cord; remarkably, the other forms muscle tissue.

Vital-staining experiments with salamander embryos appear to show that the posterior one-fifth of the neural plate is destined to form segmental muscular tissue. Extirpation of the posterior neural plate results in deficiencies in somitic mesoderm of the tail and hind end of the trunk. According to Ford (1948), the production of the tail by the tail bud does not depend on cell proliferation but is the result of stretching. The whole posterior trunk region from the tenth somite back is put in position by this stretching, with some eight or nine pairs of somites being added. The role of stretching in the process of mesodermal segmentation may be basically important. It is also notable that neural-tube formation does not necessarily imply prospective neuralization, since in the tail it forms muscle; nor, as in fish, is neuralization dependent on neural-tube formation.

In amniotes, the rapid growth in embryo length coincides with the period of regression of the primitive streak. This growth is not due to contributions from the cephalic end of the primitive streak, as was widely believed, but to growth of the embryo in front of the streak. In the chick, by the end of the second day, little more than the node and a very short portion of the streak remain. This remnant is the tail bud (Fig. 12.14*b*). At this stage there are about 24 pairs of somites. Nevertheless, the posterior portion of the embryo anterior to the tail bud continues to extend. During this extension, additional pairs of somites are added sequentially until about 50 have been formed. Both cell proliferation and tissue stretching are probably involved.

4. The renal-reproductive system

A. The renal system

On each side of the body a band of intermediate mesoderm lies between the somite-forming dorsal mesoderm and the unsegmented lateral-plate mesoderm. This intermediate zone is the *nephrotome*, or *intermediate mesoderm* (Fig. 12.15*a*), from which the *uriniferous tubules* (*nephrons*) of vertebrate excretory organs arise. If one follows the progress of kidney development in mammals and birds, a surprising sequence of events is observed: three rather distinct types of embryonic excretory systems are formed, one after the other in an anterior to posterior sequence. As will be described, this sequence of events has interesting evolutionary implications, since, at least superficially, it appears that ontogeny is recapitulating phylogeny.

At about 36 hours of development in the chick, the nephrotomic mesoderm between the fifth and sixteenth somites gives rise to a series of pronephric tubules, one per segment, which represents the *pronephros*, the most anterior and first-formed kidney unit in the vertebrate embryo. Later in development of the chick, a second and more posterior series of excretory tubules is formed, the *mesonephros*, which will serve as the excretory organ of the embryo. In birds and mammals, a third and even more posterior set develops, the *metanephros*. The pronephros is a functional organ only in some of the lower fishes, which never develop more posterior tubules. In higher fishes and amphibia, the pronephros degenerates and the mesonephros becomes the excretory organ of the adult. In birds and mammals, the mesonephros also degenerates and the metanephros, the last to appear, becomes functional toward the end of embryonic life. We will now turn to some of the morphogenetic aspects of kidney formation.

The pronephric tubules initially appear as solid outgrowths from the nephrotomes of the anterior region of the trunk (Fig. 12.16*a*). Within each nephrotome,

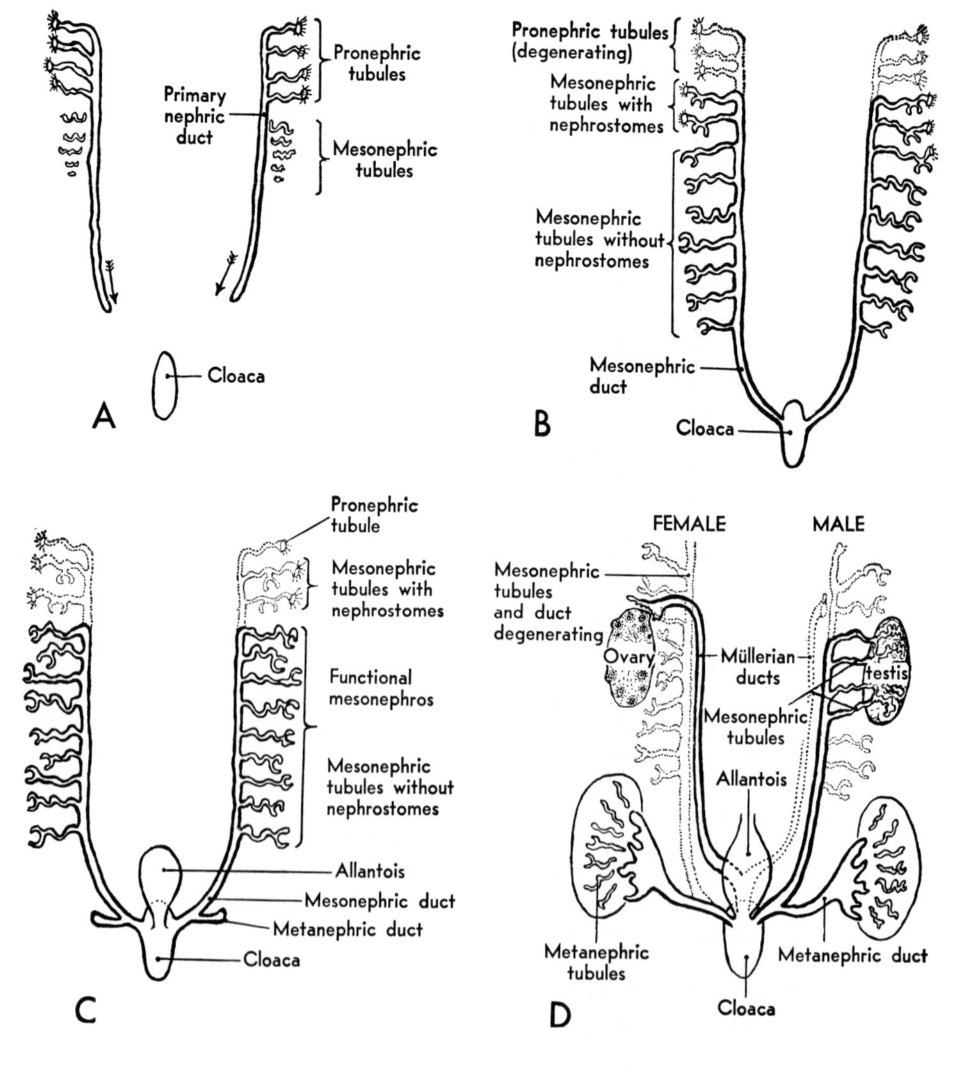

Figure 12.16 *a–c* Schematic diagrams to show the relation of pronephros, mesonephros, and metanephros at various stages of development. or the sake of simplicity, the tubules have been drawn as if they had been pulled out to the ide of the ducts. In *d* the conditions after sexual differentiation has taken lace are depicted: female, left side of the diagram; male, right side. [*From B. M. Patten and B. M. Carlson, "Foundations of Embryology," 3rd ed., McGraw-Hill, 1974.*]

a cavity termed the *nephrocoel* appears which is continuous with the coelomic cavity of the lateral plate. Although the buds of the pronephric tubules are initially solid, they soon become hollowed out to form a channel that is continuous with the nephrocoel (and thus the coelom) at one end. Meanwhile, the distal end of each tubule grows in a dorsal direction and then bends caudally to join with the adjacent tubule on its posterior side. As the ends of the tubules become linked to one another, they form a common pipeline, the *primary nephric duct* (Fig. 12.16*a*), which then grows freely in a caudal direction across a major portion of the embryo before it eventually opens into the cloaca near the end of the hindgut. The elongation of the primary nephric duct toward the cloaca can still occur under highly unusual experimental circumstances. For example, if the posterior portion of the embryo is rotated 180° dorsoventrally, the duct continues to reach the cloaca, even though it must traverse some very unfamiliar territory. This type of observation has led investigators to compare the behavior of the primary nephric duct with that of a nerve fiber growing out from the central nervous system (Section 14.4.C). In fish and amphibian larvae, the pronephric tubules have the job of transporting waste from the blood and coelom to the cloaca. It is interesting that in amniotes the pronephric tubules appear and then disappear without ever having served any apparent physiological function. If this is the case, they would represent a strictly vestigial organ of a transitory nature.

At a later stage, a second set of uriniferous tubules form nephrogenic tissue in a more posterior region of the trunk. These are the mesonephric tubules (Fig. 12.16*b*), which grow toward the preexisting primary nephric duct to which they connect. Once the lumens of the mesonephric tubules are confluent with the primary nephric duct, the pronephros disappears and the remaining portion of the channel is now referred to as the *mesonephric duct* (or *Wolffian duct*). Experimental evidence suggests that the differentiation of the mesonephric tubules depends on induction by the primary nephric duct; if the latter does not extend into the posterior region for one reason or another, mesonephric tubules tend not to appear. Unlike pronephric tubules, which open directly into the coelom, each mesonephric tubule develops a cuplike outgrowth, termed a *Bowman's capsule*, which becomes closely associated with a knot of capillaries (a *glomerulus*) from which it receives its fluid.

The formation of the metanephric kidney occurs in a very different manner from that of the previous systems. The first indication of its appearance is a pair of diverticuli (Fig. 12.16*c*) in the primary nephric duct (now the mesonephric duct) near its point of juncture with the cloaca. Each hollow epithelial outgrowth, termed a *ureteric bud*, grows in an anterolateral direction into a mass of nephrogenic tissue, termed the *metanephrogenic mesenchyme*, located in a relatively posterior region of the trunk (Fig. 12.17). After it reaches this mesenchyme, the blind end of the ureteric bud expands in diameter and branches into a number of smaller buds that will eventually give rise to the *collecting tubules* of the *pelvic* region of the kidney. The basal portion of the ureteric bud will form the *ureter*. As the tip of the ureteric bud undergoes its branching, the metanephrogenic mesenchyme differentiates into the metanephric tubules with their associated glomeruli. These tubules will eventually connect with the collecting ducts so that fluid can pass from the bloodstream of the glomeruli through the nephrons and into the cloaca. As the metanephric units become functional, the mesonephric tubules disintegrate, although a portion of the mesonephric duct is salvaged for use in an entirely different capacity, namely, as a channel for sperm passage in males (Fig. 12.16*d*). In most mammals, the metanephric kidney sees

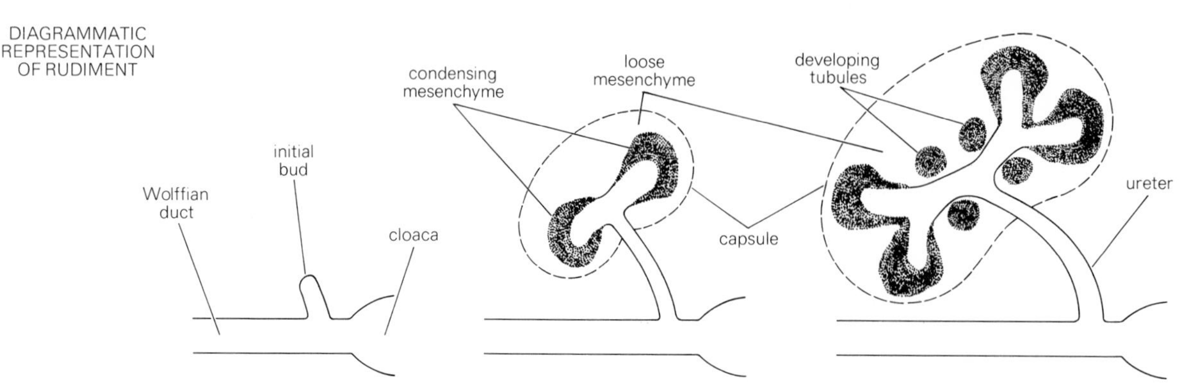

Figure 12.17 Early development of renal structure. [*After A. L. Etheridge,* J. Exp. Zool., **169:**33 (1969).]

limited use during fetal life because waste materials are carried across the placenta into the maternal circulation for elimination.

The relationship between the epithelial bud of the ureter and the metanephrogenic mesenchyme represents one of the best-studied examples of an inductive interaction. If the intact metanephric primordium (the epithelial and mesenchmal portions) is explanted and kept in culture, each component undergoes its normal differentiation into collecting ducts or coiled uriniferous tubules. If, however, the two components are separated and cultured independently, neither is capable of self-differentiation. A considerable body of evidence indicates that the interaction between the two elements is a reciprocal one. The mesenchyme induces the branching events within the epithelial bud, and the bud induces the differentiation of tubules within the mesenchyme. When the two components of a kidney unit are separated by a porous-membrane filter, inductive action still occurs. In fact, the nephrogenic mesenchyme will still produce secretory tubules if pieces of dorsal spinal cord are substituted for the ureteric bud. This is a morphogenetic response, therefore, to an inductor that is not tissue-specific. In contrast, the induction of epithelial branching in the ureteric bud does seem to specifically require nephrogenic mesenchyme. Although the inductive interaction can be obtained across a membrane filter, the means by which it occurs, i.e., as a result of long-range diffusion or short-range cell-to-cell contact, has been somewhat controversial. Whereas earlier studies suggested that inductive molecules were diffusing across the space, an extensive series of recent studies has pointed to the requirement for contact between cytoplasmic processes within the filter to accomplish the inductive stimulation. Discussion of these two opposing views can be found in the papers of C. Grobstein and of L. Saxén in "Extracellular Matrix Influences on Gene Expression," H. C. Slavkin and R. C. Greulich, eds., 1975, as well as other papers by these authors.

3. The reproductive system

The cells of the gonad, whether within a male or female amphibian, bird, or mammal, are derived from two very different embryonic sources. One group of cells, the primordial germ cells, arise far from the eventual site of gonad formation. The other group of cells, those which form the variety of nongerminal tissues in the ovary or testis, differentiate from mesodermal cells present in the region in which the gonad is formed. The gonad—whether destined to become

male or female—is represented at first by the *genital ridge*, an outgrowth of the roof of the posterior lateral plate, which projects into the dorsal coelom at each side of the midline. The ridge lies between the dorsal mesentery of the abdomen and the kidney rudiment on either side (Fig. 12.18).

The origin of the primordial germ cells (PGCs) is very different among vertebrates. In amphibians, the distinctive cytoplasm of the PGCs can be traced back to the germ plasm of the vegetal pole of the unfertilized egg (Section 7.3). After these germinal determinants become displaced upward during cleavage, they come to reside within the yolky gut endoderm, where they pass into the yolky cytoplasm of the primordial germ cells. The path taken by these cells is shown in Fig. 12.18*a*. From their location in the dorsal wall of the primary gut, the PGCs

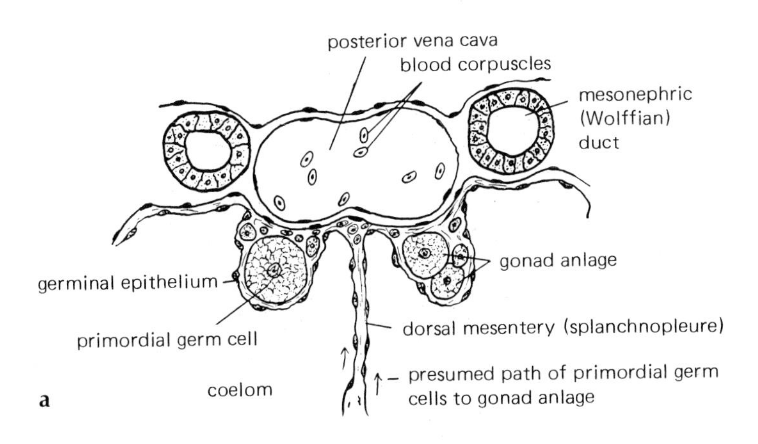

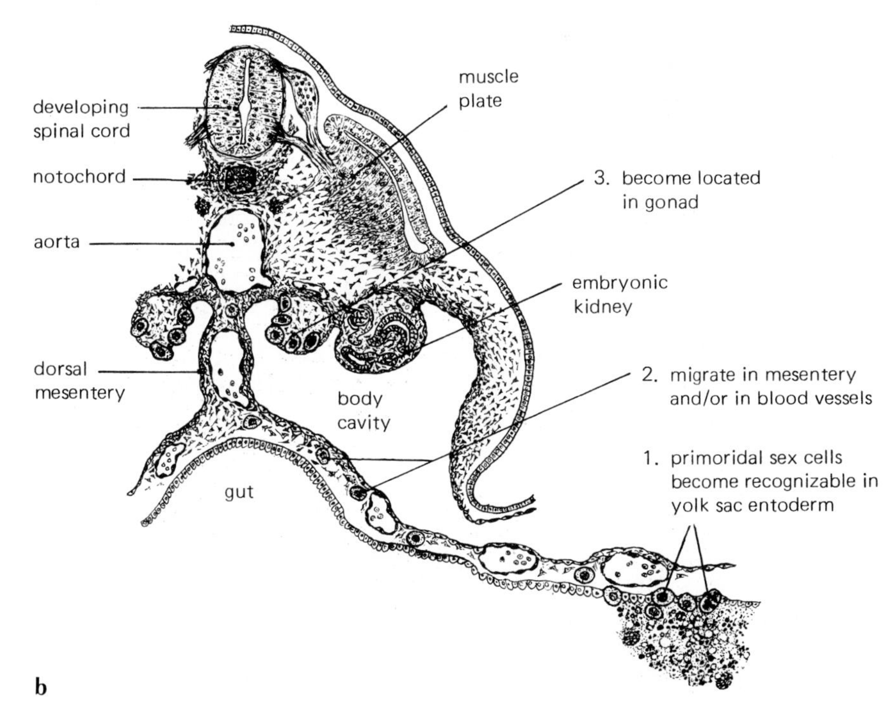

Figure 12.18 *a* Gonad primordia (anlagen) in 11-mm frog tadpole, showing probable path of germ-cell migration. [*After R. Rugh, "The Frog," McGraw-Hill, 1953.*] *b* Section through the midbody region of a chick embryo, illustrating the manner in which the primordial germ cells originate in the yolk-sac endoderm and migrate to the developing gonad. [*From B. M. Patten and B. M. Carlson, "Foundations of Embryology," 3rd ed., McGraw-Hill, 1974.*]

of the frog migrate to the dorsal mesentery, across it (Fig. 12.19), and into the formative gonad. In the chick, the PGCs, which are most readily identified cytochemically by their glycogen content (Fig. 12.20*b*) arise in the hypoblast of the extraembryonic region, specifically in an arc at the border of the areas opaca and pellucida directly anterior to the embryo proper (Fig. 12.20). From this site the PGCs enter the developing vascular system, from whence they are carried by the

Figure 12.20 *a* Diagram of a surface view of a primitive-streak stage of a chick showing the region (termed the *germinal crescent*) in which the primordial germ cells arise: h.p., head process; pr. kn., primitive knot; a.op, area opaca; a.pe., area pellucida. [*From C. H. Swift,* Am. J. Anat., **15**:*483* (*1914*).] *b* A part of the germinal-crescent area of a whole mount stained with periodic acid-Schiff to reveal the glycogen content of the large spherical primordial germ cells. [*Courtesy of T. Fujimoto, A. Ukeshima, and R. Kiyofuii.*]

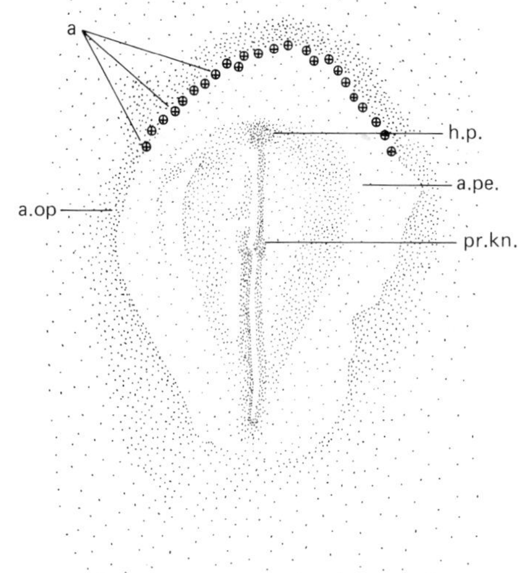

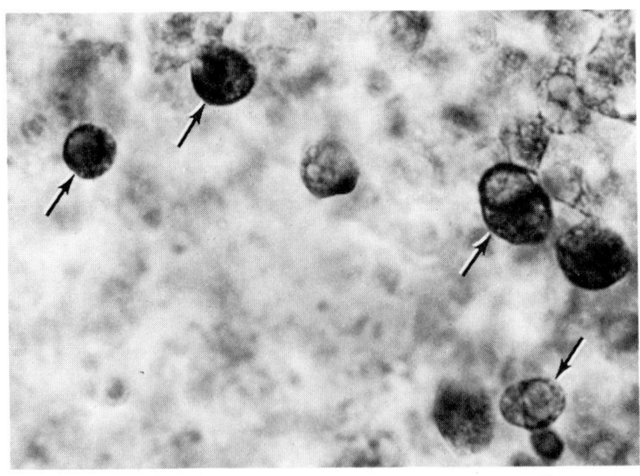

Figure 12.19 Phase contrast of a thick plastic section through the posterior body wall and mesentery of a *Xenopus* tadpole showing three large primordial germ cells in the process of movement along the dorsal mesentery. [*Courtesy of J. Heasman, T. Mohun, and C. C. Wylie.*]

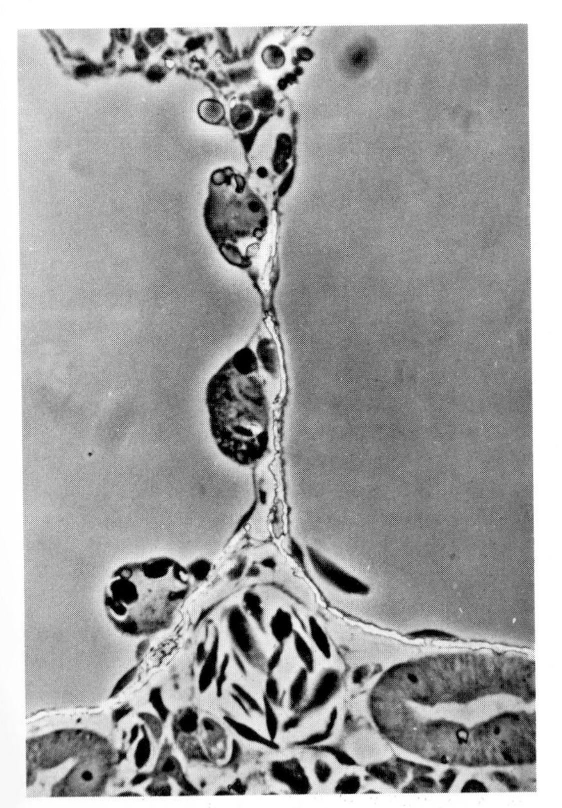

circulation to the germinal ridge. The route taken is shown in Fig. 12.18*b*. Although most of the PGCs appear to exit from the bloodstream in the vicinity of the gonad, displaced individuals are seen occasionally in other regions of the embryo, where they degenerate. In the mouse, the PGCs, which are best identified cytochemically on the basis of their high alkaline phosphatase content, arise in the endoderm of the yolk sac, from which they migrate along the dorsal mesentery to the germinal ridge. Since the PGCs are particularly sensitive to radiation, they can be selectively destroyed without affecting the remainder of the embryo to any great degree. When the PGCs are selectively killed in this manner, the gonads continue to undergo differentiation but remain sterile. It is apparent that the germ cells play no major inductive role in the development of the nongerminal structure of the ovary or testis.

Normally the PGCs become embedded in the epithelium of the germinal ridge, where they undergo proliferation while the ridge thickens and bulges out toward the coelomic cavity (Fig. 12.21). This region constitutes the outer *cortex* of the

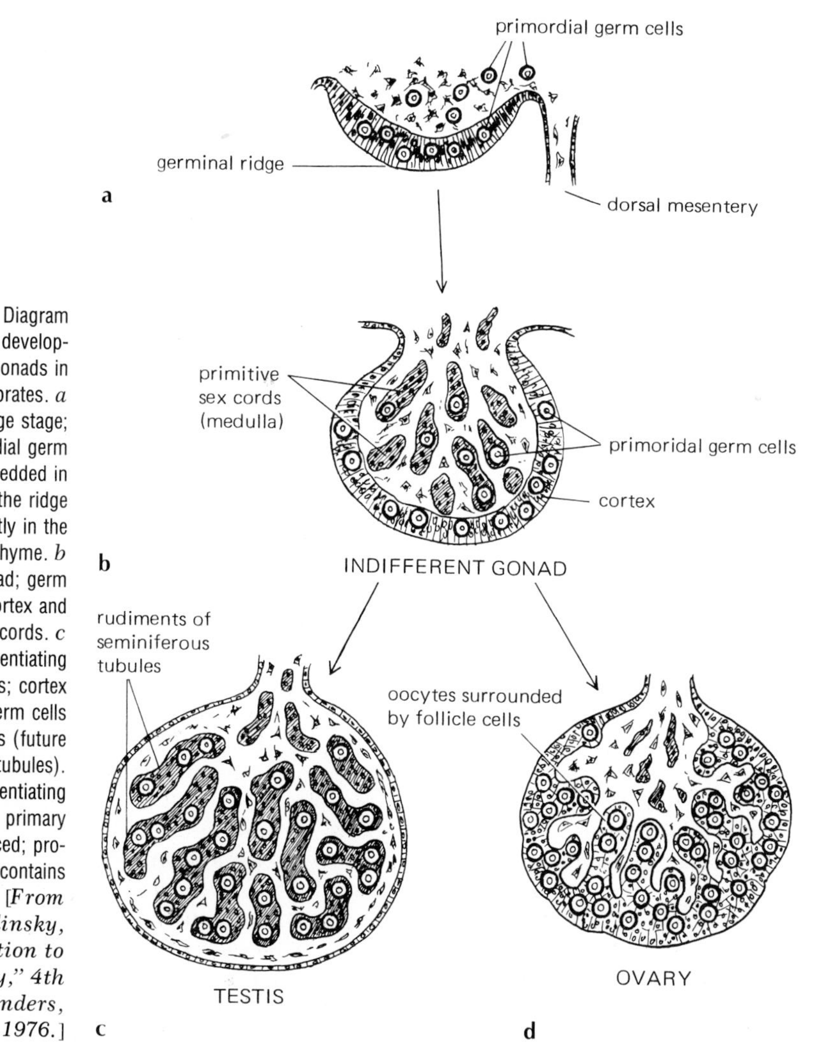

Figure 12.21 Diagram showing the development of gonads in higher vertebrates. *a* Genital-ridge stage; primordial germ cells partly embedded in epithelium of the ridge and located partly in the adjacent mesenchyme. *b* Indifferent gonad; germ cells in the cortex and in primary sex cords. *c* Gonad differentiating as testis; cortex reduced; germ cells in sex cords (future seminiferous tubules). *d* Gonad differentiating as ovary; primary sex cords reduced; proliferating cortex contains the germ cells. [*From B. I. Balinsky, "Introduction to Embryology," 4th ed., Saunders, 1976.*]

gonad. At the same time, the interior core of the ridge becomes filled with mesenchymal cells arranged in strands. These strands, or *primitive sex cords*, constitute the inner *medulla* of the gonad.

The gonad thus formed is the *indifferent gonad*, a stage that is the same in both sexes. Subsequently, in the male the primordial germ cells migrate from the cortex into the primitive sex cords, which become hollowed out and are converted into the *seminiferous tubules*. Spermatogonia arise from the primary germ cells, while Sertoli cells arise from the sex-cord tissue. The original cortex of the indifferent gonad is reduced to a thin epithelial layer covering the juvenile testis (Fig. 12.21). In the female, the medulla of the indifferent gonad becomes reduced. The primordial germ cells remain in the cortex and greatly increase its thickness. Masses of cortical cells on the inner surface of the cortex divide into groups of cells surrounding a number of primary germ cells. These then become the primary follicles which serve as a reserve throughout the fertile phase of the female life span (discussed in Section 4.3.B).

Just as the gonad passes through a stage in its development in which it is identical in both sexes, so to do the primordia of the remainder of the reproductive tract. During its initial stages, the reproductive system develops independently from the renal system; however, the two soon become closely interrelated. The formation of the primary nephric duct and its subsequent transformation into the mesonephric (or Wolffian) duct were discussed in Section 12.4.A. Following the appearance of this excretory channel, there appears a second paired duct system, the *Müllerian ducts*, which runs parallel to the former channel. The Müllerian ducts open at one end into the coelom and at the other end into the cloaca (Fig. 12.16*d*). These two pipelines represent the forerunners of the two types of reproductive tracts. In the male, a portion of the mesonephros is converted into the tubules of the *epididymis*, while the mesonephric duct is taken over for use in sperm transport, becoming the *vas deferens*. In the male, the Müllerian duct undergoes degeneration (Fig. 12.16). In contrast, in the female it is the Müllerian duct that is pressed into service for gamete transport, differentiating into the oviduct, uterus, cervix, and part of the vagina. In the female, it is the mesonephric duct that disappears. With this background behind us, we now can turn to the basis for differentiation of the gonad and reproductive tracts in a male or female direction.

The direction of differentiation of the gonad and duct system depends on several factors. In the following discussion we will confine our comments largely to observations and experiments on mammals. Although there are many differences, the basic principles seem to apply to birds and amphibians as well. Normally, the differentiation of reproductive systems into male or female structures depends on the genetic constitution of the embryo. In mammals, embryos having an XY pair of sex chromosomes develop into males, while those with an XX pair develop into females. By itself, this tells us little about the underlying chromosomal basis for sex determination. For example, is it the presence of a Y chromosome that determines maleness, or is it the presence of the single X chromosome? Consideration of various human chromosomal abnormalities provides further insight. For example, individuals having a single X chromosome and no second X or Y, i.e., XO individuals, those having Turner's syndrome, develop as immature females—vagina, uterus, and oviducts are present, but the gonad develops very abnormally. At the other extreme are individuals having a single Y chromosome together with more than one X. Persons with this type of genetic constitution, termed Klinefelter's syndrome, can range from XXY to XXXXXY. Regardless of the number of X chromosomes, these persons develop as immature males. Based

on these observations, it follows that the presence of a Y chromosome, regardless of the number of X chromosomes, dictates that the embryo develops as a male. Conversely, the absence of a Y chromosome, as occurs in Turner's syndrome, leads to differentiation along female lines. Yet there is considerably more to the story than the identity of the sex chromosomes.

One of the earliest examples of the effect of environment on sexual development came about as a result of a natural experiment. Cattle breeders have long known that when twins are born to a cow and one of the twins is a male, the other twin (termed a *freemartin*) often shows sexual abnormalities and is sterile. During the early part of the century, it was shown that this event occurs only if the two fetuses are of opposite sex and their fetal circulation becomes mixed as a result of the development of vascular anastomoses between their placentas. Under these conditions it is always the female that is rendered abnormal. It was proposed by F. R. Lillie in 1917 that some substance was passing from the male fetus and affecting the development of the female. Based on a wide variety of studies, it is clear that the presence of androgens, such as testosterone, in the fetal circulation, as occurs in the freemartin, can have a profound masculinizing effect on the development of the reproductive tract. This conclusion pushes the question of sex determination back one step to the basis for the production of sex hormones.

In 1955 an important finding was made during a study of graft rejection. It was found that when skin from a male of an inbred strain of mice was grafted to a female member of the same genetic strain, the graft was rejected. If, in contrast, the graft was made between two males or between two females, or if the skin was derived from a female and grafted to a male host, then no rejection occurred. It appeared that the male cells carried some antigen not present in cells of female members of the strain. This antigen, which is widespread among mammals, has been termed the *H-Y antigen,* a histocompatibility antigen coded for by a gene on the Y chromosome. This gene codes for an integral membrane protein that seems to be present on virtually all types of cells in the male animal. More important, it is believed that it is the presence of this Y gene that determines that the indifferent gonad will develop into a testis; i.e., it is testis-determining. In the absence of the H-Y^+ gene, the gonad develops as an ovary. In other words, ovarian development can be thought of as the passive route of differentiation, the route followed if countermanding orders from the Y are not received. If the gonad does develop into a testis, then it begins to produce substances that have secondary effects on the remainder of the reproductive tract. One of these substances is testosterone, which acts on the mesonephric duct causing it to differentiate into vas deferens. Another substance, which appears not to be a sex steroid, acts to cause the degeneration of the Müllerian ducts. This latter substance, termed *Müllerian-inhibiting hormone,* is produced by the seminiferous tubules of the developing testis. Once again, it is the presence of substances produced by the male that determines the course of differentiation. In their absence, as occurs in the normal female, the reproductive tract develops passively into uterus, oviduct, etc., and the mesonephric tubules and ducts degenerate.

Various types of experimental evidence could be cited to support this view of sexual differentiation. The following examples will suffice for our purposes. If the gonad of a fetal mammal, whether male or female, is removed, the remainder of the reproductive tract, including the external genitals, develops as female structures. In the absence of gonadal substances, the passive route to feminization of remaining structures occurs. There is a condition in humans that also provides support for this view. On occasion, women who are nonmenstrual and infertile are found to have an XY pair of chromosomes. Further analysis has

indicated that these persons, who by all external indications appear as normal females, have developed testes rather than ovaries. Why then have their secondary sexual characteristics developed in a feminine manner? One explanation might be the absence of androgen production; however, this is usually not the case. Rather, these persons have normal levels of testosterone, but their target cells are not capable of responding to the hormone. Consequently, for all practical purposes, these persons with *androgen-insensitivity syndrome* (also called *testicular-feminization syndrome*) might as well be lacking the hormone, and they develop as females. In mice, this condition has been traced to a lack of the androgen-receptor protein, the product of a gene located on the X chromosome.

As with the fetal gonad and reproductive tract, the sexual centers in the brain also pass through an indifferent stage during which they can be strongly influenced by hormonal stimulation. For example, a single injection of testosterone given to a female rat during a critical neonatal period can lead to the development of masculine sexual behavior following puberty. As with the development of the reproductive tract and secondary sex characteristics, it is the presence of male androgens that directs neuronal differentiation such that the animal expresses male behavioral patterns. In the absence of this stimulation, as occurs in the case of fetally castrated males, it appears that the brain centers passively differentiate along female lines.

5. Axial gradients

In the foregoing description of embryonic systems we have drawn attention to a time sequence in the attainment of various stages of differentiation along the embryonic or anteroposterior axis. Head structure appears before trunk and trunk appears before tail. The sequence is always from head to tail and is especially evident in the serial formation of somites. Gradients of this kind are known as *axial gradients*. Once the process of axial growth in the posterior parts comes to an end —irrespective of whether such growth is cell proliferation, cell enlargement, or tissue stretching—the more recently established embryonic regions eventually catch up with the more precociously formed regions. In other words, the term *axial gradient* is purely descriptive of the differences in the stage of development of various structures located sequentially along the main axis. It reflects the stage of development at a particular time. It denotes only differences in the time at which some process begins, first becomes evident, or completes a certain phase of development; it does not indicate differences in rate of development. The term says nothing about the process underlying the lag in initiation time along the anteroposterior axis.

6. Extraembryonic structures

Evolutionary changes in egg size, yolk content, egg envelopes, maternal association, etc. all have impacts on the course of development. The developing system itself has had to adapt to any changes that affect its circumstances. For example, a mass of condensed yolk in the egg in effect calls for the formation of an enclosing yolk sac during early development, so that the developing system comprises both an embryonic and extraembryonic structure. In fish, the yolk sac is the only extraembryonic structure. The movement of vertebrates onto the land was accompanied by major challenges to reproductive processes. Three additional extraembryonic membranes (Fig. 12.22) appeared during the evolution of an egg that could be laid by terrestrial vertebrates, such as birds and reptiles, on dry land. These extraembryonic membranes included the *allantois*, which encloses a space used for storage of nitrogenous waste, the *chorion* (or *serosa*), which be-

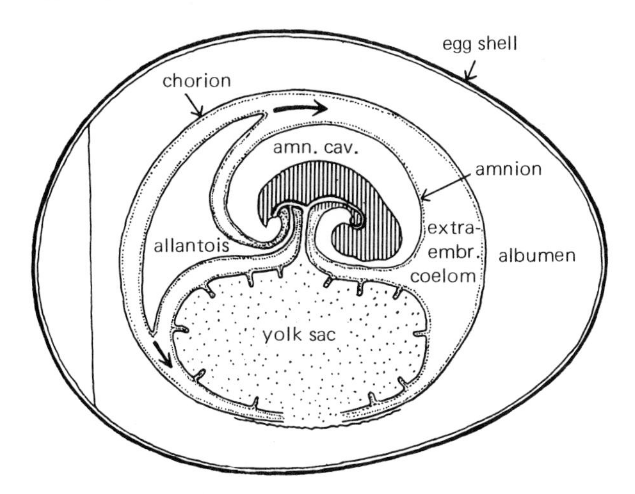

Figure 12.22 Egg of chick complete with shell, albumen, yolk, embryo, chorion, amnion, allantois, and yolk sac. [*From B. I. Balinsky, "Introduction to Embryology," 4th ed., Saunders, 1976.*]

comes highly vascularized to serve as a respiratory surface beneath the shell, and the *amnion,* which encloses a fluid-filled cavity in which the embryo is suspended. It is the amnion, from which the term *amniotic egg* is derived, that protects the embryo from dessication in its terrestrial environment and provides a watery cushion against mechanical damage.

The amnion and chorion form together in a manner not unlike the events that lead to formation of the ventral body wall (Section 12.2.B). Whereas the ventral body wall forms by folds of splanchnopleure that undercut the embryo, the amnion and chorion form from folds of somatopleure that rise above the embryo as an encompassing hood (Fig. 12.23*a*). The process begins with the formation of a head fold and, a little later, a tail fold. With further development, the body folds fuse forming two distinct membranes, an inner amnion, which consists of an inner layer of ectoderm and an outer layer of mesoderm, and an outer chorion, which consists of an outer layer of ectoderm and an inner layer of mesoderm (Fig. 12.23*b*). Between the two membranes lies a portion of the extraembryonic coelom. Fluid is secreted into the space between the embryo and the amnion, thereby forming the protective amniotic chamber. In humans, loose embryonic cells may be withdrawn from this chamber by hypodermic needle during pregnancy and examined for sex and certain other genetically determined characteristics, a procedure known as *amniocentesis.*

Unlike the amnion and chorion, the allantois has its roots in embryonic rather than extraembryonic tissue. It forms as an outgrowth of the hindgut, but it soon extends as the allantoic sac far into the extraembryonic coelom (Fig. 12.23*c*). The allantois fuses with the investing outermost layer, or chorion, of the extraembryonic tissues to form the *chorioallantoic membrane.* This structure, richly vascularized, serves the embryo as the primary respiratory surface and the embryologist as an ideal site for the culture of transplants from other sources (Section 15.4.B).

In placental mammals, the maternal system supplies all needs; there is neither yolk in the yolk sac nor accumulating wastes in the allantois. The chorioal-

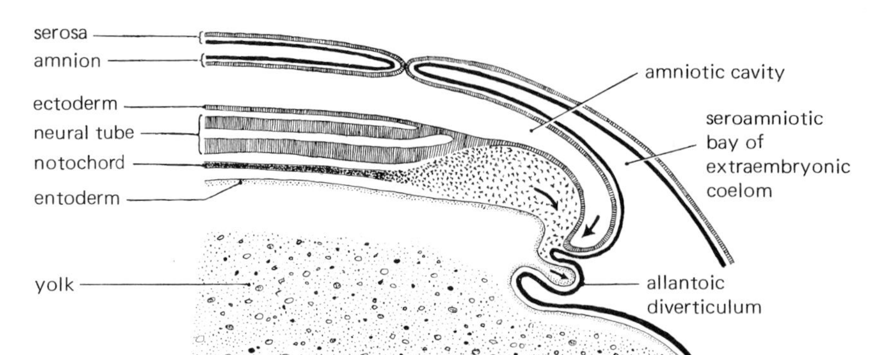

a

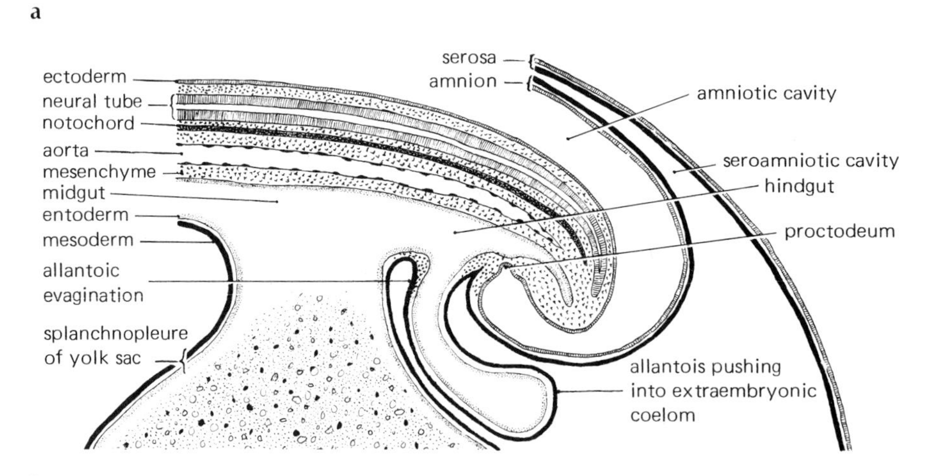

b

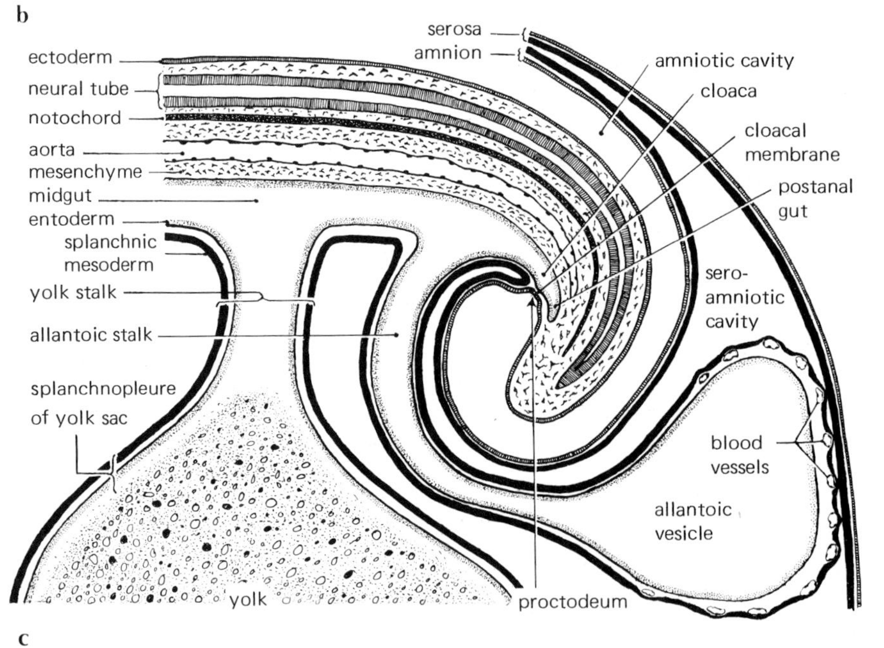

c

Figure 12.23 Three stages in diagrammatic sections of the caudal end of chick embryos to show the development of the amnion and allantois. [*From B. M. Patten and B. M. Carlson, "Foundations of Embryology," 3rd ed., McGraw-Hill, 1974.*]

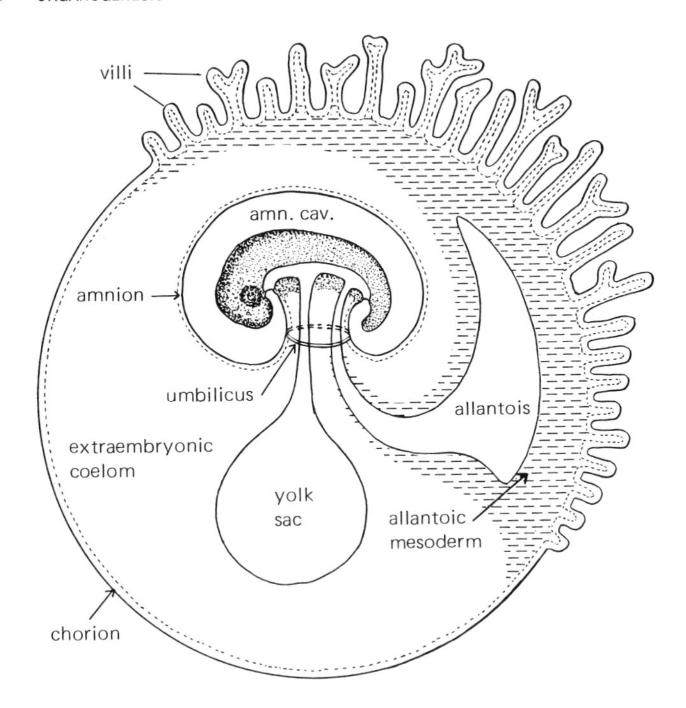

Figure 12.24 Embryo and membranes of a placental mammal, showing placental development from fused chorion and allantoic wall, extending as fingerlike villi that interlock with uterine tissue. [*From B. I. Balinsky, "Introduction to Development," 4th ed., Saunders, 1976.*]

lantoic tissue together with the trophoblast constitute the *placenta* (Fig. 12.24), which is united with the wall of the maternal uterus. The stalk of the allantois survives as the umbilical cord. It contains the blood vessels of the embryo, which supply the placenta, and returns the enriched blood to the embryo or fetus until birth do them part. The amnion persists unchanged from that seen in the chick as the fetal sac or membrane.

The amniote egg may be regarded as a remarkable evolutionary invention enabling development to proceed in a nonaquatic external environment. It primarily involves a massively increased yolk content and the confinement of the egg within a liquid-containing, nonexpansible calcareous shell that is impermeable to water.

Readings

Austin, C. R., and R. V. Short (eds.), 1972. "Embryonic and Fetal Development," Cambridge University Press.

Balinsky, B. I., 1976. "Introduction to Embryology," 4th ed., Saunders.

Bellairs, R., 1971. "Developmental Processes in Higher Vertebrates," Prentice-Hall.

Blandau, R. J., 1977. "Morphogenesis and Malformation of the Genital System," Liss.

Cooke, J. S., 1975. Control of Somite Number during Morphogenesis of a Vertebrate, *Xenopus laevis, Nature,* **254:**196–199.

Cunha, G. R., 1976. Epithelial-Stromal Interactions in the Development of the Urogenital Tract, *Int. Rev. Cytol.,* **47:**137–194.

Deuchar, E., 1975. "Cellular Interactions in Animal Development," Halsted.

DeHaan, R. L., and H. Ursprung, (eds.), "Organogenesis," Holt, Rinehart and Winston.

Eddy, E. M., 1976. Germ Plasm and Differentiation of the Germ Cell Line, *Int. Rev. Cytol.*, **43**:229–280.

Ede, D. A., J. R. Hinchcliffe, and M. Balls (eds.), 1978. "Vertebrate Limb and Somite Morphogenesis," Cambridge University Press.

Fleischmajer, R., and R. E. Billingham, 1968. "Epithelial-Mesenchymal Interactions," Williams & Wilkins.

Ford, P., 1950. The Origin of the Segmental Musculature of the Tail of the Axolotl, *Proc. Zool. Soc. (London)*, **119**:609–632.

Jacobson, A. G., 1966. Inductive Processes in Embryonic Development, *Science*, **152**:25–34.

Johnson, L. G., and E. P. Volpe, 1973. "Patterns and Experiments in Developmental Biology," W. C. Brown.

Jost, A., B. Vigier, J. Prepin, and J. P. Perchellet, 1973. Studies on Sex Differentiation in Mammals, *Rec. Prog. Hormone Res.*, **29**:1–41.

Lehtonen, E., J. Wartiovaara, S. Nordling, and L. Saxén, 1975. Demonstration of Cytoplasmic Processes in Millipore Filters Permitting Kidney Tubule Induction, *J. Embryol. Exp. Morphol.*, **33**:187–203.

Manasek, F. J., 1976. Heart Development: Interactions Involved in Cardiac Morphogenesis, in G. Poste and G. L. Nicolson (eds.), "The Cell Surface in Animal Embryogenesis," Elsevier.

Mittwoch, U., 1973. "Genetics of Sex Differentiation," Academic.

Ohno, S., 1976. Major Regulatory Genes for Mammalian Sexual Development, *Cell*, **7**:315–321.

Patten, B. M., and B. M. Carlson, 1974. "Foundations of Embryology," McGraw-Hill.

Pearson, M., and T. Elsdale, 1979. Somitogenesis in Amphibian Embryos, *J. Embryol. Exp. Morphol.*, **51**:27–50.

Rosenquist, G. C., and D. Bergsma (eds.), 1978. "Morphogenesis and Malformation of the Cardiovascular System," Liss.

Saxén, L., 1971. Inductive Interactions in Kidney Development, in "Control Mechanisms of Growth and Differentiation," 25th Symp. Soc. Exp. Biol., Academic.

Wachtel, S. S., 1977. H-Y Antigen and the Genetics of Sex Determination, *Science*, **198**:797–799.

Waterman, R. E., 1977. Ultrastructure of Oral (Buccopharyngeal) Membrane Formation and Rupture in the Hamster Embryo, *Develop. Biol.*, **58**:219–229.

Willier, B. H., P. A. Weiss, and V. Hamburger, 1955. "Analysis of Development," Saunders.

Zuckerman, S., and T. G. Baker, 1977. The Development of the Ovary, in S. Zuckerman and B. J. Weir (eds.), "The Ovary," vol. 1, Academic.

Chapter 13 Limb development

The developing limb, protruding as it does from the vertebrate body, is an obvious target for experimental analysis. The first sign of limb-bud development becomes evident in the somatic layer of the lateral-plate mesoderm (Fig. 13.1). At this stage, the somatopleure is an epithelium; it becomes thickened where the limb bud will appear. Mesenchymal cells are then released from this epithelium to migrate laterally and accumulate beneath the overlying epidermis. Together the mesodermal and ectodermal components form the rudiment of a limb bud, which first appears as a small mound of tissue on the flank at the site of a prospective limb. This disk of tissue is large enough to be subdivided, rotated, or transplanted before it develops into any sort of structure. The area involved, at least in amphibians, is roughly circular and extends over 3½ somites in diameter (Fig. 13.2*a*).

In amphibians, two separate pairs of limb disks appear from the first. In fish, they are similar, except for an initial extension along the anteroposterior axis of the embryo. In amniotes, the combined epidermal-mesenchymal thickening extends as a horizontal ridge along each side of the body (the Wolffian ridges). The intermediate part of the ridge later disappears, leaving anterior and posterior regions as the definitive limb areas.

1. Prospective limb area and the limb field

Studies conducted on amphibians and birds indicate that limb-bud determination occurs very early indeed, well before the appearance of the buds themselves. If pieces of prospective limb mesoderm of the lateral plate are cut out very soon after the closure of the neural tube and are transplanted elsewhere under the epidermis of the head or the trunk, a limb develops, although in an abnormal site, i.e., heterotopically. The epidermis everywhere responds to the presence of the underlying prospective limb mesoderm and becomes the epithelial covering of the limb bud.

Although the limbs normally form from relatively small regions of the early embryo, various types of surgical manipulations carried out in the first quarter of the century (primarily by Ross Harrison) indicated that substantially greater regions of the flank at the neurula stage had the potency to form a limb. In the salamander, if the *prospective limb area* and only that area is extirpated at an early stage (i.e., all the area that will normally give rise to a limb is removed), a limb will still develop after a short delay. If, however, a somewhat larger circular area, i.e., an additional ring of tissue surrounding the prospective limb area, is included (Fig. 13.2*b*), no limb develops. In the first case, after extirpation of the prospective area, adjoining tissue moves in to replace it and the limb disk is reconsti-

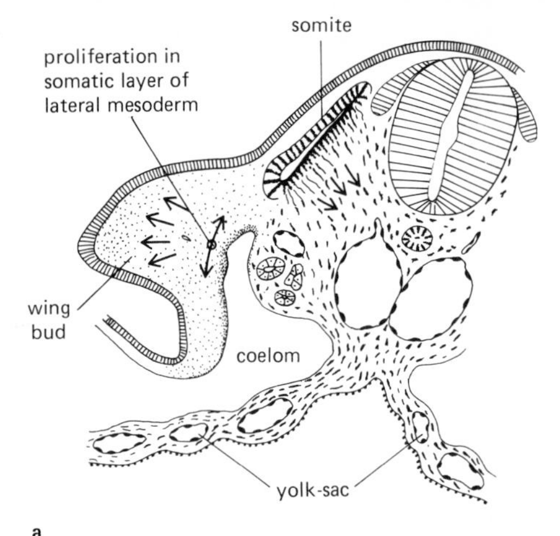

Figure 13.1 *a* Diagram showing the location of the center of mesodermal proliferation involved in forming the core of the wingbud. [*From B. M. Patten and B. M. Carlson, "Foundations of Embryology," 3rd ed., McGraw-Hill, 1974.*] *b* Section through the prospective leg field of a chick embryo showing the thickened somatopleural layer of the lateral-plate mesoderm in anticipation of the formation of the limb bud. The mesenchyme (M) is arranged in a pseudocolumnar pattern with a two-cell thick ectodermal (E) cover. Note the mesenchymal cells (*arrow*) lying subectodermally with their long axis parallel to the ectodermal plane. Spinal cord, SC; presomite, PS. [*Courtesy of E. A. Kaprio.*]

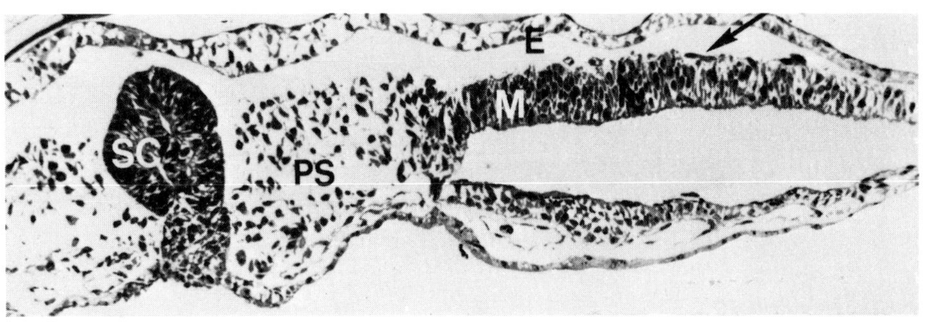

tuted; in the second, the replacement tissue also has been removed. This larger area, which represents the whole prospective limb-bud potential, is called the *limb field*.

The term *field* may best be defined as a collection of properties exhibited by a region of the embryo in which a particular structure will develop. Although the structure normally forms from the cells at the center of the field, the peripheral areas contain cells that also are capable of forming the tissues if required. The regulative powers of a field district, such as that of the limb, can be revealed in other ways (Fig. 13.2):

1. If half a limb bud is destroyed, the remaining half gives rise to a completely normal limb.
2. If a limb bud is slit vertically into two or more segments, while remaining an integral part of the embryo (Fig. 13.2*c*), and the parts are prevented from fusing again by insertion of a bit of membrane between them, each may develop into a complete limb.
3. If two limb buds are combined in harmonius orientation with regard to their axes, a single limb develops that is large at first but soon is regulated to normal size.

It is apparent from these types of experiments that the individual parts of a field (presumably the individual cells) possess some recognition of what is happening in other parts of the district; that is, they recognize their position within the

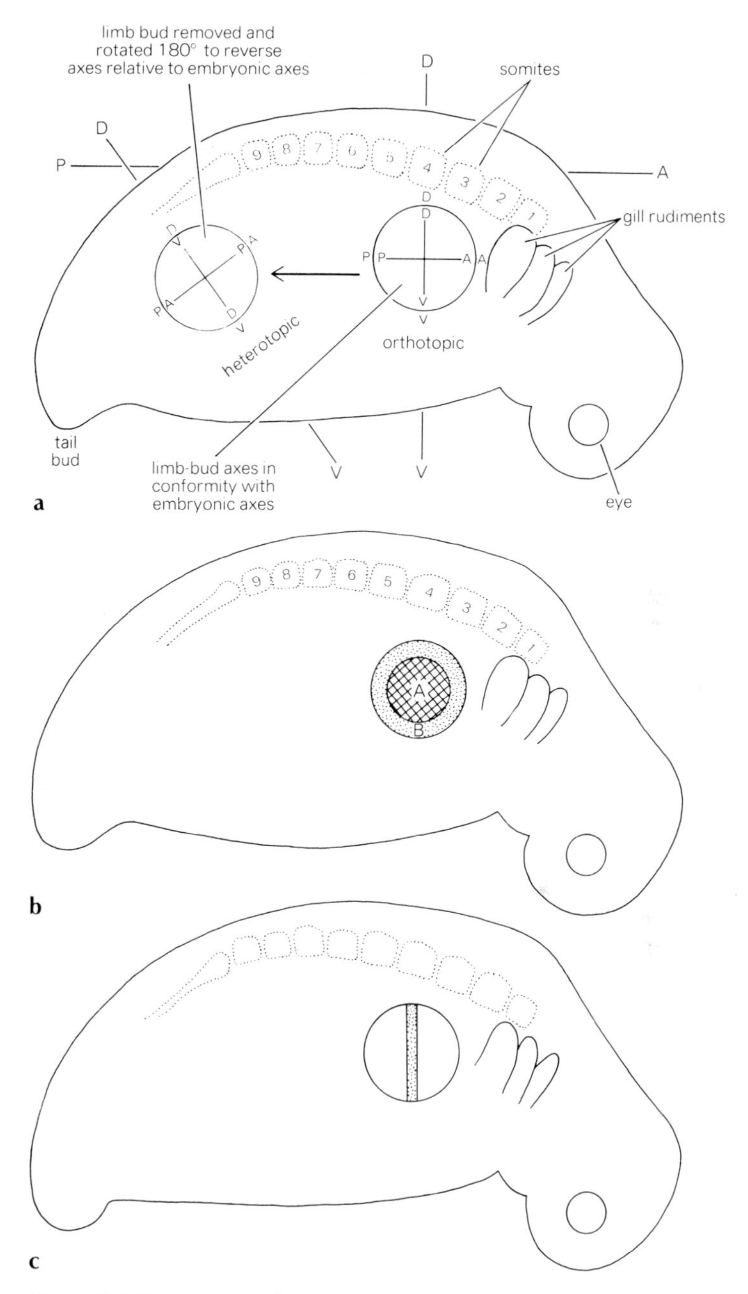

Figure 13.2 Experiments on limb-bud differentiation in the salamander (*Ambystoma*) during the tail-bud stage. *a* Transplantation of limb disk from a normal (orthotopic) site to another (heterotopic) site. The limb differentiates normally. *b* Presumptive limb area (A) and limb field ($A + B$). If the limb area is removed, peripheral-limb-field territory replaces it; if all the limb field is removed, no replacement occurs. *c* If the limb disk is split and the two parts are prevented from re-fusing, two limbs develop. [*After F. H. Swett,* Q. Rev. Biol., **12:***323* (*1937*).]

whole. If the field is split or another field is added to it, the parts recognize the disruption as some change in information about their new position; they can respond accordingly.

We have discussed at length in previous chapters the gradual, progressive nature of embryonic determination. This process is particularly evident in the behavior of organ fields. During the development of a structure such as a limb, there is an early period when small parts are totipotent and capable of forming the entire structure. Although the fate of the entire group of cells that constitutes the field is fixed with respect to the formation of a limb, the fates of individual cells remains quite undecided. A given cell, for example, might form part of the limb musculature, or one of the digits, or a tendon. In other words, the cells are still capable of substituting for one another within the context of forming a limb; the field is a sort of "harmonius-equipotential system," to use the terminology of Hans Driesch (Section 7.2). As the development of the structure proceeds, the potency of groups of cells becomes restricted such that they can form but a small part of the whole. The primary field has become subdivided into smaller fields. Eventually a point is reached at a time prior to overt cell differentiation at which the parts are fully determined and therefore capable of forming only the structures they would have if left undisturbed.

The vertebrate limb is a complex, highly asymmetrical structure with each of its axes clearly polarized. The anteroposterior axis is marked by the thumb to little finger, the dorsoventral by the upper to lower surface, and the proximodistal by the shoulder to fingertip. Although the parts of the limb field retain considerable flexibility, the various axial polarities across the field are determined quite early. The relation of limb polarities to those of the embryo as a whole is shown by inversion experiments in amphibians in which the prospective limb region is rotated 180° at various stages and the effect on subsequent limb development is monitored. If the anteroposterior and dorsoventral axes are fully determined at the time of the operation, a limb should develop that is entirely normal except that it is turned backward and upside down relative to the body. If there is no determination at this time, a limb should develop that is normal in every respect. As it turns out, the effect of limb-disk rotation in amphibian embryos is greatly dependent on the stage of the operation. If the limb field is rotated prior to the yolk-plug stage (late gastrula), the limb has a normal disposition relative to the body. If the operation is performed at the late tail-bud stage (Fig. 13.2*a*), the limb's dorsoventral and anteroposterior axes are both reversed. When the rotation is carried out between these two stages, the limb is normal along the dorsoventral axis but inverted in the anterposterior direction. It appears that the anteroposterior polarity of the limb disk is established at a time (approximately the yolk-plug stage) when the dorsoventral aspect of the field is still developmentally flexible. Other experiments of a more complicated nature suggest that determination along the third axis is delayed even further to approximately the time at which limb outgrowth begins.

2. An analysis of limb development

In the early stages of limb development, the form of the developing limb bud and the pattern of skeletal rudiments emerging within it are remarkably similar in all land vertebrates. The analysis of limb development and determination, in contrast with limb regeneration, has shifted over the years from amphibian material —in which limb rudiments, although readily accessible, are inconveniently small—to the chick (Fig. 13.3*a*). As shown in Fig. 13.3*b* for the limbs of the 2- to

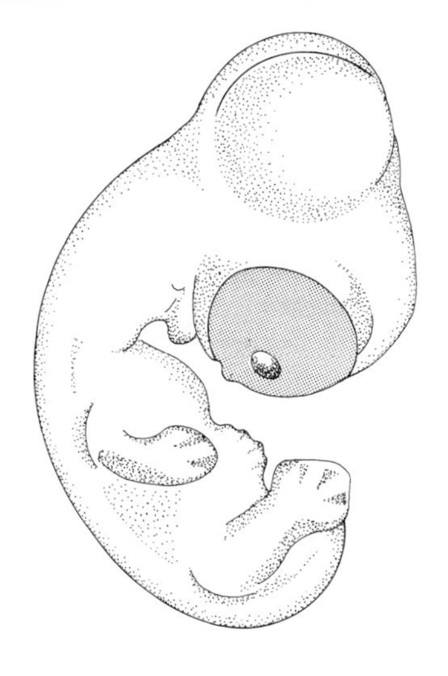

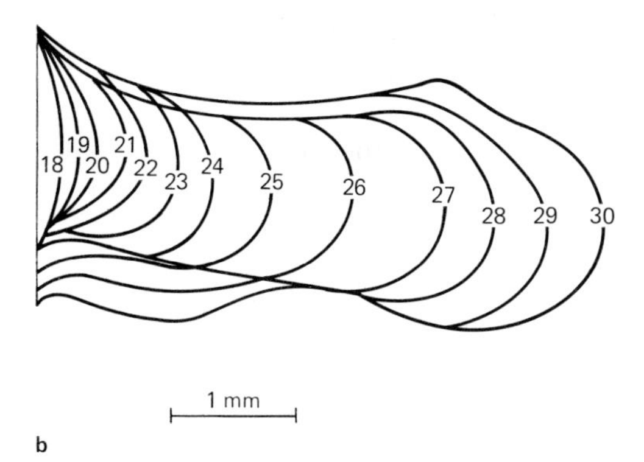

Figure 13.3 *a* Diagram of a bird embryo after 6 days of incubation showing the wing and leg buds. *b* The growth pattern of the chick wing bud. Numbers indicate the Hamburger-Hamilton stages as morphogenesis proceeds. [*From D. A. Ede, in "Cell Surface in Animal Embryogenesis," G. Poste and G. L. Nicolson (eds), Elsevier, 1976.*]

7-day chick, the disk grows out in a distal direction to give an elongated stem region, slightly narrower at its base, and an expanded distal paddle that is flattened in the dorsoventral plane. The limb grows as a result of rapid multiplication of cells within the limb disk rather than by migration of cells from outside the disk area.

The limb-forming potential is clearly localized in the mesenchymal mesoderm; no limb develops if the disk mesoderm is completely removed, leaving only the prospective limb ectoderm in place, nor does a limb develop where prospective limb ectoderm is transplanted to a new site. However, prospective limb mesoderm from as early as stages 11 to 12 (the bud appears in the chick at stages 16 to 17) is capable of promoting limb development when transplanted beneath ectoderm in other places. Limb mesoderm transplanted to the flank with a covering of flank ectoderm gives rise to a limb with normal asymmetry if the mesoderm is oriented with its anterior side forward. It gives rise to a limb with reversed asymmetry if the mesoderm is oriented with its anterior side facing backward. In other experiments, ectoderm and mesoderm were isolated separately from forelimb

and hindlimb buds and then combined and grafted to the flank of another embryo. It was found in these combinations that the limb that formed corresponded in nature to the origin of the mesoderm, forelimb mesoderm promoting wing formation and hindlimb mesoderm promoting leg formation. Similarly, when a piece of tissue is cut from the distal end of a leg bud and inserted in the distal end of a wing bud, a normal wing develops except that a toe forms at the tip of one digit. On the basis of a number of these early experiments, the concept of mesodermal determination seems to be a good working hypothesis. Nevertheless, as described below, an assumption that the prospective limb ectoderm plays no important role is by no means justified.

A. The ectodermal ridge

In the chick, the prospective mesoderm of the wing bud becomes finally localized at the 2-somite stage, although the limb-forming area is not morphologically distinguishable until the 14-somite stage. The wing rudiment is represented at this time by a slight condensation of mesenchyme lying beneath the ectoderm, which, at this early stage, consists of an outer layer of flattened cells called the *periderm* (which degenerates before hatching) and a basal layer of cuboidal cells resting on the basement membrane that covers the mesoderm. In the limb bud, under the influence of the mesoderm, the basal cells at the distal edge of the bud become tall (25 to 35 μm) and thin, forming a pseudostratified columnar epithelium that takes the shape of a ridge running along the rim of the bud in an anteroposterior direction (Fig. 13.4). This ridge, termed the *apical ectodermal ridge* (*AER*), has been shown to have a profound influence on limb development.

A variety of experiments indicate that the AER acts as an inductor of limb outgrowth and differentiation. Excision of the AER leads to the development of partial limbs, lacking distal structures (Fig. 13.5). The amount of distal deficiency

Figure 13.4 Section through the leg bud of a stage-19 chick embryo showing the apical ectodermal ridge (aer) and the underlying mesenchyme (M) [*Courtesy of E. A. Kaprio.*]

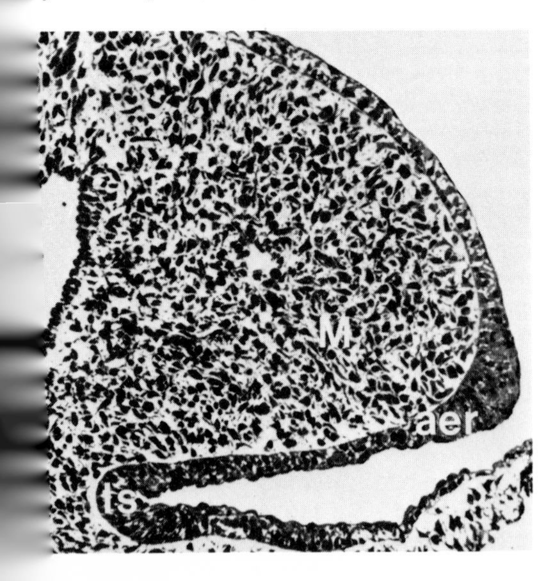

Figure 13.5 (*Left*) Differentiation of a limb bud after the removal of the apical-ectodermal cap. Only the femur and part of the tibia are differentiating. (*Right*) Graft of an apical-ectodermal cap onto the basal part of the leg after the distal half had been severed. The basal mesenchyme is induced by the apical cap to form the distal components. [*After A. Hampe,* J. Emb. Exp. Morph., **8**:247 (*1960*).]

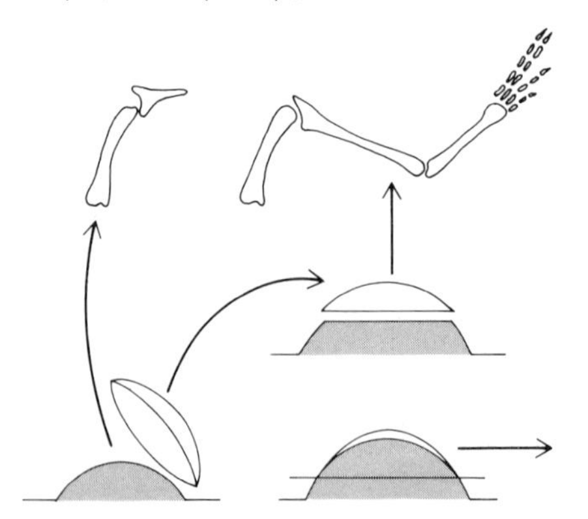

depends on the stage subjected to the procedure. Limb buds deprived of the AER at an early stage (stage 18) develop limbs lacking forearms and hands, while the same operation performed at stage 28 results in a limb that is complete except for the terminal portion of one of the digits. In the chick embryo, the only tissue that seems to foster continued limb development is the apical ectodermal cover. Ectoderm from other regions of the limb, i.e., dorsal or ventral ectoderm, or other parts of the body is incapable of influencing the limb mesoderm so that it will complete its distal development. Just as the loss of the AER stops further distal outgrowth and development, the presence of an additional AER induces the outgrowth of a second (*supernumerary*) set of distal structures. This duplication of distal structures can arise naturally, as occurs during the development of the mutant *eudiplopodia,* or it can be obtained experimentally by grafting an AER to the distal region of a host limb bud. Although the AER has a profound influence over the activities of the underlying mesoderm, this influence appears to be of a permissive rather than an instructive nature (Section 11.3). For example, replacement of an AER of a leg bud with one taken from a wing bud does not affect the direction in which the distal portion of the limb differentiates; it forms a normally proportioned leg despite the presence of wing ectoderm. Similarly, the AER does not determine the proximodistal level of differentiation achieved by the underlying mesenchyme. If ectodermal caps are exchanged between limb buds of very different ages, each composite bud proceeds to develop into a normal limb.

The importance of the AER in limb development has led to considerable speculation on its mechanism of action. It has been proposed that the AER is needed to keep the mesenchyme developmentally labile at the tip, but that it does not direct the path of progress at that site. The ridge issues a general permit rather than specific instructions. This concept of AER function is more apparent when one considers the proximodistal nature of limb differentiation. Based on a variety of cell-marking techniques, it has been shown that the early limb bud contains cells that will give rise to all parts of the limb, but that it is composed primarily of cells that will form the proximal portions. The cells whose prospective fates are to form the distal structures exist only as a thin layer beneath the AER. Once a particular stage is reached (stage 22 in the chick), the shoulder girdle and humerus begin to form, while the distal portion of the wing remains undifferentiated and greatly underrepresented in cell mass. As development proceeds and limb outgrowth continues, the remaining undifferentiated cells proliferate and become destined to form more and more distal structures. Development of overt limb structure unfolds in this manner in a distinct proximodistal sequence. Examination of the limb bud at various stages reveals the continued presence of a zone of undifferentiated cells approximately 0.4 mm thick just beneath the AER. Despite the fact that these cells are rapidly dividing (generation time approximately 11 to 13 hours), this zone of unspecialized cells remains approximately 0.4 mm thick, while the remaining cells of the limb are going about the formation of the muscular and skeletal elements. If the AER is removed, this layer of cells beneath the ridge then undergoes differentiation into the distalmost structure being formed at that time and further distal differentiation is inhibited. It is believed that the capacity for more and more distal differentiation of limb structure is programmed intrinsically within the mesoderm. It is the function of the AER to inhibit the realization of this differentiation program within the thin layer of cells in the adjacent subridge. As the cells of this distalmost layer continue to divide, the cell progeny overflow into relatively more proximal regions of the limb, where, removed from the inhibiting influence of the AER, they differentiate into the distal structures being formed at that time. It also has been proposed that the AER

exerts an influence over the mitotic index of the underlying mesenchymal cells and thus indirectly affects their capacity for differentiation. We will return to the role of the AER later.

An apical ridge has been said not to exist in amphibian limb buds, although it has been described in mammalian embryos. The scanning electron microscope, however, demonstrates its presence in *Xenopus* (Fig. 13.6). The course of the ridge is consistently related to a marginal sinus in the underlying mesenchyme; other features of limb morphogenesis, such as formation of a paddle (Fig. 13.7) and the sequence of condensation of skeletal rudiments in the mesenchyme, correspond closely to those seen in other vertebrates.

The relationship between the AER and the underlying mesoderm is clearly a reciprocal one; each depends on influences emanating from the other layer. The AER requires the close presence of limb mesoderm for its very existence. If the ectodermal ridge is separated from the mesoderm by a very thin sheet of mica, or if limb mesoderm is replaced by nonlimb mesoderm, the ectodermal ridge rapidly

Figure 13.6 Development of the apical-ectodermal ridge during limb-bud development in an amphibian (*Xenopus laevis*). *a* Scanning electron micrograph of apex preceding ridge formation. Note difference in dimension of the dorsoventral and anteroposterior transverse axes of the limb bud. *b* Scanning electron micrograph of apex of later stage with ridge well formed. *c* Section through same stage as *b*, showing thickened apical ridge ectoderm separated from underlying mesenchyme by basement membrane. [*Courtesy of D. Tarin and A. P. Sturdee.*]

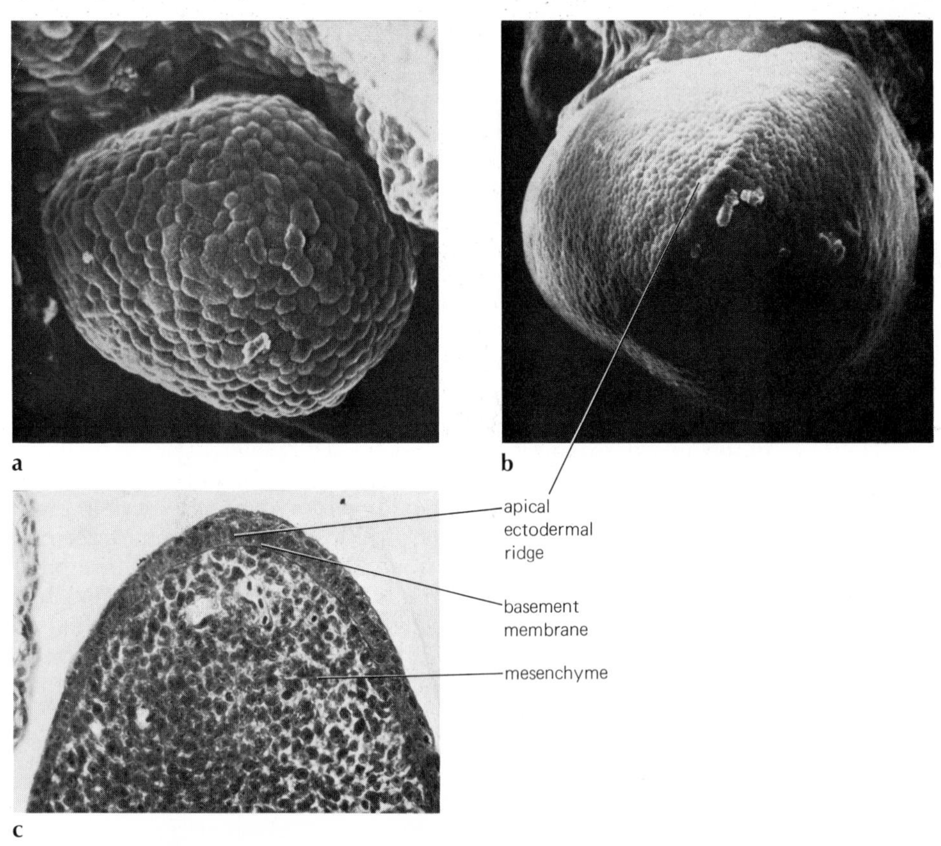

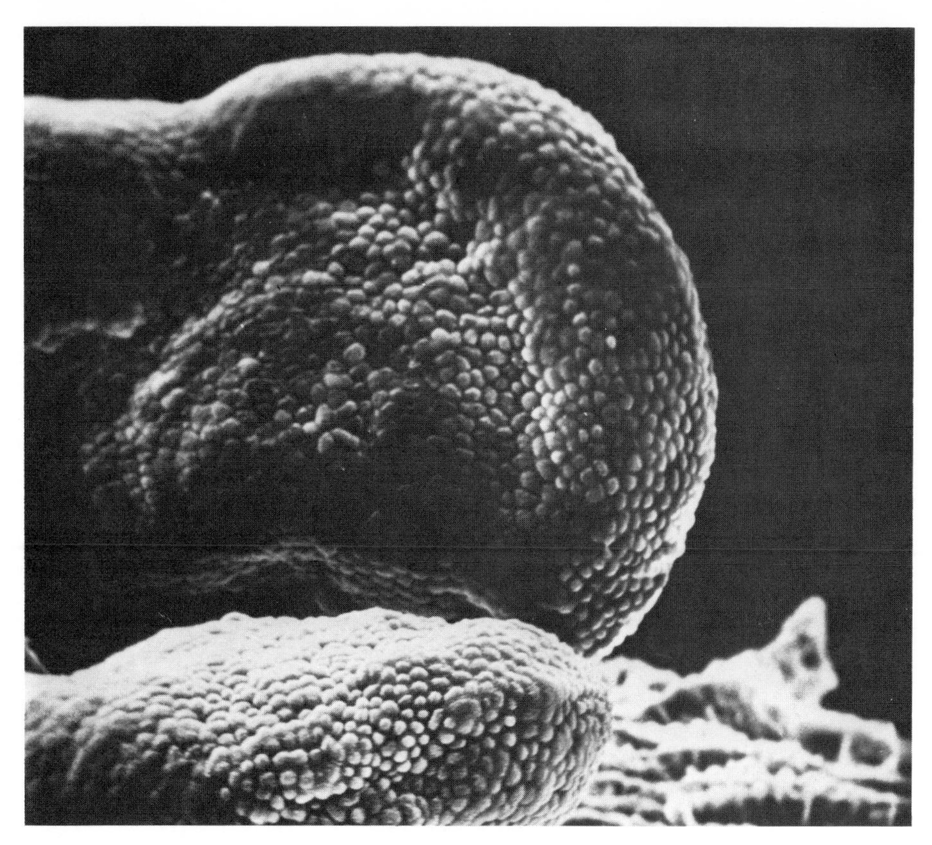

Figure 13.7 Scanning electron micrograph of the paddle stage of *Xenopus laevis*, seen broadside and edgewise. [*Courtesy of D. Tarin and A. P. Sturdee.*]

flattens and the resulting limbs are deficient in terminal structures. Conversely, a thickened AER always appears wherever there is an outpushing induced by limb mesoderm, such as occurs when limb mesoderm is grafted beneath flank ectoderm. The results of these types of experiments have led to the belief that an apical ectodermal ridge maintenance factor (AERMF) is produced within the mesoderm and acts on the overlying ectodermal epithelium.

Does the AER form independently and then induce mesodermal outgrowth, or do the thickened regions of the ectoderm form in relation to some pattern present in the mesoderm? Since the AER is itself polarized, being thickest at its posterior end, the relationship between its morphology and the organization of the underlying mesoderm can be studied using various types of surgical procedures. For example, two or three AERs can be placed in tandem along the distal edge of one limb mesoblast. Do they all remain thick and active as outgrowth inductors, or do they become modified or reorganized under the influence of the mesoderm? In every case the composite ectodermal ridge fused to form a single ridge that gradually acquired a typical configuration and gave rise to a normal limb. Much the same results were obtained when limb-bud ectoderm was rotated 180° in relation to the mesoderm. In this latter experiment, the AER became reorganized such that its thickened end once again regained its normal posterior position. In all these experiments, ectodermal-ridge pattern is reorganized so as to conform to the pattern of the underlying mesoderm. Moreover, essentially similar results were obtained when genetically normal AERs were combined with mesoderm

from limb buds of polydactyl mutants; a thickening of the ridge persisted in regions where flattening normally occurs and supernumerary distal structures developed there, as in the limb buds of polydactyl controls.

B. The role of extracellular matrix

It is apparent from the previous discussion that a dialogue takes place between the ectoderm and mesoderm that has great impact on limb morphogenesis. In order to understand this communication process it is important to consider the nature of the boundary region between these two cell layers. The following account derives from a study carried out on *Xenopus*, but very similar events occur in a wide variety of vertebrates. A basement membrane lies between the epidermis and the mesenchyme. It consists of an organized network of collagen fibrils, to which mesenchymal cells add glycosaminoglycans. This complex of cells and fibers persists throughout most of the body, but beneath the prospective limb epithelium the extracellular matrix components undergo degradation and reorganization. This is followed by a closer connection of mesenchyme with ectodermal epithelium and its associated basal lamina (Fig. 13.8). It has been proposed that the observed contact between cell surfaces and extracellular layers may mediate intercellular signals having morphogenetic significance (positional information). At the same time, specialized junctions appear between the mesenchy-

Figure 13.8 Schematic diagram illustrating progressive alterations in extracellular matrix, intercellular contacts and junctions, and epitheliomesenchymal association during stages 44-45 (*a*), 46 (*b*), and 48 (*c*) in *Xenopus laevis*. Epithelial (*above*) and mesenchymal cells (*below*) are separated by phospholipid-containing adepidermal granules (*finelined squares; a*), a basal lamina (*solid black band; b*), and the basement lamella of collagen (*lines and dots of same diameter; c*). Note progressive disorganization of basement lamella, delamination of adepidermal granules, and loss of hemidesmosomes at basal surface of epithelium. Epithelial-cell contacts (desmosomes) are stable during this period. In contrast, mesenchymal cells acquire focal tight junctions and gap junctions by stage 48, maintaining the close associations characteristic of earlier stages. Filopodia exhibiting microfilaments penetrate the disrupted matrix by stage 48 and abut the overlying basal lamina. [*After R. O. Kelly and J. G. Bluemink,* Dev. Biol., **37**:*12* (*1974*).]

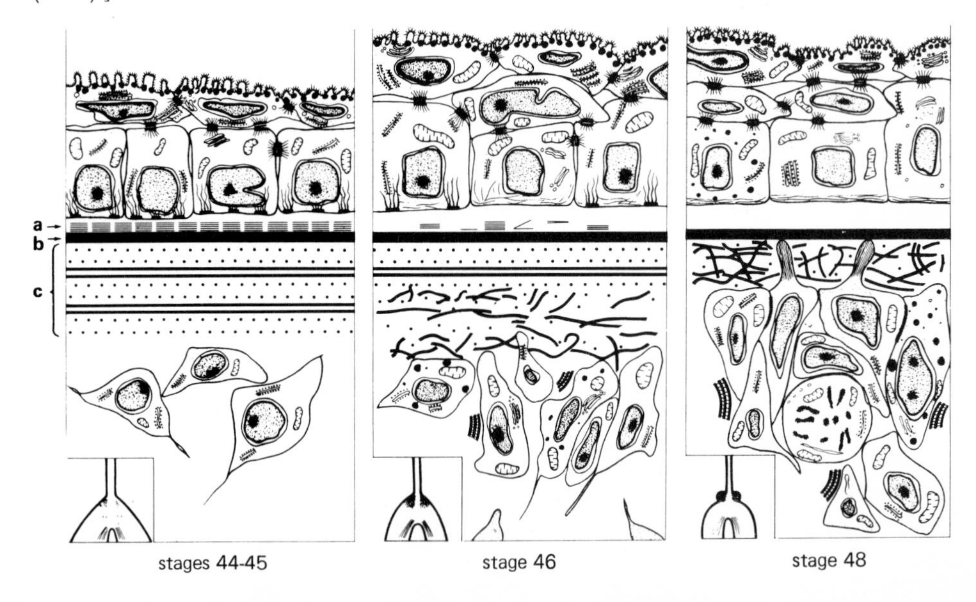

mal cells. Development of low-resistance pathways for intercellular ions and metabolite transport may functionally integrate the mesenchyme and coordinate events (such as cell division and change in cell position specific to the spatial patterns that underlie morphogenesis in the limb).

C. Models for the specification of limb structures

The vertebrate limb is a beautifully constructed mass of precisely shaped, precisely located muscles, bones, and connective-tissue elements, all of which form from a relatively homogenous assemblage of mesenchymal cells. How does a pattern of such intricacy emerge from this formless mass of limb-bud cells? For example, how is the shape of particular bones, muscles, and tendons determined? A more basic question concerns the basis for a particular mesenchymal cell differentiating into a muscle, cartilage, or connective-tissue cell. Does a cell in the core region of the limb bud become a chondrocyte because it happens to reside in that part of the developing limb (i.e., would it have differentiated into a muscle cell had it been residing in a different limb area?), or does that mesenchymal cell differentiate into a cartilage cell because it is predetermined to do so? Up until recently it was generally felt that each mesenchymal cell was capable of differentiating into a myoblast, chondroblast, or fibroblast (i.e., it was *pluripotent*), and the direction taken by a given cell was determined totally by the position in the bud at which it happened to become located. This concept of mesenchymal-cell pluripotency has received support from the results of (1) in vitro studies, where it has been reported that cultured cells can be directed into chondrogenic or myogenic phenotypes by modifications of the medium, and (2) in vivo studies, where premyogenic regions of the limb are grafted to prechondrogenic regions (and vice versa) without disrupting limb development. Recently, however, grafting experiments using quail-chick chimeras have strongly suggested that all or most of the limb's musculature is derived from somite mesoderm, while the skeletal and connective-tissue elements form from lateral plate. It follows from this observation that the cells are preprogrammed to differentiate into muscle or skeletal tissue as a consequence of their origin. If this is the case, then myogenic and chondrogenic precursors are distinct cell types from a very early stage of limb development. The two types of mesodermal cells would presumably migrate to (or proliferate in) different regions of the limb bud, the chondrogenic cells to the central core and the myogenic cells to the periphery.

Questions concerning pattern require an explanation in terms of positional information. Each cell or group of cells in the bud must receive information "telling" it of its relative position within the developing limb. Each cell must then interpret this information in terms of its own developmental background so that it can respond by differentiating in a manner appropriate for that location in the bud. The distinction between positional information and the interpretation of that information can be seen by considering an experiment in which a small piece of mesoderm from a ventral site in a leg bud is grafted to a dorsal site in a wing bud. The graft differentiates into structures characteristic of the *dorsal* portion of the *leg*. In other words, it responds to the information in its new position by differentiating in a dorsal-specific manner, but it interprets this information in terms of its own developmental background and forms structures appropriate to that position in the leg from which it came.

We considered the topic of positional information in Chapter 1 as it applied to what were essentially one-dimensional organisms, *Hydra* and *Dictyostelium*. In both of these organisms, a cell differentiates in response to its position along its

single apical-basal axis. The situation in the limb is much more complex because the cells of the mesoderm must participate in the formation of a three-dimensional pattern. Each cell must receive information telling it of its position along each of three geometric axes. In this way, each point within the limb can be assigned a unique set of positional values which can then lead to its position-specific differentiation. A cell that has received its positional information, interpreted it, and begun to differentiate in a position-specific manner is said to be *specified.*

Based largely on the results of earlier experimentation, Lewis Wolpert and coworkers have constructed a model by which mesenchymal cells of the limb can be assigned positional values along two of the three geometric axes, the proximodistal and anterposterior. Since the third, or dorsoventral, axis of the limb is much shorter and much less polarized, it is less amenable to experimental manipulation and is less well understood. It was mentioned earlier that the AER is believed to exert a negative influence over the cells of the thin subridge layer, causing them to remain unspecialized. As cells proliferated in this area, they inevitably overflow into more proximal regions of the limb, where, in the absence of the inhibitory influence of the AER, they differentiate into the appropriate structure. The Wolpert et al. model addresses the question of how the cells that leave the distal tip are directed toward differentiation in a position-specific manner. They propose that the cells beneath the AER exist in a "progress zone" in which there occurs an autonomous change with time of some metabolic variable that serves as a measure of positional value. At the beginning of limb outgrowth, the mesenchymal cells of the subridge possess some level of this variable, whatever its nature. As time passes, there occurs an autonomous rise or fall in the level of this factor within the cells of the progress zone. As its level continues to rise or fall according to its own internal program, its positional value changes accordingly such that the cells become specified to form more and more distal elements. As long as a given cell or its progeny remain in the progress zone, they stay undifferentiated and their positional value changes. The longer a cell resides in the zone, the more distal its positional value. However, cells that are displaced into a site outside the sphere of influence of the AER undergo differentiation in keeping with their positional values, which now remain frozen. According to this model, the AER controls the switch between two distinct phases, one in which undifferentiated cells change their positional value in response to an internal program and one in which this positional value is interpreted by the cells as they undergo cytodifferentiation. The time when this switch is made leads directly to the proximodistal specification of the cell in question.

Whereas information along the proximodistal axis is thought to be measured in terms of time (or actually in terms of the numbers of divisions a cell passes through while in the progress zone), it is proposed that information along the anteroposterior axis is measured in terms of the concentration of a morphogenetic substance present in gradient form. The basis for the latter part of this model derives from experiments that indicated that a small region of cells, initially located at the posterior margin of the limb bud near its junction with the body wall, had peculiar inductive properties. When this tissue was grafted to a notch in the anterior edge of a second bud, the limb developed supernumerary digits with the duplication generally showing mirror-image symmetry along the anteroposterior axis.

The chick wing is ideally suited for the analysis of aberrant patterns involving digits, because each wing normally has only three "fingers," numbered II, III,

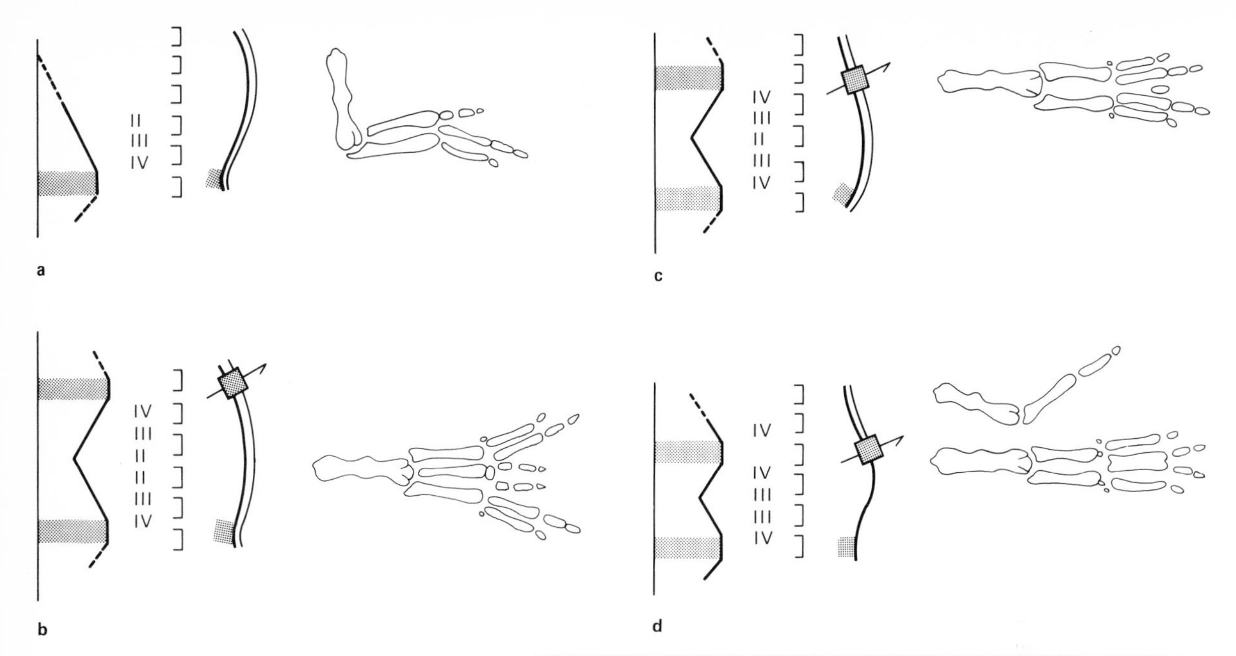

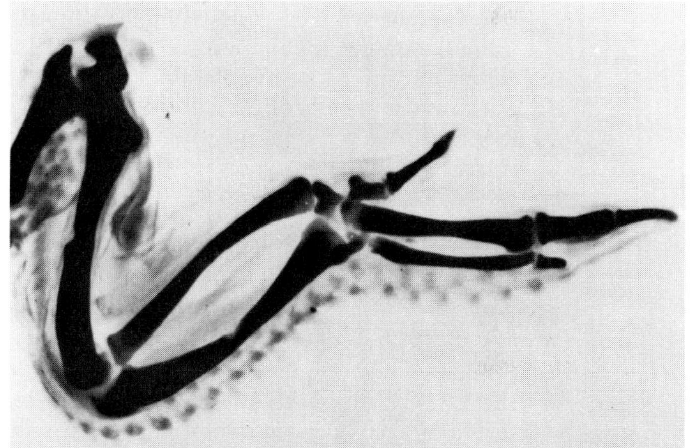

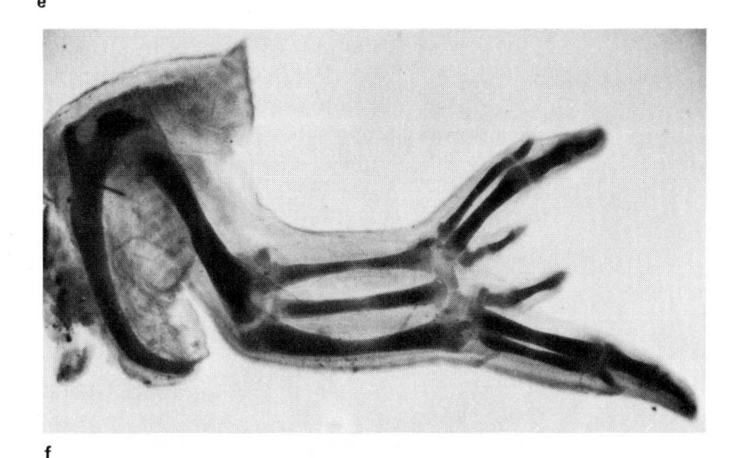

Figure 13.9 *a–d* Diagrams showing the result of grafting a zone of polarizing activity (ZPA) to different levels along the anteroposterior axis of a stage-18 host. (*Left*) Interpretation in terms of a hypothetical gradient; roman numerals indicate the digits present. (*Center*) Position of the graft on the host. (*Right*) Diagrammatic typical results. [*From L. Wolpert in "Cell Patterning," Ciba Symposium no. 29, Elsevier, 1975.*] *e–f* Photographs showing the skeletal structure of (*e*) a normal wing and (*f*) one that had developed as shown in *b* following a graft of the ZPA. [*Courtesy of D. Summerbell and C. Tickle.*]

and IV in anterior-to-posterior sequence, all of which are readily identified. The normal chick-wing pattern is shown in Fig. 13.9*a* and *e*. Analysis of the pattern of digits that develops following a graft of posterior limb-bud mesenchyme indicates that the digit developing adjacent to the graft is generally the posteriormost digit, i.e., a number IV digit. Since these grafts tend to polarize the surrounding tissue, this region of the limb bud is referred to as the *zone of polarizing activity* (ZPA). The effect of grafting the ZPA from one bud to an anterior region of another bud is shown in Fig. 13.9*b*,*c*, and *d*. In all cases, the duplication has a number IV digit closest to the graft. In the third case, the ZPA is grafted a bit more posteriorly within the bud (Fig. 13.9*d*), leaving a sufficient portion of the bud tissue anterior to the graft for a digit to develop. In this case, two supernumerary number IV digits are formed, one on each side of the graft. These types of experiments have led to the proposal that during normal development the ZPA represents a site in the young limb bud from which a morphogenetic substance is continuously released. This substance then diffuses anteriorly, where it is continually being broken down by the cells along the way. The result of the diffusion and breakdown of this substance is the establishment of a stable gradient having its highest activity adjacent to the ZPA. This graded signal serves to provide positional information along the anteroposterior axis. As the bud undergoes its outgrowth, the ZPA moves out from the body wall, maintaining itself at the posterior edge of the base of the progress zone. The concentration of the substance *within the progress zone* is what serves to specify a cell's positional value along the anteroposterior axis as it leaves the zone to differentiate into successively distal structures. Cells nearer the ZPA participate in the formation of more posterior structures. Since a region with ZPA-like properties is found in both forelimbs and hindlimbs of amphibians, reptiles, birds, and mammals, this proposal may be applicable to a wide variety of developing appendages. Evidence for and against various aspects of this model can be found in the papers by D. Summerbell (1979) and J. W. Saunders in the symposium edited by D. A. Ede et al., 1977. Another recent model can be found in the paper by S. A. Newman and H. L. Frisch (1979).

3. Limb morphogenesis and cytodifferentiation

During the first stages of limb outgrowth in the chick, up to approximately stage 22, the mesenchymal cells appear as a perfectly homogeneous cell population. Although the prospective muscle and cartilage cells appear to be derived from different types of mesoderm, they are indistinguishable on the basis of their ultrastructure, mitotic rate, or known synthetic activities. Among the various types of molecules that are synthesized by the mesenchymal cells, the sulfated glycosaminoglycans form a particularly important group because one of these species, chondroitin sulfate, is an important component of the cartilage matrix. Furthermore, the level of synthesis of these molecules at various stages in different parts of the limb is readily measured by following the incorporation of radioactive sulfate. Prior to stage 22, all limb mesenchymal cells possess roughly equivalent levels of $^{35}SO_4$ incorporation. As limb development progresses beyond this stage, sulfate incorporation in the presumptive myogenic regions (the proximal dorsal and ventral regions) decreases, while incorporation in the presumptive chondrogenic regions (the core of the bud) rises sharply. At about this time, the ultrastructure of the prospective muscle cells begins to change, primarily by the accumulation of numbers of ribosomes and glycogen particles, and the mitotic rate of the prospective cartilage cells begins to decrease. By the end of stage 25, the differences between cells of the myogenic and chondrogenic regions have become obvious. The myoblasts have begun to undergo fusion and filament formation,

while the cartilage cells have begun to secrete matrix material (primarily chondroitin sulfate and collagen) into the extracellular space. These latter changes are discussed at length in Chapter 15.

The formation of muscle and cartilage involves changes in cell-cell behavior. Chondrogenesis occurs within aggregates of cells that resemble whorls (Fig. 13.10) within the limb bud. Although the increase in cell density within these cartilage-forming aggregates is relatively small (approximately 30 percent) and the cells do not become closely pressed against one another, it is often suggested that this clustering behavior of chondrocytes plays an important role in the initiation of matrix secretion. Meanwhile, in the myogenic regions of the limb bud, contacts between the prospective muscle cells increase markedly prior to overt differentiation as these cells prepare to form dense condensations of muscle tissue.

The pattern of cartilage rudiments within the limb will determine the location and shape of the limb bones, which become ossified at a later stage (beginning at stage 32). In the intact limb, blocks of chondrogenic tissue are marked off by spaces from the limb base to its apex, indicating the upper limb, lower limb, and the metapodial and digital regions (Fig. 13.11*a*). The progress of chondrogenesis in these various regions is illustrated in Fig. 13.12 by means of the autoradiographic localization of $^{35}SO_4$ incorporation. While the skeletal elements of the limb form from the collective association of distinct cartilage rudiments, the individual muscles of the limb form from the splitting of much larger blocks of tissue located in the dorsal and ventral regions of the limb. The muscle masses are contractile even before they are split and can be activated by nerve stimulation.

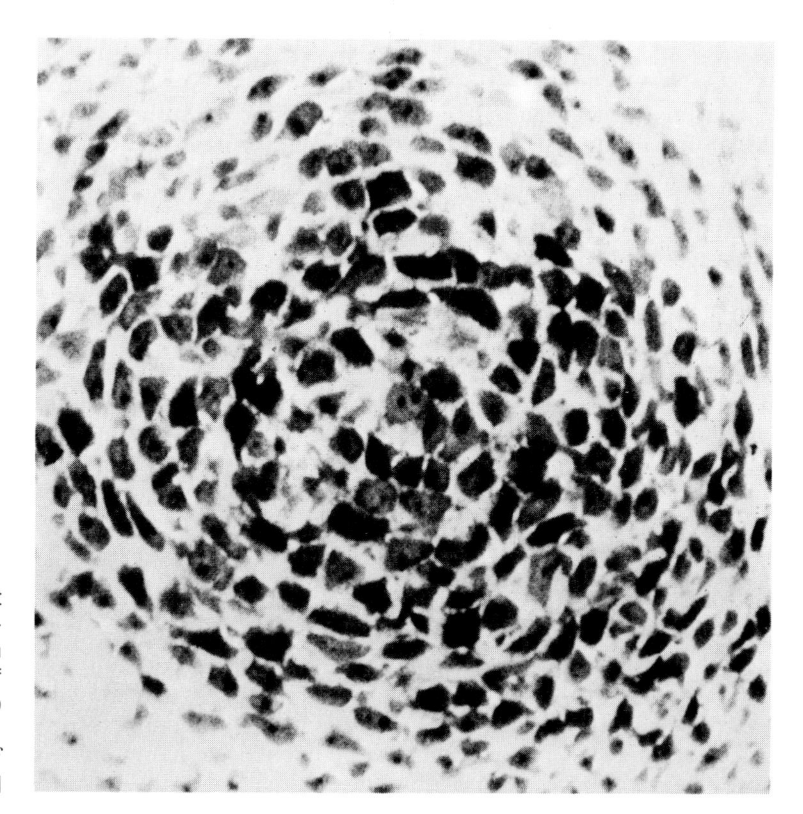

Figure 13.10 Light micrograph of a precartilage condensation in the phalangeal region of a normal chick embryo leg bud of stage 30. [*Courtesy of D. A. Ede.*]

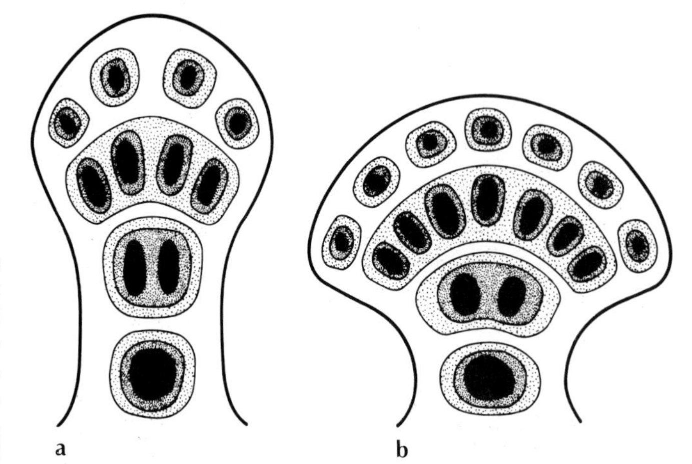

Figure 13.11 Diagrams to show chondrification centers *a* in normal limb development and *b* in talpid mutation. [*After D. A Ede,* Symp. Soc. Exp. Biol., **25**:252 (*1971*).]

As part of an attempt to understand events occurring during normal limb development, considerable attention has been paid to limb formation in mutant embryos. The mutation most exploited is the *talpid* mutant, produced by a recessive lethal gene which, when present in a homozygous state, leads to a dramatic distortion of limb morphology. In early stages, the talpid limb bud is broad and fan-shaped. In later stages, the cartilaginous skeleton of the limb shows a tendency toward fusion of the proximal elements and excessive distal elements, thereby

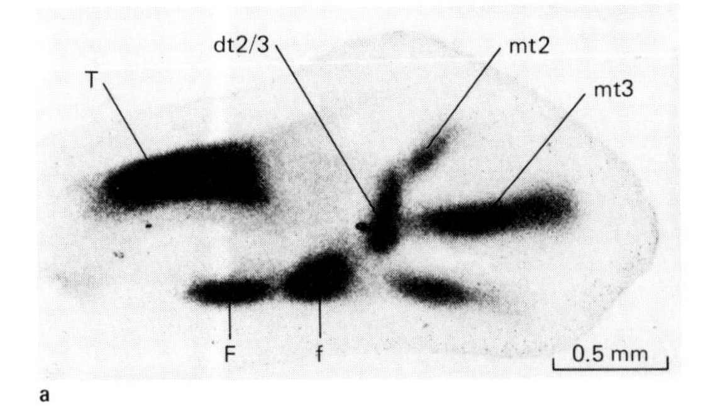

Figure 13.12 The chondrogenic pattern of the hindlimb of a chick embryo as revealed by the incorporation of $^{35}SO_4$ into chondroitin sulfate. Autoradiographs prepared at *a* stage 26, *b* stage 27, and *c* stage 30. [*Courtesy of J. R. Hinchcliffe.*]

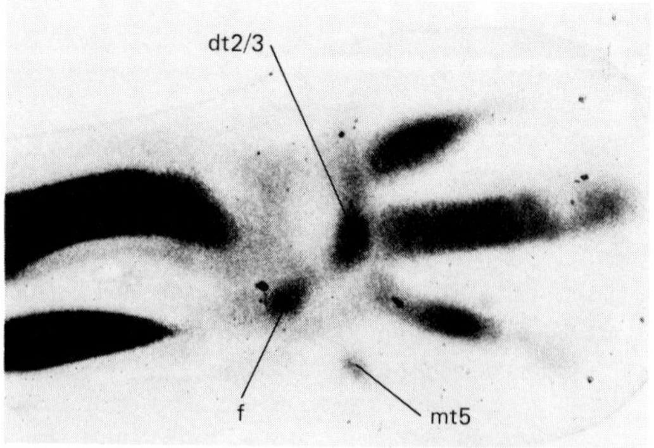

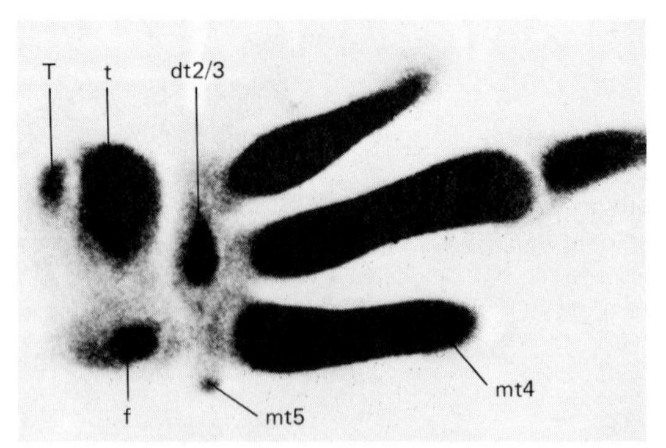

producing a broad, shovel-shaped polydactyl limb (Fig. 13.11*b*). How can these effects on morphogenesis be explained by alterations at the cellular level at which the genetic mutation must act? Changes in several cellular properties have been ascribed to the talpid mutation, and the subject remains controversial. For example, it has been proposed that the fan-shaped form of the talpid limb results from a reduced ability of the mesenchymal cells to carry out certain movements. The decreased motility of the mutant cells may be accounted for by their increased adhesion to one another, a property that also may cause less well defined aggregations and defective chondrogenesis. Talpid limbs also have a greater number of cells, a feature that could result from an increased mitotic index and/or a decreased level of cell death (see Section 13.4).

Limbs are particularly sensitive to various toxic agents, notoriously so in the case of thalidomide, a chemical commonly prescribed in Europe in the 1950s for pregnancy disturbances. It caused thousands of babies to be born without limbs. This susceptibility appears to be due to the fact that a developing limb, from the moment of its inception as a limb disk and throughout the period of elongation of the limb bud, is a relatively rapidly growing mass of tissue. Anything that interferes, directly or indirectly, with the growth process will inhibit or alter its development. Since growth of the limb bud occurs mainly at its distal end as long as the apical mesoderm remains undifferentiated, this part of the bud continues to be the most susceptible region.

4. Cell death as a morphogenetic agent

As development of the avian or mammalian limb proceeds, a sculpturing process results in the particular form of the limb. This sculpturing is not entirely a matter of relative regional growth, but includes a truly erosive aspect involving cell death, i.e., *necrosis*. In the chick wing, for example, several distinct regions (Fig. 13.13*a*) can be found in which a wave of cell death is responsible for the loss of a

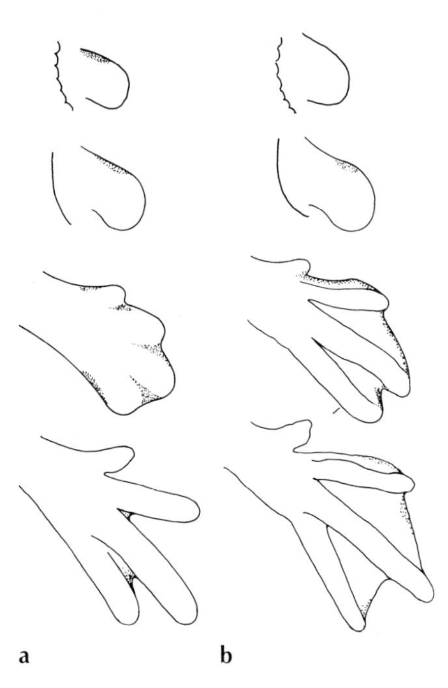

Figure 13.13 The pattern of necrosis (cell death) in the leg primordia of *a* chick and *b* duck, in relation to webbing. Shaded areas are necrotic. [*After J. W. Saunders and J. F. Fallon, 25th Symp. Soc. Dev. Biol., p. 293, 1966.*]

specific mass of cells. Cell death plays a major role in the separation of radius and ulna in the wing and tibia and fibula in the leg, as well as the separation of the digits at the distal end of both types of limbs. While the tissue between the digits of the chick foot is completely destroyed, some of these cells remain as webbing in the foot of the duck (Fig. 13.13*b*). There are also major zones of necrosis at the anterior and posterior margins of the bud. Although the reason for destruction in these regions, termed the *anterior necrotic zone* (ANZ) and the *posterior necrotic zone* (PNZ), is less clear, the underlying control of cell death in these parts of the limb is well-studied. It has been shown that the cells of the PNZ actually become committed to die at a very early stage. Even though the cells of the PNZ show no signs of ill health until stage 21 and intense necrosis does not occur until stage 24, these cells are programmed for death by stage 17, the time of limb-bud appearance. If the presumptive PNZ region of a stage-17 bud is transferred to a nonlimb region of another embryo, or if it is removed and placed in culture, the cells die at approximately the same time as they would have had they been left in place. These results suggest that a stage of differentiation is reached by stage 17 that sets in motion a condition leading to cell death approximately 30 hours later. The mechanism for this event has been referred to as a *death clock*.

Cellular death as a regional phenomenon has been widely observed in embryonic development. It is clearly a useful process inasmuch as it removes unwanted tissue whose substance can be reutilized elsewhere, as in the metamorphosis of many animals, notably amphibians and insects. In some systems, cell death seems to occur as a direct result of an increased lysosomal activity. At first cell organelles are seen to be taken up in lysosomal (autophagic) vacuoles and then general cellular autolysis occurs as the lysosomal enzymes leak out into the general cytoplasm. This type of necrosis is seen, for example, in the regression of the mesonephric tissue of the chick, the destruction of the insect salivary gland at metamorphosis, or the invasion of the uterine wall during implantation. In other cases, the primary destructive activity occurs as cells are taken up by macrophages and digested within heterophagosomes. Examination of the PNZ of the chick limb reveals the existence of both types of necrotic mechanisms. Cell death appears to begin with the appearance of autophagic vacuoles and the start of cell fragmentation. The process is then completed by the digestion of the cell remains within local macrophages.

The development of the limb as a part, or an organ, exhibits all the basic phenomena associated with the development of the egg and the organism as a whole: genomic control, morphogenetic field, sequential inductions, self-assembly, polarities, histodifferentiation, and cytodifferentiation. In one important way it is a simpler system for analysis than developing eggs. Most eggs may be assumed to be preprogrammed to proceed with at least some developmental processes, and in many cases to a very great extent. It is generally difficult to determine to what degree and in what manner an egg might develop without having undergone any preparation at all apart from the attainment of large size as a cell. Limb disks or limb buds clearly are not preprogrammed. Their axial polarities, however, are derived from those of the whole embryo, and they receive some signals designating them as "limb" and "forelimb" or "hindlimb." This last property may be regarded as some form of positional information, or it may be thought of as a component of the elusive, invisible pattern preceding and underlying the visible organization of the whole embryo. That pattern is the quintessential problem of developmental biology. Once initiated, the presumptive limb is a self-organizing system, comparable with the regeneration blastema that forms and develops from amputated salamander limbs (see Chapter 20).

Readings

AMPRINO, R., 1965. Aspects of Limb Morphogenesis in the Chicken, in R. L. DeHaan and H. Ursprung (eds.), "Organogenesis," Holt, Rinehart and Winston.

BERGSMA, D., (ed.), 1977. "Morphogenesis and Malformation of the Limb," Liss.

CHEVALLIER, A., M. KIENY, and A. MAUGER, 1977. Limb-Somite Relationship: Origin of the Limb Musculature, *J. Embryol. Exp. Morphol.*, **41**:245–258.

EDE, D. A., 1976. Cell Interactions in Vertebrate Limb Development, in G. Poste and G. L. Nicolson (eds.), "The Cell Surface in Animal Embryogenesis," Elsevier.

———, 1971. "Control of Form and Pattern in the Vertebrate Limb," 25th Symp. Soc. Exp. Biol., Academic.

———, J. R. HINCHCLIFFE, and M. BALLS (eds.), 1977. "Vertebrate Limb and Somite Morphogenesis," Cambridge University Press.

FABER, J., 1971. Vertebrate Limb Ontogeny and Limb Regeneration: Morphogenetic Parallels, *Advan. Morphog.*, **9**:127–149.

HALL, B. K., 1978. "Developmental and Cellular Skeletal Biology," Academic.

HARRISON, R., 1921. On Relations of Symmetry in Transplanted Limbs, *J. Exp. Zool.*, **32**:1–136.

HURLE, J., and J. R. HINCHCLIFFE, 1978. Cell Death in the Posterior Necrotic Zone (PNZ) of the Chick Wing Bud: A Stereoscan and Ultrastructural Survey of Autolysis and Cell Fragmentation, *J. Embryol. Exp. Morphol.*, **43**:123–136.

KELLY, R. U., and J. G. BLUEMINK, 1974. An Ultrastructural Analysis of Cell and Matrix Differentiation during Early Limb Development in *Xenopus laevis*, *Develop. Biol.*, **37**:1–17.

KOSHER, R. A., M. P. SAVAGE, and S.-C. CHAN, 1979. In Vitro Studies on the Morphogenesis and Differentiation of the Mesoderm Subjacent to the Apical Ectodermal Ridge of the Embryonic Chick Limb-bud, *J. Embryol. Exp. Morphol.*, **50**:75–97.

MACCABE, J. A., and B. W. PARKER, 1976. Evidence for a Gradient of a Morphogenetic Substance in the Developing Limb, *Develop. Biol.*, **54**:297–303.

NEWMAN, S. A., and H. L. FRISCH, 1979. Dynamics of Skeletal Pattern Formation in the Developing Chick Limb, *Science*, **205**:662–668.

RUBIN, L., and J. W. SAUNDERS, 1972. Ectodermal-mesodermal Interactions in the Growth of Limbs in Chick Embryo, *Develop. Biol.*, **28**:94–112.

SAUNDERS, J. W., 1966. Death in Embryonic Systems, *Science*, **154**:604–612.

———, 1972. Developmental Control of 3-Dimensional Polarity in the Avian Limb, *Ann. N.Y. Acad. Sci.*, **193**:29–40.

———, and J. F. FALLON, 1966. Cell Death in Morphogenesis, in M. Locke (ed.), "Major Problems in Developmental Biology," 25th Symp. Soc. Dev. Biol., Academic.

SEARLES, R. L., 1965. An Autoradiographic Study of the Uptake of S-35 Sulfate During the Differentiation of the Limb Bud, *Develop. Biol.*, **11**:155–168.

SINGER, R. H., 1972. Analysis of Limb Morphogenesis in a Model System, *Develop. Biol.*, **28**:113–122.

SLACK, J. M. W., 1977. Determination of Anteroposterior Polarity in the Axolotl Forelimb by an Interaction Between Limb and Flank Rudiments, *J. Embryol. Exp. Morphol.*, **39**:151–168.

STOCUM, D. L., 1975. Outgrowth and Pattern Formation During Limb Ontogeny and Regeneration, *Differentiation*, **3**:167–182.

SUMMERBELL, D., 1979. The Zone of Polarizing Activity: Evidence for a Role in Normal Chick Limb, *J. Embryol. Exp. Morphol.*, **50**:217–233.

SWETT, F. H., 1937. Determination of Limb Axes, *Quart. Rev. Biol.*, **12**:322–339.

TARIN, D., and A. P. STURDEE, 1971. Early Limb Development in *Xenopus laevis*, *J. Embryol. Exp. Morphol.*, **26**:169–179.

TICKLE, C., D. SUMMERBELL, and L. WOLPERT, 1975. Positional Signalling and Specification of Digits in Chick Limb Morphogenesis, *Nature*, **254**:199–203.

WOLPERT, L., J. LEWIS, and D. SUMMERBELL, 1975. Morphogenesis of the Vertebrate Limb, in "Cell Patterning," Ciba Symp. no. 29 (new series), Elsevier.

ZWILLING, E., 1968. Morphogenetic Phases in Development, 27th Symp. Soc. Dev. Biol., Academic.

———, 1961. Limb Morphogenesis, *Advan. Morphog.*, **1**:301–329.

Chapter 14 Sense organs and the nervous system

The three primary sense organs, nasal (*olfactory*), auditory (inner ear or *otic*), and eye (*optic*), develop as paired structures from or in close association with various regions of the developing brain. Their location in relation to the basic divisions of the primitive vertebrate brain is shown in Fig. 14.1. The positions of these three sense organs in the mammalian embryo is shown in Fig. 14.2. In each case we are concerned with the course of development, or developmental event, and with the nature of the agencies involved in the initiation and location of the organ rudiment.

The development of vertebrate sense organs as virtually independently forming complexes is shown in thought-provoking experiments with amphibian embryos by Johannes Holtfreter. The experiments are described under the title "Tissue affinity, a means of embryonic morphogenesis" (in English translation in "Foundations of Experimental Embryology," Willier and Oppenheimer, 1971). Anterior neural plate is induced by underlying prechordal endomesoderm. Following the initial stimulus, pieces of anterior neural plate, together with some adjoining epidermis, but without any subjacent inductor tissue, show a surprising capacity for independent and unorthodox development. Such a piece is shown in Fig. 14.3. The original piece in this case consisted, in part, of the lateral portion of the neural plate and, to a larger extent, adjacent epidermis. Located between them is a portion of the neural crest, the source of mesenchyme, pigment cells, and ganglia and cartilage as well. Shortly after isolation, the epidermis turns upward and envelops the neural material, while the latter contracts and sinks inward to form a groove. A day later, involution of the neural plate is complete. Subsequently, the epidermal vesicle expands, and mesenchymal and pigment cells develop from the neural crest. The inner neural mass of cells then develops into a hollow brain structure complete with a developing optic cup. A lens develops from the epidermis in conjunction with the optic vesicle, while an olfactory or nasal pit develops independently from the epidermis. With the exception of the olfactory pit, all the neural material is separated from the epidermis by mesenchyme. An over-

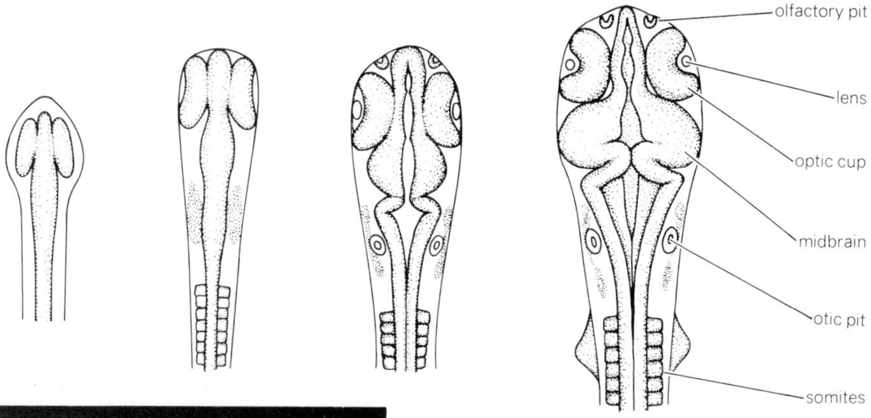

Figure 14.1 Development of brain and sense organs of a fish embryo.

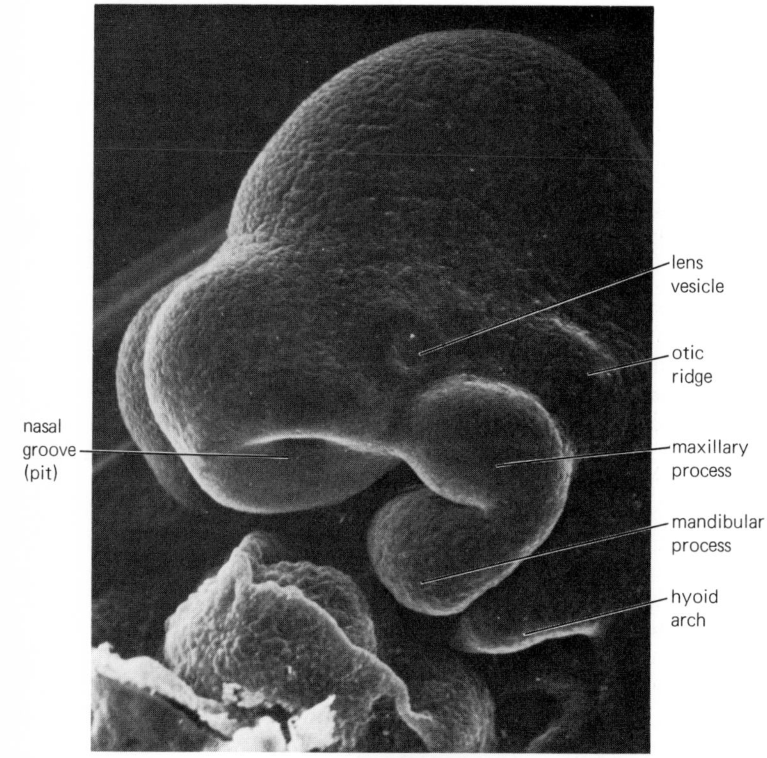

Figure 14.2 Scanning electron micrograph of a 9½-day hamster embryo showing a lens vesicle over the center of the underlying optic vesicle, together with nasal invagination (olfactory or nasal pit) and the mandibular and hyoid arches. [*Courtesy of S. M. Meller.*]

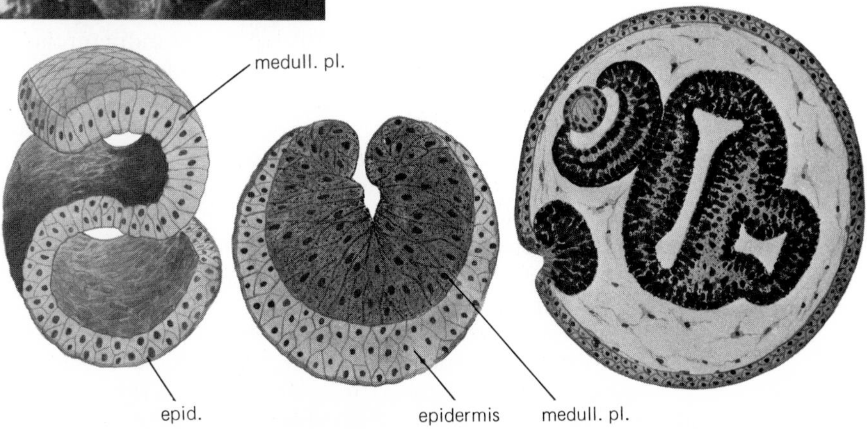

Figure 14.3 Development and differentiation of the neural tube, eye with lens, and nasal organ from a small portion of the combined ectodermal and neuroectodermal salamander embryonic epithelium. [*Courtesy of J. Holtfreter.*]

all patterning of the isolated piece of neurectodermal epithelium clearly takes place, in accordance with available territory. It involves separation of the three types of tissue initially present (neural, neural-crest, and epidermal) and regional responses between adjacent tissues. In a remarkable manner it is a self-organizing system or entity.

1. Nasal organ

Of the three sense organs, nasal organs are by far the most simple in structure and development. The nasal rudiment is first evident as a slight thickening of the ectoderm, the *nasal placode,* on each side of the anterior end of the embryo adjacent to the neural folds of the forebrain. The placode invaginates to form a *nasal sac.* The lining of the sac differentiates as olfactory neurosensory cells located among supportive epithelial cells, with axons growing inward to the forebrain. Differentiation of the nasal rudiment begins at an early stage. Transplantation experiments indicate that it does not wait until formation of the neural plate and neural folds; in amphibians it may even start during gastrulation. During and after neurulation, the rudiment acquires greater capacity for self-differentiation. Extirpation of the rudiment is followed by replacement from adjoining ectoderm, as in limb rudiments; this capacity is subsequently lost, in keeping with a general decrease in ectodermal competence.

2. Auditory organ

In higher vertebrates, the sensory machinery responsible for sound reception as well as for maintenance of equilibrium resides in a pocket within the bones of the skull. This portion of the ear, namely, the *inner ear,* derives initially from a thickening of the epidermal ectoderm, the *auditory placode* (Fig. 14.4), which lies next to the myelencephalic part of the neural tube. As in the development of the

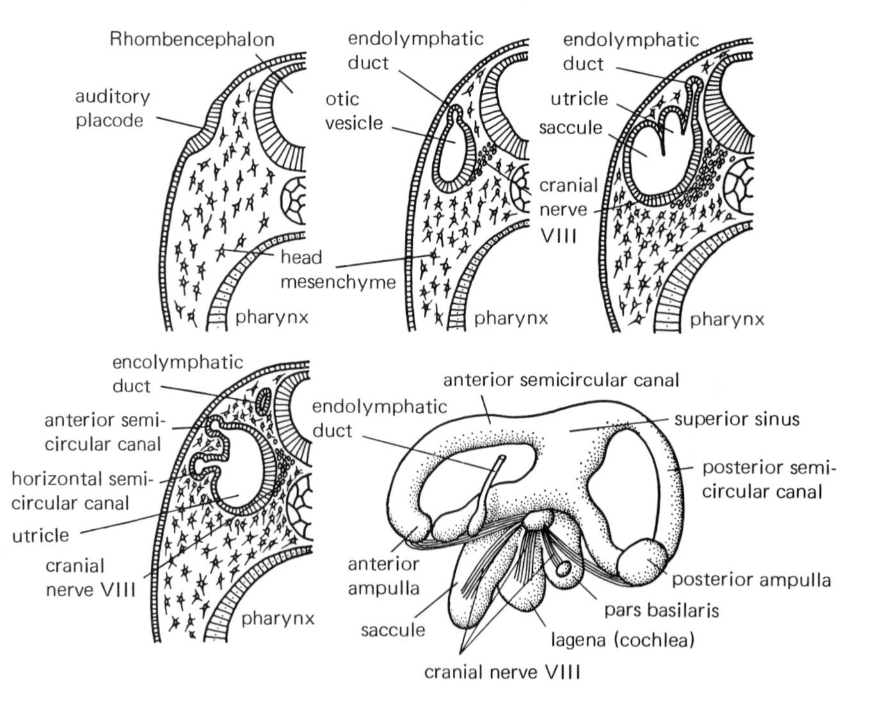

Figure 14.4 Development of the auditory apparatus of a frog from the auditory (otic) placode and vesicle. [*After Krause, from R. Rugh, "The Frog," McGraw-Hill, 1953.*]

nasal rudiment, the placode invaginates to form the *auditory pit,* which then pinches off to form the *auditory vesicle* (*otocyst*). It is from the epithelium of this vesicle that the various parts of the inner ear develop. Taken as a group, the fluid-filled chambers and canals of the inner ear comprise the *membranous labyrinth,* which consists of three *semicircular canals,* the *utriculus,* the *sacculus,* and the *cochlea.* In the chick, the auditory placode is evident at the seven-somite stage.

Soon after formation of the auditory vesicle, a section of its epithelium pushes out to form a channel, the *endolymphatic duct,* which ends in a blind sac. Later, extensions from the dorsal portion of the vesicle appear (Fig. 14.4) and develop into three looplike semicircular canals. The intermediate portion of the vesicle becomes subdivided into a more dorsal utriculus and a more ventral sacculus. Meanwhile, the epithelium of the ventral region of the otocyst develops into the auditory organ of the inner ear, the cochlea. Sensory information from the ear passes to the brain over the neurons of the eighth cranial (auditory) nerve. These sensory nerve fibers originate from the *acoustic ganglion* located close to the hindbrain. One branch of the auditory nerve grows into the cochlear duct to innervate the sensory cells involved in sound reception. The other branches grow into the vestibular organs and function in providing the brain with information on body position. With this brief description behind us, we now can reflect on some of the experimental findings related to the development of the inner ear.

In amphibians, both the auditory and nasal placodes are underlaid by an inductive layer of chordamesoderm during late gastrula stages. Determination of the rudiment at these early stages of development is not fully completed, however, since it can be reversed by being explanted to other regions. Shortly before the ear placode has formed, the prospective ear ectoderm has the characteristics of an equipotential system; a whole ear can form from two fused rudiments. By the time the auditory placode invaginates, the capacity of the parts to become a whole has been lost. The auditory vesicle has the capacity to draw mesenchymal cells toward it, as shown by transplanting vesicles to abnormal sites in the embryo. This mesenchyme, whether from a normal or abnormal source, differentiates as the cartilaginous capsule and closely envelops the epithelial compartments and tubes formed from the sensory vesicle.

The developing ear vesicle is a very strikingly asymmetrical structure. Extensive experimental analysis of the polarity determination was conducted by Ross Harrison over a generation ago. He found that before and during the determined, or mosaic, phase of development in the salamander, the rudiment becomes irreversibly polarized along both its anteroposterior and dorsoventral axes. The experiments consisted of transplanting rudiments into the ear region of donors of equivalent developmental stage in one or another of the four possible orientations relative to the axes of the host embryo. Further experiments showed that expression of polarity depends in part on the strength of the polarization within the rudiment and in part on the intensity of the polarizing factors in the host environment.

Both the anteroposterior and dorsoventral axes, which are evidently the initial guidelines for the establishment of the basic pattern of ear-vesicle development, are apparently derivative. In other words, the embryo as a whole acquires a polarization of tissue along both its anteroposterior and dorsoventral axes, which is secondarily imposed on the ear rudiment. This is shown in amphibian embryos by the direction of beat of the cilia covering the external surface; it becomes fixed during gastrulation and is very evident in the neurula, preceding the time of polarity fixation in the ear vesicle. In both ear vesicle and limb disk, however, the anteroposterior axis becomes determined earlier than the dorsoventral axis. In

both cases, reversal of the rudiment at a critical stage may result in a structure with reversed anterior and posterior features, but with normally oriented dorsal and ventral features.

The middle ear, which is responsible for transmitting sound vibrations from the outside world to the cochlea, develops by a totally independent course of events from those occurring in the inner ear. While the inner ear originates as an ectodermal thickening, the middle ear forms from an outpocketing of pharyngeal endoderm, specifically the first branchial pouch (Section 12.2.B). The blind end of the pouch, which abuts the inner wall of the ectodermal inpocketing, develops into the *tympanic cavity*, the chamber of the middle ear. This cavity is bounded laterally by the *tympanic membrane* (ear drum), which forms by the joint participation of the outer endodermal and inner epidermal layers. The passageway between the tympanic cavity and the lumen of the pharynx becomes narrowed but retains its continuity with both spaces. This channel becomes the *eustachian tube,* a passageway that continues in the adult as the connective between the mouth cavity and the middle ear. The walls of the first branchial pouch are formed by the first and second branchial arches, mesenchyme from which invades the tympanic cavity and forms the three *bony ossicles* (*incus*, *malleus*, and *stapes*) of the middle ear of amniotes. The epidermal groove between the first two arches becomes the *external auditory meatus,* the channel of the outer ear, while the cup-shaped *pinna* of the outer ear develops from cartilage of the arches themselves.

3. The eye

Of all the organs of the vertebrate body—apart from the brain—the eye is the most complex, particularly in terms of cell diversity, component tissues, and parts that are unified to form an optical instrument of amazing proficiency and efficiency. The course of development has been fairly well worked out, although it is replete with problems at every stage. The evolution of the vertebrate eye, however, and therefore the evolution of the development of the eye have hardly been considered.

Unlike the nasal and auditory organs, which originate in neurectoderm external to the neural tube, the *sensory layer*, or *retina*, of the eye develops directly as outpushings from the lateral walls of the cephalic neural tube (Fig. 14.5). These protrusions are the *optic vesicles;* they are in fact specially differentiated parts of the brain. As they constrict from the remainder of the forebrain wall, they retain a connecting *optic stalk,* which later serves for the development of the optic nerve. In the subsequent development of the eye, three very different components cooperate to become the whole organ. These are the optic vesicle, which gives rise to the layered retina; the *lens placode* and cornea, situated in the ectoderm overlying the optic vesicle; and mesenchyme adjacent to the vesicle. In brief, a large number of tissues assemble during development of the eye in such a way that their size, shape, orientation, and relative positions meet the precise geometric tolerances required by the optical function of this organ. The cells that make up these tissues are contributed by the ectoderm and the mesoderm. They become highly specialized to fulfill the diverse functional requirements of the eye. For example, both the cellular and extracellular portions of the dioptric media (cornea, lens, vitreous body) become highly transparent. The intrinsic musculature of the eye (protractor lentis, retractor lentis, or ciliary musculature) is appropriately oriented and attached to alter the position or shape of the lens and to focus the visual image in the plane of the retina. The outer tips of the visual cells are specialized to transform the light energy of these images into coded

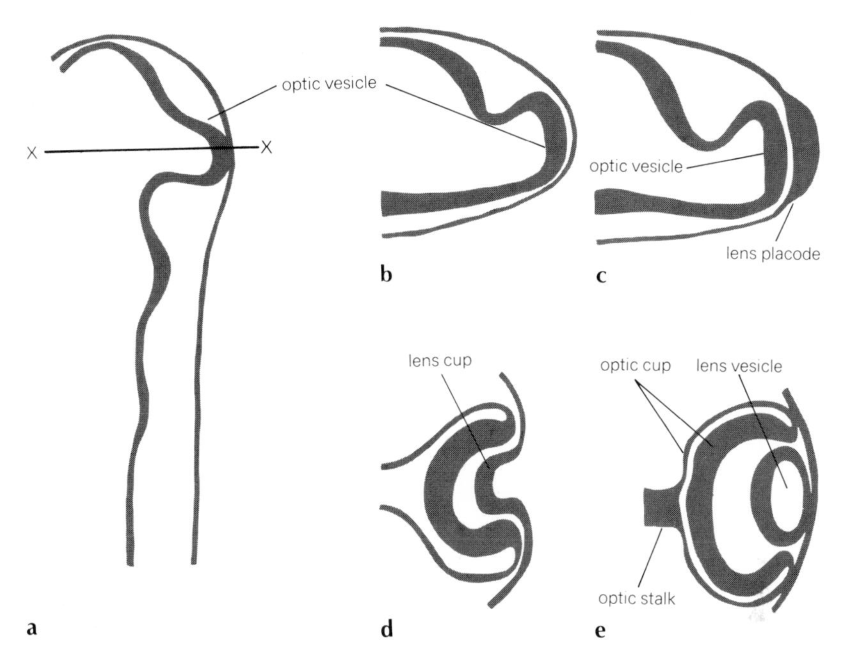

Figure 14.5 Development of the optic cup, lens, and cornea. *a* Dorsal aspect of the embryo at the stage at which the optic vesicle makes contact with the overlying ectoderm. *b* Transverse section through the level X–X at the same stage as that represented in *a*. *c* Transverse section at the same level following induction of the lens placode by the tip of the optic vesicle. *d* Invagination of the optic cup and optic vesicle. *e* Separation of the lens vesicle from the surface and reunion of the surface ectoderm to form the presumptive anterior corneal epithelium. [*After A. J. Coulombre, in "Organogenesis," R. L. DeHaan and H. Ursprung (eds.), Holt, Rinehart, and Winston, 1965.*]

trains of nerve impulses. The neural retina, which is a peripherally developed portion of the central nervous system, develops a cytoarchitectural organization that enables it to reorganize the output of the visual cells into a form suitable for transmission to the brain. It is important that the eye maintain constant shape during visual function. The scleral and corneal cells construct an outer eye wall that maintains its shape in the face of intraocular pressure and the pull of the extrinsic muscles.

The morphogenesis of the whole can be subdivided into several phases. In the first phase, inductive effects separate the specific material of a particular rudiment, such as the retinal rudiment within the neural plate and the lens and cornea within the surface ectoderm. In the second phase, cell differentiation occurs within each rudiment, e.g., the differentiation of optic-vesicle cells into pigment and retinal epithelia and the differentiation of the lens. The third phase, of a supracellular character, is concerned with the formation of the ciliary body and corneal curvature, the differentiation and growth of sclera and cornea, and the general growth of the eye.

The outpocketing of the optic vesicle usually begins before the time of closure of the neural folds. To what extent differential mitosis is involved and to what extent apical microfilaments of the cells are involved is not known. Once the anterior neuropore has closed, internal accumulation of fluid plays a part in expand-

ing the optic vesicle. If optic vesicles fail to reach the adjacent ectoderm, no lens is induced. If only the tip of the optic vesicle contacts the surface ectoderm, a perfect eye develops, but it is comparatively small. Coinciding with the period of lens induction, the optical vesicle invaginates to form the optic cup, the outer layer of which becomes the pigmented epithelium and the inner wall, the neural retina.

A. Determination of the eye rudiment

The first question concerns the events that precede the evaginative process: What determines the establishment of the pair of prospective retinal outgrowths, and when does this occur? Vital-staining experiments have shown that the material destined to become the optic vesicles and their optic stalks and chiasma region lies well forward in the neural plate. The same procedure has also shown that the area from which the future lens develops lies outside the presumptive neural plate, lateral to and in front of the presumptive-eye rudiment (Fig. 14.6*a*). Primarily, the eye develops as the result of these two areas being brought together and interacting. Identification of these areas, however, is merely a mapping process that says nothing about the state of the tissues, i.e., whether or not they are still indifferent, whether in some way they have yet to be determined.

One method that shows when retinal determination becomes irreversibly established is to constrict a developing egg in the sagittal plane—the plane of the embryonic axis. When this is done, an embryo develops as a double-headed monster, with two brains, two pairs of eyes, etc. (Fig. 14.7). This holds true in the newt up to the commencement of gastrulation; until this stage, determination of eye and lens is labile or absent, since two additional eye rudiments can still form. The determinative process may accordingly be assigned to the gastrulation or early neurulation periods.

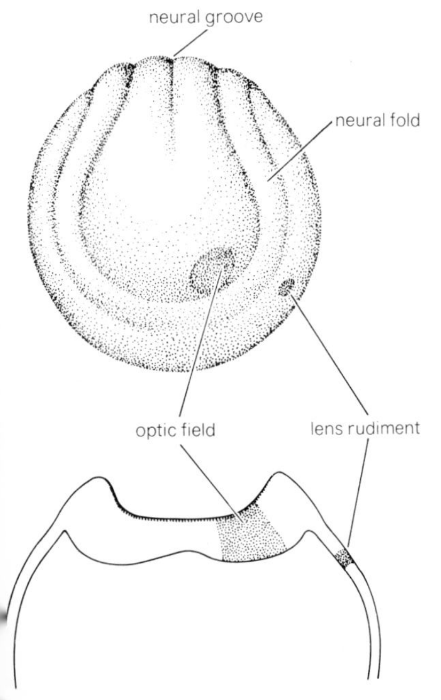

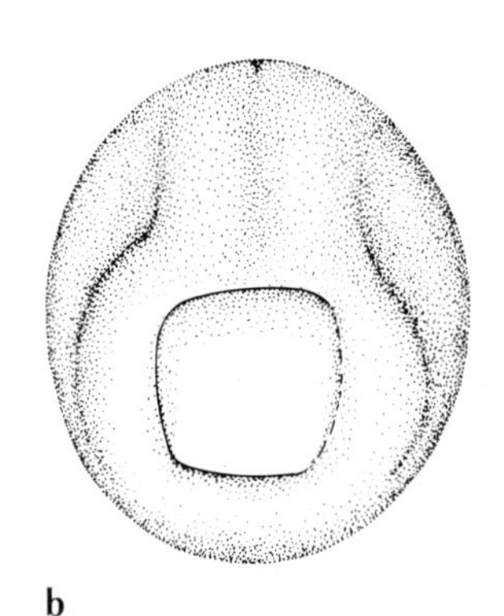

Figure 14.6 *a* The position of the presumptive optic cup and presumptive lens in the neurula stage.
b Neurula with a rectangular piece of the neural plate excised and inverted. (See text for description of results.)

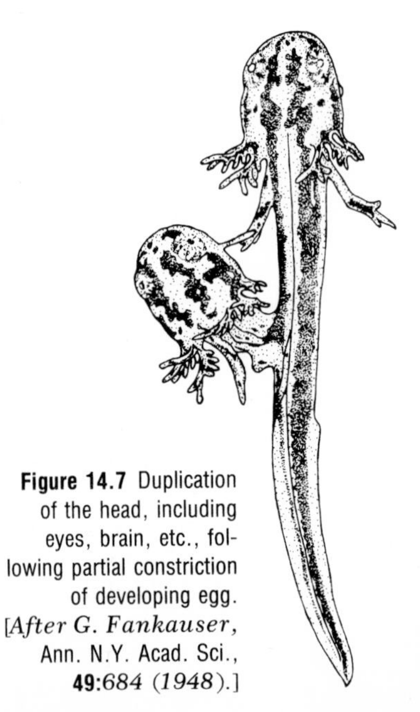

Figure 14.7 Duplication of the head, including eyes, brain, etc., following partial constriction of developing egg. [*After G. Fankauser*, Ann. N.Y. Acad. Sci., **49**:*684* (*1948*).]

A number of experiments have been made in this connection, with regard to the eye rudiment proper. One is to explant the presumptive-eye rudiment of the open-neural-plate stage into the flank of another embryo to test for its capacity for self-differentiation. When this is done, an optic cup develops in the new position.

Other experiments made with the early neurula stage consisted of excising a rectangular piece of neural plate and subjacent mesoderm and replacing it after rotating through 180°. In his classical experiment, Hans Spemann cut transversely through the prospective-eye rudiment, thereby leaving a portion in place and displacing the other portion to a more posterior location on the opposite side (Fig. 14.6*b*). Optic cups developed from each portion—the size depending on how the prospective area was divided—so that diagonal pairs resulting from a divided rudiment were complementary in size. Accordingly, the optic-cup rudiment has already been determined in the early neurula, to the extent of its general destiny to form optic cup; yet it is still labile with regard to any regional determination within it, since any part when isolated attempts to form a whole. This labile state continues for some time.

In certain circumstances one median eye (*cyclopia*) forms instead of a pair of laterally placed eyes. The presumptive-eye area, as already stated, is located in a median position in the extreme anterior part of the neural plate. In normal development, the area divides into right and left components, the definitive optic-vesicle areas. Cyclopia appears, however, if developing eggs are chemically treated (with lithium chloride, alcohol, chloretone, etc.) or morphologically altered so that the two lateral parts of the prospective-retinal area fail to become separated by an intruding wedge of median neural plate and mesoderm. Thus excision of the mesodermal substratum of the anterior region of the neural plate in the early neurula stage allows the prospective-eye rudiments to remain as one, and a single large median eye develops (Fig. 14.8).

Figure 14.8 Cyclopia in the newt, produced by mechanical defect. Excision of the mesodermal substratum of the anterior part of the medullary plate in the early neurula stage has entailed median fusion of the eyes into a single ventral eye. *a* Neurula prepared for operation by cutting around the anterior border of the medullary plate. *b* Anterior part of medullary plate is lifted. The mesodermal substratum thus exposed is then excised and the medullary flap is put back in position. *c* Cross section through the head showing the ventral eye with a single median lens. *d* Fish larva with cyclopean eye induced by magnesium (or lithium) chloride treatment of early embryo.

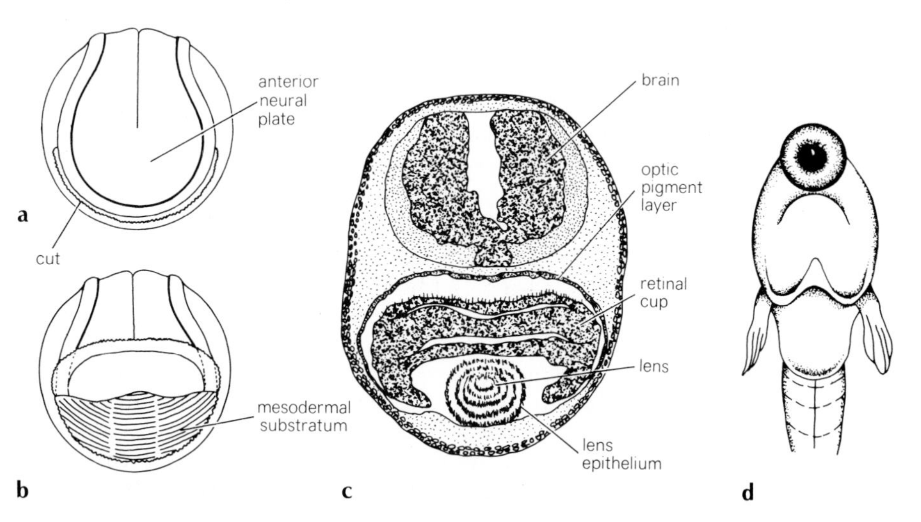

Intimate association with mesenchyme is a general condition, however, for the continuing development of the optic vesicle. This is shown when vesicles evaginating from the neural tube are isolated in vitro without access to the matrix of mesenchyme that normally surrounds them. They attain only a very rudimentary level of organization.

Under normal circumstances, the optic vesicle continues to grow; its distal region, after making contact with the lateral head epidermis, invaginates. By this means a single-layered vesicle transforms into the two-layered optic cup, the inner layer becoming the retina and the outer layer becoming the pigmented layer of the eye.

B. Lens induction

The investigation of lens induction has had a confusing history. Spemann and later investigators reported different results from experiments, depending on which species of frog was used. In *Rana sylvaticus* and *Rana palustris*, transplanted optic vesicles induce a lens from the overlying ectoderm of any part of the body. In *Rana esculentes*, only the normal prospective-lens ectoderm responds to the presence of the optic vesicle, although the *R. esculentes* vesicle evokes lens development anywhere in *R. sylvaticus* and *R. palustris;* this indicates that the vesicles are generally inductive, but that ectodermal competence is variable. In *R. catesbiani*, the lens is completely dependent on the optic vesicle, even though the ectoderm is generally responsive. In *R. esculentes* and *R. fusca*, the lens develops in the normal position even in the absence of the optic vesicle. Results such as these show that:

1. All optic vesicles are lens-inductive, even to ectoderm of other species, genus, or order.
2. A lens can develop independently of induction by optic vesicle (Fig. 14.9).
3. Ectodermal competence to form lens may be local or general.

In the salamander, studies suggest that the induction of the lens is a stepwise process involving the interaction of presumptive-lens epidermis with three successive inductors. During the early gastrula stage, presumptive-lens ectoderm is underlain by pharyngeal endoderm, which supplies the lens-inductive stimulus. As gastrulation proceeds, the mesodermal mantle sweeps into the embryo (see Fig. 10.13) such that its anterior edge, the heart-forming primordium (Section 12.2.C), comes to lie near the future lens ectoderm (see Fig. 10.14*b*). This part of the mesoderm also possesses lens-inductive activity. Later, the optic vesicles, i.e., the prospective retina, come into close proximity to the lens ectoderm and provide the final inductive stimulus that converts the overlying epidermis into the thickened lens placode. Placode formation, which occurs at the late tail-bud stage, represents the first visible sign of successful lens induction. The sequential influence of these three lens inductors is shown graphically in Fig. 14.10.

In some species, the endodermal and mesodermal influences are alone sufficient in strength and duration to initiate ectodermal lens differentiation. In others, they at least prepare the ectoderm by lowering the threshold to optic-vesicle induction. In all, the optic vesicle plays an important role—whether or not it is the principal inductor—in determining the final phase and in aligning the lens precisely with the rest of the eye. The nature of the inductive influences remains unknown; embryonic optic vesicle and guinea pig thymus, for instance, also can induce and sustain lens development in competent ectoderm. The terminal induction requires proximity between vesicle and ectoderm, although not full cell contact. In fact, an acellular layer develops between the two epithelia. It is sev-

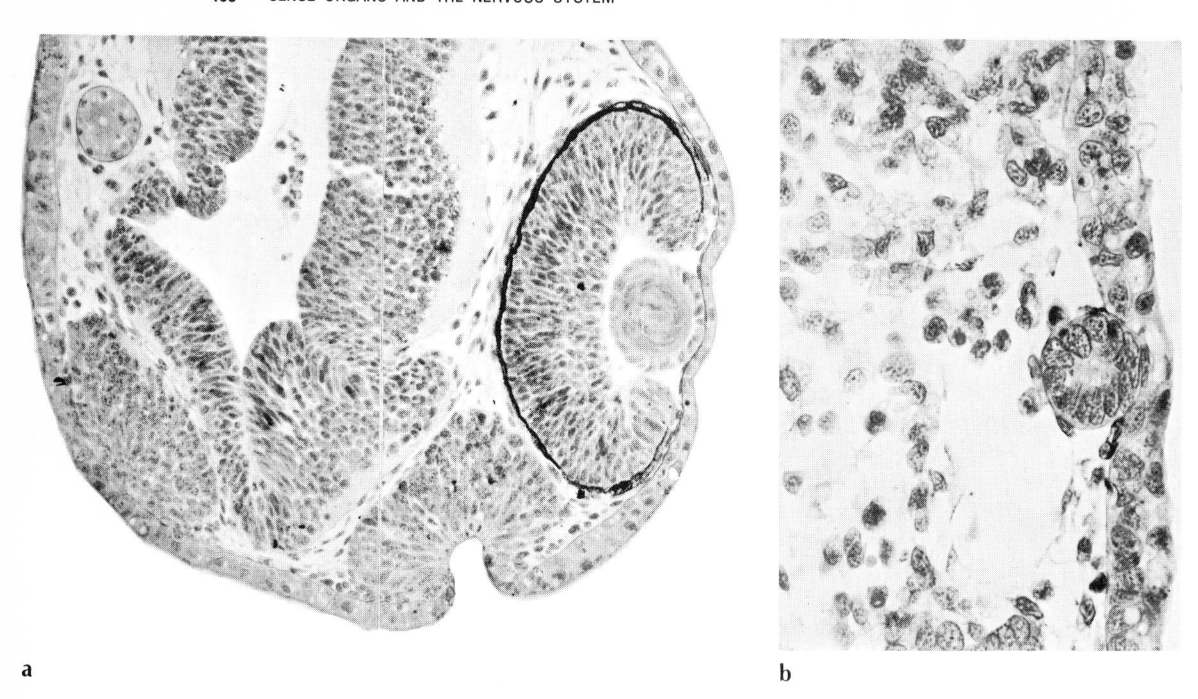

Figure 14.9 *a* Transverse section of a 12-mm salamander larva in which a lens has developed in the absence of the retina. The embryo was reared at 13°C until the early neurula stage, when the right retinal anlage was excised. *b* Section of an explant of presumptive-lens epidermis combined with the subjacent endodermal wall of the archenteron and the anterior portion of the lateral-plate mesoderm. A lens has been induced from the epidermis. [*Courtesy of A. G. Jacobson.*]

eral microns thick and results from fusion of the basement membranes of the two layers. Nevertheless, the intervening distance is very small, and the barrier has been shown experimentally to pose no difficulty to the diffusion of molecular substances.

Even when lens ectoderm has been committed to lens differentiation, the commitment is by no means an all-or-none affair. Removal of the lens at various stages of development into a neutral environment, such as the body cavity, shows that it only gradually becomes independent of the optic cup. This influence persists into the adult in some salamanders. In any case, as the lens vesicle pinches off from the ectoderm, the basement membrane of the placode envelops the vesicle and gives rise to the lens capsule. The cells constituting the back of the lens vesicle continue to elongate and, under the influence of the neural retina, produce the lens fibers, as described in the following section. As the fibers grow, they obliterate the lens cavity, while cells of the front side of the vesicle form the lens epithelium.

C. Lens differentiation

The lens of the vertebrate eye is an aggregate of fibers, each formed during terminal differentiation of an epithelial cell. At the end of this differentiation process, the lens-fiber cell lacks a nucleus and becomes filled with a distinctive collection of proteins. Lens function depends on the refractive index of its components, so that light impinging on the eye can be bent to focus on the retina. The proteins of

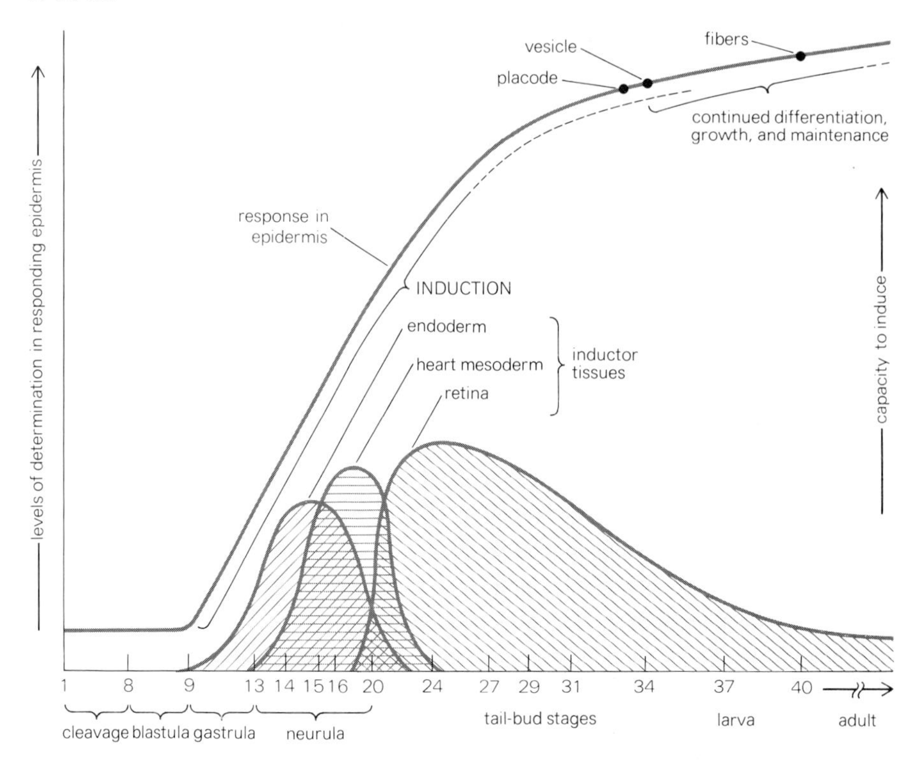

Figure 14.10 Lens induction in a salamander. The abscissa, representing time, is marked off in arbitrary stage numbers and names. The relative amount of time between stages is illustrated; actual time varies greatly with the temperature. At 17°C, the time between stages 9 and 40 is 3 weeks in the West Coast newt. The ordinate for the response curve is logarithmic, the level of response being a function of the sum of all past inductions. The ordinate representing the capacity of inductor tissue to induce is linear. [*After A. G. Jacobson,* Science, **152**:*26 (1966); copyright © by the American Association for the Advancement of Science.*]

the lens that give it this property are called *crystallins*, of which there are several types. Crystallins are classified as alpha, beta, delta, or gamma, depending on their electrophoretic mobility. They are best studied with immunologic techniques.

The lens is first induced as the lens placode (see Figs. 12.1 and 14.5*c*), a disk-shaped thickening of the ectoderm in contact with the distal surface of the optic vesicle which invaginates to form the lens vesicle. The orientation of the lens with regard to the optic axis is continuously controlled by a mechanism operating throughout development. Experimental alteration of normal lens orientation (e.g., inversion) is followed by remarkable reorganization changes that return the lens to a normal orientation. This property is controlled by an agent diffusing from the neural retina but not from the cornea or optic sclera.

Morphologically, the mature lens lacks blood vessels and is composed of an outer, single layer of epithelial cells; a zone of cellular elongation, or equatorial region, composed of cells in the process of developing into fiber cells; and the inner fiber cells. After the embryonic lens has been formed, fiber cells are continuously laid down throughout life in the zone of cellular elongation, where cuboidal epithelial cells continually transform to the elongated fiber cell characterized

by gamma-crystallin protein. Finally, since the fiber cell loses its replicative activity, it essentially enters a permanent stationary phase and can only proceed to death. In contrast with the fiber cells, the epithelial cells of the central region retain their ability to replicate and are in a reversible stationary phase (Fig. 14.11).

The time for the cell cycle of the lens epithelial cells in newborn mice has been calculated to be 56 hours, the G_1 phase taking up three-quarters of this time. In the 3-day-old mouse, most of the cells undergo DNA synthesis, but in the 12-day-old mouse, only the cells of the germinative epithelium do so. The differentiation phase, however, is very long. The time taken for cells moving from the germinative region to differentiate into an elongated fiber cell is 6 months. In the chick, the length of time required for cells of the lens ring, or annular pad, to develop into a fully differentiated fiber cell is 2 years.

The situation in the lens is typical of that in many other tissues in the body. Each kind of tissue represents a cell population of a certain sort. Each kind must in some way be maintained throughout the life of the organism. In fact, aging and death of the organism as a whole may be essentially the progressive, inevitable, collective failing of the tissue-maintenance mechanisms. In the lens, new fiber cells continually form the peripheral equatorial zone of elongating epithelial cells; these in turn derive from the adjoining germinative region of the lens epithelium. Thus fiber cells are systematically laid down, layer upon layer, throughout life. Eventually those fiber cells forming the central or nucleus fiber cells of the lens lose their cell nuclei and remain permanently entombed in the center of the lens. Inevitably the lens ages and dies at the center, although it is renewed at the edge. Its life course and destiny, therefore, depend on the relative rates of progression toward cell death and of peripheral cell birth. This situation is complicated by the fact that the aging or dying cells in the center, or their equally aging protein products, cannot be sloughed off to make way for new cells or material. As the protein ages, it loses its elasticity and becomes steadily more rigid so that accommodation decreases. Finally the protein may lose its transparent quality and the lens becomes milky white as cataract develops.

D. Lens regeneration

A lens is readily regenerated in some vertebrates, notably in salamanders, though rarely in frogs and toads. This is in keeping with the exceptional capacity of urodele amphibians, especially species of the newt *Trituris*, to regenerate whatever part may have been lost—even the front part of the head, including the jaws. The great interest in the regeneration of the lens comes from the discovery made in the late nineteenth century that this part of the eye can regenerate from

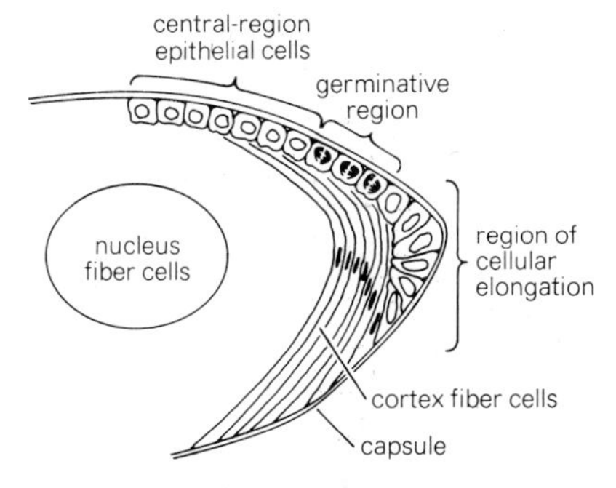

Figure 14.11 The lens of the adult vertebrate. The lens is surrounded by an external noncellular capsule. Beneath the capsule are the lens epithelial cells. In the peripheral area is the transitional region of cellular elongation, where the epithelial cells begin to elongate into fiber cells. The fiber cells that are newly laid down constitute the cortex region; those laid down during the early growth period of the lens compose the nucleus region of the adult lens. The internal, posterior side is below. [*After J. Papaconstantinou,* Science, **156**:*338 (1967); copyright © American Association for the Advancement of Science.*]

the dorsal rim of the iris (Fig. 14.12), a source remarkably different from its epidermal origin in the embryo.

The capacity to regenerate a lens in this way is generally absent during embryonic development; it first appears in the young larval stages when tissues have already attained a high degree of functional differentiation. In the newt, removal of the lens from the larval or adult stages is followed by regeneration of a lens from the iris epithelium.

As a rule, a regenerative process follows injury to the tissue. Here, lens regeneration occurs without injury to the iris, since it still occurs if the lens is removed through the roof of the mouth instead of through the cornea. Moreover, if a lens is removed and replaced, no regeneration occurs. Neither does it occur if the lens of an adult is replaced by a much smaller lens from a younger individual. Instead, the small lens grows rapidly to the size appropriate to the host. Evidently a lens has an influence, presumably chemical, that inhibits adjacent tissue from regenerating a lens as long as it is present.

The lens exerts its own influence on the development of the eye. It is principally responsible for the induction and maintenance of the cornea, although this capacity is gradually lost as development proceeds, while anterior corneal epithelium becomes progressively more independent of its inductor.

The lens also controls the accumulation of the material forming the expanding semifluid vitreous body behind the lens; this in turn generates mechanical forces important in the construction of the skeletal wall of the eye, the sclera. Scleral cartilage develops in the first place from mesenchyme that condenses around the expanding optic cup. Although the inductor of scleral cartilage has not been identified, circumstantial evidence suggests that the neural retina or

Figure 14.12 Histologic location of the proliferating zone at the representative stages of Wolffian lens regeneration. Each figure shows a section through the middorsal pupillary margin of the lens regenerate, oriented perpendicular to the main body axis. The cornea (external "anterior") side is above, the retina (internal "posterior") side below, the dorsal side toward the left, and the ventral side toward the right. *a* Regeneration stage II. *b* Stage V. *c* Stage VIII. *d* Stage XI. White circles denote cells in the proliferating zone; black circles, cells which are in proliferation. White cells are depigmented; black cells are pigmented. Lines indicate cell boundaries. [*After J. Papaconstantinou*, Science, **156**:*338 (1967); copyright © American Association for the Advancement of Science.*]

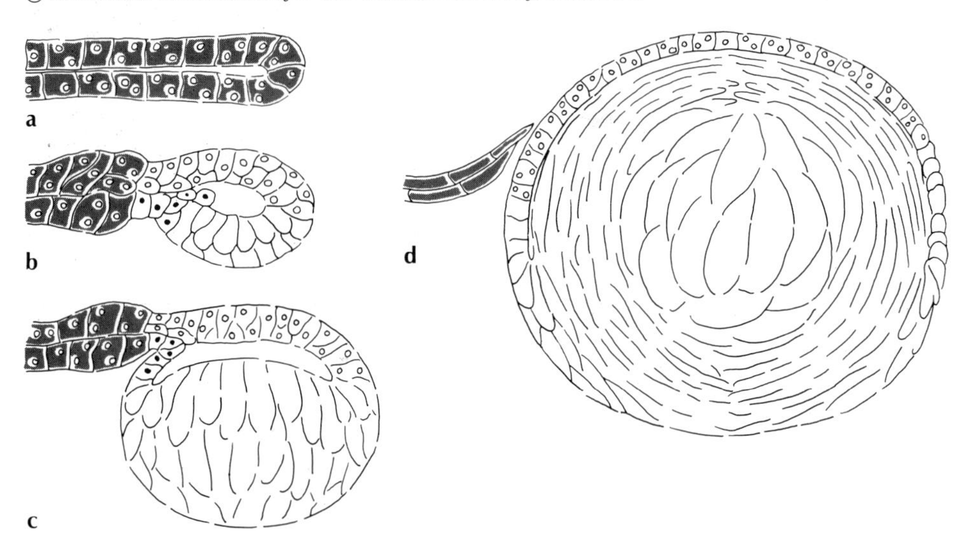

the pigmented epithelium, or both, may be the inductors, for the neural tube is known to induce vertebral cartilage, while scleral cartilage corresponds precisely in area with the underlying pigmented epithelium of the retina. It is of general interest that the cells of precartilaginous scleral mesenchyme, once their fate has been determined, form flat plates of cartilage in tissue culture after being disaggregated and reaggregated. Similary treated precartilaginous mesenchyme from chick-embryo limbs form rods of cartilage. Such properties assume great significance in connection with processes of regeneration.

E. Cornea

Competence to form corneal epithelium is present in most of the early embryonic ectoderm, although the cornea normally arises only over the embryonic eye and in a size, shape, and orientation appropriate to the eye. It is the lens vesicle that is mainly responsible for the induction and maintenance of the cornea. Recent evidence strongly suggests that it is the extracellular lens capsule (which is composed of collagen and glycosaminoglycan), rather than lens cells themselves that is responsible for inducing corneal differentiation. This can be demonstrated in several ways. Corneal epithelium can be cultured on the surface of a lens capsule in the absence of underlying lens cells. Corneal epithelium cultured in this manner responds by undergoing differentiation and the elaboration of its own extracellular product, the primary stroma (described later). A similar response is obtained when corneal epithelium is cultured on a pure collagen-containing gel; the organized structure of the lens capsule does not seem to be required for the inductive interaction. Further experiments indicate that corneal differentiation can be induced when extracellular material of the lens capsule is placed on one side of a membrane filter and corneal epithelium on the other. It can be shown in this type of experiment that cellular projections from the presumptive corneal cells will penetrate through the filter, making contact with the extracellular material on the other side. Since corneal epithelium does not differentiate when cultured alone, we can conclude that intermittent contact between these fine cellular processes and the extracellular components is sufficient to trigger the corneal response.

The cornea of the eye provides a well-studied example of a structure consisting primarily of extracellular materials. The bulk of the cornea consists of a stromal layer containing large numbers of collagen fibers embedded in a glycosaminoglycan-containing matrix with flattened fibroblasts scattered throughout the layer. The formation of the stroma has been best studied in the chick. The formation of chick-embryo stroma begins on the third day of incubation by the secretion of collagen and chondroitin sulfate, a glycosaminoglycan, by the corneal epithelium on its inner surface. These materials interact to produce the initial or primary stroma of the cornea. Most important, the collagen fibers become deposited in a lamellar organization, with the layers of fibers deposited one at a time beneath the epithelium. As each layer is deposited, its fibers become oriented in a direction perpendicular to the fibers of the previous layer, thereby forming an orthogonal ply (Fig. 14.13) with all layers roughly parallel to the corneal surface. During the sixth day of incubation, a major change occurs within the stromal layer as it swells to a much greater thickness. This swelling occurs simultaneously with the secretion of hyaluronic acid (a nonsulfated glycosaminoglycan) into the stroma, and the uptake of water is believed to result from its presence.

The sudden swelling of the primary corneal stroma sets the stage for the next step of the process, the invasion by wandering fibroblasts. Once these cells have migrated over the scaffolding provided by the primary stroma, they begin to syn-

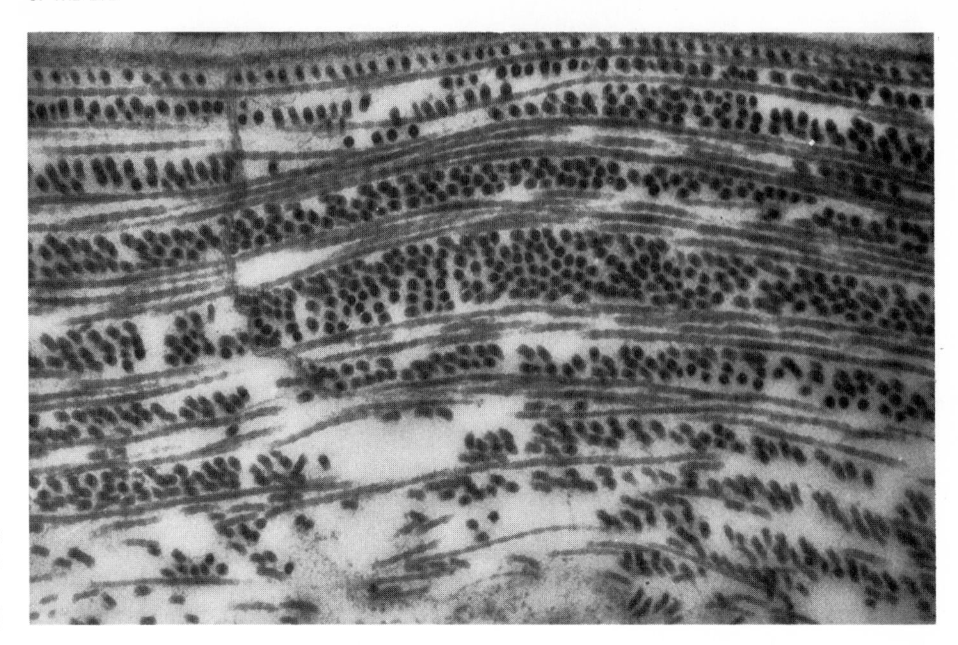

Figure 14.13 Ply layers of collagen fibrils characteristic of corneal stroma. Alternate layers appear at right angles to each other. [*Courtesy of M. Takus.*]

thesize and secrete their own collagen and chondroitin sulfate, which becomes organized into the secondary, or adult, stroma. Remarkably, the bundles of collagen fibers in the layers of the secondary stroma are organized identically to the orthogonal arrangement of the primary layer; i.e., the primary stromal architecture serves as a geometric miniature of the final layer. Either the orientation of the fibers of the primary stroma is directly involved in some template process in orienting the new fibers (which will ultimately constitute over 99 percent of the collagen fibers), or it exercises this role indirectly by orienting the cells in the proper directions before they secrete the materials. Meanwhile, during the period in which the secondary stroma is being laid down by fibroblastic cells, the corneal epithelium continues to synthesize collagen, which is deposited directly beneath the epithelium to form additional primary stromal lamellae and later thickens by secondary deposition. Beginning on day 10, an enzyme is secreted into the stroma and the hyaluronic acid produced prior to swelling is destroyed. In the wake of the removal of hyaluronate, the stromal layer shrinks by loss of water to its final characteristic composition.

These few components are assembled in time and space in such a manner that the cornea develops a large number of strikingly different functional characteristics: avascularity, tensile strength, deturgescence, a tissue-specific population of ions, regenerative capacity, a characteristic interference pattern in polarized light, an appropriate refractive index, a precisely controlled curvature that contributes to its refractory power, and transparency. Simultaneous development of such diverse properties sets strict limits on the manner and sequence in which the corneal components can be compatibly assembled during development.

F. Whole-eye regeneration

The capacity of the urodele eye to regenerate completely following removal of all except a very small remnant has long been known. As a rule, cornea, sclera, and retina are renewed from corresponding components of the eye fragment as a pro-

cess of coordinated self-assembly. The whole event is a spectacular performance, but the particular interest is in the regeneration of the sensory retina following its surgical removal. This layer readily separates from the adjoining pigmented epithelium and can be forced or flooded out of the vitreous chamber through a slit in the eyeball. A new retinal layer is then replaced by the pigmented epithelium, a phenomenon thoroughly investigated by L. Stone and his students. The pigmented epithelium not only can regenerate the neural retina early in embryonic life in most kinds of vertebrates, as long as a small piece of neural retina is present, but even when the pigmented epithelium is from one class (chick) and the retinal fragment from another (mouse).

G. Development of retinal architecture

The neural retina is one of the most complex tissues in the vertebrate body. This thin sheet of tissue is composed of several different cellular layers. They include various types of sensory cells, the cell bodies (containing the nuclei) of the neurons of the optic nerve, interconnectives between these two types, and associated structural cells. The visual image in the optic region of the brain results from the patterns of light that fall on the sensory cells of the retina, which is dependent on the development of an impressive cytoarchitecture of this layer.

During early phases of development, cell division is confined mainly to the margin of the optic cup. In embryos, autoradiographic analysis of retinas labeled with tritiated thymidine shows that immature retinal cells migrate to the outer surface of the layer as they divide; the daughter cells then migrate back to their appropriate locations within the differentiating retinal layers. Increase in thickness and complexity is accompanied by waves of cell death, which are probably related in some way to the orderly establishment of the precise neurosensory network. The innermost layers of the retina (i.e., those nearest the center of the eye) differentiate first and the sensory layer of rods and cones last. Once differentiation has begun, the intricate and interrelated processes leading to the final cytoarchitecture apparently constitute a self-directing and self-sustaining complex (Fig. 14.14).

The sequence of retinal-cell differentiations is as follows:

1. Glial cells, which become Müller's fibers and will envelop the retinal neurons, develop cytoplasmic processes connecting the inner and outer surfaces of the sensory retina; the terminals of these processes combine to form the internal and external so-called limiting membranes.
2. Ganglion cells, which form the innermost layer of neurons, send out axons which together grow to form the optic nerve by way of the optic stalk. Dendritic processes of the ganglion cells form synapses with axons from the next layer to differentiate, i.e., the inner nuclear layer, which contains at least three kinds of neurons.
3. The outer nuclear layer, which is the outermost layer of the neural retina, is the last to differentiate after local cell division ceases. There, axons synapse with the dendrites of the bipolar cells of the inner nuclear layer.
4. The light-sensitive rods and cones develop from the outer ends of the cells of the outer nuclear layer. Each cell produces a cytoplasmic bud at its outer end, which protrudes through a pore in the external limiting membrane. Here again, further differentiation is sequential and centrifugal. The inner segments of the rods and cones are first formed, and from them the outer segments later arise. Various organelles are formed (Fig. 14.15). One, a basal body or centriole, elaborates a noncontractile cilium that has nine pe-

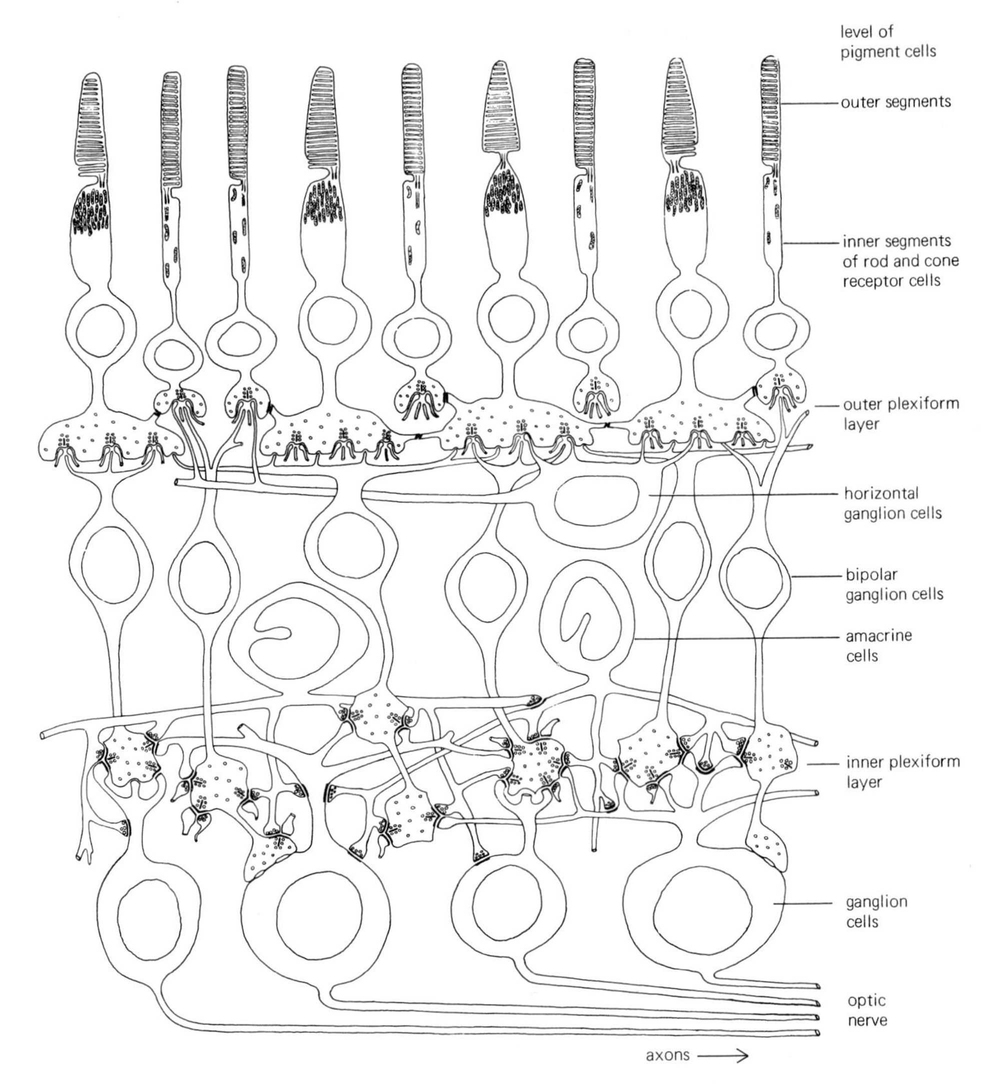

Figure 14.14 Summary diagram of synaptic contacts between the principal types of retinal cells in the human retina. The upper layer consists of rods and cones, the outermost segments of which are more or less embedded in the retinal pigment cell layer (*not shown*). These sensory cells make synaptic contact with their neighbors and with a complex layer of bipolar and horizontal ganglion cells. These in turn make synaptic contacts with the inner ganglion cells whose axons constitute the optic nerve. [*Courtesy of B. B. Boycott, 1974.*]

ripheral filaments but no central filaments and is otherwise typical. It takes part in the development of the outer segments of the rods and cones. Rod or cone sacs fold in from the plasma membrane close to the tip of the cilium, which may induce them, and become arranged like a stack of coins.

5. Finally, pigmented epithelial cells, whether differentiating from the outer layer of the embryonic optic cup or from the reconstituted layer during retinal regeneration, extend fine cytoplasmic processes. They interdigitate with

Figure 14.15 Rod and cone cytogenesis, illustrating the successive stages in the maturation of the cone or rod. *a* Cells differentiating between pigmented retinal epithelium and the external limiting membrane of the sensory retina, forming a bulbous inner segment of the prospective rod or cone. *b, c* Two later stages showing development of outer segment, the folding of the plasmalemma to form a stacked series of membrane plates, and the axial filament, which is essentially a nonmotile cilium complete with basal body. [*After A. J. Coulombre, in "Organogenesis," R. L. DeHaan and H. Ursprung (eds.), Holt, Rinehart, and Winston, 1965.*]

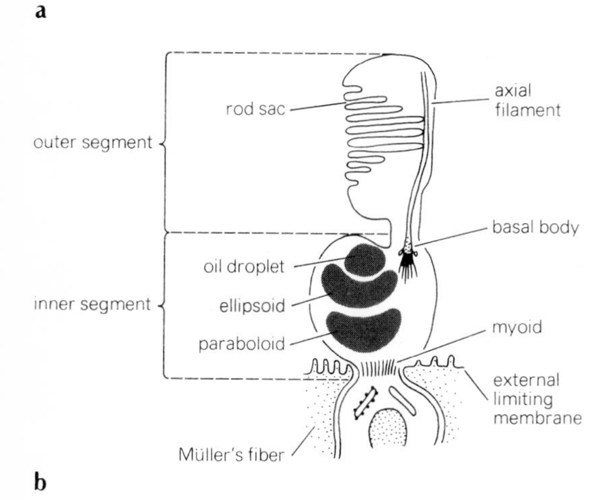

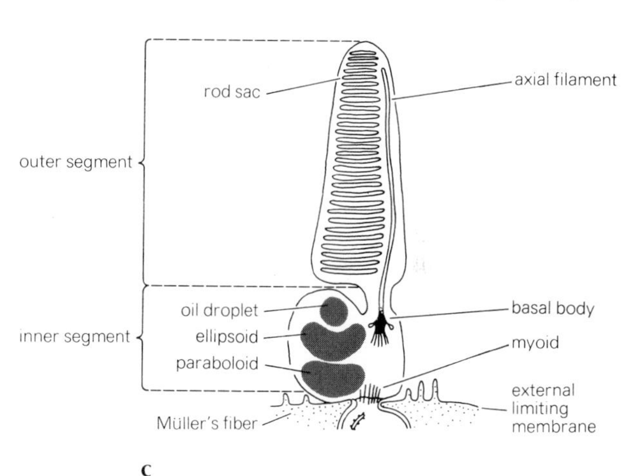

the emerging outer segments of the rods and cones, thus locking the two layers together.

6. Electron micrograph observations following injection of tritiated leucine have shown that the stacked folds of the outer segments of both rods and cones in mammals are continuously shed and re-shed from the new segments. The shed material is phagocytized by cells of the pigment layer.

4. The nervous system

Of all the organ systems of the body, we are the most ignorant of the nervous system. The problems in understanding its development are certain to be a center of research in developmental biology for many years to come. Readers seeking a comprehensive survey of the literature in this area should consult the monograph by Marcus Jacobson (1978).

A. Histogenesis of the nervous system

With closure of the neural folds, a tube is formed that is broad in front and narrow in the rear. At the time of closure, the cells of the neural tube in higher vertebrates exist as a pseudostratified epithelium, i.e., a single layer of cells whose nuclei are found at different heights, thereby giving the epithelium the superficial appearance of being composed of more than one cell layer. In actual fact, each cell extends all the way from the central canal to the basement membrane at the outer edge of the epithelium. The basis for the variation in position of the nuclei

within the neural epithelium is itself a complex and interesting story. Unlike the situation in most columnar cells, the nuclei of the neural epithelium do not remain in a fixed position, but instead move from one part of the cell to another in keeping with their stage of the cell cycle (Fig. 14.16). Autoradiographs made after exposure of embryos to [^{3}H]thymidine indicate that S-phase nuclei are found toward the external limiting membrane of the epithelium. As the cells progress toward mitosis, their nuclei move toward the lumen of the tube until they are near the apical surface by the time they actually divide.

In the following discussion we will begin by describing events as they occur within the spinal cord and progress to a consideration of brain development, where similar events are more complex and harder to follow. As the cells of the neural epithelium continue to divide, some of the cells become detached from their neighbors and move outward in a lateral direction forming a second layer that becomes thicker and thicker as cell proliferation continues in the inner layer. Soon a third layer becomes discernible lying farther from the central canal. This outer layer is not a celluar one, since it is devoid of nuclei and is composed almost entirely of cytoplasmic processes projecting from the inner cells. A cross section through a developing spinal cord is shown in Figs. 12.1*b* and 12.15. The difference in thickness between the lateral walls and the upper and lower walls is apparent; virtually all the growth of the neural tube occurs within its lateral walls. The three concentric layers just discussed are, from inside to outside, the *ependymal layer*, the *mantle layer*, and the *marginal layer*. In the embryo, the inner ependymal layer is composed of *germinal cells*, which are actively dividing stem cells (precursors) for the differentiated cells of the nervous system. In the adult, this layer is reduced to a thin strip of ependymal cells, which, among other proposed functions, play a supporting role in the spinal cord. The cells that enter

Figure 14.16 Section of the neural tube of a chick embryo showing the intermitotic migration of the nucleus of a single neuroepithelial germinal cell at approximately half-hour intervals throughout the mitotic cycle. [*From M. Jacobsen, "Developmental Neurobiology," Plenum, 1978; after F. C. Sauer,* J. Comp. Neurol., **62**:377 *(1935).*]

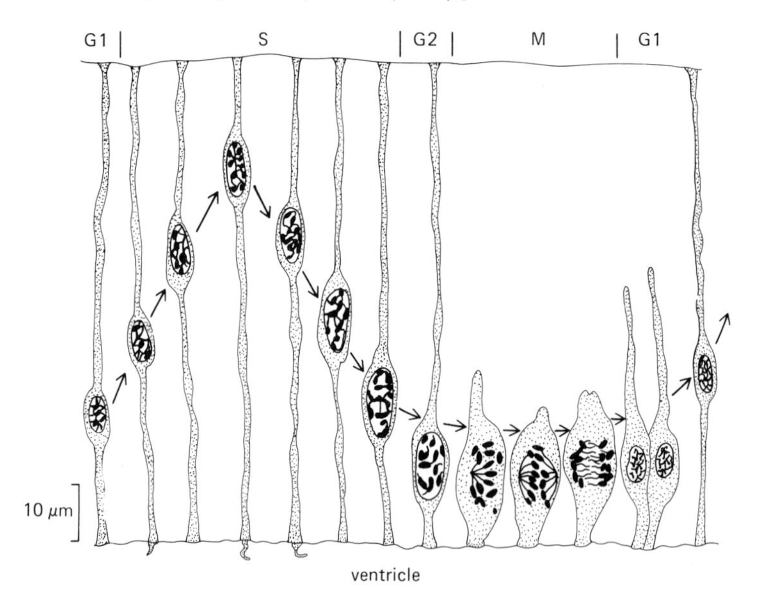

the mantle layer are postmitotic, i.e., no longer capable of mitosis. These cells proceed along two very different pathways of differentiation. One group, the *neuroblasts*, give rise to the neurons, while the other group, the *spongioblasts*, differentiate into a variety of supporting cells, most notably the *neuroglia*. It is processes from the neuroglial cells and later the axons of the neurons that course through the outer region of the spinal cord forming the marginal layer. With time, the axons of the marginal layer become covered with a sheath of myelin, giving the fibers a whitish appearance and causing the marginal layer to be referred to as the cord's white matter (Fig. 14.17). The cell bodies of the neurons, i.e., the part of the cells that contain the nuclei, remain in the mantle layer forming the gray matter of the cord.

Information, in the form of nerve impulses, comes into the spinal cord from various parts of the body via sensory (*afferent*) neurons and leaves the cord along motor (*efferent*) neurons on its way to controlling muscular activity (Fig. 14.17*b*). These two divisions of the nervous system develop in a very different way, the sensory neurons differentiating from neuroblasts of the neural crest and the motor neurons differentiating from neuroblasts of the cord's mantle layer. It was mentioned in Section 11.2 that one group of neural-crest cells collects in aggregates to the sides of the neural tube in response to the presence of the segmented somitic mesoderm. These clusters of neural-crest cells give rise to neuroblasts of the spinal (dorsal root) ganglia, i.e., cells that will form the sensory neurons of the peripheral nervous system. The neuroblasts begin their differentiation by sending out a pair of cytoplasmic processes pointed in opposite directions (Fig. 14.17*b*). One of these processes grows toward the dorsolateral region of the developing spinal cord, where it penetrates the marginal layer of the cord and eventually forms a synaptic junction with a neuron of the dorsal column. The

Figure 14.17 Simplified stereogram of the vertebrate spinal cord and nerves to show the relations of neurons involved in a simple reflex arc. [*From T. Storer et al., "General Zoology," 6th ed., McGraw-Hill, 1979.*]

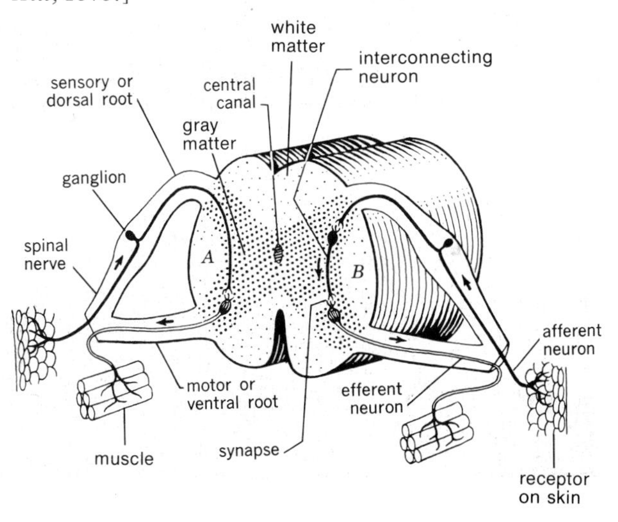

bundle of axonal processes that stretch between a spinal ganglion and the spinal cord itself is referred to as a *dorsal root* (Fig. 14.17*b*). The other process from the bipolar neuroblast grows toward the body wall, where it will innervate one of a variety of receptors. Once these axonal processes are functional, nerve impulses can travel from remote areas of the body to the spinal cord by traversing a single nerve cell. Once in the spinal cord, arriving nerve impulses are transmitted to other neurons across synaptic clefts. From this point the information can spread to other levels of the central nervous system by travelling along connecting neurons within the spinal cord or be transmitted to motor neurons, thereby completing a *reflex arc* (Fig. 14.17*b*). The motor neurons of the peripheral nervous system arise from neuroblasts whose cell bodies remain in the mantle layer of the ventrolateral region of the cord and whose axons grow out from there to innervate a skeletal muscle fiber. As the axons of the motor neurons grow out from the cord, they join up with the axons of the sensory neurons and together form the mixed spinal nerve.

Unlike the spinal cord, which is rather uniform in structure and function from one end to the other, the brain undergoes tremendous regionalization as it develops, thereby causing the various parts to take on a very dissimilar appearance. Even though many of the same basic features are evident during the development of both parts of the central nervous system, these similarities tend to become obscured or distorted in the later stages of brain differentiation. For example, the three-layered organization of the spinal cord is continued in the anterior regions of the neural tube, but the neuroblasts become very unevenly distributed within the mantle layer. Clusters of neuroblasts form discrete masses of gray matter termed *nuclei*. In two major brain regions, the *cerebellum* and *cerebrum*, the neuroblasts actually migrate through the mantle layer to form an outer *cortex* of gray matter surrounding an inner *medulla* of white matter.

The early events in brain formation are remarkably similar in all higher vertebrates (Fig. 14.18). Soon after closure of the anterior region of the neural tube, a number of distinct constrictions appear along its wall and form a series of swellings or *neuromeres*. Although most of these constrictions are transient and soon disappear, the brain region becomes divided into three distinct *primary vesicles:* the *prosencephalon* (forebrain), the *mesencephalon* (midbrain), and the *rhombencephalon* (hindbrain). The most dramatic event to occur in the newly forming brain is the outward growth of the lateral walls of the prosencephalon to form the bulbous *optic vesicles,* structures that will eventually become transformed into the eyes (Section 13.3). Soon after the optic vesicles form, the prosencephalon becomes divided into two parts, the anterior *telencephalon* and the more posterior *diencephalon,* the latter of which contains the optic vesicles. The last major division of the brain occurs when the rhombencephalon becomes constricted into a more anterior part, the *metencephalon,* and the most posterior part of the brain, the *myelencephalon.* This latter section merges imperceptibly with the anteriormost section of the spinal cord. The point of juncture between the brain and spinal cord in the developing embryo is marked by the level of the first spinal ganglion. In the following discussion we will briefly trace the fate of each of the five brain vesicles and lastly the nature of the cranial nerves.

TELENCEPHALON: In lower vertebrates, the primary role of the telencephalon is to receive and process information relating to the sense of smell. Consequently, this part of the brain in these animals is dominated by the *olfactory lobe.* In higher vertebrates, the olfactory lobe remains a very important part of the telencephalon, but it is overshadowed by the gradual evolutionary development of the *cere-*

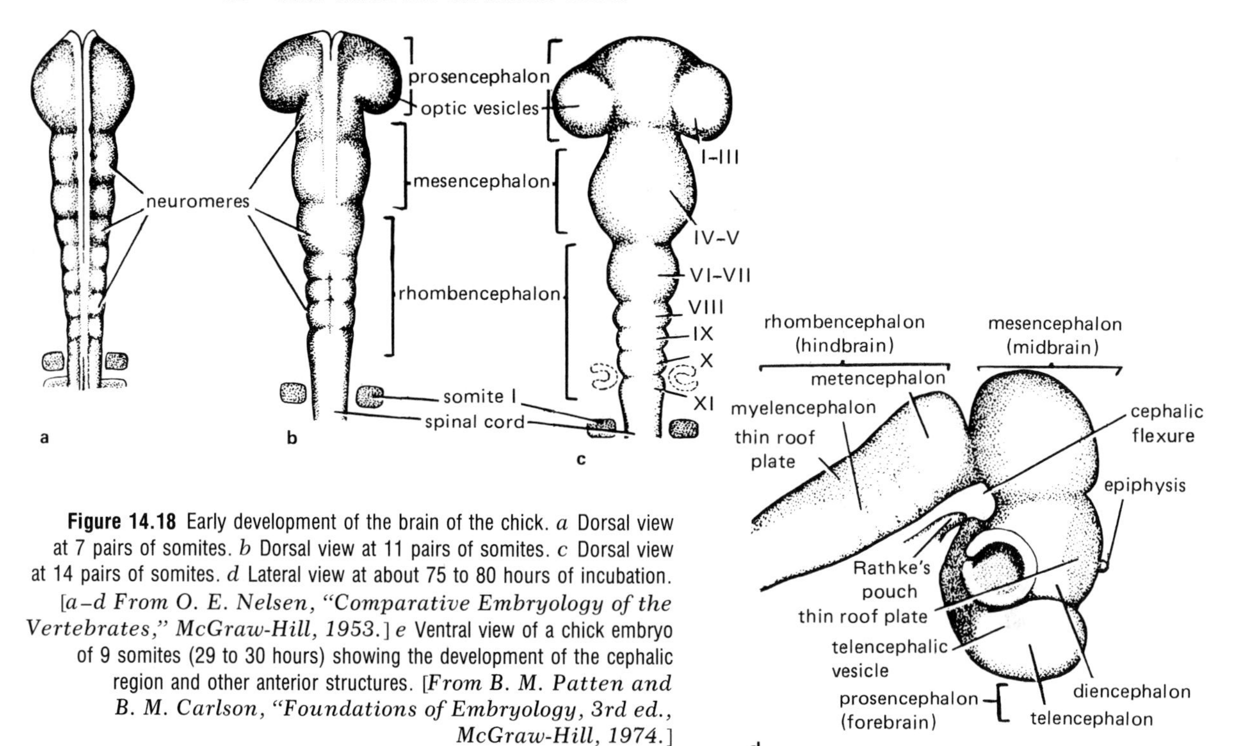

Figure 14.18 Early development of the brain of the chick. *a* Dorsal view at 7 pairs of somites. *b* Dorsal view at 11 pairs of somites. *c* Dorsal view at 14 pairs of somites. *d* Lateral view at about 75 to 80 hours of incubation. [*a–d From O. E. Nelsen, "Comparative Embryology of the Vertebrates," McGraw-Hill, 1953.*] *e* Ventral view of a chick embryo of 9 somites (29 to 30 hours) showing the development of the cephalic region and other anterior structures. [*From B. M. Patten and B. M. Carlson, "Foundations of Embryology, 3rd ed., McGraw-Hill, 1974.*]

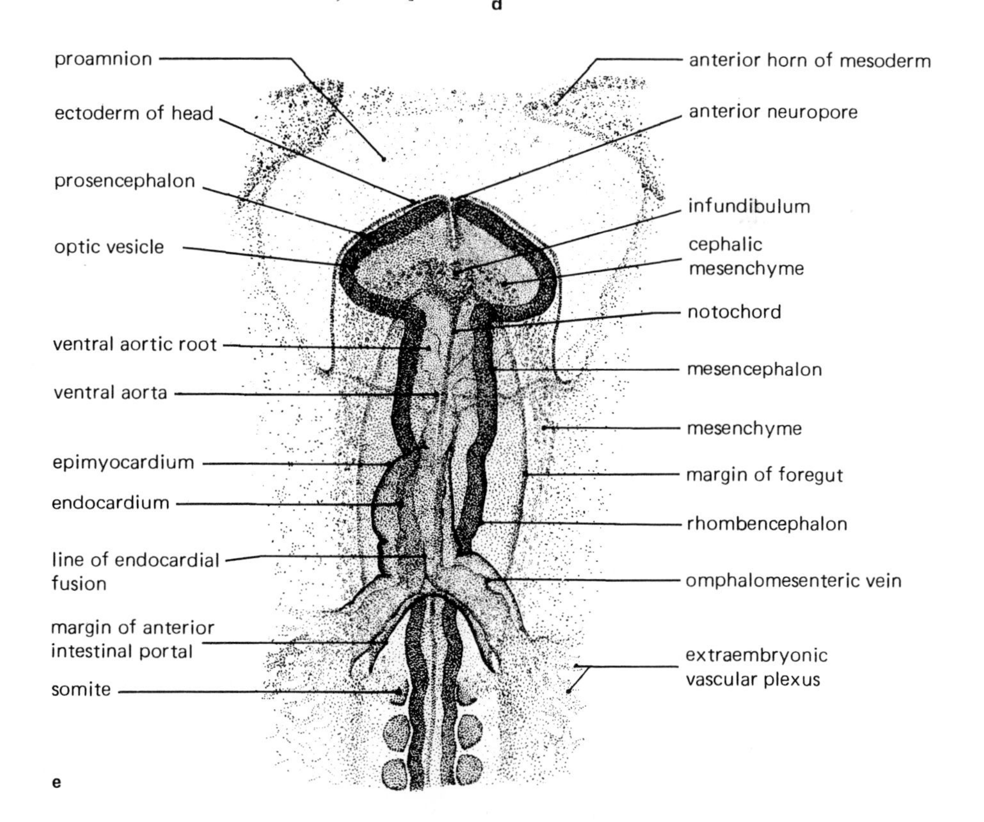

bral hemispheres, which develop from a pair of bulbous projections. The cavities of the cerebral hemispheres are the *lateral ventricles.*

DIENCEPHALON: Different parts of the diencephalon take on very different functions. The dorsal wall (roof) of the neural tube in this region is largely nonneural in function. A major portion of the roof of the diencephalon is invaded by blood vessels to form the *anterior choroid plexus,* which bulges into the *third ventricle,* the central cavity of this part of the brain. Behind the choroid plexus, an outpocketing appears in the roof that develops into the *epiphysis* (*pineal gland*), an organ that is primarily photoreceptive in lower vertebrates but possesses a poorly understood endocrine function in mammals. The dorsolateral walls of the diencephalon become sites of neuroblast aggregation, forming the *thalamic nuclei.* The primary function of the thalamus is to serve as an association center, i.e., a relay station, for impulses passing between the more primitive regions of the central nervous system, i.e., the spinal cord and brainstem, and the higher centers of the cerebral hemispheres. Another important brain center of the diencephalon is the *hypothalamus,* a structure that controls many of the most basic physiological functions and emotions. From the ventral wall (floor), there occurs an outpocketing, termed the *infundibulum,* that will give rise to the "posterior" part of the *pituitary gland.* The remainder of the pituitary is not a brain derivative, but develops from a mass of cells that grows out from the stomodeal ectoderm (Section 12.2.B). This tongue of cells loses its connection with the stomodeum and becomes intimately associated with the infundibulum.

MESENCEPHALON: The midbrain of the early embryo remains undivided and simple relative to other brain regions. The most prominent part of the mesencephalon is the *tectum,* a center for vision in lower vertebrates (Section 14.4.E). The midbrain is also the site of the greatest bend of the neural tube (Fig. 14.18), termed the *cephalic flexure.* The central cavity becomes narrowed in the midbrain and is termed the *aqueduct of Sylvius.*

METENCEPHALON: This part of the brain becomes highly specialized for controlling motor functions. Dorsally, the metencephalon forms the *cerebellum,* the center for control of coordinated movements and maintenance of body position.

MYELENCEPHALON: This final brain region, which forms the base of the brainstem, becomes the *medulla oblongata,* the center for cardiovascular, pulmonary, and other basic physiological functions. The roof of the medulla develops into a second major site of vascularization for the brain, the *posterior choroid plexus.* The cavity of the hindbrain is the *fourth ventricle,* which merges with the central cavity of the spinal cord.

CRANIAL NERVES: Unlike the spinal nerves, the cranial nerves do not emerge at regularly spaced intervals, nor do they all have roughly similar functions. There are 12 pairs of cranial nerves (Fig. 14.19): three are exclusively sensory (the olfactory, optic, and auditory), four are essentially motor (the oculomotor, trochlear, abducens, and hypoglossal), and five are mixed (the trigeminal, facial, glossopharyngeal, vagus, and accessory). The origin of the neurons of this latter group is analagous to that of the mixed spinal nerves; the sensory neurons differentiate from neural-crest cells that have migrated away from the neural tube and the motor neurons differentiate from clusters of neuroblasts in the wall of the developing brain.

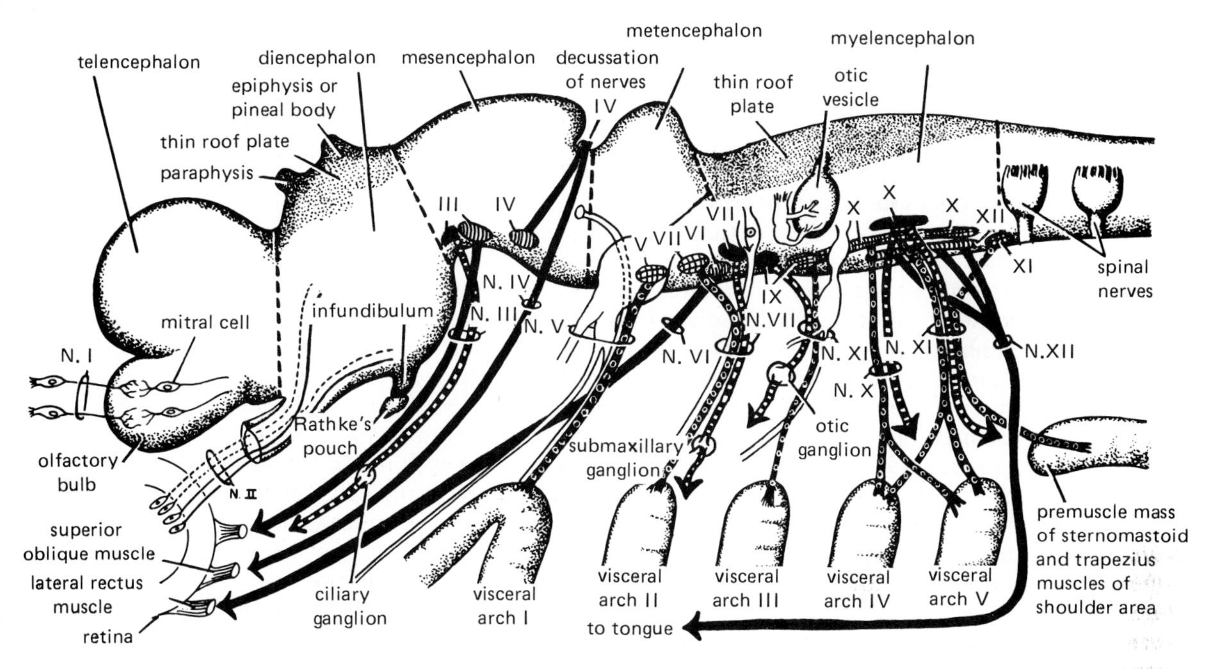

Figure 14.19 Schematic diagram showing the motor components of the cranial nerves. The nuclei of origin are shown to be located within fairly definite regions along the anteroposterior axis of the vertebrate brain. Somatic motor fibers are shown in solid black; special visceral motor fibers are indicated in black with white circles; general visceral motor fibers are black with white markings. [*From O. E. Nelsen, "Comparative Embryology of the Vertebrates," McGraw-Hill, 1953.*]

With this description of the development of brain and eye behind us, we can briefly look back in search of a few principles of general importance. In the last chapter we discussed the concept of a "field" as a region of the embryo in which an organ will develop. The cells of a field are endowed with special properties; they have become differentiated to the degree that they "understand" which organ it is that they should produce, yet they retain a flexibility as to which parts of that organ they can ultimately form. According to Johannes Holtfreter, a field can be viewed as "a transitional phase of development during which crucial decisions are made regarding the diversified trends of differentiation of the parts." In the case of the brain and eye, the anterior part of the neural plate constitutes just such a field. Prior to the establishment of the neural plate, the cells of the presumptive neural epidermis can be directed to differentiate in virtually any direction. Then comes their fateful interaction with their underlying neighbor, and they no longer possess such broad potency. At the same time, however, the fate of these cells is not completely sealed. Each cell can still differentiate into any one of a great variety of types that ultimately constitute the eye and brain. As is characteristic of fields, in general, the anterior neural plate can be demonstrated to possess extensive powers of regulation; neural-plate halves can form whole brains, just as pairs of neural plates can form single brains when fused. With time, these extensive regulatory capabilities fade; progressive determination leads to the replacement of the brain-eye field with smaller fields of more limited potency. How does the field decide which part of itself should become eye or forebrain or some other part? We do not know. Since the differentiation of the neural

plate can occur in culture after its removal from surrounding tissues, we can conclude that the assembly of structures within the field is independent of the inductive stimuli that led to its formation in the first place. The group of cells that constitute the field has become an integrated, self-determined colony in which the positions of its member cells are known to themselves and to each other. How cells communicate this positional information remains one of the foremost mysteries in developmental biology.

B. Nerve-growth factor

There is much evidence that nerves can respond to specific chemical stimulation. The best examples of such effects come from studies of hormones and of a particular protein, nerve-growth factor (NGF). The investigations into nerve-growth factor have had a long and interesting history. The first indication of the existence of a factor that could specifically affect the differentiation of certain types of neurons occurred in experiments with a mouse sarcoma, a tumor of mesodermal origin. In one early set of experiments, pieces of tumor tissue were grown on the chorioallantoic membrane of the chick, a favorite site for culturing a variety of tissues. The chorioallantoic membrane is chosen for its ease and accessibility. If a window is cut in the shell of a chick of the appropriate age, a piece of tissue from a mammalian or avian source can be dropped into the hole, where it will land on the chorioallantoic membrane and become penetrated by blood vessels that nourish the cells as the piece of tissue grows. When the mouse tissue was grown in this way, there occurred a striking enlargement of the sympathetic and sensory ganglia of the chick. A substance, NGF, had entered the circulation of the chick from the mouse tissue and had had a specific effect on these two neural-crest derivatives.

One technique for learning about the nature of an unknown factor is the determination of which enzymes it is sensitive to. One source of uncommon hydrolytic enzymes is snake venom. Surprisingly, treatment of NGF preparations with snake venom hydrolases led to the stimulation of the preparation's activity rather than its abolition. The reason for this paradoxical finding soon came to light: the venom contained the presence of a nerve-growth factor at much higher concentration than that in the malignant tissue. Further investigation turned up a still richer source of NGF, the submaxillary gland of the mouse, a homologous structure to the venom gland of the snake. Sufficient quantities of the protein have been purified from this gland to permit amino acid sequence determination. Why this particular gland contains the factor is not known. Moreover, NGF levels in the gland are not high during embryogenesis, nor does its removal have any subsequent effect on neural differentiation.

The effect of NGF is very widespread among vertebrate embryos. Injection of the purified material into newborn mammals can result in up to a twelvefold enlargement of the sensory and sympathetic ganglia. Increased size results from both increased cell number and cell size. The effect of NGF is also strikingly revealed under tissue-culture conditions. When NGF is present at high concentrations within the medium in which a sensory or sympathetic ganglion is growing, a halo consisting of large numbers of axonal processes is seen to sprout from the ganglion (Fig. 14.20). More recently, the powerful effect of NGF on the differentiation of sympathetic neurons has been revealed in a different way. It was found that immature nonneural cells of the adrenal medulla (the chromaffin cells that secrete adrenalin) could be directed to differentiate in culture into sympathetic neurons by the addition of NGF to the medium. It was subsequently found that repeated injection of NGF into rat fetuses led to a remarkable transformation of

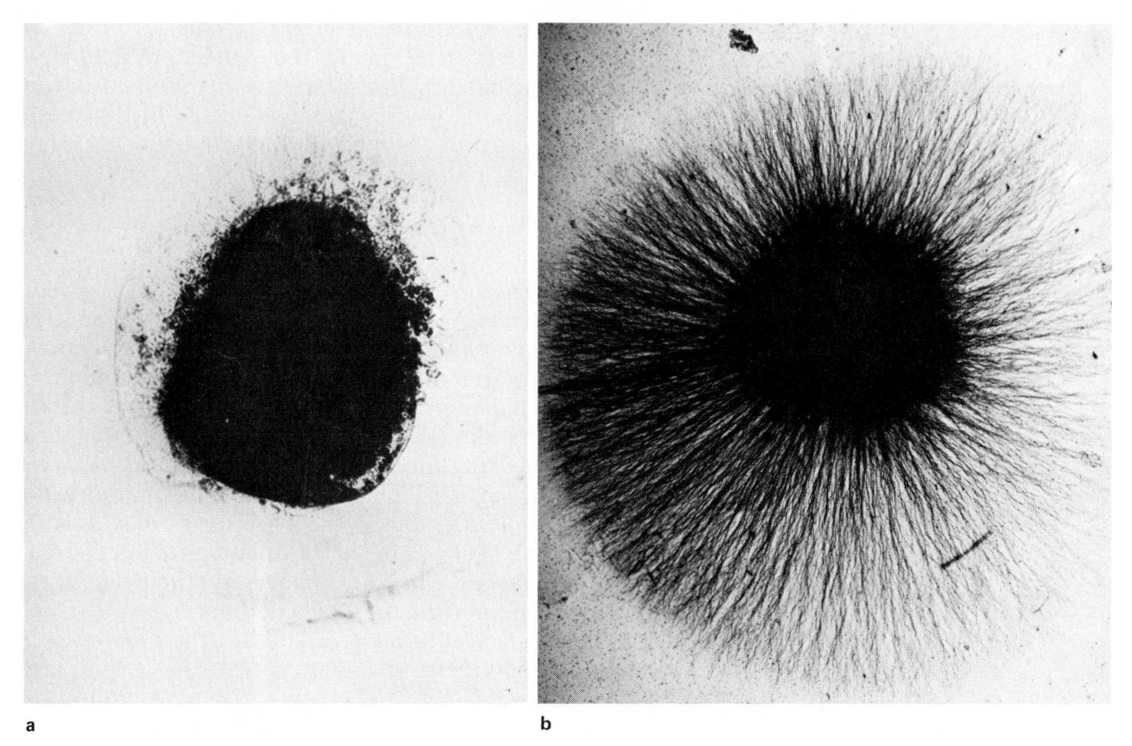

Figure 14.20 Micrograph of a sensory ganglion of a 7-day chick embryo that had been cultured in *a* the absence or *b* the presence of nerve-growth factor (0.01 μg/ml) for 24 hours. In the absence of NGF, virtually no neurite growth is observed. [*Courtesy of R. Levi-Montalcini.*]

the chromaffin cells of the adrenal medulla; they differentiated into sympathetic neurons within the core of what is normally an endocrine gland.

The importance of NGF in *normal* development is best revealed by studies employing antibodies prepared against the purified protein. When anti-NGF serum is injected into chick embryos or newborn mammals, NGF molecules are rendered inactive by combining with the antibody and widespread destruction of sympathetic and sensory neurons is brought about. Similarly, explants of both types of ganglia require the presence of NGF for their survival in culture. If the growth factor is omitted from the medium, the neurons degenerate within a short period of time. The mechanism of action of NGF remains unclear. If labeled NGF is injected intravenously (Fig. 14.21), NGF molecules bind to receptors at the outer surface of the plasma membrane of target neurons at their terminal endings and are then transported up the axon in the cytoplasm to the cell body. This process, termed *retrograde transport* to indicate its movement (Fig. 14.21) in an opposite direction (from synapse to cell body) from the general flow of materials, appears to be of vital importance to the maintenance and survival of the neuron. If the retrograde movement of NGF is blocked by the addition of agents that depolymerize microtubules, the nerve cell dies. This untimely death can be prevented by the injection of NGF. Nerve-growth factor also has been demonstrated capable of "attracting" nerve outgrowth. For example, axons of chick-embryo sensory ganglia will grow toward a source of NGF in their culture medium. With this in mind, it has been proposed that NGF concentrations in the environment may act

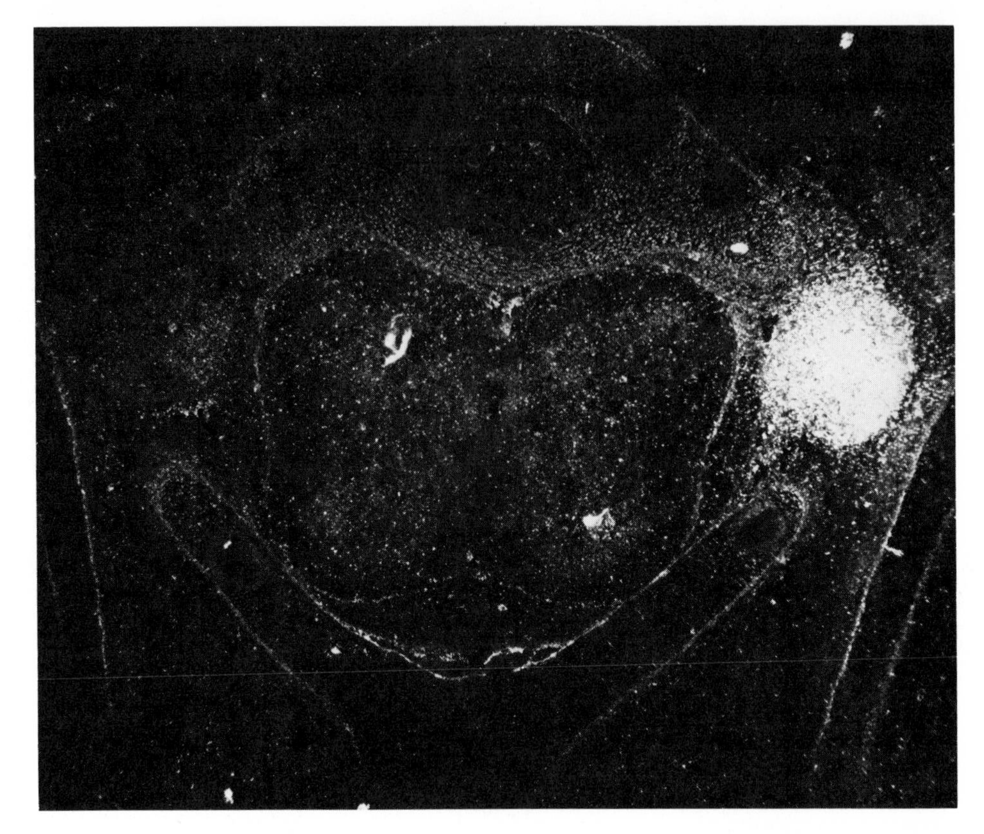

Figure 14.21 Autoradiograph (photographed in darkfield so that the silver grains appear as white sources of light) of a section through the lumbar region of the spinal cord of a chick embryo after implantation of an ^{125}I-labeled NGF pellet placed beneath the skin in the leg. Labeled NGF has been transported from the tips of the sensory neurons through the axon into the cell bodies that reside in the dorsal-root ganglion just outside the spinal cord. Radioactivity is seen only within the ganglion on the side of the implant. [*Courtesy of J. K. Brunso-Bechtold and V. Hamburger.*]

to guide nerve outgrowth through the tissues of the embryo. It is possible, for example, that somites are able to attract neural-crest cells (Section 11.2) as a consequence of their NGF production. It is also believed that NGF is released in small amounts by the tissues normally innervated by sympathetic neurons, causing the growing axon to be directed toward its target area. The general subject of nerve outgrowth is taken up in the next section.

C. Nerve outgrowth

One of the first and most important questions to be considered in the field of developmental neurobiology pertained to the mechanism whereby a single nerve cell could span the distance between the central nervous system in which its nucleus was found and the peripheral effectors that it supplied. Several opposing theories had been presented by the turn of the century, one of which suggested that nerve processes grew outward from cells that were originally present within the neural tube. The foremost proponent of this hypothesis was the 1906 Nobel Laureate, Santiago Ramon y Cajal, who came to this conclusion by examination of fixed sectioned material. The question was resolved by Ross Harrison in what

is considered the first tissue-culture experiment. Harrison removed a small part of the neural tube from a frog tail bud, placed it in a tiny drop of lymph fluid on a coverslip to provide the cells with nutrition, and inverted the coverslip over the depression on a slide to form a "hanging drop" culture in which the piece of presumptive neural tissue could differentiate. Within the next few days he found that the axons emerged from the embryonic neuroblasts and grew out into the medium just as they would have emerged from the neural tube of the embryo (Fig. 14.22). In a few favorable cases, as in certain amphibian larvae having a transparent epidermis, the outgrowth of the neuron can actually be followed as it occurs in vivo. Based on a variety of these types of observations, it appeared that the nerve tracts, which ultimately reach a length of many feet in large vertebrates, are the result of outgrowths from cells in the central nervous system. Initially, only a few pioneer fibers make the journey, but these are later followed along the established pathway by large numbers of growing fibers. At the time of their initial formation, the distance that must be traversed in the tiny embryo is only about 1 mm. Increase in axonal length occurs with increasing growth of the embryo.

What mechanism is responsible for enabling the tip of a cell to extend itself over this distance? Studies of nerve outgrowth in vivo and in vitro indicate that the end of the axon, termed the *growth cone* (Fig. 14.23), acts as a highly active ameboid terminus that "feels" its way along the substrate. Closer examination of the surface of the growth cone reveals the presence of broad lamellipodia and minute pointed filopodia, or *microspikes* (Fig. 14.23), which represent the actual motile organelles of the nerve ending. As expected from studies of other motile cells, electron micrographs show the lamellipodia and microspikes to be filled with microfilaments. Treatment of growing neurons with cytochalasin B produces a rapid disappearance of the microspikes and a cessation of axonal advance. If the main body of the growing axon is examined in the electron microscope, bundles of microtubules are found running parallel to the long axon of the neuron. These microtubules are apparently required for support of the extended young axon, since treatment of these cultured neurons with colchicine results in the disassembly of the microtubules and, after a short period, the withdrawal of the axonal process. Whereas the lamellipodia and filopodia of most cells are ac-

Figure 14.22 *a* Six successive views of the end of a growing nerve fiber, showing its change of shape and rate of growth. The sketches were made with the aid of a camera lucida at the time intervals indicated. The red blood corpuscle, shown in outline, marks a fixed point. The average rate of elongation of the nerve was about 1 μm per minute. The total length of the nerve fiber was 800 μm. The observations were made upon a preparation of frog embryo ectoderm, isolated in lymph, 4 days after isolation. [*From R. G. Harrison,* J. Exp. Zool., **9:***787–846 (1910).*]

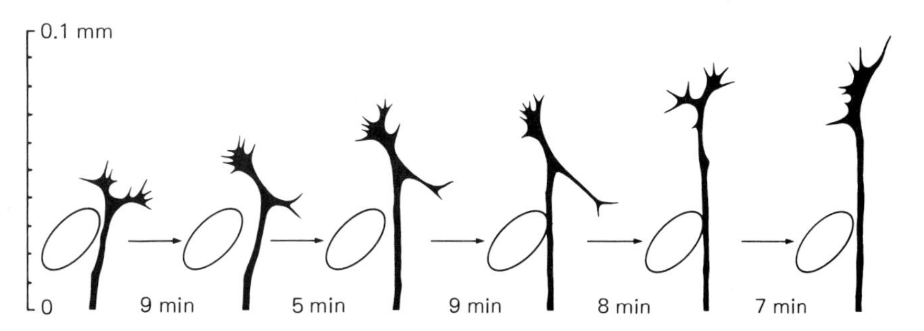

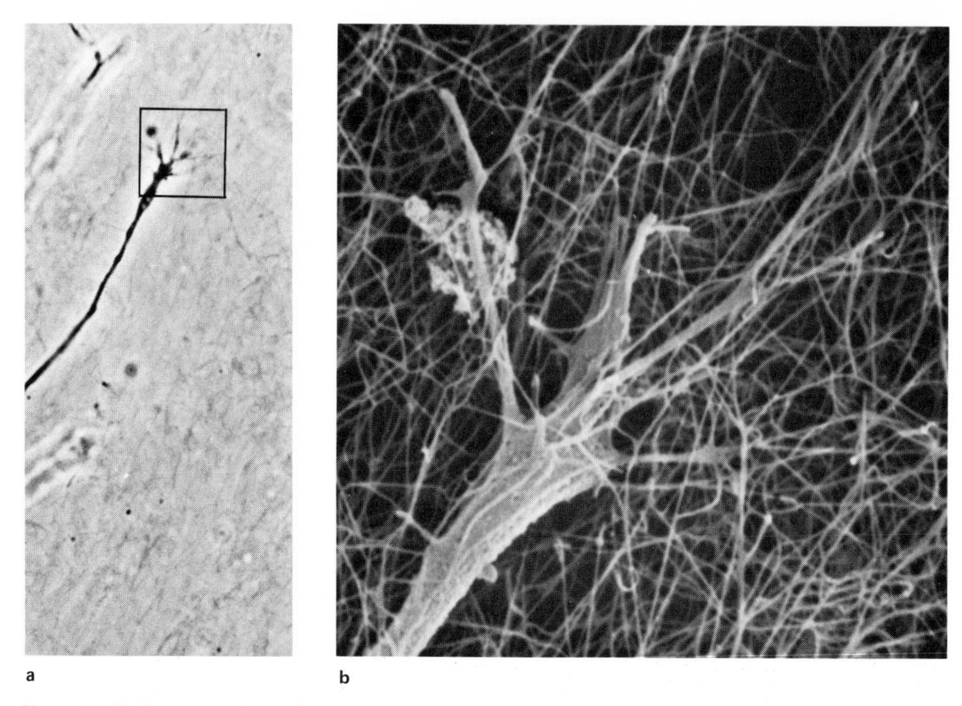

Figure 14.23 The neuronal-growth cone as seen in cell culture. The neurite bearing the growth cone extends from a dorsal-root ganglion of an 8-day chick embryo stimulated by NGF. The substratum is composed of roughly aligned collagen fibrils (oriented from lower left to upper right), which present guiding cues to the neurites. *a* The growth cone (*framed*) as seen in the living state by phase contrast. *b* The same growth cone fixed a few minutes later and prepared for scanning electron microscopy. The expanded growth cone bears several filopodia, some of which adhere to the collagenous substratum fibrils. [*Courtesy of T. Ebendal.*]

tually responsible for pulling the cell from place to place, the contractile machinery of the growth cone appears primarily to facilitate the exploratory activity of the nerve ending rather than serving to pull the axon forward. The extension of the process is accomplished by pressure from the movement of the cytoplasm, i.e., *axonal flow*, down toward the tip.

D. Nerve guidance

Now that we have described the appearance of the growing tip of the axon, we can turn to a question that has proved to be more formidable: What mechanism is responsible for guiding the original nerve fibers from their point of origin to their specific synaptic termination? This question has both anatomic and physiologic aspects.

The basic circuitry of the nervous system has a predictable and therefore overall genetic arrangement. Each nerve cell, as it emerges from a particular ganglion in the brain or spinal cord, proceeds to a predictable location in the body, where it forms a synapse with another neuron, a muscle fiber, or a sensory receptor. Are the original neurons guided to their final destinations by some inherent information within the nerve cell, or does the environment through which they migrate lead the advancing tip? If the environment does play a critical role, what types of cues does it provide the growing neuron and what mechanisms within

the neuron are operating to sense the environment? At the present time there exist a somewhat contradictory mass of experimental data and very few definitive conclusions.

One of the early observations concerning nerve outgrowth was made by Paul Weiss in his analysis of axonal extension in tissue culture. Weiss found that nerve cells did not grow unless they made contact with a solid substratum. Furthermore, the direction that the growing tip took depended on the contour of the physical substrate over which growth occurred (Fig. 14.24). A particularly well-suited substrate for culture and observation is a blood clot. It is composed of fibrous proteins that become oriented when stretched or under physical tension. When differentiating neuroblasts are cultured on this medium, the axons tend to grow along the oriented components. Similarly, when grown on a thin glass fiber, axons invariably follow the long axis of the fiber rather than the curvature of its circumference. Weiss termed the response of the nerve to the substratum "contact guidance"; he suggested that the direction axons grew in vivo reflects a similar physical pathway through which they are directed to their final, genetically

Figure 14.24 Advance of nerve fibers in fibrous media of different degrees of ultrastructural organization (randomness in center turning into prevailing horizontal orientation in left part and strict vertical orientation in right part of diagram). Along random meshes of center strip, the course of fibers *a*, *b*, and *e* is tortuous, with frequent branching; in the more orderly parts of the medium, fiber courses become correspondingly aligned; in a rigorously oriented medium, fibers (*c*, *d*) run straight and remain undivided. [*From P. Weiss, in "Analysis of Development," B. Willier, P. Weiss, and V. Hamburger (eds.), Saunders, 1955.*]

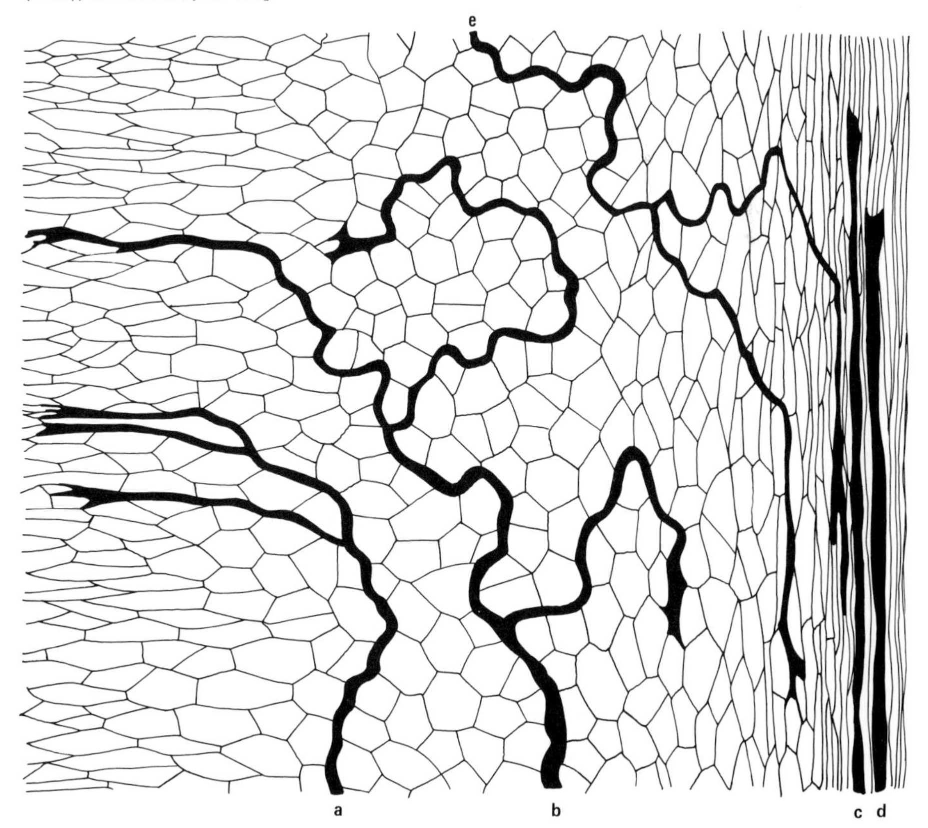

determined site. The interior of the embryo would be expected to offer a variety of surfaces whose oriented components could guide nerve outgrowth. These might include glial cell processes, blood vessels, clefts between cell layers, collagenous basement membranes, and hyaluronate-containing extracellular spaces.

Whereas contact guidance is likely to play an important role in the direction of nerve outgrowth, it seems unlikely that it is the whole story. The intricacies of the anatomic organization of the peripheral nervous system seem to demand the presence of a more specific set of guideposts for axons to follow. Furthermore, there are specific cases that do not seem totally explainable by contact guidance. For example, there are cases in which nerve fibers intermingle or cross paths and then move apart in different directions. There is a well-documented experiment in which a large nerve fiber is caused to grow out initially in an incorrect direction. After considerable extension in the wrong direction, the neuron makes a hairpin turn and proceeds to grow to its normal target location.

It has long been felt that neurons might be directed to a specific site by the chemical nature of the environment. If the tissues through which the various neurons grow are heterogeneous, so that different pathways expose the axons to different chemical stimulation, then any given neuron has the potential to be directed along any particular route. In such a mechanism, the neurons themselves must have some particular chemical identity so that each can recognize a particular set of chemical cues. This does not mean that every single neuron need be biochemically unique at its inception, but rather that the neurons of each major route would have at least a collective identity.

There are several possible types of chemically directed mechanisms. Diffusible substances may exist within the environment to which the growing tip of the axon is attracted. The target tissue itself would be a logical source of some type of "neuron-attracting" substance. The existence of chemotaxis as a directive agent for nerve outgrowth has been demonstrated in a few cases in vitro. It was mentioned earlier that sympathetic adrenergic neurons grow toward a source of NGF in the medium. Similarly, when a submandibular ganglion is cultured along with a piece of submandibular gland epithelium, its natural target tissue, parasympathetic nerve fibers grow out from the ganglion and innervate the piece of gland tissue. In contrast, when nontarget tissues are substituted for the salivary gland, no directed outgrowth is observed. Diffusible substances are not the only types of chemical cues to which nerve tips might respond. With the increasing awareness of plasma-membrane complexity, it appears very likely that specific interactions between cell surfaces might play an important role in guiding the movements of the nerve tip as well as facilitating the formation of synapses with the appropriate target tissue.

The demonstration that chemotaxis plays an important role in the formation of the peripheral nervous system is a formidable task. Given that specific nerve fibers innervate specific target tissues, it can be difficult to distinguish between (1) chemodirected outgrowth of nerve fibers to specific peripheral locations and (2) trial-and-error, nondirected outgrowth coupled with specific affinity to target cells. In the latter case, there might occur two phases of growth: first an extension of the axon in a general predetermined direction, and second, a searching phase where short-range forces guide the tip to its specific target. The alternatives are analogous to the chemotaxis versus trap-action problem discussed in Section 5.4. Unfortunately, tracks made by nerve tips within an embryo cannot be followed as easily as sperm movements within seawater.

Although various in vitro studies suggest that target organs may release specific chemotactic agents, a large body of experimental data indicates that nerve

fibers will grow to almost any potential target organ. If a limb, for example, is grafted to the head, or a tail is grafted to a limb region, nerve fibers from the central nervous system of the region will grow into the foreign tissue and, in most cases, form functional synaptic contacts. Similarly, it has been shown that neurons growing in culture are capable of forming functional synapses with a wide variety of different types of muscle cells from various species. Mouse spinal cord neurons, for example, can form synapses with muscle cells from chick or rat embryos. These types of observations raise serious questions concerning the requirement for a high degree of specificity between cell surfaces involved in synapse formation. We will return to this question of neuronal specificity once again in the next section in connection with the formation of the optic nerve.

The concept that peripheral tissues are capable of "attracting" nerve-fiber outgrowths has been based largely on a series of early experiments that showed unequivocally that the level of peripheral innervation is strongly influenced by the amount of peripheral tissue mass. If, for example, an extra limb bud is transplanted adjacent to an embryo's own limb, a much larger number of both sensory and motor neurons is ultimately found connecting the pair of limbs to the spinal cord than is found in controls. Similarly, the removal of a limb bud greatly decreases the final level of peripheral innervation from that region of the cord. The effect of the peripheral tissue mass on innervation can be explained by two different proposals. It may be that peripheral target tissue has a direct influence from a distance on the proliferation and differentiation of central neuroblasts; the greater the mass of tissue to innervate, the greater the number of differentiating neuroblasts that are generated. Alternatively, peripheral target tissue may exert its effect via neuronal survival. According to this latter explanation, nerve outgrowth is programmed within the central nervous system independent of peripheral stimulation. Regardless of the prevailing conditions, a given number of neurons grow toward the peripheral tissues. Those neurons which happen to form synaptic connections with target cells survive, while those which do not degenerate. The greater the peripheral tissue mass, the greater the number of synapses that form and the higher the percentage of cell survival.

Since these two hypotheses make very different predictions, it should be possible to distinguish between them. On casual inspection, the latter of the two explanations seems somewhat unlikely. Why should the central nervous system send out axonal processes when they may not connect with a given target but simply die? However, it has been found that even under normal conditions, a great many more neurons are produced than are capable of synapsing to available targets. In mammals, a given muscle fiber is generally capable of being innervated by only a single motor-axon terminal; consequently, the number of available sites for synapse formation is strictly limited. It appears that the strategy that prevails during neurogenesis involves the overproduction of differentiating neuroblasts, each of which sends out its axons in an indiscriminate manner, thereby guaranteeing that the peripheral targets are saturated with available fibers. Once innervation of the limited number of target cells is accomplished, those neurons which did not form connections undergo degeneration, while those which became connected are stimulated to mature and be maintained. As in the case of NGF, which is required for the maintenance of sympathetic adrenergic neurons, evidence indicates that maintenance substances are passed across the developing synapse from the postsynaptic cell to the presynaptic neuron and subsequently carried up the axon to the cell body where they have some regulatory role. Analysis of the mitotic index and level of necrosis in the central nervous system under different conditions lends support to the concept that peripheral tis-

sues do not affect proliferation and initial differentiation of neuroblasts, but rather affect the ability of neurons to survive. For example, neither amputation of existing limb buds nor grafting of supernumerary limb buds seems to affect the mitotic index of the germinal cells of the spinal cord (Section 14.4.A); however, these operations do affect the percentage of neurons that survive into the adult.

Now that we have considered how a nerve fiber might reach a postsynaptic cell, we can ponder the question of what role it plays following its arrival. Is the specific function of a neuron established before its connection with the tissue to be innervated or only after? In other words, does the program of a motor nerve growing from the spinal cord to a muscle—in the hand, for example—become determined by the muscle cell with which it makes contact, or alternatively, has the neuron already been specified as to where it was going and what it would do before it reached its destination? What happens if a limb bud is innervated by a cranial nerve? Do the limb muscles participate in normal or quasi-normal limb movements, or do they behave as if they were some type of cranial structure. In this type of experiment, it is generally found that the structure can behave normally only if supplied with the nerves that ordinarily innervate it. For example, a forelimb transplanted adjacent to an existing forelimb of an embryo will be innervated by the same nerve supply that innervates the host's own limb. Consequently, the two limbs move in a coordinated fashion. In contrast, if the same forelimb had been transplanted to the head, it would have been innervated by cranial nerves and would no longer behave in a limblike manner. Depending on the specific source of innervation, the transplanted limb may undergo noncoordinated mass contractions in synchrony with the gills, eye, or jaw. It would appear from this type of experiment that the function of a given nerve is determined, in at least a general way, by intrinsic developmental processes occurring within the central nervous system; it does not become specified as a result of its interaction with peripheral tissue or altered by sensory stimulation and use. This can be further demonstrated in these transplantation experiments in the following way. If a portion of the brachial (forelimb) region of the spinal cord from a stage 27 chick embryo is transplanted to the thoracic region of the spinal cord at the same time that the forelimb is grafted to the thoracic region of trunk, the supernumerary limb now behaves as if it were a forelimb, because it becomes innervated by neurons that normally direct forelimb movements. If the spinal-cord graft had not accompanied the limb transplantation, the resulting limb contractions would have been highly abnormal. Similar types of experiments involving sensory neurons (see the reviews by George Szekely, 1974, and Marcus Jacobson, 1978) support the concept of autonomous determination of neural function within the central nervous system.

E. Optic-nerve specificity

The central nervous system of vertebrates consists of tens of billions of neurons linked together in an exquisite intercommunicating network. Information coming into the brain and spinal cord over sensory neurons must be distributed in a differential manner so that appropriate sets of neurons in higher centers receive the information, integrate the information with that converging on it from other centers, and act on it by stimulating specific sets of motor neurons. The key to central nervous system function lies in the specific nature of its interneuronal linkages. It is difficult to probe the question of neuronal specificity using innervation of skeletal muscles or skin receptors as the basis for determination. A much better system for studying one-to-one neuronal relationships is the retina, whose ganglion cells send out axonal processes that grow to specific sites within the

brain. Most of the studies on embryos have been conducted on amphibians, whose eye and brain are more accessible to experimental manipulation during the early stages of development when the hookup between the two organs is being made. In lower vertebrates, the visual center of the brain in which the optic neurons terminate is termed the *tectum* (Fig. 14.25), and the precise location to which each part of the sensory retina sends its neurons can be mapped. Retinotectal mapping involves the simultaneous stimulation of selected sites on the retina and the measurement of electric activity in the brain. The sensory cells of the retina can be stimulated either by shining a very fine beam of light on a small retinal area or by implanting an electrode. The corresponding site, or *projection*, on the tectum is measured by determining which part of this brain region becomes electrically active. It is essential, for the sense of sight, that the spatial organization of the visual field be maintained during the transfer of this information through the eye to the brain. Consequently, it is found that there is a predictable point-to-point correspondence between the retina and the tectum. For example, a spot in the nasal portion of the visual field is "seen" by a retinal ganglion cell in the temporal region of the right retina, which projects to the rostral region of the left portion of the tectum (Fig. 14.25c).

Figure 14.25 Visual pathways of the amphibian. In *a*, the right eye with optic nerve is shown after section and regeneration. The left eye and optic nerve are not shown. The right optic pathway crosses at the optic chiasma to form the left optic tract, which is distributed to the left optic lobe or tectum, as shown in *b*. [*From R. W. Sperry,* Growth, *suppl.*, **10**:*66* (*1951*).] In *c*, the orderly maplike projections from the right and left retinas to the left and right optic tecta are shown diagrammatically. P, A, D, and V on the retinas signify posterior, anterior, dorsal, and ventral. [*After J. C. Eccles, "The Understanding of the Brain," McGraw-Hill, 1973.*]

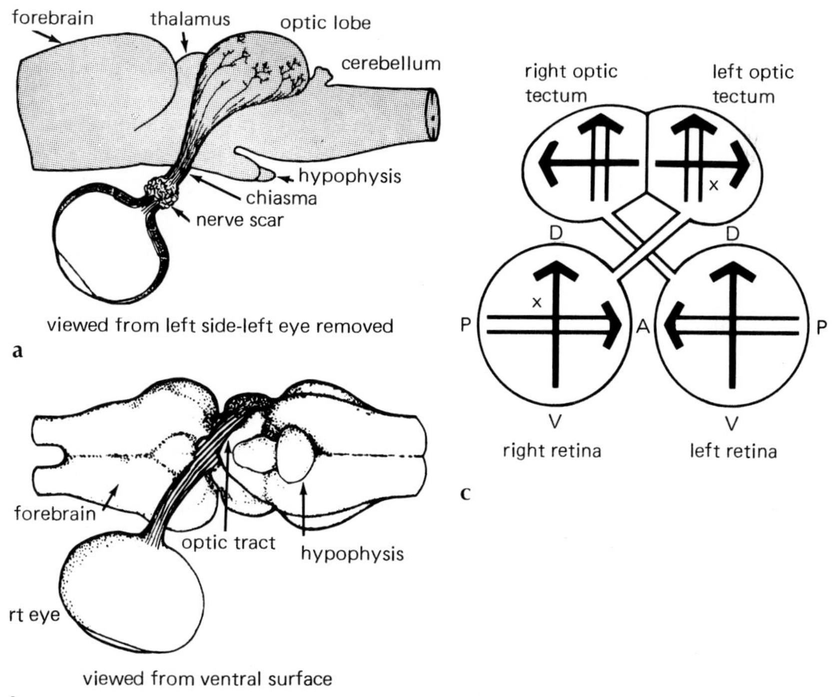

In order to better understand the mechanism by which optic axons become linked to specific brain cells, the strategy that has been most widely adopted has been to manipulate the eye in some manner at various stages before the axons reach the brain and determine the effect this has on (1) the subsequent projection to the tectum or (2) the subsequent behavior of the animal. If the retina of the salamander *Ambystoma* is rotated at a very early stage in eye development, there is no effect on the visuomotor responses of the adult salamander (Figs. 14.26). If, however, the retina is rotated at a later stage, still well before the optic neurons make contact with the brain, the visuomotor responses are inverted to the same extent as the original rotation. For example, if the retina is rotated 180° at the later stage, the adult salamander that develops will strike 180° away from a food

Figure 14.26 *a* Early optic vesicle (*central figure*) of Harrison's stages 21, 23, 28, 31, and 34 of *Ambystoma punctatum* embryos (*indicated by arrows*); in an embryo of each stage the vesicle is excised, rotated 180°, and reimplanted. D (dorsal), V (ventral), N (nasal), and T (temporal) poles are reversed. All hosts later showed normal vision through the grafted eye regardless of orientation. *b* Harrison stage 36 of *A. punctatum* in which the early optic cup is excised, rotated 180°, and transplanted to a new host. Later, the visuomotor reactions of the host are reversed. [*After L. S. Stone,* J. Exp. Zool., **145:**86 (1960).]

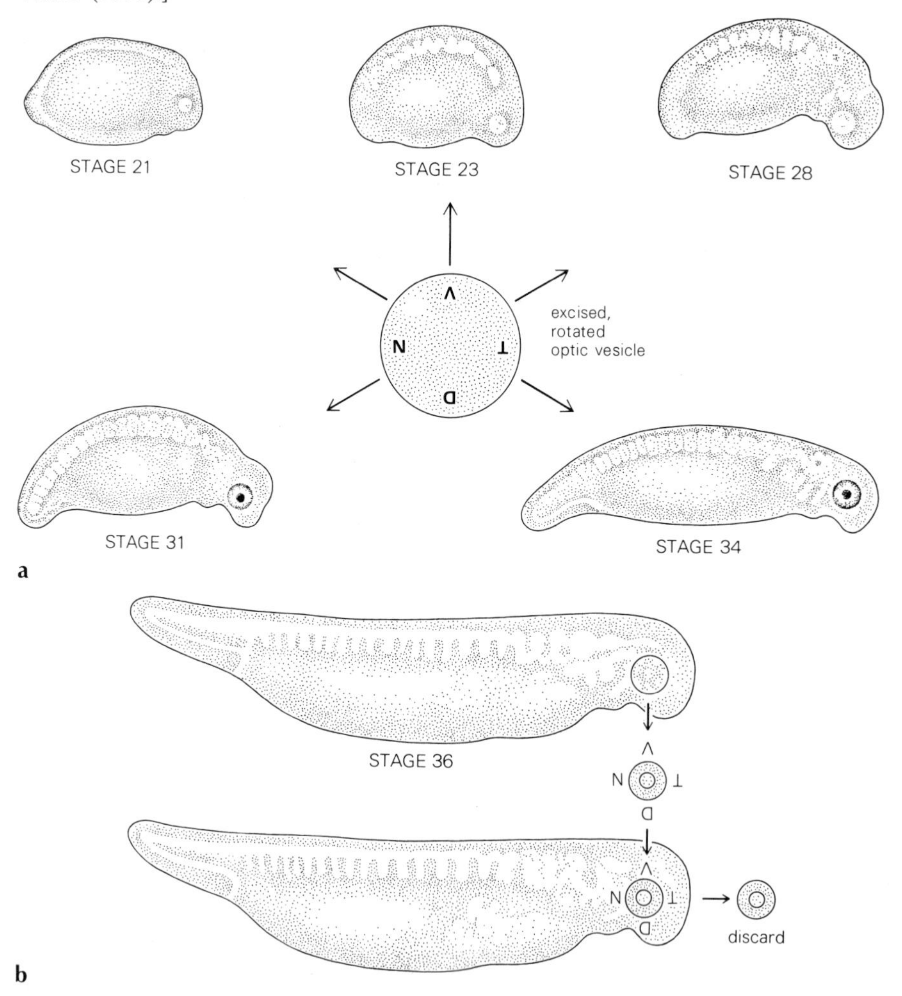

object swung into its visual field. This condition is never corrected. Responses remain completely abnormal unless the retina is rotated back to its normal orientation in adult life. Inverted visual behavior indicates that each neuron is growing out to the same site in the tectum that it would have if the retina had never been disturbed. At some point during the outgrowth process, well before axons have reached the brain, each neuron gains an identity that dictates to which part of the tectum it will proceed. This is determined by the retinal site from which it grows. If rotation is performed after this stage has been reached, the neurons continue to behave as if they were derived from the original site within the eye socket, even though this site has now been displaced.

The basis for the specification process is totally obscure. In keeping with investigations of the specification of other types of cells discussed elsewhere, it is believed that the cells of the retina exist within some type of polarized field that provides positional information to the individual cells. As with other systems, the information may exist in the form of a metabolic gradient or graded signal. Since the retina is, in essence, a two-dimensional circular screen, it is likely that each cell is specified in each of two perpendicular axes; each ganglion cell would reside at a particular latitude and longitude within the retinal field. In the case of the retinal cells, interpretation of the positional information would lead to the differentiation of cells such that each cell (or group of cells) would acquire a distinct, position-dependent surface identity. This unique identity might aid the cell in growing toward the proper site in the tectum, and it might facilitate the cell's synaptic linkage to the appropriate cell in the tectum, a cell that also has gained some type of positional identity.

These concepts have been tested in more recent studies on *Xenopus*. If the embryonic *Xenopus* retina is rotated 180° at stage 28, there is no effect on the projection of the ganglion cells to the tectum, nor is there any disturbance of the animal's visuomotor responses. This result indicates that specification has not occurred by stage 28. However, if the rotation is performed a few hours later at stage 30, the retinotectal projection is normal along the dorsoventral axis of the retina, but inverted along its anteroposterior axis, resulting in a visuomotor response correspondingly abnormal in one direction of the visual field. This result suggests that specification occurs along the two axes at different times, a finding reminiscent of determination of the amphibian limb (Section 13.1). When the retina is rotated at stage 32, the retinotectal projection is totally inverted; specification appears to be complete. Moreover, the specification process, whatever its nature, seems to occur very rapidly, because vision, if affected, is always affected to the same degree as the rotation. It is remarkable that the specification process occurs long before the optic axons reach the brain, an event that occurs about stage 40.

As mentioned earlier, rotation of the retina at stage 28 has no effect on the retinotectal projection, indicating that the retinal ganglion cells in their new location within the eye socket now project to a site in the tectum that is appropriate to that new position. If, however, a stage 28 eye is removed and cultured in vitro for a few days and is then transplanted into the socket of a suitable host, it is found that the resulting projection is normal only if the eye is grafted in the same anatomic position it had in the original donor (Fig. 14.27). The eye was originally removed at a stage when it could be influenced by rotation, yet in tissue culture its ganglion cells became specified just as they would have had the eye been left in place. This result suggests that some stable determinative influence has been felt in the retina by stage 28; otherwise a more complete specification on the same terms would not have occurred in culture. That rotation can overcome this pre-

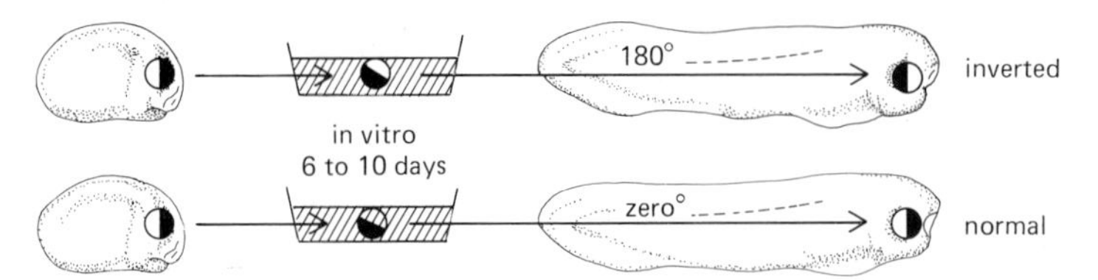

Figure 14.27 Diagram of an experiment showing that the optical axes of the *Xenopus* eye can become irreversibly specified in vitro. The eye is removed at stage 28 (a time at which specification can be reversed following experimental inversion) and cultured for several days. Specification occurs in vitro according to the original orientation of the eye within the socket. [*After R. K. Hunt and M. Jacobson,* Proc. Natl. Acad. Sci. U.S., **70**:*507 (1973).*]

liminary influence suggests that it is readily reversible. Once stage 32 is reached, specification is essentially irreversible, as measured by rotation experiments; yet other types of surgical procedures demonstrate that even after stage 32, modifications can still occur. Surgical techniques allow investigators to construct composites of retinas derived from two sources. For example, a retina can be made with the left half of one and the ventral half of another. In such a composite, one-quarter of the original retinal architecture would be absent (the right-dorsal quadrant), while another quarter (the left-ventral quadrant) would be represented twice. If retinal projection is completely fixed at stage 32, a composite such as this made after that stage should produce a tectum without connections in one quarter and doubly innervated in another. Such is not the case, and a much more normal retinotectal projection is found; specification after stage 32 is still to some extent reversible. Another example of this type of "regulation" is shown in Fig. 14.28. A similar type of result is found after the removal of half the tectum; fibers from the entire retina now project on the existing half-tectum as if the projection is compressed to meet the needs of the new situation. These latter types of regulative responses have led to the proposal that the retinal and tectal rudiments have the properties of a morphogenetic field (Section 13.1). Removal

Figure 14.28 The relationship between visual field, retina, and tectum in *Xenopus* with normal retina, double-nasal compound retina, and double-temporal compound retina. Normally, points *A, B, C* in the visual field project in the same order rostrocaudally on the tectum. In the compound retina the projection from each half-retina has spread over the whole contralateral tectum. The order of tectal projection from the compound retina is correct for the original half of the retina, but is reversed for the grafted half-retina. [*From R. M. Gaze, M. Jacobson, and G. Székely,* J. Physiol. (*London*), **165**:*484–499 (1963).*]

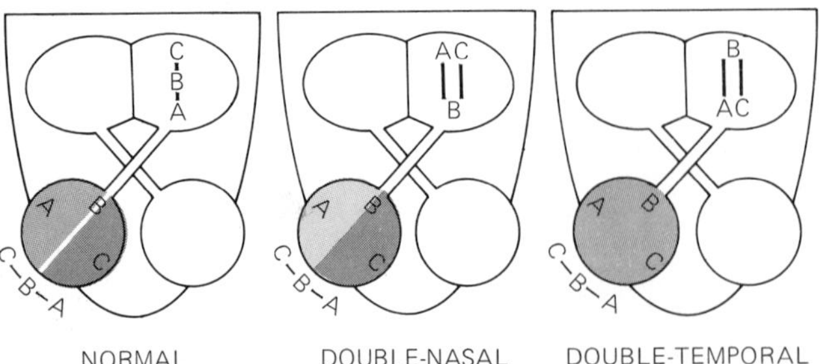

of one part of the retinal or tectal field leads to a readjustment of boundary values so that the positional information available to the remaining neurons allows them to acquire the complete spectrum of specific identities.

One of the most remarkable features of the specification of retinal ganglion cells is that it occurs at a time when the retina contains only a small percentage of its final cell number. At stage 32, when specification becomes irreversible on the basis of rotation experiments, the retina contains only a few hundred ganglion cells, compared with the approximately 50,000 of the adult eye. Growth of the amphibian retina occurs by the addition of new cells at the periphery; i.e., growth occurs concentrically. Cells in the center of the retina stop dividing and undergo terminal differentiation to become specified neuroblasts. In contrast, the cells at the periphery represent stem cells that continue to divide, thereby producing more and more cells, some of which differentiate into postmitotic neuroblasts while others continue to act as stem cells. The development of the retina poses certain thorny questions. How can rotation of the eye at stage 32 lead to the inversion of the entire retinotectal map in the adult when 99 percent of the retinal ganglion cells have not even been formed at this early stage? Clearly, the positional identity of the remainder of the ganglion cells must be obtained from those present at stage 32. How is this information passed on to the newly formed cells? Does it follow automatically from cell lineage, i.e., is the information passed on from parental cells to their daughters at division, or does it arise by some type of intercellular communication? The answer is not known. Another important question arises from the very different types of geometry of retinal and tectal growth. While the retina grows in concentric rings, the tectum extends backward (caudally) from its anterolateral pole. Despite this very different change in shape of the two tissue masses, the orderly array of neural connections that produces the retinotectal map is retained. It has been suggested (see R. M. Gaze et al., 1974) that the synapses between retinal and tectal neurons are continually shifting during the growth period (the retinal fibers would move caudally across the tectum) in order to allow the spatial organization of the visual field to be preserved.

Two aspects of retinal-cell biology have been recorded that can be correlated with the time of specification at stages 29 to 31. One of these is the cessation of cell division in the central part of the retina. The finding that cells become postmitotic just prior to their specification and outgrowth suggests that this transition may be a causal factor in the switch from the reversibly specified to the irreversibly specified state. The second change occurring during this period is the uncoupling of the cells of the central part of the retina. In the earlier stages, gap junctions are seen between cells in all regions of the retina (including gap junctions between cells of the neural and pigmented layers of the retina). However, at about the time that specification begins, the central cells are no longer in communication with each other, a phenomenon that may reflect the fact that these cells are now proceeding along separate paths of differentiation.

The point has been raised several times in this chapter that pre- and postsynaptic neurons may recognize each other as a result of some type of specific recognition and adhesion process mediated by chemical groups on their respective cell surfaces. Direct evidence for this statement has come from studies in which cell-surface affinities of retinal and tectal cells have been assayed in vitro. In one type of study ^{32}P-labeled cells dissociated from either the dorsal or ventral retina were incubated in a dish containing a ventral and dorsal half of the tectum, the brain pieces acting as collecting sites for the binding of complementary retinal cells. In this experiment, the labeled cells from the dorsal retina bind preferentially to the

ventral tectum, while the cells from the ventral retina adhere preferentially to dorsal tectal cells (Fig. 14.29). This finding is consistent with the fact that nerve fibers from the top half of the eye will connect with the bottom half of the tectum, and vice versa.

In order to determine the basis for the specific adhesion of these two types of neurons, preparations of retinal or tectal cells have been treated with various enzymes prior to their incubation with the other cell type. Treatment of ventral tectal cells or ventral retinal cells with proteolytic enzymes reduces their ability to bind to their dorsal counterparts, while treatment of dorsal tectal or dorsal retinal cells with the same protease has little effect on its adhesive properties. Conversely, treatment of dorsal retinal or tectal cells with an enzyme that removes terminal *N*-acetylhexosamine (*N*-acetylglucosamine and *N*-acetylgalactosamine) residues from the end of glycoproteins, reduces the binding of these cells to their ventral counterparts. The same enzyme has little effect on the adhesive capabilities of ventral retinal or tectal cells. Based on these types of results, it has been proposed that retinotectal adhesive specificity is mediated by the interactions of two cell-surface molecules, one a protein and the other a glycoprotein that binds via its carbohydrate moiety. It is proposed that these two molecules are each present in a gradient along the dorsoventral axis of each tissue. In both organs, the glycoprotein is present with its highest concentration dorsally, and the complementary carbohydrate-binding protein is present at greatest concentration ventrally. This double-gradient model is depicted in Fig. 14.30. Whether

Figure 14.29 Number of ^{32}P-labeled cells adhering to dorsal and ventral tectal halves of 12-day chick embryos as a function of collection time. Cell suspensions were made from 7-day dorsal half-retinas. Cells adhering to dorsal tectal halves (○); cells adhering to ventral tectal halves (●). [*From S. Roth and R. B. Marchase, in "Neuronal Recognition," S. H. Barondes (ed.), Plenum, 1976.*]

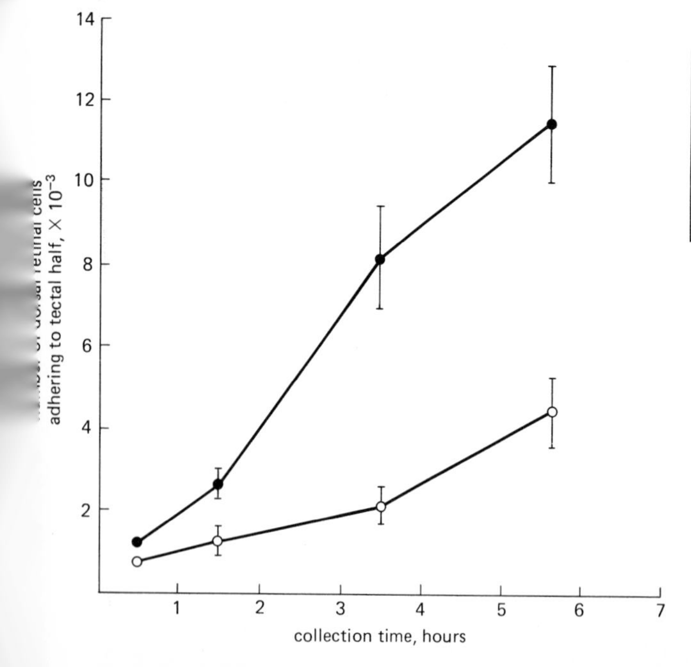

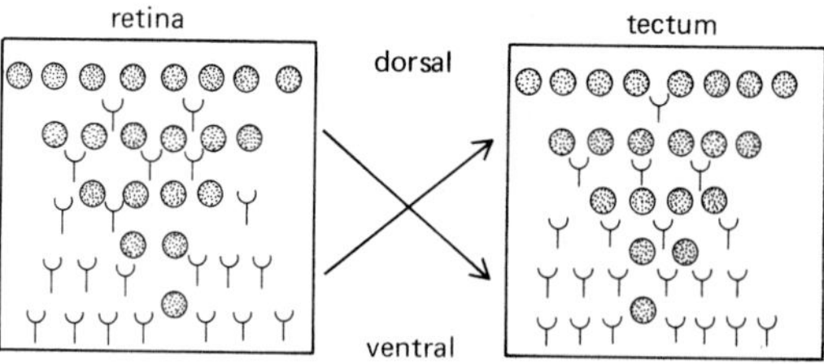

Figure 14.30 A model of retinotectal specificity based on gradients of complementary molecules in the dorsoventral axis that could provide adhesive connections corresponding to the observed retinotectal map. As a result of interactions between complementary molecules, the most stable connections would form between the dorsal tectum and ventral retina and the ventral tectum and dorsal retina. [*From R. B. Marchase,* J. Cell Biol., **75**:*244 (1977).*]

or not such a complex interaction as that between the eye and the brain can be explained on the basis of the concentration of two specific macromolecules remains to be seen.

5. Development of behavior

The study of developmental neurobiology has implications well beyond the construction of the complex circuitry of the nervous system. To what extent does an animal's behavior patterns automatically follow the generation of the proper neural connections, and to what extent is it determined by a period of trial-and-error learning experience? Stated in more familiar terms, the question seeks to measure the relative roles of genetics and environment, nature versus nurture, in the development of specific behavioral responses. During the first half of this century, the emphasis in this area was placed on the role of experience in shaping neural function. This general attitude was embodied in the concept of contact guidance, by which anonymous nerve fibers, i.e., ones whose functional identity had not been specified, were guided to their targets by mechanical directives. It was only after they had innervated their target that their function was determined. In the 1950s it became more accepted that the information coded within the genetic material, acting via the establishment of specific neuronal connections, was itself sufficient to account for behavior. The leading proponent of this view was Roger Sperry, who, for example, showed that surgical operations designed to lead to the formation of abnormal neural connections had drastic effects on behavior. It was mentioned in the last section that depending on the stage, rotation of the retina of lower vertebrates leads to permanently maladaptive visuomotor responses. A frog that has undergone this operation shows no ability to correct for its poorly aimed feeding movements.

Another approach to this question has been to block all muscular activity during the period when trial-and-error learning might be occurring. Salamander embryos raised in the presence of chlorobutanol during the formative period are immediately capable of completely normal swimming movements when the drug is removed, even in the absence of any previous muscular function. In lower vertebrates, therefore, it seems that once the proper nervous connections are made, specific behavior follows without the need for instruction to develop correct neural circuits.

In a comparable series of studies on the chick, Viktor Hamburger has analyzed the nature of the nervous system and its relation to muscular activity. The first movements in the chick embryo occur at about the fourth day of incubation with movements of the head. In human embryos, a similar initial movement occurs at about the eighth week. In the chick this activity results from the completion of motor circuits from the trigeminal nerve to cervical neck muscles. During the next few days, an increasing number of muscles become active as more and more motor innervation takes place. Until the eighth day there are no functional sensory nerves, and therefore all movements occur in the absence of sensory stimulation. In other words, nervous activity responsible for this movement originates within the central nervous system and is unrelated to environmental stimulation. Analysis of movements up to the seventeenth day suggests that there is no trial-and-error learning period in the chick. Movements through this period remain uncoordinated, and in many cases, antagonistic muscles are undergoing contraction at the same time. Only after the seventeenth day do integrated behavioral movements take place, and it appears that they do not result from previous muscular activity.

The behavioral machinery as a whole apparently works according to how it is built. For the organism to function at all, many responses have to be ready to act appropriately upon the first demand. Nevertheless, especially in mammals, much of the response system matures after birth, apparently dependent on sensory stimulation. This has been best explored in studies of the visual system, particularly in kittens. A few of these experiments will be described; the reader should keep in mind that the subject is a particularly complex one and experimentation in this area has only begun.

The fibers of the optic nerve in mammals no longer project to the tectum of the mesencephalon, but instead form synapses with neurons of the lateral geniculate nucleus (LGN) of the diencephalon. As with the tectum of lower vertebrates, the spatial organization of the retina is preserved in the pattern of connections to the cells of the LGN. From the LGN, neurons project to the visual cortex of the cerebral hemispheres, where electrodes placed into *single* cortical neurons are capable of detecting electric activity under various types of visual stimulation. There are several factors that determine whether or not a given cortical neuron will be activated by a change in the visual field in a corresponding retinal position. These factors include the size of the object, its color, the speed it is moving, the direction it is taking, its orientation, the contrast at its moving edge, and its shape. Each neuron is tuned to these various features of the environment; such neurons can be thought of as "feature detectors." As a result of the selectivity of the millions of cortical neurons, a large body of very precise information can be gleaned about an object in the visual field.

When the electric activity of cortical neurons in a young kitten is compared to that of an adult cat, marked differences are found. Although the neuronal connections are similar, there appear to be large numbers of nonresponsive neurons in the kitten cerebrum, and in many cases those which do fire on visual stimulation seem to do so in a less specific manner; i.e., various types of stimulation can activate the cells of the kitten. The fact that the brain of a newborn shows different response properties from that of the adult is no guarantee that the maturational changes are dependent on visual experience. In an attempt to determine whether or not experience plays a role in brain-cell function, kittens have been deprived of various types of stimulation and the subsequent effect, if any, monitored. These experiments have indicated that two types of deprivation can have marked effects on brain neurons. One type of deprivation affects the development of binocular vision, the other affects the development of orientation-specific cells.

Under normal circumstances, electric activity in adult-cat cortical neurons can be evoked by stimulation of corresponding sites in either retina. In other words, information from both eyes ultimately impinges on the same brain cells as a result of the complex circuitry of the brain. It is this electrical convergence that allows us to see a single image in its three-dimensional quality when observing the field with two distinct sets of photoreceptors. If kittens are deprived of vision in a single eye by lid suture or other means during a critical period from 3 weeks to 3 months of age, their cells never develop this binocularity. When single-unit measurements are made on these cats after they have grown, it is found that nearly all the cortical neurons can be driven only by stimulation of the nondeprived eye. For all practical purposes, these cats have been made blind in the closed eye.

Although these results are very informative, they do not necessarily indicate that stimulation from both eyes is normally responsible for the development of

binocular vision. It remains possible that experience is necessary to maintain neuronal function *after* it is formed by genetic activity. In this case, the deprivation would be having a destructive effect on cells that would have been binocularly driven had the eye not been closed. In other words, the development of binocularity may be experience-sensitive rather than experience-dependent. There are good reasons for believing that binocular vision would be experience-dependent, since it is hard to imagine that such a precise correspondence between points in the two retinas could develop autonomously. Slight asymmetries in the growth of the head, for example, would cause serious problems in visual alignment. It would seem more reasonable to let the brain find out which cells of the two retinas are seeing the same spot in the visual field. Since corresponding sites in the two retinal fields will almost always be stimulated in the same manner, one could conceive of various ways in which brain cells could measure the disparity between the eyes and hook up the appropriate photoreceptors with the same cortical neurons. Some of the best evidence that the development of binocularity is experience-dependent comes from studies in which kittens are caused to wear goggles that increase the disparity between the eyes, i.e., cause a relative shift of the image of the visual field on the two retinas. Under these conditions, it might be expected that single brain cells would become activated by stimuli impinging at different positions on the two eyes, the disparity reflecting the shift induced by the goggles. Preliminary evidence indicates that just such a compensatory shift as this occurs to maintain interocular matching.

As mentioned earlier, the other aspect that appears to be experience-dependent is orientation selectivity. Many cortical neurons are specific for a line or edge moving across the animal's field along a particular orientation or angle. Any one cell is specific to a particular orientation, but all angles through the entire range of 360° are covered by the population of nerve cells in this area of the brain. The development of these orientation-specific neurons can be greatly affected by visual experience. Kittens can be raised in visual environments consisting of only vertical stripes or only horizontal stripes. This is usually accomplished by placing the animals in large cylinders with one or the other type of stripe painted on the inner surface of the chamber. When cortical neurons are tested in cats that have been exposed to environments containing vertical stripes, there is a virtual absence of neurons that respond when the retina is stimulated with horizontal lines, and vice versa (Fig. 14.31). It appears that "feature detectors" do not develop for lines of the type that the animal had not seen during its early critical period. There are marked behavioral effects as well. Kittens raised with horizontal stripes were perfectly capable of jumping on a chair but, when walking on the floor, they kept bumping into the chair legs as though they were invisible. The "vertical" cats avoided the chair legs, but never attempted to jump on the chair, as though the seat did not exist.

Other workers have measured the speed at which the character of the feature detectors is specified, with the astounding discovery that as little as 1 hour of visual experience can change the nature of feature detectors, much as though it were a learning experience. All these discoveries have important bearing on human development. Infants with uncorrected squint or severe astigmatism will grow up with permanent defects because it is the brain that is affected, not the eyes. Correction must be made while the brain is still plastic, during the visual-sensitive period of the first 2 to 4 years. What is true of vision probably applies to our other senses as well. We all know that infants go through a period of intense curiosity. This may be a behavioral expression of the brain's most sensitive period

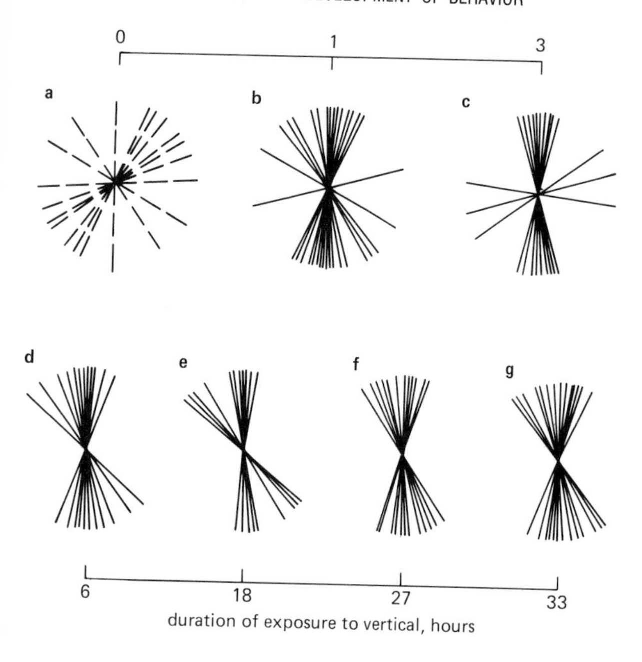

Figure 14.31 Diagrams summarizing the effects on the visual cortex of different lengths of time spent in an environment of vertical stripes for the number of hours indicated. Each line represents the optimal orientation for one cortical neuron. The control animal in *a* was totally deprived of visual experience. In this latter case, the lines are interrupted to indicate that these neurons showed only the slightest preference for the angle shown and also responded to all other orientations. It appears that as little as 1 hour of experience to the vertical stripes can create a population of cells preferring vertical stimulation. As time of visual deprivation increases, the preference for vertical stimulation increases. [*From C. Blakemore and D. E. Mitchell,* Nature, **241:***467 (1973).*]

for acquiring knowledge and learning techniques. Unless the specific sensitive periods in the development of the brain are exploited to the full at the right time, their potential may be lost forever.

Readings

ADELMANN, H. B., 1936. Problems of Cyclopia, *Quart. Rev. Biol.*, **11:**161–304.
BARLOW, H. B., 1975. Visual Experience and Cortical Development, *Nature,* **258:**199–204.
BARONDES, S. H., (ed.), 1976. "Neuronal Recognition," Plenum.
BLAKEMORE, C. B., and G. F. COOPER, 1970. Development of the Brain Depends on the Visual Environment, *Nature,* **228:**477–478.
BLOEMENDAL, H., 1977. The Vertebrate Eye Lens, *Science,* **197:**127–138.
BRADSHAW, R. A., 1978. Nerve Growth Factor, *Ann. Rev. Biochem.*, **47:**191–216.
BUNGE, M. B., 1973. Fine Structure of Nerve Fibers and Growth Cones of Isolated Sympathetic Neurons in Culture, *J. Cell Biol.*, **56:**713–735.
CHUNG, S.-H., and J. COOKE, 1975. Polarity of Structure and of Ordered Nerve Connections in the Developing Amphibian Brain, *Nature,* **258:**126–132.
COUGHLIN, M. D., 1975. Target Organ Stimulation of a Parasympathetic Nerve Growing in the Developing Mouse Submandibular Gland, *Develop. Biol.*, **43:**140–158.
COULOMBRE, A. J., 1965. The Eye, in R. L. DeHaan and H. Ursprung (eds.), "Organogenesis," Holt, Rinehart and Winston.
COWAN, M. W., 1979. Development of the Brain, *Sci. Amer.*, **241:**113–133 (Sept.).
ECCLES, J. C., 1973. "The Understanding of the Brain," McGraw-Hill.
EHRENPREIS, S., and I. J. KOPIN (eds.), 1976. "Reviews of Neuroscience," vol. 2, Raven.
GAZE, R. M., 1978. The Problem of Specificity in the Formation of Nerve Connections, in D. R. Garrod (eds.), "Specificity of Embryological Interactions," Chapman and Hall.
———, M. J. KEATING, and S.-H. CHUNG, 1974. The Evolution of the Retinotectal Map during Development in *Xenopus, Proc. Royal Soc. (London),* **B185:**301–330.
GOTTLIEB, G. (ed.), 1976. "Neural and Behavioral Specificity," Academic.
———, (ed.), 1973. "Behavioral Embryology," Academic.
GROBSTEIN, P., and K. L. CHOW, 1975. Receptive Field Development and Individual Experience, *Science,* **190:**352–358.

Hamburger, V., 1968. Emergence of Nervous Coordination, in M. Locke (ed.), "The Emergence of Order in Developing Systems," 27th Symp. Soc. Dev. Biol., Academic.
Holtfreter, J., 1939. Tissue Affinity, A Means of Embryonic Morphogenesis, in English translation, in B. H. Willier and J. M. Oppenheimer (eds.), "Foundations of Embryology," Prentice-Hall.
Hunt, R. K., 1975, The Cell Cycle, Cell Lineage, and Neuronal Specificity, in J. Reinert and H. Holtzer, "Cell Cycle and Cell Differentiation," Springer-Verlag.
———, and M. Jacobson, 1974. Development of Neuronal Locus Specificity in *Xenopus* Retinoganglial Cells after Surgical Eye Transsection or after Fusion of Whole Eyes, *Develop. Biol.*, **40**:1–15.
Jacobson, M., 1978. "Developmental Neurobiology," 2d ed., Plenum.
Kolata, G. B., 1979. Sex Hormones and Brain Development, *Science*, **205**:985–987.
Levi-Montalcini, R., and P. V. Angeletti, 1965. The Action of Nerve Growth Factor on Sensory and Sympathetic Cells, in R. L. DeHaan and H. Ursprung (eds.), "Organogenesis," Holt, Rinehart and Winston.
Levi-Montalcini, R., and P. Calissano, 1979. The Nerve Growth Factor, *Sci. Amer.*, **240**:68 (June).
Luduena, M. A., and N. K. Wessels, 1973. Cell Locomotion, Nerve Elongation, and Microfilaments, *Develop. Biol.*, **30**:427–440.
Marchase, R. B., 1977. Biochemical Investigations of Retinotectal Adhesive Specificity, *J. Cell Biol.*, **75**:237–257.
Puro, D. G., and M. Nirenberg, 1976. On the Specificity of Synapse Formation, *Proc. Nat. Acad. Sci. U.S.*, **73**:3544–3548.
Sedlacek, J., 1978. The Development of Supraspinal Control of Spontaneous Motility in Chick Embryos, *Prog. Brain Res.*, **48**:367–384.
Sidman, R. L., and P. Rakic, 1973. Neuronal Migration, with Special Reference to the Developing Brain: A Review, *Brain Res.*, **62**:1–35.
Sperry, R. W., 1965. Embryogenesis of Behavioral Nerve Nets, in R. L. DeHaan and H. Ursprung (eds.), "Organogenesis," Holt, Rinehart and Winston.
Spinelli, D. N., and F. E. Jensen, 1978. Plasticity: The Mirror of Experience, *Science*, **203**:75–78.
Stone, L. S., 1960. Polarization of the Retina and Development of Vision, *J. Exp. Zool.*, **145**:85–93.
Symposium on Synaptogenesis, 1978. *Fed. Proc.*, **37**:1999–2021.
Szekely, G., 1974. Problems of Neuronal Specificity in the Development of some Behavioral Patterns in Amphibia, in G. Gottlieb (ed.), "Aspects of Neurogenesis," Academic.
Trelstad, R. L., and A. J. Coulombre, 1971. Morphogenesis of the Collagenous Stroma of the Chick Cornea, *J. Cell Biol.*, **50**:840–858.
Weiss, P., 1955. Nervous System, in B. H. Willier, P. A. Weiss, and V. Hamburger (eds.), "Analysis of Development," Saunders.
Yntema, C. L., 1955. Ear and Nose, in B. H. Willier, P. A. Weiss, and V. Hamburger (eds.), "Analysis of Development," Saunders.

Chapter 15 Cytodifferentiation

Cytodifferentiation is the process whereby cells acquire those biochemical and morphological properties necessary to perform their specialized functions. The transformation from the *undifferentiated* to the fully *differentiated* state is a gradual one. It begins with a shift in developmental potential from the totipotent to the more restricted. As development proceeds, the repertoire of types into which a given cell can differentiate becomes gradually limited. In vertebrates there is no evidence of truly undifferentiated cells that can be directed into any morphogenetic pathway under the appropriate environmental conditions. Rather, the process of cytodifferentiation in the embryo and the continual process of tissue renewal in the adult are accomplished by a diverse population of *stem cells* (Fig. 15.1). Stem cells are ones that have simultaneously (1) an extensive proliferative capacity and (2) the ability to generate other types of cells of more restricted developmental potential. These latter cells may, in turn, have stem-cell properties, in that they are capable of both proliferation and formation of more determined cell types. From these stem cells, eventually, a large variety of specialized cells are formed.

It should be kept in mind that cytodifferentiation, which occurs within cells and leads to the formation of a specialized cytological character, is invariably accompanied by a distinct but interrelated set of *morphogenetic* activities, including cell growth, localized cell death, cell movement, change of cell shape, change of cell adhesion, and the deposition of extracellular matrices. These latter processes act at the supracellular level to sculpt the shape of an organ and its component tissues and determine the architectural arrangement of the various cell types. In this chapter we will examine the process whereby a variety of precursor cells (stem cells) undergo the terminal steps in the differentiation of each cell line. We will return in the following chapter to take up the topic of morphogenesis, which is less well understood.

Although each cell type has attained its state of differentiation via a predictable and unique series of events, developmental biologists have sought to uncover principles that can be applied to many, if not all, of these cytodifferentiations. Questions of general importance considered in this chapter include: What mechanisms are responsible for shifting a cell along one path of specialized dif-

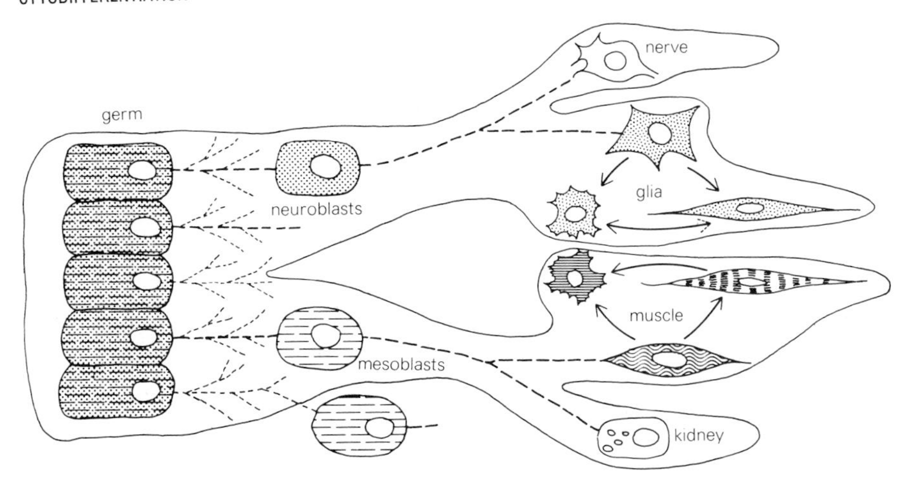

Figure 15.1 Differentiation of cells in steps, from unspecialized embryonic cells through stem cells such as neuroblasts and mesoblasts to differentiated nerve, glia, muscle, and kidney cells. [*After P. Weiss,* J. Emb. Exp. Morph., **1**:*189* (*1953*).]

ferentiation rather than another? What is the nature of the primary events leading to differentiation, and at what stage do they occur? At what stages in the differentiation process can specialized gene products, such as hemoglobin or myosin, be identified? To what extent is cytodifferentiation determined by interactions with the cellular and noncellular environment as opposed to internal factors? To what extent is the continuing cycle of mitotic divisions involved in the cytodifferentiation process? How stable is the differentiated state, and under what conditions, if any, can it be reversed? To what extent can cytodifferentiation be explained in terms of a sequential series of self-assemblies?

1. Cell specificity

An outstanding feature in a multicellular organism, whether a hydra or a mammal, is that each differentiated cell type falls into a clearly recognizable discrete category, sharply set off from other cell types.

The various types of cell specialization appear to be accentuations of the properties of the unspecialized or undifferentiated cell, such as plasma membrane, intracellular organelles, cilia or flagella, microfibrils and microtubules, etc. Secretory cells like those of the pancreas show extreme development of the endoplasmic reticulum–Golgi system, produce specific zymogen proteins, and have a sharply polarized internal organization. Nerve cells combine localized axoplasmic growth and terminal secretion with considerable development of the electrochemical features of the plasma membrane. In retinal cells, stacks of infolded plasma membrane form the pseudocrystalline rods. Schwann cells produce excessive plasma membrane that becomes wrapped around nerve axons as the myelin sheath. Contractile cells have hypertrophied systems of actomyosin microfibrils. Spermatozoa are primarily condensed cells with extreme development of the microtubule system characteristic of all cilia or flagella. All such specializations are recognizable in the general features of the undifferentiated cell, although each may be unique in having certain proteins not present in others. In a sense, the general properties of the cell represent the keyboard on which various types and patterns of selection are made.

2. Erythropoiesis

Erythropoiesis is the process by which erythrocytes (red blood cells) are formed. As with many other types of cells, erythroid cytodifferentiation (Fig. 15.2) can be studied in adults as well as embryos, although there are some important differences between the two systems, ones that will be discussed later. Erythrocytes are unusual terminally differentiated cells; they have an extremely simple internal structure and, once mature, lack virtually any synthetic activity. In fact, the red blood cells of vertebrates either lack a nucleus entirely, as in mammals, or contain one that is completely devoid of metabolic activity. They are, in essence, a membrane-bound solution of hemoglobin, a metalloprotein found only in cells of the erythroid line. Each hemoglobin molecule consists of a protein, globin, composed of four polypeptide chains—of which several different species are encoded within the genome—and four iron-containing heme groups, each capable of transporting a single molecule of oxygen through the bloodstream.

Erythropoiesis is unusual among processes of differentiation in that it occurs at successive sites within the body of the embryo. In both birds and mammals, erythropoiesis begins in the yolk sac when clusters of mesodermal cells (termed

Figure 15.2 Diagrammatic representation of the stages in the differentiation of the red blood cell. *a* The hypothetical pluripotent mesenchymal cell, which can give rise to more than one type of stem cell. *b* The hemocytoblast, the stem cell of the erythroid line. These cells are identifiable and determined but do not show overt erythroid differentiations. *c* The proerythroblast; a stage of active RNA synthesis in preparation for differentiation. *d* The basophilic erythroblast, characterized by chromosomal condensation and concomitant reduction in nuclear activity. Hemoglobin synthesis is believed to begin in these cells. *e* The polychromatophilic stage, characterized by increasing synthesis and accumulation of hemoglobin and decreasing levels of RNA synthesis. *f* The orthochromatic erythroblast, characterized by an inactive nucleus and hemoglobin-filled cytoplasm. These cells are no longer capable of cell division. *g* The reticulocyte, which has lost its nucleus and is found circulating in the bloodstream, continues to synthesize hemoglobin for a day or two. The final stage is the terminally differentiated erythrocyte, or red blood cell, characterized by the absence of protein synthesis. [*After R. A. Rifkind, in J. Lash and J. R. Whitaker (eds.), "Concepts of Development," Sinauer, 1974.*]

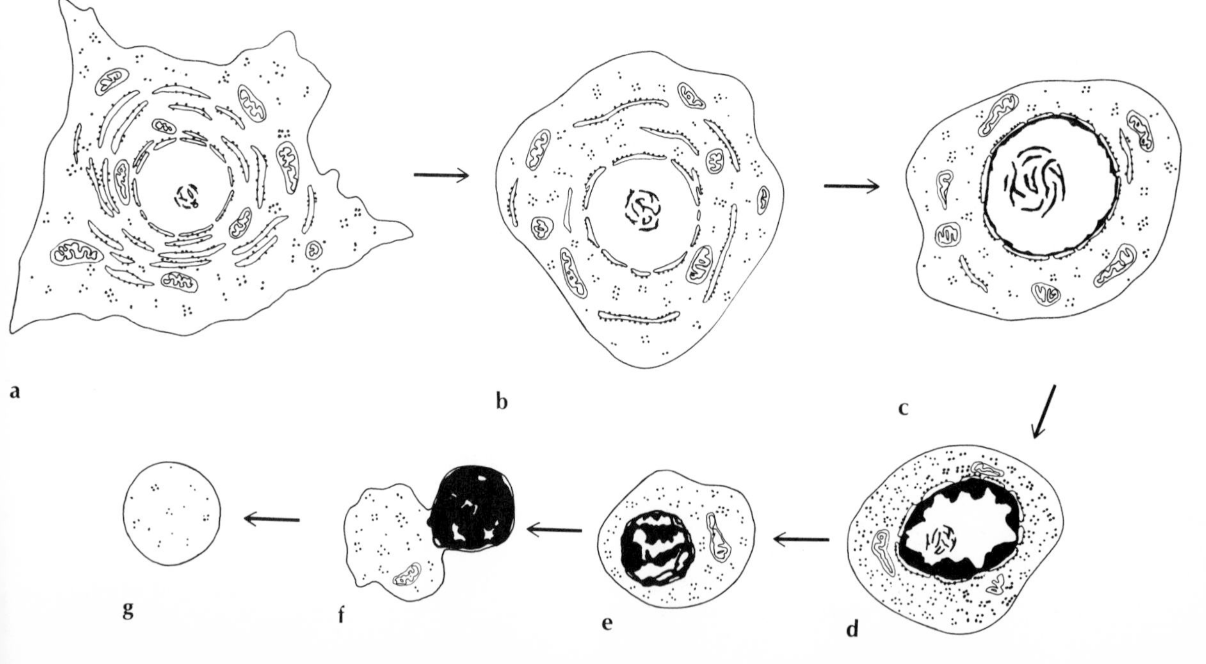

blood islands) become committed to differentiate along an erythroid pathway. These aggregations of mesodermal cells become induced by the underlying endodermal cells to form the endothelial wall of the blood vessels *and* the enclosed erythrocytes. Erythrocytes produced in the yolk sac of the early embryo are quite different from those produced at later stages of development (Fig. 15.3). The primitive nature of yolk-sac erythrocytes is reflected in their much larger size and, in mammals, the presence of a nucleus. Erythropoiesis begins in the yolk sac of the chick during the second day of incubation, and in the yolk sac of the mouse and human on the eighth and nineteenth day of gestation, respectively. Unlike that of later sites, erythropoiesis in the yolk sac is a very short-lived process, producing but a single generation of blood cells. All the erythrocyte precursors are released at about the same time from the yolk sac into the embryonic circulation, where they continue to divide several more times and complete their differentiation *in synchrony*. No reserve or self-perpetuating stem cells remain in the yolk sac; all participate in the formation of a finite number of large (*megaloblastic*) red blood cells. While the yolk-sac erythrocytes are still circulating, a new site of erythropoiesis appears, that of the liver. The erythrocytes formed in the liver are distinctly smaller and more adultlike in nature than those of the yolk sac. In many animals, erythrocytes produced in the liver are joined by similar (*normoblastic*) cells formed in the spleen. Finally, by about the twelfth day of chick development, and the fourth month of gestation in the human, erythropoiesis begins in the bone marrow, the site responsible for erythrocyte formation throughout the remainder of life. An important question in this entire matter concerns the site of origin of erythroid precursors. Some evidence suggests that cells from the yolk sac of the early embryo enter and colonize the tissues of the secondary erythropoietic organs, where they ultimately give rise to the blood cells of later stages. Other evidence suggests that the blood cells of the secondary organs arise independently within the tissues themselves. This question remains unanswered.

Not only does the site of erythropoiesis shift within the embryo, so too does the type of hemoglobin being synthesized. Once again, the situation is roughly simi-

Figure 15.3 Micrographs showing the morphologic differences between populations of erythroid cells of 7-day chick embryos separated on gradients of bovine serum albumin. *a* Definitive cell fraction. The field is chosen to include a basophilic erythroblast (*arrow*) that is normally present (in relatively low numbers) in the circulation of 7-day embryos. *b* Primitive nucleated cell fraction. [*Courtesy of* K. A. *Mahoney*, B. J. *Hyer, and* L.-N. L. *Chan.*]

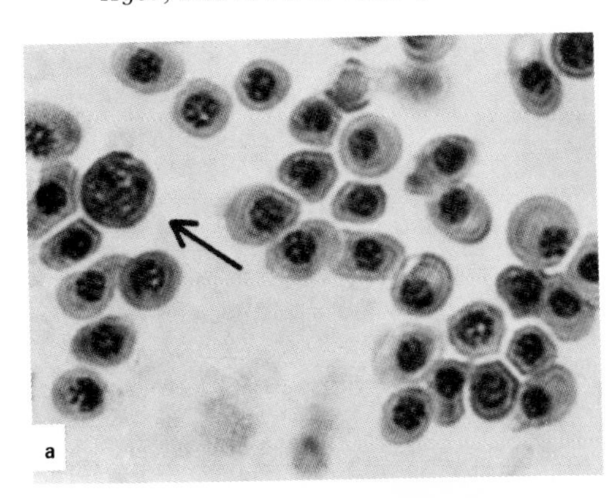

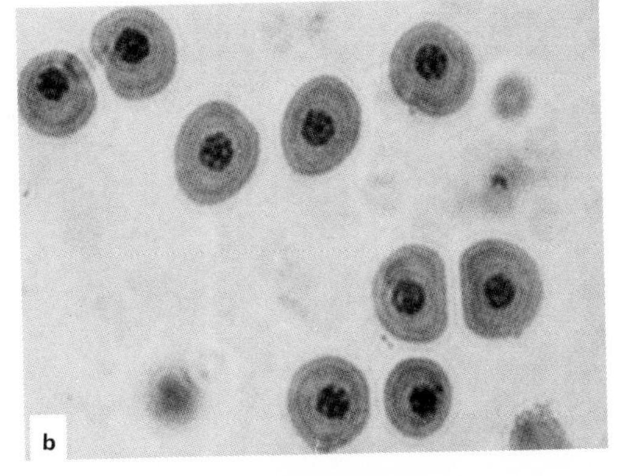

lar in birds and mammals; we will focus on events as they occur in the human organism. The various types of hemoglobins produced in humans are indicated in Fig. 15.4. The first cells found in the embryonic bloodstream contain a mixture of *embryonic* hemoglobins (Fig. 15.4). At the time of onset of erythropoiesis in the fetal liver, a new type of hemoglobin appears in the blood, namely, ***fetal*** hemoglobin (Hb F) (Fig. 15.4). It would seem likely that the switch from embryonic to

Figure 15.4 Structure and synthesis of human hemoglobins. *a* Upper row shows the various genetic loci that code for hemoglobin chains. The two γ genes code for polypeptides that differ at one amino acid residue. The two α genes code for identical chains. The lower row shows the various combinations of chains and the hemoglobins they form. [*From D. J. Weatherell and J. B. Clegg,* Cell, **16:***468 (1979)*.] *b* Developmental changes in human hemoglobin chains. [*From V. Vignon and J. Godet,* Int. Rev. Cytol., **46:***87 (1976)*.]

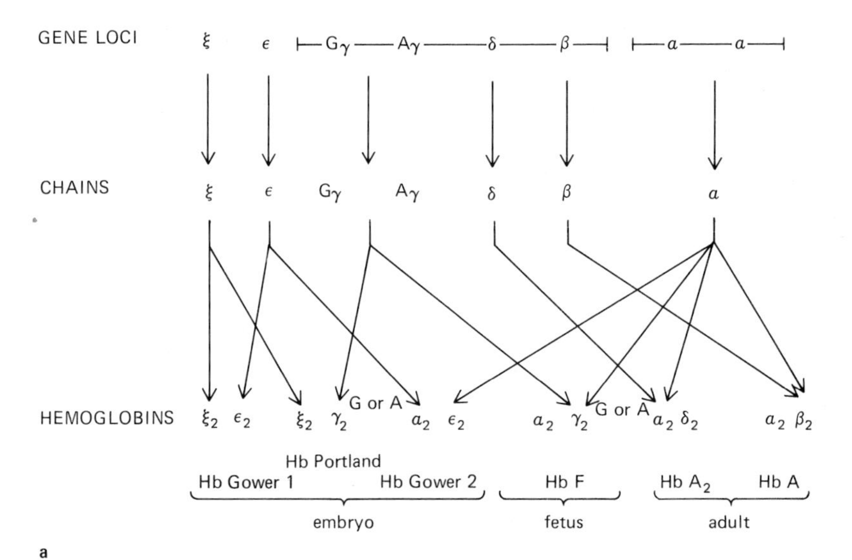

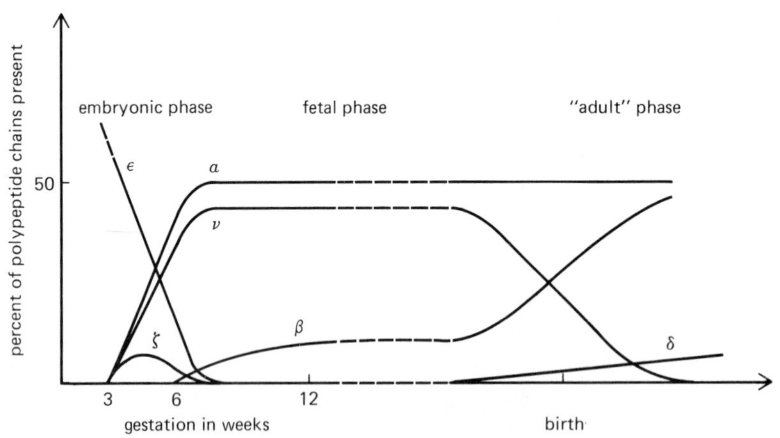

fetal hemoglobins simply reflects the switch from yolk-sac megaloblastic to liver normoblastic red blood cells. However, it has been shown that the switch in hemoglobin synthesis occurs in both populations coincidentally. Yolk-sac–derived erythrocytes, which have already been circulating for some time, begin to synthesize fetal hemoglobin, while liver-derived erythrocytes produced before the switch contain the embryonic species. That the switch actually occurs within single cells has been shown by using fluorescent antibodies; single red blood cells are found to stain with antibodies against both types of hemoglobin molecule. The genetic mechanism underlying the switch in protein synthesis is under intensive investigation.

By the end of the first trimester, the third and final type of hemoglobin, *adult* hemoglobin (Hb A) (Fig. 15.4), has appeared in the blood. As in the case of the first switch, the transition from production of Hb F to Hb A is not related to the site of erythropoiesis; it occurs synchronously throughout the liver, spleen, and bone marrow. At the end of the third month, Hb A is present at low levels in approximately 25 percent of the erythrocytes, all of them containing large amounts of Hb F. Fetal hemoglobin is characterized by a greater affinity for oxygen than the adult molecule, thereby ensuring oxygenation of the fetal blood. Within a week after birth, Hb A becomes highly predominant, and by 2½ years of age, over 99 percent of the hemoglobin circulating in the blood is Hb A, with virtually all the Hb F being restricted to a small population of cells termed *F cells*.

Now that we have surveyed the sites in which erythropoiesis occurs, we can turn to the nature of events observed during the process itself. The actual cytodifferentiation of erythrocytes is the last step in a long sequence of events occurring over several generations of cell types. It is generally believed that there exists within tissues involved in blood-cell formation, i.e., hematopoiesis, cells having striking developmental potential. These cells, referred to as *pluripotent hematopoietic stem cells* (also denoted by CFU-S in the literature), represent a self-renewing population of reserve cells which are capable of forming precursors to any one of a number of cell types. The presence of such cells can best be demonstrated by removing hematopoietic tissue from an animal, converting the tissue into a cell suspension, and injecting the cells into a recipient animal of the same strain that had been subjected to prior irradiation (a treatment that blocks division and differentiation of hematopoietic cells of the host). When this experiment is carried out, the injected cells colonize the tissues of the host animal, particularly the spleen, wherein they begin to take over the function of blood-cell formation in the irradiated animal. If the injected cells carry a detectable chromosome marker, their fate in the host can be followed. Cytologic examination of the spleens from these host animals taken at different stages after injection indicates that single graft cells form clones (i.e., colonies of cells derived from a single graft cell) containing more than one type of blood-cell precursor. Similar types of experiments have recently been carried out under in vitro conditions with similar types of results. Single hematopoietic stem cells are able to give rise to several different lines of cells; these include erythrocytes, granulocytes, megakaryocytes, eosinophils, and macrophages.

The underlying basis for the commitment of a pluripotent stem cell to one or another blood-cell line is totally obscure, but presumably depends on the microenvironment in which the cell finds itself. Regardless of the mechanism, some of these pluripotent cells are converted to erythroid precursors. The stem cell of the erythroid line is shown in Fig. 15.2*b*. After becoming committed to the erythroid pathway, these morphologically unspecialized stem cells continue to proliferate, eventually producing a population of *proerythroblasts* (Fig. 15.2*c*)

in which the first signs of erythroid differentiation are visible. The formation and subsequent differentiation of proerythroblasts is under the control of a hormone, *erythropoietin,* a glycoprotein produced by the kidney in response to lowered tissue oxygenation. Erythropoietin acts only on those cells already committed to the erythroid lineage; it is not involved in the process by which the pluripotent stem cell becomes committed to the erythroid pathway. The last steps in erythroid differentiation are ones that carry the proerythroblasts through a series of recognizable stages (Fig. 15.2) leading to the terminally differentiated red blood cell. Cytodifferentiation in the erythroid lineage is characterized by the condensation and inactivation of the chromatin, the increasing use of the protein-synthesizing machinery for globin synthesis, and the appearance in the membrane of a variety of proteins involved in the specialized function of red blood cells. One of the most important proteins of the erythrocyte membrane is spectrin, an inner peripheral protein that plays a role in immobilizing the integral proteins, thereby keeping them from diffusing through the plane of the membrane. In its early stages of differentiation, the membrane of the erythroid lineage is quite fluid, but as the density of spectrin in the membrane rises, the integral proteins of the membrane become more restricted. The final step in mammalian erythrocyte differentiation, the total enucleation of the cell (Fig. 15.2*f*), is a particularly striking event. Most interesting is the finding that the proteins of the membrane are localized in a highly asymmetric manner while the nucleus is being extruded. The spectrin becomes totally sequestered in the portion of the membrane that will be inherited by the mature erythrocyte, while the concanavalin A receptors become localized in the membrane domain surrounding the nucleus. The molecular changes occurring during differentiation will be discussed later.

3. Melanogenesis

Vertebrate pigment cells, or *melanocytes,* derive from two sources. Most pigment cells of the body arise from the neural crest after extensive migration. Melanogenesis does not begin until the precursor cells have migrated to their definitive location. The melanocytes of the retinal pigmented layer derive from a different source: cells descendant from the original optic vesicle. The coloration of melanocytes derives from a polymer of L-dopa, which is formed from the amino acid tyrosine by the enzyme tyrosinase. This polymer, termed *melanin,* is found to be associated with protein and is present in an organelle called the *melanosome* (Fig. 15.5). The formation of melanosomes by melanoblasts is the terminal step in the differentiation process by which the melanocyte is formed. The melanosomes are complex cell organelles. They are formed by dilations of the endoplasmic reticulum and are composed of subunits of fibrillar protein that form a sheetlike matrix in which tyrosinase is normally embedded. Tyrosinase arises in the Golgi complex and reaches the melanosome by fusion of the two vesicles. Melanin is then deposited on these protein structures until the melanosome is fully pigmented and enzymatic activity is lost. The number and distribution of the melanocytes, the morphology of the cell, and the color of the melanin are all influenced by identified genes, some of which act primarily within the melanoblast. Others function within the surrounding cells that make up the tissue environment.

Melanocytes, like most terminally differentiated cells, do not normally divide. Throughout the life of the animal they must be replenished through differentiation from a pool of melanoblasts. Melanoblasts do undergo mitotic replication, but with increasing age their supply diminishes, evident in the white hair of aging humans. The genetic control of pigmentation, whether acting within the

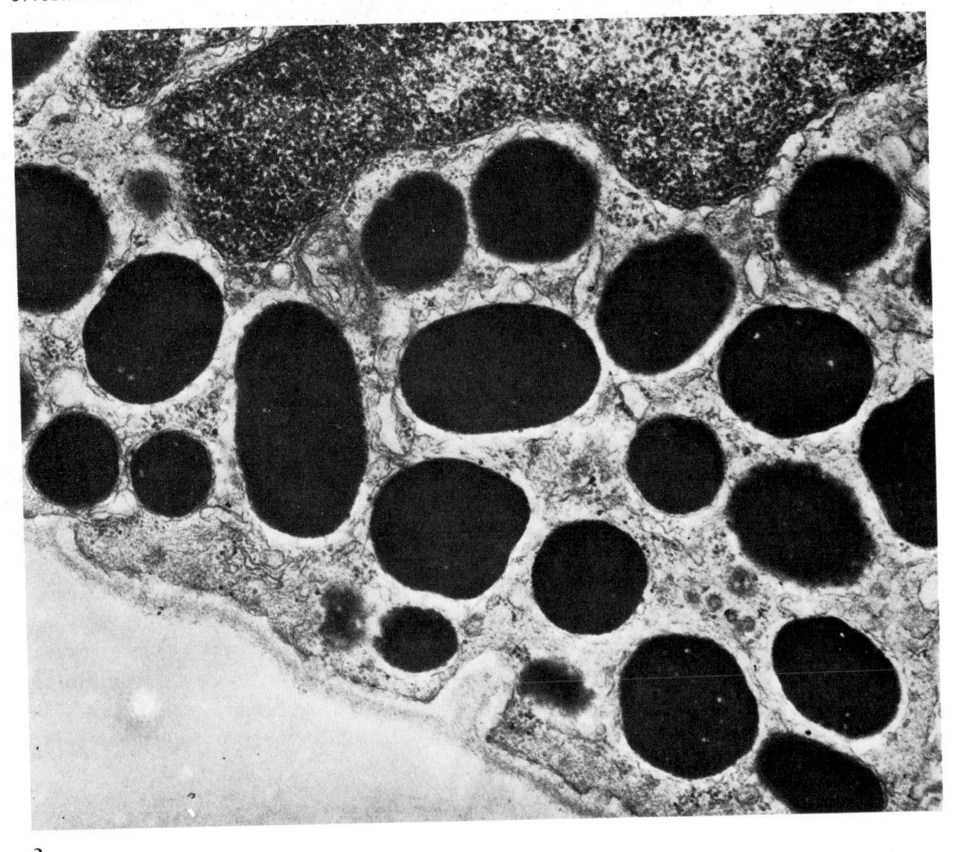

a

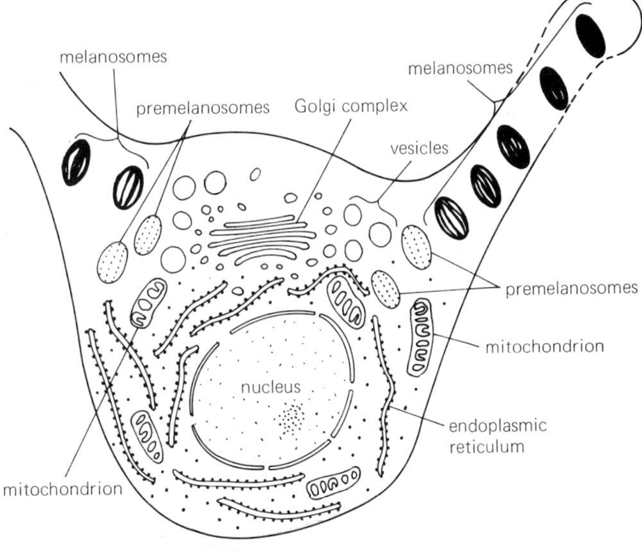

b

Figure 15.5 *a* Electron micrograph of a region of a cell from the pigmented iris epithelium. The cytoplasm of these pigmented cells is filled with melanin-containing organelles, melanosomes. [*Courtesy of J. Dumont and T. Yamada.*] *b* Development of melanosomes in a melanocyte. Vesicles containing "protyrosinase" are budded off from the Golgi. Vesicles develop into "premelanosomes" in which protyrosinase molecules have become organized into characteristic patterns. Melanin is polymerized on a protein matrix until the melanosomes are dense particles without tryosinase activity. [*Modified from M. Seiji et al.,* Ann. N.Y. Acad. Sci., **100:**497 *(1963), by permission of the New York Academy of Sciences.*]

melanocyte or through the tissue environment, is modified by the changing conditions brought about by age. This is perhaps simply another expression of the fact that the cell's genome is sensitive to the state of cell differentiation and is also the major contributor to the state of differentiation.

4. Chondrogenesis

The primary characteristic of cartilage is the presence of large amounts of extracellular matrix. It surrounds the living cells, is secreted by them, and provides rigidity and elasticity to the tissue. The study of *chondrogenesis* centers on the synthesis, secretion, and accumulation of matrix, of which there are two major components: collagen and a sulfated proteoglycan termed *chondromucoprotein*, the structure of which is shown in Fig. 15.6. The most widely used criterion for the occurrence of chondrogenesis is the incorporation of sulfate into chondroitin sulfate, a glycosaminoglycan that forms the bulk of the giant chondromucoprotein molecule. Even though other types of cells also synthesize chondroitin sulfate, cartilage cells manufacture much greater amounts of this material and can be readily identified against a background of noncartilage cells synthesizing lesser amounts (illustrated in Fig. 13.12). Moreover, recent studies indicate that the chondroitin sulfate–containing proteoglycan produced by cartilage cells is recognizably different from that produced by other cells. Differences may exist in the nature of the core protein (Fig. 15.6*a*), the ratio of chondroitin sulfate to keratin sulfate chains, and the state of aggregation of the whole complex (Fig. 15.6*b*). Similarly, the collagen made by chondrocytes is composed of a different species of polypeptide chain [an α 1 (II) chain] than that present in the collagen of other tissues. Therefore, as in the case of the contractile proteins of muscle cells discussed later, the major macromolecular markers of cytodifferentiation in cartilage cells are of a general type produced by many different cells. However, the various protein species present among the tissues of the body can often be distinguished by biochemical analysis.

Figure 15.6 *a* Diagram of a cartilage proteoglycan molecule. The formation of this molecule begins with the synthesis of the core protein, to which the sugars are added. The base of each polysaccharide chain consists of an initial xylose residue followed by two galactose residues and a glucuronic acid. To this base are added alternating sugars (*N*-acetylgalactosamine and glucuronic acid), which form the chondroitin sulfate chains. The final step in this process is the addition of the sulfate groups to the *N*-acetylgalactosamine residues. In addition to the chondroitin sulfate chains, there are a much smaller number of keratin sulfate chains. *b* The proteoglycan–hyaluronic acid complex. Each proteoglycan molecule (PG) is linked to the hyaluronate core (HA) by a noncovalent HA binding site (shown in *a*). [*From H. Muir, in "Cell and Tissue Interactions," J. Lash and M. M. Burger (eds.), Raven, 1977.*]

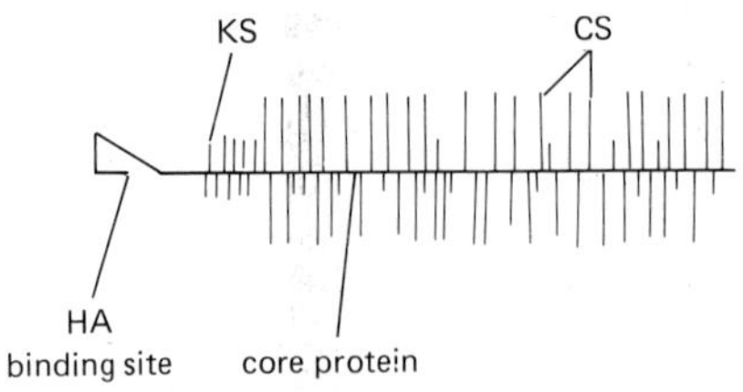

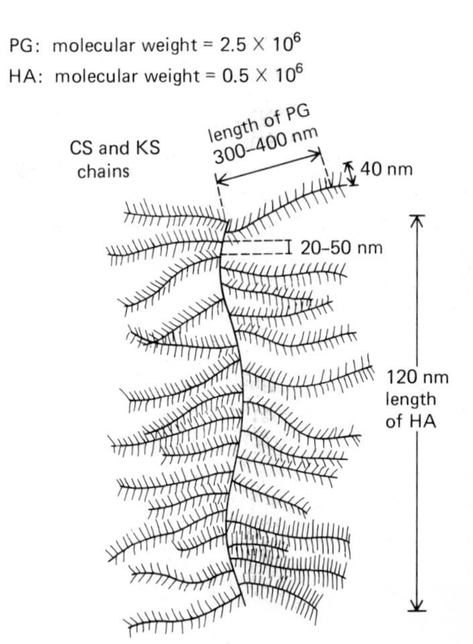

Chondrocytes in vertebrates arise from three main sources of chondroblasts: the neural crest, which form cartilaginous structures in the head and pharyngeal region (Section 11.2); the lateral plate, which gives rise to the skeleton of the limbs; and the sclerotome, which differentiates into the vertebral column. In each of these cases, the progenitors of the chondroblasts are mesenchymal cells that migrate into their definitive location prior to differentiation. This period represents the morphogenetic phase, which precedes the synthesis of cartilage-specific macromolecules and the attainment of the overt cartilage phenotype. The first histologic sign of chondrification in the precartilaginous mesenchyme of a developing embryo is the close apposition of rounding cells to form regions of greater cell density. The cell-mass effect and cell cohesion appear to be important factors among the various recognized conditions for cartilage formation. This can be demonstrated in vitro as well as within the embryo. Amphibian neural-crest cells in culture emigrate and disperse as a predominantly single layer of flattened cells. The majority differentiate into polymorphic mesenchymal cells, while others become pigment cells or neurons. However, within this layer of spreading cells, matrix-encapsulated cartilage cells appear. This cytodifferentiation is associated with an aggregation and piling up of the cells into nodules, with cell numbers varying from ten to several hundred.

Like many other types of cells, chondrocytes can be cultured in various ways. As would be expected from the previous discussion, cultures in which cell density is high are particularly favorable for chondrogenesis. However, chondrogenesis is also promoted by growing cells in liquid culture over agar (a condition in which contaminating fibroblasts do not grow and differentiate) or in low-density clonal cultures. The scanning electron micrographs in Fig. 15.7 show a clone of cells undergoing chondrogenesis in culture. The entire colony seen in this photograph has developed from a single cell, one that undergoes an initial phase of proliferation followed by a second phase of overt cytodifferentiation. The activities of a particular cell reflect to a large degree the position of the cell within the colony;

Figure 15.7 Scanning electron micrograph of a single cartilage colony (clonally derived). *a* Low-power micrograph showing the morphology of the colony. Cells most actively engaged in matrix production are found in the center of the colony. *b* Higher magnification of the central cells showing the nature of the extracellular matrix (composed of chrondromucoprotein and collagen) and its effect on cell shape on the dish. [*M. Solursh and G. Karp.*]

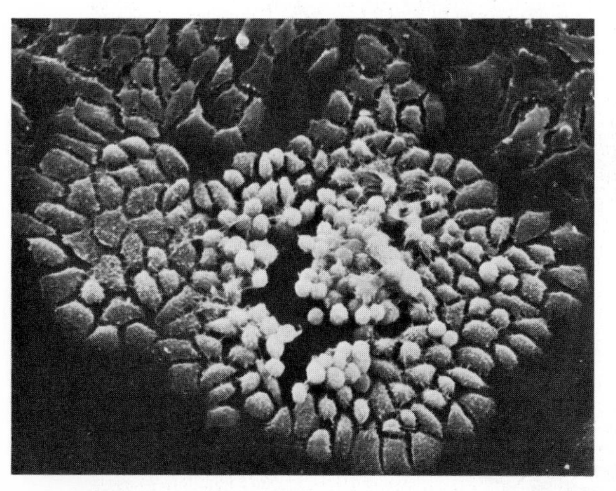

a

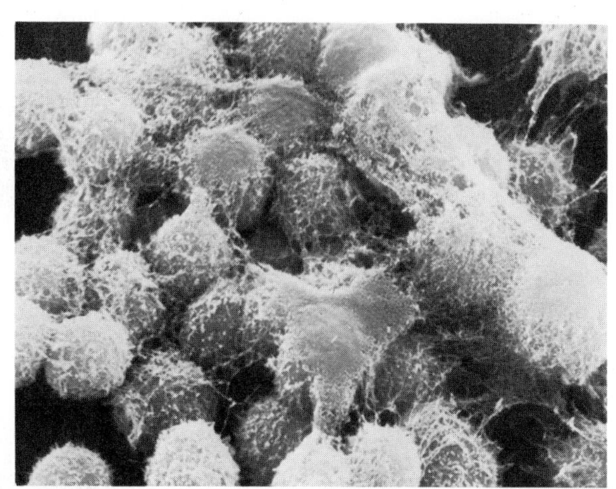

b

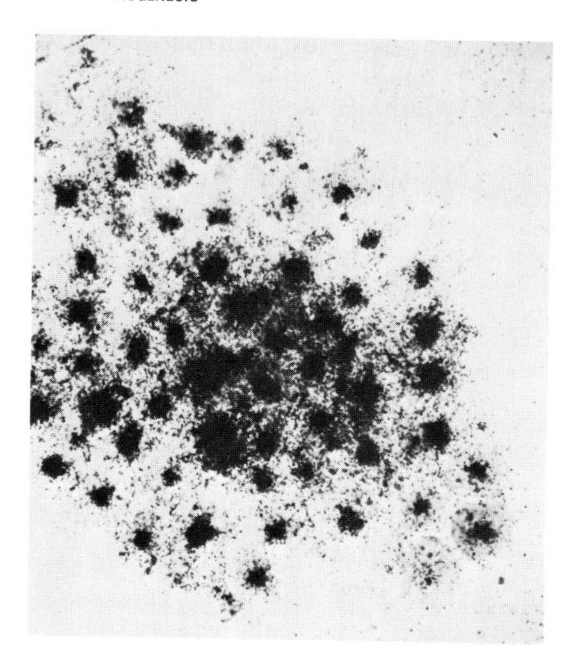

Figure 15.8 Autoradiograph of a cartilage colony (unstained) incubated in radioactive sulfate ($^{35}SO_4$) for 10 minutes. Sulfate becomes incorporated into chondroitin sulfate of the chondromucoprotein and thus provides an indicator of relative matrix production by the cells of the colony. Central cells are much more active than peripheral ones. [*M. Solursh and G. Karp.*]

those within the center are most actively engaged in matrix production. This is reflected in their increased incorporation of sulfate into chondroitin sulfate, demonstrated autoradiographically (Fig. 15.8). As a result of their biosynthetic activity, the more centrally located cells of the colony become covered with matrix, as shown in the scanning electron micrograph in Fig. 15.7*b*. Similarly, it is the cells in the interior of the colony that are found to be synthesizing type II collagen (Fig. 15.9) characteristic of differentiated chondrocytes, while the less differentiated cells of the periphery continue to synthesize the type I variety.

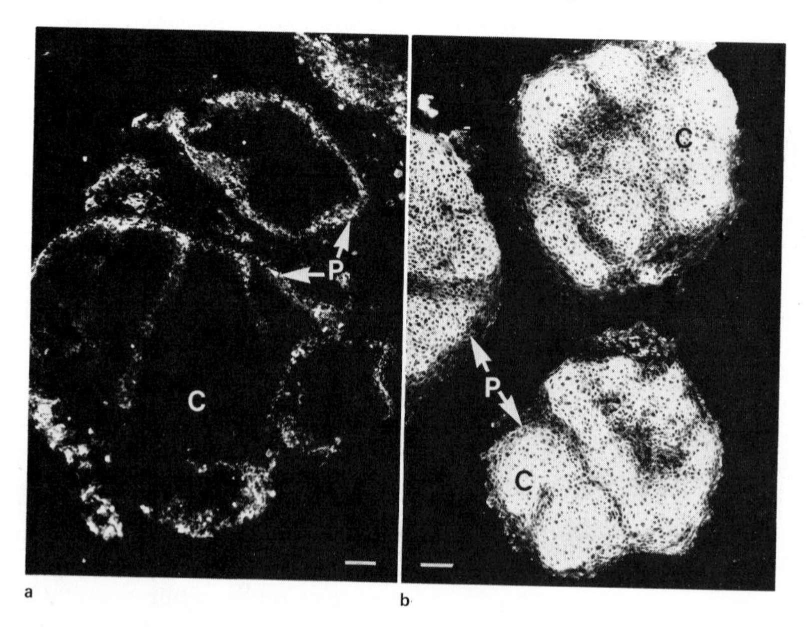

Figure 15.9 Localization of collagen types in differentiating cartilage colonies by utilization of fluorescent antibodies to type I (*a*) or type II (*b*) collagen. After 6 to 8 days in culture, a perichondrium (P) has formed around the cell aggregate which stains with anti-type I collagen antibodies (*a*). The type I collagen that had been in the center of the aggregate has disappeared and is replaced by a hyaline cartilage matrix (C) that stains with anti-type II collagen antibodies (*b*). [*Courtesy of K. von der Mark and H. von der Mark.*]

Differentiating cells exist in complex environments, the components of which can dramatically affect the metabolism within the cell. Intracellular activities can be modified by contact with the surface of other cells as well as by contact with substances of large and small molecular weight in the medium. Materials of large molecular weight can be present in a soluble or insoluble state. In the latter case, the material is termed a *matrix*, typically composed of collagen and/or sulfated proteoglycans, such as chondromucoprotein, which have increasingly been implicated as external substances capable of regulating internal events.

The importance of the extracellular matrix is clearly illustrated during chondrogenesis. Destroying its chondroitin sulfate content by the addition of the enzyme hyaluronidase serves to depress the synthesis of additional chondromucoprotein, suggesting the existence of a feedback loop between the substance on the outside of the cell and its synthesis on the inside. Furthermore, addition of chondroitin sulfate to the medium in which the cells are growing stimulates its intracellular production. Further experiments indicate that materials of the extracellular matrix not only affect the synthetic activity of chondrocytes, but also they seem to be of prime importance in calling forth the chondrocyte character in the first place. The basis for this statement comes from a long series of studies implicating the notochord and spinal cord as important inductive tissues in the differentiation of sclerotome into cartilage. It was shown in early extirpation experiments that removal of the notochord and spinal cord from a region of the embryo interfered with the formation of the cartilaginous vertebrae in that part of the embryo. Much later it was shown that either notochord or spinal cord tissue served to greatly stimulate chondrogenesis by somite tissue in vitro. Further investigation of this inductive interaction has pointed unambiguously to the extracellular materials of the inductive tissues as the source of the inductive agents. For example, removal of the collagen-proteoglycan material from around the spinal cord or notochord prior to exposure to the somite explant greatly decreases its cartilage-promoting activity. Similarly, addition of either proteoglycan or collagen to morphologically unspecialized somitic cells greatly enhances their chondrogenic potential. It is suggested that sclerotome cells migrating toward the notochord and spinal cord (Section 12.3.A) as a secondary mesenchyme are exposed to extracellular materials produced by these two embryonic structures and induced to differentiate as chondrocytes.

Although collagen and glycosaminoglycans are the most prominent materials of the extracellular (or intercellular) space, other macromolecules of importance also may be present. One of these latter types is the large glycoprotein fibronectin (also known as LETS or CSP), which is synthesized by a wide variety of mesenchymal cell types and collects at the cell surface as a fibrous-type meshwork (Fig. 15.10). Fibronectin has been implicated as a factor involved in the adhesion of cells to substrates, particularly those of a collagenous nature. In fact, the purification of this protein is accomplished by passing it through a collagen-containing matrix to which it specifically adheres. Prior to the production and accumulation of extracellular matrix, embryonic chondroblasts are characterized by having considerable amounts of fibronectin at their cell surfaces. However, as these cells begin to produce cartilage-type proteoglycan and type II collagen and amass a cartilaginous matrix, the synthesis and accumulation of fibronectin is drastically reduced (Fig. 15.10). This inverse relationship between the presence of fibronectin and cartilage matrix may be causally linked in some manner, since manipulations that affect one of these extracellular materials lead to changes in the other. For example, enzymatic digestion of proteoglycan from the matrix initiates the synthesis and secretion of fibronectin. Conversely, addition of fibronec-

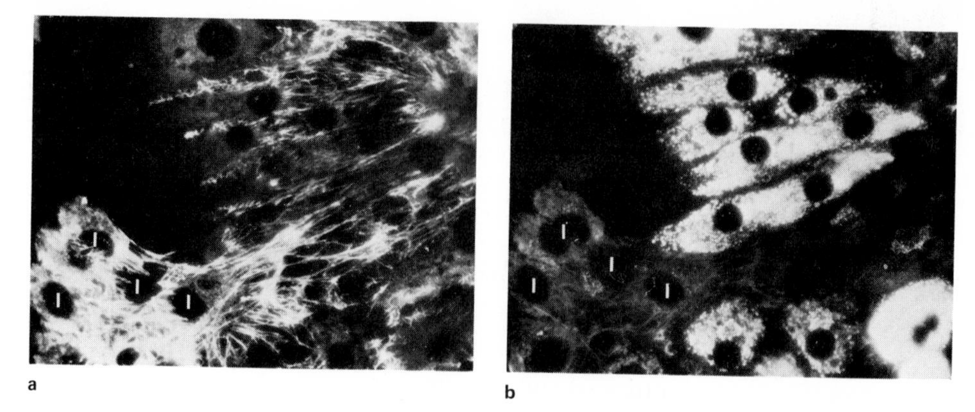

Figure 15.10 The inverse relationship between extracellular type II collagen and fibronectin as revealed by localization with fluorescent antibodies. The cells of the lower left portion of the field are less differentiated, as indicated by their synthesis of type I collagen. These cells are covered by an extensive fibronectin network (shown by antifibronectin antibodies in *a*). Conversely, the cells in the upper right portion of the field have reached a stage of differentiation at which they are synthesizing cartilage-specific type II collagen (shown by fluorescent antibody staining in *b*) and no longer contain the fibronectin cover. [*Courtesy of W. Dessau, J. Sasse, R. Timpl, F. Jilek, and K. von der Mark.*]

tin to chondroblasts maintains their "undifferentiated" fibroblastlike morphology and prevents their synthesis and accumulation of matrix material. Chondroblasts treated with fibronectin revert to the synthesis of type I collagen and stop their synthesis of the cartilage-type proteoglycan; they seem to have returned to a less differentiated state (discussed further in Section 15.9).

5. Myogenesis

Muscle tissue is well-suited for studies of cytodifferentiation. It is characterized by a highly specific morphology, centered around the presence of a well-organized fibrillar structure, and a specialized biochemistry, centered around the presence of a specific set of cytoplasmic enzymes (including myokinase and creatine phosphokinase), membrane proteins (including Ca^{2+}-dependent ATPase, acetylcholine receptors, and acetylcholinesterase), and contractile proteins (including actin, myosin, tropomyosin, tropoinin, and α-actinin). It should be noted that all these contractile proteins have counterparts in nonmuscle cells, but in each case it appears that the muscle and nonmuscle varieties are distinctly different proteins whose polypeptides are transcribed from different structural genes. In this section we will consider only the differentiation of skeletal muscle, although many of the biochemical events may be similar to those in smooth and cardiac muscle as well.

A differentiated skeletal muscle cell is large and highly elongated, with many nuclei and a marked striated or striped appearance. These cells, or *myofibers* as they are generally called, represent a type of cable made up of thousands of thinner strands, termed *myofibrils*, separated from one another by thin layers of cytoplasm containing membranous channels, mitochondria, glycogen, etc. If one examines the transverse bands of a stained muscle fiber from one end to the other, one sees a striking pattern that repeats itself many times along the length of the fiber. The basis for the banding pattern is shown in Fig. 15.11. Thin 60-A filaments made of actin (and associated tropomyosin and troponin) are found to interdigitate with thick 160-A filaments made of myosin in a highly specific man-

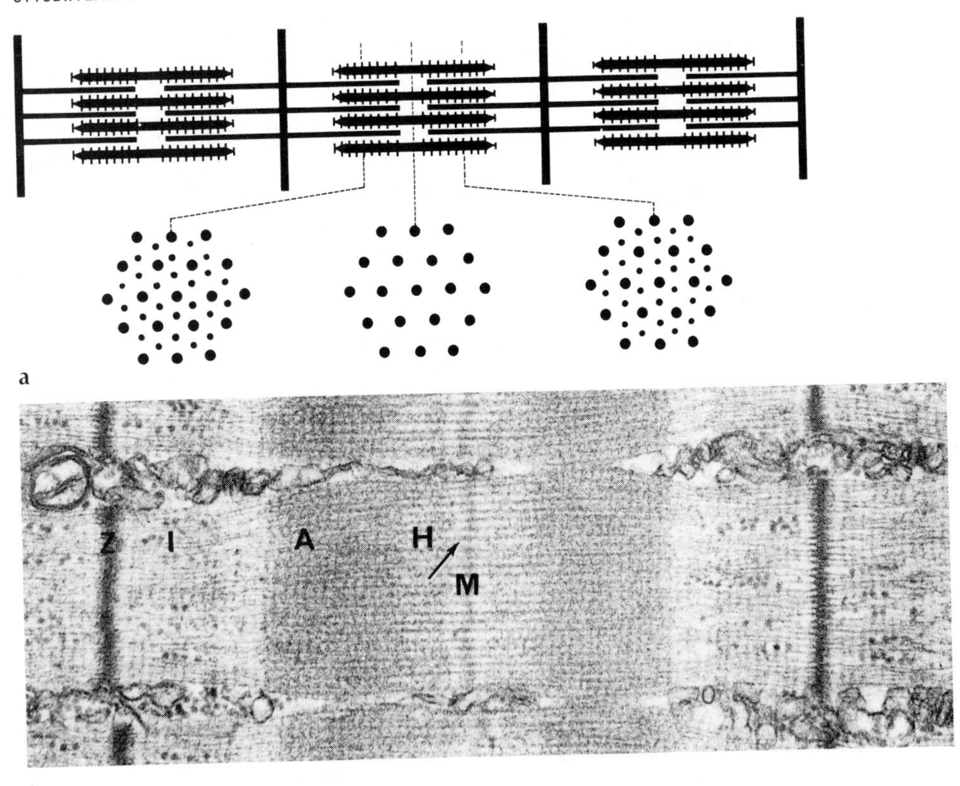

Figure 15.11 *a* Diagrammatic representation of part of a myofibril showing the overlapping array of (thin) actin- and (thick) myosin-containing filaments. The small transverse projections on the myosin fiber represent the cross-bridges. The hexagonal arrangement of the thin filaments around each thick filament in the lateral regions of the A band is apparent in the cross sections. [*After H. E. Huxley,* Science, **164**:*1357 (1969); copyright © by the American Association for the Advancement of Science.*] *b* Electron micrograph of a single sarcomere with the bands lettered. [*Courtesy of G. F. Gauthier.*]

ner, one that leads directly to the banding pattern observed in the microscope. The functional unit of the myofibril is the sarcomere, shown in the diagram and electron micrograph in Fig. 15.11.

The differentiation of skeletal-muscle tissue occurs in several stages. The *presumptive myoblast* is derived from the mesoderm of the lateral plate and the myotome region of the somite, except in the head and neck, where it is derived from nonsomitic prechordal mesenchyme. The presumptive myoblasts are believed to arise from a pluripotent mesenchymal cell capable of differentiation into muscle, skeletal, or connective tissue depending on the microenvironment. Presumptive myoblasts, although determined by this time to form muscle cells, appear in the electron microscope as typical "undifferentiated" mesodermal cells and contain little or no evidence of their future transformation. The presumptive myoblasts proliferate to form a collection of highly elongated, bipolar myoblasts (Fig. 15.12*a*) which become aligned parallel to one another and then fuse together to form an elongated cylindrical *myotube* (Fig. 15.12*b* and *c*) containing many nuclei within a common cytoplasmic compartment.

Before beginning a discussion of the cellular and molecular events found to occur during myogenesis as studied in cell culture, it is important to dispel any

morphological criteria, proceeds very rapidly. Almost as soon as the myotubes are formed, the synthesis of the muscle-specific contractile proteins and enzymes begins at rapidly increasing rates (discussed in detail in Section 15.7.B), and contractile filaments appear in the cytoplasm. The formation of myofibrils within the myotubes begins simultaneously with the appearance of cytoplasmic filaments in the cytoplasm. Myofibril formation generally begins in the peripheral regions of the cylindrical myotubes just beneath the plasma membrane (Fig. 15.14) and proceeds toward the cell interior.

In Chapter 2, the question was raised as to what extent the intracellular organelles might be generated by processes of self-assembly. Muscle differentiation provides an opportunity to examine this question, because there is considerable information on the biochemistry of the principal structural components and their highly ordered morphology. Although convincing evidence of self-assembly in vivo is difficult to obtain, it is generally believed that the geometric lattice of the myofibril can arise in this manner. It has been widely demonstrated, for example, that purified preparations of myosin can polymerize in vitro to form bipolar thick filaments remarkably similar to those present within a muscle cell. Similarly,

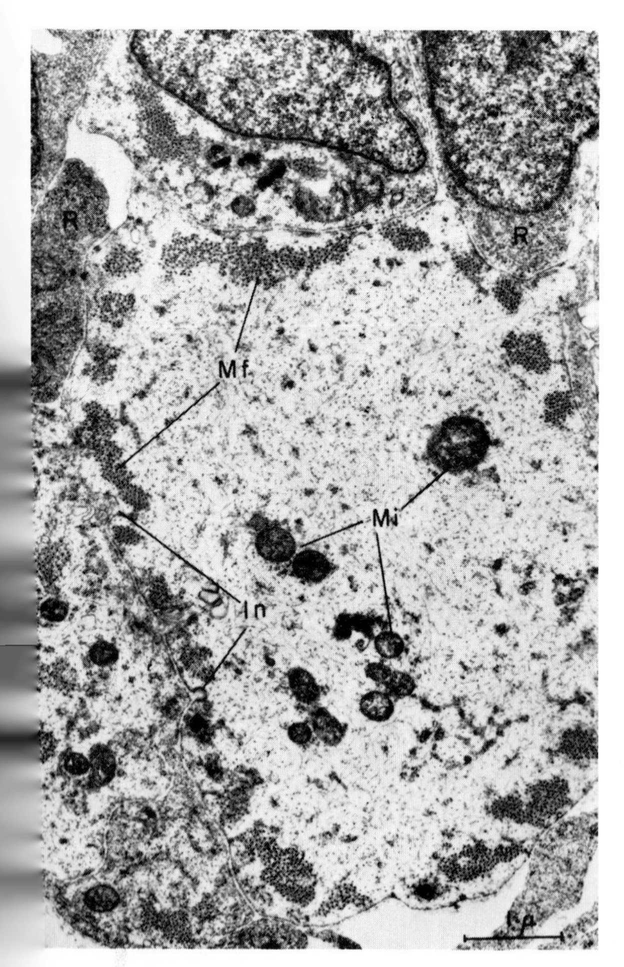

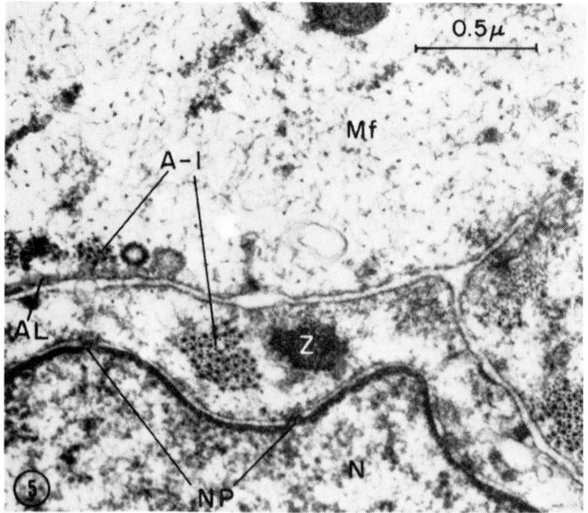

b

Figure 15.14 *a* Cross section of embryonic leg muscle from a 12-day chick embryo showing the peripheral deposition of myofibrils within each myotube. R, ribosomes; Mf, myofibrils; Mi, mitochondria; In, invaginations of the sarcolemma. *b* Higher magnification of a cross section through regions of three adjacent myofibers. Myofibrils have been cut through various sarcomere regions; sections through thin and thick filaments overlap in the A band (A-1) and the Z band are seen. All myofibrils contain myofilaments packed in the hexagonal array. An amorphous layer (AL) between the assembling myofibrils and the sarcolemma can be seen in the upper cell at the left of the micrograph. Free myofilaments (Mf) are seen in the cytoplasm. A nucleus (N) with prominent nuclear pores (NP) is visible in the lower cell. [*Courtesy of D. A. Fischman.*]

purified actin will polymerize in vitro into double-helical filament of the type seen in the cell. Moreover, when purified actin monomer is mixed with previously polymerized myosin, complexes form in which six actin filaments are bound to a central myosin filament, thereby forming a "contractile unit." The hexagonal array of thin filaments around each thick filament reflects the actin-binding cross-bridges that project from the myosin molecules to dictate a stable sixfold symmetry.

When the electron microscope is used to study the process of myofibril formation in newly formed myotubes, the presence of both thick (160 A) and thin (60 A) myofilaments is invariably seen to coexist in the same cell. In some cases, the two types of filaments are organized into the typical hexagonal lattice, which is aligned parallel to the long axis of the cell (Fig. 15.15*a*), even when only a handful of thick filaments are present in the cluster. This is apparent in some of the smaller myofibrils seen in the periphery of the myotube in Fig. 15.14. The presence of these small, ordered filament arrays suggests that both filament types may polymerize in place within the lattice, the hexagonal array being determined by the positions of the myosin cross-bridges. In other cases, thick and thin filaments can be seen to be dispersed in the cytoplasm. It would seem likely in these latter cases that the filaments are formed and subsequently organized into a myofibrillar structure. Evidence for the ability of previously formed myofilaments to become reorganized into a contractile lattice in vivo has come from studies of embryonic cardiac muscle cells that have already differentiated, but whose filaments have been secondarily dispersed in a random manner following dissociation of the cells with trypsin. In these cells, the filaments are capable of reorganizing themselves in the presence of inhibitors of protein synthesis. This study brings about the dissociation of the synthetic and assembly process; all the structural proteins needed have already been manufactured and polymerized, and the assembly step can be studied without confusion by continuing muscle-protein synthesis.

With increasing periods of time, the small myofibrils grow by the addition of thick and thin filaments at their circumference until millions of filaments are found in the mature myofibril. The first indication of sarcomeric units is seen as an appearance of electron-dense material in amorphous bands on previously formed myofilament bundles. These deposits are believed to represent forerunners of the Z bands, which probably serve as sites at which clusters of thin filaments attach to one another in an end-to-end arrangement, thereby leading to the formation of greatly extended myofibrillar cables. In addition to the myofibrils, muscle cells possess a very characteristic membranous network that also plays a vital role in contractility. Surrounding each myofibril is a network of membranous cisternae, the *sarcoplasmic reticulum,* which is seen in differentiating myotubes to be derived from the endoplasmic reticulum. The sarcoplasmic reticulum is involved primarily in the storage and release of calcium ions. Connected with the sarcoplasmic reticulum near each Z band is a membranous channel, the *T tubule,* involved in carrying the nervous impulse from the cell surface into the contractile machinery of the muscle cell. T tubules are seen in myotubes to be derived from invaginations of the *sarcolemma,* the plasma membrane of a muscle cell.

Although all these aspects of cytodifferentiation appear closely following cell fusion, they are not causally related. It has been demonstrated in many types of experiments that the inhibition of myoblast fusion in any of a variety of ways does not block the synthesis, polymerization, or assembly of contractile proteins or the differentiation of the sarcoplasmic reticulum or the T system. All these differen-

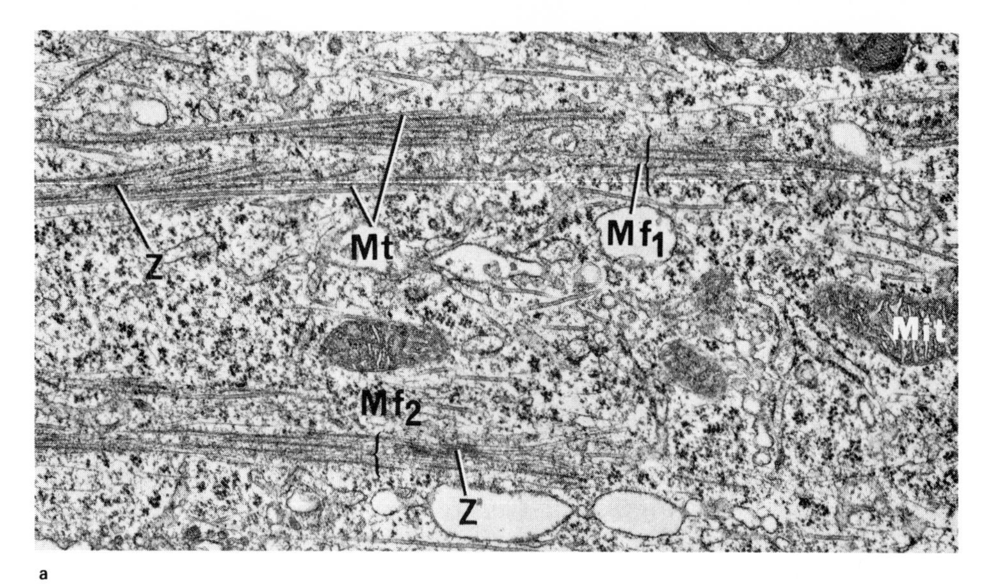

a

b

Figure 15.15 *a* Longitudinal section through a myotube within a 6-day-old muscle culture that exhibits early phases of myofibrillogenesis. Thick and thin filaments are aligned nearly parallel. Some aggregates of filaments (Mf_1) show no definite Z-band density. Other aggregates (Mf_2) show a periodic distribution of a dense material (Z) that is assumed to be a Z-band precursor. Well-defined M bands are absent. Frequently, microtubules (Mt) are seen in close association with the myofibrils. Mitochondrion, Mit. *b* Longitudinal section through a myotube in an 8-day-old culture. Developing myofibrils can be seen in which Z and M bands are clearly discernible. [*Courtesy of Y. Shimada and T. Obinata.*]

tiations are found to occur in the cytoplasm of mononucleate fusion-arrested myoblasts.

While fibrillogenesis is proceeding within the cytoplasm of the newly formed myotube, a different type of differentiation occurs within the sarcolemma. The most distinctive aspect of membrane differentiation during myogenesis is the appearance of receptors for the neurotransmitter acetylcholine. As in the case of the myofilaments, virtually no evidence of acetylcholine receptors is present in the mononucleate myoblast. With the formation of myotubes, these integral-membrane receptor proteins appear in a dispersed manner throughout the sarcolemma. These receptors are found to be synthesized within the cell on membrane-bound polyribosomes of the endoplasmic reticulum and then rapidly transported to the plasma membrane.

The distribution of acetylcholine receptors in the plasma membrane is best followed autoradiographically. In these experiments, myotubes are incubated at various stages of differentiation with radioactive neurotoxins (isolated from snake venom), which bind with high affinity and specificity to the acetylcholine receptor protein (Fig. 15.16). Although initially distributed in a homogeneous

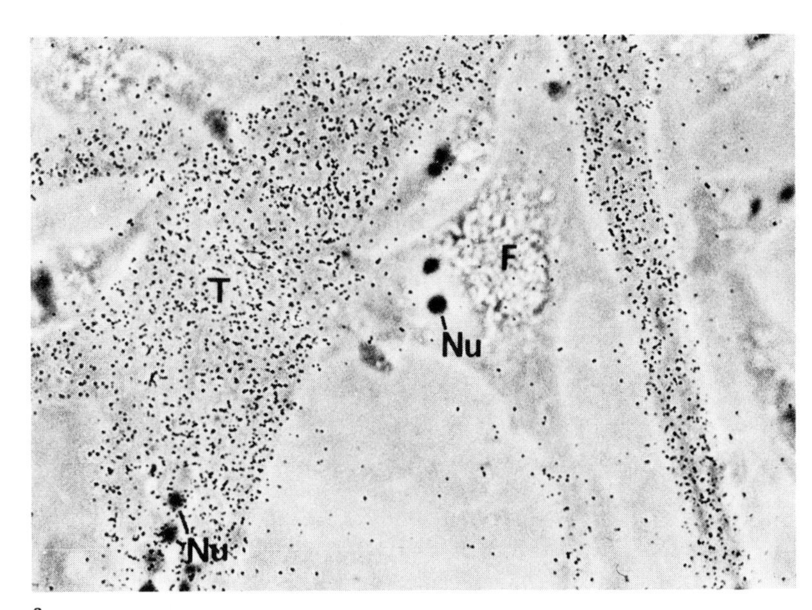

a

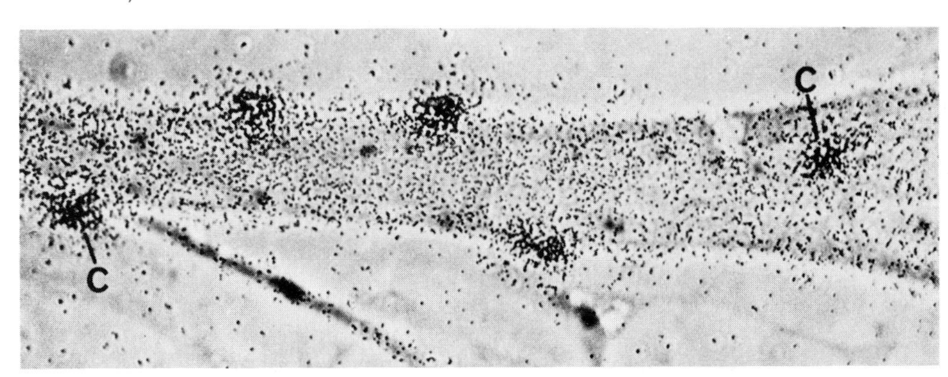

b

Figure 15.16 Autoradiographs of cultured myotubes labeled with [^{125}I]Bungarotoxin, a neurotoxin which binds selectively to acetylcholine receptors. *a* After 4 days in culture the acetylcholine receptors are distributed homogeneously over the surface of the myotubes (T) while only a few grains can be detected on the fibroblasts (F). Nu, nucleolus. *b* After 7 days in culture the acetylcholine receptors of the myotubes have become aggregated as revealed by the overlying silver grain clusters (C). (*Courtesy of J. Prives, I. Silverman, and A. Amsterdam.*)

manner throughout the sarcolemma of the myotube, the receptors subsequently become organized into clusters of much higher density than those of the remaining membrane regions. In the intact embryo, acetylcholine receptors are found only within those portions of the membrane which form synaptic associations with a nerve terminus. It has been suggested that the tips of exploratory axons associate with the membrane of the muscle cell as a result of an interaction between the nerve-cell surface and regions of high acetylecholine receptor density on the muscle fiber. Once the association has occurred, receptor proteins disappear from portions of the membrane other than the incipient neuromuscular junctions. Whether the clustering phenomenon that occurs in isolated myotubes in the absence of innervation is related to the development of neuromuscular junctions in the intact embryo remains uncertain. Regardless, the shift in localization of acetylcholine receptors in cultured myotubes provides an excellent example of the directed movement of an integral membrane protein through the lipid bilayer. A different type of redistribution of these receptors can be brought about artificially by subjecting newly formed myotubes to an electric field. In the presence of such a field, virtually all the acetylcholine receptors become localized at the edge of the cell facing the cathode.

6. The genetic control of differentiation

We are obviously a great distance from understanding the mechanism by which cells become determined and differentiate along one path or another. One of the most important aspects of this problem concerns the nature and time of appearance of tissue-specific molecules, whether they are nuclear RNAs, messenger RNAs, proteins, or some other substance. When do molecules appear that are characteristic of one type of cell relative to the stage at which overt cytodifferentiation takes place? Are some or all of the mRNA templates for cytodifferentiation produced long before the event itself? If they are, then later stages of differentiation can be thought of as being primarily controlled at the translational level. Or rather, is there progressive gene expression so that the availability of new RNA sequences limits the nature of the events that can take place? In other words, is the cell waiting for new genes to be expressed and new RNA molecules to appear? If so, differentiation is being regulated by transcriptional control mechanisms. Alternatively, control of differentiation might reside in the processing reactions by which mRNAs are selectively carved from larger nuclear transcripts. In this latter alternative, cells would synthesize similar populations of RNA molecules at all stages, but different mRNAs would be processed at various stages. These are greatly simplified and extreme alternatives; it is more likely that each type of control plays some role in the overall scheme of differentiation. The question as to what levels of control of information transfer are at work during each stage of the differentiation process is basic to our understanding of development. With the advent of molecular hybridization technology, considerable data have accumulated concerning the nature of the RNA sequences present at various stages of differentiation. Recent advances in nucleic acid cloning and sequencing methodology will hopefully provide further information in this area, particularly toward an understanding of the underlying basis of control.

Cytodifferentiation is generally accompanied by the appearance of highly specific, characteristic proteins, including myosin during myogenesis, globin during erythropoiesis, and hydrolytic enzymes and insulin during pancreogenesis. Several distinct approaches have been employed to determine whether the mRNA sequence that codes for a given protein, such as those just mentioned, has been synthesized and processed at any given stage of development.

In one of the earliest approaches, actinomycin D was administered and the synthetic activity of the tissue followed. If the mRNA for the protein being studied had already been synthesized at the time the inhibitor was added, that protein should appear at its normal time regardless of the application of the drug. If, instead, the gene coding for that protein had not been activated at the time of actinomycin D application, the given protein could not appear.

In a second technique, the presence of a specific mRNA is determined by its ability to direct the synthesis of the corresponding protein in a cell-free system. In this case, the messenger (polysomal) RNA population must be extracted from the cells at the stage in question, and this RNA must be added to the appropriate protein-synthesizing system. If the specific mRNA is present, the corresponding protein will be produced; if the mRNA is absent, there is no condition under which that protein can appear.

In a third technique, the presence of a specific mRNA is determined by its ability to hybridize to a complementary, radioactive DNA molecule, i.e., a cDNA probe. With the identification and purification of the enzyme reverse transcriptase from RNA tumor viruses, it has become possible to synthesize, in the test tube, highly radioactive DNA molecules having the complementary base sequence to any given species of RNA that can be purified. The cDNA can be made radioactive in one of two ways: the deoxyribonucleoside triphosphates used in the enzymatic polymerization can be radioactive, or alternatively, the completed cDNA molecules can be labeled to a very high specific activity using a DNA repair technique termed *nick-translation.* The cDNAs formed by the reverse transcriptase also can be amplified in amount by converting them into double-stranded DNA molecules and cloning them in bacterial cells (Section 2.2). Once the labeled cDNA is made, it can be used to probe preparations of RNA in search of the species of mRNA used as the template. For example, it becomes possible with this technique to determine at what stage of differentiation a particular mRNA appears in the cells. There are two basic approaches taken to hunt for complementary RNAs: the RNA can be extracted from cells in question and subjected to hybridization with the labeled cDNA in solution, or alternatively, the labeled cDNA can be incubated with tissue sections and the binding of radioactivity followed autoradiographically. In this latter technique, the presence of specific mRNAs can be made on mixtures of cells of various stages because the determination is made microscopically. The ability to synthesize labeled cDNA probes just described depends on techniques that allow one to isolate a sufficient quantity of one mRNA species. Since only a limited number of such mRNA species are available at this time, only a limited number of probes have been utilized. The possibility for purification of large numbers of DNA sequences prepared by fragmentation of the genome and subsequent cloning procedures should lead to considerable expansion of this type of research in the future.

Not all cDNAs are made using purified mRNAs as a template. They also may be synthesized from a diverse population of mRNAs, such as that obtained from an entire polysome preparation of any type of cell. Once synthesized, this labeled cDNA population can be used to determine the numbers of copies of the mRNAs present in the cells from which they were taken. The first step in this type of experiment is to purify the poly(A)-containing mRNA population; the second step is to use this entire population of mRNAs as templates for the production of complementary radioactive cDNAs with the reverse transcriptase; and the third step is to allow these two populations (the mRNAs and the cDNAs) to hybridize to each other under conditions of RNA excess. The greater the concentration of any given sequence in the mRNA population, the more rapidly it will hybridize. Analysis of

the kinetics of hybridization provides information on the number of different mRNA sequences present as rare, moderately abundant, or abundant copies within the polysomes.

Up to this point in the discussion we have concentrated on whether or not specific mRNAs are present in the polyribosomes. Some of the techniques described also can be used to study the sequences present in heterogeneous nuclear RNA populations. Analysis of nuclear RNAs is particularly important if one is attempting to distinguish between transcriptional and posttranscriptional levels of control (Section 2.6.B). The absence of a particular mRNA in the cytoplasm is no guarantee that this sequence is not being synthesized and destroyed in the nucleus. As discussed in Chapters 2 and 8, it can be very difficult to work with nuclear RNAs because they are so diverse in nature and present at such low concentrations. Despite the difficulties, we will be able to provide some information on hnRNAs in the following sections.

A. Genetic control of erythropoiesis

The terminal differentiation of erythrocytes is characterized by the increasing synthesis and accumulation of hemoglobin, which accounts for over 95 percent of the protein of the mature cell. For this reason, erythropoiesis is well-suited for the analysis of the control mechanisms governing protein synthesis during differentiation. In the chick, blood-island formation occurs between the definitive primitive-streak and the head-fold stage (18 to 24 hours of incubation). The time at which hemoglobin can be detected in the developing chick embryo depends on the sensitivity of the technique by which it is measured. Earlier studies using a fairly sensitive staining assay found the first indications of the presence of hemoglobin in the six- to eight-somite stage. This was approximately 10 hours after the head-fold stage and the formation of the blood islands and approximately 10 hours after its appearance became insensitive to actinomycin D. Actinomycin D added prior to the head-fold stage will block the appearance of hemoglobin at the six- to eight-somite stage, while application after the head-fold stage does not block subsequent hemoglobin synthesis. These results suggest that the mRNA for hemoglobin is made by the head-fold stage, after which the inhibition of RNA synthesis does not affect the appearance of the corresponding protein. The question raised by these results concerns the apparent delay between the synthesis of the mRNA and its translation into the polypeptide chains of the globin protein. Does the mRNA lie dormant in the cytoplasm for 10 hours before some translational-control mechanism causes it to be translated? More recent studies used isotopic-tracer methods capable of detecting very small quantities of newly synthesized hemoglobin. They indicate that translation of the globin mRNAs begins soon after their transcription and, therefore, that complex translational-control mechanisms need not be invoked. These results illustrate an ever-present difficulty: the interpretation of data depends on the limitations of the methods used. Some methods of detection are more sensitive than others; all have a defined limit below which detection cannot be made and conclusions cannot be drawn.

Hemoglobin is not the only protein characteristic of erythrocytes; so too are a number of membrane components. In fact, the proteins of the red-blood-cell plasma membrane are the best characterized of any membrane and their appearance during differentiation is of general interest. The synchronous nature of red-blood-cell formation in the yolk sac allows one to obtain pure populations of cells at each of the various stages of megaloblastic erythropoiesis. When membrane proteins are extracted from cells at various stages and fractionated on SDS-polyacrylamide gels, a developmental profile of the changes occurring in the mem-

brane during differentiation is revealed (Fig. 15.17). Several features are evident in the profile. Most of the important changes in the membrane occur during the early stages of erythroid differentiation, i.e., prior to the sixth day. A steady-state membrane-protein composition seems to be established by the late polychromatophilic erythroblast stage (Fig. 15.2). During this early period, there are three components (120,000, 45,000, and 29,000 daltons) that gradually decrease in relative amount and then disappear. Meanwhile certain of the major proteins of the mature erythrocyte, bands 1, 2.1, and 3.1, increase manyfold during differentiation.

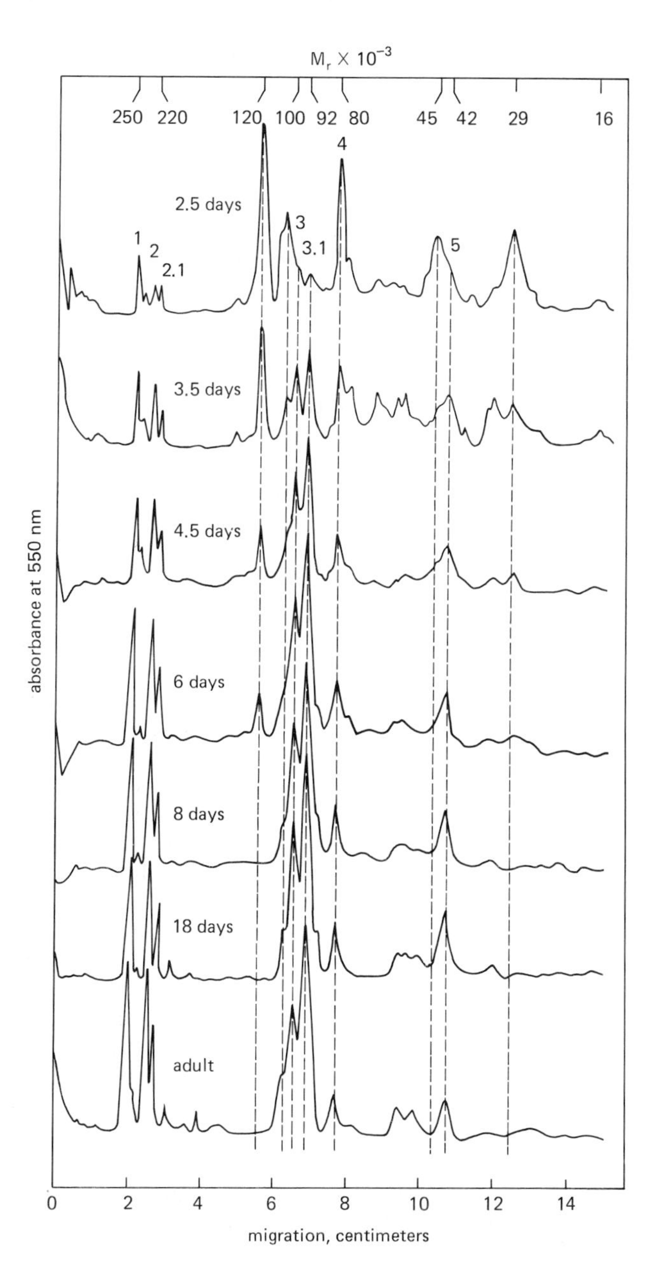

Figure 15.17 Results of SDS polyacrylamide gel electrophoresis of proteins extracted from erythrocyte plasma membranes at various stages of early chick development. The molecular weights (which are inversely related to the distance of migration) are indicated at the top. The nomenclature used to identify erythrocyte protein bands is given in association with the scan of the 2.5-day preparation. [*From L.-N. Chan*, Proc. Natl. Acad. Sci. U.S., **74:***1064 (1977)*.]

Although the initial formation of erythrocytes in the yolk sac of the embryo has the advantage of being synchronous, the availability of such limited amounts of material make it difficult (or impossible in the case of mammals) to carry out analyses of nucleic acid sequences. Consequently, most studies of genetic control of erythropoiesis have utilized adult systems of differentiation. By the use of certain experimental procedures, populations of cells at a given stage of differentiation can generally be obtained. For example, mixtures of cells at various stages of erythropoiesis often can be sorted out according to their size. Or alternatively, bursts of erythropoiesis can be initiated by treatments that stimulate the formation of red blood cells. In addition, the development of better tissue-culture techniques in recent years offers promise for the study of the entire process in vitro.

From a molecular point of view, the analysis of circulating erythrocytes in anemic birds has proven to be one of the most revealing. Consider the following facts. Circulating erythrocytes at a late stage in the process of differentiation are found to contain approximately 4,000 different species of poly(A^+)-nuclear RNA (total complexity of about 5×10^6 nucleotides) and fewer than 100 species of poly(A^+)-polysomal mRNA. The preoccupation of the erythrocyte with hemoglobin synthesis is reflected in the composition of the mRNA population; there are approximately 1,500 copies of each of the globin mRNA species and only about 7 copies, on the average, of each nonglobin mRNA species.

To what extent is this difference in mRNA frequencies attributable to transcriptional-level as opposed to posttranscriptional-level control? In the first place, the fact that there are only about 4,000 nuclear RNAs, by itself, points to an extremely high level of transcriptional control. This is a very small variety of RNA molecules, much smaller than most populations of cytoplasmic messenger RNA. These are unusual cells on their way to becoming totally inactive in transcription and capable primarily of transporting oxygen for a limited period of time. Regardless of the physiological reasons, some mechanisms must be operating at the transcriptional level to inhibit the expression of nearly all the genome. As discussed in Chapter 2, the concept of transcriptional-level control carries with it the mechanism of selective gene activation, i.e., the transcription of different DNA sequences in different cell types. A comparison of nucleic acid populations in different types of cells provides support for regulation at the transcriptional level. Liver cells of the adult chicken contain approximately 14,000 different mRNA species. Molecular-hybridization studies indicate that of the 4,000 nuclear RNA species of the "nearly mature" chicken erythrocyte, only about 100 are shared with the liver mRNA population. This can be contrasted with the results obtained when comparisons of nuclear and messenger RNA populations in different sea urchin tissues or embryonic stages were compared (Section 8.3.C).

Further support for transcriptional-level control during erythropoiesis comes from studies with isolated chromatin. Two types of experiments will be mentioned. It has been shown that genetic loci actively involved in transcription are more sensitive to degradation by DNase I than inactive loci when isolated chromatin is subjected to treatment with this enzyme. With this criterion for genetic activity in mind, it has been shown that the globin genes in isolated chromatin are sensitive to DNase I digestion when taken from a hemoglobin-producing tissue, such as bone marrow, but are insensitive to the same treatment when isolated from a nonerythropoietic organ. Furthermore, sensitivity of the globin DNA sequences to DNase I digestion appears at an early stage in the differentiation sequence, well before the appearance of globin mRNA can be detected (discussed later). This finding suggests that conformational changes in specific sites within the chromatin precede the actual transcription of that genetic region.

The differential template activity of chromatin isolated from different tissues has been widely used as evidence in support of transcriptional-level control (this technique is discussed critically by P. Chambon in Cold Spring Harbor Symp. **42**:1209, 1977). Experiments carried out in several laboratories indicate that chromatin isolated from erythropoietic tissues will serve as a template for the synthesis of the globin mRNA sequence in vitro, while chromatin from nonerythropoietic tissues will not. The results of one of these experiments is shown in Fig. 15.18. In this case, transcription was accomplished with a mammalian polymerase, and no endogenous globin mRNA was detected in the preparation prior to addition of the polymerase. In this particular experiment it was found that approximately 0.05 percent of the RNA synthesized in vitro would hybridize to globin cDNA when rabbit bone-marrow chromatin was used as the template. Presumably, this percentage reflects the fact that the globin genes represent only 1/2,000 of the active fraction of the genome. Despite the fact that the globin genes represent such a small fraction of the genes being transcribed in these cells, they account for over 90 percent of the protein being synthesized. How can this be accomplished?

Kinetic studies of various types suggest that the globin sequence receives preferential treatment in erythroid cells all along the way. In anemic chickens it is calculated that each species of globin mRNA arrives on polyribosomes at a rate of about one molecule per minute per cell, approximately 50 times that of nonglobin mRNA species. Most of its difference is believed to reflect a greater rate of transcription of globin as opposed to nonglobin genes, but some of the enhanced flow of globin mRNA onto the polysomes seems to reflect its greater rate of processing from larger nuclear precursors. Once in the cytoplasm, the globin message is generally found to be more stable than those of other species. Consequently, a given globin mRNA will be translated a greater number of times than a nonglo-

Figure 15.18 Hybridization of globin cDNA to RNA transcribed from chromatin isolated from an erythropoietic organ, i.e., rabbit bone marrow (○), and a nonerythropoietic organ, i.e., rabbit liver (●), using a eukaryotic polymerase prepared from sheep liver to carry out the transcription. Triangles (△) represent a control in which RNA polymerase is deleted. [*From A. W. Steggles et al.,* Proc. Natl. Acad. Sci. U.S., **71**:*1221 (1974)*.]

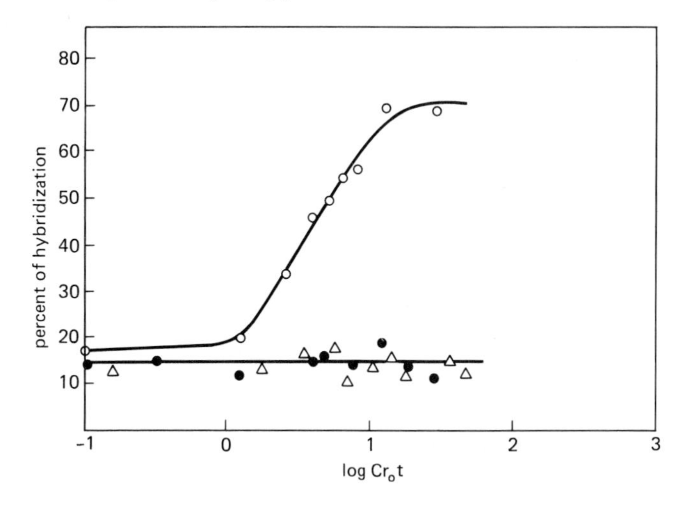

bin message. Taken together, these various figures illustrate the point that two genes per cell can, under the appropriate conditions, account for a cell's becoming packed with one type of protein.[1] In addition to the controls just discussed, the synthesis of globin chains is also regulated by a mechanism operating at the translational level. It has been known for a number of years that the synthesis of globin does not occur in the absence of heme groups with which it must combine to form active hemoglobin molecules. Recent studies have revealed the manner in which hemin levels serve to regulate protein synthesis within reticulocyte cytoplasm (reviewed in Ochoa, 1979). Regulation occurs via phosphorylation and dephosphorylation of specific proteins, much like that during hormone-dependent control of sugar metabolism in liver cells (Section 3.4). One of the substances required for protein synthesis is the initiation factor eIF-2, which normally binds to the methionyl tRNA during the initiation step of polypeptide formation. This protein can be phosphorylated by a protein kinase, a modification that renders the factor inactive and unable to participate in the initiation process. In the presence of hemin, it is the protein kinase that is inactivated, leaving the initiation factor fully functional and able to carry out its work.

Figure 15.2 showed the various stages of differentiation found within the erythroid lineage. Where in this pathway are the globin mRNAs first synthesized, and when do globin polypeptides first appear on the scene? Studies of a variety of erythropoietic systems suggest that it is during the proerythroblast stage that the globin genes are first transcribed, most likely as part of the response of the cell to erythropoietin. Given the normal condition in which hemin groups are present in the cytoplasm, the synthesis of the globin message is followed by its immediate translation; there is no storage of untranslated message under normal conditions. By the time differentiation has progressed to the basophilic erythroblast stage, large amounts of globin message have accumulated in the cells, and globin polypeptides can be detected.

One system that has been widely exploited as a model for the study of erythropoiesis is a transformed (malignant) cell line, the murine erythroleukemia (MEL) cell. These cells are derived from normal erythroid precursors (probably ones at the proerythroblast stage) that have been infected (in vitro or in vivo) by a group of tumor viruses, the Friend leukemia virus complex. Once transformed, these cells have several properties not found in normal erythroid precursors which make them ideal for analysis. Most important, one can maintain long-term cultures of these unspecialized precursors, obtain various genetic variants, and induce the synchronous differentiation of large numbers of cells by treatment with one of a variety of inducing agents, including dimethylsulfoxide (DMSO), fatty acids, and purines. Once transformed, these cells no longer require erythropoietin for their maintenance or differentiation. It should be noted that differentiation of MEL cells does not proceed to completion. The final stages of differentiation resemble the orthochromatic erythroblast (Fig. 15.2*f*), but enucleation steps are not initiated. During the several days over which differentiation of MEL cells occurs, the appearance and/or accumulation of a number of erythrocyte-specific substances is found to occur. These include globin mRNA and globin chains, hemin, carbonic anhydrase, and the membrane-specific proteins.

[1] The lack of gene amplification in erythropoietic cells is discussed in Section 8.4.A. Analysis of molecular hybridization data, genetic variants, and restriction endonuclease fragments has revealed the exact number of globin genes in a number of organisms. There are two alpha genes in the human genome, both on chromosome 16 and very close to each other. These two loci are identical in nearly all persons. The β, γ, and δ genes are also located close to each other, but on a different chromosome, number 11. There are two copies of the γ gene that code for polypeptides that are different at one amino acid residue. There appears to be only one copy of the β and δ genes.

Analysis of the molecular events occurring during MEL-cell differentiation serve as an interesting contrast to those described earlier for the terminal steps of erythrocyte formation in the anemic chicken. Unlike the nearly mature chicken erythrocyte, the MEL cell has a large variety of mRNA species in its polysomes. It is estimated that over 10,000 different mRNAs are being translated in the uninduced Friend cell, and significantly, there is no major change in the variety of mRNAs occurring after induction with DMSO. In this case, the switch from a committed but unspecialized erythroid cell to one that is overtly expressing the erythroid program does not seem to require a major qualitative overhaul of the mRNA population. For example, the globin message can be demonstrated to be present in the uninduced MEL cell, but its concentration greatly increases after DMSO treatment (Fig. 15.19). These results suggest that the committed precursor, a proerythroblastlike cell in this case, is already engaged at a low level in the synthetic activities characteristic of the cytodifferentiated cell, despite the fact that the cell's morphology remains unspecialized. We will return to this concept in the discussion in Section 15.6.C.

B. Genetic control of myogenesis

Biochemical analysis of muscle cell differentiation has concentrated on a number of contractile proteins and enzymes characteristic of muscle tissue. When are these proteins, and the mRNAs that code for them, first detectable? The answers to these questions may well vary somewhat depending on the particular muscle tissue undergoing differentiation. As discussed earlier, one of the key steps during myogenesis is myoblast fusion. This event occurs rapidly, is clearly definable, and takes place at a stage between proliferation and cytodifferentiation. At the same time, the fusion event itself is not required for the ensuing morphological transformation to occur. Since fusion represents only a convenient developmental marker with which other events can be compared rather than a trigger for cy-

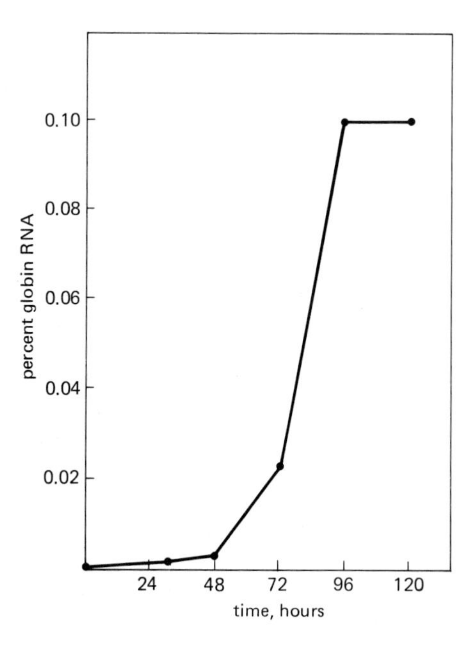

Figure 15.19 Time course of accumulation of globin mRNA in erythroleukemic cells after treatment with dimethylsulfoxide. [*From S. H. Orkin,* In Vitro, **14:***151 (1978)*.]

todifferentiation, it is not surprising that some variability exists with respect to the time of appearance of muscle-specific gene products relative to fusion itself. For example, mononucleate myoblasts present within the somites of a stage 16 chick embryo may contain myosin, as detected by its binding to antimyosin antibodies and the presence of thick filaments in the electron microscope. In contrast, studies of other myogenic systems in vivo or analysis of cultured myoblasts undergoing fusion and differentiation in vitro suggest that myosin and other contractile proteins do not appear in significant amounts until after myotube formation. Since it is virtually impossible to prepare cultures in which all the myoblasts are at approximately the same stage of differentiation, it can always be questioned as to whether the presence of low levels of a given protein is due to (1) the existence of this protein within all cells or (2) the presence in the culture of a small percentage of developmentally advanced cells containing larger amounts of the protein being studied.

The results of one of the more extensive recent studies of the synthesis of muscle-specific proteins is shown in Fig. 15.20. In this case, proteins were extracted from quail myoblasts whose differentiation had been synchronized by transfer to a medium lacking growth-promoting factors, as discussed in Section 15.5. Newly synthesized proteins, labeled with [^{35}S]methionine, were fractionated by two-dimensional gel electrophoresis. The identity of the spots were determined by comparison with known polypeptides purified from adult muscle. In this study, no evidence for the synthesis of any of eight different polypeptides of contractile proteins was detected prior to myoblast fusion. Soon after fusion, however, the presence of all eight polypeptides became apparent and the rates of synthesis of all eight increased at least 500-fold over the next 24 hours. Moreover, the increased rate of synthesis of all eight polypeptides showed exactly the same time course (Fig. 15.21), suggesting that their synthesis may occur in a coordinated manner. The concept of coordinate regulation of the synthesis of contractile proteins is reinforced by the finding that the synthesis of the noncontractile muscle-specific proteins follow several very different time courses, which suggests that they are subject to different patterns of regulation.

Of the noncontractile proteins, the best-studied with respect to differentiation is creatine phosphokinase, the enzyme that serves to transfer phosphate groups from an energy-storage compound, creatine phosphate, to ADP, thereby forming ATP, which can fuel further muscular activity. Enzymes often exist in multiple molecular forms, termed *isoenzymes* or *isozymes*. The different isozymic forms of a given enzyme catalyze the same reaction, but generally have different kinetic properties, different sensitivities to low molecular weight modulators, different subcellular localization, and/or different distribution within the animal. Creatine phosphokinase (CPK) is a dimer, i.e., made of two independent subunits, each of which by itself is catalytically inactive. Furthermore, there exists two distinct types of subunits, M and B species, coded by two different genes. Three possible enzymes can be formed as a dimer containing two of these subunits (BB, BM, MM), and all are active and can be found within various tissues. Since the M and B subunits have differing electric charge, the various isozymes can be separated from one another and the relative amounts of each of the three varieties of enzyme determined in a given tissue preparation. The preponderance of one or another isozyme varies markedly among various tissues. Skeletal muscle of adult birds and mammals has only the MM isozyme, while brain tissue has only the BB form.

It has been found that the isozymic pattern of CPK shifts dramatically during muscle differentiation, a feature that can be demonstrated both in culture (Fig.

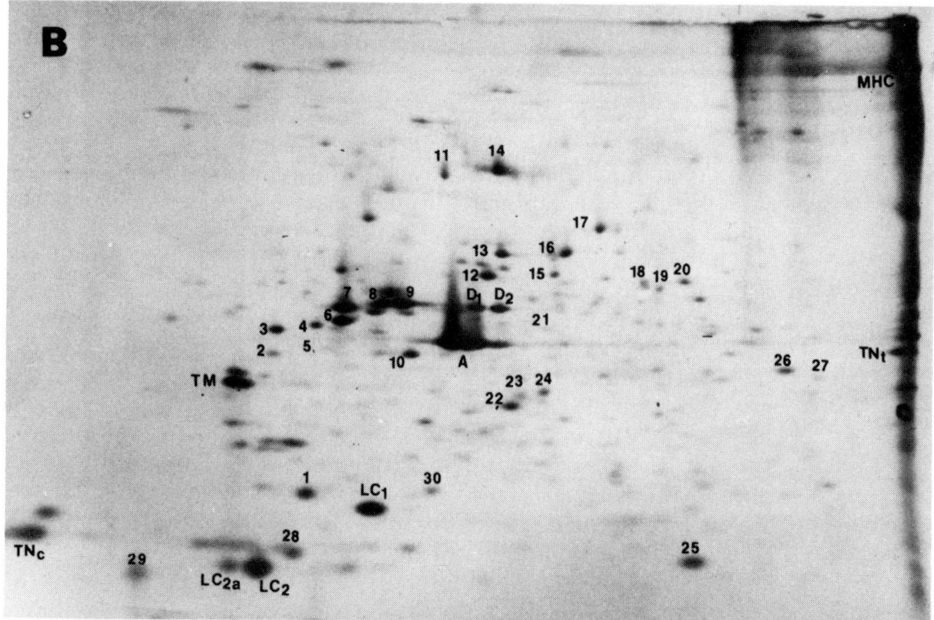

Figure 15.20 Autoradiograms of ^{35}S-labeled proteins synthesized by dividing myoblasts (*a*) or muscle fibers (*b*) after separation by two-dimensional gel electrophoresis. The arrows shown in *a* indicate the positions of proteins that are not synthesized by myoblasts but are synthesized by muscle fibers. The positions of the contractile proteins shown on the autoradiogram in *b* were identified by their comigration with stained markers of purified contractile proteins and are designated by the following abbreviations: myosin heavy chain, MHC; myosin light chains, LC_1, LC_2, and LC_{2a}; tropomyosin, TM; troponin, TN_t and TN_c; and actin, A. Other muscle-cell proteins are designated by numbers. [*Courtesy of R. B. Devlin and C. P. Emerson, Jr.*]

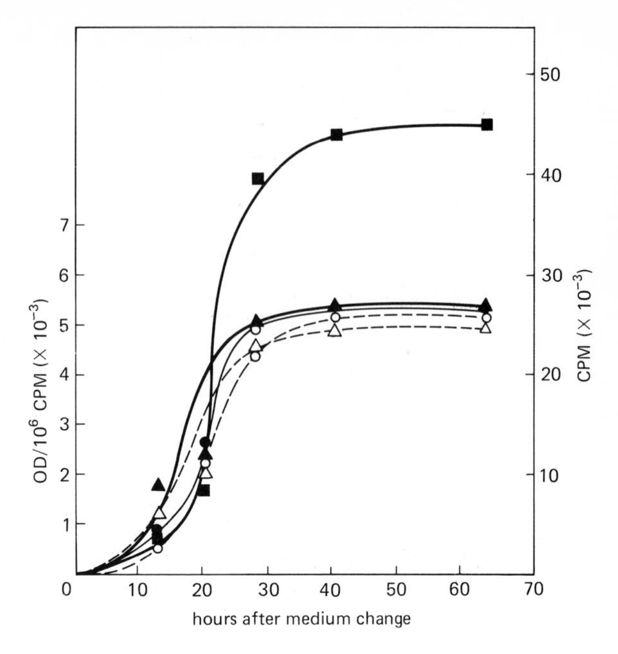

Figure 15.21 Kinetics of synthesis of contractile proteins during myoblast differentiation. Cultures of dividing myoblasts in growth medium were shifted to a medium that initiates myogenesis. Cultures were labeled with [^{35}S]methionine for 2-hour periods at various stages during the process of differentiation. Labeled proteins from each time period were separated by two-dimensional gel electrophoresis and the density of the spots determined. LC_1 (●); LC_2 (○); TN_t (△); TN_c (▲); MHC (■). The ordinate for the much larger MHC is on the right, the ordinate for the other proteins is on the left. Similar curves are found for TM_1, TM_2, and actin. [*From R. B. Devlin and C. P. Emerson, Jr.,* Cell, **13**:*603 (1978).*]

15.22*a*) and within the intact embryo (Fig. 15.22*b*). Under culture conditions, proliferating mononucleate chick skeletal myoblasts show no evidence of the presence of the MM isozyme, and only low levels of the MB form, which may represent a contaminant from the small percentage of more advanced cells in the culture. However, with the onset of fusion and subsequent cytodifferentiation, the MM form rapidly becomes predominant (Fig. 15.22*a*). A similar type of shift from BB to MM isozyme occurs in the rat between the seventeenth day of gestation and birth (Fig. 15.22*b*). Shifting isozyme patterns during differentiation re-

Figure 15.22 Creatine phosphokinase isozyme changes during myogenesis in chick embryonic cell cultures (*a*) and rat skeletal muscle (*b*). Activity of each isozyme is expressed as a percentage of the total activity. [*a From J. Lough and R. Bischoff,* Dev. Biol., **57**:*337 (1977). b From F. A. Ziter,* Exp. Neurol., **43**:*542 (1974).*]

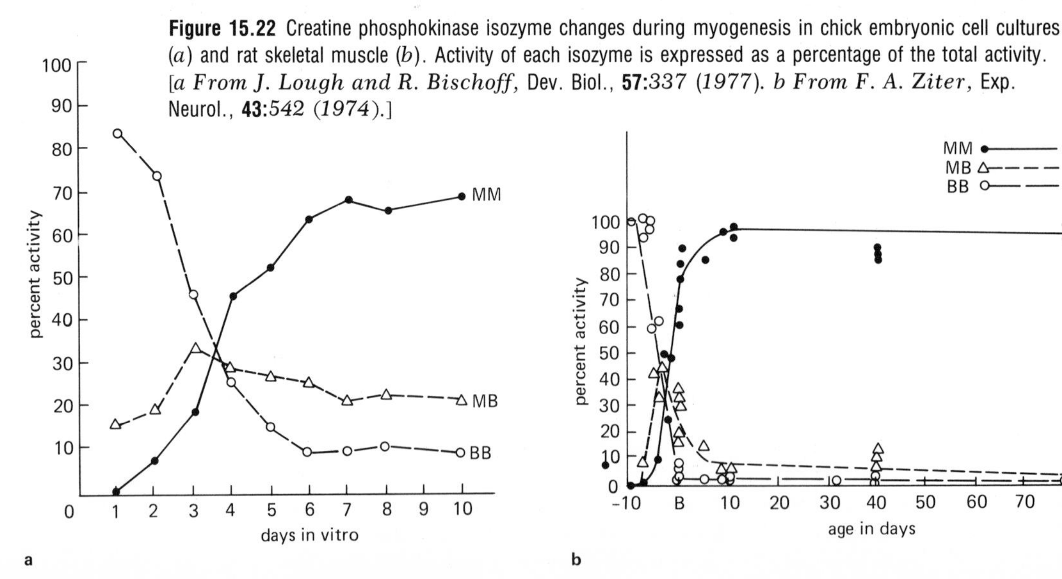

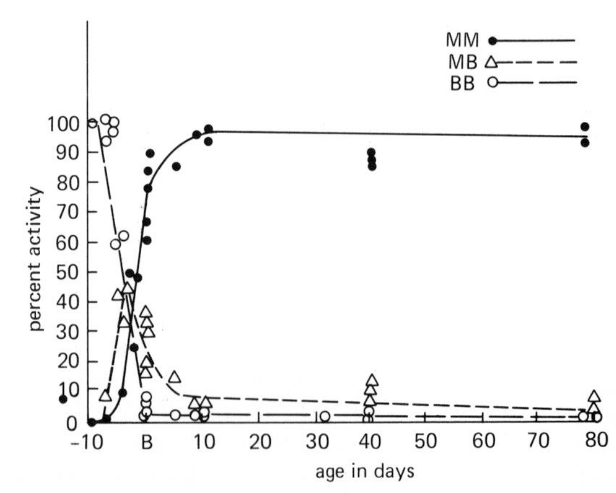

flect changes in gene expression, in this case the presumed shift from the expression of the B gene to that coding for the M subunit.

Questions concerning the relationship between the transcription of muscle-specific genes and the translation of their mRNAs have generated contradictory sets of data; the answers may well vary with the particular system of myogenesis under study. Some investigators find evidence for the early synthesis of muscle-specific mRNAs, such as that which codes for the very large polypeptide of the myosin molecule, and claim that these messages are stored in an inactive state in the cytoplasm of proliferating myoblasts. Following fusion, these mRNAs are released for recruitment into polyribosomes by some translational-control mechanism. The situation is analogous to the storage of mRNAs in the unfertilized sea urchin egg (Section 8.3.A). It is more widely believed that the synthesis of muscle-specific proteins is limited by the availability of mRNA sequences. According to one proposal, the synthesis of mRNAs for these proteins occurs well before the time of myoblast fusion, but the molecules are unstable and do not accumulate. In this view, the fusion process is accompanied by a major stabilization of muscle-specific messengers and the subsequent accumulation of their respective polypeptides. Alternatively, there is a body of evidence that suggests that mRNA sequences (as measured by translation in an in vitro cell-free system) for muscle-specific proteins are not present in dividing myoblasts and it is only with the appearance of these mRNAs at the onset of cytodifferentiation that their respective polypeptides can be produced.

Regardless of whether or not muscle-specific mRNAs are present in myoblasts prior to fusion, their concentration increases markedly as cytodifferentiation begins. While this special collection of mRNAs is increasing in abundance, the overall complexity of the RNA population of differentiating muscle cells is being greatly reduced. It has been found that total cell RNA from proliferating myoblasts saturates approximately twice as much single-copy DNA as does a comparable amount of RNA extracted from differentiated myotubes (see appendix to Chapter 8 for discussion of methodology). It would appear that gene expression is most extensive in the proliferating myoblast population and decreases progressively as muscle differentiation proceeds. In this case at least, the acquisition of specialized characteristics is accompanied by a large-scale repression of previously active genes and a marked increase in transcription of a small number of genes of particular importance to muscle cell development and function.

One of the major handicaps in the study of the molecular biology of myogenesis has been the inability to obtain highly purified mRNA sequences for muscle-specific proteins. Unlike erythropoiesis, in which one particular protein greatly dominates the cell's synthetic activity, no single protein constitutes more than about 5 percent of the total synthetic picture during myogenesis. It is hoped that cloning techniques soon may make available purified mRNA sequences for some of the major muscle-specific polypeptides.

C. Genetic control of pancreas formation

The pancreas is also an excellent system for study at the biochemical level. Its differentiation from a primitive rudiment will occur in vitro accompanied by the synthesis of a variety of proteins that have been very well characterized. The proteins of the pancreas are of two types, produced in different cells and secreted into different vessels. One group of proteins, a wide variety of hydrolytic enzymes secreted into the ducts leading to the intestine, is the product of the bulk of the pancreatic cells, the *exocrine acinar cells*. This component of the pancreas consists of a single convoluted layer of secretory cells. The other proteins are hor-

mones, insulin and glucagon, products of islets of endocrine cells embedded within the acinar tissue of the pancreas.

The exocrine acinar cells have a distinct ultrastructural morphology reflecting their high rates of protein synthesis and their extensive secretory activity. Fully differentiated pancreas cells have an extensive rough endoplasmic reticulum where the enzymes are synthesized, an elaborate Golgi complex where they are packaged, and a great number of secretory, or zymogen, granules in which the enzymes are prepared for release. Insulin is produced by the B (or beta) cells, which can be readily distinguished from the glucagon-producing A (or alpha) cells. Beta cells, which differentiate several days after the precocious alpha cells, are characterized by their dense cytoplasm and the presence of beta granules.

In the mouse, pancreas morphogenesis begins approximately halfway through the gestation period by the bulging of a group of cells of the dorsal endoderm (roof of the archenteron) into the surrounding mesenchyme. The intimate association of an epithelium, in this case derived from the gut, and a mesenchymal tissue is important in the formation of a number of organs including the salivary gland, lung, kidney, thyroid, mammary gland, pituitary, and liver. It is generally found that a mesenchymal factor acts as an inducing agent to promote the differentiation of the epithelium. As a result of this epithelial-mesenchymal interaction, a bulbous pancreatic rudiment is formed.

Biochemical analysis of the cells of the pancreatic rudiment on the twelfth day of mouse development indicates that the enzymes characteristic of the mature exocrine pancreas, as well as insulin, are present at a low level. The low level of these proteins is maintained until the fourteenth day, after which it begins to rise to reach the high levels characteristic of the fully differentiated pancreas cell. Insulin and lipase, for example, are present in at least a 10,000-fold greater concentration in the cells of the pancreas rudiment compared with the cells of the embryo in general. The presence of these proteins prior to the fourteenth day is clear indication that the cells have achieved a tissue-specific differentiated state; yet morphologically they remain relatively unspecialized. The acinar cells lack the highly differentiated rough endoplasmic reticulum, Golgi, and zymogen granules; the beta cells lack the beta granules. Rutter has termed this the *protodifferentiated state* to distinguish it from an earlier period where these proteins are absent and a later period where they are present at much greater concentration (Fig. 15.23).

The first transition (to the protodifferentiated state) is accompanied by the activation of a large battery of pancreas-specific genes. The primary transition, therefore, must be considered a major transcriptive event leading to cytodifferentiation, which is delayed for several days. The primary transition can be considered a change from a precursor cell to one with a pancreatic phenotype, even though the phenotype is not that of the fully differentiated cell.

The protodifferentiated phase is accompanied by a great change in the appearance of the pancreatic rudiment. At its start, the epithelial-mesenchymal rudiment is little more than a bulge on the wall of the intestine. At its termination, cell proliferation has resulted in a great increase in the cell number and morphogenesis has produced a convoluted epithelium surrounding an intricate system of ducts. The cytologic appearance of the cells, however, remains unspecialized.

A second transition step (occurring after day 14 in the mouse) converts the protodifferentiated cells to the fully cytodifferentiated state. Enzyme levels after the second transition range from 1,000 to 10,000 times that of the protodifferentiated state. Since many, or all, of the specific messenger RNAs must already be available within the cells to account for the presence of all the pancreatic en-

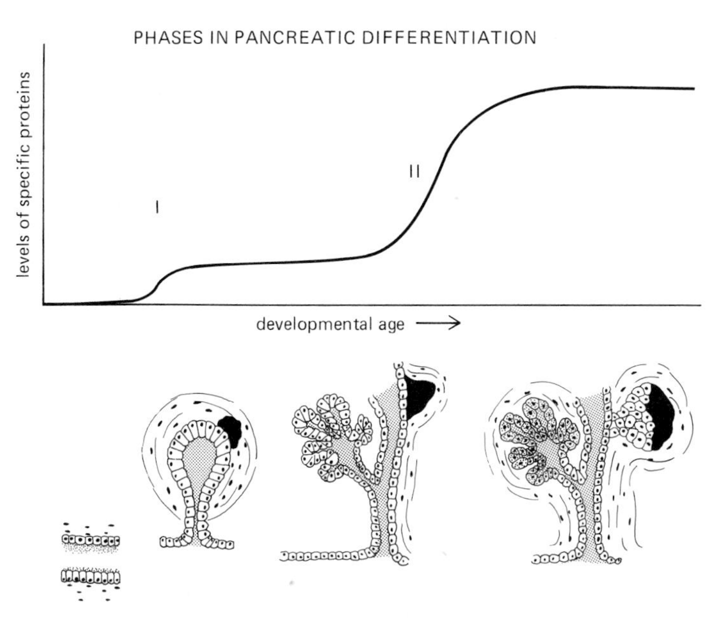

Figure 15.23 Diagrammatic representation of the stages in the accumulation of various proteins during the differentiation of the pancreas. Three recognizable phases are evident: an early stage, in which pancreas-specific proteins are lacking; a second protodifferentiated stage, when these proteins are present at low concentration; and a third stage, when the proteins are present at high levels. Roman numerals I and II indicate the transitions. The morphologic correlates of the synthetic stages are shown below. [*From W. J. Rutter et al., in "Differentiation and Development," F. Ahmad (ed.), Academic, 1978.*]

zymes, the secondary transitional step may reflect primarily a translational-control mechanism.This does not mean that additional transcriptional regulation is not required for this second transition as well. Actinomycin D administered during the protodifferentiated phase will block the development of the fully differentiated state, suggesting that additional RNA synthesis is required. Once the protodifferentiated state has ended, however, actinomycin D is no longer effective in blocking the subsequent cytodifferentiation.

More recently, the direct analysis of mRNA levels during pancreas development has provided support for the conclusions reached using actinomycin D. Using a labeled cDNA prepared from a purified amylase mRNA, it was found that amylase mRNA levels increase approximately 600-fold between the fourteenth and twentieth day of gestation. This increase roughly parallels that found for the enzyme itself, suggesting that the availability of the message serves as the rate-limiting step in the control of protein synthesis. It is estimated that cells at the protodifferentiated stage contain approximately 20 molecules of the amylase mRNA, approximately 0.03 percent of their eventual content. Indications have been found that the biphasic accumulation of specific proteins during cytodifferentiation occurs in other types of cells, including cartilage, thyroid, and mammary gland tissue, although the general applicability of this model remains to be determined.

The differentiation of the pancreatic epithelium into a secretory cell layer occurs in response to a protein produced by the surrounding mesenchymal cells. This protein, termed *mesenchyme factor,* or MF, has been extracted from the membrane fraction of the mesenchymal cells in high salt solution and has been greatly purified for use in studies of growth and differentiation of the pancreas. In the absence of MF, the pancreas rudiment develops into a structure having only endocrine function. Evidence suggests that MF exerts its inductive effect via interaction with the responding epithelial cell surface. Mesenchymal factor can be covalently linked to inert beads of sepharose, which blocks MF entry into the epithelial cells, and it will still promote growth and differentiation of the pancreatic epithelium. In fact, in response to the presence of these MF-coated beads, the responding cells become tightly pressed against the bead surface, attached by the bridges of mesenchymal factor. Autoradiographs of these cells after exposure to [^{3}H]thymidine indicate that nearly all the cells that are synthesizing DNA are directly attached to the coated beads (Fig. 15.24). It appears that the interaction of MF with the cell surface greatly stimulates the DNA synthetic capacity of these cells. Not only are the attached cells stimulated to divide, they can undergo cytodifferentiation into mature pancreas cells, with the basal side of the cell attached to the bead and the apical microvillar surface in the opposite direction. The results of this last experiment are particularly important because they indicate that cytodifferentiation can be dissociated from the morphogenetic activities that normally accompany pancreatic acinar formation. It would appear that the differentiation of a specialized pancreatic exocrine cell can occur independently from the development of the normal histologic architecture of the organ.

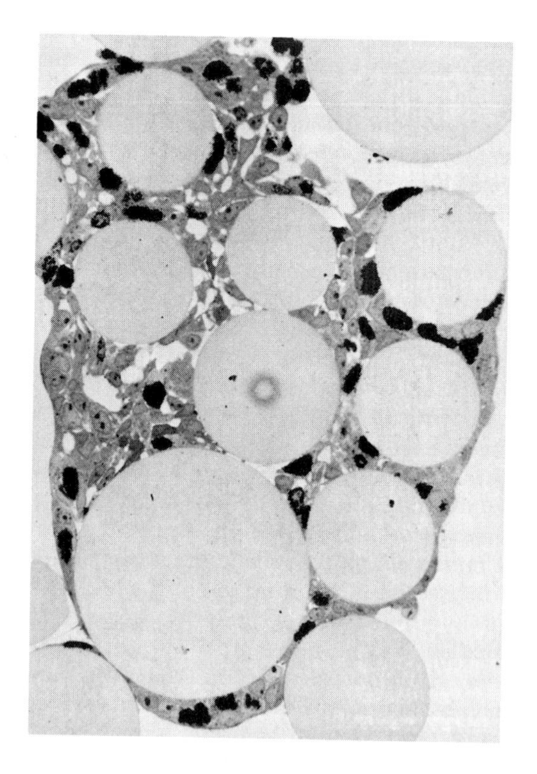

Figure 15.24 Autoradiograph of a section cut through pancreas cells undergoing replication (and subsequent differentiation) in response to contact with MF-coated beads. Blackened nuclei are those which have incorporated [^{3}H]thymidine into DNA. [*Courtesy of S. Levene, R. Pictet, and W. J. Rutter.*]

7. Embryonic determination

In the first part of this chapter a few events were described that take place during the terminal steps in the differentiation of some cell types that have been studied. During the terminal stages, the cells become visibly differentiated; they have attained a phenotypic expression. These stages have received the most attention because they are the most readily examined. It is the earlier stages, of which little is known, in which a cell's fate becomes fixed. We have paid considerable attention to the change in prospective potency of cells in vertebrate embryos during gastrulation. Following gastrulation, cells transplanted to new environments would differentiate in the foreign surroundings into the same tissues they would normally have formed if left undisturbed. The onset of determination marks the first step in the differentiation process; the cells of the embryo are no longer equivalent with respect to the tissues each can form. The term *differentiated*, which expresses the fact that cells have become different from one another, is as suitably applied to the earlier stages as it is to those cells in which visible morphological evidence can be found.

It is generally assumed that the transition from the undifferentiated to the fully differentiated state is a gradual one with many steps. Indirect evidence for this comes from many sources. It is inherent in the concept of the embryonic field that the cells of the field are capable of forming other parts of the organ but not parts of some tissue of a foreign field. As later and later stages are reached, the ability of a given part of the field to form peripheral structures becomes progressively limited. During the early development of the hemopoietic tissues, cells can be found that are *pluripotent:* they can form a variety of specialized cells of the lymphoid, erythroid, and granulocytic series. At a later stage, the pluripotent cell is no longer found, and each of these individual lines has its own precursor. The pluripotent cell has become *unipotent*. Similarly, the exocrine, endocrine, and duct cells of the pancreas are believed to differentiate from a common precursor cell in the initial pancreatic rudiment. Another example of progressive differentiation is the somite cells. They differentiate into cartilage, muscle, and connective tissue. As development proceeds, each of these tissues has its own precursor cell—the chondroblast, myoblast, and fibroblast, respectively. In the initial stages of somite formation, all cells appear indistinguishable; there is no compelling reason to suspect that any given cell cannot form all three fully differentiated cell types.

The intermediate in the sequence from the undifferentiated to the cytodifferentiated is the stem cell. A given stem cell is committed to differentiate along a path from which one or a few closely related cell types can emerge. Stem cells appear in the embryo, but in many tissues they remain throughout the life of the animal to provide the pool of precursors from which the terminally differentiated cells will arise.

In the context just discussed, embryonic development can be likened to a tree with its singular trunk and circle of limbs that branch progressively to form an increasing number of twigs and leaves. At the base there exists a single cell, the egg, which divides in a binary fashion producing an increasing number of cells. In this analogy, the developmental tree can be viewed as a complex series of cell lineages; the closer to the base, the more multipotent the cell and the broader the lineage. Near the base, for example, one might identify ectodermal, mesodermal, and endodermal cell lineages. As one moves farther up the tree, the limbs increase in number and decrease in size, a reflection of the progressive restriction of the stem cells in their prospective potency. At each branch point, a decision must be made that restricts the possibilities for differentiation. In some cases, branch points occur at times of cell division such that each of the two daughter

cells become committed to differentiate along diverse paths. Alternatively, a given division might produce two equivalent cells that diverge following exposure to different environmental influences. While there are many branch points, there are also many straight portions between the forks. Along these unbranched segments, cells may undergo bursts of proliferation to produce a large population of cells of similar commitment. We have been concerned in this chapter with the most distant portions of the tree past the points of the last branch. In many cases, these last segments are of considerable length, if length can be equated to time passed or numbers of cells produced. These straight portions can represent periods in the lineage when dramatic morphological changes can occur. This is evident from the steps that occur in the erythroid lineage (Fig. 15.2) following the time of the last branch. With this concept of cell lineages in mind, we can briefly reexamine the types of inductive interactions that might be responsible for the branch points themselves.

8. Nature of the inductive interaction

In Chapter 11, considerable attention was paid to the process of embryonic induction as a means whereby a responsive tissue was directed along a particular path of development. The nature of the inductive interaction has remained one of the least understood events despite a long history of investigation. The intervention of an inducing molecule into the activities of a reacting tissue has been considered in two ways. The earlier view emphasized an "instructive" role of the inducing substances, in which the inducer carried some specific information that was directly utilized by the responding tissue to cause it to differentiate in one direction as opposed to another. In this type of induction, the responding tissue is viewed as being flexible with regard to the opportunities available for its differentiation. In this concept, the direction ultimately taken by the responding tissue is dependent on the nature of the inducing substance. In other words, in the case of an instructive interaction, the induction stimulus enters into the decision-making or determinative process that precedes overt cytodifferentiation. Inducers of this type might be expected to interact directly or indirectly with the genome in some specific manner as to direct its expression in some tissue-specific direction.

An alternate concept gives the inducing substances a "permissive" role, one that provides a favorable environment in which the phenotypic expression is promoted. In this view, the competent state of the responding tissue at the time of induction is emphasized, suggesting that the inducer releases a preprogrammed sequence of events rather than supplying information needed for the differentiation process. This latter view stresses the background of the responding tissue and its previously determined potential. It has come primarily from recent work in tissue-culture differentiation in which the importance of the environment has been emphasized.

Cartilage, for example, has been a favorite tissue for in vitro culture and can be used to illustrate these points. The early evidence, gained from both in vivo and in vitro studies, suggested that the differentiation of somite tissue into cartilage required an induction by substances emanating from either notochord or spinal-cord tissue. Extracts from these, and only these, cells were found to be active among a wide variety of types that were tested. In other words, there appeared to be a high degree of specificity associated with these inducing tissues; specificity is generally interpreted as reflecting an instructive interaction.

Several lines of evidence, however, suggest that even in this case, the inducing substances are having a permissive role. With the development of more favorable culture media, it was discovered that somite tissue could differentiate into

cartilage by itself, in culture isolation. Such differentiation is termed *spontaneous cartilage*. It indicates cartilage formation does not require some extraneous piece of information that must be provided from its environment. Rather it suggests that the notochord or spinal cord provides a particularly favorable environment, but not an essential one, for the expression of the cartilage phenotype. The inducing substances are believed to enhance a chondrogenic bias already preprogrammed into these cells. Cells not having this bias are not induced. Even cartilage-forming cells from other parts of the embryo, such as the limb, are not stimulated by notochord or spinal-cord tissues. The specificity is as much a property of the responding tissue as it is of the inducing cells.

The induction of cartilage by notochord is like the induction of myogenesis by collagen, neurogenesis by nerve-growth factor, pancreas formation by a mesenchymal factor, erythropoiesis by erythropoietin, etc. In each of these cases, the responding cells are already determined, at least partially, toward their definitive courses. In cases such as these, a permissive interaction is more understandable than the inductive interactions occurring earlier in development that are involved in laying down the major organ primordia of the embryo. Prior to gastrulation, the roof of the archenteron has far-reaching prospective potency; it is only after its interaction with the underlying chordamesoderm that its choices become restricted to that of forming sensory and neural tissue. The question of the roof of the archenteron as an inductor versus an evocator has already been discussed in Chapter 11 and need not be reconsidered here.

It is apparent from this discussion, and others in this book, that we know very little about the nature of inductive interactions, whether they be primary, secondary, tertiary, etc. Over the years, three general mechanisms have been proposed to account for the passage of a stimulus from inductive to responding tissue: via freely diffusible substances, via direct cell-to-cell contact, and via an action of intercellular matrix materials interposed between the two interacting tissues. Some evidence has been gathered which supports each of these mechanisms in one or another example of induction. There is some consensus of opinion that primary induction occurs as a result of the free diffusion of chemical inducers (Section 11.4), that kidney induction involves cellular contact between apposing cell membranes (Section 12.4.A), and that the induction of chondrogenesis and corneal differentiation are mediated by materials of the extracellular matrix (Sections 15.4 and 14.3.E). However, in none of these cases is the evidence overwhelming for the contention presented, and considerable controversy remains. For example, much of the work on induction rests on the use of membrane filters, and investigators still argue whether or not cellular processes are extending deeply into these filters and whether or not the contact of such processes can account for the induction observed.

9. Modulation of the phenotype in culture

The effect of the culture medium on differentiation can be shown with a wide variety of cells. It has been generally found that if a differentiated tissue is removed from the body, dissociated into single cells, and cultured in vitro, these cells rapidly assume an undifferentiated appearance. Cartilage cells, for example, cease the production of matrix materials, pigment cells lose their melanosomes, etc. In the earlier literature, these cells are considered to have *dedifferentiated*. This term has generally been avoided in recent years, or at least carefully defined, since it carries the implication that the differentiated cell has in some way reverted to a more primitive, undifferentiated state. Such is not the case. Analysis of the capability of these cells reveals that in many respects they are no less dif-

ferentiated than the state from which they were derived. Cartilage cells, for example, when placed into culture, retain the enzymes needed for matrix production, even though matrix is no longer synthesized; some other factor has intervened to shut production down.

Whether previously differentiated cells remain undifferentiated in appearance or regain their original phenotype depends to a great extent on the culture conditions. The phenotype of cartilage cells, for example, depends on numerous factors including the nature of the substrate the cells grow on, the density at which they are growing, and the chemical nature of their environment. Under certain conditions, the cells become polygonal-shaped and are seen to be incorporating large quantities of sulfate and accumulating matrix, such as those shown in Fig. 15.8. If the medium is changed by the addition of fibronectin (Fig. 15.25) or substances of high molecular weight from an embryo extract, the cells undergo a change in shape to a stellate, fibroblastlike cell with a much greater surface area. The synthetic activities of these cells also change dramatically; the production of chondroitin sulfate drops to low levels and the collagen being manufactured changes from the type II form characteristic of cartilage cells to the type I form. In other words, they lose the characteristics of the cytodifferentiated state. If the substances from the embryo extract are removed in a suitable period, the cells will regain the cartilage phenotype; they can revert to the fully differentiated cartilage state, but no other. Cells that have lost their overt differentiation are in a state of covert, or hidden, differentiation. The oscillation between the overt and covert state is termed *modulation*.

Changes in the overt state of differentiation of a cell also can be brought about by abruptly changing the cell's cytoplasmic content, a feat that can be brought about by cell fusion. One of the earliest and best-studied examples of this experimental phenomenon occurs in hybrids formed between mouse or hamster melanoma cells and mouse L cells. The malignant melanoma cell is normally very darkly pigmented, but following fusion to a nonpigmented L cell, it loses its coloration and its ability to produce melanin. The differentiated state of the cell is said

Figure 15.25 The morphological effects of fibronectin on chondroblasts. *a* Control cells. *b* After 144 hours of exposure to low concentrations of fibronectin the cells have lost their polygonal morphology. [*Courtesy of C. West.*]

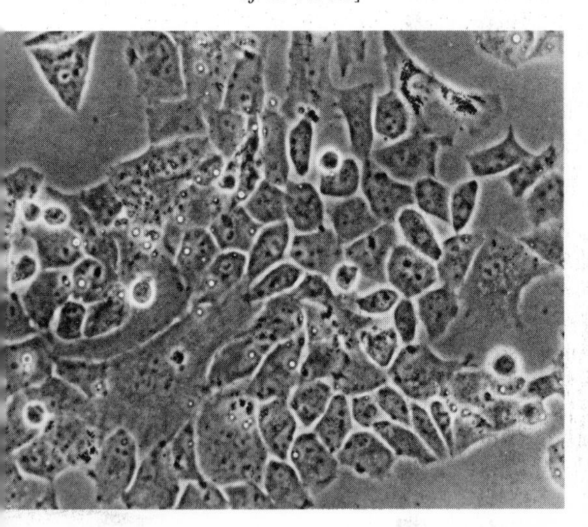

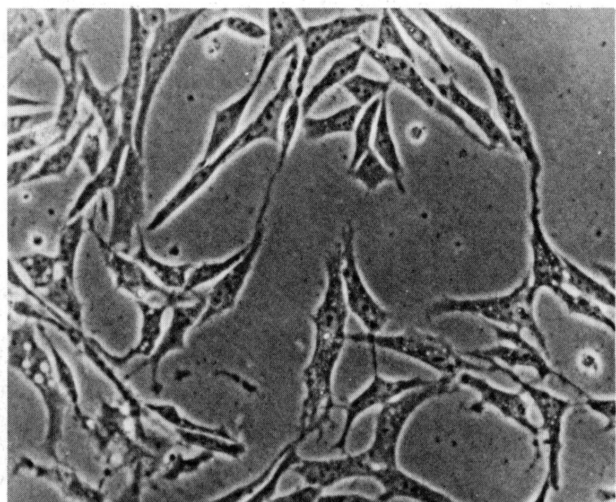

b

to have been extinguished. Presumably, cytoplasmic factors in the L cell act at some level to block the expression of those genes, such as that for tyrosinase, which are essential for the phenotypic expression of the pigmented state. Analysis of the proteins in these hybrid cells indicates that extinction is selective for the differentiated traits; the "housekeeping" functions of both genomes continue to be expressed. Later studies indicated that the chromosome content of the pigmented melanoma cell also was an important factor in the outcome of the experiment. Cancer cells having very different numbers of chromosomes can often be isolated and propagated in culture. If melanoma cells having a near-tetraploid number of chromosomes, rather than the usual diploid cells, were fused with L cells, many of the hybrids retained their pigmentation. It is apparent that the *gene dosage* of the hybrid, i.e., the ratio of chromosomes of one parent to that of the other parent, can also be important in the expression of the pigmented phenotype. Further investigation has revealed that the extinction of the differentiated state depends on the continued presence of particular chromosomes donated by the undifferentiated parent. It has often been found that the chromosome content of cell hybrids is unstable and, as cell division continues, that chromosomes from one or the other species are preferentially lost. In the case of the hamster melanoma–mouse L cell hybrid, pigmentation reappears as mouse chromosomes are selectively lost from the karyotype. Even though a given hybrid clone might remain unpigmented for many cellular generations, the cells retain their commitment to being a pigment cell. With the disappearance of certain mouse chromosomes, the environment within the cell becomes conducive once again to the expression of the differentiated state and the genes of the hamster chromosomes respond in a stable, preprogrammed manner.

10. The stability of the differentiated state

One of the most important questions in the analysis of differentiation is whether the events that lead to the mature terminal cell are such that the fate of that cell is forever restricted. In the previous section, the modulation of a cell between its fully expressed phenotypic state and some covertly differentiated state was described. The underlying difference between these two states may be quite subtle even though their phenotypes are highly diverse. It was already demonstrated in Chapter 8 that nuclei could be taken from differentiated cells and reprogrammed such that they could support the development of an entire new organism. To what degree can this reversibility be demonstrated for the entire cell itself? The question presently under consideration is whether, a cell that once was fully differentiated and has lost these properties can be shifted into any other path of differentiation. This question is a complex one and has had a long and turbulent history; unambiguous examples in vertebrates are very difficult to find. This reflects either the infrequency with which it occurs or the difficulty in arranging for the proper conditions under which it can be observed. The most generally cited example of a change in the state of differentiation of a group of cells (often termed *transdifferentiation,* or *metaplasia*) occurs in Wolffian lens regeneration (shown in Fig. 14.12) as studied in adult urodeles. In this event, the dorsal part of the iris (a neural-ectoderm derivative) loses its differentiated properties and redifferentiates into a lens (an epidermal-ectoderm derivative). The essence of the transformation during Wolffian lens regeneration is the extrusion from the iris cell of all its melanosomes, the activation of DNA and RNA synthesis and cell division, and the reprogramming of the "dedifferentiated" cell along the entirely different phenotypic pathway of the lens cell.

In more recent years this type of transformation from pigmented epithelial cell of the eye cup to lens-fiber cell has been studied in culture using eye tissues taken from 8- to 9-day chick embryos. It was found initially that lenslike structures, termed *lentoid bodies,* would appear after several weeks in mass cultures of cells derived from the chick pigmented retina, i.e., the outer wall of the optic cup (see Fig. 14.5). These lentoid bodies contained lens-specific antigens and a lenslike ultrastructure, but one could not be certain that these cells were actually derived from differentiated pigmented cells in the original inoculum. These reservations were subsequently removed when it was shown that lentoid bodies could arise under clonal conditions in which the entire colony was begun by the proliferation of a single pigmented epithelial cell. In these cases, certain progeny of the pigmented progenitor cell lost their melanosomes after a period of extensive proliferation and redifferentiated into lens-fiber cells just as was demonstrated to occur in the case of Wolffian lens regeneration in vivo. Does the existence of this clear-cut case of transdifferentiation point to the general reversibility of the differentiated state, or does this event simply reflect some peculiarity of this particular tissue? One might argue that cytodifferentiation in the cells of the pigmented retina and iris occurs at a state prior to the terminal determinative stages typical of most tissue. The resolution of this question awaits more examples.

The eye has been the source of several other examples of metaplasia, included among them the interconversion of the pigmented and sensory layers of the retina. Although these are much more closely related with respect to their embryonic origin than are lens and iris, they are very different in the components that make up their cytoplasm, and each is a highly complex cell. The interconversion of the two layers of the retina appears to be responsible for a disease in humans. Retinitis pigmentosa is characterized by the pigmentation of the sensory layer of the retina, suggesting that some defect in the stabilization of the differentiated state has occurred in these cells.

The greatest claims for metaplastic transformations have been made for regenerating tissues, although the extent to which this occurs is highly debatable. Many investigators in this area would agree that at least with respect to the cells of muscle, cartilage, bone, and connective tissue, interconvertibility can occur. Each of these cells is highly specialized, with many tissue-specific molecules; yet, after a process of "dedifferentiation," redifferentiation into another type of cell may well be able to occur. There is one last example of metaplasia that should be mentioned owing to its widespread occurrence and extensive analysis. This case concerns the skin. The application of vitamin A compounds to differentiating skin cells can rescue these cells from their presumptive pathway leading to keratinization and death and convert them into glandular, mucous-secreting epithelial cells.

11. Cell division and differentiation

There is a background of controversy about the relationship between cell division and cytodifferentiation. Tissue differentiation, both in vivo and in vitro, is characterized by an initial period of proliferation followed by the cytodifferentiation of a nondividing population of cells. It has been argued that the transition to the overtly differentiated state is accompanied by a loss of proliferative capacity. For a while mitosis and cytodifferentiation were thought to be mutually exclusive phenomena. Although a highly structured cytoplasm would be expected to interfere at least mechanically with division, there are numerous examples of a cyto-

differentiated cell capable of DNA synthesis and subsequent cytokinesis. For example, cartilage cells with surrounding matrix are readily shown to incorporate both [^{3}H]thymidine into DNA and $^{35}SO_4$ into matrix-bound chondroitin sulfate. Similarly, embryonic cardiac muscle cells, containing well-ordered myofibrils, continue to divide until a very advanced stage of cytodifferentiation. In this latter case, there is a mitosis-dependent degradation of the Z bands of the myofibril, thereby releasing the myofilament bundles for distribution into the two daughter myocytes.

The role of mitosis in differentiation is unclear at this time. In the studies by Howard Holtzer of a variety of differentiating systems, including chondrogenesis, myogenesis, and erythropoiesis, mitosis as a causative event for differentiation is stressed. In this concept, mitosis facilitates the differentiation process whereby two daughter cells attain reactive states not in the repertoire of the mother cell. The cell cycle, in this view, can lead either to duplication of the mother's phenotype or to one or two daughter cells with pathways very different from those active in the mother cell. The former is a *proliferative cell cycle;* the latter is a *quantal cell cycle*. Proliferative cell cycles lead to increased numbers of similar cells. Quantal cycles are postulated to be the means whereby diversity via genetic reprogramming is introduced into replicating systems. If mitosis is suppressed either by adding inhibitors or by culturing cells at high cell density, differentiation does not occur. Accordingly, myoblasts just before cell fusion are considered "postmitotic" and no longer members of the pool of dividing cells. Some event has occurred that has rendered them incapable of division and ready for cytodifferentiation. The contrasting view is that a terminal myoblast, i.e., one with full potential for fusion, is still capable of further rounds of division provided the proper environment can be created. According to Holtzer, the transition to the terminal generation capable of cytodifferentiation can be coupled to one particular round of DNA synthesis. The actual expression of the quantal cycle leading to full phenotypic differentiation may occur immediately. Alternatively, the altered state may be covertly transmitted through several subsequent proliferative cycles, to be expressed many generations later in response to environmental factors. If mitosis is suppressed before the quantal cycle, by inhibitors or by high cell density, differentiation cannot occur. Studies aimed at resolving this question are now in progress.

A. The effect of 5-bromodeoxyuridine

One compound, 5-bromodeoxyuridine (BrdU or BUdR), has been found to have an anomalous effect on a wide variety of differentiating cells. This compound generally blocks the development of a cytodifferentiated phenotype, although the precise effects vary with the cell type under study and the stages at which the compound is applied. BUdR is incorporated into DNA. Its incorporation can be blocked by excess thymidine, with which it competes. In certain cases, the presence of BUdR in the genome is mutagenic. This is not the basis for its effects on differentiation, since in many cases its effects can be readily reversed by removing it from the medium. Reversibility would not occur if stable genetic changes (mutations) were being promoted.

The mechanism of action of BUdR remains a mystery, although several theories have been put forth. In some manner it appears to repress the expression of genetic information that is characteristic of the specialized cell, i.e., the nonessential information, without disrupting the genetic expression for the metabolism required for life itself. As a result, RNA synthesis, protein synthesis, cell di-

vision, etc. generally continue in a normal manner. The effect of BUdR generally requires its incorporation into DNA during an S phase. One such exposed cycle is sufficient, though evidence exists that the BUdR effect may also be directed at the cell surface in some cases. Examples of inhibition by BUdR include myosin synthesis and fusion during myogenesis, matrix production during chondrogenesis, zymogen granule formation during pancreas development, antibody synthesis by lymphocytes, hemoglobin synthesis during erythropoiesis, and numerous others. The ability of BUdR to selectively suppress specialized functions with minimal disruption to "housekeeping" activities has suggested that the development of these specialized differentiating properties involves a separate regulatory mechanism, one that may have many common features among all the cell types. As such, BUdR is an important probe into the nature of differentiation. Its mechanism of action is being intensively investigated.

In summary, we have tried in this chapter to bring together information about a variety of aspects relating to the transformation of cells from the undifferentiated to the fully differentiated state. The processes of differentiation remain poorly understood. Analysis with the electron microscope has provided an in-depth description of the events of cytodifferentiation in a great number of cell types, but very little is known concerning the underlying controlling mechanisms. The study of the molecular biology of cytodifferentiation is providing increasing descriptive information about which macromolecules are present at each stage in a variety of cell types but very little information about the events responsible for their synthesis or the nature of the determination process. The development of more suitable in vitro culture conditions has provided developmental biologists with the opportunity to study differentiation independently of the multitude of systemic influences to which a cell is normally subjected in vivo. The environment in which the cultured cell is growing and differentiating can be manipulated at will, and the importance of such factors as cell density, cell adhesion, and cell division can be studied. In the first section we posed a variety of questions. Some of these have been answered, at least partially, in the pages between; others cannot be answered now, but they are being investigated currently.

Readings

BAGNARA, J. T., 1979. Common Origin of Pigment Cells, *Science*, **203**:410–415.

BENZ, E. J., J. E. BARKER, J. E. PIERCE, P. A. TURNER, and A. W. NIENHUIS, 1978. Hemoglobin Switching in Sheep, *Cell*, **14**:733–742.

BISCHOFF, R., 1978. Myoblast Fusion, in G. Poste and G. L. Nicolson (eds.), "Membrane Fusion," Elsevier.

BUCKINGHAM, M. E., 1977. Muscle Protein Synthesis and Its Control during the Differentiation of Skeletal Muscle Cells in vitro, in J. Paul (ed.), "Biochemistry of Cell Differentiation II," vol. 15, University Park Press.

BURGESS, A. W., D. METCALF, and S. M. WATT, 1978. Regulation of Hemopoietic Cell Differentiation and Proliferation, *J. Supra. Struct.*, **8**:489–500.

CLINE, M. J., and D. W. GOLDE, 1979. Cellular Interactions in Haematopoiesis, *Nature*, **277**:177–181.

DAVIDSON, R. L., 1974. Gene Expression in Somatic Cell Hybrids, *Ann. Rev. Genet.*, **8**:195–218.

DEVLIN, R. B., and C. P. EMERSON, Jr., 1979. Coordinate Accumulation of Contractile Protein mRNAs during Myoblast Differentiation, *Develop. Biol.*, **69**:202–216.

DIENSTMAN, S. R., and H. HOLTZER, 1977. Skeletal Myogenesis: Control of Proliferation in a Normal Cell Lineage, *Exp. Cell Res.*, **107**:355–364.

DUMONT, J. N., and T. YAMADA, 1972. Dedifferentiation of Iris Epithelial Cells, *Develop. Biol.*, **29**:385–401.

EGUCHI, G., and T. S. OKADA, 1973. Differentiation of Lens Tissue from the Progeny of Chick Retinal Pigment Cells Cultured in Vitro: A Demonstration of a Switch of Cell Types in Clonal Cell Culture, *Proc. Nat. Acad. Sci. U.S.*, **70**:1495–1499.

ELLISON, M. L., and J. W. LASH, 1971. Environmental Enhancement of *in vitro* Chondrogenesis, *Develop. Biol.*, **26**:486–496.

FISCHMAN, D. A., 1970. The Synthesis and Assembly of Myofibrils in Embryonic Muscle, *Curr. Topics Dev. Biol.*, **5**:235–280.

———, 1967. An Electron Microscopic Study of Myofibril Formation in Embryonic Chick Skeletal Muscle, *J. Cell Biol.*, **32**:557–575.

FLEISCHMAJER, R., and R. E. BILLINGHAM (eds.), 1968. "Epithelial-Mesenchymal Interactions," Williams and Wilkins.

GARROD, D. R., (ed.), 1978. "Specificity of Embryological Interactions," Halsted.

GEIDUSCHEK, J. B., and S. J. SINGER, 1979. Molecular Changes in the Membranes of Mouse Erythroid Cells Accompanying Differentiation, *Cell*, **16**:149–163.

HARRISON, P. R., 1976. Analysis of Erythropoiesis at the Molecular Level, *Nature*, **262**:353–356.

HOLTZER, H., 1976. Lineages, Quantal Cell Cycles, and Generation of Cell Diversity, *Quart. Rev. Biophys.*, **8**:523–559.

KAFOTOS, F. C., 1972. The Cocoonase Zymogen Cells of Silk Moths: A Model for Terminal Cell Differentiation for Specific Protein Synthesis, *Curr. Topics Dev. Biol.*, **7**:125–191.

KALDERON, N., and N. B. GILULA, 1979. Membrane Events in Myoblast Fusion, *J. Cell Biol.*, **81**:411–425.

KARKINEN-JAASKELAINEN, M., and L. SAXÉN (eds.), 1977. "Cell Interactions in Differentiation," Academic.

KONIGSBERG, I. R., and P. A. BUCKLEY, 1963. Clonal Analysis of Myogenesis, *Science*, **140**:1273–1284.

KONIGSBERG, I. R., P. A. SOLLMANN, and L. O. MIXTER, 1978. The Duration of the Terminal G_1 of the Fusing Myoblast, *Develop. Biol.*, **63**:11–26.

KOSHER, R. A., and J. W. LASH, 1975. Notochordal Stimulation of *in vitro* Somite Chondrogenesis before and after Enzymatic Removal of Perinotochordal Materials, *Develop. Biol.*, **42**:362–378.

LASH, J. W., and M. M. BURGER, 1977. "Cell and Tissue Interactions," Raven.

LASKEY, L., and A. J. TOBIN, 1979. Transcriptional Regulation in Avian Erythroid Cells, *Biochem.*, **18**:1594–1598.

LEDOUARIN, L. (ed.), 1979. "Cell Lineage, Stem Cells, and Cell Determination," Elsevier.

LEVITT, D., and A. DORFMAN, 1974. Concepts and Mechanisms of Cartilage Differentiation, *Curr. Topics Dev. Biol.*, **8**:103–150.

LIPTON, B. H., 1977. A Fine Structural Analysis of Normal and Modulated Cells in Myogenic Cultures, *Develop. Biol.*, **60**:26–47.

LOUGH, J., and R. BISCHOFF, 1977. Differentiation of Creatine Phosphokinase during Myogenesis: Quantitative Fractionation of Isozymes, *Develop. Biol.*, **57**:330–344.

MARKS, P. A., and R. A. RIFKIND, 1978. Erythroleukemic Differentiation, *Ann. Rev. Biochem.*, **47**:419–448.

MINTZ, B., and W. B. BAKER, 1967. Normal Mammalian Muscle Differentiation and Gene Control of Isocitrate Dehydrogenase Synthesis. *Proc. Nat. Acad. Sci. U.S.*, **58**:592–598.

NADAL-GINARD, B., 1978. Commitment, Fusion, and Biochemical Differentiation of a Myogenic Cell Line in the Absence of DNA Synthesis, *Cell*, **15**:855–864.

NATHANSON, M. A., S. R. HILFER, and R. L. SEARLES, 1978. Formation of Cartilage by Non-Chondrogenic Cell Types, *Develop. Biol.*, **64**:99–117.

NIGON, V., and J. GODET, 1976. Genetic and Morphogenetic Factors in Hemoglobin Synthesis during Higher Vertebrate Development: An Approach to Cell Differentiation Mechanisms, *Int. Rev. Cytol.*, **46**:79–176.

OCHOA, S., 1979. Regulation of Protein Synthesis, *CRC Critical Revs. Biochem.*, **7**:7–22.

OKADA, T. S., K. YASUDA, M. ARAKI, and G. EGUCHI, 1979. Possible Demonstration of Multipotential Nature of Embryonic Neural Retina by Clonal Cell Culture, *Develop. Biol.*, **68**:600–617.

ORDAHL, C. P., and A. I. CAPLAN, 1976. Transcriptional Diversity in Myogenesis, *Develop. Biol.*, **54**:61–75.

RUMYANTSEV, P. P., 1977. Interrelations of the Proliferation and Differentiation Processes during Cardiac Myogenesis and Regeneration, *Int. Rev. Cytol.*, **51**:187–273.

RUTTER, W. J., R. L. PICTET, and P. W. MORRIS, 1973. Toward Molecular Mechanisms of Developmental Processes, *Ann. Rev. Biochem.*, **42**:601–646.

RUTTER, W. J., et al., 1978. Pancreas Specific Genes and Their Expression during Differentiation, in F. Ahmad et al. (eds.), "Differentiation and Development," Academic.

SEARLS, R. L., 1973. Chondrogenesis, in S. J. Coward (ed.), "Developmental Regulation: Aspects of Cell Differentiation," Academic.

SPOONER, B. S., H. I. COHEN, and J. FAUBION, 1977. Development of the Embryonic Mammalian Pancreas: The Relationship between Morphogenesis and Cytodifferentiation, *Develop. Biol.*, **61**:119–130.

STAMATOYANNOPOULOS, G., and A. W. NIENHAUS (eds.), 1978. "Cellular and Molecular Regulation of Hemoglobin Switching," Grune and Stratton.

SYTKOWSKI, A. J., Z. VOGEL, and M. W. NIRENBERG, 1973. Development of Acetylcholine Receptor Clusters on Cultured Muscle Cells, *Proc. Nat. Acad. Sci. U.S.*, **70**:270–274.

TOBIN, A. J., 1979. Evaluating the Contribution of Posttranscriptional Processing to Differential Gene Expression, *Develop. Biol.*, **68**:47–58.

WEATHERALL, D. J., and J. B. CLEGG, 1979. Recent Developments in the Molecular Genetics of Human Hemoglobin, *Cell*, **16**:467–479.

WEINTRAUB, H., 1975. The Organization of Red Cell Differentiation, in J. Reinert and H. Holtzer (eds.), "Cell Cycle and Cell Differentiation," Springer-Verlag.

WESSELS, N. K., and W. J. RUTTER, 1969. Phases in Cell Differentiation, *Sci. Amer.*, vol. 269 (March).

WHITTAKER, J. R., 1974. Aspects of Differentiation and Determination in Pigment Cells, in J. Lash and J. R. Whittaker (eds.), "Concepts in Development," Sinauer.

WILT, F. W., 1974. The Beginnings of Erythropoiesis in the Yolk Sac of the Chick Embryo, *Ann. N. Y. Acad. Sci.*, **241**:99–112.

URSPRUNG, H. (ed.), 1968. "Stability of the Differentiated State," Springer-Verlag.

YAMADA, T., 1977. "Control Mechanisms in Cell Type Conversion in Newt Lens Regeneration," Karger.

Chapter 16 Morphogenesis and tissue assembly

Nowhere is the complexity of the living organism better revealed than by examination of a section through a vertebrate organ. How does the histological architecture of an organ originate; i.e., what is the basis of *histogenesis*? We do not know the answer to this question, but we can presume that the development of specific three-dimensional cellular patterns must depend heavily on the nature of the cell-cell interactions that occur between members of each cell type. Intercellular interactions involve such phenomena as cell motility, cell communication, the elaboration of extracellular matrices, cell recognition, and cell adhesion. Some of these topics have already been discussed in Chapter 3 and in other places in the book; we will concentrate in this first section of this chapter on the topic of cell adhesion, particularly as it might relate to the determination of histological patterns.

1. Tissue sorting and assembly

Much of our current view concerning the formation of organ-specific cellular patterns derives from the work of Johannes Holtfreter in the 1930s and 1940s, some of which was discussed in Section 10.1.B.ii. It was Holtfreter's belief that the positions that cells occupy within embryonic organs was largely a result of their selective affinities to other cells. Given the ability of cell populations to proliferate, differentiate, and move around within a cell mass, it might be expected that cells would accumulate neighbors to which they most strongly adhered. The experiments described in Fig. 10.17 utilized small fragments taken from different parts of the gastrula. It was demonstrated in these experiments that small parts of the embryo were capable of undergoing many of the same types of morphogenetic movements in isolation that they would normally carry out in the intact embryo. The parts of the embryo can be thought of as capable of a considerable degree of self-assembly. To what degree can this morphogenetic capability for self-assembly be demonstrated by the component cells? Investigators seeking to answer this question have studied the capacity of cells obtained from dissociated embryonic tissues to reassemble and differentiate. In the initial experiments with amphibian embryos, cell dissociation was accomplished by brief exposure of the tissues to a medium of alkaline pH. Amphibian material has been notably suited to this

investigation, as it has for experimental embryology in general. Amphibian cells are relatively large, and different kinds vary greatly in degree of pigmentation. Consequently, cell movements can be followed with comparative ease. When cells dissociated from ectoderm, mesoderm, and endoderm of gastrulas and neurulas were mixed in various combinations, the different cell types in a composite aggregate became sorted into distinct homogeneous layers whose stratification corresponded to the normal germ layer arrangement (Fig. 16.1). Thus in an aggregate of cells from the three germ layers, Holtfreter was able to see lightly pigmented mesodermal cells vanish before his eyes as they moved into the depths of the tissue mass while darkly pigmented ectodermal cells moved to the periphery to replace them.

gure 16.1 Segregation and differentiation of arious combinations of dissociated tissues of amphibian embryos. : Mesoderm and endoerm mixtures result in inward migration of sodermal cells forming uscle and notochordal sue. *b* Epidermal cells, when combined with esodermal cells, move outward, and mesodermal cells move inrd. *c* Mixtures of axial nesodermal, medullary (neural) plate, and pidermis cells sort out and re-form respective tissue types. [*After P. L. Townes and Holtfreter, 1955.*]

In mixed aggregates of ectoderm and mesoderm (Fig. 16.1*b*), the ectodermal cells at first mingle with the mesoderm but are later expelled to the surface where they form a separate epidermal layer; i.e., an initial affinity is later lost. In mixtures of endoderm and mesoderm, the mesoderm likewise moves to the interior and differentiates, leaving the endodermal epithelium at the surface (Fig. 16.1*a*), not unlike the organization of the exogastrula shown in Fig. 11.18. Prospective neural plate cells and prospective epidermal cells separate to form neural tube and epidermis, respectively. Neural and epidermal tissue can be kept from separating, however, if some mesodermal cells are present to form an intermediate layer adhesive to both.

The directed movements appear to be of two kinds: (1) inward movements, as manifested by the cells of the neural plate, or of the mesoderm, which have been combined with either epidermal or endodermal cells, and (2) outward movements and peripheral spreading, as manifested by epidermal cells. Tissue segregation becomes complete because of the emergence of a selectivity in cell adhesion. When they meet, homologous cells remain permanently united to form functional tissues, whereas a cleft develops between certain nonhomologous tissues, such as between neural or endodermal tissues and adjacent epidermis. It would appear from these experiments that cells preferentially adhere to others of their same germ layer. Moreover, given the opportunity to undergo morphogenetic movement, dissociated cells are capable of rearranging themselves without the intervention of outside agents in order to reconstitute the basic organization of the embryo from which they were isolated.

A. The basis for cell adhesion

We will return to questions of morphogenetic importance in subsequent paragraphs, following consideration of some of the experimental approaches to the study of how cells can recognize each other and selectively adhere. The first experiments on dissociated cells were carried out by H. V. Wilson in the first decade of this century. Sponges are ideal for these types of experiments because they are composed of only a few cell types. The cells are relatively loosely attached to each other, and once dissociated, they rapidly reassociate in simple isotonic salt solutions. Wilson found that whole sponges could be forced through a silk mesh and converted to a suspension of cells in seawater. The cells are said to be *mechanically* dissociated. If these cells were allowed to remain together, they would aggregate to form large clusters which would ultimately become new sponge individuals. Further research on the phenomenon revealed that the re-formation of new individuals depended on the movement of cells within the cluster so that their appropriate position within the aggregated mass was regained. As in the case of the amphibian embryo, cells that were lining the outer surface of the original organism would take up a corresponding residence in the aggregate. Although not interpreting his data in such a way, Wilson had discovered the ability of different types of cells to self-assemble into structures having a higher level of organization. In subsequent experiments it was found that if cells dissociated from two different species of sponge were mixed together, the cells of each species would become grouped with their own type. The sponges that re-formed were found to be exclusively of one or the other species' cells. Since the cells of the two species used were of different colors, red and yellow, one could actually watch like cells associate with each other and "sort out" of the mixed cell mass.

The simplicity of the sponge system has made it of continuing value in the study of recognition and adhesion, and various investigators have continued to explore it. Attention turned to the molecular nature of the cell surface when it

was found that antibodies prepared against the cells of one sponge species would selectively block the aggregation of dissociated cells prepared from that same species. Another important finding was made when the means of dissociating the cells was changed. Instead of using natural seawater, sponges were placed in artificial seawater (made in the laboratory from ordinary salts), in which Ca^{2+} and Mg^{2+} had been left out. When this was done, the sponges fell apart into their component cells, just as if they had been mechanically dissociated. However, there is an important difference between the "chemically" dissociated cells and a mechanically prepared suspension. If both are washed and placed in normal seawater (containing Ca^{2+} and Mg^{2+}) and the temperature kept very low (for example, 4°C), only the mechanically dissociated cells will reaggregate; the chemically dissociated cells remain as single cells or small, loose clusters. If, however, the supernatant in which the chemical dissociation took place is added back to the cells, then their reaggregation also can occur. It appears that in the process of chemical dissociation in Ca^{2+}-Mg^{2+}–free seawater, something is lost from the cells that is required for their reaggregation. As long as the temperature is kept low, the cells cannot resynthesize this aggregation factor (their metabolism is too sluggish at cold temperatures) and they cannot adhere. Analysis of the supernatant demonstrated that the factor responsible for promoting reaggregation was a gigantic multimeric proteoglycan composed of over 50 percent carbohydrate. This factor is sufficiently large to be visualized in the electron microscope after negative staining. In some cases the aggregation factors have a "sunburst" appearance with radiating fibrous arms extending from an inner circle, while the comparable factor from other species exists as a linear backbone with projecting side chains. Where it has been investigated, destruction of either the protein or carbohydrate part of the molecule renders the aggregation factor inactive.

To determine if the presence of the aggregation factor was involved in the species-specific nature of reaggregation, the supernatant in which a red sponge was chemically dissociated was added to a mixture of chemically dissociated red and yellow sponge cells kept at low temperatures. In the presence of the factor from red sponges, only the red *Microciona* sponge cells would aggregate, leaving dissociated yellow *Haliclona* sponge cells behind. The results on the sponge system suggest that species-specific intercellular adhesion is mediated by an intercellular cement (aggregation factor) that is complexed to the sponge-cell surface, probably by a receptor protein or base plate, as shown in Fig. 16.2. Somewhere in this complex of cell surface, receptor protein, and aggregation factor is the Ca^{2+} ion, whose presence is required to hold the multicomponent complex together.

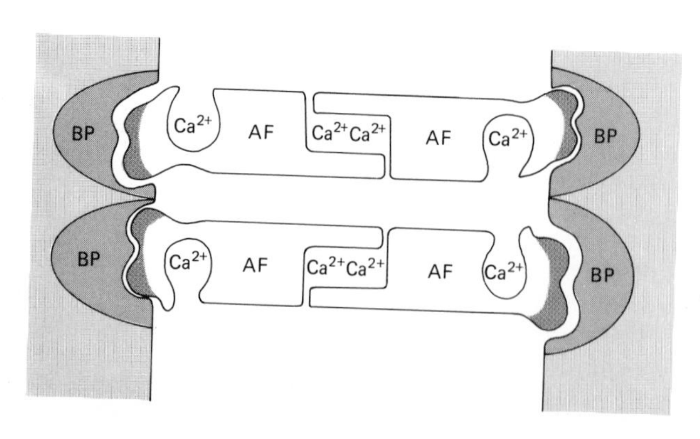

Figure 16.2 A tentative model for specific cell-cell interaction in tissue construction. Two macromolecular aggregation factors (AF) are illustrated, each consisting of at least two subunits. The black termini at each pole carry the carbohydrates, which are recognized by the base plate (BP) anchored in the adjoining cell surfaces. [*From G. Weinbaum and M. M. Burger*, Nature, **244:***510 (1973).*]

Studies of the reaggregation of cells dissociated from embryonic chick and mouse tissues have added new dimensions to our present concept of selective cell adhesion. Dissociation of cells from vertebrate embryonic tissues requires more stringent measures than those employed for sponges. In the method routinely chosen for tissue dissociation, brief trypsin treatment is used to disrupt intercellular adhesions followed by mechanical agitation to completely separate the cells. Although the treatment with trypsin temporarily affects the adhesive properties of the dissociated cells, following a period of recovery during which cells resynthesize cell surface proteins, these cells can be used in a wide variety of experimental procedures for determining their surface recognition properties. In one type of experimental approach, the suspension of dissociated cells is rotated in a flask, causing the cells to collide with each other and thereby providing the opportunity for them to adhere. Aggregates can then be observed at various times and their number and size recorded as a measure of the affinity of the cells for each other (Fig. 16.3). Moreover, these aggregates can be followed over extensive periods of time in culture to determine their potential for morphogenetic movements and histogenetic differentiation (discussed below).

As in the case of the sponge system, cultured embryonic cells have been found to release aggregation-promoting factors into the medium. Several laboratories working with numerous types of tissues have isolated a variety of different factors having a potential role in cell adhesion. In many cases, the approaches used to isolate these materials as well as to determine their biological activity are quite different and comparisons can be difficult to make. Rather than trying to survey this body of data, we will hold to a more general discussion using specific examples to illustrate points. Aggregation-promoting factors are often released into the supernatant when cells are grown in a monolayer on the bottom of a culture dish containing serum-free medium. Similar, if not identical, factors can be extracted

Figure 16.3 *a* Suspension of skin cells obtained by trypsin dissociation of 8-day chick embryo skin. *b* The appearance of an aggregate of dissociated skin cells after 24 hours of incubation. [*Courtesy of M. H. Moscona and A. A. Moscona.*] *c* Scanning electron micrograph of an aggregate of 8-day chick embryo neural-retina cells originally dissociated by trypsin. [*Courtesy of Y. Ben-Shaul and A. A. Moscona.*]

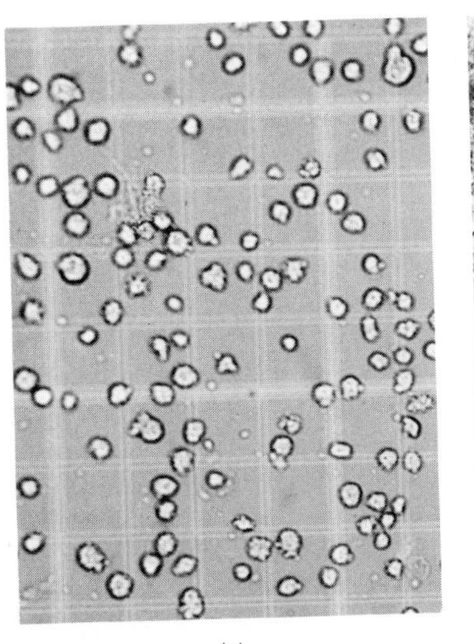

(*a*)

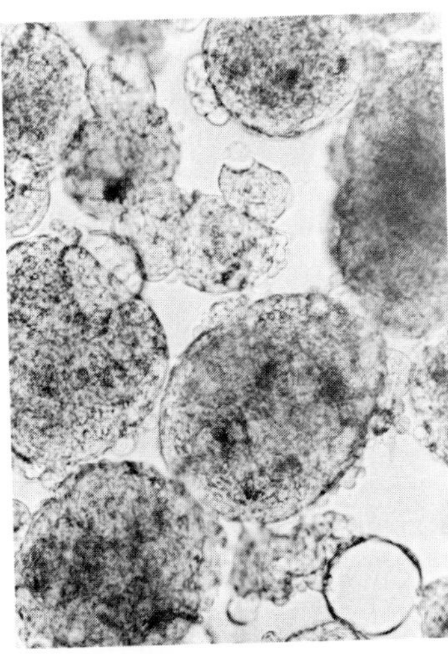

(*b*)

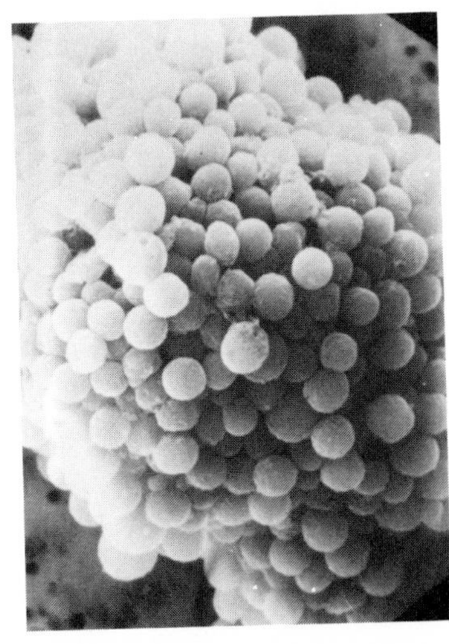

(*c*)

directly from the cells' plasma membranes. One of the best-studied of these cell surface components is a 50,000-dalton glycoprotein isolated from the membranes (or cell culture supernatants) of chick embryonic neural retina cells. When purified, this material will greatly enhance the aggregation capacity of dissociated neural retina cells but has little or no effect on the aggregation of cells of other organs. Similarly, antibodies prepared against this molecule will inhibit the reaggregation of chick embryonic neural retinal cells without blocking the adhesive properties of other types of cells. The localization of these antibodies on the surface of embryonic neural retina cells (which have been allowed to recover from trypsin dissociation) confirms the belief that they represent proteins normally present within the external leaflet of the plasma membrane. Treatment of the neural retinal factor with glycosidases or periodate, agents that destroy its carbohydrate moieties, has no effect on its aggregation-promoting capacity, suggesting that in this particular case, it is the protein component of the molecule that carries the specificity with regard to intercellular adhesion.

Another approach to the study of cell affinities takes advantage of the fact that cells can be pulled from suspension by adsorption to a complementary surface. In most cases, the collecting surface consists of a layer of unlabeled cells present as a coating on the bottom of a culture dish, on small beads, or nylon fibers. Several examples are given in Fig. 16.4. It is generally assumed that the rate at which radioactively labeled cells of type A are collected onto type B cell-coated surfaces is a measure of the affinity of the two types of interacting cells. Consequently, these types of experimental procedures can be used to compare affinities between cells in an attempt to gauge the degree of selectivity of a given cell population. For example, it is nearly always the case that cells adhere preferentially to those of similar type, i.e., *homotypic cells*, than to those prepared from a different type of tissue, i.e., *heterotypic cells* (Fig. 16.5). Cells are therefore capable of discriminating behavior with respect to cell recognition and adhesion in vitro. It is reasonable to assume that preferential adhesion of cells to others of the same tissue type is a major factor in the formation of tightly adherent cell layers in vivo.

Cellular affinity is not an all-or-none phenomenon. Although homotypic adhesions may be preferred, cells are clearly capable of interacting with varying degrees of intensity with other cell types; otherwise, the formation of complex cell patterns during histogenesis would hardly seem possible. Among heterotypic encounters, the rates of collection of labeled cells onto cell-coated surfaces provides some measure of relative cell affinities. These affinities may result from qualitative and/or quantitative differences. For example, the preferential adhesion of A cells to B cells over C cells may reflect the fact that A cells have a greater affinity for B cell adhesion sites than for those on the C cell surface, or alternatively, B and C cells may have identical contact sites, but those on the B surface may be present in greater quantity or may be more appropriately positioned to interact with the A cell surface than those of the C cell. All these factors, and many others, are probably involved in tissue and organ construction in the embryo.

The development of procedures for collecting cells on surfaces has led to further analysis of the nature of the binding sites. It has been found that some types of cells will bind to surfaces to which simple chemical groups, specifically sugars, have been covalently bound. For example, cells prepared from chicken liver tissue (hepatocytes) will bind to insoluble matrices to which the sugar *N*-acetylglucosamine has been attached. Conversely, cells from rat liver tissue will bind selectively to surfaces on which galactose has been attached. These results suggest that specific sugar groups present at terminal positions on oligosaccharide chains of membrane glycolipids and/or glycoproteins (Section 3.1.A) play

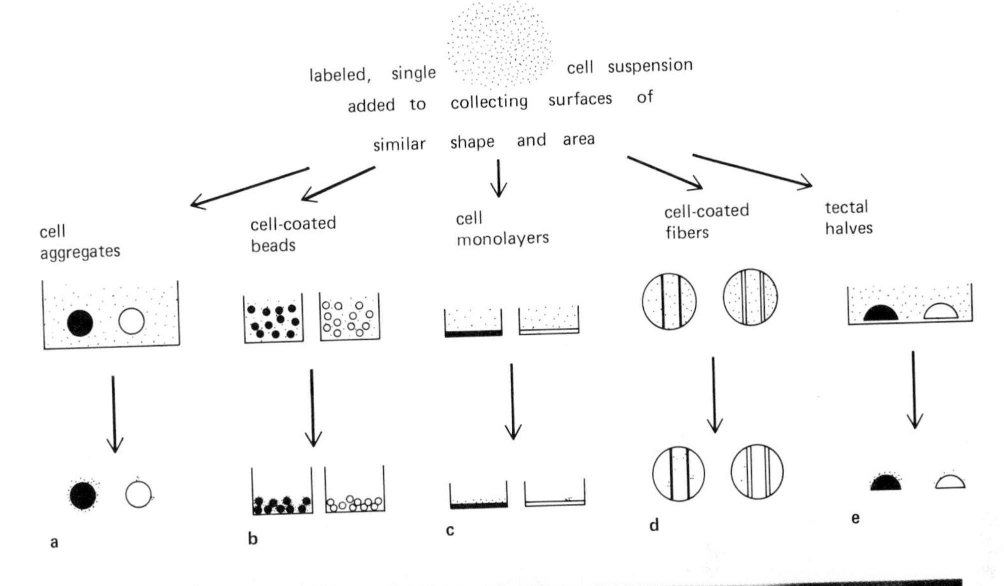

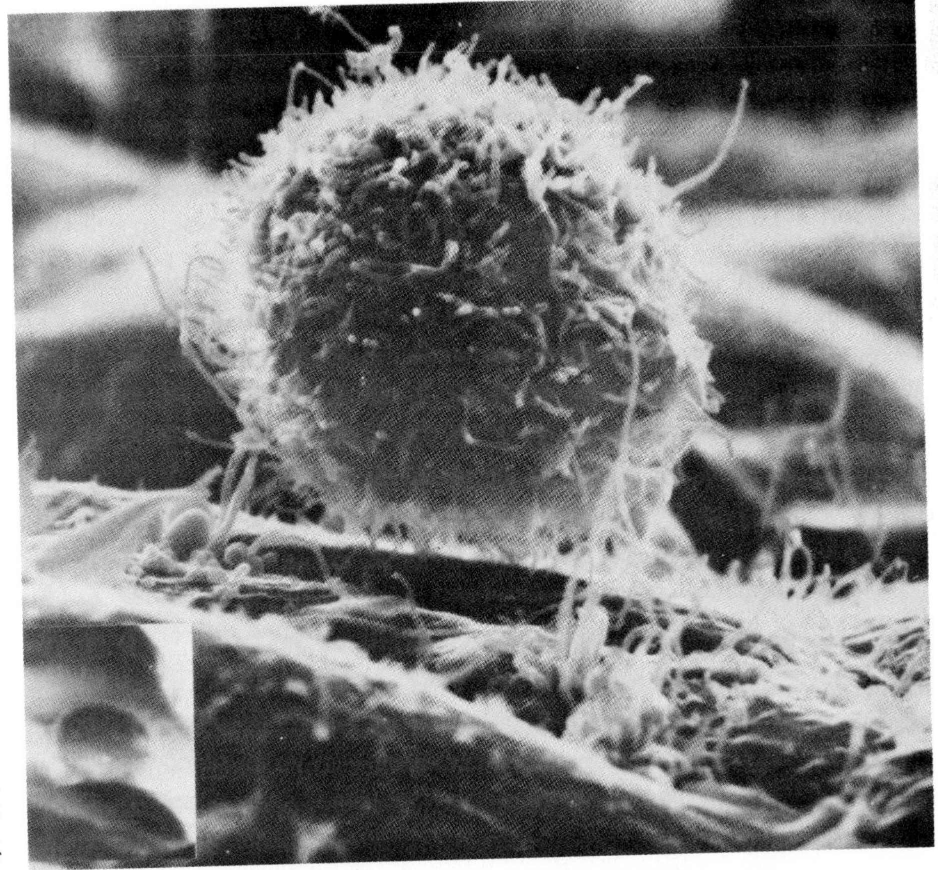

Figure 16.4 *a–e* Protocols for collection assays. A single-cell suspension, prepared from one tissue and labeled with a radioactive isotope, is exposed to various types of collecting surfaces of high or low affinity. Since the collision frequencies between the surfaces and the labeled cells are identical, the number of radioactive cells adhering to a surface gives a direct estimate of the relative adhesivity between the cells in suspension and the cells making up that surface. [*From B. Marchase, K. Vosbeck, and S. Roth,* Bioc. Biop. Acta, **457**:*404 (1976).*] *f* Scanning electron micrograph of a 3T12 fibroblast adhering to a monolayer of 3T12 cells. [*Courtesy of S. Roseman.*]

settes, and so on. It is apparent that embryonic mouse and chick cells derived from homologous tissues at comparable stages of development are able to form interspecific mosaics and to participate jointly in the histogenesis of such chimeric structures. Regardless of the taxonomic distance, intermingled cells become assorted and matched selectively in accordance with their functional similarity. This result is particularly evident when pairs of tissues from both chick and mouse are mixed together and sorting out is followed. For example, when embryonic mouse cartilage and liver cells are mixed with embryonic chick cartilage and liver cells, the cartilage cells sort together to form a chimeric structure and so do the liver cells. If we assume, based on the types of studies previously discussed, that adhesion does require the interaction of complementary stereospecific molecules, it would appear that these molecules have been highly conserved over considerable periods of evolutionary time.

The experiments with dissociated cells of the amphibian gastrula (Fig. 16.1) indicated that cells of a mixed aggregate can sort out in such a way that one type of cell, such as those of ectodermal origin, moves to the periphery of the aggregate and another type of cells, such as those of mesodermal origin, will move into the central region of the aggregate. It has been proposed that the relative positions that masses of cells occupy after sorting has occurred is determined by the strengths of the contacts that cells make with each other. This hypothesis is not concerned with the specific chemical interactions involved in adhesion, but is compatible with any molecular theory. Rather, it concerns itself with the physical aspects of cells capable of moving around and sorting into predictable configurations. When cells of two types of embryonic tissues are mixed together, one of the types tends to move inward, forming a central core surrounded by cells of the other type. The final equilibrium configuration that the cells maintain is the thermodynamically favored one, i.e., the one that minimizes the interfacial free energy of the system, with the cells having the strongest contacts being located in the interior. The strength or tenacity of adhesion between cells will be dependent on the number of bonds formed and their average stability. For example, if procartilage cells of the embryonic limb bud are mixed with cells taken from embryonic heart ventricle, at equilibrium the latter cells will have surrounded the former (Fig. 16.8*a*). If these same ventricle cells are mixed with cells of the em-

Figure 16.8 *a* Section through an aggregate formed by intermixed precartilage and heart-ventricle cells. The precartilage is reassembled internally to the heart tissue. *b* Section through an aggregate formed by intermixed heart-ventricle and liver cells. Heart-ventricle tissue is reassembled internally to liver tissue. [*Courtesy of M. S. Steinberg.*]

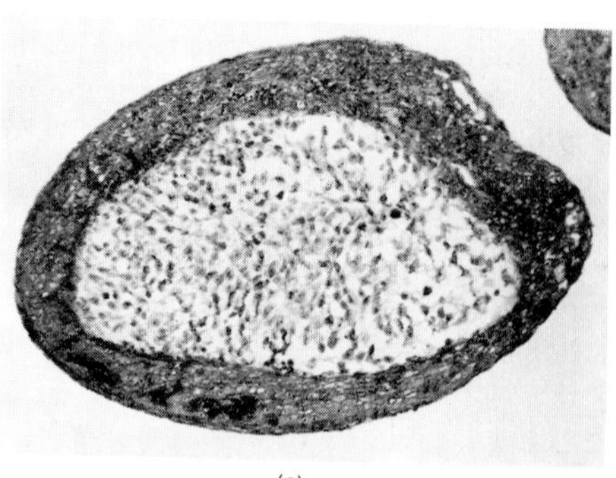

(*a*)

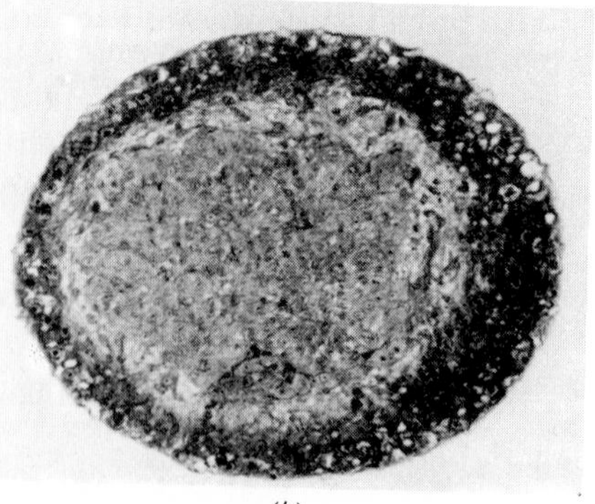

(*b*)

bryonic liver, they will move to the interior of the combined mass, being surrounded by the liver cells (Fig. 16.8*b*). The behavior of cells from a given type of tissue can change markedly with age. For example, ventricle tissue taken from embryonic chick hearts between days 2 and 5 becomes increasingly more internalizing when tested in combination with other tissues. This is paralleled in its own self-recognition behavior; when younger and older ventricle cells are mixed, younger cells spread over older ones. The differential adhesion hypothesis explains these predictable relationships by proposing that in any combination of cells, those with the greater cohesiveness will be found in the interior, having squeezed out the less cohesive cells to the periphery. Each type of cell-cell interaction can be assigned a relative cohesiveness and a scale or hierarchy of adhesiveness can be established.

In summary, control of tissue arrangements is a central problem of developmental biology. The reestablishment of normal architecture of tissues and organs by cell sorting is a counterpart, at the tissue and organ level, of the phenomenon of molecular self-assembly. Both in molecular self-assembly and in cell sorting, normal structure is regenerated by initially disordered populations of subunits. As in the case of the molecules, the various cellular components of a tissue and organ have their normal mutual relations and functional activity. The results reemphasize that internal self-organization is one of the most basic problems in the study of development, in contradistinction to recent preoccupation with external inductions.

C. Dermal-epidermal interactions

The skin of vertebrates serves well as a model system of a tissue formed from two components of very different origin. It consists of two layers, the epidermis, derived from ectoderm, and the mesenchymal dermis, which arises from mesoderm. The epidermis consists of a basal layer of columnar cells which are loosely united as an epithelium resting on the basement membrane, together with a number of layers of progressively keratinized cells externally. All cell division takes place in the basal layer; cells above the basal neither divide nor synthesize the DNA that must precede mitosis. Following division, both daughter cells remain basal cells for an indefinite period; yet every time a basal cell divides, a nearby cell begins to move toward the skin surface. Once free of the basal layer, such cells change from columnar to cuboidal to flat and progressively produce the mixture of proteins that constitute keratin. At the same time they form stable connections, desmosomes, with their immediate neighbors. Consequently, the epidermal cells external to the basal layer migrate toward the surface bonded together as a continuous sheet.

Differentiation of embryonic skin and the other specialized derivatives of the integument—such as hairs, feathers, and scales—is based on specific interactions between dermal and epidermal components. Separation of the two interacting skin components has provided insights into the relative contributions of the two tissues to the initiation and maintenance of special structures. In normal development of feathers (Fig. 16.9), four stages are generally recognized:

PREPLACODAL STAGE: In the 5-day embryo, the ectoderm is a simple epithelial layer of cuboidal cells overlaid by flat peridermal cells, the underlying mesoderm being a loose mesenchymal network.

PLACODAL STAGE: In the 7- and 8-day embryo, the feather germs (primordia) first appear as localized nodes, or centers of dense cell populations, along

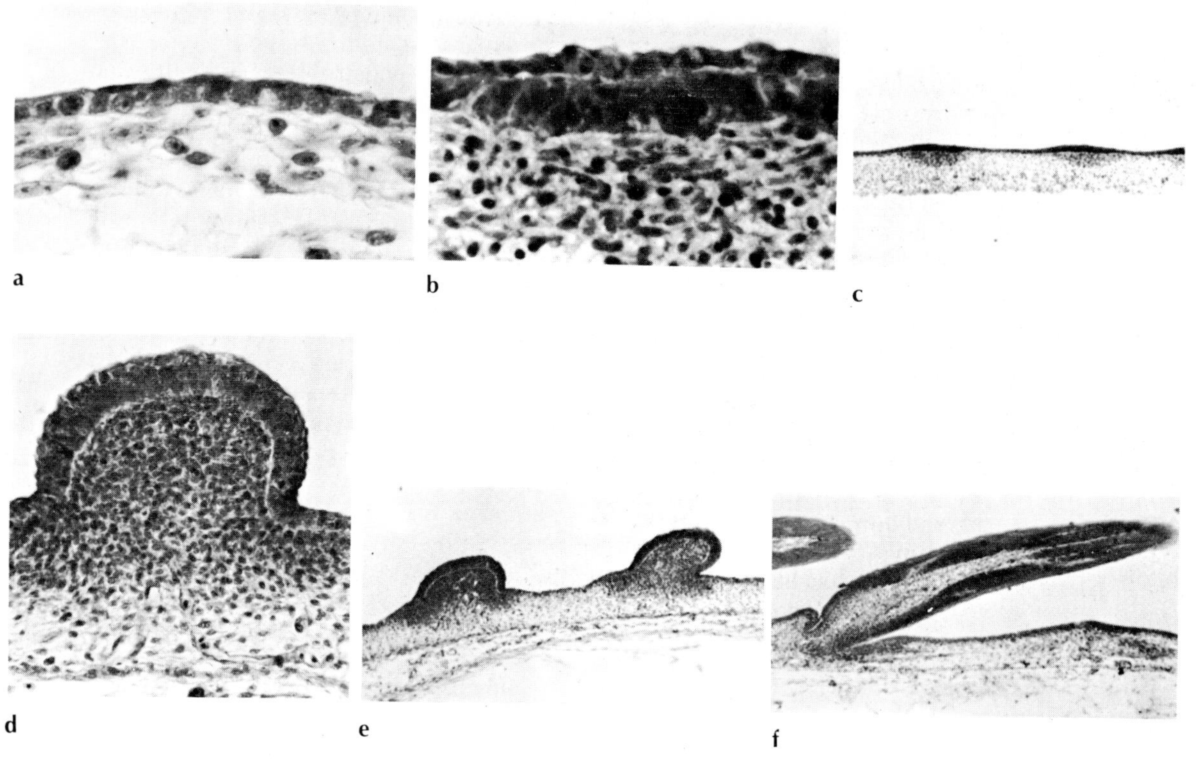

Figure 16.9 Stages of normal embryonic feather development; high-power details showing epidermal differentiation and dermal papilla. *a* Preplacodal stage, 6-day skin. Two-layered epithelium, cuboidal cells covered by flat peridermal cells; mesenchyme forms a loose subjacent network; feather papillae are not yet defined. *b, c* Placodal stage, 8-day skin. Epithelium has palisaded into a thickened placode of columnar cells covered by a peridermal layer; dermal condensation beneath the placode forms a discrete dermal papilla. *d, e* Feather bud in hump stage, 10-day skin. Dermal papilla has greatly increased in volume and has elevated the placodal epithelium; blood vessels invade the papilla from the underlying connective tissue. *f* Elongation phase. [*Courtesy of B. Garber.*]

the middorsal line of the embryo, the ectodermal cells now becoming columnar and forming placodes while the underlying mesenchymal dermis condenses into dermal papillae. Other rows of feather germs later appear successively on either side of the primary row; i.e., the feather germs originate progressively both posteriorly and bilaterally.

HUMP STAGE: In the 8- to 10-day embryo, the dermal component grows rapidly and raises the ectodermal component to form the feather bud.

ELONGATION STAGE: The feather germ elongates, and condensed dermal papilla cells are evident at the feather base.

The capacity of dissociated and subsequently reaggregated embryonic cells to develop into organized structures changes with the age of the cells at the time of their dissociation; i.e., it changes as their differentiation progresses. Thus aggregates of cells dissociated from 5- to 8-day old embryonic skin (preplacodal and placodal stages) form well-developed, typical feathers enclosed in large, thin-

walled keratinized epithelial vesicles in the chorioallantoic membrane. Aggregates of cells dissociated from 10-day embryonic skin (hump stage) form imperfect feathers and skin, normal in cell types but poorly constructed. No feathers of any sort, only dense and extensive sheets of dermis and keratinized epidermis, form from aggregates of 12- and 14-day-old skin cells.

Feathers are exceedingly complicated structures that vary in morphology and function on different parts of a bird's body. Although feathers are made almost entirely by ectodermal cells, the mesoderm controls the type of feather that will form from the ectoderm. Three types of experiments demonstrate this fact: (1) When a block of mesoderm from the thigh of a chicken embryo is inserted beneath the ectoderm covering the proximal portion of an embryonic wing, the wing ectoderm forms leg feathers. Both the feather morphology and the arrangement of feathers on the wing surface are characteristic of leg. (2) A combination of feather-forming mesoderm and ectoderm located in an area that does not form feathers results in feather development. (3) If a piece of ectoderm destined to form feathers is combined with mesoderm from the lower leg, where "scales" normally form, the ectoderm forms scales.

It is apparent that the differentiation of uncommitted ectodermal cells can be directed along any of a number of different pathways. Results of these experiments in which ectoderm and mesoderm from various regions are combined are consistent with the conclusion that *mesoderm controls the kind of specialized structure* (scale or type of feather) *that will form from ectoderm.* This conclusion, however, has to be qualified, for a remarkable experiment shows that the corneal epithelium (ectoderm) of a 17-day-old chick embryo, when supplied with dermis (mesoderm) from the flank of a mouse embryo, develops feathers (Fig. 16.10). The *stimulus* to form feathers in place of corneal differentiation is sup-

Figure 16.10 Induction of feathers in chick cornea by flank dermis cells of mouse embryo. *a* Axial section of the epithelium in the corneal region of a 17-day-old-chick embryo. At 5 days of incubation the lens of this eye was replaced by a block of flank dermis from a mouse embryo. A group of mouse cells lies just beneath the chick epithelium. *b* Feathers are forming from the corneal epithelium. Mouse cells are present in the feather pulp. [*Courtesy of J. L. Coulombre and A. J. Coulombre.*]

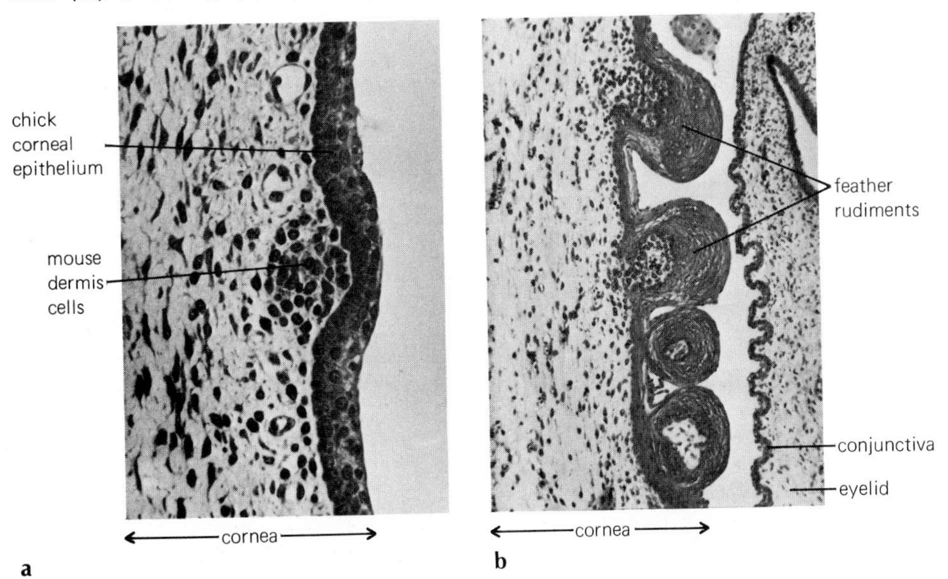

plied by mouse mesoderm, which in a mouse would have induced hairs. The *capacity* to develop feathers rather than hairs is part of the ectodermal repertoire of chick ectoderm.

2. Tissue-pattern formation

A. Feather patterns

Sheets of skin from the middorsal area of 6-day-old chick embryos, maintained in organ culture, show that the first change in the dermis in connection with feather development is the appearance of cells elongated in an anteroposterior direction, while small clusters of cells appear marking the sites of the first row of dermal papillae. At the 8-day stage, the entire dorsal feather area (field) is organized into a latticelike system of oriented dermal cells linking the sites of the dermal papillae. Epidermis alone shows a generalized meshwork, but dermis shows an additional fibrous pattern. It first appears as a midline streak and then extends laterally, i.e., corresponding to the order of appearance of the feather germs. The pattern of the fibrous material and the pattern of feather buds are the same, both spatially and temporally. The fact that the enzyme collagenase, which breaks down collagen, inhibits the development of the fibrous material and feather germs, as well as other evidence, indicates the collagenous nature of the fibrous lattice and its importance.

A birefringent, fibrous dermal lattice has a strikingly gridlike organization and, as already stated, is collagenous. The development of dermal papillae coincides with a characteristic distribution pattern of dermal cells in relation to this fibrous lattice: elongated cells align along the tracts of the lattice; rounded cells form clusters at the intersections of the lattice, which represent the sites of the future dermal papillae (Fig. 16.11). Cell clusters at the lattice intersections are the precursors of the dermal papillae and arise by aggregation of cells that migrate along the fibrous tracts. Whether the fibrous material is directly involved in cell guidance, or whether this is due to nonfibrous materials associated with the fibers, is an open question. The fibers might conceivably provide a visible indication of some organization at a finer level.

That the fibrous lattice plays a role in skin development is indicated by the following results: Treatment of skin with collagenase, which disrupted the organized structure of the lattice, caused a reversible inhibition of skin development. In skin of the "scaleless" chicken mutant, both the birefringent lattice and the dermal condensations are absent. Although collagen synthesis proceeds normally, an organized dermal lattice does not form. Evidence suggests that the absence of a fibrous lattice is not the result of a reduced synthesis of collagen but is probably the result of failure of synthesized collagen to become organized into a lattice. Further experiments suggest that the site of gene action in the scaleless mutant resides in the epidermis rather than the dermis, which is actually responsible for production of the collagen. For example, mutant dermis can be combined with normal epidermis and the composite structure will develop a normal lattice and feather pattern. In the reverse combination, employing mutant epidermal tissue, normal development does not take place. It would appear that it is the epidermis which is responsible for the proper alignment of the underlying dermal cells, which, in turn, then produce and secrete collagen molecules in the required orientation.

Feather-forming skin of the chick thigh tract is capable, in isolation, of autonomous pattern formation. Separated lateral and medial halves of the tract can each initiate the development of spatially ordered feather primordia (Fig. 16.11*b*). Therefore, no one row serves as a specific "initiator row" in whose absence the entire area of skin remains devoid of feathers. Apparently all the skin

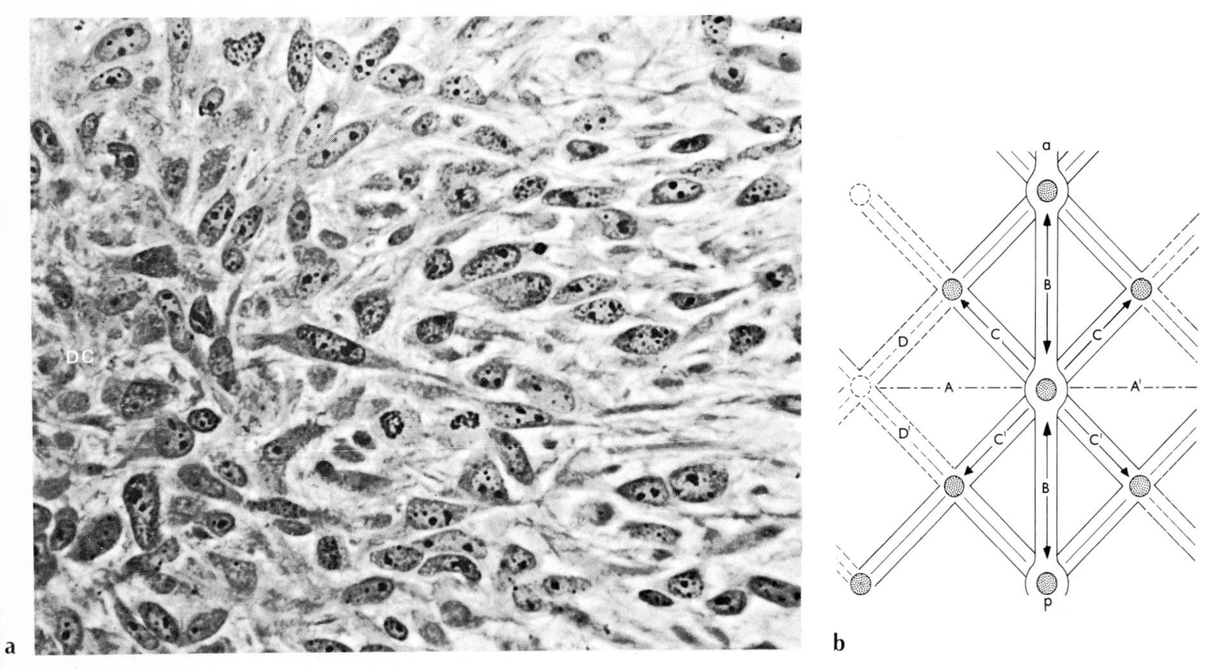

Figure 16.11 Pattern of dermal condensations relating to feather-germ formation in dorsal skin of chick embryo. *a* A dermal-cell condensation on the left and oriented cells pointing toward a neighboring condensation. [*Courtesy of N. K. Wessells and J. Evans.*] *b* Diagram of a portion of an embryonic feather tract to show the relative positions and sequential appearance of aligned fibrous material and dermal cells during development of papillae. The primary row contains major goups of aligned fibers and cells (B–B) extending anteroposteriorly between successively forming papillae; this midline region contains a full, continuous spectrum of the stages in fiber and cell alignment, as related to papilla formation. In the lateral regions, the major bands of aligned fibers and cells (C–C^1, D–D^1) extend diagonally from the more mature medial papillae toward the prospective lateral papillae; thus a cross section (A–A^1) through the lateral skin does not coincide with the directions of fiber and cell alignment. [*After A. A. Moscona and B. Garber.*]

A–A^1 hypothetical plane of cross section
B–B^1 primary row of fibers and cells between papillae
C–C^1, D–D^1 } bands of oriented cells and aligned fibers
a–p anteroposterior axis (midline axis) through the primary row of feather germs
Stippled circles—developing papilla
Open circles—prospective papilla
Solid lines—aligned fibers and dermal cells
Broken lines—prospective bands of aligned fibers and cells

within the feather tract is capable of forming a first row of equally spaced primordia and of adding secondary rows in the normal pattern.

The mechanism controlling pattern formation must act, accordingly, over a shorter distance rather than over the whole tract, probably a distance not greater than the width of one primordium. Both in vivo and on the chorioallantoic membrane, a misplaced primordium is propagated as a whole new row within the pattern. Thus a single, misplaced primordium can act as a whole new frame of reference within the pattern. This strongly suggests that each individual primordium not only is a component of the pattern but acts directly in the actual extension of the pattern. The simplest explanation of pattern generation, therefore, is that the place where each primordium forms is determined by some influence from the

adjacent primordia, once they themselves are established. This would account for the change in the pattern caused by a single, misplaced primordium.

B. Retinal pattern

A very different example of lattice pattern is seen in the retina. The retina of fish and birds and possibly other classes of vertebrates exhibits a lattice pattern in the distribution of rods and cones, and particularly in the distribution of a variety of types of cones. In 1900, Shafer mapped the position of double and single cones in the anterior region of the retina of the fish *Micropterus salmonides*. He found a remarkable correlation between the distribution and orientation of single and double cones on the one hand and a system of intercrossing coordinates on the other. Single cones occur in the middle of each lattice space; double cones are midway on each coordinate section between crossovers. Further, the separation line in each double cone is at right angles to the coordinate line.

The system is somewhat more complex in the chick (Fig. 16.12). Rods, double cones (a principal and an accessory cone), and two types of single cones appear in

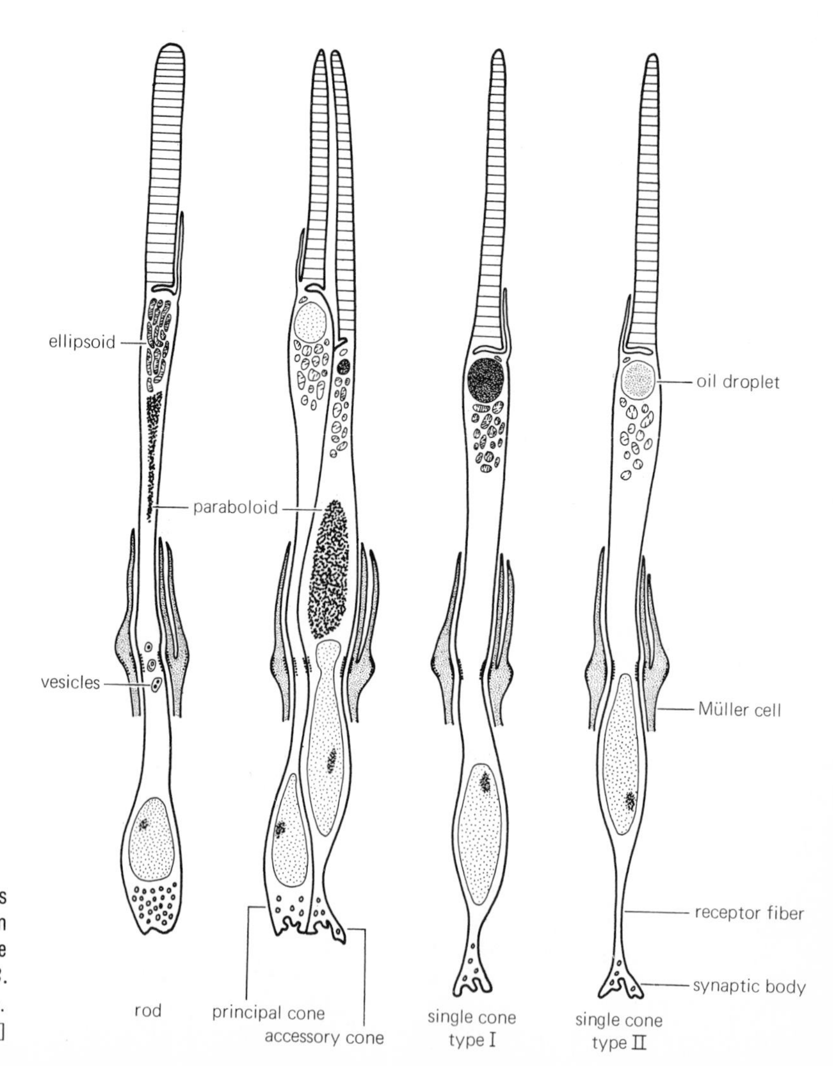

Figure 16.12 Differentiation of four types of visual receptors in the chick retina. From left to right: rod, double cone, single cone type 1 and type 2. [*After V. B. Morris and C. D. Shorey*, J. Comp. Neurol., **129**:*315* (*1967*).]

the chick retina, with associated differences in the synaptic structures. The types of receptors are arranged in a hexagonal-lattice pattern. The pattern in the mosaic is regarded as the outcome of an evenly spaced distribution of receptor types (Fig. 16.13). The pattern appears to be formed, during the development of the retina, from an array of stem cells. Each stem cell gives rise, by sequential division, to the four receptors in the repeating group of the pattern, according to one hypothesis.

This hypothesis has been tested in a study of the times at which the receptor types complete their final S phase. Tritiated thymidine is injected at a series of incubation times and the embryos are then reared to hatching. Analysis of the results lends support to the contention that the receptor types depend on cell lineage, but not from one stem cell. Each receptor cell in a repeating group is probably formed, in the final cell cycle, from a different precursor cell. The problem of diversified differentiation remains acute, and so does that of the periodic patterning of the several cell types in a complex mosaic.

C. Pigmentation patterns

One of the most interesting developmental systems, and one that has received relatively less attention in recent years, concerns the basis for the differentiation of melanocytes and the formation of pigmentation patterns. It was pointed out in Chapter 11 that all the pigment cells of the body are derived from neural crest progenitors, except those of the pigmented retina, which arise directly from the

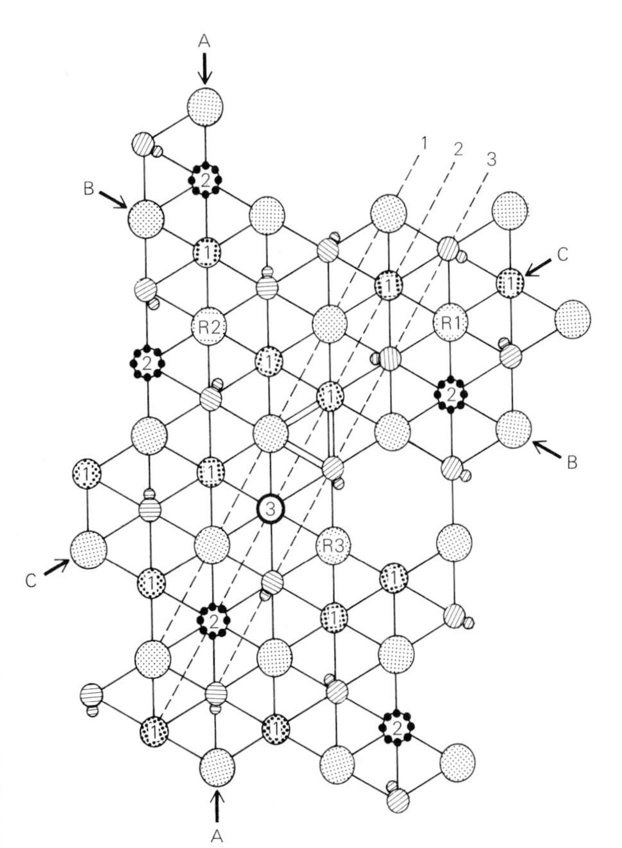

Figure 16.13 Pattern of the receptors in the peripheral retina of the chick. Rows of receptors (AA, BB, and CC) intersect at angles of 60°. Each of these rows is composed of repeating sequences of one rod and two other types of receptors. Another pattern is indicated along the broken lines: broken line 1 is a row of rods, broken line 2 is a row of single cones. A third regular pattern, visible throughout the lattice, is a triangle of receptors consisting of one rod, one single cone, and one double rod. (An example is indicated with double lines.) [*After V. B. Morris,* J. Comp. Neurol., **140:**378 (1970).]

optic cup. Consequently, melanoblasts that appear throughout the body occur in these locations only after an extensive migration during early embryonic development. Most melanoblasts migrate into the superficial regions of the embryo, where they participate in formation of melanocytes of the skin and its epidermal derivatives (hair, horn, feathers, scales). Numerous other sites, including the brain, iris, thymus, etc., also contain some pigmentation.

Considerable evidence (Section 11.2) indicates that neural crest cells emigrating from the trunk region of the neural tube have an equivalent prospective potency at the time they begin their journey. It is influences from the environment along their migratory path and in their definitive locations that ultimately determine what type of cell they become. Consequently, analysis of melanocyte differentiation provides information on the broader question of the influence of tissue environment on cell differentiation. Melanocytes in mammals are particularly well-suited for studying these questions because their state of differentiation is clearly revealed by superficial examination of the surface of the animal.

It has been found that pigmentation of the coat of mammals is under the direct or indirect control of a wide variety of genetic loci, many of which may exist in several different allelic forms. As will be seen later, the state of differentiation of these cells is very sensitive to gene activity occurring within cells of the environment, as well as genes acting within the melanoblast itself. The determination of the site of action of a given genetic locus is readily investigated in studies of pigmentation by conducting experiments in which melanoblasts of one genotype are caused to appear in an environment having a genotype of different constitution. This is best accomplished in mammals by exchanging small pieces of skin between newborn individuals carrying different alleles of the genes in question. When grafts are made as shown in Fig. 16.14, melanoblasts of the host epidermis will wander into the peripheral regions of the graft and some melanoblasts of the graft will move out into the host epidermis, thereby allowing the relative influences of genes acting in the pigment cell versus the cellular environment to be determined.

Figure 16.14 Grafting experiment to test the relative roles of the genotype of cells present in the environment of the melanoblast versus that of the melanoblast itself. In this case (discussed toward the end of the section) intensely pigmented black and yellow mice, grafted at birth with histocompatible skin from newborn genetically yellow (but nonpigmented) and genetically black (but nonpigmented) mice, respectively, exhibit some intensely pigmented hairs within the graft margin when grown. These hairs are pigmented by host melanocytes, which have migrated into hair bulbs of the graft, where their functional behavior is dictated by the milieu (agouti-locus genotype) of the graft. [*From W. K. Silvers,* Science, **134**:*370 (1961); copyright © by the American Association for the Advancement of Science.*]

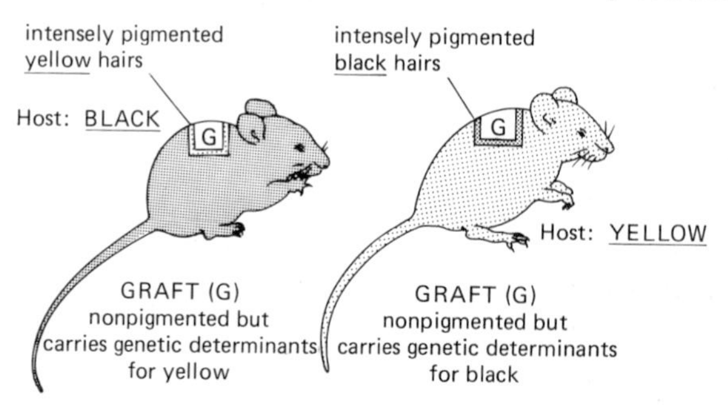

Before beginning a discussion of the genes themselves, a few remarks on the mechanism for pigmentation of epidermal derivatives will be useful. The growing basal end of a hair resides within a complex hair follicle in which living cells become filled with keratin before dying and contributing to the growth of the nonliving hair fiber. Melanocytes involved in pigmentation of a hair exist as highly extended cells with long dendritic processes containing melanosomes (pigment granules; Section 15.3). A number of these cells reside within the generative region of each hair follicle. During the formation of the hair, pigment granules are actually passed from the cytoplasm of the melanocyte process into that of the keratinocyte prior to its entombment in the body of the hair.

Genetic loci known to affect pigmentation in the laboratory mouse are shown in Fig. 16.15. Two major types of genes are indicated, those which appear to function in the melanoblast itself (lower part of figure) and those which are expressed in the cellular environment and influence pigmentation by a secondary action on the pigment cell (bottom part of figure). Of this latter group, some genes are believed to act in early stages of melanoblast formation and differentiation (left-hand part of figure), others at steps in the terminal stages of cytodifferentiation and pigment formation (right-hand part of figure). Alleles of a given locus are indicated in Fig. 16.15 as entries above and below each other in a row. We will begin by describing a few of the genes that act within the melanoblast at one or another stage of differentiation. One of the best-studied genetic loci is the *C* locus, the structural gene for tyrosinase, an enzyme (Section 15.3) that is absolutely required for the formation of any type of coloration in a pigment cell. Mutations of the *C* locus that lead to the production of an inactive tyrosinase molecule, when present in the homozygous state, are responsible for albinism. The only de-

Figure 16.15 Probable times and places of gene action during melanoblast differentiation in the mouse. Alleles at the same locus are shown in rows. [*From C. L. Markert and W. K. Silvers,* Genetics, **41**:*446 (1956)*.]

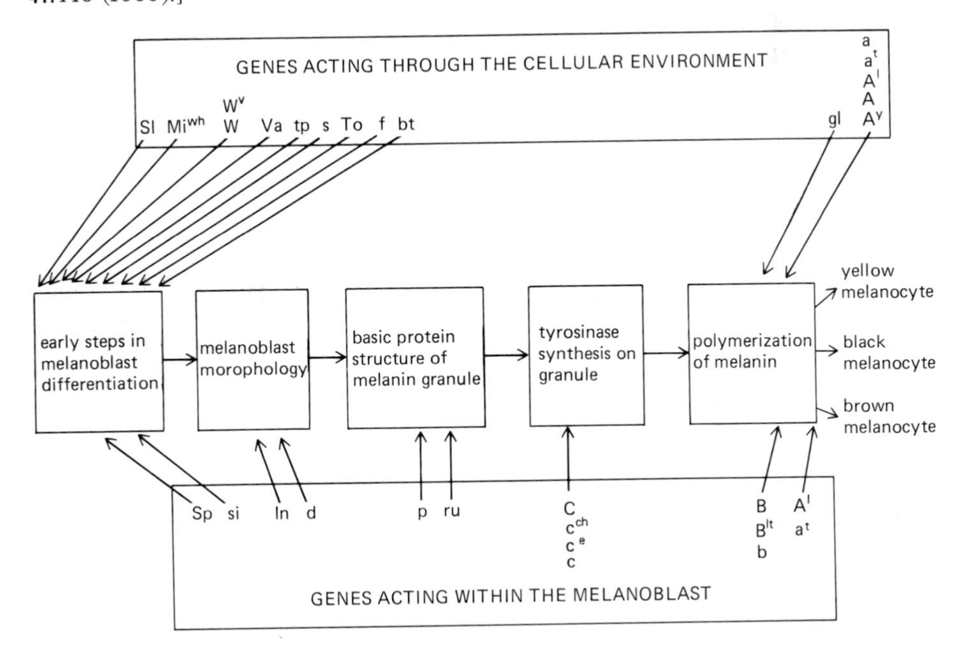

fect in the albino condition is the lack of pigmentation; melanosomes of a nonpigmented nature are readily demonstrated within melanocytes in the generative regions of a growing hair. Grafts of skin from a normal mouse made into an albino host release melanoblasts into the albino regions of the epidermis where they become fully pigmented. It is evident that skin tissue of the albino animal is fully capable of supporting melanocyte coloration given the presence of pigment cells having the proper enzymatic constitution. The gene at the P locus operates within the melanoblast and affects the structure of the melanosome (Section 15.3) rather than its pigmentation. The genetic information at the B locus acts in some manner to control the polymerization process by which melanin is produced. Mice having the BB genotype contain black melanin granules, those of the bb genotype are brown. Granules with either black or brown coloration are said to contain *eumelanin*.

Among genes that affect epidermal pigmentation but operate outside the melanoblast, the best-studied are the nine or so alleles of the A (agouti) locus. It was just mentioned that BB mice produce black melanocytes and bb mice produce brown ones. In actual fact, this statement must be greatly qualified, since the differentiated state shown by a melanocyte of BB, Bb, or bb genotype depends on (1) the location of the cell within the body and (2) the nature of the alleles present at the agouti locus. Assuming no other genetic deficiencies in pigmentation genes, melanocytes of a BB individual that exist outside of hair follicles, such as in the skin of the ear or in the brain or eye, will be black in color. However, those melanocytes residing in the hair follicles of the skin may be black or yellow depending on which alleles exist at the agouti locus. The exact nature of the yellow pigment, termed *pheomelanin* to distinguish it from *eumelanin*, is unclear. If there is an A^y allele at *either* of the homologous agouti loci, all melanocytes of hair follicles will be yellow. However, pigment cells of the skin itself, as revealed in the ear, will be black or brown depending on the alleles of the B locus. This shows that cells containing an A^y gene are capable of synthesizing eumelanin. Whether or not pheomelanin or eumelanin is produced depends entirely on tissue location. If there is an A^w allele at the agouti locus, both yellow and black melanocytes will be present and hairs will be produced that contain regions of yellow pigment and other regions of black pigment. This is precisely the case in the wild-type population. If the animal is aa or a^ea^e, the hairs will be completely black (or brown) owing to the absence of melanocytes containing yellow pigment. It appears that the agouti gene is somehow able to override the effects of the B gene, but only within melanocytes exposed to the environment of the hair follicle. Where does the gene at the agouti locus act? This question has been clearly answered using the grafting procedure described earlier. When a graft from a mouse of BB,A^y- genotype is made onto the back of a mouse with BB,aa genotype, the following type of pigmentation pattern is seen. Melanocytes from the graft (potentially yellow) that find their way into neighboring host follicles are black in color; the aa genes of cells of the hair follicle are determining the pigmentation of the melanocyte. Conversely, melanocytes of the host (potentially black) are found to produce yellow melanosomes under the influence of the A^y gene in the surrounding follicle cells. The nature of the stimulus provided the melanoblast by the gene product of the follicle cells is unknown.

The influence of the agouti gene is felt on the terminal steps of differentiation within the melanoblast. Although the evidence is less convincing, other genes are believed to act from the outside environment on one or more of the earlier steps of melanoblast differentiation. There exists a common condition among mammals in which spots of white appear within the fur. In mice these nonpig-

mented regions can be restricted to a particular patch or may extend to the entire coat of the affected animal depending on its genetic constitution. Examination of the hair follicles of *white-spotting* phenotypes indicates that nonpigmented regions are completely devoid of cells recognizable as melanocytes. Unlike the albino condition, where melanocytes lacking pigment (*amelanotic* melanocytes) are readily observed, no such cells are found in the the white spots of this other group. Why should melanocytes be lacking in one region of the epidermis, while they are present and fully pigmented in a neighboring region having the same genotype? The same question can be asked of genetically determined pigmentation patterns such as that of the zebra. There are no clear answers to these questions. The question is complicated by the presence of relatively large numbers of genetic loci (at least 14 in mice) capable of producing white-spotting phenotypes.

Given the complex lifestyle of cells that serve as ancestors to melanocytes, the absence of these pigment cells in a particular body region could be explained in many different ways. It has been suggested that there are a very small number of cells set aside early in development which serve as the progenitors of all melanocytes. If one or more of these cells were to die at an early stage prior to undergoing proliferation, corresponding embryonic regions might be expected to be devoid of pigmentation. Alternatively, the white-spotting phenotype might involve a migratory defect; cells heading for a particular region of the epidermis may die, for some reason, enroute to their final destination. Another possibility is that conditions are such in the skin of the white region that differentiation of the neural crest cell into recognizable melanoblasts cannot occur. In this explanation, it is proposed that the white-spotting genes are required to act within the cellular environment of the melanoblast in support of some early step in the differentiation sequence (Fig. 16.15). It is apparent that nonpigmented regions of white-spotted mice can support the later stages of melanocyte differentiation because grafts made into these regions are capable of donating nonmutant melanoblasts that differentiate into black pigment cells within the otherwise white environment.

3. Epithelial-mesenchymal interactions and branching morphogenesis

A variety of organ rudiments first appear in the embryo as a rounded epithelial bud surrounded by a mass of loosely packed mesenchyme. With time there occurs a progressive branching and lobulation of the epithelial layer in response to the presence of mesenchyme. This type of morphogenesis has already been briefly described in connection with the development of the metanephric kidney (Section 12.4.A) and pancreas (Section 15.7.C). Similar events occur during the formation of the lung, mammary gland, salivary gland, and thymus. In all these cases, morphogenesis can be studied in vitro following explantation of the embryonic rudiment into organ culture. In each case, the branching of the epithelial component and its subsequent differentiation require the presence of associated mesenchyme; if the two are separated and cultured independently, neither undergo morphogenesis or differentiation.

The repetitive branching of the epithelial layer of these embryonic organs is accomplished by a combination of mitotic growth within the layer and a change in cell shape. Initially the bud is a rounded structure whose branches are brought about by the formation of deep clefts (Fig. 16.16). The clefts form from the infolding of the cell layer, an event initiated by the constriction of the basal end of the cell at the leading edge of the furrow. As in the analogous events of neurulation (Section 11.1.A), the change in cell shape results from the contractile action of a band of microfilaments located at one end of the cell (Fig. 16.16). If developing glands are exposed to cytochalasin B, microfilaments disappear and clefts in the

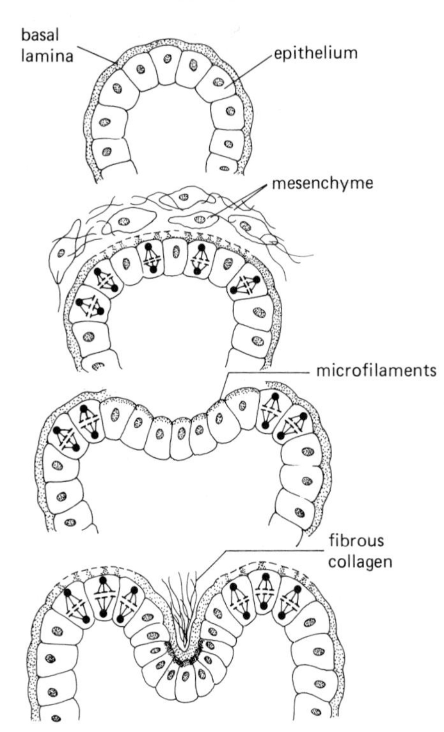

Figure 16.16 Schematic model depicting a possible sequence of events in branching morphogenesis as discussed in the text. [*After M. R. Bernfield, in "Birth Defects," J. W. Littlefield and J. de Grouchy (eds.), Excerpta Medica, 1978.*]

process of formation or those which have only recently been produced are seen to regress. Older clefts remain as such in the presence of cytochalasin B, indicating their stabilization by components other than intracellular microfilaments. When the drug is washed from the tissue, the reappearance of microfilaments is particularly evident in the basal end of those cells which will form the bottom of the recovery clefts. In addition to changes in cell shape, the expansion of the epithelium by cell division contributes to the pressure leading to infolding. The tight association of the surfaces of the cells ensures that the layer will move inward as a cohesive sheet.

Given that a number of organs develop in a similar manner, how specific are the requirements of a particular type of epithelium for a particular type of mesenchyme? What happens, for example, if one separates the epithelial and mesenchymal components from mammary gland and salivary gland rudiments, recombines them with the opposite partner (heterotypically), and then follows their development in culture? When mouse submandibular salivary gland mesenchyme is combined with mouse mammary gland epithelium, the morphological development of the composite structure closely follows that characteristic of salivary gland. The lobules are unusually broad and closely packed compared with the narrow tubular structure of mammary epithelium grown with its own mesenchyme. In other words, morphogenesis follows the pattern dictated by the source of mesenchyme, but this is not the end of the story. If this rudiment formed by experimental manipulation is grafted under the kidney capsule of an adult female mouse and this host animal is allowed to become pregnant and give birth, the graft responds to the new hormonal environment by synthesizing various proteins needed for lactation and it becomes distended with milk. It is evident that despite undergoing a salivary type of morphogenesis, the epithelium "remem-

bers" its roots as a mammary gland and differentiates as mammary tissue. Similar results are obtained when pancreatic epithelium is combined with salivary mesenchyme; the composite produces enzymes characteristic of pancreatic function, indicating that the biosynthetic activities of the epithelium are predetermined by its own developmental origin. As in the example discussed in Section 15.7.C, morphogenesis can be dissociated from cytodifferentiation.

Until recently it was believed that the ability of an epithelium to respond to a heterotypic mesenchyme was dependent on the particular source of the components. Pancreas, lung, and mammary epithelia were felt to be less specific in their requirements, since they would readily respond to other types of mesenchyme by undergoing branching morphogenesis. Parotid and submandibular salivary gland epithelia were believed to be more fastidious in their requirements and would develop only in the presence of homotypic mesenchyme. Recent experiments appear to have demonstrated that salivary gland epithelia can develop with other types of mesenchyme, although generally not as well as with their own. Results of recombination experiments indicate that the factors involved in these types of inductive interactions are widespread among mesenchymal tissue taken from organ rudiments that normally undergo branching. If the mesenchyme is obtained from the tracheal portion of the developing respiratory tract (which does not become branched) and combined with salivary gland epithelium, branching of the epithelium does not occur.

Now that we have established the need for mesenchyme in epithelial morphogenesis and the role of intracellular contractility in branching, we can turn to a more speculative topic, namely, the manner in which the mesenchyme is able to affect events within the cells of the epithelium. Most of the attention given this question has focused on the extracellular material present between the apposing cellular components. As is typically the case, the space between the epithelium and mesenchyme is composed primarily of collagen and glycosaminoglycan (GAG). The best-studied of the organs is the salivary gland, in which the extracellular material is organized into two distinct layers. One is a loosely organized layer of collagen fibers and amorphous material, probably glycoprotein and proteoglycan. The other is a thin, well-ordered basal lamina that is pressed tightly against the epithelium, which is responsible for its production. The basal lamina of the salivary gland rudiment is unusual in that it contains a much higher proportion of GAG than that of adult epithelia or embryonic epithelia that do not undergo branching. Basal lamina of these other types of epithelia are primarily collagenous. In fact, the basal lamina of the *adult* salivary gland is also primarily made of collagen, indicating that a transition from one type to the other occurs following glandular morphogenesis.

A. A proposed role for the extracellular matrix

It has been suggested on the basis of an extensive series of experiments that the basal lamina is of particular importance in epithelial morphogenesis. If a developing salivary gland epithelium is isolated with its surrounding basal lamina, it will retain its lobular morphology and, when combined with mesenchyme, will continue with its branching morphogenesis. However, if the epithelium is isolated, its basal lamina subsequently removed by treatment with hyaluronidase to destroy its GAG, and then combined with mesenchyme, the epithelium loses its lobular morphology and becomes a rounded mass. Eventually the epithelium resynthesizes its basal lamina and branching begins again. It would appear that the basal lamina is required for maintenance of the branched morphology, and the mesenchyme is required for the branching process to continue.

It is generally assumed that extracellular layers such as basal laminae act as tight-fitting investments around an epithelium, much like an exoskeleton acts on an arthropod. It has been suggested that the high GAG content of the basal lamina of the *embryonic* salivary gland causes the structure to be pliant and malleable, properties that facilitate the dramatic changes in organization that occur in the underlying cell layer. Owing to its high GAG content, the basal lamina stains intensely with dyes such as ruthenium red and alcian blue. An alcian blue-stained section of a salivary gland rudiment at an early stage in morphogenesis is shown in Fig. 16.17*a*. It can be seen that the basal lamina is not of uniform thickness across its length. Rather, the lamina is thickest in the clefts and thinnest at the rounded distal ends of the lobules (arrow) which represent the sites of subsequent branching activity. The distribution of collagen, which lies outside the basal lamina closer to the mesenchyme, is also much thinner at the distal tips of the lobules and densely packed in the clefts and along the stalk regions (Fig. 16.17*b*). Consequently, there is an inverse relationship between extracellular material and morphogenetic activity. It is suggested that the thinning of the basal lamina at the distal tips of the lobules promotes the formation of clefts at that site. We will return to this point shortly.

What is the role of the mesenchyme in branching morphogenesis? The answer is by no means certain, but at least one major effect of the mesenchyme appears to be directed at the basal lamina, leading indirectly to subsequent changes in the epithelial layer. If an intact salivary gland rudiment is incubated briefly in $[^3H]$glucosamine, a precursor that is readily incorporated into GAG of the basal

Figure 16.17 *a* Micrograph showing a 13½-day mouse embryo submandibular salivary gland stained with Alcian blue so as to better reveal the location of glycosaminoglycans. The basal lamina is apparent between the lobulated epithelium and surrounding mesenchyme. The arrow points to the distal aspect of the lobule, the C indicates a cleft. [*Courtesy of M. R. Bernfield, R. H. Cohn, and S. D. Banerjee.*] *b* Location of collagen fibers around a developing salivary-gland unit. Collagen fibers are believed to stabilize the branching by accumulating in the indentations. Collagen is not accumulated in the morphogenetically active areas, i.e., those that will soon indent.

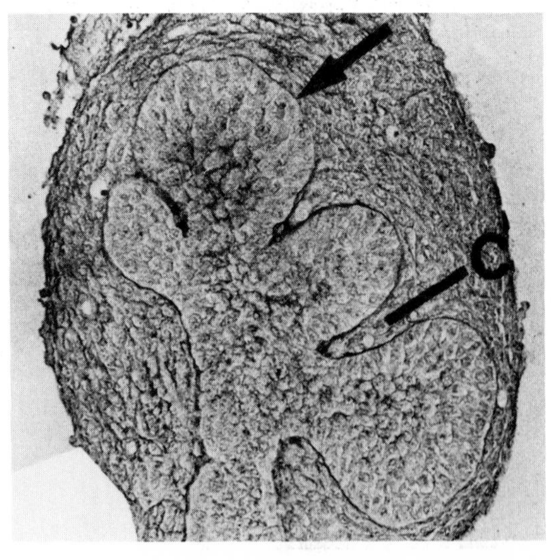

a

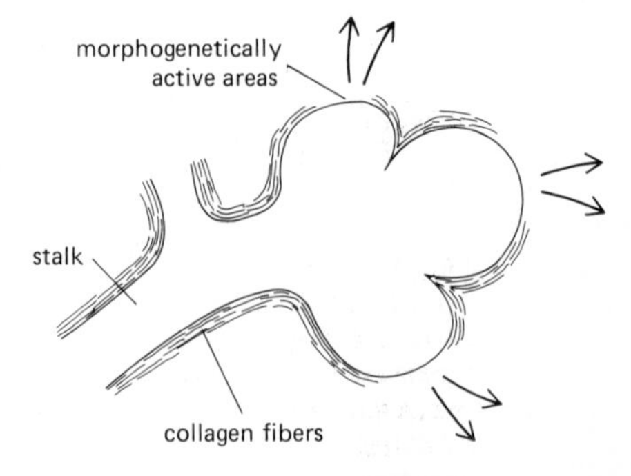

b

lamina, and then chased by transfer to unlabeled medium, the fate of the newly synthesized GAG can be followed autoradiographically. When this experiment is performed, it is found that radioactivity incorporated into the basal lamina at the distal tips of the lobules is rapidly lost during the chase period. This indicates there is a rapid turnover of extracellular GAG in these morphogenetically active regions. In contrast, label incorporated into the basal lamina of the clefts and stalk is retained to a much greater degree during the chase period, indicating the relatively slow turnover of GAG in regions where branching does not occur.

Further experiments indicate that GAG turnover in the basal lamina is dependent on the presence of associated mesenchyme. If an intact embryonic salivary gland is incubated for 2 hours with [^{3}H]glucosamine, the mesenchyme then removed, and the epithelium transferred to unlabeled medium for 5 additional hours, all parts of the basal lamina become radioactive in a relatively uniform manner (Fig. 16.18*b*). If however, mesenchyme is added to an isolated epithelium that has been previously labeled, radioactivity is preferentially lost from the basal lamina in the region outside the rounded ends of the lobules (Fig. 16.18*a*). These results suggest that a major role of the mesenchyme is to selectively degrade material of the basal lamina in morphogenetically active regions. How does such localized destruction facilitate a branching response at those sites? It may be that the localized thinning of the basal lamina promotes the direct interaction of processes from the two apposing cell layers. Electron micrographs do show an interaction between epithelial and mesenchymal cells at these sites. Alternatively, destruction of extracellular material may lead to an activation of the contractile apparatus on the other side of the membrane. Evidence for transmembrane control of cytoskeletal elements (Section 3.1.A) appears to be well es-

Figure 16.18 Effect of mesenchyme on the distribution of [^{3}H]glucosamine incorporated at the basal surface of submandibular epithelia. Intact glands were labeled for 2 hours, and following isolation by collagenase treatment, the epithelia were chased in the presence (*a*) and absence (*b*) of mesenchyme for 5 hours. In the absence of mesenchyme, label is distributed uniformly along the basal lamina, while in the presence of mesenchyme, radioactivity in the basal lamina of the distal tips of the lobules is relatively low, a condition believed to be due to increased degradation of GAG in that region. [*From M. R. Bernfield, in "Birth Defects," J. W. Littlefield and J. de Grouchy (eds.), Excerpta Medica, 1978.*]

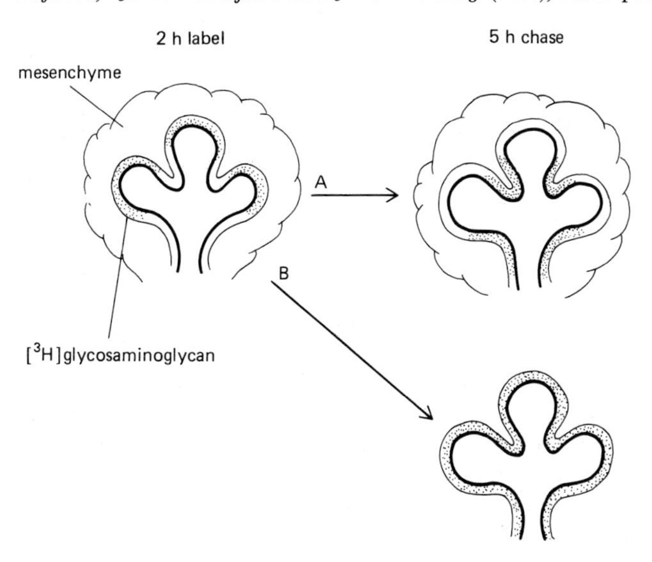

tablished. It may be that the effect is mediated via localized changes in Ca^{2+} concentration. Glycosaminoglycans are highly negatively charged and capable of complexing large amounts of calcium. If these ions were to be released locally by the degradation of GAG and pass through the plasma membrane, they might trigger the contractile response on the cytoplasmic side. A very different proposal suggests that changes in the extracellular matrix may have an effect on the mitotic activity of adjacent cells, causing them to proliferate in the morphogenetic areas of the layer. As in the case of contractility, cell division also may be widely regulated by a transmembrane control mechanism. At the present time, very little evidence exists for any of these postulated mechanisms; thus there is little basis to choose among them.

4. Bone morphogenesis

The importance of the extracellular matrix in sclerotome chondrogenesis was stressed in Section 15.4. Another series of experiments has revealed the potential that extracellular matrices possess for shaping the cartilaginous and bony structures that develop in their midst. Bone is a tissue composed of three components: scattered cells or osteocytes, an extensive extracellular organic matrix composed primarily of collagen and glycosaminoglycan, and mineral deposits of calcium phosphate. It has been found that the extracellular organic matrix, by itself, is very active in inducing chondrogenesis and subsequent osteogenesis in cells that normally would not have differentiated in that direction. In these experiments, bone or tooth matrix is demineralized and implanted intramuscularly, where fibroblastic (mesenchymal) connective-tissue cells abound. These mesenchymal cells, which were previously part of the connective-tissue framework of the adjacent muscle, invade old vascular channels and begin restriction toward bone-cell differentiation. In the presence of the morphogenetic substratum provided by the bone matrix, these cells now begin to synthesize cartilage matrix and then become ossified into bone. The quantity of woven bone deposited by cells is proportional to the mass of preimplanted matrix. The evidence suggests that the total three-dimensional structure of the bone matrix imposes the bone morphogenetic pattern on proliferating mesenchymal cells.

A wide variety of extracellular substances may act as morphogenetic substrates and as cues for subsequent differentiation. In this light, the genetic control of collagen synthesis may be an important intermediate step, or means, for determining specific patterns, e.g., those of the feather or retina. At the very least, the nature of the substratum imposes an orientation on cells migrating over it (Fig. 16.19) which may, in turn, lead to a defined orientation of materials produced by these cells as they differentiate. In addition, extracellular matrices may provide soluble, diffusible substances that act as inducers within adjacent cells. The conversion of fibroblasts to cartilage-bone cells by implantation of bone matrix is believed to be mediated by a substance released by the matrix and transferred to the population of mesenchymal cells. Although nonliving, the extracellular matrix need not be a static structure. As in the case of the basal lamina of the developing salivary gland, neighboring cells may alter the structure and properties of these external layers, thereby giving them new morphogenetic capabilities. In a very real sense, the extracellular matrices represent a continuous supracellular agency, integrating populations of cells into organized and differentiating patterns. At the same time, each cell within the matrix as an individual entity contributes to the pattern. This area of investigation is still in its infancy; yet it shows considerable promise of helping to place abstract concepts of morphogenetic fields and positional information into a tangible context.

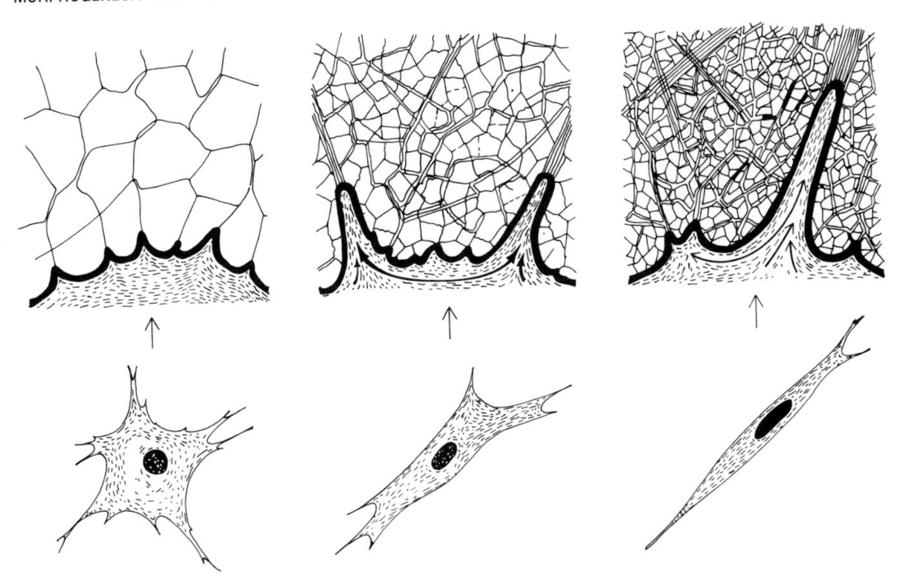

Figure 16.19 Structure of mesenchymal cells in three media of (from left to right) increasing plasma fibrous concentrations. Showing protoplasmic flow directed along main paths of fibrous gel structure. [*After P. Weiss and B. Garber,* Proc. Nat. Acad. Sci. U.S., **38**:270 (1952).]

Discussions of morphogenesis have been scattered throughout this book, which is as it should be in a text on embryonic development. Taken as a whole, it is evident that quite a variety of interrelated activities are involved in shaping the organization of the tissues and organs of the early embryo. These include changes in cell adhesion, as occur during epithelial-mesenchymal interconversions (Chapter 10) or histogenesis (Section 16.1.B); localized deposition and/or degradation of extracellular matrices, as occurs during branching morphogenesis (Section 16.3.A) or corneal differentiation (Section 14.3.E); localized cell death, as occurs during limb development (Section 13.4) or metamorphosis (Section 19.3.D); changes in cell shape, as occur during the folding (Section 11.2.A) or spreading (Section 10.1.B) of an epithelium; and localized mitotic activity, as occurs during limb formation (Section 13.2.C) and cleft formation (Section 16.3). Most important of all aspects is cell movement, for it is only by the relative displacement of cells that a predetermined spatial pattern can emerge. Movements on a localized scale are believed to be of primary importance in histogenesis, allowing cells of particular types to become arranged in their proper spatial relationships to each other. Morphogenetic movements of this type may depend on changes in cell adhesion and/or cell shape as well as the locomotory activity of the cells involved. Large-scale movements, such as occurs during gastrulation (Section 10.1.B), formation of the gonad (Section 12.4.B), nerve outgrowth (Section 14.4.C), neural crest dispersal (Section 11.2), etc., require extensive locomotion and are of obvious importance in morphogenesis. The major long-range movements of cells which occur in the vertebrate embryo are shown in Fig. 16.20.

There is one major aspect of the subject of morphogenesis that has been ignored in this chapter—its genetic control. It is evident that even the most subtle aspects of morphogenesis are somehow determined by the genetic inheritance. Consider how similar the features of a parent and their offspring can appear, or the symmetry of human facial structure, even though each side forms in a spatially independent manner. Since genes consist of a one-dimensional informa-

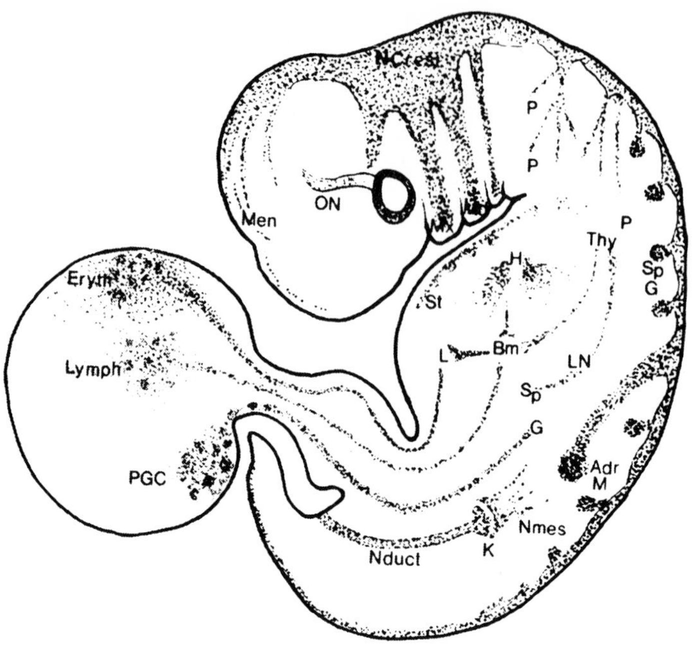

Figure 16.20 Diagrammatic representation of some of the cellular traffic involved in the formation of tissues and organs during mammalian embryogenesis. The diagram combines events taking place at different stages of development. The "pathways" connect the original sites of cells with their destinations; they do not accurately depict the actual routes of the cells. The progenitors of erythroid cells (Eryth) originate in the yolk-sac endoderm, enter blood vessels, and their progeny are transported by the circulation to the subsequent sites of erythropoiesis (L, liver; Bm, bone marrow). The progenitors of lymphoid cells are transported to the bone marrow (Bm), thymus (Thy), lymph nodes (LN), and spleen (Sp). The primordial germ cells (PGC) migrate actively from the yolk sac to the germinal ridges and colonize them to form the gonads (G). Cells from the neural crest (NCrest)—a temporary structure along the dorsal axis of the early embryo—migrate in swarms or individually, and their progeny give rise to a variety of structures, among them meninges (Men), cartilages of the embryonic skull (Mx, Md), pigment cells (P), spinal ganglia (SpG), and adrenal medulla (AdrM). Axons that grow out of retinal-ganglion cells form the optic nerves (ON), which advance toward the visual centers in the brain. The nephric ducts (Nduct) elongate toward aggregations of nephric mesenchymal cells (Nmes) to jointly form the kidney rudiments (K). The sternum (St) is formed by mesenchymal cells that migrate from the dorsal region of the embryo. The heart (H) arises from aggregations of cells originating in the "heart-forming territory." The intricate translocations of cells involved in histogenesis of the nervous system are not included in this diagram, nor are other examples of morphogenetic cell migrations and aggregations (e.g., those resulting in the formation of hair follicles, teeth, limb rudiments, etc.). [*From A. A. Moscona and R. E. Hausman in "Cell and Tissue Interactions," J. W. Lash and M. M. Burger (eds.), Raven, 1977.*]

tional system serving to code for amino acid sequences, their role in morphogenesis must be indirect. The genome bears the responsibility of producing the appropriate RNA transcripts; all else follows from there. We have discussed the role of selective adhesion at length in this chaper. The appearance of particular cell-surface proteins might be expected to lead to particular adhesive interactions involved in histogenesis. The existence of such genes becomes apparent when the genes undergo mutation and normal intercellular contacts cannot occur. The reeler mutation mentioned in Section 16.1.B may be of this type. Other types of morphogenetic mutations were described in connection with development in

Drosophila. Mutations in homeotic genes lead to the entire substitution of one morphogenetic pathway with another. Unfortunately, we know very little about the nature of these genetic loci. We are just as ignorant about the nature of genes whose mutations can have drastic effects on human development. It may be that certain of these genes act in a similar manner to those required for the construction of viruses (Chapter 2). It is for this reason that these studies on molecular assembly are of such importance to developmental biologists. Once again, much remains to be discovered.

Readings

BANERJEE, S. D., R. H. COHN, and M. R. BERNFIELD, 1977. Basal Lamina and Embryonic Salivary Epithelium, *J. Cell Biol.*, **73**:445–463.

BERNFIELD, M. R. and S. D. BANERJEE, 1978. The Basal Lamina in Epithelial-Mesenchymal Morphogenetic Interations, in N. A. Kefalides (ed.), "Biology and Chemistry of Basement Membranes," Academic.

COULOMBRE, J. L., and A. J. COULOMBRE, 1971. Metaplastic Induction of Scales and Feathers in the Corneal Anterior Epithelium of the Chick Embryo, *Develop. Biol.*, **25**:464–478.

EBERT, J. D., and T. S. OKADA, 1979. "Mechanisms of Cell Change," Wiley.

FRAZIER, W., and L. GLASER, 1979. Surface Components and Cell Recognition, *Ann. Rev. Biochem.*, **48**:491–523.

GARBER, B. B., and A. A. MOSCONA, 1972. Reconstruction of Brain Tissue from Cell Suspensions. II. Specific Enhancement of Aggregation of Embryonal Cerebral Cells by Supernatant from Homologous Cell Cultures, *Develop. Biol.*, **27**:235–243.

GARROD, D. R., (ed.), 1978. "Specificity of Embryological Interactions," Chapman & Hall.

GLASER, L., 1978. Cell-Cell Adhesion Studies with Embryonal and Cultured Cells, *Rev. Physiol. Pharmacol.*, **83**:89–122.

GOETINCK, P. F., and M. J. SEKELLICK, 1972. Observations on Collagen Synthesis, Lattice Formation, and Morphology of Scaleless and Normal Embryonic Skin, *Develop. Biol.*, **28**:636–648.

GROBSTEIN, C., 1967. Mechanism of Organogenetic Tissue Interaction, *Nat. Cancer Inst. Monogr.*, **26**:279–299.

HAUSMAN, R. E., and A. A. MOSCONA, 1979. Immunological Detection of Retinal Cognin on the Surface of Embryonic Cells, *Exp. Cell Res.*, **119**:191–204.

HAY, E. D., 1973. Origin and Role of Collagen in the Embryo, *Amer. Zool.*, **13**:1085–1107.

HUMPHREYS, T., 1963. Chemical Dissolution and *in vitro* Reconstruction of Sponge Cell Adhesions, *Develop. Biol.*, **8**:27–47.

KRATOCHWIL, K., 1969. Organ Specificity in Mesenchymal Induction Demostrated in the Embryonic Development of the Mammary Gland of the Mouse, *Develop. Biol.*, **20**:46–71.

LAWRENCE, I. E., 1971. Timed Reciprocal Dermal-Epidermal Interactions between Comb, Mid-dorsal, and Tarsometatarsal Skin Components, *J. Exp. Zool.*, **178**:195–210.

LINSENMAYER, T., 1972. Control of Integumentary Patterns in the Chick, *Develop. Biol.* **27**:244–271.

MARCHASE, R. B., K. VOSBECK, and S. ROTH, 1976. Intercellular Adhesive Specificity, 1976. Intercellular Adhesive Specificity, *Biochem. Biophys. Acta*, **457**:385–416.

MARKERT, C. L., and W. K. SILVERS, 1956. The Effect of Genotype and Cell Environment on Melanoblast Differentiation in the House Mouse, *Genetics*, **41**:429–450.

MARKERT, C. L., and H. URSPRUNG, 1971. "Developmental Genetics," Prentice-Hall.

MORRIS, V. B., 1970. Symmetry in a Receptor Mosaic Demonstrated in the Chick from the Frequencies, Spacing and Arrangement of the Types of Retinal Receptor, *J. Comp. Neurol.*, **140**:359–397.

———, and C. D. SHOREY, 1967. An Electron Microscope Study of the Types of Receptor in the Chick Retina, *J. Comp. Neurol.*, **129**:313–339.

MOSCONA, A. A., 1974. "The Cell Surface in Development," Wiley.

———, 1962. Studies on Cell Aggregation and Demonstration of Materials with Selective Cell-Binding Activity, *Proc. Nat. Acad. Sci. U.S.*, **49**:742–747.

PHILLIPS, H. M., L. L. WISEMAN, and M. S. STEINBERG, 1977. Self versus Nonself in Tissue Assembly, *Develop. Biol.*, **57**:150–159.

RAWLES, M. E., 1963. Tissue Interaction in Scale and Feather Development as Studied in Dermal-Epidermal Recombinations. *J. Embryol. Exp. Morphol.*, **11**:765–789.

REDDI, A. H., and W. A. ANDERSON, 1976. Collagenous Bone Matrix-Induced Endochondrial Ossification and Hematopoiesis, *J. Cell Biol.*, **69**:557–572.

SAWYER, R. H., 1972. Avian Scale Development. I. Histogenesis and Morphogenesis of the Epidermis and Dermis during Formation of the Scale Ridge, *J. Exp. Zool.*, **181**:365–384.

SENGEL, P., 1971. The Organogenesis and Arrangement of Cutaneous Appendages in Birds, *Adv. Morphol.*, **9**:181–230.

———, and D. DHOUAILLY, 1977. Tissue Interactions in Amniote Skin Development, in M. Karkinen-Jaaskelainen, L. Saxen, and L. Weiss (eds.), "Cell Interactions in Differentiation," Academic.

SILVERS, W. K., 1961. Genes and the Pigment Cells of Mammals, *Science*, **134**:368–373.

SPIEGEL, M., 1955. The Reaggregation of Dissociated Sponge Cells, *Ann. N.Y. Acad. Sci.*, **60**:1056–1078.

SPOONER, B. S., 1975. Microfilaments, Microtubules, and Extracellular Materials in Morphogenesis, *Bioscience*, **25**:440–450.

———, 1974. Organogenesis in Vertebrate Organs, in J. Lash and J. R. Whittaker (eds.), "Concepts of Development," Sinauer.

STEINBERG, M. S., 1970. Does Differential Adhesion Govern Self-Assembly Processes in Histogenesis, *J. Exp. Zool.*, **173**:395–434.

STUART, E. S., B. GARBER, and A. A. MOSCONA, 1972. An Analysis of Feather Germ Formation in the Embryos and *in vitro*, in Normal Development and in Skin Treated with Hydrocortisone, *J. Exp. Zool.*, **179**:97–118.

SYMPOSIUM, 1973. Factors Controlling Cell Shape during Development, *Amer. Zool.*, **13**:941–1129.

TOOLE, B. P., 1973. Hyaluronate and Hyaluronidase in Morphogenesis and Differentiation, *Amer. Zool.*, **13**:1061–1065.

TOWNES, P. L., and J. HOLTFRETER, 1955. Directed Movements and Selective Adhesion of Embryonic Amphibian Cells, *J. Exp. Zool.*, **128**:53–120.

URIST, M. R., 1970. The Substratum for Bone Morphogenesis, 29th Symposium, *Develop. Biol. (suppl.)*, **4**:125–163.

URSPRUNG, H., 1966. The Formation of Pattern, *Develop. Biol.*, **25**:251–277.

WEISS, P., 1961. Ruling Principles in Cell Locomotion and Cell Aggregation, *Exp. Cell Res.*, **8**:260–281.

———, and A. C. TAYLOR, 1960. Reconstitution of Complete Organs from Single-Cell Suspensions, *Proc. Nat. Acad. Sci. U.S.*, **46**:1177–1185.

WESSELS, N. K., 1977. "Tissue Interactions and Development," Benjamin.

———, 1970. Mammalian Lung Development: Interactions in Formation and Morphogenesis of Tracheal Buds, *J. Exp. Zool.*, **175**:455–466.

———, and J. EVANS, 1968. The Ultrastructure of Oriented Cells and Extracellular Materials between Developing Feathers, *Develop. Biol.*, **18**:42–61.

WISEMAN, L. L., M. S. STEINBERG, and H. M. PHILLIPS, 1972. Experimental Modulation of Intercellular Cohesiveness: Reversal of Tissue Assembly Patterns, *Develop. Biol.*, **28**:498–517.

Chapter 17 Development of the immune system

At the present time, research on the development, nature, and function of the immune system is producing a tremendous body of fact and theory. In this chapter we will focus on the developmental questions that have arisen and on the background of antibody structure and formation necessary to understand them.

The immune system is considered primarily as a system of defense at the molecular level. The targets are molecules recognized as foreign to the individual; the weapons are proteins called *antibodies* (or *immunoglobulins*). The cells responsible for antibody production are those of the lymphoid tissue. In mammals this includes the thymus, bone marrow, lymph nodes, spleen, certain tissues of the gut (Peyer's patches), and the fetal liver.

1. Antibody structure and function

Each animal has within its body macromolecules that distinguish it from other members of the same species or a different species. The identification of one's own materials as "self" and those of an outside source as "nonself" forms the basis for the immune system's ability to react against foreign substances. For example, if proteins extracted from one animal, such as a mouse, are injected into another animal, such as a rabbit, the latter will recognize these proteins as being foreign, i.e., distinct from its own proteins. It will respond to them by the production of antibodies. Any substance (whether protein, carbohydrate, nucleic acid, or other) that evokes the production of antibodies is termed an *antigen*. A closer look at a large-molecular-weight antigen indicates that the molecule is composed of a number of distinctive subregions, each having a defined spatial and electronic configuration. These smaller sections of a large molecule are termed *antigenic determinants* and provide the actual stimulus for the eventual production of a particular antibody. In other words, the combining sites of antibody molecules are directed against small sections of a macromolecule rather than the entire molecule itself.

Once produced, the antibody is capable of reacting chemically with the antigen responsible for its production or any other molecule having the same determinant. One of the most remarkable features of the immune system is its speci-

ficity. If a particular antigen, such as measles virus, is injected into an animal, the antibodies formed in response to that injection are highly specific: they will combine only with measles-virus protein and not with protein obtained from any other source. If one tests the combining powers of antibodies present in an immune serum, one often finds that two proteins having only a single amino acid difference can be distinguished. Several very important questions are raised by these observations. How can certain antibody molecules combine with one antigen and not with another? That is, what is the basis of antibody specificity? This question can be answered by considering antibody structure and variability, as discussed in subsequent paragraphs. How does a given antigen evoke the production of an antibody that can specifically combine with it? This latter question will be considered in a following section.

Antibodies are proteins (termed *immunoglobulins*) built of two types of polypeptide chains. These two chains, referred to as heavy (H) chains (molecular weight of 50,000 to 70,000) and light (L) chains (molecular weight of 23,000) occur in pairs linked to one another by disulfide bonds. The most widely studied class of immunoglobulin, IgG, is composed of two light chains and two heavy chains arranged in the manner shown in Fig. 17.1. The light chains of IgG (or any class of Ig) are of two distinct types, kappa (κ) and lambda (λ). In contrast, there is only one type of H chain found in all IgG molecules.

Five major classes of immunoglobulin (IgA, IgD, IgE, IgG, and IgM) have been identified, each having a somewhat different structure and, presumably, a different set of functions. Each of these types of immunoglobulin is built from pairs of heavy and light chains. All classes utilize the same two types of light chains, although each class has its own distinct heavy chain. The three classes representing the bulk of the antibody content of serum are IgM, IgG, and IgA. IgM is a very large complex containing five pairs of L-H subunits held together by a totally different polypeptide termed the *J chain*. IgM molecules appear in the serum most rapidly after contact with a given antigen. With the passage of time, the IgM molecules are replaced by IgG species having the same antigen-combining sites. IgA molecules, which also have the basic subunit structure shown in Fig. 17.1 (with the gamma H chain replaced by an alpha H chain), can exist as monomers, dimers, trimers, or even higher-order complexes. Although IgA is also found in the serum, it is present as the primary immunoglobulin in the secretions of various exocrine-gland cells. These secretions include nasal mucus, tears, and milk.

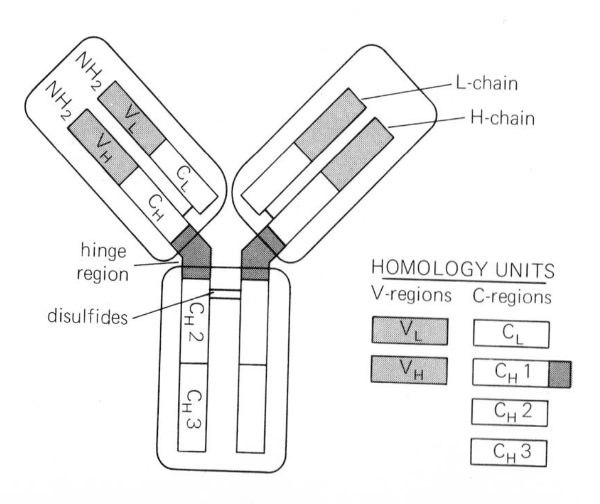

Figure 17.1 A model of an immunoglobulin (IgG) molecule. V, variable portion of a polypeptide chain; C, constant portion of a polypeptide chain; subscript H represents heavy chain; subscript L represents light chain. [*Reproduced with permission from G. P. Smith, L. Hood, and W. M. Fitch*, Ann. Rev. Biochem., **40;** copyright © *1971 by Annual Review, Inc.*]

The basis of antibody specificity is revealed upon consideration of the amino acid sequences of the component polypeptide chains. The first step in amino acid sequence analysis is to obtain a preparation of purified protein. However, under normal conditions it is impossible to obtain a purified preparation of a given antibody since each individual produces many different antibody molecules, all of which are similar enough in structure to be very difficult to separate. Even after the injection of one antigen (or a molecule containing only one antigenic determinant), antibodies with many different combining sites are produced, all of which are capable of combining with the antigen. A way out of this dilemma has been found by taking advantage of a particular disease condition.

Amino acid sequence data have been obtained from immunoglobulins of patients with multiple myelomas, tumors of lymphoid cells. In patients with these tumors, large amounts of a single class and molecular species of antibody are produced and secreted. The particular species of immunoglobulin produced depends on the particular cell that became malignant (as described later, a given lymphocyte is destined to make only one species of antibody). Different patients produce distinct immunoglobulin molecules. When the amino acid sequences of several light and heavy chains of IgG from different myeloma patients were compared, an important pattern was revealed (Fig. 17.2). In the group of kappa light chains that were studied, it was found that one-half of each chain was constant in amino acid sequence among all the kappa chains, while the other half was variable from patient to patient. Similarly, comparison of the amino acid sequences of several lambda chains from different patients revealed that they too consisted of a section of constant sequence and a section whose sequence varied from immunoglobulin to immunoglobulin. Further analysis indicated that the heavy chains of these antibodies also contained a variable (V) and a constant (C) portion. In the case of the heavy chain, approximately one-quarter (110 amino acids at the amino end) has a variable amino acid sequence, the remaining three-quarters being constant for all species of IgG. Regardless of the Ig class, the constant portion of the heavy chain for that class can be divided into sections of approximately equivalent length which are clearly homologous to one another. There are three such *homology units* in the H chain for an IgG molecule (Figs. 17.1 and 17.2) designated C_H1, C_H2, and C_H3. It would appear that the component sections of the C part of the heavy chains arose during evolution by the duplication of an ancestral gene. Similarly, the amino acid sequence of the constant portion of light chains bears sufficient similarity to the sequences of the homology units of the heavy chain to indicate an evolutionary relationship as well. A closer look (in

Figure 17.2 The general structure of the heavy and light chains of an IgG molecule. Each half of the light chains contains approximately 110 amino acid residues. The heavy chain has approximately 450 amino acid residues, about one quarter of which compose the variable region. The constant portion of the heavy chain in IgG molecules is composed of three homology units. The location of the hypervariable regions in this diagram (indicated by black bands in V regions) has no relationship to their positions in the actual amino acid sequence.

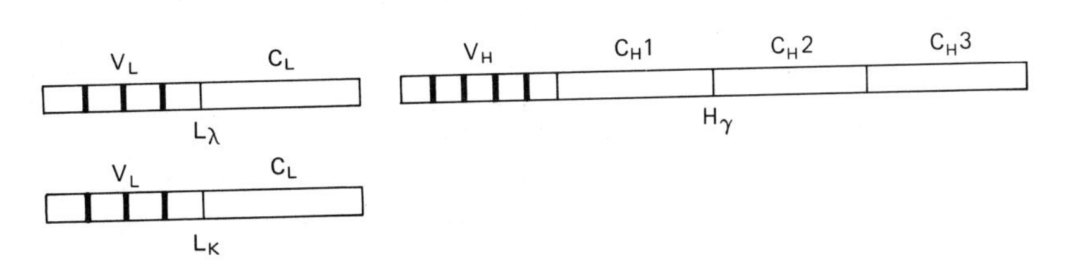

both humans and mice) has revealed that the variable portions of both the heavy and light chains contain subregions that are highly variable, i.e., *hypervariable*, from Ig to Ig. The hypervariable portions of the chains (three in the L and four in the H) contain deletions and insertions of amino acids as well as substitutions.

As shown in Fig. 17.1, each IgG molecule contains two *pairs* of polypeptide chains, each consisting of one H and one L. Various types of analyses have indicated that in the complete immunoglobulin molecule, the V part of each L chain is associated with the V part of one of the H chains. The specificity of a particular antibody molecule is determined by the amino acids present in the antigen-combining sites (two identical sites per IgG molecule). Each combining site is made of the variable portion of both a heavy and a light chain. Given the availability of large amounts of unique myeloma immunoglobulin molecules, the question was raised as to whether or not the immunoglobulins secreted by these tumors were capable of combining with any known substance; i.e., were they indeed antibody molecules? If such an antigenic determinant could be found, then the opportunity for studying the interaction between a given antigen and antibody would be greatly improved. The search for "antigens" complementary to purified immunoglobulins has been successful in a few cases. For example, one human myeloma protein was found to bind vitamin K in a highly specific manner. The discovery of suitable "antigens" has recently led to crystallographic studies that have defined the precise three-dimensional organization of the combining sites of antibody molecules. As expected, it is the hypervariable regions of each chain that make up the walls of the combining site and are therefore most intimately involved in antigen binding.

A particularly important advance in the field of immunology has occurred in recent years with the development of cell lines that divide indefinitely and secrete large amounts of a single antibody. These cells, termed *hybridomas*, are produced by the fusion of lymphocytes and malignant myeloma cells. The lymphocyte brings to the hybrid the property of production and secretion of a specific antibody, while the tumor cell brings to the hybrid the property of immortality, i.e., indefinite cell division. The soluble antigen (or preparation of foreign cells) is simply injected into a mouse to cause the proliferation of specific antibody-forming cells. The spleen is then removed and the lymphocytes fused with a population of malignant myeloma cells. Hybrids are selected from nonfused cells and then individually screened for production of antibody against the antigen in question. Lymphocytes containing the appropriate receptors can then be cloned in vitro or in vivo (by growing as tumor cells in a recipient animal), and huge amounts of monoclonal antibody can be prepared. Surprisingly, with this technique it can be easier to obtain purified antibody than purified antigen. Consequently, the antibody can be employed in subsequent steps to isolate purified antigen by use of affinity chromatography.

2. Two genes, one polypeptide

An important question in immunology concerns the presence in a given H or L polypeptide chain of two regions (a V and a C) that appear to have separate genetic origins. Analysis of mRNAs for the L and H chains of immunoglobulins indicate that a single mRNA molecule contains the coding regions of both the V and C regions. In other words, the L or H polypeptide chain is *not* formed by the joining of completed V and C polypeptides. Therefore, if all the light chains (of a given type, κ or λ) produced by an individual have identical C portions but diverse V portions, then one or the other of the following conditions presumably exist.

1. There are many copies of each type of C gene so that each V gene is present next to one of these copies. In this way, an mRNA containing both sequences in tandem can be synthesized.
2. There is one (or a few) copy of each type of C gene and it can become associated with different V genes in different cells. In this case, some type of DNA-joining mechanism must exist whereby two separate DNA sequences are brought together in such a way that a single mRNA containing both their information can be transcribed.

The evidence obtained to date on this question points unequivocally to the second alternative. Earlier hybridization studies suggested that the number of C genes within the genome is very limited and, therefore, that the same segment of DNA must be associated with different V genes in different antibody-producing cells. It was then shown that the V and C genes complementary to a particular kappa-chain mRNA appear to be separated from each other in the DNA of the embryo, while they are much closer in the DNA of the lymphoid tumor cells responsible for secreting this polypeptide. This was the first direct evidence that a movement of DNA sequences occurs during the generation of antibody-forming cells. As a consequence of the shift, a single polypeptide can be formed from information that was originally present at two distinct genetic loci. Although the mechanism by which recombination occurs is not understood, the precise nucleotide sequences involved in the formation of a few light chains has been worked out. The organization of DNA sequences involved in the formation of mouse kappa chains is shown in Fig. 17.3. In this case, a variety of V genes are located in a linear array separated from the single C gene by some distance. Nucleotide-sequence analysis of these V genes indicated that they were shorter than that required to code for the corresponding portion of the polypeptide comprising the V region of the light chain. The reason for this foreshortened condition became clear when other segments in the region were sequenced. The stretch of nucleotides that codes for the 13 amino acids of the carboxyl end of the V region is present at a distance from the remainder of the V-gene sequences. This small portion at the end of the V gene is termed the *J segment*. As shown in Fig. 17.3*a*, there are five distinct J segments (of similar nucleotide sequence) arranged in a linear array. The cluster of 5 J segments is then separated from the C gene of the kappa chain by an additional stretch of over 2,000 nucleotides. The actual translocation event (Fig. 17.3*b*) occurs when a specific V gene is caused to recombine with one of the five J segments, thereby forming a complete and specific V-gene sequence that is still separated from the C gene by 2,000 to 4,000 nucleotides. No further DNA movement is necessary prior to transcription; the entire genetic region is transcribed into a large primary transcript from which noncoding sections are excised. The translocation of the V gene is believed to serve as the event that leads to the commitment of that cell and all its progeny to the production of antibodies containing that specific light chain; the cell is said to be specified.

It is interesting that during the maturation of immunocompetent cells, the cells switch from one immunoglobulin class to another, such as from IgM to IgG or IgD. These classes differ from each other with respect to the type of heavy chain. However, all antibodies of all classes produced by a given immune-cell line contain the same exact combining sites and thus the same specificity. In other words, the V regions of all heavy chains produced by a cell line are identical, despite the shifts in the C regions as one class of Ig is replaced by another. The manner in which this shift of genetic components is accomplished remains obscure.

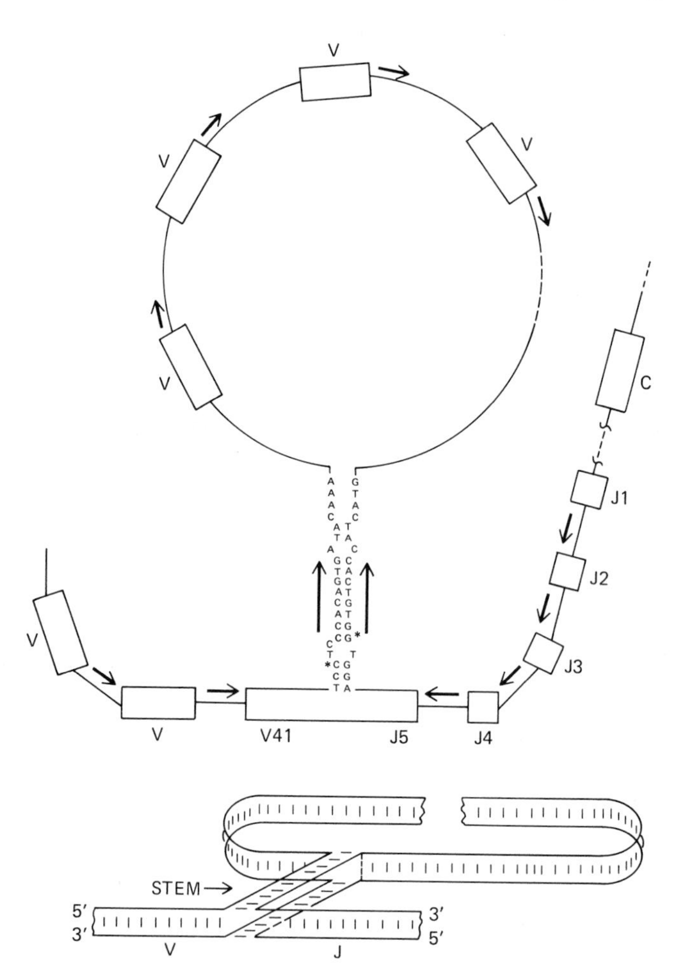

Figure 17.3 *a* Diagrammatic representation of the organization of sequences in the kappa region of the genome. *b* Hypothetical stem structure formed between inverted repeats located next to germ-line V- and J-region genes. Above is a diagram of hypothetical intermediate in the chromosomal rearrangement κ V- , J- , and C-region genes. Distances are not to scale. Each V- and J-region gene is bordered by a palindrome that is also an inverted repeat sequence (*boldface arrow*) located on the 3′ side of the V genes and on the 5′ side of the J genes. Each of these sequences can be written as a complementary stem structure, as shown in the figure, in which the actual sequence of MOPC-41 recombinant is used as an example. The bases marked with asterisks are those actually joined to form the recombinant. The resulting amino acid codon sequence is indicated. Similar stem structures which draw V and J regions together can be written for each κ and λ V and J sequence now known. Below is a three-dimensional representation of the hypothetical stem intermediate. V- and J-coding sequences interact with their opposite strands in a normal DNA duplex, but the inverted repeat adjacent to the V gene is drawn as a stem interacting with its complement located on the same strand, presumably many thousands of bases away adjacent to a J sequence. [*b. From E. E. Max, J. G. Seidman, and P. Leder,* Proc. Nat. Acad. Sci. U.S., **76:** *3454 (1979).*]

The movement of DNA sequences within the genome is of particular importance in a developmental context because it results in a limitation of the potency of the cell in which it occurs. From the time of translocation, that cell becomes committed in a stable, inheritable manner. The evolution of a mechanism for somatic rearrangement, as occurs during immune differentiation, could have general significance for embryonic development. It is conceivable that the progressive determination seen to occur during development rests on a mechanism of genetic recombination. Although no evidence has been found to support such a concept, little evidence exists to negate it. Of all the experiments performed to date, it is the nuclear-transplantation experiments that are the most difficult to reconcile with a mechanism of widespread genomic rearrangement, since it is these studies that suggest that changes occurring during development are reversible.

3. The genetic basis of antibody diversity

Another important question in the study of the immune system concerns the number of antibodies with different combining sites that an individual can make. Although there is no current answer to this question, it is generally estimated that a person might be able to synthesize over 1,000 different light or heavy chains. If we assume that any H chain can combine with any L chain, we could conclude that a person might be able to synthesize in the neighborhood of 10^6 to 10^7 different species of immunoglobulin. Are the large number of genes required for immune function present at the beginning of development in the DNA of the sperm and the egg, or does this great diversity arise during the ontogeny of each animal? Since the amino acid sequences of the V regions of both the L and H chains are similar to each other, it is believed that the genes are related and that DNA sequences which code for them arose as a result of duplication, mutation, and recombination. The question centers on the events surrounding the generation of the large number of variable sequences. In the *germ-line theory*, the variable-gene diversity has arisen during evolution, causing the entire spectrum of sequences to be present in the DNA of the gametes. In the *somatic-recombination theory*, the diversity arises by some special process in the immune tissues during development and differentiation. There is considerable evidence that now suggests that both theories have some validity, i.e., that the answer lies somewhere between the two extremes. Genetic analysis of the mouse kappa region discussed in the previous section suggests that all of the DNA sequences needed to produce kappa chains are present within the germ line. Additional variability is achieved by having the V region split into two parts, since any one of the five different J segments can be joined with a given V gene, thereby generating several times as many V regions as might be expected from this amount of DNA. Moreover, it has been proposed that each J segment may have more than one site at which recombination can occur. If V genes can be linked with J segments at various sites within the J DNA sequence, a large amount of additional variability can be produced. In contrast to these findings, results from studies of other genetic regions (such as that coding for the lambda chains in the mouse) suggest that a relatively small number of variable genes are inherited through the gametes and that this small collection of genes is then amplified during development by some type of somatic process involving duplication and mutation to form a much larger number in the adult. It may be, for example, that the hypervariable stretches of the V portion of the H and L chains arise by somatic generation, while the differences in the remaining parts of the variable regions (forming subgroups) reflect the actual diversity in the germ-line DNA.

4. The clonal-selection theory

Regardless of the means by which cells of the immune tissue acquire a great number of different genes that code for immunoglobulins, some mechanisms must exist whereby appropriate antibodies are produced in response to specific antigens. The theory presented in detail by F. M. Burnet in 1957 to explain the general basis of antibody production, the *clonal-selection theory*, has gained virtually complete acceptance. The theory is based on several premises.

1. As the cells of the immune system differentiate, they become capable of producing only one species of antibody molecule, i.e., Igs with *particular* V regions in both their H and L chains.
2. The entire spectrum of possible antibody-producing cells is present within the lymphoid tissues *prior to stimulation by antigen* and, therefore, independent of the presence of foreign materials (Fig. 17.4). In other words, the preliminary step in the differentiation of specific lymphocytes, i.e., that step in which they become specified to produce only one type of antibody molecule, occurs in the absence of a potential antigen for that antibody.
3. The ability of an antigen to elicit the production of a complementary antibody molecule results from the recognition, i.e., the *selection*, by the antigen of the appropriate antibody-producing cell. Those cells selected by antigen respond by proliferation (Fig. 17.4) to produce a clone of cells capable of producing the suitable antibody. The basis for the selection process resides in the cell membranes of the lymphocytes. Receptor molecules (discussed later) specific for a particular antigenic determinant are embedded in the membrane and are capable of interaction with antigen.

Figure 17.4 A model of the clonal-selection theory. In the first phase of differentiation, a large variety of lymphocytes are generated, each capable of synthesizing a single antibody type (indicated by the particular number). The diversity of DNA sequences could arise by a somatic process (which this diagram was initially intended to illustrate) or by an evolutionary one. The second phase of differentiation involves the selection of specific lymphocytes (B cells in this case) which are induced to proliferate and form antibody-secreting clones. [*From G. M. Edelman,* Cold Spring Harbor Symp. Quant. Biol., **41:**892 (1976).]

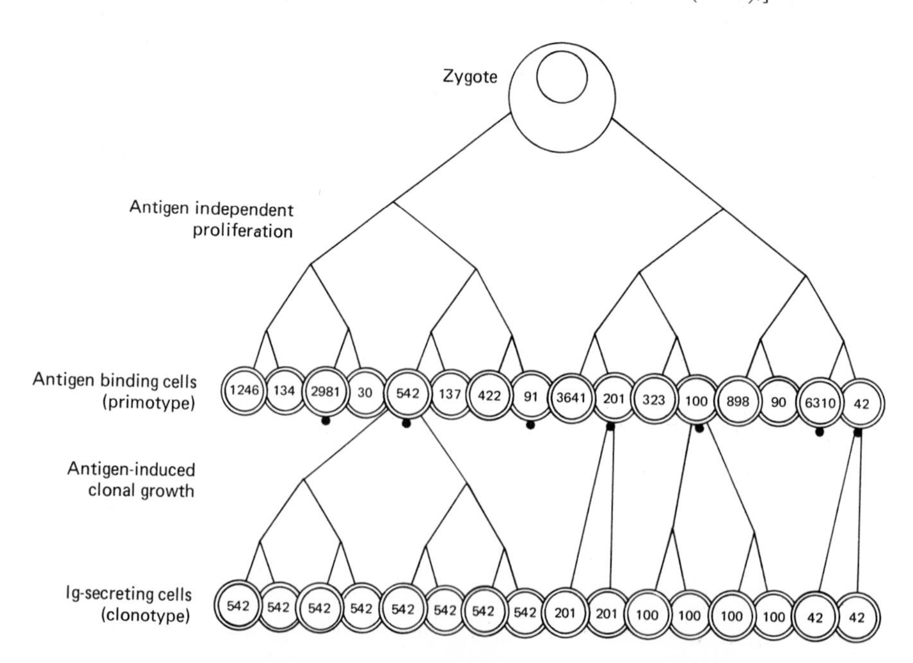

5. Differentiation of lymphoid tissues in mammals

The development of the immune system begins with the appearance of a group of large hematopoietic (blood-forming) stem cells in the blood islands of the yolk sac. The potential for migration and differentiation of these cells was discussed in Section 15.3. The original concept of stem-cell migration via the bloodstream came from the analysis of twin cattle in which blood cells from both individuals populated the tissues of each twin. The stem cells of the yolk sac are capable of differentiating into a wide variety of different types of cells of both the myeloid (red blood cells, polymorphonuclear leukocytes, monocytes, megakaryocytes) and the lymphoid (lymphocytes) line. The direction of differentiation taken by a given stem cell depends on the influences it receives after it reaches its specific destination. Those which migrate into lymphoid tissues become lymphocytes. Cells destined to form lymphocytes are first detected in the embryonic thymus, which forms from the third and fourth branchial pouches and plays a critical role in the development of the entire avian and mammalian immune systems. In the mouse, if the thymus rudiment is removed at the tenth day of gestation and is cultured in vitro, no evidence is found of lymphoid cells in the culture. If, however, the rudiment is explanted at day 12, lymphocytes are produced in large numbers, suggesting that the thymus has been "seeded" by migrating stem cells between the tenth and twelfth days of development (days 7 to 8 in the chick). These large basophilic (readily stained by basic dyes) cells can be seen in sections of the thymus rudiment as they penetrate through the surrounding mesenchyme heading toward the epithelium. Once these stem cells are present, the rudiment is capable of self-differentiation in vitro.

Once in the thymus, the large lymphoid stem cells proliferate in the cortex of the thymus to produce a population of smaller lymphocytes (present by day 16 in mice), called *thymocytes*, which now bear distinctive markers on their surfaces. Two of the surface markers are called *Thy 1* (formerly *theta*) and *TL* (*thymus leukemia*) *antigens*. The TL antigen disappears during maturation. The term *antigen* is used because the markers are detected by injecting these cells into another animal, which then produces antibodies against these particular surface components. The final steps in the development of thymus cells involve the acquisition of the ability to interact specifically with antigen. Cells from the thymus that have become mature and are now capable of synthesizing antibody molecules with only one antigen-combining site are said to be *immunocompetent* and are called *T lymphocytes*, or simply *T cells*.

The main role of the thymus is to receive and send out lymphocytes (Fig. 17.5). The greatest flow of cells through the thymus occurs during the period of early development of the immune system, but the flow continues at a diminishing rate into adult life. In the initial stages of thymus development, the cells are derived from the yolk sac. Then they are derived from the liver; and after birth, cells from the bone marrow are found to enter the thymus. Presumably all these cells can be converted into mature T cells having characteristic surface structures and functions. From their maturation in the thymus, T cells migrate through the lymphatic and blood vessels into a variety of lymphoid tissues, including the spleen and lymph nodes. If the thymus of a newborn mouse is removed (a *thymectomy*), T cells, as detected by their surface antigens, are absent in the circulation, and the spleen and lymph nodes have a severely depleted lymphocyte population. In the mouse, the first week after birth is a time of active seeding of peripheral lymphoid tissues by T cells. Removal of the thymus in the newborn mouse precedes the appearance of a significant number of circulating lymphocytes. In humans, congenital defects in the thymus are found on occasion, and the symptoms are similar to those of thymectomy in the mouse.

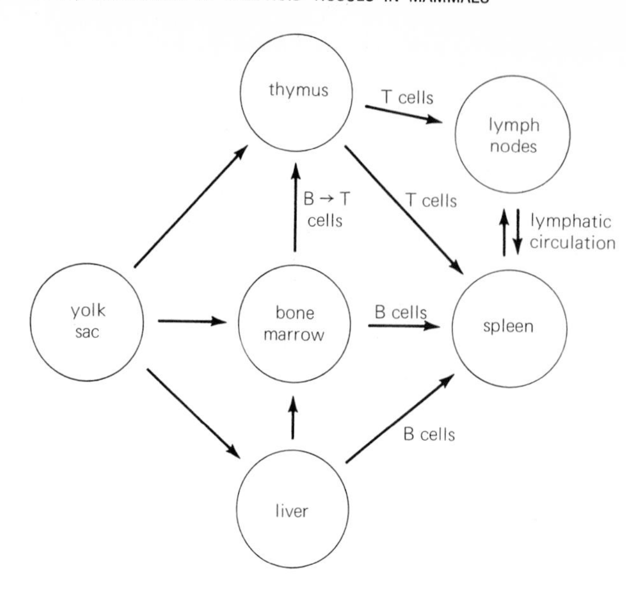

Figure 17.5 Cell-migration pattern during the development of the immune system. [*After R. Auerbach, "Concepts in Development," Sinauer, 1974.*]

Under normal conditions, it is difficult to determine the extent to which cells enter and leave a particular organ, such as the thymus. To measure thymus-cell turnover, the thymus can be removed and in its place can be grafted a thymus whose cells have been labeled with [^{3}H]thymidine. Under these conditions, there is a rapid loss of labeled cells from the thymus and an appearance of radioactivity in lymph-node and spleen tissue. In a matter of days there is essentially a completely new population of host cells in the thymus, and the old population has been replaced. In addition to the seeding of lymphocytes, the thymus is responsible for the secretion of a hormone, *thymosin*, which appears to act in some manner on the differentiation of the thymus-dependent lymphoid tissues. If, for example, a thymus graft into a thymectomized mouse is kept in a container that allows macromolecules out but holds cells back, there is considerably greater development of the peripheral lymphoid tissues, even though none of the implanted cells can leave the graft.

Thymus-derived cells (T cells) are only one of two main types of lymphocytes; the others are thymus-independent and are called *B lymphocytes*, or *B cells*. In birds, the differentiation of B lymphocytes occurs within the *bursa of Fabricius*, an organ that develops from a dorsal evagination of the wall of the gut in the region of the cloaca. Multipotent blood-borne stem cells that originate in the yolk sac enter the developing bursa of the chick at about the fourteenth day of incubation and are soon converted into mature lymphocytes, which, like the T lymphocytes, can be detected by the presence of cell-surface molecules. Unlike the T cells, B cells contain immunoglobulin molecules in their plasma membranes that can be readily identified. The differentiation of the B cell is accompanied by the synthesis of IgM and its rapid incorporation into the plasma membrane. With time, additional classes of immunoglobulin are synthesized (IgD, IgG, and IgA), and each becomes included within the cell surface. The site of initial differentiation of B lymphocytes within the mammalian fetus, i.e., the mammalian analogue of the avian bursa, has not been identified. Of all the lymphoid tissues of the body, the fetal liver appears to be the most likely site at which the original multipotent stem cells are caused to differentiate into immunocompetent B cells.

As in the case of T lymphocytes, the differentiation of B lymphocytes occurs in the absence of antigen. The specification of the lymphoid precursor cells is believed to involve the translocation of a given V-gene sequence, as described in a previous section. Once this specification event occurs, all immunoglobulins produced by this clonal cell line, regardless of Ig class, will have the same combining-site specificity. In the early stages of immune function, the immunoglobulins exist primarily within the membrane, where they serve as potential receptors for antigenic stimulation. As predicted by the clonal-selection theory, the interaction of receptor and antigen triggers the proliferation of antibody-producing cells and the secretion of antibody with the same specificity (and thus guaranteed combining ability) as the interacting receptor. This is described in greater detail in Section 17.6.

6. Role of T and B lymphocytes

Before the effects of thymectomy and the subsequent deficiency in T cells can be understood, it is necessary to briefly examine the types of immune response that an animal can mount. There are two broad categories of immunity: cell-bound and humoral. In *cell-bound immunity,* the antigen-combining sites are carried at the surface of a lymphocyte and the destruction of the antigen involves the direct participation of the lymphocyte. *Humoral immunity* refers to the secretion of immunoglobulin molecules into the blood as soluble antibody. These two types of immune response can be dissociated to a large extent because they are mediated by the two different classes of lymphocytes just discussed. In humans, for instance, there is a disease (congenital agammaglobulinemia) in which humoral antibody is deficient and cell-bound immunity is normal. In contrast, congenital thymus deficiencies have greatly impaired cell-bound immunity with relatively high serum antibody levels.

A. Cell-bound immunity

Cell-bound immunity is completely within the province of T cells and is implicated in several functions. The best-studied immune response utilizing T cells is graft rejection. Many of the proteins and glycoproteins located on all cell surfaces are highly specific markers to which the immune system can respond. One group of macromolecules, the histocompatibility antigens, is particularly important in the recognition of foreign cells. Histocompatibility antigens are coded for by a number of different genes for which many different alleles of each are present in the population. The probability that two individuals will have the same alleles for all the histocompatibility genes is very unlikely, even among nonidentical siblings. Graft rejection is therefore a virtual certainty after tissue transplantation in all but identical twins and highly inbred strains of laboratory animals. Differences between individuals at certain of the histocompatibility loci result in very rapid rejection, while others are considered weak antigens and rejection is a more prolonged process.

The basis of graft rejection is an attack by T cells that have proliferated in response to the presence of foreign tissue containing antigenic determinants. In mice that have been deprived of thymus tissue from birth, graft rejection does not occur and such grafts will remain in place. One special case of tissue rejection by immunocompetent T cells is the graft-versus-host reaction, which has been widely studied. If spleen cells are injected into an animal that has been irradiated or thymectomized and is thus unable to reject foreign cells from another individual, a gradual process of host rejection by the descendants of the injected cells will occur. The spleen cells will seed the lymphoid tissues of the host; being im-

munocompetent themselves, they will be stimulated by the presence of the host tissue, which is recognized as foreign. The tissues of the host will gradually be destroyed. The graft-versus-host reaction has been observed in humans under certain clinical conditions. In one report, an infant was born with a severe thymus deficiency that had prevented the development of his immune system. In an attempt to provide him with a population of functioning lymphocytes, spleen cells from a very young sister were injected into him. These would be expected to seed his lymph nodes, spleen, and other tissues and to provide him with a basis to attack foreign molecules. In this case, however, the injected cells had already been specified, and they contained cells capable of reacting against their new host. In the ensuing months, host tissues were destroyed by cell-bound lymphocytes that found the tissues foreign. The infant died from what is termed "wasting disease."

Although the study of graft rejection is an important experimental system and has significant consequences for the future of clinical transplantation, it does not help explain the role of T cells or histocompatibility antigens in normal body function. It is generally found that the targets of T lymphocytes are cells themselves. Foreign graft cells are one type of target cell, but there are altered cells that appear within the body under nonexperimental conditions that also can serve as targets of T cells. For example, numerous types of pathogens, both viruses and bacteria, are capable of infecting host cells. In these cases, T lymphocytes are capable of recognizing the infected cell by the presence of the foreign antigen and destroying it. Further analysis of the interaction between sensitized T cells, i.e., T cells that have responded to the presence of the pathogenic antigens, and their target cells has revealed an important role of the histocompatibility antigens. The histocompatibility antigens serve as recognition factors for the sensitized T cell. In order for the T cell to destroy a cell infected with a particular sensitizing virus, that infected cell must share identical histocompatibility antigens. The T cell is said to be *restricted* by its histocompatibility antigens. It appears that the T cell must recognize two distinct types of molecules on the surface of the cell with which it interacts, one the sensitizing antigen and the other the cell-surface histocompatibility protein it must match.

The basis for the destruction of grafts or infected cells is an attack by T lymphocytes that have proliferated in response to the presence of foreign antigens. In mice that have been deprived of thymus tissue from birth, graft rejection does not occur and such grafts will remain in place. The ability of T lymphocytes to destroy cells containing foreign surface molecules has led to the proposal that these cells play an important role in the detection and destruction of potential tumor-forming cells. The conversion of a normal cell to the malignant state is accompanied by changes in the nature of the cell surface (discussed in Chapter 18), changes that should make the cell vulnerable to attack by immunocompetent T lymphocytes. Although the topic remains controversial, T lymphocytes are believed by many investigators to carry out a process termed *immunological surveillance,* whereby the surfaces of cells are continually examined for the appearance of antigenic molecules. Cells bearing inappropriate antigens would be destroyed. In recent years, evidence has mounted against the direct participation of T cells in tumor-cell destruction. It is now generally accepted that mice with a severely deficient T-cell response do not develop an inordinately high incidence of spontaneous tumors, as would be expected by the hypothesis. Attention has shifted in the past few years to a newly discovered type of lymphocyte termed a *natural-killer* (*NK*) *cell.* These cells are present in spleen tissue and have the capability of lysing a variety of cells, most notably those from malignant tissue. Although these cells resemble conventional lymphocytes, they are clearly distinct

in their function, since their ability to lyse target cells is independent of prior sensitization and does not appear to involve antibody molecules. The basis for the recognition and destruction of the tumor cell by these nonspecific natural killer cells is presently obscure, as is their role, if any, in the control of tumor formation.

The mechanism whereby T cells are able to destroy foreign cells also remains obscure. Presumably, it relates to the ability of T cells to produce a variety of highly active substances, one of which has a cytotoxic action on other cells and could lead to their destruction. These substances, termed *lymphokines*, inhibit cell migration, chemotactically attract other cells, cause other cells to divide, and have lytic action. It is obvious that many of the important steps in lymphocyte activity remain to be worked out.

7. Tolerance

The ability of an individual to secrete specific antibodies into the bloodstream is one of the bases of our defense against invading pathogenic organisms, whether as a response to an infection itself or an immunization. The cells responsible for the production of circulating immunoglobulins are the plasma cells, descendants of the B cells. The plasma cells are the end products of the B-cell line, cells that have differentiated (in response to the presence of antigen) in order to secrete large quantities of antibody. Although the precise nature of the steps is not known, the combination of antigen with Ig receptors on the surface of the B lymphocyte results in the proliferation of the lymphocyte to form a clone of cells (called *blast* cells), all of which are specified to make the same antibody. Some of these blast cells will then differentiate into plasma cells and begin to secrete antibody, while others will remain in the lymphoid tissues as "memory" cells to respond rapidly at some later date if that antigen becomes reintroduced. It is the memory aspect of the immune response to which booster immunizations are geared; the reintroduction of an antigen can cause a much more rapid production of antibody than occurred after the initial injection.

Although the appearance of immunocompetent B cells occurs independently of the thymus, it was found that thymectomy greatly reduced the ability of an animal to produce circulating antibodies against a variety of antigens. Similarly, if B cells alone are injected into a mouse whose immune system has been destroyed by irradiation, the injected B cells are not able to react strongly to many antigens. If, however, T cells are injected along with B cells, a full response by the B cells can result. Even though it is the B cell that is responsible for antibody production, the T cell is required in some "helper" role. T cells that have this function are termed *helper T cells* and represent a distinct subpopulation having unique cell-surface properties. As with the interaction between cytotoxic T cells and virus-infected target cells discussed earlier, the interaction between helper T cells and B cells is also restricted by histocompatibility antigens and mediated by surface receptors. Before this intercellular response can occur, there must be a match between the antigen-combining specificity and the histocompatibility antigens on the B and T cells involved. Not all antigens require the intervention of a T cell prior to evoking a B-cell response; some antigens are said to be thymus-independent and able to directly and independently stimulate the B lymphocyte toward differentiation as an antibody-forming cell.

In addition to helper T cells and cytotoxic T cells, there is another subpopulation (also demarcated by its cell-surface components) that functions to inhibit the response by the B lymphocyte, thereby blocking the production of soluble antibody. In this case, the T lymphocyte is termed a *suppressor T cell*. In these various interactions with other cells, there is evidence for the actual contact between cells, as well as evidence for the secretion of active substances by the T lympho-

gators believe that both these properties of malignant cells, i.e., their loss of growth control and their detachment from other cells of the tumor, reflect certain basic changes in the cell membrane. We will return to this point later, but first we will consider the types of stimuli that can cause a cell to become malignant.

A. Carcinogenic stimuli

Various types of agents—chemical, physical, and biological—have been shown to be capable of converting a normal cell in culture into a malignant one, a process termed *transformation*. Since the same agents can be used to induce tumors in the organism, it would appear that the study of the transformation of cells in vitro by these various agents has relevance for the development of tumors in the body. Similarly, in most cases, cells transformed in culture will cause tumors when injected into an appropriate host. Usually, the test recipient is a *nude* mouse (one carrying a genetic defect that prevents the development of a thymus gland), which does not reject the injected cells. The problem in the study of "spontaneous" human malignancy is the unknown nature of the carcinogenic stimulus. It is generally believed that various chemicals present in our complex environment are responsible for a large percentage of human cancers.

The tremendous diversity in molecular structure of the chemicals known to be carcinogenic has led to a search for some unifying principle. The common denominator that has emerged in recent years is the finding that in order to be carcinogenic, a chemical must be either directly or indirectly *mutagenic*. Direct mutagenesis is easy to understand; compounds such as alkylating agents or acridine result in alterations of the nucleotide sequence of the DNA. However, it was found in the early studies that many of the most potent carcinogens, such as the polycyclic aromatic hydrocarbons, were not capable of inducing mutations in susceptible bacterial cells. After considerable study, it was found that these chemicals become carcinogenic only upon activation within the body. The endoplasmic reticulum of the liver contains various enzymes that serve to detoxify drugs and other compounds. Ironically, in the process of modifying chemicals, various cellular oxidases actually convert them into carcinogenic species. The actual carcinogens are believed to be electrophilic epoxide stages that interact covalently with DNA. It is difficult to consider any target in the cell other than DNA that would allow the effect of chemical exposure to cause cancer to appear many years later, or allow the malignant phenotype to be retained through each successive cell cycle during the development of the tumor. One of the advances in the area of chemical carcinogenesis in recent years has been the development of an in vitro system to screen large numbers of compounds for potential carcinogenicity. The assay involves subjecting sensitive bacterial cells to chemicals in the presence of rat-liver microsomal enzymes. If a given chemical proves to be mutagenic under these circumstances, it has a very high likelihood of being carcinogenic in further tests.

Other agents capable of converting a normal cell to a malignant one include radiation and infection by tumor viruses. Radiation presumably exerts its effect via direct mutagenesis. The role of tumor viruses in the etiology of cancer is a very controversial one. As will be discussed later in some detail, a number of different types of viruses are know that can transform cells.

What is the relationship among these various cancer-causing agents? Does a cell transformed by different chemicals have the same defect, and is this defect the same as the one induced by radiation or tumor viruses? Or rather, are there various genetic sites that can be altered in different malignant cells, all of which will produce a cell with a similar loss of growth control and the ability to form

tumors? These are very important questions which cannot be answered at present with certainty. One possibility that has been raised is that all forms of cancer involve the presence of a tumor virus and that the role of the chemical or physical carcinogen is simply to activate the viral agent.

B. The phenotype of the transformed cell

It is generally assumed that in order to understand the cancer cell, its properties relative to the normal cell must be determined. Toward this end, a large number of differences, both structural and biochemical, have been catalogued. In fact, it is the diversity of the observations that poses one of the major problems. Which of the changes that occur when a normal cell becomes malignant is a primary change and which is a secondary alteration? Similarly, are any of the differences that have been detected directly responsible for the altered behavior of the cells themselves? One point should be kept in mind in the following discussion: nearly all the observations are made on cells transformed by oncogenic (tumor-causing) viruses, and there are questions as to the general nature of the properties of virally transformed cells. Viruses are used in these studies for the ease with which they transform a wide variety of cells growing in culture, for the predictable nature of the events that follow addition of the virus at well-defined times, and for the availability of mutants, particularly temperature-sensitive ones. By the use of temperature-sensitive viruses, transformation and reversion can be studied by simple temperature shifts in one direction or the other. In addition, transformation by viruses occurs in response to the integration of a small, well-studied genetic element. In at least a few cases, it has been shown that the transformation process, with all of its attendant changes, is the result of the activity of a *single* viral gene. Scattered evidence from spontaneous or chemically transformed cells has corroborated much of the observations on the altered properties of virally transformed cells. Another very important point to consider is the extreme variation that exists with respect to the properties of one type of malignant cell or another. Although one would hope that all malignant cells might share some basic property that reflected their altered growth control, no such universal property has yet been discovered. At best, there appear to be tendencies that are revealed by a greater or lesser variety of tumor cells.

Within the cell nucleus, the most striking and widespread alterations occur to the chromosomes. Normal cells are generally fastidious in maintaining their normal diploid chromosomal complement as they grow and divide, whether in vivo or in vitro. In contrast, transformed cells often have highly aberrent chromosome complements, a condition termed *aneuploidy*. These very abnormal karyotypes are presumably a *result* of the type of growth of the transformed cells rather than a cause of it. Regardless, it is evident that transformed cell growth is much less dependent on a normal chromosome content than is normal cell growth. In some cases, rather characteristic chromosome changes are seen in malignant cells, as evidenced by the presence of an extra band in one of the chromosomes of most cells from a Burkitt's lymphoma. The most widespread alterations involving the cytoplasm of the transformed cell concern its enzymatic properties, particularly those related to glycolysis. Malignant cells generally have a much greater level of glycolytic activity (regardless of the oxygen levels) than do normal cells. The importance of this observation is not understood. The most striking morphological changes that occur in the cytoplasm following transformation center around the cytoskeletal elements. Whereas the normal cell generally contains a well-organized and oriented network of these microfilaments and microtubules, the transformed cell tends to have a reduced and/or disorganized cytoskeleton. This point

will be brought up again later with regard to the possible role of this change in the loss of growth control. One of the consequences of the alterations in the cytoskeleton is a change in the overall morphology of the transformed cell and the colonies it forms (Fig. 18.1). The most extensive alterations that have been found to occur upon transformation concern the cell surface, an observation not altogether surprising considering the role of the cell surface in growth control and cell-cell interaction.

At the biochemical level, the changes at the cell surface are marked by the appearance (or increase) and disappearance (or decrease) of particular components. Although there is little evidence of widespread changes in the phospholipid nature of the bilayer, there are reports of various changes in the glycolipids, the glycoproteins of the membrane, and the glycosaminoglycans of the extracellular coat. Some of the surface changes may result from the increased activity of proteolytic enzymes in the medium. One of the widespread reports on a variety of transformed cells is the disappearance (or marked reduction) from the cell surface of the extracellular protein, fibronectin. In addition to the loss of this protein, transformed cells generally possess new cell-surface proteins, proteins that are generally referred to as antigens because they can induce the formation of antibodies directed against the cell. In some cases these antigens are coded by the transforming virus, while in other cases they appear to be host-cell derived. In some cases, proteins typical of embryonic cells appear on the surface of transformed adult cells.

In addition to the biochemical changes, numerous physiological or behavioral differences involving the cell surface have been reported. One of the most characteristic features of transformed cells in culture is their lack of contact inhibition of movement. As pointed out in Chapter 3, the membrane of a normal cell, when contacted by another cell, ceases its activity and movement in the original direction. When normal cells become surrounded on all sides by other cells, their

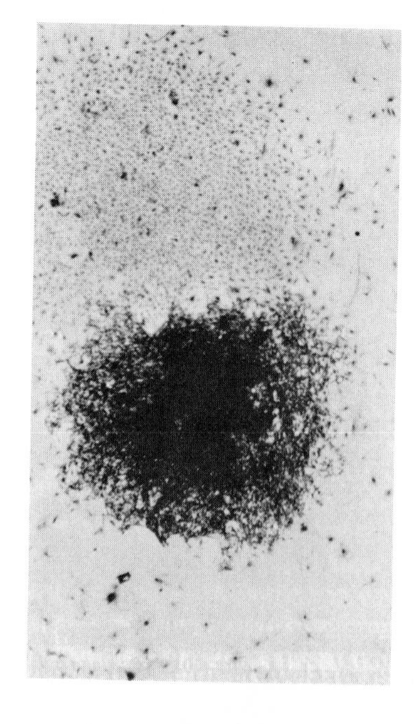

gure 18.1 A compari- ı of the morphologies f two colonies, one of polyoma-transformed hamster embryo cells tom) and one of nor- hamster embryo cells (top), at 7 days after ting. [*Courtesy of L. Sachs.*]

motility ceases; they are not capable of moving over one another and they form a monolayer on the bottom of the dish. Another characteristic that generally accompanies contact inhibition of movement is density-dependent inhibition of growth. When normal cells reach a particular density in culture, their growth potential drops to a low value, and the number of cells on the dish remains relatively constant (Fig. 18.2). In contrast, the movement of a transformed cell is generally not obstructed or diverted when contacted by another cell, nor is its division potential reduced nearly as much as the cell density increases. In addition, malignant cells are much less dependent on serum in the medium. As a result of these properties, cultures of malignant cells form clumps rather than monolayers and are capable of proliferating to many times the cell density of normal cultures (Fig. 18.3). There are other important differences in the growth characteristics between normal and transformed cells. Whereas normal cells seem to have a limited capacity for cell division, cancer cells are seemingly immortal in the sense that they continue to divide indefinitely. In addition, cancer cells, unlike normal cells, can generally grow in a state of suspension in a medium made of soft agar or methyl cellulose. In this regard, cancer cells are said to have lost their anchorage dependence on which the growth of normal cells depends. Other widespread properties of transformed cells relative to the normal counterparts are an increased agglutinibility by lectins, increased mobility of cell-surface receptors, increased transport of various compounds, decreased adhesiveness to the substrate, and lack of gap junctions between cells (Fig. 18.4).

When a population of cells is transformed in culture and their progeny grown over successive generations, the properties of the cells remain highly uniform. In

Figure 18.2 Normal cells growing (*a*) at low density making contacts that cause them to adhere to one another and (*b*) at high density where they have formed a mosaiclike arrangement in a monolayer and are contact-inhibited. [*Courtesy of R. Dulbecco.*]

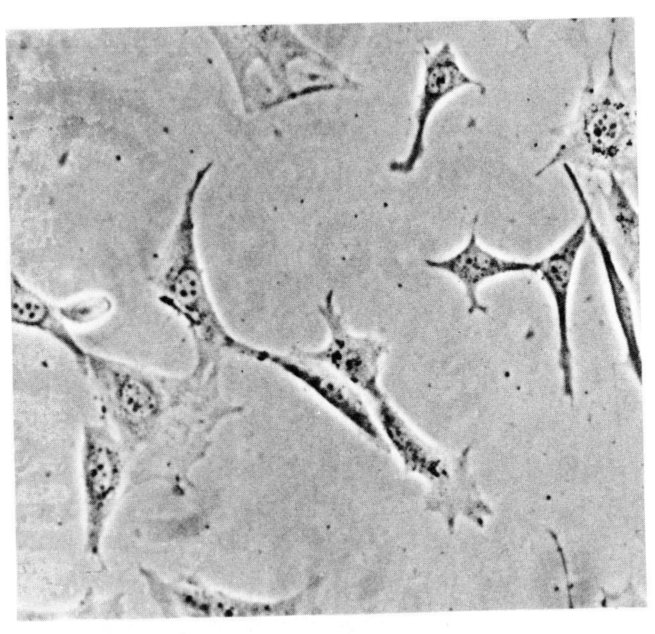

a

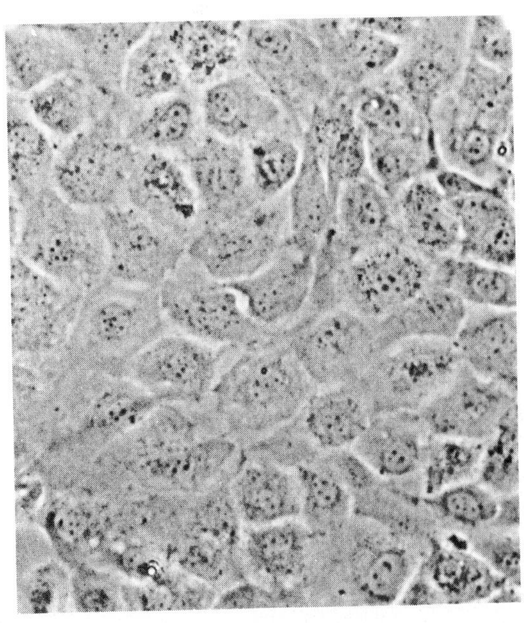

b

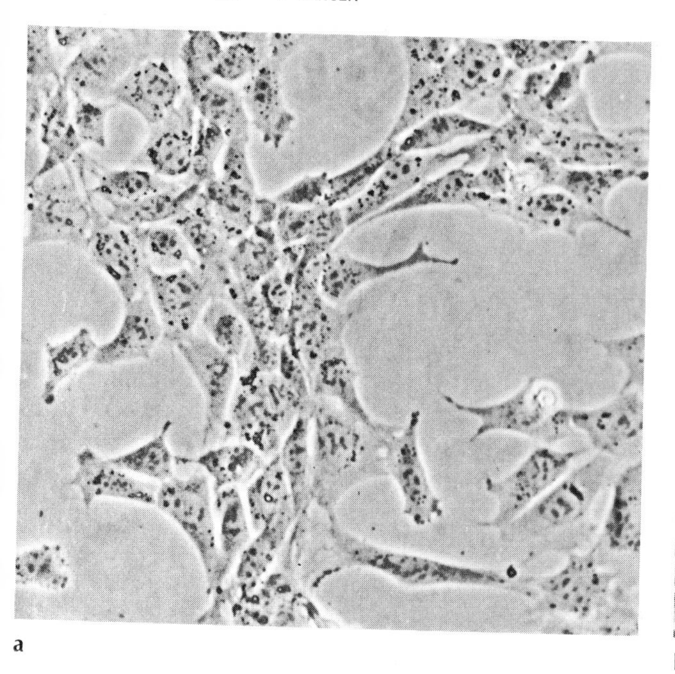

a

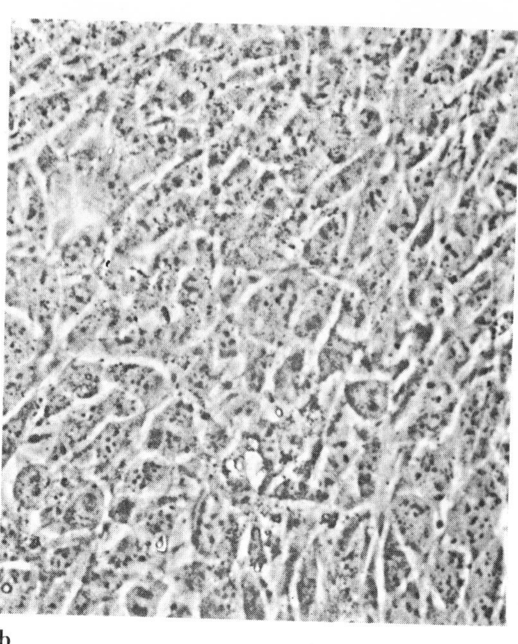

b

Figure 18.3 Cells transformed by viruses generally overlap each other (*a*) and form multilayered clumps and irregular patterns (*b*). [*Courtesy of R. Dulbecco.*]

contrast, when cells proliferate within a malignant tumor, their properties often undergo significant change, a phenomenon termed *tumor progression.* Unlike the normal cell, the genetic composition of the malignant cell appears to possess an inherent instability, one that causes it to undergo spontaneous alteration. As a result, the phenotype of tumor cells—their surface properties, chromosome com-

Figure 18.4 Lack of coupling between a cancer cell and the adjoining normal cells. The middle cancer cell was manipulated into the gap in the line of normal cells. *a* The seven cells in a line; the middle cell is malignant. Microelectrodes for measuring electrical coupling are seen penetrating the cells. *b* Tracing of the cells; the cancer cell is black. *c* Dark-field photograph after injection of a fluorescent dye into the last of the normal cells in the line. The fluorescence rapidly penetrates the two adjoining normal cells but does not move into the cancer cell because of lack of coupling. If a normal cell had been manipulated into this line of cells, the dye would have passed through it. [*Courtesy of R. Azarnia and W. Loewenstein.*]

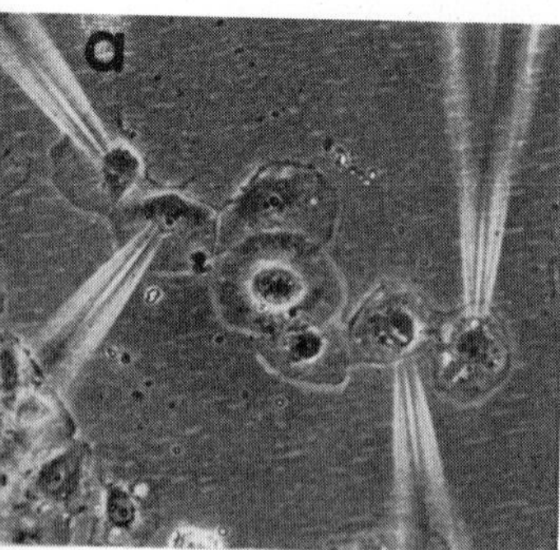

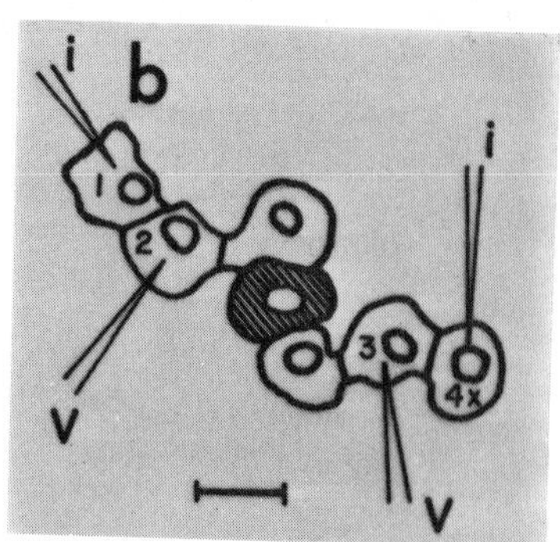

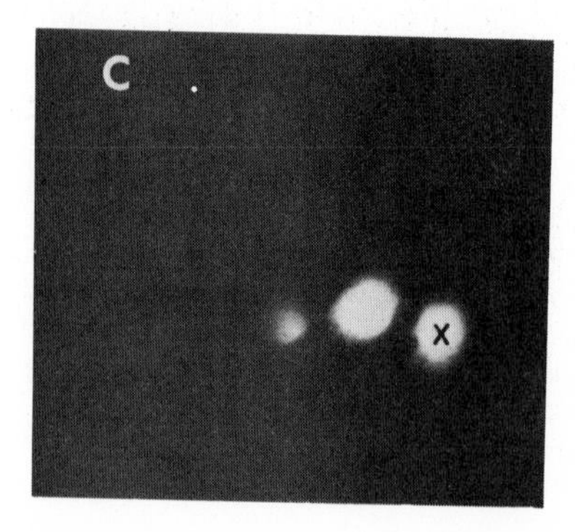

position, drug resistance, hormone dependence, potential for metastasis, etc.—becomes highly variable. Those variant cells which have an increased potential for uncontrolled growth possess a selective advantage and tend to overgrow the remaining cells. For example, cells that appear in the tumor whose surfaces are less immunogenic would be expected to be favored since they would be less susceptible to attack by the body's defenses. As a result of progression, the tumor becomes even more autonomous and more malignant.

C. The underlying basis

In previous sections we have mentioned an extensive, though by no means exhaustive, number of alterations (Fig. 18.5) that occur upon transformation. The most important question concerns the relationship between these properties and the underlying causes of tumor formation. The changes listed earlier are simply correlations without any mention of cause and effect. A particular property of transformed cells is discovered, often by chance, and various investigators begin to study it without knowing the underlying genetic basis for the change. In a sense we are dealing with several levels at which the malignant phenotype is expressed. At one level, we speak of the tumor cell's loss of growth control, or its invasiveness, or its ability to be attacked by the immune system, or the ability of tumors to metastasize. At another level, we speak of the behavior of transformed cells in culture, for example, their immortality, their anchorage independence, or their loss of contact inhibition. At the biochemical level, specific alterations in cytoplasmic elements or membrane proteins have been reported. Although these observations are generally thought to be related, the greatest challenge in this field is attempting to assign causes and effects among these various properties.

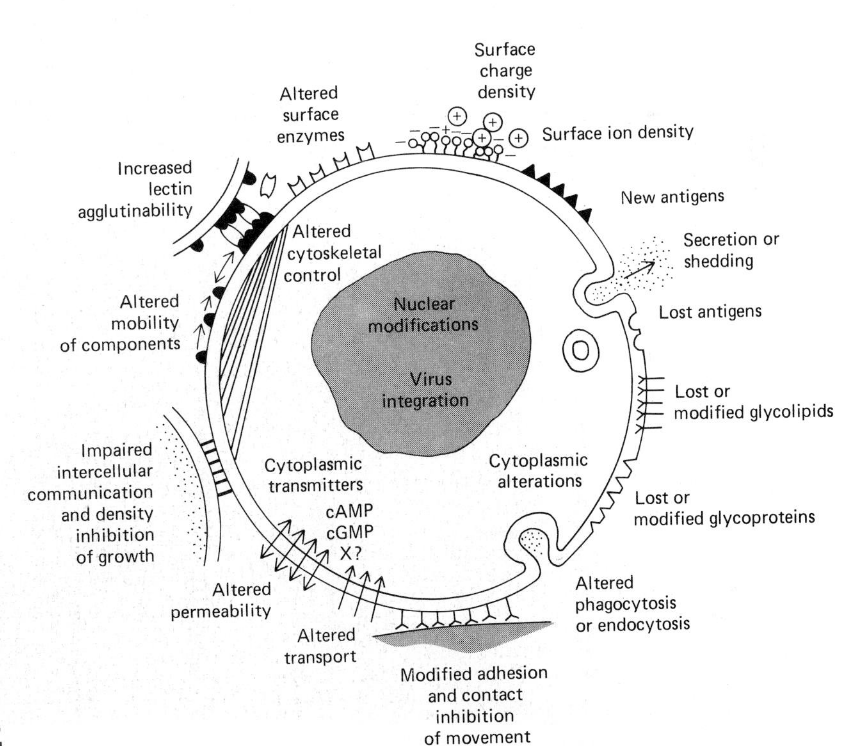

Figure 18.5 Diagrammatic illustration of various cell-surface alterations found after neoplastic transformation. [*From G. L. Nicolson,* Biochim. Biophys. Acta, **458:***16* (*1976*).]

Even though we may not yet understand the basis for all the alterations that occur after transformation, it is still important to search for the basis of the loss of growth control in the malignant cell. It was suggested in Chapter 3 that the cell surface was an important mediator of the growth potential of the cell. Two general hypotheses have been presented in an attempt to explain how events at the edge of the cell might dictate policies of growth and division within the cell itself. In one theory it is suggested that the levels of the cyclic nucleotides cAMP and cGMP are of the utmost importance in determining the cell's growth potential, while in the other theory the mobility of cell-surface receptors is of primary importance. In this latter theory, the information from the cell surface is carried inward by a cytoplasmic network of cytoskeletal and/or contractile proteins. Since cyclic nucleotide levels may be an important determinant of the state of the cytoskeletal machinery, the two theories may be closely interrelated.

Evidence has been obtained which supports the concept that alterations in the levels of cyclic nucleotides and/or the cytoskeletal-contractile elements of the cell are involved in the transformation process. For example, cyclic AMP levels tend to be low in the transformed cell, as they are in the mitotic cell. Furthermore, the treatment of transformed cells with cyclic AMP causes them to temporarily develop a more normal phenotype: they become more adhesive to their substrate, they lose their increased agglutinibility, they regain their normal morphological appearance, and in some cases they regain their density-dependent inhibition of growth. The dramatic effect of cyclic AMP on cell behavior is shown in Fig. 18.6. Under certain conditions, *normal* Chinese hamster ovary cells do not exhibit contact inhibition, and they grow in multilayered clumps. Cells in these clumps are compact and randomly oriented (Fig. 18.6*a*). The addition of cyclic AMP to these cells converts them to elongated, fibroblastlike cells, which are contact inhibited and grow as a monolayer (Fig. 18.6*b*). This morphological change presumably reflects an underlying change in the distribution of cytoplasmic skeletal elements. With regard to the second theory, it has already been noted that transformed cells generally have an increased surface-receptor mobility and a disorganized cytoskeletal-contractile apparatus. These are conditions one would expect to lead to a loss of growth control, on the basis of this theory.

D. Oncogenic viruses

Viruses were first reported in tumors as early as 1908, when cell extracts were found to transmit leukemia through successive passages from fowl to fowl. In 1910, Peyton Rous began his pioneering work on the propagation of virus-induced sarcomas. Although the initial studies were met with great skepticism, the role of viruses in animal cancers was gradually established, and more recently, attention has turned to their probable involvement in some human tumors. Oncogenic viruses can be broadly divided into two heterogeneous groups based on the presence of DNA or RNA as their genetic material, i.e., the nucleic acid in the virus particle released from one cell and capable of infecting another. The transformation of a cell by an oncogenic virus is analogous in many ways to the temperate infections of bacteria by bacteriophage. The viral genome becomes hidden within the host-cell DNA, and the genomes undergo replication together as the virus is passed from each cell to its progeny at cell division. Whereas the typical lysogenic bacteriophage remains transcriptionally inactive (other than producing its own repressor), the oncogenic virus produces RNAs that become translated into proteins that profoundly disturb the host cell's metabolic machinery. The presence of the oncogenic virus is manifested as the transformation of a normal cell to the malignant state.

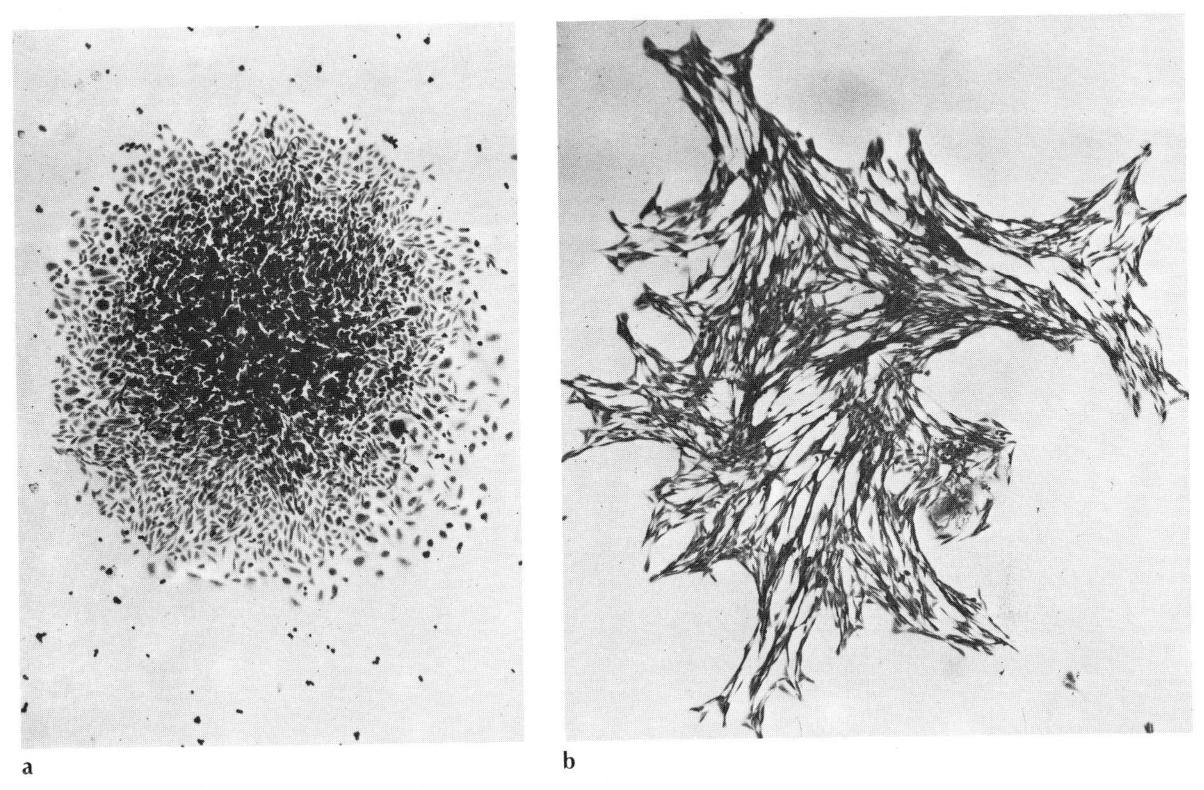

Figure 18.6 Effect of cyclic AMP on the morphology of a colony of Chinese hamster ovary cells. *a* Grown on standard medium where contact inhibition does not occur. *b* Cells grown in the presence of dibutyryl cAMP are now contact-inhibited. [*Courtesy of A. W. Hsie and T. T. Puck.*]

i. DNA tumor viruses. Most of the DNA tumor viruses that have been studied contain very small genomes and are not believed to be a natural factor in tumor formation. These viruses are highly infectious; cells growing in culture as well as cells growing within the body are susceptible to transformation by the addition of virus particles. The two best-studied viruses are polyoma and simian virus 40 (SV40). These two viruses, which are very similar in nature, are icosahedral types of approximately 450-A diameter, having a circular DNA molecule (3.5×10^6 daltons) containing as few as three genetic functions and associated with a few histone molecules derived from the host cell. Two types of host cell are studied in tissue culture in connection with each virus. In the case of the permissive host cell, the virus multiplies unchecked until the cell is killed. Certain types of monkey kidney cells are permissive for SV40, while various mouse cells are permissive for polyoma. In the other case, that of the nonpermissive cell, the viral DNA becomes integrated into the host DNA and does not undergo independent replication. Rather than producing a productive infection, as in the permissive cell, the virus converts the cell to the malignant state.

Of the three genes present in the SV40 genome, two code for viral capsid proteins, while the third, the A gene, codes for a protein termed the *T (tumor) antigen*, which accumulates in the cell nucleus (Fig. 18.7*a*) and appears to be responsible for the transformation process. In fact, the only gene expressed in the

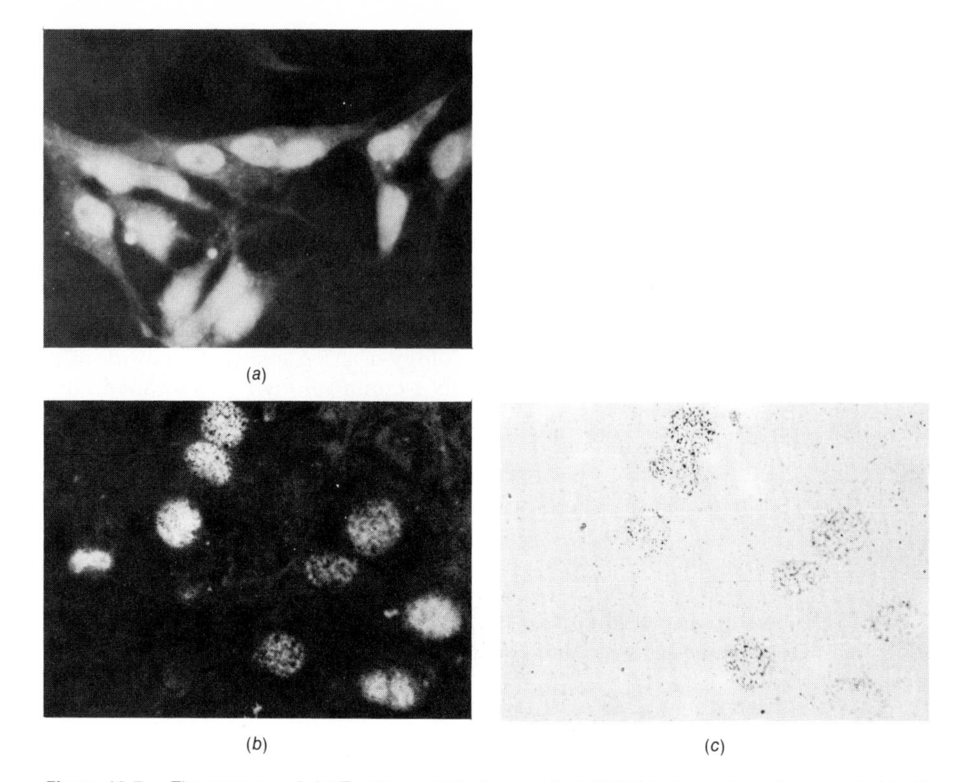

Figure 18.7 *a* The presence of the T antigen within the nuclei of SV40-infected cells as demonstrated by the binding of fluorescent antibody. [*Courtesy of B. Steinberg, R. Pollack, W. Topp, and M. Botchan.*] *b, c* Results of an experiment which demonstrates the importance of the viral T antigen in the activation of DNA synthesis. Mouse kidney cells in a nondividing, confluent culture were injected with very small volumes of purified T antigen. The cultured cells were then incubated in [^{3}H]thymidine for 18 to 20 hours before fixation. Those cells which incorporate [^{3}H]thymidine (*c*) are found to be the ones into which the T antigen had been injected, as demonstrated by immunofluorescence (*b*). [*Courtesy of R. J. Tjian, G. Fey, and A. Graessmann.*]

transformed cell is that which codes for the T antigen; its presence is both necessary and sufficient for the change. It would appear that this one protein is responsible for all of the behavioral changes catalogued in Section 18.1.B. Since it is unlikely that a single protein could directly interfere with so many functions, we can conclude that most of the phenotypic changes that occur during transformation are secondary responses to some more basic change(s), possibly in some aspect of the cell's regulatory machinery.

The primary response to infection by polyoma or SV40 is the initiation of DNA synthesis. The role of the T antigen in this response is clearly shown by the injection of purified T protein into the cytoplasm of *normal* nondividing cells. The protein rapidly accumulates within the nucleus of the cell and soon activates DNA synthesis, as revealed by its incorporation of [^{3}H]thymidine (Fig. 18.7*b* and *c*). Activation of DNA synthesis in a previously nondividing cell may involve the production of a variety of required enzymes (such as thymidine kinase) that precede the actual synthesis of DNA. If the cell is a permissive one, the activation of the *A* gene is followed by a later activation of the remainder of the genome and the subsequent production of capsid proteins. The synthesis of the components of the

virus is followed by their assembly and the lysis of the infected cells. In contrast, if the cell is a nonpermissive one, the viral DNA becomes integrated into the chromosomes of the host cell, only the *A* gene is expressed, and the cell becomes transformed.

Another type of oncogenic DNA virus that has been widely studied is the adenovirus. These agents, which cause influenzalike infections in humans, are capable of forming tumors in a variety of animals. Although the genome of the adenovirus is approximately seven times that of polyoma or SV40, it has been found that only a very small portion of that genome (approximately 2,400 nucleotides) is actually required for transformation itself. Apart form its ability to induce tumors, adenovirus has been most valuable as a research system in providing information on the molecular biology of DNA expression in mammalian cells.

The other major group of DNA-containing viruses implicated in tumor formation, including a number of human malignancies, consists of the herpes viruses. The first indication that a herpes-type virus was involved in cancer came in 1964. Tumor cells from patients with Burkitt's lymphoma, a rare form of cancer endemic to Africa, were found to contain a virus, since called the Epstein-Barr virus (EBV) after its discoverers. As with other herpes viruses, a large percentage of the population has had contact with them or even harbor them, evidenced in this case by the presence of serum antibodies against this virus in most healthy individuals. It would appear that some factor in addition to the presence of the virus is responsible for the development of the cancer.

ii. RNA tumor viruses. Although there are several different groups of RNA tumor viruses, all of them are basically similar and are believed to be evolutionarily related. RNA tumor viruses are constructed as shown in Fig. 18.8. They consist of an inner core in which the viral RNA resides, together with a few internal proteins, surrounded by an outer shell, which is in turn surrounded by an outer envelope. The outer envelope, which contains certain virally coded glycoproteins, is involved in the attachment of the particle to a new host cell. The envelope forms from the plasma membrane of the host cell by a budding process (Fig. 18.9) after that part of the membrane is occupied by viral proteins and essentially cleared of host proteins. Although it was initially believed that the RNA genome consisted of a single 70-S RNA molecule, it is now known that the core contains several smaller RNAs present together in a tightly held 70-S aggregate. The individual

Figure 18.8 Diagram of a cross section of a typical RNA tumor virus. The diameter of the virion is about 100 to 150 nm. The diameter of the ribonucleoprotein particle is about 50 nm. [*Reproduced with permission from H. M. Temin,* Ann. Rev. Gen., **8**; *copyright © 1974 by Annual Reviews Inc.*]

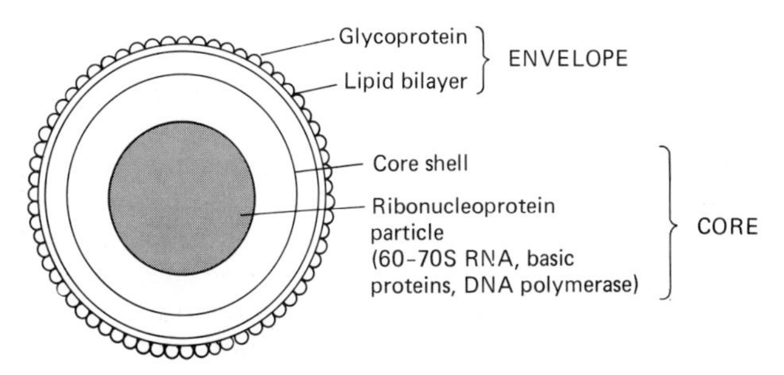

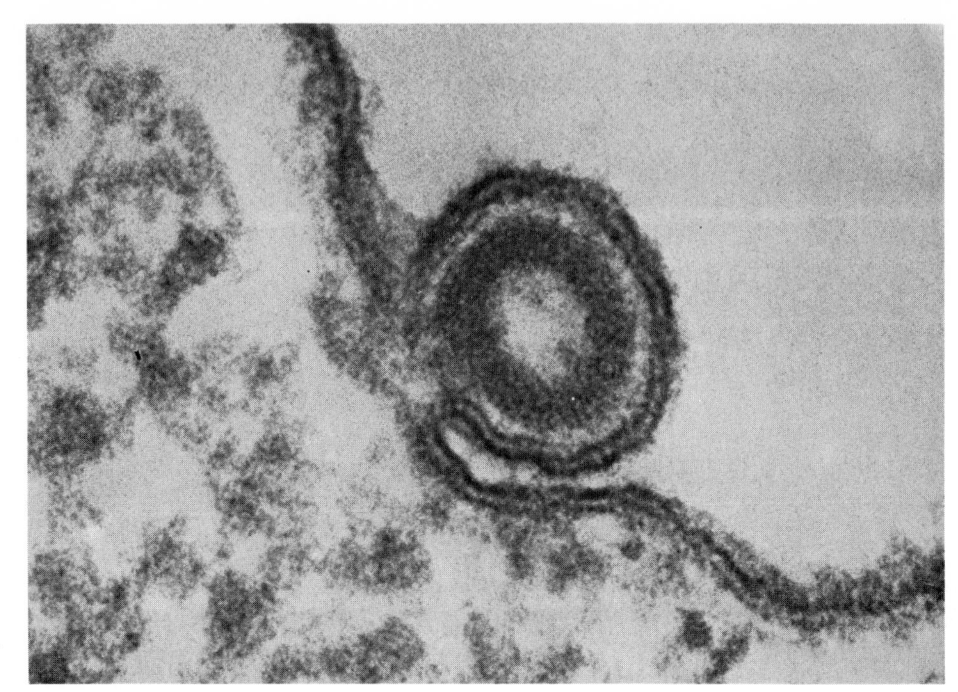

Figure 18.9 Budding of a murine Friend leukemia virus from the surface of a cultured leukemic cell. [*Courtesy of E. de Harven.*]

molecules are believed to be identical 35-S copies that contain all the information necessary for either transformation or productive cell infection. The best-studied core protein is the RNA-dependent DNA polymerase, commonly known as the "reverse transcriptase." After the infectious particle has entered a host cell, this enzyme functions to make a complementary copy of the RNA genome in the form of DNA, which is then inserted into the DNA. Although the details have not been completely worked out, it appears that this DNA polymerase uses a tRNA molecule present in the core as a primer and produces a single-stranded DNA complement. The tRNA is then digested, a complementary DNA strand is synthesized to form a double-stranded DNA form of the viral genome, and it becomes circularized and then integrated into the host DNA. Although the genetics of RNA viruses are not as well understood as those of polyoma or SV40, it is apparent that the number of genetic functions is very limited. There are four main genetic elements known to exist in the 35-S (3×10^6 daltons) RNA:

1. The gene (*env*), which codes for the viral-specific glycoprotein of the envelope.
2. The gene (*pol*), which codes for the reverse transcriptase.
3. The gene (*gag*), which codes for a precursor polypeptide that is subsequently cleaved to form several proteins of the inner core.
4. The gene (*src* or *onc*), which codes for a protein responsible for the transformation process. As in the case of the small DNA tumor viruses, a single genetic function appears responsible for the entire spectrum of changes in the transformed cell. Unlike the *A* gene of SV40, the *src* product is not required for replication of the viral genome; it seems required only for transformation itself. Moreover, the product of the *src* gene does not become localized within the host-cell nucleus, but is concentrated out at the cell surface, possibly associated with proteins of the cytoskeleton. Recent evidence strongly

suggests that the *src* gene codes for a protein kinase, i.e., an enzyme that serves to phosphorylate other proteins. It may be that the alteration in growth control associated with transformation by this virus results from a primary interaction and enzymatic alteration of the cytoskeletal components.

iii. The role of RNA viruses in cancer. The role of RNA-containing viruses in the etiology of cancer has become a focus of research activity. The involvement of these viruses as *natural* agents of certain types of tumors in animals, particularly rodents, is well established. One of the most interesting aspects of the topic concerns the manner in which the virus is transmitted from one individual animal to another. Viral infections are normally thought of as being caused by the release of particles from an infected cell and their subsequent infection of other susceptible cells. Transmission of this sort, termed *horizontal transmission*, does appear to be responsible for the transmission of certain types of animal cancers, such as feline leukemia. However, there is a very different type of transmission that may be more important. The early studies on mice indicated that certain tumor viruses could be passed from mother to offspring in the milk. In these cases, the virus particles could be detected in the milk and the offspring become susceptible to development of the tumor at a later age. In other cases it has been clearly shown that the virus can be passed from one generation to another in the DNA of the sperm and/or the egg, just as the bacteriophage can be passed from one bacterium to its progeny. The inheritance of a genetic element containing information for virus formation adds a whole new perspective to the concept of viral infection and has stimulated a great deal of research and speculation. Some investigators believe these viral genomes may actually have evolved from cellular DNA, while others believe they represent the descendants of viruses that had infected the species at some point in the evolutionary past. Regardless of their origin, the consequences to a cell harboring the information for one or more of these viruses can be quite variable. In some cases, the virus appears to result in the transformation of the cell in which it exists and the subsequent death of the individual. Inbred strains of rodents have been bred that inherit various viruses, some of which can lead to production of tumors in the individuals. In other cases, however, the virus does not appear to affect the metabolic activities of the cell. In some cells, there is no evidence of the virus other than the presence of viral DNA sequences; in other cells, the presence of viral antigens can be demonstrated. In many cases, mature viral particles are also found. All these conditions, including the production of a complete virus, can occur in a "normal" cell, one that is neither malignant nor in the process of being destroyed. In some cases, embryonic cells are characterized by the production of virus particles which then stops later in life without any sign of malignancy. Generally, the viral particles released by the nonmalignant cell are not infectious to the cells of the strain from which they are derived. Whether these viruses have a role in the cell or simply represent the remnants of some evolutionary infection that is no longer harmful remains unclear.

The point of most obvious importance in the topic of tumor viruses is their role in the etiology of human cancer. Although the subject remains controversial, RNA-containing viruses have been directly implicated in specific types of human malignancies. The first indication in this regard came when virus particles were identified in certain samples of human milk. It was soon established that human breast *tumor* cells can contain RNA in their polyribosomes that is complementary to labeled DNA probes prepared from mouse mammary tumor virus. Since normal cells did not seem to contain evidence of viral RNA, it was strongly sug-

gested that the presence of a human virus very similar to that causing tumors in mice was responsible for human breast cancer. Subsequent experiments on human leukemias and related malignancies also have suggested that the affected cells contain viral information, both in the genome (as integrated DNA) and in the cytoplasm (as viral mRNA). Once again, the results from hybridization studies suggested that the viral nucleotide sequences were similar to those known to be responsible for causing leukemia in other animals. Most important, it has been found that normal cells do not appear to contain complete copies of these particular viral sequences, and therefore, that the agents responsible do not seem to be passed from parents to offspring in the germ line. The extent to which tumor viruses, both DNA and RNA, are responsible for human cancer remains one of the topmost priorities of research in this field.

. Genetic aspects of malignancy

If one considers the types of agents capable of transforming cells in culture, namely, chemicals, radiation, and viruses, it becomes apparent that the primary alterations induced by the agents are genetic. Although it is possible that radiation and chemical carcinogens cause the transformation by activating an endogenous tumor virus, it is more likely that they act by damaging, i.e., mutating, the DNA. If this is the case, then we could conclude that the transformed cell is one that is deficient in a particular genetic function that secondarily manifests itself in the altered behavior of the malignant cell. In contrast, treatment of the culture with the tumor virus would appear to *add* a genetic factor, one whose products somehow interfere with the cell's activities, thereby causing it to become malignant. To some degree, these are alternative concepts, one suggesting that cancer results from a negative genetic basis, the other suggesting a positive genetic basis. One possibility that can be eliminated is that tumor viruses act to produce cellular mutations that secondarily cause the transformation. Results with temperature-sensitive tumor viruses clearly show that the viral genome must continue to produce active protein if the transformed state is to be maintained.

It has been shown that spontaneous tumor formation can be associated with the presence of a mutant genetic condition. For example, persons with the genetic disease Fanconi's anemia are much more susceptible to the development of tumors. This condition is not simply due to some systemic problem, such as a defective immune system, since the cells from persons with this or certain other genetic diseases are much more susceptible to transformation by various agents in vitro. A similar condition was mentioned in the case of xeroderma pigmentosum in Chapter 14. Cells from these persons are known to have deficient genetic repair mechanisms, a condition that causes them to be much more susceptible to agents such as ultraviolet light. The correlation between a tendency toward cancer and the inability to repair DNA damage can be considered independent evidence for the concept of the cancer cell as genetically deficient.

Another approach to determining whether cancer is a result of positive or negative genetic alterations is to fuse a normal cell with a cancer cell and observe the phenotype of the hybrid cells. Superficially at least, one might expect that if the malignant state were due to a genetic defect, the presence of the normal genes should correct the deficiency and restore the normal condition. Experiments of this type have generally tended to support the concept that the tumor cell is one suffering from the loss of genetic function. The hybrid formed between the normal and malignant cell is generally found to have greatly reduced malignant properties in culture and to be much less tumorigenic if injected into a suitable host. Regardless of these experiments, we know that the introduction of a

viral genome can transform the host cell, so we are still left with the same basic question. The answer to the question may be that either type of genetic alteration can cause the transformation; the underlying similarity in all cases is the loss by the cell of growth control.

F. Teratomas and teratocarcinomas

Of all the various types of malignancies, one is the most important to developmental biologists, the *teratocarcinoma*. In their usual solid form, these tumors consist of two classes of cells, one an undifferentiated type termed an *embryonal carcinoma cell* and the other a wide variety of differentiated cell types. Some teratocarcinomas may be mostly embryonal, either entirely undifferentiated or with neural or myoblastic tissue. Most of these tumors contain notochord, together with respiratory, alimentary, and glandular epithelia, as well as cartilage, bone, and marrow. Many even have hair and teeth. Various degrees of tissue organization may be seen within the mass, and microenvironmental influences seem to be responsible for the diversity of differentiation. These influences may be no more than local variations in electrolyte concentrations within each population of cells.

Of the two classes of cells in a teratocarcinoma, the embryonal carcinoma (EC) cells are malignant, i.e., will upon transplantation give rise to another tumor, while the differentiated cells are not. Even though the bone, muscle, nerve, or other types of cells are derived from EC cells, once they have undergone their respective cytodifferentiation, they no longer retain their malignant capability. The EC cells can be thought of as pluripotent stem cells that can divide to produce other malignant EC cells or differentiate into benign derivatives. Before continuing this aspect of the story, it will be useful to describe how these tumors are obtained initially under laboratory conditions. Using mice, which are the best studied, there are three procedures that predictably lead to the formation of teratocarcinomas:

1. There are two strains of mice in which these tumors arise spontaneously at very high frequency. In one strain, 129, the tumors arise within the seminiferous tubules of male animals, while in the other strain, LT, they arise from oocytes that are retained within the ovary of females. In this latter case, the oocytes, for some unknown reason, become activated parthenogenetically, undergo the first meiotic division, and proceed to form a small embryolike structure. After a few days, however, the mass of "embryonic" cells becomes increasingly disorganized and begins to develop into one of these tumors.
2. If the genital ridge of a 12-day male mouse embryo (a stage at which primordial germ cells have moved into the ridge) is grafted to any of a variety of sites in an adult animal (e.g., kidney, liver, testis), the mass also develops into one of these tumors.
3. If a normal mouse embryo prior to the eighth day of development is removed from the uterus and grafted to an extrauterine site, the embryo becomes disorganized and progresses toward the formation of these tumors.

In most cases, the tumors stop growing after a period and are no longer capable of forming additional malignancies upon transplantation. Nontransplantable tumors have become benign because all their EC stem cells have undergone differentiation to form nonmalignant derivatives. Benign tumors of this type are termed *teratomas*. In a smaller percentage of these solid tumors, the mass retains a population of malignant stem cells, and the tumor (now specifically re-

ferred to as a *teratocarcinoma*) continues to be capable of growth in additional hosts.

One question that has received considerable attention in this area concerns the origin of teratomas. It is generally believed that these tumors can arise from two distinct types of pluripotent cells: germ cells or somatic cells of an early embryo. Of the three procedures just described, the first and second appear to be examples of germ cells giving rise to teratomas, while the third is an example of the formation of teratomas from embryonic cells. To some degree, the distinction is an arbitrary one, since cells of the early embryo are destined to give rise to germ cells at later stages and germ cells seem to divide and form embryolike complexes before they grow out of control and form teratomas. These matters are discussed at length in the paper by B. Mintz et al. (1978). We will simply conclude that teratomas and teratocarcinomas arise from totipotent cells that retain this totipotency as long as they contain undifferentiated EC stem cells.

Over the years, several types of culture techniques have been worked out for the study of EC cells. If as few as one of these cells is injected under the skin of a host, a solid teratocarcinoma may develop. If, however, the EC cells are cultured in suspension, either within a flask or within the peritoneal cavity of a host mouse (termed an *ascites* in this case), the EC cells proliferate and form highly characteristic masses of cells, termed *embryoid bodies*. These structures are of considerable importance to developmental biologists, for, as the name implies, they closely resemble an early embryo. These free-floating cellular aggregates contain an internal mass of cells surrounded by a single-celled epithelium. (Fig. 18.10) and have been likened in appearance to the internal contents (the ICM) of a 5-day mouse blastocyst composed of a mass of inner ectodermal cells surrounded by a thin layer of endoderm (see Fig. 7.36). In the case of the embryoid body, the internal contents consist of a small population of malignant embryonal carcinoma

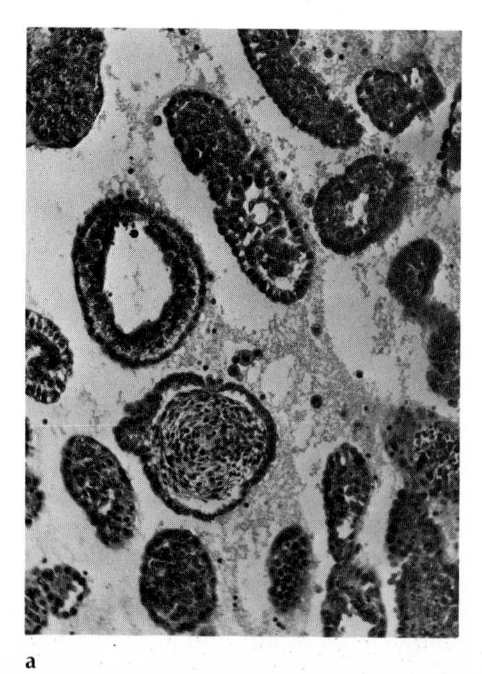

a

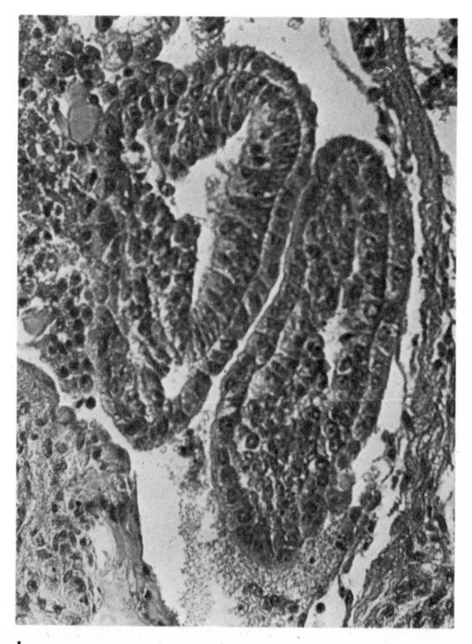

b

Figure 18.10 Embryoid bodies derived from a transplanted mouse testicular teratocarcinoma. [*Courtesy of L. C. Stevens.*]

cells. If these initial embryoid bodies are allowed to develop further in the culture flask or peritoneal cavity, they begin to appear like a more advanced embryo, one characterized by the formation of a variety of differentiated cells. The more advanced structure is referred to as a *cystic embryoid body*.

The formation of embryoid bodies bears a striking morphologic similarity to the normal development of the embryo and, as such, has been thought of as a model system of mammalian development. The similarities extend beyond the morphological parallels and the ability of the EC cells to undergo differentiation. Biochemical properties of the malignant pluripotent cells indicate that they too are embryonic in character. This has best been shown by examination of their cell surfaces. One cell-surface marker, the F9 antigen, is found to occur on EC cells prior to their differentiation. The striking observation has been made that this antigen also exists on embryonic cells during the period of cleavage but is lost by about the ninth day, when the cells of the embryo begin to lose their developmental potency and become committed to the formation of specific embryonic and extraembryonic parts. The F9 antigen is also present on the surfaces of male germ cells. There is one very important property that these three types of cells —EC, blastomeres, and germ cells—have in common; they are all very pluripotent, if not totipotent. It would appear that the presence of this cell-surface protein is correlated with the potency of the cell that bears it. Moreover, in both normal development and in teratocarcinomas, the F9 antigen disappears from the surfaces as the cells undergo differentiation.

Although the pluripotency of the EC cell is indicated by its potential for differentiation in vitro, it is much more strikingly illustrated in the following experiment, the technique for which is shown on the front cover. Individual EC cells have been injected into the blastocoel of genetically marked mouse blastocysts and their fate during subsequent development closely followed. Embryos of this composite nature are found to proceed with normal development when placed within the uterus of a foster mother, where they develop to term. Histological examination of the tissues of these mice indicates that they are often highly chimeric, i.e., are composed of large numbers of both types of genetically marked cells. Organs normally derived from all three germ layers are generally found among the products of the single injected EC cell.

We can conclude from this experiment that an EC cell that would otherwise have differentiated into a disorganized mass of tissue within a teratocarcinoma can participate in the development of normal embryonic tissues and organs when allowed to express its potential in a normally constituted embryo. Not only can these injected EC cells differentiate into a wide variety of somatic tissues, they are fully capable of forming germ cells that subsequently give rise to fully fertile spermatozoa. Normal mice have been raised from eggs fertilized by spermatozoa bearing the genetic marker of the ancestral embryonal carcinoma cell. The totipotency of the EC cell is thus established. Furthermore, all signs of malignancy are lost. Tissues from chimeric mice can, for example, be grafted to extrauterine sites, but teratomas never result. It would appear that the environment in which these malignant cells are placed is of utmost importance in determining whether or not they retain their malignant potential.

In addition to demonstrating the potency of the embryonal carcinoma cell, these experiments bring into focus the more general problem of the nature of the malignant state. It is argued that if a malignant EC cell is capable of reversion to what appears to be a completely normal state, one that allows for controlled proliferation and differentiation, then the original malignant ancestor could not have become such as a result of a loss of genetic information required for normal

growth control. Rather, it is argued that the malignant state of the EC cell must result from its presence within a disorienting environment. This type of reasoning is further strengthened when one considers that the grafting of a normal early embryo to an extrauterine site is a near certain guarantee of causing that embryo to grow as a teratocarcinoma with its full complement of malignant EC cells. Is there a close relationship between the EC cell and the other types of malignancies? The answer to this question remains unclear, although increasing study of the varieties of tumors that can form is leading investigators more and more to the conclusion that the differences in properties among malignancies far outweighs their similiarities. Although the EC cell can certainly be a killer under many internal environments, it is clearly capable of acting in a civilized manner within the environment of the embryo. Another property that distinguishes the EC cell from most other malignant cells is its ability to retain its normal diploid genome over long periods of growth and culture. EC cells that have been serially cultured for periods of years are still capable of participation in normal development following their injection into embryos. In contrast, cells of most tumors become increasingly aberrant in chromosomal composition, a property that apparently leads to their continued ability to adapt and survive.

2. Aging

As an animal ages, the outward signs are obvious; yet the underlying basis for the deteriorative processes that occur in all animals is very poorly understood. Physiological measurements of organ function, nerve conduction velocity, muscle power, etc. indicate a decreasing capacity with age. It is generally believed that these more easily observed alterations reflect a progressive deterioration in the elements of which the tissues are composed. In this brief discussion, we will consider two sites, the cellular and the extracellular, in which age-related events occur. There is no doubt that progressive changes take place both within cells and in their surrounding environments. The principal controversy in this regard concerns which of these two sites within the tissues is the primary one from which the aging of the whole animal results. This problem is similar to that previously discussed on the underlying basis of cancer. Not only is there controversy over the basic causes of aging, there is considerable disagreement over many of the observations themselves. Various laboratories working with different types of aging systems have not always agreed on even the more basic findings. In the following discussion we will simply present some of the basic observations that have been made and a little of the speculation about their significance.

A. Cellular aging

Theories that attempt to explain the aging process as a result of defective intracellular processes are based on several assumptions. The underlying assumption is that there is a finite frequency of error in the biochemical operations of every cell. Over a period of time, these errors might accumulate to produce a cell with a defective function or they might result in the death of the cell. Biochemical mistakes or damage can occur at numerous sites and manifest themselves in numerous ways. For example, in one theory of aging, damage is believed to result from the formation of highly reactive free radicals, particularly those of oxygen. In aerobic cells, certain enzymatic reactions are responsible for generating a superoxide radical (O_2^-) as a free intermediate. In the presence of water, this species is converted into the highly reactive hydroperoxyl radical $HO_2\cdot$). Although there are numerous macromolecules with which this radical might react, one of its prime targets would be the unsaturated fatty acids of the phospholipids of cell-

ular membranes. The interaction of fatty acids in the membrane with a free radical can lead to a chain reaction in which considerable lipid autoxidation can occur. In order to cope with the prospect of this type of damage, aerobic cells possess an enzyme termed *superoxide dismutase* that is responsible for the destruction of the superoxide radical in the following reaction.

$$2O_2^- + 2H^+ \rightarrow H_2O_2 + O_2$$

Although lipid autoxidation may be an important consequence of aerobic metabolism, it is generally believed that the most sensitive site in the cell for the accumulation of age-related damage is the genetic material itself. Any change in the linear base sequence in the DNA will result in its altered regulatory or template properties and subsequent effect on all processes in which that gene is involved. Alteration of the DNA template can result from damage (known to occur after most types of radiation) or from a mistake during replication by insertion of an incorrect nucleotide. Alterations in the genome of cells outside the germ line are called *somatic mutations*. Such mutations would be expected to be random, and each cell would have its information content affected in a different way.

The best evidence that individual cells undergo deterioration has been obtained with cells growing in culture. During the early 1960s, it was reported that when human lung fibroblasts are followed in tissue culture, the number of divisions these cells can undergo is limited. To perform this experiment, a small piece of tissue is removed, the cells are dissociated, and a certain number of these single cells are transferred to a culture dish and allowed to attach to the surface and divide. After a number of divisions, the cells are removed from the dish and placed into suspension; a small percentage is used to inoculate new culture dishes, just as the first had been done. These cells will continue to divide and cover the new dish, after which the process can be repeated. Each time the cells are removed from the dish and plated on a new dish at a lower concentration, they are said to be *subcultured*. If cells are subcultured numerous times, a point is reached at which they stop dividing and the cell strain dies out. This has been found for many types of cells. It is generally reported that the number of divisions the strain goes through determines its life span, rather than the time that has passed during the experiment. This is clearly an example of age-related cellular death.

Tumor cells provide a significant exception to the rule that cells undergo a limited number of doublings in culture. Malignant cells have no such restriction; they continue to divide in essentially an immortal way. Tumor cells are often characterized by an abnormal chromosome number, which may be related to their immortality in vitro. Interestingly, occasional cells derived from a normal donor will continue to divide rather than become senescent, as in most cases. When the chromosomes of these peculiar cells are examined, they are generally found to be aneuploid, like the malignant cells, and they often behave in other respects as if they were tumor cells. Cells that undergo this change are said to be a *cell line* as opposed to a *cell strain*, as in cells that have a restricted division potential. The frequency with which cells become a cell line is related to the species from which they were derived. Mouse cells will frequently become altered in this way, human cells only rarely.

If aging and ultimate age-related death of an animal result from intracellular damage, it would be expected that cells from a short-lived species would have less potential for cell division than those from a long-lived animal. Similarly, cells from a young animal should be more capable of extended division than cells from an older individual. This has become a controversial topic. It appears that

human fetal lung fibroblasts are capable of approximately 50 divisions before they lose this ability; those from an adult at maturity are capable of approximately 30; and those from more aged individuals are capable of less. These numbers, however, are averages, and in one case, an 87-year-old man was reported to have cells capable of an average of 29 further divisions. In other words, there is convincing evidence that cells age, but it is equally clear that we do not run out of all types of cells and then die as a result. If we take the value of 50 to 100 divisions for all cells of the fetus, sufficient cells for many lifetimes could be produced. If the exhaustion of the capacity of cells to divide is responsible for age-related processes, a search must be made for a tissue that is particularly sensitive that could secondarily affect the other tissues of the body in a destructive way. A candidate would be one or more of the endocrine glands, whose effects on the function of the whole body are well known. Another is the lymphoid system which has received a great deal of attention as a potential aging pacemaker.

Up to this point we have considered the accumulation of simple, randomly occurring mutations. To understand the potential for cellular damage inherent in the occurrence of somatic mutation, we must consider the concept of an "error catastrophe." In this theory, it is pointed out that certain types of errors are likely to produce a great number of subsequent errors. Consider, for example, a mistake occurring in a DNA polymerase gene that results in an enzyme that will make further mistakes during replication. Such a mistake would be considered an error catastrophe, and examples of altered enzymes that cause an increased number of mistakes are known from studies of bacterial viruses. Replicational enzymes are only one place where an error catastrophe could occur. Proteins are needed for transcription and translation, and altered proteins that serve these functions could similarly be expected to rapidly fill the cell with an entire spectrum of defective proteins. At first glance, the somatic-mutation theory of aging seems incompatible with the observed differences in life span among organisms. If senescence were due to the accumulation of random mutations, all cells and organisms might be expected to show a similar aging time course. However, our knowledge of mutation rates and genetic-repair mechanisms among organisms is far too meager to justify this conclusion. The degree to which the accuracy of replication, or transcription, or DNA repair can vary from species to species remains to be determined.

Is there evidence that as cells of an organism (or a culture) become older they have an increasing content of deficient proteins? Over the past decade, hundreds of observations have been made with regard to this question, although no definitive answer has been forthcoming. In many cases it has been shown that during the senescence of an organism (or a cell culture), certian of its enzymes undergo a decrease in specific activity. Similarly, it has been shown that aging is accompanied by an increase in the presence of inactive or unstable enzymes. More recently, specific enzymes have been purified from young and old animals and differences in activity have been noted. These types of reports provide evidence for a theory of aging based on somatic mutation, although one cannot rule out the possibility that the deterioration of these proteins results from posttranslational modification. Even though there are many reports of changes in enzyme activity with age, there are many studies in which no such changes have been found. One might conclude that different enzymes undergo deteriorative changes at different rates, a conclusion that is difficult to reconcile with a proposal based on random genetic damage. Another finding that argues strongly against the somatic-mutation theory (and particularly the error-catastrophe theory) comes from studies of viral production in young versus old cells. If the protein-synthesizing

machinery of a cell deteriorates with age, older cells should be less able to support a viral infection. This does not seem to be the case.

If the aging process does result from the effects of somatic mutation, then it should be possible to accelerate aging by increasing the mutation frequency. Once again, results in this area are controversial, although evidence on behalf of this prediction has been reported. For example, it has been shown that *Drosophila* larvae that have been fed compounds that increase the content of defective proteins experience a markedly shortened life span. Similar findings have been reported for cells growing in culture—compounds that increase translational errors can reduce the doubling capacity of these cells.

One of the main lines of evidence cited in support of theories of cellular aging has been obtained by irradiation. It is a well-established fact that irradiation, in proportion to the dose, has life-shortening effects. Remarkably, animals whose life expectancy has been shortened as a result of such exposure undergo the signs of aging prematurely. Figure 18.11 shows two groups of litter mates of mice. The group exposed to X-radiation shows the debilitating signs of age at a chronological age well in advance of the normal time for these animals. The means by which the radiation is administered and whether the effects are truly mimicking the natural aging process are debatable and beyond the scope of this book. If we assume that radiation is accelerating the natural aging events, the most likely explanation for its action is via somatic mutation, i.e., changes in DNA.

The aging of cells as they grow in vitro was described earlier. By *serial transplantation,* whereby cells from one animal are repeatedly transplanted from one host to another, the longevity of cells in vivo can be estimated. With this technique, the descendants of the cells of the original transplant can be maintained in the bodies of animals of any age rather than be exposed to a continually aging environment. In other words, the technique of serial transplantation allows one to dissociate the cell from its normal environment, yet maintain it within an appropriate physiological container. Several types of tissues, including skin, mammary gland, and spleen cells, have been treated in this way. In all these cases, the proliferation of cells is limited, just as in the in vitro experiments. However,

Figure 18.11 Two groups of mice that were originally identical. Mice on the left are control animals; mice on the right received a large but nonlethal dose of radiation. Only three mice of the irradiated group remain alive; these are "old" and senile. [*Courtesy of H. J. Curtis.*]

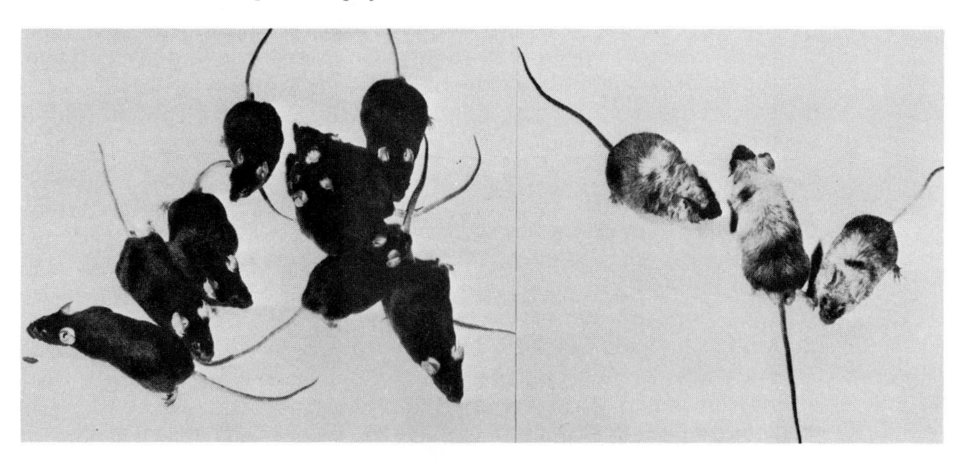

these transplanted tissues will outlive, to a considerable extent, the animal they were originally taken from. The results of these experiments, therefore, are inconclusive. They confirm that cells undergo aging in vivo but do not establish if this phenomenon is responsible for the age-related death of the animal.

Before questions of cellular aging within the mammalian body can be considered, a distinction must be made as to the nature of the cells involved. A certain percentage of the cells of an animal, including neurons and muscle cells, is formed during the developmental period of the life cycle and are not replaced during the entire life of the organism. The study of a postmitotic cell for signs of aging allows one to separate aging processes from mistakes made during division. Neurons or muscle cells, once formed, might be expected to accumulate errors during their long life. These errors might be expected to reduce the capabilities of each cell and to cause deteriorating changes in the physiological properties of the tissue. Both nervous and muscle tissue undergo age-related physiological changes.

Another class of cells produced continually throughout the life of the animal includes circulating blood cells, lymphoid and bone-marrow cells, skin cells, mucous membrane epithelial cells, etc. These cells are produced with a finite lifetime, from days to months, and are destroyed. Use of radioactively labeled cells indicates that in many cases these cells are not removed on a random basis; rather, as a cell's lifetime in the body increases, its chance of becoming destroyed also increases. This is an important observation because it indicates that there are mechanisms by which aging cells can be marked and mechanisms whereby they can be destroyed. These short-lived cells can be shown to age in very short periods, while longer-lived cells show no such changes in these short time periods. Different cells seem to undergo different rates of age-related deterioration.

Another class of cells can be distinguished that either continue to divide or retain the capacity, to some degree, throughout the life of the organism. Examples are hepatocytes, osteoblasts and chondrocytes. In this group are the many types of stem cells that are partially differentiated, at least to the extent that they can give rise to only one or a limited variety of differentiated cells. Aging to this group of cells would be reflected in their inability to renew the tissues they are responsible for as well as in their decreased ability to function physiologically. Each type of cell must be considered individually; each may have its own error frequency, its own capacity for DNA repair, its own molecular turnover rate, etc., and each may contribute in a different way to the overall aging process of the animal.

As stated earlier, theories of cellular aging must explain the entire age-related spectrum of deteriorative changes. Can an increased likelihood of malignancy, increased heart disease, etc. be accounted for by underlying cellular damage? At the present time, this question cannot be answered.

B. Extracellular aging

A significant percentage of the dry weight of an animal resides outside the cell in the extracellular space. The extent to which a tissue is composed of extracellular material varies greatly, reaching a maximum in the supportive tissues of the body. The primary components of the extracellular space are glycosaminoglycans and fibrous proteins, particularly collagen and elastin, which are secreted out of the connective-tissue cells in which they are synthesized. Collagen is estimated to account for up to 40 percent of the body protein, is present in extracellular spaces of virtually all tissues, and has been suggested as the primary site for age-related changes.

Several properties of collagen suggest it may be responsible for the aging process. That collagen undergoes molecular modification with age is primary to the theory. The fibrous protein molecules are polymers of collagen monomers, each composed of three polypeptide chains. In the newly polymerized molecule, the collagen monomers are held together by noncovalent bonds. As collagen is maintained under physiological conditions, covalent cross-linking takes place both within the collagen monomer (among the three polypeptides) and between the monomers. These cross-linked reactions have profound effects on both physicochemical and biological properties of collagen. For example, the ease with which collagen can be extracted and dissociated drops markedly with age, until it is essentially insoluble by the time of maturity in mammals. From that point through old age, the cross-linking continues and its effect can be measured. One study of its changing properties has utilized collagen fibers from rat tail and their resistance to shrinkage. If collagen fibers are heated to 65°C, they shrink in length unless maintained at their original length with attached weights. The amount of weight required to prevent shrinkage is a measure of the extent of cross-linking and, therefore, of aging of the collagen molecules (Fig. 18.12). In one study it was found that 2-month-old rat-tail collagen can be maintained by 1.5 g, 5-month by 3 g, and 30-month by approximately 10 g. However, comparative studies suggest that cross-link density, by itself, cannot be used as a measure of the degree to which an animal has aged. While a 10-g weight prevents the shrinkage of collagen fibers of a 30-month rat, it takes approximately 50 g to accomplish the same task when the collagen being tested is obtained from a cat of comparable age (for example, 15 years). It would appear that the accelerated aging experienced by rats as compared with cats (or humans) is not a direct result of the accelerated aging of those collagen molecules which can be readily tested, i.e., those of tails,

Figure 18.12 Age change in isotonic thermal contraction. The plot shows the amount of weight that is required to prevent shortening by rat-tail tendon fibers (collagen fibers) when the fibers are taken from various aged animals. [*From F. Verzar,* Int. Rev. Connect. Tissue Res., **2**:257 (*1964*).]

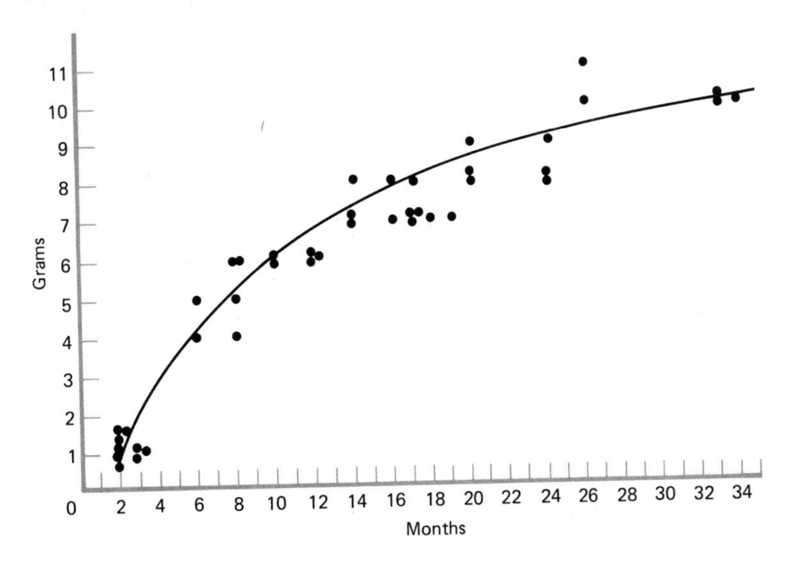

tendons, etc. Much less is known about the more physiologically important collagen of the extracellular matrices of the various internal organs.

The theory of aging based on collagen is analogous to the theory based on intracellular changes in that the evidence of age-related modification is indisputable. The question is whether or not all the other age-related events are explainable as a consequence. The primary chemical modifications are believed to directly affect a wide variety of physiological activities, as a result of the widespread distribution of collagen. For a cell to maintain a healthy intracellular composition, it must receive oxygen, nutrition, ions, hormones, etc. from its environment and release carbon dioxide and other waste products. The occurrence of collagen in spaces between cells suggests that it could play a critical role in the exchange activities between a cell and its environment. The presence of large amounts of collagen in the lining of all blood vessels would similarly affect movement from the blood into the extracellular space as well as affect the distensibility of the arterial wall.

Readings

ADLER, W. H., 1975. Aging and Immune Function, *Bioscience,* **25**:652–657.

BALTIMORE, D., 1976. Viruses, Polymerases, and Cancer, *Science,* **192**:632–636.

BURNET, F. M., 1974. Intrinsic Mutagenesis: A Genetic Basis of Aging, *Pathology,* **6**:1–11.

BURNET, F. M., 1976. "Immunology, Aging, and Cancer," Freeman.

COMFORT, A., 1964. "Ageing: The Biology of Senescence," Holt, Rinehart and Winston.

CRISTOFALO, V. J., 1974. Aging, in J. Lash and J. R. Whittaker (eds.), "Concepts in Development," Sinauer.

DAMJANOV, I., and D. SOLTER, 1974. Experimental Teratoma. *Curr. Topics Pathol.* **59**:69–130.

DANIEL, C. W., 1972. Aging of Cells During Serial Propagation *in vivo, Advan. Gerontol. Res.,* **4**:167–199.

DULBECCO, R., 1976. From the Molecular Biology of Oncogenic Viruses to Cancer, *Science,* **192**:437–441.

GRAHAM, C. F., 1977. Teratocarcinoma Cells and Normal Mouse Embryogenesis, in M. I. Sherman (ed.), "Concepts in Mammalian Embryogenesis," MIT Press.

HAYFLICK, L., 1965. The Limited *in vitro* Lifetime of Human Diploid Cell Strains, *Exp. Cell Res.,* **37**:614–636.

HAYFLICK, L., 1975. Cell Biology of Aging, *Bioscience,* **25**:629–637.

HOLLIDAY, R., and G. M. TARRANT, 1972. Altered Enzymes in Aging Human Fibroblasts, *Nature,* **238**:26–30.

HSIE, A. W., and T. T. PUCK, 1971. Morphological Transformation of Chinese Hamster Cells by Dibutyryl Adenosine 3′,5′ Monophosphate and Testosterone, *Proc. Natl. Acad. Sci. U.S.,* **68**:358–361.

HYNES, R. O., 1976. Cell Surface Proteins and Malignant Transformation, *Biochem. Biophys. Acta,* **458**:73–107.

IKAWA, Y., and T. OKADA (eds.), 1979. "Oncogenic Viruses and Host Cell Genes," Academic.

ILLMENSEE, K., and L. C. STEVENS, 1979. Teratomas and Chimeras, *Sci. Amer.,* **240**:87–98 (April).

JACOB, F., 1977. Mouse Teratocarcinoma and Embryonic Antigens, *Immunol. Revs.,* **33**:3–32.

KERBEL, R. S., 1979. Implications of Immunological Heterogeneity of Tumors, *Nature,* **280**:358–360.

KLEIN, G., 1976. Analysis of Malignancy and Antigen Expression by Cell Fusion, *Fed. Proc.,* **35**:2202–2204.

KOHN, R. R., 1971. "Principles of Mammalian Aging," Prentice-Hall.

LAMB, M., 1977. "Biology of Aging," Halsted.

MARTIN, G. R., 1977. The Differentiation of Teratocarcinoma Stem Cells in Vitro: Parallels to Normal Embryogenesis, in M. Karkinen-Jaaskelainen et al. (eds.), "Cell Interactions in Differentiation," Academic.

MINTZ, B., 1979. Teratocarcinoma Cells as Probes of Mammalian Differentiation, in J. Ebert and T. S. Okada, (eds.), "Mechanisms

———, C. CRONMILLER, and R. P. CUSTER, 1978. Somatic Cell Origin of Teratocarcinomas, *Proc. Natl. Acad. Sci. U.S.*, **75**:2834–2839.

ORGEL, L. E., 1973. Ageing of Clones of Mammalian Cells, *Nature*, **243**:441–445.

PIMENTAL, E., 1979. Human Oncovirology, *Biochem. Biophys. Acta*, **560**:169–216.

SETLOW, R. B., 1978. Repair Deficient Human Disorders and Cancer, *Nature*, **271**:713–717.

SHOPE, R. E., 1966. Evolutionary Episodes in the Concept of Viral Oncogenesis, *Perspectives in Biol. and Med.*, **9**:258–274.

STEVENS, L. C., 1975. "Spontaneous Parthenogenesis and Teratocarcinogenesis in Mice," 33rd Symp. Soc. Dev. Biol. Academic.

———, 1973. Developmental Approach to the Study of Teratocarcinogenesis, *Bioscience*, **23**:169–172.

STREHLER, B. L., 1976. "Time, Cells, and Aging," 2d ed., Academic.

TANZER, M. L., 1973. Cross-Linking of Collagen, *Science*, **180**:561–566.

TEMIN, H. M., 1977. The Relationship of Tumor Virology to an Understanding of Nonviral Cancers, *Bioscience*, **27**:170–176.

"Tumor Viruses," 1975. *Cold Spring Harbor Symp. Quant. Biol.*, **39.**

VERZAR, F., 1963. The Aging of Collagen, *Sci. Amer.* (April).

WEINBERG, R. A., 1977. How Does T Antigen Transform Cells? *Cell*, **11**:243–246.

WIENER, F., G. KLEIN, and H. HARRIS, 1974. Analysis of Malignancy by Cell Fusion, *J. Cell Sci.*, **15**:177–183.

WILLIAMSON, A. R., and B. A. ASKONAS, 1972. Senescence of an Antibody-Forming Cell Clone, *Nature*, **238**:337–339.

Chapter 19 Transformative development: metamorphosis

Transformation during development is typical of most animals. When the change is dramatic, it is known as *metamorphosis*. In many, where no radical change in lifestyle occurs, the course of development may be a gradual, progressive change from one condition to another, or the change may be sudden but not so superficially remarkable that it goes by the name of metamorphosis. Such is the shift from fetal structures and adaptations to postnatal life in human beings and other mammals. Metamorphic change during the developmental cycle is an acceleration or condensation of essentially the same basic processes characteristic of most forms of development. Primarily it consists of the differential destruction of certain tissues, accompanied by an increase in growth and differentiation of other tissues. This is seen in spectacular form in the development of the starfish (Fig. 19.1). A bilaterally symmetrical larval form, (the bipinnaria larva) designed for living as a plankton organism, is replaced by a radially symmetrical bottom-living type of organism that has developed within the larva.

1. Critical mass and metamorphosis

Every particular example of specific structural organization, whether that of a whole organism or of a part thereof, requires a minimal or critical mass of tissue for its visible expression. This critical value may be seen as volume, surface area, or number of cells. It is critical in the sense that full expression of a pattern is wholly or partially inhibited at lower values. Larger values, however, permit developmental patterns to be initiated on a larger scale. For example, tadpoles of the common toad metamorphose when about half an inch long and form toadlets even smaller, about as small a size as the terrestrial creature can be expressed in. On the other hand, bullfrog tadpoles, developing from eggs no larger, grow to about 6 inches as tadpoles before metamorphosing.

Even the smallest egg, generally of the order of 0.1 mm in diameter, is a relatively very large cell compared with most tissue cells, and simply by successive cleavages without growth it can become a mass of some 2,000 cells. By that time

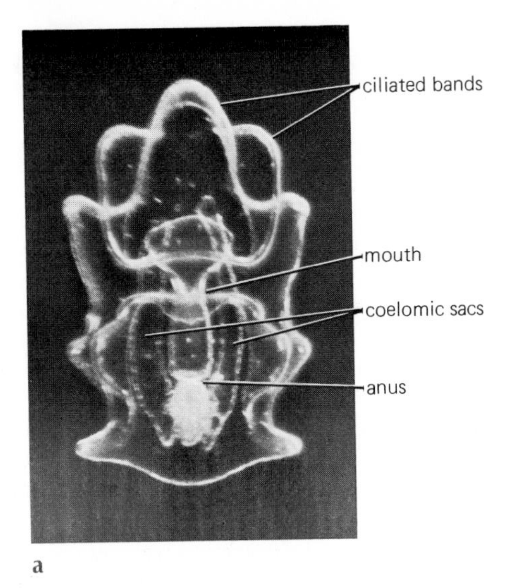

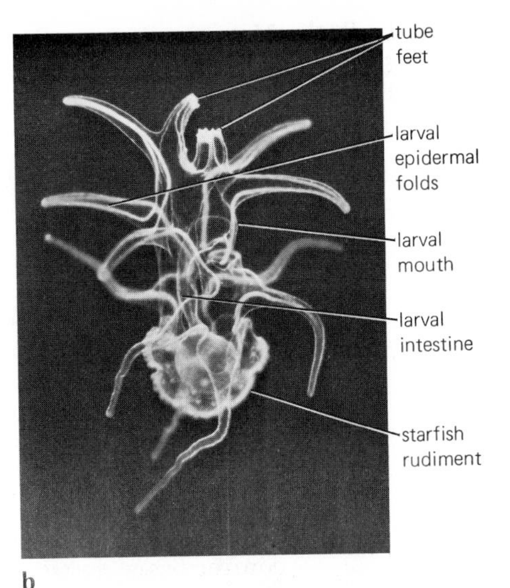

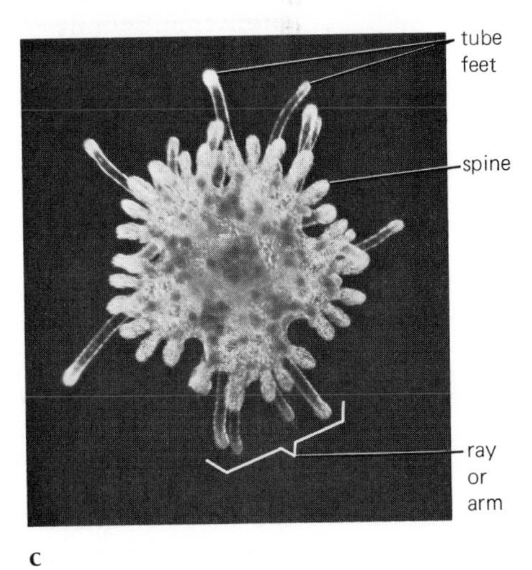

Figure 19.1 Metamorphosis during development of the starfish *Asterias*. *a* Bipinnaria larva with extensive ectodermal ridges bearing locomotory cilia and intestine, all exhibiting a bilateral symmetry. *b* Older bipinnaria larva with rudiment of juvenile starfish forming from portions of internal tissues and exhibiting a radial symmetry; the two organizations coexist for a while as a dual planktonic organism. *c* Fully metamorphosed stage, as a benthic (bottom-living) juvenile starfish with arms yet undeveloped, larval tissues resorbed or sloughed off, and clearly radially symmetrical. [*Courtesy of D. P. Wilson.*]

it must be a viable organism capable of feeding and growing, or it must be already supplied with an alternative nutritive system, either in the form of yolk reserve or a placental system. In most cases, growth is necessary before the final organization can be expressed even on the smallest scale. When eggs are comparatively large, especially when a large blastoderm forms over a large mass of yolk, a miniature adult, or juvenile, is laid out directly. The contrast is seen, for example, in the development of marine mollusks. Those with the smallest eggs develop into minute trochophore larvae that must grow and metamorphose. In the relatively much larger and yolky squid egg, the adult-type organization forms like a developing photograph in the extensive blastoderm of the egg as a whole, with no need for subsequent metamorphosis or any change in lifestyle (see Chapter 7).

2. Dual developmental systems

Metamorphosis presents a developmental-evolutionary problem. In many marine invertebrates and in the insects, the nature of the egg and the genome have been changed from that associated with a relatively more direct development to a more complicated course. The development of the egg is forced into a new path to yield a viable organism strikingly different from the adult; this is accomplished without harming the capacity to develop the adult organization. In the amphibians, the genome (and the egg as a whole) has evolved so as to produce a terrestrial tetrapod in place of an ancestral lobe-finned fish. It has done so without changing the particular character of the early freshwater vertebrate type of egg, which accordingly follows the old developmental path for a considerable developmental distance.

In both invertebrates and amphibians, a dual system of development is very evident. In both groups, the egg exhibits a specific early organization that relates to the development of the first of the two organismal stages. In terms of development, the two situations have much in common, although they are different enough to call for separate discussion.

Several fundamental developmental and evolutionary problems accordingly confront us. The development of an organism adapted to two very different environments at different times during its life span implies the operation of two sets of environmental-selection pressures. Whatever such selection for functional phenotypes may have been, an internal developmental selection of genotypes may well have been the primary factor in determining evolutionary direction. In the examples of metamorphic life cycles just mentioned, at least two effective genotypes, i.e., two virtually independent sets, exist together in the same genome, although expressed at different times.

A. Transformative development

The egg of the common jellyfish *Aurelia*, like that of virtually all hydrozoan and scyphozoan coelenterates, hydra included, develops into a small elongate, cilia-covered planula larva. In *Aurelia* and other scyphozoans, the planula attaches by one end and develops a mouth and ring of tentacles at the other; i.e., it becomes a hydralike polyp or scyphistoma. The polyp grows and produces buds laterally from the body wall in essentially the same way as hydras do, the buds separating and becoming attached alongside the parent polyps which they replicate. Extensive polyp colonies are thus produced, as in hydras, that persist throughout most of the year. Each polyp grows to a certain maximum size, with individual growth otherwise being siphoned off, so to speak, through the process of budding.

In nature, during spring months, polyps undergo a segmentation, or strobilation, process. That is, transverse epidermal constrictions mark off a series of segments of the trunk, forming in sequence beginning at the distal end and eventually consuming all but the basal part of the polyp (Fig. 19.2). During the process of constriction, each segment develops the basic organization of a medusa, complete with radial symmetry, gastric filaments, statocysts, and pulsating musculature. The young medusa, set free as an ephyra, grows massively to become a sexually mature aurelia. This life cycle is of course well known, but it exemplifies again that diverse forms—polyp and medusa—can be expressed by the same genome under different circumstances, and also that here there is an actual transformation of polyp (hydranth) into multiple medusae. Segmentation initiates the process and is followed in each segment by a true metamorphosis. The residual polyp, following strobilation, regrows and repeats the process year after year.

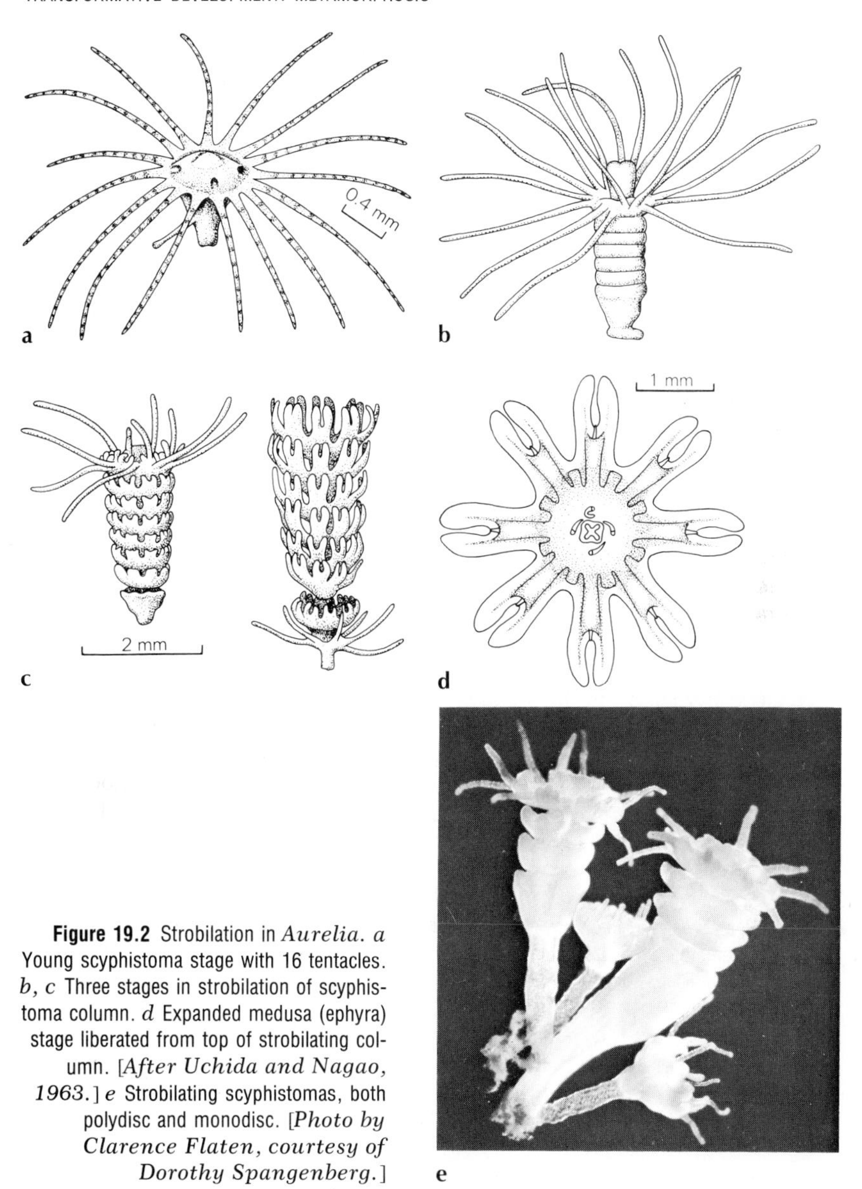

Figure 19.2 Strobilation in *Aurelia*. *a* Young scyphistoma stage with 16 tentacles. *b*, *c* Three stages in strobilation of scyphistoma column. *d* Expanded medusa (ephyra) stage liberated from top of strobilating column. [*After Uchida and Nagao, 1963.*] *e* Strobilating scyphistomas, both polydisc and monodisc. [*Photo by Clarence Flaten, courtesy of Dorothy Spangenberg.*]

Many attempts have been made to identify the environmental factor or factors responsible for triggering the strobilation process. Several have been implicated, particularly temperature; a preconditioning by exposure to comparatively low temperatures appears to be generally necessary. The only dependable effective agent so far discovered is iodine. It is effective as iodide or as thyroxine when added to artificial seawater used as a culture medium for laboratory-maintained *Aurelia* polyps. Natural seawater normally contains iodine in trace amounts, and since the iodine content undoubtedly fluctuates with the seasonal growth and decay of iodine-binding seaweeds, iodine may well be the trigger in nature as well. In any case, it is striking that iodine and thyroxine are capable of initiating

a profound metamorphosis in such unrelated organisms as jellyfish and amphibians. What the significant change may be that iodine brings about in the tissue metabolism of the responsive organisms remains a question for the future.

B. Metamorphic and direct development in ascidians

Among lower-chordate animals, such as ascidians, metamorphosis occurs in dramatic form, particularly with regard to absorption of the tail. At the time of fertilization, the ascidian egg is essentially programmed to develop into a tadpole larva. When the period of larval activity comes to an end and larval tissue is resorbed (Fig. 19.3), the residual tissue takes a great leap forward, so to speak. It is released from the dominating presence of tadpole organization and at the same time receives a new nutrient supply in the form of autolyzed tadpole-tail tissue. The factors that trigger tail resorption of the tadpole, which is the most striking feature of ascidian metamorphosis, are still unknown. The development of the ascidian egg was described earlier in this book, with regard to the organization of the egg and cleavage pattern and with regard to gastrulation and neurulation. The duality of such egg development is evident. On the one hand, the egg develops into a chordate larva, and the organization of the egg, cleavage pattern, and gastrulation appear to serve this end primarily. The swimming, nonfeeding tadpole larva, with its neurosensory-musculature equipment, functions as a special site selector for the permanent settlement of the organism. On the other hand, from the start, but more slowly, the egg develops the permanent organization of the ascidian, which continues to develop through the metamorphic period into the juvenile and later adult condition, that of a sessile filter feeder.

Even in egg development, however, the tadpole stage is not an essential step in the process of becoming a young ascidian. Ascidians generally occupy the multiple attachment sites available along rocky shores, lagoons, reefs, etc., where site selection is a vital need. Some ascidians, however, inhabit the vast mud and sand flats of the continental shelf, where there is little or no need for site selec-

Figure 19.3 Metamorphosis in an ascidian. *a* Metamorphosing tadpole showing resorbing tail. *b* Attached metamorphosing tadpole, with tail nearly completely resorbed. *c* Fully metamorphosed ascidian with functional internal filter system, showing inhalant siphon (mouth) and exhalant siphon.

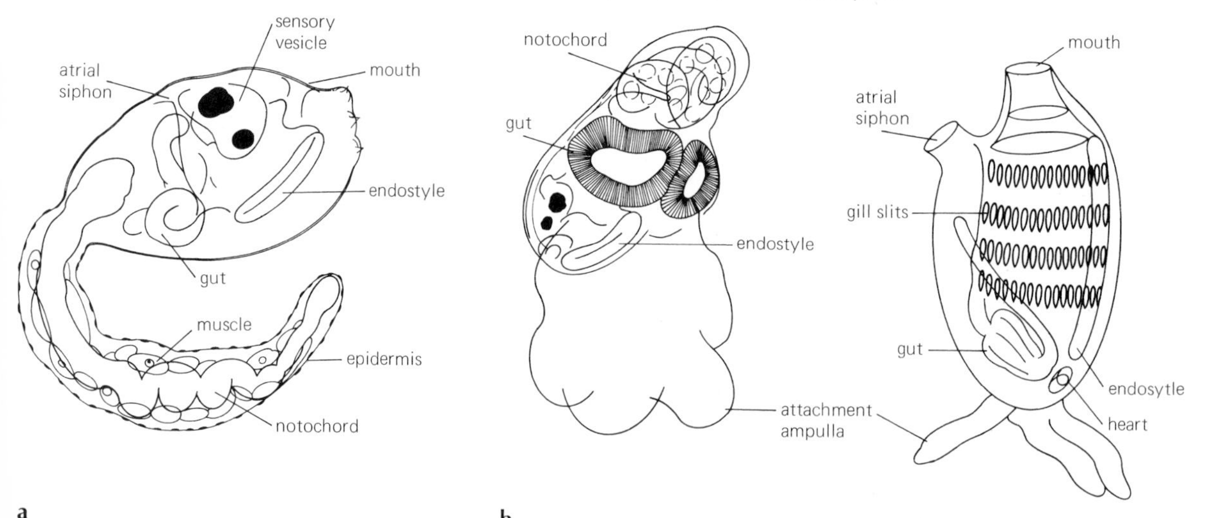

tion. In nearly all such species, the egg develops directly to form a miniature ascidian without forming a chordate embryo or tadpole larva. Although egg size, developmental rate, gastrulation, and time of hatching all remain unaltered, only traces remain of the embryonic enzymes characteristic of notochord, tail muscle, and sensory differentiation in the tadpole larva.

C. Development of nonsexual rudiment

This concept of the developing ascidian egg as a special development contained within another slower and more durable developing system is further established by the phenomenon of asexual bud development in ascidians. Buds are essentially small fragments of tissue isolated from the parental organism either physically or physiologically in circumstances that support developmental growth.

Buds of different species differ widely in their constitution, depending on what part of the parental organism is included. Some are minute disks or spheres consisting of very few cells. Whether as a disk or a sphere, such a primordium consists of an outer epidermal layer of simple epithelium and of an inner layer, also a simple epithelium. As such it can be compared in organization with the two-layered, invaginated gastrula stage of embryonic development, although it is much smaller. In all cases, the outer layer is derived from the specialized parental epidermis. The inner layer consists of unspecialized cells which, according to the species, are derived from parental tissue that may be ectodermal, endodermal, or mesodermal in origin. This layer gives rise to a whole, sexually mature ascidian individual, except for the epidermis.

It is evident, therefore, that unspecialized tissue cells, unrelated to specifically reproductive tissue, have the potentiality to develop into the complete organism. It is equally evident that the intricate course of egg development leading to tadpole formation and subsequent metamorphosis is a consequence of the special differentiation of the egg as an egg, in addition to its developmental potential as a cell.

The early determination process with ascidian eggs has already been described in earlier chapters (Chapters 7 and 9), particularly the yellow-crescent region of myoplasm in the undivided egg, which is eventually found within the muscle-lineage cells of the tadpole larva in association with acetylcholinesterase. This has been shown to be a cytoplasmic determinant found within the egg before the onset of cleavage and the initiation of nuclear lineages, while other cytoplasmic determinants relating to specifically larval cells and tissues appear to be equally precocious, altogether seeming to comprise a set produced by genetic transcription at or close to the onset of development. It is this complex of tissues that undergoes destruction at metamorphosis. At the same time, however, a determinant relating to intestinal development and function (which is complete very much later than the sensorimotor system of the nonfeeding tadpole) has been identified in an even earlier stage of egg development. In other words, following fertilization, development is initiated and leads immediately, although at different rates, to both the sessile, filter-feeding adult-type organization and the motile, nonfeeding tadpole larva. The basic problem is how two such greatly differing systems are activated or established within the same genome.

The two systems, however, do have much in common compared with the contrast seen in ascidian egg and bud development. Here, the egg develops to basic vertebrate organization and this persists throughout life, but in structure, function, and cellular differentiation, the larval and adult systems, respectively, represent special modifications of the basic plan in virtually every detail. Significantly, the relatively large and yolky eggs of certain tropical toads of the genus

Eleutherodactylus are laid out of water and develop directly to the adult form; that is, no aquatic gill-breathing stage with its concomitant aquatic structures and tissues is produced; no metamorphosis is necessary; and the differentiations characteristic of the postmetamorphic amphibian develop without prelude.

3. Metamorphosis in amphibians

The amphibian egg and larva are, in their general organization, clearly a retention of an ancient ancestral type, serving only while development proceeds in the original freshwater environment. Frogs, toads, and in a less spectacular way, urodeles transform from an aquatic gill-breathing larva to a lung-breathing terrestrial animal. The general course of development through this life cycle is well known (Fig. 19.4). Two transformations take place:

1. Certain adaptive structures formed during embryonic development—namely, the ventral suckers and the external gills of the anuran tadpole and the balancers of the urodele larva—are resorbed during early functional life. These are merely resorptions of precociously formed structures; they are not a part of the general metamorphic event, which occurs much later and in anurans includes the resorption of the tail. These earlier structures disappear when they have served their purpose.
2. Almost every organ system of the frog undergoes alterations during metamorphosis. Before the onset of metamorphosis, the tadpole, a vegetarian, is a fully aquatic creature with well-developed gills, a long flattened tail, lidless eyes, horny rasping teeth, and a long coiled intestine. After the completion of metamorphosis, the froglet, no heavier and usually somewhat lighter than the tadpole that gave rise to it, is a lung-breathing creature with well-developed limbs, eyelids, teeth, gut, and other structures associated with its carnivorous habits, and no tail. Even structures that persist into the adult undergo changes. For instance, the skin thickens, becomes more glandular, attains an outer keratinized layer, and acquires a characteristic pattern of pigmentation. The brain becomes more highly differentiated. At the cellular level, cell modifications are evident in eyelids, limbs, lungs, tongue, eardrum, operculum, skin, liver, pancreas, and intestine. Horny teeth are lost, and the tail and the larval cloaca atrophy. Probably no cell, tissue, or organ remains entirely unaffected.

A. Biochemical changes

Many physiological and biochemical changes take place during metamorphosis. The biochemical alterations may be considered to have direct adaptive value or to serve as a basis for morphological, chemical, or other changes that have adaptive

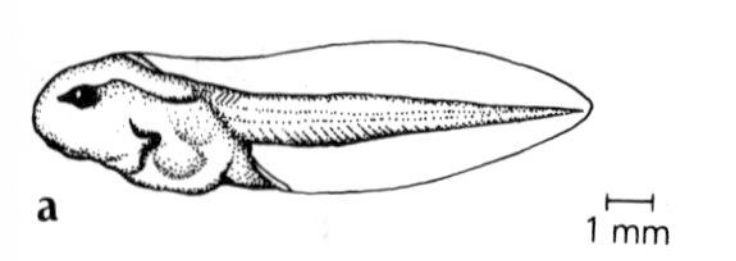

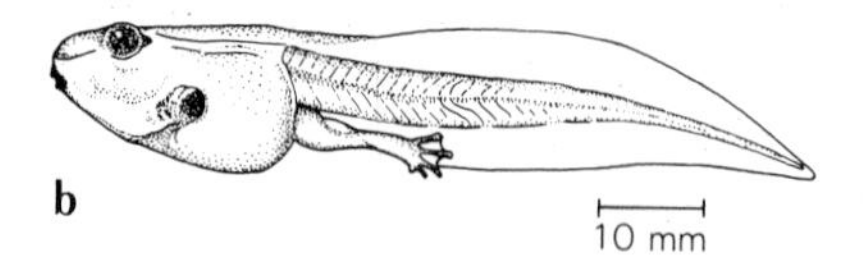

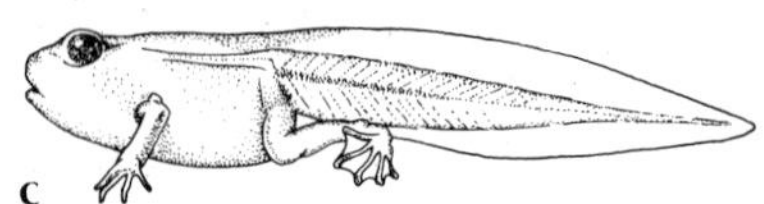

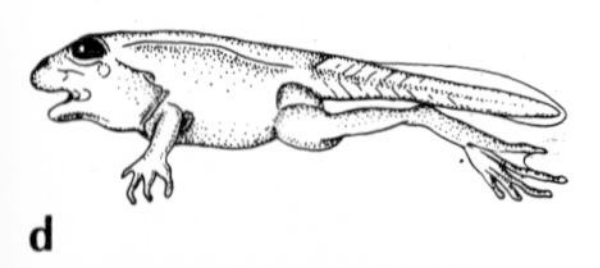

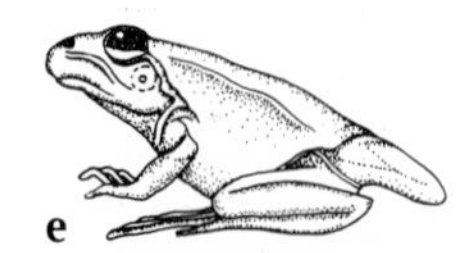

Figure 19.4 Typical anuran metamorphosis (*Rana pipiens*). *a* Premetamorphic tadpole. *b* Prometamorphic tadpole (growth of hindlimbs). *c* Onset of the metamorphic climax (eruption of forelimbs, retraction of tail fin). *d, e* Climax stages, showing gradual appearance of the froglike organization.

value relating to the change from freshwater to land. For example, among the most important adaptive changes are the shift from ammonotelism to ureotelism (i.e., from the excretion of ammonia to the excretion of urea), the increase in serum albumin and other serum proteins, and the alteration in the properties and biosynthesis of hemoglobins. The development of certain digestive enzymes and the augmentation of respiration also contribute to the success of the differentiation process. During metamorphosis, there are many additional important chemical developments, which may be secondary to the primary morphologic or cytologic transformations that aid in the adjustment to land. These include alterations in carbohydrate, lipid, nucleic acid, and nitrogen metabolism. Major modifications in water balance, visual pigments (vitamin A), pigmentation, and tail metabolism are also observed. Finally, there is a partial mobilization of the enzyme machinery to promote the metamorphic process and the colonization of the land.

It is not yet certain whether the frog and the tadpole enzymes differ in their basic subunits or whether they are made up of the same subunits but in different proportions. The synthesis of what seems to be a new enzyme during metamorphosis of the tadpole to the frog, with each enzyme serving in what appears to be the same metabolic role, indicates that biochemical differentiation from the stage of the fertilized ovum to the tadpole stage involves a different, or modified, genetic expression from that in the biochemical differentiation of the tadpole to the adult frog.

B. Hormonal control

The sequence of metamorphic changes relates to the buildup of thyroid activity. At first the hypothalamic-pituitary-thyroid system begins activity very slowly, accelerating during premetamorphic growth. During this period, early metamorphic events, such as leg growth, are completed before the climax intervenes. The thyroid hormone, acting on the hypothalamic system, builds it up to higher and higher levels, finally resulting in an explosive release of thyroid hormone at the beginning of climax. The moment in the lifespan when this positive-feedback system first starts varies with each species depending on the time its hypothalamic system becomes sensitive to thyroid hormones. The time of acquisition of this sensitivity is genetically determined differently for each species according to the length of its larval period. This is the basic clock of metamorphosis. The rate of feedback from the thyroid serves as the subsidiary clock regulating the pattern of metamorphic change.

Frogs metamorphose after various periods of growth, according to the species. It was recognized early that the attainment of a critical species-specific size, rather than the duration of growth, is crucial. Thus, in nature, bullfrog tadpoles metamorphose at the end of the third summer-growth season in the North, but at the end of the second in the South, after having attained a certain size. The time required depends on the mean environmental conditions. It has also long been known that adding iodine to the water or feeding with thyroid-gland tissue causes metamorphosis to occur earlier, at a smaller size (Fig. 19.5). Elimination of iodine from the diet postpones it. Conversely, elimination of iodine and therefore of thyroxine by thyroidectomy of the tadpole results in failure of metamorphosis to take place. Nevertheless, such a tadpole may continue to grow and become sexually mature, although still a tadpole. This phenomenon is known as *neoteny*. From the first, therefore, the iodine-containing thyroxine of the thyroid gland has been studied intensively with regard to the timing of metamorphosis and the metabolic effect of the hormone. Any vertebrate hormone, however, is a component

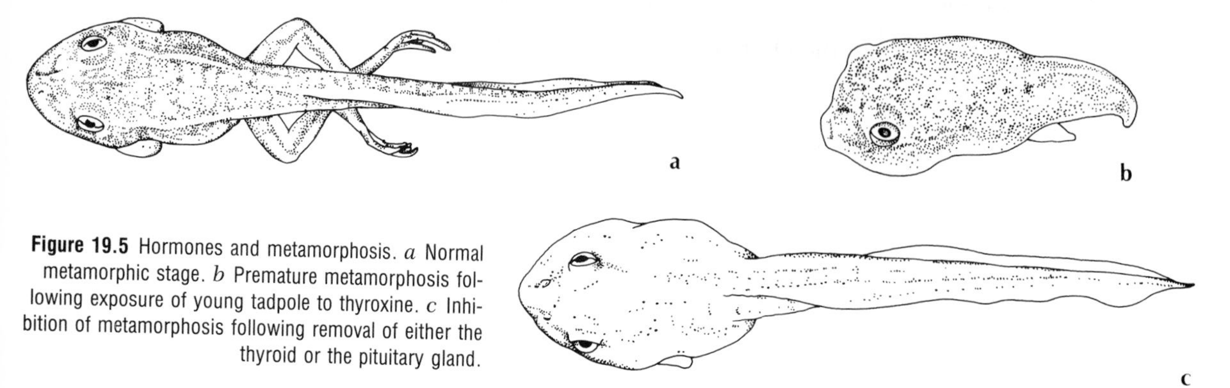

Figure 19.5 Hormones and metamorphosis. *a* Normal metamorphic stage. *b* Premature metamorphosis following exposure of young tadpole to thyroxine. *c* Inhibition of metamorphosis following removal of either the thyroid or the pituitary gland.

in a complex interacting hormonal system, in which pituitary hormones are particularly involved.

In nature, the perennibranchiate, or permanently gilled, salamanders grow to sexual maturity as permanently aquatic creatures that retain larval features, such as external gills, and do not undergo metamorphosis. Typically they have relatively inactive thyroid glands, and they do not respond to thyroxine or iodine treatment. They are also remarkable in having tissues that consist of comparatively enormous cells, as though the larva had grown to adult size by means of cell enlargement rather than cell proliferation. The phenomenon calls for further analysis.

At the level of endocrine action, we again see that developmental control is effected by a balance between inhibition and activation rather than by simple stimulation. The pituitary-thyroid axis is kept at a low level of activity in the growth phase of the tadpole's development by negative feedback, and at metamorphosis its activation is brought about by hypothalamic action. The complexity of this push-pull type of interaction in governing metamorphosis is further emphasized by the discovery of the role of prolactin as a thyroid antagonist in amphibian development. Whether this substance acts at the peripheral level or at a higher one, or in both ways, its role again emphasizes that development is controlled by a dynamic balance of plus and minus factors. The thyroid-hormone signal appears able to be locally modulated, either amplified or attenuated. The molecular basis of such local modulation is unknown.

The developmental system, seen as a continuously maintained although progressively elaborating and changing organization, the very molecules of which are in perpetual flux as long as life persists, is a reality whose essence is elusive. Yet any change in the genetic information appears as some sort of change in the developmental outcome, and any subtance to which cells and tissues are sensitive can alter the course in some degree. Hormones, which are themselves the products of development, have such a role and act on cells and tissues in various ways to modify or modulate the timing and direction of events. Yet no more than the so-called inductors, organizers, and organ-forming substances of the embryo can they be considered to be truly developmentally instructive. It should be kept in mind that hormonal activity in a developing system is possible only in organisms already sufficiently developed to produce hormones. Hormones are unlikely to be operative at the time of metamorphosis of the small larval organisms of marine invertebrates such as sea urchins and ascidians.

C. Differential tissue sensitivity

Metamorphosis involves alteration of larval tissues that normally acquire sensitivity to thyroid hormones long before significant quantities of such hormones are released into the circulation. In general, structures that metamorphose early are more sensitive than those which undergo change later. Successive metamorphic events have different and somewhat higher thresholds from preceding ones. Thus limb development, inhibited during most of the growth period of the tadpole, apparently needs a progressive booster at a certain stage. This may be associated with the enormous growth that anuran hind limbs finally undergo during metamorphosis compared with urodele limbs, which remain relatively small and require only developmental release. In the frog, however, as hormone concentrations increase, tissue responses become progressively more rapid, until maximum rates of change are attained. In effect, at high concentrations, all metamorphic events become crowded together, and the time sequence is disturbed; the tail begins to resorb before limbs become well developed, so that without providing means of locomotion and other essential factors, the transformation leads to death.

Difference in tissue sensitivity is seen in a striking form in an early experiment in which an eye cup was transplanted to a tadpole tail, where it differentiated. During metamorphosis, the eye moved forward as the tail resorbed and finally came to rest in the sacral region when metamorphosis was complete (Fig. 19.6). Transplanted limbs and transplanted kidney tumor are likewise unaffected by the degenerative processes in the surrounding tail tissue.

Perhaps the most striking single feature of the hormonal control of amphibian metamorphosis is that a single hormone, the low-molecular-weight compound thyroxine, evokes multiple responses from diversified tissues. Responses are specific, although the inducer varies only quantitatively. Moreover, they are both constructive and destructive, depending on the target tissue. Thus, in response to triiodothyronine, a companion hormone of thyroxine, biosynthesis of nucleic acid is decreased in the tail but increased in the liver. Similarly, thyroxine induces rapid aging and destruction of the red blood cells of the metamorphosing tadpole, which carry tadpole-type hemoglobin; simultaneously (or later), it stimulates the

Figure 19.6 Organ specificity of metamorphic responses in tadpoles. *a–c* Tail tip transplanted to the trunk region undergoes atrophy simultaneously with the host's tail. [*After Geigy*, Rev. Suisse Zool., **48**:*483 (1941)*.] *d–f* Eye cup transplanted to the tail remains unaffected by the regressing tail tissue. [*After Schwind, 1933.*]

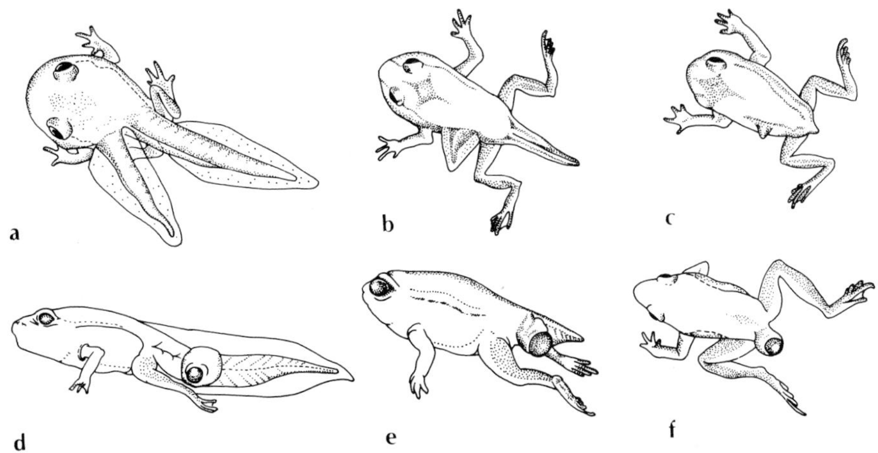

development of cells that synthesize the adult-frog type of hemoglobin exclusively. The agent is the same, but the response is cell death and cell proliferation, respectively, in the two erythrocytic-cell populations.

D. Tail resorption

Changes in the fine structure of the tail muscle, involving myofibrils, mitochondria, and sarcoplasm, occur before any appreciable shortening of the tail takes place. Digestion of tissue responsible for tail resorption is accomplished by lysosomal enzymes (Fig. 19.7). Experimental evidence indicates that thyroid hormone acts directly on peripheral tissues to accelerate metamorphic changes at both a morphological and a biochemical level.

Tail-isolation culture has been used as a model to study the direct metamorphosing action of thyroid hormone at the biochemical level. The process of tail resorption has been investigated in isolated tails or pieces thereof of *Xenopus* in particular. Isolated tails respond to thyroxine. Involution always begins at the very tip, and the typical pattern of metamorphosis is as follows: thickening of the epidermis, migration of pigment cells, and involution of notochord, neural tube, and muscle cells.

Inhibition of RNA synthesis with actinomycin D or of protein synthesis with puromycin or cycloheximide completely abolished regression that had been induced in tail-organ cultures by triiodothyronine. This indicates that additional

Figure 19.7 Regression of tadpole tail, in metamorphosis of *Xenopus laevis*, is accomplished by lysosomal digestion of cells. As metamorphosis proceeds, the enzyme concentration increases (the absolute amount of enzyme remaining constant). Eventually the stub contains almost nothing but lysosomal enzymes, and it falls off. [*Partly after W. Etkin and R. Weber.*]

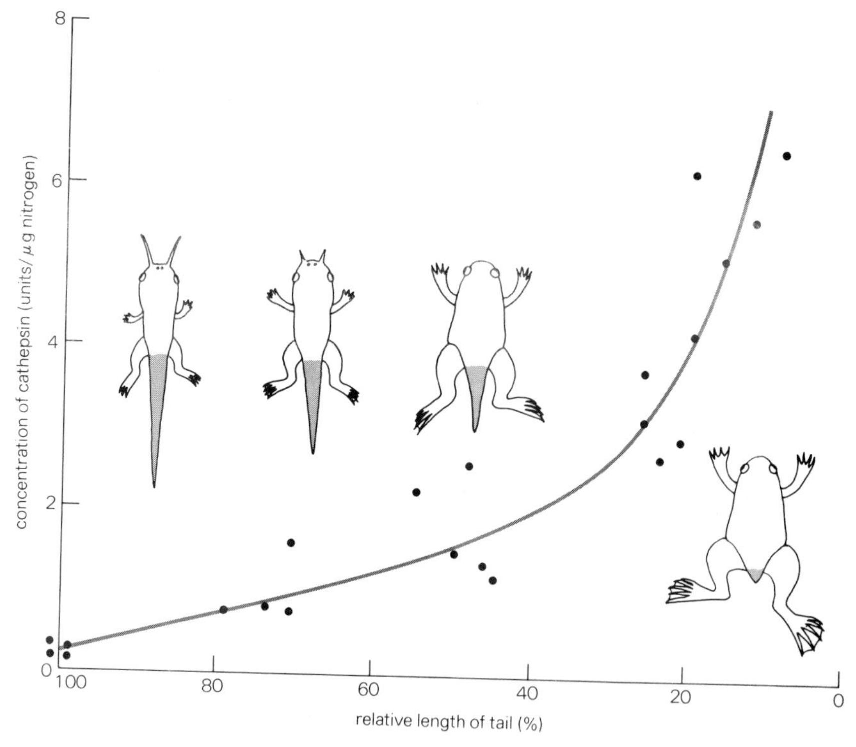

RNA and protein synthesis are essential for the process at the onset of regression in cultured tails. A basic question remains: Does the general patterned process of tissue regression and cell death trigger the release of the new phenotypic development, or is the differential destruction of the "juvenile" structure a response to the activation of the new phenotype?

4. Insect metamorphosis

Insects, in both egg and adult form, have become fully terrestrial. The most primitive insects lay small but yolky eggs on land, which develop directly to the adult state without passing through special larval stages or through a process of metamorphosis. Special larval forms, especially those of holometabolous insects (those which undergo pupation), are relatively new evolutionary inventions interpolated into the originally direct course of development. This holometabolous transformation is surely as radical an alteration of form as graces the animal kingdom. The extraordinary fact is that a caterpillar hatches from the egg of a butterfly; the subsequent change of the caterpillar into the butterfly is merely the return of the metamorphosed young to the form of its parents. The transformation of the caterpillar is a visible event reenacted with each generation. The legs of a caterpillar, for example, are not primitive structures representing an early step in the development of the adult legs, because the legs of the adult exist within the larva as distinct structures, segregated off as imaginal discs. The larval legs serve as legs, but they are relatively new structures evolved for the special conditions of larval life and are sacrificed in the return toward the ancestral form.

A. Developmental cycle

The eggs of the large *Cecropia* silkworm moth hatch as larvae after 10 days; they subsequently molt four times as they grow some 5,000-fold to mature as fifth-instar larvae, all the while transforming leaves into silkworm. They then enter pupation; the pupa is enclosed in a cocoon (Fig. 19.8).

The completion of the cocoon signals the beginning of a new and even more remarkable sequence of events. On the third day after a cocoon is finished, a great wave of death and destruction sweeps over the internal organs of the caterpillar. The specialized larval tissues break down, but meanwhile, certain more or less discrete clusters of cells, tucked away here and there in the body, begin to grow rapidly, nourishing themselves on the breakdown products of the dead and dying larval tissues. These are the imaginal discs, already discussed in Chapter 7, which throughout larval life have been slowly enlarging within the caterpillar. Their spurt of growth now shapes the organism according to a new plan. New organs arise from the discs. Also, some less specialized larval tissues, such as the epidermal layer of the abdomen, are transformed directly into pupal tissues. Pupation is followed by a developmental standstill, a diapause, lasting 8 months. This device allows the pupa to survive the winter and emerge at an appropriate time the following spring.

Then comes a second period of intense morphogenetic activity. The result is a predictable pattern of death and birth at the cellular level as the specialized tissues of the pupa make way for the equally specialized tissues of the adult moth. Spectacular changes occur throughout all parts of the insect: in the head, the formation of compound eyes and featherlike antennae; in the thorax, the molding of legs, wings, and flight muscles; in the abdomen, the shaping of genitalia and, internally, the exorbitant growth of ovaries and testes. In the newly formed skin we can witness the strangest behavior of all—the extrusion and transformation of

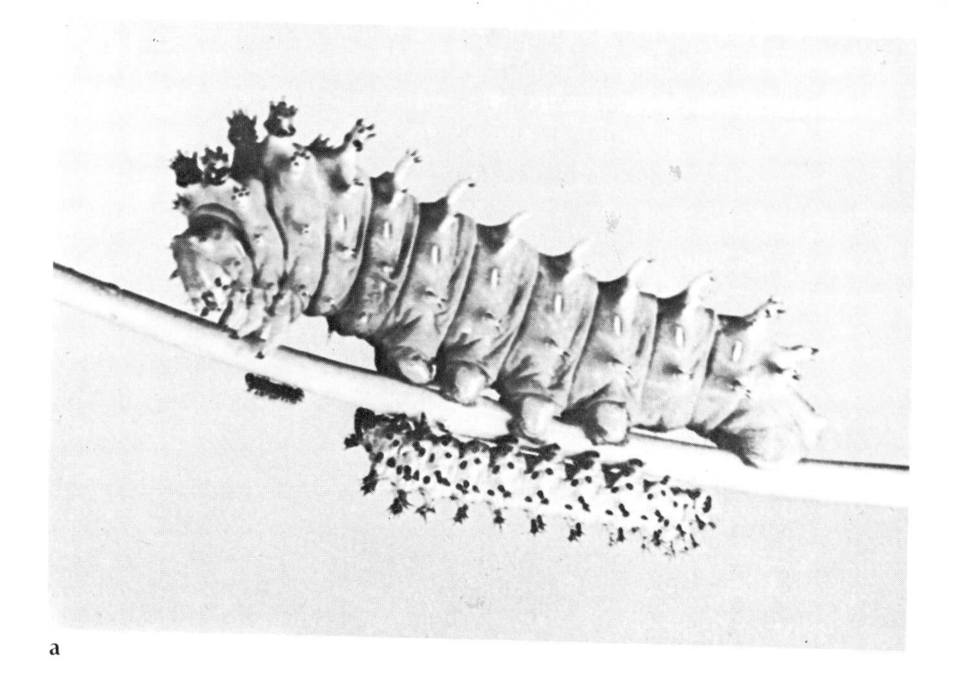

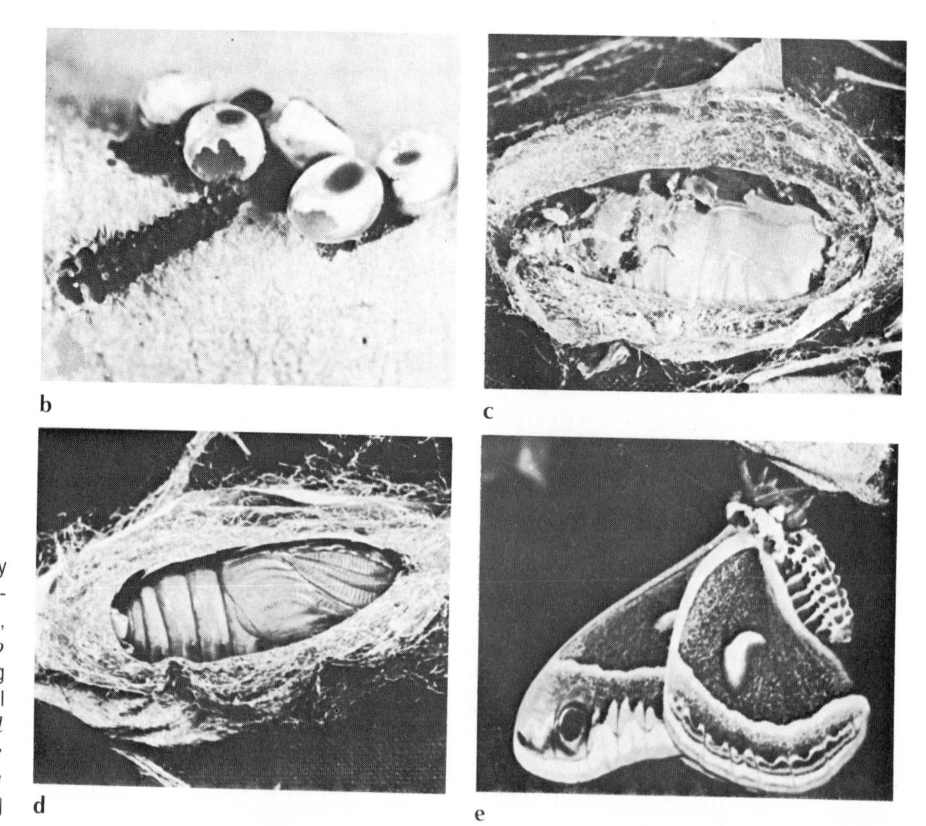

Figure 19.8 Life history of the *Cecropia* silkworm. *a* First-, third-, and fifth-instar larvae. *b* First-instar larva hatching from egg. *c* Larval-pupal molt within cocoon. *d* Pupa within cocoon. *e* Adult male. [*Courtesy of L. I. Gilbert.*]

tens of thousands of individual cells into the colorful but lifeless scales so typical of moths and butterflies. After 3 weeks of adult development, the process is complete. The full-fledged moth escapes from the cocoon and unfurls its wings.

B. Extent and timing of metamorphosis

The many orders of insects show a great variation in life cycles. Of the winged orders, the exopterygote insects have wings that develop internally, as in stone flies, termites, dragonflies, cockroaches, and locusts. All these insects have a series of larval stages that lack wings and have immature reproductive systems. They exhibit a gradual transition to the winged, sexually mature adult, a process called *incomplete metamorphosis*. Each nymphal stage molts into the next stage, which may have slightly more developed wing buds and reproductive system. The greatest change even in this general type, however, is seen at the last, or nymphal-adult, molt, when a dramatic growth of wings and reproductive systems takes place.

The other group of winged insects are the endopterygotes, with wings and other structures developing internally in invaginated imaginal epidermal pockets. They include moths and butterflies, bees and wasps, flies, and others. Metamorphosis is described as complete. Larval stages lack any external evidence of wings and reproductive organs. The transformation into the adult is a two-step process, the fourth or fifth instar, or larval stage, molting into a pupa and the pupa into an adult. In all members of this group of insect orders, not only are reproductive systems, wings, and compound eyes present in the adult and absent in younger, larval stages, but with changes in feeding habits, extensive changes occur in mouth parts and intestine, while muscle requirements may be drastically different in larva and adult. No system, in fact, remains unaffected by the metamorphic changes occurring during the pupal stages (Fig. 19.9).

a

b

Figure 19.9 *a* Female pupa of the silkworm *Antherea polyphemus*. *b* Adult of the same species prior to wing expansion. [*Courtesy of L. I. Gilbert.*]

C. Molting and juvenile hormones

Insects have been studied mainly with regard to their external features and structures, which consist of cuticle laid down by a single layer of epidermal cells. Internal organ systems are as important to the insect as to any other animal but are even less amenable to experimental, developmental, or genetic studies. It seems, therefore, that in the insect the epidermis is the chief agent of morphogenesis, although our ignorance may be bliss and seeing is believing. However this may be, the general characters of the cuticle change in successive stages of growth. Since the cuticle is a mosaic made up of the contributions of each individual epidermal cell, the growth activity of each cell, at each stage of the whole developmental history, is firmly registered in the characters of the little patch of cuticle it lays down. And since the cells constituting the single epidermal sheet not only lay down cuticle, a process incompatible with cell division, but are also responsible for growth and change of form, which requires cell multiplication, these two processes do not occur at the same time. Accordingly, the growth of insects takes place in cycles, in which mitosis and cellular growth alternate with the deposition of a new cuticle and the shedding of the old, i.e., in molting cycles. All insects are subject to this pattern of growth and its restraints.

The sequence of events is under precise hormonal control. In brief, certain stimuli associated with the state of nourishment cause the brain to discharge the "brain hormone" from the neurosecretory cells of the pars intercerebralis. This hormone, in turn, activates the endocrine organ known as the *prothoracic gland,* which then secretes the molting hormone *ecdysone.*

Ecdysone is a cyclic compound of small molecular size. Just as the same thyroid hormones are produced in all classes of vertebrates, the hormone ecdysone is produced by all insects and other arthropods that have been investigated. Molting cycles are characteristic of all classes of arthropods. Ecdysone extracted from insects causes molting in shrimps, while ecdysone from shrimps causes molting in insects. It is an activator of a nonspecific type. The growth and molting cycles of an insect are more specifically controlled, however, by a third hormone, the *juvenile hormone* (Fig. 19.10), secreted by the corpora allata, which are endocrine glands located near the brain. A general pattern of hormonal control in insects accordingly emerges, although there are differences among insect groups.

In *Drosophila melanogaster,* during the prepupal period, the imaginal discs are converted in a matter of hours into the basic form of the adult insect. There are extensive gross morphological changes involving the evagination of the appendages, fusion of the discs into a continuous sheet of tissue, and formation of the cuticle. Discs invaginated in pockets during larval life evaginate at the onset of metamorphosis. Evagination and RNA synthesis in the discs can be evoked in vitro by ecdysone.

Differential response must be a property of the target tissues. For example, many of the cells of the epidermis are highly differentiated and are no longer capable of division or of renewal of the cell cycle; other cells lying between the groups of specialized cells remain dormant or undifferentiated.

Under the impact of ecdysone, at each molt the old cuticle is loosened and thrown off, the specialized epidermal cells are resorbed or discarded, and the ordinary epidermal cells that have been stimulated to grow and divide give rise to new cuticle and to new groups of specialized cells. In fact, ecdysone acts directly on those cells which are in a state of dormancy. Within a few hours, the nucleolus is enlarging, RNA begins to accumulate in the cytoplasm, and mitochondria enlarge and multiply by subdivision. By the time the old cuticle is thrown off, the renewed epidermal growth has been virtually completed, with locally expanded

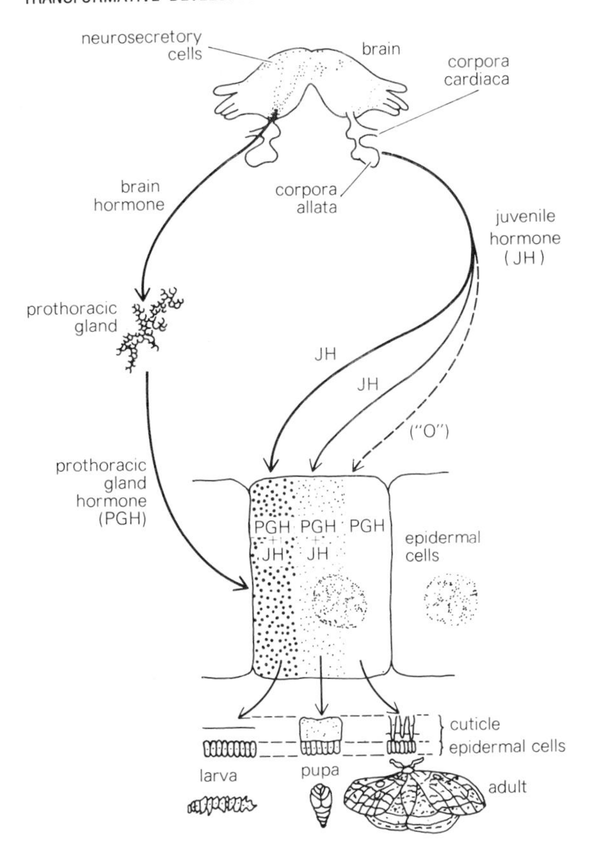

Figure 19.10 The endocrine control of growth and molting in the *Cecropia* silkworm. The nature of the molt is determined by the relative levels of the two hormones, PGH (ecdysone) and JH, the latter dropping steadily through the life cycle. [*After H. Schneiderman and L. I. Gilbert, 1959.*]

regions folded compactly while awaiting release from cuticular confinement. Remarkably, the muscles of the larva, e.g., of *Rhodnius*, dedifferentiate between molts and redifferentiate muscle fibers shortly before the next molt occurs. To a degree, the organism is renewed on each occasion and may therefore be said to undergo a degree of metamorphosis. As a general but not absolute rule, molting occurs each time the volume of the growing animal doubles. Such is the cyclic basis of growth, which underlies the more spectacular changes commonly seen in the life history of many insects and crustaceans.

Growth regulation in a developing system requires that the component cells communicate with each other. In insects, the epidermal cells are ionically coupled, and in the larva of the beetle *Tenebrio*, for example, gap junctions make up to 20 to 30 percent of the functional membrane. Functional-membrane conductivity is elevated by ecdysone, apparently either by the formation of new gap-junctional channels or by increasing the bore size of existing channels. Both cyclic AMP and calcium ions reverse the ecdysone-stimulated changes. It is possible that changes in intercellular communication before metamorphosis may reflect the timing of the signals that trigger proliferation and the generation of new spatial patterns in the epidermis.

In holometabolous insects, whether fruit flies or giant saturniid silk moths, the life cycle consists of usually four or five larval molts, a larval-pupal molt, and

a pupal-adult molt. The larval-pupal-adult transformations constitute an extensive metamorphosis. In the silkworms, metamorphosis is usually interrupted soon after pupation by a prolonged pupal diapause, during which development usually ceases. Sexual maturation, which is a part of metamorphosis, occurs during the construction of the adult during pupation. Six stages follow the hatching of the embryo: four (in some, five) larval stages (or instars), the pupation phase, and the mature adult. The intermittent molting, with its discard of the old and replacement with the new, allows control both of phase duration and of remodeling during the span of individual life. In the absence of ecdysone, for instance, the insect lapses into a state of developmental standstill. When ecdysone is again secreted, this state, the diapause, is terminated and growth is resumed.

The dual nature of the development of the egg is evident and important. On the principle of first come, first served, the embryonic development leads directly to the formation and function of the insect larva—whether it be the so-called nymph of a grasshopper (resembling its parent except dimensionally and in the absence of wings), the dragonfly nymph (remarkably different from the adult in being adapted to aquatic life), or the grub or caterpillar (which must be entirely remodeled as a pupa).

In every case, the developing system, having been diverted to some degree during embryonic development, moves toward expression of the adult organization. The adult organization is latent but ready to go, so to speak. The hormonal controls operate to determine when the major event occurs, to what extent development can be temporarily arrested, and how long juvenile states may be prolonged. For example, in the presence of the prothoracic-gland hormone (PGH), juvenile hormone (JH) promotes larval development or maintains the status quo and so prevents metamorphosis. The presence of JH in an immature insect, whether larva or pupa, ensures that when the immature insect molts, it retains its larval or pupal characters and does not differentiate into an adult. When the insect molts in the absence of JH, it differentiates into an adult. In other words, withdrawal of JH initiates metamorphosis. Conversely, implantation of the JH-secreting gland from a second-stage larva to a fourth-stage larva maintains the larval state, in effect inhibiting metamorphosis, so that larval growth continues and a giant larva is produced (Fig. 19.11).

Quantitative differences are as significant as the extremes of presence and absence, as is seen in the response of epidermal cells to JH. When the larval epidermis molts in response to ecdysone (PGH), the response varies as follows:

1. In the presence of a high concentration of JH, the cells secrete larval cuticle.
2. In the presence of a low concentration of JH, the epidermal cells secrete a pupal cuticle.
3. In the absence of JH, the epidermal cells secrete the adult, or imaginal, cuticle directly and omit the pupal molt.

Moreover, if the corpora allata are removed in the last larval stage of the honeybee or the *Cecropia*, for instance, instead of a normal pupa appearing, a monstrous form intermediate between a pupa and an adult is developed. However, if these glands are surgically removed during one of the earlier stages, the larvae undergo precocious pupation at the very next stage and eventually develop into midget-sized adult insects. The larval form is developed in the presence of a high concentration, the pupa in the presence of a very low concentration, and the adult when the hormone is completely absent.

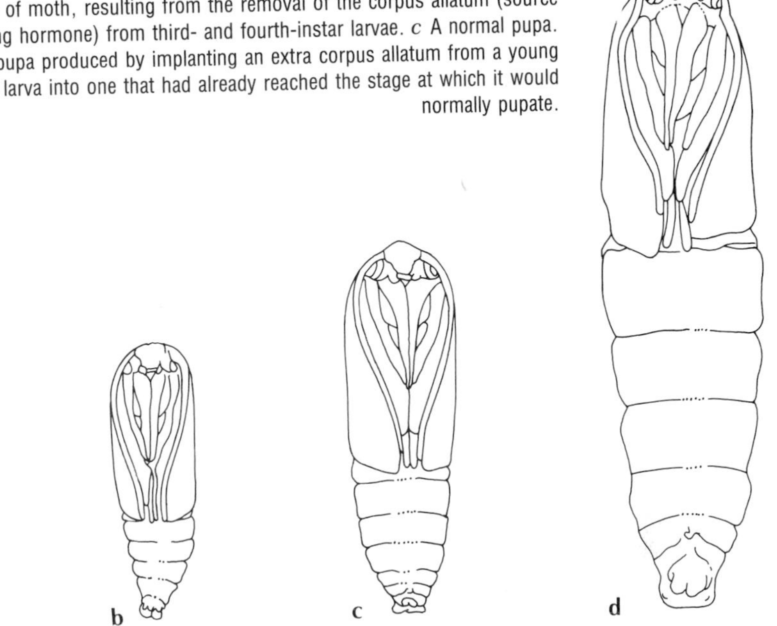

Figure 19.11 Experiments on effect of molting hormone on pupa size. *a*, *b* Dwarf pupae of moth, resulting from the removal of the corpus allatum (source of the molting hormone) from third- and fourth-instar larvae. *c* A normal pupa. *d* A giant pupa produced by implanting an extra corpus allatum from a young larva into one that had already reached the stage at which it would normally pupate.

D. Pupal reorganization

Metamorphosis in the insect in its complete form is complex indeed, for there is a relatively great disruption of old tissues and a correspondingly great degree of new development. Moreover, not only is there much variability among different insects, but in every case each tissue in the assembly that constitutes a larva undergoes changes particular to its own character. The cells of some tissues commit suicide. In others, they may multiply, for example, to give rise to a new intestine in place of the old, or they may have persisted throughout larval life as undifferentiated, though not undetermined, cells of imaginal buds or discs and be permitted at last to fulfill their destiny. Within the confines of the pupal epidermis, the broken-down tissues yield a veritable nutritive soup. This promotes massive proliferation and reconstructive processes by the less specialized cells that survive the cellular massacre or mass suicide, whichever it may be.

Two main types of larval tissue have been recognized with regard to growth and metamorphosis. Many tissues grow by increase in cell size and undergo disintegration at metamorphosis; the corresponding adult tissue is formed from imaginal discs that were not functional parts of the larval tissue. In the second class, tissues grow by cell division and are carried over, with or without modification, into the adult. Some other tissues behave in a manner intermediate between these two methods.

The muscle system is a case in point. In some of the more primitive orders, the majority of larval muscles are carried over into the adult stage. However, in the thorax of the honeybee, for example, no larval muscles remain unchanged, although most of them are associated with the development of adult muscles. In general, metamorphosis involves the destruction of some larval muscles, the rebuilding of others, and the formation of muscles that were never represented in the larva. In muscles that disappear, the cross-striations are lost, fiber bundles lose their connections and separate, nuclear membranes disappear, and nuclei

degenerate. Only when the tissue is disintegrating do phagocytes pick up the pieces. The rebuilding of a muscle from larva to adult involves the replacement of large nuclei (which have multiplied by amitosis) by smaller nuclei (which multiply by mitosis and have been sheltered in the larval muscle). At metamorphosis the large nuclei degenerate, and each of the small nuclei becomes enclosed in a small bag of myoplasm, thus forming the myoblasts that result in new adult muscles.

A comparable situation is seen in the reconstruction of the intestine in the mosquito, for example. Larval growth is accomplished solely by increase in cell size. At metamorphosis the adult tissue is formed by the simultaneous division, reduction in size, and increase in number of these same larval cells. The great increase in size of the intestinal cells associated with larval growth as a whole is accompanied by replication of chromosomes in the individual cell, with the largest cells having as many as 32 sets of chromosomes. During metamorphosis, these cells divide until daughter cells are produced having the normal diploid sets. The new cell population gives rise to the new intestine of adult character.

Other types of tissues have cells that grow large but do not divide during larval growth, although the tissues may be augmented by growth of associated small reserve cells. These tissues also undergo excessive chromatin replication but of a different sort. Such cells do not subdivide at metamorphosis; characteristically they suffer cell death, and they may or may not be replaced, depending on organ function.

E. Hormones and gene activation

As described in previous sections, the rising level of ecdysone and the falling level of juvenile hormone bring about a sweeping change in the organization of the metamorphosing insect. These dramatic events are believed to result from a far-reaching activation of gene expression. Although it is the imaginal discs that are primarily responsible for the construction of the adult organism, the cells of these structures are not well-suited for biochemical analysis. It is known, however, that inhibition of RNA synthesis by α-amanitin or actinomycin D does act to inhibit the eversion and differentiation of imaginal-disc tissue following the rise in ecdysone concentration at the end of larval development. Much more is known about the transcriptional events that occur in certain differentiated larval tissues at the time of pupation and metamorphosis, since these cells possess giant interphase chromosomes (see Fig. 2.1).

The giant chromosomes of specialized dipteran larval and adult tissues are enclosed in a typical but large nucleus and attain a length of ten times and a cross-section of a thousand times that of normal univalent interphase chromosomes. These chromosomes are polytene; that is, they are multivalent in the fashion of a rope consisting of thousands of strands, as a result of repeated chromatid replication (typically 8 to 9 times) without mitotic condensation and separation (Fig. 19.12). Under the microscope, stained preparations of these chromosomes appear banded, the bands reflecting the fact that the strands are in perfect "register." The banding pattern itself results from differences in the density of the components from one end to the other. The banding pattern is essentially constant from fly to fly and from tissue to tissue, although very great differences are observed between the chromosomes from flies of different species.

The giant chromosomes provide a striking visual portrait of the genetic map of the individual species under examination, and over the years, the genetic function of a large number of bands in the giant *Drosophila* chromosomes has been determined. Moreover, the giant chromosomes are not simply a stable, unchang-

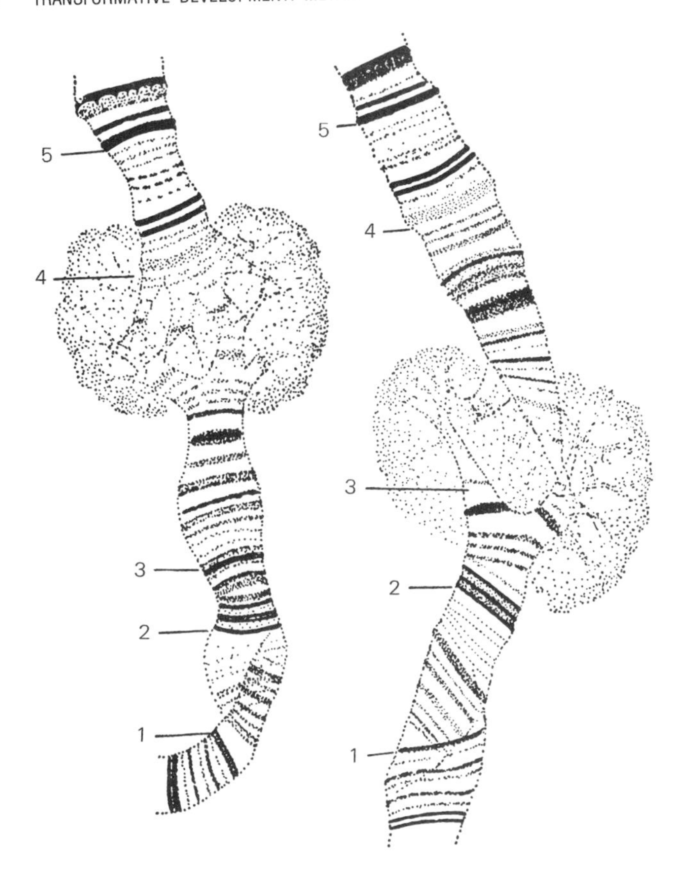

Figure 19.12 Puffing in insect chromosomes. A short section of one of the chromosomes of a fly, *Trichocladius*, from a cell of a salivary gland. Five prominent bands (*numbered*) can be recognized in each chromosome. The same chromosome region is seen from different lobes of the salivary gland, showing a strikingly different pattern. [*After W. Beerman, 1966.*]

ing refelection of the composition of the chromosomes, but they undergo important physiological changes as well. These changes involve the decondensation of the chromatin fibers of specific bands such that bulges appear on the surface of the chromosome. These bulges, termed *puffs* (Fig. 19.12), represent sites of gene transcription, as is readily demonstrated by providing the cells with radioactive RNA precursors and determining the sites of incorporation by autoradiography (Fig. 19.13).

The giant chromosomes of the salivary glands have opened the door to a visualization of the process of gene activation, particularly in relation to metamorphic events. If one examines the polytene chromosomes of the salivary glands of a *Drosophila* larva at about 15 hours before pupation, approximately 10 prominent puffs can be seen. Approximately 10 hours before pupation occurs, a dramatic change in the puffing pattern is seen. From this point to about 2 hours after pupation, approximately 125 different bands undergo puffing, each appearing and disappearing at specified times in the process (Fig. 19.14). Remarkably, the entire series of puffing changes can be induced to occur in isolated salivary glands incubated in vitro with ecdysone. The two puffs that appear first in vivo are the same two puffs to appear first after the administration of ecdysone in vitro. Altogether, a handful of genes appear to become activated within the first hour in the presence of elevated levels of ecdysone. The precise mechanism by which gene expression is activated at these loci is not understood. The hormone

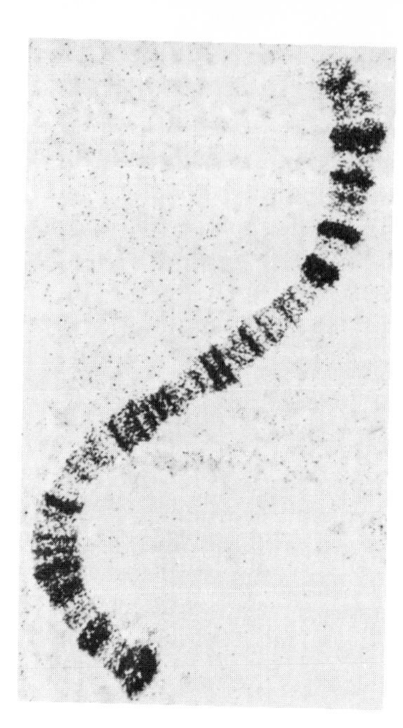

Figure 19.13 Autoradiograph of a *Chironomus* chromosome showing that the puffs are the sites of the major incorporation of [^{3}H]uridine into RNA. [*Courtesy of C. Pelling.*]

Figure 19.14 Puffs appearing and disappearing during the third larval instar and the prepupal stage at the base of chromosome arm 111 L of *Drosophila melanogaster* salivary glands. Numbers indicate hours before or after puparium formation. [*After H. Becker,* Chromosoma, **10**:*654 (1959), by permission of Springer-Verlag.*]

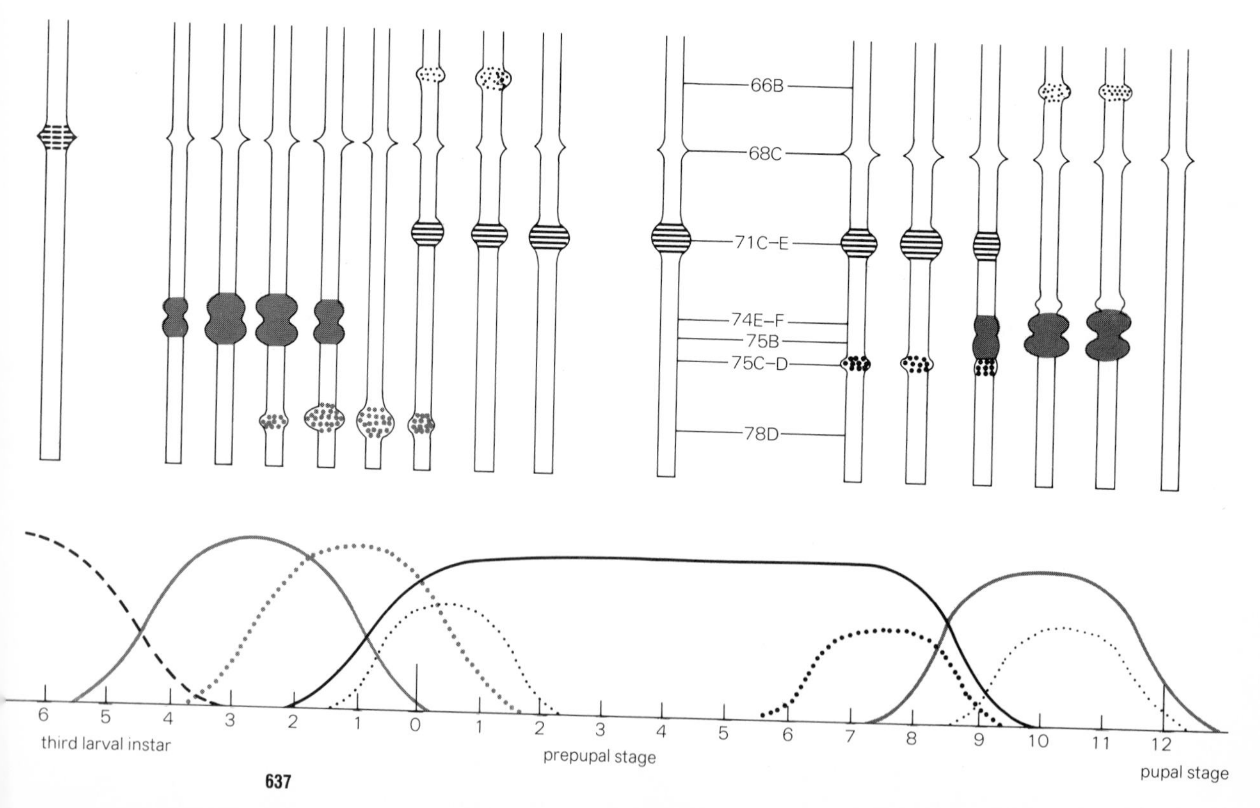

is known to enter the cytoplasm of the target cells and becomes complexed to a specific proteinaceous receptor. This hormone-receptor complex then enters the nucleus, where it is believed to become bound to specific sites within the chromatin, acting to initiate localized transcription.

The rate of transcription of the first puffs to appear seems to be dependent on the concentration of ecdysone, since the size of the puffs is proportional to hormone levels over a considerable range. If the hormone is withdrawn from the medium during the first hour or two, the "early" puffs soon regress, the DNA strands seeming to fold back into the chromosome to re-form compact bands. The formation of the early puffs is followed, after a lag of several hours, by the successive appearance of a much more diverse set of "late" puffs. The appearance of the late puffs does not require the continued presence of ecdysone but does require that the early set of puffs had previously appeared. It has been proposed that among the gene products of the early puffs are substances responsible for the activation of the secondary set of genetic loci. Although puffing patterns have been best studied in the larval-pupal salivary gland, other tissues also have been examined and found to respond quite differently to the same hormonal stimulus. This would be expected, since the morphological responses by the various tissues to increased ecdysone levels are diverse in nature.

Readings

Barrington, E. J. W., 1968. Metamorphosis in the Lower Chordates, in W. Etkin and L. Gilbert (eds.), "Metamorphosis," Appleton-Century-Crofts.

Clever, U., 1965. Chromosomal Changes Associated with Differentiation, in "Genetic Control of Differentiation," Brookhaven National Laboratory, Symposium in Biology, no. 18.

Cloney, R. A., 1978. Ascidian Metamorphosis, in F. Chia and M. Rice (eds.), "Settlement and Metamorphosis of Marine Invertebrate Larvae," Elsevier, North-Holland.

Cohen, P. P., 1970. Biochemical Differentiation during Amphibian Metamorphosis, *Science*, **168**:533–543.

Etkin, W., 1978. Searching for the Clocks of Metamorphosis, in "Pioneers in Neurobiology," vol. 2, Plenum.

———, 1968. Hormonal Control of Amphibian Metamorphosis, in W. Etkin and L. Gilbert (eds.), "Metamorphosis," Appleton-Century-Crofts.

Frieden, E., 1961. Biochemical Adaption and Anuran Metamorphosis, *Amer. Zool.*, **1**:115–150.

Gilbert, L. I., and H. A. Schneiderman, 1961. Some Biochemical Aspects of Insect Metamorphosis, *Amer. Zool.*, **1**:11–52.

Hensen, H., 1946. The Theoretical Aspect of Insect Metamorphosis, *Biol. Rev.*, **21**:1–14.

Kollros, J. J., 1961. Mechanisms of Amphibian Metamorphosis, *Amer. Zool.*, **1**:107–114.

Kroeger, H., 1968. Gene Activities during Insect Metamorphosis and Their Control by Hormones, in W. Etkin and L. Gilbert (eds.), "Metamorphosis," Appleton-Century-Crofts.

Lambert, C. C., 1971. Genetic Transcription during the Development of the Tunicate *Ascida callosa*, *Exp. Cell Res.*, **66**:401–409.

Lash, J. R., A. Cloney, and R. R. Minor, 1973. The Effect of Cytochalasin B upon Tail Resorption and Metamorphosis in Ten Species of Ascidians, *Biol. Bull.*, **145**:360–372.

Lisk, R. D., 1971. The Physiology of Hormone Receptors, *Amer. Zool.*, **11**:755–768.

Lynn, W. G., 1961. Types of Amphibian Metamorphosis, *Amer. Zool.*, **1**:151–162.

Schneiderman, H. A., 1967. Insect Surgery, in F. W. Wilt and N. K. Wessells (eds.), "Methods in Developmental Biology," Crowell.

Schwind, J. L., 1933. Tissue Specificity at the Time of Metamorphosis in Frog Larvae, *J. Exp. Zool.*, **66**:1–14.

Subtelney, S., and I. R. Konigsberg (eds.), 1979. "Determinants of Spatial Organization," 37th Symp. Soc. Dev. Biol. (1978), Academic.

WEBER, R., 1967. Biochemistry of Amphibian Metamorphosis, in R. Weber (ed.), "Biochemistry of Animal Development," vol. 2, Academic.
WITTAKER, J. R., 1979. Cytoplasmic Determinants of Tissue Differentiation in the Ascidian Egg, in S. Subtelney and I. R. Konigsberg (eds.), "Determinants of Spatial Organization," 37th Symp. Soc. Dev. Biol., Academic.
WHITTEN, J., 1968. Metamorphic Changes in Insects, in W. Etkin and L. Gilbert (eds.), "Metamorphosis," Appleton-Century-Crofts.
WHYTE, L. L., 1960. Developmental Selections and Mutations, *Science,* **132:**954.
WIGGLESWORTH, V. B., 1966. Hormonal Regulation of Differentiation in Insects, in W. Beermann (ed.), "Cell Differentiation and Morphogenesis," North-Holland.
———, 1959. Metamorphosis and Differentiation, *Sci. Amer.,* **200:**100–110, February.
WITSCHI, E., (ed.), 1961. Metamorphosis in the Animal Kingdom, *Amer. Zool.,* **1:**3–171.

Chapter 20 Reconstitutive development

The capacity of cells and multicellular organisms to restore missing structure so as to reconstitute the whole is so widespread that it must be regarded as a general property of living matter. Where the capacity is absent, the lack is therefore the result of special inhibiting organismic conditions rather than a primitive state.

1. Reconstitution

Reconstitution may take one of two paths, or even a combination of both. In *epimorphic reconstitution*, the most common form of regeneration, new structure grows out from the old surviving part until what was missing is fully restored—never more, though sometimes less. In *morphollactic reconstitution*, no outgrowth occurs, but the surviving part of the organism undergoes reorganization to form a new whole within the confines of the old. An example of the former is the regeneration following amputation in many worms, such as *Clymenella* (Fig. 20.1*a*), where whatever the number of segments missing from a cut anterior surface, that number is re-formed from a zone of growth. An example of the latter is seen in the hydroid *Tubularia*, a well-known form much studied in this conncection for almost a century. Here a distal cut surface quickly heals over and a process of reorganization of the stem proximal to the new surface reconstitutes a new hydranth or polyp (Fig. 20.1*b*) without growth as such.

Another form of reconstitution is *intercalary regeneration*, where new tissue is formed where two cut edges not normally in contact are brought together and fuse. If the flatworm *Dugesia* is cut transversely near the middle of the body, a head forms from the cut surface of the posterior piece and a tail forms from the anterior piece, as in *Clymenella*. If the two pieces are immediately brought together, they fuse, and no regeneration occurs. If, however, a transverse piece is cut from the middle region and the surviving anterior and posterior pieces are similarly brought together, fusion occurs and is followed by intercalated growth of new tissue between the anterior and posterior parts.

A comparable situation is seen in regeneration in the imaginal discs of insects. If a quadrant, for instance, is removed from a wing disc, the two radial cut

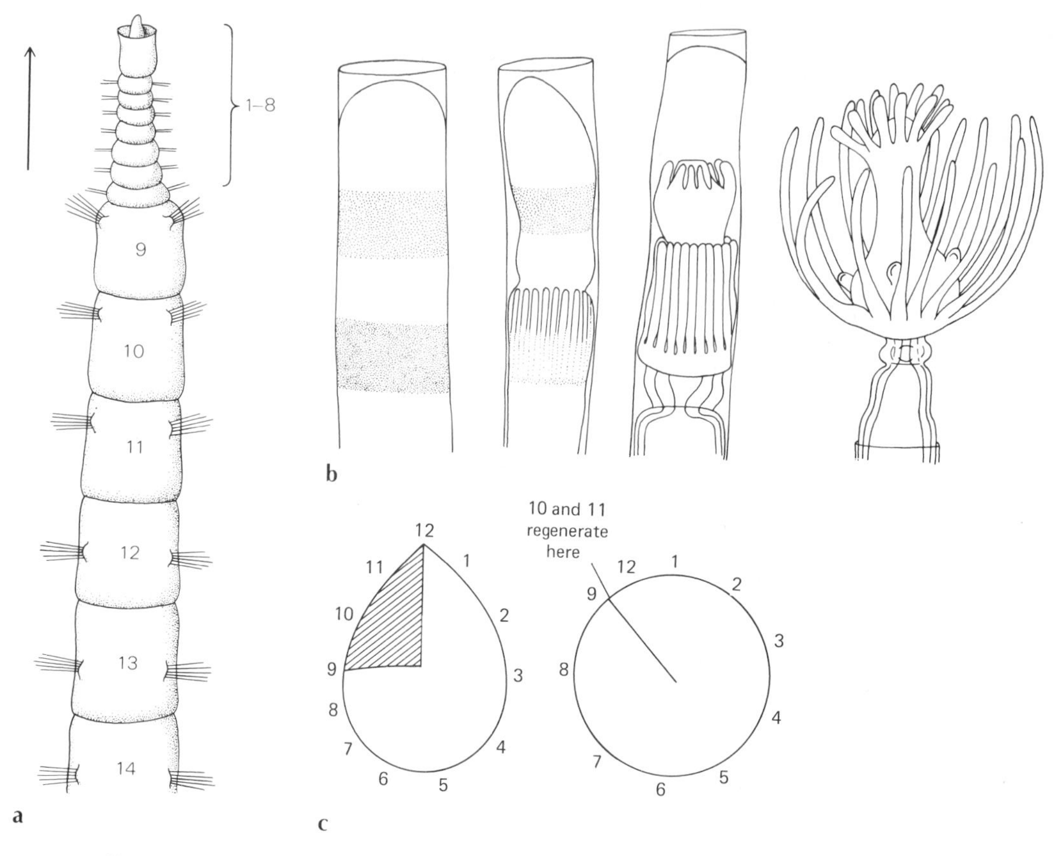

Figure 20.1 *a* Polarity and reconstitution in the polychaete *Clymenella,* showing polarized (anteroposterior axis) structure and reconstitution, anteriorly or posteriorly, of whatever number of segments are required to restore the original total of 22 segments. In this instance, segments 1 to 8 are regenerated from the anterior surface of segment 9. *b* Reconstitution of a new head (hydranth) from the distal part of a piece of *Tubularia* stem. Four stages, from left to right: pigment-band stage representing future tentacle zones, later stage with tentacle ridges in proximal but not in distal zone, completed hydranth still in original position within cuticle, and functional hydranth extended beyond cuticle. [*After Davidson and Berrill,* J. Exp. Zool., **107**:*473 (1948*).] *c* The wing disc of *Drosophila.* The cut edges of the fragments come together and intercalary regeneration occurs between them, indicated by arrow.

edges come together and intercalary regeneration occurs (Fig. 20.1*c*) at the junction.

In all cases of reconstitution, of whatever sort, we are confronted with the phenomenon and problems of morphogenetic fields, with other difficult problems relating to the source of the participating cells, and with the processes of differentiation. These are all features of development associated with the development of eggs, but here are seen in another context.

2. Limb regeneration in vertebrates

Virtually all the phenomena and problems associated with development generally, especially those associated with the vertebrate embryo, are inherent in the regeneration of a limb. The development of limb buds, as we have seen, more or

less parallels the early development of the embryo as a whole. It begins with an initial and relatively undetermined stage, i.e., the spherical egg or the circular limb disc, with little more than primary polarity, and develops mainly as the result of interactions between epithelial (epidermal) and mesenchymal (mesodermal) tissues, leading to tissue differentiation and pattern elaboration. A comparable event is seen in the reconstitution or regeneration of a part from the whole at later stages of growth and development, even in fully grown and mature organisms. The process of reconstitution of the vertebrate limb accordingly offers further opportunity to analyze the morphogenetic and histogenetic events responsible for the creation of such a structure. The situation differs from embryonic limb development inasmuch as the stump from which a new limb, following amputation, is produced already consists of differentiated cells and tissues. The following account, therefore, extends the analysis of limb development already given and also serves as an introduction to the general phenomenon of regeneration widely encountered among lower animals.

Limb regeneration has been studied mostly in amphibians, particularly in salamanders of various ages. In these forms, the limbs are readily regenerated throughout life, although more rapidly when the amphibian is young and small. In fact, no other vertebrates exhibit such a wide capacity for regeneration of missing parts, including tail and snout and even the eye to some extent.

A. The blastema

When a limb is amputated, a process of restoration begins immediately, in which three phases are recognized: a period of wound healing, a period of blastema formation, and a period of differentiation (Fig. 20.2). The present discussion of the blastema is limited to that of the limb. Three phases are evident in the regeneration process as a whole:

WOUND HEALING: This process consists essentially of epidermal cells migrating from the basal layer of the adjacent epidermis toward the center of the wound. Active migration of this epithelial layer of cells continues until the wound is closed.

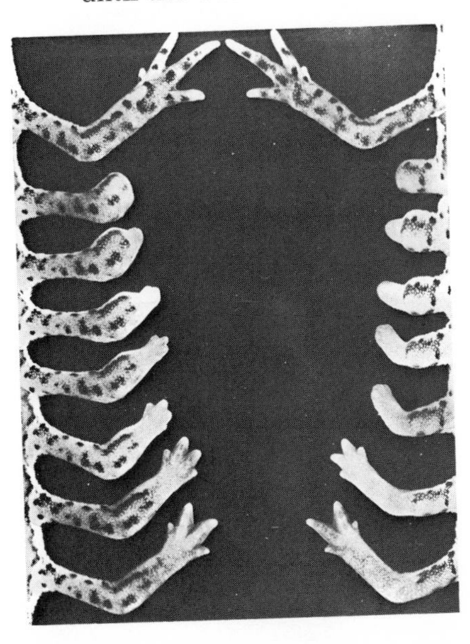

Figure 20.2 Regeneration of forelimb of salamander. (*Left*) Amputation below elbow. (*Right*) Amputation through upper arm. [*Courtesy of R. J. Goss.*]

BLASTEMA FORMATION: Cells accumulate beneath this newly formed epithelial covering, and the combined cap of epithelial and subjacent cell mass is known as the *blastema*. As such, it has the general properties associated with a limb bud or an egg insofar as it has the potential of growing and differentiating into a highly organized unit structure. The production of the blastema continues for some time before developmental events become discernible through cell division and cell migration. Epidermal cells continue to accumulate at the apical region, possibly through further migration, but certainly in part through mitotic division of cells initially forming the cap. Mesenchymal cells progressively accumulate beneath the cap, so that an epithelial-mesenchymal reacting system is established resembling that of a limb bud or a feather bud but on a much larger scale.

DIFFERENTIATION: The limb blastema passes through several definable stages during the course of regeneration. Morphologically, in the adult newt at 20°C, 15 days after amputation, the blastema is filled with undifferentiated cells; by 20 days the blastema has become a *cone;* the *palette* stage (a flattened cone) is reached at 25 days; the *notch* stage, representing the first sign of digits, is reached at 30 to 35 days. From then on, digital pattern becomes progressively in evidence, the precartilaginous skeleton condenses, and a complete limb is present by 75 days (Fig. 20.3).

The general questions that arise are already familiar. What is the role of the newly formed ectoderm, and what is the role of the mesodermal elements in the subsequent processes of differentiation and organization? What is the develop-

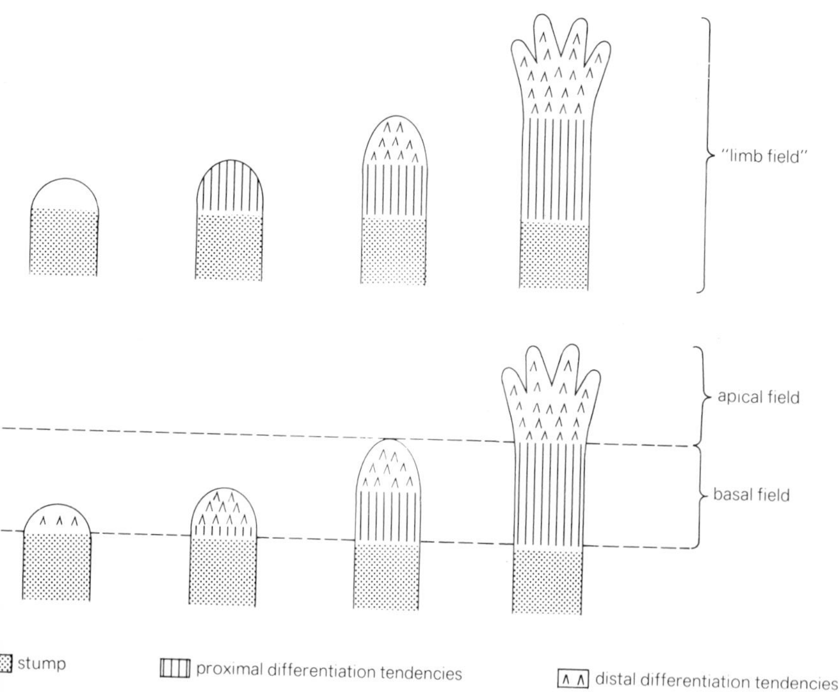

Figure 20.3 Two concepts of regional organization in the limb regenerate. Four successive stages of regeneration are shown, from left to right. *a* The youngest blastema is depicted as being morphogenetically neutral. Limb pattern arises as a result of determinative events proceeding from stump distalward. *b* The youngest blastema has distal differentiation tendencies. Proximal differentiation tendencies appear later than distal ones. [*After Faber, 1965.*]

mental sequence? And so on. The new questions relate to the fact that the blastema cells derive from mature epidermis and from internal tissue that contained fully differentiated muscle, nerve, and cartilage together with connective-tissue cells (fibroblasts or mesenchyme).

B. Dedifferentiation

The source of the mass of mesenchymal cells constituting the bulk of the blastema has been a problem plaguing analysis of limb regeneration for many years. Following amputation, the residual differentiated tissue of the stump, mostly cartilage and muscle, undergoes extreme disintegration and apparent dedifferentiation, e.g., cartilage cells lose their matrix, muscle cells their myofilaments. At the same time, the mesenchymal cells beneath the epidermal cap increase in number enormously (Fig. 20.4). The problem has been to ascertain the connection between these two events. Opposing interpretations, not entirely exclusive of each other, have been variously adopted. The points at issue have been questions not of plausibility, but of fact. Possible interpretations are

1. Is the apparent dedifferentiation of mature muscle and cartilage cells a true return to an undifferentiated state? If so, they can obviously give rise to a mass of undifferentiated blastema cells. It has been extraordinarily difficult, however, to establish whether or not true dedifferentiation and subsequent redifferentiation may actually take place.
2. Are the disassociating and otherwise changing cartilage and muscle cells degenerating and in fact dying, allowing their products to stimulate and support rapid multiplication of any unspecialized mesenchymal cells present? A dependent population replacement of this sort would be effective and is known to occur in regenerative processes in some other forms (e.g., ascidians). In this event, no transformation of cells is called for, only differential multiplication.
3. Do the cartilage and muscle cells merely appear to dedifferentiate? That is, do they lose their distinctive histologic character, but in actuality retain their fundamental character and as such contribute to the blastema? This phenomenon of pseudodedifferentiation is commonly seen in tissue culture as *modulation* (Section 15.9). If this is the case, however, in a blastema, then we are faced with another problem, because the implication is that modulated cartilage, muscle, and other cells, having been mixed together, subsequently sort out according to their kind, i.e., like to like, and also become reassembled in the new configurations of the developing limb structures.

The long-standing difficulty has been the failure in following the history of individual cells during the crucial period, that is, in tracing individual cartilage or muscle cells from the original stump tissues to whatever their destiny may be, or in determining the source of the mesenchymal cells with any precision.

C. Redifferentiation

When the stumps of amputated limbs are treated with tritiated thymidine and the tissues are fixed at later intervals and autoradiographed, the pattern of incorporation of the thymidine indicates that DNA synthesis begins 4 to 5 days after amputation in all differentiating tissues within 1 mm of the wound. In animals into which thymidine has been injected before amputation, only the epidermis of the limb shows incorporation of the tracer. When this is followed by amputation, the labeled epidermis migrates over the wound surface and forms a labeled apical

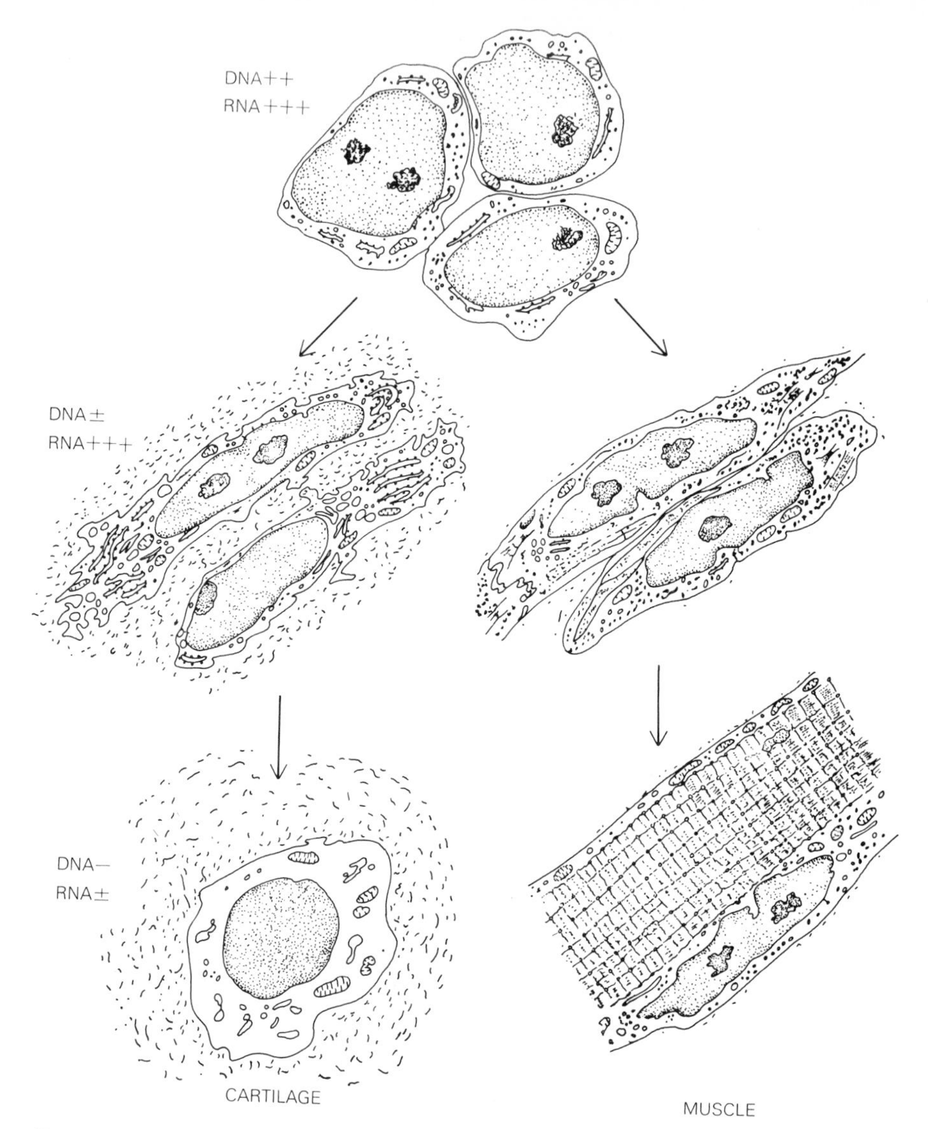

Figure 20.4 Changes in fine structure of mesenchymal cells as they aggregate within the limb blastema *(top)* and differentiate into cartilage *(left)* and muscle *(right)*. In cartilage, the newly formed differentiated products are destined for the extracellular compartment. They are synthesized and secreted by the rough endoplasmic reticulum and Golgi complex. In muscle, the differentiated proteins (myofibrils) are intracellular in location. They are synthesized by free ribosomes in the cytoplasm without the intervention of the rough endoplasmic reticulum. The muscle cell contains only a few profiles of rough reticulum and has a smaller Golgi apparatus than the cartilage cell. It later acquires a special kind of smooth-surfaced reticulum. Also depicted in the diagram are the relative amounts of DNA and RNA synthesized by the cells in the developmental stages illustrated. [*After Hay, 1965.*]

cap, which remains labeled throughout blastema formation. None of the labeled epidermal cells contribute to the internal blastema.

Other evidence shows that cells from the body proper do not contribute to the mesenchymal blastema. Blood cells, for instance, that appear at an early stage in the limb do not derive from the blood-forming organs of the body. The dedifferentiating tissues of the stump, more or less by default, give rise to the unspecialized blastema cells. Evidence that cartilage cells can dedifferentiate and subsequently redifferentiate into muscle cells, or vice versa, is supplied by labeling experiments of a different kind. The question is a general one of vital significance: Can the daughter cells of a highly differentiated vertebrate cell, following mitosis, exhibit a variety of differentiations, or must they perpetuate the parental type? Can they truly dedifferentiate and redifferentiate, or only modulate?

One crucial experiment consisted of grafting triploid salamander cells of a known differentiated type into the limbs of diploid hosts, amputating the limbs, allowing limb regeneration to occur, and then examining the regenerates to see whether the triploid cells had given rise to cell types other than those of the original graft cells. The grafts consisted of pure cartilage, pure muscle, cartilage plus perichondrium, and epidermis, i.e., the various types typically present in the stump of an amputated limb, but introduced separately into the hosts. Host limbs were previously exposed to x-irradiation to discourage host cells from participating in the regeneration process when the host limb was amputated. Triploid cells are recognizable as having three nucleoli in the interphase nucleus instead of two.

When the graft tissue consists of pure cartilage or of cartilage plus perichondrium, the marker appears subsequently in the regenerated limb in the following cell types: cartilage, perichondrium, the connective tissue of joints, and in fibroblasts, but not in muscle. When pure muscle is grafted, the marker appears in all the preceding cell types and in muscle cells as well; no regeneration occurs when pure epidermis is grafted. Similar experiments made independently but employing both tritiated thymidine and triploidy (used either independently to mark graft and host cells differentially or simultaneously to label the same cell) have provided cross-checks on any uncertainties in either method alone. Again, clean cartilage grafts give rise to morphologically dedifferentiated blastema cells, which redifferentiate almost exclusively into chondrocytes that retain their specific character through at least five divisions; and again, muscle tissue contributes to both muscle and cartilage, *although it is highly significant* that muscle tissue consists of undifferentiated connective-tissue cells as well as contractile elements. The situation is even more unclear because muscle tissue also contains a population of undifferentiated *satellite cells*, which have long been thought to represent a type of reserve stem cell from which myoblasts can differentiate. If these satellite cells are capable of undergoing myogenesis, they may well be the ones that give rise to the muscle tissue of the regenerating limb rather than the muscle fibers of the stump adjacent to the cut. The question of whether real dedifferentiation takes place, with subsequent redifferentiation along another line, still receives a somewhat dusty answer.

D. Neural trophic factor

Regeneration of the limb blastemas of vertebrates, particularly amphibians, and regeneration blastemas of many invertebrates are normally dependent on a critical supply of nerves at a very early stage. Many nerve endings in fact enter the epidermal cells. However, if a limb is first denervated and then amputated, or if nerves are by any means blocked from penetrating the epidermis, no regenera-

tive outgrowth occurs. The stump tissues merely undergo degenerative changes. Yet once the process of dedifferentiation and blastema formation has taken place, interference with the nerve supply no longer has any effect. The influence of nerves is something other than nerve function as such and is not due to acetylcholine. The agent is known as the *neural trophic factor*. What it is and how it works are still unsettled, although the establishment of neural-epidermal junctions seems to be an essential condition in such nerve-dependent regenerations.

Dependence of regeneration on neural factor, however, is not universal. Embryonic limb buds, for instance, comparable to though smaller than regeneration blastemas, develop without initial neural stimulation. Similarly, salamander limbs that have been denervated and maintained nerveless for a month or more, by repeated denervation at 5-day intervals, acquire the capacity to regenerate amputated limbs. In other words, the dependency is not absolute. This situation may be compared with that seen in *Hydra* (described in Chapter 1).

Experimental morphologic and cytologic studies suggest that the neutrophic agent is a peptide and that it affects the rate of regenerative events and not the kind of events, whether in limb regeneration or in other cases of nerve-dependent growth. Increase in rate of molecular synthesis results in an increased cell population, which accordingly has formative capabilities that do not exist in a smaller cell population. This is a hypothesis.

Mammals do not significantly regenerate lost parts. The inaccessibility of mammalian embryos makes it impossible to be sure of this point in fetal mammals. An exception is the marsupial. It is born at a relatively very early stage, after which it migrates to the maternal pouch and attaches to the nipples. The newborn North American opossum has proved to be uniquely suited for studies of replacement of the mammalian limb. Since this marsupial is born without lymphocytes, xenoplastic as well as homoplastic transplants are tolerated. Supplementary nerve tissue was transplanted to newborn hindlimbs and the limbs were amputated immediately above the ankle. The implant remained in place, and distal-limb regeneration resulted. Control experiments indicate that neither the trauma of simple amputation and implantation nor the implantation of other tissues can evoke the response that results after nerve tissue is implanted. These studies demonstrate that young opossum limbs are capable of regenerating when additional nervous tissue is supplied.

Adult frogs also are ordinarily unable to regenerate limbs, but limb regeneration can be stimulated by the implantation of small electrogenic bimetallic couplings in the amputation stump. Regeneration occurs only if the orientation of the coupling enhances the distal negativity and is lacking or much decreased with all other orientations. The initial or primary stimulus for this process may be a relatively simple electric field requiring only a minimal current level acting on a population of susceptible cells.

It is now known that the direct-current potentials measurable on the intact surfaces of all living animals demonstrate a complex field pattern that is spatially related to the anatomical arrangements of the nervous system (Fig. 20.5). The surface potentials appear to be directly associated with some element of the nervous system. They can be measured directly on peripheral nerves, where they demonstrate polarity differences depending on whether the nerve is primarily motor or sensory in function. The existence of standing electric potentials in a conducting medium implies the existence of a steady current flow. Experiments have demonstrated that such a current exists, longitudinally, in the neural elements.

Three general possibilities are that (1) bulk ionic currents and fields may directly modify the surface-charge pattern on the membrane of sensitive multipo-

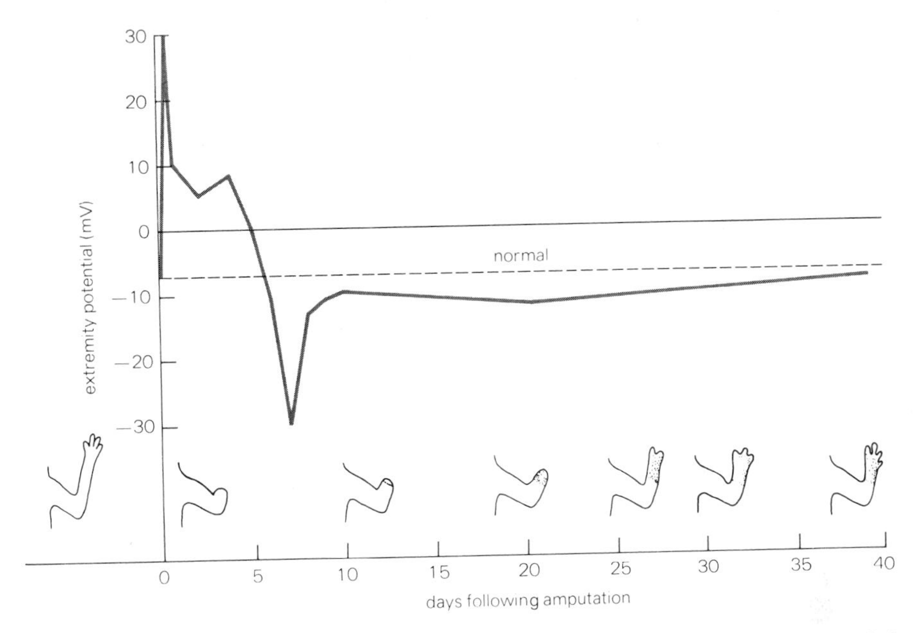

Figure 20.5 Voltage changes in adult-newt forelimb during regeneration. Zero point on horizontal scale immediately precedes amputation. Very high peak in positive voltage curve is concurrent with limb amputation. Secondary positive peak is observed on the fifth day. Negative maximum is reached on the eighth day. Negative bias is maintained until limb regeneration is complete. Normal limb voltage is about −10 mV (*interrupted line*) at point of amputation. Lower figures show schematically the stages of regeneration. [*After Becker, 1961.*]

tent cells and thus transmit information to the cell machinery, (2) cells can be directly influenced by changes in the electrochemical environment at or near an electrode interface, and (3) cells can be indirectly influenced by changes in enzyme concentrations or configurations produced by electrochemical reactions at the electrodes.

Recent experiments show that the response to cathodal current is mediated by electric fields rather than by electrode products, which raises the question of what the target or targets of this minute current may be. The two main possibilities are the initial wound epithelium and the underlying nerve, the latter being the more likely. A minimum ratio of neuroplasm to stump surface area is normally needed for amphibian regeneration. Moreover, at least 20 times more nerve is found in the distal part of a cathodally stimulated stump than in controls, and nerve processes, with their dilated terminal growth cones regenerating toward the stump surface, should be exceptionally sensitive to external fields.

E. Blastema as a self-organizing system

The cone stage is of particular interest morphogenetically because not only do the various tissues and structures differentiate within the regeneration outgrowth, but they form in continuity with the stump tissues to produce an anatomically and functionally complete limb. Only those structures distal to the plane of amputation are formed. A long-standing explanation has been that the differentiated tissues of the limb stump induce the undifferentiated blastema to redif-

ferentiate the lost parts in conformity with the stump organization. The failure of young blastemas to continue their differentiation when transplanted to foreign sites and the ability of older blastemas to do so have supported this interpretation.

The validity of this has been tested by removing cone- , palette- , and notch-stage blastemas and grafting them to the body (the dorsal fin) of the larval salamander, with or without a stump. Whether the stump is included or not, whole blastemas are able to self-organize into all the skeletal and muscular components of the lost limb distal to the level of amputation. Distal parts of blastemas grafted in the same way develop hand structure; proximal parts develop the more proximal parts (Fig. 20.6*a*). In other words, the limb blastema as early as the cone stage is not continuously influenced by the surviving stump tissue but is a *self-organizing system* already imbued with a pattern representing the prospective regional structure.

Figure 20.6 Regeneration of salamander limb in different circumstances. *a* Leg and arm amputated and regrafted on to arm and leg stump, respectively, then cut off again, leaving disc of arm or leg on stump of other kind; result is regeneration according to type of disc. *b* Regeneration of limb from stump devoid of epidermis; result is complete regeneration including new epidermis. *c* Amputation of limb longitudinally, resulting in no regeneration, followed by transverse amputation; result is complete regeneration, followed by transverse regeneration of whole distal portion from reduced base. [*After Needham, 1942.*]

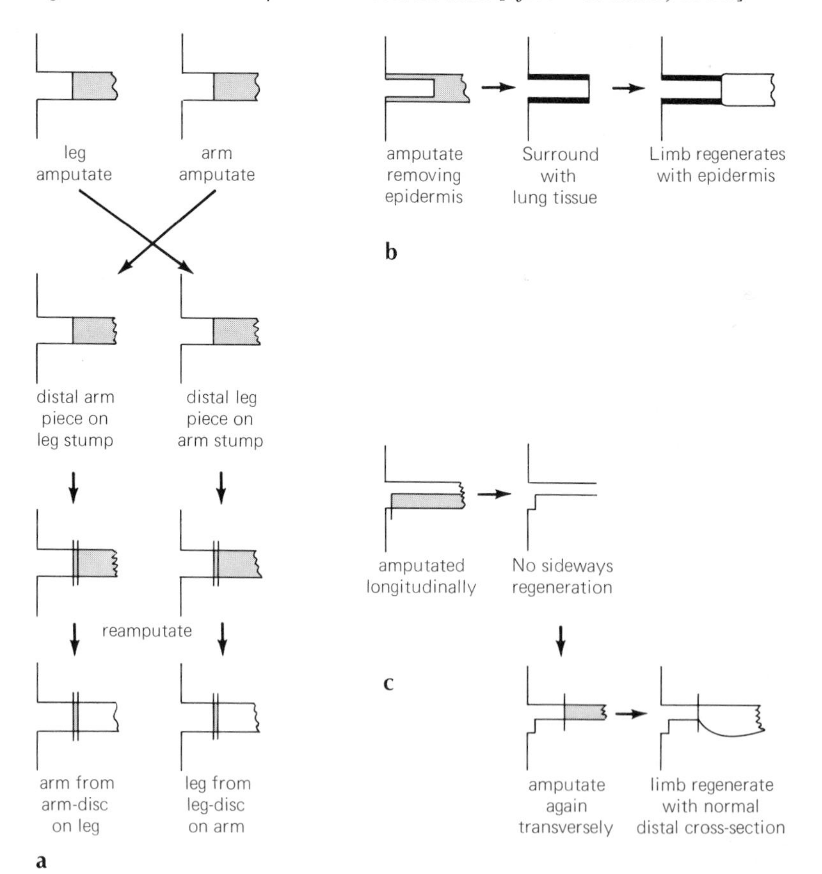

Blastemas of all three stages are able to differentiate precartilage and striated muscle in vitro in complete absence of the stump. Inclusion of stump in an explant does not enhance the frequency of differentiation of precartilage and muscle in the blastema, and it actually appears to be detrimental to development of the regenerate. It has been concluded that by the time the blastema has reached the cone stage, it is not dependent on inductive messages from the stump for differentiation but is a self-differentiating system.

Nevertheless, a blastema has received an imprint of regional character by the time amputation of the blastema becomes possible, so that in some way it already represents that portion of the limb that has been amputated, neither more nor less. If a forelimb of a salamander larva is transplanted to the region of the hindlimb, or vice versa, and the transplant, after healing, is partly reamputated, what develops depends on the original character of the stump rather than on the site to which it is transplanted. If a thin transverse slice of forelimb, for example, is healed onto the stump of a hindlimb, a new forelimb regenerates.

Epidermis derived from stump epidermis, moreover, is not essential. If the epidermis from the stump is removed following amputation and the stump is then surrounded by lung tissue, the limb that regenerates is well provided with normal epidermis (Fig. 20.6*b*). There is, however, one striking limitation to the regenerative capacity of a mutilated limb. If a limb is amputated longitudinally, rather than transversely, the injured surface heals but does not regenerate as a whole. If such a split limb is then cut transversely, a distal regenerate forms that is normal in cross section and structure (Fig. 20.6*c*).

A model for regenerating limbs has been proposed in which the cells are considered to respond to two components of positional information. One component corresponds to position around the circumference of the limb, i.e., to a circular sequence of positional values, while the other component, relating to the proximodistal axis of the limb, consists of the sequence of values along the radius of the circular sequence. Such is the *polar-coordinate*, or *clock, model* (Fig. 20.7). According to this model, amphibian limb regeneration is governed by two rules for cellular interaction. One is the rule of shortest intercalation, which states that missing values in either the radial or circular sequences are filled in by intercalary regeneration. The complete-circle rule states that distal transformation of blastema cells takes place only if these cells are derived from a limb stump possessing a complete-circle sequence of positional values. The complete-circle sequence exists after simple transverse amputation of a limb, as well as when two normally separated half circles are experimentally opposed (grafted together).

The organization of the morphogenetic field of the limb-regeneration blastema has been described as a three-dimensional set of cell-surface–oriented properties that define pattern boundaries while permitting changes in enclosed positional values when a discontinuity arises. These hypothetical properties (of both the mesodermal and epidermal tissues) of the limb are considered to form gradients of positional value along each of the three axes (anteroposterior, dorsoventral, and proximodistal in the mesodermal cells, but to remain constant everywhere in the limb epidermal cells. In the blastema, the surface states of the epidermal and mesenchymal (mesodermal) cells, considered to be inherited from the mature limit cells, are mixed up during wound healing and dedifferentiation. Mesenchymal cells recognize incongruous, or disparate, states among themselves and respond by continually changing their surface states until a complete gradient of states is restored up to the boundary values. This hypothesis of a molecular mechanism (cell-surface states) for the specification of positional infor-

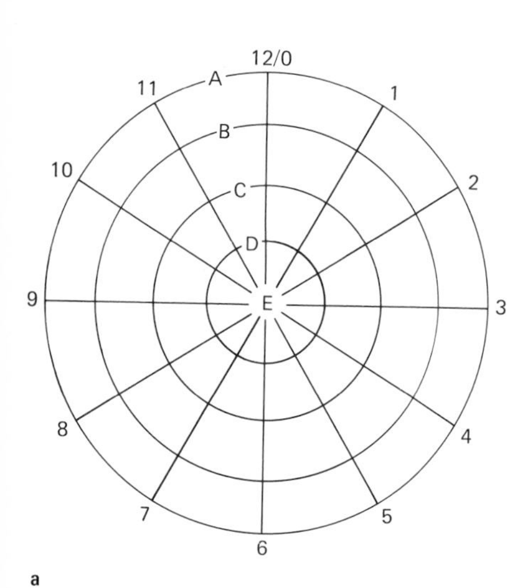

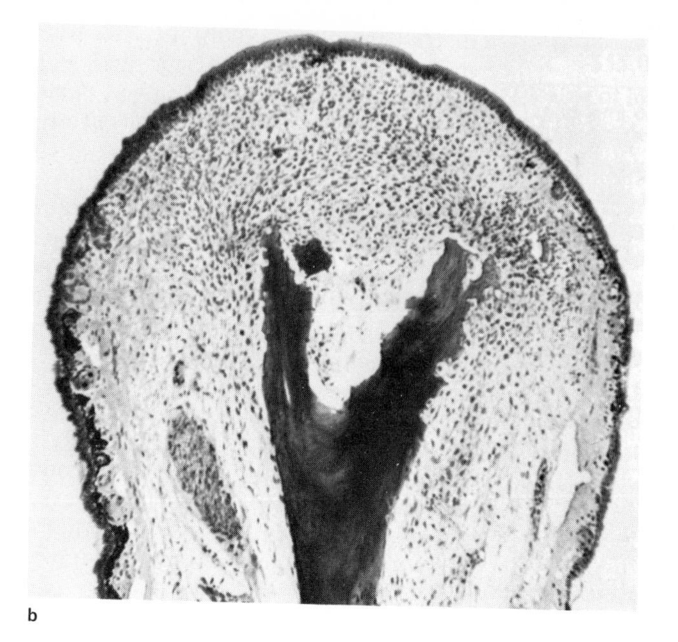

Figure 20.7 *a* Positional information in the polar coordinate model. Each cell is assumed to have information with respect to its position on a radius (A through E) and its position around the circle (0 through 12). Positions 12 and 0 are identical, so that the sequence is continuous. *b* Longitudinal section through an early regenerating forelimb bud and stump showing the mound of mesenchymal cells forming the blastema. The central dark shaft is the humerus. Muscle and connective tissue in the stump have been broken down in preparation for the formation of blastema cells. [*Courtesy of M. Singer.*]

mation by the blastema field, with or without an involvement of extracellular matrix, is untested. It does, however, suggest avenues for exploration.

Distinguishing positional information from the cells' responses to it is justified on the grounds that genetic mutations can affect the two events separately and that different patterns of cytodifferentiation can apparently result from the same underlying map of positional values because of differences in the interpretation event. The whole topic of morphogenetic field organization in this context is fully reviewed by Stocum (1978).

3. Regeneration in imaginal discs

Imaginal discs of *Drosophila* have been used extensively for regeneration studies. Fragments of wing discs that have been cut in various ways are injected either into larval hosts, where they differentiate according to what is already determined, or into adult hosts, where the hormonal conditions do not induce metamorphosis but do allow growth to occur, these last then being transferred to host larvae for metamorphosis.

Disc development varies strikingly according to the size of disc fragments: (1) If a district fragment is too small at a critical time of metamorphosis of the host, no organ is formed at all. (2) If the size of a fragment passes a critical level, a defective organ forms, although the gross shape may be "regulated." (3) If a disc district has time to proliferate so that it becomes much larger than normal, the

district usually splits up into two districts of similar size and shape rather than producing an abnormally large organ. (4) When either two organs are formed from an overlarge district or two partial organs are formed, as in cases where excess growth is more restricted, the two organs, or two parts thereof, are mirror images of each other.

When discs are cut into three pieces by two parallel cuts and the pieces cultured separately, many of the central-fragment pieces regenerate the missing parts of the disc in both directions, whereas each of the two edge pieces usually produce a mirror-image duplication of the pattern normally produced by that part of the disc. The regulative behavior of duplicating fragments, however, can be altered by mixing them with fragments from different disc regions. Fragments from opposite edges of a disc, when grafted together, are able to produce the missing pattern elements from the central part of the disc by intercalary regeneration. In other words, imaginal-disc tissue can undergo intercalation in response to a positional-information discontinuity. Although the concept is generally plausible, however, both in amphibian and insect regeneration, some experimental results are anomalous, and the universality of the model as an explanation has not been established. The application of the polar-coordinate model to regulation in *Drosophila* imaginal discs is reviewed by Bryant (1978).

4. Regeneration in a cell

The giant unicellular green alga, *Acetabularia*, long the subject of morphogenetic investigation, consists of a distinctly shaped cap, long stalk, and a basal rhizoid containing the single nucleus. Removal of the cap results in regeneration of a new cap. Polarity seems to reside in the membrane sheath. When the cytoplasm of an isolated section of stalk is removed, homogenized, and returned, the cap regenerates with the same polarity as before. Cap regeneration occurs even if the nucleus is also removed.

Cap regeneration is a complex process involving the synthesis of cap-specific enzymes, structural proteins, and polysaccharides, assembly of these macromolecules into particular structures, and the appearance of a sharp morphologic discontinuity between stalk and cap, all of which can occur in the absence of the nucleus. Nuclear products are essential for the process, but all the processing of nuclear information, synthesis of specific substances in specific places, and morphogenesis of species-specific structures occur without the benefit of any gene-switching network. In other words, the "field" makes all the temporal and spatial decisions, whereas gene products are the raw materials of the morphogenetic process. Neither field nor genome alone can sufficiently account for morphogenesis. In the study of regeneration, therefore, field phenomena predominate, with special regard to polarities, symmetrics, scale, and experimental modifications.

For nearly a century, investigations of reconstitutive development, both regenerative and reorganizational, have been conducted on a wide assortment of organisms, mostly invertebrates. The literature is vast and has been reviewed at considerable length in another publication (Berrill, 1961). In the present context, space permits only a brief sampling of the rich diversity of material and experiments, although it is perhaps sufficient to serve as a general indication of the scope of this field of enterprise and to give minimal leads to the literature.

Lund (1921) showed that in the hydroid *Obelia*, polarity has an electrical basis; there is an electric-potential difference between the two ends of a regenerating piece of hydroid, and this can be neutralized or reversed by sending an electric current through a stem in the direction opposing its own bioelectric polarity. The imposed electric field has to be only slightly greater than the inherent

field. He reasoned that since an applied field of the same order of magnitude as the natural inherent field, but of opposite sign, could change the site and reverse the direction of differentiation, the inherent field must under natural conditions determine the site and direction of differentiation. Tests were made on stems of *Tubularia*, another hydroid, to ascertain whether conditions that determine the site and direction of regeneration do so by producing a polarized (bioelectric) field prior to visible regeneration. The tests show that conditions favoring regeneration at the ends of stems do so by first making those ends electronegative. Conditions that retard regeneration locally tend to make that region electropositive. This and other experiments on *Tubularia* by S. M. Rose are described at length in his book "Regeneration" (1970).

5. Inductive action of blastemas

In most cases of regeneration (other than in hydroids), undifferentiated cells, whatever their source, gather as a group at the wounded area and constitute the blastema. Here regeneration is initiated. In flatworms (planaria), these undifferentiated cells are termed *neoblasts*. They are generally considered to be totipotent; i.e., they are capable of forming any and all of the missing tissues and organs. There is at present some dispute as to their origin. There is also evidence that blastema formation depends initially on the presence of cells already available, inasmuch as (1) a high mitotic activity does not necessarily cause blastema formation, and (2) blastema formation is sometimes possible when mitoses are inhibited. Blastema formation begins with the determination of the distal part, and this occurs as soon as wound healing is over.

A. Heads and tails

Pieces of flatworm possess an inherent polarity. Transversely cut pieces retain the original anteroposterior polarity of the whole. A head is reconstituted at the anterior region and a tail at the posterior region. This polarity is associated with the presence of nerve fibers, directly or indirectly. Lateral pieces containing nerves also form a new head and tail. In the absence of all nerve fibers, a lateral piece can reconstitute a head, complete with brain and sense organs, but there is no elongation of the piece to form the remainder of the body. "Head humps" develop, but a typical reconstituted worm develops only if a fragment of nerve tissue is present. The head alone, therefore, appears to be a self-organizing system.

Induction phenomena are well documented in planarian regeneration. It has been demonstrated that the brain induces eye formation. In a series of experiments in the 1950s, Lender and his group worked with *Polycelis nigra*, which has numerous eyes forming a rim around the anterior region. If the brain and part of the eye rim are removed, subsequent removal of the "brain" blastema every 2 days prevents regeneration of eyes. Eyes regenerate before optic-nerve connection occurs. Therefore, Lender concluded that the brain was capable of inducing eyes "at a distance" through the diffusion of chemical substances. If "brainless" planaria were raised in an extract of planarian brains, the eyes also regenerated.

Transplantation of the head into the postpharyngeal region caused a supplementary pharynx to be regenerated. Removal of the induced pharynx caused another one to be regenerated. A prepharyngeal fragment differentiates a pharynx inside the old tissues. A postpharyngeal fragment regenerates a pharynx within the blastema but not before the brain and eyes develop. From experiments such as these, the investigators conclude that the brain induces a prepharyngeal zone, which induces a pharyngeal zone capable of forming a pharynx.

Based on these demonstrations of inductions and inhibitions in planarian regeneration, the Lender-Wolffian school has presented a model to explain regeneration in planaria (Fig. 20.8). The brain induces the formation of eyes and a prepharyngeal zone. It also inhibits the formation of supernumerary brains. The prepharyngeal zone induces a pharyngeal zone, followed by the formation of the pharynx. Inhibitory substances from the pharynx prevent additional pharynges from forming. The anterior region also induces the gonads and copulatory apparatus.

The concept of a "time-graded regeneration field" to explain interactions occurring in planarian regeneration has been fully presented by Brønstedt in his book "Planarian Regeneration" (1970). This concept is that there is a regenerative "high point" located anteriorly and centrally in the worm where the neoblasts begin to differentiate first. These neoblasts would differentiate into a brain. As these neoblasts differentiate into nerve cells, they prevent other cells from becoming brain. These neoblasts then differentiate according to the next choice open—that of eyes. Therefore, there exists a sort of hierarchy of differentiation possibilities open to a neoblast. What it actually becomes is determined by gradients of time and space as well as by interactions with its fellow cells.

Another group of worms, also capable of remarkable regenerative and reorganizational performance, are the annelids, or segmented worms, especially the more diverse and even beautiful marine polychaete annelids. Anatomically the annelids are more advanced than flatworms and nemerteans. They are divided into a series of body segments, each of which has much in common with the other segments and at the same time has morphological and physiological features distinctively its own. Therefore, the segmented worms offer several advantages to the investigator of regenerative phenomena:

1. An anteroposterior polarity or axis is present, comparable to that of others.
2. A regional differentiation exists along the primary axis.
3. New body segments form from the posterior zone of growth, whether in normal posterior growth or regenerative posterior growth.
4. Blastema formation occurs at the onset of anterior regeneration.
5. Reorganization of old body structure takes place under the influence of new anterior regenerating tissue.

Figure 20.8 *a* Sequence of induction in regenerating flatworm from left to right: brain first differentiates, which induces differentiation of eyes along anterior and anterolateral margin, to constitute head structure; head induces prepharyngeal region and then the pharyngeal region posteriorly, which induces reproductive organs posteriorly. [*After Wolff, 1962.*] *b* Window is cut out of forepart of flatworm, and head regenerates within it, whether or not original head is amputated. [*After Bromstedt, 1955.*]

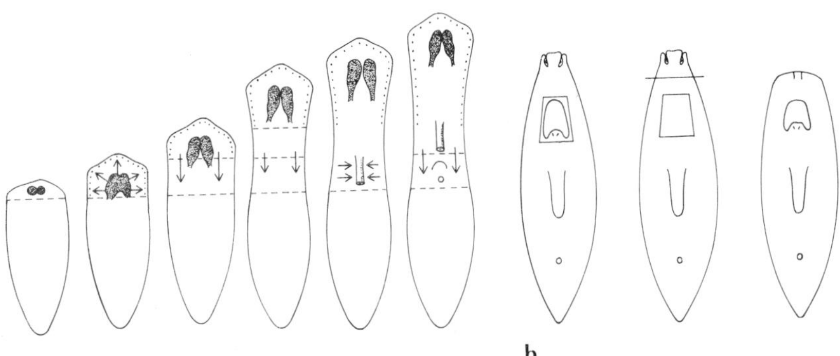

These phenomena are recognized in flatworm regeneration but in polchaete regeneration they are more clearly defined. Body segmentation makes every region along the axis numerically definable. This is immediately apparent in the anterior and posterior regeneration of pieces of certain polychaetes such as *Clymenella*. A common species of this genus consists of 22 segments. These are formed very early during the juvenile stage; further growth occurs through enlargement of individual segments.

Any segment or group of segments separated from the rest will regenerate the missing anterior and posterior parts and will do so exactly. If the anterior eight segments are amputated, eight are replaced. In a piece consisting of original segments 13 and 14, twelve will be regenerated anteriorly and eight posteriorly, so that the two old segments come to be numbers 13 and 14 in a new worm consisting altogether of 22 segments. The problem is the same as in the specific determination of the number of body segments (including tail) in vertebrate embryos and adults. No satisfactory explanation has yet been given for the basic nature of the posterior rhythmic growth responsible for segment production, or for the precise limitation of such growth among the many creatures that exhibit it.

In essentials, an annelid consists of a self-organizing entity at each end of the organism, together with an intervening zone of polarized, elongate, growth-embodying, segment-producing tissue. An annelid is also a tube (body wall) enclosing a tube (intestine). It is bilaterally symmetrical about an anteroposterior axis, has dorsoventral polarity, and has a segmented nerve cord that is dorsal anteriorly and ventral throughout the remainder of the body. Each end of the worm appears to be dominant in a morphogenetic sense: two fields in potential conflict ranging from overlapping influence to a state where a neutral zone lies between.

The most spectacular example of reconstitution in polychaetes is seen in sabellids, where both epimorphic and morphollactic phenomena occur together. The body organization is important: an anterior head bearing a crown of tentacles, a thorax usually of about seven segments, all bearing a distinctive arrangement of parapodial hooks (ventral) and bristles, or setae (dorsal), and an abdomen of indeterminant length and segment number, the segments bearing dorsal hooks and ventral setae (Fig. 20.9). At the junction between thorax and abdomen, the appearance is as if the worm had been twisted through 180°, with consequent inversion of the parapodia of the two regions relative to each other. At the posterior end of the worm is a terminal structure surrounding the anus, the *pygidium,* immediately in front of which is a minute zone of growth that forms new abdominal segments successively from its anterior part (Fig. 20.9*c*).

Anterior regeneration of a new head, consisting of the crown, collar, imperfect thoracic segment, and a perfect thoracic segment, can take place from an anterior cut surface from all levels of the body. The same head organization develops from the blastema irrespective of the type of segment from which it forms (Fig. 20.9*b*). It is a self-organizing system comparable to the head in the "head hump" of a flatworm or the limb blastema of an amphibian. Posterior regeneration occurs from a posterior cut surface, again from any level along the anteroposterior axis. The unusual feature is that thoracic segments are rarely regenerated, and a new thorax is attained by transformation of abdominal segments adjoining a regenerated head, i.e., by morphollaxis.

Transformation of abdominal segments to thoracic type takes place successively, beginning with the most anterior segment, which is in contact with the basal (thoracic) segment forming in the blastema. The number of segments thus reorganized is usually about the same as in an intact worm, typically seven. It may be significant, in this respect, that the nervous system is dorsal in the head

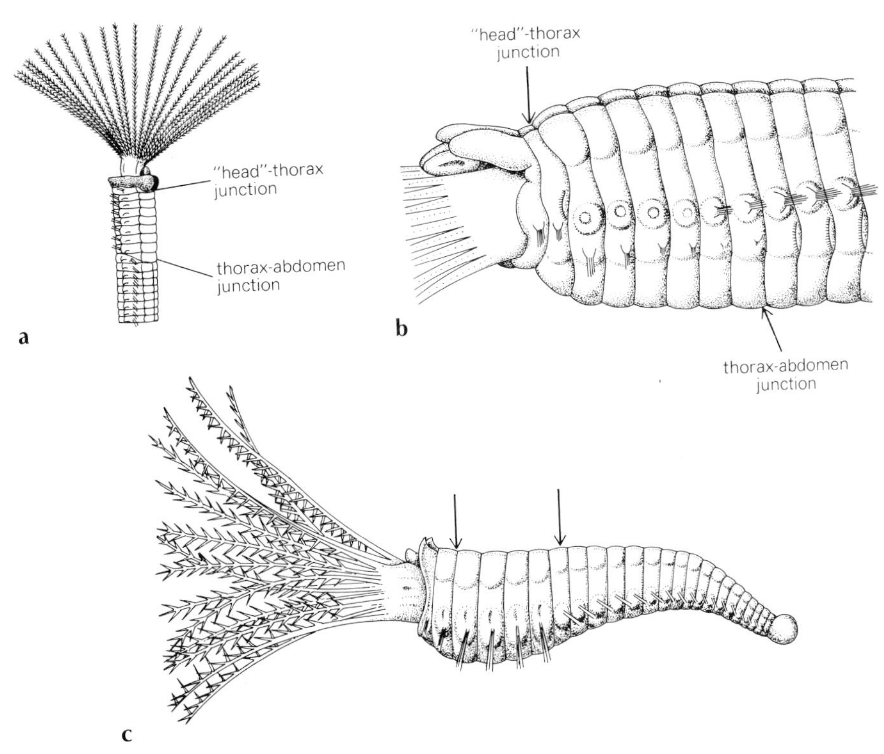

Figure 20.9 Reconstitution in the polychaete *Sabella*. *a* Anterior end of worm showing crown of tentacles, junction between "head" and thorax, and junction between thorax and abdomen. The bristle tuft and row of hooks characteristic of each segment are reversed in position in thoracic segments compared with abdominal segments. *b* Stage in reconstitution of abdominal piece of worm, showing regenerating "head" in front of the junction with the thorax and a gradation in the stage of transformation of abdominal-type segments into thoracic-type segments. *c* Completed reconstitution of an individual from a piece consisting of four originally abdominal segments. A new functional "head" has regenerated anterior to the left arrow, the four originally abdominal segments have transformed into thoracic segments, and 19 new abdominal segments have regenerated posterior to the arrow on the right.

but ventral from the collar backward, suggesting the possible presence of two opposing fields, a dorsoventral and a ventrodorsal. This is in keeping with a long-standing concept, based on regeneration studies, that a worm consists of two ends in morphogenetic confrontation, with a neutral or an overlapping zone in between.

The process of reorganization, particularly the number of segments or extent, is experimentally modifiable, at least in the species *Sabella pavonina* and probably in the very closely related *Sabella spalanzini*. In total darkness, although head regeneration is normal, an average of only four segments becomes reorganized, over a wide range of temperature. In augmented visible light, with no ultraviolet, and again irrespective of temperature, the number of segments transforming is greatly increased, to as many as 80, even when only the regenerating blastema is exposed (Fig. 20.10*a*). This is literally an extension of a morphogenetic field, and the effects of both darkness and light indicate that the field is light-sensitive and may have a photochemical basis.

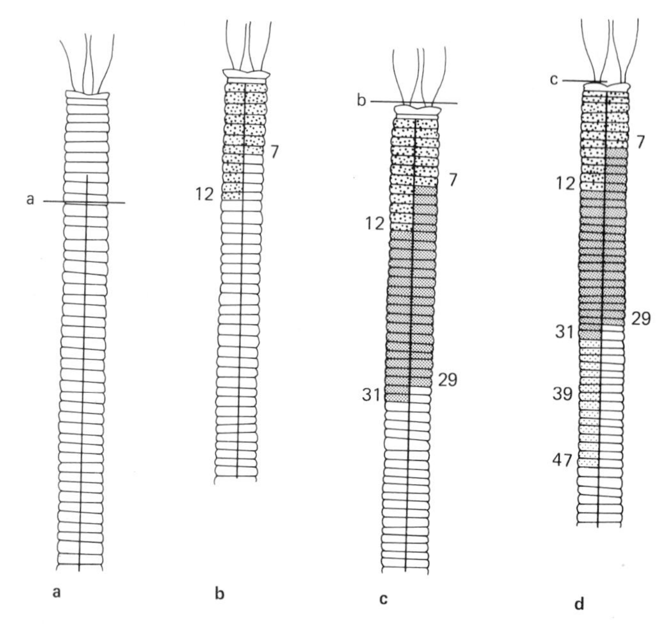

Figure 20.10 Summation effect and bilateral independence in *Sabella*. (Numbers refer to segments posterior to original amputation level *a*.) *a* Intact worm from ventral side, showing turning of ventral groove and level of amputation. *b* Extent of reorganization, different on two sides, at end of regeneration period. *c* Extension of reorganization of abdominal to thoracic segments after amputation of branchiae *b*. *d* Extension on one side only after amputation of branchia *c* of that side. [*After Berrill and Mees.*]

Another aspect of the field is a summative quality, particularly evident in augmented thoracic cases. The developing sabellid crown is bilaterally distinct, the right and left halves developing independently of each other. If the new crown is removed from a worm with an induced excessive reorganization, an additional reorganization of comparable extent proceeds from the posterior limit of the first reorganization. This can be repeated. If only one side of such a crown is removed, the additional reorganization occurs along that side alone (Fig. 20.10*b*).

Throughout the affected region, a dorsoventrality is imposed on a preexisting ventrodorsality in segmental organization. The evidence indicates that this is the result of an enhancement of dorsal-field influence emanating from the anterior end. The related question is: What is the source of the ventral field that becomes overwhelmed or superceded? This appears to be established in the zone of growth at the posterior end of the abdomen. Each minute segment segregated from the anterior region of the zone of growth (Fig. 20.10*c*) is initially endowed with the ventrodorsal polarity, each such segment subsequently growing to full size without change in typical pattern. In each segment is a bilateral pair of ventral ganglia with lateral segmental nerves extending dorsally in the body wall. If, however, a length of the ventral nerve cord is excised from a piece of the abdomen consisting of a number of segments, all the abdominal-type segments transform to the thoracic-type with inversion of polarity.

These sensitivities to light and to reactivations of a dorsal morphogenetic influence extending from the blastema, together with the influence of the ventral

nervous system in establishing and maintaining an inverted polarity, offer as promising a prospect for investigation of field properties as may exist.

6. Total reconstitution

It has long been known that very short pieces of a stem of the hydroid *Tubularia* and, similarly, short pieces of a flatworm generally reconstitute "head" structure at both ends. This was explained by Child in terms of his theory of metabolic gradients: When the two cut surfaces are close together, the metabolic rates at the two ends are not sufficiently different to establish an effective polarizing gradient, and both ends accordingly develop as dominant regions. Normal head reconstition (regeneration) and bipolar development are shown in Figs. 20.1*a* and 20.11*a*).

Bipolar regeneration also occurs in very short pieces of the abdominal region of the ascidian *Clavelina*. The anatomy of this ascidian is not important here, except that a nerve cord is not present. In this ascidian, the timing of cuts is as important as the distance between them. If short pieces are isolated by cuts made almost simultaneously, anterior (thoracic) structure forms from both surfaces. If the posterior cut is made an hour, more or less, *after* the first cut, typical anterior and posterior regeneration occurs at the two surfaces respectively; i.e., the original polarity of the tissue persists. The *orientation*, presumably of macromolecular components, may be more significant than concentration gradients of either ions or molecules in determining and maintaining tissue polarity. The DC electric current of injury may well be responsible for establishing the initial direction of polarization. That is, in short pieces resulting from simultaneous cuts, the polarizing current produces its effect in opposite directions from the two ends.

Bipolar regeneration has been induced in the polychaete worm *Sabella* by means of exposure to colchicine. Colchicine inhibits all regeneration, but when colchicine-treated pieces are returned to normal seawater, a head may regenerate from each end. When such a two-headed piece is bisected, however, the anterior portion regenerates a tail posteriorly, and the posterior portion regenerates a head anteriorly (Fig. 20.11*b*). The original polarity is regained. The interpretation offered is that colchicine reversibly disrupts the normal array of microtubules in the nerve cord (i.e., it causes a temporary depolarization), while the production of neurosecretory granules is not inhibited.

B. Bipolar head regeneration

The term *regeneration* is customarily used to denote replacement or reconstitution of a missing structure by a substantial surviving fragment of the original whole. In extreme cases, the surviving tissue may be so minute that what is reformed is virtually the complete organism. This is commonplace among plants, where even a single somatic cell may develop into a large multicellular plant by virtue of photosynthetic growth of itself and its descendants. Single animal cells, excluding eggs, are unable to perform such a feat because of the difficulty in obtaining nourishment from parental tissues, directly or indirectly. Budding, as is seen in *Hydra*, for instance, is rarely possible in more mobile creatures. Somatic fragments consisting of remarkably few cells, however, are able to do so in special circumstances. Cloning of somatic individuals is in fact highly developed among invertebrates, particularly by coelenterates and ascidians and their closest relatives. This capacity has been exploited to produce either extraordinary large numbers of dispersed individuals or to form colonial organizations in which many individuals are organically joined to form a single superorganism, comparable to most plants.

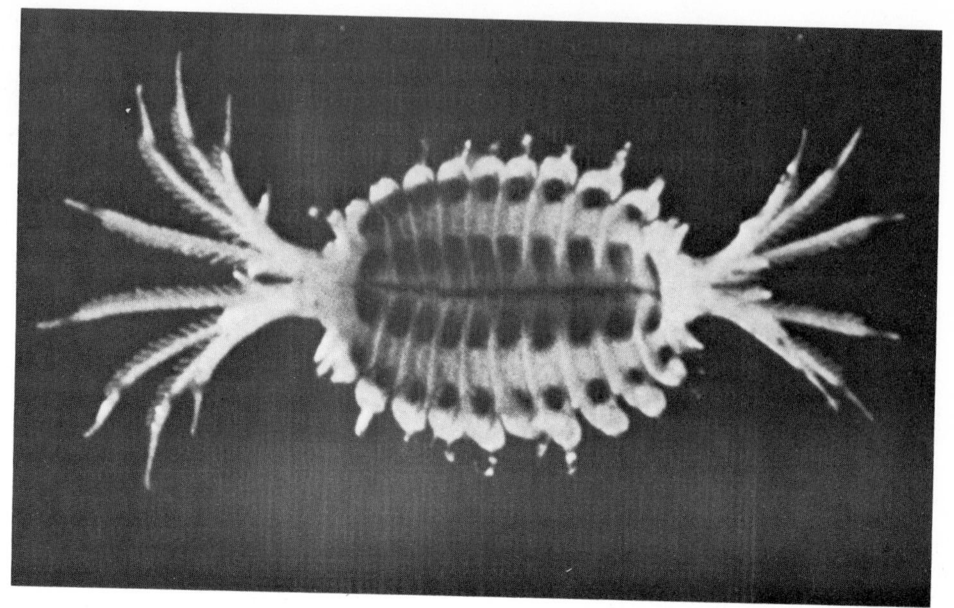

a

b

Figure 20.11 *a* Bipolar head regeneration in *Sabella*. Following exposure to colchicine, a typical three-segment head regenerates from the healed posterior cut surface in place of a tail as in normal regeneration. Reorganization of abdominal segments to thoracic segments takes place under the influence of each head. If bisected, the original anterior portion, however, now regenerates a tail, while the original posterior portion regenerates a head. [*Courtesy of T. P. Fitzharris.*] *b* Independence of apical region in hydroid *Tubularia*. Partial regeneration occurs in short-stemmed pieces of stalk, which may be uniaxial; it gives rise to as much of the rest of the organism as can be formed from the available material, with apical structure forming at the expense of more proximal structure. Short pieces also commonly produce whole or partial hydroid structure at each end. [*After Child, 1915.*]

In certain ascidians, a microscopic assembly of somatic cells, in aggregate much smaller in mass or volume than the smallest egg, develops into a complex, filter-feeding, sexually mature ascidian. In *Botryllus*, the form most studied, the rudiment first appears as a small, disclike thickening of unspecialized atrial epithelium overlain by a matching area of differentiated epidermis. The two layers of the disc may each consist of as few as three or four cells in optical section when first visible, the two layers being close together but not in contact. The epi-

dermal layer in every case is already a specialized tissue whose cells secrete the external tunicin that forms a supporting extracellular coat or tunic for the organism and from which the phylum (Tunicata) gets its name. The epidermis of the bud inherits the character of the parental epidermis. It continues to secrete tunicin. As a boundary epithelium, it plays a role in shaping the body as a whole, particularly since the body wall consists only of the one-cell-thick epidermis and some extracellular material. The inner layer of bud tissue is therefore responsible for developing the whole internal organization and the cytodifferentiations associated with it. The origin of the internal layer in terms of the three primary germ layers recognized in embryonic development apparently has no significance. In every case, the internal layer involved in bud formation has remained a simple limiting membrane, the cells of which presumably have undergone no histologic or cytologic specialization. Such cells may be regarded as units of no less potency than eggs and with nothing of the egg's elaborate specialization. Some mesenchymal cells may be included between the two basic layers constituting a bud.

In the two chosen examples, the colonial organism contains numerous individual ascidiozooids several millimeters long. In *Botryllus*, the zooid is ovoid; in *Distaplia*, it is slender and much longer with a very long posterior vascular extension. Body form is a specific character that is very readily perceived. In *Botryllus*, the bud primordium is initially a disc of enlarged atrial (ectodermal) epithelial cells associated with a corresponding area of overlying epidermis. It shortly converts into a two-layered vesicle (Fig. 20.12) with unmodified epidermis externally surrounding an inner sphere of enlarged epithelial cells. Such primordia arise from the lateral body wall of the parent. In *Distaplia*, buds arise from the distal end of the long posterior epidermal extension of the body containing a tubular epithelial extension known as the *epicardium*. The buds form by constriction of the epidermal structure and are pinched off as minute, two-layered vesicles resembling those of *Botryllus* formed from the bud disks. Apart from the different sources of the buds in these two genera and the need in *Botryllus* to convert a disc into a vesicle, the starting point for development is virtually identical: a double-layered epithelial sphere with some resemblance to the late gastrula stage of an amphioxus and various invertebrates.

Starting with the closed-sphere stage (Fig. 20.12) and considering mainly the inner formative layer, two folds divide it into a median and two lateral divisions, thereby establishing the right and left atrial cavities and the central pharyngeal sac, i.e., the primary spatial divisions of the organism-to-be. Then evaginations from the central chamber establish the intestinal loop posteriorly, the heart ventrally, and the nerve tube dorsally. Somewhat later, the walls of the central sac become perforated simultaneously by rows of minute prospective gill slits. Before this, the external form has been more or less completely acquired. Further development consists mainly of expansion and histologic elaboration. Throughout the whole development, the epidermal layer plays a relatively minor role, conforming to or developing the shape of the whole and producing as a structure of its own only the stolonic outgrowth that unites the bud to the system of vessels of the parental colony substituting for the original stalk. The fact that it is possible to describe the developmental processes so briefly even in outline indicates to some extent the monumental directness of the phenomenon.

Several points merit further emphasis. One is the rigorous correlation of the features characterizing any given stage, so that if the developmental stage of one feature is known, the stage as a whole can be accurately defined. The separation

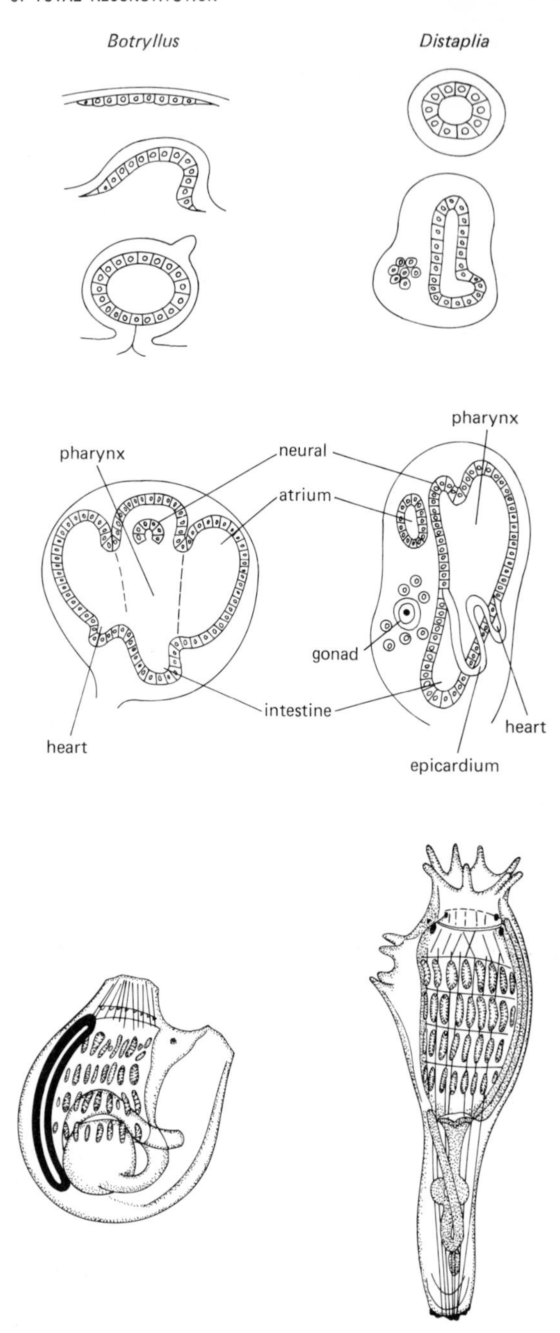

Figure 20.12 Direct development in the buds of two types of ascidians, showing development of double-layered disk of *Botryllus* bud into stage with six primary organization divisions by invaginative and evaginative foldings, and comparable stages of bud of *Distaplia*.

of the heart evagination indicates a definite degree of development of, for example, the digestive loop or neural complex. In the development of pattern there is, accordingly, an invariable relationship at a given moment. There is no shifting in the first appearance of a character to an earlier or later stage. Such constant relationships, while obvious and possibly universal, gain rather than lose significance by virtue of their general occurrence.

Another point is the general relationship of a stage of development to its spatial dimensions. The sphere stage already mentioned is less than one-tenth of a millimeter in diameter, consisting of on the order of 100 cells. It is no more than a tenth of a millimeter long when the atrial divisions and neural, heart, and intestinal evaginations are all discernible, and the characteristic shape of the organism is already emerging. It is hard to conceive how the gross architecture of the organism could be expressed in a smaller entity or by significantly fewer cells.

Apart from the morphogenesis associated with the inner layer of the bud rudiment, the preceding observations concerning the extremely precocious appearance of characteristic body shapes, expressed by the simple epidermal layer with no contact with another tissue, suggest that an overall field of determination may reside in the epidermis and that the inner layer responds to that field. This conclusion is strongly supported by various observations and experiments on *Botryllus*, particularly with regard to diverse origins of buds, where only the epidermis is a constant feature.

7. Initiation and determination

Just as removal of a piece from an organism initiates regenerative or reorganizing processes leading to its replacement, so isolation of a minute fragment allows that fragment to redevelop into a new whole. The two phenomena appear to be variables of a common basic response. Except when tissues have been disaggregated and then reaggregated into small masses, the anteroposterior and left-right polarities of the new structures or organism are inherited from the old. A small part may become isolated from parental control physically (as in bud production in the ascidian *Distaplia*) or in some manner physiologically (as in *Botryllus*, where the rudiment remains a physically integrated part of the parental epidermis in the beginning). The question here is: In what way has the rudiment become isolated, or independent, of its adjoining tissue, to the extent that it acts as though it were alone? The question persists.

Among coelenterates, hydroids in particular, the problem is compounded. With some exceptions, including *Hydra* and *Tubularia*, hydroids generally are united by stolons or stalks, slender tubular structures consisting of epidermis and endodermis (gastrodermis) united back to back. New individuals are constantly produced from small, more or less circular areas of the stolon wall, as in *Hydra*, which grow and develop into a functional organism. These may originate from the growing end of a stalk (see *Tubularia*, Fig. 20.1*b*) or from an area on the side (see *Hydra*, Fig. 1.5). The general question is essentially the same as in *Botryllus*, namely: What initiates an area of local tissue to develop into a new organism? The additional feature, seen in many hydroids, is that the initiate may develop into a hydranth (like a hydra) or it may develop as a medusa (Fig. 20.13). In some cases, one and the other sprout close together from along the same short length of stolon. What determines an initiated unit to take one path or the other? In our present state of ignorance, even speculation seems idle.

Production of buds in hydras and in those medusae which produce medusa buds from the manubrium follows a characteristically sequential course. New buds are continually produced at a critical distance from the distal end of the col-

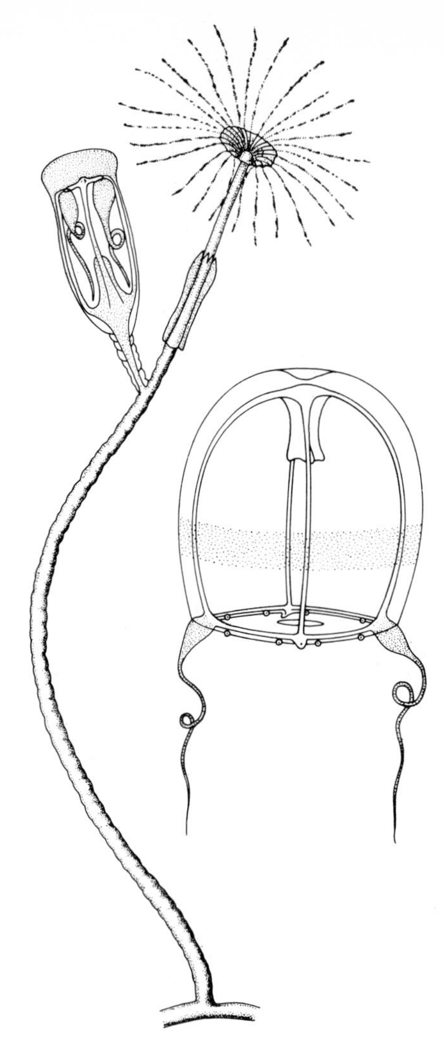

Figure 20.13 Polymorphic development in hydroid, showing fully formed hydranth with a fully developed medusa growing from stalk, and a liberated, expanded medusa. [*After Rees,* J. Mar. Biol. Assoc. U.K., **23:***1* *(1939)*.]

umn or manubrium, as the case may be. Older buds and their adjoining parental tissue are seen to shift proximally accordingly. Bud rudiments appear initially to be of essentially the same size, and the number of buds present at any one time depends on the circumference and extent of the total area involved. In other words, irrespective of the details of the budding process in any particular case, buds are initiated and exist in direct relation to available space. In essence, the procedure is meristematic and may be compared to the situation in the vegetative and floral meristems of the higher plants. However, the investigation of the development of meristematic units is still in its infancy.

A. The plant meristem

The shoot or the flowering apex, as the case may be, shows two phenomena:(1) a series of well-defined territories becomes established in sequence immediately below the tip of the meristem; and (2) each territory is the site of a more or less independent development leading to leaf, shoot, or flower formation, depending

on various circumstances. The active growth centers are characterized by increased metabolism, cell division, and growth, resulting in the outgrowth of a leaf or bud. The fundamental problems are the same as those associated with local asexual development of buds of colonial animals, particularly hydroids, bryozoans, and tunicates. These are essentially threefold:

1. The initiation or establishment of a local territory as an independent growth center within the general organization of the parental organism, which includes the question of positioning within the meristematic system.
2. The properties of such a growth center, or the territory involved, that determine what type of developmental path will be followed.
3. The course of subsequent development of the unit in terms of morphogenesis, histogenesis, and cytodifferentiation.

The term *phyllotaxis* has long been used to denote the regular spacing of successive leaf rudiments or other primordia on the surface of a meristem and, accordingly, the arrangement of the developed structures on the mature stem. It refers not only to the regularity of the spacing, but to the particular character of the spacing. The growth centers formed on the flank of the meristem are started near the apex a certain initial distance apart in series and at a certain distance from those already forming the ring of centers below. As a rule, the series thus formed are arranged either in a close spiral or as a number of whorls extending from the meristem apex to its base. Accordingly, there are both primary and secondary growth centers. The primary center is the distal meristem itself, from which growth of the shoot apex as a whole proceeds. This region exhibits a rhythmic growth, which is seen as a regular change in shape correlated with the initiation of each secondary center. During this growth, the apex passes from a minimum to a maximum volume. Each secondary center initiated successively from the apical region, whatever its manner of formation may be, represents a local region of increased growth rate relative to the rate of growth of its immediate surroundings. A rhythm in the frequency of mitosis is evident in the meristematic zone immediately distal to the zone of initiation of the primordia.

It becomes increasingly evident from the studies of the plant meristem, and equally from studies of the development of eggs and other developmental units, that the great technical problem arises from the fact that in all cases the most important developmental events occur in the beginning. This, of course, is far from surprising; yet it means that the primary analysis concerns events taking place either in single cells (although they may be large, as in the case of the apical cell of a meristem and all animal eggs) or in a group of apical-meristem cells and the buds of colonial animals. So much must go on in so little space, or in so little mass of living material, that analytical progress has been severely retarded by technological difficulties. The need, with regard to experimental analysis, is to isolate the growing systems so that they can be studied under various conditions in the laboratory. Animal eggs, of course, are ready-made for such a procedure and have therefore been exploited in this way for nearly a century. Excised shoot meristems of Norway spruce and other plants under laboratory conditions are making the meristematic system, normally hidden and almost inaccessible, available for experimental study (Fig. 20.14). The domelike meristem of the spruce is about 0.3 mm in diameter and somewhat less in height. To culture it, one must, while maintaining sterility, expose it to view, cut it off with a microtool (causing as little damage as possible), and transfer it to a suitable nutrient medium. The culture must then be kept sterile and be nurtured under a controlled

a

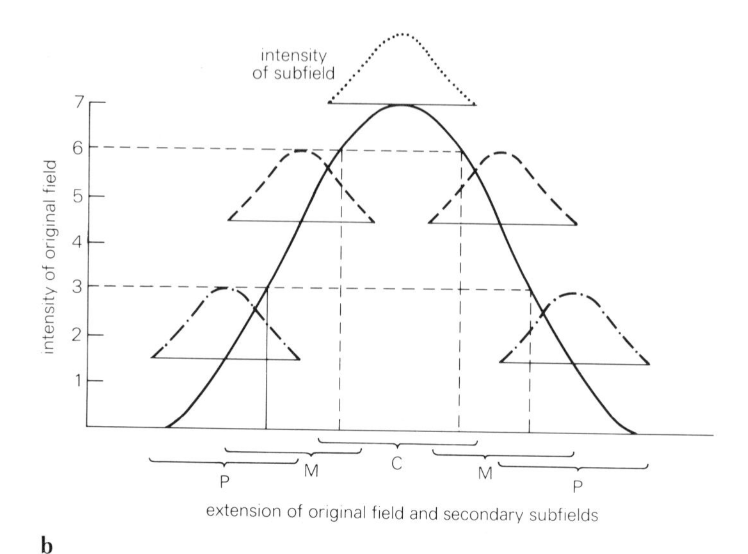

b

Figure 20.14 *a* Lateral view of whole-shoot meristem from a dormant vegetative bud of spruce, showing domelike apical meristem at the top, and helical whorls of primordia (secondary fields). [*U.S. Forest Service Photograph No. 520097, by J. A. Romberger, reproduced by permission.*] *b* Extension and subdivision of morphogenetic fields. [*After Nieuwkoop, 1967.*]

environment while its growth response to various treatments is periodically observed. The practical problems are legion. For plant scientists, the problem is comparable to that of growing test-tube mammalian embryos, although the difficulties are not as great.

The meristematic process, the strobilating process, and the morphogenetic process as a whole—as seen in the initiation and development of bud primordia—remain challenging subjects for analysis. These phenomena are to be pondered, and they call for an understanding of the nature of morphogenetic fields far beyond anything we now have. The distance from the concept of gene regulation seems almost infinite, and the prospect is wide open.

8. Points of view

"A theory of development would effectively enable one to compute the adult organism from the genetic information in the egg."

L. Wolpert and J. Cairns, 1975, *Ciba Symp.* **29**:95–130.

"In conclusion, we can see that development occurs as a series of complex and inter-related events under the control of physiological and molecular conditions which are initiated and directed by gene action. The complexities of the developmental system, which encompass the entire span from oogenesis to final differentiation, are kept in balance by the wild-type-genotype so that development is normal. This balance is both molecular and temporal. Normal development is governed by molecular reactions and interactions whose rates and sequences are of primary importance. The intracellular molecular organization, controlled and directed by gene action, results in a balanced supramolecular organization which in turn results in normal development."

R. Hillman, 1977, *Amer. Zool.* **17**:531.

"At this point one sometimes encounters the proposition that the organism is 'nothing but' genes and gene products, and that talk about a field or any other type of spatial and temporal organizing influence is to introduce a mysterious, not to say vitalistic, element into the argument. The simple answer to this, however, is that composition does not determine form. Just as there are allotropic crystalline forms of elements such as carbon (diamond and graphite) or sulphur (rhombic and monoclinic) whose composition is identical but whose forms are distinctly different, so there are different ways in which protomers can be assembled into higher-order structures (e.g., flagellin assembly into wavy or curly flagella). Despite many contrary but incorrect claims, primary structure does not in general determine quaternary form. Thus a knowledge of an organism's genome does not allow one to predict its morphology, so that the adult organism is not computable from the genetic information in the egg. This is why Weismann's (and Wolpert's) position is logically untenable, as is any similar proposition concerning a genetic programme of development. The assumption that the switching on or off of specific genes is a sufficient condition for the generation of a particular structure is simply wrong, so that any theory of development which reduces to a description of the position and the temporal order of gene switching is logically incomplete. Such a theory is equivalent to trying to predict the motion of a planet from a knowledge of its initial position and velocity only, without knowing the laws of motion of bodies under the action of gravitational forces. Our position is that one must attempt to define these laws of motion of the developmental process; and, since we are dealing with a process occurring in space-time, these laws must be expressible in terms of field concepts and field equations. The division of the organism into a controlling and a controlled part plays no useful role in this analysis. The genes, their products, and cytoplasmic organization constitute an indissoluble unity in the generation of organic form."

B. C. Goodwin, 1978.

"We are only beginning to get a glimpse of the mechanisms of field action. It is unrealistic to expect any quick solutions in the problems posed in the relationship between the whole and its parts; nevertheless, these problems are the most fascinating ones in developmental biology and will continue to be a challenge for a long time to come."

D. L. Stocum, 1978, *Amer. Zool.* **18**:894.

In any case, there are no final answers, no more in biology than in physics. Moreover, the great variety of phenomena and organisms available for study, although for the most part difficult to obtain and often hard to cultivate, holds much promise for students of developmental biology for generations to come. It is this ongoing activity that is most meaningful rather than the attainment of dogma. The philosopher A. N. Whitehead wrote that "the process itself is the reality," and this phrase applies equally and cogently both to developmental processes and to the activity of developmental biologists.

Readings

BERRILL, N. J., 1961. "Growth, Development and Pattern," Chaps. 11 and 12, Freeman.

———, 1978. Induced Segmental Reorganization in Sabellid Worms, *J. Embryol. Exp. Morph.*, **47**:85–96.

BERRILL, N. J., and D. MEES, 1936*a*. Reorganization and Regeneration in Sabella. I. Nature of gradient, summation and posterior regeneration, *J. Exp. Zool.*, **73**:67–83.

———, 1936*b*. Reorganization and Regeneration in Sabella. II. The Influence of Temperature. III. The Influence of Light, *J. Exp. Zool.*, **74**:61–89.

BORGENS, R. B., J. W. VANABLE, JR., and L. F. JAFFE, 1977. Bioelectricity and Regeneration. I. Initiation of Frog Limb Regeneration by Minute Currents, *J. Exper. Zool.*, **200:**403–416.

BRØNSTEDT, H. V., 1970. "Planarian Regeneration," Pergamon; also *Biol. Rev.*, **30:**65–126 (1955).

BRYANT, P. J., 1978. Pattern Formation, Growth Control and Cell Interactions in *Drosophila* Imaginal Discs, in S. Subtelny and I. R. Konigsberg (eds.), "Determination of Spatial Organization," 37th Symp. Soc. Dev. Biol., Academic, pp. 295–316.

CAMPBELL, R. D., 1978. Development of *Hydra* Lacking Interstitial and Nerve Cells ("Epithelial *Hydra*"), in S. Subtelny and I. R. Konigsberg (eds.), "Determination of Spatial Organization," 37th Symp. Soc. Dev. Biol., Academic, pp. 267–293.

FITZHARRIS, T. P., 1973. Control Mechanisms in Regeneration and Expression of Polarity, *Amer. Sci.*, **61:**456–462.

FRENCH, V., P. J. BRYANT, and S. V. BRYANT, 1976. Pattern Regulation in Epimorphic Fields, *Science* **193:**969–981.

GOODWIN, B. C., 1978. A Field Approach to Biological Pattern, *Proc. Netherland Soc. Dev. Biol.*, p. 8.

GOSS, R. J., 1968. "Principles of Regeneration," Academic.

JAFFE, L. F., and R. NUCCITELLI, 1977. Electrical Controls of Development, *Ann. Rev. Biophys. Bioeng.*, **6:**455–476.

ROSE, S. M., 1970. "Regeneration," Appleton-Century-Crofts.

SHOSTAK, S., 1974. The Complexity of *Hydra:* Homeostasis, Morphogenesis, Controls, and Integration, *Quart. Rev. Biol.*, **49:**287–310.

SINGER, M., 1978. On the Nature of the Neurotrophic Phenomenon in Urodele Limb Regeneration, *Amer. Zool.*, **18:**829–841.

SPERRY, P. J., D. A. KRYSTINA, and F. K. TITTEL, 1973. Inductive Role of the Nerve Cord in Regeneration of Isolated Post-Pharyngeal Body Sections of *Dugesia dorotocephala*, *J. Exp. Zool.*, **186:**159–174.

STEEVES, T. A., and I. M. SUSSEX, 1972. "Patterns in Plant Development," Prentice-Hall.

STOCUM, D. L., 1978. Organization of the Morphogenetic Field in Regenerating Amphibian Limbs, *Amer. Zool.*, **18:**883–896.

SYMPOSIUM, 1974. R. L. Miller, (ed.), The Developmental Biology of the Cnidaria, *Amer. Zool.*, **14:**437–866 (32 articles on *Hydra*, Hydroids, and Scyphozoa).

Chapter 21 Methods in cell biology

1. Cell fractionation

A section of Chapter 2 dealt with the nature of the ribosome, a particle so small that it is seen as a mere dot in the electron microscope. One question that must be satisfied is how investigators have learned so much about something so small. The ribosome exists as a part of a cell, and to perform biochemical studies of its nature, the integrity of the cell must first be destroyed. In higher organisms, the cell is a fragile unit whose outer membrane can be broken or dissolved by numerous treatments including pressure, shearing or grinding forces, ultrasonic vibration, detergents, or osmotic shock. Once the cells are broken and have formed an homogenate, some procedure must be used to obtain a preparation containing only the part of the cell under study. The term *cell fractionation* refers to the separation of the contents into purified fractions. The fractionation of the cell's components is desired for many reasons, and many techniques have been designed to meet this need. Typically, cells are fractionated by differential centrifugation, which describes the fact that different parts of the cell will sediment toward the bottom of a centrifuge tube at different rates of centrifugal force.

Numerous factors determine whether or not a given component will settle through a liquid medium, including the size, shape, and density of the structure and the density and frictional resistance, i.e., viscosity, of the medium. In order for a substance to sediment toward the bottom of the tube it must (1) have a greater density than the surrounding medium and (2) be centrifuged at sufficient speed to overcome the tendency for the components to be redistributed uniformly as a result of diffusion. In a eukaryotic organism, if a nuclear preparation is sought, the homogenate is centrifuged at low centrifugal force (below $1{,}000 \times g$), the nuclei are spun to the bottom, and the supernatant is removed. If mitochondria are sought, the nuclei are first removed, the supernatant is then centrifuged at greater speeds (approximately $12{,}000 \times g$), and the mitochondria are spun out. If ribosomes are sought, as in this example, the nuclei and the mitochondria are first removed, the supernatant is subjected to ultracentrifugation at approximately $100{,}000 \times g$, and the ribosomes pellet at the bottom of the tube. The post-ribosomal supernatant now consists of the soluble phase of the cell's protoplasm and those particles too small to be easily removed. In actual practice, many more procedural steps would be required to obtain a purified organelle preparation, but the principle is that just outlined. Technological advances in cell biology con-

tinue to allow greater sophistication in all methodology, cell fractionation included. For example, techniques of subnuclear purification include procedures for the isolation of nucleoli, nuclear membranes, specific types of chromosomes, and even the nuclear-pore complex itself.

2. Sucrose density gradient centrifugation

In ribosome assembly it is clear that each of the two subunits of the bacterial ribosome, the 30S and 50S particles, can be isolated from each other. To begin to separate cell components as similar as these two particles, a more sophisticated technique must be employed. One can still use the centrifuge, but an environment within the tube must be constructed that takes advantage of differences in properties of the components. The major difference between the 30S and the 50S subunit is the greater mass of the latter. These two particles are easily separated in a sucrose density gradient. To perform this operation, one must begin with a centrifuge tube filled so that the bottom contains a high concentration of sucrose, and the concentration decreases linearly or nonlinearly toward the top, producing a gradient of sucrose. A typical gradient to separate ribosomal subunits might have 40 percent sucrose at the bottom and 10 percent sucrose at the top. To begin the separation, a small volume containing the ribosomes is carefully layered on top of the gradient. Since its density is less than 10 percent sucrose, it will form a layer without mixing. The tube is then centrifuged at high speed (generally above 30,000 $\times$ g), which drives the particles down into the gradient. If centrifugation is done at very low Mg^{2+} levels, the two subunits of each ribosome come apart from each other. The velocity with which a particle sediments depends on its molecular weight. The greater the molecular weight, the faster it moves; therefore, these subunits become separated within the gradient. As a particle sediments through the tube, the increasing gravitational forces it experiences cause the particle to accelerate, while the increasing viscosity of the sucrose produces a counteracting effect.

This same technique can be used to fractionate larger-size particles, such as polyribosomes. Each polyribosome consists of a messenger RNA thread and ribosomes, along with the various factors needed for protein synthesis and the nascent polypeptide chain. The bulk of the polyribosome is composed of ribosomes; the number of ribosomes attached to a given mRNA will determine how rapidly that polyribosome will sediment in a sucrose gradient. Generally, larger mRNAs will have a greater number of ribosomes in their polyribosomes and will be found closer to the bottom of the tube after centrifugation. The procedure generally used is to homogenize the cells, centrifuge the homogenate to pellet the nuclei and the mitochondria, layer the supernatant (containing the polyribosomes) over the sucrose gradient, and then centrifuge the contents. If membrane-bound polyribosomes are to be examined, they must first be dissociated from their membrane attachments, a feat generally accomplished with detergents, such as sodium deoxycholate. An example of a polyribosome gradient is shown in Fig. 8.16. As in the separation of the ribosomal subunits, density-gradient centrifugation allows one to spread the various elements of a mixture into their component parts, in this case, the various-sized polyribosomes. The use of labeled amino acids or uridine prior to homogenization allows one to analyze which polyribosomes are most actively synthesizing protein and which polyribosomes contain newly synthesized RNA.

Density-gradient centrifugation can be used to separate macromolecules as well as particles. For example, if the ribosomal pellet was first extracted for its RNA, a mixture of 18S and 28S RNA would be present together. If this mixture

were layered over a sucrose gradient (in this case 5 to 20 percent), these two rRNAs could be separated as easily as the 30S and 50S subunits in the previous sucrose gradient. The S values of a particle or molecule refer to its sedimentation velocity, which, for a given type of substance, is a measure of relative molecular weight. Since the density of the medium is less than that of the nucleic acid molecules even at the bottom of the tube, these molecules will continue to sediment as long as the tube is being centrifuged, i.e., equilibrium is never reached.

We have arrived at the place where the components of the original mixture have been separated, but they remain in the tube and somehow must be analyzed. The problem is to displace the contents from the tube for analysis without disturbing the separation. This is generally accomplished by carefully puncturing the bottom of the tube and collecting the drops into different tubes. The first drops to emerge are from the bottom of the tube, and soon drops will fall that contain the 50S subunit (or 28S rRNA in the last example). Eventually drops will emerge that contain the small subunit (or the 18S rRNA). Collection of sample from the top of the tube also can be made. The amounts in each collection tube, or fraction, can be determined with the spectrophotometer or radiation counter.

3. Sedimentation equilibrium centrifugation

In the previous section, we discussed the use of a sucrose density gradient to facilitate the separation of a mixture of two particles or macromolecules. In that example, the longer the tubes were centrifuged, the farther into the gradient the materials would move. If left long enough, they would pellet at the bottom. We will consider another type of gradient centrifugation, this time using DNA extracted from germinal vesicles of *Xenopus* oocytes. We begin with a centrifuge tube containing a solution of cesium chloride. To this solution we mix in the sample of DNA, place the tube in the proper holder, and centrifuge it at high speeds for a few days in the ultracentrifuge. During the centrifugation, the cesium ions are so heavy that they can be influenced by the centrifugal force, and a gradient of cesium chloride is formed. To shorten centrifugation, one can begin with a preformed gradient as for sucrose. This gradient of cesium chloride is also a gradient of density, owing to the different concentration of cesium along the length of the tube. As a result, the DNA, in this case, sediments to a position in the gradient having a CsCl density equal to its own. If DNAs of different buoyant density (different guanine plus cytosine content) are present, they become separated. If proteins are present, they are less dense and will go to the top of this particular gradient. If RNA is present, it is more dense and will go to the bottom. Single-stranded DNA is more dense than double-stranded DNA. In these latter examples, the CsCl gradient can be used to purify as well as to separate these molecules. The contents of the tube can be removed and the amounts determined, as in the sucrose gradient.

Sedimentation equilibrium (or isopycnic) centrifugation has been an important technique in determining that the synthesis of rRNA during amphibian oogenesis takes place on selectively amplified rDNA (Section 4.3.A). If DNA is extracted from somatic cells of *Xenopus*, such as liver cells, and is centrifuged on a cesium chloride gradient to equilibrium, only one DNA peak, with a density of 1.699 g/cm^3, is found. If, however, DNA is extracted from germinal vesicles collected from *Xenopus* oocytes and is then centrifuged in a similar manner, two distinct peaks are found. One of these peaks has a density of 1.699 g/cm^3; the other peak has a higher density of 1.729 g/cm^3. Since the bulk chromosomal DNA of *Xenopus* has a guanine plus cytosine content of 40 percent while the ribosomal DNA has 63 percent, it would be expected that the high-density peak contained the DNA that codes for rRNA and was present within the many nucleoli of the

germinal vesicle. In the somatic-cell DNA, no high-density peak is seen, since in these cells having only two nucleoli, the rDNA should account for only about 0.1 percent of the genome (450 copies of ribosomal RNA genes per haploid set of chromosomes). This is too low a value to detect in these gradients. In the gradients of oocyte DNA, however, the rDNA accounts for a minimum of nearly 50 percent of the chromosomal DNA and represents a striking selective gene amplification.

4. Autoradiography

Autoradiography takes advantage of the ability of a particle emitted from a radioactive atom to activate a photographic emulsion, much like light or x-rays activate the emulsion that coats a piece of film. If such a photographic emulsion is brought into close contact with a radioactive source, the particles emitted by the source leave tiny, black-silver grains in the emulsion after processing. Autoradiography is used to analyze radioisotopes within cells and tissues, or sections of tissues, that have been immobilized on a slide or electron-microscope grid. The emulsion can be purchased in two forms. In the liquid form, the slide containing the tissue is dipped into the fluid and withdrawn, leaving a thin film over the tissue. In the other type, the emulsion is already in the form of a thin film, which is floated on water to a position on top of the slide. In either case, these operations are performed in the darkroom and the slides are then put aside in a lightproof container to allow the film to be exposed by the emissions. The longer the slide is left, the greater will have been the number of emissions, and therefore the greater will be the number of silver grains. After the desired exposure time, the slides are developed in the darkroom, in the same manner a piece of film is developed, and examined under the microscope. Wherever radioisotope was present in the tissue, as in a section of a fly ovary (Fig. 4.20), silver grains will be found in the layer of emulsion directly over those sites.

Autoradiography, therefore, allows one to see where in a cell or tissue the radioactive molecules that were administered have become localized. In addition, since the number of silver grains can be estimated or even counted, the quantity of radioactive molecules can be determined.

5. Inhibitors

The use of inhibitors to probe function is widespread. There are specific inhibitors for the disruption of a multitude of biological processes. Some inhibitors are analogs, which are effective because they resemble a biological molecule and the cell cannot distinguish between the inhibitor and the proper molecule. Once the analog is used in place of the proper molecule, either the further synthesis of the product is halted or the product is nonfunctional. For example, cordycepin is an analog of adenosine and blocks poly-A metabolism as a result. Puromycin resembles an amino acyl–tRNA molecule and is added to the growing polypeptide chain; the entire nascent chain then falls off the ribosome without elongation. Many inhibitors work to block specific enzymes. For example, there is a potent RNA polymerase inhibitor isolated from mushrooms, alpha-amanitin, which blocks RNA synthesis by binding to the enzyme. Actinomycin D blocks RNA synthesis in an indirect manner by attaching to the DNA and stopping the movement of the polymerase at that point. Other inhibitors have particular effects on biological molecules, such as the ability of colchicine to cause the depolymerization of microtubules and therefore the dissolution of the spindle apparatus.

The general method for the use of an inhibitor is to add the compound at one specific time and watch for a disruptive effect at a later time. In the example of meiosis (Section 4.2), an inhibitor of DNA synthesis is added at the stages of

meiosis and abnormalities are recorded. If the inhibitor is added during zygotene, the ability of the chromosomes to properly synapse and form the synaptonemal complex is blocked. The conclusion is that the synthesis of DNA at zygotene is somehow needed for this physiological process to occur; in other words, a causal relationship is implied.

Interpretation of inhibitor studies must be made with caution. First, it is important to show that the inhibitor is working, by measuring the effect on the specific synthetic process. Inhibitors are often large molecules that may or may not penetrate a given cell. If no effect is seen, the reason may be the lack of penetration and subsequent inhibition rather than that the synthesis is unnecessary. Comparsion of the effects of actinomycin D in different systems illustrates the permeability problem. Mammalian cells growing in tissue culture are inhibited at less than 1 μg/ml, while a comparable inhibition in sea urchin embryos requires over 20 μg/ml. The difference, presumably, reflects the different penetration rates rather than differences in the sensitivity of RNA synthesis to this inhibitor.

Another difficulty that is harder to evaluate is the possibility that the observed effect is not due to the specific inhibition but to a side effect. For example, actinomycin D is known to have an effect on respiration; it is therefore difficult to assign its blocking of a certain process, such as cleavage in mammalian eggs, to its inhibition of transcription rather than to its inhibition of respiration or some other event.

6. Electron microscopes

Clearly the most important tool for the analysis of cell structure is the electron microscope. The theoretical limit of resolution of a given type of microscope is restricted by the wavelength of the illuminating source. In the light microscope, using light of the shortest wavelength, the theoretical limit of resolution is approximately 0.2 μm. In other words, if two structures being examined are not separated by at least that distance, one could not see them as distinct entities in the light microscope. Instead they would be so close together that they would appear as one structure. In the electron microscope, illumination is by means of electrons, whose wavelength is vastly shorter than that of visible light; the limit of resolution is therefore vastly greater. As a direct result of the electron microscope, the tremendous body of knowledge of the complexity and structure at the subcellular level has been gained. We are all accustomed to the sight of the elongate mitochondria with its numerous membranous elements, the ribosome-studded membranes of the rough endoplasmic reticulum, or the many other features in common to many cell types or characteristic of only one or a few types of cells. In the world of the cytologist of only a few decades ago, the structures within a cell were either unknown or recognized at a poorly defined level.

Two types of electron microscopes have been developed and both have been used to study a multitude of processes that occur in developing systems. The most familiar type is the *transmission electron microscope*, which is used to analyze the structure of very thin sections of tissue. Before sections can be prepared, the tissue must be subjected to a prescribed series of treatments. The first step in tissue preparation is fixation. The tissue is immersed in a fixative that has the following properties: It causes the cessation of all the cell's activities and its rapid death, and it maintains the structure of the cell as close as possible to that of the living state. It is of obvious importance, when one examines cell structure at the termination of the procedure, to have confidence that what is being observed is a reflection of the true structure of the cell rather than an artifact produced after

the cell was fixed. The most common fixatives in use for electron microscopy are formaldehyde, glutaraldehyde, and osmium tetroxide. These may be used singly, in sequence, or in combination.

After fixation, the tissue is dehydrated by transfer through a series of alcohols of increasing concentration. It is finally placed in a solution of liquid plastic, which is allowed to penetrate into the spaces of the cells by replacement of the previous solvent. This tissue is then placed in the oven, which causes the molten plastic to harden as a supporting medium of the tissue. Blocks of tissue are then prepared and very thin sections of plastic-embedded tissues are cut. Sections for the electron microscope are typically about 500 A in thickness (1 cm equals 10^8 A). These sections float on water as they come off the knife and are then picked up on small grids, which act as a support in a manner analogous to the slide for a light microscope section. The sections on the grids are then stained by floating the grids on drops of solutions of heavy metals, typically uranium and lead, which become complexed with the tissue. Because these heavy metals interfere with the passage of electrons through the tissue, those parts of the cell which have bound more of the stain will appear relatively darker on the viewing screen.

One point should be kept in mind when examining a series of electron micrographs. Each micrograph represents the state of the cell at only one point in time, and any suggestions concerning the dynamic processes of that cell must be based on conclusions that compare many different micrographs at different stages in the process.

The freeze-fracture technique uses the transmission electron microscope in an entirely different way. The first step is to place the tissue on a small metal disk and freeze it by dropping it into liquid freon or a similar freezing fluid. This disk is then placed in a special freeze-fracturing apparatus, which maintains its frozen state and cracks the tissue in half. The fracture planes characteristically run along surfaces and within membranes, and these parts of cells have been studied best by this technique. Following the fracture, the open surfaces of the tissue are still held in place by the apparatus, and heavy metals are used to coat these surfaces with a very thin layer. Once the coating has been made, the tissue that served as the template is discarded and the metallic cast, or replica, is then examined in the electron microscope. Since the molecular metallic layer was able to fit itself to the contours of the fractured surface, we are provided with an elegant replica of that contour. That the technique works is demonstrated by the micrographs of Figs. 3.6*c* and 4.7.

The other type of electron microscope is the *scanning electron microscope*, which is used to observe the nature of the cell surface. In this procedure, tissue is fixed but not sectioned. Instead, the specimen is dried under special conditions and covered with a very fine layer of heavy metal, which again acts to block the penetration by electrons. The resulting image in the scanning electron microscope reveals the intricate three-dimensional nature of the outer boundary layer. This tool has recently been used to explore processes where surface events play a significant role in the activity. These include the changes in egg surface at fertilization, the surface of migratory embryonic cells, and the surfaces of cells during cleavage.

7. Purification and fractionation

The need to purify a particular protein or other macromolecule is not the only reason one would use fractionation procedures. Information on the variety of proteins, RNAs, or polysaccharides can be of great importance in understanding cellular activity. For example, one can ask how many different proteins make up

a microtubule or a chromosome, or how many RNAs are present in the ribosomes, or how great a variety of sulfated polysaccharides exist at the cell surface. Answers to these questions require a means of separating the molecules so that their variety can be displayed. A related question to the preceding ones concerns the synthesis of a given species. For example, one may inquire whether the proteins of microtubules are being synthesized by the embryo and also about their variety. Questions of synthesis are best answered with isotopes. The embryo can be provided with radioactively labeled amino acids for incorporation into proteins made prior to homogenization. After incubation, the microtubules are purified and fractionated to determine which of the proteins under study are labeled. The question has been answered.

Each type of macromolecule—RNA, DNA, protein, polysaccharide, or fat—has properties that enable it to be prepared free of contaminating impurities. The preliminary steps in most isolation and purification procedures involve the differential solubility of these different macromolecules. Cells are generally homogenized, and the homogenate or cell fraction treated with solutions that selectively dissolve or precipitate the desired material. For example, if one were extracting DNA, the homogenate might first be shaken with phenol, possibly in the presence of a detergent and high salt; these are conditions that remove the protein from the DNA and precipitate it while the DNA goes into solution in the aqueous phase. The addition of ethanol to the aqueous DNA solution causes the DNA to come out of solution (along with any RNA), and the fibrous DNA can be wound on a glass rod (leaving the finely precipitated RNA behind). The next step in a purification procedure is often the treatment of the material with enzymes that will digest any remaining contaminant (RNA, protein, or polysaccharide in this case). By this time we should have a relatively pure preparation of that macromolecule, but we might want a particular species and must continue the purification. We might be purifying one RNA species (e.g., 18S rRNA), one protein species (e.g., collagen), or one glycosaminoglycan (e.g., chondroitin sulfate). To reach this goal, more sophisticated procedures must be employed to separate a mixture of molecules into its spectrum of individual components.

Certainly one of the greatest gifts to the developmental biologist from the biochemist has been the tremendous number of materials made available that allow the fractionation of these macromolecules. Two species of macromolecule with a different linear sequence of components will inevitably have different physical properties. The dramatic effect of a single amino acid substitution in one polypeptide chain of the hemoglobin molecule of a sickle-cell anemia victim is a clear illustration of this fact. The most common differences between species are molecular weight, ionic charge, solubility, shape, and density. A given fractionation procedure generally depends on more than one of these properties. In the technique of electrophoresis, for example, the population of macromolecules is placed in an electric field, with the positive electrode on one side and the negative electrode on the other. Under these conditions, a macromolecule with an overall negative charge—i.e., it has more negatively charged groups than positively charged ones—will migrate toward the positive electrode. The greater the negative charge, the faster will be the migration; therefore, this mixture of species will begin to separate out into its components on the basis of charge. Although they all began at one spot, those species carrying the greatest negative charge will move farthest toward the positive electrode. The mobility of these molecules in an electric field, however, is affected by other properties as well. Larger molecules, as might be expected, move more slowly. Similarly, certain shapes are more conducive to rapid migration than others. The distance a given species will migrate in a

period of time is a complex matter, and two different molecules are not likely to be the same.

In the most commonly employed electrophoretic technique, mixtures of proteins (or nucleic acids) are caused to move through a semisolid column of polymerized acrylamide, i.e., polyacrylamide gel. To begin a fractionation on polyacrylamide gel, small tubes are filled with solutions of acrylamide, which then solidifies into a gel when exposed to ultraviolet light. The tube is inserted between two compartments containing buffer in which electrodes are immersed and the protein solution is layered in the tube over the gelated acrylamide. Current is applied across the column of the gel and the proteins begin their migration as disks moving at different rates. The rate of movement is dependent on charge and molecular weight. In many cases, polyacrylamide gel electrophoresis (PAGE) is carried out under conditions in which separation occurs solely on the basis of molecular weight. This is usually accomplished by conducting the PAGE in the presence of the ionic detergent sodium dodecyl sulfate (SDS). In the SDS technique, the negatively charged detergent molecules become complexed with the protein to an extent that is proportional to the protein's molecular weight. Consequently, each protein species, regardless of its molecular weight, will have an equivalent charge per unit molecular weight and will tend to move at an equivalent rate in an electric field. However, as a result of the cross-linked nature of the acrylamide in the tube, the larger the molecular weight of the migrating molecule, the more it will be held up, and the slower it will move. As a result, proteins become separated on the basis of their molecular weight.

Another variation of electrophoresis, that of isoelectric focusing, has become an extremely powerful means to separate a mixture of proteins into very fine bands. In this technique, the tube to be used for separation is filled with a substance that forms a pH gradient when subjected to the applied electric field. As proteins migrate through the gel, they are exposed to a continually changing pH, which produces a continuous change in their ionic charge. At some point within the tube each protein encounters a pH that is equal to its isoelectric point, thereby converting the protein to a neutral molecule and causing its migration to stop. Each protein comes to equilibrium to form a very sharp band at a predictable position along the tube's length.

Different techniques utilize different physical properties; therefore, fractionation procedures often involve more than one method to ensure maximum separation. In the preceding example, consider that the maximum separation by electrophoresis has been obtained and the protein being isolated is migrating closely with a contaminating species. This pair of proteins can be removed and subjected to a different technique that depends on a different basis of separation—for example, one that depends most heavily on molecular weight, such as gel filtration with Sephadex. Sephadex consists of tiny beads perforated by holes of very specific diameters. Different Sephadex grades have different-size holes for use with various molecular-weight ranges. These beads allow molecules up to a certain molecular weight inside and exclude molecules with a larger molecular weight. Consider that the protein being isolated has a molecular weight of 100,000 and the contaminant, 50,000. In this case, a column of Sephadex G-75 beads can be made within a tube. This grade of Sephadex allows molecules of 50,000 to penetrate the beads, but those of 100,000 remain outside. When a solution containing these two proteins is poured through the column, the solution that first drips out the bottom contains the protein being studied. All the contaminant remains held up in the beads. Other major fractionation procedures take advantage of differences in solubility or sedimentation. Studies in immunology have provided some

of the most powerful fractionation techniques, based on the specific interactions of antibodies and their antigens, but these are beyond the scope of this book.

In 1975, a new technique for protein fractionation was introduced that has proved invaluable in the separation of large numbers of different proteins. The technique involves the separation of proteins in two dimensions across a flat gel surface. As is generally the case for two-dimensional fractionation procedures, it is most desirable that the separation in each direction take advantage of independent properties of the species in the mixture, thereby maximizing the distribution of the components over the surface of the plate. In this technique, proteins are separated in one direction according to their isoelectric point using isoelectric focusing. Proteins are then fractionated in a second direction according to their molecular weight by SDS PAGE. The resolution of this technique is so great that virtually all the proteins present in a cell (in detectable amounts) can be distinguished. It is estimated that up to 5,000 different proteins could be theoretically displayed by this procedure. Examples of the use of this technique are shown in Fig. 15.20.

8. Cell-free protein synthesis

One property of a population of RNA that can be measured is its ability to direct amino acid incorporation into protein, i.e., its template activity. To measure this property, a technique has been developed that uses the protein-synthesizing machinery of one type of cell and the mRNA of the cells under study. For example, a preparation can be made from reticulocytes or wheat germ that contains all the necessary ingredients for protein synthesis (the ribosomes; initiation, elongation, and termination factors; the tRNAs and their activating enzymes; an energy-generating system; etc.) except that they lack mRNA. When mRNA, from whatever source, is added, this added RNA can be used by the protein-synthesizing system to translate into protein. The system is termed *cell free* because the cells that donate the ribosomes and other ingredients have been homogenized and no intact cells are used. Since rRNA and tRNA are not capable of serving as templates for protein synthesis because of the changes in their nucleotide bases (such as methylation), the amount of amino acid incorporation is proportional to the added template-active RNA. To arrive at the percentage of the total RNA that can be used as a template, the RNA being studied is compared with that of an RNA known to be pure mRNA, such as viral RNA. In the case of RNA present in the sea urchin egg, the viral RNA causes 20 to 25 times greater amino acid incorporation per milligram of RNA; therefore, only one-twentieth to one-twenty-fifth of the total-egg RNA is template-active.

The use of this technique is particularly valuable because it allows one to determine the variety of mRNAs present in a cell independent of their normal translation. For example, there are cases in which cells contain mRNA molecules that are not being translated to their full template activity; i.e., the cells are operating to some extent under translational-level control. This is best determined by fractionating proteins that have been synthesized in vitro by two-dimensional gel electrophoresis and comparing this pattern to that produced by labeled proteins synthesized within the cells. Since the cell-free system uses as a template all abundant mRNAs present in the cell, those missing from the pattern of cellular synthesis must be selectively kept from translation within the cells.

9. Determination of rate of RNA synthesis

In this section, a general approach to measure the amount of RNA synthesized per unit of time is described. The same general procedure would apply to the determination of any product using a radioactive precursor with appropriate modifi-

cations. The availability of a precursor relatively specific to the macromolecule to be measured greatly simplifies the determination procedure. For the measurement of RNA synthesis, the most widely used precursors are [^{3}H]uridine or [^{3}H]guanosine. Before being incorporated, these molecules must be converted to the triphosphate form (UTP or GTP), which is utilized by the RNA polymerase in RNA synthesis. The triphosphates are soluble under many conditions where RNA is not, such as in 70% ethanol or cold 5% trichloroacetic acid or 0.6 *M* perchloric acid. To measure the amount of labeled precursor incorporated into RNA, the cells are homogenized; the homogenate is treated with a solution that renders the unincorporated precursor soluble and the RNA (containing the *incorporated label*) insoluble and therefore precipitated. The radioactivity in the precipitate is a measure of the added uridine or guanosine that was incorporated during the period of incubation. To be certain that all the label was actually incorporated into RNA and not, for example, into DNA after conversion into a DNA precursor, the susceptibility of the labeled product to RNase can be measured. If all the radioactivity becomes soluble after enzyme treatment, it had all been incorporated into RNA.

Up to this point, the only consideration has been the value of incroporation of precursor. This is not necessarily a measure of synthesis. When a labeled precursor is used, the rate of synthesis is reflected by three values: incorporation, uptake, and pool size. Assume that between two stages, *A* and *B*, there is no actual change in the synthesis of RNA. Assume also that between these stages there is a twofold increase, for some reason, in the permeability of precursor into the embryo. In other words, if the same number of embryos is placed in the same concentration of [^{3}H]guanosine, the embryos of stage *B* will *take up* twice as much isotope. This means that twice as many GTP molecules in the cell will be radioactive; and when the incorporation value is measured, it will be twice as high for *B* as for *A*, although no actual change had occurred in the synthetic rate. A changing precursor uptake pattern is a common observation during development, and therefore uptake must be measured to establish synthetic rate. Uptake can be measured as the amount of radioactivity present as precursor, i.e., soluble in 5% trichloroacetic or by other means.

The other value necessary to measure the actual synthetic rate is pool size. The term *pool* refers to the amount of precursor actually available for synthesis at a given time. In the previous example, no change in synthesis occurred between stages *A* and *B*. Now assume there is no change in uptake or actual synthesis, but there is a change in the amount of uridine or guanosine that the embryo is making. If in stage *B* the embryo has made enough guanosine so that its pool has doubled (there are twice as many GTP molecules available for synthesis), then the percentage of labeled GTP molecules in the cells relative to the total number will be one-half, and this will result in half the value if incorporation. To measure the size of the pool, the actual quantity of precursor in the embryo must be measured. If both uptake and pool size are known, then incorporation values can be converted into true synthetic rates. This type of procedure still assumes that there is no compartmentation of precursor pools within the cell; i.e., all molecules taken up by the cell have an equal chance of being incorporated.

Index